Trigonometric Table

Angle in degrees	Angle in radians	Sine	Cosine	Tangent	Angle in degrees	Angle in radians	Sine	Cosine	Tangent
0°	0.000	0.000	1.000	0.000					
1°	0.017	0.017	1.000	0.017	46°	0.803	0.719	0.695	1.036
2°	0.035	0.035	0.999	0.035	47°	0.820	0.731	0.682	1.072
3°	0.052	0.052	0.999	0.052	48°	0.838	0.743	0.669	1.111
4°	0.070	0.070	0.998	0.070	49°	0.855	0.755	0.656	1.150
5°	0.087	0.087	0.996	0.087	50°	0.873	0.766	0.643	1.192
6°	0.105	0.105	0.995	0.105	51°	0.890	0.777	0.629	1.235
7°	0.122	0.122	0.993	0.123	52°	0.908	0.788	0.616	1.280
8°	0.140	0.139	0.990	0.141	53°	0.925	0.799	0.602	1.327
9°	0.157	0.156	0.988	0.158	54°	0.942	0.809	0.588	1.376
10°	0.175	0.174	0.985	0.176	55°	0.960	0.819	0.574	1.428
11°	0.192	0.191	0.982	0.194	56°	0.977	0.829	0.559	1.483
12°	0.209	0.208	0.978	0.213	57°	0.995	0.839	0.545	1.540
13°	0.227	0.225	0.974	0.231	58°	1.012	0.848	0.530	1.600
14°	0.244	0.242	0.970	0.249	59°	1.030	0.857	0.515	1.664
15°	0.262	0.259	0.966	0.268	60°	1.047	0.866	0.500	1.732
16°	0.279	0.276	0.961	0.287	61°	1.065	0.875	0.485	1.804
17°	0.297	0.292	0.956	0.306	62°	1.082	0.883	0.469	1.881
18°	0.314	0.309	0.951	0.325	63°	1.100	0.891	0.454	1.963
19°	0.332	0.326	0.946	0.344	64°	1.117	0.899	0.438	2.050
20°	0.349	0.342	0.940	0.364	65°	1.134	0.906	0.423	2.145
21°	0.367	0.358	0.934	0.384	66°	1.152	0.914	0.407	2.246
22°	0.384	0.375	0.927	0.404	67°	1.169	0.921	0.391	2.356
23°	0.401	0.391	0.921	0.424	68°	1.187	0.927	0.375	2.475
24°	0.419	0.407	0.914	0.445	69°	1.204	0.934	0.358	2.605
25°	0.436	0.423	0.906	0.466	70°	1.222	0.940	0.342	2.748
26°	0.454	0.438	0.899	0.488	71°	1.239	0.946	0.326	2.904
27°	0.471	0.454	0.891	0.510	72°	1.257	0.951	0.309	3.078
28°	0.489	0.469	0.883	0.532	73°	1.274	0.956	0.292	3.271
29°	0.506	0.485	0.875	0.554	74°	1.292	0.961	0.276	3.487
30°	0.524	0.500	0.866	0.577	75°	1.309	0.966	0.259	3.732
31°	0.541	0.515	0.857	0.601	76°	1.326	0.970	0.242	4.011
32°	0.559	0.530	0.848	0.625	77°	1.344	0.974	0.225	4.332
33°	0.576	0.545	0.839	0.649	78°	1.361	0.978	0.208	4.705
34°	0.593	0.559	0.829	0.675	79°	1.379	0.982	0.191	5.145
35°	0.611	0.574	0.819	0.700	80°	1.396	0.985	0.174	5.671
36°	0.628	0.588	0.809	0.727	81°	1.414	0.988	0.156	6.314
37°	0.646	0.602	0.799	0.754	82°	1.431	0.990	0.139	7.115
38°	0.663	0.616	0.788	0.781	83°	1.449	0.993	0.122	8.144
39°	0.681	0.629	0.777	0.810	84°	1.466	0.995	0.105	9.514
40°	0.698	0.643	0.766	0.839	85°	1.484	0.996	0.087	11.43
41°	0.716	0.656	0.755	0.869	86°	1.501	0.998	0.070	14.30
42°	0.733	0.669	0.743	0.900	87°	1.518	0.999	0.052	19.08
43°	0.750	0.682	0.731	0.933	88°	1.536	0.999	0.035	28.64
44°	0.768	0.695	0.719	0.966	89°	1.553	1.000	0.017	57.29
45°	0.785	0.707	0.707	1.000	90°	1.571	1.000	0.000	∞

PHYSICS

Theodore P. Snow ⋯ J. Michael Shull
University of Colorado at Boulder

West Publishing Company
St. Paul □ New York □ San Francisco □ Los Angeles

Copyediting: Tech Edit
Design/Art Direction: Wendy Calmenson
Composition: The Clarinda Company
Text Illustrations: John and Jean Foster, Art by AYXA, Donna Salmon
Photo Research: Roberta Spieckerman

A STUDENT SOLUTIONS MANUAL

A solutions manual that contains full solutions to all odd-numbered numerical problems is available to all students from the local bookstore under the title *Student Solutions Manual to Accompany Physics*, prepared by J. Michael Shull. An examination copy is available to instructors by contacting West Publishing Company.

COPYRIGHT (c) 1986 By WEST PUBLISHING COMPANY
50 West Kellogg Boulevard
P.O. Box 64526
St. Paul, MN 55164-1003

All rights reserved

Printed in the United States of America

Library of Congress Cataloging-in-Publication Data

Snow, Theodore P. (Theodore Peck)
   Physics.

   Includes index.
   1. Physics. I. Shull, J. Michael. II. Title.
QC21.2.S625   1986   530   85-26989
ISBN 0-314-93515-0

Part Opening Photos

Part One Courtesy Six Flags over Texas. **Part Two** Courtesy Fundamental Photographs, NY **Part Three** Courtesy Pennsylvania Power and Light Company. **Part Four** © Allan Birnback. **Part Five** Courtesy TRW, Inc. **Part Six** Courtesy General Electric Company.

Chapter Opening Photos

Chapter 1 American Institute of Physics, Niels Bohr Library. **Chapter 2** © Pete Saloutos Photography. **Chapter 3** NASA. **Chapter 4** Photo Researchers, Inc., N.Y. **Chapter 5** CSX Corporation. **Chapter 6** Utah Power and Light. **Chapter 7** NASA. **Chapter 8** L. E. Weiss, Department of Geology and Geophysics, University of California at Berkeley. **Chapter 9** © Fundamental Photos, NY. **Chapter 10** © David Krathwehl/Stock, Boston. **Chapter 11** COMSAT. **Chapter 12** © Peter Menzel/Stock, Boston. **Chapter 13** Photo courtesy of AMCA INTERNATIONAL. **Chapter 14** © Michael Philip Manheim/Photo Researchers, Inc., NY. **Chapter 15** © Richard Wood/Taurus Photos. **Chapter 16** Fermilab, Batavia, IL. **Chapter 17** Del Webb Corporation. **Chapter 18** Courtesy of National Radio Astronomy Observatory. **Chapter 19** Doug Johnson. **Chapter 20** © Fundamental Photographs. **Chapter 21** © Brown Brothers, Sterling, PA. **Chapter 22** NASA. **Chapter 23** Corning Glassworks, Corning, NY. **Chapter 24** © OMIKRON/Photo Researchers, Inc. **Chapter 25** Fermilab, Batavia, IL. **Chapter 26** Courtesy of Fermilab, Batavia, IL.

Figure Credits

Chapter 1

Fig. 1.1 Library of Congress. **Fig. 1.2** City Museum, Torun, Poland. **Fig. 1.3** The Bettman Archive. **Fig. 1.4** The Granger Collection. **Fig. 1.5** © Alinari/Art Resource. **Fig. 1.6** The Granger Collection. **Fig. 1.7** The Granger Collection. **Fig. 1.8** National Bureau of Standards, Gaithesburg, MD.

Chapter 2

Fig. 2.13 Dr. Harold E. Edgerton, Massachusetts Institute of Technology. **Fig. 2.1.1** NASA photograph.

Chapter 3

Fig. 3.4 © Bettman Newsphotos, UPI-Reuters Photo Libraries. **Fig. 3.1.2** Stickfigure diagram and related data courtesy John A. Miller, Jr., University of Iowa.

Chapter 4

Fig. 4.1 Russell Thompson/Taurus Photos. **Fig. 4.13** © 1984, John Zimmerman/LPI.

Chapter 5

Fig. 5.8 © Wide World Photos.

Chapter 6

Fig. 6.9a © Lester V. Bergman. **Fig. 6.9b** The United States Figure Skating Association, Colorado Springs, CO. **Fig. 6.12** Sandia National Laboratories. **Fig. 6.1.1** © Fundamental Photographs.

Chapter 7

Fig. 7.8 Palomar Observatory, California Institute of Technology. **Fig. 7.10** NASA photograph. **Fig. 7.15** Lowell Observatory photograph.

Chapter 8

Fig. 8.1 © Fundamental Photographs. **Fig. 8.10** *Powerlifting USA*. **Fig. 8.1.1** American Museum of Natural History.

(continued following index)

To Our Parents

Contents

Preface, xiii

Part One Motion and Energy 1

Chapter 1 Introduction to Physics 2

1.1 The Philosophy of Science, 3
1.2 The Method of Science, 4
1.3 A Historical Perspective, 5
1.4 Physics Today and This Textbook, 10
1.5 The Elements of Quantitative Physics, 11
Perspective, 18
Problem Solving: Overall Strategy, 19
Thought Questions, 21
Problems, 21
Special Problems, 23
Suggested Readings, 25

Chapter 2 Kinematics 26

2.1 Distance and Displacement, 28
2.2 Speed and Velocity, 29
2.3 Acceleration, 34
2.4 Gravitational Acceleration, 39
Focus on Physics: Launching Sounding Rockets, 44
Perspective, 45
Problem Solving: Solving Equations in Sequence, 46
Problem Solving: Graphical Analysis and Integration, 47

CONTENTS

Summary Table of Formulas, 50
Thought Questions, 50
Problems, 51
Special Problems, 54
Suggested Readings, 55

Chapter 3   Dynamics   56

3.1 Force, 57
3.2 Newton's Laws of Motion, 58
Focus on Physics: Biomechanics, 64
3.3 Tension and Transmission of Force, 66
Focus on Physics: A Biomechanical Analysis of a World Record, 68
3.4 Friction and Air Resistance, 69
Perspective, 74
Problem Solving: Coordinates and Force Relations, 75
Summary Table of Formulas, 76
Thought Questions, 77
Problems, 77
Special Problems, 82
Suggested Readings, 83

Chapter 4   Forces in Balance   84

4.1 The Nature of Equilibrium, 85
4.2 Center of Gravity and Torque, 88
Focus on Physics: Equilibrium in Architecture, 95
Perspective, 99
Summary Table of Formulas, 100
Thought Questions, 100
Problems, 101
Special Problems, 109
Suggested Readings, 111

Chapter 5   Momentum, Work, and Energy   112

5.1 Momentum, 113
5.2 Work, 121
5.3 Energy, 124
5.4 Power, 133
5.5 Simple Machines, 134
Focus on Physics: Work, Energy, Exercise, and Your Weight, 138
Perspective, 140
Problem Solving: Shortcuts with Energy Conservation, 141
Summary Table of Formulas, 142
Thought Questions, 142
Problems, 143

CONTENTS v

Special Problems, 150
Suggested Readings, 151

Chapter 6 Circular and Rotational Motion 152

6.1 Measures of Angular Quantities, 153
6.2 Circular Motions of Particles, 160
6.3 Inertia, Momentum, Work, and Energy of Rotation, 168
Focus on Physics: Angular Momentum and Gyroscopes, 174
Perspective, 176
Problem Solving: Problems in Rotational Motion, 177
Summary Table of Formulas, 178
Thought Questions, 179
Problems, 179
Special Problems, 185
Suggested Readings, 186

Chapter 7 Gravitation and Orbits 188

7.1 The Law of Gravitation, 189
7.2 Orbits and the Two-Body Problem, 193
7.3 Escape Velocity and Orbital Energy, 200
7.4 Artificial Satellites, 208
7.5 Differential Gravitational Forces and Tides, 210
Focus on Physics: The Secrets of Double Stars, 214
Perspective, 216
Problem Solving: Gravitation and Orbit Problems, 216
Summary Table of Formulas, 217
Thought Questions, 218
Problems, 219
Special Problems, 223
Suggested Readings, 224

Part Two Vibrations and Waves 225

Chapter 8 Solids and Elasticity 226

8.1 The Nature of Matter, 227
8.2 Restoring Forces and Hooke's Law, 230
8.3 Elastic Deformations: Stress and Strain, 232
Focus on Physics: Rocks Under Stress and Earthquake Prediction, 245
Perspective, 247
Summary Table of Formulas, 248
Thought Questions, 249
Problems, 249
Special Problems, 252
Suggested Readings, 253

Chapter 9   Oscillations   254

9.1   The Restoring Force and Simple Harmonic Motion, 255
9.2   Analyzing Simple Harmonic Motion, 258
9.3   The Simple Pendulum, 266
Focus on Physics: Oscillations and Clocks, 270
9.4   Oscillations that are not SHM, 273
9.5   Forced Vibrations and Resonance, 274
Perspective, 276
Problem Solving: Frequencies, Displacements, and Sinusoidal Motion, 276
Summary Table of Formulas, 278
Thought Questions, 279
Problems, 279
Special Problems, 282
Suggested Readings, 283

Chapter 10   Wave Phenomena   284

10.1   The Nature of Waves, 285
Focus on Physics: Waves in the Earth, 294
10.2   Interactions of Waves, 297
10.3   Stationary Waves, 306
10.4   Sound Waves, 309
Focus on Physics: Musical Harmony and Scales, 316
10.5   Shock Waves and Sonic Booms, 317
Perspective, 319
Summary Table of Formulas, 319
Thought Questions, 320
Problems, 321
Special Problems, 325
Suggested Readings, 327

Part Three   Fluids and Gases   329

Chapter 11   Fluids   330

11.1   Pressure and Buoyancy in Stationary Fluids, 331
11.2   Surface Tension and Capillary Action, 340
11.3   Fluids in Motion, 343
Focus on Physics: Drag and Fuel Economy in Cars, 355
Perspective, 357
Summary Table of Formulas, 357
Thought Questions, 358
Problems, 359
Special Problems, 361
Suggested Readings, 363

CONTENTS

Chapter 12 Temperature and Heat 364

12.1 Temperature, 365
12.2 Thermal Expansion, 368
12.3 Equations of State, 372
12.4 Heat and Energy, 382
12.5 Heat Transport, 389
Focus on Physics: Convection and the Earth, 396
Perspective, 398
Problem Solving, 398
Summary Table of Contents, 400
Thought Questions, 401
Problems, 402
Special Problems, 405
Suggested Readings, 407

Chapter 13 Gas Kinetics, Thermodynamics, and Heat Engines 408

13.1 Kinetic Theory, 409
13.2 Diffusion and Osmosis, 418
13.3 Heat Capacity and the First Law of Thermodynamics, 422
13.4 Heat Engines and the Second Law of Thermodynamics, 426
Focus on Physics: Throwing Away Heat, 434
Perspective, 436
Summary Table of Formulas, 436
Thought Questions, 437
Problems, 438
Special Problems, 440
Suggested Readings, 441

Part Four Electricity and Magnetism 443

Chapter 14 Electrical Charge and Field: Electrostatics 444

14.1 Charge, 445
14.2 Electrical Force and Field, 447
14.3 Electric Energy and Potential, 455
14.4 Storage of Electrical Energy, 461
Focus on Physics: Lightning and Thunder, 466
Perspective, 468
Summary Table of Formulas, 468
Thought Questions, 469
Problems, 470
Special Problems, 473
Suggested Readings, 475

CONTENTS

Chapter 15 — Current and Circuits 476

15.1 Electromotive Force, 477
15.2 Current, 478
15.3 Resistance, 479
15.4 Elements of Circuits, 487
15.5 Experimental Analysis of Circuits, 497
Focus on Physics: Household Circuits, 502
Perspective, 503
Problem Solving: Strategies for Analyzing Circuits, 504
Summary Table of Formulas, 506
Thought Questions, 506
Problems, 507
Special Problems, 512
Suggested Readings, 513

Chapter 16 — Magnetism and Electricity 514

16.1 Magnetism, 515
16.2 Moving Charges and Magnetism, 517
16.3 Magnetic Forces on Moving Charges, 520
16.4 Analyzing Magnetic Fields, 527
16.5 Magnetic Induction, 531
Focus on Physics: The Earth's Magnetic Dynamo, 537
Perspective, 540
Problem Solving: The Cross Product and EM Forces, 540
Summary Table of Formulas, 542
Thought Questions, 542
Problems, 543
Special Problems, 547
Suggested Readings, 549

Chapter 17 — Applied Electricity and Magnetism 550

17.1 Generators and Motors, 551
17.2 Transmission and Transformation of Power, 556
17.3 Time-Dependent Circuits, 559
17.4 Vacuum Tubes and Semiconductors, 570
Focus on Physics: Computer Memory and Data Storage, 578
Perspective, 580
Problem Solving: LRC Circuit Problems, 581
Summary Table of Formulas, 582
Thought Questions, 583
Problems, 583
Special Problems, 587
Suggested Readings, 588

Part Five Electromagnetic Radiation 589

Chapter 18 Electromagnetic Waves 590

18.1 The Origin of Electromagnetic Waves, 591
18.2 The Nature of EM Waves, 596
Focus on Physics: Measuring the Speed of Light, 599
18.3 Transmission of Energy, 601
18.4 Information Transport, 605
Perspective, 607
Summary Table of Formulas, 608
Thought Questions, 608
Problems, 609
Special Problems, 611
Suggested Readings, 613

Chapter 19 Geometric Optics 614

19.1 General Principles of Optics, 615
Focus on Physics: Optical Fiber Communications, 623
19.2 Image Formation by Mirrors, 625
19.3 Image Formation with Lenses, 631
19.4 Optical Instruments, 637
Focus on Physics: Optical Defects of the Eye, 640
Perspective, 653
Summary Table of Formulas, 653
Thought Questions, 654
Problems, 655
Special Problems, 658
Suggested Readings, 659

Chapter 20 Wave Phenomena in Light 660

20.1 The Evidence for Light Waves, 661
20.2 Diffraction and Interference, 666
20.3 Reflection and Interference, 674
Focus on Physics: Space Geodesy and Satellite Ranging, 680
20.4 Polarization and its Applications, 682
Perspective, 688
Summary Table of Formulas, 688
Thought Questions, 689
Problems, 689
Special Problems, 692
Suggested Readings, 694

CONTENTS

Part Six Atomic Structure and Modern Physics 695

Chapter 21 Relativity 696

21.1 Galilean Relativity, 697
21.2 Special Relativity, 699
21.3 General Relativity, 712
Focus on Physics: Testing Relativity, 717
Perspective, 718
Problem Solving: Special Relativity, 719
Summary Table of Formulas, 721
Questions, 721
Problems, 722
Special Problems, 725
Suggested Readings, 727

Chapter 22 Particles and Waves 728

22.1 The Particle Nature of Light, 729
Focus on Physics: Solar Cells and Energy, 737
22.2 The Wave Nature of Particles, 741
22.3 The Uncertainty Principle, 746
Perspective, 750
Summary Table of Formulas, 750
Thought Questions, 751
Problems, 751
Special Problems, 754
Suggested Readings, 755

Chapter 23 Light and the Atom 756

23.1 Experimental and Observational Clues, 757
23.2 The Bohr Model, 761
23.3 Practical Spectroscopy, 769
Focus on Physics: Natural Lasers in Space, 778
Perspective, 780
Summary Table of Formulas, 780
Thought Questions, 781
Problems, 782
Special Problems, 784
Suggested Readings, 785

Chapter 24 Essentials of Quantum Mechanics 786

24.1 Wave Mechanics, 787
24.2 Atomic Structure, 791
24.3 Changes in Quantum Numbers: Atomic Spectra, 803

CONTENTS

24.4 Molecular Bonds and Molecular Spectra, 810
Focus on Physics: Bonds in Biology, 812
Perspective, 820
Summary Table of Formulas, 821
Thought Questions, 821
Problems, 822
Special Problems, 824
Suggested Readings, 825

Chapter 25 The Nucleus and Nuclear Energy 826

25.1 Properties of the Nucleus, 827
25.2 Radioactivity, 833
Focus on Physics: The Effects of Low-Level Ionizing Radiation, 847
25.3 Nuclear Reactions, 850
25.4 Particle Accelerators and Nuclear Reactors, 857
Perspective, 866
Summary Table of Formulas, 867
Thought Questions, 867
Problems, 868
Special Problems, 871
Suggested Readings, 873

Chapter 26 Elementary Particles and Cosmology 874

26.1 Particle Exchange and Field Forces, 876
26.2 Particle Interactions, 882
26.3 Quarks and Gluons, 891
26.4 Matter in the Universe, 894
Focus on Physics: The Cosmological Redshift and Universal Expansion, 895
Perspective, 904
Summary Table of Formulas, 904
Thought Questions, 904
Problems, 905
Special Problems, 907
Suggested Readings, 908

Appendix 1 Fundamental Constants A1

Appendix 2 Review of Mathematical Techniques A2
Numerical Accuracy and Scientific Notation A2
Algebra A3
Trigonometry A5
Logarithms A7
Exponential Functions A9
Binomial Expansion A11

CONTENTS

Appendix 3 Units and Conversion Factors A12

Appendix 4 Table of Planetary Data A14

Appendix 5 Periodic Table of the Elements A15

Appendix 6 Elemental Abundances A16

Appendix 7 Selected Isotopes A18

Appendix 8 Answers to Odd-Numbered Problems A22

Index, I1

Preface

In writing this text, we have endeavored to bring home to the student the connection between everyday experience and physics. The text is designed for the science student pursuing a field other than physics, who is therefore ultimately most interested in applying physics to another field. At the same time, we have written a book that is basically about physics; its fundamental concepts, and how those concepts are applied to understand the interactions of matter and forces. Thus, we have tried to meet a dual goal. We have, we believe, succeeded in this by including both the basic physics and applications to other fields, with distinct boundaries between. The fundamental material is treated in the main text, with applications being discussed in insert boxes entitled *Focus on Physics,* and in special problems. This allows the student to concentrate without distraction on the principles of physics, while reading separately about the applications to fields of interest. Applications discussed include biophysics, medicine, biomechanics, geophysics, geodesy, astrophysics, architecture, computer science, music theory, and others related to everyday life.

Our aim was to write a concise book, one not overly long or burdened with too much detail, particularly in specialized applications. Thus, we have only 26 chapters, each covering a broad area in physics. We hope that this will allow instructors to cover the entire book in a full year course, something often impossible with longer texts. At the same time, we have left out no basic physics; our economy has been gained by avoiding undue emphasis on secondary topics, not by excluding fundamental ones. Because of the many special features and pedagogical aids that it contains, the book is substantial in size, even though the written text itself is briefer than most of the competitors.

The use of mathematics in the text is limited to algebra and trigonometry, although in a few places it proves useful to mention the notions of limits and graphical integration. An extensive appendix on mathematical methods refreshes the student on the techniques and explains some special types of functions (logarithms, exponentials, binomial expansion) that arise in the text.

Great emphasis is placed on teaching the students how to solve problems. This is done by example; there are many worked problems in each chapter. It is also done by including 12 special boxed inserts called *Problem Solving* throughout the book, wherever new physical concepts lead to new methods for solving problems. Thus, the student receives continual assistance from the text in learning how to approach problems. Because problem-solving skill is perhaps the most important goal of the student using this book, we have placed strong emphasis on the problems for the student to solve, at the end of each chapter. There are 1715 original problems, keyed both for difficulty and for reference to the section of the chapter they are based on. In addition to the problems, there are 529 challenging thought questions, aimed at reinforcing the comprehension of concepts, and there are 110 "Special Problems," which require unusual imagination and understanding on the part of the student, but which are also creative and interesting. Many of these are applied problems in fields where physical principles from the text are needed.

The opening chapter has the first problem-solving insert, as well as explanations in the text about experimental values and significant figures, and about vectors and vector operations. The *Systeme International* (SI) system of units is explained there also, and used throughout the book. In worked examples and in the problems, the SI units are used predominantly, but it is our belief that students should gain some experience in the use of other units, since they cannot be avoided in real life. Thus, some of the examples and problems require the use of other units, or, more commonly, their conversion to the SI standards.

We have included over 700 drawings and photographs to enhance the text and make it more understandable. One 32-page signature of the book (in the section on light) has been printed by a four-color process, allowing us to use full color illustrations in that section. This not only beautifies the book, but aids in the explanation of concepts related to light.

Included in the book are many reference tables giving accurate and up-to-date values for various parameters. We have gone to great lengths to ensure that we have the best and most recent values from the literature, rather than simply relying on standard tables that have been reproduced for years. In the Appendices are equally current and thorough tables of fundamental constants, conversion factors, astronomical data, and data on the elements and their abundances, and on atomic isotopes. We believe that these tables will prove useful to students and instructors alike.

The organization of the book is traditional, beginning with classical mechanics. It then moves on to elastic properties of solids and a treatment of oscillations, which is followed by a section on thermodynamics and gas kinetics, and then two on electricity and magnetism and on electromagnetic radiation. The final section covers modern physics, starting with relativity and then delving into the world of atoms, nuclei and elementary particles. This sequence of topics has proven to be pedagogically sound and self-consistent, allowing students to build on the sometimes difficult concepts of force and energy developed in the early chapters as they move on to the other topics later in the book.

The opening section on mechanics (Part One) contains seven chapters; an opening chapter which is really a basic introduction to the entire science of physics (including a brief overview of historical developments), and then five chapters in which traditional topics are developed in the standard sequence, starting with kinematics (chapter 2), then dynamics (chapter 3), statics (chapter 4), energy, momentum, work, and conservation laws (chapter 5), and circular motion (chapter 6). The final chapter of this section is unique among texts at this level; it contains the essentials of the two-body problem, orbits, and tidal forces, topics of increasing importance in the space age.

Some may prefer to follow the study of classical mechanics directly with a treatment of relativity, although our preference is to reserve the latter for the final section on modern physics. Even so, the relativity chapter (chapter 21) can be taught directly following the section on classical mechanics.

Part Two of the text covers solids and elasticity (chapter 8), oscillations (chapter 9), and waves (chapter 10), including discussion of sound waves. Although many of the principles developed in these chapters are applicable to light, we defer the treatment of light waves until after the section on electricity and magnetism.

Part Three, also containing three chapters, covers fluids (chapter 11), heat and temperature (chapter 12), and thermodynamics and heat engines (chapter 13). The next section (Part Four) has four chapters in which electricity and magnetism are thoroughly but succinctly treated: there are chapters on electrostatics (chapter 14), current and circuits (chapter 15), magnetism (chapter 16), and, finally, on applied electricity and magnetism (chapter 17), including both DC and AC applications, along with modern electronics. Part Five, three chapters long, contains our discussions of electromagnetic waves (chapter 18), geometric optics (chapter 19), and interference phenomena (chapter 20).

The final section of the book (Part Six) is a collection of chapters on modern physics. We view this section as very important for students using this text, because so many of today's applications of physics to other fields involve modern physics concepts, and because these students, unlike those majoring in physics, may not get another good chance to become acquainted with the concepts of modern physics. For these reasons we have tried to make the book compact enough so that the final section can be covered in a normal course. The first chapter in this section (chapter 21) is on special and general relativity, with application to cosmology. The next are on waves and particles (chapter 22), atomic spectra (chapter 23), essentials of quantum mechanics (chapter 24), the nucleus (chapter 25), and on elementary particles and the early universe (chapter 26). We believe we have written the most up-to-date and complete summary of these topics to be found in any text at this level (examples are the discussions of quark theory and quantum chromodynamics in Chapter 26), and we hope that students and instructors alike will enjoy them.

Several people assisted us, in various ways, in preparing this text. Among them were John Taylor, our colleague at the University of Colorado, who subjected much of the manuscript to intense scrutiny and advised us on both accuracy and strategy; and a number of people who provided information and advice for the applications inserts: Stefanie Baker, Peter Bender, Edward Benton, Ron Canterna, Robert Gallawa, Scott Shull, Peter Stoner, and Ken Zweibel. Larry Esposito provided invaluable assistance in locating current data on the planets for Appendix 4. Help with photographs and illustrations was provided by Bruce Bohannan, Mary Bonneville, Mircea Fotino, Doug Johnson, Jeremy Pickett-Heaps, Uriel Nauenberg, Gary Rottman, Ron Thomas, Bill

McClintock, and John Cumalat. We are also grateful to the Astronomy Department at UCLA, for providing hospitality to one of us (T.P.S.) while the first draft was written.

A number of reviewers, all of them teachers of physics at this level, also assisted in the preparation of this text, particularly in helping to ensure its accuracy. We acknowledge their help, with gratitude:

Paul A. Bender, Washington State University,
Roger N. Blais, University of Tulsa,
Steve Brooks, Diablo Valley College, California,
John Brunn, Chabot College, California,
Robert Clark, Texas A & M University,
Harold Cohen, California State University, Los Angeles,
J. P. Davidson, University of Kansas,
Dean Dragstorf, Kansas State University,
Harry T. Easterday, Oregon State University,
Robert J. Endorf, University of Cincinnati,
David Ernst, Texas A & M University,
Rex A. Freeman, North Hampton Community College, Pennsylvania,
James B. Gerhart, University of Washington,
Christoph K. Goertz, University of Iowa,
Marvin Goldberg, Syracuse University,
Walter Gray, University of Michigan,
H. James Harmon, Oklahoma State University,
Richard Harris, McGill University,
Stanley Hirschi, Central Michigan University.
Gordon E. Jones, Mississippi State University,
Michael Lieber, University of Arkansas,
David Markowitz, University of Connecticut,
Raymon W. Mires, Texas Tech University,
Eugene R. Niles, Antelope Valley College, California,
Michael J. Nolan, Millersville University, Pennsylvania,
Kenneth D. O'Dell, Northern Arizona University,
Michael Paesler, North Carolina State University,
T. R. Palfrey, Jr., Purdue University,
Hans Plendl, Florida State University,
Kwangjai Park, University of Oregon,
Roger W. Rollins, Ohio University,
Lee Rutledge, Oklahoma State University,
Manual Schwartz, University of Illinois,
John Shelton, College of Lake County, Illinois,
John L. Stanfor, Iowa State University,
John P. Wefel, Louisiana State University,
Stan A. Williams, Iowa State University,
John G. Wills, Indiana University,
Timothy M. Young, Schenectady Community College.

All of the drawing concepts were reviewed by Kwangjai Park, to whom we are indebted. The problems and their solutions were all checked by Murali Ramanathan and Dave Rost. Roger Culver (Colorado State University) assisted with the preparation of the thought questions. Marsha Allen provided invaluable assistance in the preparation of the reading lists and the tables, both in the text and in the Appendices. Wendy Calmenson designed the book and ably oversaw the acquisition of all of the illustrations.

Finally, we recognize with thanks the efforts of the editorial staff at West Publishing Company: Denise Simon, who originated the book and saw it through its early development; Peter Marshall, who oversaw the creation of the manuscript; and Pam Rost, the production editor who turned it all into a real book. Theresa O'Dell coordinated the reviews. Without these people, the book not only would have been impossible; it never would have been attempted.

PART ONE

MOTION AND ENERGY

Chapter 1
Introduction to
Physics 2

Chapter 2
Kinematics 26

Chapter 3
Dynamics 56

Chapter 4
Forces in Balance 84

Chapter 5
Momentum, Work,
and Energy 112

Chapter 6
Circular and
Rotational
Motion 152

Chapter 7
Gravitation and
Orbits 188

Chapter 1

Introduction to Physics

To begin a study of physics, it is helpful first to understand what physics is, and the role it plays in our lives. These aspects of physics are intermingled with the nature and role of science in general, so in a sense the first part of this chapter could have been written for a text in many other fields.

It is also important to start our study with the appropriate tools and background information. Therefore, following our discussions of the philosophy and historical development of physics, we will conclude this chapter with some of the important techniques to be used throughout the book, particularly the mathematical representation of physical quantities.

1.1 The Philosophy of Science

To undertake a study of physics is to study science itself. Indeed, for most of recorded history no distinction was made; natural philosophy, the general effort to understand all natural phenomena, embraced all the sciences. There was no such thing as a physicist or a geologist or an astronomer. Today we have such things, but we recognize that physics is the modern embodiment of the ancient concept of natural philosophy; it is the foundation for understanding all natural phenomena.

When we speak of physics and why it is studied, we speak inevitably of humankind's urge to explore, to seek out the workings of the universe. "Physics" is, after all, only a word, but what it entails is more than a discipline or a methodology; it is the basic concept that natural phenomena are capable of being understood in terms of a few simple laws and that the universe operates according to certain rules. In this sense physics (or more generally science) may not seem any different from any adopted system of belief such as a religion, but there is a vital distinction: in science a rule of the game is not accepted unless it can be tested and verified repeatedly, and in science modifications of the rules are always accepted if new tests or new observations indicate that changes are needed. Unlike a dogmatic follower of a faith, a scientist can never believe that he or she understands perfectly the ways of the universe.

The challenge for the scientist is to uncover the underlying order in a complex cosmos. If anything is taken on faith by a scientist, it is the belief that such order exists and can be understood. The seeker of the rules of the game is tantalized by the knowledge that the universe exists and operates, and that the rules being sought operate continuously all around him. The understanding that is sought is literally at one's fingertips.

1.2 The Method of Science

The pursuit of science is often said to follow the **scientific method**, a sort of recipe alleged to lead inevitably to new discoveries. From elementary school on, we learn the steps in the scientific method as though they were gospel. There is certainly some accuracy in the description of the scientific process embodied in the classical scientific method, but is incorrect to assume that scientists succeed by simple adherence to it. Above all, a scientist must be capable of recognizing and accepting new ideas. A scientist cannot merely follow a recipe.

Despite these cautions, it is generally fair to characterize the discovery process as having certain distinct stages. First there must be some input, some information on natural phenomena that provides the basis for asking and attempting to answer a question. This input may be the result of a deliberate effort to obtain data, or it may occur accidentally. Second, a hypothesis must be formed. This is a guess at the underlying rule or principle that explains the input information, and the scientist advancing the hypothesis is supposed to regard it as merely tentative until the next stage, testing the hypothesis, is carried out. Some scientists retain this remoteness from their newborn ideas better than others, but this does not matter, so long as they are willing to change or drop the ideas in the face of contrary evidence. The third stage in the development of a new concept is most crucial, and is where science and faith diverge. The hypothesis must be tested by observation or experiment, and the tests must be repeatable. A new rule of the game must always apply to phenomena other than the observation or experiment that led to the hypothesis in the first place, and new observations or experiments must be carried out to see whether the rule accurately predicts the outcome. If it does, then the new rule can continue to be regarded as possibly correct. If it does not, then the new rule is clearly not a rule at all, but only an inadequate suggestion. A rule can never properly be regarded as proved. At best it can only be recognized as having passed all the tests applied to it so far. Hence scientists tend to refer to their current concepts of the universe as theories, no matter how well established they may be.

It is worthwhile to reconsider in more detail the three steps toward scientific discovery. Each seems simple in concept, but each can be quite complex in practice. Gathering data, for example, is a very complicated, perilous business.

One must be certain that the observation or experiment is properly planned, so that the information gathered will be as unambiguous as possible. In addition, there are many possible sources of uncertainty in any experiment or observation, and care must be taken to understand and allow for all possible errors. These can be quite subtle, and can be introduced not only by the complex nature of the phenomenon being studied, but also by the very instruments used to carry out the study. One of the most important aspects of any analysis is to ferret out all the effects introduced by instrumentation. Failure to find them all is one of the principal reasons for erroneous results.

The second step in the development of new concepts, the formation of a hypothesis, is the most difficult to explain. The ability to make new hypotheses varies greatly among individuals, and is the factor that determines a scientist's success. To gather data and to test a hypothesis are processes for which a person can be trained, to invent new ideas is not. The word "intuition" is generally used to describe the mysterious mechanism by which hypotheses are born.

Finally, the testing of a hypothesis involves at least two potential pitfalls. All of the difficulties of gathering data are encountered again as new observations or experiments are carried out, and there is the added difficulty of designing experiments that truly test the hypothesis. It is necessary to predict correctly the consequences of the hypothesis, and then devise a test that will unambiguously confirm or contradict that hypothesis.

1.3 A Historical Perspective

Despite the difficulty of describing in rigorous terms how it happens, science does go forward. There are many false turns along the way, and no doubt more will be encountered, but certainly humankind's understanding of the universe has grown vastly. It is perhaps instructive to learn a little about the course of this growth.

Early Developments

Today scientific research is carried out vigorously throughout the world, and it probably always has been. Because few ancient records have survived, our summary of historical developments is confined to those cultures best known to us today. Much of the earliest scientific speculation was astronomical because ancient peoples sought to explain the many cyclical events they observed in the heavens, and how those cycles were related to events on the earth, such as the seasons. We know that astronomy developed in the first two millennia B.C. in several locations. We also know that it was the early Greek science, probably influenced by even earlier cultures, such as the Babylonian, that led more or less directly to our modern astronomical science. Other early astronomies either died out or were eventually merged into the Greek.

As it was in astronomy, so it was in other areas of natural philosophy. The Greek civilization provided the earliest framework for comprehending the

Figure 1.1

Aristotle.

workings of natural phenomena, set the stage for the Renaissance and later Western developments, and gave rise to all modern science. The philosopher usually credited with the first rational inquiry was Thales (ca. 624–547 B.C.), who lived in Miletus, on the eastern shore of the Aegean Sea. His principal contribution was the idea that inquiry can lead to understanding, that is, that the cosmos is knowable. Another influential figure was Pythagoras (ca. 570–500 B.C.), widely known today for his geometrical theorems. His major contribution was more fundamental, however; it was Pythagoras and his followers who introduced the notion that natural phenomena can be described mathematically, a concept at the very heart of all modern science.

The Greek culture reached perhaps its most refined development in the city state of Athens, and it was there that the philosopher Plato (428–347 B.C.), who had studied with Socrates as a youth, established his school. Plato's fundamental precept was that what we can observe of the natural world is only an imperfect representation and that the universe consists of perfect bodies governed by idealized laws. Plato adopted several natural laws, which he derived more by intuition than by observation. Furthermore, he believed that observation and experimentation were unnecessary, since it was impossible to actually see the true nature of things. Hence, Plato and his followers established laws of nature without ever taking the next step, verification of those laws.

The most influential of Plato's followers was Aristotle (ca. 384–322 B.C.; Fig. 1.1), who attempted to explain the workings of the universe in the context of a complex set of physical laws. He taught that all natural bodies, including the earth, are perfect spheres, and that in the heavens all motions are circular. Even the universe itself had to be spherical, or else one was confronted with the untidy possibility that it had edges. On the earth, the only natural motion was downward; one of Aristotle's laws of physics was that every body naturally sought the center of the universe, identical with the center of the earth (he even used this law to confirm that the earth is spherical, since only on the surface of a sphere can all lines pointed towards the center be perpendicular to the surface). Another tenet of Aristotle's was that horizontal motion occurs only while a force is applied; the natural tendency of any moving object was to stop.

Although we have spoken of only astronomy and physics, the attitude towards inquiry that led the way in these sciences also prevailed in others. Early studies of anatomy, for example, were guided by direct observation of dissected corpses, but many of the theories that were adopted were based principally on the Platonic tradition of deduction by reason.

The Renaissance

Following the decline of the Greek civilization, there was a long period when little scientific advance took place. Fortunately, extensive records and verbal traditions survived the Dark Ages, for much of the rebirth of the Renaissance was stimulated by the discovery of ancient Greek teachings. Copernicus (1473–1543; Fig. 1.2), for example, was well acquainted with the astronomical teachings of Aristotle and later Greeks such as Hipparchus and Ptolemy. Co-

1.3 A HISTORICAL PERSPECTIVE

Figure 1.2

Nicolaus Copernicus.

pernicus developed his sun-centered model of the universe in part because he felt that it more closely adhered to the principles set forth by Aristotle than did the earth-centered theory, which had become rather complex. Hipparchus and Ptolemy had departed from earlier Greek tradition by insisting that theories of the heavens agree with observations, and had been forced to develop a model of the universe in which the planets moved on small circles, called epicycles, which themselves orbited the earth. Even worse, the orbits of the epicycles were not centered precisely on the earth, further violating the teachings of Aristotle (offsets were required because the planets actually move in elliptical orbits around the sun and their orbital speeds are not constant). Copernicus found it much more elegant and intuitively satisfying to adopt the view that each planet circles the sun. It was especially pleasing to him that the distances of the planets from the sun could then be derived by simple geometrical arguments, and the orbits of the plants made a nice pattern of widening concentric circles. Unfortunately, Copernicus adhered to the notion that all motions must be circular, and he was forced to add epicycles to make theory agree with observation. In the end, his revolutionary theory was neither much simpler nor any more accurate than the earlier earth-centered models.

By 1514, the time of Copernicus's earliest known reference to his theory, the ancient Greek teachings had acquired the force of religious dogma. Copernicus was reluctant to publish, until very late in his life. There were no immediate repercussions, but the academic world took notice, and later philosophers were strongly influenced. The spirit of inquiry that characterized the Renaissance led to continued investigations, many of them still primarily in astronomy. Several European monarchs engaged court astronomers, most of whom were assigned to keep track of the heavens for astrological purposes. One who went far beyond this was Tycho Brahe (1546–99), appointed by the king of Denmark. Tycho systematically observed planetary positions with considerably more diligence and accuracy than earlier astronomers. Although he could not fully accept the sun-centered view of Copernicus (he suggested that the sun orbited the earth while the other planets orbited the sun), he never developed his own detailed theory of the universe.

Just before his death, Tycho hired a young assistant named Johannes Kepler (1571–1630; Fig. 1.3), who was given the task of deriving a law of planetary motion from the wealth of observational data Tycho had accumulated. This Kepler did, formulating and rejecting many hypotheses suggested by geometrical figures. Kepler's work well illustrates the principle that a law must not be accepted until it can be verified repeatedly. He developed intuitive hypotheses to which he felt strongly committed, but he was always willing to abandon them if the observations failed to bear them out. For example, he once developed an epicyclic earth-centered model for the motion of Mars that closely matched the observations, missing the measured position of the planet by only 8 minutes of arc at most (typical observational accuracy was then not much better than this). Nevertheless, Kepler rejected this model, knowing that Tycho's measurements were so accurate that even so small an error was unlikely.

Kepler eventually arrived at three laws of planetary motion. He found that the planets move around the sun in elliptical orbits, that the orbital speed of each planet varies inversely with its distance from the sun, and that the size

Figure 1.3

Johannes Kepler.

Figure 1.4

Galileo Galilei.

of the orbit is systematically related to the time each planet takes to orbit the sun. These were empirical laws, deduced by fitting hypotheses to the data, with no deeper understanding of orbits and planetary motions (Chapter 7). Kepler did speculate about why the universe works as it does, and he realized that somehow the sun causes the planets to move, but he did not stumble onto the concept of gravity.

While Kepler worked in northern Europe, Galileo Galilei (1564–1642; Fig. 1.4) pursued his own investigations in Italy. Galileo's accomplishments are often cited as the beginning of modern science, although others of his era should share the credit. Certainly Galileo helped establish the general principles now known as the scientific method, and he deserves recognition as a founder of experimental science. Argumentative and egotistical, Galileo was always ready to undertake debate, and he delighted in trapping opponents into logical fallacies.

Galileo's early work was in the science now known as mechanics, the study of moving objects. He carried out many simple experiments with balls and inclined planes (Fig. 1.5), as well as weights and pendulums, and quickly overthrew some cherished ancient Greek tenets which were still held in Galileo's time. Galileo found that a moving body tends to stay in motion unless acted on by a force, quite contrary to the understanding of Aristotle, and that objects of unequal weight fall at the same rate, contradicting another widespread idea. To reach the first of these conclusions, now known as the principle of inertia, Galileo performed experiments, such as rolling a ball, in which friction was reduced and established that in most instances a force acts to stop the motion of an object. To understand the rate of fall of objects, Galileo allowed balls to roll down inclined planes, thus slowing the fall to permit careful observation. By gradually inclining the plane he showed that the rate of fall approached that of freely dropped objects. These experiments not only showed that the rate of fall does not depend on weight, but also led to the concept of acceleration, or rate of change of speed. He found that falling objects accelerate at a constant rate, increasing their speed downward by equal amounts in equal time intervals. Galileo also discovered that a weight suspended from a string swings with a constant frequency (that is, the time required for each swing is the same), even though the length of the swing decreases gradually due to friction. The first pendulum clocks were constructed on this principle.

The work of Galileo established physics as a science, and from that time physics and astronomy have been pursued independently (although certainly not in isolation, for modern astronomy is regarded simply as the application of physics to phenomena outside of the earth). Galileo himself actively pursued astronomy in later years, and applied his principles of observation, experimentation, and verification to it. He was the first person known to have studied the nighttime sky with a telescope, and he made many discoveries that conflicted with the established earth-centered doctrine. He did not develop any new theories of the cosmos, but rather found many logical and observational arguments to support the sun-centered theory introduced by Copernicus and refined by Kepler. Moreover, Galileo brought his ideas before the public in an aggressive manner that made him unpopular with Church and government officials, but which also permanently overthrew the ancient teachings. His most famous publication was the *Dialogue on the Two Chief World Systems*, in

1.3 A HISTORICAL PERSPECTIVE

Figure 1.5

Galileo analyzing gravitational acceleration using inclined planes.

Figure 1.6

Isaac Newton.

which he ridiculed the Church while demonstrating the merits of the sun-centered theory. He attempted to conceal his own beliefs by posing his arguments in the form of a conversation among three fictional characters. It was a thin disguise, however, and Galileo found himself in trouble for the rest of his days, for he had earlier promised not to publish radical ideas. He was under house arrest for the last decade of his life, a relatively quiet period for him except for the publication of a book on his earlier discoveries in mechanics.

Newton and the Development of Modern Physics

The stage was now set for the entrance of Isaac Newton (1643–1727; Fig. 1.6), one of the most influential scientists of all time. After his education at Cambridge, Newton, an Englishman, spent two years (1665–67) on his family's farm while the Plague was raging in London. During this brief interval, surely one of the most productive periods of individual discovery in human history, Newton developed unprecedented understandings of mathematics and physics. He stated the laws of motion which are the basis for the science of mechanics, developed the calculus, realized that the gravitational force exists and deduced its mathematical description, applied his laws of motion and gravitation to a wide range of phenomena (deriving, among other things, Kepler's laws of

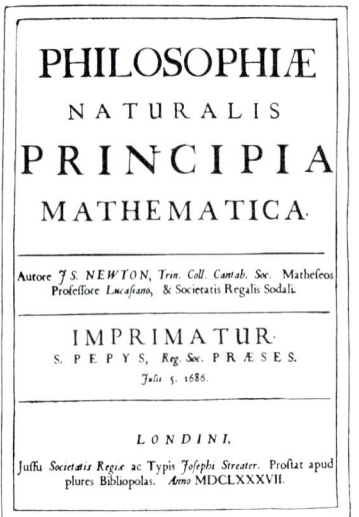

Figure 1.7

The title page from an early edition of the Principia.

planetary motion), and even undertook an analysis of light and optics which included inventing the reflecting telescope.

Most of what Newton accomplished while sequestered on the farm went unpublished for years. His massive book *Philosophiae Naturalis Principia Mathematica* (now known simply as the *Principia*), did not appear until 1687 (Fig. 1.7). In the intervening years, some of Newton's discoveries had been repeated independently by others; most notable was Leibnitz's development of the calculus (which led to a distasteful dispute between the two men and their followers). In his later life Newton became a government official and ceased scientific research, except for publishing two revisions of the *Principia*, in 1713 and 1726, and a book on his discoveries in optics, in 1704.

The work of Newton in mechanics and gravitation has survived almost intact to the present, and is still considered a valid representation of nature in many circumstances. The modern science of mechanics is based on Newtonian principles, which are superseded only in cases of extremely high velocities, strong gravitational fields, or large distances, when special and general relativity theory must be brought to bear, and in atomic interactions, where quantum mechanics applies.

In modern physics we distinguish only four fundamental forces in nature. If we were to discuss the history of physics in terms of these four fundamental forces, then we have begun properly with the events leading to the work of Newton, for the gravitational force, the weakest of the four, was also the earliest known. The electromagnetic force was not discovered until two hundred years after Newton, and the weak and strong nuclear interactions were unknown until the twentieth century. Other fundamental phenomena, such as light and electromagnetic radiation, the structure of atoms and subatomic particles, and the properties of gases and fluids, have also been understood in the past century. The developments that led to these discoveries will be mentioned in later chapters. With our present background on the development of scientific thought and of mechanics, we have seen physics safely on its way into the modern era, where progress is made in ever-diversifying fields and at a vastly increasing rate.

1.4 Physics Today and This Textbook

The science that is known today as physics has broadened rapidly from its humble beginnings in astronomy and mechanics. A person planning a career in physics is pressed, early in graduate training or perhaps even before, to decide upon a specialty. One can enter solid-state physics, nuclear physics, chemical physics, astrophysics, high-energy physics, biophysics, or any of several others. The choice of careers is just as broad as the choice of disciplines. Today's graduates enter industry, government research, the academic realm (advanced degrees are usually required in the latter two areas), or related sciences such as astronomy, geophysics, chemistry, biology, and medicine, or seemingly unrelated fields such as law. Engineering is usually treated as a separate field, but it really can be viewed as applied physics, since there is a wide overlap between engineering and physics curricula. Even the growing field of computer science employs physicists as well as engineers and mathematicians; the development and design of memory devices and of new electronic components fall under the heading of solid-state physics.

This textbook is designed for students of sciences other than physics. The goal is to develop in the student a working familiarity with the physics principles that come into play in these other sciences. It is clearly impossible to tailor the descriptions and examples for each of the possible applications, but it is feasible to demonstrate the principles so that the student will be able to see their relevance by using everyday examples. It is vital that the student follow and understand the examples, and work out the problems. Physics is a participatory science.

1.5 The Elements of Quantitative Physics

To make quantitative analyses of physical phenomena, we must have a mathematical framework. Like Pythagoras, we must have a system of units of measure and techniques enabling us to represent mathematically the aspects of the universe that we wish to understand.

Dimensions and Units

Since the time of Pythagoras, science has been based on mathematics. To express quantities and the relationships among them with numbers and equations, we must understand the dimensions of a quantity. When we want to discuss the size of an object, for example, we must decide whether its linear dimensions, its volume, or its mass are of interest. Linear dimensions such as diameter, width, or length are measured in linear units such as inches, feet, or meters. Volume is measured in cubic inches, cubic feet, or cubic meters. Mass is measured in grams, kilograms, or the relatively obscure unit, the slug.

Communication requires a common language. Communication among scientists, especially to compare results, is simplified if everyone uses the same set of units. If two physicists perform experiments to measure the same quantity, but each expresses it in different units, then the results cannot easily be compared. In most cases, conversions between units are straightforward, but not always, and the use of different systems certainly introduces unnecessary confusion. Hence a standardized unit system called the *Système International d'Unités* (International System of Units or SI system) has been adopted by scientists around the world. Many of the units used in the SI system are becoming familiar to us as units of the metric system: the unit of length is the meter, for example, and the unit of mass is the kilogram. As a rule, we will use the SI units, but in some cases we also express quantities in the units that are more familiar to most students in the United States.

A standard system of units implies a set of standard weights, lengths and so forth, by which those units are defined. Historically, the standards have been nonuniform (the foot, for example, has been defined in many different ways, including its original definition, the length of the foot of an English

Figure 1.8

The standard kilogram meter at the U.S. Bureau of Standards, similar to the one in France.

king), but gradually a set of uniform standards has been developed. At first, the standards actually were material objects, such as a platinum meter stick kept in France, but today most standards are expressed in terms of quantities measurable in the laboratory. For example, the meter is now defined in terms of the speed of light, which has been very accurately measured. Because time, through the known vibration rates of certain kinds of atoms, can be measured more accurately than length, a length standard, the meter, is defined as the distance light travels in a certain time. The kilogram, on the other hand, is based on the mass of a specific block of metal that is kept in France (Fig. 1.8).

There are only seven *basic* quantities: length (measured in meters in the SI system), mass (kilograms), time (seconds), electric current (amperes), temperature (degrees Kelvin, called simply Kelvins), amount of substance, or the number of atoms or molecules (moles); and luminous intensity (candela). There are in addition many *derived* quantities, such as speed (meters per second), which are combinations of basic units. For convenience in avoiding the use of extremely large or small numbers, various multiples of the standard units are commonly used. These are usually factors of 10^3 larger or smaller than the base unit (but not always—the centimeter is 10^2 smaller than the meter). We speak of kilometers, for example, or grams (one gram is 10^{-3} kilogram).

Appendix 1 summarizes the units in the SI system, outlines other commonly used systems, and gives useful conversion factors. Very frequently in solving physics problems it is necessary to convert quantities from one system to another. The conversion of units is very simple, but nevertheless care must be taken to do it correctly, for improper conversion is a very common source of errors in problem solving. Generally, a conversion factor is known in terms of how many of one unit equal one of another unit; for example, 0.0254 m equals one inch. The conversion of a measurement from one unit to another requires only multiplication by the conversion factor. For example, if we wish to find how many meters are in a length of 14 inches, we write

$$14 \text{ in} = (14 \text{ in}) \times (0.0254 \text{ m/in}) = 0.36 \text{ m}.$$

Note that in this example, care was taken to write the units for each quantity *including the conversion factor,* which was given in terms of the ratio of meters to inches. When the multiplication was carried out, the inches cancelled out, leaving only meters as the unit in the result. Including the units in this way is very helpful, because had there been an error, incorrect units would have been left after cancellation.

Accuracy in Measurements and Calculations

There is always a limit to how accurately we know the value of any quantity in physics. When we perform an experiment, for example, we can measure a value only within some range. We may improve the accuracy by using better techniques for measurement, but, nevertheless, some uncertainty persists. When we express values, therefore, we must be careful to keep their intrinsic accuracy in mind. For example, we may roughly measure a person's height to be 1.7 m, or we may find more accurately that it is 1.700 m. The difference between the way we express the value in these two cases is important. In the first case, we have measured the value to two **significant figures**; in the second, we have measured it to four significant figures.

In dealing with practical problems in any science, we should never speak of measured quantities without implicitly allowing for the uncertainty of the measurements. It is nonsense to calculate the density of an object to six significant figures, for example, when its volume and mass may be known only to two or three significant figures. The uncertainty of the measurement is almost as important to know as the measurement itself, and the uncertainties are always stated with the results. In doing calculations, it is important to keep the accuracy of the numbers in mind, so that the result does not appear more accurate than the data were. In general, the accuracy of a result should be stated with no more significant figures than the *least* accurate measured number. Throughout this book, the examples carefully adhere to this principle. Equal care should be taken by the student doing the exercises.

Scalars and Vectors

Most of the mathematics in this textbook is familiar to anyone who has had high school algebra. Hence algebra and trigonometry will not be reviewed, although the elements of trigonometry are summarized in Appendix 2 for those who wish to review them. There is one mathematical formalism, however, that may not be familiar to many students and that is discussed here.

Throughout our study of physics we will encounter quantities that have a direction associated with them. A force is not properly specified, for example, unless we say which way it is acting; a displacement likewise means little to us unless we know its direction. This becomes especially apparent when we consider the cumulative effect of two or more forces or displacements. Suppose we drive a car a distance of 3 km in one direction, and then 2 km in the opposite direction. We do not finish 5 kilometers from our starting point, as you would conclude if you simply added up the distances travelled, but only 1 km away. To compare and add such quantities as force and displacement it is necessary to consider direction, and the addition requires geometrical or trigonometric considerations. A quantity with both a magnitude and a direction, and which obeys certain rules for addition and subtraction (described below) is called a **vector**. Velocity, force, acceleration, and electric field are examples of vectors. In this book, as is customary in printed materials, vector quantities are represented by boldface letters: **v**. In handwritten equations, a small horizontal arrow is usually written just above symbols for vector quantities ($\vec{v}$).

Some quantities can be completely expressed by stating the magnitude without reference to direction. The mass of an object is such a quantity; masses can be added without regard for geometry or directionality. Quantities like this, which can be expressed by a magnitude only, are called **scalars**. Other examples of scalars are distance, time, and electric charge. Most of us become familiar with scalar quantities at a young age, although we may not know the term for them. All our early training in arithmetic and algebra involves strictly scalar quantities—quantities that can be added, subtracted, multiplied, or divided.

14 CHAPTER 1 INTRODUCTION TO PHYSICS

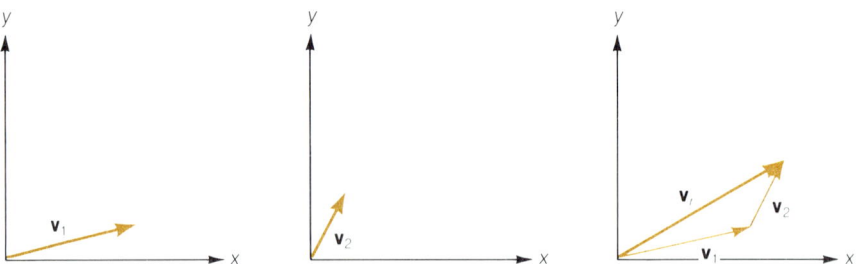

Figure 1.9

Vector addition. To add vectors V_1 and V_2, we place the tail of V_2 at the head of V_1, then draw the resultant vector V_r. The magnitude and direction of V_r can be measured directly on the graph or it can be calculated by component analysis.

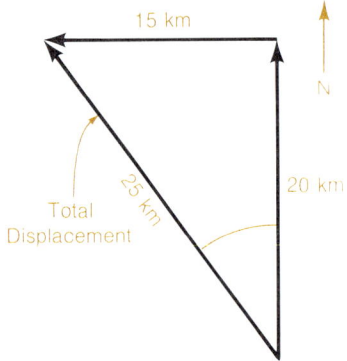

Figure 1.10

Adding displacement vectors. A ship sails 20 km north, then 15 km west. The final displacement is found by placing the two vectors as shown, and drawing the resultant vector.

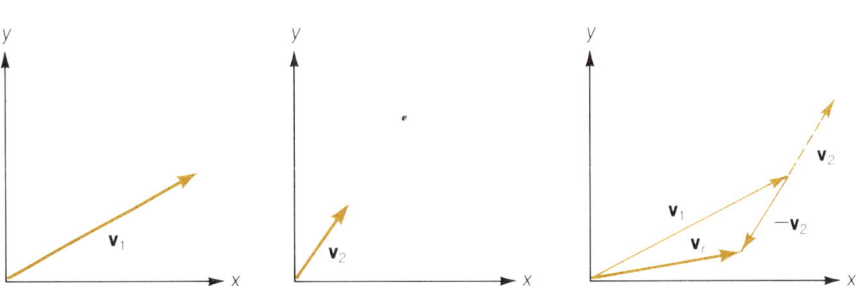

Figure 1.11

Vector subtraction. Here V_2 is to be subtracted from V_1. This is accomplished graphically by reversing the sign, hence the direction, of V_2 and then adding by placing V_2 head to tail with V_1.

In physics, scalars can be added and subtracted like ordinary numbers (but the scalars being added or subtracted must have the same units!). However, for vectors the sum depends on the directions of the vectors. One way to add or subtract vectors is to plot them on graph paper, placing the tail of one vector at the head of the other, then drawing a new vector from the tail of the first one to the head of the second (Fig. 1.9), and measuring this vector. This vector, called the **resultant**, represents the sum of the two vectors. For example, in Fig. 1.10, arrows represent the two displacements as a ship sails 20 km due north, then 15 km to the west. The other arrow indicates the final displacement of the ship from the starting point; the length and direction of the arrow indicates that the ship is now 25 km, in a direction about 37° west of north from the starting point. (In navigation terms, directions are measured east from north. The direction of the ship from its starting point is 323°. To subtract vectors, simply reverse the sign of the one being subtracted from the other (it now has the same length but the opposite direction), and add them (Fig. 1.11).

1.5 THE ELEMENTS OF QUANTITATIVE PHYSICS

Example 1.1 What is the resultant of the three vectors shown in Fig. 1.12?

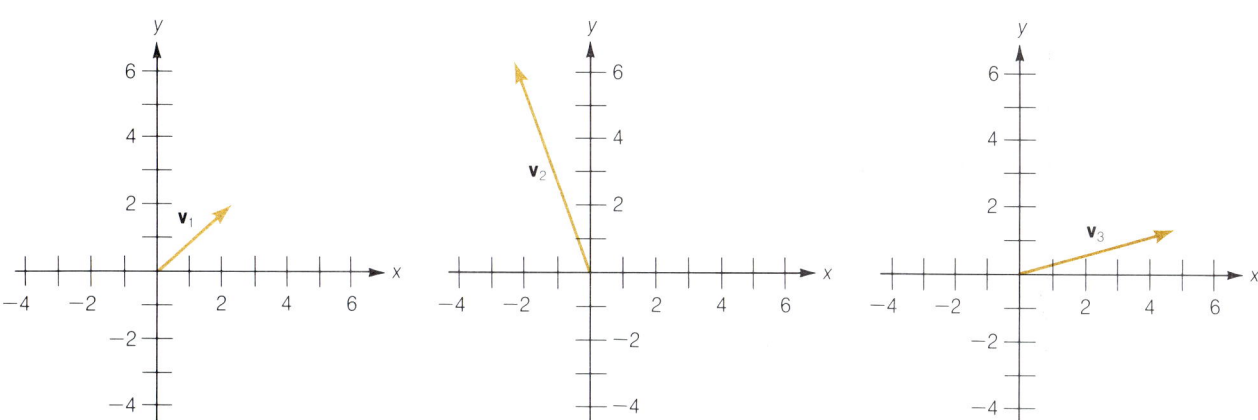

Figure 1.12

We make another sketch (Fig. 1.13) in which the three vectors are drawn with the tail of the second touching the head of the first, and the tail of the third touching the head of the second. Then we draw a new vector connecting the tail of the first with the head of the third. This is the sum or resultant. We simply measure its length and direction on our scale drawing, as indicated. We find that the length, or magnitude, of the resultant is 10.5 units, and its direction is 65° from the x-axis (you may wish to make your own drawing, and see how well your answer agrees. The differences will give you an appreciation of experimental and measurement uncertainties).

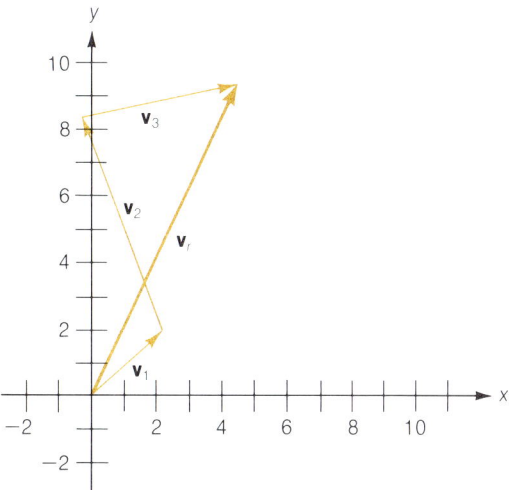

Figure 1.13

Note that the three vectors could have been drawn head to tail in any order with the same resultant.

16 CHAPTER 1 INTRODUCTION TO PHYSICS

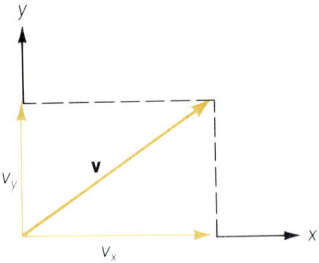

Figure 1.14

*Vector components. The projections of vector **V** along the x- and y-axes are its components V_x and V_y.*

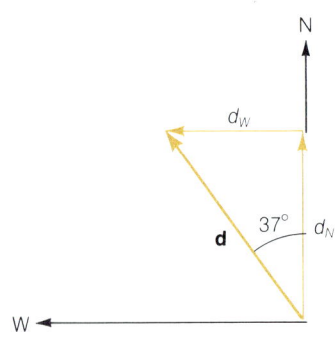

Figure 1.15

*Finding vector components. Here a ship has sailed along vector **d**. The north and west components of its displacement can be found by direct measurement or by trigonometry: $d_N = d \cos(37°)$, and $d_W = d \sin(37°)$, where d is the magnitude of **d**.*

Whenever we speak of quantities having a direction, we must specify a frame of reference (indeed, we have already been doing this in describing vectors and in Example 1.1). For example, it is impossible for one person to tell another how to drive somewhere unless he uses a frame of reference that the other person understands. On the surface of the earth we normally use compass directions (and some starting point) as our reference frame, but in physics we often use special frames. Hence whenever we deal with vector quantities, we must specify a coordinate system to give the magnitudes and directions of the vectors a frame of reference.

Very often we express directions (or quantities having directions) in terms of graphical coordinates. For example, in order to represent the ship's motion and find its displacement we make a scale drawing to represent the actual displacement on the sea. Whether we solve vector problems by making a drawing or by other techniques, we must always have a frame of reference in which to express and compare the vector quantities.

With a frame of reference, it is possible to analyze the sum or difference of vector quantities without resorting to graphs. In some cases, the graphical method becomes impractical; consider, for example, the final displacement of a helicopter. It would be very difficult to represent the flight of a helicopter on a two-dimensional sheet of graph paper, yet in physics it is often necessary to add three-dimensional vectors. This requires isolating the **components** of the vectors; that is, resolving each vector into component vectors representing its extent along each axis of the reference frame. If the frame of reference consists of coordinate axes at right angles to each other (that is, a rectilinear coordinate system), any vector can be expressed as the sum of its components, which are simply the projections of the vector along the coordinate axes (Fig. 1.14).

The components of a vector can be calculated by simple trigonometry. Suppose the ship (Fig. 1.10) had simply sailed 25 km in a direction 37° west of north. We draw a right triangle whose sides represent the ship's displacement d (the hypotenuse of the right triangle) and the north and west components d_N and d_W (the legs; Fig. 1.15) of its displacement. The cosine of one of the acute angles in a right triangle is equal to the length of the adjacent side divided by the length of the hypotenuse, so we see that the cosine of 37° is equal to the north component of the ship's displacement divided by the total displacement; that is, $\cos(37°) = d_N/d$. Solving, we find that the north component is $d_N = d \cos(37°) = (25 \text{ km}) \times \cos(37°) = 20$ km. The sine of an acute angle in a right triangle is equal to the opposite side divided by the hypotenuse, so we have $\sin(37°) = d_W/d$. Thus the west component is $d_W = d \sin(37°) = (25 \text{ km}) \times \sin(37°) = 15$ km.

The major benefit of finding the components of vectors is that the components are scalars. Vectors can then be added or subtracted by adding or subtracting their components. The result gives the components of the sum, or resultant vector, which can then be reconstructed. For example, suppose a ship sails 12.0 km 30° east of north, and then sails 40.0 km 10° east of north (Fig. 1.16). To find its total displacement, we must *resolve*, that is, find the components of, the vectors representing the two legs of the ship's path. The north component of the first leg is $(12 \text{ km}) \times \cos(30°) = 10.4$ km, and the north component of the second leg is $(40 \text{ km}) \times \cos(10°) = 39.4$ km; hence

1.5 THE ELEMENTS OF QUANTITATIVE PHYSICS

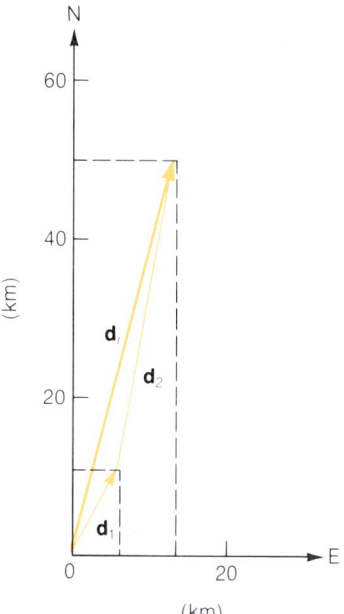

Figure 1.16

Addition of components to find resultant. Here two displacement vectors d_1 and d_2 are resolved into components which are now added to yield the components of the resultant vector d_r.

the north component of the resultant vector is 10.4 + 39.4 = 49.8 km. The two east components are (12 km) × sin(30°) = 6.0 km and (40 km) × sin(10°) = 7.0 km, resulting in a total eastward displacement of 13.0 km. Now we use the total north and east components to reconstruct the resultant vector. Its length, given by the Pythagorean theorem, is $\sqrt{(49.8)^2 + (13.0)^2}$ = 51.5 km, and its direction is the angle whose tangent is 13.0/49.8; that is, the direction is arc tan 0.26 = 14.6° east of north. Note that we could have found the angle first, then used trigonometry to find the length of the vector: 49.8/cos 14.6° or 13.0/sin 14.6°.

The same principles apply when working in three dimensions; one simply calculates the components in all three dimensions, adds all the components in each dimension, and then reconstructs the resultant vector. The calculations are more complicated when frames of reference that are not rectilinear must be used. For example, polar coordinates are more convenient for spherical or axial symmetry. Our examples of the ship at sea were oversimplified because the earth is a sphere, but for these small distances, treating the sea surface as a flat plane is quite adequate. Discussions and problems involving vectors in this book will generally use rectilinear frames of reference.

It is essential that students learn to manipulate vectors. The worked examples accompanying this section should be examined, and sample problems at the end of the chapter should be worked until they become easy.

Example 1.2 Find the resultant of the three vectors from Example 1.1 by component analysis rather than by making a graph.

The vectors are redrawn in Fig. 1.17, which shows the *x*- and *y*- components.

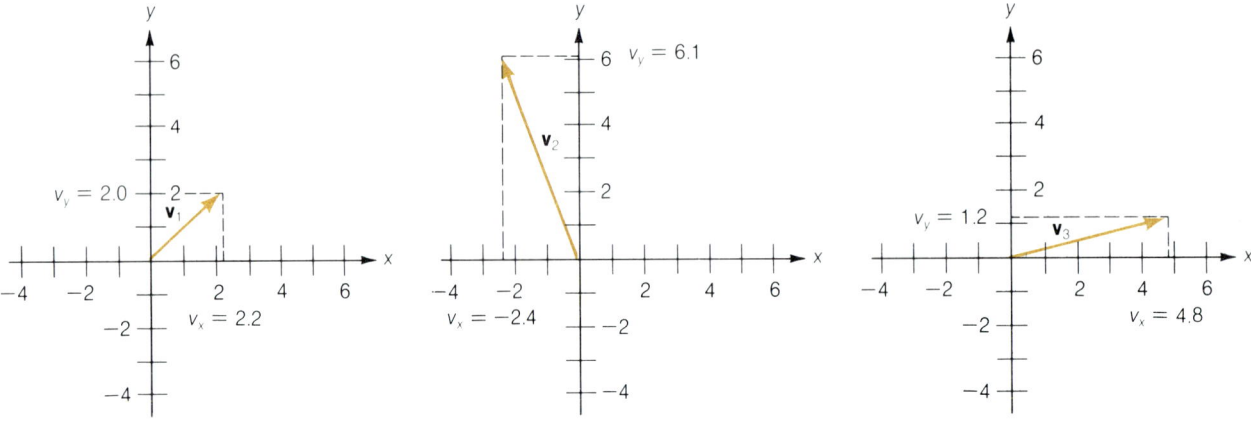

Figure 1.17

To find r_x, the *x*-component of the resultant, we add the *x*-components of the three vectors.

$$r_x = 2.2 - 2.4 + 4.8 = +4.6$$

Similarly the *y*-component of the resultant is

$$r_y = 2.0 + 6.1 + 1.2 = +9.3.$$

The magnitude *r* of the resultant is

$$r = \sqrt{r_x^2 + r_y^2} = \sqrt{(4.6)^2 + (9.3)^2} = 10.4$$

Measured from the *x*-axis, the direction of the resultant is the angle whose tangent is equal to r_y/r_x.

$$\theta = \text{arc tan } (r_y/r_x) = \text{arc tan } (9.3/4.6) = \text{arc tan } 2.02 = 63.7°$$

Note this agrees, within the measurement uncertainties, with the graphical result in Example 1.1.

Perspective

We now have most of the preliminary knowledge to allow us to begin to learn physics. The material on units and on vectors and scalars will be essential to our comprehension of later chapters. We begin with motion and forces: the science of mechanics.

PROBLEM SOLVING

Overall Strategy

Physics is an experimental science. Physical theories must be tested by observations, experiments, and analytical problems. In this section we will describe several general strategies for solving problems, and we will add a few tricks of the trade. More detailed techniques for specific problems are provided in later chapters. Our strategy follows six themes:

1. Units and dimensions
2. Diagrams and coordinate axes
3. Input and output variables (knowns and unknowns)
4. Relations among variables
5. Sequential reasoning and complex reasoning
6. Proportional reasoning, scaling, and shortcuts

The first theme, units and dimensions, is important for communication, as well as for checking the steps in a calculation. To describe a quantity, one must first understand the units in which it is measured (many times there are several ways of measuring the same quantity). The relations among physical quantities are often expressed in algebra, without dimensions, because we wish to find a general formula for solving all problems of a certain type. In a particular solution, though, we will insert actual quantities into the general formula. Units should be attached to each quantity, unless it is a dimensionless constant like π. Note also that some constants will have dimensions. The proportionality constant in the equation $F = kx$, relating force F to distance x is a good example. When you arrive at a result, check that it has the correct units. If it does not, then an error was made in an earlier step. Perhaps a factor whose dimensions were required for proper cancellation was omitted.

Example 1 An object with velocity v moves in a circular orbit of radius R. What is the period of motion T, if $R = 3$ m and $v = 10$ m/s?

Since for a circular path, $v = 2\pi R/T$, a general expression for the period is $T = 2\pi R/v$. Now we insert actual quantities with units.

$$T = \frac{2\pi R}{v} = \frac{(2\pi)(3 \text{ m})}{(10 \text{ m/s})} = 1.88 \text{ s}$$

Notice that the meters, m, cancel in numerator and denominator, and that the 1/s in the denominator becomes s when raised to the numerator. The dimensions are treated algebraically, just like numbers.

Converting units is a related problem. A single measurement may be expressed in a variety of unit systems (for example, meters, feet, miles, and kilometers are all distances). When doing problems, these units need to be converted to one system, usually SI. This conversion is done by multiplying by a conversion factor as shown in the next example.

Example 2 The Empire State Building is 102 stories tall. A story averages 10 feet high. How tall is the building in meters?

$$H = (102 \text{ stories})\left(\frac{10 \text{ feet}}{1 \text{ story}}\right)\left(\frac{1 \text{ m}}{3.28 \text{ ft}}\right) = 311 \text{ m}$$

(continued)

There are two conversion factors here, one to convert stories to feet, and a second to convert feet to meters.

A good way to check the units conversion is to make sure that units cancel, as shown in Example 2. If they do not, then you may have inverted the conversion factor. When converting from unit system A to system B, the conversion factor must have the form (B units/A units). It may not be obvious that all units have cancelled, because some physical quantities are actually combinations of more basic SI units (kg, m, s). For example, we will see later that the unit of energy is a Joule, where 1 Joule equals 1 kg m^2/s^2. Similarly, the units of force are Newtons, where 1 Newton equals 1 kg m/s^2. It is generally advisable to leave these derived quantities, such as Newtons or Joules, intact unless you wish to check units after the problem is in final form.

We will now quickly discuss the other five themes, leaving examples to later chapters, after the necessary material has been covered. The second theme emphasizes the importance of drawing a diagram of the problem. Label all the known quantities (input variables) and unknown quantities (output variables). In force problems, it is useful to make separate sketches, showing forces acting on each body. Choose a set of rectangular axes for resolving the forces into x- and y-components.

Themes 3 and 4 address the actual mechanism of solving problems. Once you have set up axes, made diagrams, and labelled the variables, you need to identify which variables you seek. Once the problem is put into algebraic terms, it is easy to forget which quantities are known and which are unknown. A valuable technique is to think of the problem as a machine. You have some input variables, and you are looking for output variables. To solve for the unknowns, the machine needs fuel (relations or equations among the variables). The basic rule of all problem solving is

For each unknown variable, you must have one equation which relates it to the other variables (known or unknown).

Thus, if the problem has 5 variables (3 known and 2 unknown), then you must find two relations involving the unknown variables. These relations may be physical laws, force balance equations, or simply direct relations among the variables (for instance, in a circle, you know circumference equals 2π times radius).

Themes 5 and 6 concern the logical steps that go into solving a problem, once it is set up with diagrams, variables, and equations. This is really just a summary of the mathematics of solving simple algebraic equations. If the problem has only one unknown variable, the solution simply requires substituting values into a single formula. Problems with several unknowns are more difficult. The easier of these may be solved sequentially: first solve one equation for one unknown, then substitute that expression into another equation. In more advanced problems, three or more equations must be solved for three or more unknown variables. The strategy is to solve the simplest equation first, then substitute into the other equations. This procedure requires some trial and error and some ingenuity; practice helps.

Finally we come to theme 6, which involves more advanced techniques in problem solving and that nebulous ability called physical intuition. Intuition comes mainly from experience, but there are a few aids. Proportional reasoning and scaling arguments are two of them. In many problems, you will derive a general formula from which it should be obvious that the answer depends in a straightforward manner on a variable. For example, the unknown variable y may be linearly proportional to the known variable x ($y = cx$, where c is a constant of proportionality). Thus, if you double x, you double y; y is said to scale linearly with x. More rapid variations are possible. Suppose $y = cx^2$. Then y is proportional to the square of x. If you double x, then y increases by a factor of 4. In problem solving, you should be aware of these proportionalities, so that once you have done the problem with one set of input variables, you know what will happen if those variables change. This is a useful shortcut. For instance, in Example 1, if we wanted to know the orbital period for an object with $R = 3$ m and $v = 30$ m/s, we could simply divide the answer $T = 1.88$ s by 3, since

T is inversely proportional to velocity v ($T = 2\pi R/v$).

Other shortcuts involve noticing that a problem does not depend on one variable, so that you can omit it from the calculations. As we will see in later chapters, there are often several ways to do a problem. Choosing a method that takes advantage of a conservation law (mass, momentum, energy, for example) or a certain symmetry (usually a geometric symmetry) can save you time and algebra. These techniques are best learned by doing problems in later chapters, but this summary should be helpful. Return to it as you progress through this text. Good luck.

THOUGHT QUESTIONS

1. What are the "rules of the game" for the modern scientific method of experimental and theoretical science?
2. Why do scientists insist on repeatability of their experiments?
3. Are there reasons why certain sciences developed before others? Consider, for example, astronomy, biology, chemistry, geology, medicine, and nuclear physics.
4. What are the links between Newtonian physics and astronomical studies of planetary motions? Mention Galileo, Tycho, and Kepler.
5. What are the relationships between mathematics, numbers, and dimensions of measurement? Discuss, for example, length, volume, and mass.
6. Is it a coincidence that the sizes of standard units of mass (kilogram), length (meter), and time (second) are common in everyday experience? For example, humans are about 2 meters tall, bricks have masses about 1 kilogram, human reaction times are about 1 second.
7. Algebra has been described as a technique for understanding and working with relationships between known and unknown quantities. Why is such a mathematical system useful in physics?
8. What is the difference between a scalar and a vector?
9. Decide which of the following are scalars and which are vectors.
   (a) The distance from Toronto to Winnipeg.
   (b) The velocity of an airplane flying due west.
   (c) The gravitational force on a ball falling to the ground.
   (d) The distance of San Francisco relative to Los Angeles.
   (e) The air temperature on a hot day.
10. Why is the Pythagorean theorem useful in finding distances between two points?

PROBLEMS*

•• 1 A 10-meter ladder leans against a vertical wall. The base of the ladder is 3 meters away from the wall. How high on the wall does the ladder touch? What angle does the ladder make with the wall?

• 2 A mountain climber trudges up a steep slope tilted 30° from horizontal. If the total vertical ascent is 3 km, how far must the climber walk along the slope?

• 3 A knight on a chessboard moves two squares left and one square forward. What is the displacement of the knight if the squares are 3 cm on a side?

• 4 An airplane flies at 600 km/h in a direction 62° west of north. Find the components of the velocity vector in the north and west directions.

• 5 A 400-meter race is run on a circular track (one lap). What is the runner's displacement vector from the start halfway through the race? At the finish?

*Problems are ordered by difficulty: One bullet (•), two, or three to indicate the order of increasing difficulty. In later chapters, problems are also numbered by sections.

- **6** An ice block slides down a ramp at a constant velocity of 1 m/s. If the ramp is tilted 35° from horizontal, find the horizontal and vertical components of the velocity vector.
- **7** A regiment marches 30 km northwest, then 40 km 30° east of north. What is the resultant displacement?
- **8** A lonely sailor walks 100 m in a straight line, turns left at a right angle and walks 50 m, then turns left again at 90° and walks 25 m. How far from his original position is he when he stops?
- **9** The 3 vectors shown in Fig. 1.18 can be added in various orders. Show that the resultant vector is always the same.

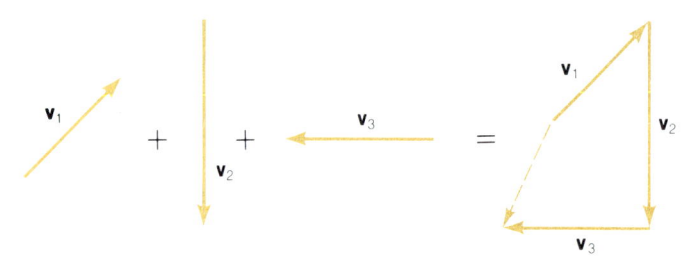

Figure 1.18

- **10** In a mythical city of giants, the streets are laid out in a rectangular grid of blocks 1 km in length. A giant walks 10 blocks east, 6 blocks north, then 8 blocks east. What is his final displacement from his starting point? What was the total distance he walked?
- **11** Point B lies at an angle 30° above the x-axis (see Fig. 1.19) on a circle of radius 1 m. Compute the x- and y-coordinates of point B.
- **12** Compute the length AB on the circle (Fig. 1.19). (The entire circumference has length $2\pi r$.)
- **13** A diver descends 100 m vertically from a boat, then swims horizontally 50 m west and 60 m south. What is the shortest distance back to the boat?
- **14** A blimp hangs over the Los Angeles Coliseum during a football game. At the kickoff the blimp moves north 40 m, then west 10 m. It then moves straight up 100 m. What is the magnitude of the displacement from its starting point?
- **15** A regiment marches 40 m northwest. What are the x- and y-displacements on a rectangular coordinate system?
- **16** A hiker walks 100 m NE, then 100 m SE. What is his displacement from his starting point?
- **17** On a radar screen, an air controller sights a UFO 30° east of north and 5 miles away. One minute later, the UFO is 60° west of north and 3 miles away. How far did it move during the minute? What is its speed in miles per hour?
- **18** Give the (x, y) coordinates of the following points in polar coordinates (r, θ).
  - (a) (1, 30°)
  - (b) (2, 45°)
  - (c) (5, 245°)
- **19** Points A and B have polar coordinates (r_1, θ_1) and (r_2, θ_2). Find a general formula for the distance d between them.
- **20** A geometric curve has equation $r = a\theta$, where a is a constant. Graph this curve on polar graph paper and describe how it is formed.
- **21** Track and field meets have recently converted from the 100-yard dash to the 100-m dash. How many meters are in a yard? What fraction longer is the 100-m race?
- **22** How many seconds are in a day? In a year? (One Julian year is 365.25 days; the Gregorian year is slightly different.)
- **23** Using the definitions of the prefixes, express the following lengths in meters with scientific notation: megameter, kilometer, centimeter, millimeter, micrometer, and nanometer.
- **24** A common unit of length in the atomic scale is the Ångstrom (1 Å = 10^{-10} m). How many Ångstroms wide is a human hair (25 micrometers)?
- **25** The diameter of the earth is 12,576 km. How many meters is this? How many miles?
- **26** If the speed limit for cars in Canada is 100 km/h, what is this in m/s? In miles/h?
- **27** A typical adult male has a mass of 70 kg. Since his body is mostly water, his density is about 1,100

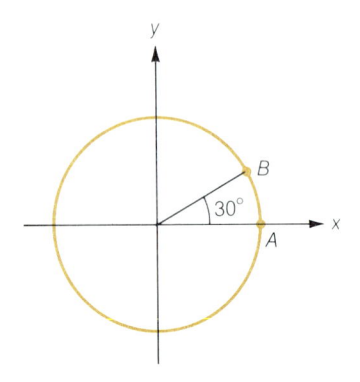

Figure 1.19

kg/m³. What is the man's volume (m³)? What would be his mass if he were made of iron (density 5,500 kg/m³)?

•• 28 A playing field measures 60 yards × 100 yards. What are its dimensions in meters? If sod costs $1/m², what does it cost to sod the field?

•• 29 A spherical tank holds 1,000 gallons of water. One gallon is about 0.004 m³ and the density of water is 1,000 kg/m³. One kg weighs 2.2 lbs on earth. What is the tank's volume in m³? What is the sphere's radius? How much does the water weigh (in lbs)?

•• 30 A furlong is an ancient unit of length equal to ⅛ mile; a fortnight is two weeks. If an ant has a speed of 1 furlong/fortnight, what is its speed in m/s?

SPECIAL PROBLEMS

1. Units in the Universe

According to best estimates, the universe is about 15 billion years old. The edge of the visible universe is therefore 15 billion light-years away. One light-year is the *distance* light travels in one year (the nearest star, Alpha Centauri, is about 4.3 light-years away). Light travels at a speed of 3×10^8 m/s in a vacuum.

(a) How many seconds are in a year?
(b) How many meters are in a mile? (One meter equals 39.37 inches.)
(c) Using the relation distance = velocity × time, show that one light year is about 10^{16} meters or 6 trillion miles (6×10^{12} miles).
(d) How many miles away is the edge of the visible universe?

2. Volumes and Masses

The mass of an object equals its volume times its density. In SI units, volume is measured in m³, and density in kg/m³. However, volumes and densities are not always given in SI units; you must first convert units. You must also be able to compute the volume from the appropriate geometric formula (Fig. 1.20).

(a) If 1 cm³ of water has a mass of 1 gram, show that its density is 1000 kg/m³.
(b) What are the masses of the following containers of water: a sphere with a 5-meter radius, a cube with 10-meter sides, a cylinder of radius 5 meters and height 10 meters.
(c) A town council wishes to build a water storage tank, and is debating over whether to build a sphere, a cube, or a cylinder. The tank cannot be more than 10 meters tall or 10 meters wide. Which type of tank should they choose in order to hold the most water?
(d) What is the maximum mass of water that the tank chosen in part c will hold? How many gallons will it hold? (One gallon has a volume of about 3,785 cm³.)

3. Office Arithmetic

One floor of an office building is laid out with 20 identical, rectangular offices (Fig. 1.21) with desks at the center of each office. The office manager wishes to install carpeting ($20/m²) and phone cables between each desk ($0.10/m).

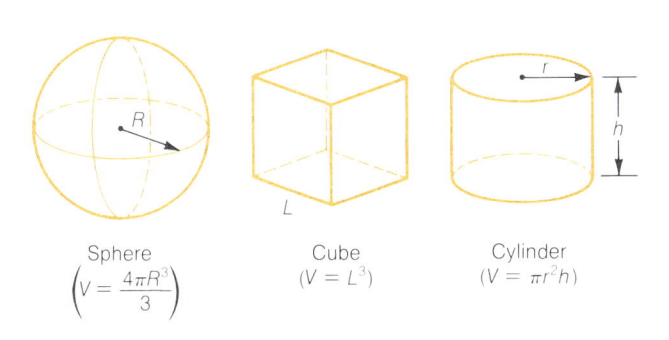

Sphere $\left(V = \dfrac{4\pi R^3}{3}\right)$

Cube $(V = L^3)$

Cylinder $(V = \pi r^2 h)$

Figure 1.20

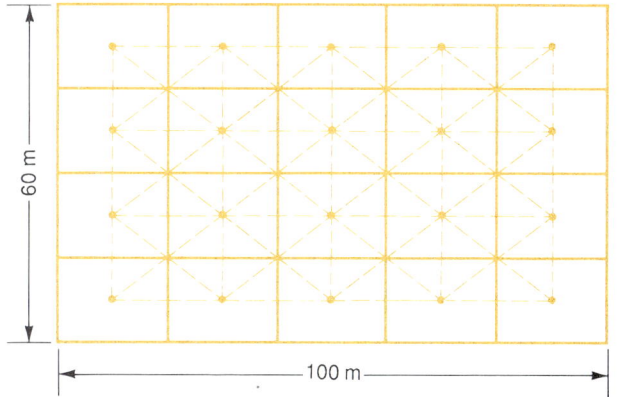

Figure 1.21

24 CHAPTER 1 INTRODUCTION TO PHYSICS

(a) What is the cost of carpeting one office?
(b) What is the cost of wiring all the offices?
(c) What are the three lengths into which the phone cables must be cut?

4. Wind Drag on Cars

A car travels north at $V_c = 25$ m/s (about 55 mi/h) while a crosswind blows from the west at $V_w = 10$ m/s. The force of wind drag is proportional to the square of the magnitude of velocity, $F = kv^2$, where $k = 0.80$ N · s²/m² is a constant (N stands for Newtons, a unit of force.) Note that this is a nonlinear force law (force is not proportional to velocity, but to velocity squared).

(a) Find the wind velocity vector V_{tot} relative to the car. In Fig. 1.22, V_{tot} is the vector difference $V_w - V_c$.
(b) Find $F_\parallel$ and $F_\perp$, the components of wind drag parallel and perpendicular to the car's motion. Then find the total force F_{tot}.
(c) What are the angles Θ_1 and Θ_2? Notice that they are not the same. The net wind force is not aligned with the net velocity. Explain why this effect is a result of the nonlinear force law.

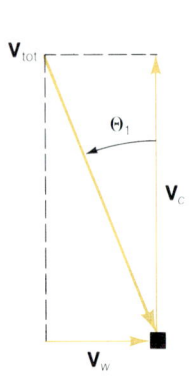

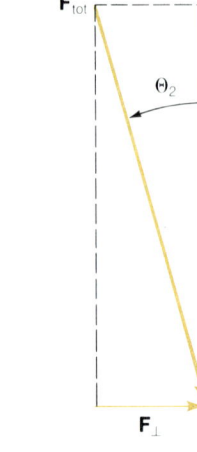

Figure 1.22

SUGGESTED READINGS

"The standards of measurement." A. V. Astin, *Scientific American*, Jun 68, p 50, **218**(6)

"Intuitive physics." Michael McCloskey, *Scientific American*, Apr 83, p 122, **248**(4)

"Is physics human?" Victor F. Weisskopf, *Physics Today*, Jun 76, p 23, **29**(6)

Two hundred years of American physics." *Physics Today*, Jul 76, p 23, **29**(7)

"Can physics develop reasoning?" Robert G. Fuller, Robert Karplus, Anton E. Lawson, *Physics Today*, Feb 77, p 23, **30**(2)

[Fifty years of physics in America—special issue]. *Physics Today*, Nov 81, **34**(11)

"Solving physics problems-how do we do it? Robert G. Fuller, *Physics Today*, Sep 82, p 43, **35**(9)

"Research on conceptual understanding in mechanics." Lillian C. McDermott, *Physics Today*, Jul 84, p 24, **37**(7)

"Math anxiety and physics: Some thoughts on learning 'difficult' subjects." Sheila Tobias, *Physics Today*, Jun 85, p 60, **38**(6)

[Special issue-Twenty discoveries that changed our lives]. *Science 84,* Nov 84, **5**(9)

"Foundations of the international system of units (SI)." Robert A Nelson, *The Physics Teacher*, Dec 81, p 596, **19**(12)

The history of physics. Isaac Asimov, Walker, New York, 1984

From x-rays to quarks. Emilio Segrè, Freeman, New York, 1980

From falling bodies to radio waves. Emilio Segrè, Freeman, New York, 1984

Science and unreason. Michael Radner, Daisie Radner, Wadsworth, Belmont, CA, 1983

The powers of ten. Philip and Phyllis Morrison, Charles and Ray Eames, Freeman, New York, 1982

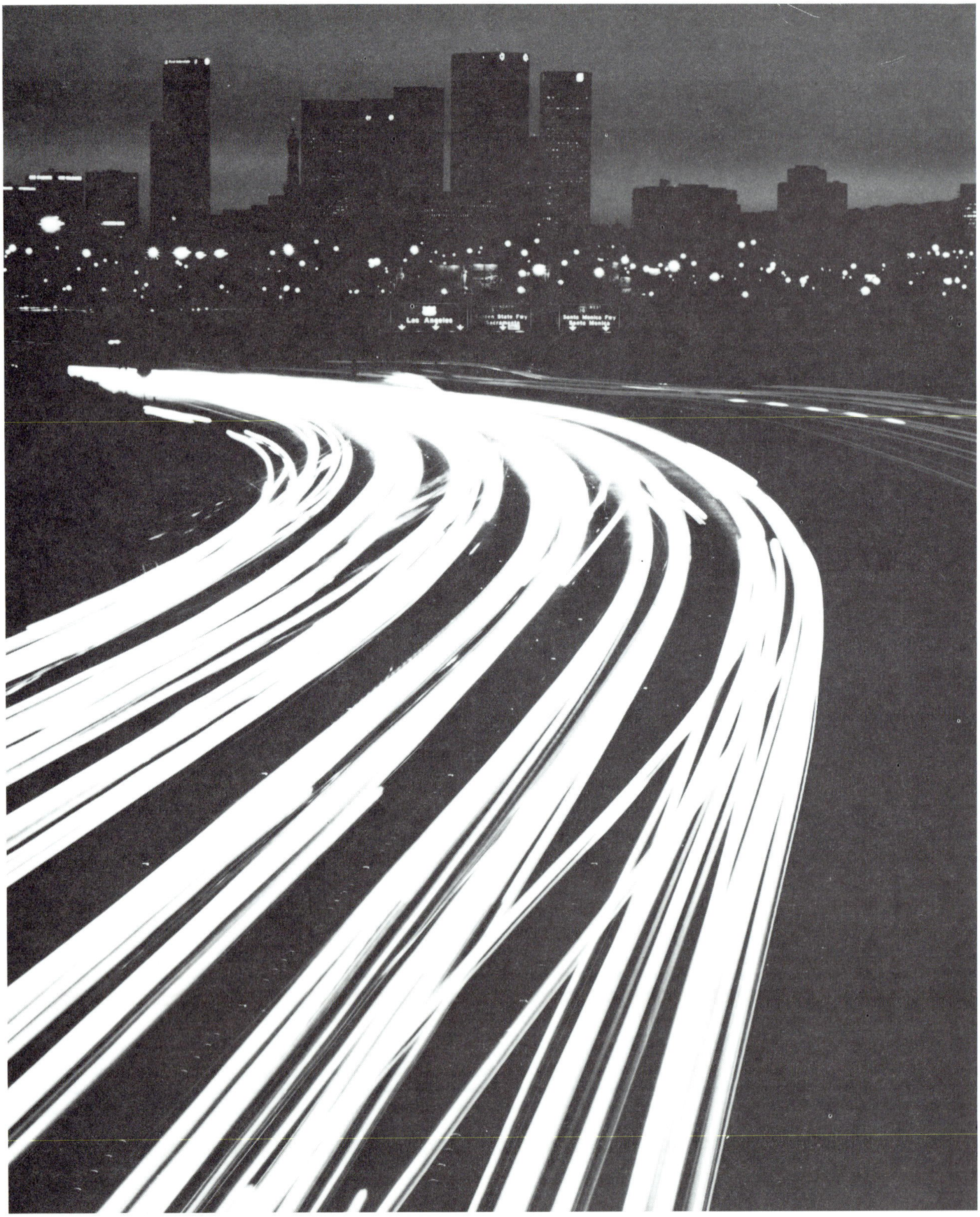
Los Angeles
Santa Monica Fwy
Santa Monica

Chapter 2

Kinematics

The word "kinematics" refers to the description of motion. Therefore our purpose in this chapter is to understand what we mean by motion and how we may analyze it. In Chapter 3 we will discuss dynamics, which includes not only motions but also forces (which cause changes in motion).

There are many kinds of motion. A spacecraft moving at constant speed through a vacuum far from any massive bodies that would exert gravitational forces represents perhaps the simplest form of motion, with no variations or deviations from a straight path. An object sliding along a surface on the earth experiences friction and slows. A falling object speeds up under the influence of gravity, but this increase in speed is moderated by air resistance. The pendulum of a clock moves back and forth, speeding up as it approaches the bottom of its arc, and slowing as it reaches the top. The moon orbits the earth in a nearly circular path. Electrons near a quasi-stellar object spiral around magnetic field lines at velocities near the speed of light. All of these motions are subject to mathematical analysis by a few simple physical laws.

To begin this discussion, we will confine ourselves to the simplest cases and will disregard forces. The most basic motion we can describe is uniform motion in a straight line, but first we must discuss the concepts of distance and displacement which were mentioned in Chapter 1.

2.1 Distance and Displacement

All of us have sound ideas of what the words "distance" and "displacement" mean, but our common usages are not very precisely defined. To a physicist, however, these terms have very specific meanings. When we speak of **distance**, we mean a scalar quantity that indicates the length of the path between two points, whereas **displacement** refers to the vector that describes both the magnitude of the separation and the direction. If a person walks around the block, returning to his starting point, the distance travelled is the total path-length around the block, but the displacement is zero. The displacement is the separation between two points, expressed as a vector whose magnitude is the straight-line distance between them. Thus, the displacement of point B from point A in Fig. 2.1 is 7 cm in the direction 35° from the x-axis, whereas the distance an object travels from A to B would depend on the path followed.

The concept of displacement is independent of how the displacement occurs; your displacement when you go from your dormitory to the student union is the same, no matter whether you got there by following the shortest path across campus, or went to the bank on the way. It is quite possible to make a long excursion but end up with zero displacement when you return to your starting point. To find the displacement mathematically, we must add the individual displacements that occur as we journey, and this is done following the techniques for vector addition described in Chapter 1.

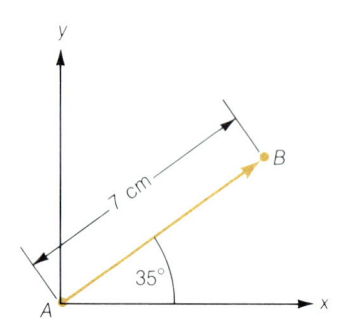

Figure 2.1

The displacement of B from A is represented by a vector whose magnitude is equal to the straight-line separation of the points, and whose direction is specified in some reference frame. Here it is 35° from the x-direction.

Example 2.1

Fig. 2.2 shows three different routes you might take in going from your dormitory to the student union, which is 2.5 km due east. If you take the direct route, it is easy to see that the displacement is 2.5 km due east. Suppose, however, that you go to the bank, 1 km due north of the dormitory, on the way and then proceed to the student union by going 2.5 km east from the bank, and then 1 km south. Since you have followed streets that parallel the reference axes, adding the vectors representing your displacements is very easy, because the displacements are readily resolved into their north-south and east-west components. You have a pair of components in the north-south direction, one of +1 km (the plus sign representing north), and one of −1 km (the minus sign representing south). Adding these two yields a zero displacement in the north-south direction. There is only the single +2.5 km displacement in the east-west direction left, and that is your resultant displacement vector. Another route is to travel 1 km due north to the bank, but then take a diagonal path from the bank to the student union, a distance of 2.69 km in the direction 21.8° south of due east. Now to find the resultant we must find the components of this diagonal vector. The north-south component is $(2.69 \text{ km}) \times \sin(-21.8°) = -1$ km, and the east-west component is $(2.69 \text{ km}) \times \cos(-21.8°) = +2.5$ km. Adding the north-south components again yields zero displacement in that direction, while the east-west component is +2.5 km.

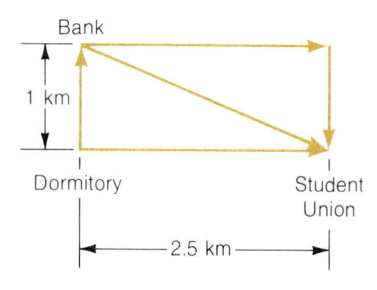

Figure 2.2

2.2 Speed and Velocity

As we continue our discussion of kinematics, we will be interested in rates of change with time. A rate of change is the amount of change of a given quantity divided by the time during which the change took place.

We now consider the rate at which displacement may change with time. The **velocity** is the vector quantity representing the rate of change of displacement. The **speed** is the scalar representing the magnitude of the velocity at a given moment. In common usage, these words are often interchanged, but to a physicist there is a clear distinction. If you wish to specify only how fast an object is moving, then it is the speed that is to be measured. If, however, the direction is to be considered as well, then it is the velocity that must be measured. The speedometer in a car measures only the speed.

If you want to calculate where you are or how far you are from your starting point after moving for some period of time, then knowing your speed alone will not provide this information. You could travel for an hour at 80 km/h, but end where you started, or at any other point within 80 km of your origin. If, however, you knew that you had travelled for an hour at 80 km/h towards the south, then you would know that you finished 80 km south of your starting point.

In most quantitative analyses of motions, we will want to know the direction, hence we will generally work with velocities, not speeds. The units for both speed and velocity are units of length divided by units of time; in the SI system, we use meters/second (m/s).

Uniform Velocity in a Straight Line

We have already noted that the simplest form of motion is straight-line motion with no deviations. In this case it is particularly easy to calculate the velocity, because all we need to know is the displacement and the time during which this displacement occurred. The velocity vector **v** is given by

$$\mathbf{v} = \frac{\mathbf{d}}{t} \tag{2-1}$$

where **d** is the displacement vector, and t is the time. The vectors **v** and **d** are in the same direction. Note that t is a scalar; when a vector is multiplied or divided by a scalar, the direction of the vector remains unchanged but the magnitude is altered.

Eq. 2–1 represents the actual velocity only if the velocity is constant during time interval t. This is rarely the case, but for now we make the simplifying assumption that it is so. As we will see, the equation holds in general if **v** is the average velocity.

Suppose you drive your car at a constant speed to the west along a straight stretch of highway, starting at 6:00 A.M. and stopping at 9:00 A.M. for coffee at a point 210 km away from where you started. Your speedometer does not work, and you want to know what your (constant) velocity was. Your displacement **d** is 210 km due west, so your velocity **v** is

$$\mathbf{v} = \frac{\mathbf{d}}{t} = \frac{210 \text{ km due west}}{3 \text{ hrs}} = 70 \text{ km/h due west}.$$

Knowledge of your velocity and the time during which you travel at that velocity can give you your displacement by solving Eq. 2–1 for **d**:

$$\mathbf{d} = \mathbf{v}t,$$

or

$$\mathbf{d} = \mathbf{d}_0 + \mathbf{v}t, \qquad (2\text{–}2)$$

where $\mathbf{d}_0$ represents the initial displacement, which can often be defined as equal to zero.

If, in the preceding example, you knew that your velocity was a constant 70 km/h due west, but you did not know how far you had gone when you stopped for coffee after 3 hours, Eq. 2–2 could be used to determine your displacement, assuming your initial displacement to be $\mathbf{d}_0 = 0$:

$$\mathbf{d} = (70 \text{ km/h due west})(3 \text{ h}) = 210 \text{ km due west}$$

Relative Motion

Uniform motion in a straight line can occur in a moving frame of reference. Consider, for example, the case of a ship sailing against an ocean current (Fig. 2.3). The ship is moving at a velocity of 30 km/h due east relative to the water, but the current is flowing at a uniform velocity of 12 km/h due west relative to an observer on land (that is, its velocity is −12 km/h due east, if we take the eastward direction as positive). Clearly, the ship's resultant velocity is 18 km/h due east, relative to the land.

This example illustrates how vector addition is used to find the resultant

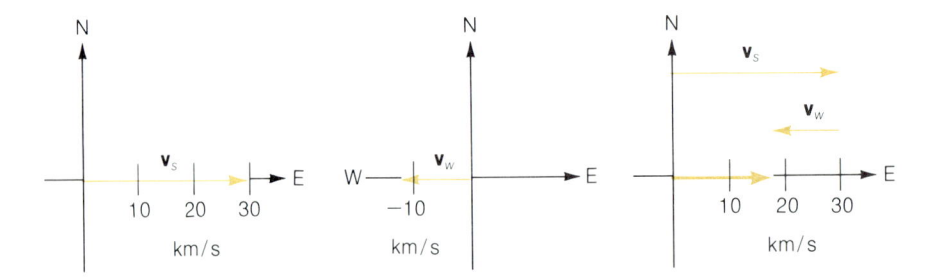

Figure 2.3

Resultant velocity. The vector at left represents the velocity of a ship travelling east at 30 km/h with respect to the water, and the vector at center represents the velocity of the current at 12 km/h west, with respect to land. At right, these two velocity vectors are added graphically to yield the resultant velocity of 18 km/h to the east, with respect to land.

2.2 SPEED AND VELOCITY

velocity when the motion results from two or more velocities. Since the velocities to be added were along the same east-west direction, the addition was particularly simple. In general the vector addition procedure described in Chapter 1 is used.

Example 2.2

A commercial jet is cruising at 900 km/h airspeed (that is, with respect to the air), in the direction 70° east of north (Fig. 2.4), and is passing through a jetstream (high-altitude air current) whose velocity is 300 km/h on a heading of 100° east of north. We want to find the resultant velocity of the airliner with respect to the ground.

We must begin by finding the components of the velocity vectors representing the plane's motion and that of the air current. In the north-south direction, defining north as positive, the plane's velocity component is (900 km/h) × cos(70°) = +308 km/h, and in the east-west direction, with east positive, it is (900 km/h) × sin(70°) = +846 km/h. The components of the jetstream are (300 km/h) × cos(100°) = −52 km/h north, and (300 km/h) × sin(100°) = 295 km/h east. Addition tells us that the components of the resultant vector are +256 km/h north and 1,141 km/h east. The resultant vector has the direction arc tan (1,141/256) = arc tan (4.46) = 77.4°, and its magnitude is 1,141/sin(77.4°) = 1,169 km/h. Hence the combination of motions leads to a velocity for the plane of 1,169 km/h in the direction 77.4° east of north.

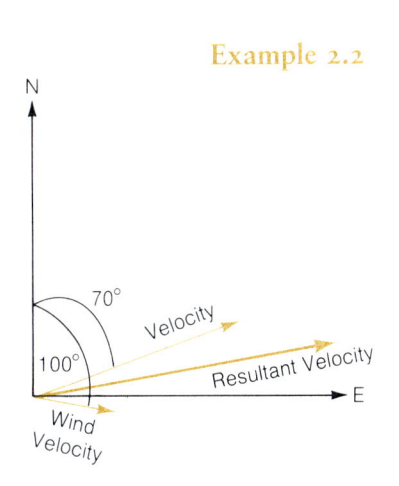

Figure 2.4

Average Velocity

Perfectly uniform velocity is quite rare. Hills, traffic, and the whims of the driver usually cause a car to slow down and speed up irregularly. In nearly every situation there are forces that cause changes in velocity. Even so, it is still possible to calculate the velocity that represents a given displacement during a given time interval, but now it is an **average velocity**, which can be quite different from the actual, or **instantaneous**, velocity.

The average velocity, commonly represented by $\bar{\mathbf{v}}$ (a bar over the top of a symbol generally is used to represent the average value) is the total displacement divided by the total elapsed time:

$$\bar{\mathbf{v}} = \frac{(\mathbf{d} - \mathbf{d}_0)}{(t - t_0)} = \frac{\Delta \mathbf{d}}{\Delta t} \qquad (2\text{–}3)$$

where $\mathbf{d} - \mathbf{d}_0$ represents the net displacement vector, and t and t_0 the final and initial times, respectively. The Greek delta is generally used to refer to a change in a quantity, as shown at right.

Suppose that you start from home at 6:00 A.M. and drive 210 km due west. You stop for 30 minutes to have coffee, then continue your journey, and stop for lunch 140 km further west at 11:30 A.M. You want to know what your average velocity has been. From Eq. 2–3, we see that

$$\bar{v} = \frac{350 \text{ km}}{(11{:}30-6{:}00 \text{ h})} = \frac{350 \text{ km}}{5.5 \text{h}} = 64 \text{ km/h}.$$

The velocity while in motion was always 70 km/h, but because of a stop along the way, the average velocity was lower. If you finish lunch at 1:00 P.M. and then drive straight back home, arriving at 6:00 P.M., your net displacement would be zero, and so would your average velocity, even though you had been driving for most of the day, and your total distance was 700 km. Your average speed, which is the distance divided by the time interval, would be 58.3 km/h (you travel 700 km in 12 hours).

Graphical Representation of Displacement and Velocity

It is often useful to represent motions graphically, because visual information is readily understood. Fig. 2.5a shows the displacement of a car moving at a uniform velocity of 70 km/h. The graph consists only of a straight line whose slope, which gives the velocity, is 70 km/h. Fig. 2.5b shows how the velocity varies as a function of time. This plot is a horizontal line because the velocity is constant. Fig. 2.6 describes a car which travels for 3 hours at 70 km/h, then stops for 30 minutes, then resumes travel at 70 km/h for 2 more hours. The displacement curve has three distinct segments, and two different slopes, and the velocity curve is discontinuous.

It is important to realize that the graphs shown in Fig. 2.5 and Fig. 2.6 represent motion in a single direction; that is, such graphs can easily represent only one component of a displacement or velocity. If the car had changed

(a) Displacement

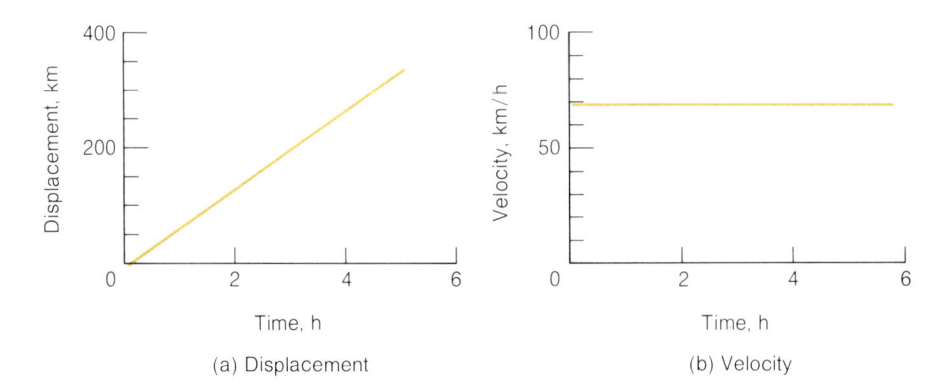

(b) Velocity

Figure 2.5

The displacement (a) and velocity (b) vary as a function of time for a car travelling at a constant velocity of 70 km/h. The slope of the displacement plot is equal to the velocity.

2.2 SPEED AND VELOCITY

direction, it would be much more difficult to show the displacement as a function of time on such graphs. It would be possible to plot each component separately, however, and then reconstruct the total displacement by calculating the resultant of the components.

If displacement is measured as a function of time and then plotted, it is easy to see not only what the total displacement is, but also how the velocity may have varied during the measurement interval. Fig. 2.7a shows the displacement for a maintenance train moving along a straight track, slowing and stopping, then speeding up, even reversing its direction for awhile. In Fig. 2.7b we see the velocity of the train as a function of time. This plot does not consist of simple straight-line segments, but is instead quite complex.

(a) Displacement

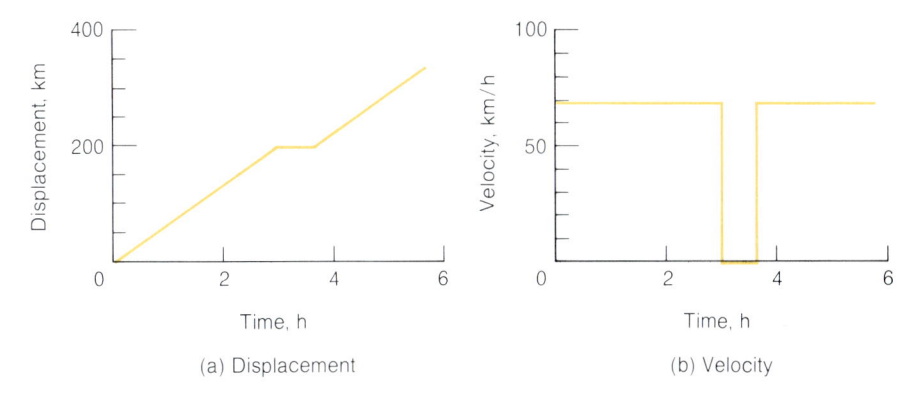

(b) Velocity

Figure 2.6

The velocity was constant except for the interval from 3 to 3.5 hours, when the velocity was zero. The velocity at any time is equal to the slope of the displacement graph.

(a) Displacement

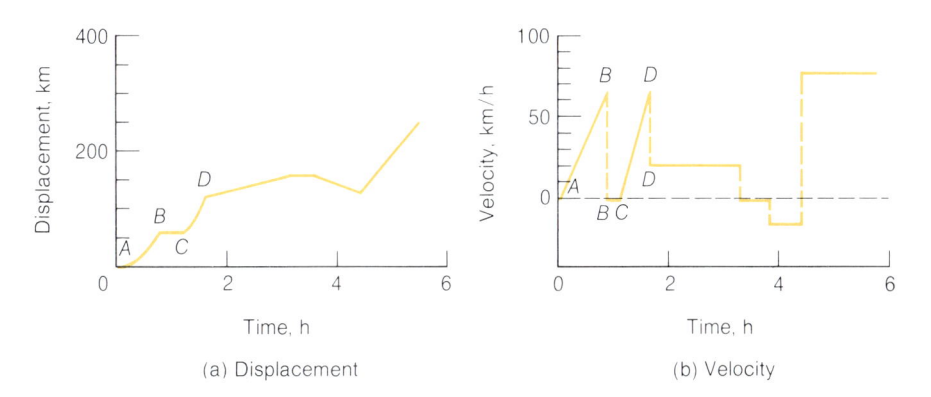

(b) Velocity

Figure 2.7

Displacement and velocity for a train that changed its velocity several times during 6 hours. Note that the displacement graph is curved in time intervals corresponding to changing velocity (i.e., in intervals A to B and C to D).

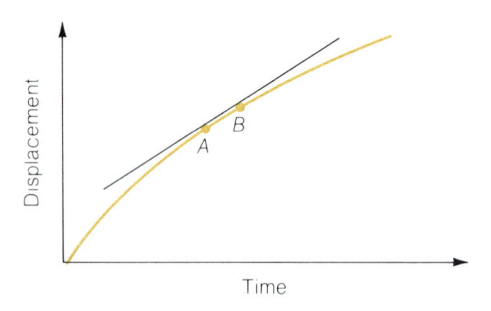

Figure 2.8

Instantaneous velocity. As the time interval between points A and B is decreased so that A and B approach a single point, the slope is the tangent to the curve at that point, and is the instantaneous velocity at that time.

Instantaneous Velocity

We have seen that the average velocity can be quite different from the actual velocity. The **instantaneous velocity** is a term we have already mentioned but not yet precisely defined. The correct definition of this quantity requires the concept of a **limit**, in which we speak of the value the velocity approaches as we consider an ever-smaller time interval. The word "instantaneous" implies an infinitesimally short time interval, but we cannot measure a change in displacement over an infinitesimally short time.

The concepts of limits and instantaneous velocity are perhaps most easily understood by considering a graph of displacement versus time (Fig. 2.8). We already know that the slope of the line represents the velocity. Now consider a small portion of the line, between points A and B. The velocity for that small segment of the graph is simply the displacement between points A and B divided by the time difference between the two points; that is, the average velocity for that interval is the slope of a straight line between points A and B. Now consider what happens as the separation of points A and B decreases. The slope will approach a certain value, corresponding to the slope of a straight line that is tangent to the plotted curve at the point in question. This value is the limit of the ratio $\Delta d/\Delta t$, as Δt approaches zero; that is, it is the instantaneous velocity.

2.3 Acceleration

Uniform velocity does not really occur in many situations. For instance, in our earlier example, we assumed that the car instantaneously went into motion at 70 km/h, and just as instantaneously stopped. Actually, a car requires some time to reach a speed, and some time to come to rest.

2.3 ACCELERATION

The rate of change of velocity is called the **acceleration**. This is a vector quantity, since it involves a change in velocity, which is a vector. Either an increase or a decrease in the magnitude of a velocity is called an acceleration; the commonly used word "deceleration" is avoided. Furthermore, a change of direction, even if the speed is constant, is also an acceleration because the vector **v** has changed. Whether an acceleration results in an increase in speed, a decrease, or a change of direction depends entirely on the direction of the acceleration vector relative to the initial velocity vector.

In mathematical terms, the acceleration is the change in velocity during a time interval divided by the time interval, that is

$$\mathbf{a} = \frac{(\mathbf{v} - \mathbf{v}_0)}{(t - t_0)} = \frac{\Delta \mathbf{v}}{\Delta t} \qquad (2\text{–}4)$$

where **v** and $\mathbf{v}_0$ are the final and initial velocities during the time interval, and t and t_0 are the final and initial times. As in our discussion of velocity, we are assuming that the acceleration is constant; otherwise Eq. 2–4 does not give the actual acceleration, but instead represents the average acceleration (discussed below).

The units of acceleration are those of velocity divided by time, which is the same as displacement divided by time, then divided by time again. This may be clear if we say that a car that accelerates from rest to 60 mi/h in 30 seconds has an acceleration of (60 mi/h)/30 s = 2 (mi/h)/s. In the SI system we use seconds and meters, so the car's final speed is 26.8 m/s, and the acceleration is (26.8 m/s)/30 s = 0.89 (m/s)/s, which would normally be written 0.89 m/s^2.

We stress that acceleration is a vector quantity, having a direction associated with it. For motion in a constant direction, we commonly treat that direction as positive, so that as an object increases its velocity in that direction, the acceleration is positive. Then if the object decreases its velocity, the acceleration is negative. Thus, if the same car, travelling at a velocity of 60 mi/h or 26.8 m/s, is now brought to a stop in 3 seconds, its change in velocity is −26.8 m/s, and the acceleration is (−26.8 m/s)/3 s = −8.9 m/s^2. A body may undergo velocities and accelerations in two or three dimensions, so a coordinate system must be carefully defined and vector addition techniques must be used.

Average and Instantaneous Acceleration

Average acceleration and **instantaneous acceleration** are analogous to average and instantaneous velocity. Eq. 2–4 defines the average acceleration over the same time interval Δt, and in general could have been written with a horizontal bar over the **a** representing the acceleration vector. Fig. 2.9 shows the velocity as a function of time for a car that travels in a straight line but varies its speed, starting from rest, accelerating smoothly to 40 km/h, then slowing to 20 km/h (undergoing a negative acceleration) before accelerating to 80 km/h,

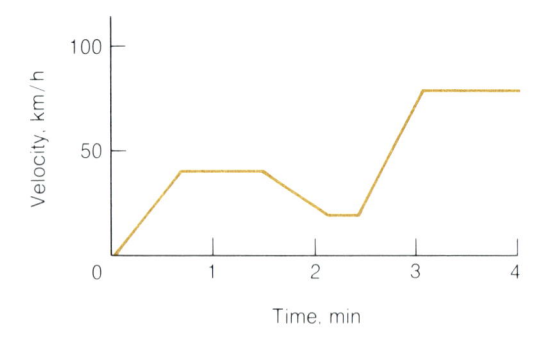

Figure 2.9

Average acceleration. This is a velocity graph for a car that started from rest and reached a velocity of 80 km/hr. after 3 minutes. Its average acceleration after 3 minutes was 80 km/hr/ 3 min. = 0.12 m/s².

reaching that speed 3 minutes after starting. The average acceleration during the 3 minutes is

$$\bar{\mathbf{a}} = \frac{(80 \text{ km/h})}{180 \text{ s}} = \frac{(22.2 \text{ m/s})}{180 \text{ s}} = 0.12 \text{ m/s}^2.$$

The instantaneous acceleration is defined as the limit of $\Delta \mathbf{v}/\Delta t$ as Δt approaches zero. On a plot of velocity versus time (as in Fig. 2.9), this limit is the slope of the tangent to the curve.

Uniform Acceleration

In some circumstances, acceleration is uniform, that is, the acceleration is constant in a fixed direction, so that the velocity changes by the same amount in every equal interval of time. A plot of velocity versus time for uniform acceleration is shown in Fig. 2.10. If a car accelerates from rest at a steady rate of $a = 0.5$ m/s², it will reach a velocity of 0.5 m/s after 1 second; after 2 seconds, it will be 1.0 m/s; after 3 seconds, it will be 1.5 m/s, and so on. If it continues to accelerate at this rate for a full minute, it will be travelling at 30 m/s, or 108 km/h (67 mi/h). In equation form, we write

$$\mathbf{v} = \mathbf{v}_0 + \mathbf{a}t \tag{2-5}$$

where $\mathbf{v}_0$ is the initial velocity (which is zero if a body starts from rest), and $\mathbf{a}$ is the acceleration.

Very large velocities can be achieved if a uniform acceleration is applied over a long period of time, even if the acceleration is small. A proposed form of propulsion for interplanetary or even interstellar spacecraft is the ion drive, which would produce an acceleration comparable to that of a car, but could achieve enormous velocities over periods of years, as a spacecraft journeyed

2.3 ACCELERATION

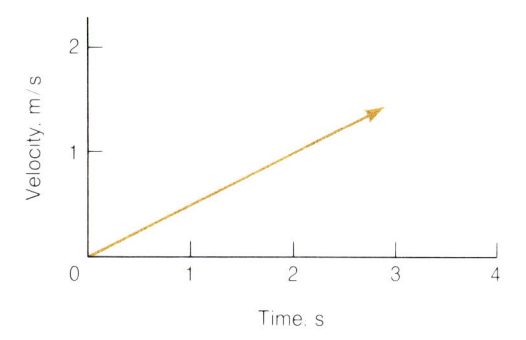

Figure 2.10

Constant acceleration. Here we see that the velocity increases at a constant rate when the acceleration is constant. In this case the acceleration is 0.5 m/s².

toward a neighboring stellar system. Suppose a spacecraft is accelerated from rest (so that v_0 is zero) by an ion drive that imparts an acceleration of 0.1 m/s². After one year, or 3.15×10^7 s, the velocity is

$$\mathbf{v} = (0.1 \text{ m/s}^2) \times (3.15 \times 10^7 \text{ s}) = 3.15 \times 10^6 \text{ m/s}$$

or over 3000 km/s (roughly 7 million mi/h). This is over one percent of the speed of light!

Calculating the Average Velocity and the Displacement

To calculate the displacement of a body is straightforward when the acceleration has been uniform, but not so simple when it has not. If the acceleration is uniform, the interval average velocity is halfway between the initial and final velocities, and is found by dividing the sum of the initial and final velocities by 2.

$$\bar{\mathbf{v}} = \frac{1}{2}(\mathbf{v} + \mathbf{v}_0), \tag{2-6}$$

where $\mathbf{v}$ and $\mathbf{v}_0$ are the final and initial velocities. The average velocity of the car that accelerates uniformly from rest to 60 mi/h is 30 mi/h, and the average velocity of a car that slows from 60 mi/h to 10 mi/h is 35 mi/h. This simple calculation does not depend on the time interval over which the acceleration occurs, but it does require that the acceleration be constant.

To calculate the displacement after a period of constant acceleration, we rewrite Eq. 2–2 for the displacement in terms of the average velocity $\bar{\mathbf{v}}$,

$$\mathbf{d} = \mathbf{d}_0 + \bar{\mathbf{v}}t \tag{2-7}$$

and substitute Eq. 2–6 into it.

$$\mathbf{d} = \mathbf{d}_0 + \frac{1}{2}(\mathbf{v}_0 + \mathbf{v})t \qquad (2\text{-}8)$$

Now we substitute for the velocity **v**, using Eq. 2–5.

$$\mathbf{d} = \mathbf{d}_0 + \frac{1}{2}(\mathbf{v}_0 + \mathbf{v}_0 + \mathbf{a}t)t$$

$$\mathbf{d} = \mathbf{d}_0 + \mathbf{v}_0 t + \frac{1}{2}\mathbf{a}t^2 \qquad (2\text{-}9)$$

This equation gives the displacement for any uniformly accelerated motion, provided the initial displacement $\mathbf{d}_0$, the initial velocity $\mathbf{v}_0$, the acceleration $\mathbf{a}$, and the time t are known. By solving the equation for t, the time required to travel a given distance under a known uniform acceleration can be found.

In most applications of these equations, the displacement, acceleration, and velocity will be in one dimension. Therefore, we can drop the vector notation, but we must keep using plus and minus signs to denote opposite directions. Using this convention, we can derive another useful relation between velocity, acceleration, and displacement. If we solve Eq. 2–5 for t, then substitute into Eq. 2–7, and use the expression (Eq. 2–6) for $\bar{v}$, the result is

$$v^2 = v_0^2 + 2a(d - d_0). \qquad (2\text{-}10)$$

Example 2.3

How far does the ion-drive spacecraft described earlier travel in one year?

We can define its starting point as $d_0 = 0$, and its initial velocity as $v_0 = 0$. The acceleration is a = 0.1 m/s^2. From Eq. 2–9, we see that

$$d = \frac{1}{2}(0.1)(3.15 \times 10^7)^2 = 5.0 \times 10^{13} \text{ m}.$$

This is over 300 times the distance from the earth to the sun.

How long it would take to reach the nearest star, a distance of about 4.3 light-years or 4.1×10^{16} m, if the ship continues to accelerate at 0.1 m/s^2? Solving Eq. 2–9 for t (with $d_0 = 0$ and $v_0 = 0$) and substituting for the distance yields

$$t = \sqrt{\frac{2d}{a}} = \sqrt{\frac{2(4.1) \times 10^{16} \text{ m}}{.1 \text{ m/s}^2}} = 9.1 \times 10^8 \text{ s} = 29 \text{ years}.$$

The velocity at the end of this time would be

$$v = at = 9.1 \times 10^7 \text{ m/s}$$

just about one third of the speed of light. (Actually, relativistic effects modify the results when velocities are this large. See Chapter 21.)

2.4 GRAVITATIONAL ACCELERATION

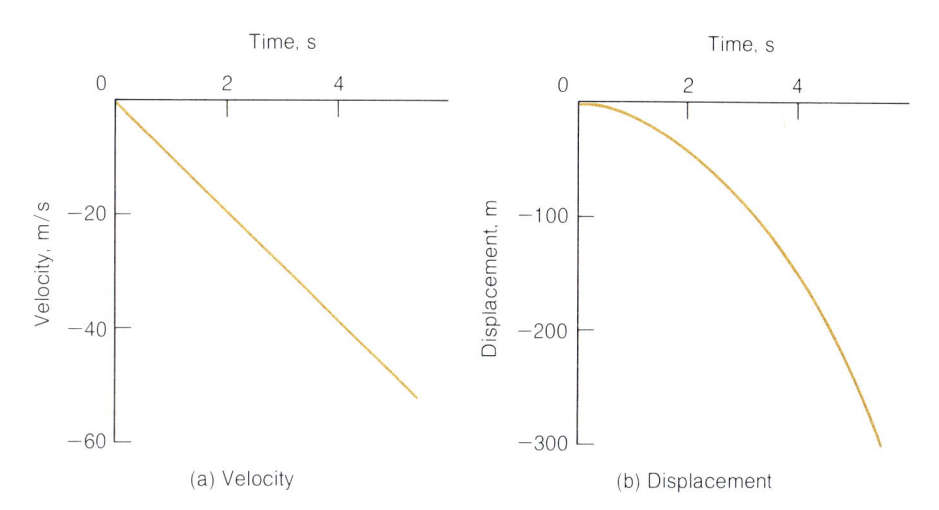

Figure 2.11

Gravitational acceleration. The acceleration of gravity is −9.8 m/s². At left is a graph showing how the velocity of an object falling from rest varies with time, and at right is a graph showing the displacement as a function of time.

2.4 Gravitational Acceleration

We now have all the tools to discuss the motions of falling bodies. It was shown by Galileo, and later demonstrated mathematically by Newton (see Chapter 3) that the acceleration of a falling object at the earth's surface is both constant and independent of its mass (provided air resistance is negligible). Thus, in principle we have a very simple situation, one already described in our discussion of uniform acceleration.

In many gravitation problems, we need consider only the vertical components of the displacement, velocity, and acceleration. Thus we can drop vector notation, and note directions by using plus and minus signs. Later we will see that combined horizontal and vertical motions are easily treated by considering the components in each dimension separately.

Experimental data show that the acceleration due to gravity at the earth's surface, usually designated as g, is 9.8 m/s² downward; that is, the speed of a falling object increases by 9.8 m/s for every second of fall (Fig. 2.11). We can adopt a one-dimensional coordinate system in which the upward direction is defined as positive (this is the usual convention, although it could just as easily have been defined so that the downward direction is positive).

Example 2.4   A pebble is dislodged from the roof of a building that is 100 m tall. How long does it take to reach the ground, and how fast it is moving when it gets there?

To find the time, we use Eq. 2–9, with the initial displacement and velocity set equal to zero, and solve for t:

$$t = \sqrt{\frac{2d}{g}} = \sqrt{\frac{2(-100 \text{ m})}{(-9.8 \text{ m/s}^2)}} = 4.52 \text{ s}$$

Now we can find the velocity from Eq. 2–5, with $v_0 = 0$.

$$v = gt = (-9.8 \text{ m/s}^2)(4.52 \text{ s}) = -44.3 \text{ m/s}$$

If the height of the building were not known, but the time of the pebble's fall were measured, then the height could be deduced directly from Eq. 2–9.

Although we have used examples in which the initial velocity and displacement are zero, describing other motions is not much more complicated. It is always possible to define the coordinates so that the starting point is zero. In principle, it is also possible to choose a coordinate system in which the initial velocity is zero as well, but this may make the problem more complicated because then the coordinate system itself may have to move. It is much easier to work with a nonzero initial velocity.

Suppose a ball is thrown straight up with an initial velocity v_0 (taking the positive direction upward). If air resistance can be neglected (see Chapter 3), then the ball accelerates at $g = -9.8 \text{ m/s}^2$, eventually stops its upward motion, and begins to fall down. Now it speeds up, since its acceleration is in the same direction as its velocity, until it hits the ground at the point from which it was originally thrown. The height the ball reaches is determined by its displacement when the velocity has decreased to zero, which we can find from Eq. 2–10 by setting $d_0 = 0$ and $v = 0$ and solving for d_{max}.

$$d = d_{max} = \frac{v_0^2}{2g} \tag{2-11}$$

The time required for the ball to reach its maximum height is found by solving Eq. 2–5 for t, with $v = 0$ and $a = g$.

$$t = \frac{-v_0}{g} \tag{2-12}$$

Now let us calculate the time it takes for the ball to fall back to earth, and the velocity it reaches. For simplicity, we can redefine our coordinate system so that downward displacement and acceleration are positive, since both are in the same direction. Then the starting displacement is $d_0 = 0$, and the final displacement is $v_0^2/(2g)$ from Eq. 2–11 (the sign is now positive because g is positive). What we seek is the time for the ball to fall from the initial displacement $d_0 = 0$ to the final displacement $d = d_{max} = v_0^2/(2g)$, where v_0 is the final velocity, equal to the initial upward velocity. From Eq. 2–9, we solve for t to find

2.4 GRAVITATIONAL ACCELERATION

$$t = \sqrt{\frac{2d}{g}}. \tag{2-13}$$

If we substitute d_{max} for d, we find

$$t = \sqrt{\frac{2(v_0^2/2g)}{g}} = \frac{v_0}{g} \tag{2-14}$$

which is equal to the time it took the ball to reach the height d_{max} (the signs in the two expressions for t are opposite because the sign of g was changed).

The velocity of the ball when it returns to the ground is given by Eq. 2–5, with the initial velocity $v_0 = 0$ and the time given by Eq. 2–12.

$$v = gt = g\left(\frac{-v_0}{g}\right) = -v_0$$

Hence when the ball reaches the ground on its descent, it has a velocity equal in magnitude to its initial upward velocity, but opposite in direction.

We have rigorously proved something that may have seemed intuitively obvious, that the motion of a projectile thrown or launched upwards is symmetric. The projectile takes as much time to rise to its greatest height as it does to fall from that point, and its velocity when it returns to the ground is the same as its initial velocity.

Example 2.5 Suppose a football is punted straight up, and has a "hang time" (the total time the ball is in the air) of 2.8 s. How high did it go, and what was its initial velocity?

We must use half this time to find the initial velocity and the height reached. We can use Eq. 2–13 which relates the time and the maximum displacement, and solve for the displacement.

$$d = \frac{-gt^2}{2} = -\left(\frac{-9.8}{2}\text{ m/s}^2\right)(1.4\text{ s})^2 = 9.6\text{ m}$$

The initial velocity can be found from Eq. 2–14.

$$v_0 = -tg = -(1.4\text{ s})(-9.8\text{ m/s}^2) = 13.7\text{ m/s}$$

Now let us add a second dimension to the motion. Suppose we throw a ball upward at an angle to the vertical, so that the ball returns to the ground at some distance from its starting point. This kind of motion can be depicted

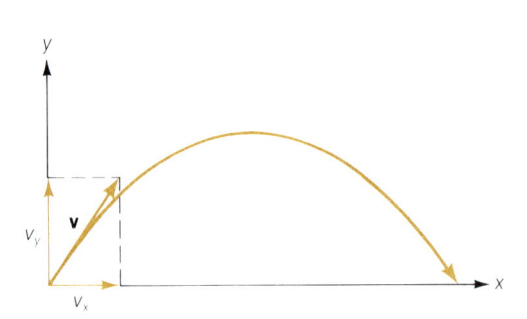

Figure 2.12

*Two-dimensional motion under the acceleration of gravity. Here an object is thrown with velocity **V**, at some angle away from the vertical. The motion in the vertical direction is the same as if the object were thrown straight up with speed V_y, and the horizontal motion is that of a body moving at constant speed V_x. This separation is possible because gravity only acts in the vertical direction.*

Figure 2.13

Vertical and horizontal motion. The ball falling vertically accelerates downward at the same rate as the ball thrown horizontally. The horizontal component of the ball at right is uniform.

graphically, as shown in Fig. 2.12. We must now consider the components of the displacement, velocity, and acceleration vectors. This need not be complicated if we choose our coordinate system wisely. It is best to use the vertical as one axis, and to define the horizontal axis so that it coincides with the horizontal component of the initial velocity. That is, if the ball is thrown at a 45° angle to the ground in a westerly direction, let west be the positive horizontal axis. Then the vertical motion is the same as for one-dimensional motion under the acceleration of gravity, and the horizontal component is the same as for uniform velocity. Note that gravity acts only in the vertical direction, leaving the horizontal velocity component unchanged (Fig. 2.13).

Example 2.6   Suppose a ball is thrown up at an angle of 30° from the vertical, with an initial velocity of 10 m/s. How far from the starting point will it land?

The first step is to calculate the time it will be in the air, and this is the same as if the ball were thrown straight up. This time is twice that given in Eq. 2–12 (allowing for the ascent and the descent), or

$$t = \frac{-2v_{0y}}{g}$$

where v_{0y} is the vertical component of the initial velocity; that is,

2.4 GRAVITATIONAL ACCELERATION 43

$$v_{0y} = (10 \text{ m/s})\cos(30°) = 8.66 \text{ m/s}.$$

(The time is positive despite the minus sign, because g is negative in this frame of reference. Then t is

$$t = \frac{-2(8.66 \text{ m/s})}{9.8 \text{ m/s}^2} = 1.77 \text{ s}.$$

Now to find the horizontal displacement component d_x of the ball at the end of its flight, we use Eq. 2–2, with $d_0 = 0$.

$$d_x = v_{0x} t = v_0 \sin(30°)t = (10 \text{ m/s})(\sin(30°))(1.77 \text{ s})$$

Thus the ball travels 8.9 meters in the horizontal direction and takes 1.77 seconds to return to the ground.

Example 2.7 Refer to the Example 2.5 (the football that was kicked and stayed in the air 2.8 s). Now suppose that the ball is kicked at a 35° angle to the vertical with the same vertical velocity component, so that its maximum height and hang time are the same, but now there is a horizontal component. How far downfield will the ball travel before it hits the ground?

We first must find the horizontal component of the initial velocity. The vertical component is 13.7 m/s, and the angle from the vertical is 35°. Then

$$13.7 \text{ m/s} = v_0 \cos(35°)$$

which yields

$$v_0 = 16.7 \text{ m/s}.$$

The horizontal component of the velocity (see Fig. 2.12) is

$$v_x = v_0 \sin(35°) = (16.7 \text{ m/s})\sin(35°) = 9.6 \text{ m/s}.$$

Now it is simple to find the horizontal displacement.

$$d_x = v_x t = (9.6 \text{ m/s})(2.8 \text{ s}) = 27 \text{ m}$$

This was a poor punt—30 yards down the field.

FOCUS ON PHYSICS

Launching Sounding Rockets

Even in the era of the *Space Shuttle,* which provides easy and frequent opportunities to launch scientific instruments into orbit, an older method is still used for making measurements from above the earth's atmosphere. Sounding rockets (Fig. 2.1.1) are used to launch payloads to altitudes of several hundred kilometers, where they have a few minutes to collect data before falling back to the ground. The descent is slowed by parachute, and, with luck, the instrument survives to be launched again.

The first sounding rocket experiments were carried out in the 1940s, with captured German V-2 rockets. Later, United States rockets, with names like *Aerobee* and *Nike,* were developed, and by the 1950s the country had an active research program in atmospheric science and astronomy. The need for making measurements at high altitudes is clear for the study of the atmosphere, where this method provides *in situ* measurements of conditions from the base of the atmosphere to its top. For astronomy, the importance lies in the access that can be gained to wavelengths of light that do not penetrate the atmosphere. The first astronomical observations made by sounding rocket were ultraviolet measurements of the sun. Today rocket instruments measure solar and stellar radiation at ultraviolet and other wavelengths (such as x-ray) that do not reach the earth.

Sounding rocket experiments are still profitable, despite the short measurement time, because they are relatively inexpensive (compared to launching satellites, although the *Space Shuttle* may become even less expensive) and because instrument preparation and scheduling are so flexible that rockets can be efficiently used to develop new instruments. Furthermore, there are still scientific problems in atmospheric science and in astronomy that can be solved with five minutes' data collection.

Figure 2.1.1

Sounding rocket on launching pad.

We already know enough about motions of bodies under uniform gravitational acceleration to appreciate the delicacy of predicting the trajectory of a sounding rocket. It is important to do this correctly, because after the rocket is launched the scientists and technicians on the ground cannot correct its course. Many sounding rockets are launched from the U.S. Army's White Sands Missile Range in southern New Mexico. The rockets are launched to the north from the southern end of the range, which is about 100 miles long and 50 miles wide. Although the rockets can reach altitudes of over 200 miles, they must fall back to earth within the narrow corridor established by the range boundaries. Crosswinds that may carry a rocket many miles off course must be compensated for by the direction and angle of the launch. Often the sideways displacement caused by winds is greater than the width of the range, meaning that the rocket is pointed towards a spot outside of the boundaries, so that the wind will bring it back in. If a miscalculation or a sudden shift in the wind puts a rocket in a path that would end outside the range, the rocket is destroyed and the payload is dropped with a parachute without making its scientific measurements.

A rocket is typically accelerated at an average rate of 5 times the gravitational acceleration for about 30 seconds. Then the rocket motor shuts down. The vehicle is then more than 20 km above ground. The maximum velocity is about 1,500 m/s (well over 3,000 mi/h). Now the payload is released and about 150 seconds later coasts up to the highest point, 200 miles. Thus, because it falls back to the same altitude in 150 seconds more, there are about five minutes for doing useful scientific work. Then the parachute opens and the payload falls to the ground slowly enough that it is not destroyed on impact (although some damage may occur).

All of the figures given so far are approximately correct, and could have been calculated using equations presented in this chapter. Real trajectories are a bit more complicated, however. First, air resistance always retards acceleration. Second, the rocket may not accelerate uniformly on the upwards boost; sometimes two-stage rockets are used, so that there are two distinct acceleration stages. Third, calculating the effect of crosswinds is not simple because the winds vary with height (and time, for that matter), and compensating for wind must allow for its speed at each height. For several hours before a launch, radar techniques and small balloons make wind speed measurements. To compensate for winds and other effects, a sophisticated computer system aims the rocket by tilting the launch tower to the correct angle and direction. This aim is checked and adjusted as wind data are fed into the computer right up to the moment of launch.

Perspective

We have now developed a complete formal mechanism for describing motions of various types. We can write and solve equations which describe uniform motions and accelerated motions, and we have learned graphical techniques for analyzing motions that are not simple. We know how to solve problems about bodies falling under the acceleration of gravity, and we have the rudiments for calculating the trajectories of projectiles.

We are now ready to discuss the effects of forces on motions; that is, we are ready to study dynamics.

PROBLEM SOLVING

Solving Equations in Sequence

Problem solving can be easy if one follows the guidelines set out in Chapter 1. There, Theme 5 discussed how to solve a problem once you have drawn a figure, identified the known and unknown variables, and found several relations among the unknowns. The task then is how to solve these relations for the unknowns. The easiest problems can be done sequentially.

Example 1 Consider a body undergoing constant acceleration a, starting from rest at time $t = 0$. The distance and velocity after time t are given by the relations

$$d = \frac{at^2}{2} \text{ and } v = at.$$

Thus there are four variables (d,v,a,t). If d and a are known variables, we can derive general formulas for the two unknowns v and t. This is a two-step problem, which may be worked by solving the first equation for t to get $t = \sqrt{2d/a}$, then substituting this expression into the second equation and solving for v, to get

$$v = at = a\sqrt{\frac{2d}{a}} = \sqrt{2da}.$$

Notice that we have derived a general expression for v in terms of d and a. This method is usually preferable, since it allows us to find the unknowns algebraically from the values of the known variables d and a.

In more advanced problems, two or more equations must be solved for the unknowns. A good strategy is to solve the simplest equation first, then substitute into the other equations. This procedure requires some trial and error and some ingenuity. Practice helps.

Example 2 A boy throws a rock at angle $\theta = 40°$ from the horizontal toward a building 10 m tall and 30 m away. What initial velocity v_o is needed to barely clear the building?

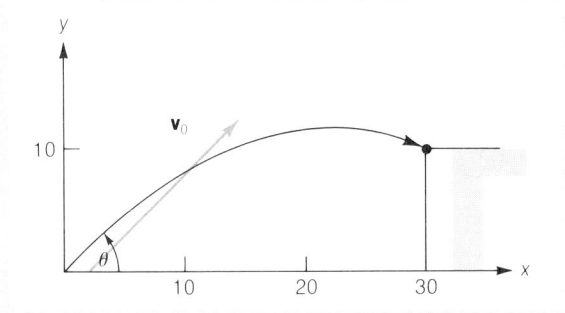

Figure 2.2.1

We first make a diagram with x- and y-axes (Fig. 2.2.1). The initial velocities in the x- and y-directions are $v_o \cos\theta$ and $v_o \sin\theta$. Gravitational acceleration only affects the motion in the vertical (y) direction, and the rock's trajectory is given by the relations

$$x = (v_o \cos\theta)\, t \text{ and } y = (v_o \sin\theta)\, t - \frac{1}{2} gt^2.$$

We know θ and the coordinates $(x,y) = (30 \text{ m}, 10 \text{ m})$ of the building top, and we wish to solve for the unknown v_o. The first equation is simplest, so we solve it for the time t and substitute that expression into the second relation.

$$t = \frac{x}{v_o \cos\theta}$$

$$y = (v_o \sin\theta)\left(\frac{x}{v_o \cos\theta}\right) - \frac{1}{2}g\left(\frac{x}{v_o \cos\theta}\right)^2$$

$$= (\tan\theta)\,x - \left(\frac{g}{2v_o^2 \cos^2\theta}\right)x^2$$

We recognize this as the equation of a parabola. Finally, we solve for v_o^2,

$$v_o^2 = \left(\frac{gx^2}{2\cos^2\theta}\right)\frac{1}{(x\tan\theta - y)}$$

Substituting for x, y, θ, and g, we find $v_o = 22.3$ m/s (about 50 mi/h).

PROBLEM SOLVING

Graphical Analysis and Integration

We have seen that when an object travels at a given velocity for some period of time, it is possible to calculate the displacement by knowing the velocity and the time. Similarly, it is possible to calculate the velocity after a period of known acceleration. The equations we introduced in the text, however, apply only to uniform or average velocity (for calculating a displacement) or to uniform or average acceleration (for calculating a velocity).

To calculate the displacement or the final velocity when the acceleration is not uniform is more difficult, but can be done graphically. First consider uniform velocity. Figure 2.3.1 shows plots of displacement versus time and of velocity versus time. The cumulative displacement that has occurred at any time is the product of velocity and time, and is represented by the shaded area under the velocity curve. Fig. 2.3.2 shows the curves for uniform acceleration in which the velocity increases at a constant rate as a function of time. Now the displacement at any time, which could also be found from Eq. 2–9, is represented by the triangular shaded area. The legs of this triangle are t and v since $v = at$ when the acceleration is uniform and the initial displacement and velocity are zero. The area of a right triangle is one half the product of the legs, or $\tfrac{1}{2}at^2$, hence the area of the triangle is $\tfrac{1}{2}at^2$. Thus, measuring this area is equivalent to calculating the displacement from Eq. 2–9.

Now suppose the acceleration is not uniform (Fig. 2.3.3). The area under the velocity curve up to any time represents the total displacement up to that time. Therefore we can find the displacement by measuring the area, even though there is no simple algebraic equation that would allow us to calculate the displacement easily.

A special tracing device called a planimeter can precisely measure the area within a closed curve, however irregular, but a good approximation can also be made by dividing the time axis into many small divisions (Fig. 2.3.3). Then each division is the width of a rectangle whose length is equal to the height of the curve at the center of the division. Therefore, the entire area under the curve is approximated by the sum of the areas of the small rectangles. The smaller the division, the more accurate this approximation becomes (indeed, the limit of this approximation as the width of the rectangles approaches zero is the precise area under the curve). Since a large number of rectangles means more laborious calculation, the widths are determined by the shape of the velocity curve (the

(continued)

Figure 2.3.1

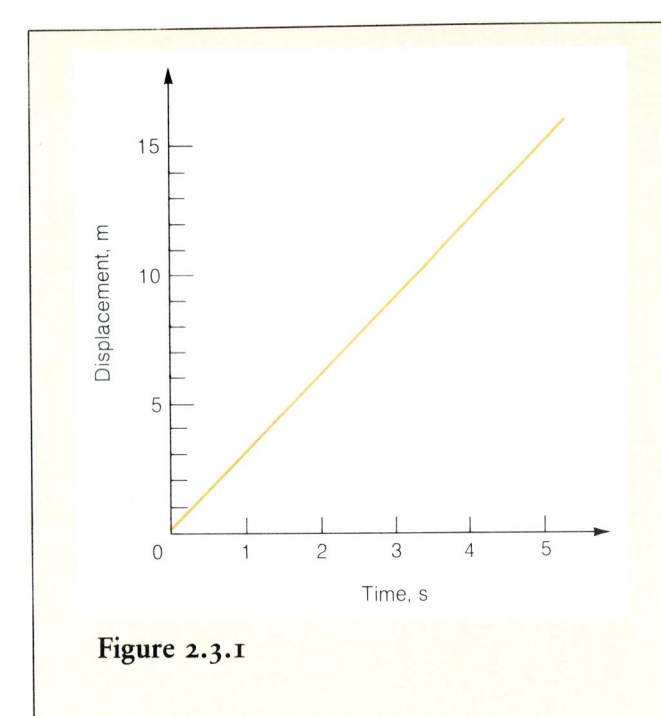

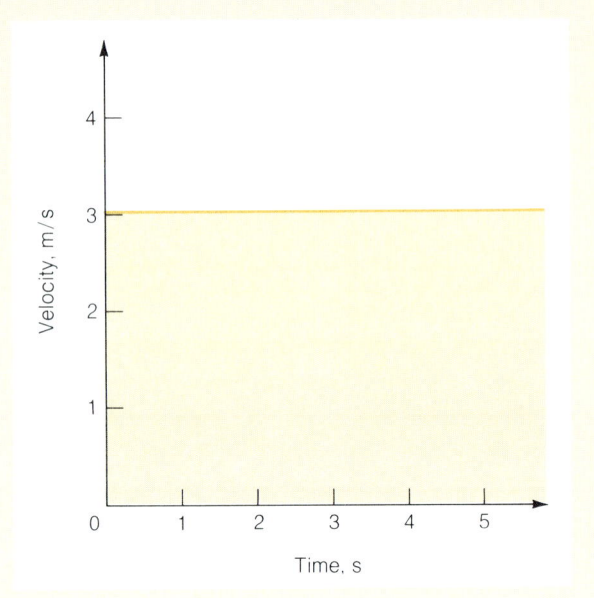

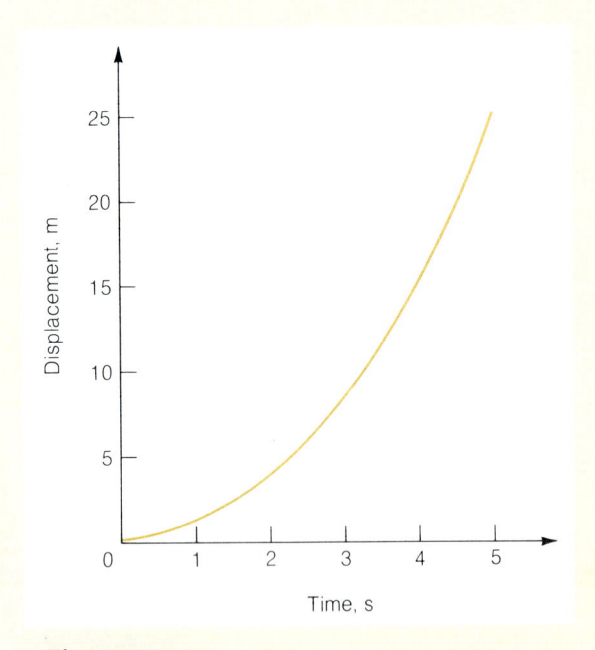

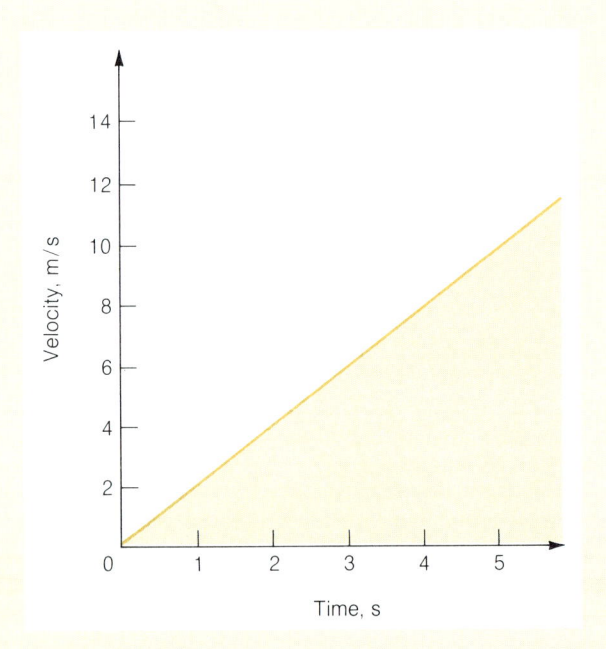

Figure 2.3.2

more wiggles it has, the more divisions are needed to represent it accurately) and by the accuracy of the input data.

We will often find in physics that the area under a curve has important applications. Finding such areas is called integration in calculus, and if the equation of the curve is known, the area can be found without resorting to measuring the graph. There are still many problems for which the graphical method must be used, however, and often com-

PERSPECTIVE

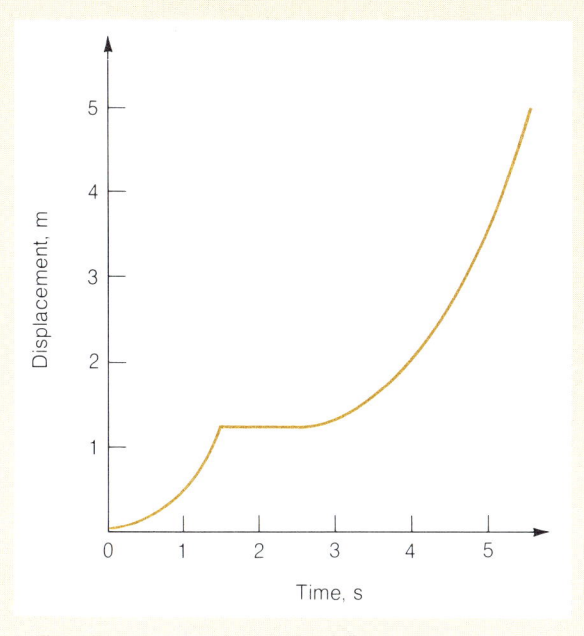

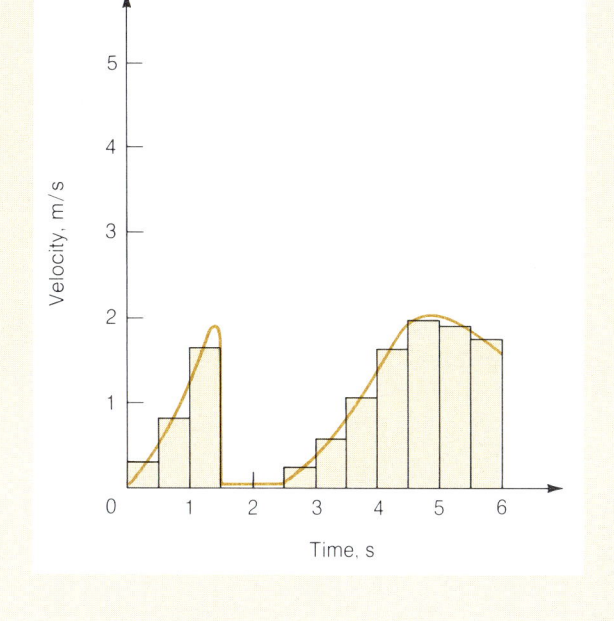

Figure 2.3.3

puters are employed to do the work. Coordinates of points on the curve are supplied to the computer which measures the area under the curve by dividing it into rectangles and summing their areas. Since computers do repetitive calculations very quickly, they can efficiently find and sum the areas of many very small rectangles. This technique is called numerical integration.

Example A train on a straight track slows down when it encounters road crossings and speeds up while it traverses open country. The train stops at a broken trestle spanning a river 6 h 24 min after its previous stop. The engineer would like to report the problem, but he doesn't know his location. He needs to calculate how far he has travelled since his last stop. Fortunately, his fireman had been recording the train's velocity every few minutes and was able to make the plot shown in Fig. 2.3.4.

The engineer can find the total displacement of the train by measuring the area under the curve. Because the fireman measured the velocity in km/h and the time in minutes, the engineer first redraws

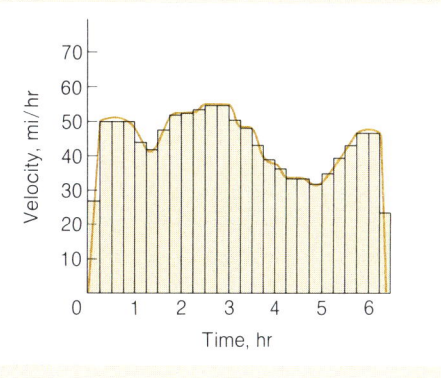

Figure 2.3.4

the horizontal scale in units of hours (this has already been done in the figure). He decides that rectangles of width 0.25 h are sufficiently narrow to represent the shape of the curve. Now he measures the height of each rectangle in km/h and multiplies it by 0.25 h to find its area. Then he sums the areas of all the rectangles. He finds that the train has travelled 283 miles from its last stop. The map shows the track crossing a river at just this distance from the town where the train last stopped, so the engineer is able to correctly report the location of the broken trestle.

SUMMARY TABLE OF FORMULAS

Velocity:
$$\mathbf{v} = \frac{\mathbf{d}}{t} \tag{2-1}$$

Displacement:
$$\mathbf{d} = \mathbf{d}_o + \mathbf{v}t \tag{2-2}$$

Average velocity:
$$\bar{\mathbf{v}} = \frac{(\mathbf{d} - \mathbf{d}_o)}{(t - t_o)} = \frac{\Delta \mathbf{d}}{\Delta t} \tag{2-3}$$

Acceleration:
$$\mathbf{a} = \frac{(\mathbf{v} - \mathbf{v}_o)}{(t - t_o)} = \frac{\Delta \mathbf{v}}{\Delta t} \tag{2-4}$$

Velocity after uniform acceleration:
$$\mathbf{v} = \mathbf{v}_o + \mathbf{a}t \tag{2-5}$$

Average velocity:
$$\bar{\mathbf{v}} = \frac{1}{2}(\mathbf{v} + \mathbf{v}_o) \tag{2-6}$$

Displacement calculated from average velocity:
$$\mathbf{d} = \mathbf{d}_o + \bar{\mathbf{v}}t \tag{2-7}$$

$$\mathbf{d} = \mathbf{d}_o + \frac{1}{2}(\mathbf{v} + \mathbf{v}_o)t \tag{2-8}$$

Displacement after uniform acceleration:
$$\mathbf{d} = \mathbf{d}_o + \mathbf{v}_o t + \frac{1}{2}\mathbf{a}t^2 \tag{2-9}$$

Velocity after uniform acceleration (in one dimension):
$$v^2 = v_o^2 + 2a(d - d_o) \tag{2-10}$$

Height reached by a body thrown upward:
$$d = d_{\max} = \frac{v_o^2}{2g} \tag{2-11}$$

Time for a body to reach its maximum height (positive up):
$$t = \frac{-v_o}{g} \tag{2-12}$$

Time for a body to fall from a height (positive down)
$$t = \sqrt{\frac{2d}{g}} \tag{2-13}$$

$$t = \frac{v_o}{g} \tag{2-14}$$

THOUGHT QUESTIONS

1. What is the key distinction between displacement and distance? Between speed and velocity?
2. Why does adding two vectors to form a resultant automatically take into account the different directions associated with those vectors?
3. Are the components of a vector scalars? Can they be added and subtracted directly?
4. Why does a particle moving in a circular orbit experience acceleration even if its speed is constant?
5. The analysis of relative motion, for example a boat moving in a current, is best done with vectors. Why?
6. In general, the average and instantaneous velocities of a moving object are not the same. Why?
7. At the instant that an apple falls from a tree, you throw a second apple horizontally outward from the tree at the same height as the first. In the absence of air resistance, why do both apples strike the ground at the same time?
8. The speed of an object falling from rest increases by 9.8 m/s each second it falls. That is, speed is linearly proportional to time, $v = gt$. Why is the distance fallen proportional to time squared ($d = \frac{1}{2}gt^2$)?
9. If one knows the velocity of a train at many closely-spaced intervals in time, then one can graphically determine the distance the train has travelled. What law of physics is used to do this? How could one graphically determine the average acceleration over each time interval?
10. Consider an object moving in a circular orbit at constant speed, as shown in Fig. 2.14. By considering two closely spaced points along the circle, separated by a time interval Δt, show that the velocity difference $\Delta \mathbf{v} = \mathbf{v}_2 - \mathbf{v}_1$ points toward the center of the circle. Since acceleration is defined as the limiting value of $\Delta \mathbf{v}/\Delta t$ as Δt becomes smaller and smaller, which way does the acceleration point?

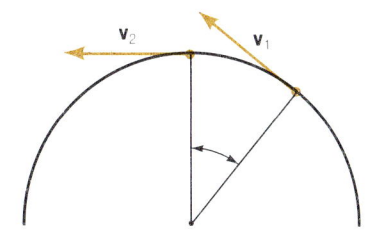

Figure 2.14

11 Suppose speed limits had to be posted in mi/h, km/h, and m/s. How would a 55 mi/h sign read?

12 Displacement **d**, velocity **v**, and acceleration **a** are all vectors. Describe a situation in which **d** and **v** are parallel, but **a** points in an opposite direction.

13 Why is the time for a mass m to fall from height h independent of the mass m?

14 If the gravitational acceleration g were reduced by a factor of 2, how much farther and higher could you jump?

15 What is the difference between: acceleration, deceleration, negative acceleration. Explain the significance of acceleration as a vector.

PROBLEMS

Sections 2.1–2.2 Displacement and Velocity

- 1 If you want to drive from San Francisco to Denver, 2,000 km away, in 22 hours, what must your average speed be?

- 2 At Mach 2 (2,400 km/h) how far can a jet fly in 2.5 hours?

- 3 At a track meet, the winning times for the 100 m, 400 m, and 800 m races are 10.0 s, 44.1 s, and 1 min 44 s respectively. What are the average speeds?

- 4 A thoroughbred horse covers 2.1 miles (3.4 km) in 2 min 10 s. How much faster is the horse than a human who runs a four-minute mile?

- 5 Two cars, each moving at 80 km/h, approach one another on a straight highway. If they are initially 1 km apart, after how many seconds will they meet?

- 6 If one meter equals 39.37 inches and one mile is 5,280 feet, what is the conversion factor between miles/h and km/h?

- 7 If it takes three days for a letter to be delivered from Los Angeles to New York (5,000 km), what is the letter's average speed?

- 8 You hear a thunderclap 9 seconds after you see lightning strike. If the speed of sound is 350 m/s and the speed of light is 300,000 km/s, how far away was the strike?

- 9 On a bicycle trip lasting three days and three nights, you cover 60 km, 100 km, and 115 km on the three consecutive days. What was your average speed?

•• 10 In a relay race, each swimmer covers his lap 0.5 s faster than the preceding person. If the total time for the four-lap race was 53.0 s, what was the time of the first lap?

•• 11 The earth has a circumference of 40,000 km and rotates once each 23 h 56 min. What is the average speed of a point on the surface at the equator?

•• 12 A spider crawls forward at 1 cm/s across a rug, as a housekeeper pulls the rug slowly forward at 2 cm/s (relative to the floor). What is the spider's velocity relative to the floor?

•• 13 A boat, whose speed in still water is 10 km/h, heads directly across a 1 km wide river with a 5 km/h current. How long does it take to cross? How far downstream does the boat land?

•• 14 An airliner can fly 1,000 km/h in still air. In a 200 km/h headwind the plane flies three hrs, then turns around and flies back with the wind. How many hours does the return trip last? What is the total distance travelled?

•• 15 A wise old bird, whose airspeed is 100 km/h, wishes to fly south for the winter, but a 50 km/h wind is blowing from a direction 45° west of north. In what direction should the bird head? What will be its resultant velocity relative to the ground?

••• 16 A boy riding on a 100 km/h train throws a rock at 30 km/h perpendicular to the train's velocity, hoping to hit a window on a house. What is the rock's velocity relative to the ground? If the rock strikes the house 3 s later, how far in advance (in meters) did the boy throw the rock?

••• 17 Two cars approach an intersection on perpendicular streets. *A* moves at 50 km/h and is initially 4 km from the intersection. *B* moves at 40 km/h and is

52 CHAPTER 2 KINEMATICS

initially 3 km away. How far apart are the two cars initially? How long until they are 1 km apart? Which car arrives at the intersection first?

••• 18 In Problem 17, if the cars continue through the intersection at the same speeds, when will they be 5 km apart?

••• 19 A train moves at 150 km/h, and a baseball pitcher on board throws a fastball in the opposite direction, toward the rear of the train, at 150 km/h. What is the velocity of the ball as seen by an observer at rest outside the train? When does the ball hit the rear wall of the car, initially 15 m away from the pitcher?

• 20 A plane flying east with airspeed 500 km/h drops a parcel when the plane is above the target. The wind is blowing toward the northeast at 100 km/h. What is the plane's ground speed? If the pilot wishes the parcel to hit the ground directly southeast of the target, with what velocity perpendicular to the plane (south) must the parcel be thrown out?

Section 2.3 Acceleration

• 21 A stroboscopic camera which flashes at intervals of 1 second photographs a speeding falcon. Four successive images show the bird has moved 50 m, 55 m, and 60 m in the three intervals. What is the falcon's acceleration? What is its average speed over the three seconds? Assume the falcon flies at constant acceleration in a straight line.

• 22 A sprinter accelerates from rest to 8 m/s in 3.1 s. What is her average acceleration?

• 23 A racecar accelerates from rest to 150 km/h in 10 s. What is its acceleration? How far does it travel from the start?

•• 24 As a car decelerates smoothly from 100 km/h to 20 km/h, it travels 300 m. Find its average velocity $\bar{v}$, the time elapsed, and the average acceleration.

•• 25 Two cars in a drag race approach an intersection on perpendicular streets. A starts from rest, 300 m from the intersection, accelerates at 3 m/s^2 for 10 s, then maintains constant speed. B also starts from rest, 400 m from the intersection, accelerates at 6 m/s^2 for 6 s, then maintains constant speed. What are the final velocities of each? How long does each car take to reach the intersection?

•• 26 A baseball pitcher accelerates a fastball to 90 mi/h (about 40 m/s) in 0.1 s. What is the average acceleration of the ball? How far does the pitcher's hand move during this time? Assume constant acceleration.

•• 27 A 90 mi/h (40 m/s) fastball stops in a catcher's glove, depressing the glove 2 cm. What is the deceleration of the ball? How long does the ball take to stop? How many g's is the deceleration? (1 g = 9.8 m/s^2) Assume that the catcher does not move. (In practice, the catcher recoils from the collision.)

•• 28 During an earthquake, a fault moves 5 m at an angle tilted 20° from vertical. If the slippage occurred over a 2 s interval, what are the horizontal and vertical components of the velocity?

•• 29 Two children play tag. Annie is at rest; Bobby runs by at 5 m/s and tags her, then continues at the same speed. How fast must Annie accelerate from rest to catch Bobby in 10 seconds? Assume Annie starts immediately after being tagged.

•• 30 A frightened rabbit runs along a 10 m × 10 m square path. It reaches the corners 10 s, 22 s, 36 s, and 52 s after it started. Draw a graph showing the displacement vectors relative to the starting point at each of the corners. Draw a second graph showing the average velocity vectors during the time intervals corresponding to the four sides.

•• 31 The height of the *Space Shuttle* as a function of time after launch is shown in Fig. 2.15. During what periods is its vertical velocity negative? Plot the vertical velocity as a function of time. During which periods did it accelerate?

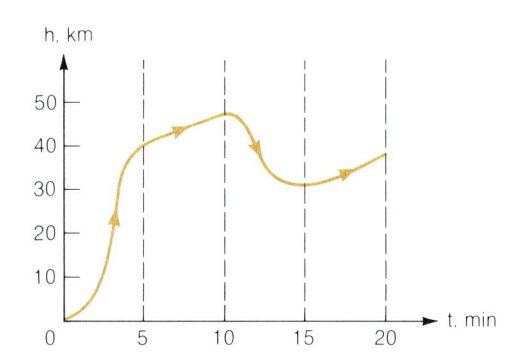

Figure 2.15

•• 32 From Fig. 2.15 estimate the average vertical velocity during the intervals 0–5 min, 5–10 min, and 10–15 min after launch.

•• 33 Draw a graph of distance versus time for a bus whose velocity is shown in Fig. 2.16.

•• 34 In a trip between two commuter towns, a train accelerates at 0.1 m/s^2 for half the trip, then decelerates at 0.1 m/s^2 for the second half. If the towns are 20

PROBLEMS

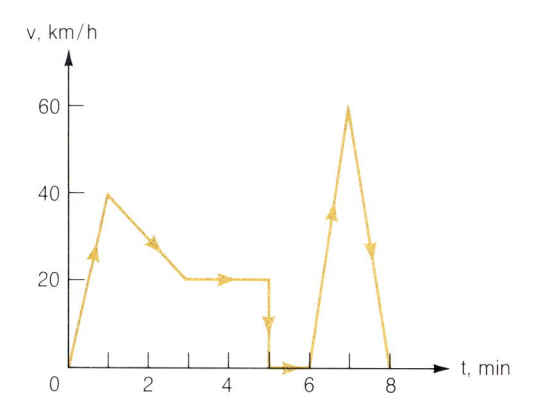

Figure 2.16

km apart, how long does the trip last? What was the train's maximum velocity?

••• 35 The earth moves in a circular orbit 3×10^{11} m in diameter. What is its average speed? How fast must a rocket accelerate (m/s^2) from rest across the orbit in order to meet the earth six months later? How fast would the rocket be moving at this time?

••• 36 A police officer in a specially-equipped squadcar is tailing a getaway car. A new on-board computer records the speeds at one-second intervals: 0 km/h (start), 5 km/h, 12 km/h, 19 km/h, 30 km/h, 41 km/h. What are the car's average accelerations over each 1-second interval? What is the average acceleration over the 5-second interval? Why do these accelerations differ?

Section 2.4 Gravity

• 37 A baseball is thrown straight up and lands 10 s later. How high did it travel? What was its initial velocity?

• 38 How fast must a baseball be hit straight up in order to hit the top of a 100 m domed stadium?

• 39 Wandering through the woods, you encounter an abandoned mineshaft. You drop a stone down the shaft and hear it hit bottom 5 s later. How deep is the mine? Neglect sound travel time.

•• 40 A water sprinkler shoots out two streams at angles of 30° and 60° from vertical at speeds of 10 m/s. How far out do the streams land? Neglect air resistance.

•• 41 A baseball player hits a four hundred-foot (122 m) homerun. Assuming that the ball left the bat at an angle of 45° from horizontal and neglecting air resistance, calculate how long the ball was in the air.

•• 42 If you can throw a ball 50 m straight up on earth, how high could you throw the same ball on the moon, where gravity is ⅙ that on earth?

•• 43 In a Hollywood movie stunt, a car plunges off a cliff, initially moving horizontally at 100 km/h. If the cliff is 400 m high, how long does the car take to hit the bottom?

•• 44 An athlete can jump a horizontal distance of 7 m on earth. How much farther can he jump on the moon, where gravity is ⅙ that on earth?

••• 45 An object is released at angle θ_0 from horizontal with initial velocity v_0.
   (a) What is the horizontal distance D travelled downrange when the object strikes level ground again? Ignore air resistance.
   (b) Show that the maximum distance D is obtained when $\theta_0 = 45°$. You may find the trigonometric relation $\sin(2\theta) = 2\sin(\theta)\cos(\theta)$ of use.
   (c) Considering air resistance, why should realistic trajectories begin with angles θ_0 somewhat *less* than 45°?

••• 46 A bird flying at 50 m/s wishes to get the attention of a man sitting on a park bench 100 m below. If the bird drops an object vertically in flight, how soon and how far in distance before flying directly over the man should the bird release its object?

••• 47 A baseball pitcher stands 60 ft 6 inches (18.44 m) from home plate and throws a ball with initial velocity v_0 *horizontally*. What is the minimum velocity required to make it to home plate, assuming that the point of release is 1.2 m above the level of the plate?

••• 48 A field goal kicker wishes to make a field goal from the 55-yard line. He kicks the ball 40° from horizontal. The goal's crossbar is 3 m high and 60 m from the ball's initial position. What minimum initial velocity must the ball have in order to clear the crossbar?

••• 49 A cannonball is shot at velocity v_0 at angle θ_0 from horizontal. Neglecting air resistance, how high does it travel vertically? How long is it in the air? How far does it travel horizontally before hitting level ground? Show that the ball travels farthest when $\theta_0 = 45°$.

••• 50 A diver leaves a springboard 10 m above water with an initial velocity of 10 m/s at an angle 30° from vertical. How high above her starting point does she rise? How long has she been in flight and what is her horizontal distance when she hits water?

••• 51 A ball is thrown vertically upward at 30 m/s. It strikes a 30 m ceiling and reflects downward at the same velocity with which it hits the ceiling. With what velocity does the ball hit the ceiling? What is the total time until the ball hits the floor?

••• 52 Superman is trying to throw a ball over Mt. Everest (elevation 8,848 m), standing at a horizontal distance of 10 km at an elevation of 4,000 m. If the ball just clears the peak at the maximum height of its trajectory and one neglects air resistance, what was the initial velocity (speed and angle from horizontal) of the ball?

••• 53 A student jumps out a fifth story window, falls 20 m onto a trampoline, and bounces completely elastically. That is, his upward velocity off the trampoline equals the downward velocity with which he hits the trampoline. How long after he jumped does he reach the window again? If he rebounds at only 0.8 of his downward velocity, how high will he rise?

SPECIAL PROBLEMS

1. Monkey Problem

Standing at a horizontal distance d from a tree, a hunter takes aim with a peashooter at a monkey sitting at a height h in the tree (Fig. 2.17).

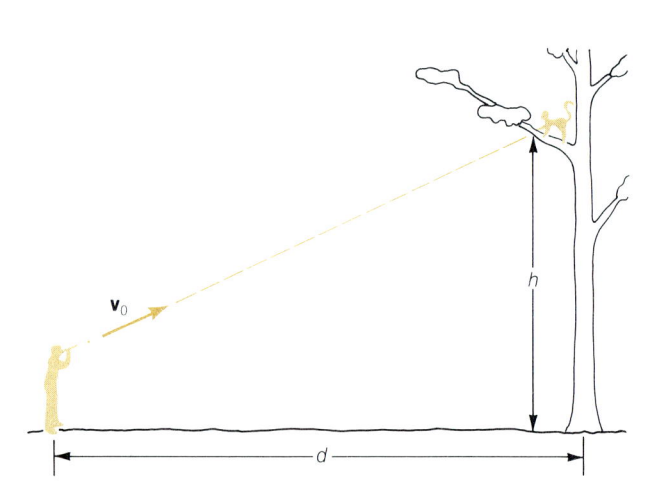

Figure 2.17

At $t = 0$, the hunter shoots the pea with initial velocity $\mathbf{v}_0$. However, the monkey, thinking itself clever, drops vertically from the tree the moment it sees the pea released.
(a) How long does the pea take to reach the tree?
(b) How far below the monkey's initial height does the pea pass the tree?
(c) How far has the monkey dropped at that time?
(d) Show that the pea always hits the monkey.

2. Nervous Reactions

Signals travel at a rapid speed along nerves in the human body. If human reaction times are 0.1 s across the length of the body (about 2 m), estimate the average signal speed along the nerve in km/h and mi/h.

3. Juggling Problem

A clown can juggle five balls at once. Each one is in the air for 1 s.
(a) How high must she toss the balls?
(b) With what initial vertical velocity must she toss them?
(c) If she juggled on the moon (where gravity is ⅙ that on earth), how many seconds would the balls stay in the air? How many more balls could she juggle? Assume the initial vertical velocity remains the same.

4. Geology Problem

Two geologists wish to measure the depth d of a special layer of rocks that reflects sound waves back to the surface (see Fig. 2.18). Mica bangs her hammer on the ground, while Flint puts his ear to the ground at a point 5,000 m away. He hears the hammer sound twice, separated by an interval of 3 s. How deep is the rock layer, if sound travels at 5,000 m/s in rock?

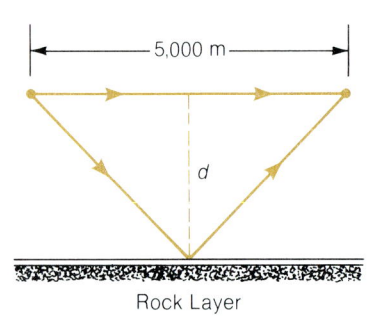

Figure 2.18

5. Space Flight Problem

The formula for the height reached by an object shot with initial velocity v_0 is $h = v_0 t - \frac{1}{2}gt^2$. However, the acceleration due to gravity g is not constant if the object moves far above the earth. Figure 2.19 shows the orbital trajectory calculated using this formula (constant g). But gravity weakens with height above the earth. Sketch the actual trajectory. Ignore air resistance.

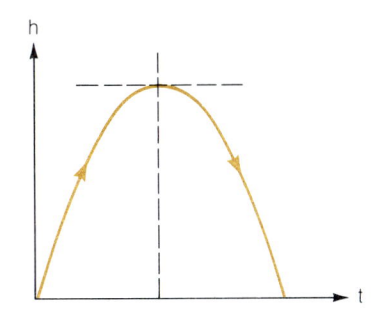

Figure 2.19

SUGGESTED READINGS

"The physics of the follow, the draw and the massé (in billiards and pool)." Jearl Walker, *Scientific American*, Jul 83, p 124, **249**(1)

"Thinking about physics while scared to death (on a falling roller coaster)." Jearl Walker, *Scientific American*, Oct 83, p 162, **249**(4)

"Deep think on dominoes falling in a row and leaning from the edge of a table." Jearl Walker, *Scientific American*, Aug 84, p 122, **251**(2)

"Galileo's discovery of the parabolic trajectory." Stillman Drake, James MacLachlan, *Scientific American*, Mar 75, p 102, **232**(3)

"Sir Isaac and the rising fastball." Peter J. Brancazio, *Discover*, Jul 84, p 44 5(7)

Newton at the bat: The science in sports. Eric W Schrier, William F. Allman, eds, Scribners, New York, 1984

Sport science. Peter J. Brancazio, Simon and Schuster, New York, 1984

A source book in physics. William F. Magie, Harvard University Press, Cambridge, 1965

The flying circus of physics with answers. Jearl Walker, Wiley, New York, 1977

Science from your airplane window. Elizabeth Wood, Dover, New York, 1975

Great scientific experiments. Rom Harre, Phaidon Press, Oxford, 1981 (Galileo's experiments with inclined planes)

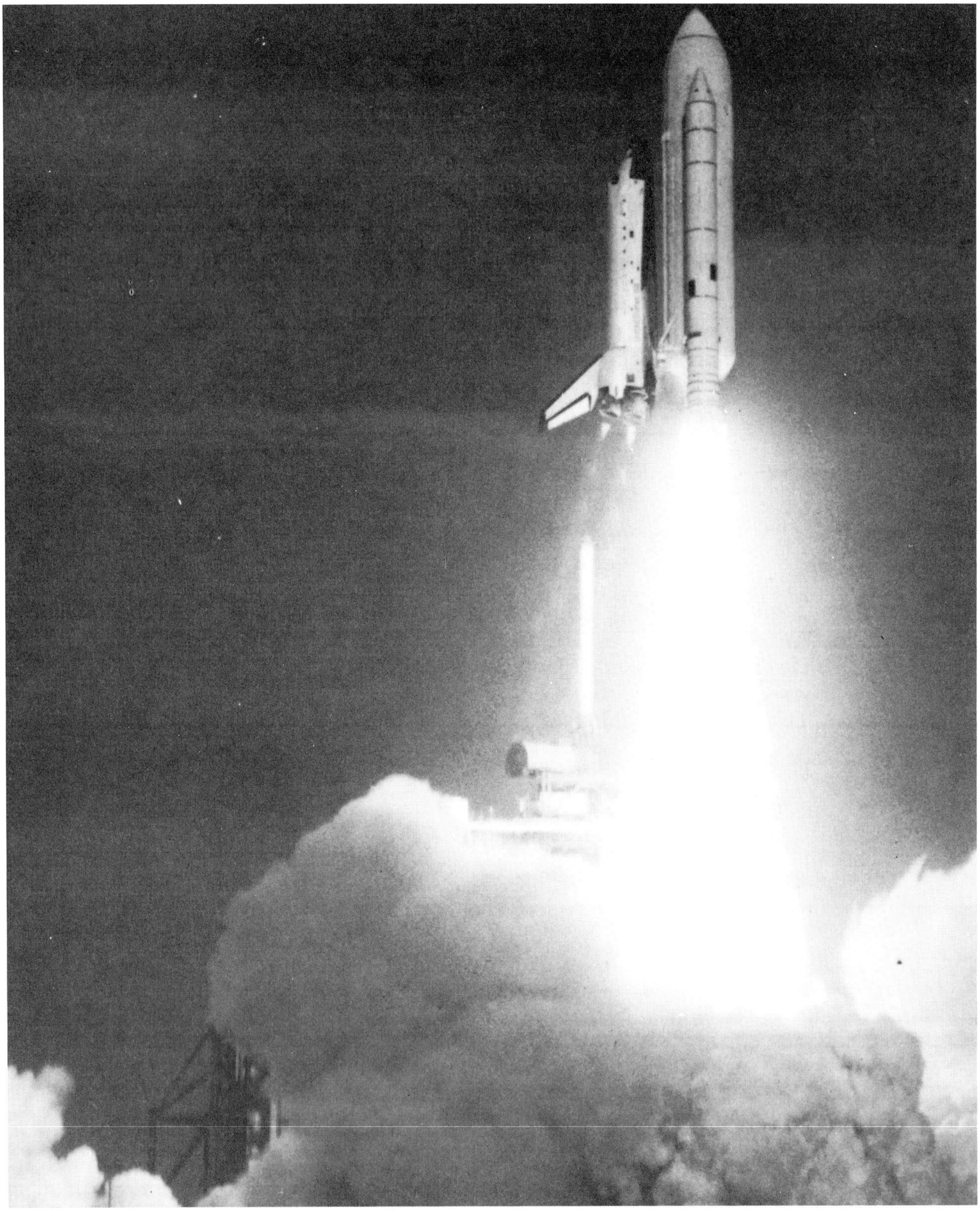

Chapter 3

Dynamics

Now that we have some background in kinematics, the descriptive study of motions, we are ready to ask *why* things move as they do; that is, we are ready to study **dynamics,** the study of motions and their relation to forces. This chapter will discuss dynamics as understood by Newton. Modern modifications, such as those imposed by special and general relativity, will be deferred to later chapters.

3.1 Force

Aristotle had the misconception that objects could move or stop without any intervention, whereas Galileo realized that a *force* was always necessary to bring about any change in motion. Most of us have an intuitive idea of force, although it is more difficult to define it precisely. We may think of a force as any kind of push or pull. Many forces are most readily recognized by the motions that result. This is a limited view, however, because forces do not always cause motion. A book lying on a table exerts a force on it, but there is no motion as a result (because, we shall see, the table exerts a balancing force on the book). A change in motion can occur only when an unbalanced force is exerted, but the exertion of a force does not always cause a change in motion.

The Nature of Forces

Because a force does not always cause motion, we cannot simply define force as something that causes motion, even though we know that a force *can* cause motion. It is more useful to define a force as that which can cause a change in motion, or, more precisely, a force is that which can cause the acceleration of an object. We use the imprecise phrase "that which" because we are defining force in the operational sense, that is, by saying what it does rather than saying precisely what it is or how it works. We use such a phrase when we cannot say what it really is that we are talking about, when we define something by stating its effect.

We classify forces as **mechanical forces,** those brought about by the physical contact of material objects; or as **field forces,** those exerted at a distance by force fields. Actually, mechanical forces, at some fundamental level, are always derived from field forces, so the question of what force really is becomes one of understanding how field forces work. We have already mentioned one field force, gravity, and we will spend considerable effort on another, the electromagnetic force. We will deal with the effects of such forces, but not with their causes.

There are four known field forces: the two just mentioned, and the strong and weak nuclear forces, which govern the interactions of subatomic particles. Although the nuclear forces have profound impact on the nature of matter, they have little direct impact on how we view our environment. The question of field forces and their fundamental nature, as well as how they are related to each other, is a major area of modern research (Chapters 25–26). For our study of mechanics, we will ignore the fundamental questions about the nature of forces and focus on mechanical forces and their effects.

Net Force

Whether or not acceleration results from the exertion of a force depends on the **net force,** the sum of all forces acting on the body. A force has both a magnitude and a direction, so when two or more forces act on a body, the net force is found by vector addition.

It can be very useful in dynamics problems to make a diagram in which the body is isolated and all forces acting on it are represented by vectors. A **space diagram** shows the object with arrows representing the forces acting on it [Fig. 3.1(a)]. A **force diagram** is a vector drawing of the forces placed head to tail to find the net force [Fig. 3.1(b)]. When we deal with an object consisting of several parts, such as, for example, a pair of weights connected by a rope, we will find it useful to construct a **freebody diagram,** in which each part is separately depicted with the forces acting on it [Fig. 3.1(c)]. We will use space and freebody diagrams in examples, and force diagrams to find net forces in problems.

We will find, in the discussions that follow, that forces can create motions. The resultant motion always depends on the net force. Motion occurs when the net (or unbalanced) force is not zero. Later, especially in our discussions of equilibrium in the next chapter, we will study balanced forces.

3.2 Newton's Laws of Motion

The following discussion is organized by treating Newton's three laws of motion in sequence. Newton himself considered these three laws so self-evident that he mentioned them only in an introductory section of the *Principia,* preferring to place the emphasis of his work on applying them to various situations.

3.2 NEWTON'S LAWS OF MOTION

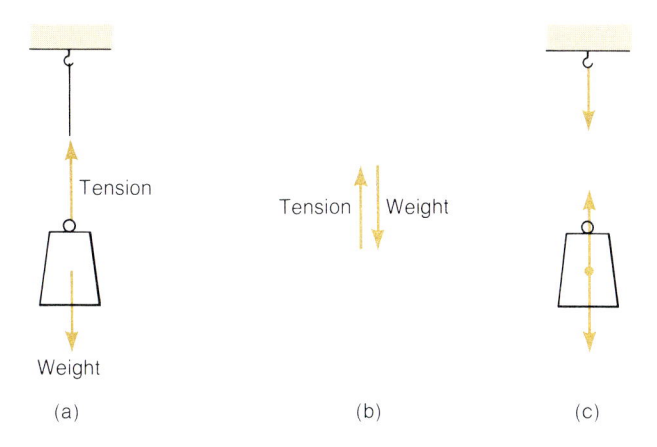

Figure 3.1

Illustrating forces. At left is a space diagram, showing the forces acting on a body. At center is a force diagram which shows with vectors the forces acting on an object so that the resultant could be derived by adding the vectors. At right is a freebody diagram, in which the forces on individual parts of the object are depicted.

Inertia and the Mass

Galileo realized that an object in motion will continue in motion unless a force acts to stop it. **Inertia** was his name for this tendency of an object to resist a change in its motion. Newton later stated his first law of motion more generally, and today we often adopt the following wording:

A body in a state of rest or uniform motion remains in its state of rest or uniform motion unless it is acted on by an unbalanced external force.

Thus, when a shuffleboard disk comes to a stop, it does so because a force (which we call friction) acts on it, not because, as Aristotle would have said, its natural tendency is to stop unless a force acts to keep it going. To see this more clearly, imagine that the shuffleboard disk is moving along an ice rink. The force of friction is greatly reduced, and the disk travels much farther before stopping, even though its initial velocity may have been the same as it was on the concrete shuffleboard court. In the vacuum of space, there is no friction, and an object in motion will never stop, so long as no forces act on it.

The inertia a body has is closely related to its **mass**. The greater the mass of a body, the greater its resistance to any change in its state of rest or motion. To properly understand inertia, as well as the other laws of mechanics, we must have a concept of what we mean by mass. Indeed mass and Newton's laws are so closely related that a full discussion of mass requires knowledge of Newton's laws.

For now we can give an intuitive definition of mass: it is the quantity of matter a body contains. This is distinct from the volume or the density of an object (but is related to both), and it is also distinct from the weight (which is a force). The weight of an object depends on the gravitational acceleration, whereas the mass is an intrinsic property that does not depend on external factors. Thus, the mass of an astronaut is the same whether he is on the earth or on the moon, but his weight is much less on the moon. The SI unit of mass is the kilogram. A quart of water, which weighs a little more than two pounds at the earth's surface, has a mass of almost one kilogram.

Mass, Force, and Acceleration

Newton elucidated something that had been discussed by Galileo, namely, that the amount of acceleration a body undergoes as the result of an applied force depends on the mass of the body. This is just a more general statement of the meaning of inertia. Whereas Newton's first law of motion dealt with the case of zero net force, hence no acceleration, the second law allows for nonzero forces and accelerations, and describes their relationship.

The acceleration of a body is directly proportional to the net force applied to it, is inversely proportional to the mass of the body, and is in the same direction as the net force.

The greater the force applied to a body, the greater its acceleration. The greater the mass of a body, the less the acceleration when a given force is applied (Fig. 3.2). In order to accelerate from 0 to 100 km/h in 10 seconds, a large sedan needs a greater force applied to it than does a small sports car. Given the same force, the sports car will accelerate at a greater rate.

The second law is written mathematically as

$$\mathbf{a} = \frac{\mathbf{F}}{m} \tag{3–1}$$

or, more commonly,

$$\mathbf{F} = m\mathbf{a} \tag{3–2}$$

where **F** represents the net force acting on a body of mass m, and **a** is the acceleration of the body.

3.2 NEWTON'S LAWS OF MOTION

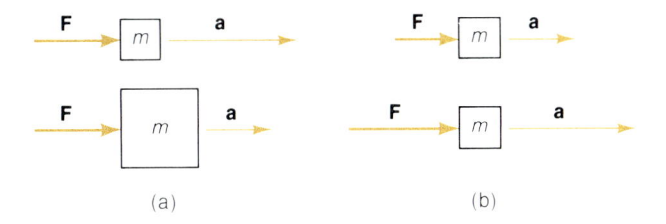

Figure 3.2

Newton's second law. In (a) equal forces are applied to objects of different mass. Their accelerations are inversely proportional to their masses. In (b) different forces are applied to equal masses. The accelerations are proportional to the forces.

Example 3.1 Suppose a spaceship of mass 10^5 kg is accelerated at 5g's or 5×9.8 m/s² = 49.0 m/s². How much force must be exerted on the spaceship?

This is a straightforward application of Eq. 3–2. The force and the acceleration are in the same direction, so we can drop the vector notation:

$$F = ma = (10^5 \text{ kg})(49.0 \text{ m/s}^2) = 4.9 \times 10^6 \text{ kg} \cdot \text{m/s}^2$$

We will soon discuss the units in which force is measured; for now, we note that 4.9×10^6 kg · m/s² = 1.1×10^6 pounds, or about 550 tons.

Mass

Like force, mass is defined by its effects. The most easily measured effect is the acceleration of a body under a known force. The relationship among mass, acceleration, and force is stated in Newton's second law, and we use this law to formulate our definition of mass. We define the mass of a body by measuring how it is accelerated when a known force is applied. If we push two masses with identical forces and measure the resultant accelerations, the ratio of accelerations is inversely proportional to the ratio of the masses of the objects. Thus, if one of the objects is a known standard mass, we can measure the mass of the other. The universal standard in the SI system is the kilogram, which by definition is the mass of a standard block of metal stored in Paris. In practice we commonly measure mass by making comparisons with objects of known mass, which are ultimately calibrated by comparison with standard blocks such as the one in Paris.

Another way to compare masses is to use a known acceleration to compare forces. Living on the earth's surface, we have available a convenient, constant

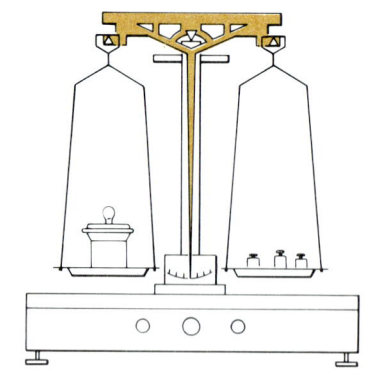

Figure 3.3

A balance used to determine masses by comparing weights.

source of acceleration, namely gravity. The gravitational force on an object in terms of Newton's second law can be expressed as

$$\mathbf{W} = m\mathbf{g} \qquad (3\text{–}3)$$

where **W** is the force that we know as the weight of the object, and **g** is the acceleration of gravity at the earth's surface. If we want to compare the masses of two objects, then, we simply compare their weights. One way to do this is by using a balance (Fig. 3.3).

Because the gravitational acceleration does not vary over the surface of the globe, it is easy to confuse weight and mass. Even in many technical applications, no distinction is made and people often speak of "weights" in terms of grams or kilograms. It is important, however, to keep in mind the difference between weight and mass.

Masses are sometimes expressed in units other than the kilogram, most commonly in **grams,** where 1 gram (g) is equal to 10^{-3} kilograms. The gram is one of the basic units in the *cgs* (centimeter-gram-seconds) system of units, which is a widely used metric system related to the SI system by simple power-of-ten conversion factors. Historically, the gram was defined as the mass of water (at its maximum density, which occurs at 4°C) contained in one cubic centimeter. The kilogram was simply 1000 grams, but, as we have noted, in modern usage the kilogram is regarded as the fundamental unit, and it is defined in terms of a standard block of metal.

Mass is rarely expressed in English units. When it is, it is in units called **slugs.** A force of 1 lb acting on a mass of 1 slug produces an acceleration of 1ft/s². Thus, the mass of an object in slugs is its weight (in pounds) divided by the gravitational acceleration (which in English units is 32 ft/s²). An object whose mass is one slug weighs 32 lbs at the earth's surface. Slugs will not be used in this text.

The Units of Force

We have discussed the concept of force at some length, but have said little about how it is quantified. Newton's second law offers a means to do this: we define our units of force in terms of units of mass and acceleration. From Eq. 3–2, we see that force has the dimensions of mass times acceleration, or, in the SI system, units of kg · m/s². A 1 kg mass accelerated at 1 m/s² by definition experiences a force of 1 **newton** (N); that is, a newton is the quantity of (net) force that will accelerate a mass of 1 kg at the rate of 1 m/s². The weight of a 1 kg mass at the earth's surface is 9.8 N, since the acceleration of gravity is 9.8 m/s².

In the cgs system, the unit of force is the **dyne,** the force required to accelerate a mass of one gram at a rate of 1 cm/s². A dyne is a relatively small unit, equal to only 10^{-5} newtons. The acceleration of gravity in the cgs system is 980 cm/s², so the weight of a single gram of material (about the mass of an aspirin tablet) is 980 dynes. In the English system, we find that a kilogram weighs about 2.205 pounds, so 1 newton is equal to 0.2248 pounds. If your weight is 140 pounds, it can be expressed as 623 newtons. Your mass in this case is just under 64 kilograms (or 4.4 slugs).

Example 3.2   A baseball pitcher throws a fastball at a speed of 90 mi/h, taking 0.2 s to accelerate the ball from rest. The mass of the ball is 200 grams. What is the force exerted by the pitcher as he throws the ball (Fig. 3.4)?

First, we need to calculate the acceleration. We consider only the acceleration in the direction toward home plate (which we define as positive), because we are concerned only with the net force in this direction. The ball accelerates from 0 to 90 mi/h, or 40 m/s, in 0.2 s, so the acceleration is

$$a = \Delta v/\Delta t = \frac{(40 \text{ m/s})}{(0.2 \text{ s})} = 200 \text{ m/s}^2.$$

(We assume that this is uniform acceleration for the duration of the horizontal throwing motion.) The force applied by the pitcher to the ball is therefore

$$F = ma = (0.2 \text{ kg})(200 \text{ m/s}^2) = 40 \text{ N}.$$

This force equals about 9 pounds. Since this is not a very large force, we realize that factors other than the pitcher's own strength govern how fast he can throw the ball. Physiological factors prevent a person from applying larger forces in very short times. We have assumed the acceleration was constant, but a proper description would consider the rate of change of acceleration.

Figure 3.4

Action and Reaction

Newton's third law of motion is perhaps less intuitively obvious than the first two:

Whenever one body exerts a force on a second body, the second body exerts a force of equal magnitude but opposite direction on the first body.

Figure 3.5

Action and reaction. When the person exerts a force on the anchor, the anchor exerts an equal and opposite force on the person.

Newton's own statement used the words *action* and *reaction*: for every action, there is an equal and opposite reaction. In either wording, the essence of this law is that forces always occur in pairs. There is a force only if it is exerted by one body on another.

Sometimes the relationship between an exerted force and its reaction force is obvious. If a person standing in a small boat throws an anchor horizontally out of the boat, the anchor is accelerated in one direction, and the boat is accelerated in the opposite direction (Fig. 3.5). The action-reaction force pair

FOCUS ON PHYSICS

Biomechanics*

Human motion, a complex interaction between the body's internal musculo-skeletal structure and external forces, has fascinated and stimulated intellectual curiosity from the times of the early Egyptians and Greeks to the present. The modern discipline of **biomechanics,** part of a more general field called **kinesiology,** deals with aspects of human motion where mechanics plays an important role. Our study of the laws of motion in this chapter prepares us to understand the basic principles of biomechanics.

The design and use of prosthetics, or artificial moving parts, is as an excellent illustration of the importance of biomechanics. For example, a widely accepted treatment for osteoarthritis of the hip joint in elderly people is a total hip replacement. Orthopedic surgeons, biomechanists, and biomedical engineers have many mechanical problems to consider before the operation is performed. One of the most important questions concerns the design, structure, and material used in the hip joint prosthetic, which must withstand the pressures and twisting forces generated by the patient during normal walking. Consider the complexity of this task. During normal gait, the support leg starts its function when the heel impacts the ground, and at midstride takes on a vertical force equal to the weight of the patient. The hip joint, to which this load is transferred, goes through a normal range of motion of about 30° to 40°, until the other leg takes over. To design a prosthetic hip, one needs to analyze the forces acting on the hip, and to know how these forces are distributed over the ball and socket. It is also necessary to find materials that will have sufficient strength as well as being lightweight, durable, and physiologically adaptable to the body's natural defense system. Furthermore, the forces exerted by the prosthetic on surrounding structures such as the pelvis and femur must be understood, so that damage to these bones can be prevented. The detailed explanation of how all of this is accomplished is quite complex and beyond the scope of this text, but the basic study of mechanics is an important first step for future surgeons and doctors.

Biomechanics has found wide application to sports as well as medicine. Often major competitions are decided by very thin margins, so even subtle refinements in equipment or technique can be crucial. The great emphasis placed on sports today by athletes, fans, team owners, and even governments has given biomechanics a permanent place in modern sports. The main objective of sports biomechanics research is to investigate improvements in technique and apparatus that influence performance. Many examples of gains made through sports biomechanics analysis can be cited.

In 1983 the United States lost the America's Cup, one of the premier prizes in yacht racing, for the first time in 132 years. The U.S. crew lost to an Australian yacht that used an innovative new keel design (Fig. 3.1.1) that enabled the boat to tack with extreme speed. This was a clear example of a better, physically sound mechanical design improving the performance of the equipment and hence influencing the outcome of a race.

In the 1984 Olympic Games, held in Los Angeles, the United States cycling team won an unprecedented number of medals. The U.S. riders wore peculiar-looking helmets and clothing and rode newly designed, futuristic bicycles. The helmets, clothing, bike frames, and wheels were all tested in wind tunnels for minimum air resistance. The extensive research program that was

*Contributed by Dr. Ron Canterna, Colorado College

3.2 NEWTON'S LAWS OF MOTION

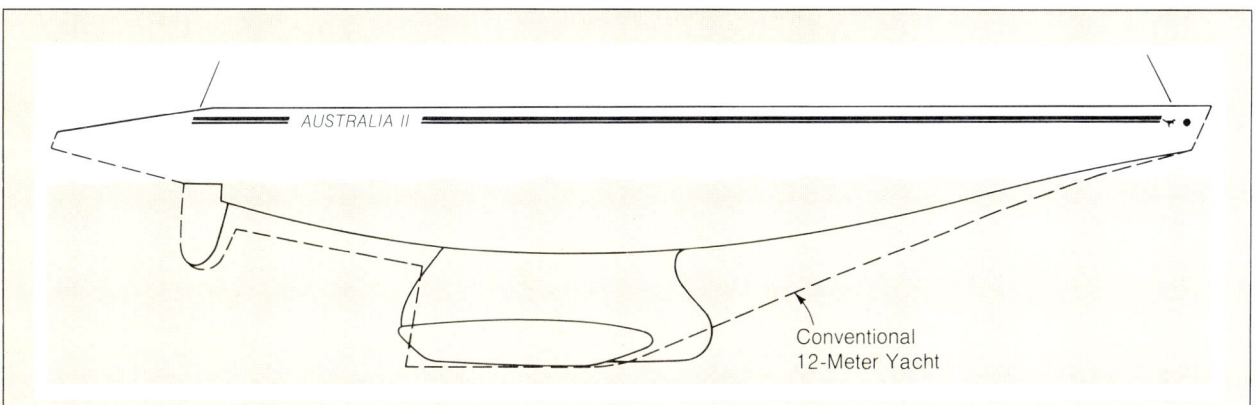

Figure 3.1.1

responsible for these innovative features contributed greatly to the team's success.

In both cases, mechanical principles were applied to the design of equipment to improve the aerodynamic or hydrodynamic performance. Many other examples of improvements in sports equipment can be found. Each one uses the laws of mechanics discussed in this chapter.

Biomechanics also plays a role in the refinement of techniques used in various sports. For example, in the late 1960s and early 1970s, swimming techniques were optimized through the use of wind tunnel analyses. This was an era of unprecedented improvement of swimming records, and the new techniques were partly responsible. (It was also an era of great improvement in endurance training techniques.) Other examples of biomechanical analyses of athletic techniques can be cited. In throwing events, such as the shot put, simple considerations of projectile motions (as discussed in Chapter 2) are important. In jumping events not only projectile motions, but also complex forces, must be analyzed. Later in this chapter, another *Focus on Physics* article illustrates in some detail how a jumping event is analyzed.

consists of the force exerted by the person on the anchor and the force exerted by the anchor on the person, which was transmitted through his body to the boat. Another common example is the kick of a rifle; in this case the bullet exerts a force on the gun, and the gun exerts an equal and opposite force on the bullet. It is important to keep in mind that the accelerations of the two objects can be quite different, even though the forces are equal. According to Newton's second law, the acceleration is inversely proportional to the mass. The gun is accelerated much less than the bullet, and the boat much less than the anchor thrown overboard.

Example 3.3 Let us reconsider the baseball pitcher (Example 3.2). Now let him stand on a frictionless surface (an ice rink may be a reasonable approximation), and calculate his velocity as a result of the throw. The pitcher weighs 185 lbs.

We know that the ball exerts a force of 40 N on the pitcher. We simply calculate the acceleration on a mass equivalent to 185 lbs under this force.

First, a weight of 185 lbs at sea level corresponds to a mass of 84 kg. The acceleration of the pitcher is therefore

$$a = F/m = -40 \text{ N}/84 \text{ kg} = -0.48 \text{ m/s}^2$$

where the negative sign is used to show that the acceleration is in the opposite direction from that of the ball, which was taken as positive in Example 3.2. This acceleration is applied to the pitcher for the duration of his throwing motion, which is 0.2 s, so his velocity is

$$v = at = (-0.48 \text{ m/s}^2)(0.2 \text{ s}) = -0.096 \text{ m/s}.$$

This is less than one fourth of one percent of the velocity imparted to the ball.

Note that when two objects exert forces on each other in a frictionless environment, their accelerations and velocities are inversely proportional to their masses. We could have greatly simplified the solution by noting that the mass of the pitcher is 420 times the mass of the ball, so his acceleration (and therefore his final velocity) will be 1/420 times those of the ball. This simpler method will be explored more fully in Chapter 5.

The action-reaction pair is often not easily recognized. We have already mentioned the balance of forces on a book that lies on a table. Now we see that the book exerts a downward force on the table, and at the same time the table exerts an equal upward force on the book. How do we know this? Imagine that the table did not exert a force on the book. There would be a net downward gravitational force on the book, and the book would be accelerating downward in accordance with Newton's second law. Because the book is stationary, it has no acceleration and, according to Newton's second law, the net force must be zero. We conclude that there must be a second, upward force, equal and opposite to the downward gravitational force supplied by the table.

Let us now imagine that the table disappears, leaving the book to fall. Clearly a force is being exerted on the book, creating a downward acceleration. We have said that for every force there must be an equal and opposite reaction force. What is the reaction force exerted by the book? It exerts a gravitational force on the earth that is equal and opposite to the gravitational force exerted by the earth on the book. This force is not apparent. Because of the great mass of the earth, its acceleration is too small to be easily detected.

3.3 Tension and Transmission of Force

We mentioned earlier that a person who throws an anchor out of a boat transmits to the boat the force exerted on him by the anchor, causing the boat to accelerate in the direction opposite to the motion of the anchor. Let us look a little more closely at the notion of transmitting force. As the person's hand pushes the anchor, the anchor pushes back on the hand. The force transmitted

3.3 TENSION AND TRANSMISSION OF FORCE

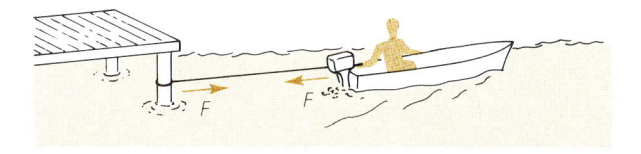

Figure 3.6

Transmission of force. The force exerted by the boat on the rope is transmitted to the dock by many microscopic action-reaction pairs in the rope.

through his body to his feet is exerted against the boat, while the boat exerts an equal and opposite force against the person's feet. If the person as a whole experiences no net force, then all the force exerted on the person by the thrown anchor is transferred to the boat. The transmission of forces through solid bodies is a very complicated matter, ultimately related to the field forces that hold atoms and molecules together. The force is actually transmitted to the boat by means of a series of action-reaction pairs of molecules in the person's body.

The transmission of force is perhaps easier to see in a simpler case. If the boat is moored to a dock, and the boat's motor is on, the rope is pulled taut (Fig. 3.6). The boat exerts a force on the water, and the water exerts an equal and opposite force on the boat, which is transmitted by the rope to the dock. The dock exerts an equal and opposite force on the rope, which is transmitted to the boat. There are no accelerations, because the net force on each body is zero. If we view the rope as an object that simply transmits force, we speak of the **tension** in the rope, and we identify two action-reaction pairs: the action of the boat on the rope and the tension of the rope on the boat; and the action of the rope on the dock and the reaction of the dock on the rope. We ignore the multitude of microscopic action-reaction pairs along the rope.

Example 3.4

Suppose we want to tow a disabled car by using a rope whose tensile strength is 300 lbs (that is, the rope can withstand a tension of up to 300 lbs without breaking). If we accelerate the car, whose mass is 1000 kg (about 2200 lbs in weight), from rest to 30 mi/h in 20 seconds, will the rope break?

We first convert to SI units. The tensile strength of the rope is 1330 N, and the car's final speed is 13.4 m/s. We note that the acceleration, velocity, and force are all in one dimension, so we drop the vector notation. Then the acceleration is

$$a = \Delta v/\Delta t = (13.4 \text{ m/s})/(20 \text{ s}) = 0.67 \text{ m/s}^2$$

and the force applied to the car, which is equal to the tension in the rope, is

$$F = ma = (1000 \text{ kg})(0.67 \text{ m/s}^2) = 670 \text{ N}.$$

FOCUS ON PHYSICS

A Biomechanical Analysis of a World Record*

On Sunday, June 16, 1985, at Indianapolis, Indiana the American triple jumper Willie Banks broke the world record by three inches, achieving a new standard of 58 feet, 11½ inches. Banks had broken the American record two weeks earlier. These achievements were made especially remarkable by the fact that Banks had performed poorly in the 1984 Olympics, and some experts had considered his athletic career to be over.

For the past three years the Biomechanics Laboratory at the University of Iowa has had the opportunity to analyze the techniques of Willie Banks and other U.S. triple jumpers as part of the United States Olympic Committee's Elite Athlete program. These analyses are complicated and require patience and time. More importantly, they also require a solid foundation and understanding of mechanics. In this brief article, we describe the procedures for a typical biomechanical analysis of a triple jump.

Two fixed movie cameras record the major phases of the triple jump. One camera is positioned to record the last two strides of the run-up and the *hop*, the first part of the triple jump. The second camera records the *step* and *jump* phases, as well as the landing. Placement of the cameras is critical because all phases of the jump must be recorded, and there must be overlap between the fields of view of the two cameras, so that the data can be correctly combined. Scaling markers, separated by known distances, are placed as close to the runway as possible to gauge actual distances. High-speed cameras with accurately known filming rates are used with film that has timing marks along the edges. To provide a back-up in case the film timing indicators fail, a crew member is filmed as he drops a ball from rest next to a meter stick. This allows an independent calibration of the rate of film advance of the camera (Chapter 2). Once the cameras are calibrated and placed in position, the competition jumps are recorded (sometimes this is made difficult by interference from bystanders).

More than 100 frames of developed film are studied. A measuring device called a digitizer is used to determine the location of the athlete at many points during the jump. In order to understand the importance of these positional data, it is necessary first to know the major mechanical factors that influence triple jump performance. The objective of any horizontal jump is to achieve a maximum distance. This distance can be separated into three parts: the *take-off position,* where the position of the body and placement of the support foot relative to the foul board are important factors; the *in-flight distance,* which is governed primarily by the velocity and angle of the athlete's center of gravity at take-off; and the *landing distance,* which depends on the rate of rotation of the jumper and his ability to place himself in an advantageous position. For the triple jumper the task is especially demanding because at the end of each phase the jumper must be in position for the next jump.

In order to analyze how well the jumper meets the demands outlined above, detailed data are needed on the positions of the feet and the jumper's center of gravity. The center of gravity is located by a complex analysis of twenty-one body points whose positions are measured from the film. The result of this analysis is a stick figure diagram (Fig. 3.2.1), showing the positions of the twenty-one body points and the center of mass.

*Contributed by Drs. Ron Canterna and James Hay, University of Iowa

3.4 FRICTION AND AIR RESISTANCE

WILLIE BANKS
World Record Jump

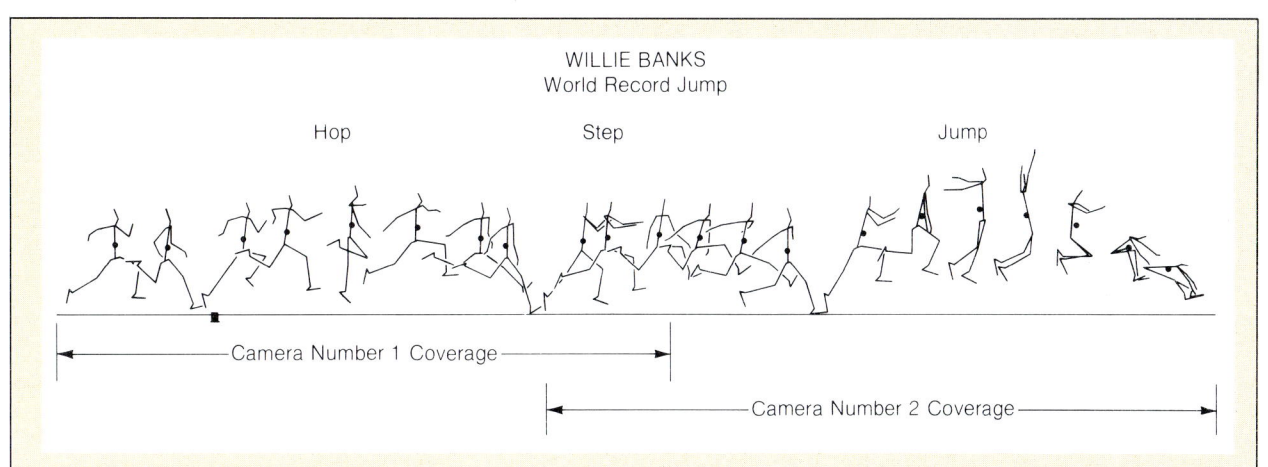

Hop — Step — Jump

Camera Number 1 Coverage
Camera Number 2 Coverage

Figure 3.2.1

Next the measured positions of the center of gravity and of the feet at contact points with the ground are used to determine the athlete's velocity, angle of take-off, distance achieved, and rotation rate during each phase of the jump. The forces exerted by the jumper can also be calculated from the changes in velocity at impact with the ground. Then the biomechanist can tell the athlete and his coach which phases of the jump the athlete should emphasize in his training and which the athlete is doing correctly, or answer specific questions the coach or the athlete may have regarding the technique.

The analysis of Willie Banks' 1985 world-record jump shows two significant results. First, his takeoff velocities on the last stride and the step and jump phases were 1 ft/s faster than in previously analyzed efforts. Second, his jump distance was longer, in relative and absolute terms, than in any previous world-record jump. Banks was able to maximize his total distance by adopting the Polish technique of putting maximum effort into the jump phase.

This is less than the tensile strength of the rope, so the towing job should be successful. Notice how critically this conclusion depends on the magnitude of the acceleration: if you towed the car from 0 to 30 mi/h in 10 seconds instead of 20, the force would exceed the tensile strength of the rope, and the rope would break. Also, if the acceleration were not smooth (short bursts of increase in speed), the force could briefly exceed the tensile strength of the rope, even though the average acceleration was lower than the critical value.

3.4 Friction and Air Resistance

So far we have unrealistically ignored the forces of nature that tend to resist motions. Few descriptions of real motions are correct if these forces are neglected.

Friction

Whenever two surfaces are in contact and exert forces on each other, they tend to stick together, and resist sliding. This resistance, called **friction,** arises because no surface is perfectly smooth on a microscopic scale, and very fine bumps and irregularities catch on each other. Details of this interaction depend on the atomic and molecular structure of the two surfaces, but the net effects can be represented in simple terms.

The nature of the friction force depends on the type of motion that occurs between the two surfaces. If there is no motion, then we are dealing with **static friction,** which is greater in magnitude than **kinetic friction,** the simplest kind when there is sliding motion. (If one of the surfaces is curved and rolls along the other, then we are dealing with **rolling friction.**) Static friction, the friction that resists you when you try to push a heavy object, is the strongest; once the object is in motion, the friction usually decreases. It is important to understand that static friction always arises in reaction to a force. For example, the static friction is zero for a book lying at rest on a table, unless a horizontal force is exerted on the book (Fig. 3.7). As the horizontal force increases, the static friction increases until the applied force overcomes friction and the book finally starts moving. Hence, static friction has a maximum value, which cannot be exceeded without motion.

Many experiments (some as early as the sixteenth century) have revealed several properties of all friction.

1. It is proportional to the net perpendicular force (called the net *normal* force) between two bodies (for static friction, it is the maximum value that is proportional to the normal force).
2. It is approximately independent of the surface area of contact between the two bodies.
3. It is approximately independent of the relative velocity of the two bodies.
4. Its direction always opposes motion or an external force that could cause motion (that is, it always acts along the surface of mutual contact and in the direction opposite to the motion or to the net external force parallel to the surface).
5. Its magnitude depends strongly on the nature of the two surfaces.

These statements can be expressed mathematically by the simple relation

$$F_f = \mu F_N \tag{3-4}$$

where F_f represents the force of kinetic friction or the maximum force of static friction, μ is the **coefficient of friction,** and F_N is the normal force. The coefficient of friction depends on the two surfaces, and is tabulated in scientific handbooks. Note that F_f and F_N have units of force, while μ is a dimensionless constant of proportionality which represents either sliding friction or static friction (Table 3.1).

Sliding friction is a constant force, independent of velocity and surface area; static friction depends on the magnitude of the applied horizontal force.

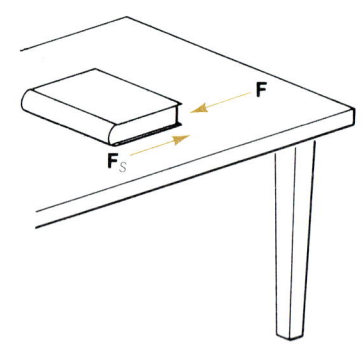

Figure 3.7

Static friction. This force balances a horizontal force until the applied force exceeds the maximum value of the static friction and the book moves.

Table 3.1
Coefficients of friction

Materials	Static μ_s	Sliding μ_k
Rubber on solids	1.-4.	1.
Copper on cast iron	1.05	0.29
Glass on glass	0.94	0.40
Copper on glass	0.68	0.53
Steel on steel	0.58	0.3
Wood on wood	0.4	0.2
Wood (waxed) on wet snow	0.14	0.1
Teflon on steel	0.04	0.04
Human joints	0.01	0.01

3.4 FRICTION AND AIR RESISTANCE

For kinetic friction we write

$$F_{kf} = \mu_k F_N \qquad (3\text{–}5)$$

where F_{kf} is the kinetic friction force and μ_k is the coefficient of kinetic friction. Because static friction depends on the applied force, and has a specific maximum value, we write

$$F_{sf} \leq \mu_s F_N \qquad (3\text{–}6)$$

where F_{sf} and μ_s represent the static friction force and the coefficient of static friction. This is an inequality because the term $\mu_s F_N$ is the maximum value of the friction. The actual value of the static friction is always equal to the net horizontal force until motion starts.

After motion has started, then the friction is the kinetic friction given by Eq. 3–5. Kinetic friction is less than static friction because the motion depresses the microscopic bumps and irregularities that resist motion.

Example 3.5 The coefficient of static friction for the contact between two wooden surfaces is $\mu_s = 0.35$, and the coefficient of kinetic friction is $\mu_k = 0.15$. Suppose a 100 kg crate of wood is standing on a wooden floor. What horizontal force is required to start sliding it along the floor (Fig. 3.8)? What force is required to keep it sliding at constant velocity?

First we need to calculate the maximum force of static friction because that is the force required to start the crate moving. The magnitude of the normal force, due entirely to gravity, is

$$F_N = mg = (100 \text{ kg})(9.8 \text{ m/s}^2) = 980 \text{ N}.$$

Then the maximum force of static friction is

$$F_{sf} = \mu_s F_N = (0.35)(980 \text{ N}) = 343 \text{ N}.$$

Hence, a horizontal force of 343 N is required to start this crate moving.

After the crate is in motion the kinetic coefficient of friction applies. The force required to keep the crate moving is

$$F = F_{kf} = \mu_k F_N = (0.15)(980 \text{ N}) = 147 \text{ N}.$$

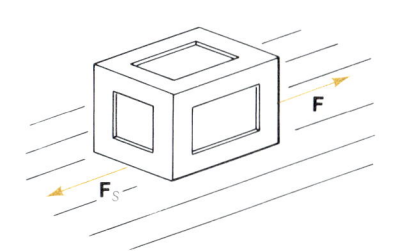

Figure 3.8

In our examples so far the normal force is equal to the force of gravity because the surfaces were horizontal. We stress, however, that the *net* normal force is the relevant parameter. If forces other than gravity act on the surfaces, or if the surfaces are not horizontal, (as when an object is on an inclined plane), we must calculate the net normal force by vector addition before computing the friction force.

Example 3.6

Suppose a roofer is reshingling the roof of a barn with a peaked roof inclined 35° to the horizontal (Fig. 3.9). The roofer wants to know whether his bundles of shingles, which weigh 100 lbs each, will slide off the roof. The bundles are wrapped in cardboard, and the existing roof is wood. The static coefficient of friction for cardboard on wood is 0.25.

In the space diagram for a bundle of shingles (Fig. 3.9), we see that the normal force exerted by the shingles on the roof is equal to the component of the gravitational force that is normal to the roof, so that

$$F_N = (100 \text{ lbs})\cos(35°) = 82 \text{ lbs}$$

Similarly, the net force parallel to the roof is the component of the gravitational force on the shingles in that direction, which is

$$F_P = (100 \text{ lbs})\sin(35°) = 57 \text{ lbs}$$

The maximum force of static friction is

$$F_{sf} = \mu_s F_N = (0.25)(82 \text{ lbs}) = 20.5 \text{ lbs}.$$

This is considerably less than the component of the bundle's weight (57 lbs) that is parallel to the plane of the roof, so the bundles of shingles will certainly slide off the roof unless the roofer anchors them.

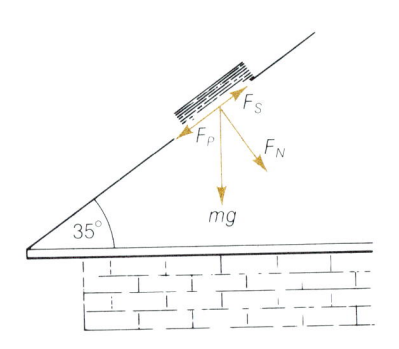

Figure 3.9

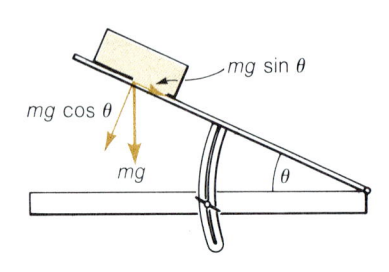

Figure 3.10

Measurement of μ_s. The board is tilted until the block just begins to move. At that point the maximum value of the static friction has been reached, and is equal to the component of the block's weight that is directed downhill, parallel to the board.

Inclined planes are used to measure coefficients of friction. An object of known weight is placed on a plane surface such as a board, and the surface is tilted until the object begins to slide. The angle of inclination is then measured, and the components of the weight parallel and perpendicular to the surface are calculated. If θ is the angle of inclination (Fig. 3.10), then the normal force is

$$F_N = mg \cos\theta$$

where mg is the weight of the object. The component of force parallel to the inclined plane, which equals the maximum static friction force F_{sf} when the angle is set so that the object is just starting to slide, is

$$F_{sf} = mg \sin\theta.$$

When the object is just starting to slide because the force parallel to the plane is equal to the maximum force of static friction, then Eq. 3–6 becomes an equality, and can be solved for μ_s.

$$\mu_s = \frac{F_{sf}}{F_N} = \frac{\sin\theta}{\cos\theta} = \tan\theta \qquad (3\text{--}7)$$

3.4 FRICTION AND AIR RESISTANCE

To measure the kinetic coefficient of friction is more complicated because in principle we would need to measure the acceleration of an object sliding down an inclined surface, so that the net force parallel to the surface could be deduced and the coefficient of friction calculated. In practice, it is easier to tilt the plane just enough to start motion and then lower the tilt until the velocity is uniform, with no acceleration. Then there is no net force parallel to the plane, and the kinetic coefficient of friction can be found from Eq. 3–5

$$\mu_k = \frac{F_{kf}}{F_N} = \tan\theta \qquad (3\text{–}8)$$

(because $F_{kf} = mg\sin\theta$ and $F_N = mg\cos\theta$, and $\frac{\sin\theta}{\cos\theta} = \tan\theta$).

Example 3.7 A toboggan manufacturer wants to test new waxes to find the ones that best minimize friction. He needs to know both the static and kinetic coefficients of friction. A measurement is made in which a 20 kg block of wood coated with the test wax is placed on a surface covered with hard-packed snow, and the surface is tilted. When the tilt is 23° to the horizontal, the block begins to slide. When the angle is decreased, it is found that the block slides without acceleration when the tilt is 7°.

From Eq. 3–7, we see that the coefficient of static friction is

$$\mu_s = \tan(23°) = 0.42.$$

Similarly, from Eq. 3–8, the kinetic coefficient of friction is

$$\mu_k = \tan(7°) = 0.12.$$

Air Resistance

Objects moving through air meet another kind of resistance force, **air resistance**. Like friction, air resistance always acts in the direction opposite to the motion. Unlike friction, however, air resistance depends on the size and shape of the object and on its speed. Although a quantitative treatment of air resistance would be quite complex, we can make a few useful remarks about its behavior.

Two falling objects of equal mass would have identical accelerations if gravity were the only force acting, but if they have different shapes or different sizes, one will encounter a stronger air resistance force than the other. The object experiencing the stronger air resistance will have a smaller net downward force and will therefore be accelerated at a slower rate.

As an object falls and accelerates, the air resistance increases, because this force depends on speed. Eventually the air resistance force becomes equal to

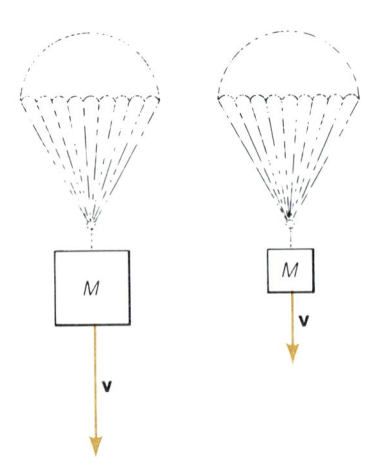

Figure 3.11

Air resistance and terminal velocity. For the same shape and size, larger masses have larger terminal velocities.

the gravitational force, and the net force on the falling object becomes zero. After this, the object falls at a constant speed. This **terminal velocity** is therefore determined by the mass of the object and its shape. The greater the mass, the greater the weight, and the greater the air resistance force required to overcome the weight (Fig. 3.11). Therefore, for objects of the same shape, the terminal velocity increases with increasing mass. Hence, the famous experiment supposedly performed by Galileo, in which he dropped balls of unequal weight from the Leaning Tower of Pisa to show that objects of unequal mass are accelerated equally by gravity, probably would have failed (for this and other reasons, it is doubtful whether Galileo actually performed this public demonstration).

Since air resistance can substantially distort our intuitive understanding of projectile motion we see again the importance of designing experiments that isolate the phenomena being studied. In dynamics, this means that useful experiments either are performed in a vacuum or on so small a scale that air resistance is negligible.

Perspective

We have now developed an understanding of the causes of changes in motions, namely forces, and we know how forces and motions are related to each other. We have discussed the basic laws of motion and their applications to everyday life, we have seen how to treat some kinds of friction, and we have developed a qualitative understanding of air resistance.

Next we will study forces in balance, the state of equilibrium where there is no acceleration. This is really just a special case of Newton's second law, but an important one, particularly for structural design.

PROBLEM SOLVING

Coordinates and Force Relations

In Chapter 1 we discussed six themes of problem solving. The second theme emphasized the importance of making a drawing of the problem, labelling the known and unknown quantities. In force problems it is useful to make separate sketches (free body diagrams) that show all forces acting on each body. Be careful to include only forces that act on that body, gravity, friction, or reaction forces. Choose a set of rectangular axes for resolving the forces into x- and y-components. Often a convenient choice is to place these axes parallel and perpendicular to a plane of motion—for example, an inclined plane on which a mass is sliding. At the same time, you must decide on where the origin of the coordinates is; a common choice is at the point where most of the forces are applied.

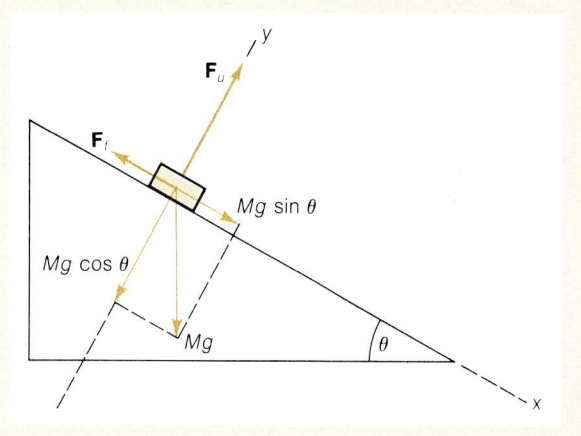

Figure 3.3.1

Example 1 A mass M rests on a plane inclined at angle θ. Figure 3.3.1 shows the gravitational force Mg, resolved into components parallel ($Mg \sin\theta$) and perpendicular ($Mg \cos\theta$) to the plane. Notice that we have chosen tilted coordinate axes passing through the mass and aligned with the plane. In equilibrium, the upward force F_u and the frictional force F_f must balance the components of gravity along the two axes. Thus, we have

$$F_u = Mg \cos\theta \qquad F_f = Mg \sin\theta$$

Three forces act on the mass—gravity, friction, and the upward reaction force from the plane. We resolved gravity into its two components along the coordinate axes. This is the key to solving for the net force along the x-axis. For example, if the mass is moving and the coefficient of kinetic friction is μ, then $F_f = \mu\, Mg \cos\theta$, and the net force along the plane is $Mg \sin\theta - \mu\, Mg \cos\theta$. The net acceleration would be $g(\sin\theta - \mu \cos\theta)$.

The third theme addressed the actual technique of solving problems. The rule is, you must have one relation with the known variables for each unknown variable. In the previous example, we were seeking the net acceleration down the plane. The known quantities were the mass M, the angle θ, and the coefficient of friction μ. The one unknown, a, was derived by finding a single relation, Newton's law $F = ma$, where F is the net force along the plane.

Example 2 Two masses are suspended from a frictionless, massless pulley, with $m_2 > m_1$ (Fig. 3.3.2). What are the tension T and the net acceleration a of the system?

Figure 3.3.2 is a freebody diagram, showing the two unknowns, T and a. Thus we need two relations, which are just Newton's laws, $F = ma$, for each mass. Both masses move together with the same acceleration. The net force on mass 1 is $T - m_1 g$, and the net force on mass 2 is $m_2 g - T$. The two relations are

(continued)

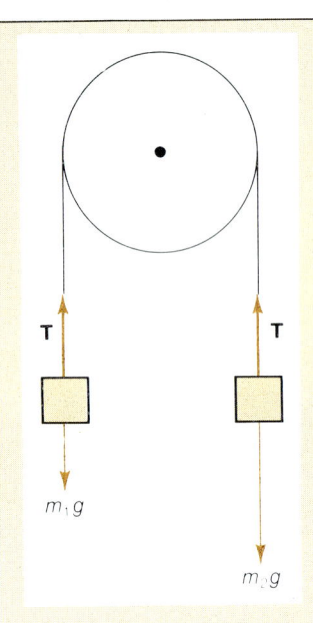

Figure 3.3.2

$$T - m_1 g = m_1 a \quad \text{and} \quad m_2 g - T = m_2 a$$

which are easily solved:

$$a = \frac{(m_2 - m_1) g}{(m_1 + m_2)} \quad T = \frac{2 m_1 m_2 g}{m_1 + m_2}$$

Theme 6 mentioned looking for shortcuts, such as noticing that a problem's solution does not depend on one variable. In most force problems involving gravity, the net acceleration does not depend on the mass m. In Example 1 mass cancels in Newton's Law, $F = ma$, because both the gravitational force $mg \sin\theta$ and the frictional force $\mu mg \cos\theta$ are proportional to m.

Example 3 At what critical angle θ_c will the mass in Example 1 begin to slide? The slippage begins when friction can no longer balance the component of gravity along the plane.

$$\mu \, \cancel{mg} \cos\theta_c = \cancel{mg} \sin\theta_c$$

Thus $\theta_c = \arctan(\mu)$. Both m and g cancel, and the angle depends only on μ. Once you have solved a number of problems, you will develop an intuition about whether such cancellations make sense. You might have made an error in algebra, so it always pays to ask yourself whether your answer is reasonable physically. Does it scale proportionally as certain variables are increased or decreased? In this example, as the coefficient of friction increases, θ_c also increases, which makes intuitive sense.

SUMMARY TABLE OF FORMULAS

The second law of motion:

$$\mathbf{a} = \frac{\mathbf{F}}{m} \quad (3\text{-}1)$$

$$\mathbf{F} = m\mathbf{a} \quad (3\text{-}2)$$

Weight:

$$\mathbf{W} = m\mathbf{g} \quad (3\text{-}3)$$

Friction:

$$F_f = \mu F_N \quad (3\text{-}4)$$

Kinetic friction:

$$F_{kf} = \mu_k F_N \quad (3\text{-}5)$$

Static friction:

$$F_{sf} \leq \mu_s F_N \quad (3\text{-}6)$$

Coefficient of static friction:

$$\mu_s = \frac{F_{sf}}{F_N} = \tan\theta \quad (3\text{-}7)$$

Coefficient of kinetic friction:

$$\mu_k = \frac{F_{kf}}{F_N} = \tan\theta \quad (3\text{-}8)$$

THOUGHT QUESTIONS

1. When all forces on an object balance, why may the object still move?
2. How does one measure a force in the laboratory? Give examples.
3. Explain the difference between a gram, a kilogram, a pound, and a newton.
4. A man standing on the moon experiences an upward force from the surface one sixth of the force he would feel standing on earth. Why? How is it that the moon pushes upward just enough so that the man does not move?
5. If one pushes twice as hard on an object with twice the mass, why is the acceleration the same?
6. To keep a tetherball moving in a circle at constant speed, a force must be applied inward along the string. Describe the role tension plays. Why doesn't the ball's speed change, since it is experiencing a force?
7. How do ice skaters accelerate from rest on a rink?
8. A boy accelerates a wagon from rest along a road. Draw freebody diagrams showing all forces on the boy, on the cart, and on the ground. Explain how he is able to move the wagon, despite the law of action and reaction.
9. Explain the role of inertia in the following situations:
    (a) A hockey puck sliding along a frictionless ice rink.
    (b) The slow acceleration of a large truck compared to a car.
    (c) Hitting a baseball with a bat.
    (d) A whiplash head injury in a car accident.
10. An astronaut on the moon is struck by a speeding moon buggy. Since this vehicle's weight is only ⅙ that on earth, is the accident as serious?
11. Why is the coefficient of static friction always greater than the coefficient of kinetic friction? Why is the frictional force greater for heavier objects?
12. A block resting on a wooden inclined plane does not begin to slide until the plane is tilted at an angle from horizontal larger than a critical value. Explain why.
13. As the plane in Problem 12 is tilted even more, the block's acceleration down the plane increases. Why?
14. As a car accelerates from rest, its passengers are pushed back against their seats. While this force is often called *g*-force by jet pilots, explain why it has nothing to do with gravity. Would the force be the same on the moon?
15. When you jump up inside a train, you come down in the same spot. Yet when you toss an object out the train's window, the object is quickly left behind. Explain why in each case.
16. Why does a rocket illustrate Newton's third law of action and reaction? Explain why a rocket still moves forward in empty space.
17. As a ship sails, its prow pushes aside the water. How is an icebreaker different?
18. Why is it easy to walk on the ground, but difficult to walk through a puddle?
19. Devise an experiment that could measure the coefficient of static friction between two bodies in contact.
20. Explain the role of friction in the following situations:
    (a) A car coasting down a hill.
    (b) A bowling ball rolling without slipping down an alley.
    (c) A rock climber standing on a slope of 60°.
    (d) A phonograph stylus staying in the groove.
    (e) Disk brakes on a car.

PROBLEMS

Sections 3.1–3.3 Forces, Newton's Laws

- 1 Calculate the force required to accelerate a 2,000 kg car at 1 m/s².
- 2 A bicycle and its rider (total mass 100 kg) accelerate uniformly from rest to 10 m/s in a time of 10 s. What is the acceleration?
- 3 What is the mass of an object that weighs 100 N on the earth's surface?
- 4 What is the mass of an alien who weighs 100 N on the surface of Mars, where gravity is 0.38 times that on earth?
- 5 On the surface of the Sun, the gravitational acceleration is 274 m/s². How many *g*'s is this? What would a 60 kg human weigh (in N and lbs)?

- 6 A 20 kg block sits at rest on a flat, frictionless surface. What horizontal force is needed to accelerate the block to 5 m/s in 3 s?
- 7 In Problem 6, suppose that a 6 N frictional force acts horizontally on the block in the direction opposite to the applied force. What force is now needed to accelerate the block at 2 m/s^2?
- 8 A child slides down a frictionless plane inclined at angle θ from horizontal. What is her acceleration?
- 9 Two children tug back on a 30 kg wagon while another pulls forward (Fig. 3.12). What is the resultant force on the wagon? Its acceleration?

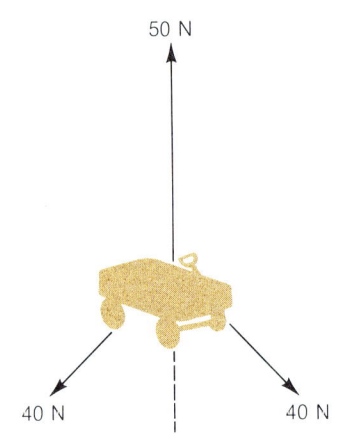

Figure 3.12

- 10 A 10 kg mass rests on the floor.
  (a) Draw a space diagram showing all forces acting on the mass.
  (b) Draw a second diagram showing the forces if the mass is accelerated horizontally at 1 m/s^2 by a rope. A 2 N frictional force acts against the acceleration.
- 11 For the force diagram in problem 10(b), what is the magnitude and direction of the resultant force?
- 12 What is your mass in kg and your weight in newtons on earth and on the moon (where gravity is one-sixth as strong)?
- 13 A 200 kg motorcycle accelerates from rest to 100 mi/h in 10 s. What is its acceleration? What force was exerted on the cycle?
- 14 In a race, two motorcyclists accelerate at constant rates. One travels 400 m in 10 s; the other travels 100 m in 8 s. What were the forces on the cycles, if the mass of each, including riders, is 150 kg?

- 15 A 400 kg boat is propelled by a motor exerting a 100 N force. A stream exerts a 20 N resistance force. What is the resultant force on and acceleration of the boat?
- 16 A racecar crashes into a solid wall at 30 m/s. If the crash lasts 0.1 s, calculate the average deceleration of the car and the force exerted on the 1,500 kg vehicle.
- 17 A 100 kg chandelier hangs from the ceiling by a thin wire. What is the wire's tension?
- 18 A car towing a 1,000 kg wagon accelerates away from a stoplight at 1 m/s^2. What is the tension in the rope connecting the two vehicles?
- 19 A 1,000 kg elevator is accelerated upward by a cable whose tensile strength is 10^4 N. What is the largest acceleration possible?
- 20 Two 1 kg blocks hang over a massless, frictionless pulley (Fig. 3.13). What is the tension T in the massless cord connecting the weights?

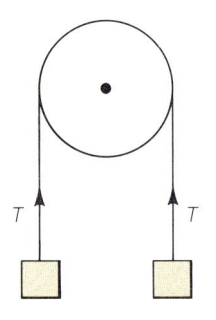

Figure 3.13

- 21 A 1,000 kg elevator is pulled upward by a force of 2 × 10^4 N. How many 60 kg passengers can ride this elevator if an acceleration of 0.1g is required?
- 22 Before opening his parachute, a skydiver falls against a wind resistance of 200 N. What is his initial acceleration if his mass is 70 kg?
- 23 A rollercoaster in an amusement park descends a 45° slope, beginning from rest. What is its speed after a distance of 60 m (neglect friction)?
- 24 During an earthquake, a 100,000-ton layer of granite is accelerated at 3 g's for 0.5 s. How far does the layer move? If the same force were applied to a 20,000-ton layer for the same time, what would be its acceleration and how far would it move?
- 25 As you step from a stationary platform onto a train moving at 5 m/s, you feel a force for an interval of

1 second. What is the magnitude of this force if your mass is 50 kg?

·· 26 A 30 kg child is accelerated by gravity down a slide inclined at 30° to horizontal. If the slide is frictionless, what is the child's speed after 3 s? How far along the slide will the child have moved?

·· 27 In a severe auto crash, passengers decelerate at 20 g's for 0.1 s. How far will a passenger move before halting? What force is exerted on a 70 kg passenger to achieve this deceleration?

·· 28 A truck tows a 2,000 kg car with a rope having tensile strength of 150 N. If the driver wishes to attain a final speed of 20 m/s, what is the shortest interval of time to achieve this speed without breaking the rope?

·· 29 A person of mass m stands on the floor of an elevator which is being accelerated downward with acceleration a. What force does the person feel? If $a = g$, what reaction will the person experience?

·· 30 A 200,000 kg locomotive pulls two 100,000 kg cars, each joined by identical couplings. If the train's acceleration is 0.4 m/s², what are the tensions in the couplings?

·· 31 In Fig. 3.14, two blocks of mass m_1 and m_2 move on a frictionless surface. They are connected by a massless cord and a massless frictionless pulley; the tension in the cord is T.
(a) Find the net force accelerating each block.
(b) If the cord does not stretch, show that $T = [m_1 m_2/(m_1 + m_2)]g$.
(c) What is the acceleration of the system?

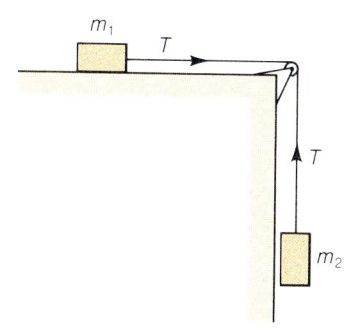

Figure 3.14

·· 32 A weightlifter pulls upward on a rope attached to a 50 kg mass. If the rope has a tensile strength of 400 N, what is the shortest interval of time in which the system may be smoothly accelerated to a height of 1 m?

·· 33 In a TV tube, an electron ($m_e = 9.11 \times 10^{-31}$ kg) is smoothly accelerated from rest at the cathode to the anode 0.2 m away. If its terminal velocity at the anode is 10^7 m/s, what was its acceleration? What force was exerted on the electron by the electric field in the tube?

·· 34 A 2 gram bullet moving at 300 m/s strikes a maple tree and penetrates 5 cm. If the deceleration was constant, what force was exerted on the bullet? How long did the force act?

·· 35 The force of wind resistance, or wind drag, is approximately proportional to velocity squared: $F = k_2 v^2$. Show that the terminal velocity of a body falling under gravity is $v_t = \sqrt{mg/k_2}$.

·· 36 The terminal velocity of a 70 kg skydiver is about 100 mi/h (44.7 m/s). What force of wind drag must be encountered at that speed? Use the law $F = k_2 v^2$ to evaluate the constant k_2.

·· 37 Using the proportionality constant k_2 of Problem 36, compute the downward acceleration of a 60 kg sky diver who falls at velocity 10 m/s (not yet at terminal velocity).

··· 38 A luxury train advertises that its maximum deceleration will not exceed 0.01 g, so that its passengers can sip their coffee in peace. If the train reaches a maximum speed of 100 mi/h (44.7 m/s), how far before a stop should the train apply its brakes? How many minutes does this correspond to? What is the force of compression C on the coupling between the locomotive and the first car? Assume the locomotive pulls four identical cars of mass 10^5 kg each (Fig. 3.15).

Figure 3.15

·· 39 An athlete throws a 2 kg mass vertically upward, accelerating it at 10 m/s² for 0.2 s. What force was exerted during this interval? With what velocity does it strike the ground? What force is exerted by the ground on the weight, if the weight is brought to a halt in 0.3 s?

80 CHAPTER 3 DYNAMICS

••• 40 Draw a force diagram for the two masses in Fig. 3.16 (neglect friction). What is the acceleration of the system? What is the tension T in the cord? Assume the cord and pulley are massless.

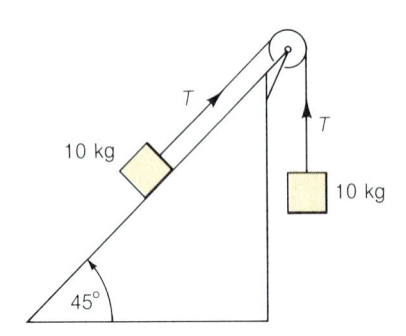

Figure 3.16

•• 41 A 60 kg human riding a 10 kg bicycle can coast down a 7° hill at 10 m/s. If the force of wind resistance is $F = k_1 v^2$, calculate the constant k_1 assuming force balance. Find the force that must be exerted on the cycle in order to descend the hill at 20 m/s.

•• 42 A 1 kg ball falls from rest under gravity through a viscous oil. The viscous drag force in newtons is linearly proportional to velocity: $F = 10v$, where v is the ball's speed in m/s. What is the ball's initial acceleration? What is its terminal velocity?

•• 43 A 0.1 kg rubber ball falls vertically 2 m from rest, bounces off the floor, and returns to its original height. Make a graph showing the total forces acting on the ball as a function of time. Assume that the bounce lasts 0.2 seconds.

•• 44 A fisherman in a boat tosses a rotten fish horizontally away from the boat. The 1 kg fish is accelerated to a speed of 5 m/s in 1 s. What is the force on the fish? What is the final velocity of the boat? Assume that the boat has a mass 80 kg and the fisherman 60 kg.

••• 45 Suppose in Problem 44 that the 80 kg boat contains two 60 kg fishermen who start an argument. One throws a 1 kg fish at the other, accelerating it for 1 s to a speed of 5 m/s. The other man throws a 2 kg fish in the opposite direction with the same acceleration. The fish miss their targets and fly off in opposite directions. What is the final velocity of the boat?

••• 46 On either side of a massless, frictionless pulley, two masses of 2 kg and 4 kg are connected by a massless cord. Draw a space diagram of the forces on each mass and calculate the acceleration of the system.

••• 47 A 0.1 kg hockey puck strikes a goalie's mitt, which recoils 0.2 m and brings the puck to rest. If the puck was initially moving at 40 m/s, find the force on the puck and the deceleration. How long did the impact last?

••• 48 Two teams of five persons each have a tug-of-war. Each team member pulls with a 100 N force. Draw a diagram showing the tension forces in the rope between each member. How strong must the rope be?

••• 49 In a vertical standing jump, a 50 kg athlete bends her knees to lower her body 0.25 meters, then accelerates upward at a constant rate for 0.25 m. If she wants to raise her body level 0.75 m above its original position, what force must she exert on the ground? How long must she exert this force?

••• 50 A mass $m_1 = 50$ kg and a platform with mass $m_2 = 100$ kg are suspended by a rope below a massless, frictionless pulley (Fig. 3.17). One end of the rope is attached to a rigid vertical wall. If the rope has a tensile strength of 5,000 N, how many 60 kg people may sit on the platform without breaking the rope? What then are the tensions in the rope above and below m_1?

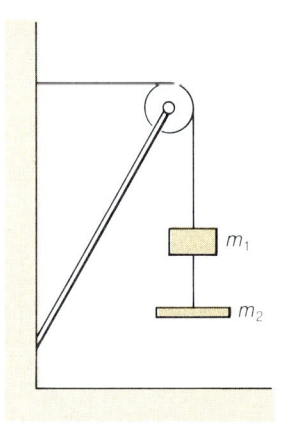

Figure 3.17

Section 3.4 Friction

• 51 A 5 kg mass slides at constant speed down a plane inclined at an angle θ from the horizontal. If its velocity is 5 m/s and the coefficient of moving friction is 0.3, what is θ?

• 52 A 20 kg block of ice begins to slide down a plane when its angle from the horizontal is 20°. What is the coefficient of static friction?

PROBLEMS

- 53 A force of 100 N is required to start a 50 kg mass moving across a surface. What is the coefficient of static friction?

- 54 A 20 kg dog sits in front of a door, refusing to budge. If the coefficient of static friction is 0.4, what force is required to move him?

- 55 Waiting impatiently in line at an airport, you kick your suitcase. It moves off at 2 m/s. How far will the suitcase travel if the coefficient of friction is 0.2? Does it matter how much the luggage weighs?

- 56 A 70 kg rock climber hopes to ascend a steep slope, inclined at an angle θ from the horizontal. If the coefficient of friction between the climber's shoes and the slope is 0.7, what is the steepest ascent he can attempt without slipping?

- 57 A roller skater is persuaded to buy some expensive new skates by a salesman who claims they have "a coefficient of friction with the road of only 0.1." Why was this not such a good idea?

- 58 You wish to accelerate a 200 kg mass from rest to 8 m/s in 4 s. An engine purchased at a flea market will deliver a force of 1,000 N. What is the maximum acceptable coefficient of kinetic friction?

- 59 Suppose in problem 31 that mass m_1 now moves with a coefficient of friction μ. Draw a force diagram and derive the acceleration of the system.

- 60 Suppose in Problem 40 that the sliding mass experiences a coefficient of friction 0.3 with the plane. What is the acceleration of the system?

- 61 If the coefficient of kinetic friction between a sliding block and the floor is 0.2, compute how far the block moves if its initial velocity is 10 m/s.

- 62 A sled lies motionless on a slope inclined 15°. The sled's mass is 10 kg, the coefficient of static friction is 0.2, and a line of children are waiting to pile on top. How many children are needed on the sled to start it accelerating down the slope?

- 63 A man steps on a bar of soap, exerting a horizontal force of 100 N and a vertical force of 600 N. If the coefficient of static friction between soap and floor is 0.2, will he slip? Explain your reasoning.

- 64 What is the maximum acceleration that a drag racer can reach on a flat surface, starting from rest, if the coefficient of static friction is 0.6?

- 65 Two objects with equal velocity have coefficients of kinetic friction μ and 2μ on the same surface. They coast to a halt. Which object travels farther? By what factor?

- 66 A 50-ton truck (4.5×10^4 kg) starts to roll down a 10% grade (that is, the slope drops 1 m for every 10 m travelled along the road). Calculate the speed of the truck after it has travelled 1 km:
  - (a) If there were no friction.
  - (b) If there were an effective friction coefficient $\mu = 0.05$ (accounting for wind resistance and internal friction in the wheel bearings).

- 67 A custodian pushes a 50 kg floor waxer at constant speed. He applies a force of 100 N along a handle inclined at 45° from horizontal.
  - (a) What is the horizontal frictional force acting on the waxer?
  - (b) What is the coefficient of kinetic friction between floor and waxer?
  - (c) Calculate the force the custodian would need to exert to accelerate the waxer from rest to 2 m/s in 5 s.

- 68 A car travelling at velocity v applies its brakes and decelerates smoothly to rest. The coefficient of static friction is μ. Show that its minimum stopping distance is $(v^2/2\mu g)$.

- 69 What is the maximum acceleration that a drag racer can achieve up an incline, starting at rest on a track sloped at 20° from horizontal, if the coefficient of static friction is 0.6?

- 70 If the coefficient of kinetic friction between a car's tires and the road is 0.45, what is the car's stopping distance if it was initially moving at 55 mi/h? That is, what is the length in which the car can decelerate smoothly to rest without slipping?

- 71 In Problem 70, what is the maximum speed on a wet surface (friction coefficient 0.2) for the car to stop in less than 100 meters?

- 72 A large plane needs a speed of 200 mi/h (90 m/s) to take off. If its mass is 30,000 kg, and if the coefficient of friction between tires and runway is 0.5, what is the minimum length of runway needed?

- 73 Spiderman clings to the front vertical surface of an accelerating train car (Fig. 3.18). The coefficient of friction between his boots and the surface is 0.8. If he does not fall, what is the minimum acceleration?

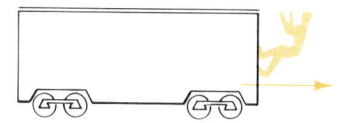

Figure 3.18

••• 74 A 10 kg crate rests on the flat surface of a pickup truck, which accelerates away from a stoplight. The coefficient of friction on the floor is 0.2.
(a) What acceleration can the crate tolerate before slipping?
(b) If the acceleration is 3 m/s², how far will the crate slip in 2 s?

••• 75 Some out-of-town physicists are worried that their small rental car may not make it up a steep San Francisco hill. Their 1,000 kg car (including them) can accelerate at 2 m/s² on a flat surface. What is the maximum slope they can climb without slowing down?

SPECIAL PROBLEMS

1. The Jupiter Effect

Several years ago, some astronomers suggested that the gravitational forces of the large planets Jupiter and Saturn might have an effect of the earth when they were aligned on the same side of the Sun. As we will learn in Chapter 7, the gravitational force between two masses M_1 and M_2 separated by distance R is $F = GM_1M_2/R^2$, where $G = 6.67 \times 10^{-11}$ N·m²/kg² is Newton's constant of gravitation. In Figure 3.19 the Sun, Earth, Jupiter, and Saturn are aligned.

Planet	Mass, kg	Distance from Sun, m
Earth	5.98×10^{24}	1.50×10^{11}
Jupiter	1.90×10^{27}	7.78×10^{11}
Saturn	5.69×10^{26}	1.43×10^{12}

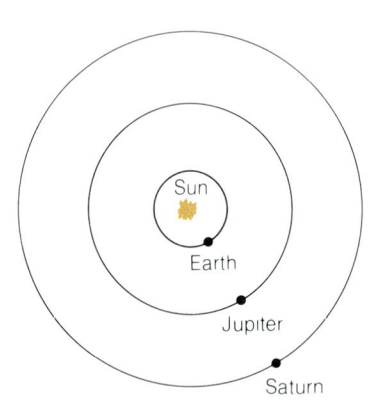

Figure 3.19

(a) Use Newton's formula to compute the forces exerted on the earth by the Sun (mass 1.99×10^{30} kg) and by the two planets Jupiter and Saturn.
(b) What is the total (vector sum) force exerted on the Earth by all three? What fraction of the Sun's force are the net forces of Jupiter and Saturn?
(c) What is the ratio of the force of the moon to the force of Jupiter? The moon has a mass of 7.35×10^{22} kg and is 3.84×10^8 m from the Earth.

2. Spacecraft Acceleration

A spacecraft's rocket engines are capable of exerting a thrust (force) of 5 million newtons. The craft's initial mass is 5×10^5 kg (mostly fuel).
(a) What is the spacecraft's initial acceleration (in m/s² and in g's)?
(b) Assume that the rockets maintain constant thrust for 4.1 min, burning and expelling fuel at a rate of 10^5 kg/min. At intervals of 1, 2, 3, and 4 minutes after launch, compute:
(1) the total mass of the spacecraft;
(2) the instantaneous acceleration.
(c) Assume that the change in velocity Δv during each interval Δt of 1 minute is given by $\Delta v = \bar{a}(\Delta t)$, where $\bar{a}$ is the average acceleration over the previous one-minute interval. Estimate the rocket's velocity after 4 minutes and compare it to the escape velocity from the Earth's surface, 11.2 km/s.

3. Slopes of Volcanoes

The steepness of a hill of loose rock or gravel is described by its angle of repose. Suppose a loose rock of mass m rests on a hill inclined at θ from the horizontal and that the coefficient of static friction between the loose rock and the hill is μ.
(a) Draw a force diagram for the rock. Show gravitational and frictional forces along the slope.
(b) Show that the mass will remain at rest for angles less than a critical angle θ_{cr}, the angle of repose. Derive a formula for θ_{cr}, and explain why it does not depend on the size of the mass m or the value of the gravitational acceleration g.
(c) If volcanoes have a steepness angle $\theta = 30°$, what is the coefficient of friction between their rocks and the slope?

4. Terminal Velocities

It was shown in the nineteenth century that the drag force on a sphere of radius R, falling smoothly through a fluid without turbulence, is given by Stoke's formula: $F_{dr} = 6\pi R\eta v$, where η is the coefficient of dynamic viscosity of the fluid and v is the sphere's velocity. The viscosity coefficient at 20°C is 0.001 kg/m · s for water and 1.8×10^{-5} kg/m · s for air.

(a) Derive a formula for the terminal velocity of a sphere of radius R and density ρ, falling under gravity through a fluid of viscosity η. Note that all forces balance at the terminal velocity. Check your units.

(b) Estimate the terminal velocities of the following:
  (1) A raindrop of radius 0.5 mm falling through air. Assume the raindrop has a density of 1,000 kg/m^3.
  (2) Buckshot of radius 1 mm and density 5,500 kg/m^3 falling through water.
  (3) A human being in tuck position falling through air. Approximate the human as a sphere of radius 0.25 m and density 1,100 kg/m^3. Why is your answer much greater than the observed terminal velocity of about 100 mi/h?

SUGGESTED READINGS

"Boomerangs! How to make them and also how they fly." J. Walker, *Scientific American*, Mar 79, p 162, **240**(3)

"More on boomerangs, including their connection with the dimpled golf ball." J. Walker, *Scientific American*, Apr 79, p 134, **240**(4)

"The essence of ballet maneuvers is physics." J. Walker, *Scientific American*, Jun 82, p 146, **246**(6)

"In which simple equations show whether a knot will hold or slip." J. Walker, *Scientific American*, Aug 83, p 120, **249**(2)

"Newton." I. B. Cohen, *Scientific American*, Dec 55, p 73, **193**(6)

"Friction." F. Palmer, *Scientific American*, Feb 51, p 54, **184**(2)

"Stick and slip." E. Rabinowicz, *Scientific American*, May 56, p 109, **194**(5)

"Form and function in fish swimming." P. W. Webb, *Scientific American*, Jul 84, p 72, **251**(1)

"Jet propelled swimming in squids." J. M. Gosline, M. E. DeMont, *Scientific American*, Jan 85, p 96, **252**(1)

"The art and physics of soaring." L. Hunter, *Physics Today*, Apr 84, p 34, **37**(4)

"Sir Isaac and the rising fastball." Peter J. Brancazio, *Discover*, Jul 84, p 44, **5**(7)

Newton at the bat: The science in sports. Eric W. Schrier, William F. Allman, eds, Scribners, New York, 1984

Sport science. Peter J. Brancazio, Simon and Schuster, New York, 1984

A source book in physics. William F. Magie, Harvard University Press, Cambridge, 1965

The flying circus of physics with answers. Jearl Walker, Wiley, New York, 1977

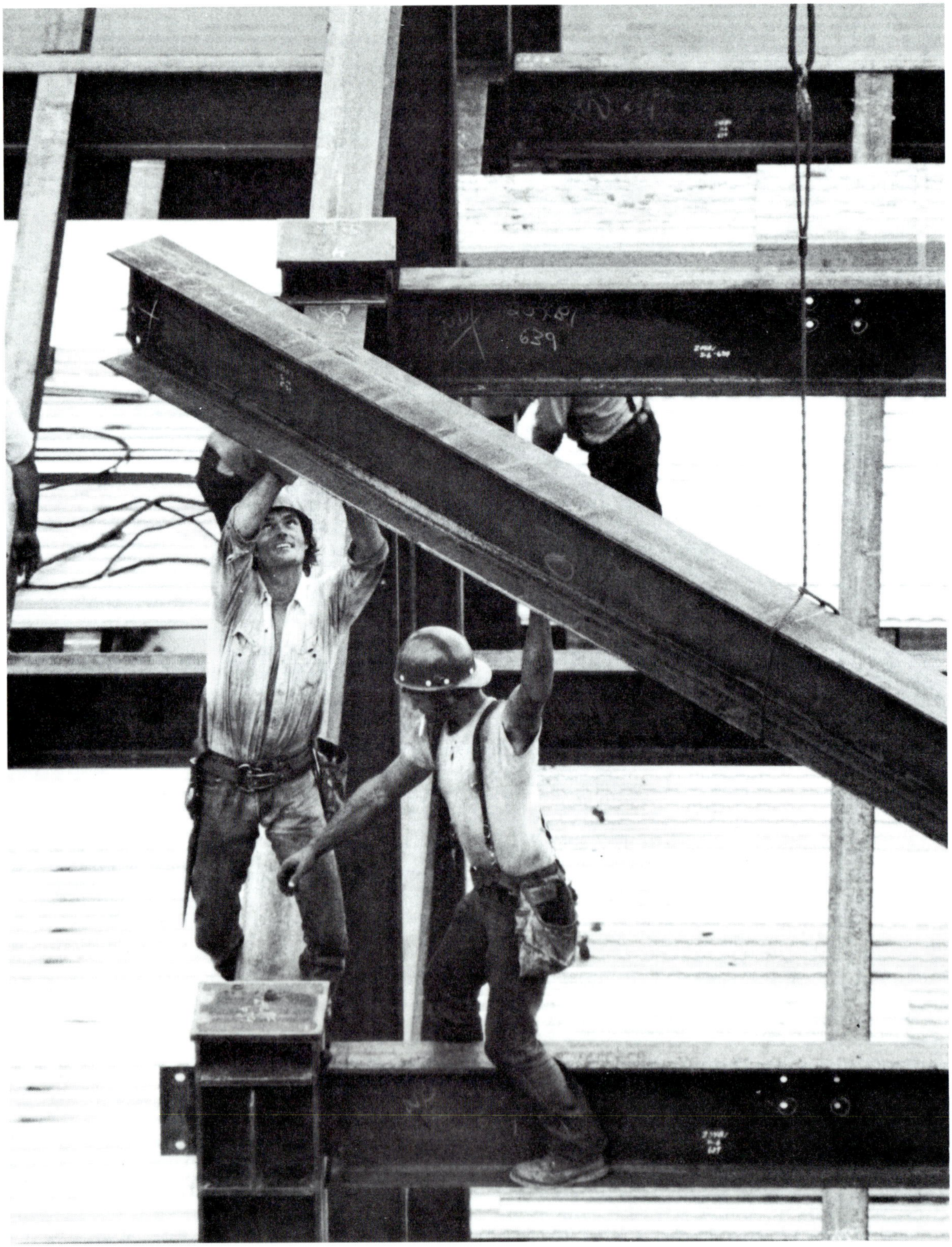

Chapter 4

Forces in Balance

In this chapter we discuss situations where the net force on a body is zero; that is, there is no acceleration. It might seem at first that such circumstances would be rare; why should it happen that all the forces on a body happen to cancel? The answer is that a balance of forces is quite common because one force often naturally develops to counteract another. When a person sits in a chair, exerting a downward gravitational force on it, the chair exerts an equal upward force, and the two forces balance so that there is no net force on the person and no acceleration. The force of static friction acts in a similar fashion, always equalling and opposing the external sideways force on an object resting on a surface (unless the maximum static friction force is exceeded, in which case the balance is destroyed and the object moves). A body in free fall reaches a condition of balance when the air resistance equals its weight and it stops accelerating. It is nearly impossible to imagine a situation where no forces act on a body, yet there are countless instances of objects that are not undergoing acceleration.

4.1 The Nature of Equilibrium

A state of balance is said to be an **equilibrium** state. When there is no motion, by far the most common kind of equilibrium, it is **static equilibrium**; when there is uniform motion (no acceleration), it is **dynamic equilibrium**. In equilibrium, forces are still present, and they can have significant effects. A building is in static equilibrium (Fig. 4.1), because the engineer who designed it calculated the forces acting on its arches and walls so that sufficiently strong materials would be selected. An airplane in flight is in dynamic equilibrium, even though immense forces act on its wings. The study of equilibrium even has important applications to the human body (consider the forces acting on joints, for example).

86 CHAPTER 4 FORCES IN BALANCE

Figure 4.1

Static equilibrium. This structure is in equilibrium, so that the net force at all points is zero. Evidently this has not always been so, because partial collapse has occurred in places.

Mathematically, the condition for equilibrium, that the net force be zero, is written

$$\Sigma \mathbf{F} = 0 \tag{4-1}$$

where $\Sigma \mathbf{F}$ means the vector sum of the forces. If the forces sum to zero, the components in each direction must sum to zero; therefore

$$\Sigma F_x = 0, \quad \Sigma F_y = 0 \tag{4-2}$$

The condition of equilibrium (Eq. 4–2) is often used to find unknown forces. To do so requires writing the equation in terms of the components of the forces involved, and then solving for the forces.

4.1 THE NATURE OF EQUILIBRIUM

Example 4.1 A tightrope walker is at rest on the rope (Fig. 4.2a). The mass of the performer is 70 kg. The angle of one rope segment is 15° below horizontal, and the angle of the other is 10° below horizontal. Find the tension in each segment of the rope.

If the x-axis is defined as horizontal (Fig. 4.2b), then the condition of equilibrium in the x-dimension is

$$T_{X1} + T_{X2} = 0$$

where T_{X1} and T_{X2} are the x-components of the tension. The weight W is not shown because its x-component is zero. By substituting values, we find

$$-T_1 \cos(15°) + T_2 \cos(10°) = 0$$

or

$$T_1 = \frac{T_2 \cos(10°)}{\cos(15°)} = 1.02 T_2.$$

Similarly, the y-components must add to zero:

$$T_{Y1} + T_{Y2} + W_Y = 0$$

Hence

$$T_1 \sin(15°) + T_2 \sin(10°) - (70 \text{ kg})(9.8 \text{ m/s}^2) = 0.$$

Combining our two expression relating T_1 and T_2 leads to

$$T_1 = 1599 \text{ N}$$

$$T_2 = 1568 \text{ N}$$

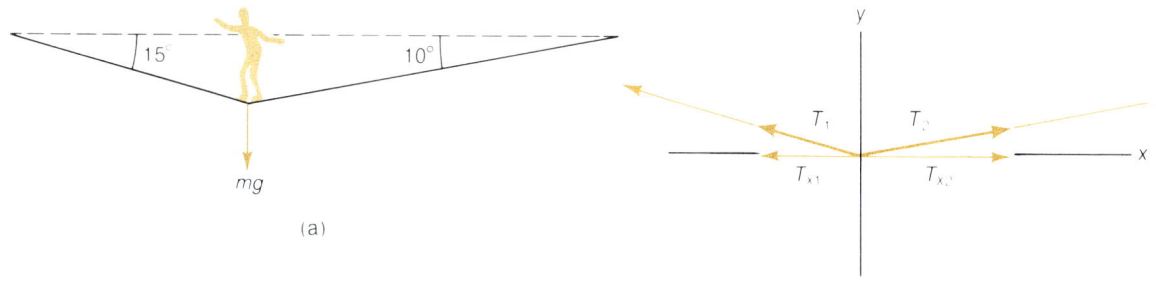

Figure 4.2

4.2 Center of Gravity and Torque

The equilibrium condition was sufficient to solve the last example because all the forces acted at the same point. In the example of the tightrope walker, for instance, we assumed without really saying so that all forces acted at the point where the performer's feet pressed down on the rope. But forces in equilibrium commonly act on different points, and we shall have to broaden our conditions of equilibrium.

Motions of Extended Bodies

When objects have spatial extent, that is, when they are not simple points, forces applied to different points cause rotation. Even if the net force is zero, the forces do not balance at all points on the body. The pinwheel (Fig. 4.3) is a familiar example. In general, we recognize two types of motion: **translational motion,** in which all parts of a body are displaced together, and **rotational motion,** in which a body spins about some center.

A rotating body must have an axis of rotation, an imaginary line about which it spins. For example, the earth rotates on an axis that joins the poles. If a body is fixed at some point, then that point will act as the pivot, locating the axis of rotation. If, on the other hand, the body is not rigidly attached to anything, then if unbalanced forces are applied, it will tend to rotate about the **center of gravity,** the point in a body through which the net downward force of gravity appears to act (Fig. 4.4). It is the point where the net gravitational force is equivalent to the gravitational force that would act on a single particle whose mass is equal to that of the body. For a symmetric, uniform object, the center of gravity is at the geometrical center, but it can also be found for a nonsymmetric or nonuniform object. We sometimes speak of the **center of mass,** which has essentially the same meaning but is defined independently of the earth's gravitational field.

In the absence of any other forces, a force applied at the center of gravity (or the center of mass) of an object will produce only translational motion, moving the body as a whole, with all of its parts on parallel paths. A force applied at any other point will cause rotational motion about the center of gravity.

The condition of equilibrium in Eq. 4–1 and Eq. 4–2 applies only when no forces cause rotation. Now we can say that Eq. 4–1 and Eq. 4–2 apply only when all forces act on the center of mass. These equations treat extended bodies as single point particles.

Torque

If a force is applied to an object at some point displaced from its axis of rotation, this force may create a **torque.** The dimensions of torque are force times length, and it is measured in the SI system in newton-meters (N · m). Dyne-centimeters (dyn · cm) are used in the cgs system, and foot-pounds (ft · lb) in the English system.

Figure 4.3

Rotational motion. The net force applied to the pinwheel is zero, but unbalanced forces due to wind strike different points, causing rotation.

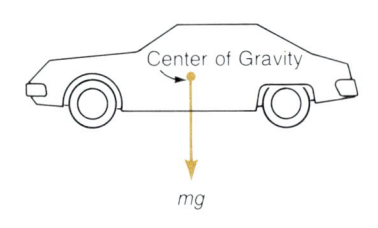

Figure 4.4

Center of gravity. The center of gravity is the point through which the downward force of gravity acts. A force applied to the car at the center of gravity can create only translational motion; a force applied elsewhere creates rotation.

4.2 CENTER OF GRAVITY AND TORQUE

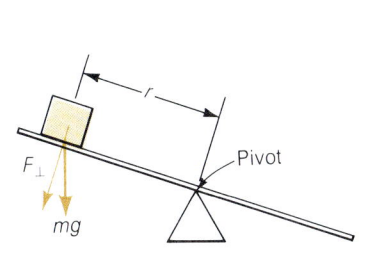

Figure 4.5

Torque. The torque created by the weight of mass m is the product $F_\perp r$, where $F_\perp$ is the component of the weight perpendicular to r. The axis of rotation is perpendicular to the plane of the figure.

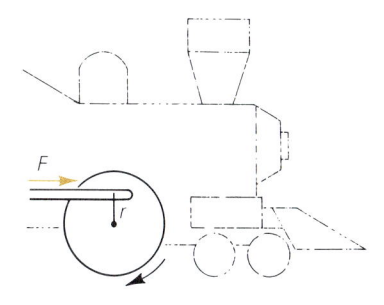

Figure 4.6

Lever arm. The force applied by the pushrod is always horizontal, and the torque is the product Fr, where r is the lever arm, the shortest perpendicular distance from the pivot point to a line drawn along the direction of **F**. Because r varies as the wheel turns, the torque also varies.

Torque may be defined as the product of the tangential component of an applied force times the distance from the axis of rotation to the point of application of the force. By "tangential component" we mean the component of the force that is perpendicular to a line connecting the axis of rotation with the point of application of the force (Fig. 4.5). This definition is equivalent to a simpler one: torque is the product of an applied force times the **lever arm,** which is defined as the shortest (perpendicular) distance from the axis of rotation to a line drawn along the direction of the force (Fig. 4.6). Thus, we can define torque in terms of the component of the force perpendicular to the line between the point of application and the axis of rotation:

$$\tau = rF_\perp \qquad (4\text{--}3)$$

where r is the distance from the axis of rotation to the point of application of the force, and $F_\perp$ is the component of the force perpendicular to r. We can also define torque in terms of the applied force and the (perpendicular) distance between the axis of rotation and the line of action of the force:

$$\tau = r_\perp F \qquad (4\text{--}4)$$

where $r_\perp$ is the perpendicular distance from the axis of rotation to the line along which the force acts. The definitions given by Eqs. 4–3 and 4–4 are equivalent.*

---

*The definition of torque can be expressed as the **cross product** of the two vectors **r** and **F**, where the magnitude of the cross product is $\tau = |\mathbf{r} \times \mathbf{F}| = rF \sin\theta$. The magnitudes of the vectors **r** and **F** are r and F and θ is the angle from **r** *to* **F**. The direction of the torque vector is given by the right-hand rule. Note that it is important to specify that θ is measured from **r** to **F** and not the other way, because the direction of **τ** would be changed.

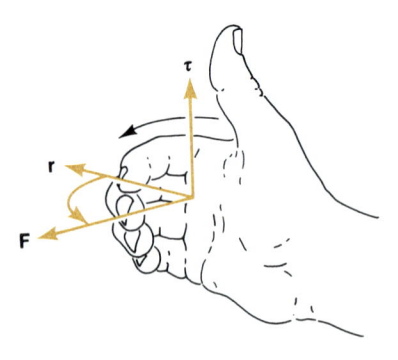

Figure 4.7

The right-hand rule.

Torque is a vector quantity, but the direction of the vector may not seem intuitively obvious. By convention, its direction is taken to be along the axis of rotation (that is, perpendicular to both the force and the lever arm) and in the direction dictated by the **right-hand rule**: this is the direction in which the thumb points if we rotate the fingers of our right hand, in the direction from **r** to **F**. (Fig. 4.7).

Example 4.2 What is the maximum torque exerted by a child on a seesaw, if the child has a mass of 35 kg and sits 2 m from the pivot?

The force applied to the seesaw is the child's weight, and the maximum torque occurs when the seesaw is horizontal, because then the full weight acts perpendicular to the lever arm. At that point the torque is simply

$$\tau = rF_\perp = (2 \text{ m})(35 \text{ kg})(9.8 \text{ m/s}^2) = 686 \text{ N} \cdot \text{m}$$

Example 4.3 A construction crane with a 40 m boom lifts a pallet loaded with bricks. The mass of the loaded pallet is 1200 kg, and the mass of the boom is 5000 kg. What is the torque exerted on the base of the crane when the boom is set at an angle 40° from vertical (Fig. 4.8)?

We can solve this using either Eq. 4–3 or Eq. 4–4. For the purpose of illustration, we do it both ways here. According to Eq. 4–3, the torque is

$$\tau = r_1 W_1 \sin(40°) + r_2 W_2 \sin(40°)$$

where r_1 is the length of the boom, $W_1 \sin(40°)$ is the component of the pallet's weight that is perpendicular to the boom, and r_2 is the distance along the boom to its center of gravity (equal to 20 m, if we assume that the boom is symmetric about its center), the point through which its weight W_2 acts. Substituting, we find

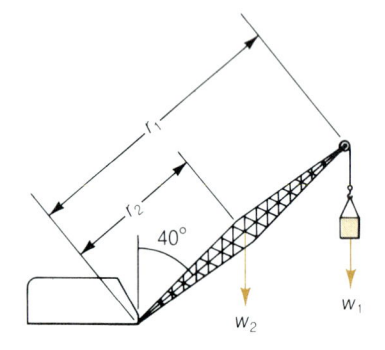

Figure 4.8

4.2 CENTER OF GRAVITY AND TORQUE

$$\tau = (40 \text{ m})(1200 \text{ kg})(9.8 \text{ m/s}^2)\sin(40°) + (20 \text{ m})(5000 \text{ kg})(9.8 \text{ m/s}^2)\sin(40°)$$
$$= 9.3 \times 10^5 \text{ N} \cdot \text{m}.$$

Or, if we use Eq. 4–4, we write

$$\tau = r_1\sin(40°)W_1 + r_2\sin(40°)W_2$$

where $r_1\sin(40°)$ is the perpendicular distance between the axis of rotation and the line along which the force W_1 acts, and $r_2\sin(40°)$ represents this quantity for W_2, the weight of the boom. Clearly, this will yield the same answer as we found above, using Eq. 4–3. For this example, both equations are equally convenient.

Center of Gravity

Now that we have discussed torques, we can calculate the location of the center of gravity of a body. We recall that the center of gravity is the point through which gravity acts, the point whose translational motion is governed by Newton's laws. Thus the weight of a body acts at the center of gravity to create a torque about any other point. If the body is in equilibrium, then the total torque due to the weight at all points in the body must be the same as the torque due to the body's total weight applied at the center of gravity. Therefore, for an axis of rotation at an unknown distance from the center of gravity, we can set the sum of the individual torques caused by the individual weights acting at different points on the body equal to the torque produced by the total weight acting at the center of gravity.

Consider, for example, two masses m_1 and m_2 attached by a massless rod of length L (Fig. 4.9). If we choose one end of the rod (the location of mass m_1) as the pivot, the torque due to the total weight of the system is

$$\tau = g(m_1 + m_2)x$$

where x is the distance from the pivot to the center of gravity. Since m_1 is at the pivot, the sum of the torques due to individual masses is

$$\tau = (0 \times gm_1) + Lgm_2 = Lgm_2.$$

Setting the total torque due to the weight of the entire system equal to the sum of the individual torques leads to

$$x = L\frac{m_2}{m_1 + m_2}.$$

If the masses m_1 and m_2 are equal, the center of gravity is at a distance $x = L/2$ from the end of the rod; that is, the center of gravity is in the center, as we would expect.

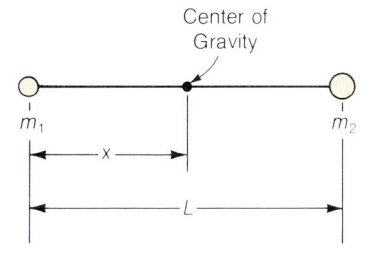

Figure 4.9

Finding the center of gravity. To find the c.g. of a body such as this massless rod with attached masses, choose a pivot point and set the sum of the torques about that point equal to zero. The resulting equation can be solved for the unknown distance x.

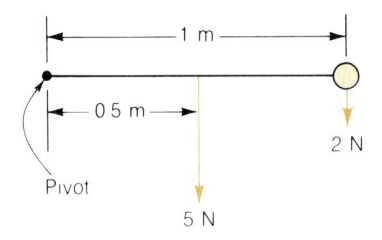

Figure 4.10

Center of gravity of an extended object. In this case, the rod has mass, creating an additional torque that must be included when calculating the location of the c.g.

Let us now find the center of mass for a slightly more complicated system, a uniform rod of known weight and length, with a known mass attached to the end. We choose a pivot point at an arbitrary distance x from the center of gravity, then calculate the individual torques due to the weight of the mass attached to the end of the rod and to the weight of the rod itself. We set the sum of these individual torques equal to the torque due to the total weight of the rod and its attached mass, acting through the center of gravity. For example, if the rod is 1 m long, its weight is 5 N, the attached mass weighs 2 N, and we choose the unweighted end as our pivot (Fig. 4.10), then

$$(0.5 \text{ m})(5 \text{ N}) + (1 \text{ m})(2 \text{ N}) = (5 \text{ N} + 2 \text{ N})x$$

or

$$x = 0.64 \text{ m}$$

that is, the center of gravity is 0.64 m from the unweighted end of the rod.

Because the center of gravity is the point about which the sum of the torques due to the weights of the individual parts of an extended body is zero, it is the balance point for an extended object.

Example 4.4

Suppose a crane operator wants to lift a girder with extra masses welded to it (Fig. 4.11). He needs to know where the balance point is. The girder's mass is 300 kg and its length is 12 m. A mass of 500 kg is attached 3 m from one end, and a mass of 700 kg is attached 2 m from the other end.

Take as a pivot point the left end of the girder and set the torques due to the girder and its attached masses equal to the torque due to the total weight acting through the center of gravity at a distance x from the left end.

$$(500 \text{ kg})(9.8 \text{ m/s}^2)(3 \text{ m}) + (300 \text{ kg})(9.8 \text{ m/s}^2)(6 \text{ m}) \\ + (700 \text{ kg})(9.8 \text{ m/s}^2)(10 \text{ m}) = (500 + 300 + 700 \text{ kg})(9.8 \text{ m/s}^2)x$$

Solving for x gives

$$x = 6.87 \text{ m}$$

The center of gravity for the girder with its extra loads is 6.87 m from the left end of the girder.

Figure 4.11

We may calculate the torque due to the weight of any body by finding the center of gravity first. Then the torque will be simply the torque due to the total weight acting on center of gravity. For extended bodies, calculating the torque in this way is very much simpler than calculating the sum of the individual torques due to the weights of all the parts of the body.

4.2 CENTER OF GRAVITY AND TORQUE

Example 4.5

Suppose the girder whose center of gravity was calculated in the previous example is set on the roof of a building, so that the center point of the girder is right at the edge of the roof, with (naturally) the more massive half on the roof (Fig. 4.12). A man whose mass is 70 kg walks along the girder. How far beyond the edge of the roof will he get before the girder pivots and the man falls?

The problem is to find the man's position at the point where the torque due to his weight is equal to the torque due to the other forces acting on the girder. The edge of the roof is the pivot. We could calculate all the individual torques due to the man, the weight of the girder and the weights of the extra masses attached to it. It may be easier, however, to set the torque due to the man equal to the torque due to the weight of the girder and its attached masses, acting at the center of gravity. Then the problem reduces to two steps: calculating the center of gravity, which we have already done, and then setting the torque due to the man's weight equal to the torque due to the weight of the girder and its loads and solving for the position of the man. Setting the torques equal gives

$$W_m x_m = W_g x_g$$

where W_m refers to the weight of the man, x_m is his distance from the edge of the roof, W_g is the total weight of the girder and its attached masses, and x_g is the distance from the edge of the roof and the girder's center of gravity. We know already that the total mass of the girder and its attachments is 1500 kg, that its center of gravity is 6.87 m from one end, and that it is 12 m long. Hence the torque due to the total weight of the girder and its attached masses acts at a point 0.87 m from the edge of the roof, and our equation becomes

$$(70 \text{ kg})(9.8 \text{ m/s}^2)x_m = (1500 \text{ kg})(9.8 \text{ m/s}^2)(0.87 \text{ m})$$

or

$$x_m = 18.64 \text{ m}.$$

Hence the man can walk safely all the way to the end of the girder, which is only 6 m beyond the edge of the roof.

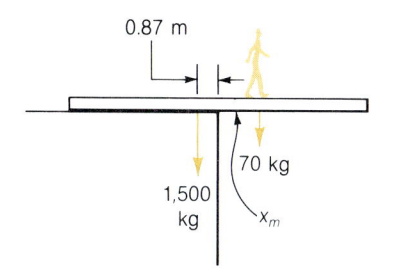

Figure 4.12

Equilibrium Revisited

We have learned that a body can satisfy the equilibrium condition in Eq. 4–1 and Eq. 4–2 yet not be in equilibrium because of net rotational forces. Now that we understand torque, we can revise our definition of equilibrium to include forces which do not act at the same point. We now say that, in addition to the condition that the net force must be zero, we must also have a net torque that is zero. In equation form, we write

$$\Sigma \tau = 0. \tag{4-5}$$

94   CHAPTER 4   FORCES IN BALANCE

Figure 4.13

Rotational equilibrium. When the diver is motionless on the board, the torques are balanced.

Thus, for a three-dimensional body, we may have as many as six equations: an equation for each component of the net force and an equation for each component of the net torque. Usually, however, we will have only one or two dimensions. For example, it will often be possible to define a coordinate system so that the axis of rotation is in the z-direction, perpendicular to the x, y plane; then we need only the two equations in Eq. 4–2 and a third to represent the net torque, which is due to forces acting in the x, y plane.

There are many examples of rotational equilibrium, where the torques add to zero. A diver standing on the end of a diving board exerts a torque about the board's pivot point that is exactly balanced by the torque due to the board's rigid attachment at its other end (Fig. 4.13). A drawbridge raised over a moat is in equilibrium because the torque due to the weight of the raised bridge is balanced by the torque due to the tension in the cables that hold it up. Cantilevers, projecting beams whose torque due to gravity is balanced by a strong fastening at the end are often used in building construction. The design engineer must calculate the tensile strength needed in the bolts that hold down the fixed end of a cantilever.

FOCUS ON PHYSICS

Equilibrium in Architecture*

An important component of the history of architecture is the chain of physical discoveries that has allowed us to move stepwise from straw houses to skyscrapers. The diversity of building types and forms that we have created over the last 10,000 years provides an eloquent record of our cultural development because buildings must be able to stand up to be of any use, and are frequently built so well that they far outlast the society that created them. Whatever the intended use of a building may be, whether purely a symbol such as the Washington Monument or simply a warehouse enclosing space, the architect's understanding of materials and structure is critical to the success of the design.

In the past, structural systems had to be advanced by the unforgiving method of trial and error. The builders of the Gothic cathedrals knew they had exceeded the limits of their structural system when tons of stone arches crashed to the ground. Experiment is still the proof of all science, as witness the occasional collapse of a modern structure such as the Hartford Civic Center after a heavy snow storm, but the theory of structures has advanced to the point that buildings are limited more by the strength of materials than the engineer's ability to analyze the structure on paper. Computers have greatly enhanced our ability to apply physics theory to actual problems and to design structures that were unthinkable only because the calculations would have taken a lifetime.

The objective of the architect and engineer is to design a building that will perform its desired functions and remain in static equilibrium indefinitely. This must now be done without recourse to trial and error in the field, so the forces that will act on the structure must be thoroughly understood. Space stations are now on the drawing boards, but since all of our buildings to date are earthbound we must somehow direct all the forces acting on them into the earth in such a way that the net force is zero. If we are successful the building will not move except within the elastic limits of the materials chosen to carry the loads. What forces must be considered before a structure is designed?

Gravity is both an obvious and important part of the problem, but it is also constant and nonvariable. If there were no other forces acting on our building the problem would be relatively simple; compute the force of gravity acting on the mass of the building (the dead load) and on the mass of the occupants, contents, and snow on the roof (the live load), add a safety factor, and channel the forces cumulatively down to the ground. A sixty-story building for instance might exert a force of 100,000 tons due to gravity, and directed straight down. If the compressive strength of the earth beneath the foundation were adequate to resist this force, the building would stand up. Taken together, the dead load and the live load constitute the static loads on the building; that is, they change slowly or not at all. The live load can never be known exactly, but reliable, conservative quantities for different uses are contained in building codes, eg. a hallway in a public building must be designed to support one hundred pounds per square foot. The downward direction of the force due to these static loads has various implications on the design of structures. The height of vertical structures, for instance, is limited not by the action of gravitational forces, but by the other forces acting on the structure. In the absence of other forces, granite blocks could be stacked to a height of over three miles without structural failure of the material. In the case of horizontal structures, which must span a distance perpendicular to the force of

*Contributed by Peter B. Stoner, AIA

(continued)

gravity, the static loads become dominant. The longest spans now exceed a mile in the case of suspension bridges and are approaching the limit of the available materials, the limit of the attraction between the molecules in a high-strength steel cable. Long span roofs designed as arches or domes, in which the roof loads are carried in compression to the ground, have reached 700 ft in the New Orleans Superdome.

Unfortunately, gravity-induced static loads are not the only forces which must be considered when designing a structure to meet the conditions for equilibrium. Buildings must now be designed to at least survive, not necessarily undamaged, all but the most severe windstorms and earthquakes. Both of these forces act horizontally, (perpendicular to the force of gravity), and may be directed at a building from any point of the compass. Since buildings are designed not to move, these horizontal forces will tend to rotate them around a point on the surface of the earth unless an attachment can be made to the earth's crust. The equilibrium problem is essentially the same whether the wind is pushing the building with respect to the ground, or an earthquake is accelerating the earth's crust and taking the building with it. The difference will be that a wind of uniform velocity will exert force on the windward face of the building and act through the centroid of that area, whereas the mass of the building will resist the acceleration caused by an earthquake, thereby inducing a force through the center of gravity of the structure. With the axis of rotation at ground level, buildings must act as cantilevered beams to resist these horizontal forces. Thus a gust of wind hitting the top of the Sears Tower will be transferred to the ground through a lever arm equal to the height of the building, 1,450 feet, and will generate a significant torque in the process.

Unlike the force of gravity, winds and earthquakes vary over time, often very quickly, and cannot always be considered as static loads, loads which vary slowly, if at all. Whether strong gusts of wind and earthquakes produce static or dynamic loads depends on the form and structural characteristics of a given building. Remember that we can only hope to maintain static equilibrium within the elastic limits of the materials used for construction. All materials will shorten when compressed and elongate when in tension, and vary widely in their strength and stiffness. Wood and masonry are still the most commonly used structural building materials, but lack the strength to build highrise or long-span structures. Our sixty story building for instance would be framed with structural steel or reinforced concrete, materials whose elastic deformations are a tiny fraction of their length. The building will bend like a tree in the wind, but to a much smaller degree. For the comfort of the occupants, high-rise buildings are usually designed so that the top floors will not be displaced more than the height of the building divided by 500. Thus if our building were 800 ft high this displacement, or wind drift would be less than two ft. Has the building experienced a static or a dynamic load?

The answer depends on the time which elapses during the gust of wind, compared to the period of the building. A body is in stable equilibrium if it returns to its original position after a slight displacement. Suppose a child on a swing is given one push. The swing will oscillate back and forth until it comes to rest, and each oscillation will take the same period of time. Our 800 ft building will also swing back and forth after a displacement, until it returns to static equilibrium. The duration of each swing, the period of this building, would be about six seconds. If the gust of wind lasted less than six seconds it would be a dynamic load, whereas a wind which reached the same intensity and subsided over more than six seconds would be a static load. Dynamic loads are potentially the most dangerous because buildings, like the child on the swing, move, however slightly, in response to forces even though they are basically in static equilibrium. Some of the most spectacular structural failures occur when wind or seismic forces impact a building repeatedly in rhythm with its natural period, causing the displacement to increase with each swing until materials fail and the building collapses.

In practice then, there are many forces which

4.2 CENTER OF GRAVITY AND TORQUE

must be considered simultaneously when designing any kind of structure. As an example we will consider only two forces at once and find out if our sixty story building will tip over in a strong wind. The weight of the building is 100,000 tons or 2×10^8 lb, it is 800 ft tall, and has a square plan 100 ft on a side. Assume a wind load of 60 lb per square ft, 60 psf.

The weight of the building is a downward force through the center of the building and acts through a lever arm of 50 ft, half the width of the building, when resisting the torque caused by the wind (Fig. 4.1.1).

$$(2 \times 10^8 \text{ lb})(50 \text{ ft}) = 1 \times 10^{10} \text{ lb-ft}$$

The total force of the wind will be the area of one face of the building, 80,000 sq. ft, times the unit force of 60 psf, or 4.8×10^6 lb. This force acts through the centroid of the building face so the lever arm is 400 ft, half the height of the building. The torque due to the wind is,

$$(4.8 \times 10^6 \text{ lb})(400 \text{ ft}) = 1.92 \times 10^9 \text{ lb-ft}$$

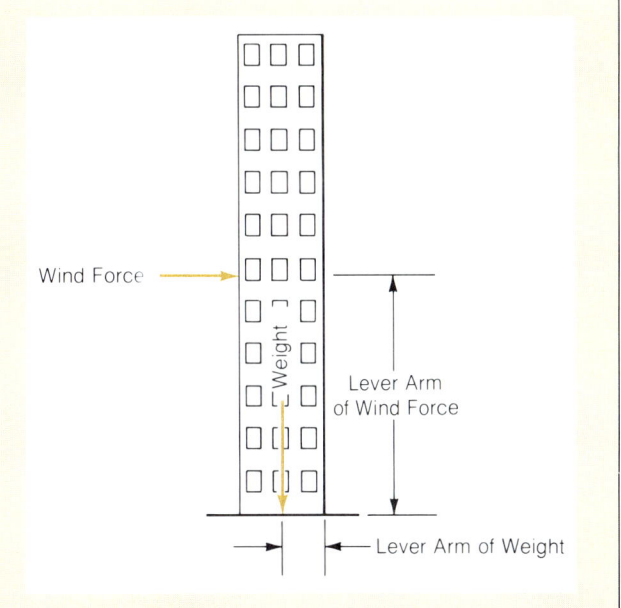

Figure 4.1.1

In this case the mass of the building is sufficient to resist the torque caused by the wind and the building will remain in static equilibrium.

In equilibrium problems, we set the sum of the torques acting in one rotational direction (for example, clockwise) equal to the sum acting in the opposite direction (counterclockwise). We can arbitrarily choose the pivot, but often one particular choice may simplify the problem.

Example 4.6

Let us consider a crane that has a torque of about 9.3×10^5 N · m exerted on it by the weight of its boom and a loaded pallet it is lifting. The crane is in equilibrium because it has enough mass on the opposite side of its axis of rotation from the side the boom is on. If the center of gravity of the crane is 3 m from the axis of rotation, how much mass must the crane have in order to balance the loaded pallet?

Because the net torque must be zero, the torque due to the weight of the body of the crane must be 9.3×10^5 N · m, in the counterclockwise direction (Fig. 4.14). Since the crane is resting on a horizontal surface, its weight W acts at a right angle to the line connecting the point of application (the center of gravity) with the axis of rotation. We write

98   CHAPTER 4   FORCES IN BALANCE

$$9.3 \times 10^5 \text{ N} \cdot \text{m} = (3 \text{ m})W$$

or

$$W = \frac{9.3 \times 10^5 \text{ N} \cdot \text{m}}{3 \text{ m}} = 3.1 \times 10^5 \text{ N}.$$

Hence the mass of the crane's body must be

$$m = \frac{W}{g} = \frac{3.1 \times 10^5 \text{ N}}{9.8 \text{ m/s}^2} = 3.2 \times 10^4 \text{ kg}.$$

Notice how much more massive the crane's body must be than the load it lifts. This is because of the much shorter lever arm between the crane's body and the axis of rotation.

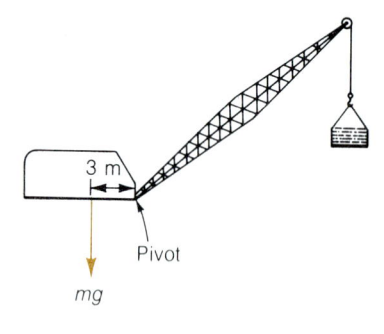

Figure 4.14

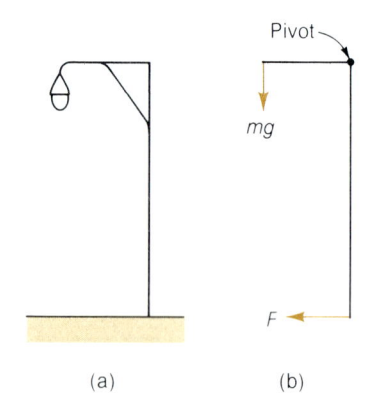

Figure 4.15

Torques in balance. The torque due to the weight of the lamp and its support arm is balanced by the torque due to the horizontal force exerted at the base of the pole by the ground.

Sometimes it is difficult to identify all the forces that act in equilibrium. The greatest difficulty is isolating the body and the forces acting *on* it from the forces exerted *by* it. The examples discussed so far have been carefully designed to make this easy, but it can be difficult unless systematic steps are followed. It is helpful to draw a space diagram of the body and the forces acting on it and their points of application. Then choose a pivot and identify the torques.

Fig. 4.15a shows a street light fixture with a lamp suspended from a pole. Since there is no acceleration, we know it is in equilibrium, but it may not be clear where the torque arises that counteracts the torque due to the weight of the lamp and its support arm. Fig. 4.15b shows a space diagram of the forces. It is often easiest to adopt as the pivot point the place where there is a bend in the structure. Here we may choose to assume that the top of the vertical pole is the axis of rotation, and consider the torques acting about that point. We see that there are torques in the counterclockwise direction due to the weight of the lamp acting at the end of the horizontal arm and to the weight of the arm acting at its center and a clockwise torque due to the horizontal force exerted at the foot of the pole. Without drawing such a diagram it might not have been clear what force counteracts the weight of the lamp and its support arm, or in which direction it acts. Once we draw a space diagram, it is simple to solve this kind of problem.

Example 4.7   A person weighing 500 N climbs a ladder leaning against a wall. The ladder weighs 100 N and is 2 m long, and is set at an angle of 25° from the vertical. The wall and the floor are both made of smooth wood, and the coefficient of static friction for the ladder against them is 0.30. As the person climbs the ladder, at some point the bottom starts to slip. How high does the person get before this begins to happen?

PERSPECTIVE

The ladder itself and the forces acting on it are shown in the space diagram (Fig. 4.16). At the top of the ladder are a horizontal force H exerted by the wall and an upward friction force equal to $\mu_s H$. At the bottom of the ladder are a vertical force N and a horizontal friction force to the left $\mu_s N$ and vertical forces W_P and W_L due to the weight of the person and the ladder, acting at point x and at the middle of the ladder, respectively. We must use the first condition of equilibrium (Eq. 4–1) to find the forces. In the horizontal direction

$$H = \mu_s N.$$

(we write an equality because the ladder is about to slip so that the static friction equals its maximum value.) In the vertical direction,

$$\mu_s H + N = W_P + W_L.$$

Solving for H and N yields

$$N = \frac{W_P + W_L}{\mu_s^2 + 1} = 550 \text{ N}$$

$$H = \frac{(W_P + W_L)\, \mu_s}{\mu_s^2 + 1} = 165 \text{ N}$$

Now we choose the bottom of the ladder as the pivot and set the clockwise torques equal to the counterclockwise torques.

$$H\,(2 \text{ m})\cos(25°) + \mu_s H\,(2 \text{ m})\sin(25°) = x\, W_P \sin(25°) + W_L\,(1 \text{ m})\sin(25°)$$

$$(165 \text{ N})(2 \text{ m})\cos(25°) + (.3)(165 \text{ N})(2 \text{ m})\sin(25°) = (x)(500 \text{ N})\sin(25°) + (100 \text{ N})(1 \text{ m})\sin(25°)$$

Solving for x yields

$$x = 1.41 \text{ m}.$$

The ladder will start to slip when the person has climbed 1.41 m.

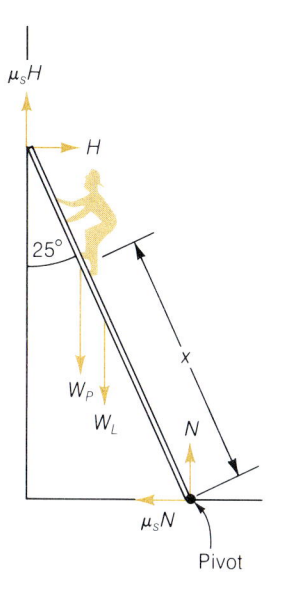

Figure 4.16

Perspective

Now that we understand forces and equilibrium, we can solve many problems in everyday life. We appreciate how engineers design buildings, and we can calculate the stresses acting on structural elements. We are ready to study energy, work, momentum, and power, and their conservation laws in the next chapter.

SUMMARY TABLE OF FORMULAS

Dynamic equilibrium:
$$\Sigma \mathbf{F} = 0 \tag{4-1}$$
$$\Sigma F_x = 0, \quad \Sigma F_y = 0 \tag{4-2}$$

Torque:
$$\tau = rF_\perp \tag{4-3}$$
$$\tau = r_\perp F \tag{4-4}$$

Rotational equilibrium:
$$\Sigma \tau = 0 \tag{4-5}$$

THOUGHT QUESTIONS

1. Give two examples of bodies that move yet are in equilibrium.

2. Suppose forces balance on an object at rest inside a train. Why doesn't the train's constant velocity affect the equilibrium?

3. When an inclined plane is tipped beyond a critical angle, gravity overcomes friction, and an object resting on the plane begins to slide. Why is the critical angle independent of the object's mass?

4. At the instant when an object is dropped, are the forces in balance? In air, will the object accelerate indefinitely?

5. Force is transmitted to a mass by means of tension in a rope. What is the microscopic basis for tension?

6. A rope wound around a pulley changes the direction of a force. Explain the role of friction.

7. What is the microscopic basis for friction? Why is μ_s usually greater than μ_k? Why is the friction of steel on ice less than that of steel on wood?

8. Friction can be both a hindrance and an aid. Give examples of both.

9. Why is the center of gravity of an object not necessarily at its geometric center? Give an example.

10. Explain the importance of the center of gravity for the following: the Leaning Tower of Pisa, a gymnast on a balance beam, a backpacker, a bicyclist leaning into a turn, a shopper loaded with an armful of boxes.

11. Archimedes said, "Give me a lever long enough, and I'll move the world." Explain what he meant.

12. Explain the importance of torque and leverage in the following: crowbars, diving boards, hammers, screwdrivers, bottle openers, staplers.

13. In athletics, forces are applied to objects (balls, bats, pieces of wood or metal). Explain why knowing the length of bones and the distance away from a joint at which muscles attach to the bones could be useful.

14. In architecture, the arch and the dome provide efficient means of spanning a space. With a space diagram, show how the force of gravity is resolved into compression and extension forces in the arch.

15. Why do rowboats have two oars and triremes have an even number of crew members?

16. A suspension bridge can be thought of as a series of weights hanging from a cable. Explain, with a diagram, how tension forces in the cable support the weight of the bridge.

17. Explain the concept and design of counterweights in terms of torques.

18. Explain how force is transmitted to the road in a bicycle and a car. Discuss the role of gears, wheels, and friction.

19. When one pushes a tall object which is free to slide against friction, why is it best to push near the ground? Explain how this applies to pushing heavy boxes and blocking football players.

20. Why do anthills almost all have the same slope?

PROBLEMS

Section 4.1 Equilibrium

1. Forces of 3 N, 5 N, 8 N and 10 N act on an object (Fig. 4.17.) Compute the force (magnitude and direction) required to keep the object stationary.

2. A 100-ton airplane flies at constant speed and altitude. Draw a freebody diagram showing the forces of gravity, lift, engine thrust, and air drag acting on the plane. Compute the magnitude of the lift required.

3. A 60 kg rock climber supports herself on a steep slope, inclined at 70° from horizontal, by a rope attached at the top of the slope. What rope tension is required to keep her stationary if the coefficient of static friction is 0.7?

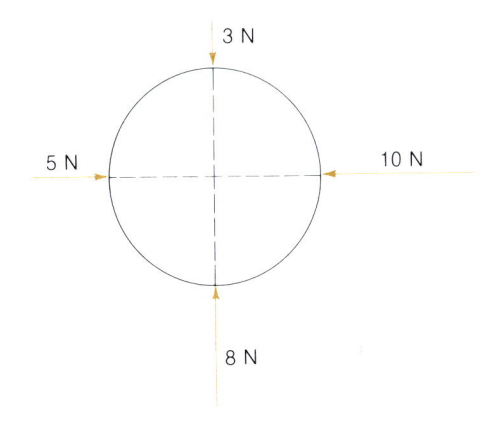

Figure 4.17

4. A 1,500 kg car lies midway across the bottom of a canyon 50 m deep and 200 m wide. Two ropes are attached to the car and strung over pulleys fixed at the top of the canyon rims, and equal forces F are applied horizontally to the ropes. What force F is needed to lift the car?

5. A 200 kg mass is placed directly over the center of gravity of a 50 kg symmetric table top. Compute the upward forces that the four support legs must exert on the table top.

6. Two zookeepers push on a 200 kg gorilla in perpendicular directions with forces of 200 N each. What minimum coefficient of static friction will ensure that the gorilla does not budge?

7. Compute the tension, T, in the cord which supports the 25 kg beam in Fig. 4.18, and find the force F exerted by the wall on the beam. Give the horizontal and vertical components, F_x and F_y.

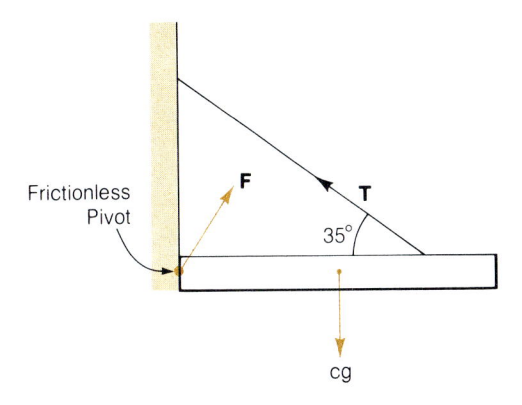

Figure 4.18

8. Find the tensions in the two wires shown in Fig. 4.19.

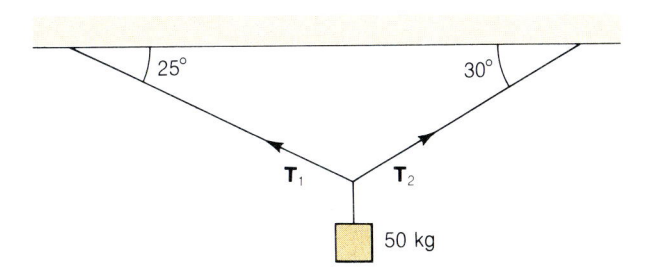

Figure 4.19

9. What force is needed to support the 10 kg mass shown in Fig. 4.20? Neglect the mass of the rope and assume the pulley is frictionless.

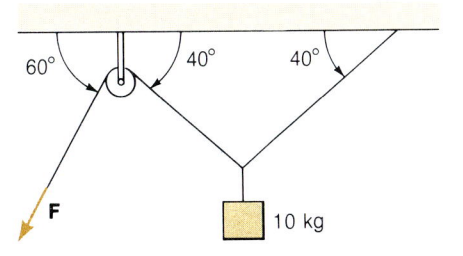

Figure 4.20

10 Suppose a tightrope walker of mass M is at rest on a rope (Fig. 4.21) with the angles and rope lengths θ_1, θ_2, L_1, and L_2. The rope tensions are T_1 and T_2.

(a) From the conditions of equilibrium described in Example 4.1, show that

$$T_2 = T_1 (\cos\theta_1 / \cos\theta_2)$$
$$T_1 = Mg/[\sin\theta_1 + \sin\theta_2(\cos\theta_1/\cos\theta_2)]$$

(b) Using the trigonometric identity, $\sin(\theta_1 + \theta_2) = \sin\theta_1 \cos\theta_2 + \cos\theta_1 \sin\theta_2$, show that T_1 and T_2 can be rewritten in the convenient forms

$$T_1 = Mg[\cos\theta_2/\sin(\theta_1 + \theta_2)]$$
$$T_2 = Mg[\cos\theta_1/\sin(\theta_1 + \theta_2)].$$

11 Use the general formulas in Problem 10 to obtain the answer to Example 4.1, where $M = 70$ kg, $\theta_1 = 15°$, $\theta_2 = 10°$. Do your results agree with the text?

12 For the tightrope walker in Problem 10 (Fig. 4.21), show that the lengths of the rope segments 1 and 2 are given by

$$L_1 = D[\sin\theta_2/\sin(\theta_1 + \theta_2)]$$
$$L_2 = D[\sin\theta_1/\sin(\theta_1 + \theta_2)].$$

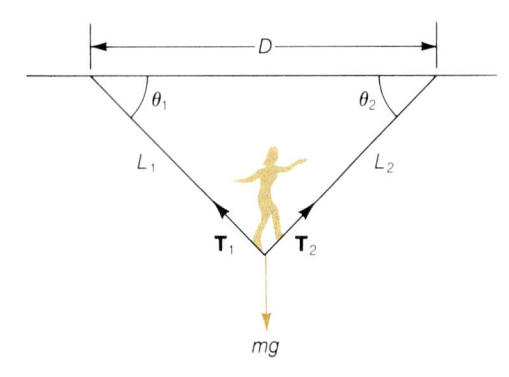

Figure 4.21

Examine horizontal and vertical components of L_1 and L_2 and use the trigonometric identity in Problem 10.

13 Suppose the tightrope walker in Problems 10 and 12 is exactly in the middle of the rope ($L_1 = L_2$). If his mass is 70 kg and the rope is twice the length D, calculate
(a) The angles θ_1, θ_2.
(b) The tensions T_1 and T_2.

14 To keep your 20 kg backpack away from bears who have a 3 m reach, you hang the pack from the midpoint of a rope strung between two trees. If the rope is attached to points 6 m up on the trees and the trees stand 10 m apart, what tension is required in the rope so that the bears cannot touch your food?

15 Compute the tensions T_1, T_2, T_3, and T_4, in the ropes shown in Fig. 4.22. From the symmetry of the situation, can you make any statements about the relationship between pairs (T_1, T_4) and (T_2, T_3)?

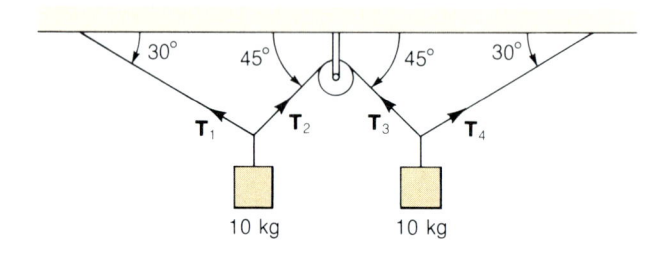

Figure 4.22

Section 4.2 Torque

16 What force F_1 is needed to balance a 100 kg mass in Fig. 4.23?

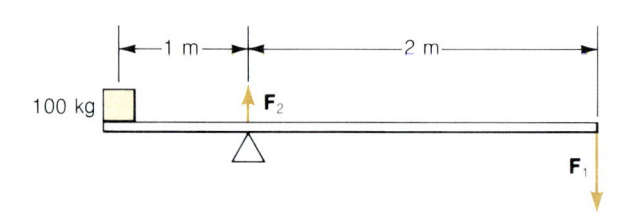

Figure 4.23

17 For the system in Problem 16, what force F_2 is exerted by the fulcrum?

18 A 200 kg birthday cake for a king rests on a 100 kg beam (Fig. 4.24).

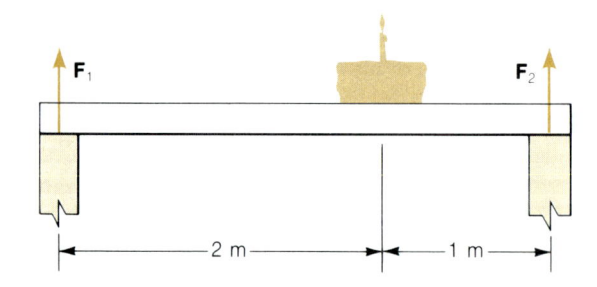

Figure 4.24

(a) How far from the left support leg is the center of gravity of the beam alone?

(b) Compute the forces F_1 and F_2 exerted on the beam.

• 19 In Problem 18, what would F_1 and F_2 be if the cake were placed directly over the center of gravity of the beam? Would their sum, $F_1 + F_2$, change?

• 20 A 10 m beam with 100 kg mass overhangs a cliff, as shown in Fig. 4.25. Two masses, M_1 and M_2, are placed at each end.

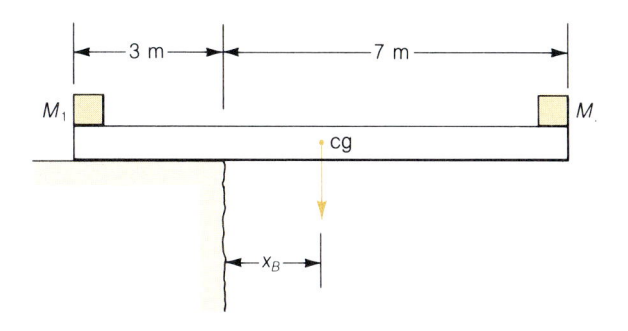

Figure 4.25

(a) What is the overhang distance, x_B, of the beam's center of gravity, excluding the masses?

(b) If $M_2 = 50$ kg, what must M_1 be to counterbalance?

• 21 If, in Fig. 4.25, $M_1 = 500$ kg, how large can M_2 be before tipping the beam?

• 22 Calculate the forces F_1 and F_2 on the supports of a diving board (Fig. 4.26) when a 70 kg diver stands at one end. Neglect the mass of the board.

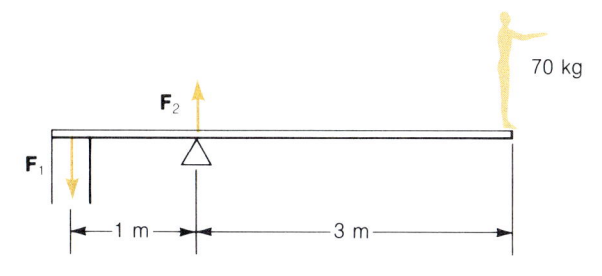

Figure 4.26

• 23 Suppose, in Fig. 4.26, the diver jumps in the air and lands on the end of the diving board with force 1,500 N. What instantaneous force F_2 acts at the fulcrum?

•• 24 A uniform 8 m beam of mass 10 kg, supported at each end, is loaded with masses of 2 kg, 4 kg, and 6 kg at points 2 m, 4 m and 6 m from the left end.
(a) How far from the left end is the center of gravity of the system?
(b) What are the vertical forces exerted by the two supports on the beam?

•• 25 In Problem 24, suppose the order of the three masses is changed, so that the 2 kg and 4 kg masses switch places. What are the forces on the supports now?

•• 26 A 3 m × 3 m square table having of a mass 10 kg supports four stone mugs (Fig. 4.27), placed at the table edges, midway between corners.

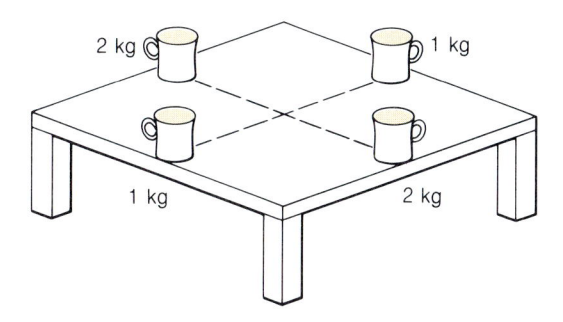

Figure 4.27

(a) Find the center of gravity of the system (Find it separately in x- and y-directions).
(b) What upward force F is exerted by each of the support legs?

•• 27 Suppose each of the four table legs in Fig. 4.27 can support a weight of 200 N. Find the mass of the largest dancer who can stand at the center of the table with the mugs still in place.

••• 28 A ladder of mass m and length L rests at angle 60° against a smooth wall (Fig. 4.28), supported by a friction force $F_1 = \mu_s F_2$ along the floor, and by perpendicular forces from the wall (F_4) and floor (F_2). Since the wall is frictionless, $F_3 = 0$. A painter of mass M wishes to stand a distance d up the ladder. For the ladder standing alone (set $M = 0$), find the forces F_1, F_2, and F_4 in terms of the ladder's weight mg.

••• 29 For the ladder alone ($M = 0$) in Fig. 4.28, what coefficient of friction, μ_s, at the floor is needed to prevent the ladder from slipping?

•• 30 Suppose a $M = 80$ kg painter in Fig. 4.28 wishes to stand at a distance d = 6 m up a 10 kg ladder.
(a) Calculate the forces F_1, F_2, and F_4.

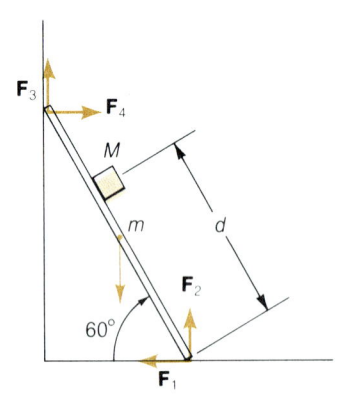

Figure 4.28

(b) What coefficient of friction, μ_s, is needed at the floor so that the ladder does not slip?

•• 31 What is the highest distance d up the 10 kg ladder the painter in Fig. 4.28 can climb without the ladder slipping, assuming the coefficient of friction at the floor $\mu_s = 0.4$?

••• 32 Suppose the 10 kg ladder in Fig. 4.28 rests against a wall with the same coefficient of friction as the floor (F_3 is nonzero now). Calculate the maximum height d the 80 kg painter can climb ($\mu_s = 0.4$). Why is d greater than in Problem 31?

•• 33 A 60 kg rock-and-roll star is tugged on by three fans. One fan pulls forward with 400 N. The other two pull backward from opposite sides with 200 N each, at 120° angles from the first. If the coefficient of friction along the ground is $\mu_s = 0.3$, how much extra mass in metal bracelets should the star wear to avoid being moved?

•• 34 At what distance x should a fulcrum be placed under the loaded beam (Fig. 4.29) for support and balance?

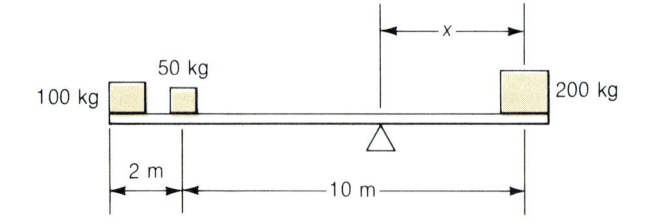

Figure 4.29

• 35 A 40 m boom of mass 4,000 kg lifts a 2,000 kg pallet (Fig. 4.30). What is the net torque exerted by the boom and pallet alone about the base B when the boom is set at angle $\theta = 60°$? Neglect the mass of the base.

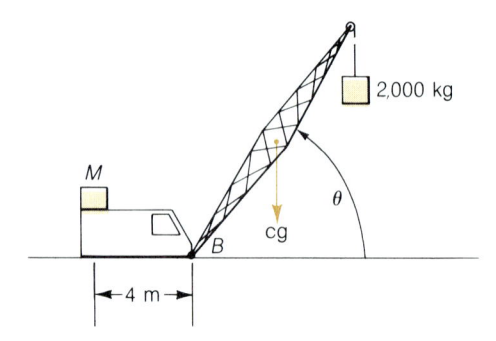

Figure 4.30

• 36 In Fig. 4.30 what mass M is needed as a counterbalance if $\theta = 60°$?

•• 37 In Fig. 4.30 at what angle θ would the torque about the base be greatest? What mass M would be needed as a counterbalance at that angle?

•• 38 A fisherman's whopper with 80 kg mass hangs from a wooden pole, supported by a separate guidewire attached to the top of the pole (Fig. 4.31).

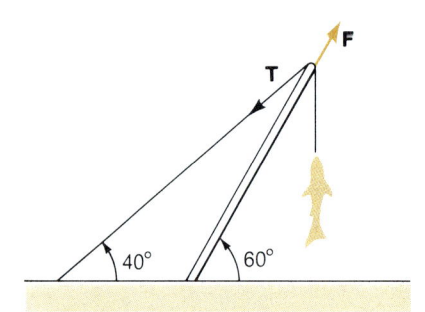

Figure 4.31

(a) Calculate the tension T in the guidewire (neglect the mass of pole and wire).
(b) What force F is exerted by the pole on the wire?

• 39 Find the center of gravity of a 2 m uniform 10 kg sled, loaded at front and back by two children of mass 40 kg and 20 kg.

•• 40 In a movie filmed on Mt. Rushmore, an 80 kg stuntman is prevented from sliding down the 70° slope by the handgrasp of an actor above, who pulls along the same 70° angle. What force must the actor exert before the stuntman can regain frictional support ($\mu_s = 0.3$)?

PROBLEMS

- **41** Calculate the torques about the center exerted on a bicycle wheel by (a) 100 N tangential force applied at the rim (70 cm radius); (b) 500 N tangential force exerted at the gear cluster (4 cm radius).

- **42** Force F is applied to a bicycle wheel at the inner gear cluster (Fig. 4.32). If the coefficient of friction at the ground is $\mu_s = 0.3$, and the weight on the wheel is W, what is the maximum force F that can be applied without the wheel slipping?

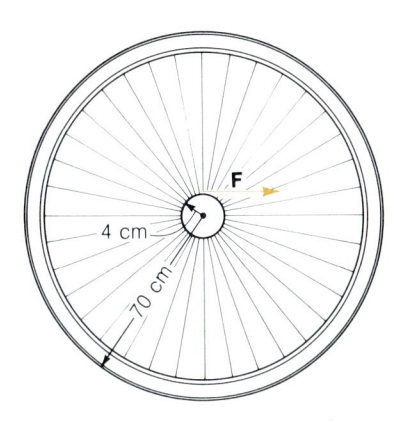

Figure 4.32

- **43** Two children of mass 30 kg and 50 kg sit on opposite sides of a see-saw. The smaller child is 2 m from the fulcrum. How far from the fulcrum must the larger child sit to keep the smaller child in the air?

- **44** A thief tries to push a 50 kg gum machine along the floor (friction coefficient $\mu_s = 0.2$) by applying a force F at height $h = 50$ cm. The base $b = 30$ cm as shown in Fig. 4.33.

  (a) What force F is required to start the machine moving?
  (b) Will the machine slide or tip over?

- **45** What is the maximum coefficient, μ_s, at the floor for which the gum machine in Fig. 4.33 will slide and not tip?

- **46** How much force, F_m, must the biceps muscle apply to a 37 cm forearm (Fig. 4.34) to support a 4 kg mass held at 90°? Assume that the forearm mass is 2.2 kg and that the muscle attaches to the bone 5 cm from the joint.

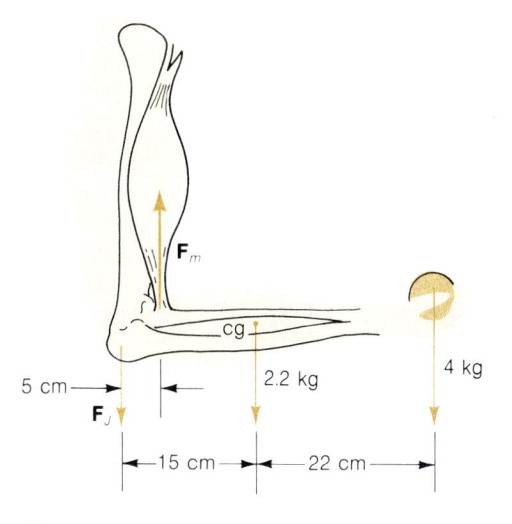

Figure 4.34

- **47** Suppose an athlete's biceps muscle attaches to the bone in Fig. 4.34 at 6 cm from the joint. What force would the athlete's muscle need to support 4 kg at 90°?

- **48** If the biceps muscle in Fig. 4.34 can exert a 1,000 N force and attaches to the bone 5 cm from the joint, what is the maximum mass it can support at 90°?

- **49** When a 70 kg ice skater pushes with his blade against the ice, the ice pushes back with a force F_1 (Fig. 4.35), allowing him to skate at constant velocity against a frictional force F_{fr}. If the coefficient of friction between ice and blade is 0.1, and if no one pushes the skater ($F_2 = 0$), what is the force F_1?

- **50** If the skater in Fig. 4.35 is pushed at height $h = 60$ cm with a force F_2, and $F_1 = 0$, what is the maximum force that can be exerted without tipping him over?

- **51** A pickup truck carrying a 1,000 kg box full of cream pies suddenly accelerates at 3 m/s². If the coefficient

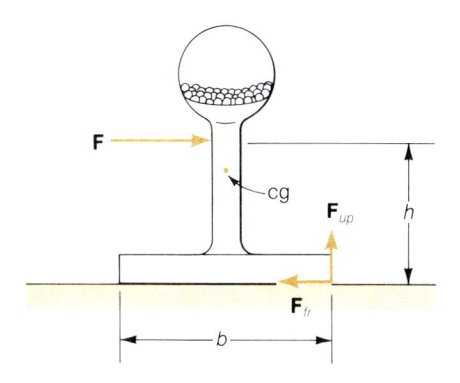

Figure 4.33

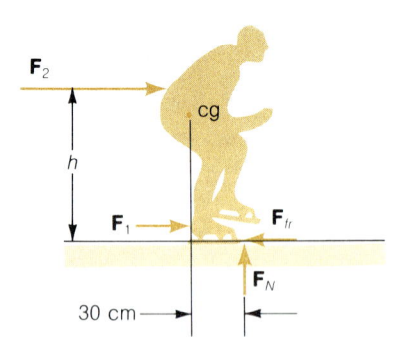

Figure 4.35

of friction between the box and the floor of the truck is $\mu_s = 0.35$, is the friction great enough to prevent the box from sliding off the truck?

•• 52 What is the minimum mass of the box in Problem 51 that would stay on the truck if $\mu_s = 0.5$? Be careful here.

•• 53 A patient recovering from a ski accident wishes to wave hello to the doctor by raising her 10 kg leg with 5 kg cast (Fig. 4.36) by means of a pulley held rigidly to the ceiling.

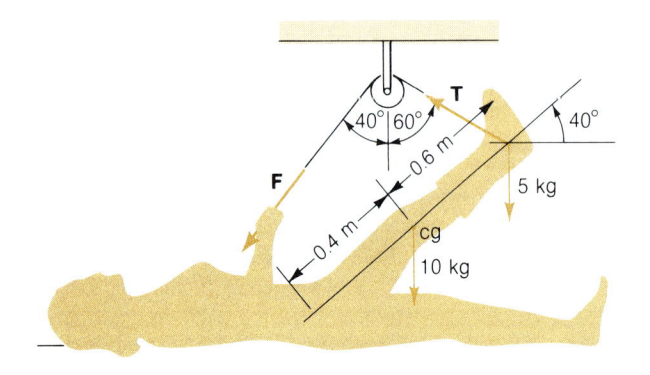

Figure 4.36

(a) What force F must she apply?
(b) What is the tension in the rope?

•• 54 In Problem 53, what is the relation between F and T? Explain the role of external forces on the pulley.

• 55 A 10 m uniform 1,000 kg steel beam overhangs a building top by 3 m (Fig. 4.37). What is the vertical force F exerted by the edge of the building on the beam?

•• 56 Three types of levers (Fig. 4.38) apply a force F to a mass M with a fulcrum. The lever arms are d_1 and

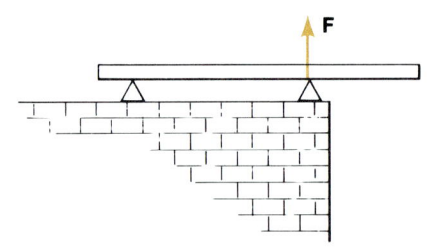

Figure 4.37

d_2. For each type, solve for F in terms of M, d_1, d_2, and g. Give an example of each type from everyday life.

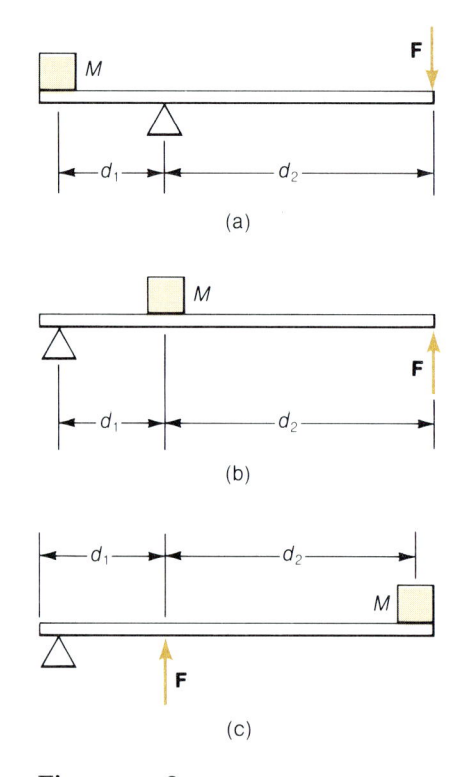

Figure 4.38

•• 57 For each type of lever in Fig. 4.38, compute the ideal mechanical advantage, defined as the ratio of the weight applied to force applied, Mg/F.

•• 58 For the three types of levers in Fig. 4.38, assume that $d_1 = 1$ m and $d_2 = 2$ m and assume the levers are initially horizontal.
(a) Which lever can lift the most weight for the same force?

PROBLEMS

(b) Which can lift a weight the farthest vertically, for the same small distance of travel by the end of the lever arm?

59 An exhibitionistic waiter wishes to impress his customers by pulling their tablecloth out from under their plates. The coefficient of friction between the 300 g plates and the 100 g tablecloth is $\mu_s = 0.3$. With what minimum force must the waiter pull horizontally on the tablecloth?

60 A tower of uniform composition is 60 m tall with an 8 m wide base. At what angle can the tower lean from vertical before it becomes unstable?

61 The rafter beams of a building (Fig. 4.39) have mass 20 kg each, length $l = 10$ m, and angle $\theta = 30°$. If each beam supports a uniform weight of 3,000 N, what is the compression force F at their junction?

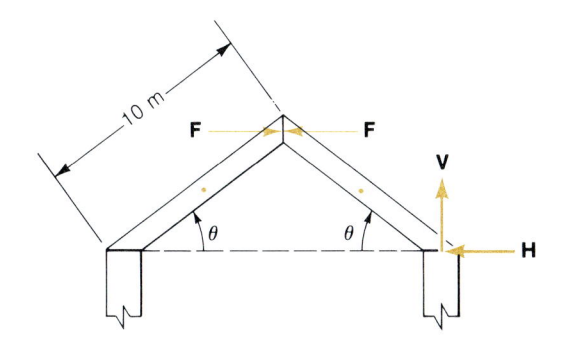

Figure 4.39

62 For the rafters in Fig. 4.39, what are the horizontal and vertical forces, H and V, exerted by the vertical supports on the beam?

63 A policeman, 2 m in height, leans against a frictionless wall while inclined at an angle θ from vertical. The floor has friction coefficient $\mu_s = 0.2$, and his center of gravity is 1.2 m from his feet. What is the steepest angle at which he can lean?

64 Compute the force that must be applied perpendicularly to the crank in Fig. 4.40 to produce a tension in the rope of 600 N.

65 A loose piece of rock rests on the slope of a volcano which makes an angle θ from the horizontal. If the coefficient of static friction is μ_s, show that the steepest possible angle, the angle of repose, is given by $\tan\theta = \mu_s$. If $\mu_s = 0.4$, what is θ?

66 When a hockey player takes a slapshot, he holds the top end of his stick fixed and applies a horizontal force $F_2 = 200$ N to the middle of the stick (Fig.

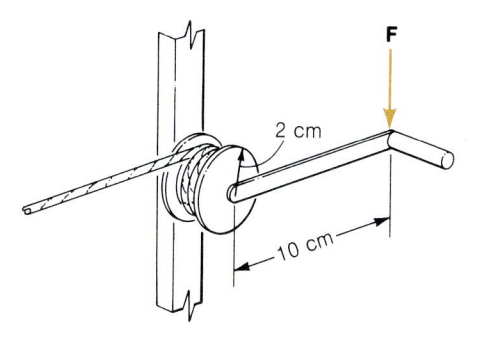

Figure 4.40

4.41). Unfortunately, some prankster has anchored the puck to the ice, so that it cannot move. What is the horizontal force delivered to the puck at impact?

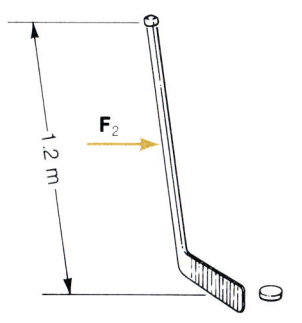

Figure 4.41

67 Two firemen carry a 70 kg unconscious smoke victim on a 2 m long stretcher. The victim is 2 m tall and his center of gravity is 0.8 m from his head. What upward forces must the two firemen exert on the ends of the stretcher?

68 A 100 kg block rests on a table with friction coefficient $\mu_s = 0.4$. Two ropes are attached to pulleys and masses M_1 and M_2 (Fig. 4.42). If $M_1 = 50$ kg, what is the range of masses M_2 for which the system remains at rest?

69 For the system in Fig. 4.42, when M_2 has been increased 1 kg more than the critical amount needed to get the block moving, what is the block's acceleration? Assume $\mu_k = 0.3$.

70 What force must your friends exert on your car with you inside to push it out of a ditch, inclined at a 15° angle from horizontal? The car has 1,500 kg mass, your mass is 70 kg, and the effective coefficient of friction is 0.05.

108   CHAPTER 4 FORCES IN BALANCE

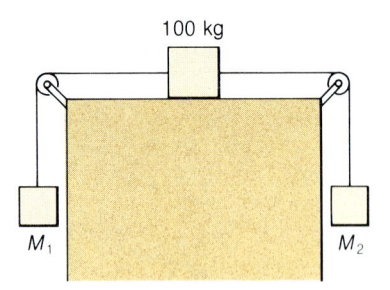

Figure 4.42

71 A boy chases an iceblock which has fallen out of a truck and is sliding ($\mu_k = 0.1$) down a street inclined at 10° from horizontal. When he reaches top speed, he sees that both he and the iceblock have the same velocity. Will he be able to catch the iceblock?

72 Two skiers race down a hill inclined at angle $\theta = 15°$. The coefficients of kinetic friction between their skis and the snow are $\mu = 0.09$ and 0.10, respectively.
  (a) Show that the net acceleration of a skier is $g(\sin\theta - \mu\cos\theta)$.
  (b) At the end of the run, a distance of 400 m, how much farther ahead is the first skier? Neglect wind resistance.

SPECIAL PROBLEMS

1. Whiplash Injuries

In a whiplash accident a passenger's head is wrenched backward by the sudden acceleration of the car. Suppose a car of mass 1,500 kg (including all passengers) is struck from behind with a force F = 44,100 N (Fig. 4.43). A passenger's mass is 70 kg, and his head's center of mass (4 kg) is located a height h = 17 cm above the point of support by the seat.
  (a) Calculate the acceleration of the car upon impact. Give your answer in m/s² and in g's.
  (b) Calculate the torque exerted by force F_1, about the center of mass of the head if no headrest is present.
  (c) How much less torque would be exerted if the total car mass were 2,500 kg?
  (d) How much less torque would be exerted if the passenger had a shorter neck, h = 14 cm?
  (e) For the force and dimensions of parts (a) and (b), calculate what force, F_2, a headrest would need to exert to prevent whiplash.

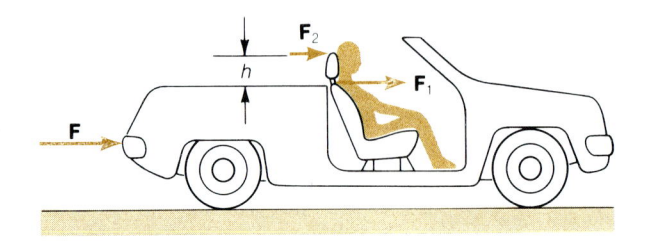

Figure 4.43

2. Ski Lift

A ski lift carries a gondola of mass m up a 40° slope, using cables supported by towers of height h = 10 m (Fig. 4.44). The tensions in the cables are T_1 and T_2, and the distances and angles are as shown. Suppose during an accident the gondola is just scraping the ground.

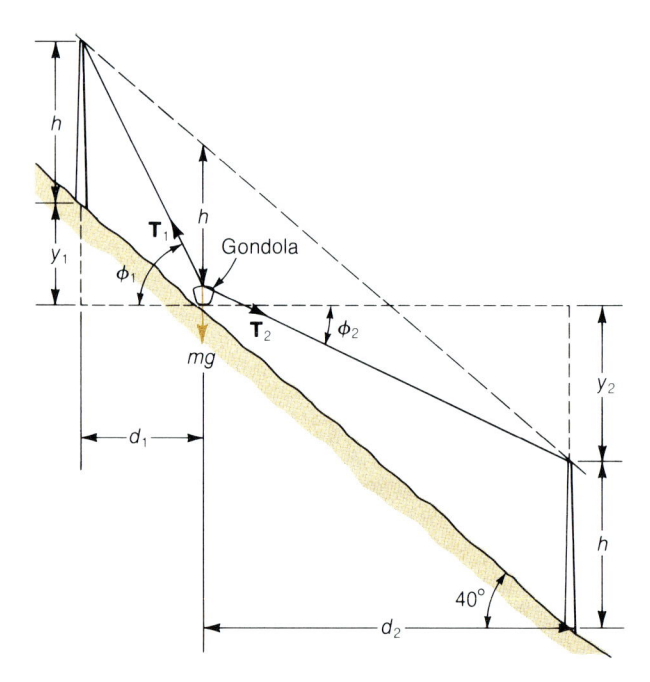

Figure 4.44

SPECIAL PROBLEMS

(a) Using trigonometry, show that the angles ϕ_1 and ϕ_2 are given by the formulas

$$\tan\phi_1 = [d_1 \tan(40°) + h]/d_1$$
$$\tan\phi_2 = [d_2 \tan(40°) - h]/d_2.$$

First solve for the distances y_1 and y_2.

(b) Solve for ϕ_1 and ϕ_2, assuming that $d_1 = 40$ m, $d_2 = 60$ m, and $h = 10$ m.

(c) If the forces on the gondola are in balance, show that

$$T_1 = mg \cos\phi_2/\sin(\phi_1 - \phi_2)$$
$$T_2 = mg \cos\phi_1/\sin(\phi_1 - \phi_2).$$

See Problem 10 and use the relation $\sin(\phi_1 - \phi_2) = \sin\phi_1 \cos\phi_2 - \cos\phi_1 \sin\phi_2$.

(d) If the gondola alone has a mass of 1,000 kg, how many 70 kg passengers can the cable carry in the position shown? Assume that the cable has a tensile strength of 50,000 N.

3. Mountain Roads

In the mountains, residents often refer to rental car roads, jeep roads, and winch roads, each progressively steeper. This problem illustrates how friction and engine force limit cars in hill climbing.

(a) Suppose a 2,000 kg rental car can accelerate from rest to 20 km/h in 5 s on level ground in low gear. What is the steepest hill it can climb at constant speed in low gear (angle θ)?

(b) If the coefficient of friction between the tires of a rental car and the road is $\mu_s = 0.3$, will its tires be able to apply maximum force to the road without slipping?

(c) Suppose a 1,500 kg jeep, in low gear, can exert six times more force on the road through its tires than the rental car. What is the steepest angle the jeep can climb?

(d) On extra steep roads, a jeep uses a winch and cable, which when wrapped around a tree (Fig. 4.45) can exert a force F on the jeep. If $F = 3,700$ N, what is the steepest hill the jeep can climb with a winch if the jeep's engine applies maximum force to the road as in part (c)?

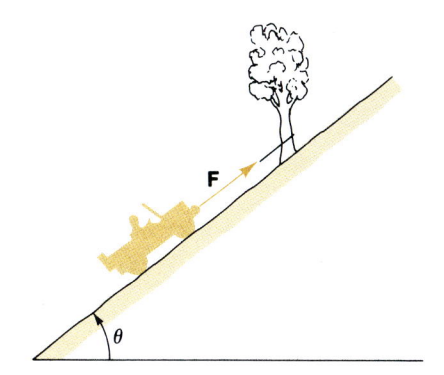

Figure 4.45

4. Pulleys

Systems of pulleys can be used to lift heavy loads with less applied force. The block and tackle in Fig. 4.46a, allows one to lift a mass M with force F which is less than Mg. The mechanical advantage is defined as the ratio (Mg/F).

(a) If the left end of the rope is pulled down with force F, what is the tension T in the rope and the mechanical advantage, Mg/F, in static equilibrium?

(b) If the left end of the rope is pulled a distance L_1, what distance L_2 does the mass rise?

(c) For the system in Fig. 4.46b, compute the tension T in the rope and the mechanical advantage, (Mg/F). Note that the weight is carried by four ropes, indicated by tickmarks.

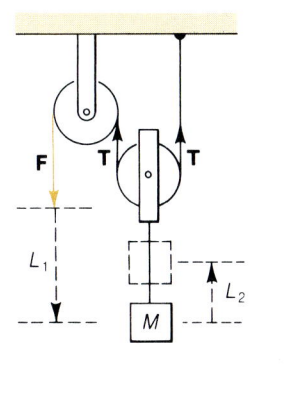

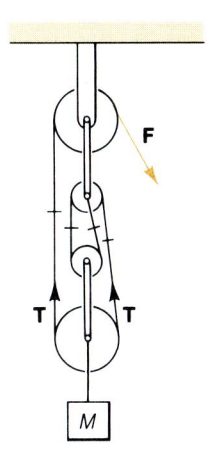

Figure 4.46

5. Foot Forces

When running, jumping, or standing on tiptoe (Fig. 4.47), one's foot is subject to an upward force W from the floor, a frictional force F along the floor, gravity acting downward through a compressional force J on the joint, and a tension force T from the Achilles tendon.

(a) If the foot is stationary, W is the weight supported by the foot. By balancing horizontal and vertical forces and torques about the base, find T and J in terms of W and θ. Assume that $d_2 = 1.3\, d_1$.

(b) What is the frictional force F in terms of W and θ?

(c) At what steepest angle θ can a runner lean without slipping? Assume a coefficient of friction, μ_s. Show that for any realistic road surface ($\mu_s < 1$) the angle θ must be less than 45°.

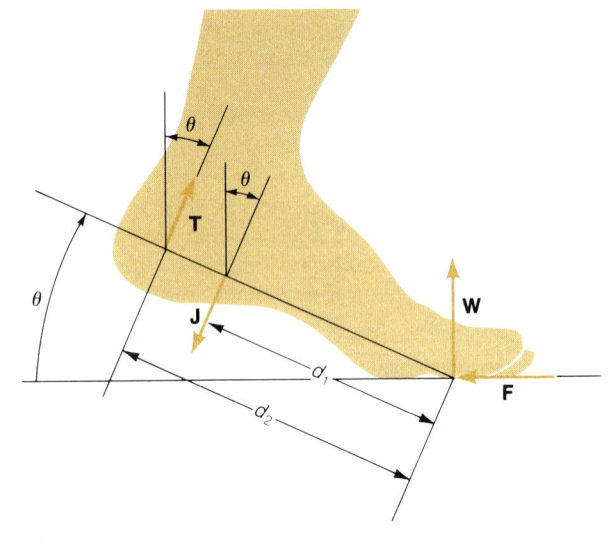

Figure 4.47

SUGGESTED READINGS

"In judo and aikido, applications of the physics of forces makes the weak equal to the strong." J. Walker, *Scientific American,* Jul 80, p 150, **243**(1)

"The physics of spinning tops, including some far-out ones." J. Walker, *Scientific American,* Mar 81, p 182, **244**(3)

"A field formula for calculating the speed and flight efficiency of a soaring bird." J. Walker, *Scientific American,* Mar 85, p 122, **252**(3)

"Structural analysis of Gothic cathedrals." R. Mark, *Scientific American,* Nov 72, p 90, **227**(5)

"The physics of karate." M. S. Feld, R. E. McNair, S. R. Wilk, *Scientific American,* Apr 79, p 110, **240**(4)

"Gothic structural experimentation." R. Mark, W. W. Clark, *Scientific American,* Nov 84, p 176, **251**(5)

"The physics of dance." K. Laws, *Physics Today,* Feb 85, p 24, **38**(2)

The flying circus of physics with answers. Jearl Walker, Wiley, New York, 1977

Newton at the bat: the science in sports. Eric W. Schrier, William F. Allman, eds, Scribners, New York, 1984

Sport science. Peter J. Brancazio, Simon and Schuster, New York, 1984

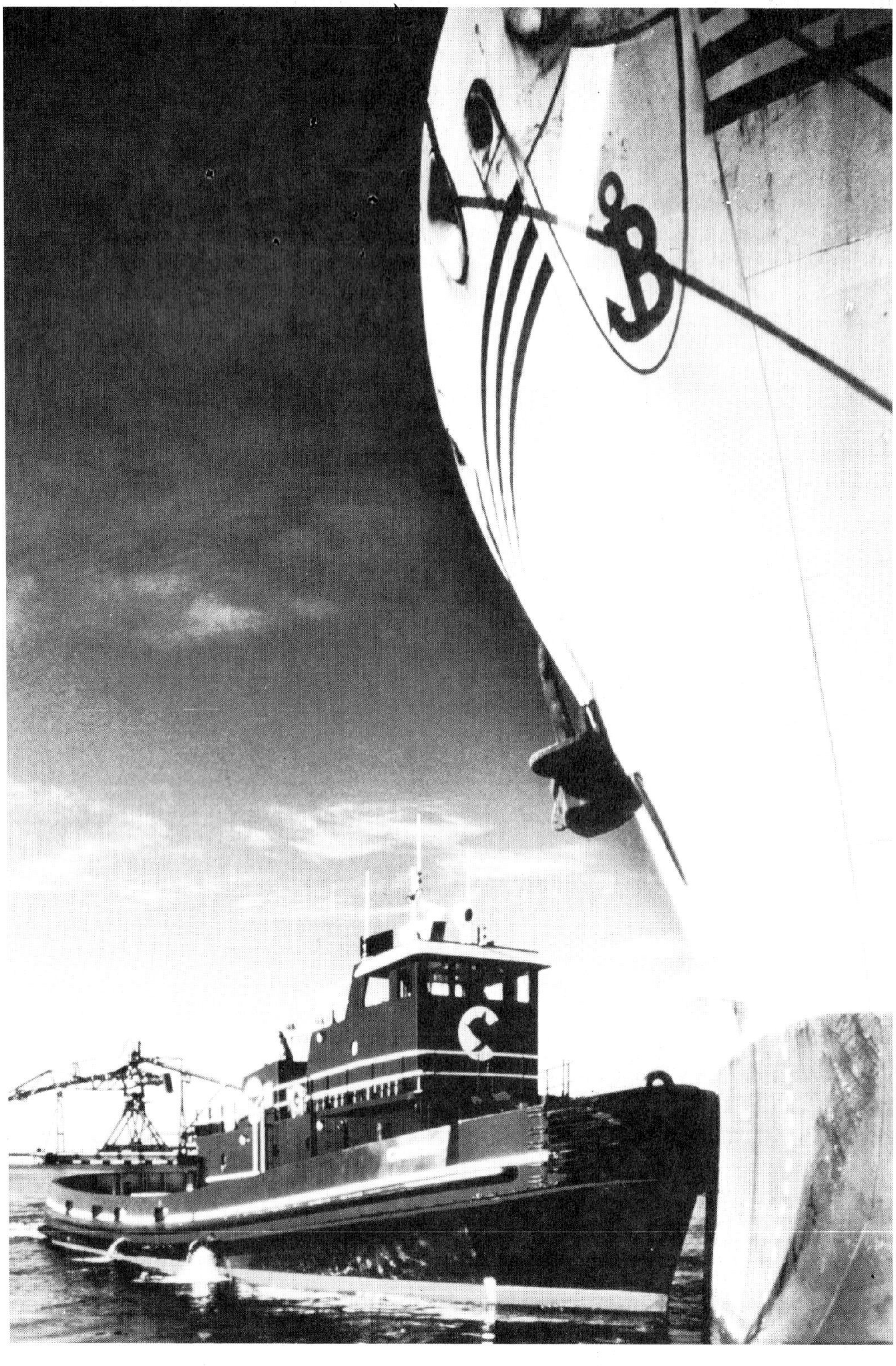

Chapter 5

Momentum, Work, and Energy

In this chapter we discuss in specific terms several concepts that are intuitively familiar to most people. We speak often of momentum, we occasionally act on impulse, we all are familiar with the notion of work, we may be concerned with various kinds of power, many of us worry about the energy crisis, and we deal with machines every day. Despite this widespread familiarity, however, few of us attempt to specifically define and use the physical principles these commonplace words describe. In this chapter, we will see that the precise definitions of these words in physics can be quite distinct from their everyday meanings.

We will also come to appreciate one of the most elegant principles in the world of science: the conservation law. We will see the problem-solving power that derives from the simple statement that momentum or energy cannot be lost during a physical interaction.

5.1 Momentum

Many of us have the idea that an object in motion carries with it a certain quality, called momentum, that depends on both its speed and its mass. In sports and politics the concept of momentum is often transferred into the realm of psychology or metaphysics, when we say that a player, a team, or a candidate that is doing well has momentum.

The Nature of Momentum

In physics the concept of **momentum** is quantified in terms of velocity and mass: the momentum of a body is the product of its velocity and its mass. In equation form, the momentum **p** is given by

$$\mathbf{p} = m\mathbf{v} \tag{5-1}$$

which shows that momentum is a vector quantity whose direction is parallel to the velocity vector. Rather than define a special unit for momentum, we just use the combination of the units for mass and velocity, and express momentum in terms of kilogram-meters per second (kg · m/s).

This technical definition of momentum is consistent with the popular notion: the heavier and faster a body is, the greater its momentum. A large truck travelling at a high speed has more momentum than a small car travelling at a lower speed. A small speedboat may have more momentum than a much larger vessel that is moving very slowly.

Usually the mass of an object is constant, so the only way to change its momentum is to alter its velocity. From Newton's first law we see that this requires a force. We can relate the change of momentum to the force that causes the change by Newton's second law,

$$\mathbf{F} = m\mathbf{a} = m\frac{\Delta \mathbf{v}}{\Delta t}$$

but since m is constant,

$$\Delta \mathbf{p} = m\Delta \mathbf{v}$$

so we conclude that

$$\mathbf{F} = \frac{\Delta \mathbf{p}}{\Delta t}. \qquad (5\text{--}2)$$

The rate of change of momentum is equal to the applied force. Actually, this was the way Newton himself expressed his second law. The more familiar form was then derived from Eq. 5–2.

In deriving Eq. 5–2 we implicitly assumed that we were dealing with a constant force, so that

$$\mathbf{a} = \frac{\Delta \mathbf{v}}{\Delta t}$$

is a valid expression. The relationship between force and rate of change of momentum is, however, valid even for a variable force. This can be intuitively understood if we recall the concept of a limit, as discussed in Chapter 2. As the time interval over which a constant force acts grows smaller, Eq. 5–2 still holds; if we allow the time interval to approach zero, and we consider infinitesimally short time intervals, we see that the equation is still valid, but now it relates the instantaneous force to the instantaneous rate of change of momentum. We remind ourselves what is meant by the instantaneous force. Fig. 5.1 shows a plot of the rate of change of momentum versus time, where we see that the instantaneous force at any time is the tangent to the momentum curve.

The mathematical relationship between change of momentum and applied force tells us something that we already knew from experience: the ease of stopping an object depends not only on its mass but also on its velocity. A

5.1 MOMENTUM

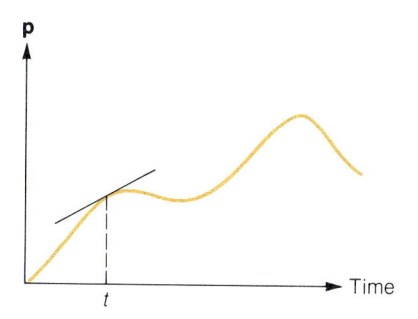

Figure 5.1

Instantaneous force. The force at time t is equal to the instantaneous rate of change of momentum at time t; that is, the slope of the tangent at t.

football player running at high speed is more difficult to tackle than one of equal mass who is walking. Much more force is required to stop a speeding car within a certain time than to stop a slowly-moving car, and more force is needed to stop a large car than a small one moving at the same speed. A large ship moving very slowly is more likely to damage a pier than a small boat, even if the small boat is moving faster, if mv is smaller for the small boat.

Example 5.1 A ship whose mass is 20,000 tons rams into a pier at a speed of 2 mi/h, taking 5 s to stop. What is the force exerted on the ship by the pier?

First we convert the known quantities to SI units: the ship's mass is 1.8×10^7 kg, and its initial speed is 0.89 m/s. Then we use Eq. 5–2, taking the ship's initial direction of motion as positive, so that the change in velocity is -0.89 m/s.

$$\mathbf{F} = \frac{\Delta \mathbf{p}}{\Delta t} = \frac{m\Delta \mathbf{v}}{\Delta t} = (1.8 \times 10^7 \text{ kg}) \frac{(-0.89 \text{ m/s})}{(5 \text{ s})}$$
$$= -3.2 \times 10^6 \text{ N} = -7.2 \times 10^5 \text{ lb} = -360 \text{ tons}$$

Impulse

The total change in momentum that is caused by a force depends on how long the force is applied. This leads us to define a new quantity: the **impulse** is the product of the force and the time interval over which it acts. We can rearrange Eq. 5–2 to get

$$\mathbf{F}\Delta t = \Delta \mathbf{p} \qquad (5\text{–}3)$$

where the quantity $\mathbf{F}\Delta t$ is the impulse. Its units in the SI system are newton-seconds (N · s).

From Eq. 5-3 we see that the same change in momentum can be caused by a strong force acting over a short time, or by a weaker force acting over a longer time. A car will be brought to rest quickly if a strong braking force is applied, and slowly if the brakes are applied gently, but the change in momentum will be the same.

Often we must find the force needed to accomplish a known change of momentum; we now see that this depends on how quickly the change of momentum is brought about. In an example in Chapter 3, we noted that a tow rope might break if it were used to accelerate a disabled car too quickly. Now we can use Eq. 5-3 to see how quickly we can tow the car from rest to a given speed without exceeding a certain force (such as the known tensile strength of the rope).

Example 5.2 Turkeys, not very intelligent creatures, lay their eggs in a standing position, so that the eggs fall about a foot. A turkey farmer wants to minimize the number of broken eggs. He realizes that he needs to reduce the force on a falling egg as it hits the ground to less than a critical value, the value at which it breaks. He tries padding the floor of the turkey coop, so that a 0.2 lb falling egg changes its momentum slowly enough that the force does not exceed the critical value, 3 lb. He finds a material that slows a falling egg to rest in 0.02 s. Is this sufficient to prevent breakage?

To calculate the change in momentum we first find the velocity at which an egg is falling when it reaches the ground. From Eq. 2-13, we see that the time of fall for an object starting at rest and travelling 1 ft is

$$t = \sqrt{\frac{2d}{g}} = \sqrt{\frac{2(-1 \text{ ft})}{-32 \text{ ft/s}^2}} = 0.25 \text{ s}$$

taking the upward direction positive. The velocity of the egg when it reaches the ground is

$$v = gt = (-32 \text{ ft/s}^2)(.25 \text{ s}) = -8 \text{ ft/s} = -2.44 \text{ m/s}.$$

If a turkey egg weighs 0.2 lb, its mass is 0.09 kg, and its momentum when it reaches the ground is -0.2 kg · m/s. Therefore the change of momentum, equal to the **impulse**, is 0.2 kg · m/s as the egg comes to rest.

Since the force must not exceed 3 lb (about 13 N), then from Eq. 5-3 we see that the time required to stop the egg must be greater than

$$\Delta t = \frac{\Delta p}{F} = \frac{(0.2 \text{ kg} \cdot \text{m/s})}{(13 \text{ N})} = 0.015 \text{ s}.$$

The material the farmer has found will save the eggs from breaking.

5.1 MOMENTUM

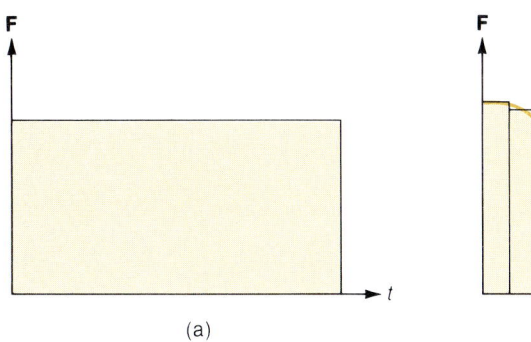

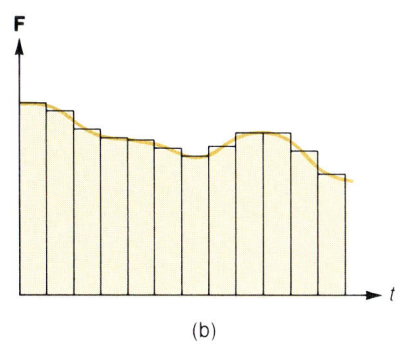

Figure 5.2

Graphical calculation of impulse. The area under a graph of force versus time (t) is equal to the product Ft. This technique is useful when F varies with time.

It is possible to calculate the impulse when the force varies with time. This is useful if we want to know the total change in momentum of a body subject to a variable force. Fig. 5.2a is a plot of force versus time for a constant force. The impulse is simply the product of force and time, which is equal to the area of the shaded rectangle. Fig. 5.2b is a plot of force versus time for a variable force. The time axis is divided into a number of small intervals. The impulse during an interval is approximated by the area of the rectangle. Just as in Chapter 2 we found the total displacement from a velocity plot, we now find the total impulse, which is equal to the net change in momentum, by adding the areas under the curve. For two- or three-dimensional motion, we must perform the summation for each component of force separately, then calculate the vector sum of the components of the impulse in order to find the total impulse and thus the net change of momentum.

Example 5.3 A three-stage rocket is launched from a space station. It accelerates a payload and then releases it. Tests of the types of engines used show that they exert a net force over time as shown in Fig. 5.3. Find the final velocity of the payload, and thus its final momentum. The mass of the payload is 2,000 kg.

The motion is in a straight line, and we can represent the total impulse by a single plot. Summation of the small rectangles in the figure yields a total impulse of 4.9×10^6 kg · m/s. Hence the final momentum of the payload is equal to this value, and the change in velocity is

$$v = \frac{(4.9 \times 10^6 \text{ kg} \cdot \text{m/s})}{(2,000 \text{ kg})} = 2.45 \times 10^3 \text{ m/s} = 2.45 \text{ km/s}.$$

This is the final velocity of the payload with respect to the space station.

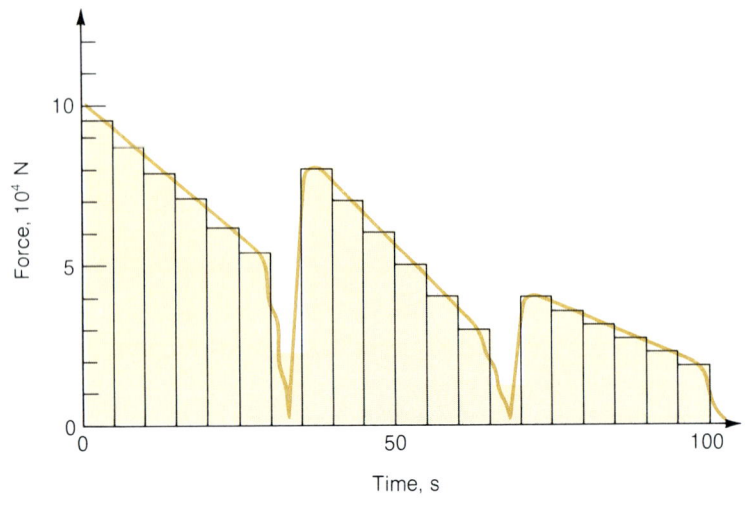

Figure 5.3

Conservation of Momentum

It was observed long ago that momentum can be transferred from one body to another. Pool players often observe that when the cue ball strikes another, the cue ball may come to rest while the other ball begins moving (Fig. 5.4). Careful observation shows that the other ball, if equal in mass to the cue ball and hit squarely by it, will have the same velocity as the cue ball had. Since the masses are equal, the momentum of the struck ball is equal to the momentum of the cue ball before the collision. If the collision is a glancing one, so that both balls are in motion afterwards, it is found that their total momentum after the collision is equal to the momentum the cue ball had before the collision. The total momentum of the two balls remains constant.

The simple conclusion of the experiment with billiard balls has been expanded to a much more general statement:

The total momentum of an isolated system is constant.

That is, in a system of bodies with no external forces acting on the system, the total momentum of the bodies is not changed by any forces the bodies exert on each other. We say that momentum is conserved.

We can prove this easily for a pair of bodies. We know from Newton's third law that the two bodies must exert equal and opposite forces on each other; that is,

$$\mathbf{F}_1 = -\mathbf{F}_2$$

5.1 MOMENTUM

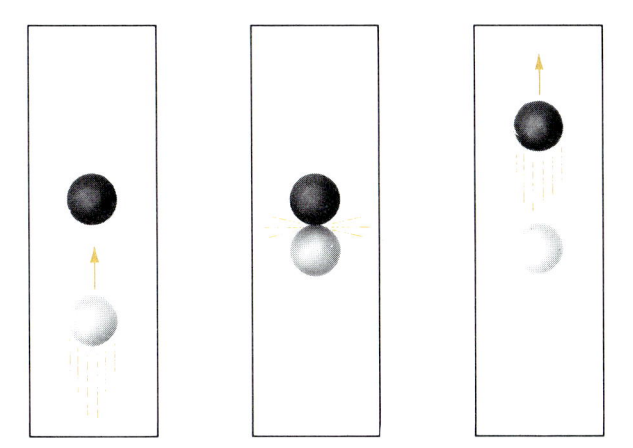

Figure 5.4

Transfer of momentum. If the cue ball (white) strikes a resting ball squarely, the second ball gains the momentum lost by the cue ball.

where $\mathbf{F}_1$ is the force exerted on body 1 by body 2, and $\mathbf{F}_2$ is the force exerted on body 2 by body 1. If these are the only forces acting on the bodies, then from Newton's second law, we write

$$m_1 \mathbf{a}_1 = -m_2 \mathbf{a}_2$$

or

$$m_1 \frac{\Delta \mathbf{v}_1}{\Delta t} = -m_2 \frac{\Delta \mathbf{v}_2}{\Delta t}$$

where Δt is the time interval during which the two bodies exert forces on each other. Since Δt is the same for both bodies, we can write

$$m_1 \Delta \mathbf{v}_1 = -m_2 \Delta \mathbf{v}_2$$

or

$$\Delta \mathbf{p}_1 + \Delta \mathbf{p}_2 = \Delta(\mathbf{p}_1 + \mathbf{p}_2) = 0.$$

The total change of momentum is zero; that is, momentum is conserved. We could have done this for any number of bodies, reaching the general conclusion that the sum of all their individual changes of momentum is zero.

The statement that momentum is conserved provides us with a powerful tool for solving problems in which initial velocities and masses are known and we want to find the final velocities. In an earlier chapter we cited the example of a baseball pitcher standing on a frictionless surface (so that there would be no external force) and throwing a fastball at 90 mi/h. We calculated the force, hence the acceleration, acting on the pitcher, and used the time duration of the acceleration to find his velocity after the throw. Now we see that we could have found his final velocity quite directly by noting that the total momentum of the ball plus the pitcher before the throw is zero, so it must be zero after the throw. Therefore the momentum of the pitcher is equal in magnitude (but opposite in direction) to the momentum of the thrown ball, which means that the ratio of the pitcher's velocity to that of the ball is simply the inverse ratio of the masses of the pitcher and the ball.

Example 5.4 Suppose a railroad car of mass 15,000 kg is at rest on a straight track, and another car of mass 12,000 kg rams it at a velocity of 15 mi/h (6.7 m/s). The two cars link together. What is the velocity of the combined mass after the collision?

We need not use vector notation, because all the motions are in the same direction. To assume that the momentum of the two cars is conserved, we must ignore any external forces, such as friction, exerted on the cars in the direction of their motion. Conservation of momentum dictates that

$$m_1 v_1 + m_2 v_2 = (m_1 + m_2) v_f$$

where the two masses are m_1 and m_2, their initial velocities are v_1 and v_2, and the final velocity is v_f. If m_1 is the mass of the 15,000 kg car, then $v_1 = 0$, and solving for v_f yields

$$\begin{aligned} v_f &= \frac{m_2 v_2}{m_1 + m_2} \\ &= \frac{(12{,}000 \text{ kg})(6.7 \text{ m/s})}{(15{,}000 + 12{,}000) \text{ kg}} \\ &= 3.0 \text{ m/s} = 6.7 \text{ mi/h}. \end{aligned}$$

Example 5.5 A battleship fires a shell directly ahead. The ship was moving at 15 mi/h, and its mass is 2×10^7 kg. The mass of the shell is 100 kg, and its velocity is 150 m/s. By what fraction is the ship's velocity changed by the kick of the gun that fired the shell?

We again assume that the momentum of ship and shell is conserved, and ignore external forces such as water or air resistance. Since the shell is fired along the same direction as the ship's course, we can drop the vector notation. We set the sum of the final momenta of ship and shell equal to the initial momentum of the ship and shell travelling together.

$$(m_1 + m_2) v_1 = m_1 v_1' + m_2 v_2'$$

The masses of the ship and the shell are m_1 and m_2, v_1' and v_2' are the final velocities of the ship and of the shell, and v_1 is the initial velocity of the ship. Since the mass of the shell is negligible compared to the mass of the ship, we may say

5.2 WORK

$$m_1 + m_2 \approx m_1.$$

We know that $m_1 = 2 \times 10^7$ kg, $v_1 = 15$ mi/h $= 6.7$ m/s, and that $v_2' = 150$ m/s. Solving for the ship's final velocity yields

$$v_1' = v_1 - v_2' \frac{m_2}{m_1}$$

$$= 6.7 \text{ m/s} - 150 \text{ m/s} \frac{100 \text{ kg}}{2 \times 10^7 \text{ kg}} = 6.7 - 7.5 \times 10^{-4} \text{ m/s}.$$

The ship's final velocity is reduced by only one part in 10^4, a factor equal to the inverse ratio of the masses. Notice that the ratio of final velocities equals the inverse ratio of masses only if the initial velocity is zero. If the ship had been at rest when it fired its shell, then v_1 would have been zero, and the result would have been

$$\frac{-v_1'}{v_2'} = \frac{m_2}{m_1}$$

as it was for the pitcher throwing the fastball.

5.2 Work

In physics the definition of work is quite restricted compared with the colloquial meaning. We often call any physical or mental effort work, but in physics work is done only when a force is applied to a body and moves it. **Work is the product of an applied force and the distance through which it is applied.** More specifically, work is the product of the component of the force in the direction of the displacement and the displacement. If **F** and **d** represent a force and the displacement caused by the force, then the work is

$$W = Fd \cos\theta \tag{5-4}$$

where F and d are the magnitudes of the vectors **F** and **d**, and θ is the angle from **F** to **d** (so that $F \cos\theta$ is the component of **F** that is parallel to **d***).

Work is a scalar quantity, having a magnitude but no direction (see Eq. 5–4). Its dimensions are force times displacement, and its units are the newton-meter (N · m) in the SI system, or **joule** (J), and the dyne-centimeter

---

*The product $Fd \cos\theta$ is called the scalar product of the vectors **F** and **d** because the result is a scalar. It is also called the dot product. In general the dot product of **A** and **B** is written **A** · **B** $= AB \cos\theta$, where θ is the angle between **A** and **B**.

CHAPTER 5 MOMENTUM, WORK, AND ENERGY

(dyn · cm) or **erg** in the cgs system. One joule is equal to 10^7 ergs. In the English system, the unit of work is the foot-pound (ft · lb).

Example 5.6 A tractor pushes a bale of hay 50 m across a field. The bale weighs 500 lb, and the coefficient of friction is 0.8. How much work does the tractor do on the bale of hay?

The force applied by the tractor is equal to the force of sliding friction, since work is done only when the bale is in motion, and we assume there was no acceleration after the bale was moving. The direction of displacement is parallel to the direction of the force, so the $\cos\theta$ factor equals one. Hence the work performed is

$$W = Fd = \mu_k mgd$$

where μ_k is the coefficient of kinetic friction, mg is the weight of the bale (the normal force), and d is the displacement. Substituting the values gives

$$W = (0.8)(227 \text{ kg})(9.8 \text{ m/s}^2)(50 \text{ m}) = 8.9 \times 10^4 \text{ J}.$$

It is important to realize that the work, like the force that creates it, is done *by* something *on* something. In the example of the tractor pushing the bale of hay, the work done on the bale by the tractor was calculated. Work was also performed on the bale by the ground, however, because friction was acting on the bale during the time it was moving. Since friction force always acts against the motion, the angle between the force and the motion was 180°. Because $\cos(180°) = -1$, the work performed on the bale by the ground was negative. Its magnitude was equal to that of the work done by the tractor on the bale, so the net work done on the bale was zero. This is because the net force was zero.

When calculating the work done, it is important to distinguish which force acts on which body. The net work is often zero, but substantial work may be performed by some forces on a body. A common example is work done by vertical forces or forces with vertical components because gravity affects the net vertical force. If an object is lifted steadily there is zero net force, but work is done on the object by the applied force and an equal but negative amount of work is done on the object by gravity.

Example 5.7 A person pushes a 200 lb sled at a constant speed and direction up a steep snow-covered hill, with an angle of inclination of 20° (Fig. 5.5). How much work has the person done when he reaches the top of the hill, which is 500 ft above his starting point? Assume the coefficient of kinetic friction is $\mu_k = 0.05$.

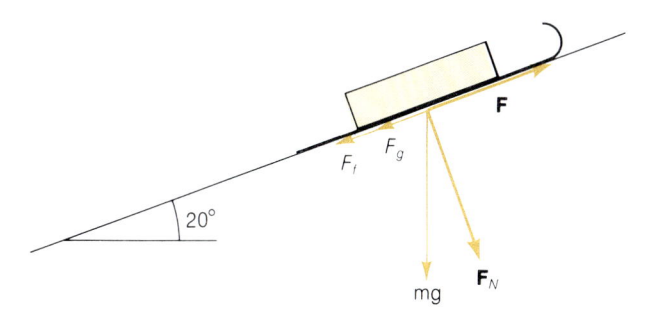

Figure 5.5

Even though the motion is in one dimension, this is a two-dimensional problem because the force applied by the person to the sled must be derived from known forces acting on the sled in two dimensions. The forces that the person must counteract are gravity and friction. The gravitational force is 200 lb in the downward direction, so its component in the direction of motion is

$$F_g = -(200 \text{ lb})\sin(20°) = -68.4 \text{ lb}.$$

The minus sign indicates that the force is opposite to the direction of motion. The friction force, which acts parallel to the hillside and opposite to the motion, is

$$F_f = \mu_k F_N = -(0.05)(200 \text{ lb})\cos(20°) = -9.4 \text{ lb}$$

where μ_k is the coefficient of kinetic friction and F_N is the component of the weight of the sled that is perpendicular to the hillside. The total force $F_g + F_f$ exerted by the sled against the person parallel to the motion is -77.8 lb, so the person must exert an equal force to push the sled uphill. The distance along the hillside is 500 ft/sin(20°), so the work done by the person on the sled is

$$W = 77.8 \text{ lb} \frac{500 \text{ ft}}{\sin(20°)} = 1.1 \times 10^5 \text{ ft} \cdot \text{lb}.$$

By setting the net force equal to zero, we found the force exerted on the sled by the person. The net work done on the sled was therefore zero.

When the applied force varies while it is causing a displacement, a graphical method can be used to find the total work (Fig. 5.6). The area under a plot of force versus distance is measured. Note that the component of the force that is parallel to the displacement, given by $F\cos\theta$, is the y-coordinate in the graph. As usual, we can divide the area under the curve into a number of small rectangles and add their areas.

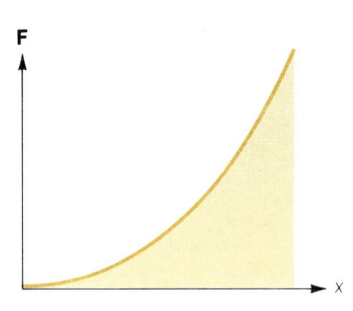

Figure 5.6

Work done by a variable force. The work done is equal to the area under the curve.

5.3 Energy

Energy is one of the most important concepts we will encounter. As with the definition of force, we resort to saying what energy does, rather than what it is. **Energy** is the ability to do work. Now that we have a clear and precise definition of work, energy has a more definite meaning for us.

Energy can exist in many forms, for example, we speak frequently of electrical energy, which operates most modern machinery, we use chemical energy which is released by chemical reactions in a car engine, and we think of heat as a form of energy. All forms of energy come under two headings; kinetic energy and potential energy.

Kinetic Energy

A moving object carries energy. We know this because it can perform work on another body that it hits or pushes. We might get some intuitive feeling for this by considering, for example, a bowling ball striking a group of pins. If the ball is moving slowly when it hits the pins, they tip over but do not scatter. If, on the other hand, the ball is rolling very rapidly when it strikes the pins, the pins fly about. Clearly, the ball has imparted more energy to the pins when the ball was moving rapidly.

The term for energy of motion is **kinetic energy,** after the Greek word for motion. Kinetic energy is closely related to momentum, since the momentum of a moving object is a measure of the force it will apply if it hits something, and this force is related to the amount of work performed.

We can quantify these relations. Suppose a football player rams into a blocking dummy and moves it a short distance. The dummy's rate of change of momentum is equal to the force applied to it by the player, and the work done on the dummy is this force times the distance it is moved. If we adopt a one-dimensional coordinate system with positive as the direction in which the player was moving, then the work done on the dummy is

5.3 ENERGY

$$W = Fd.$$

From Newton's second law, we can substitute for F.

$$W = mad$$

Here m is the mass and a is the acceleration of the dummy.

From Chapter 2, we know that the displacement d is

$$d = v_0 t + \frac{1}{2}at^2$$

where v_o is the initial velocity, a is the acceleration, and t is the time during which the acceleration occurs. Also from Chapter 2, we have

$$v = v_0 + at.$$

If we solve this latter expression for t and substitute into the expression for d just above, we find

$$d = \frac{v^2 - v_0^2}{2a}.$$

We now substitute this for d in our expression for the work done on the dummy:

$$W = \frac{m(v^2 - v_0^2)}{2} = \frac{1}{2}mv^2 - \frac{1}{2}mv_0^2$$

Since the dummy was initially at rest, we set $v_0 = 0$ and find

$$W = \frac{1}{2}mv^2.$$

We see that the player's work on the dummy has changed the dummy's state of motion; it was initially at rest, and now is moving at velocity v. Furthermore, the work done on the dummy is equal to the quantity $\frac{1}{2}mv^2$, which we call the **kinetic energy** KE.

$$\text{KE} = \frac{1}{2}mv^2 \tag{5-5}$$

Kinetic energy is a scalar quantity, and is always positive if there is any motion at all. We need only know the magnitude of the velocity, that is, the speed, to compute the kinetic energy.

We also note that the units of energy and of work are the same. We have said that energy is the ability to do work; now we see that energy and work are somehow equivalent. This equivalence will be explored further in a later section.

Example 5.8   What is the kinetic energy of a football player whose mass is 100 kg, running at a speed of 6 m/s? How does this compare with the kinetic energy of a speeding bullet, whose mass is 0.01 kg and whose speed is 1000 m/s?

The kinetic energy of the football player is

$$KE = \frac{1}{2}mv^2 = (0.5)(100 \text{ kg})(6 \text{ m/s})^2 = 1800 \text{ J}.$$

The kinetic energy of the bullet is

$$KE = (0.5)(0.01 \text{ kg})(1000 \text{ m/s})^2 = 5000 \text{ J}.$$

It is no wonder that a bullet is more likely than a football player to damage its target. (Of course, damage in a collision depends on the area and duration of impact, which depend on the shape and composition of the moving object.)

Potential Energy

Kinetic energy always involves motion, but the other general form of energy does not. **Potential energy** is commonly called stored energy, and exists due to the relative positions of the parts of a system. Some examples of potential energy are gravitational potential energy, chemical energy, and electrical energy. The height of an object above the earth determines the gravitational potential energy stored in the object. Chemical potential energy is stored in the electrons which are bound to the nuclei of atoms by chemical binding forces (which are electromagnetic in nature).

An object gains potential energy when work is done on it that alters its position or shape (that is, the relative positions of its molecules). When we raise a book from the floor to a table and leave it there, the work we have done on the book is stored in it in the form of gravitational potential energy. Every object on the earth's surface has an amount of gravitational potential energy determined by its distance from the earth's center of gravity. When we wind up a watch spring, we store energy in it because we do work on it that changes its shape. The molecules in a volatile liquid or gas have chemical potential energy because chemical reactions can release energy.

Energy can be converted from one form to another. When a watch spring is wound, then released, potential energy is converted into kinetic energy. When vaporized gasoline burns in the cylinders of a car engine, chemical potential energy is converted into the kinetic energy of rapid motions of gas molecules whose pressure pushes the pistons which turn the drive shaft. In the next section we will discuss the conversion of energy from one form to another in more detail.

No single mathematical expression describes all forms of potential energy. Instead we must deal with different expressions for different forms. Here we will describe several common forms. Others will be introduced later.

5.3 ENERGY

We often need to calculate gravitational potential energy. We have already said that all objects at the earth's surface have gravitational potential energy, because of their distance from the earth's center of gravity. We do not usually measure the energy objects have at the earth's surface; we usually choose a point where we define the gravitational potential energy of an object as zero, and then calculate the energy gained if it is raised above this point. We have seen that work done on an object changes the energy of the object and that the work done on an object can be converted into kinetic energy. If, instead of changing the motion, the work done results in a change in potential energy, then the change in potential energy is equal to the work done. Thus, if we raise an object, the potential energy gained is

$$\text{PE} = W = Fd = mgh \qquad (5\text{--}6)$$

where the force F is the weight mg of the object, and h denotes the height to which the object is raised. We will see (Chapter 7) that Eq. 5–6 is valid only for values of h that are small compared to the radius of the earth; for very large displacements we need a more general expression for gravitational potential energy.

We have mentioned that a wound spring has potential energy because of its deformation. Whenever a body is deformed, it tends to return to its original shape, and there is potential energy stored in the body by virtue of its deformation. In Chapter 8 we will discuss the elastic properties of solids, but for now we note that the restoring force is proportional to the distance x that an object is deformed; that is, the restoring force is $F = kx$, where k is a constant. The work required to deform a body is therefore $W = \bar{F}x$, where $\bar{F}$ is the average force as the deformation occurs. Since the force is zero at first, and becomes kx at the end of the deformation, the average force is $\frac{1}{2}kx$. Hence the work done is equal to the potential energy given to the object.

$$\text{PE} = \bar{F}d = \frac{1}{2}kx^2 \qquad (5\text{--}7)$$

In Chapter 8 we will be more specific about deformation, and we will explore applications of elasticity that are beyond the scope of the present discussion.

Conservative Forces and the Work-Energy Principle

The gravitational potential energy gained by an object that is raised, or the energy lost by an object that is lowered, does not depend on the path taken. Eq. 5–6 shows that the energy depends only on the vertical displacement h. This can be shown rigorously by calculating the work required to lift an object by different paths. The reader may find it a useful exercise to do this. (For example, the work can be calculated for a straight vertical displacement, or for an inclined plane. The answer will be the same provided friction is ignored.)

Sometimes the amount of work required to move an object from one place to another *does* depend on the path taken. An example is the friction force between two surfaces. Obviously, the amount of work you do to push an

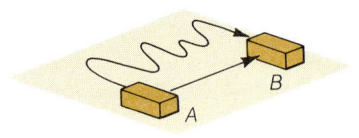

Figure 5.7

Work against a nonconservative force. The block can be pushed from A to B following either of the paths shown. Because friction always opposes motion, the amount of work depends on the path length.

object from one place to another depends on friction, and hence on the length of the path taken (Fig. 5.7). Some kinds of forces, such as gravity, are independent of the direction of motion of a particular body, and for these forces the work done is independent of path. Most of these forces always act in the same direction, regardless of the direction of motion of the body on which they act. Friction and air resistance, on the contrary, always act against the motion, no matter what direction it takes. Forces such as gravity for which the work done is independent of path are called **conservative forces**. We will encounter several other kinds of conservative forces in our study.

We have commented on the close relationship between work and energy. In calculating the kinetic energy associated with any moving body (Eq. 5–5), we found that the kinetic energy of a body is equal to the work that body can perform. We then defined potential energy in terms of the work done on a body. Now that we understand conservative forces, we can be more specific about the relationship between work and energy.

We can define potential energy as equivalent to work *only* when work is done against a conservative force. Because potential energy is stored energy, work put into a motion or deformation can be gotten out again; we say that the potential energy can be released. For work done against a force that is not conservative, work put in cannot be gotten out again because the energy cannot be stored. You can push a heavy crate around all day, doing immense quantities of work against the force of friction, but at the end of the day the crate has not gained any potential energy. Only if the crate is lifted, so that it can later fall back down, does it gain potential energy.

We can now state the *work-energy principle:*

The work done on a body by the net external force is equal to the total change in energy of that body.

The net external force must be a conservative force. In equation form, we write

$$W = \Delta KE + \Delta PE. \qquad (5\text{–}8)$$

It is important to note that the work is done by a conservative force, and that the PE term must include any other forces, such as gravity, that have acted. An extension of this principle, which leads to a very powerful problem-solving technique, is discussed in the next section.

Conservation of Energy

We have mentioned the transformation of energy from one form to another, and we have discussed the work-energy principle. We recognize that when work is done on an object by some force, the object does an equal and opposite amount of work on the agent that exerted the force. In the preceding sections

5.3 ENERGY

we showed in a few examples that the sum of the net work on a body and by the body is always zero. The total work done in any interaction is always zero, even though the energies of the bodies may change. Since the work-energy principle states that work and energy changes are equivalent, energy that is given to one body must be lost by another because the total work must remain zero. This leads to the principle of *conservation of energy*.

The total energy in an isolated system cannot be increased or decreased by any process.

This is the second of the great conservation laws that we will encounter. We stress again that this is strictly true only if all forces are conservative. Energy can be changed from one form to another, but the total energy of a system is constant. Mathematically, we can write

$$\Delta KE + \Delta PE = 0 \qquad (5-9)$$

$$KE + PE = \text{constant.} \qquad (5-10)$$

We have used a brief argument to demonstrate this principle in terms of Newton's laws. It is possible to prove it more rigorously for any mechanical system, but it is not easy to do so for nonmechanical forms of energy such as electrical or chemical energy. Nevertheless the principle is valid for all these forms of energy. The only exceptions arise in cases where relativistic effects become significant, for then we find that mass and energy are interchangeable and the conservation law must become more general.

We can find innumerable simple examples to demonstrate the conservation of energy (Fig. 5.8). If a weight is dropped from a height h to the ground, for example, its initial potential is mgh, and its final potential energy is zero. Equation 5–9 leads us to believe that its kinetic energy when it reaches the ground should be mgh; let us see if it is. From Eq. 2–13 we see that the time it takes for the weight to reach the ground is

$$t = \sqrt{\frac{2h}{g}}.$$

Then its velocity when it reaches the ground is

$$v = gt = \sqrt{2gh}.$$

The kinetic energy when it reaches the ground is

$$KE = \frac{1}{2}mv^2 = \frac{1}{2}m(2gh) = mgh.$$

Figure 5.8

Conservation of energy. As the athlete bends the pole, his kinetic energy is converted into potential energy stored in the pole. When the pole then straightens, the athlete regains kinetic energy.

The kinetic energy at the end of the fall is equal to the potential energy that disappeared during the fall.

The knowledge that energy is constant provides us with a very powerful means of solving problems. For example, if we want to know the velocity of a falling object when it has fallen from some height, we can simply equate the loss in potential energy to the gain in kinetic energy, and solve for the velocity. If we measure the speed of a falling object as it reaches a certain point, we can use that to calculate the original potential energy of the object, hence its initial height.

Example 5.9   A BB gun uses a spring to shoot small metal pellets which weigh 0.01 lb. If cocking the gun requires compressing the spring with an average force of 20 lb over a distance of 3 in, how high will a pellet go if shot straight up? Ignore air resistance.

The potential energy gained by the spring under compression is equal to the work put into it, which is

$$\text{PE} = W = \bar{F}d = (20 \text{ lb})(0.25 \text{ ft}) = 5 \text{ ft} \cdot \text{lb}$$

The initial kinetic energy is zero, and we want to find the height the pellet reaches when the kinetic energy is again zero; that is, when the pellet is at its highest point. We could have solved this problem in Chapter 2 by calculating the acceleration of the pellet, its velocity when it leaves the gun, and the distance it would travel upward against the deceleration due to gravity, but to do this would have been laborious and would have required more information, such as properties of the spring. Now that we understand the principle of conservation of energy, however, we find this problem extremely simple. We write

$$\text{KE}_1 + \text{PE}_1 = \text{KE}_2 + \text{PE}_2$$

where the subscripts indicate time, before the spring is released and when the ball is at rest at the top of its excursion. In both cases the KE is zero, so we write

$$\text{PE}_1 = \text{PE}_2$$

or

$$5 \text{ ft} \cdot \text{lb} = mgh = (0.01 \text{ lb}) h$$

so that

$$h = 500 \text{ ft.}$$

It would also be easy to find the velocity of the pellet when it left the gun, because at that point all of the initial potential energy has been converted into kinetic energy.

5.3 ENERGY

$$\frac{1}{2}mv^2 = \text{PE}$$

then

$$v = \sqrt{\frac{2(\text{PE})}{m}} = \sqrt{\frac{2g\,(\text{PE})}{W}}$$

where $W = mg$ is the weight of the pellet. We find

$$v = \sqrt{\frac{2(32\text{ ft/s}^2)(5\text{ ft}\cdot\text{lb})}{.01\text{ lb}}} = 179\text{ ft/s}.$$

We may ask what happens when we deal with nonconservative forces. A person may push a crate across the floor, doing work and expending energy, yet when he is done the crate has not gained any energy. Where did the energy go that was created as the crate was moved? The answer is that the energy was converted into other forms and lost to the surroundings. Most of it went into internal thermal energy; we are familiar with the heat friction creates (a form of internal kinetic and potential energy of moving atoms and molecules). The floor and the crate both gained thermal energy in the process. The person moving the crate also gained energy, mostly internal heat generated by biochemical reactions that enabled him to push the crate. Some small amounts of energy may also have been converted to sound waves while the crate was being moved. Whenever a nonconservative force does work, energy is lost to the environment rather than conserved by the moving body.

In some collisions, energy is transferred between bodies. In a real collision, some energy is lost to the surroundings in the form of heat and sound waves. However, sometimes these losses are very small and we can neglect them. We then say that the collision is **elastic**, which means any deformation is perfectly restored as the objects bounce apart. In an elastic collision, we can assume that the kinetic energy is constant. We also know that the momentum is constant, so we have two conditions that we can apply in problem solving.

Example 5.10 Consider a billiard table on which two balls are in play. If ball 2 is at rest and is hit squarely by ball 1 (so that there is no torque and no rotation), what is the speed of ball 2 after the collision?

Here we can assume the collision is elastic. We need to work in only one dimension, since the motion of ball 2 will be in the same direction as the original motion of ball 1. We write the equations for conservation of momentum and kinetic energy.

$$m_1v_1 = m_1v'_1 + m_2v'_2$$

$$\frac{1}{2}m_1v_1^2 = \frac{1}{2}m_1v'^2_1 + \frac{1}{2}m_2v'^2_2$$

The primes indicate velocities after the collision. Ordinarily there would have been terms for the initial momentum and kinetic energy of ball 2, but these are both zero since that ball is initially at rest. We solve the first equation for the velocity of ball 1 after the collision.

$$v'_1 = v_1 - \frac{m_2}{m_1}v'_2$$

If we square this expression and substitute it for v'^2_1 in the second equation, we get (after some algebraic manipulation)

$$\frac{v'_2}{v_1} = \frac{2m_1m_2}{m_1m_2 + m_2^2} = \frac{2m_1}{m_1 + m_2}.$$

Now let us consider some examples. If the two balls have equal mass, then we find

$$\frac{v'_2}{v_1} = 1$$

that is, the second ball after the collision has a velocity equal to that of ball 1 before the collision. Conservation of momentum (or conservation of kinetic energy) quickly tells us that ball 1 must be at rest after the collision, because ball 2 now has all the momentum and all the kinetic energy that ball 1 had.

If ball 2 has half the mass of ball 1 (that is, $m_2 = m_1/2$), then

$$\frac{v'_2}{v_1} = \frac{4}{3}.$$

This indicates the (perhaps) surprising result that ball 2 after the collision is actually moving faster than ball 1 was before the collision. The German philosopher Nietzsche once claimed to prove that God exists by declaring that this extra speed had to come from divine intervention. Actually the "extra" speed is simply a consequence of conserving momentum in the collision. We can also solve for the velocity v'_1 of ball 1 after the collision, by solving the energy equation for v'_2 and substituting into the equation of momentum conservation:

$$\frac{v'_1}{v_1} = 1 - \frac{2m_1m_2}{m_1 + m_2} = \frac{m_1 - m_2}{m_1 + m_2}.$$

This verifies that the final velocity of ball 1 is zero when the two masses are equal, and shows that when $m_2 = m_1/2$

5.4 POWER

$$\frac{v_1'}{v_1} = \frac{1}{3}.$$

Thus when ball 2 has half the mass of ball 1, ball 2 moves away with one-third greater speed than ball 1 had initially, and ball 1 continues moving after the collision with one third of its original speed.

5.4 Power

When we speak of doing work, we often mean the instantaneous rate at which it is done, rather than the total amount done over a long time. The work done per unit of time is called the **power**, that is

$$P = \frac{\Delta W}{\Delta t}. \tag{5-11}$$

We may equally well think of power as the amount of energy transformed per unit time. In either case, power is a measure of the capability of a person or a machine to do work because it includes the important notion of the time period in which a task will be performed. The units of power are units of energy per time. In the SI system, one joule per second (J/s) is equal to one **watt** (W). In the cgs system, we speak of erg/s, and in the English system, ft · lb/s. A common unit called the **horsepower**, equal to 550 ft · lb/s, is often used in the English system instead of ft · lb/s. (In modern metric usage, one horsepower is defined as 750 W, whereas a straightforward conversion of 550 ft · lb/s to watts yields 1 hp = 746 W, which is the modern British value.)

The power of a person or machine is usually governed by intrinsic characteristics related to the internal conversion of energy. A person may be able to walk 10 km quite comfortably, but may be unable to run that far because his body cannot convert the potential energy stored in his food into usable kinetic energy fast enough. The process of conditioning a human body for rigorous exercise is really a process of optimizing the body's efficiency for converting energy.

Similarly, the power of a car depends on the rate at which it converts chemical potential energy into kinetic energy, which depends on how much fuel it can burn in a given time. The greater the capacity of its cylinders, and the greater the rate of fuel supplied by the gas pedal, the more power can be generated.

Example 5.11 Suppose a car climbs 10 miles along a 5% grade. If the car weighs 3,000 lb and travels at a steady speed of 60 mi/h, how much power is needed, in watts and horsepower? Neglect air resistance, which would, of course, increase the power required.

There is a mixture of units in the problem as stated, so it is wise to convert them all to SI units. The distance covered is 1.609×10^4 m, the car's weight is 1.33×10^4 N, and its speed is 26.8 m/s.

We need to calculate the work done on the car as it goes up the hill, and this requires knowing the force applied to the car. The velocity of the car is constant, so we know that the net force must be zero. Therefore, if we ignore friction, the force applied to the car to make it go up the hill must be equal and opposite to the component of the car's weight that is parallel to the road. For a 5% slope, the angle of inclination is 2.87°, and the component of the weight parallel to the road is $(1.33 \times 10^4 \text{ N}) \sin(2.87°) = 666$ N. The work done on the car is therefore

$$W = Fd = (666 \text{ N})(1.609 \times 10^4 \text{ m}) = 1.07 \times 10^7 \text{ J}.$$

This work is done over a time $t = (1.609 \times 10^4 \text{ m})/(26.8 \text{ m/s}) = 600$ s, so the power is

$$P = W/t = \frac{1.07 \times 10^7 \text{ J}}{600 \text{ s}} = 1.78 \times 10^4 \text{ W} = 23.8 \text{ hp}.$$

We could have arrived at the same answer a bit more simply by noting that the rate at which work is done in this simple case is just the force times the velocity of the car, since velocity is the rate of change of displacement. In short, the power is

$$P = Fv$$

and all we needed to solve the problem was to multiply the force required to keep the car going uphill by the speed of the car.

$$P = (666 \text{ N})(26.8 \text{ m/s}) = 1.78 \times 10^4 \text{ W} = 23.8 \text{ hp}$$

5.5 Simple Machines

Machines are devices that perform work (that is, they apply forces over distances). The simplest mechanisms transmit force. Such a machine cannot reduce the amount of work needed to perform a task, because the task will consist of applying the same force over the same distance, no matter what applies the force. Thus a machine does not save work.

A machine is still useful, because, by increasing the displacement, it can reduce the amount of force needed to perform a task. Since work is the product of a force and a displacement, a given amount of work can be performed by reducing the force and increasing the displacement. Thus for example, when a lever is used to lift a heavy object (Fig. 5.9), a force is applied to one end of the lever, the lever rotates about the pivot, and the other end of the lever exerts the force to do the work. If the pivot is close to the heavy object, the

5.5 SIMPLE MACHINES

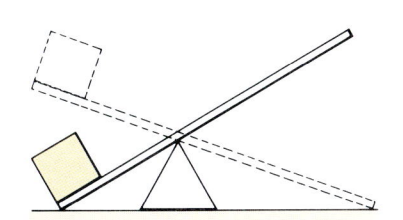

Figure 5.9

A lever. A heavy object can be lifted by a lever, so that a smaller force is applied over a greater displacement to accomplish a given amount of work.

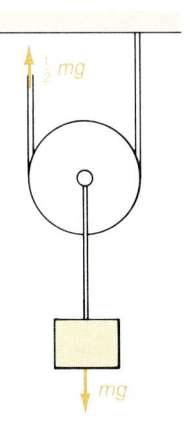

Figure 5.10

A simple pulley. One half the force is applied over twice the displacement to lift the weight.

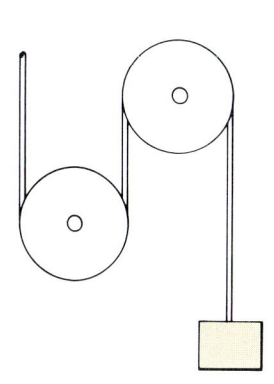

Figure 5.11

Multiple pulleys. Multiple pulleys reduce the force required and increase the displacement required to do the same amount of work.

force applied to the other end is much smaller than the force the lever exerts on the object, because the applied force acts over a much larger distance.

A moving pulley also reduces the required force by increasing the distance over which force is applied (Fig. 5.10). The pulley wheel divides the weight over two ropes. To lift the load, the pulley exerts on each rope a force equal to half the weight. A specified vertical displacement of the load, that is, a specified quantity of work, now requires twice as much displacement of the one rope that is pulled. Additional pulleys can further reduce the force required while proportionally increasing the displacement required (Fig. 5.11).

We measure the effectiveness of machines by calculating the **mechanical advantage**, which is the ratio of the force required without a machine to the force required with a machine to do the same work. For a lever, if the load is one fourth as far from the pivot as the point where the force is applied, the net torque would be zero so that the force applied to the end of the lever would be one fourth of the weight of the object, and the mechanical advantage would be 4 (Fig. 5.9). For a single pulley, the load is spread over two ropes, so the mechanical advantage is 2 (Fig. 5.10). These are **ideal mechanical advantages** (abbreviated **IMA**), because they ignore the effects of friction and the weight of the machines themselves. The IMA is always equal to the ratio of the displacements over which the input and output forces act, that is

$$\text{IMA} = \frac{d_i}{d_o}. \tag{5-12}$$

In reality, friction and the weight of the machine reduce the mechanical advantage. The **actual mechanical advantage** (**AMA**) is the ratio of the actual output force to the input force

CHAPTER 5 MOMENTUM, WORK, AND ENERGY

$$\text{AMA} = \frac{F_o}{F_i}. \tag{5-13}$$

The AMA measures the effectiveness of a machine.

Example 5.12 A very simple machine is the inclined plane, which raises an object with less force than the weight of the object. Suppose we have a plane whose angle of inclination is 5°, and we want to raise an object weighing 500 N to a platform 0.5 m above the floor. Calculate the IMA and the AMA of this machine.

The IMA is simply the ratio of input distance to output distance, or

$$\text{IMA} = \frac{(0.5 \text{ m})/\sin(5°)}{0.5 \text{ m}} = \frac{1}{\sin(5°)} = 11.5.$$

The AMA is the ratio of output force F_o to input force F_i and is calculated by taking friction into account. If the coefficient of kinetic friction is 0.2, then the friction force acting against the motion is

$$F_f = \mu_k F_N = (0.2)(500 \text{ N}) \cos(5°) = 99.6 \text{ N}.$$

The output force is the weight, 500 N. The actual input force is the sum of F_f and the component of weight along the plane, $(500 \text{ N}) \sin(5°)$, and the AMA is

$$\text{AMA} = \frac{F_o}{F_i} = \frac{500 \text{ N}}{43.6 \text{ N} + 99.6 \text{ N}} = 3.49$$

Thus, friction created a significant difference between the ideal and the actual mechanical advantage, and we see the need for reducing friction in machines.

We can also characterize a machine in terms of the work that must be done *to* it for a certain amount of work to be done *by* it. The ratio of input work to output work is called the **efficiency**. For an ideal machine the efficiency is always precisely 1. In any real machine, extra work must be done to counteract friction or the weight of the machine itself, and the efficiency is always less than 1.

The efficiency can be expressed in terms of the IMA and the AMA. In an ideal machine, the ratio of input force to output force is always equal to the inverse ratio of the displacements over which the input and output forces are applied, so we can write

$$\text{IMA} = \left(\frac{F_o}{F_i}\right)_{\text{ideal}} = \frac{d_i}{d_o}$$

5.5 SIMPLE MACHINES

where d_i and d_o represent the distances over which the input and output forces are applied. The AMA is still

$$\text{AMA} = \frac{F_o}{F_i}$$

so if W_i and W_o represent the input and output work, the efficiency is

$$\text{efficiency} = \frac{W_o}{W_i} = \frac{F_o d_o}{F_i d_i} = \frac{\text{AMA}}{\text{IMA}}. \qquad (5\text{--}14)$$

The efficiency is simply the ratio of the actual and ideal mechanical advantages.

Example 5.13 Calculate the efficiency of the inclined plane whose IMA and AMA were calculated in Example 5.12.

To illustrate the principles, we will do this two ways. First, we can calculate the work done to push the weight up the plane and the net work that would have been done if there were no friction, and take the ratio. The work against friction can be found in the same way as in Example 5.7. If we substitute the values in the present example (that is, an inclination of 5°, a load of 500 N, a kinetic friction coefficient of 0.2, and a net vertical displacement of 0.5 m), we find

$$F_g = (500 \text{ N}) \sin(5°) = 43.6 \text{ N}$$
$$F_f = .2(500 \text{ N}) \cos(5°) = 99.6 \text{ N}$$
$$W_i = (43.6 + 99.6) \text{ N} \frac{0.5 \text{ m}}{\sin 5°} = 821.5 \text{ J}.$$

The work done in the absence of friction would have been

$$W_o = (500 \text{ N})(0.5 \text{ m}) = 250 \text{ J}.$$

(You can also calculate W_o by neglecting friction and multiplying the force required to counteract gravity by the length of the inclined plane; you get the same answer.) The efficiency is

$$\text{efficiency} = \frac{W_o}{W_i} = \frac{250}{821.5} = 0.30.$$

The second way to solve this problem is just to write the expressions for IMA and AMA, and divide. The IMA is the ratio of the input to the output displacements.

$$\text{IMA} = \frac{d_i}{d_o} = \frac{0.5 \text{ m}/\sin(5°)}{0.5 \text{ m}} = \frac{1}{\sin(5°)} = 11.47$$

The AMA is the ratio of output to input forces. The output force is simply the weight of the object, and the input force is the component of the required force parallel to the inclined plane.

$$F_i = F_g + F_f = (500 \text{ N}) \sin(5°) + (0.2)(500 \text{ N}) \cos(5°) = 143.2 \text{ N}$$

The AMA is then

$$\text{AMA} = \frac{F_o}{F_i} = \frac{500 \text{ N}}{143.2 \text{ N}} = 3.49$$

and the efficiency is

$$\text{efficiency} = \frac{\text{AMA}}{\text{IMA}} = \frac{3.49}{11.47} = 0.30.$$

This is the same answer as before.

FOCUS ON PHYSICS

Work, Energy, Exercise, and Your Weight*

You have been overeating during the past few months and have gained a few pounds. You think that you can lose those extra pounds by running two or three miles for a few days. You will have to work and expend energy to shed that excess weight, but will a four-mile run or a thirty-minute swim satisfy the energy requirements to lose those extra pounds? What are the energetics of weight control by diet and exercise?

Physiologists and dieticians use units of energy measurement that are not in the SI system. Probably the most familiar unit to most of us is the Calorie. One Calorie (note that it is capitalized) is equal to 1,000 of a smaller unit known as the calorie (lower case), which in turn is defined according to its equivalent quantity of heat (these units are defined more fully in Chapter 12). Hence one Calorie is equal to one kilocalorie (kcal), and is the equivalent of 4,180 J. The caloric content of foods is determined by a device called a **bomb calorimeter**, in which a food substance is burned and the heat content determined by the temperature rise in water that circulates around the combustion chamber (Fig. 5.1.1).

All foods that we take in provide our bodies with energy that is stored within our skeletal muscles and fat cells, and gives us the capacity to perform work and to exercise. How do we measure the quantity of energy used by an individual during a typical exercise program? What information is necessary for us to plan an adequate exercise program and diet for weight control?

*Contributed by Dr. Ron Canterna.

5.5 SIMPLE MACHINES

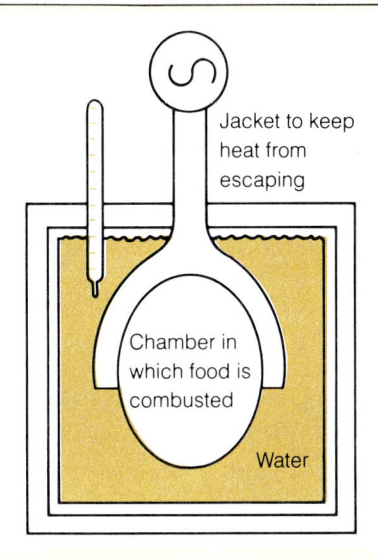

Figure 5.1.1

A simple form of the principle of conservation of energy states that there is a constant ratio between a fixed amount of energy expended by the body and the resulting production of heat. Physiologically, this principle was first shown by Rubner in the late 1800s using a bomb calorimeter with canines as the subjects. The dogs were placed in a chamber that monitored the intake of oxygen and the production of heat from the dogs' metabolic functions, by measuring the temperature of water circulating around the chamber. The results of these studies and others show that the energy expended by an individual, which produces heat during an activity, is exactly equal to the energy released by metabolism, determined from the oxygen consumption.

The caloric content of food is measured directly in a bomb calorimeter, as described above. For human and animal subjects, the heat production is measured indirectly from the oxygen consumption, which ultimately depends on the types of food metabolized by the body. Carbohydrates yield approximately 4 kcal per gram (kcal/g), protein 5.2 kcal/g, and fats 9.0 kcal/g. Physiologists can determine the caloric equivalent of the consumed oxygen by measuring the respiratory exchange ratio R, which is the ratio of the volume of expired CO_2 per minute (abbreviated $\dot{V}CO_2$) to the volume of consumed O_2 per minute ($\dot{V}O_2$). If carbohydrates are oxidized completely, an R value of 1.0 is found, and approximately 5.1 kcal of energy are released for each liter of oxygen consumed by the body. If fats are metabolized completely, an R value of 0.70 is found, with only 4.7 kcal of energy released per liter of oxygen. (Notice that consumption of carbohydrates releases more energy per unit of consumed oxygen than fats, even though fats store almost twice the amount of energy per unit mass than carbohydrates. What does this say about the amount of energy it takes to break down fats and carbohydrates?) Oxidizing proteins will not be discussed here, because of the complexity of this subject and the fact that generally, protein is not significantly utilized by the body as a fuel for exercise.

To determine the energy cost of a specific exercise activity, the respired CO_2 and consumed O_2 are measured, an R value is determined, and the total amount of energy released as heat is determined. A typical human will consume between 0.2 and 0.3 liters of oxygen per minute during normal resting activities (a liter is a measure of volume equal to 1,000 cubic centimeters, the equivalent of a little more than a quart). This amounts to an average of 360 liters of oxygen per day. If we assume that approximately 60 percent carbohydrates and 40 percent fats are metabolized during rest ($R = 0.85$, which yields 4.85 kcal/liter of consumed O_2), then approximately 1,800 kcal are needed. When you include energy requirements for digestion, the normal daily energy consumption by a human is approximately 2,000 kcal (or 2,000 Calories).

For activities such as running, walking, cycling and other exercise routines that are easily monitored in a laboratory, the energy requirements can be easily measured. However, for exercise programs that are not easily monitored in the laboratory, the physiologist will use a relationship between the heart rate and $\dot{V}O_2$ for a given subject. This heart rate-$\dot{V}O_2$ relationship is determined for each subject using a treadmill or **bicycle ergometer**

(continued)

(a device that measures kinetic energy). Then during the exercise activity of interest, the subject's heart rate is monitored by **telemetry,** and the heart rate-$\dot{V}O_2$ relationship is used to determine the total amount of oxygen consumed by the subject. For a given exercise or sport, the average energy cost can be determined in this manner. A list of some popular activities and their energy costs is given in the accompanying table.

Now let us get back to our main objective, to see how much weight we lose during exercise. As an example, let us calculate how much energy will be used during a four-mile run that is completed in 30 minutes (which is a 7½ minute/mile speed). From the table, we see that we expend 10.6 kcal/minute for this run, or approximately 320 kcal total. Assuming 3,500 kcal of expended energy is needed to release one lb of fat, we find that only 0.1 lb of fat weight would be lost. (Generally carbohydrates are consumed in larger quantities at the start of an exercise activity, but the preference for fat metabolism increases as the activity continues. This means that exercise must be prolonged in order to consume even small amounts of fat.) This does not seem to be a significant loss in weight for such a strenuous activity. What is usually forgotten is that you put on the extra weight over a period of weeks or months. If you continued the running for one month, every other day, you would lose 1.5 pounds, which is significant. Exercise over an extended period of time will definitely result in loss of weight. From energy considerations, however, you can easily see the impracticality of losing permanent weight in a few days.

Approximate energy costs of various physical activities

Sport or activity	Kilocalories/minute
Climbing	10.7—13.2
Cycling at 5.5 mph	4.5
Cycling at 9.4 mph	7.0
Cycling at 13.1 mph	11.1
Dancing	3.3—7.7
Football	8.9
Golf	5.0
Gymnastics	
Balancing	2.5
Abdominal exercise	3.0
Trunk bending	3.5
Arm swinging, hopping	6.5
Rowing	
51 strokes/min	4.1
87 strokes/min	7.0
97 strokes/min	11.2
Running	
Sprint	13.3—16.6
Cross-country	10.6
Tennis	7.1
Skating (fast)	11.5
Cross-country skiing, fast	10.8—15.9
Cross-country skiing, fast uphill	18.6
Squash	10.2
Swimming	
Breaststroke	11.1
Backstroke	11.5
Crawl (55 yards/min)	14.0
Wrestling	14.2

*Data from *Nutrition for the Athlete,* a pamphlet published by the American Association for Health, Physical Education, and Recreation (Washington D.C.), 1971.

Perspective

This has been a lengthy chapter, made so by the significance of the concepts introduced and the need to discuss them thoroughly. With the information on momentum, work, and energy presented here, it is possible to solve a very wide range of problems, and applications of some of the principles from this chapter will be made repeatedly throughout the book.

We are not quite finished with our general discussion of mechanics, however, because we still have not fully treated rotational and circular motions. We are well prepared to do so next, because we will apply many of the concepts we have just learned.

PROBLEM SOLVING

Shortcuts with Energy Conservation

The laws of energy and momentum conservation provide some of the best shortcuts in problem solving. Problems in Chapters 2 and 3 which required a fair amount of algebra can now be done in a few lines. However, one must understand how to take advantage of these conservation laws. For example, one should choose a convenient reference point for zero potential energy (PE). Also, it is often useful to separate the kinetic energy (KE) into two parts, $\frac{1}{2}mv_x^2 + \frac{1}{2}mv_y^2$, corresponding to motion in the x and y directions.

Example 1 A stone is thrown straight up with velocity v_0. How high does it go? If we choose PE to be zero at the ground, then at height y, $PE = mgy$. The sum KE + PE must remain constant. Since the KE is zero at maximum height ($v = 0$),

$$mgy = \tfrac{1}{2}mv_0^2$$

so that $y = v_0^2/2g$, independent of mass m.

Example 2 Suppose the rock is thrown at angle θ from horizontal with initial velocity v_0. What is the maximum height and horizontal range on level ground? In this example, we must separate the x and y components of motion. The horizontal velocity $v_x = v_0 \cos\theta$ remains constant, but the vertical velocity decreases from its initial value $v_0 \sin\theta$ to zero at height y. Energy conservation requires

$$\tfrac{1}{2}mv_x^2 + \tfrac{1}{2}m(v_0 \sin\theta)^2 = \tfrac{1}{2}mv_x^2 + mgy.$$

The kinetic energy $\frac{1}{2}mv_x^2$ cancels, and we can solve for the height.

$$y = \frac{(v_0 \sin\theta)^2}{2g}$$

This is similar to Example 1, except that only the vertical velocity appears in the final solution. The time t_{up} for the rock to reach this height is the initial vertical velocity divided by the gravitational acceleration g. The horizontal distance is twice the value of x at this point.

$$x_{max} = 2v_x t_{up} = \frac{2(v_0 \cos\theta)(v_0 \sin\theta)}{g}$$
$$= \left(\frac{v_0^2}{g}\right) \sin(2\theta)$$

Example 3 A mass m has initial velocity v_0 at the bottom of a plane inclined at angle θ (Fig. 5.2.1). What distance d up the plane does the mass reach if the coefficient of kinetic friction is μ? Energy conservation must now include the work done against the frictional force, $F_f = \mu mg \cos\theta$, as well as the gravitational potential energy = mgh.

$$\tfrac{1}{2}mv_0^2 = mgh + F_f d$$
$$= mgd \sin\theta + \mu(mg \cos\theta)d$$

The mass cancels out and we can solve for the distance.

$$d = \frac{v_0^2}{2g(\sin\theta + \mu \cos\theta)}$$

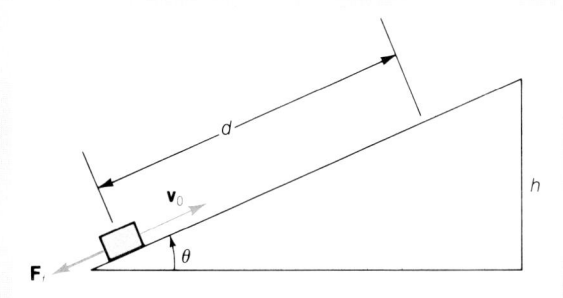

Figure 5.2.1

SUMMARY TABLE OF FORMULAS

Momentum:
$$\mathbf{p} = m\mathbf{v} \tag{5-1}$$

Force and rate of change of momentum:
$$\mathbf{F} = \frac{\Delta \mathbf{p}}{\Delta t} \tag{5-2}$$

Impulse:
$$\mathbf{F}\Delta t = \Delta \mathbf{p} \tag{5-3}$$

Work:
$$W = Fd \cos\theta \tag{5-4}$$

Kinetic energy:
$$KE = \frac{1}{2}mv^2 \tag{5-5}$$

Gravitational potential energy at earth's surface:
$$PE = mgh \tag{5-6}$$

Potential energy of a spring:
$$PE = \frac{1}{2}kx^2 \tag{5-7}$$

Work-energy principle:
$$W = \Delta KE + \Delta PE \tag{5-8}$$

Conservation of energy:
$$\Delta KE + \Delta PE = 0 \tag{5-9}$$
$$KE + PE = \text{constant} \tag{5-10}$$

Power:
$$P = \frac{\Delta W}{\Delta t} \tag{5-11}$$

Ideal mechanical advantage:
$$IMA = \frac{d_i}{d_o} \tag{5-12}$$

Actual mechanical advantage:
$$AMA = \frac{F_o}{F_i} \tag{5-13}$$

Efficiency of a machine:
$$\text{Efficiency} = \frac{W_o}{W_i} = \frac{AMA}{IMA} \tag{5-14}$$

THOUGHT QUESTIONS

1. Why is a rocket propelled forward, even in a vacuum with no air to push on?
2. If every person in China climbed up a 2 m ladder, then jumped off at the same time, what would happen to the Earth?
3. Is there any difference in the motion of a mass acted on by a force F for a time Δt, and a force $2F$ acting for $\Delta t/2$?
4. When two objects of widely different masses collide elastically, which carries away more energy?
5. Can an astronaut in free space change his state of motion?
6. A Canadian fisherman who is trapped on a perfectly frictionless frozen lake wants to get home. How can he use some of his fish to get to shore?
7. Explain the transformation of energy as a high jumper makes his approach, jumps, and lands.
8. Explain why a car or bicycle needs friction to move forward.
9. Every mile, as a train rolls down a flat, frictionless track, several new passengers jump aboard. What happens to the train's velocity?
10. After two billiard balls collide, we measure their outgoing velocities and the angle between their paths. Is this enough information, with the laws of energy and momentum conservation, to determine the initial velocities?
11. In a race between a tortoise and hare, which exerts more force? More energy? More power?
12. When a car brakes, where does its KE go?
13. Why does the velocity of an object falling down a slope not depend on its mass, if friction is small? How does friction affect the velocity of heavy objects compared to light ones?
14. Of the common energy sources, which ultimately come from the sun?
15. Explain the energy source for tides and their gravitational PE.

PROBLEMS

Section 5.1 Momentum

1. If momentum $p = mv$, show that kinetic energy $KE = \frac{1}{2}mv^2$ may be written as $KE = p^2/2m$. If two particles have the same KE, which has more momentum?

2. What is the momentum of a 200 g baseball moving at 98 mi/h?

3. A strong tennis serve imparts a velocity of 130 mi/h to a 60 g ball. If the ball is in contact with the racket for 0.025 s, what is the average force on the ball?

4. On a football team, a fullback (235 lb), a split end (180 lb), and a lineman (275 lb) can run the 20-yard dash in 3.4 s, 3.2 s, and 4.6 s, respectively. Which player attains the highest momentum, if they accelerate at a constant rate for the whole distance?

5. A 100 kg rugby player, running downfield at 6 m/s, is tackled and brought to a complete stop in 1 s. Compute
   (a) the initial momentum of the runner
   (b) the average force exerted in the tackle.

6. A jet expels hot gases out its rear engine with velocity 50,000 m/s at a rate 10 kg/s. Show that in one second it ejects 10 kg of gas. What is the momentum of the gas? What is the force on the gas? Hence, what is the force on the plane? Recall Newton's third law of action and reaction.

7. An empty train car moving at 15 m/s strikes a similar, but loaded train car which is at rest. The two cars stick together and roll at 3 m/s. If each car weighs 4 tons empty, what was the load of the stationary car?

8. A 1 kg bird flying at 20 m/s collides head-on with the radiator of a 1,000 kg car travelling at 30 m/s and is crushed in 0.05 s. If the bird sticks to the car, calculate
   (a) The impulse given to the car.
   (b) The average force on the bird.
   (c) The change in the car's velocity.

9. Suppose the car in Problem 8 collides with many such birds. How many head-on collisions are required to slow the car's speed by 1%? As the car slows down, why do the later collisions have a slightly less effect?

10. The approximate density of air at sea level is 1.3×10^{-3} g/cm^3.
    (a) Calculate the mass of air contained in a cube 1 m on a side.
    (b) Calculate the momentum carried by this air when the wind is 40 mi/h.

11. For the air density in Problem 10, consider a 100 mi/h hurricane. The wind is completely stopped by a wall.
    (a) Calculate the momentum in a 1 m^3 volume.
    (b) Calculate the force exerted on a 1 m^2 area of wall by the wind impact during one second.
    (c) Find the total force on a 4 m × 20 m wall.

12. Two masses, $m_1 = 5$ kg and $m_2 = 10$ kg, approach each other at $v_1 = 10$ m/s and $v_2 = 20$ m/s. They collide and stick. Calculate the final velocity of the combined mass.

13. Suppose, in Problem 12, the two masses are initially moving in the same direction, with mass 2 overtaking mass 1. What is the velocity of the combined mass after they collide and stick?

14. A 70 kg stockbroker jumps out of a two-story building, landing on the street 7 m below. If he bends his knees, the impact lasts 0.04 s.
    (a) What is his velocity when he hits the ground?
    (b) What is the vertical force exerted on him by the ground during the impact?

15. Suppose in Problem 14 that the force of impact is absorbed by the bones of both legs (cross-sectional area of bone in each leg is about 10 cm^2). Bones fracture under a compressive stress (force per area) of about 10^8 N/m^2. Will his bones break? Include the downward force of gravity.

16. A 150 kg astronaut (including all gear) throws a 1 kg wrench away from herself with force 50 N applied for 0.02 s. What is her recoil speed?

17. Suppose that in a 10 minute cloudburst, 1 inch (2.54 cm) of rain falls. Each drop has a volume 0.07 cm^3, density 1 g/cm^3, and terminal velocity of 25 m/s (about 56 mi/h).
    (a) How many drops fall per second on a 1 m^2 area?
    (b) What is the force of the rain on a person's head (350 cm^2), assuming that the drops are brought completely to rest?

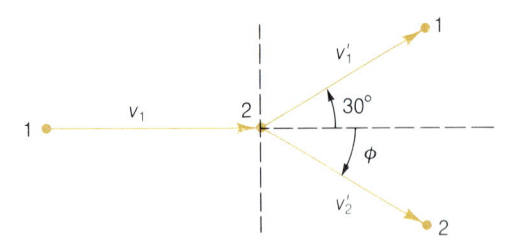

Figure 5.12

••• 18 Two identical 500 g balls collide inelastically, as shown in Fig. 5.12. Initially, ball 1 moves at velocity $v_1 = 1$ m/s; ball 2 is at rest. After the collision, ball 1 has been deflected 30° and has speed $v_1' = 0.6$ m/s. Write the equations of conservation of momentum in the x- and y-directions separately. Solve these equations for the speed v_2' and the angle ϕ.

•• 19 A 100,000 kg spacecraft moving in outer space, far from any gravitational forces, accelerates from 5 km/s to 10 km/s. Its rear rockets expel hot gases at 50 km/s relative to the rocket. How much mass in hot gas must the rocket expel? Neglect the velocity of the rocket in computing the momentum impulse.

•• 20 After chasing each other all over Rome, four cars collide in the square at the same instant and stick together (Fig. 5.13). What is the final velocity of the wreck, neglecting friction? Instead of numerical values, use the masses and velocities in the figure.

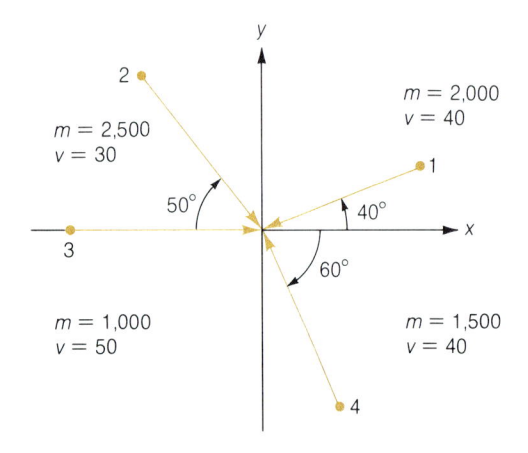

Figure 5.13

••• 21 A 40 g bullet, fired at 500 m/s from ground level strikes a 1 kg wood block sitting on a fencepost 1 m high and 4 m distant horizontally. The bullet buries itself in the block, which flies off the post.

(a) What is the block's initial velocity (speed and direction)?
(b) How far horizontally from the post does the block land when it hits ground? Neglect friction and air drag and neglect the tiny change in the bullet's velocity before it hits the block.

••• 22 A gas consists of electrons and protons in random motion. Electrons are light particles of mass m and velocity v; protons are slow heavy particles of mass $M = 1,836\,m$, and velocity $V = v/42.85$. If we neglect their electric forces, we can study collisions between these particles.
(a) What are the ratios of momenta of the two particles?
(b) If a light particle collides with a heavy particle at an angle of 90° (Fig. 5.14) and sticks, what is the angle θ at which the composite particle flies away?
(c) If the light and heavy particles collide repeatedly from 90° angles, how many such collisions are required to alter the heavy particle's velocity by a total angle of 45°?

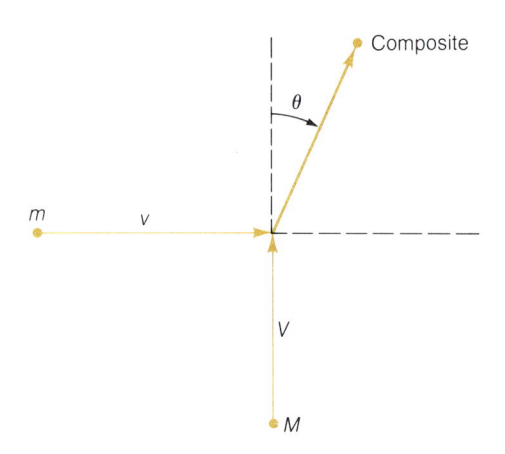

Figure 5.14

•• 23 A 1,000 kg race car, starting from rest, experiences a variable force shown in Fig. 5.15. What is the total impulse of the car? What is the car's final velocity? Neglect friction.

•• 24 To escape from a kidnapper's car, which is travelling at 15 m/s, you jump sideways, perpendicular to the velocity of the car. If you push off with a horizontal force of 1,000 N for 0.2 s, and if your mass is 60 kg, what is your total velocity (three-dimensional) when you strike ground after a 1 m vertical drop?

PROBLEMS

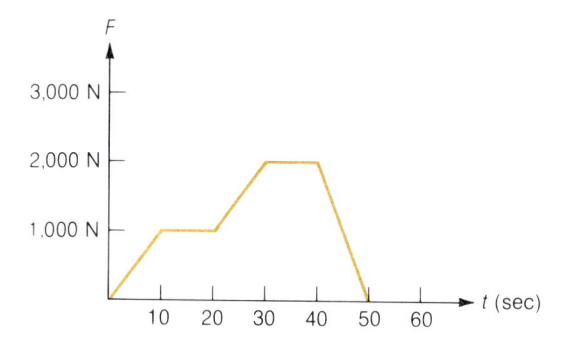

Figure 5.15

- **25** As you stand on bathroom scales, you jump up, exerting a force of 800 N for 0.3 s. If your mass is 60 kg, what do the scales read? (1 N = 4.45 lbs).
- **26** Suppose that every hundred years, the Earth is struck from behind at a 30° angle at 40 km/s by an asteroid (swarms of these asteroids sometimes intersect the Earth's orbit). After how many years would the Earth's orbit change by 1% from its present shape? (That is, the Earth obtains a velocity component perpendicular to its circular velocity which is 1% of its orbital velocity.) Assume the asteroid's mass is 10^{16} kg, the Earth's mass is 6×10^{24} kg, and the Earth's velocity is 30 km/s.
- **27** A pitcher throws 200 g baseballs at a 30 kg target on wheels. The coefficients of friction between wheels and ground are $\mu_s = 0.2$ and $\mu_k = 0.1$. In each 0.2 s collision, the ball reflects backward with 0.5 of its forward velocity.
    (a) How fast must the pitcher throw the ball to start the target moving?
    (b) If the pitcher throws a 40 m/s fastball (about 90 mi/h), how far does the target slide?
- **28** A policeman shoots a stream of water moving at 22 m/s and carrying 15 kg/s of water at a 1,500 kg vehicle moving toward him at 13 m/s. The water stops on impact (assume that it sticks to the car).
    (a) What is the force of the water stream on the car?
    (b) Approximately how long would the stream take to bring the vehicle to a halt, neglecting friction? (This part is difficult. One must graphically compute the impulse due to the variable force of the water stream as the car slows down.)

Section 5.2 Work

- **29** How much work is done against gravity when a 60 kg person climbs 6 flights of stairs (25 m)?
- **30** How much work is required to push a 10 kg vacuum cleaner horizontally across a 4 m floor? The friction coefficient is 0.3 between floor and bottom of vacuum.
- **31** While climbing a mountain pass, a 2,000 kg car will gain 7,000 feet in vertical elevation, but has the choice of various routes with different steepness and different distances. Calculate the work required for a constant velocity climb
    (a) With no friction.
    (b) With friction ($\mu = 0.2$) on a 4° grade.
    (c) With friction ($\mu = 0.2$) on a 6° grade.
- **32** How much work is required for a 100,000 lb airplane to climb from the ground to 35,000 feet? Neglect friction and air drag.
- **33** How much energy is required to pound a nail 2 inches into wood, if the force of resistance is 800 N?
- **34** A 5 kg bucket is lowered into a 50 m well, where it fills with 20 kg of water and is pulled back up. Both steps are done slowly at constant velocity. Calculate the work done by the operator during each step.
- **35** A child pulls a 20 kg sled with 43 N force, applied at an angle 20° above horizontal. If the friction coefficient is 0.2, how much work is done pulling the sled 100 m along level ground?
- **36** A child pulls a loaded sled of total mass M at constant velocity up a slope of angle ϕ. A force F is applied to the sled at angle θ (Fig. 5.16), and μ, the coefficient of friction, is 0.2.

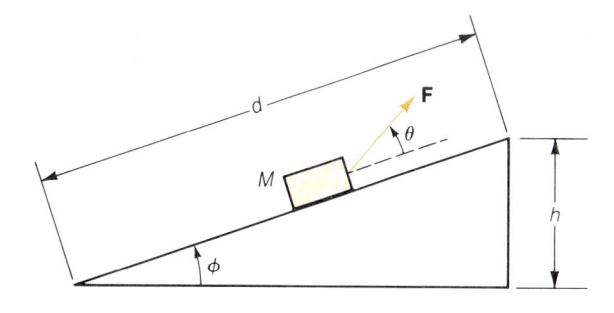

Figure 5.16

(a) By solving the equations of force balance along and perpendicular to the slope, show that $F = Mg [\sin\phi + \mu \cos\phi]/[\cos\theta + \mu \sin\theta]$.

(b) If $\phi = 10°$ and $M = 20$ kg, calculate the work done in pulling the sled 100 m for two different angles, $\theta = 5°$ and $10°$.

(c) If ϕ is fixed, find the angle θ for the child to pull the sled and have the least force. From part (a), notice that the force F is least when the denominator, $(\cos\theta + \mu \sin\theta)$ is largest. Assume $\mu = 0.2$ as before, and use a trial and error method to find θ.

37 A librarian wishes to stack 100 books on the floor. Each book has mass 1 kg and width 3 cm, and must be lifted from floor level to the top of the stack. How much work is required to lift all 100 books, one by one? (Use the mathematical result that the sum of the first N integers equals $(N + 1)(N/2)$.)

38 Suppose, in Problem 37, that the 100 books are already stacked in piles of 10. Calculate the work needed to pile these 10 stacks into one single stack. Why does the work differ from that in Problem 37?

39 A 2,000 kg car brakes uniformly from 25 m/s to rest in 5.5 s. How much work was done?

40 A 100 kg elevator lifts ten 70 kg people to the top of a 60-story building (3.7 m/story). How much work was done, if the elevator moved at constant velocity?

41 Suppose the elevator in Problem 40 accelerated from rest at 0.1 g for 30 stories, then decelerated to rest at the same rate. How much work was done by the motor on the elevator?

42 The force of resistance in compressing a spring a distance x is given by Hooke's law, $F = kx$, where k is a constant. Show that the work done in compressing the spring a distance x is $W = \frac{1}{2}kx^2$.

43 The force of gravity on a mass M (in kilograms) a distance r (in m) from the center of the Earth is not constant, but is $F = (4 \times 10^{14} \text{ N} \cdot \text{m}^2/\text{kg}) M/r^2$. How much work is done against gravity in launching a 10,000 kg rocket from ground level ($r = 6,380$ km) to an elevation of 3,000 km? Plot F versus r and determine the work from the graph.

44 Two 1,500 kg cars race up a 10 km, 5° hill. The vehicles have coefficients of road friction $\mu_1 = 0.1$ and $\mu_2 = 0.12$. Travelling at constant speeds, the first car finishes the race in 10 minutes, the second in 12 minutes. Calculate the work done against gravity and friction by each car. Are the finishing times needed?

Section 5.3 Energy

45 What is the KE of the following:
(a) A 0.01 gram fly moving at 1 m/s.
(b) A 100 gram bird flying at 10 m/s.
(c) A 20,000 kg plane flying at 300 mi/h (134 m/s).
(d) The Earth moving around the Sun at 30 km/s.
(e) The Sun moving around the Milky Way at 250 km/s.
(f) The Milky Way (5×10^{11} solar masses) falling toward the Virgo cluster of galaxies at 450 km/s.

46 A speeding bullet (100 g, 1,000 m/s) strikes Superman (100 kg), and stops inelastically. If momentum is conserved, what are the energies of the bullet initially and of Superman afterward? Neglect friction.

47 Air at room temperature has an average energy of 6×10^{-14} ergs per molecule. If the average mass per molecule is 29 atomic mass units (1 u = 1.66×10^{-24} g), calculate the average velocity of the air molecules.

48 How many nutritionist's Calories are burned up when a 60 kg person climbs a 10 story building (3.7 m per story)? Compute only work against gravity. (1 Cal = 4,180 J)

49 How fast is a 70 kg sprinter moving if his KE is 3500 J?

50 The energy put into accelerating a 100 kg object 100 m along a frictional track is 60,000 J. The coefficient of friction is 0.3. What is the final velocity?

51 A 50 kg child sleds down a 20° hill, 50 m long. The sled's mass is 20 kg and its final velocity is 8 m/s. How much energy was dissipated as heat? Neglect other dissipation.

52 In Problem 51, use the amount of frictional heat dissipated to evaluate the average coefficient of friction along the hill.

53 How much work must be done to stop a 2,000 kg car moving at 60 mi/h?

54 A 20 kg bicycle (and 70 kg rider) accelerate by gravity down a 15° hill. Calculate the final velocity after 400 m, without friction and with friction ($\mu = 0.2$).

55 Calculate the gravitational potential energy of a 80 kg climber atop Mt. Everest (29,000 feet above sea level). Express your answer in both Joules and Calories (1 Cal = 4,180 J).

PROBLEMS

56 Suppose a spring-loaded catapult obeys Hooke's law, $F = kx$ (see Problem 42). If the catapult can shoot a 50 kg rock 30 m straight up in the air, calculate the rock's initial velocity when shot at 45°. If the catapult was compressed a distance $x = 30$ cm before release, what was the spring constant k?

57 Artillery observers noticed that certain cannons were capable of firing 100 kg shells 1,000 m vertically.
(a) What was the muzzle velocity of the shell?
(b) What would be the maximum horizontal range of the shell?

58 How fast must a fountain spray water to reach a 20 m height? Neglect friction or drag.

59 A six-foot tall high jumper can clear a bar 7'7" above ground, raising his center of gravity from 1.0 m to 2.31 m. How high could he jump on the Moon (where gravity is one sixth that on Earth)?

60 A 65 kg swimmer uses up 3,000 J of energy during a 100 m race in which her final speed is 2 m/s. How much energy was expended against fluid friction with the water?

61 A 70 kg parachutist accelerates from rest to a terminal velocity of 120 mi/h during free fall from a height of 4,000 m to 3,000 m. How much energy was lost to air friction?

62 A projectile is released at angle θ from horizontal with initial velocity v. From energy considerations, derive general formulas for
(a) The maximum height h during the flight. (Neglect air drag.)
(b) The horizontal range on level ground. (Neglect air drag.)

63 A boy can throw a stone over a building of height h. How far can he throw the same stone horizontally on level ground? Ignore air drag.

64 A roller coaster (Fig. 5.17) has a speed of 1 m/s at point A. What are its speeds at points B, C, and D, neglecting friction?

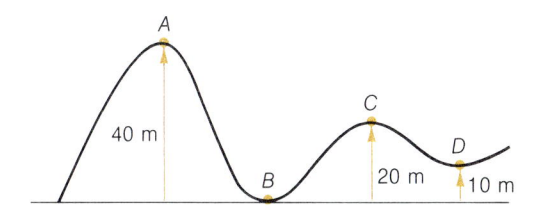

Figure 5.17

65 A mass m slides from rest at height h down a frictionless track with a "loop-the-loop" of radius r (Fig. 5.18) before being released horizontally a distance d above ground. From energy conservation, find
(a) The horizontal velocity v_0 and the velocity at the top of the loop.
(b) The minimum release height h required to keep the mass on the track at the top of the loop. (To keep on track, the outward centrifugal force, $F_c = mv^2/r$, must exceed the downward force of gravity, $F_g = mg$.)

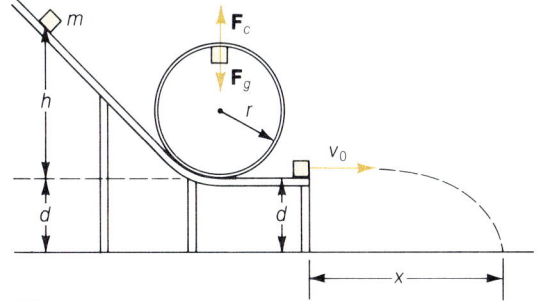

Figure 5.18

66 In Fig. 5.18, if $h = 3r = 3$ m, and $d = 2$ m, how far does the mass travel horizontally, x?

67 A bullet of mass m and velocity v is stopped inelastically by a pendulum of mass M, supported by a massless wire (Fig. 5.19). Using conservation of momentum and energy where valid, derive an expression for the maximum height of the pendulum h in terms of m, M, v, and g.

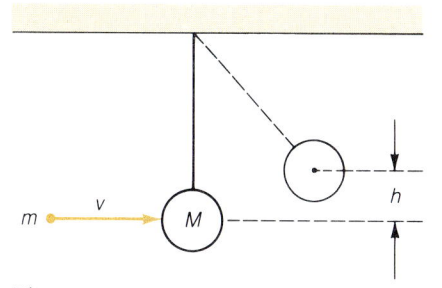

Figure 5.19

68 Masses of 2 kg and 4 kg hang at rest on opposite sides of a frictionless pulley, then the masses are released. Calculate the speed of the 4 kg mass when it has fallen 0.4 m from its initial position. Use energy conservation.

148 CHAPTER 5 MOMENTUM, WORK, AND ENERGY

69. How many meters can a 125 lb person climb by utilizing the energy contained in 10 g of sugar? Assume 10% of the energy (1 g of sugar contains about 4 Calories) is converted into usable form.

70. The heat of combustion of 1 g of wood is about 4 Calories. How much wood must be burned in order to accelerate a 2,000 kg steam engine (20% efficiency) from rest to 50 mi/h in 1,000 m against friction ($\mu = 0.3$) on level ground?

71. The heat of combustion of gasoline is 11.5 Calories per g. One gallon contains 3,786 cm^3 and gasoline has a density of 0.68 g/cm^3
    (a) Compute the energy contained in one gallon of gasoline.
    (b) If a 2,000 kg car is 30% efficient, how many gallons of gasoline are needed to climb a 3,000 m, 6° hill, if $\mu_{eff} = 0.05$?

72. Every second, the Sun's rays, shining through the atmosphere on a 1 m^2 area, deposit about 500 J. If solar panels are 10% efficient, compute how long a 3 m × 4 m panel must collect energy to equal the equivalent of 1 gram of coal, which has a heat of combustion of eight Calories/g.

73. A motorcyclist starts at rest on one side of a 50 m deep ravine, descends to the bottom under gravity, then rolls up the other side (Fig. 5.20). Cycle and rider have a mass of 300 kg. How high up the other side does he get
    (a) Assuming no friction.
    (b) With a frictional force 0.3 times his weight.

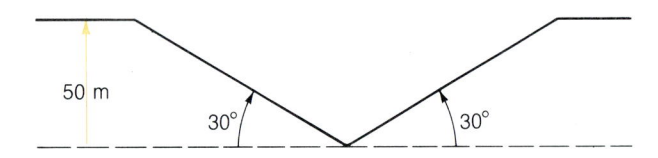

Figure 5.20

74. In the collision of particles described in Problem 18, calculate how much energy was lost in the inelastic collision.

Section 5.4 Power

75. Show, by conversion of units, that the British horsepower (550 ft-lb/s) equals 746 W, different from the metric horsepower of 750 W.

76. What power (in both W and hp) is required to push a 50 kg crate at 1 m/s across a floor with friction coefficient $\mu = 0.3$?

77. If a car requires 30 hp to cruise at 60 mi/h, what is the average force of resistance due to friction and air drag?

78. How many watts are equivalent to 1 Calorie per hour?

79. A 1,500 kg car can generate 200 hp, of which 60% is applied to the road.
    (a) What is the steepest hill the car can climb at 30 mi/h, if frictional forces average 1,000 N?
    (b) At that angle, is the normal component of weight sufficient to maintain frictional support?

80. The flow in a 150 m waterfall is 1,000 kg/s. Calculate the velocity of water at the bottom of the falls, and the power available, in principle, for a hydroelectric plant.

81. A 60 kg woman runs up a 10 story building (3.7 m per story) in 1 minute. Calculate her average power output in W and hp.

82. The winning race car in the Pike's Peak hill climb had a time of 12 minutes. If the road rises 8,000 ft, with average slope 10°, and if the frictional forces average 800 N, calculate the horsepower required for a 1,000 kg car with 70% efficiency.

83. Electric companies measure energy in kilowatt-hours. How many Joules in 1 kWh? How many Calories?

84. A fountain pumps 100 gallons of water per second 20 m in the air. If one gallon weighs 8.35 lbs, how many horsepower are required for the pump?

85. A 4 hp motor is attached to a pulley system, with AMA = 3. What force is required to lift a 500 lb piece of machinery a distance of 5 stories (19 m)? How long will the lift take?

86. A steam engine burns coal at 100 kg/h. What power is it capable of producing at 30% efficiency? Coal has a heat of combustion of 8 Calories per gram.

87. If the maximum Caloric intake of an active person's diet is 8,000 Cal/day, what is the person's effective power output in watts?

88. A 100-ton rocket reaches escape velocity (about 25,000 mi/h) four minutes after launch. Compute the average horsepower, neglecting friction, air drag, and work against gravity.

89. For the rocket in Problem 88, estimate the power consumed in overcoming gravity. Assume constant acceleration in a vertical flight, and take the force of gravity to be constant.

90. A lawnmower battery runs a 2 hp motor and two 100 W bulbs. What is the drain of chemical potential energy during 20 minutes?

PROBLEMS

91 To prepare for the next day's power needs, a hydroelectric plant wishes to pump 10 million gallons of water overnight (10 h) from a reservoir into a storage lake 200 m in elevation. How many hp are required?

92 For the power plant in Problem 91, what is the power output generated during the day, if the turbines are 70% efficient and if water flows back to the reservoir at 1,000 kg/s?

93 The human heart pumps about 5,000 cm^3 of blood through the circulatory system in about 15 s. For a 5'8" person, estimate the work rate of the heart (in W) against gravity. Neglect fluid friction and assume that blood has a density about 1 g/cm^3.

Section 5.5 Machines

94 For each of the three classes of levers shown in Fig. 4.38, calculate the IMA.

95 Calculate the IMA of a pair of scissors, whose length is 20 cm. The force is applied at one end, the blades intersect midway along their length, and the paper being cut is placed 5 cm from the intersection.

96 A crowbar is placed so that 2 cm of the total 100 cm length lies under a rock. Calculate the IMA.

97 A wedge may be thought of as a lever by which a lesser force F, applied to the wedge top, can transmit a greater force F' laterally. Calculate the IMA of a wedge with opening angle θ. (Fig. 5.21).

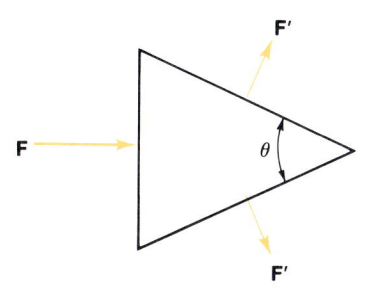

Figure 5.21

98 To climb a 1,000 m hill, a 6,000 kg truck travels along a road which maintains a 6° incline. If the effective friction coefficient $\mu = 0.05$, calculate the AMA, IMA, and efficiency of the grade.

99 To ascend 2 m, piano movers slide a 600 kg piano up a 10 m ramp. If the coefficient of friction is 0.25, what are the IMA, AMA, and efficiency?

100 Calculate the IMA of each pulley system in Fig. 5.22.

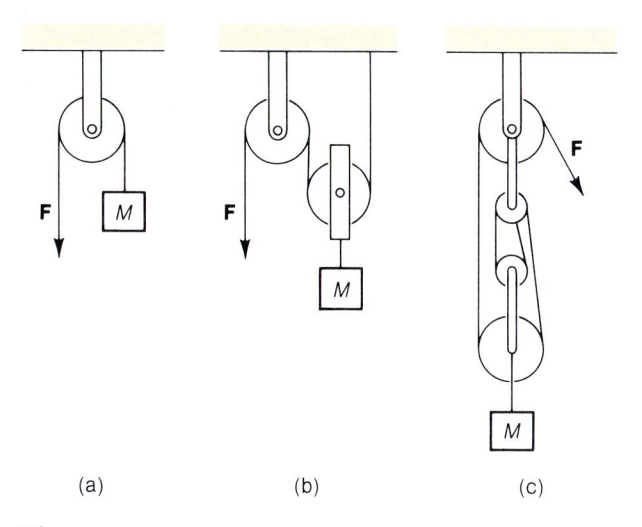

(a)   (b)   (c)

Figure 5.22

101 Show that for an inclined plane, with angle θ and friction coefficient μ, in which the force is applied along the plane, IMA = $1/\sin\theta$, AMA = $1/(\sin\theta + \mu\cos\theta)$, and efficiency = $1/(1 + \mu/\tan\theta)$.

102 What is the IMA from chain to road of a bicycle wheel, in which force is applied by the chain, 4 cm from the axle, and exerted at the rim, 70 cm from the axle?

103 What is the IMA of the biceps muscle shown in Fig. 4.34?

104 What are the IMA and AMA for the child pulling the sled in Fig. 5.16? Give your answer in terms of angles θ and ϕ and the friction coefficient μ.

105 For the sled in Problems 36 and 104, what angle θ gives the best AMA if $\mu = 0.2$? Your answer should be independent of the slope angle ϕ.

106 A farmer uses a set of pulleys to hoist a 100 kg bale of hay 4 m into the barn window. The bale rises 1 cm for every 50 cm pull on the pulley rope. If the farmer expends 6,000 J of energy, calculate the IMA, AMA, and efficiency of the pulleys.

107 A prehistoric architect wished to construct a stone monument by lifting 1 m cubes of granite (density 2.7 g/cm^3) a vertical distance of 3 m. If workers of that era could exert a maximum 1,000 N force on pulleys of 60% efficiency, what minimum IMA must the pulleys produce?

SPECIAL PROBLEMS

1. Radioactive Nuclear Decay

When radioactive nuclei decay, they often split apart into smaller particles. A common mode of decay emits alpha particles, which are helium nuclei. Consider the alpha-decay of radioactive uranium-238, which splits into thorium-234 and an alpha particle (Fig. 5.23).

The masses of the particles are given in atomic mass units (1 u = 1.6606×10^{-27} kg).
(a) For the decay of a nucleus of uranium-238 at rest, write equations describing conservation of momentum and energy, assuming that the total energy released (sum of energies of thorium and alpha nuclei) is 6.87×10^{-13} J.
(b) Compute the velocities v_1 and v_2, and explain why they are oppositely directed (180° from each other).
(c) Why does the alpha particle move faster than the thorium nucleus?

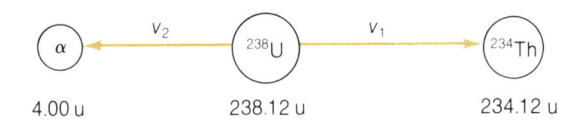

Figure 5.23

2. Rocket Problem

For a rocket to move forward, it must shoot mass at high velocity in the opposite direction. Suppose a rocket alone has a mass $m = 10^6$ kg and carries an initial mass of fuel $M_o = fm$ (f is the fuel-to-load ratio and is usually larger than 1). The rocket boosters eject hot gas at 10^5 m/s at 10^3 kg/s.
(a) Compute the upward thrust of the rocket.
(b) What is the maximum mass of fuel that the rocket can carry to get off the launchpad? What is f?
(c) If $M_o = 5 \times 10^6$ kg ($f = 5$), what is the initial acceleration?
(d) Using Newton's law, $F = (M + m)a$, estimate and plot the acceleration of the rocket versus time, from launch ($t = 0$) until the time of fuel exhaustion ($t = 5,000$ s). What is the rocket's final velocity? (The fuel mass M decreases with time, so you will need to compute a new acceleration for each time step after launch. Fairly accurate results may be obtained for time steps of 500 s.)

3. Calorie Intake

A typical diet for an active person is about 3,000 Calories per day (1 Calorie = 4,180 J). Calculate the energy output for the following activities. (Give your answer in W and in Calories per hour.)
(a) *Normal Rate at Rest*
    How many watts does the body use to keep itself at 98° F, resulting in heat losses of about 60 W/m² over the body? Compute the surface area of a typical human body.
(b) *Walking at 4 Miles per Hour*
    Compute the energy expended in work against gravity and friction. Assume 65 kg mass and average effective friction coefficient $\mu = 0.3$. For the gravitational work, assume that during each step of 0.8 m, one lifts the center of gravity of each 25 lb leg a distance of 10 cm. Why doesn't one regain the potential energy on the downstep?

4. Pyramid Construction

To construct the pyramids, the Egyptians used levers, inclined planes, and many humans. To appreciate the amount of human labor involved, consider a small pyramid, with a square base 25 m × 25 m and four triangular sides inclined at 45° (Fig. 5.24). The pyramid is constructed of 1 m³ stone blocks of 2,500 kg mass laid in 13 levels. Each level is a square, with two fewer blocks per side than the one below.
(a) Show that the pyramid contains 2,925 blocks and calculate the mass of the pyramid in kg.
(b) Calculate the gravitational potential energy of the pyramid. That is, how much work against gravity was required to lift the 2,300 blocks above the first level? (Note that the first 625 blocks required no lifting.) You may need a calculator here.
(c) If an Egyptian worker was capable of expending 1,000 Calories worth of work per day (1 Cal = 4,180 J), how many slave-days were needed to build this small pyramid?

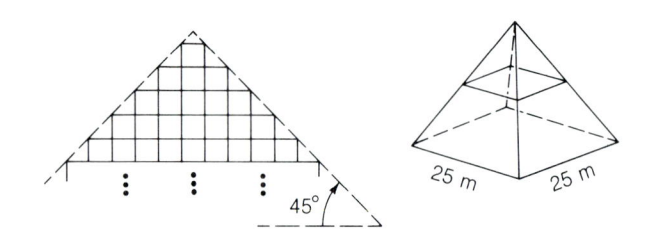

Figure 5.24

5. Automobile Crashes

It is well known that head-on collisions are far more damaging than collisions with stationary objects. Consider (Fig. 5.25) two crashes: (I) a 2,000 kg car collides at 55 mi/h with a stationary car of the same mass, (II) two 2,000 kg cars collide head-on at 55 mi/h. Suppose that the collisions are totally inelastic (the cars stick together).

(a) For each case, draw a diagram of the relative velocities before and after the collision, as viewed in the frame of one of the moving cars. (Show the relative velocity of approach as seen by one of the drivers.)

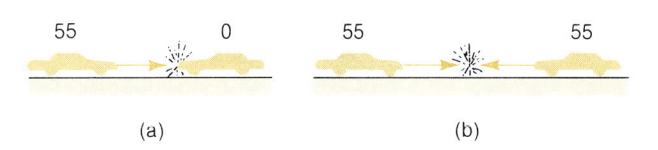

Figure 5.25

(b) Using conservation of momentum, compute the total KE, before and after the collision.
(c) Show that the KE lost in (II) is four times that lost in (I). Why? Where does the KE go?

SUGGESTED READINGS

"In judo and aikido, applications of the physics of forces makes the weak equal to the strong." J. Walker, *Scientific American*, Jul 80, p 150, **243**(1)

"The crossbow." V. Foley, G. Palmer, W. Soedel, *Scientific American*, Jan 85, p 104, **252**(1)

"The mechanics of swimming and diving." R. L. Page, *The Physics Teacher*, Feb 76, p 72, **14**(2), *errata*, May 76, p 309

The flying circus of physics with answers. Jearl Walker, Wiley, New York, 1977

Newton at the bat: The science in sports. Eric W Schrier, William F. Allman, eds, Scribners, New York, 1984

Sport science. Peter J. Brancazio, Simon and Schuster, New York, 1984

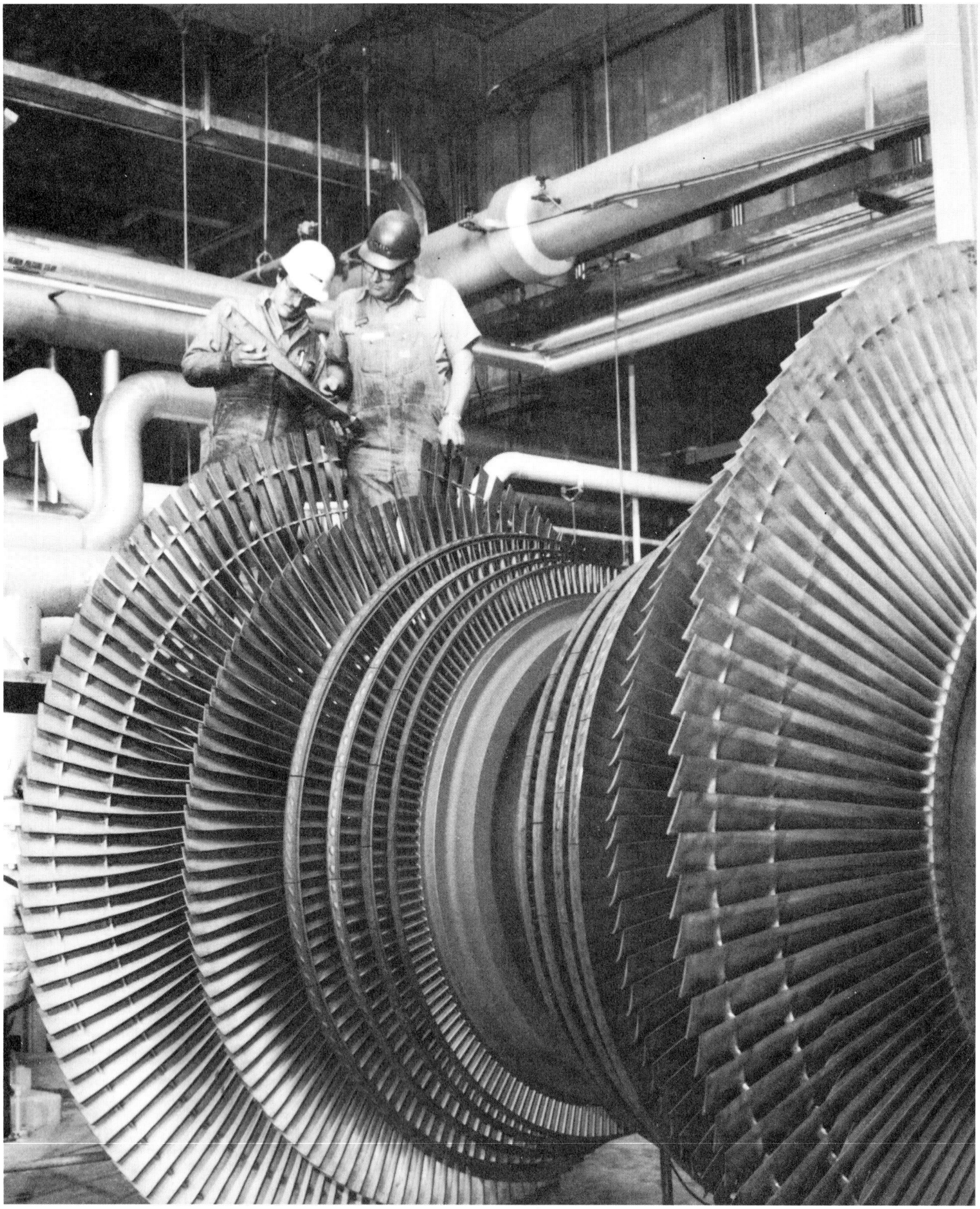

Chapter 6

Circular and Rotational Motion

We have learned quite a bit about how to describe motions and solve problems, but there remains a large class of motions that we have avoided. We know that acceleration occurs whenever an object's direction of motion changes, yet we have not discussed this, nor the forces that cause it. Curved motion is a very complex subject, but sometimes it can be treated quite simply. In this chapter we will discuss the simpler cases. In circular or rotational movement, symmetries make the mathematical treatment rather straightforward compared to nonlinear accelerated motion.

Most of the circular motions in nature involve the rotation or orbits of celestial bodies, which are discussed in the next chapter. However, countless artificial situations involve such motions because most machines have rotating parts. Wheels and bearings are good devices for reducing friction, and rotating shafts, gears, propellers, and flywheels commonly store or transmit forces or energy.

6.1 Measures of Angular Quantities

Just as we had to define certain quantities related to linear motions, such as displacement, velocity, acceleration, force, and momentum, we must now introduce corresponding quantities for circular motion. Like their linear counterparts, the quantities that describe circular motion are also vectors. The direction of the angular vectors is given by the right-hand rule: the vector representing an angular quantity is along the axis of rotation, in the direction your right thumb points when you make a fist and rotate it, fingertips first, in the direction of the angular quantity. We will generally not use vector notation in this chapter, however, because for most applications here, angular quantities are aligned, so that addition and subtraction do not require vector techniques.

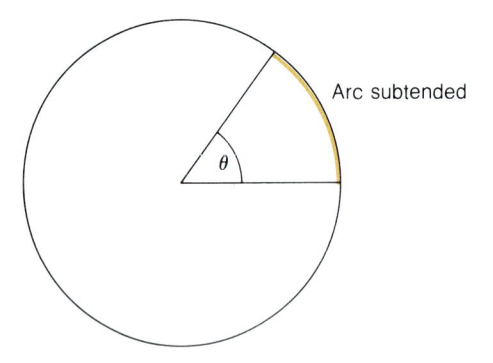

Figure 6.1

Arc and angle. The angle θ subtends the arc.

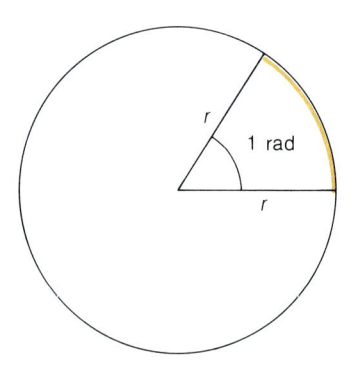

Figure 6.2

The radian. The angle subtended by an arc equal to the radius of the circle is one radian. Thus, there are 2π radians in a full circle.

Angular Displacement

To begin, we must address the question of measurement of angles. We will often express angular quantities in terms of the radius of curvature, and this leads us to a unit of angle that is much more convenient than the degree. The degree is, after all, quite arbitrary; it was introduced by ancient Babylonian astronomers who based it on their lunar year of 360 days (and a few extra days as needed to reconcile the calendar with the actual year).

The circumference of a circle is $2\pi r$, where r is its radius. A segment of arc along the circumference of a circle corresponds to a certain angle which that arc subtends (Fig. 6.1). If the circumference is divided into segments equal to the radius of the circle, we define our unit of measure, called the **radian** (rad), as the angle subtended by one such arc (Fig. 6.2). Since the circumference of a circle is $2\pi r$, there are 2π radians in a full circle, corresponding to 360°, so 1 rad = 360°/2π = 57.3.°

If we now express in radians the angle θ subtended by a segment of arc s along the circumference, then we have the simple relation

$$\theta = \frac{s}{r}. \tag{6-1}$$

This expression will be very useful when a distance and an angle are known, and a size is wanted, or a size and angle are known, and the distance is wanted.

Example 6.1   A golfer is preparing to hit his ball towards a distant green, but doesn't know how far it is. He has a device for measuring angles, though, and finds that the flagstick on the green has an angular height of 0.64° (Fig. 6.3). He knows that the actual height of the flagstick is 6 ft, and he wants to find the distance.

6.1 MEASURES OF ANGULAR QUANTITIES

Actually, we are asking for the value of s in Eq. 6–1, with r representing the distance from the golfer to the flag, and θ the angular height of the flag (this is only approximately correct, since we are assuming that the actual height of the flag is equal to the segment of a circle whose radius is the distance from the golfer to the flag. We express the angular height of the flagstick in radians, finding that $\theta = 0.011$ rad. Then the radius of the circle is

$$r = \frac{s}{\theta} = \frac{(6 \text{ ft})}{0.011} = 545 \text{ ft}$$

or about 180 yards.

Devices which do exactly what we have done in this example are commercially available to golfers.

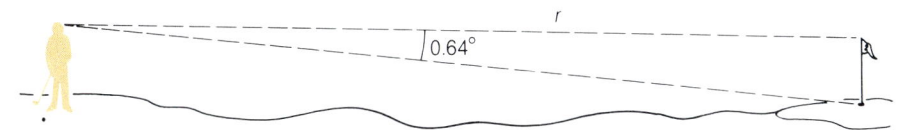

Figure 6.3

Angular Velocity

The circular equivalent of linear displacement is expressed as an angular measure, as we have just seen. Similarly, the equivalent of linear velocity is **angular velocity,** expressed in terms of rate of change of angle. In mathematical form, when the rate of change is constant, we write

$$\omega = \frac{\theta}{t} \quad (6-2)$$

where ω is the Greek letter omega, used by convention to represent angular velocity; and θ/t is the time rate of change of the angle θ. Angular velocity is measured in units of radians per second (rad/s).

For a simple example, suppose a car wheel rotates 10 times per second. The angular displacement of a point on the wheel is 2π radians for each rotation, so the angular velocity is $(10/s)(2\pi \text{ rad}) = 62.8$ rad/s.

The angular velocity and displacement are related to each other in the same way that linear displacement and velocity are.

$$\theta = \theta_0 + \omega t \quad (6-3)$$

Eq. 6–3 can be used to calculate the final angle after a period of rotation at a known constant angular velocity (compare to Eq. 2–2).

Eq. 6–2, strictly speaking, applies only when the rate of change of angle is constant, or it represents the average angular velocity as θ changes over

156   CHAPTER 6   CIRCULAR AND ROTATIONAL MOTION

time. Therefore, we need to define the instantaneous angular velocity when the angular velocity is variable.

$$\omega = \frac{\Delta\theta}{\Delta t} \quad (6\text{--}4)$$

Here the concept of a limit is useful: we say that the instantaneous angular velocity is the limit of $\Delta\theta/\Delta t$ as Δt approaches zero. Graphically, this limit is represented by the tangent to a plot of θ versus t (Fig. 6.4). The total change in angle as the result of a variable angular velocity acting over time is represented by the area under a curve (Fig. 6.4), just as total linear displacement can be found by finding the area under a curve of linear velocity, as we saw in Chapter 2. We will often find that in real situations the angular velocity is constant, so we will not have to deal with graphical methods.

Two new quantities that we did not encounter in our earlier discussion of linear motions arise in rotational and circular motion. One is the **period**, which is the time required for a full circle or revolution to be completed. The period is the time T required to undergo an angular displacement of 2π rad. We set $\theta = 2\pi$ rad in Eq. 6–2, and solve for T.

$$T = \frac{\theta}{\omega} = \frac{2\pi}{\omega} \quad (6\text{--}5)$$

Closely related to period is the **frequency**, the number of revolutions per unit time. The relationship between period and frequency is very simple; consider the car whose wheels have a frequency of 10 revolutions per second. If the wheels rotate ten times in a second, each revolution must take one tenth of a second. Hence we see that the frequency and the period are simply reciprocals of each other, and we can write

$$\nu = \frac{1}{T} = \frac{\omega}{2\pi}. \quad (6\text{--}6)$$

Very often when we discuss machines with rotating parts, we speak of the frequency, rather than the angular velocity or the period. Frequency is expressed in revolutions per second (or cycles per second), which has the special name **hertz**. One hertz (Hz) is equal to one revolution per second.

When a rigid body rotates, every point on the body undergoes translational motion (Fig. 6.5). The relationship between the speed of the point and the angular velocity can be found by combining Eq. 6–1 and Eq. 6–2. If we solve Eq. 6–1 for s and then divide by t, we find

$$\frac{s}{t} = \frac{r\theta}{t}$$

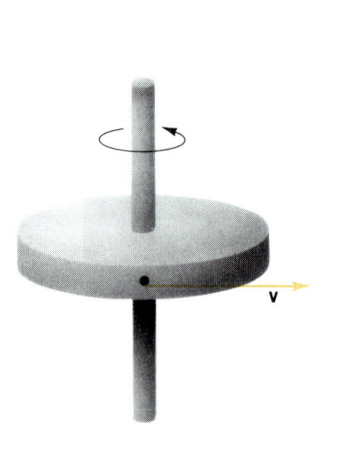

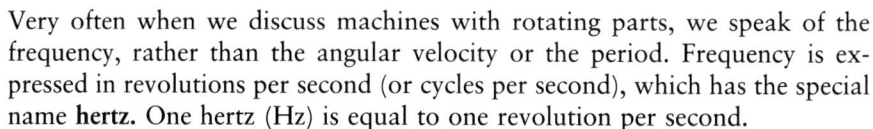

but s/t is the **linear** speed of a point at r from the center or axis of rotation, and θ/t is the angular velocity, so we have

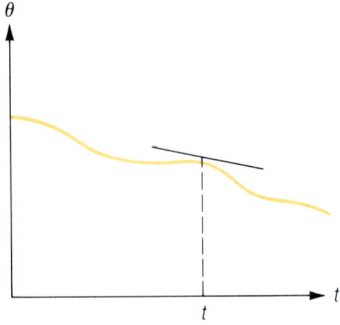

$$v = r\omega. \quad (6\text{--}7)$$

Figure 6.4

Instantaneous angular velocity. The instantaneous angular velocity at time t is equal to the slope of the tangent to the curve.

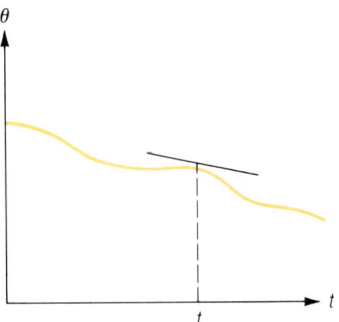

Figure 6.5

Linear velocity of rotation. The vector v represents the instantaneous linear velocity of a point on the rotating disk.

6.1 MEASURES OF ANGULAR QUANTITIES

This expression tells us how fast a point on a rigid rotating object is moving. Here the direction of v is always tangent to the circular path of the point. In terms of the period, the linear velocity is

$$v = \frac{2\pi r}{T}. \tag{6-8}$$

That is, a point on a circle moves through a linear displacement equal to the circumference of the circle in time T. In terms of the frequency, the linear velocity is

$$v = 2\pi r \nu. \tag{6-9}$$

For example, the earth spins once in 24 hours (86,400 s), and has an average radius of 6,368 km. From Eq. 6–8 we see immediately that a point on the equator has the linear velocity

$$v = \frac{2\pi r}{T} = \frac{2\pi(6.368 \times 10^6 \text{ m})}{(86,400 \text{ s})} = 463 \text{ m/s} = 1036 \text{ mi/h}.$$

It is interesting that the linear velocity varies as a function of distance from the axis of rotation. Thus, for example, a person north or south of the equator has a smaller linear velocity than we have just calculated, because the distance from the axis of rotation is smaller. For this person, the value of r to use in Eq. 6–8 is the earth's radius times the cosine of his latitude.

Example 6.2

An engineer is designing a speedometer for a new model car. He knows the tire radius, and he needs to establish a calibration for the car's speed in terms of wheel frequency, which is directly related to the frequency of the shaft that turns the speedometer. Suppose the tire radius is 0.3 m, and the engineer wants to know the wheel frequency that corresponds to a linear speed of 40 mi/h.

Using Eq. 6–9, we substitute for r and v (40 mi/h is 17.88 m/s), and solve for the frequency ν.

$$\nu = \frac{v}{2\pi r} = \frac{(17.88 \text{ m/s})}{2\pi(0.3 \text{ m})} = 9.5 \text{ Hz}$$

The engineer now sets up the speedometer so that the gauge reads 40 mi/h when the wheels are rotating at 9.5 revolutions per second.

Example 6.3

Suppose that for whimsical reasons we want to know how fast the stylus on a phonograph turntable runs in the groove on a record, and how long the groove is. Assume that the record is 12 in in diameter, and that the continuous groove starts at an outer radius of 5.7 in and ends at an inner radius of 3.2 in. This is a long-playing record, designed to play at 33.3 revolutions per minute for about 40 minutes.

158 CHAPTER 6 CIRCULAR AND ROTATIONAL MOTION

The frequency in hertz is $\nu = 0.56$ Hz. The speed of the record relative to the stylus is a function of the radius, so we calculate the speed separately at the outer and inner limits of the play area, and then we can take the average. The radii at these points are 0.14 m and 0.08 m. The linear velocity at the outer point is

$$v = 2\pi r \nu = 2\pi(0.14 \text{ m})(0.56 \text{ Hz}) = 0.49 \text{ m/s}$$

and at the inner point it is

$$v = 2\pi(0.08 \text{ m})(0.56 \text{ Hz}) = 0.28 \text{ m/s}.$$

In miles per hour, these velocities are 1.1 mi/h and 0.63 mi/h. Note that, once we find the velocity at any radius r, we can find it at any other radius simply by scaling our answer proportionally to the change in r, since v is proportional to r for a rigid object.

The average speed is 0.39 m/s = 0.87 mi/hr. The length of the groove is then

$$L = vt = (0.39 \text{ m/s})(40 \text{ min})(60 \text{ s/min}) = 936 \text{ m} = 0.58 \text{ mi}.$$

Angular Acceleration

Once we have the concept of angular velocity, we can easily define **angular acceleration** as the rate of change of angular velocity, the change in angular velocity divided by the time over which the change occurred. If an object is accelerated from rest to an angular velocity ω in time t, then the angular acceleration α, equal to the change in ω divided by the time interval, is

$$\alpha = \frac{\omega}{t} \tag{6–10}$$

where α is in rad/s². More generally, the angular acceleration is

$$\alpha = \frac{\Delta \omega}{\Delta t} \tag{6–11}$$

where $\Delta \omega$ is the change in angular velocity that occurs in time Δt. Although the angular acceleration may vary so that graphical methods (such as we have described in discussing variable linear acceleration) would be needed, we will confine ourselves to constant angular acceleration.

We can write an equation like Eq. 2–5 for the angular velocity after angular acceleration α has occurred over time t.

$$\omega = \omega_0 + \alpha t \tag{6–12}$$

We also have (compare with Eq. 2–9)

6.1 MEASURES OF ANGULAR QUANTITIES

$$\theta = \theta_0 + \omega_0 t + \frac{1}{2}\alpha t^2. \tag{6-13}$$

If the origin of either of these equations is unclear, it might be useful to derive them. Follow the steps used in Chapter 2 to derive Eq. 2–5 and Eq. 2–9, but substitute the angular quantities.

Just as a point on a rotating body has linear velocity related to the angular velocity of the body, it also has linear acceleration if the body is undergoing angular acceleration. The linear acceleration consists of a tangential component and a component directed inward, toward the center of rotation. We will discuss the inward component, which changes only the **direction** of motion, in a later section; here we consider the tangential component, which changes the speed. The angular acceleration is related to the tangential linear acceleration a at some distance r from the center of rotation by

$$a = r\alpha. \tag{6-14}$$

Like the linear velocity due to circular or rotary motion, the linear acceleration varies with distance r from the center of rotation.

Example 6.4 A windmill is accelerated by a gust of wind from rest to a frequency of 2 Hz in a time of 20 s. What is its angular acceleration? If each blade is 3 m long, what is the linear acceleration of a point at its tip? How many revolutions does the windmill undergo during this time?

Here we assume that the acceleration was uniform, so that Eq. 6–10 can be used, with a substitution for ω from Eq. 6–6.

$$\alpha = \frac{\omega}{t} = \frac{2\pi\nu}{t} = \frac{(2\pi \text{ rad})(2/\text{s})}{20 \text{ s}} = 0.63 \text{ rad/s}^2$$

The linear acceleration at the tip of one of the blades is

$$a = r\alpha = (3 \text{ m})(0.63 \text{ rad/s}^2) = 1.89 \text{ m/s}^2.$$

To find the number of revolutions of the windmill during the 20 s acceleration period, we use Eq. 6–13 (with θ_0 and ω_0 both set equal to zero) to calculate the angle through which it turned.

$$\theta = \frac{1}{2}\alpha t^2 = (0.5)(0.63 \text{ rad/s}^2)(20 \text{ s})^2 = 126 \text{ rad}$$

The number of revolutions is this angle divided by 2π, the number of radians per revolution. We find that the windmill rotated 20 times. This could also have been found by noting that the average rotational velocity during the start-up period was 1 Hz, so in 20 s the windmill rotates 20 times.

6.2 Circular Motions of Particles

Most of the equations we have developed so far can be applied to either rotating, extended bodies or to discrete particles that circle about some fixed point. Some considerations, however, apply only to a detached body moving about a center. Now we will consider a body in **uniform circular motion**; that is, it circles at constant angular velocity, with zero angular acceleration.

Centripetal Acceleration

Even though the angular acceleration of a body in uniform circular motion is zero, it is still undergoing linear acceleration. Recall that in Chapter 2 we were very careful to specify that *any* change in velocity constitutes acceleration. Since velocity is a vector quantity, this means that if the *direction* of motion is changing, even though the speed is not, the body is undergoing acceleration. Clearly an object in uniform circular motion is constantly changing its direction and must therefore be undergoing acceleration.

We can see intuitively what the direction of this acceleration is by imagining a velocity vector attached to the body whose velocity it represents (Fig. 6.6). As this vector travels around the circle, its direction constantly changes so that it is always tangent to the circle. Hence the direction of the change is always towards the center of the circle. Because of this, the acceleration is called **centripetal acceleration**; centripetal means "seeking the center."

To quantify the centripetal acceleration, we consider the vector difference between the vectors **v** and **v'**, which represent the velocity at two times. The acceleration is given by Eq. 2–4

$$\mathbf{a}_c = \frac{\Delta \mathbf{v}}{\Delta t} = \frac{(\mathbf{v}' - \mathbf{v})}{\Delta t}$$

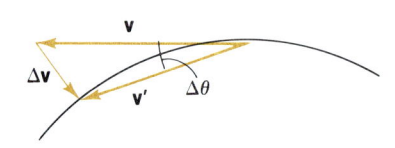

where $\mathbf{a}_c$ represents the centripetal acceleration.

In order to evaluate this expression, we consider Fig. 6.7, in which the vectors **v** and **v'** have been placed together so that the dashed vector $\Delta \mathbf{v}$ represents the difference between them. The velocity change $\Delta \mathbf{v}$ appears approximately equal to the segment of arc connecting the end of vector **v** with **v'** (this becomes exactly true as $\Delta \mathbf{v}$ approaches zero). Application of Eq. 6–1 to Fig. 6.7 shows that

$$\Delta \theta = \frac{\Delta v}{v}.$$

We note that the change of angle $\Delta \theta$ here is the same as the angle through which the body moves along its circular path. In Fig. 6.8 we see that as the body moves from *A* to *B*, the radial lines connecting it to the center of the circle change by $\Delta \theta$, and that this is the same angle by which the velocity vector changes.

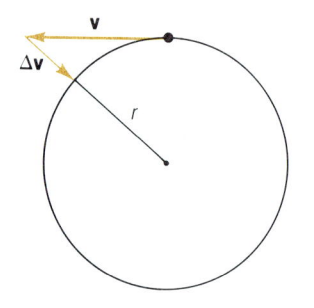

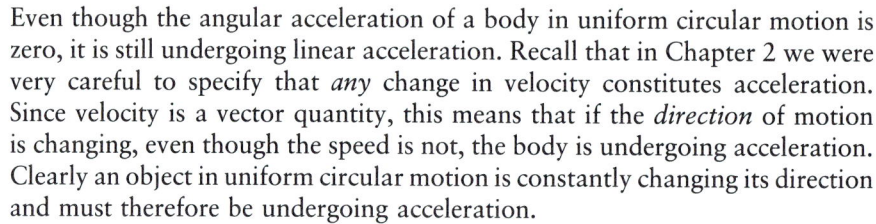

Figure 6.6

Centripetal acceleration. The change of velocity required to maintain circular motion is directed toward the center of the circle.

Figure 6.7

Change in velocity corresponding to centripetal acceleration. The small vector $\Delta \mathbf{v}$ represents a small change in velocity for a body moving in a circular path.

6.2 CIRCULAR MOTIONS OF PARTICLES

Rearranging and dividing by Δt, the small time interval when the angle changed by $\Delta\theta$, we find

$$\frac{v\Delta\theta}{\Delta t} = \frac{\Delta v}{\Delta t}.$$

In the limit as Δt approaches zero, we substitute the instantaneous linear acceleration a_c for $\Delta v/\Delta t$ and the instantaneous angular velocity ω for $\Delta\theta/\Delta t$.

$$a_c = v\omega \qquad (6\text{--}15)$$

In this derivation, ω represented the angular velocity of the velocity vector as it changed direction. We note now, however, that this is the same as the angular velocity of the body as it follows its circular path. Hence Eq. 6–15 relates three quantities associated directly with the uniform motion of a body in a circular path. We derived this equation for the instantaneous acceleration, velocity, and angular velocity, but it is valid for all uniform motion because the magnitudes of these quantities are constant.

If we use Eq. 6–7 to substitute for either v or ω in Eq. 6–15, we find

$$a_c = r\omega^2 \qquad (6\text{--}16)$$

and

$$a_c = \frac{v^2}{r}. \qquad (6\text{--}17)$$

We can also take advantage of Eq. 6–5 through Eq. 6–8 to write the centripetal acceleration in terms of the period and the frequency of the circular motion.

$$a_c = \frac{4\pi^2 r}{T^2} \qquad (6\text{--}18)$$

$$a_c = 4\pi^2 r v^2 \qquad (6\text{--}19)$$

All of these expressions for centripetal acceleration are useful, depending on which quantities are known and which are sought.

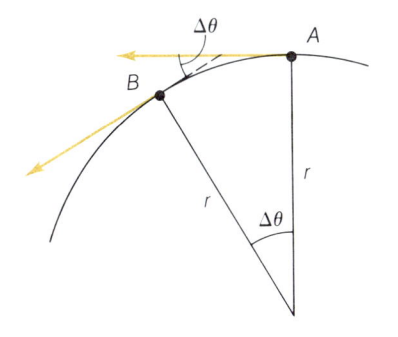

Figure 6.8

Angular velocity and circular motion. The change of angle about the rotational point as a body moves in a circular path is equal to the change in direction of motion during the same interval.

Example 6.5 What is the centripetal acceleration of a car travelling at 100 km/h around a circular track whose radius is 0.5 km?

Here we are given values for v and r, so Eq. 6–17 is easiest to use (after converting the speed and radius to SI units).

$$a_c = \frac{v^2}{r} = \frac{(27.8 \text{ m/s})^2}{(500 \text{ m})} = 1.55 \text{ m/s}^2$$

(a)

(b)

Figure 6.9

Central force. A body can remain in circular motion only if a force (supplied by the string or by the other skater in these cases) acts toward the center of motion.

Central Force

We now get to the "punch line" of this section: the application of what we have learned about centripetal acceleration to the forces implied. From Newton's first law we know that whenever there is an acceleration, there must be a force to create it. We now know that there is acceleration when there is uniform circular motion, so we conclude that there must be a net applied force to keep a body in uniform circular motion. This is something that Galileo realized and Newton eventually quantified; it is, in fact, what led Newton to understand the properties of gravity.

Consider an ice skater being spun around by another, or a child whirling a ball over his head, using a string tied to the ball (Fig. 6.9). The ball follows a circular path, because the string exerts a constant force on it. If the string breaks, the ball will fly off in a straight line tangent to the circle at the point where the ball was when the string broke. Clearly the force applied to the ball is directed inward along the string, or the string would not be taut. The force on the ball is simply

$$F_c = ma_c = mv\omega \tag{6-20}$$

if we use Eq. 6–15 to substitute for a_c. We could, of course, use any of the other expressions for a_c. If we put in values for the mass of the ball, the length of the string, and the angular velocity (or, perhaps more likely, the frequency—then we would use Eq. 6–19 for a_c), we could solve for the force, which is the tension in the string.

In many practical applications, when a body is to move in a circular path, a source of sufficient centripetal force must be found. For example, the child whirling the ball overhead must have a string strong enough to supply the required force.

6.2 CIRCULAR MOTIONS OF PARTICLES

Example 6.6 An engineer who designs a windmill needs to know the required strength of the bolts that will hold the rotor blades to the hub. The windmill is designed to rotate at frequencies as high as 10 Hz, and the mass of each blade is 100 kg. The radius of the hub to which the blades are attached is 0.3 m.

The engineer must calculate the centripetal force that the bolts exert on the rotor blades using Eq. 6–19 for the centripetal acceleration.

$$F_c = ma_c = 4\pi^2 r \nu^2 m = 4\pi^2 (0.3 \text{ m})(10 \text{ Hz})^2 (100 \text{ kg}) = 1.2 \times 10^5 \text{ N}$$

The engineer chooses bolts that have a tensile strength safely in excess of this value.

Example 6.7 For a movie stunt, Tarzan is required to swing from tree to tree on a vine. Being cautious, he wants to check whether the vine will support him. He makes tests that tell him the tensile strength of the vine is 600 lb, and he finds that the length of the vine is 45 ft. Tarzan weighs 220 lb.

His question is whether the tensile strength of the vine is sufficient to offset the combination of his weight and the force required to keep him in circular motion as he swings (Fig. 6.10). Hence the problem reduces to finding out whether the central force required for circular motion exceeds 380 lbs, the residual tensile strength after subtracting Tarzan's weight.

Tarzan needs to estimate what his velocity is as he swings. He knows that he will start from rest, and that gravity will cause him to accelerate downward. To be safe, Tarzan assumes that the maximum velocity he will reach is equal to what his free-fall velocity would be if he dropped straight down to the ground. He chooses an initial height of 50 ft. Aware that energy is conserved, so that his initial potential energy is equal to his final kinetic energy, Tarzan writes

$$\frac{1}{2}mv^2 = mgh$$

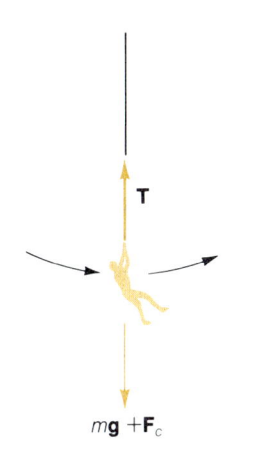

Figure 6.10

and solves for v.

$$v = \sqrt{2gh} = \sqrt{2(32 \text{ ft/s}^2)(50 \text{ ft})} = 56.6 \text{ ft/s}$$

Now Tarzan is ready to calculate the maximum tension in the vine by assuming that at most he will have a velocity of 56.6 ft/s, and by noting that his mass is 6.88 slugs. He uses Eq. 6–17 for the centripetal acceleration to find the maximum centripetal force.

$$F_c = ma_c = \frac{mv^2}{r} = \frac{(6.88 \text{ lb} \cdot \text{s}^2/\text{ft})(56.6 \text{ ft/s})^2}{45 \text{ ft}} = 490 \text{ lb}$$

The vine will break. Tarzan decides he'd be better off as a physics professor.

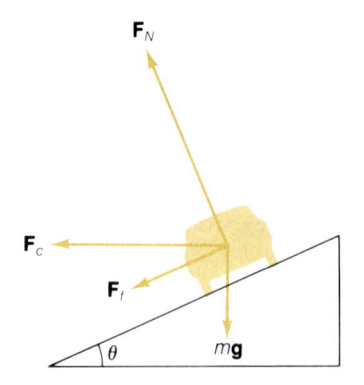

Figure 6.11

The forces acting on a car moving on a banked curve are gravity, the normal force F_N exerted by the road, and friction F_f. Both the normal force and friction have horizontal components, creating the centripetal force F_c.

Every day, car drivers encounter centripetal force. When a car is to be driven around a curve, the driver must decide whether the car will follow the curve or skid. Most of us develop intuition that tells us how fast we can go without skidding, but it is a straightforward matter to calculate this. Furthermore, road engineers who decide how sharply to curve highways or how much to bank them must make calculations of the centripetal force.

Suppose, for example, that a roadway has a curve that is a circular arc with a known radius of curvature. We want to decide on the speed limit to post, so that drivers will go slowly enough not to skid. Here we need to calculate the centripetal force required to keep a car moving in a circle, and then set it equal to the maximum force of friction, which is the force of static friction. Thus we write

$$\frac{mv^2}{r} = \mu_s mg$$

where m is the mass of a car, v is the speed of a car that is just starting to slide, r is the radius of curvature of the road, and μ_s is the coefficient of static friction for tires on pavement. This expression could be solved for v, and values substituted for the known quantities. For example, suppose the radius of curvature is 50 m, and the coefficient of static friction is $\mu_s = 0.5$. The m's cancel out, and we have

$$v = \sqrt{r\mu_s g} = \sqrt{(50 \text{ m})(0.5)(9.8 \text{ m/s}^2)} = 15.7 \text{ m/s} = 56.3 \text{ km/h}.$$

Thus a speed of about 35 mi/h is on the verge of being too fast, and we probably should choose a conservative speed limit of 25 or 30 mi/h.

The maximum safe speed can be increased if the road is banked (Fig. 6.11). Then the normal force that the road applies to the car has a horizontal component, and this component supplies a centripetal force. As the diagram shows, there are three forces acting on the car: the normal force, friction, and gravity. Only the normal force and friction contribute to the centripetal force, and we write the net horizontal force

$$F_h = F_N \sin\theta + F_f \cos\theta = F_N \sin\theta + \mu_s F_N \cos\theta$$

where θ is the angle of inclination of the roadway, F_N is the normal force, and F_f is the friction force which, for the case of the car's maximum speed without slipping, is directed parallel to the road and downhill. This expression incorporates the horizontal component of the normal force, but we must express it in terms of the known vertical component, which is the weight mg plus the vertical (downward) component of the friction. Hence we have

$$mg + F_f \sin\theta = F_N \cos\theta$$

or

$$mg + \mu_s F_N \sin\theta = F_N \cos\theta.$$

6.2 CIRCULAR MOTIONS OF PARTICLES

Solving for F_N yields

$$F_N = \frac{mg}{(\cos\theta - \mu_s \sin\theta)}.$$

Now we substitute this expression for the normal force into the expression for the net horizontal force.

$$F_h = F_N(\sin\theta + \mu_s \cos\theta) = \frac{mg(\sin\theta + \mu_s \cos\theta)}{\cos\theta - \mu_s \sin\theta}$$

We set this equal to the centripetal force mv^2/r, and solve for v.

$$v = \sqrt{\frac{rg(\sin\theta + \mu_s \cos\theta)}{\cos\theta - \mu_s \sin\theta}}$$

We check this result by noting that when $\theta = 0$, it reduces to the same answer we got previously for a road that was not banked.

If we put our previous example values into this expression and assume that the angle of inclination is $\theta = 15°$, we find that

$$v = 20.8 \text{ m/s} = 74.9 \text{ km/h}.$$

The maximum speed is just over 45 mi/h, so a safe speed limit to post might be 35 or 40 mi/h.

It is interesting to note that, even if there were no friction (that is, if $\mu_s = 0$), there is still a velocity (about 26 mi/h) for which no slippage will occur. This is because the normal force still has a horizontal component that can supply centripetal force. Thus, on a very icy curve, a car can travel safely if it is driven at just the right speed (if it is driven too slowly, then it will slip down the incline, towards the inside of the curve).

It is also interesting that whether the road is banked or not, or whether there is friction or not, the speed at which slippage will occur is independent of the car's mass. Thus the posted speed limit is valid for all vehicles.

Centrifugal Force

We have seen that, from the point of view of someone observing a body that is moving in a circle, there is an inward centripetal force. From Newton's third law, we know that there must be an opposite but equal force, and this is exerted by the moving body in the direction away from the center of rotation. The ball whirled about on a string exerts a force on the string, and the car driven around a curve exerts a force on the road.

If we now take the point of view of someone rotating or travelling with the moving body, we see that such an observer feels an apparent force in the

Figure 6.12

A large centrifuge used to simulate high acceleration.

outward direction. The outside observer knows that this is not really a force; it is just the tendency of the person in the rotating reference frame to travel in a straight line. The constant centripetal force creates an inward acceleration which, to the inside observer, produces the same sensation as an outward force. In relativity, this kind of force is called an **inertial force,** because it is due to the inertia of a body as it resists some applied external force. There is no discernable difference to the observer between an inertial force and an applied external force, so the person in the rotating frame simply feels an outward-directed force.

The name **centrifugal force** is used for this force. The word centrifugal means "fleeing the center." Even though it is not a real force in the external frame of reference, we can calculate its strength in the rotating frame. There are even practical uses of the centrifugal force, particularly in devices called **centrifuges,** which use rotation to separate fluids of differing density. Centrifugal force is used in many kinds of machines, including centrifugal clutches (such as those used in the reels of seat belts), in which rotation causes a latch inside a cylindrical drum to expand and lock, and in large centrifuges used in training pilots and astronauts to withstand extreme accelerations (Fig. 6.12).

We live in a rotating frame of reference because the earth is rotating, and it is instructive to see what effect this has. Our inertia makes us tend to fly off

6.2 CIRCULAR MOTIONS OF PARTICLES

into space in a straight line, and to us here on the earth's surface, the result is a reduction in the effective gravity we feel. This effect is maximized at the equator, and reaches zero at the poles. We can calculate the reduction in the gravitational acceleration g for a person at the equator.

$$g' = g - a_c = g - \frac{4\pi^2 r}{T^2}$$

The centripetal acceleration a_c is expressed in terms of the radius r and the rotation period T, as in Eq. 6–18. By substituting $r = 6.4 \times 10^6$ m and $T = 24$ h $= 8.64 \times 10^4$ s, we find

$$g' = 9.8 \text{ m/s}^2 - 0.034 \text{ m/s}^2.$$

The reduction in effective gravity is negligibly small (about 0.35%), from a practical point of view.

Example 6.8 Suppose a test pilot wants to simulate zero gravity by flying his plane in a circular arc so that the centripetal acceleration is in the upward direction and equal in magnitude to the downward acceleration of gravity (Fig. 6.13). If the plane has an airspeed of 1,000 km/h, what is the radius of curvature of the circular arc that he should follow?

The problem is simply one of solving for r, given a_c and v, so again we use Eq. 6–17 (after converting 1,000 km/h to 278 m/s).

$$r = \frac{v^2}{a_c} = \frac{(278 \text{ m/s})^2}{9.8 \text{ m/s}^2} = 7{,}890 \text{ m} = 7.89 \text{ km}$$

The plane would have to follow a circular path whose radius is nearly 8 km. Because the direction of the centripetal acceleration would be changing constantly as the plane flew, exact zero acceleration would occur only for an instant at the moment when the plane was flying at the top of its arc. Maneuvers such as this have actually been used to prepare astronauts for conditions in orbit, and approximate weightlessness can be achieved for periods of a few minutes.

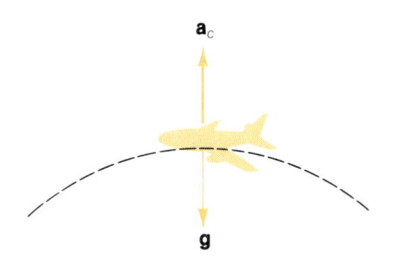

Figure 6.13

Example 6.9 Suppose NASA is designing a space station that will consist of a circular tube structure that rotates (Fig. 6.14), so that occupants in the outer tube will experience a centrifugal force that simulates that of gravity on the earth. If the outer radius of the tube is 100 m, how fast must it rotate so that the centrifugal acceleration equals g?

The rotational frequency may be the most practical value to calculate, so let us use Eq. 6–19 for the centripetal acceleration, and set this equal to g.

$$4\pi^2 r \nu^2 = g$$

Solving for the frequency ν yields

$$\nu = \sqrt{\frac{g}{4\pi^2 r}} = \sqrt{\frac{9.8 \text{ m/s}^2}{4\pi^2 (100 \text{ m})}} = 0.05 \text{ Hz}.$$

The space station must complete one revolution every 20 s. Note that the occupants of the outer tube will stand with their feet against the outer wall, and "up" to them will be the inward direction, toward the hub of the station. It is amusing to imagine living in an environment like this, for the floor curves uphill in either direction, yet walking along the tube will cause no feeling of walking uphill.

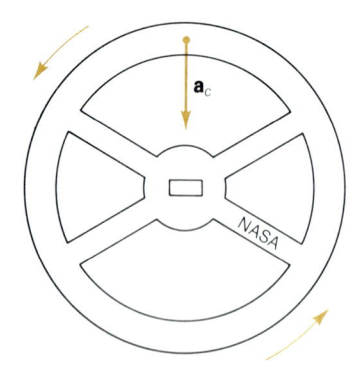

Figure 6.14

6.3 Inertia, Momentum, Work, and Energy of Rotation

We have seen that circular or rotational motion has an analog for each of these linear quantities: displacement, velocity, and acceleration. We now consider the analogs for inertia, momentum, and energy, and we will use conservation of momentum and energy.

Rotational Inertia

Any body resists changes in its rotational velocity, just as it will resist changes in its linear velocity. A torque is required to overcome this resistance, just as a force is required to change a body's linear velocity.

In order to put this idea into mathematical terms, suppose a single point mass m is rotating at a constant speed at radius r from the center of rotation. In this case a torque is required to bring about an angular acceleration, so we write

$$\tau = Fr$$

where F here represents the net force tangent to the circular path. We substitute for F using Newton's second law.

$$\tau = mar$$

Here a is the tangential acceleration. Now we use Eq. 6–14 to express a in terms of the angular acceleration α and substitute.

$$\tau = mr^2 \alpha \qquad (6\text{–}21)$$

6.3 INERTIA, MOMENTUM, WORK, AND ENERGY OF ROTATION

This is Newton's second law for a point mass in circular motion. We infer from this that the rotational analog to mass is the quantity mr^2, since the torque is the product of this quantity and angular acceleration, just as the linear force is the product of mass and linear acceleration.

For an extended body, the rotational inertia is the sum of the rotational inertias for all points in the body, and for each point, Eq. 6–21 is valid. Hence the total torque required to induce angular acceleration is

$$\tau = (\Sigma mr^2)\,\alpha = I\alpha \qquad (6\text{–}22)$$

where I is called the **moment of inertia**, representing the sum of the quantity mr^2 for all points in a body. Eq. 6–22 is the analog of Newton's second law, for rotation of a body while Eq. 6–21 is the form for a single point mass moving at a radius of curvature r.

The units in which the moment of inertia are expressed are simply a combination of the units for mass m and displacement r. In the SI system, moment of inertia is therefore expressed in $kg \cdot m^2$.

Calculation of the moment of inertia is simple for point masses or for bodies that can be represented by point masses, but for extended bodies, an integration must be performed to sum the contributions mr^2 for all points. This is usually difficult even by graphical integration so calculus must be brought into play. Hence the calculation of moments of inertia is largely beyond the scope of this text. It is always possible, of course, to measure the moment of inertia experimentally by subjecting a body to a known torque and measuring the angular acceleration. Table 6.1 contains expressions for the moment of inertia of bodies of various shapes and uniform density.

When the value of I and the torque are known, they can be used to find the angular acceleration, or vice versa.

Example 6.10 A person exerts a 15 lb force on the edge of a roulette wheel for 1 s. The mass of the wheel is 20 kg, and its radius is 0.5 m. What is the frequency of rotation of the wheel after the force is applied? Ignore friction.

The moment of inertia is approximated by the value for a uniform disk, which is $I = \tfrac{1}{2}mr^2$, so we have

$$I = (0.5)(20\text{ kg})(0.5\text{ m})^2 = 2.5\text{ kg} \cdot \text{m}^2.$$

A 15 lb force is equivalent to 66.8 N. When this force is applied at radius 0.5 m the torque is

$$\tau = Fr = (66.8\text{ N})(0.5\text{ m}) = 33.4\text{ N} \cdot \text{m}.$$

Equating this to the torque given by Eq. 6–22 and solving for α yields

$$\alpha = \frac{\tau}{I} = \frac{(33.4\text{ N} \cdot \text{m})}{(2.5\text{ kg} \cdot \text{m}^2)} = 13.4\text{ rad/s}^2.$$

Combining Eqs. 6–6 and 6–10 yields

$$\nu = \frac{\alpha t}{2\pi} = \frac{(13.4 \text{ rad/s}^2)(1 \text{ s})}{2\pi} = 2.13 \text{ Hz}.$$

The roulette wheel is spinning just over two times per second after the force is applied.

Angular Momentum

If there is a rotational moment of inertia, this suggests that there is a rotational momentum, just as the existence of linear inertia suggests linear momentum. We could derive an expression for angular momentum by following the same steps as in Chapter 5 for linear momentum, but here we will simply state the result: the **angular momentum** (L) is the product of the moment of inertia and the angular velocity.

$$L = I\omega \tag{6-23}$$

In the applications of angular velocity and acceleration that we discussed, the vectors happened to be aligned. In the case of angular momentum, however, it is important to recall that these quantities have a direction given by the right-hand rule, because we will want to add angular momentum vectors.

The action of a gyroscope or top helps us see the importance of the directionality of angular momentum. We know that a straight vertical rod set on its end is very likely to fall; that even a small torque will tip it over. If it is spinning, however, then it has angular momentum, directed vertically along the axis of rotation. Typically, this angular momentum is much greater than that caused by any torques that might act to rotate the spinning body lengthwise, that is, tip it over. Furthermore, if a small torque acts on the body to tip it over, the angular momentum vector is horizontal, in the direction given by the right-hand rule. The net angular momentum is the sum of the large vertical vector representing the angular momentum of the body's spin plus the relatively small horizontal vector representing the applied torque. The resultant vector representing the sum of the torques is directed at a small angle to the vertical (Fig. 6.15). This means that the body does not fall, but retains an angular momentum vector at a constant angle to the vertical. Because the angular momentum vector does rotate about the vertical, the object wobbles or **precesses**. The greater the angular momentum of the spinning rod, the more it resists changes in direction, and the smaller the angle through which precession will occur when a torque is applied. Therefore it is advantageous to increase the moment of inertia by giving it an extended shape, such as that of a typical top or gyroscope.

We can discuss another useful property of angular momentum. There is a very important analog to the conservation of linear momentum: in an isolated system (that is, one with no external torques acting on it), angular momentum is conserved. Thus, when angular momentum is lost by one body in a system or when the moment of inertia is changed, there must be a compensating change in other bodies so that the total angular momentum does not change.

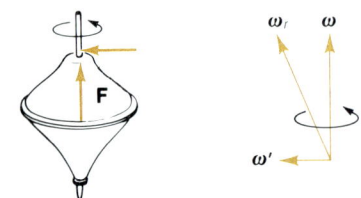

Figure 6.15

Angular momentum of a top. The top's angular momentum vector ω is directed vertically upward. If a horizontal force F, tending to push the top over, is exerted toward the axis into the page, it creates angular momentum ω' directed to the left. The resultant angular momentum vector is ω_r, which rotates about the vertical.

6.3 INERTIA, MOMENTUM, WORK, AND ENERGY OF ROTATION

Table 6.1
Moments of inertia

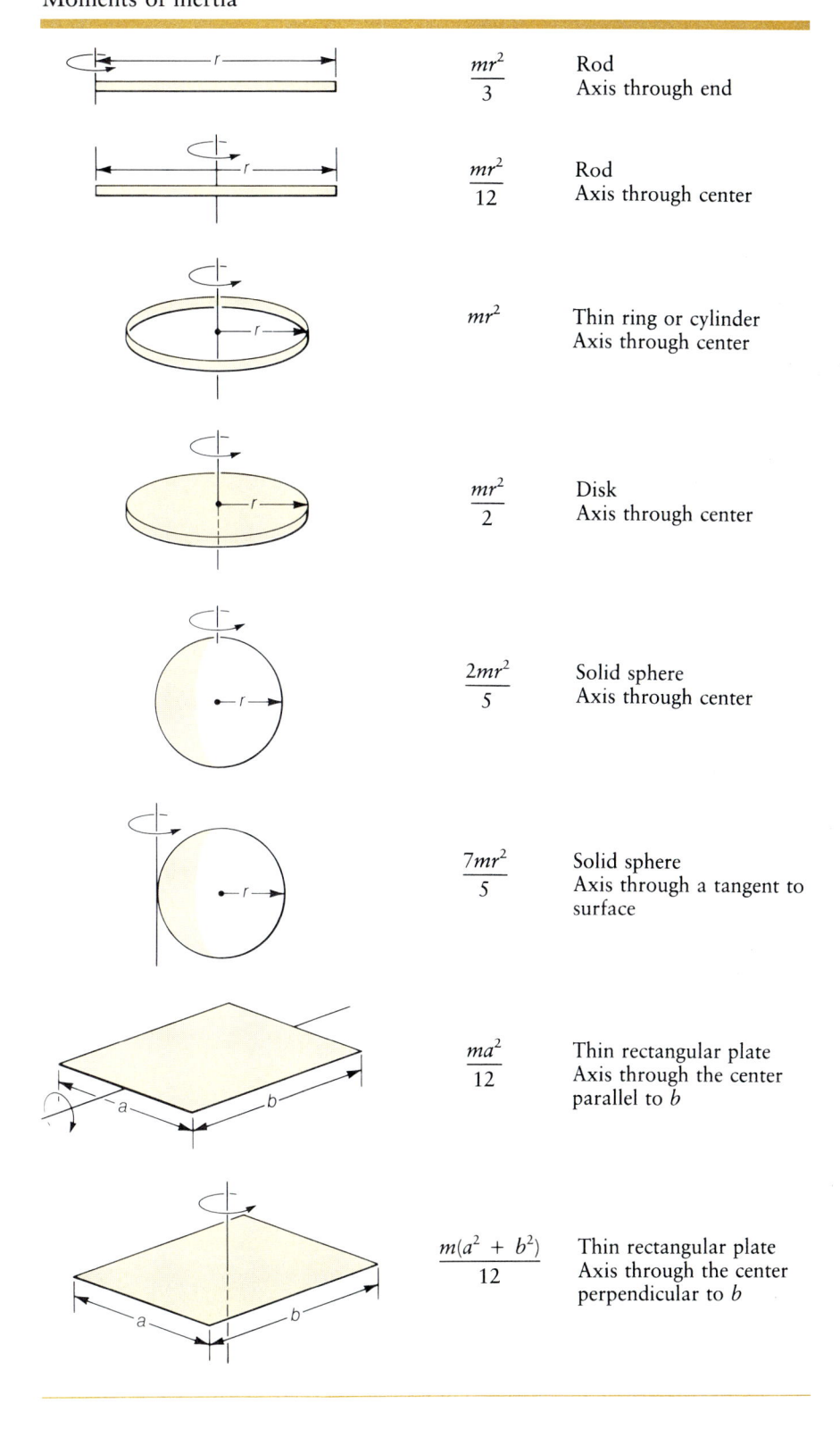

Shape	Moment of Inertia	Description
	$\dfrac{mr^2}{3}$	Rod Axis through end
	$\dfrac{mr^2}{12}$	Rod Axis through center
	mr^2	Thin ring or cylinder Axis through center
	$\dfrac{mr^2}{2}$	Disk Axis through center
	$\dfrac{2mr^2}{5}$	Solid sphere Axis through center
	$\dfrac{7mr^2}{5}$	Solid sphere Axis through a tangent to surface
	$\dfrac{ma^2}{12}$	Thin rectangular plate Axis through the center parallel to b
	$\dfrac{m(a^2 + b^2)}{12}$	Thin rectangular plate Axis through the center perpendicular to b

CHAPTER 6 CIRCULAR AND ROTATIONAL MOTION

For example, the earth-moon system has a certain total angular momentum due to the spins of the earth and the moon, and to the motion of the moon in a nearly circular orbit around the earth. It is known that the spin rate of the moon was once much faster than it is now, but that it was slowed due to gravitational torques exerted on it by the earth. As this happened, the moon moved farther from the earth, increasing its orbital radius and angular momentum to compensate for the spin angular momentum that it lost. As we will see in the next chapter, when the moon was closer to the earth, its orbital period was shorter, that is, the lunar month was shorter than it is now.

Angular momentum conservation has a striking effect for which there is no analogy in linear momentum conservation. We do not normally have situations where the mass of a body changes, so when we discussed conservation of linear momentum, we assumed that the momentum of a body could change only if its velocity changed. Now, however, we find that the angular momentum of a body can be altered if its mass distribution changes, because this causes a change in the value of I. The most widely-cited example of this is the ice skater who begins a spin with arms outstretched, then pulls them in. Because the radius is shorter now, the skater's angular velocity increases dramatically to maintain constant angular momentum $I\omega$. Astronomers have discovered stars that collapsed into small, compact objects and in so doing sped up to enormous angular velocities. Neutron stars, for example, which are collapsed remnants of stars whose initial radii were about 10^{10} m, now have radii of about 10^4 m. A compression factor of 10^6 on a sphere increases its angular velocity by a factor of 10^{12}, because I is proportional to r^2 for a sphere. Some of these stars rotate several hundred times per second!

Example 6.11

A cylindrical artificial satellite is telescoped lengthwise (by equal amounts at both ends), so that its length is increased from 7 m to 12 m. Its mass is 2,000 kg. Before the length was increased, it was spinning end-over-end at 0.6 Hz. What is its new rotational frequency?

We assume that the mass is uniformly distributed before and after the extension. From Eqs. 6–23 and 6–6, we can write an expression for the angular momentum in terms of ν, the spin frequency.

$$L = I\omega = 2\pi\nu I$$

Now, solving for ν and comparing the frequencies before and after the cylinder is extended yields

$$\frac{\nu'}{\nu} = \frac{L'I}{LI'} = \frac{I}{I'}$$

where the primes indicate values after the extension. The conservation of angular momentum requires that $L' = L$; hence the angular momentum has cancelled.

For a uniform cylinder of mass m and length l rotating about its center, the moment of inertia is $I = ml^2/12$, so the new rotational frequency is

6.3 INERTIA, MOMENTUM, WORK, AND ENERGY OF ROTATION

$$v' = v\left(\frac{I}{I'}\right) = v\left(\frac{ml^2/12}{ml'^2/12}\right) = (0.6 \text{ Hz})\left(\frac{7 \text{ m}}{12 \text{ m}}\right)^2 = 0.204 \text{ Hz}.$$

As in many of our simple examples, it would have been much more efficient to note that the rotational frequency is inversely proportional to the square of the length, and begin with our final expression here.

Work and Energy of Rotation

We can write more expressions for angular quantities by analogy with their linear counterparts. The expression for work is

$$W = \tau\theta \qquad (6\text{--}24)$$

where τ is a torque and θ is the angle through which it is applied.

Similarly, the kinetic energy of rotation or circular motion is

$$\text{KE} = \frac{1}{2}I\omega^2 \qquad (6\text{--}25)$$

where I is the moment of inertia and ω is the angular velocity. As in the linear case, both work and energy are scalar quantities.

The expression in Eq. 6–25 for kinetic energy and the conservation of energy provide us with a tool for solving circular or rotational motion problems. For example, we have already referred to the changes in the orbit of the moon and in the spin rate of the earth that occurred when the moon's spin was slowed; now we could solve the energy and momentum equations to find how much the moon's distance increased and how much the length of the day decreased. Other equations would relate the changes in potential energy and kinetic energy.

Example 6.12 An old-fashioned weight-driven clock is wound up by turning a crank that activates a system of gears which winds up a chain attached to a 5 lb weight. How high is the weight lifted, if a force of 1 lb is applied through 12 full turns of the crank, whose radius is 2 in? What is the ideal mechanical advantage of this clock? If this is sufficient to run the clock for 24 h, how much electric power would be required to run the clock?

Ordinarily to find how far the weight would be lifted would require knowing the gear ratios so that the vertical displacement of the weight corresponding to a given angular turn of the crank would be known. By using conservation of energy, however, we can actually find the effective gear ratios; that is, we can find the vertical displacement that corresponds to a given turn of the crank.

The work-energy principle tells us that the work required to turn the crank must equal the energy gained by the system. If the initial energy is defined to be zero, then the input work equals the gravitational potential energy of the weight after the lifting has been done.

$$Fr\theta = mgh$$

Solving for the height h gives us

$$h = \frac{Fr\theta}{mg} = \frac{(1 \text{ lb})(.17 \text{ ft})(24\pi \text{ rad})}{5 \text{ lb}} = 2.56 \text{ ft}.$$

Note that we have idealized the problem by ignoring friction. In reality, some of the input work would be used in overcoming friction and would not contribute to the potential energy gain. Thus, h would be less than 2.56 ft.

The IMA is equal to the ratio of distances over which the input and output forces act, which is the same as the ratio of the output and input forces when friction is neglected.

$$\text{IMA} = \left(\frac{F_o}{F_i}\right)_{\text{ideal}} = \frac{5 \text{ lb}}{1 \text{ lb}} = 5$$

The power is the energy expended per unit of time. Over 24 h, the clock gradually releases the gravitational energy stored when the crank was turned, so the power is

$$\text{power} = \frac{mgh}{t} = \frac{(2.27 \text{ kg})(9.8 \text{ m/s}^2)(0.78 \text{ m})}{(8.64 \times 10^4 \text{ s})} = 2.0 \times 10^{-4} \text{ W}.$$

This is a very low rate of energy consumption, and electric clocks are not significant energy users (although internal friction increases the actual power required).

FOCUS ON PHYSICS

Angular Momentum and Gyroscopes

In this chapter we have learned that a rotating body has angular momentum, and that angular momentum is conserved in any closed system. This has many useful applications.

Many machines that use motors contain **flywheels** for storing angular momentum. A flywheel is simply a massive disk mounted on a shaft, using bearings as free of friction as possible. A rather

6.3 INERTIA, MOMENTUM, WORK, AND ENERGY OF ROTATION

small motor can gradually increase the spin rate of the flywheel until it contains a large amount of angular momentum. This stored angular momentum can then be transferred to the machinery as needed through the use of a clutch system which brings the flywheel in contact with a belt or gear. Many amusement park rides work in this manner. A small motor builds up angular momentum in a flywheel; when the operator brings the belt or gears into contact with the flywheel, angular momentum is transferred to the machinery of the ride, and the passenger cars move.

Other examples of angular momentum storage in flywheels are common. The small "friction" cars that children play with use the flywheel principle. The car is rubbed against the floor, to turn the wheels to build up angular momentum in the flywheel. When the car is set down, the angular momentum is transferred to the wheels and the car moves rapidly by itself.

The **gyroscope** employs another characteristic of angular momentum conservation. As we discussed in the text, a body containing angular momentum tends to maintain a constant orientation of its spin axis. This has many novel consequences (Fig. 6.1.1), but also many practical applications. Everyone knows that it is easier to keep a bicycle upright when it is moving rapidly than when it is stopped or moving very slowly. This is because the angular momentum vector of the wheels tends to maintain constant orientation. In the 1984 Olympic Games, the U.S. cycling team introduced bicycles with special disk rear wheels, designed to store more angular momentum than ordinary spoked wheels, thereby increasing the stability of the bikes (the increased angular momentum capacity of these wheels also helps them to act as flywheels, giving the rider greater momentum when coasting).

Gyroscopes are used for stabilization and navigation of vehicles of various kinds. When a torque is applied to the spin axis of a gyroscope, the gyroscope will precess, or wobble, as discussed in this chapter. It is virtually impossible to avoid precession in a toy gyroscope because of gravitational torques; The earth itself precesses (with a

Figure 6.1.1

26,000 year period) due to unbalanced torques created by the gravitational attraction of the sun and the planets. The rate (frequency) of precession is proportional to the applied torque; that is, the harder you push sideways on the spin axis of a gyroscope, the more rapidly it precesses (this is a straightforward consequence of conservation of angular momentum).

A massive gyroscope can take advantage of precession to stabilize a rolling ship. A ship will roll with a reasonably constant frequency. If this frequency is measured, then a torque can be applied to the gyroscope so that it will precess at the same frequency. Because of angular momentum conservation, a precessing gyroscope exerts a torque against anything that acts to inhibit the precession. Therefore the torque exerted by the gyroscope can act against the roll of the ship if its precession is adjusted to have the same period as the ship's roll. Then, every time the ship rolls to one side, the gyroscope exerts a torque that counters the roll, resulting in a more comfortable ride for the passengers.

The most widespread use of gyroscopes is in

(continued)

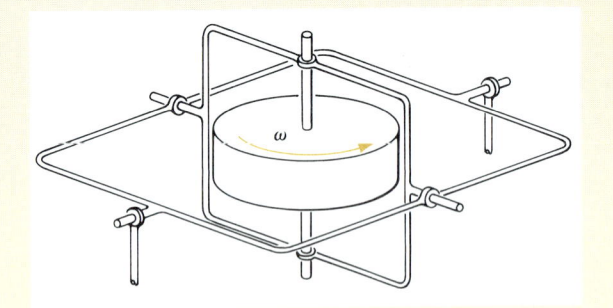

Figure 6.1.2

guidance systems for aircraft and missiles. These systems take advantage of two characteristics of gyroscopes: their tendency to maintain constant orientation, and the relationship between applied torque and precession rate. A typical **inertial guidance** system consists of two main components. The first is a platform attached to a large gyroscope which is mounted on a special gimbal system that allows freedom of motion in all three dimensions (Fig. 6.1.2). When the airplane or missile moves, this gyroscope will maintain its original orientation in an external frame of reference.

Mounted on this inertial platform are three **accelerometers**, devices using gyroscopes to measure the total acceleration of the vehicle in each dimension. An accelerometer consists of a gyroscope with a small mass attached to an extension of its spin axis. When the vehicle accelerates, this mass exerts a force on the spin axis that is proportional to the acceleration. The perpendicular component of this force creates a torque, causing the gyroscope to precess at a frequency that is proportional to the torque. Because the torque is proportional to the acceleration, and the precession rate is proportional to the torque, the precession rate of the gyroscope is a measure of the acceleration the vehicle has undergone. As the acceleration continues, the precession rate continues to increase. Therefore at any time, the precession rate is a measure of the total acceleration of the vehicle in one dimension. The accelerometers mounted in mutually perpendicular directions allow the total cumulative acceleration in all three dimensions to be measured at all times. Given this, and the original orientation of the platform (provided by the stability of the main gyroscope), the displacement of the vehicle from its initial position can be calculated (recall from Chapter 2 how velocity is derived from a known acceleration, and displacement from a known velocity).

Inertial guidance systems can be very accurate. Modern aircraft equipped with such systems can make transoceanic flights with virtually no reference to external navigation beacons (although external beacons are normally used as a precaution). Missiles using inertial guidance systems can be aimed with pinpoint accuracy at targets halfway around the world (the accuracy of these missiles depends also on photographs taken by surveillance satellites). Inertial systems maintain orientation in spacecraft and satellites, and will doubtless be used in future space explorations.

Perspective

Much of what we have discussed in this chapter was not derived in detail, because we quickly established the analogies between rotational and linear quantities. Now we have equipped ourselves to handle a much broader assortment of practical problems. There are still complex types of motion that can only be treated with the methods of calculus, but the tools now in our hands will allow us to analyze quantitatively many kinds of real motions, and to understand qualitatively many others.

We move on in the next chapter to a particular application of many principles we have learned, namely, the study of gravitation and orbits. We will become familiar with the properties of one of the four fundamental forces of nature and its effects on the interaction of bodies.

PROBLEM SOLVING

Problems in Rotational Motion

Problems in rotational motion have a direct relation to problems in linear translational motion. In Chapter 2 we derived a set of equations relating a particle's displacement d, velocity v, and acceleration a at time t, given the initial values d_0 and v_0. These equations carry over to rotational motion, with angle θ, angular velocity ω, and angular acceleration α (Table 6.2). Linear momentum $p = mv$, and kinetic energy $\frac{1}{2}mv^2$ also have rotational analogs: angular momentum $L = I\omega$ and rotational KE $= \frac{1}{2}I\omega^2$, where the moment of inertia I takes the place of mass. Even Newton's force law, $F = ma$, translates to a relation, $\tau = I\alpha$ between torque and angular acceleration. These relations should be applied only after first setting up a coordinate system and choosing a rotation axis. The torque, moment of inertia, angular momentum, and rotational KE are all computed with respect to this axis.

Table 6.2

Linear motion	Rotational motion
$d = d_0 + v_0 t + \frac{1}{2}at^2$	$\theta = \theta_0 + \omega_0 t + \frac{1}{2}\alpha t^2$
$v = v_0 + at$	$\omega = \omega_0 + \alpha t$
$v^2 = v_0^2 + 2a(d - d_0)$	$\omega^2 = \omega_0^2 + 2\alpha(\theta - \theta_0)$
$p = mv$	$L = I\omega$
$KE = \frac{1}{2}mv^2 = \frac{p^2}{2m}$	$KE = \frac{1}{2}I\omega^2 = \frac{L^2}{2I}$
$F = ma$	$\tau = I\alpha$

Example 1 A cylinder of mass m and radius r rolls without slipping down a hill of vertical height h inclined at angle θ to horizontal. What is the cylinder's velocity when it reaches the bottom?

We will use energy conservation to find the velocity v at any point a vertical distance y above the bottom. The total energy is

$$\frac{1}{2}mv^2 + \frac{1}{2}I\omega^2 + mgy$$

where we include rotational KE. The moment of inertia of a cylinder about an axis through its center is $I = \frac{1}{2}mr^2$. Because the cylinder rolls without slipping, its angular velocity $\omega = v/r$. This relation is the key to doing the problem and should be fully understood. Because the cylinder does not slip, the speed v of the cylinder center relative to the contact point is the same as the speed of the contact point relative to the cylinder center. At the top of the hill, v and ω are zero, while at the bottom $y = 0$. Using energy conservation, we set the total energy at the top equal to the total energy at the bottom.

$$0 + 0 + mgh = \frac{1}{2}mv^2 + \frac{1}{2}I\omega^2 + 0$$

$$mgh = \frac{1}{2}mv^2 + \frac{1}{2}\left(\frac{1}{2}mr^2\right)\left(\frac{v}{r}\right)^2$$

The mass m and radius r cancel and we find

$$v^2\left(\frac{1}{2} + \frac{1}{4}\right) = gh \quad \text{or} \quad v = \sqrt{\frac{4gh}{3}}.$$

This is less than if the cylinder slides down the plane without rotating, $v = \sqrt{2gh}$, since the extra rotational kinetic energy results in a lower translational velocity. Notice that static friction is required to make the cylinder roll without slipping. There is no kinetic friction and no frictional work, because the contact point is instantaneously at rest at every moment.

(continued)

Example 2 A stick of mass M and length L is nailed through its center to a frictionless surface and is free to pivot. A smaller mass m travelling at velocity v strikes the stick at one end and sticks. What is the stick's angular velocity ω?

We choose the rotational axis through the nail and conserve angular momentum about that axis. The initial angular momentum of the moving mass is $mv(L/2)$. The final angular momentum is $I\omega$, where the moment of inertia I includes contributions from the stick and mass m.

$$I = \frac{ML^2}{12} + m\left(\frac{L}{2}\right)^2$$

The angular velocity is then

$$\omega = \frac{\frac{mvL}{2}}{I} = \frac{mvL}{\frac{ML^2}{6} + \frac{mL^2}{2}} = \frac{6mv}{L(M + 3m)}.$$

SUMMARY TABLE OF FORMULAS

Angle subtended by an arc:
$$\theta = \frac{s}{r} \tag{6-1}$$

Angular velocity:
$$\omega = \frac{\theta}{t} \tag{6-2}$$

Angular displacement:
$$\theta = \theta_o + \omega t \tag{6-3}$$

Angular velocity:
$$\omega = \frac{\Delta\theta}{\Delta t} \tag{6-4}$$

Period of circular motion:
$$T = \frac{\theta}{\omega} = \frac{2\pi}{\omega} \tag{6-5}$$

Frequency:
$$\nu = \frac{1}{T} = \frac{\omega}{2\pi} \tag{6-6}$$

Linear speed of point moving in circular path:
$$v = r\omega \tag{6-7}$$
$$v = \frac{2\pi r}{T} \tag{6-8}$$
$$v = 2\pi r\nu \tag{6-9}$$

Angular acceleration:
$$\alpha = \frac{\omega}{t} \tag{6-10}$$
$$\alpha = \frac{\Delta\omega}{\Delta t} \tag{6-11}$$

Angular velocity after angular acceleration:
$$\omega = \omega_0 + \alpha t \tag{6-12}$$

Angular displacement after angular acceleration:
$$\theta = \theta_0 + \omega_0 t + \frac{1}{2}\alpha t^2 \tag{6-13}$$

Linear acceleration of a point undergoing angular acceleration:
$$a = r\alpha \tag{6-14}$$

Centripetal acceleration:
$$a_c = v\omega \tag{6-15}$$
$$a_c = r\omega^2 \tag{6-16}$$
$$a_c = \frac{v^2}{r} \tag{6-17}$$
$$a_c = \frac{4\pi^2 r}{T^2} \tag{6-18}$$
$$a_c = 4\pi^2 r\nu^2 \tag{6-19}$$

Centripetal force:
$$F_c = ma_c = mv\omega \tag{6-20}$$

Torque and angular acceleration for point mass in circular motion:
$$\tau = mr^2\alpha \qquad (6\text{--}21)$$

Torque and moment of inertia:
$$\tau = (\Sigma mr^2)\alpha = I\alpha \qquad (6\text{--}22)$$

Angular momentum:
$$L = I\omega \qquad (6\text{--}23)$$

Work:
$$W = \tau\theta \qquad (6\text{--}24)$$

Kinetic energy:
$$KE = \frac{1}{2}I\omega^2 \qquad (6\text{--}25)$$

THOUGHT QUESTIONS

1. Can a mass be accelerated if its speed is constant?
2. What happens to the rotation of the Earth when the trees drop their leaves to the ground in the fall? When the tree's sap rises in the spring?
3. If a body moves in a circle, why is no work done, despite the fact that a force acts to accelerate the body?
4. When a gymnast does a flip, where does the torque come from?
5. Thorn-proof bicycle tires are considerably heavier than normal tires. Is it more difficult for the cyclist to pedal?
6. When hiking, one pound extra weight on one's boots is more fatiguing than one pound on the back. Why? Would this be true if there were no friction in the muscles?
7. If a plant grows on a rotating platform it will lean inward. Why?
8. Explain why a diver goes into tuck position to do faster somersaults.
9. In a tetherball game, explain why the ball rotates faster as it wraps the rope around the pole.
10. Why is the velocity of a cylinder greater when it slides without friction down an inclined plane than when it rolls without slipping?
11. Imagine a small circular piece cut out of the edge of a large circular pie rotating about its center on a platform. Is its KE translational, rotational, or both?
12. Explain why a yo-yo can climb vertically up the string.
13. Explain why a top precesses.
14. Explain why rotational KE, $\frac{1}{2}I\omega^2$, is equivalent to the translational KE, $\frac{1}{2}mv^2$, of all the parts of a rotating body.
15. Why do a rod, a cylinder, and a sphere of the same length or diameter have different moments of inertia about their lengthwise axes?

PROBLEMS

Section 6.1 Angles and Angular Velocity

- 1 Express the following angles in radians: 30°, 45°, 90°, 180°, 215°.
- 2 Express the following angles in degrees: $\pi/3$ rad, $\pi/2$ rad, π rad, 0.147 rad, 6π rad.
- 3 How many rotations correspond to 5 radians?
- 4 A 12 in diameter record is played at 33.3 rev/min for 20 min.
    - (a) How many revolutions does it make?
    - (b) Through how many radians did it turn?
- 5 How far does a point on the circumference of the record in Problem 5 move?
- 6 A discus thrower holds a disc at arm's length (1 m) and spins through an angle of 4.5π radians before throwing it. How far does the disc travel before release?
- 7 What is the linear velocity of a point on the rim of a wheel of radius 30 cm rotating 10 times per second?
- 8 The tip of an airplane's propeller spins with a linear speed of 500 mi/h. If the propeller's radius is 1.2 m, what is its angular frequency?

9. Calculate the angular velocity (rad/s) of
   (a) The second hand of a clock.
   (b) The minute hand of a clock.

10. During a solar eclipse, the Moon just covers the Sun (angular diameter 0.5°). The Earth has radius 6,378 km, the Moon 1,738 km, and the Sun 696,000 km. If the Sun is 93 million miles away from Earth, calculate the distance to the Moon.

11. The distance from the Sun to the Earth (1 astronomical unit) is 1.496×10^{11} m. Calculate the angular and linear velocity of the Earth in orbit around the Sun.

12. A red giant star has a radius 300 times that of our Sun. With telescopes, astronomers can measure angles as small as 0.3 arc-seconds. How far away (in light-years) can they distinguish the disk of such a star? One arc-second equals 1/3600 of a degree, and 1 light-year equals 10^{16} m.

13. The human eye can distinguish angles as small as 1 arc-minute (1/60 degree). How far away can one read 5-in letters on a sign?

14. If the top of the Empire State Building is 1,100 feet above ground, and a person with eye level 5 feet high stands 5 miles away, what is the height of a fence 50 feet away from him which just blocks the building from view?

15. A lighthouse beam makes a turn every 10 s. What is its angular velocity (radians per second)?

16. For the lighthouse in Problem 15, compute the linear velocity of the spot of light as it moves across a ship 5 miles from shore.

17. Pulsars are rapidly spinning neutron stars with beams of radiation which rotate with the star like a lighthouse. The most rapid pulsar spins 642 times per second. With what linear velocity does the pulsar beam sweep across the Earth if the pulsar is 10,000 light-years away? Compare your answer to the speed of light (3×10^8 m/s). One light year is 10^{16} m.

18. A cyclist's top speed is 40 mi/h. If the radius of the wheels is 70 cm, what is their angular velocity?

19. Two gears of radius 5 cm and 8 cm (Fig. 6.16) are meshed together. The 8 cm gear turns at 100 rev/min.

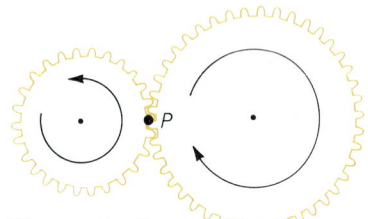

Figure 6.16

(a) What is the linear velocity of the teeth near the point of contact, P?
(b) What is the angular velocity of the 5 cm gear?

20. Two gears of radius R_1 and R_2 are connected by a chain, which moves with constant linear velocity v. Assume that $R_2 > R_1$.
    (a) Compute the angular velocities ω_1 and ω_2 of the gears.
    (b) Compute the gear ratio, the fraction ω_1/ω_2, in terms of v, R_1, and R_2.

21. A car with 32 cm wheels accelerates smoothly from rest to 60 mi/h in 10 s. Compute the angular acceleration of the wheels and their final angular velocity.

22. A ten-speed bicycle has 66 cm radius wheels and a gear ratio of 3.5:1 (front cog to back cog; see Problem 20). At 40 mi/h, how many revolutions per second are made by the front (pedaling) gear?

23. Calculate the average linear velocity of the inner and outer rims of a record (radii 3 in and 6 in) played at 33.3 rpm, 45 rpm, 78 rpm.

24. For the record in Problem 23, compute the playing times if the total groove length is 0.5 km.

25. A 4-in radius bowling ball rolls without slipping down a 50 ft alley. How many revolutions does it make?

26. A basketball player shoots at a 12 ft high rim. He releases the ball 8 ft above ground and 30 ft away from the basket at an angle 45° from horizontal. If the ball has a backspin of 2 rev/s, how many revolutions does the ball undergo before hitting the rim?

27. A record turntable accelerates smoothly from rest with angular acceleration of 0.5 rad/s². How long does it take to reach 33.3 rev/min? How many revolutions did the turntable make?

28. A boat propeller slows down from 500 rev/min to 300 rev/min during 50 rotations. What was its angular acceleration?

29. A fan blade, initially rotating at 100 rpm, accelerates smoothly for 10 s, during 30 rotations.
    (a) What is the final rotation rate (rev/min) of the fan?
    (b) What was the angular acceleration?

30. A car accelerates from rest to 60 mph in 10 s. What is the angular acceleration of its 30 cm radius wheels?

31. A fan rotates with angular velocity ω (Fig 6.17). Compute the total number of rotations.

32. For the fan in Problem 31, what was the average angular acceleration during the first 20 s, the first 30 s, the first 50 s?

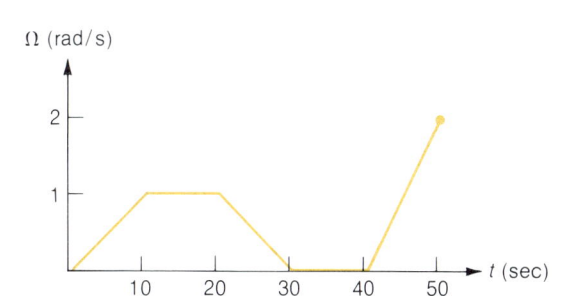

Figure 6.17

Section 6.2 Circular Motion and Centrifugal Force

- **33** A 40 kg child sits on a horse 10 m from the center of a carousel platform, which rotates once each 15 s. What is the child's linear velocity? What is the net force on the child?

- **34** A centrifuge rotates 10,000 times per minute. At what radial distance from the rotation axis does the outward centrifugal acceleration equal 50,000 g's?

- **35** A 50 mi/h train goes around a circular bend, changing direction by 45° over a distance of 1 mile. Calculate the radius of curvature and the centripetal acceleration of the train.

- **36** Suppose a train accelerates from 30 mi/h to 50 mi/h in 1 minute while going around a bend with a 1 mile radius of curvature. What is the total acceleration when the speed is 40 mi/h? (Combine tangential and radial or centripetal acceleration vectors.)

- **37** Compute the centripetal acceleration of the Earth orbiting the Sun. What force must be exerted on the Earth? Consult Appendix 4 for Earth's mass and orbital radius.

- **38** How fast would a planet the size and mass of the Earth have to rotate so that the net force on a body at the equator is zero? (Centrifugal force balances gravity.) How long is a day on such a planet?

- **39** In a carnival centrifuge ride, a vertical cylinder 5 m in radius rotates at 30 rev/min, throwing riders outward against the walls. How many g's do they experience?

- **40** For riders on the centrifuge in Problem 39, what minimum coefficient of static friction will ensure that they do not slip down the wall?

- **41** If the coefficient of friction between a car's tires and the road is $\mu = 0.5$, compute the smallest radius of curvature of an unbanked road for a car moving at 50 mi/h without slipping.

- **42** A roller coaster goes around a vertical loop 10 m in radius. What is the minimum velocity required at the top of the loop to prevent the car from falling downward off the track?

- **43** A donkey stands on a rotating circular platform, 5 m from the rotation axis. If the coefficient of friction at the floor is 0.4, how fast (rev/min) must the platform rotate in order to start the donkey sliding outward?

- **44** A 70 mi/h racecar makes a turn of 200 m radius. If the track has friction coefficient 0.4, what minimum banking angle is required?

- **45** On an icy curve of radius 500 m, bank angle 5°, and friction coefficient 0.05, what is the range of velocities for which a car may make the turn without slipping outward or down the incline?

- **46** A cycle and its rider, with a total mass 90 kg, go around an unbanked curve with 50 m radius at 15 m/s. If the friction coefficient is 0.5, at what angle from vertical should the rider lean to keep from tipping over?

- **47** A roller coaster goes over a crest with radius of curvature 25 m. What is the maximum velocity possible without flying off the track?

- **48** A child on a 5 m swing reaches an angle of 30° from vertical on the backswing. What is the acceleration (in g's) at the midpoint of the swing?

- **49** A 600 mi/h jet flies in a vertical loop. What radius of curvature is needed to simulate weightlessness at the top of the loop?

- **50** You wish to twirl a 5 kg rock horizontally at the end of a 1 m string. If the string has a tensile strength of 100 lb, what is the fastest (rev/s) that you may spin the rock?

- **51** Could you spin the rock in Problem 50 faster if you twirl it in a vertical direction? Why or why not?

- **52** After a power dive, a pilot turns upward at 500 mi/h along a circular arc with 1 km radius. How many g's does the pilot experience?

- **53** A 70 kg ski jumper starts at height $h = 40$ m, then encounters a stretch with radius of curvature 20 m (Fig. 6.18). What upward force F is exerted at the end of the curve?

Figure 6.18

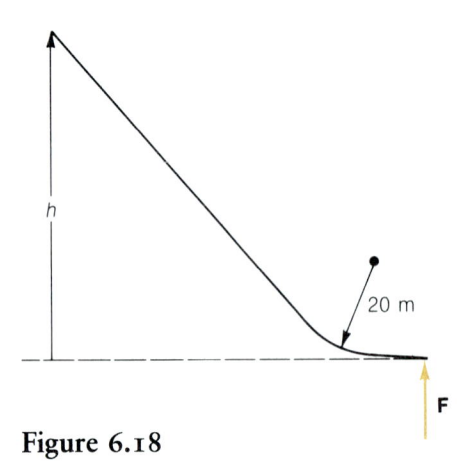

Section 6.3 Rotational Inertia, Momentum, and Energy

- 54 What torques are exerted on a door by perpendicular forces of 200 N and 100 N applied at horizontal distances of 40 cm and 60 cm from the hinge axis?

- 55 Calculate the moment of inertia about its base of a uniform 10 m ladder with 25 kg mass.

- 56 Calculate the moment of inertia of a 10 kg uniform, solid sphere about an axis through its center. The radius is 50 cm.

••• 57 A bulb with mass 10 g and radius 5 cm rests on a socket with kinetic friction coefficient 0.5 (Fig. 6.19). Calculate the twisting force that must be exerted at the bulb's 5 cm radius to accelerate the bulb to 1 rev/s in 0.5 s. Assume the bulb is a solid sphere (neglect the mass of the neck).

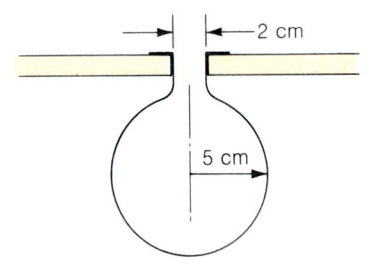

Figure 6.19

•• 58 Calculate the moment of inertia of the set of 1, 2, and 3 kg masses shown in Fig. 6.20, all of which are attached to the same vertical axis.

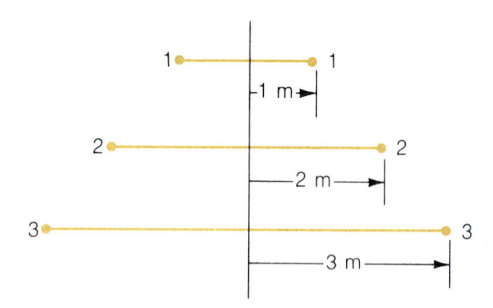

Figure 6.20

• 59 Two wheels each have a mass of 20 kg. Wheel A is a solid, uniform disk, 40 cm in radius. Wheel B is a hollow cylinder 30 cm in radius. Calculate the moments of inertia. Which would be easier to accelerate?

•• 60 A 10 m radius Ferris wheel consists of a 1,000 kg circular rim, 8 radial spokes of mass 500 kg each, and 8 small gondolas of mass 200 kg when loaded. Calculate the total moment of inertia about the center.

•• 61 If a motor exerts a torque of 1,000 N · m on the Ferris wheel in Problem 60, how long does it take to accelerate the wheel to 1 rev/min?

•• 62 A horizontal circular platform with moment of inertia 3 kg · m^2 is struck from the side by a 1 kg mass travelling at 10 m/s. The mass sticks to the platform 20 cm from the center axis. Calculate the platform's final angular velocity.

• 63 What is the rotational kinetic energy of a 1 kg, 60 cm baton twirled twice per second? What is the angular momentum about its center?

•• 64 A 1 m radius uniform sphere can be accelerated about its central axis from rest to 10 rev/s in 10 s by a torque of 27 N · m. What is the sphere's mass?

•• 65 A bicyclist exerts a force of 800 N along a chain to the 4 cm radius rear axle. If the rear tire has 66 cm radius and supports a weight of 400 N, find the minimum coefficient of static friction between the tire and the road so that the tire accelerates without slipping.

•• 66 A diver leaps horizontally off a 10 m platform, exerting a torque of 500 N · m about his center of gravity for 0.2 s. His moment of inertia is 7 kg · m^2 about an axis through the center of gravity. How many turns can be completed before he strikes the water? Assume that his moment of inertia stays the same while turning (in practice it can change).

PROBLEMS

67 A 2 m long hockey stick with mass 600 g lying on the ice is struck at one end by a 170 g puck moving at 50 mi/h perpendicular to the stick. If the collision is totally elastic, calculate the angular velocity ω of the stick. (Use the impulse equation to calculate Δω.)

68 From conservation of energy (rotational and translational), compute the translational velocity of the hockey stick in Problem 67.

69 A spherical ball of mass m and radius r rolls without slipping down an inclined plane with translational velocity v.
  (a) Show that the sphere rotates with angular velocity $\omega = v/r$.
  (b) Show that the translational and rotational kinetic energies are $mv^2/2$ and $mv^2/5$.

70 An 80 kg sphere, 20 cm in radius, rolls without slipping down a plane inclined 10° from horizontal. Use conservation of energy to compute the velocity of its center of gravity 10 m down the plane. Refer to Problem 69 and neglect frictional losses.

71 A cylindrical log of radius r, length l, and mass m rolls without slipping down an inclined plane with velocity v. Show that the translational and rotational kinetic energies are $mv^2/2$ and $mv^2/4$, respectively.

72 A 200 ft redwood tree falls to the ground, but does not slip at its base. Using energy conservation, calculate the linear velocity with which the tree's top end strikes the ground. Assume the tree to be a uniform cylinder.

73 In slamming a 10 kg, 1m × 2m door, you exert constant, perpendicular force of 100 N at a horizontal distance of 1 m from the hinge axis.
  (a) Calculate the moment of inertia by considering it to be a number of horizontal strips each 1 m long.
  (b) How long does the door take to close (rotate through an angle of 90°)?

74 An angry child delivers a horizontal kick of 0.2 s duration to a 2 kg, 1 m long vertically balanced stick, at a point 10 cm below its center of gravity. What force kick is needed for the stick to land flat after a rotation of 90°?

75 A supercentrifuge rotates at 50,000 rev/min on a platform 1.5 m above ground. If a piece breaks off at the edge (10 cm radius), how far will it travel horizontally? Neglect friction.

76 A rod of length l and mass M floating in outer space is struck elastically at one end by an object of mass m moving perpendicular to the rod with speed v. From conservation of angular momentum and energy, find the velocity of its center of gravity and the rotational velocity about the center of gravity (Fig. 6.21). Assume $m \ll M$.

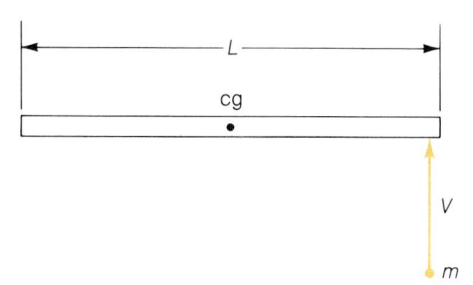

Figure 6.21

77 A spherical ball of mass m and radius r rolls without slipping down a plane inclined at angle θ from horizontal. The coefficient of static friction is μ. After a distance l along the plane, the ball's translational velocity is v. From energy conservation and the no-slip condition ($v = \omega r$), show that $v^2 = (10gl/7)(\sin\theta)$.

78 Suppose the ball in Problem 77 slides down the plane without rolling. What is v^2 after a distance l? Why is v^2 greater than when the ball rolls?

79 A 10 kg sphere of radius 40 cm spins at 1 rev/s. Suddenly it collapses to 10 cm radius, preserving its uniform composition. How fast will it spin?

80 A uniform circular carousel platform of mass 1,800 kg carries 10 children of mass 50 kg each standing at the rim, 5 m from the axis. If the loaded carousel rotates once each 10 s, how fast will it rotate if the children all jump off radially, exerting no torque?

81 A 70 kg fisherman trapped on a frictionless frozen lake in the Yukon tries to reach shore by throwing his 1 kg fish at 5 m/s in the opposite direction. He releases the fish with his right arm, at a point 50 cm from the center of his body, which has a moment of inertia 0.9 kg · m² about a vertical axis passing through his center of gravity. Using conservation of angular momentum and energy, calculate
  (a) How fast his center of gravity moves toward shore.
  (b) How fast he rotates (rev/min).

82 A circular disk of mass 10 kg and radius 1 m is set rotating by two guns which shoot 40 g bullets at 500 m/s in tangential directions from positions on the rim 180° apart. If the firing rate is 1 bullet/s, after what time will the disk rotate at 1 rev/s? Neglect the velocity of the wheel in computing the torque of the bullets.

83 A satellite orbits the Earth in an elongated orbit (Fig. 6.22). If its velocity $v_1 = 3{,}160$ m/s, what is v_2 if angular momentum about the Earth's center is conserved?

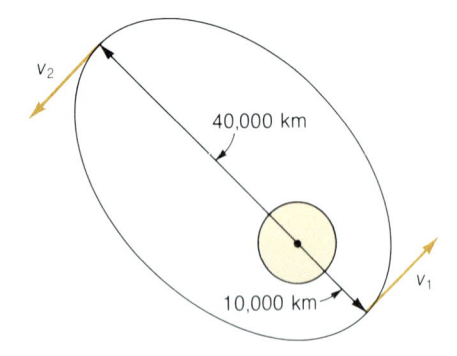

Figure 6.22

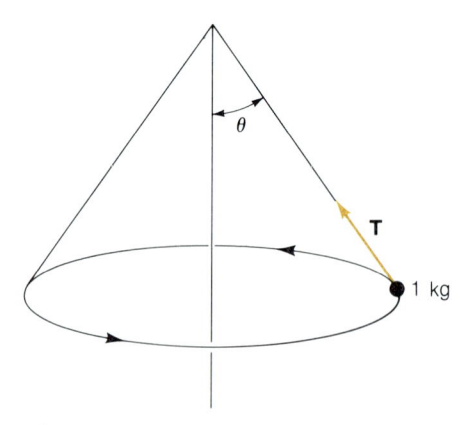

Figure 6.23

•• 84 A 100 kg astronaut with $I = 1$ kg·m² floating motionless in space is handed a 1 kg spherical gyroscope (5 cm radius) spinning at 100 rev/min. If he clasps the gyroscope toward his body center, how fast will he spin?

•• 85 An ice skater spins at 1 rev/s, with each of her 4 kg arms extended 1 m from a vertical axis through her center of gravity. If she pulls in her arms, how fast will she spin? Treat her arms like uniform rods 1 m long, and assume that the moment of inertia of her body (excluding her arms) about the spin axis remains constant at 0.8 kg·m².

•• 86 A gyroscope with $I = 10$ kg·m² spinning at 1,000 rev/min slows uniformly to rest in 1 h. What frictional torque was experienced? How much energy was dissipated?

•• 87 A shrewd investor hits upon the idea of buying gold by weight at the equator, then selling it by weight at the North pole, where gravity is slightly stronger. If the price of gold is $400 per ounce, how many kg of gold must the investor sell to cover the plane fare to the North pole, $10,000? One troy ounce has a mass of 31.1 g.

••• 88 A string is attached at radius 4 cm to a 50 g cylindrical yo-yo with $I = 700$ g·cm². The top of the string is held fixed, and the yo-yo falls while unrolling down the string without slipping. Find the yo-yo's acceleration. What is the tension in the string?

• 89 A 500 g rock is whirled in a horizontal circle on a 100 cm cord. Neglect gravity. If the tension in the cord is 12.5 N, what is the rock's velocity? Energy?

•• 90 Find the angle from vertical of a 1 kg mass held out on the end of a 1 m string and rotating at 1 rev/s in a conical path (Fig. 6.23). What is the tension T?

• 91 A cylindrical, uniform flywheel of mass 50 kg and radius 1 m rotates at 6,000 rev/min. How much rotational energy is stored?

• 92 A cylindrical, uniform flywheel with $m = 500$ kg and $r = 2$ m rotates at 5,000 rev/min. How long can it supply energy at 10 hp?

•• 93 A model airplane of mass m attached to a string is whirled in a circular path of radius r at velocity v. The airplane is then pulled in to a new radius $r/2$ without exerting any torque. Find the new velocity. What is the difference in rotational energies before and after? Where did this energy come from?

• 94 In the Bohr model of the hydrogen atom, an electron with mass 9.11×10^{-31} kg moves around the nucleus in a circular orbit of radius 5.29×10^{-11} m. If the orbital period is 1.51×10^{-16} s, what are the orbital velocity, orbital angular momentum, and kinetic energy?

•• 95 If the Earth's radius expanded by 1%, what would be the percent change in the length of the day? Assume $I = 2mr^2/5$ for the Earth and assume the expansion exerted no torque and that the Earth remains uniformly dense.

••• 96 Calculate the downward acceleration of the mass M_2, attached to the uniform cylinder of mass M_1 (Fig. 6.24). Neglect friction.

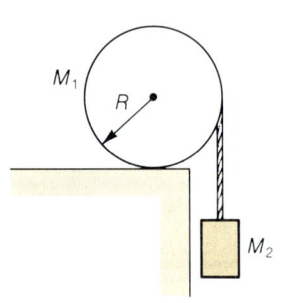

Figure 6.24

SPECIAL PROBLEMS

1. Platform Rotation

A circular platform of mass $M = 200$ kg and radius $R = 10$ m rotates on frictionless bearings at 5 rev/min (Fig. 6.25). A person with mass $m = 60$ kg hops on at the rim, exerting no torque, then walks radially inward. The coefficient of friction between the platform and the person's feet is $\mu = 0.5$. Assume that the person's velocity inward is very small.
(a) What is the new rotation rate after the person hops on?
(b) What is the rotation rate when the person has moved inward to radius r?
(c) At what radius does the person slip? You may do this part graphically.

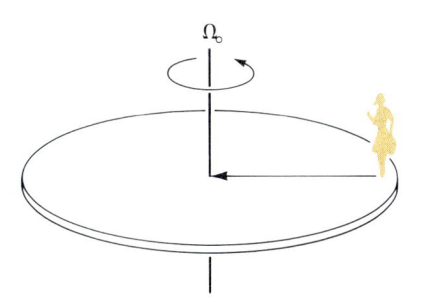

Figure 6.25

2. Pulleys with Mass

Consider the pulley system shown in Fig. 6.26, in which the pulley's mass and moment of inertia may not be neglected. Assume the pulley has mass M and radius r, that $M_2 > M_1$, and that the pulley rope does not slip.
(a) Write three equations of motion for the two masses and pulley, involving tension, gravity, and torques.
(b) Solve these equations for the downward acceleration, a, of the system, and for the forces T_1 and T_2.

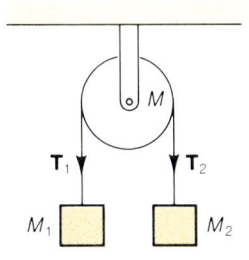

Figure 6.26

3. Ice Dancing

Two 60 kg ice skaters moving at 5 m/s approach one another from opposite directions, separated by a perpendicular distance of 10 m (Fig. 6.27). When directly opposite each other they grab onto a 10 m rope.
(a) From conservation of angular momentum, find their rotation rate about the center of the rope.
(b) If each pulls on the rope until they are separated by 5 m, what is their linear speed?
(c) If the rope has a breaking strength of 800 lbs, how close can they get by pulling on the rope?

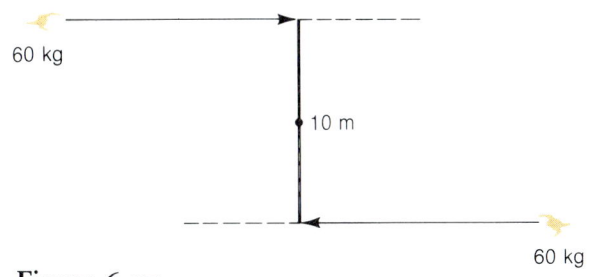

Figure 6.27

4. Jelly Sandwich

A 100 g, 10 cm × 10 cm square jelly sandwich topples off the tip of one's hand (Fig. 6.28) from a height of 1.2 m. The coefficient of friction between one's fingers and the bread is $\mu = 0.3$.
(a) Compute the angle θ at which the sandwich slips off the fingers.
(b) Using energy conservation, calculate the bread's angular velocity ω_s about the point of finger support at the time it slips. Then find the angular velocity ω_{cg} about the center of gravity. (Notice that the total KE equals the translational energy of the center of gravity plus the rotational energy about the center of gravity.)
(c) How many revolutions has the bread made when it strikes ground? Does it land jelly side up?

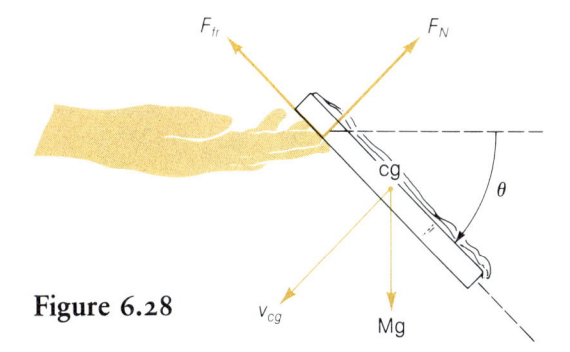

Figure 6.28

SUGGESTED READINGS

"The physics of spinning tops, including some far-out ones." J. Walker, *Scientific American,* Mar 81, p 182, **244**(3)

"The essence of ballet maneuvers is physics." J. Walker, *Scientific American,* Jun 82, p 146, **246**(6)

"The physics of somersaulting and twisting." C. Frolich, *Scientific American,* Mar 80, p 154, **242**(3)

"The tornado." J. T. Snow, *Scientific American,* Apr 84, p 86, **250**(4)

"The physics of dance." K. Laws, *Physics Today,* Feb 85, p 24, **38**(2)

"Why do things spin? Why not?" K. C. Cole, *Discover,* Mar 85, p 83 Vol. 6, No. 3

"The moment of inertia of a tennis racket." H. Brady, *The Physics Teacher,* Apr 85, p 213, **23**(4)

"Year of the tornado." S. Olson, *Science 84,* Nov 84, p 174, **5**(9)

The flying circus of physics with answers. Jearl Walker, Wiley, New York, 1977

Newton at the bat: The science in sports. Eric W. Schrier, William F. Allman, eds, Scribners, New York, 1984

Sport science. Peter J. Brancazio, Simon and Schuster, New York, 1984

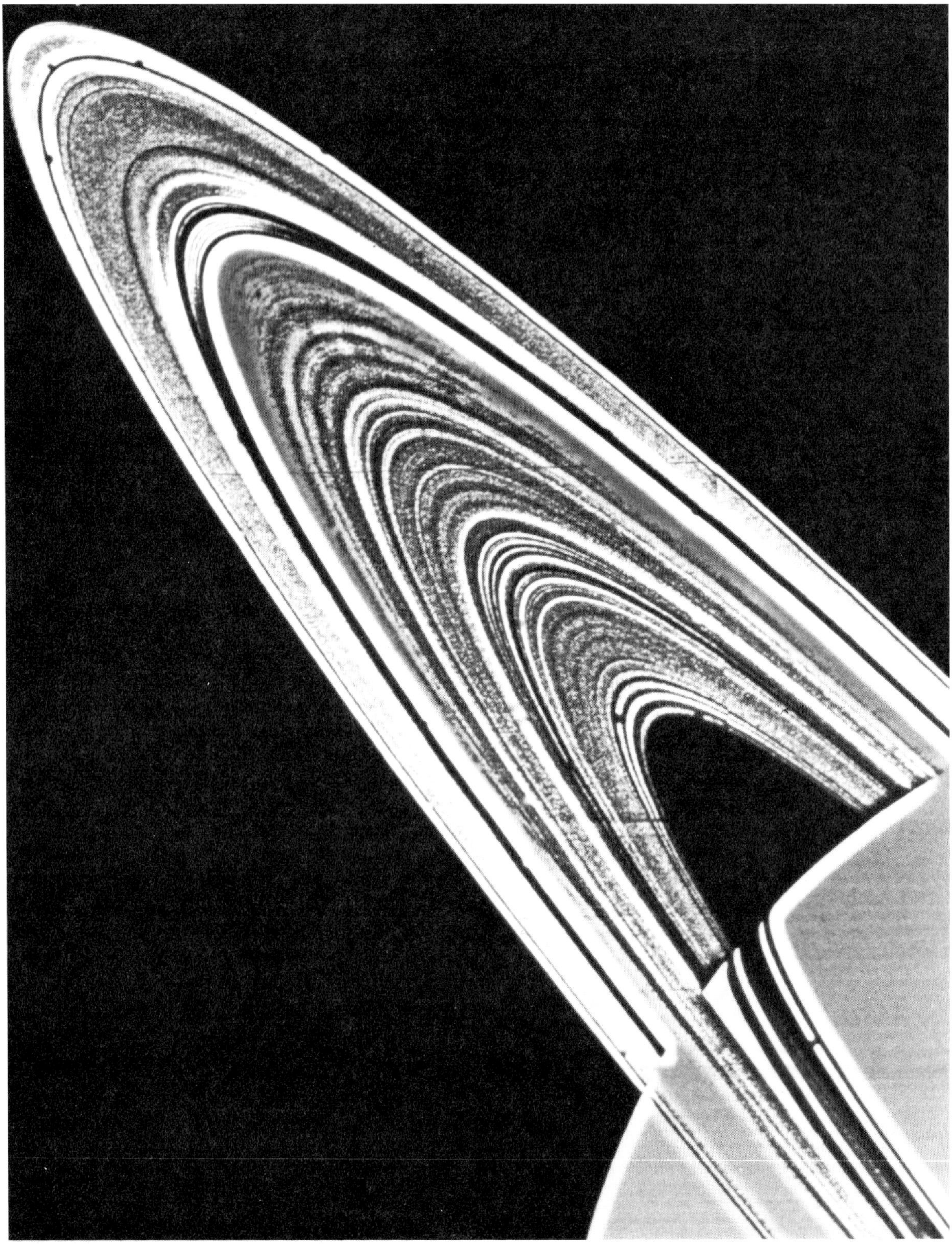

Chapter 7

Gravitation and Orbits

We have spent considerable effort getting familiar with moving bodies, and we have discussed motions influenced by gravity. Now we are ready to understand gravity in a more general way, and to investigate orbital motions. This will help us not only to understand our natural environment better, but also to appreciate the difficulties of deploying artificial satellites, which play a growing role in our culture.

7.1 The Law of Gravitation

Kepler and Galileo realized that the sun must be an important factor in governing the motions of the planets, but it was Isaac Newton who first deduced the properties of the attractive force that the sun exerts on the other bodies in the solar system. In this section we will trace Newton's reasoning and discuss the nature of the law of gravitation that he derived.

The Deductive Reasoning of Newton

Newton knew that there must be a central force; that is, a centripetal force, to keep the planets moving in nearly circular orbits around the sun. He was also aware of the work of Kepler, who had discovered three laws of planetary motion:

1. The orbits of the planets are ellipses with the sun at one focus.
2. A line connecting a planet to the sun sweeps out equal areas in equal intervals of time.
3. The square of the period of a planet is proportional to the cube of its semimajor axis.

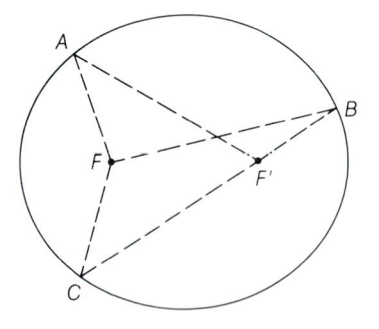

Figure 7.1

The ellipse. In an ellipse the distances FAF', FBF', FCF' are all equal.

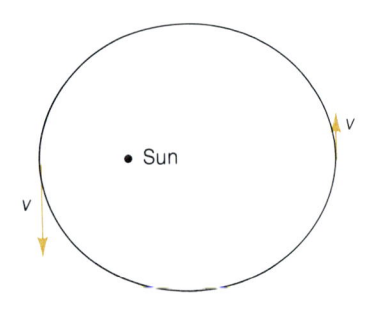

Figure 7.2

Kepler's second law. The orbital speed v of a planet is greatest when the planet is closest to the sun and smallest when it is farthest from the sun. This law is a statement of conservation of angular momentum.

An **ellipse** is a geometrical figure such that the sum of the distances from any point on it to two fixed points called **foci** is constant (Fig. 7.1). Kepler had found, after exhaustive experimentation, that the planetary orbits for which data existed were elliptical. The sun is located at one focus, and the other one is empty.

The second law was Kepler's statement that the orbital speed of a planet varies. It is greatest when the planet is closest to the sun, and slowest when it is farthest away (Fig. 7.2). This is actually a statement that angular momentum is conserved.

The third law, probably the most pleasing to Kepler because it ties all the planetary orbits into a common system, relates the size of each planet's orbit to its period. The long axis of an ellipse is called the **major axis**; half of this length is the **semimajor axis**. Kepler had found that the square of the orbital period was proportional to the cube of the semimajor axis. If the period T is expressed in years and the semimajor axis a in units of the sun-earth distance (called the **astronomical unit**, AU, 1.496×10^{11} m, or about 93 million miles), then the proportionality becomes an equality.

$$T^2 = a^3 \tag{7-1}$$

These laws, as developed by Kepler, were strictly empirical laws, that is, they were deduced from experiment or observation, and were not derived from some underlying physical laws. Newton sought the physical basis for these laws.

Newton took the point of view that the sun must be exerting some force on the planets, and attempted to find how this force depended on distance. Kepler's third law provided information on this. The relationship between the period of a planet and its distance from the sun led Newton to realize that the centripetal acceleration varies as the inverse square of the distance from the sun. We can easily see this from Eq. 6–18, which expresses the centripetal acceleration a_c in terms of the radius of circular motion r and the period T.

$$a_c = \frac{4\pi^2 r}{T^2} \tag{7-2}$$

If we assume that we are dealing with a circular orbit, so that $r = a$, we find

$$a_c \propto \frac{a}{T^2}.$$

(The symbol $\propto$ means "is proportional to.") Now, using Eq. 7–1, we find

$$a_c \propto \frac{1}{a^2}.$$

Newton also calculated the centripetal acceleration required to keep the moon in its orbit around the earth, based on its known period and orbital radius. When he compared this with the acceleration of falling objects at the earth's surface, he realized that the acceleration on the moon was less than that at the earth's surface by a factor of $(R/r)^2$, where R is the earth's radius

7.1 THE LAW OF GRAVITATION

and r is the moon's distance from the earth. Hence, whether the sun acts on the planets or the earth acts on the moon, he found the acceleration depended on the inverse square of the distance. Newton concluded that a universal force acts between any pair of bodies. He knew that the sun's attraction for the planets dominated their motions, and reasoned that this must be due to the sun's greater mass. He knew from his third law that there must be a reaction force for every action, so that a planet must exert a force on the sun that is equal and opposite to the force exerted on the planet by the sun. He summarized all these relationships in one statement:

$$F = \frac{Gm_1m_2}{r^2} \tag{7-3}$$

Here F represents the force acting on either of a pair of bodies whose masses are m_1 and m_2 with separation r, and G is the **gravitational constant,** whose value had to be derived experimentally. **F** is a vector whose direction is along the line between the two masses.

The Nature of the Gravitational Force Law

Newton's law of gravitation is usually called the **law of universal gravitation,** for every body in the universe attracts every other body with a force given by Eq. 7–3. Gravitation is one of the four fundamental forces of nature (see Chapter 3), and one of several physical laws that express an inverse square dependence on distance.

While the nature of a field force such as gravitation still eludes intuitive understanding, it is not so hard to see why the force should have this dependence on distance. If we imagine that the force is caused by some kind of field that radiates from a body, the strength of this field must be spread over an ever-growing area as we move farther from the mass. The force must act uniformly in all directions, so its strength must be spread uniformly over a spherical surface of area $4\pi r^2$, and its intensity at any point on this spherical surface is therefore inversely proportional to r^2. The same reasoning explains why any radiant quantity that is emitted uniformly in all directions has a $1/r^2$ dependence. The apparent brightness of a star, for example, shows this dependence.

We were already familiar with the dependence of the gravitational force on mass from knowledge of motion on the earth's surface, although we may not have realized the whole story. We knew that the weight of an object is proportional to its mass, but it is not so obvious that it is also proportional to the mass of the earth. If we wish, we can solve Eq. 7–3 for the weight W of an object of mass m, on the surface of a planet whose radius is R and whose mass is M:

$$W = \frac{GmM}{R^2} \tag{7-4}$$

We have dropped the vector notation because the direction of this force is always downward.

Earlier we said that the weight of an object at the earth's surface is given by

$$W = mg$$

where g is the acceleration of gravity. Substituting into Eq. 7–4 and solving for g leads directly to

$$g = \frac{GM}{R^2}. \tag{7-5}$$

The mass m of the object has cancelled because the acceleration of a falling body is independent of its mass, as Galileo first observed. This expression gives the acceleration of gravity at the surface of any spherical body of mass M and radius R.

The only body other than the earth that has been visited by humans is the moon. It is easy to calculate how a person's weight there compares with his weight on the earth. Using subscripts m and e to represent quantities on the moon and on the earth, we find

$$\frac{W_m}{W_e} = \frac{mg_m}{mg_e} = \frac{g_m}{g_e} = \frac{GM_m/R_m^2}{GM_e/R_e^2} = \frac{M_m}{M_e}\left(\frac{R_e}{R_m}\right)^2.$$

Since the moon's mass is 0.0123 times the mass of the earth, and the moon's radius is 0.273 times that of the earth, this yields

$$\frac{W_m}{W_e} = \frac{(0.0123)}{(0.273)^2} = 0.165.$$

Thus, a person weighs about one sixth as much on the moon as on the earth.

Example 7.1   How much is a person's weight decreased as he travels from sea level on the earth to the summit of Mt. Everest, 29,028 ft above sea level?

Here both m and M in Eq. 7–4 are constant, and the weight varies as the inverse square of the distance from the earth's center. The earth's radius at sea level is $R = 6.378 \times 10^6$ m, and the height of Mt. Everest is 8,850 m $\approx$ 0.009 $\times$ 10^6 m, so the weight W' at its summit (R') compared to the weight W at sea level is

$$\frac{W'}{W} = \left(\frac{R}{R'}\right)^2 = \left(\frac{6.378 \times 10^6}{6.387 \times 10^6}\right)^2 = 0.997.$$

The decrease in weight is very small, because the height of Mt. Everest is very small compared to the radius of the earth. The mass of the mountain makes an additional contribution to the weight that we have not estimated.

7.2 Orbits and the Two-Body Problem

With our knowledge of the law of gravitation and the properties of circular motion (Chapter 6), we are ready to analyze the orbits of bodies. This can be done analytically (that is, the equations can be solved exactly) only for two bodies in orbit about each other and in certain other special circumstances. Furthermore, simple algebraic treatment applies only to circular orbits, so we will derive equations for these restricted cases, but we will argue that the results are actually more general. Hence we will deduce the properties of orbits in a somewhat idealized situation, since we know that there are more than two bodies in the solar system, but we will often find that the influence of other bodies is small.

Center of Mass

In Chapter 4 we noted that the center of gravity is the point in a body at the earth's surface through which the weight acts as a force on a point mass. Now we must be more general, because we will discuss bodies in space, where weight is meaningless. The term we now use is **center of mass.** In the presence of a uniform gravitational field, the center of mass is the center of gravity.

Conservation of momentum tells us that the total momentum of an isolated system must not change; that is, all changes of momentum within such a system are balanced. This implies that there is a point in an isolated system that is not accelerated, regardless of any internal motions within the system; this point is the center of mass. For a pair of bodies in mutual orbit, neither body is fixed in space, but instead both orbit the center of mass, which is fixed or moves uniformly, with no acceleration (the center of mass is sometimes called the **barycenter**). Hence the orbit of any planet is not about a fixed sun, but about a center of mass between that planet and the sun; similarly, the moon orbits the center of mass of the earth-moon system. The location of this point for a circular orbit is found by using Eq. 6–18 for the centripetal accelerations a_c of the two bodies, and setting their centripetal forces ma_c equal (because these are equal to the mutual gravitational force).

$$\frac{4m_1\pi^2 r_1}{T^2} = \frac{4m_2\pi^2 r_2}{T^2}$$

Here r_1 and r_2 are the distances of the two bodies from the center of mass. After cancelling common terms, this becomes

$$\frac{r_1}{r_2} = \frac{m_2}{m_1}. \tag{7-6}$$

This result might have been expected; we know that a see-saw balances when the two riders are at distances from the pivot that are inversely proportional to their masses, because then the torques balance.

CHAPTER 7 GRAVITATION AND ORBITS

The concept of center of mass applies equally well to noncircular orbits and to many-body systems. In a cluster of stars, for example, the stars orbit about a center of mass that is fixed with respect to other clusters.

Example 7.2 Where is the center of mass of the sun-earth system?

From Eq. 7–6 we know that the distances of the sun and the earth (r_s and r_e) from the center of mass are in inverse proportion to the ratio of their masses.

$$\frac{r_s}{r_e} = \frac{m_e}{m_s} = \left(\frac{5.976 \times 10^{24} \text{ kg}}{1.991 \times 10^{30} \text{ kg}}\right) = 3.002 \times 10^{-6}$$

We also know that the sum of r_s and r_e must equal the sun-earth distance, one AU. Hence we can also write

$$r_s + r_e = 1 \text{ AU} = 1.496 \times 10^{11} \text{ m}.$$

Solving, we find that the distance of the sun from the center of mass is

$$r_s = 4.5 \times 10^5 \text{ m}.$$

This is less than the radius of the sun, which is about 7×10^8 m, so the center of mass of the sun-earth system is inside the sun, very near its center. Hence the sun's orbital motion about this center of mass is difficult to observe (and, of course, is vastly complicated because it actually orbits the center of mass of the entire solar system).

Newton's Derivation of Kepler's Laws

The laws of planetary motion discovered by Kepler were later derived by Newton directly from his own laws of motion and the law of universal gravitation. This requires calculus to deal with the instantaneous velocities and accelerations of the planets and to find comprehensive expressions for their motions. Without extensive formal calculus, we must take the first law on faith for now. We can use our knowledge of limits and of angular momentum to derive the second law, and we can derive the third law for the case of circular motion with simple algebra and our knowledge of circular motions (Chapter 6).

To derive the second law, we assume strictly circular orbits, and consider the triangle in Fig. 7.3, which represents the area swept out by the planet in a small interval Δt. The area of the triangle, if $\Delta \theta$ is small, is

$$\Delta A = \frac{1}{2} r \Delta s$$

7.2 ORBITS AND THE TWO-BODY PROBLEM

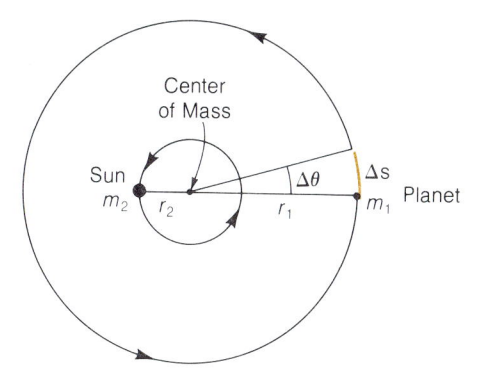

Figure 7.3

Orbits about the center of mass. In a two-body system, each body orbits the center of mass. Here the orbits are circular for simplicity, and the size of the sun's orbit relative to that of the planet is greatly exaggerated. This figure is used to derive both Kepler's second and third laws for circular orbits.

where s is the segment along which the planet moved in time Δt. We know $\Delta s = r\Delta\theta$, so we write

$$\Delta A = \frac{1}{2} r^2 \Delta\theta.$$

Conservation of angular momentum tells us that $I\omega$ is constant where $\omega = \Delta\theta/\Delta t$. For a point mass at r from the center of mass, the moment of inertia is $I = mr^2$, so we know that $mr^2\omega$ is constant. The mass of the planet must be constant, so we conclude that the product $r^2\omega$ is constant.

The rate of change of the area A of the triangle is

$$\frac{\Delta A}{\Delta t} = \frac{1}{2}\frac{r^2\Delta\theta}{\Delta t} = \frac{1}{2}r^2\omega.$$

We have just argued that $r^2\omega$ is constant, so we conclude that

$$\frac{\Delta A}{\Delta t} = \text{constant}.$$

That is, equal areas are swept out in equal intervals of time. Thus Kepler's observation was equivalent to the fact that angular momentum is conserved.

Kepler's third law can be easily derived for circular motion. We just set the gravitational acceleration of the planet equal to the centripetal acceleration (Eq. 6–18).

$$\frac{Gm_2}{(r_1 + r_2)^2} = \frac{4\pi^2 r_1}{T^2}$$

Here m_2 is the mass of the sun (1 M$_\odot$ = 1.99 × 10^{30} kg), and r_1 and r_2 represent the distances of the planet and the sun, respectively, from the center of mass. We can write a similar expression equating the gravitational and centripetal accelerations of the sun in terms of the planet's mass m_1.

$$\frac{Gm_1}{(r_1 + r_2)^2} = \frac{4\pi^2 r_2}{T^2}$$

Adding the two expressions leads to

$$\frac{G(m_1 + m_2)}{(r_1 + r_2)^2} = \frac{4\pi^2(r_1 + r_2)}{T^2}$$

or

$$(m_1 + m_2)T^2 = \frac{4\pi^2}{G}(r_1 + r_2)^3 \qquad (7\text{-}7)$$

which is the mathematical form of Kepler's third law. Here we see some dependencies of which Kepler was unaware: first, the masses of the two bodies come into play, and second, the distance that is represented is actually the sum of the semimajor axes of their orbits.

The reason Kepler's form of this law was simpler is that even the most massive planet, Jupiter, is very much less massive than the sun. Therefore the factor $(m_1 + m_2)$ is essentially a constant for all planet-sun pairs, and the observational data that Kepler analyzed showed no dependence on planetary mass. Likewise, the semimajor axis of the sun's orbit about the center of mass is negligibly small compared to that of the earth, and the sum $r_1 + r_2$ is dominated by the semimajor axis of the planet's orbit.

Astronomers define a system of units so that the constant term $4\pi^2/G = 1$. In this system, masses are measured in units equal to the sun's mass, the period is in years, and the separation is in units equal to the sun-earth distance. Then Newton's general form of Kepler's third law becomes

$$(m_1 + m_2)T^2 = a^3 \qquad (7\text{-}8)$$

where $a = r_1 + r_2$ now represents the sum of the semimajor axes of the two bodies. Eq. 7-8 (and Eq. 7-6 as well) applies to any pair of bodies orbiting their center of mass, and is widely used by astronomers to find stellar masses in binary systems, where two stars orbit a common center of mass (many stars are members of binary or larger systems). The period and the separation of the stars are observable so Eq. 7-8 can be solved for the sum of the two masses. The ratio of the masses may also be found, in some cases, by careful observation of the orbits or velocities of the two stars, so that the stationary center of mass

7.2 ORBITS AND THE TWO-BODY PROBLEM

can be located relative to the distance between the stars. From the sum and the ratio of the masses, it is possible to solve for each mass. This is the only direct technique astronomers have for measuring the masses of stars.

Example 7.3

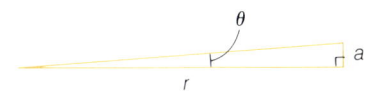

Figure 7.4

The small angle approximation for θ in radians. If θ is small θ ≈ a/r. Also sin θ ≈ θ for small angles.

A pair of stars in mutual orbit have a period of 23.4 yr and an angular separation of 3 seconds of arc. The center of mass is one fourth of the distance from star 1 to star 2. The double star is 29.6 light-years (2.8×10^{17} m) away. What are the masses of the two stars?

We are given the period, but not the separation of the stars, although we know the angular separation. We use the small-angle approximation (Fig. 7.4) (First convert 3 seconds of arc to 1.45×10^{-5} radians).

$$a = \theta r = (1.45 \times 10^{-5} \text{ rad})(2.8 \times 10^{17} \text{ m}) = 4.06 \times 10^{12} \text{ m} = 27.14 \text{ AU}$$

Now we solve Kepler's third law, (Eq. 7–8), for the sum of the masses.

$$m_1 + m_2 = \frac{a^3}{T^2} = \frac{(27.14)^3}{(23.4)^2} = 36.5 \text{ solar masses}$$

From the location of the center of mass, the ratio of the masses is

$$\frac{m_1}{m_2} = \frac{r_2}{r_1} = 3$$

The individual masses are $m_1 = 27.4$ solar masses, and $m_2 = 9.1$ solar masses.

We made a tacit assumption in deriving the actual linear separation of the two stars from their angular separation: we assumed that we viewed the orbit face-on, so that there was no projection effect. In reality, the tilt of the orbit is very difficult to determine, and this and other complications make the measurement of stellar masses a very uncertain business.

Conic Sections and Orbital Shapes

The ellipse is one of a family of shapes that orbits can take in a two-body system with an inverse square law of attractive force. These orbits are the **conic sections**, the figures formed by the intersection of a plane and a cone (Fig. 7.5). We can derive equations defining the shapes and other properties of these figures, but will do so in detail only for the ellipse.

For an ellipse, we can write

$$r + r' = 2a = \text{constant}$$

CHAPTER 7 GRAVITATION AND ORBITS

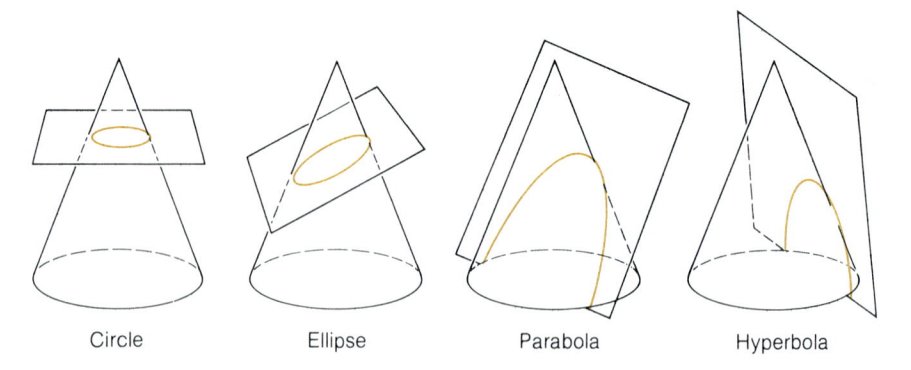

Circle Ellipse Parabola Hyperbola

Figure 7.5

The conic sections.

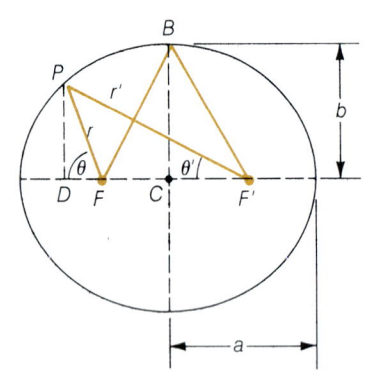

Figure 7.6

Elliptical orbits.

because by definition the sum of distances r and r' from any point on the ellipse to the two foci is constant (Fig. 7.6). As before, we define a as half the major axis, and we define b as half the minor axis. The **eccentricity** e of the ellipse is the ratio of the separation of the foci to the major axis, so that the distances CF and CF' in Fig. 7.6 are each equal to ae. For point B on the minor axis, the lines BF and BF' are each equal to a, the semimajor axis. The Pythagorean theorem then tells us

$$b^2 = a^2 - a^2e^2 = a^2(1-e^2).$$

Kepler's first law says that the sun is located at one of the foci; let us say that it is at focus F. The maximum distance on the ellipse from the sun, called the **aphelion** distance, is then $a + ae = a(1 + e)$, and the minimum distance, or **perihelion** distance, is $a - ae = a(1 - e)$. The *average* distance is simply a.

It is useful to find an expression for the distance r of any arbitrary point P on the ellipse from the focus F. We can define a **polar coordinate system** (one in which positions are defined by an angle θ and a distance r from a fixed location and direction) such that θ is the angle between r and the major axis. Then r' is the distance to focus F' and θ' is the angle from the major axis to r'. The Pythagorean theorem applied to triangle $F'DP$ tells us that

$$r'^2 = (2ae + r\cos\theta)^2 + (r\sin\theta)^2$$

which, after some manipulation, leads to

$$r'^2 = (2ae)^2 + 4r(ae)\cos\theta + r^2(\cos^2\theta + \sin^2\theta).$$

But $\cos^2\theta + \sin^2\theta = 1$, so we have

$$r'^2 = (2ae)^2 + 4r(ae)\cos\theta + r^2.$$

7.2 ORBITS AND THE TWO-BODY PROBLEM

We also know that

$$r' = 2a - r$$

and substituting for r' gives us finally

$$r = \frac{a(1 - e^2)}{1 + e \cos\theta}. \tag{7-9}$$

This is the equation for an ellipse in polar coordinates.

For some values of e, this equation leads us into difficulties. The orbit is an ellipse when $0 < e < 1$. We see that if $e = 0$, then $r = a$, and we have a circle. But what if e is 1 or greater? This possibility gives rise to new classes of orbits. If $e = 1$, then the orbit is a **parabola**, which may be viewed as an ellipse with one focus at infinity. The product $a(1 - e^2) = 2p$ remains constant as $a \to \infty$ and $e \to 1$. The equation for a parabola is then

$$r = \frac{2p}{(1 + \cos\theta)} \tag{7-10}$$

where p is the point of closest approach to the focus that is not at infinity.

If e is greater than 1, then the orbit has only one focus and it is a **hyperbola**, whose equation is

$$r = \frac{a(e^2 - 1)}{1 + e \cos\theta}. \tag{7-11}$$

The two most common types of conic sections are the ellipse and the hyperbola, and, as we shall see in the next section, these correspond to the two general possibilities of orbits that are gravitationally bound or unbound. In a bound orbit the two bodies cannot escape from each other; we will discuss this in terms of orbital energy later in the chapter.

The equations shown here give the distance from a body to the focus of the conic section representing its orbit, in terms of the angle θ. To fully define an orbit requires several additional parameters, because Eq. 7-8 and Eq. 7-9 give only the size and shape, along with an angle representing the position of the orbiting body on the ellipse. Other important quantities such as the inclination of the orbit and its orientation with respect to some reference direction and plane (usually the **ecliptic,** the plane of the earth's orbit) remain unspecified. To completely describe an orbit so that the position of the orbiting body at any time can be found requires seven parameters known as **orbital elements.** It is beyond the scope of this text to present these elements in further detail.

Example 7.4 A new asteroid is discovered, and its orbit is calculated after its motion has been observed for some time. The orbital period is 6.8 years, and the eccentricity of the orbit is $e = 0.09$. How close does the asteroid come to the sun at its closest approach, and how far does it recede from the sun at its farthest approach?

Since an asteroid is a minor body whose mass is much less than a planet, we can use the simplest form of Kepler's third law, from Eq. 7–1, and solve for the semimajor axis.

$$a = (T^2)^{1/3} = 3.59 \text{ AU}.$$

We know that the perihelion distance is $a(1 - e)$, so the asteroid's closest approach to the sun will be 3.27 AU. Its greatest distance, the aphelion, will be $a(1 + e) = 3.91$ AU. This is a typical orbit for an asteroid.

7.3 Escape Velocity and Orbital Energy

We have discussed the force of gravity between two bodies, and from our earlier analysis of work and energy we can see that energy must be associated with orbital motions. Furthermore, the conservation laws that we have developed in previous chapters have many important applications to orbits. In this section we will define orbital potential and kinetic energy, then see how these concepts can be used in various applications.

Gravitational Potential Energy

If we recall the work-energy principle, which says that the work done on a body subjected to a conservative force equals the potential energy gained by the body, we can derive an expression for gravitational potential energy. In our previous discussions of gravitational potential energy, we have usually chosen some arbitrary point where the potential energy is zero. In most cases, we set the potential energy equal to zero at the starting height, and calculated the energy gained by an object as it was lifted higher. Now, however, we need to establish a more general zero of potential energy because we will be speaking of total energies in a two-body system where gravitational acceleration is not constant.

We say that the potential energy is zero when two bodies are separated by an infinite distance. If the two bodies are brought closer together, potential energy is lost, just as it is when an object is lowered at the earth's surface. Hence the potential energy in general is negative. We refer to a potential energy well surrounding a massive object in space (Fig. 7.7), because the gravitational field of the object creates an imaginary hole that requires work in order to escape from it.

To derive an expression for the gravitational potential energy, we calculate how much work it would require to bring two bodies initially separated by an infinite distance to a separation of distance r, then invoke the work-energy principle, which says that the potential energy change is equal to this amount of work. The work required is the force times the displacement from infinity

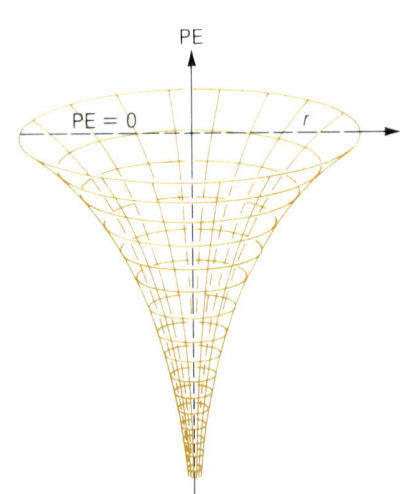

Figure 7.7

Potential well around a massive object. The potential energy is zero for an infinite separation between two bodies, and increasingly negative as they approach each other.

7.3 ESCAPE VELOCITY AND ORBITAL ENERGY

to r, but the force varies as a function of r, and the computation requires calculus. The result is

$$\text{PE} = -\frac{Gm_1m_2}{r}. \tag{7-12}$$

This is the potential energy of the two-body system, not of either body individually. The potential energy is negative because the force exerted on one body as it is brought closer to the other is exerted in the direction of decreasing displacement; this is like calculating the work done as a weight is brought from one height to a lower one at the earth's surface.

It will be useful to keep in mind that the absolute value of the gravitational potential energy of one object in the gravitational field of another represents the amount of work required to move the first object to an infinite distance.

The Total Energy

When two bodies orbit a center of mass, each has kinetic energy. In addition, there is system potential energy given by Eq. 7–12, so the total energy TE is

$$\text{TE} = \text{PE} + \text{KE} = -\frac{Gm_1m_2}{r} + \frac{1}{2}m_1v_1^2 + \frac{1}{2}m_2v_2^2.$$

To evaluate this, we recall that the total momentum of the system is constant. If we arbitrarily set this equal to zero (this is equivalent to saying that the system center of mass is at rest, and we can always define our frame of reference so that this is true), then

$$m_1\mathbf{v}_1 + m_2\mathbf{v}_2 = 0.$$

This implies that the magnitudes of the momenta are equal; that is

$$m_1v_1 = m_2v_2.$$

It also implies that the velocities must always be in opposite directions, so the net speed of the two bodies with respect to each other is

$$v = v_1 + v_2.$$

Solving this for v_1 and substituting $v_2 = v_1 m_1/m_2$ yields

$$v_1 = \frac{m_2 v}{m_1 + m_2}.$$

In a similar way, we find

$$v_2 = \frac{m_1 v}{m_1 + m_2}.$$

We now substitute these speeds into our expression for the total energy.

$$\text{TE} = -\frac{Gm_1m_2}{r} + \frac{1}{2}m_1\left[\frac{m_2 v}{(m_1+m_2)}\right]^2 + \frac{1}{2}m_2\left[\frac{m_1 v}{(m_1+m_2)}\right]^2$$

$$= m_1 m_2\left[\frac{v^2}{2(m_1+m_2)} - \frac{G}{r}\right] \quad (7\text{–}13)$$

Let us examine this expression for an elliptical orbit. From Eq. 7–9 and the conservation of energy and angular momentum, we can show that the velocities at perihelion, $r_p = a(1-e)$, and aphelion, $r_a = a(1+e)$, are given by

$$v_p = \left[\frac{G(m_1+m_2)}{a}\right]^{1/2}\left[\frac{(1+e)}{(1-e)}\right]^{1/2}$$

$$v_a = \left[\frac{G(m_1+m_2)}{a}\right]^{1/2}\left[\frac{(1-e)}{(1+e)}\right]^{1/2}$$

If we set $v = v_p$ and $r = r_p$ in Eq. 7–13, we find that the total energy simplifies to the expression

$$\text{TE} = -\frac{Gm_1m_2}{2a}. \quad (7\text{–}14)$$

This expression for TE contains no information on the relative values of the potential or kinetic energy, but it does tell us something very important: the total energy of an elliptical orbit is less than zero. We say that such an orbit is *bound*, because the two bodies cannot escape each other. We also note that the energy increases (that is, becomes less negative) as a increases. In the next section we will explore what happens when the total energy becomes zero or greater.

Continuing our examination of elliptical orbits, we can substitute Eq. 7–14 for the TE into Eq. 7–13 to yield

$$-\frac{Gm_1m_2}{2a} = \frac{m_1m_2 v^2}{2(m_1+m_2)} - \frac{Gm_1m_2}{r},$$

or, upon rearranging terms,

$$v^2 = G(m_1+m_2)\left[\frac{2}{r} - \frac{1}{a}\right]. \quad (7\text{–}15)$$

This expression, sometimes known as the *vis visa* equation, has many important applications. For example, we can use it to study how the kinetic energy of an orbiting body increases as the orbital size increases. We see this by regarding the eccentricity e as fixed and substituting Eq. 7–9 for r into the *vis visa* equation:

7.3 ESCAPE VELOCITY AND ORBITAL ENERGY

$$v^2 = \frac{G(m_1 + m_2)}{a}\left[\frac{2(1 + e\cos\theta)}{(1 - e^2)} - 1\right] \quad (7\text{--}16)$$

We could reduce this further, but there is no need; we simply note that v^2 decreases if a increases. This explains why the outer planets move more slowly than the inner planets, since all the planetary orbits have nearly the same shape.

Eq. 7–15 has other implications that we might have expected. For example, in an orbit with semimajor axis a, the speed v is smaller for greater values of r. This is consistent with what we have said previously about conservation of angular momentum and Kepler's second law; namely, that a planet moves slower where its orbit is farthest away from the sun.

Because the *vis visa* equation was derived from conservation of energy and angular momentum, we can use it and Eq. 7–9 to calculate the orbital velocity of a planet in any point in its orbit. This is especially simple at perihelion and aphelion. For example, the earth's orbital speed at perihelion, where $r = a(1 - e)$ is

$$v = \sqrt{G(m_1 + m_2)\left(\frac{2}{a(1-e)} - \frac{1}{a}\right)} = \sqrt{\frac{G(m_1+m_2)}{a}\left(\frac{1+e}{1-e}\right)}$$

The eccentricity of the earth's orbit is $e = 0.0167$ and its semimajor axis is 1 AU or 1.496×10^{11} m, so we find (setting $m_1 + m_2 \simeq M_\odot = 1.99 \times 10^{30}$ kg)

$$v = 3.03 \times 10^4 \text{ m/s} = 30.3 \text{ km/s}$$

(about 68,000 mi/h). Similarly, at aphelion, where $r = a(1 + e)$, the speed is

$$v = \sqrt{\frac{G(m_1 + m_2)}{a}\left(\frac{1-e}{1+e}\right)} = 29.3 \text{ km/s.}$$

Example 7.5

Halley's comet has period 76.08 years and semimajor axis $a = 17.95$ AU (Fig. 7.8). The eccentricity of its orbit is 0.967. What is its speed with respect to the sun at perihelion and aphelion?

At perihelion we have $r = a(1 - e)$. Substituting into Eq. 7–15 yields

$$v^2 = G(m_1 + m_2)\frac{(1 + e)}{a(1 - e)}.$$

Neglecting the mass of the comet in the sum of masses and substituting the sun's mass, 1.99×10^{30} kg, then yields

$$v = 54.3 \text{ km/s.}$$

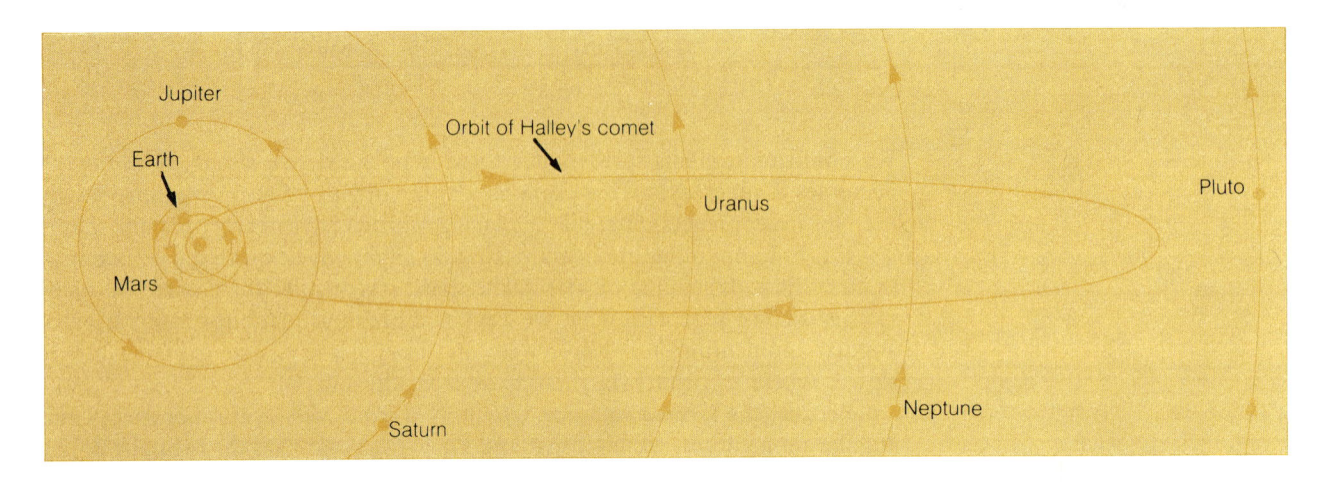

At aphelion, $r = a(1 + e)$, and we get

$$v = 0.91 \text{ km/s}.$$

Because of its elongated orbit, the comet has extreme variations in speed.

Figure 7.8

Example 7.6 Suppose we wish to launch a space probe that will reach Jupiter. What velocity must this probe have as it leaves the earth?

The most efficient way to send a probe to an outer planet is to give it the proper speed so that it enters an elliptical orbit whose aphelion is equal to the planet's distance from the sun (Fig. 7.9), with the earth at the perihelion of the probe's orbit. We first derive the semimajor axis of the orbit by noting that the major axis is equal to the sum of the sun-earth distance and the Jupiter-sun distance: this is 6.2 AU, so $a = 3.1$ AU $= 4.64 \times 10^{11}$ m. We then substitute this value for a in Eq. 7–15, and for r we substitute 1 AU or 1.496×10^{11} m. In the equation, the probe's mass is insignificant, so $m_1 + m_2 = 1.99 \times 10^{30}$ kg, the sun's mass. Substituting these values leads to

$$v = 3.86 \times 10^4 \text{ m/s} = 38.6 \text{ km/s}.$$

Since the earth is moving at about 30 km/s in its orbit around the sun, the problem of launching a probe to Jupiter is greatly simplified if we launch in the direction of the earth's motion. Of course, the timing must be precise, for

7.3 ESCAPE VELOCITY AND ORBITAL ENERGY

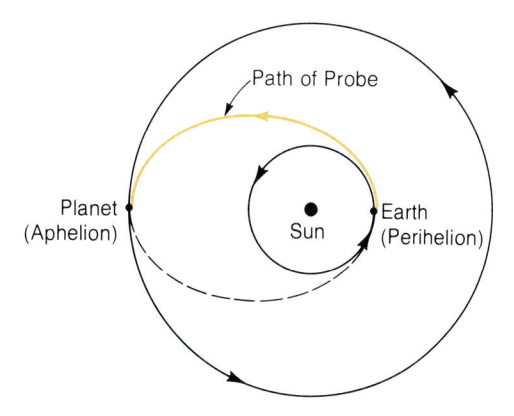

Figure 7.9

Jupiter must be at the same spot in its orbit as the probe when the probe intercepts Jupiter's orbit.

If a probe is to be sent to an inner planet, then the earth is at the aphelion of the probe's orbit and the planet is at perihelion, and we find that the probe's velocity is less than the earth's orbital speed. Hence a probe to an inner planet is launched in the direction opposite the earth's motion.

Escape Speed and Unbound Orbits

We note from Eq. 7–14 that the total energy of an orbit becomes zero when $a = \infty$; this result is expected because we have already said that the total energy is zero when the two bodies are separated by an infinite distance. An infinite value for a substituted into Eq. 7–15 yields

$$v_e = \sqrt{\frac{2G(m_1 + m_2)}{r}} \qquad (7\text{--}17)$$

where v_e is called the **escape speed,** the minimum relative velocity required for the two bodies to escape each other entirely. The relative velocity means the speed of the two bodies away from each other; for a rocket leaving the earth, this would be the vertical component of the velocity. There is no bound orbit when the relative velocity exceeds the escape speed.

In most practical applications, we need to find the speed for a small body to escape from the surface of a larger one, such as a planet. Then the value of r in Eq. 7–17 is the planetary radius, and the sum of the masses is approximately equal to the mass of the planet alone, that is, the escape velocity is independent of the mass of the small body. Thus each planet or moon has an escape velocity that is often cited as one of its basic characteristics. This is useful not only for determining the speed necessary should we ever wish to

Figure 7.10

The first launch of the Space Shuttle Columbia.

leave a planet by rocket, but also because it governs the atmospheric composition of a planet. Some gases may escape from the planet because the average speed of the molecules exceeds the escape velocity, while other gases may not.

Substituting the earth's mass (5.976×10^{24} kg) and radius (6.378×10^6 m) into Eq. 7–17 yields $v_e = 11.2$ km/s. This is the minimum vertical launch velocity required for a space probe to escape the earth. This is a difficult technological achievement, but sufficiently powerful rockets (Fig. 7.10) were developed in the early 1960s. If staged rockets are used, the escape speed is achieved through a series of outward accelerations. Each boost in acceleration puts the probe at a higher altitude with a smaller escape velocity. One variation on this is to use large rockets to place the probe in earth orbit, then launch it with the lower escape velocity required from there.

Example 7.7 Suppose an interplanetary spaceship is assembled on the moon, and is launched to the outer solar system. Ignoring, for the moment, the moon's gravity, what is the velocity required for the spaceship to escape the earth's gravitational field? How does this compare with the velocity required to escape from the surface of the moon? What velocity is required to leave the solar system entirely?

7.3 ESCAPE VELOCITY AND ORBITAL ENERGY

To calculate the velocity needed to escape the earth at the moon's distance, we simply substitute the earth's mass and the radius of the moon's orbit into Eq. 7–17. The earth's mass is 5.976×10^{24} kg, and the semimajor axis of the moon's orbit is 3.84×10^8 m; these values yield

$$v_e = 1.44 \times 10^3 \text{ m/s} = 1.44 \text{ km/s}.$$

This is smaller than the escape velocity from the earth's surface by nearly a factor of eight. This could have been found immediately by calculating the square root of the ratio of the earth's radius to the radius of the moon's orbit.

The escape velocity from the surface of the moon is found by substituting the moon's mass and radius into Eq. 7–17; these values are $m = 7.35 \times 10^{22}$ kg and $r = 1.74 \times 10^6$ m. We find

$$v_e = 2.37 \times 10^3 \text{ m/s} = 2.37 \text{ km/s}.$$

Thus the moon's escape velocity is greater than the speed needed to escape the earth's gravitational field at the moon's distance from the earth, but it is still much smaller than the escape velocity from the earth's surface.

Finally, if we wish to escape the solar system, we find the required speed by substituting into Eq. 7–17 the sun's mass and the sun-earth distance; that is, $m = 1.99 \times 10^{30}$ kg and $r = 1.5 \times 10^{11}$ m. We find

$$v_e = 4.2 \times 10^4 \text{ m/s} = 42 \text{ km/s}.$$

This is far greater than the escape velocity from either the moon or the earth. Thus, not so much is gained by launching from the moon. The most efficient method for reaching the outer solar system or escaping it is to launch from the earth in the direction of its orbital motion, to take advantage of the earth's 30 km/s velocity.

If the vehicle's velocity just equals the escape velocity, so that the total energy is exactly zero, then the orbit is a parabola, and Eq. 7–10 would be used to calculate the separation of the two bodies. If, as is much more likely, the total energy has a positive value, then the orbit is a hyperbola, and Eq. 7–11 is invoked. Parabolic orbits are rare because it is unlikely for the total energy to exactly equal zero, but some orbits are nearly parabolic. Comets are small sun-orbiting bodies that enter the inner solar system from a cloud of debris that orbits the sun at about 10^5 AU. It is thought that these bodies ordinarily orbit far from the sun, in roughly circular paths, but that occasionally random forces due to collisions among them or to gravitational perturbations of nearby stars cause them to fall toward the sun. Thus, they approximate the idealized case of an object falling from rest at an infinite distance. In the absence of any other forces, the orbits of comets would be such elongated ellipses that they would be nearly parabolic. Many comets do have such orbits, with eccentricities near 1 and periods measured in the millions of years. (It often happens, however, that an incoming comet is subjected to acceleration by the gravitational field of one of the giant outer planets, either gaining speed so that its total energy becomes positive and it escapes the solar system after passing near the sun, or losing speed so that the orbit becomes more tightly bound, with a smaller semimajor axis and shorter period.)

A positive total energy corresponds to a hyperbolic orbit. In such an orbit, the two bodies pass near each other once and never return. Thus a hyperbola describes the paths followed by any pair of free bodies.

In two-body systems, if the total energy exceeds zero, a bound orbit cannot be achieved without the intervention of some force in addition to the gravitational attraction of the two bodies. Thus, a planet cannot capture another body that passes near it unless some force decreases the relative speed. But free bodies have been captured in the solar system and it is believed that a third body must have interfered or that the relative speed was decreased by the drag of an atmosphere or a cloud of debris surrounding one of the bodies. In general, though, we are forced to conclude that planetary satellites and double stars must have formed as bound systems, because captures are naturally rare.

7.4 Artificial Satellites

Large numbers of man-made moons now circle the earth and play an important role in our culture. These satellites are used for scientific and technical purposes, for military applications, and, perhaps of greatest direct impact, for communications. It is therefore worthwhile to discuss some aspects of the orbits and other characteristics of artificial satellites.

First, we address weightlessness. It is common knowledge that astronauts in orbit are able to float about inside the spacecraft because they are weightless. Many important technical applications of artificial satellites, such as materials manufacturing techniques, depend on weightlessness. But Newton's law of universal gravitation tells us that there is indeed a gravitational force between an astronaut in orbit and the earth, so how is it that he does not feel this force?

The answer is that the spacecraft and its contents are in free fall toward the earth, so there is no acceleration of the contents with respect to the spacecraft. Recall that the centripetal acceleration of an orbiting body is the gravitational acceleration; hence the orbiting body is constantly falling inward, toward the center of the orbit. Its tangential velocity prevents it from crashing into the earth. For most artificial satellites, the tangential velocity is chosen so that the orbit will be circular or nearly so.

Hence it is not correct, technically speaking, to say that an astronaut is weightless. The proper statement is that there is no acceleration with respect to the spacecraft, even when an object is left unsupported.

The tangential velocity required for circular orbit is found easily from the *vis visa* equation (Eq. 7–15), by setting $r = a$, since the radius of the circle is equal to the semimajor axis when the eccentricity is zero and the two foci of

7.4 ARTIFICIAL SATELLITES

the ellipse merge into a single point. Making this substitution leads to

$$v_c = \sqrt{\frac{G(m_1 + m_2)}{r}} \qquad (7-18)$$

where r is the radius of the orbit. When we discuss artificial satellites orbiting the earth, the sum of the masses in this expression is approximated by the earth's mass. Note that the circular velocity is less than the escape velocity from orbit (from Eq. 7–17), by a factor of $\sqrt{2}$ (that is, $v_c = 0.707 v_e$).

Most artificial satellites, including all the manned earth-orbit missions so far, have altitudes above the earth's surface of about 150 to 200 miles, and typical orbital periods of about 90 minutes. Such orbits are high enough above the earth's atmosphere to be stable. Atmospheric drag will not immediately slow the spacecraft and cause them to spiral in and burn up. (This does eventually happen because of the trace of residual atmosphere in these altitudes. There was a great deal of public concern, for example, when the *Skylab* spacecraft reentered the earth's atmosphere a few years ago.) The circular velocity for such an orbit is about 7.8 km/s. Therefore, the job of launching a satellite into earth orbit becomes one of getting it to an altitude of about 150 miles and giving it a tangential velocity of 7.8 km/s, or about 17,000 mi/h.

Because a satellite orbit must be centered on the earth's center (approximately the center of mass of the two-body system), all orbits follow great circles, that is, their planes pass through the center of the earth. This means that a satellite launched from some latitude other than the equator will be inclined to the earth's equatorial plane (Fig. 7.11), and will pass back and forth over the equator as it circles the earth. This can be avoided only by diverting the satellite's velocity vector after it is in orbit so that the orbital plane coincides with the earth's equatorial plane.

Any of a variety of orbits can be chosen for different applications. One of the most widely used is the **geosynchronous orbit,** in which the orbital period is equal to 24 hours, so that the satellite remains fixed over a given meridian on the earth. Communications satellites are generally placed in synchronous orbits over the equator, so that the orbit is not inclined to the equatorial plane, and there is no north-south motion with respect to the earth's surface. Such a satellite stays fixed over one spot on the earth (Fig. 7.12).

The radius of a geosynchronous orbit can be calculated from Kepler's third law (Eq. 7–7), for a period of 24 hours. The result is $r = 4.22 \times 10^7$ m (from the earth's center), or about 22,300 mi above the earth's surface. This value of r is about one tenth of the earth-moon distance. The circular velocity at this altitude is about 3.1 km/s, or 6,900 mi/h. The total energy of such an orbit is significantly greater than for a low-earth orbit, so it is more difficult to place a satellite in geosynchronous orbit.

At an altitude of over 22,000 miles, more than five times the earth's radius, a satellite has direct line-of-sight contact with a large portion of the earth's surface. Three equally spaced satellites hovering over the equator can broadcast directly to the entire earth, except for the extreme polar regions. A single satellite can reach all of the United States. Hence telephone communi-

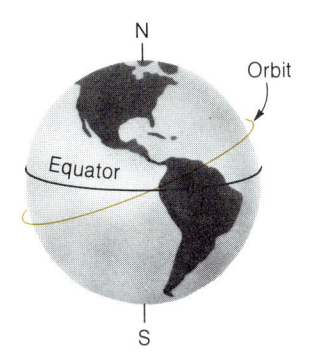

Figure 7.11

Orbit of an artificial satellite. The orbital plane passes through the earth's center, and must therefore be inclined to the equator, if the satellite is launched from any latitude north or south of the equator.

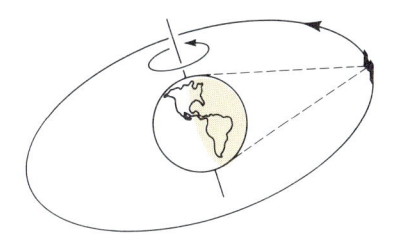

Figure 7.12

Geosynchronous satellite. At an altitude of 22,300 miles, this satellite orbits once every 24 hours. It therefore remains fixed with respect to the rotating earth and has line-of-sight contact with a large area (shaded region on earth's surface).

cations or television broadcasts can be sent directly into receiving stations all over the country. Such satellites are used to broadcast cable television; cable companies purchase the right to receive the satellite signals and sell them to customers who pay for direct links to the ground station. (Tremendous legal controversy has arisen over the issue of whether private individuals have the right to circumvent the cable companies by purchasing their own receiving equipment.)

The use of artificial satellites will undoubtedly continue to grow, and we can expect to see humans in orbit more and more frequently as the Space Shuttle ferries people and materials into orbit. The United States is even making plans to establish a permanent manned space station in a low-earth orbit. This would be capable of supporting a few people at a time, with new crews and supplies brought periodically by the Shuttle.

7.5 Differential Gravitational Forces and Tides

Before we complete our discussion of gravity and orbits, we consider the effect of gravity on an extended body. So far we have dealt only with the motions of bodies with respect to the center of mass, and we have treated the bodies themselves as point masses. This is not strictly correct, because the gravitational force experienced by an extended body varies with distance from the body that exerts the force. For example, the side of the earth facing the moon experiences a stronger gravitational force from the moon than does the opposite side. Although the earth's center of mass behaves as we have discussed, the so-called **differential gravitational force** (sometimes called the **tidal force**) can have indirect effects that modify orbits, as well as direct effects that we can observe on earth and in the solar system.

Consider a sphere of radius R, located a distance r from a large mass M (Fig. 7.13). If we consider two points on opposite sides of that body at distance $(r + R)$ and $(r - R)$ from the second body, we can calculate the differential gravitational acceleration for the two points. The difference between the gravitational accelerations is

$$\Delta a = \frac{GM}{(r - R)^2} - \frac{GM}{(r + R)^2}.$$

After finding the lowest common denominator and simplifying, this is

$$\Delta a = \frac{4RrGM}{(r + R)^2(r - R)^2} = \frac{4RrGM}{(r^2 - R^2)^2}.$$

If R is much smaller than r, which is generally the case for astronomical bodies, then this reduces to

$$\Delta a = \frac{4GMR}{r^3}. \tag{7-19}$$

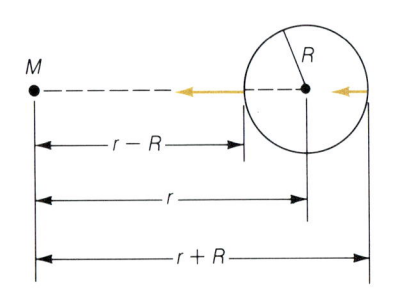

Figure 7.13

Tidal forces. The gravitational force on the near side of the sphere (separation r − R) *is larger than the force on the far side* (separation r + R).

7.5 DIFFERENTIAL GRAVITATIONAL FORCES AND TIDES 211

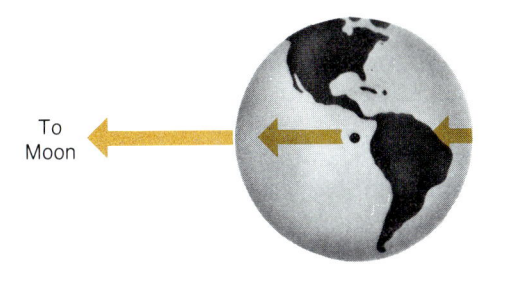

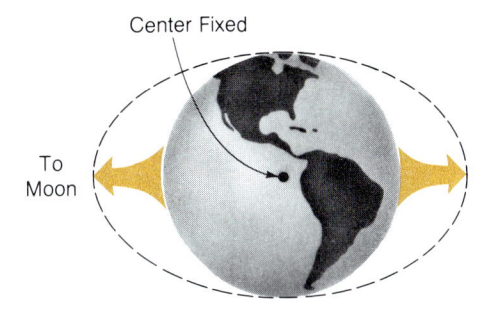

Figure 7.14

Tidal forces on earth. The upper figure shows the gravitational forces on the near and far sides of the earth. The lower figure shows the differential forces, relative to the center of the earth. Tidal bulges are produced on both sides (dashed lines).

Hence the differential gravitational acceleration (or the differential force, if the mass distribution is uniform) varies as the size of the extended body and as the inverse cube of the separation between the two bodies. Properly speaking, this expression should be derived by summing the differential forces at all points in the extended body, but the result has the same form.

The direction of this differential acceleration is very important: it acts to separate the two points on the extended body along the line connecting this body with the distant one. Thus the earth feels a force that acts to stretch it along the earth-moon line (Fig. 7.14). The earth's center of mass moves as dictated by orbital mechanics, but all other points on the earth tend to be pulled away from the center of mass. A surprising consequence is that the side of the earth opposite the moon is stretched just as the near side is; in an intuitive sense, this is due to the lesser gravitational acceleration on the far side compared with the acceleration at the center of mass.

Because of the moon's gravitational force on the earth, the earth has two bulges on opposite sides, one facing the moon and one away from it. Because of the rigidity of the earth, these bulges are not very pronounced, and furthermore, they are not quite aligned with the moon's direction, because they are pulled ahead by the earth's rotation. The two bulges pass by a point on the earth's surface at intervals of a little more than 12 hours (it is not exactly

12 hours because the moon moves along in its orbit while the earth spins).

The earth's oceans bulge much more than the solid earth, and they experience tides. A full cycle at any location takes just over 12 hours. Thus, high and low tides at a shoreline occur at intervals of just over 6 hours (the detailed timing as well as the amount of rise and fall depends on the local topography). The sun also exerts a tidal force on the earth, but it is not as strong as the moon's. (This is easy to show, by noting that M/r^3 for the moon is 2.2 times greater than for the sun, because the sun's great distance weakens its larger gravitational force.) The sun does have some effect, however, and the tides are maximized (spring tides) when the sun and moon are aligned (as at full moon or new moon), and minimized (neap tides) when they are at right angles from the earth. The strength of the tides also depends on where the moon is in its elliptical orbit; small variations in distance have a significant effect on the tides because of the r^3 in the denominator.

The effect of the earth's differential gravitational force on the moon is interesting. The earth's greater mass means that its stretching force on the moon is much stronger than the moon's force on the earth, and in fact the smaller rigid body of the moon is distorted by this force. The moon has significant tidal bulges, even though it has no water. We have said that the moon once rotated more rapidly than it now does, but the rotation has slowed so much that now the moon keeps one side facing the earth. The reason for this can now be understood. As the moon rotated, the earth's gravitational force acted on its bulge to create a torque that worked to slow the rotation. In addition, internal stress created by the distortion of the moon caused the dissipation of energy (in the form of internal heating) as the moon rotated, and this loss of energy was supplied by a decrease in the rotational energy. The result was that the moon's spin slowed until the torque due to the earth acting on its bulge became zero, and no internal stress was created by motion of the bulge over the moon's surface. These conditions are met when the spin period equals the orbital period, so that the bulge on one side points directly towards the earth at all times. No net torque is created by the earth's gravitational force. We have already discussed the effects on the moon's orbit due to this slowing of its spin.

Since all real bodies are extended, it should happen that in all two-body systems, tidal forces will eventually cause both bodies to slow their spins until each perpetually faces the other. The only reason that this has not happened everywhere is that the tidal forces are often so small and synchronization comes about so slowly that it has not yet occurred in many systems. Eventually the earth's spin will slow to equal the moon's orbital period, but not during the expected lifetime of the solar system. It is true, however, that many cases of synchronous rotation are known, involving satellites and planets in the solar system as well as double-star systems in which the two stars are very close together.

Tidal forces can also destroy orbiting bodies that come too close to very massive objects. If the tidal force exceeds the binding force holding a body together, the body will be torn apart. If the only force holding a body together is the mutual gravitation of its own material, then it can be shown that, as

7.5 DIFFERENTIAL GRAVITATIONAL FORCES AND TIDES 213

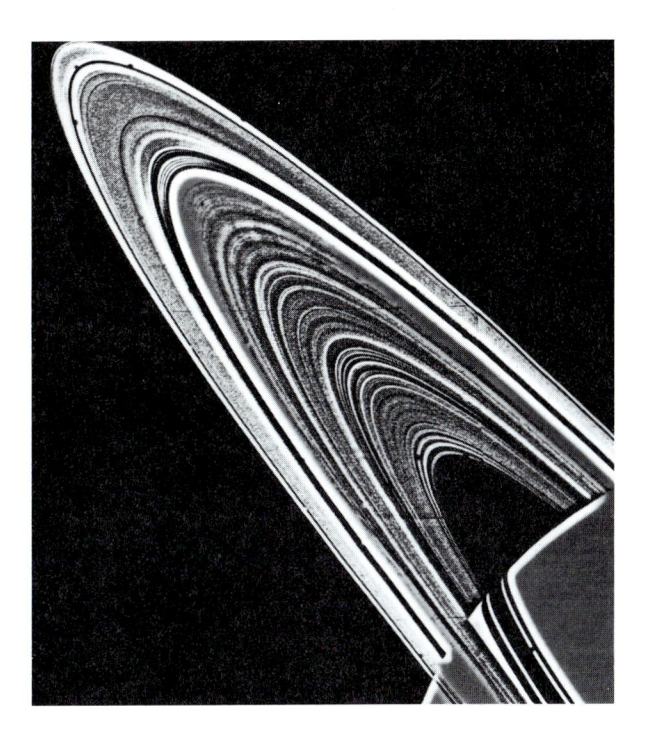

Figure 7.15

The rings of Saturn. The rings consist of countless tiny particles of ice and rock, which were prevented by tidal forces from forming a satellite.

one body approaches another of greater mass, at a certain separation the tidal force acting to stretch the lesser body will equal the gravitational force holding it together. The distance, known as the **Roche limit**, has the form

$$r_R \propto \left(\frac{\rho_M}{\rho_m}\right)^{1/3} R \qquad (7\text{--}20)$$

where R is the radius of the massive body, and ρ_M and ρ_m are the densities of the greater and less massive bodies, respectively.

It has been suggested that the ring system girdling Saturn is the debris of a former moon that was broken apart because it wandered inside the Roche limit. More likely, the countless particles of the ring system were fragments from the process by which Saturn formed, and were prevented from forming a moon by tidal forces. Whatever the explanation, differential gravitational forces have created a spectacular phenomenon (Fig. 7.15).

FOCUS ON PHYSICS

The Secrets of Double Stars

It is very fortunate for astronomers that many stars are members of double, or binary, systems. If this were not so, our knowledge of the universe and of many physical processes would not be as advanced as it is.

It is not usually easy to tell whether a star is double. The two stars are often too close together to be seen as separate stars, even when viewed through a telescope, or one star is very much fainter than the other. But evidence of orbital motion may show that the star is double. Carefully measuring the position of the visible star for many years may show its slight wobbling motion, or analyzing the light from the star may show shifts in wavelength characteristic of motion along the line of sight. (This is the Doppler effect. See Chapter 10.) Once the orbital motion is detected, it is possible to determine the orbital period and to gain some information on the separation of the two stars. Usually the tilt of the orbit is not known, so the separation of the stars is often difficult to measure accurately. Even so, a great deal of information about the two stars may be gained by analyzing the orbit.

We have discussed in this chapter measuring masses in double stars using Kepler's third law. The most accurate values of stellar masses are found for star systems where the separation of the two stars and the tilt of the orbit are well known. From the masses of a number of stars, astronomers in the early decades of this century determined the internal structure of stars. Using simple laws of gravitation and heated gases, scientists developed comprehensive theories of stellar structure. Along the way nuclear reactions became well understood, because it became evident that these reactions were the only possible source of the tremendous quantities of energy that stars radiate (nuclear reactions are discussed in Chapter 25). Thus, an entire field of physics, very important to us today, arose from astronomical research.

Binary stars tell us more than the masses of stars. In certain binaries, where the plane of the orbital motion is viewed edge-on, the two stars periodically eclipse each other, causing periodic variations in the total observed brightness (Fig. 7.1.1). If the orbital velocities are also known, then the diameters of the two stars can be determined from the duration of the eclipses, using the simple formula distance equals speed times time. If both the masses and the diameters are known, then the densities, the ratios of mass to volume, can also be calculated. Surface temperatures can be derived from the changes in brightness during eclipses, because surface brightness is a very sensitive function of temperature (this is discussed in Chapter 22). In short, virtually *all* that we know about the properties of stars is learned from analyses of binary stars.

Most of the fundamental work on stellar properties derived from binary stars took place in the first half of this century. For some time after this, many astronomers thought the analysis of binary orbits was no longer interesting. Starting in the late 1960s, however, there has been a strong resurgence of interest, because new observing techniques unveiled new and mysterious kinds of stars. Observations from above the earth's atmosphere revealed stars that emit x-rays, and periodic eclipses of the x-ray emission showed that many of these stars are binary. During the same period of time, the existence of highly condensed stars that were the collapsed remnants of more normal stars was being hypothesized, for both theoretical and observational reasons. Theories of stellar evolution failed to explain what happened to certain very massive stars when their nuclear reactions ended, and radio observations revealed a class of objects called **pulsars,** which were soon found to be very rapidly rotating, highly compact objects. Other theoretical work using general relativity theory (Chapter 21) showed that some stars might collapse and become **black holes** at the end of their lifetimes. The study of binary systems was quickly recognized as the best means of testing the theories of these bizarre objects.

Even if a star has collapsed to a size so small that it cannot emit enough light to be seen, or has become a black hole from which light literally cannot escape, its gravitational effects are still present. Thus, if such a star is orbiting with a normal star in a double system, analysis of the orbit of the visible star can reveal some of the characteristics of the unseen companion. From analyses of such systems, many of them x-ray emitters, has arisen strong evidence for two kinds of collapsed stars. The pulsars have been identified as **neutron stars**, objects so condensed that a star with the mass of our sun has a diameter of only 10 km and a density of 10^{18} kg/m^3 (ordinary rock has a typical density of 3×10^3 kg/m^3). The existence of black holes is now considered nearly certain, because binary systems have been found in which the unseen star has a mass far too great to avoid infinite collapse. In both cases the mass determination was critical to determining the nature of the invisible star, and the masses were estimated using techniques described in this chapter. The x-ray emission, from both neutron stars and black holes in binary systems, arises in gas that is compressed and heated to millions of degrees under the intense gravitational forces of the collapsed stars.

Other fantastic phenomena have been inferred from the analysis of binary star orbits. We know now that one star can lose material to its companion, altering the masses of both, and changing the course of their evolution. In systems where the two stars are quite close to each other, they are invariably locked into synchronous orbits, so that the same side of each star constantly faces the companion, and is therefore much hotter than the dark side. In some binaries, the two stars are so close to each other that they literally touch, and their shapes are distorted by tidal forces into ellipsoids rather than spheres. Today the study of binary stars is one of the most interesting and challenging areas of research in astrophysics, and the material covered in this chapter provides a basis for appreciating how this research is done.

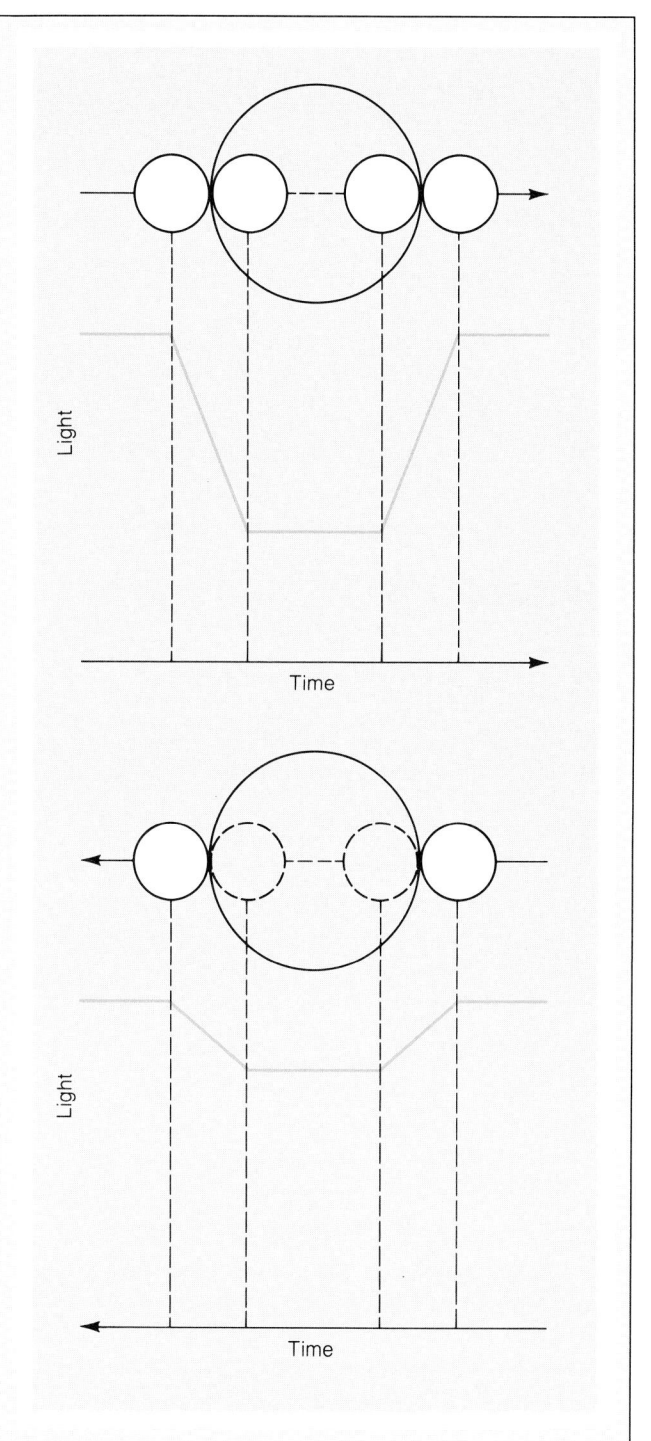

Figure 7.1.1

Perspective

The discussion of gravitation and orbits in this chapter completes our treatment of mechanics. We now have sufficient background to understand in principle, and in many cases to analyze quantitatively, many types of motion that we observe. We are now familiar with certain kinds of forces, and we have constructed a framework in which we can easily deal with others as they arise in later chapters.

In the next section, we will study a special type of motion: oscillation and vibration. We will discuss the properties of solids that lead to such motions, and the waves that can result from them.

PROBLEM SOLVING

Gravitation and Orbit Problems

Gravitation introduces a new set of laws and problem types. Newton's law defines the gravitational constant $G = Fr^2/m_1 m_2$ in terms of the force F between two masses m_1 and m_2 separated by a distance r. The SI units of G are therefore $N \cdot m^2/kg^2$. However a Newton is not a basic unit. Force equals mass times acceleration (1 N = 1 $kg \cdot m/s^2$) so G has units

$$G = \frac{kg \cdot m}{s^2} \cdot \frac{m^2}{kg^2} = \frac{m^3}{kg \cdot s^2}.$$

In doing gravity problems, it is useful to check your answer's units to make sure they make sense.

Example 1 A small mass m moves in a circular orbit of radius r about a large mass M. There are three variables, v, R, and M. If two are known, say M and R, then we need one relation to find v. That relation is the equation of force balance. The attractive gravitational force must equal the centripetal acceleration, which for a circular orbit is mv^2/r. Thus, we solve

$$\frac{GMm}{r^2} = \frac{mv^2}{r}$$

for the orbital velocity.

$$v = \sqrt{\frac{GM}{r}}$$

Note that the answer does not depend on the mass m, which cancels. If we substitute the units of G, we easily see that the final units are m/s.

Example 1 illustrates a simple, but common situation, in which the orbiting mass is much smaller than the large mass, so one can consider the large mass as fixed; the orbital radius r_1 therefore is zero in Eq. (7–7). Kepler's laws follow directly from the velocity in Example 1. The period T is just the orbit's circumference divided by the velocity.

$$T = \frac{2\pi r}{v} = 2\pi r \sqrt{\frac{r}{GM}}$$

Squaring this, we find

$$T^2 = \frac{4\pi^2 r^3}{GM}$$

which agrees with Eq. (7–7) when m is much less than M.

Now $T = 1$ yr and $r = 1$ AU for the earth orbiting the sun. There is nothing special about the SI units of period (s) and distance (m), so if we choose to measure periods in years, distances in astronomical units (1 AU = 1.496×10^{11} m), and mass in solar masses (1 $M_\odot$ = 1.99×10^{30} kg), we may write Kepler's law in the convenient form

$$T^2 = \frac{r^3}{M}.$$

For the solar system, $M = 1$.

Example 2 If Neptune has a mean distance of 30.11 AU from the sun, what is its period? Next, find the orbital distance of a satellite with period 35.4 yrs orbiting a star with 10 times the mass of the sun.

For Neptune orbiting the sun, we set $M = 1$ and $r = 30.11$ and solve for $T = (30.11)^{3/2} = 165$ yr. For the star, we set $M = 10$ and $T = 35.4$ and solve for

$$r^3 = MT^2 = (10)(35.4)^2 = 12{,}532.$$

If we take the cube root, we find $r = 23.2$ AU.

Not all orbits are circular, but an elliptical orbit introduces only one extra parameter, the eccentricity e. The velocity in an elliptical orbit can be found from the *vis visa* equation (Eq. 7–15 or Eq. 7–16). At aphelion and perihelion, the radius and velocity are particularly easy,

$$r_a = a(1 + e) \qquad v_a = \left(\frac{GM}{a}\right)^{1/2}\left(\frac{1-e}{1+e}\right)^{1/2}$$

$$r_p = a(1 - e) \qquad v_p = \left(\frac{GM}{a}\right)^{1/2}\left(\frac{1+e}{1-e}\right)^{1/2}$$

where a is the mean radius (semimajor axis).

Example 3 What are the ranges in distance and velocity for the planet Pluto, whose period is 248.5 yr and eccentricity is $e = 0.25$?

Kepler's law gives a semimajor axis $a^3 = T^2 = (248.5)^2$, or $a = 39.5$ AU. The mean velocity is thus $(GM/a)^{1/2} = 4.74$ km/s, where $M = 1\,M_\odot = 1.99 \times 10^{30}$ kg. The ranges are

$$r_a = (39.5\text{ AU})(1 + 0.25) = 49.4\text{ AU}$$
$$r_p = (39.5\text{ AU})(1 - 0.25) = 29.6\text{ AU}$$
$$v_a = (4.74\text{ km/s})\left(\frac{1-0.25}{1+0.25}\right)^{1/2} = 3.67\text{ km/s}$$
$$v_p = (4.74\text{ km/s})\left(\frac{1+0.25}{1-0.25}\right)^{1/2} = 6.12\text{ km/s}$$

Thus an eccentricity of 25% ($e = 0.25$) leads to comparable changes in radius and velocity at the two extremes.

SUMMARY TABLE OF FORMULAS

Kepler's third law:
$$T^2 = a^3 \tag{7–1}$$

Centripetal force:
$$a_c = 4\pi^2 r/T^2 \tag{7–2}$$

Law of universal gravitation:
$$F = \frac{Gm_1 m_2}{r^2} \tag{7–3}$$

Weight of body of mass m on planet of mass M and radius R:
$$W = \frac{GmM}{R^2} \tag{7-4}$$

Acceleration of gravity:
$$g = \frac{GM}{R^2} \tag{7-5}$$

Relative displacements from center of mass in two-body system:
$$\frac{r_1}{r_2} = \frac{m_2}{m_1} \tag{7-6}$$

Complete form of Kepler's third law:
$$(m_1 + m_2)T^2 = \frac{4\pi^2}{G}(r_1 + r_2)^3 \tag{7-7}$$

Kepler's third law for solar units:
$$(m_1 + m_2)T^2 = a^3 \tag{7-8}$$

Equation for ellipse in polar coordinates:
$$r = \frac{a(1 - e^2)}{1 + e\cos\theta} \tag{7-9}$$

Equation for parabola, polar coordinates:
$$r = \frac{2p}{(1 + \cos\theta)} \tag{7-10}$$

Equation for hyperbola, polar coordinates:
$$r = \frac{a(e^2 - 1)}{1 + e\cos\theta} \tag{7-11}$$

Potential energy of a two-body system:
$$PE = -\frac{Gm_1 m_2}{r} \tag{7-12}$$

Total energy of a two-body system:
$$TE = m_1 m_2 \left[\frac{v^2}{2(m_1 + m_2)} - \frac{G}{r} \right] \tag{7-13}$$

Total energy for elliptical orbit:
$$TE = -\frac{Gm_1 m_2}{2a} \tag{7-14}$$

Energy equation (*vis visa* equation):
$$v^2 = G(m_1 + m_2)\left[\frac{2}{r} - \frac{1}{a} \right] \tag{7-15}$$

Orbital velocity and orbital size for fixed eccentricity:
$$v^2 = \frac{G(m_1 + m_2)}{a}\left[\frac{2(1 + e\cos\theta)}{(1 - e^2)} - 1 \right] \tag{7-16}$$

Escape speed:
$$v_e = \sqrt{\frac{2G(m_1 + m_2)}{r}} \tag{7-17}$$

Tangential velocity for circular orbit:
$$v_c = \sqrt{\frac{G(m_1 + m_2)}{r}} \tag{7-18}$$

Tidal acceleration:
$$\Delta a = \frac{4GMR}{r^3} \tag{7-19}$$

Roche limit:
$$r_R \propto \left(\frac{\rho_M}{\rho_m} \right)^{1/3} R \tag{7-20}$$

THOUGHT QUESTIONS

1. If Galileo had dropped two objects in a vacuum, they would have fallen with the same acceleration. Why?
2. When two people stand near one another, why doesn't their mutual gravitational attraction pull them together?
3. Would two people in space move together because of gravity?
4. A tunnel is drilled on a straight line through the earth, from New York to Los Angeles. Would water flow into or out of the tunnel?
5. Because of drag in the atmosphere, a satellite spirals in and speeds up. Explain why this happens. (What happens to the velocity of a circular orbit as it contracts?)
6. Why do the outer planets have slower velocities around the sun than the inner planets?
7. If an astronaut steps out of the Space Shuttle, she would not fall behind. Why?
8. Why are astronauts in orbit weightless?

9 Two objects, one in an eccentric orbit and one in a circular orbit with the same apogee (maximum distance from the earth's center) have the same energy. Which has the greater angular momentum about the center?

10 Newton's version of Kepler's laws involves the sum of both masses, m_1 and m_2. Explain why this dependence was not noticed by Kepler. How could this dependence be perceptible?

11 For two bodies in orbit, a measurement of their period and separation determines the sum of their masses. What additional information is needed to find each mass?

12 How is the eccentricity of an orbit related to its apogee and perigee?

13 If the sun suddenly doubled its mass, what would happen to the earth's angular momentum and the shape and size of the earth's orbit?

14 Give several observational effects of differential gravitational forces (tides).

PROBLEMS

Section 7.1 Gravitational Force Law

1 From the gravitational force law and data in Appendix 4, show that on the Earth's surface, $g = 9.8$ m/s^2.

2 What is the gravitational force between two 10 kg spheres 2 m apart?

3 The density of iron is about 7.9 g/cm^3. Calculate the gravitational force between two iron balls, 5 cm in diameter, separated by 15 cm.

4 Compute the gravitational force between the proton ($m_p = 1.67 \times 10^{-27}$ kg) and the electron ($m_e = 9.11 \times 10^{-31}$ kg) in the hydrogen atom. Assume the electron travels around the proton in a circular orbit of radius 5.29×10^{-11} m.

5 Calculate the centripetal force on the electron in circular orbit with period 1.52×10^{-16} s (Problem 4). Is the force of gravity sufficient to keep it in orbit?

6 How much weaker is the force of gravity on the Earth's surface than on the surface of the Sun?

7 Two 60 kg people stand almost touching, 50 cm between their centers of gravity. What is the ratio of their mutual gravitational force to their weight?

8 How many steel cables would be needed to replace the gravitational force that holds the Moon in its orbit about the Sun? Assume that each cable can withstand a force of 100,000 N.

9 Two 80 kg bear cubs, trapped on a frictionless lake, sit 1 m apart, measured center to center. How long would they have to wait until they each move 1 cm toward one another?

10 Two spheres have equal mass M, radius R, and mass density ρ (kg/m^3). What is the gravitational force between them when they touch?

11 Calculate the gravitational acceleration on the surfaces of Mars, Jupiter, and Saturn. Refer to the table of astronomical data in Appendix 4.

12 Compute the net gravitational force on the Moon during three phases (Fig. 7.16): (a) full moon, (b) new moon, (c) half moon. Assume that the Earth remains at a constant distance of 1 AU from the Sun.

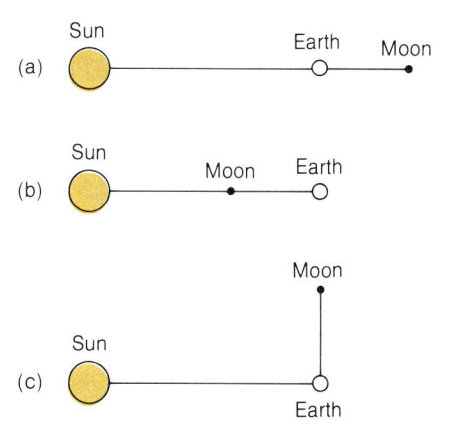

Figure 7.16

13 How much bigger are the tidal forces on the Earth during phase (b) than phase (a) in Fig. 7.16? (Use Eq. (7–19) and convert from acceleration to force.)

14 Knowing that $g = 9.8$ m/s^2 on the Earth's surface ($R = 6378$ km), compute the mass and average mass density (g/cm^3) of the Earth. Compare to water (1 g/cm^3). (Recall that $g = GM/R^2$ for the Earth.)

15 A newly discovered planet has an average density one third that of Earth, but the gravitational acceleration on its surface is the same as Earth's. What is the planet's radius in Earth radii?

16 The Sun rotates once each 30 days. Calculate the orbital radius of a synchronous satellite (one which stays in the same position above the Sun). Compare this radius to that of Mercury's orbit.

17 Suppose astronomers discover Planet X, beyond Pluto, with an orbital radius of 50 AU. What is its orbital period (in earth years)?

18 A periodic comet was seen in 1910 A.D. and next in 1986. How far from the sun (in AU) did the comet travel? Assume its eccentricity is 0.97.

19 Several astronomers recently proposed the existence of a very distant, low-mass stellar companion to the sun, orbiting once every 26 million years. How far from the Sun (in light-years) would such a hypothetical star travel? One light-year is 6.32×10^4 AU. Neglect the small star's mass and assume its orbit has $e = 0.5$.

20 If you weigh 125 lbs on Earth, what would you weigh on Mars? On Jupiter?

Section 7.2 Orbits

21 Using Kepler's Laws, find the orbital period of two stars of the mass of our sun ($1 M_\odot$), rotating about one another in circular orbits with centers 1 AU apart.

22 Where is the center of mass of a binary star system which consists of a 1 $M_\odot$ and a 3 $M_\odot$ star?

23 A three-body system consists of two stars of 0.5 $M_\odot$ separated from each other by 1 AU and orbited by a distant star of 1 $M_\odot$, 2,000 AU away from the others.
  (a) Where is the center of mass of the system?
  (b) What is the approximate period of the 1 $M_\odot$ star about the two smaller stars? (Treat the two smaller stars as a single object of 1 $M_\odot$.)

24 Three stars are, for the moment, travelling on the unstable circular orbit shown in Fig. 7.17. Where is the center of mass, and what is the magnitude of the net force of gravity on each star? What is the velocity V (in km/s)?

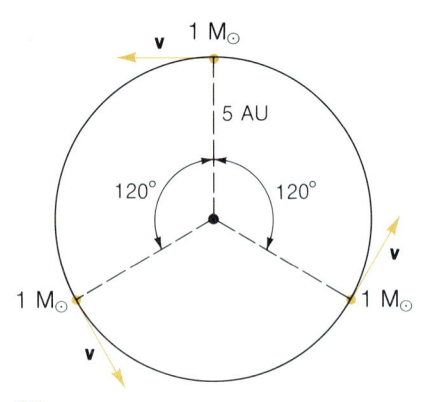

Figure 7.17

25 A satellite orbits a much larger mass M in a circular path of radius R. By balancing "centrifugal" force with gravity, derive formulas for the orbital velocity V and the period T.

26 A sleek satellite moves in a circular orbit just above the Earth's surface. Neglect atmospheric drag and find the satellite's velocity and period.

27 How many times faster is the satellite in Problem 26 moving than the ground below it due to the Earth's rotation?

28 Using Kepler's Laws and the values 1 AU = 1.496×10^{11} m and 1 year = 3.156×10^7 s, find the mass of the Sun. If the solar radius $R_\odot = 6.96 \times 10^8$ m, what is the average density of the Sun? Compare to water (1 g/cm^3).

29 Two stars just like our Sun (same mass, same radius) orbit each other on circular paths. What is the shortest period of orbit they can have without touching? One solar radius is 6.96×10^8 m.

30 What is the orbital velocity of the two stars in problem 29?

31 Two stars of mass m_1 and m_2 and circular velocities v_1 and v_2 have a period T and semimajor axis $a = (r_1 + r_2)$. Show that
  (a) $m_1/m_2 = v_2/v_1$
  (b) $(v_1 + v_2) = 2\pi a/T$
  (c) $(m_1 + m_2) = (a^3/T^2)(4\pi^2/G)$.

32 A binary star system is observed to have $T = 30$ d, and the individual stars are found to have orbital velocities $v_1 = 50$ km/s and $v_2 = 70$ km/s. Assuming circular velocities and using the results of Problem 31, find their separation, a, and their masses m_1 and m_2. Give your answer in AU and $M_\odot$.

33 The Sun moves around the Milky Way galaxy in a nearly circular orbit, with $v = 250$ km/s and with

$r = 30{,}000$ light-years from the galactic center. What is the approximate mass of the galaxy inside the Sun's orbit? Assume the mass is distributed spherically and give your answer in solar masses ($1\,M_\odot = 1.99 \times 10^{30}$ kg). Neglect the mass outside the Sun's orbit.

··34 A planetary system is discovered surrounding a nearby star. The orbits have periods of 50 d, 10 d, and 300 d, and the star is found to have a mass of $0.4\,M_\odot$. How far from the star are the planets (in AU)?

··35 For the planets in Problem 34, compute the angular radii in arc-seconds of their orbits as seen from Earth, if the star is 5 light-years away. Recall that 1 AU is 1.5×10^{11} m and 1 light-year is 9.5×10^{15} m.

··36 A comet has period 50 years and eccentricity 0.2. What are its perihelion and aphelion (in AU)?

··37 If a comet has a period of 10 years, what is the largest eccentricity it can have without crashing into the Sun?

··38 For very distant comets, with eccentricity $e = 0.5$, what period would take them to the distance of the nearest star ($a = 4.2$ light-years)?

··39 Pluto is not always the most distant planet from the Sun. Pluto has an orbital period $T = 248.5$ yr and eccentricity $e = 0.250$, while Neptune has $T = 164.8$ yr and $e = 0.0086$. Calculate the perihelion and aphelion of each planet.

···40 Jupiter and Saturn are 5.20 AU and 9.55 AU from the Sun, respectively.
(a) What are the periods of their nearly circular orbits (in years)?
(b) If their positions are aligned, how many years will pass before they are aligned again? (Consider the angular velocities, $\omega = 2\pi/T$ and use $\theta = \omega t$.)

··41 Halley's comet has a period of about 76 years and passes within 0.59 AU of the Sun at perihelion. What are its orbital eccentricity e and its aphelion?

Sections 7.3–7.4 Escape Velocity, Satellites

·42 Calculate the escape velocity of a small mass ejected from the surface of the sun.

·43 Calculate the escape velocity from the surface of a neutron star, which has a mass of $1\,M_\odot$ and a radius of 10 km.

··44 What is the radius of a star of mass M at which the escape velocity equals the speed of light, c? If $M = 1\,M_\odot = 1.99 \times 10^{30}$ kg, what is this radius? Use $c = 3 \times 10^8$ m/s. (This radius is the size of a black hole of one solar mass.)

··45 The mass of the Milky Way galaxy is about $10^{12}\,M_\odot$ within a sphere of radius 300,000 light-years (about 3×10^{21} m). Calculate the escape velocity from the edge of the galaxy.

··46 A space station orbits the Earth at a distance 10,000 km from Earth center. Calculate the velocity needed to launch the station from the Earth's surface into that orbit.

··47 For the space station in Problem 46, what is the velocity required to launch a probe from the station that will escape the Earth's gravity completely?

··48 NASA wishes to launch a space probe to Neptune (30.11 AU from the Sun). What velocity is required when the probe leaves Earth?

··49 A small mass m orbits a much larger mass M on an orbit of eccentricity e. The velocities at farthest radius (r_a) and closest radius (r_p) are v_a and v_p. From conservation of angular momentum and total energy, show that:
(a) $v_a r_a = v_p r_p$
(b) $\tfrac{1}{2}v_a^2 - GM/r_a = \tfrac{1}{2}v_p^2 - GM/r_p$

···50 Using the definitions $r_a = a(1+e)$ and $r_p = a(1-e)$, where a is the semimajor axis, with the results of Problem 49, show that:
(a) $v_a^2 = (GM/a)(1-e)/(1+e)$
(b) $v_p^2 = (GM/a)(1+e)/(1-e)$

···51 A comet on an elliptical orbit passing the Sun is found to have a velocity of 65 km/s at perihelion ($r_p = 0.4$ AU). What are the eccentricity, the period, and the semimajor axis of the orbit? Use the results of Problem 50.

··52 Calculate the velocity a probe must have as it leaves Earth for Venus. (Venus is 0.72 AU from the Sun.)

··53 Each of the four orbits shown in Fig. 7.18 passes through the same perihelion, r_p from the Sun. The eccentricities are 0.0, 0.5, 1.0, and 2.0 for orbits 1 through 4, respectively. If $r_p = 1$ AU, compute the velocities v_p at perihelion for each orbit.

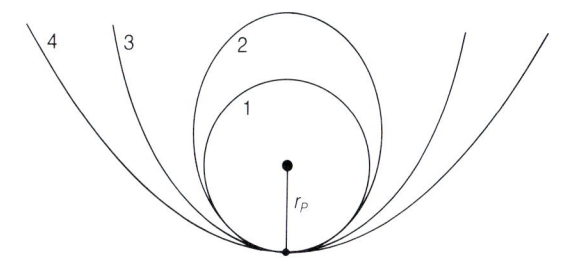

Figure 7.18

54 Using Eqs. 7–9 and 7–15, show that, for an elliptical orbit of semimajor axis a and eccentricity e, the velocity at angle θ (see Fig. 7.6) is $v^2 = [G(m_1 + m_2)/a](1 + 2e\cos\theta + e^2)/(1 - e^2)$.

55 Use the result of Problem 54 to show that at perigee ($\theta = 0°$) and at apogee ($\theta = 180°$), the velocities are
$v_p^2 = [G(m_1 + m_2)/a](1 + e)/(1 - e)$
$v_a^2 = [G(m_1 + m_2)/a](1 - e)/(1 + e)$.

56 For an elliptical orbit, show that the eccentricity is related to the apogee and perigee by the equation $e = (r_a - r_p)/(r_a + r_p)$.

57 Using polar graph paper or a suitably marked diagram, plot the shape of ellipses with $e = 0.1$ and 0.5.

58 For a mass m travelling in an elliptical orbit around a much larger mass M,
(a) Write down formulas for the total energy E and the angular momentum L. (Evaluate L at apogee or perigee.)
(b) Using the results of Problem 50, show that $(1 - e^2) = -(2EL^2/G^2M^2m^3)$.

59 A small mass m is released from rest at a distance r from a much larger mass M. What is the time it takes mass m to fall into mass M? (Consider the fall to be one-fourth an eccentric orbit, in which the eccentricity e approaches 1.)

60 Suppose the Earth suddenly stopped. How long would it take to fall into the Sun? See Problem 59.

61 A space station circles the Earth in a geosynchronous orbit. With what velocity must a rocket on the Earth's surface be launched to reach it?

62 A piece of space debris falls radially from rest at $r = 20{,}000$ km from the Earth's center. Use energy conservation to find its velocity at $r = 10{,}000$ km.

63 A small mass m is raised from the Earth's surface, $r = R$, to a height h.
(a) Show that the work done against gravity is $W = GMmh/R(R + h)$.
(b) Show that if h is much smaller than R, $W \approx mgh$, where g is the acceleration of gravity at $r = R$.

64 A gun shoots a bullet vertically upward with $v = 500$ m/s from the Moon's surface. From energy conservation and neglecting any friction, find the maximum height of the bullet.

65 A probe is launched from Earth to Jupiter, as described in Example 7.6. After swinging by Jupiter, the probe returns to the Earth orbit.
(a) How many years have passed?
(b) How far in its orbit has the Earth moved since launch?

66 At closest approach, a UFO is travelling horizontally at 10.9 km/s, 100 km above the Earth's surface. Assuming that the object does not use its engines, will it escape the Earth's gravity? Calculate the eccentricity of its orbit. [Use the results of Problem 50 to show that $v_p^2 = (GM/r_p)(1 + e)$].

Section 7.5 Tidal Forces

67 The planet Mercury, 2,439 km in radius, orbits the Sun at 0.39 AU radius. What is the difference Δa in gravitational acceleration between the two sides of the planet?

68 For Mercury (Problem 67), what is the ratio $\Delta a/g$ of the gravitational tidal force across it compared to the gravitational acceleration g at its surface (due to its own mass)?

69 Several years ago, some people predicted disastrous effects when Jupiter aligned with the Earth on the same side of the Sun. In that position, how much smaller is Jupiter's tidal force on the Earth than the Moon's?

70 Show that the average tidal force of the Sun on the Earth is 2.2 times smaller than that of the Moon on the Earth.

71 Refer to Problem 12 for the three phases of the Moon. Calculate the differential tidal force between the opposite sides of Earth for phases a, b, and c. (Notice that the tides in a and b are the same. For c, compute the net gravitational force vector.)

72 A spherical body of mass m and radius r comes within the Roche distance r_R of a large body of mass M.
(a) Show that the differential tidal acceleration between the center and radius of mass m is $\Delta a = (2GMr/r_R^3)$.
(b) By equating Δa to the gravitational acceleration at the surface of mass m, show that $r_R = (2M/m)^{1/3}r$.

SPECIAL PROBLEMS

1. Cavendish Experiment

In 1798, Henry Cavendish designed an experiment to measure G, the constant of universal gravitation. He used a torsion pendulum (Fig. 7.19) with two lead balls (5 cm diameter) attached to a 50 cm horizontal rod, which responds to twists in angle θ by a restoring torque, $\tau = k\theta$ which tries to return the rod to its normal position, $\theta = 0$. Two additional lead balls (20 cm diameter) are placed 15 cm (measured center to center) from the 5 cm balls. The gravitational attraction between the balls twists the pendulum by an angle θ_c until the torques balance.
  (a) Calculate the masses of large and small balls and the gravitational force between them. Assume lead has a density 11.3 g/cm³ and assume that θ_c is small, so that the distance between centers stays essentially constant at 15 cm.
  (b) Find θ_c in terms of G and k. Neglect the mass of the rod and assume again that θ is small.
  (c) Solve for G in terms of k, θ_c, and the other quantities. If $k = 1.3 \times 10^{-5}$ N · m/rad, as determined by experimental calibration, what angle θ_c must one measure to perform the experiment?

2. Communication Satellites for Other Planets

From the astronomical data below, calculate the altitude above the surface for surface-stationary orbits (orbit period equal to rotation period) for the following six planets. Is such an orbit possible for all six planets?

Planet	Mass/M_{Earth}	Rotational Period
Mercury	0.056	58.65 d
Venus	0.815	243.01 d
Earth	1.000	23.93 h
Mars	0.107	24.62 h
Jupiter	318.05	9.93 h
Saturn	95.15	10.67 h

3. Center of Mass and Angular Momentum of Solar System

Most of the mass in the solar system resides in the Sun, but most of the angular momentum is in the planets, mainly in the four large ones (Jupiter, Saturn, Uranus, and Neptune). Neglect the five smaller planets.
  (a) Compute the shift in the Sun's position when the four large planets are aligned on one side of the Sun and then the other. (Compute the center of gravity of the system in each position.)
  (b) Compute the angular momentum of the planets (four large ones) and compare it to that in the Sun's rotation. Use $I = 2MR^2/5$ and $T_{rot} = 30$ days.

4. Black Hole Candidate

Astronomers suspect that some x-ray emitting binary star systems may contain a compact object which could be a black hole. But to justify this speculation, they must first show that the compact object has a mass in excess of 3 $M_\odot$ (otherwise it could be a neutron star). From the following data, find the mass m_1 of the compact x-ray-emitting object in the hypothetical binary system Orion X-1 (The orbits are illustrated in Fig. 7.20.)

Orbital Period $T = 5.6$ days
Orbital Velocity $V_1 = 52$ km/s
Orbital Velocity $V_2 = 208$ km/s

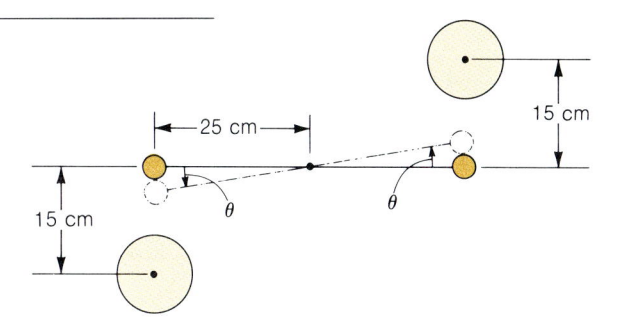

Figure 7.19

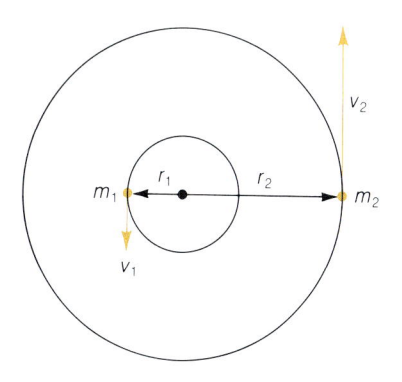

Figure 7.20

5. Roche Limit

The Roche limit is quite interesting in astronomy, since it determines the conditions under which certain bodies are torn apart by tidal gravitational forces. As derived in Problem 72, the Roche radius is given by $r_R = (2M/m)^{1/3} r$, where M is the mass of the large body, and where m and r are the mass and radius of the small body to be torn apart. (A more detailed calculation, including the effects of centrifugal forces, yields $r_R = 2.45 (M/m)^{1/3} r$.)

(a) If the average mass densities of masses M and m are ρ_M and ρ_m, show that

$$r_R = 2.45 (\rho_M/\rho_m)^{1/3} R_M$$

where R_M is the radius of the large body.

(b) Evaluate r_R for each of the following situations:
1. An icy comet ($\rho_m = 1$ g/cm^3) passing near the Sun (give your answer in solar radii).
2. A rocky asteroid ($\rho_m = 5.5$ g/cm^3) passing near the planet Saturn (give your answer in Saturn radii).
3. A solar type star ($\rho_m = 1.4$ g/cm^3) passing near a 10^8 $M_\odot$ black hole of radius $R_M = 3 \times 10^{11}$ m (give your answer in km).

SUGGESTED READINGS

"The Coriolis effect." J. E. McDonald, *Scientific American*, May 52, p 72, **186**(5)

"Gravity." G. Gamow, *Scientific American*, Mar 61, p 94, **204**(3)

"How did Kepler discover his first two laws?" C. Wilson, *Scientific American*, Mar 72, 92, **226**(3)

"Tides and the earth-moon system." P. Goldreich, *Scientific American*, Apr 72, p 42, **226**(4)

"Newton's apple and Galileo's dialog." S. Drake, *Scientific American*, Aug 80, p 150, **243**(2)

"Newton's discovery of gravity." I. B. Cohen, *Scientific American*, Mar 81, p 166, **244**(3)

"Rings in the solar system." J. B. Pollack, J. N. Cuzzi, *Scientific American*, Nov 81, p 104, **245**(5)

"The Earth's orbit and the ice ages." C. Covey, *Scientific American*, Feb 84, p 58, **250**(2)

Constructing the universe. David Layzer, Freeman, New York, 1984

Experiencing Science. Jeremy Bernstein, Basic Books, New York, 1978, (chapter on Kepler)

The flying circus of physics with answers. Jearl Walker, Wiley, New York, 1977

PART TWO

VIBRATIONS AND WAVES

Chapter 8
Solids and Elasticity
226

Chapter 9
Oscillations 254

Chapter 10
Wave Phenomena
284

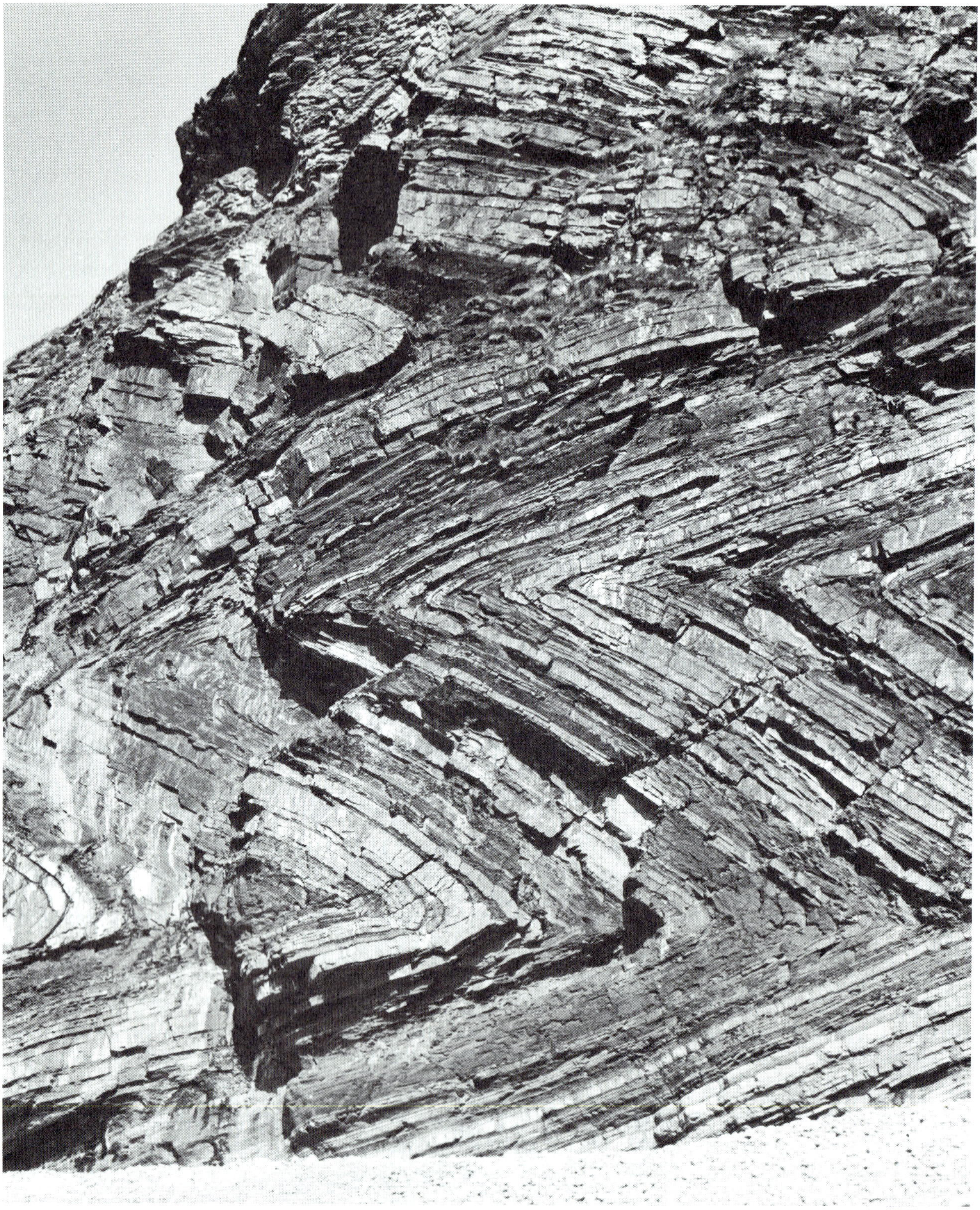

Chapter 8

Solids and Elasticity

We have spent considerable effort so far in developing an understanding of the motions and interactions of rigid bodies. Much of the formalism we have worked with is based on the assumption that objects can be treated as point masses; when that assumption is inadequate, we have added the concepts of torque and rotational quantities in order to discuss motions of extended bodies. We still have not considered internal motions and forces, which play an essential role in all mechanical interactions.

When two balls collide, for example, they rebound because of internal forces caused by the deformation of the balls (Fig. 8.1). When an object rests on a table, the table sags, and the upward force that counteracts gravity arises from the elasticity of the table, which tends to return to its original shape. When an object falls, its elastic properties determine whether or not it will break. No object is truly rigid; all contact between bodies involves internal elastic forces.

The nature of solids and their elastic properties depends on the structure of matter at the molecular level, so it is appropriate to begin this chapter with a discussion of the properties of matter and the internal forces that govern its behavior.

8.1 The Nature of Matter

Ever since the times of the early Greek philosophers, it has been theorized that matter is composed of a small number of basic substances or particles. The ancient notion of four basic elements (earth, water, air, and fire; later a fifth, the aether, was added) has evolved into the modern concepts of atomic and molecular structure. Today we generally speak of the atom as the basic particle of which all matter is composed, but atoms are composed of subatomic particles which themselves consist of elementary particles.

Figure 8.1

An elastic collision. When two balls collide, both balls are momentarily deformed. Elastic forces restore the balls to their original shapes and cause them to rebound.

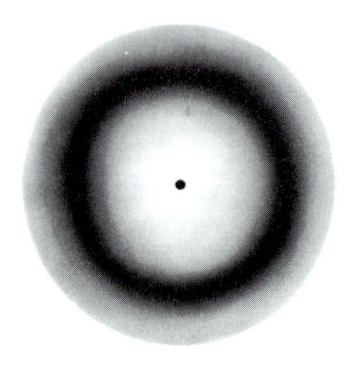

Figure 8.2

Electron probability distribution. An electron does not orbit the nucleus of an atom in a precisely defined orbit, but instead is described by a probability distribution (Chapter 24).

We need not begin our present discussion at quite so fundamental a level. We note, however, that the elementary and subatomic particles, as well as atoms and molecules, act upon each other exclusively through field forces (Chapter 3). The subatomic particles within an atomic nucleus are bound to each other by the strong nuclear interaction force, and the electrons are bound to the nucleus by the electromagnetic force. Since electrons orbit the nuclei, atoms are also bound to each other in molecules by the electromagnetic force. The detailed nature of the interactions between electrons and nuclei in atoms and molecules will be discussed in Chapter 24; for now, we begin by stating a few general properties of solids that result from electromagnetic bonds.

Atoms and Elements

Even though many details will be deferred, it is necessary to include here a brief description of the nature of atoms. Extensive experimentation in the late nineteenth and early twentieth centuries on the interactions of matter with light and of matter with matter laid the groundwork for an understanding of the structure of atoms during the first four decades of this century. Atoms were found to consist of three types of subatomic particles: **protons** and **neutrons**, which together form the nucleus; and **electrons**, which orbit the nucleus. The nucleus of an atom contains nearly all its mass. The neutrons and protons, collectively called **nucleons**, have nearly identical masses, but differ in electrical properties: the proton has a positive charge, while the neutron is without charge. Thus the nucleus has a net positive charge.

An electron, on the other hand, has a very much smaller mass than either type of nucleon, and carries a negative charge which is equal in magnitude to the positive charge carried by a proton. A normal atom is electrically neutral, because the number of electrons in orbit is equal to the number of protons in the nucleus.

The typical size scale of an atom is 10^{-10} m, while that of a nucleus is of order 10^{-15} m. The volume of an atom is mostly empty space. While it is tempting to think of an atom as a miniature solar system, with electrons orbiting the nucleus like planets orbiting the sun, this picture is vastly oversimplified. It is better to view the electrons as a cloud of electrical charge with a more-or-less smooth distribution about the nucleus, with no sharp limit at the outer boundary (Fig. 8.2).

Each element is identified by the number of protons in the nucleus, which is called the **atomic number Z**. For simple elements, the number of neutrons N is normally comparable to the number of protons, whereas for the more massive and complex atoms, the number of neutrons may be quite a bit larger. Some heavy elements with a large disparity between the numbers of neutrons and protons are unstable and spontaneously emit particles; such nuclei are said to be **radioactive**.

A given element may exist in different **isotopes**, which are forms with differing numbers of neutrons. Carbon, for example, always has six protons ($Z = 6$), but is found in nature with six (the most common form), seven, or eight neutrons. A given isotope is characterized by its **atomic mass**, which is related to the total number of nucleons it has. The atomic mass is represented by the **mass number** A, so that $A = Z + N$. The atomic mass is not simply equal to the mass number, but is instead defined as the number of **atomic mass units** (amu, usually abbreviated as simply u) the nucleus contains. The u is defined as one twelfth of the mass of the common form of carbon (that is, one-twelfth the mass of a carbon nucleus consisting of six protons and six neutrons). In standard units, 1 u = 1.6606×10^{-27} kg. Thus, the atomic mass of this isotope of carbon is precisely 12.000 u, and the mass number is 12. There are three isotopes of hydrogen: normal hydrogen, whose nucleus consists only of a single proton, with mass number 1 and atomic mass 1.007276 u; **deuterium**, which has one proton and one neutron, so that the mass number is 2 and the atomic mass is 2.014102 u; and **tritium**, which has one proton and two neutrons, with mass number 3 and atomic mass 3.016049 u. In the standard notation used by physicists, these three isotopes are designated ^1_1H, ^2_1H, and ^3_1H. In this notation, the letter (or letters) designate the atom, the superscript gives the mass number A, and the subscript gives the atomic number Z.

There are nearly a hundred naturally-occurring elements. Several more have been created in nuclear particle accelerators, but are so unstable that they exist for very short periods, then spontaneously emit particles and change into other elements. Each isotope of each element is characterized by its atomic number and its atomic weight. Appendix 5 lists all the known elements and gives their chemical symbols, atomic numbers, and atomic masses, and Appendix 6 gives their relative abundances in the universe.

Atoms in Combination

Atoms generally are not found as free particles, but are instead bound together into complexes of many atoms. In a gas or a liquid, atoms may become bonded by electromagnetic forces into structures called **molecules** (Fig. 8.3). The manner of bonding, the particular kinds of atoms that comprise molecules, and the size of the molecules are determined by the chemical properties of the atoms (which are determined in turn by their electrical properties). The study of the processes by which molecules form and interact is the basis for the branch of physical science that we call chemistry.

Atoms are also combined in solids. The key contrast between a liquid or gas and a solid is that in a solid the atoms are constrained to stay in the same location relative to their neighbors, rather than being able to move as free particles or as parts of free molecules.

We have said that electrons are bound to nuclei by the electromagnetic force, and that atoms in solids are also held together by this force. In the very close-range and complex interaction of electrons and protons within a solid, the electromagnetic force is referred to as the **elastic force**. The elastic force, of particular interest in this chapter, may be viewed as the force that holds the atoms and molecules of a solid in a specific relationship to each other. The exact nature of this relationship depends on a complicated interaction of electromagnetic forces due to the ensemble of charged particles in the atoms and

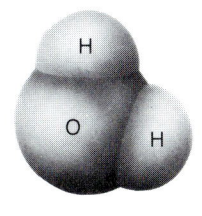

Figure 8.3

A molecule. Atoms bond together in specific arrangements to form molecules. This shows the structure of water molecules.

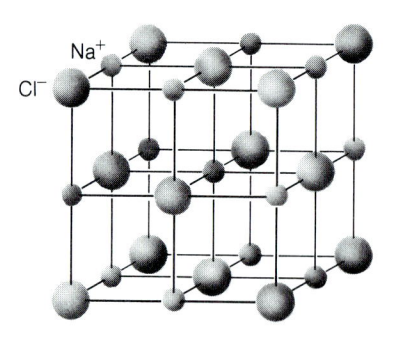

Figure 8.4

Crystal structure. The atoms in a crystal are arranged in a regular lattice.

molecules. The net effect is that a solid of a particular composition has a particular atomic or molecular structure which the elastic force acts to restore if the solid is deformed.

When the arrangement of atoms within a solid forms a lattice with a regular, repeating sequence of atoms (Fig. 8.4), we call the substance a **crystal**. We often think of crystals as minerals with regular external shapes, but most solid matter is composed of microscopic volumes that are crystalline. Whether crystalline or not, all solids can be characterized by properties related to the strength of elastic forces.

8.2 Restoring Forces and Hooke's Law

When an object is stretched or compressed, it will often return to its original size and shape when the force that deformed it is removed. When a diver leaves the board, the board returns to its original horizontal position; when a tennis ball is compressed by the impact of the racket, it returns to its initial shape (and the racket strings return to their initial alignment); when a spring is stretched and released, it returns to its original length. The bonds holding the atoms of a solid in place always have this property as well; if an atom is displaced from its position relative to its neighbors, it will return to that position when the distorting force is removed. A force that acts to return a body to its original shape or position is called a **restoring force**.

As a solid is stretched or compressed, the amount of force that is required changes. Experimentation has shown that for many materials the force required to displace a point on a solid is proportional to the displacement of that point from its equilibrium position (so long as the displacement is small). This means, for example, that as we stretch a spring, the force required to stretch it increases

8.2 RESTORING FORCES AND HOOKE'S LAW

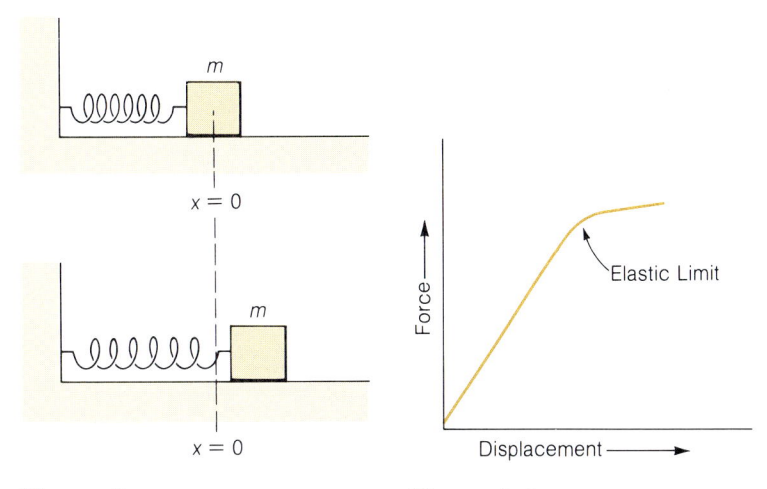

Figure 8.5

Restoring force. When the mass is displaced from its equilibrium position the spring exerts a force on it that is proportional to the displacement.

Figure 8.6

Elastic limit. After the deformation exceeds the elastic limit, the restoring force is no longer proportional to the displacement.

in direct proportion to the amount by which it has already been stretched (Fig. 8.5). In mathematical form, this property is written

$$F = kx, \tag{8-1}$$

where F is the force on the spring, x is the amount of displacement of the point from its equilibrium position, and k is a constant known as the **force constant.** This relationship, first discovered experimentally in the seventeenth century by Robert Hooke, is known as Hooke's law. The value of k depends on the nature of the material that is being deformed, and can even depend on whether the spring is stretched or compressed.

Under most circumstances, the body will return to its original shape when the deforming force is removed. There is a limit, however, beyond which this will not occur. If the deformation is too great, Hooke's law no longer applies. The displacement x is no longer proportional to the deforming force (Fig. 8.6). The internal structure of the solid is permanently altered, and we say that the **elastic limit** has been exceeded. If we bend a metal rod far enough, for example, it will reach a point where it no longer springs back to its original straight alignment, but will instead stay bent. Like the force constant, the elastic limit depends on the nature of the solid and may even depend on external factors such as temperature. If a solid is deformed well beyond the elastic limit, it may reach the **breaking point** and fracture.

In the next chapter we will fully explore the role Hooke's law plays in vibrations. For now, we further explore the relationship between deforming force and deformation.

8.3 Elastic Deformations: Stress and Strain

A solid can be deformed in a variety of ways. Besides being folded, stapled, spindled, or mutilated, as the classical warning on computer cards reads, a solid can also be compressed, stretched, squeezed, or twisted. All of these deformations fall into three general categories: (1) changes in length (either a stretch or a compression); (2) changes in internal angular orientation (a twist or **shear deformation**); or (3) changes in volume (a **bulk deformation**). We have mentioned only solids in connection with deformations, but liquids and gases can undergo bulk deformations.

The term **stress** refers to the force that causes a deformation, and **strain** refers to the resulting change in dimension of the solid. The stress is the force applied per unit area, and it has units N/m². To see why the force per unit area is relevant, consider a simple length deformation. Suppose a weight hangs from a strap, so that the strap is stretched (Fig. 8.7). If we have two identical weights and straps, the amount of stretch is the same for each. The total weight is doubled, as is the total cross-sectional area of the two straps. The result would be the same if we had a single weight equal to both of the individual ones, and a single strap with twice the cross section of the original one. Hence, the amount of stretch is a function of the weight divided by the cross-sectional area of the strap. For shear and bulk deformations, the cross-sectional area in question will be specified differently, but stress will retain essentially the same meaning. Strain, however, must be defined differently for each type of deformation, and we will do so in the following sections. Strain in all cases represents the fractional deformation of an object (or a fluid, in the case of bulk deformation).

Stress always has units of force divided by area, or N/m² in the SI system. Strain is always dimensionless, since it is a change in dimension divided by the original dimension.

In all three types of deformation, we will find that strain is proportional to stress, so long as the elastic limit of the material is not exceeded. Thus Hooke's law may be regarded as a special case.

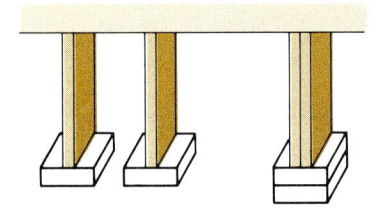

Figure 8.7

Stress and strain. At left, two identical weights hang from matching rubber straps. At right, the same two weights are attached to a single strap with twice the cross-sectional area. The force per unit area (the stress) is the same, and so is the lengthening of the rubber straps (the strain).

Length Deformations

When an object is stretched or compressed, the cross-sectional area that defines the stress is simply the area of the object in the plane perpendicular to the direction of the deforming force (Fig. 8.8). Mathematically, we say that the stress σ is

$$\sigma = \frac{F_\perp}{A} \tag{8-2}$$

where $F_\perp$ is the perpendicular applied force, and A is the area over which it is applied. The force $F_\perp$ can either be tension or compression.

For a length deformation, the strain is the fractional change in length. If

8.3 ELASTIC DEFORMATIONS: STRESS AND STRAIN

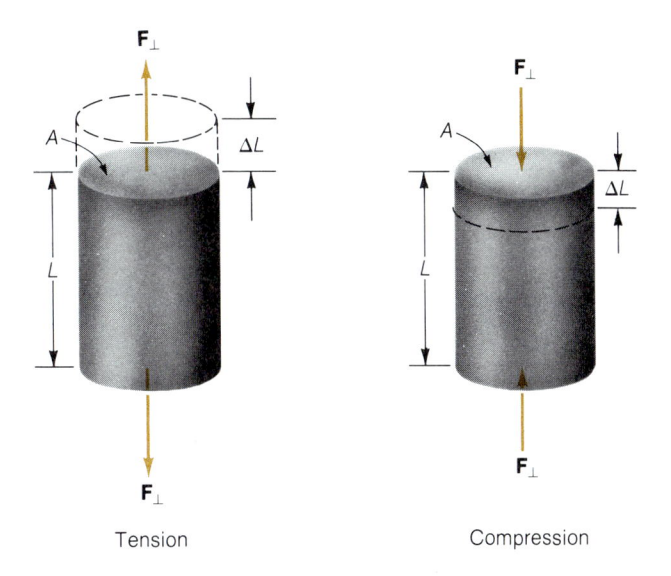

Figure 8.8

Stress. Under a length deformation, ΔL, the stress is the perpendicular force $F_\perp$ *(tension or compression) divided by the cross-sectional area* A. *Strain equals* ΔL/L.

a body has an initial dimension L, and it is stretched or compressed by an amount ΔL when a stress is applied, then the strain ϵ is

$$\epsilon = \frac{\Delta L}{L}. \tag{8-3}$$

We can find a relationship between stress and strain by generalizing Hooke's law. Experiment shows that the fractional deformation of a solid is indeed proportional to the deforming force, but inversely proportional to the cross-sectional area (in the plane perpendicular to the direction of the deformation); hence the fractional deformation $\Delta L/L$, which we have just defined as the strain, is proportional to the stress. The constant of proportionality k in Eq. 8–1, which depended on the cross-sectional area, is replaced by one that does not, and is therefore a more fundamental property of the material. The new constant is called **Young's modulus** or the **elastic modulus** and is designated Y. The units of Y in the SI system are N/m^2 (Table 8.1). The relationship between stress and strain can be written

$$\sigma = Y\epsilon \tag{8-4}$$

from which we see that the elastic modulus is

$$Y = \frac{\text{(Stress)}}{\text{(Strain)}} = \frac{(F_\perp/A)}{(\Delta L/L)} \tag{8-5}$$

CHAPTER 8 SOLIDS AND ELASTICITY

Table 8.1
Moduli for solids

Material	Elastic modulus $Y(N/m^2)$	Shear modulus $G(N/m^2)$	Bulk modulus $B(N/m^2)$
Aluminum	7.0×10^{10}	2.5×10^{10}	7.0×10^{10}
Brass	10.0×10^{10}	3.5×10^{10}	8.0×10^{10}
Copper	11.0×10^{10}	4.2×10^{10}	$14. \times 10^{10}$
Glass	5.5×10^{10}	2.3×10^{10}	3.7×10^{10}
Iron, cast	$10. \times 10^{10}$	4.0×10^{10}	9.0×10^{10}
Lead	1.6×10^{10}	0.56×10^{10}	0.77×10^{10}
Nickel	$21. \times 10^{10}$	7.7×10^{10}	$26. \times 10^{10}$
Steel	$20. \times 10^{10}$	8.0×10^{10}	$14. \times 10^{10}$
Tungsten	$36. \times 10^{10}$	$15. \times 10^{10}$	$20. \times 10^{10}$
Concrete	$2. \times 10^{10}$		
Brick	1.4×10^{10}		
Marble	5.0×10^{10}	2.0×10^{10}	7.0×10^{10}
Granite	4.5×10^{10}	2.0×10^{10}	4.5×10^{10}
Wood (pine)			
∥ to grain	1.00×10^{10}		
⊥ to grain	0.1×10^{10}		
Nylon	0.5×10^{10}		
Bone	1.5×10^{10}	8.0×10^{10}	
Water			0.2×10^{10}
Ethanol			0.1×10^{10}
Mercury			0.25×10^{10}

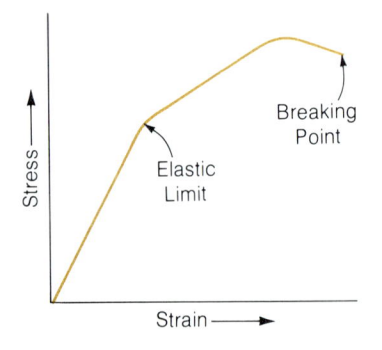

Figure 8.9

Stress-strain curve. The strain is proportional to the stress until the elastic limit is exceeded. The shape of the curve beyond that point depends on the substance.

Hence the elastic modulus of a substance can be measured experimentally by determining how the strain varies as a function of stress. The result of such an experiment is called a **stress-strain curve** (Fig. 8.9). As long as the strain is proportional to the stress, such a curve is a straight line whose slope is Y.

As we would expect from our discussion of Hooke's law, there is a point where a stress-strain curve becomes nonlinear; that is, there is a limiting stress beyond which Eq. 8–4 no longer applies, and the curve is no longer a straight line. This is the point where the solid has exceeded its elastic limit, and permanent deformation takes place (a material may only partially return to its original shape in this region of the stress-strain curve). If stress continues to be applied, strain will occur at a faster rate. Eventually, at a critical point, the stress required to produce additional strain stops increasing and actually begins to decrease. Finally, as stress continues, the material fractures. The point where the curve becomes nonlinear is called either the elastic limit, as before, or the **yield point**. The maximum value of the stress, that is, the point beyond which the stress lessens as deformation continues, is called the **ultimate tensile strength**, denoted S_t, or the **ultimate compression strength**, designated S_c (see Table 8.2). Deformations that exceed the elastic limit and hence are permanent (at least partially) are called **plastic deformations**. A material that fractures quickly once the ultimate tensile strength has been reached is said to be **brittle**; one that will continue to deform substantially is **ductile.**

8.3 ELASTIC DEFORMATIONS: STRESS AND STRAIN

Table 8.2
Ultimate strengths

Material	Tensile N/m²	Compressive N/m²	Shear N/m²
Iron, cast	170×10^6	550×10^6	170×10^6
Steel	500×10^6	500×10^6	250×10^6
Brass	250×10^6	250×10^6	200×10^6
Aluminum	200×10^6	200×10^6	200×10^6
Concrete	2×10^6	20×10^6	2×10^6
Brick		35×10^6	
Marble		80×10^6	
Granite		170×10^6	
Wood, pine			
∥ to grain	40×10^6	35×10^6	5×10^6
⊥ to grain		10×10^6	
Nylon	500×10^6		
Bone	130×10^6	170×10^6	150×10^6

Example 8.1 Human bone has an elastic modulus of about $Y = 1.5 \times 10^{10}$ N/m² for compression, and an ultimate compression strength of $S_c = 1.7 \times 10^8$ N/m². Suppose the total cross-section of the leg bones of a weightlifter is 1×10^{-3} m², and their length is 0.5 m. By what fraction is the length of these bones decreased when the weightlifter picks up a weight of mass 100 kg (Fig. 8.10)? What is the maximum weight he can lift before permanently deforming his leg bones?

We consider the equilibrium length of the bones to be the length they have when the lifter is standing; hence we ignore his body weight in calculating the deformation of the bones when he picks up the weight. From Eq. 8–2 we find that the stress is

$$\sigma = \frac{mg}{A} = \frac{(100 \text{ kg})(9.8 \text{ m/s}^2)}{(1 \times 10^{-3} \text{ m}^2)} = 9.8 \times 10^5 \text{ N/m}^2.$$

From Eq. 8–4 the strain is

$$\epsilon = \frac{\sigma}{Y} = \frac{(9.8 \times 10^5 \text{ N/m}^2)}{(1.5 \times 10^{10} \text{ N/m}^2)} = 6.5 \times 10^{-5}.$$

Hence the fractional change in length is only about a hundredth of one percent. From Eq. 8–3 we calculate the actual change ΔL.

$$\Delta L = L_0 = (6.5 \times 10^{-5})(0.5 \text{ m}) = 3.3 \times 10^{-5} \text{ m} = 0.033 \text{ mm}$$

To see how much the lifter can pick up without permanent damage to his leg bones, we substitute the ultimate compression strength S_c into Eq. 8–2, and solve for $F_\perp$.

Figure 8.10

$$F_\perp = S_c A = (1.7 \times 10^8 \text{ N/m}^2)(1 \times 10^{-3} \text{ m}^2) = 1.7 \times 10^5 \text{ N}$$

In the English system, this weight is just over 38,000 lb, which is far more than any human has ever lifted. Bones do get broken, however, but most commonly through bending or shearing, not compression.

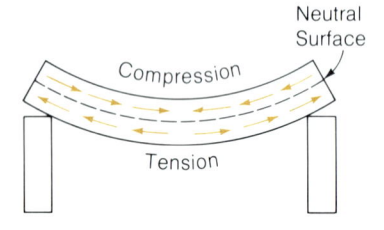

Figure 8.11

Bending. A bar bends due to its own weight, creating compression in the upper surface and tension in the lower surface. The neutral surface is neither compressed nor stretched.

The principles used to analyze length deformations can also be applied when an object is bent. At first it may seem that this kind of deformation is unrelated to the simple stretches and compressions we have been discussing, but in fact it is not. When an object is bent (Fig. 8.11), it is stretched along one side and suffers compression along the other, with a surface somewhere inside that is neither compressed nor stretched (this is called the **neutral surface**). The cross-sectional shape of the object determines how much compression or stretching occurs along surfaces displaced from the neutral surface; thus, two bars of equal length and cross-sectional area but different cross-sectional shapes can have different responses to the same bending force.

It is helpful to consider the forces acting on a bar that is being bent. A bar supported at both ends, but subject to bending due to its own weight (Fig. 8.11) is in equilibrium, but only because the internal elastic force produces torques that counteract those caused by gravity acting downward at the bar's center and the supports pushing upward at its ends. The elastic forces counteract the compression of the upper part of the bar and the stretching of the lower part. The deformation is greatest at the upper and lower surfaces, so the elastic forces are greatest along these two surfaces. The farther these surfaces are displaced from the neutral surface, the greater the effective lever arm through which the elastic force acts, and the greater the torque that is produced to counteract the bending force. Therefore, for a rod of a fixed quantity of material, the maximum rigidity is created by maximizing the displacement of these surfaces. Hence a hollow metal tube is stronger (against bending) than a solid metal rod of equal cross section (Fig. 8.12).

Although its derivation is beyond the scope of the present discussion, we can write an expression relating the stress and the strain for a bar being bent. The equivalent to the stress is the internal torque τ divided by the **area moment of inertia** I_A (note that this is not the same as the moment of inertia I that was defined in our discussions of mechanics). The strain is related to the radius of curvature R of the deformed object. The equivalent of the stress-strain relation is

$$\frac{\tau}{I_A} = \frac{Y}{R} \tag{8-6}$$

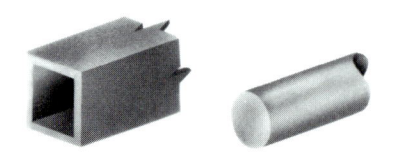

Figure 8.12

Rigidity. Here are cross-sectional sketches of two metal beams of equal cross-section. The one at left is far stronger against bending, because the long lever arm resists bending torques.

where Y is the elastic modulus as before. The value of I_A depends sensitively on the shape of the cross section. For example, as shown in Table 8.3, I_A for a cylindrical rod equals $\pi r^4/4$, proportional to the fourth power of the cross-sectional radius of the rod. Distributing mass at the outside of a cross section therefore increases its area moment of inertia. This is why beams with an I-shaped cross section have greater rigidity than beams of other shapes, such as cylinders or rectangles, with the same cross section.

Example 8.2 Two steel beams of equal length are constructed of equal quantities of metal, so that their cross sections of metal are the same. One has a cylindrical shape, and the other is a hollow tube whose inside radius r is equal to the outer radius of the other (Fig. 8.13). Each is supported at its ends and has a weight of equal mass attached at the center. What is the ratio of the radii of curvature of the two bent beams?

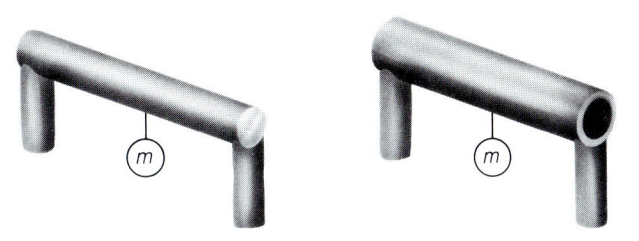

Figure 8.13

The total weight supported by each beam is the same, so the torques exerted on each by their own weight and by the supports at the ends must be equal. Therefore the internal torques in the two beams must be equal. We use Eq. 8–6 to write expressions for the internal torque in each beam, and then set these two expressions equal to each other. We then solve for the ratio of the radii of curvature R_1/R_2.

$$\frac{R_1}{R_2} = \frac{(YI_A)_2}{(YI_A)_1} = \frac{(I_A)_2}{(I_A)_1}$$

The elastic modulus Y cancels because it depends entirely on the nature of the material, which is the same for both beams. Table 8.3 shows that if the cylindrical beam is beam 1, then

$$(I_A)_1 = \frac{\pi}{4}r^4$$

and for the hollow beam,

$$(I_A)_2 = \frac{\pi}{4}(a^4 - r^4)$$

where a is the outer radius. Hence the ratio of radii of curvature is

$$\frac{R_1}{R_2} = \frac{(a^4 - r^4)}{r^4}.$$

We can find the relationship of a and r by recalling that their cross-sectional areas are equal. Therefore

$$\pi r^2 = \pi a^2 - \pi r^2$$

Table 8.3
Area moments of inertia

Rectangular solid

$$I_A = \frac{a^3 b}{12}$$

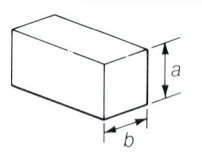

Solid cylinder

$$I_A = \frac{\pi r^4}{4}$$

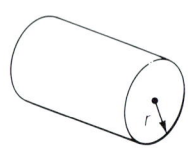

Hollow cylinder

$$I_A = \frac{\pi(a^4 - b^4)}{4}$$

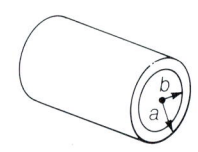

I-beam

$$I_A = \frac{a^2 bt}{2} + \frac{a^3 t}{12} \quad (t \ll a,b)$$
(t = thickness of each member)

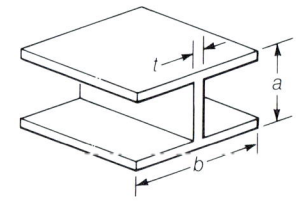

---

or

$$a^2 = 2r^2.$$

Substituting this into our expression for R_1/R_2 yields

$$\frac{R_1}{R_2} = \frac{(4r^4 - r^4)}{r^4} = 3.$$

The cylindrical beam bends three times more than the hollow tube. Thus we have shown quantitatively what we have already said qualitatively; namely, that it is better to make structural beams with an open cross-section than to make solid rods or rectangular bars.

8.3 ELASTIC DEFORMATIONS: STRESS AND STRAIN

A body subjected to bending deformation can buckle and collapse if the torque applied to it exceeds the maximum internal torque that can be produced by elastic forces. The key to this is the ratio of the length of a beam or rod to its cross section, because the external torque is a function of the length, while the internal torque is a function of the cross section, as we have just seen. To visualize what happens, imagine that a beam supported only at its ends is lengthened. The torque due to its weight acting at its center and to the supports at the ends is increased because of the increased lever arm, so the internal torque that resists this external torque must also increase. As the beam is lengthened, the elastic limit of its surface material is eventually exceeded. A little more lengthening causes the ultimate tensile strength (at the lower surface) or the ultimate compression strength (at the upper surface) to be exceeded, and then the beam quickly bends further, soon reaching its breaking point.

If we now consider an upright cylinder, we see that a small displacement from vertical will produce a torque due to the weight of the cylinder acting at its center of gravity (Fig. 8.14). As the cylinder grows taller while the radius is constant, eventually this torque will exceed the maximum internal torque, and the cylinder will collapse. The height at which this occurs is called the **critical height,** and it is a function of the elastic modulus as well as the radius. It can be shown that the critical height l_c has the form

$$l_c \propto r^{2/3} \tag{8-7}$$

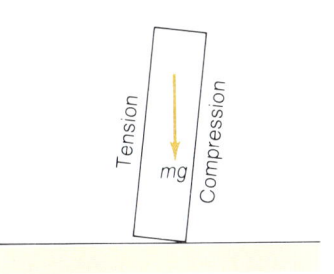

Figure 8.14

Bending deformation of a cylinder. If a cylindrical body is displaced from the vertical, it is bent by a torque due to its weight acting at its center of gravity. The cylinder will buckle if the torque exceeds the maximum internal torque for its material.

where r is the radius of the cylinder. The constant of proportionality that would make this expression an equality is a function of the elastic modulus and of the shape of the cylinder; the general form of Eq. 8–7 applies to tapered or hollow cylinders as well as uniform ones.

It is interesting to note that many biological structures apparently have evolved so that they have nearly the maximum length allowed before exceeding the critical height. The height and diameter of trees, for example, follow Eq. 8–7. This appears to be true of animal and human limbs also.

Shear Deformations

A shearing deformation occurs whenever an applied force acts to displace adjacent planes within a solid with respect to each other (Fig. 8.15). The displacement can be linear or rotational in character; for example, shear occurs when a book is pushed across a table (Fig. 8.16a), and it also occurs when the opposing covers of a book are twisted with respect to each other (Fig. 8.16b). The general mathematical treatment of shearing deformations can be very complex, but there are simple circumstances that we can discuss quantitatively.

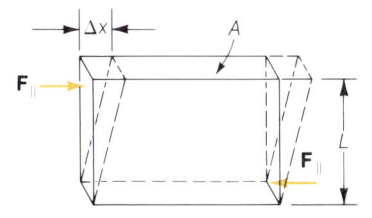

Figure 8.15

Shear stress and strain. The force $F_\parallel$ parallel to area A creates a shear stress $F_\parallel/A$. The shear strain is $\Delta x/L$.

If we push sideways on the top cover of a book lying on a horizontal surface, the book is deformed as the top cover is displaced. Planes within the book parallel to the cover are also displaced, but not as much, and the bottom cover is displaced the least. The **shear stress** σ_s is defined as the applied force

CHAPTER 8 SOLIDS AND ELASTICITY

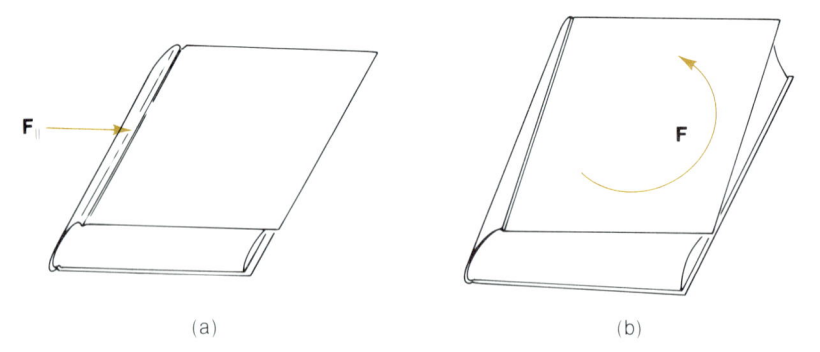

Figure 8.16

Linear and rotational shear.

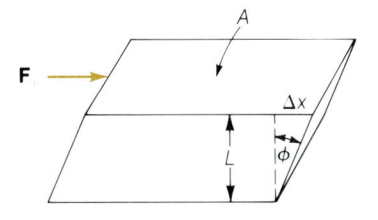

Figure 8.17

Shear stress and strain. The force $F_\parallel$ creates a shear stress $\sigma_s = (F_\parallel/A)$, which produces a shear strain $\epsilon_s = (\Delta x/L) = \tan \phi$.

divided by the area of the plane parallel to the direction of the force; that is,

$$\sigma_s = \frac{F_\parallel}{A} \tag{8-8}$$

where $F_\parallel$ is the force parallel to the plane of the top cover of the book, and A is the area of the cover (Fig. 8.17). The area parallel to the deforming force governs how strongly the internal parallel planes within the solid resist being displaced (ignoring the book's binding). The molecular bonding forces or friction forces that resist displacement of adjacent planes depend on how much area of contact there is between planes.

The resulting **shear strain** ϵ_s may be defined as displacement per unit height; that is, the horizontal displacement of the top cover of the book with respect to the bottom cover, divided by the thickness of the book. The shear strain is

$$(\text{Shear Strain}) = \epsilon_s = \frac{\Delta x}{L} \tag{8-9}$$

where Δx is the horizontal displacement and L is the thickness of the book. An alternative expression is

$$\epsilon_s = \tan \phi \tag{8-10}$$

where ϕ is the angle through which the deformation takes place (see Fig. 8.17).

In analogy with the elastic modulus in length deformations, the **shear modulus** G is a constant of proportionality defined so that

$$\sigma_s = G\epsilon_s \tag{8-11}$$

or

8.3 ELASTIC DEFORMATIONS: STRESS AND STRAIN

$$G = \frac{\text{(Stress)}}{\text{(Strain)}} = \frac{\sigma_s}{\epsilon_s} = \frac{(F_\parallel/A)}{(\Delta x/L)}. \qquad (8\text{–}12)$$

The value of the shear modulus is usually about one third to one half the value of the elastic modulus. Some representative values are given in Table 8.1. The **ultimate shear strength** S_s is defined to be the shear stress at which an object breaks (Table 8.2).

Example 8.3 A heavy piece of machinery being transported on a flatbed truck is mounted on rubber pads to reduce vibrations. There are four square pads, each 0.2 m on a side and 0.015 m thick. The machinery they support has a mass of 5,000 kg. The truck, moving at 30 mi/h (13.4 m/s) goes around a curve with radius of curvature 50 m. How much horizontal displacement of the load occurs?

We know the values of A (the total area of the four pads) and L (their thickness), and we want to find Δx, the displacement of the top of a pad with respect to its bottom, which is fixed to the truck bed. We need a value for $F_\parallel$, the force parallel to the plane of the truck bed. Recalling Chapter 6, we assume that this force is the centrifugal force, which is equal to the centripetal force. From Eq. 6–17, the centripetal acceleration is

$$a_c = \frac{v^2}{r}$$

so the force is

$$F_\parallel = ma_c = \frac{(5{,}000 \text{ kg})(13.4 \text{ m/s})^2}{(50 \text{ m})} = 1.8 \times 10^4 \text{ N}.$$

Now we use Eq. 8–8 to find the shear stress.

$$\sigma_s = \frac{F_\parallel}{A} = \frac{(1.8 \times 10^4 \text{ N})}{4(0.2 \text{ m})^2} = 1.1 \times 10^5 \text{ N/m}^2$$

Finally, the displacement is found by using this result with Eqs. 8–9 and 8–11 and a representative value of 5×10^6 N/m² for G.

$$\begin{aligned}
\Delta x &= \text{(Shear Strain)} \times L \\
&= \text{(Shear Stress)} \times \frac{L}{G} \\
&= \frac{(0.015 \text{ m})(1.1 \times 10^5 \text{ N/m}^2)}{5 \times 10^6 \text{ N/m}^2} = 3.3 \times 10^{-4} \text{ m}
\end{aligned}$$

The load is displaced sideways by only 0.33 mm.

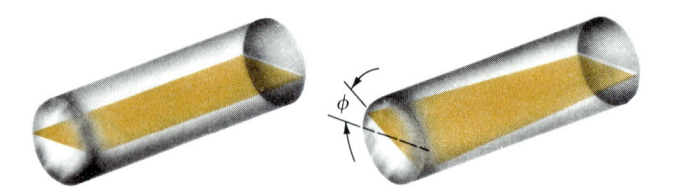

Figure 8.18

Angular deformation. At left an internal plane is shown in an undeformed cylinder. At right the cylinder is twisted, and the internal plane is bent by an angle φ over its length.

The analysis of twisting shears is a bit more difficult, but we can summarize what is found and how the result may be applied. Suppose a cylinder is twisted so that the planes formed by its ends are displaced by an angle φ (Fig. 8.18). The cylinder's resistance to twisting is characterized by the **polar moment of inertia** I_p, which is $\pi r^4/2$ for a solid cylinder of radius r, and the analog to stress is the torque divided by this moment of inertia. The strain is characterized by ϕ/l, where l is the length of the cylinder. Then the relationship between stress and strain becomes

$$\frac{\tau}{I_p} = \frac{G\phi}{l}. \tag{8-13}$$

From this relation, the amount of twisting deformation that will result from a torque can be calculated.

Example 8.4 A flywheel on an engine is a circular disk of radius 0.5 m and mass of 50 kg. When the engine is started, the flywheel reaches a rotational frequency of 5,000 rpm ($\nu = 83.3$ Hz) in 5 s. The flywheel is attached to a cylindrical metal shaft of radius 0.03 m, which is 0.5 m long. How much is this shaft twisted when the engine starts?

We are given the value for l, and we can find I_p from the expression for a solid cylinder.

$$I_p = \frac{\pi}{2}r^4 = 1.3 \times 10^{-6} \text{ m}^4$$

The value of G for a typical metal is 3×10^{10} N/m². To solve for φ, we now need to calculate the torque τ. To do this, we use Eq. 6–22.

$$\tau = I\alpha$$

Here I is the moment of inertia (not to be confused with I_p), which for a uniform disk is $\frac{1}{2}mr^2$, and α is the angular acceleration. In this example, the

8.3 ELASTIC DEFORMATIONS: STRESS AND STRAIN

angular acceleration is

$$\alpha = \frac{\Delta\omega}{\Delta t} = \frac{2\pi\nu}{\Delta t} = \frac{2\pi(83.3 \text{ Hz})}{(5 \text{ s})} = 105 \text{ rad/s}^2.$$

Hence the torque is

$$\tau = {}^1\!/_2 mr^2 \alpha = (0.5)(50 \text{ kg})(0.5 \text{ m})^2(105 \text{ rad/s}^2) = 656 \text{ N} \cdot \text{m}.$$

Now we can solve for the angular displacement of one end of the shaft with respect to the other.

$$\phi = \frac{\tau l}{GI_p} = \frac{(656 \text{ N} \cdot \text{m})(0.5 \text{ m})}{(3 \times 10^{10} \text{ N/m}^2)(1.3 \times 10^{-6} \text{ m}^4)} = 0.0084 \text{ rad}$$

Hence the shaft is twisted by about 0.48°.

A sufficiently large twisting deformation will cause a body to fracture. This type of fracture is fairly common in human bones, particularly for people who risk large polar torques. A skier, for example, has a long lever arm rigidly attached to his foot, so that a relatively small force applied to the end of the ski can produce a large torque in the bones of his leg.

Bulk Deformation

When stress is applied uniformly to all surfaces of a body, its volume will change (Fig. 8.19). Here we refer to the **volume stress** σ_v, defined as

$$\sigma_v = \frac{\Delta F}{A} = \Delta P \tag{8-14}$$

where $\Delta F/A$ is a change in force per unit area. Fluids tend to exert uniform force per unit area, and the term **pressure** is normally used; hence the equality with ΔP.

The **volume strain** ϵ_v is defined as the fractional change in volume V, or

$$\epsilon_v = \frac{\Delta V}{V}. \tag{8-15}$$

Then the relationship between stress and strain when volume is reduced is

$$\epsilon_v = \frac{-\sigma_v}{B} \tag{8-16}$$

where B is the **bulk modulus**. The minus sign is used because an increase in stress causes a negative strain (that is, a decrease in volume).

Solids and liquids alike can undergo bulk deformations, whereas only solids

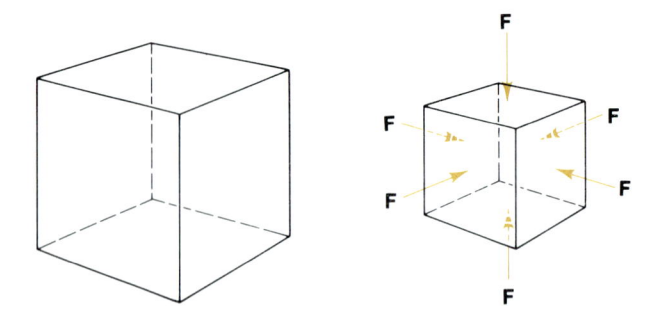

Figure 8.19

Bulk deformation. Here a uniform force of compression reduces the volume of a solid.

are subject to length or shear deformations. Thus a few liquids are included in Table 8.1, which lists representative values of B. The large values indicate that most liquids are difficult to compress. This is the reason that fluids in sealed containers transmit force well in hydraulic systems.

A quantity called the **compressibility** is sometimes used to describe how easily a liquid can be compressed. The compressibility is the fractional change in volume per unit change in pressure, and this turns out to be simply the reciprocal of B, the bulk modulus.

Example 8.5 Suppose a baseball with a bulk modulus of 1×10^8 N/m^2 is dropped into the ocean and sinks to the bottom, where the pressure is 100 times the sea-level atmospheric pressure. If the ball's initial radius is .045 m, what is its radius when it reaches the sea floor?

Atmospheric pressure at sea level is about 1×10^5 N/m^2, so the change in pressure as the ball sinks, which is approximately equal to the pressure at the bottom, roughly is 1×10^7 N/m^2. The initial volume of the ball is

$$V = \frac{4}{3}\pi r^3 = 3.8 \times 10^{-4} \text{ m}^3.$$

By using Eq. 8–14, Eq. 8–15, and Eq. 8–16, the change in the volume of the ball can be written

$$\Delta V = \frac{-V\Delta P}{B} = -\frac{(3.8 \times 10^{-4} \text{ m}^3)(1 \times 10^7 \text{ N/m}^2)}{(1 \times 10^8 \text{ N/m}^2)} = -0.38 \times 10^{-4} \text{ m}^3.$$

The new volume is therefore about 3.4×10^{-4} m^3, so the new radius is

$$r = \left(\frac{3V}{4\pi}\right)^{1/3} = 0.043 \text{ m}.$$

FOCUS ON PHYSICS

Rocks Under Stress and Earthquake Prediction

We normally think of rocks as hard, brittle objects that cannot be bent or squeezed. Indeed, the expression "hard as a rock" is commonly used to indicate complete rigidity. Nevertheless rocks can be bent or compressed, and when the ultimate strength of rock is exceeded the consequences are disastrous.

Geological forces acting over long periods of time can create large-scale deformations of rock in the earth's crust. Many of the world's mountain chains were created by such forces, which can bend and uplift layers of rock that may originally have been horizontal. Sometimes striking evidence of this process is seen in the form of curved, tilted layers of sedimentary rock (Fig. 8.1.1). Immense pressure applied over periods of tens or hundreds of millions of years has gradually changed the shapes of rocks. Forces of comparable strength may be applied over shorter times, straining rocks beyond their breaking point, as in earthquakes.

The stress that creates the strain leading to earthquakes is created by large-scale motions of the earth's crust. The crust consists of several plates (Fig. 8.1.2) that gradually move about, sometimes colliding or sliding along each other. Most earthquakes occur along boundaries where adjacent plates move relative to each other. As they do, rocks on opposing sides of the boundary catch on each other so that they resist the motion. The rocks become increasingly strained as the crustal plates, driven by motions deep inside the earth, try to continue moving. Eventually the strain exceeds the

Figure 8.1.1

(continued)

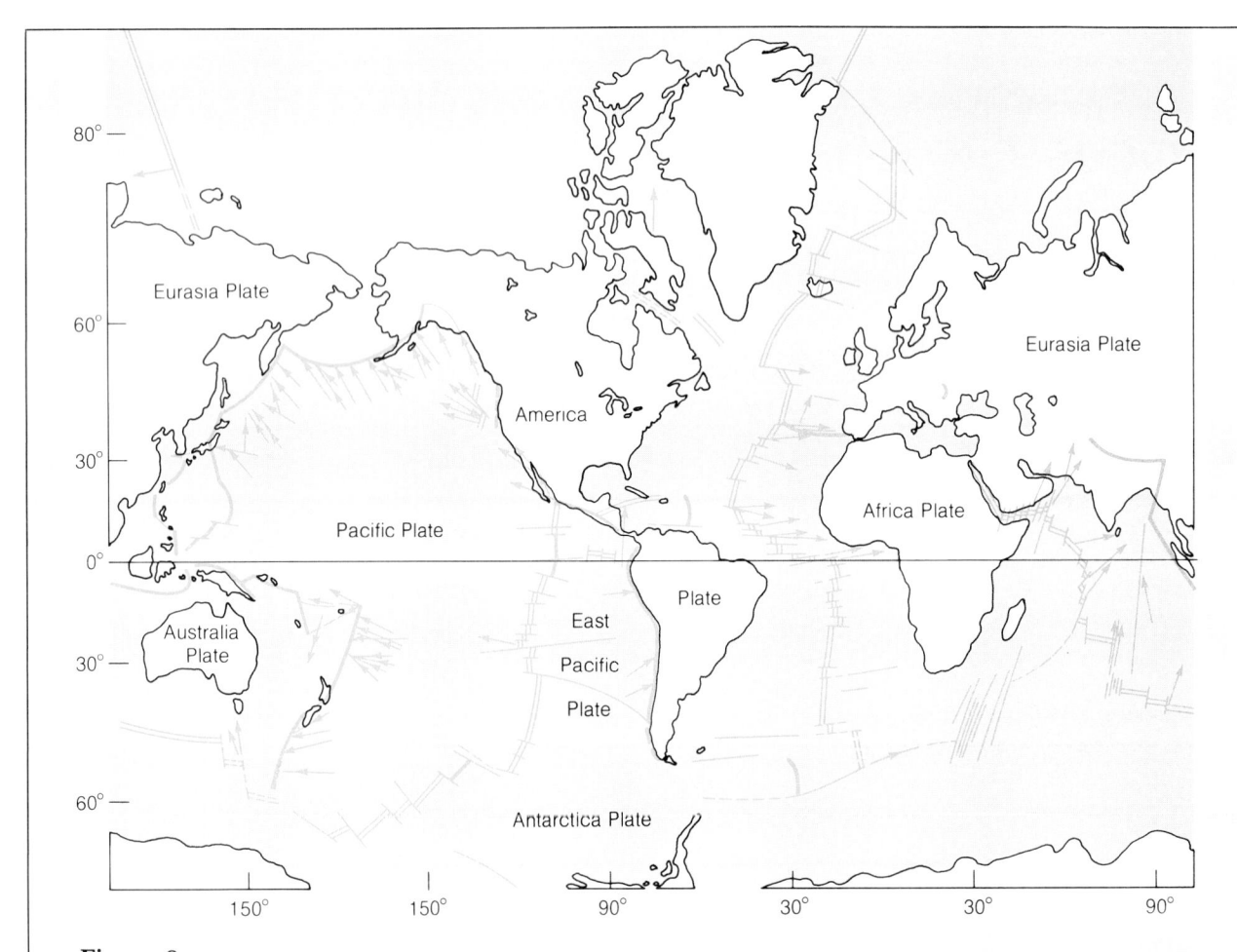

Figure 8.1.2

ultimate compressive strength of the rock and it breaks. The crustal plates suddenly move several feet, and an earthquake occurs. Enormous amounts of energy may be released in the process, resulting in widespread destruction.

Rocks under strain can produce interesting and useful effects, even if they do not reach the breaking point. One of the most important has to do with earthquake prediction, a science still in its infancy. Clearly, earthquake prediction becomes possible if the strain in rocks along faults (lines where crustal motion occurs and earthquakes take place) can be measured. There are at least three potential methods for detecting strain in these situations.

The most direct method is simply to measure the deformation that takes place as stress is applied. This may be done by placing sighting devices on opposing sides of a fault, and measuring their relative motion over long periods of time. Very accurate measurements must be made, because the motions are usually only a few centimeters per year. Sophisticated devices using **lasers** (Chapter 23) have been developed for this purpose. In some cases, vertical motion accompanies the build-up of strain. For many years it was said that such motion was occurring in a region north of Los Angeles, California, resulting in a local uplift known as the "Palmdale bulge." It was expected that this deformation would lead to a major earthquake in the next several years, but now the measurements of the bulge have been questioned, so the prediction is not taken so seriously. Nevertheless, measurements of small motions along faults, either vertical or horizontal, may predict earthquakes.

A second manifestation of strain in rocks has led to an earthquake prediction method that many scientists discounted for a long time. It has been observed that animals such as livestock often behave unusually just before an earthquake occurs. This has led to the prediction of earthquakes based on observations of animal behavior, a technique exploited particularly in China. Now a reason for this peculiar behavior has been suggested, and it may lead to a more direct and objective prediction technique. All rocks contain minute quantities of various trace gases, and these gases are released very gradually under normal conditions. When a rock experiences strain, the release of trapped gases is accelerated. Hence when rock along a fault has passed its elastic limit and is approaching the breaking point, unusually large quantities of trace gases may be released. It is possible that these gases are detected by animals, leading to the unusual behavior that has been observed. It should be possible to directly measure the concentrations of the released gases without relying on observations of animals. One potentially useful gas in this regard is **radon,** which is naturally radioactive and therefore easy to detect and measure. (It should be pointed out that another, entirely different, interpretation of animal behavior preceding earthquakes has been suggested. Animals may be sensitive to low-frequency, low-amplitude "microquakes" that occur before major earthquakes, and their strange behavior may be reactions to these vibrations instead of to released gases. Regardless, the detection of released radon is potentially a powerful method for measuring strain in rocks.)

Finally, a third manifestation of strain in certain rocks is the **piezoelectric effect.** In some kinds of crystals, strain causes small electric fields to be established, so that free electrons flow. There are many practical applications of this; for example, motions in the crystal stylus of a record player are converted into electrical signals by the piezoelectric effect, thus allowing the bumps in the groove on the record to be converted into a signal that produces sound in the speakers. It is thought that piezoelectric currents are created in rocks under strain along geologic faults, so that it may become possible to predict impending earthquakes by monitoring these currents. (Piezoelectric currents in rocks under strain have also been suggested as an explanation for certain UFOs. Repeated sightings of strange lights along a section of railroad track in New Jersey have been tentatively attributed to electrical discharges caused by these currents in the rock underneath the tracks.)

Perspective

We have dealt in this chapter with practical situations that arise because of the elastic properties of matter, and, along the way, we have learned a good deal about those properties as well. Much of the problem-solving ability that we have gained relies on the quantitative characteristics of different materials, something that has had little influence on our previous discussions. The stress-strain relations are summarized in Table 8.4. Whereas we could usually calculate quantities, the complexity of the internal forces in solids dictates that we now must rely on quantities measured in experiments.

So far we have not much explored the consequences of the fact that deformations under the elastic limit are always reversed when the deforming force is removed. In the next chapter we will learn how this causes vibrations in solids.

Table 8.4
Stresses and strains

Stress	Form	Strain	Elastic Modulus
Tension	$\sigma_t = \dfrac{F_\perp}{A}$	$(\Delta L/L)$	$Y = \dfrac{F_\perp/A}{\Delta L/L}$
Compression	$\sigma_c = \dfrac{F_\perp}{A}$	$(\Delta L/L)$	$Y = \dfrac{F_\perp/A}{\Delta L/L}$
Shear	$\sigma_s = \dfrac{F_\parallel}{A}$	$(\Delta x/L)$	$G = \dfrac{F_\parallel/A}{\Delta x/L}$
Bulk	$\sigma_v = \Delta P$	$(\Delta V/V)$	$B = \dfrac{-\Delta P}{\Delta V/V}$

SUMMARY TABLE OF FORMULAS

Hooke's law:
$$F = kx \tag{8-1}$$

Linear stress:
$$\sigma = \frac{F_\perp}{A} \tag{8-2}$$

Linear strain:
$$\epsilon = \frac{\Delta L}{L} \tag{8-3}$$

Linear stress and strain:
$$\sigma = Y\epsilon \tag{8-4}$$

Elastic modulus:
$$Y = \frac{(F_\perp/A)}{(\Delta L/L)} \tag{8-5}$$

Stress and strain for bending deformation:
$$\frac{\tau}{I_A} = \frac{Y}{R} \tag{8-6}$$

Critical height of a cylinder:
$$l_c \propto r^{2/3} \tag{8-7}$$

Shear stress:
$$\sigma_s = F_\parallel/A \tag{8-8}$$

Shear strain:
$$\epsilon_s = \Delta x/L \tag{8-9}$$

$$\epsilon_s = \tan \phi \tag{8-10}$$

Shear stress and strain:
$$\sigma_s = G\epsilon_s \tag{8-11}$$

Shear modulus:
$$G = \frac{(F_\parallel/A)}{(\Delta x/L)} \tag{8-12}$$

Stress and strain for twisting shear:
$$\frac{\tau}{I_p} = \frac{G\phi}{l} \tag{8-13}$$

Volume stress:
$$\sigma_v = \frac{\Delta F}{A} = \Delta P \tag{8-14}$$

Volume strain:
$$\epsilon_v = \frac{\Delta V}{V} \tag{8-15}$$

Volume stress and strain:
$$\epsilon_v = \frac{-\sigma_v}{B} \tag{8-16}$$

THOUGHT QUESTIONS

1. Suppose that a spring is stretched beyond its elastic limit. Will a force produce more or less deformation than the same force on the spring within its elastic limit?
2. Give three examples of restoring forces.
3. Sometimes science fiction movies portray giant insects. Is it possible for such creatures to exist on earth? Explain.
4. Give an everyday example of a substance which is brittle and one which is ductile.
5. Describe how you would experimentally measure the elastic modulus of a substance.
6. Why is the shear modulus smaller than the elastic modulus?
7. What would you estimate the ratios of elastic, shear, and bulk modulus to be for gelatin?
8. In the text, the structure of the atom is compared to that of the solar system. Can you think of any basic differences?
9. Why do you suppose the atomic mass unit, u, is defined in terms of the carbon atom rather than of the hydrogen atom?
10. Do any of the classical "fold, staple, spindle, or mutilate" activities which one can perform on a computer card fall into the three general categories of deformation listed in the text?
11. Explain in terms of atomic structure and forces why substances reach an elastic limit.
12. Place an encyclopedia and a paperback novel on a table and exert the same shearing force on each. Describe how they respond. Are your observations consistent with Eq. 8–8?
13. Explain how you would experimentally determine the value of the force constant for a spring or rubber band.
14. Why is it harder to bend a hollow tube than a solid bar, if both have the same length and are made of the same amount of metal?
15. Two tubes are made from identical sheets of paper. The first tube has a large radius and a thin wall. The second has a small radius and a thick wall. Which will support the larger load?
16. Measure the height and circumference of several species of grown trees and see if Eq. 8–7 is verified.
17. Do you think mountains and mountain ranges would obey Eq. 8–7? Explain.
18. Why are liquids hard to compress?
19. What sort of substances would you expect to have large compressibilities?
20. Is it possible to change the physical characteristics of a substance by subjecting it to extremely high pressures?
21. Give an example of plastic deformation.
22. Two bones of the same radius are subjected to identical twisting shears. If one bone is twice as long as the other, which bone will shatter first?
23. Why do beams with an I-shaped cross section have greater rigidity than beams of similar cross-sectional area but different shape?

PROBLEMS

Section 8.1 Atoms and Elements

- 1. If the spherical nucleus of a hydrogen atom has radius 1.2×10^{-15} m and mass of 1 u, find its density (kg/m^3).
- 2. If an atom has a diameter of 10^{-10} m, how many such atoms can be contained in a cube 1 cm on a side?
- 3. What are the atomic numbers and mass numbers of the following isotopes: $^{13}_{6}C$, $^{16}_{8}O$, $^{238}_{92}U$?
- 4. From the periodic table (Appendix 5), find the ratio of the number of protons to the number of neutrons for carbon, iron, and uranium.
- 5. Estimate the number of neutrons in a typical nucleus of gold.
- 6. How many silver nuclei are contained in one gram?
- •• 7. Estimate the number of water molecules (H$_2$O) in the ocean. Assume that the earth has a radius of 6,378 m and is three-fourths covered by oceans which have an average depth of 1 mile. Water is nearly incompressible, $\rho = 1{,}000$ kg/m^3.

8 How many nucleons (protons or neutrons) are contained in the earth? Assume the earth has a mass 5.976×10^{24} kg.

9 The earth's atmosphere is 78% nitrogen (N_2), 21% oxygen (O_2), and 1% trace gases, by mass. If 22.4 liters of air at standard temperature and pressure contain 6.023×10^{23} molecules, what is the mass of that volume?

10 Calculate the mean number of atoms in a 70 kg human body. Using typical dimensions for the atom and nucleus, estimate what fraction of the body is empty space.

11 Compute the density of the H atom if all its mass were spread uniformly throughout the volume of the electron orbit.

12 Calculate the atomic mass of one $^{13}_{6}C$ atom. How many such atoms are in a 1 kg piece of graphite? Assume that the isotopic abundance of $^{13}_{6}C/^{12}_{6}C$ equals 1/90.

13 The sun has a radius 7×10^8 m, and Mercury has an orbital radius 6×10^{10} m. How does the ratio of these dimensions compare with that of an atom and its nucleus?

Section 8.2 Restoring Forces

14 A spring stretches 10 cm when a force of 5 N is applied. How much force is needed to stretch the spring 1 meter?

15 A spring has force constant 150 N/m. How far will it be stretched by a force of 20 N?

16 A spring having mass 0.1 kg is hung from the ceiling. If the spring doubles its initial length, 20 cm, when hanging in this position, finds its force constant.

17 A 100 N force extends a spring 2 cm beyond its normal position. What is its force constant k?

18 What force on the spring in Problem 17 is required to extend the spring 3 cm? Assume it remains elastic.

19 An experimenter applies forces of 10 N, 20 N, 30 N, 40 N, and 50 N to a spring and compresses it distances of 1.0 cm, 2.0 cm, 3.0 cm, 3.8 cm, and 4.5 cm. What is the spring's force constant? What is its approximate elastic limit?

20 An object of mass m hangs from a spring with force constant k and equilibrium length x_0. Derive an expression for the displacement x in terms of m, g, and k.

21 A 1 m, 1 kg spring increases in length by 1 cm due to gravity, when hung vertically. What is its force constant?

22 A spring with force constant $k = 10^4$ N/m is loaded with 100 kg masses. If its breaking point = 20 cm, what is the maximum number of masses it can hold?

23 A table is supported by four springs, each with constant $k = 10^3$ N/m. If the table is to be used as a weight scale, and the elastic limits of the springs are 10 cm, what is the maximum weight the scale can measure reliably? Give your answer in pounds.

24 A football player tugs on a rope which obeys Hooke's law, with $k = 10^4$ N/m. If the rope's breaking point is 2,200 N, how far can he stretch the rope? Assume the rope remains elastic until it breaks.

25 A 1 kg ball on a 1 m string is twirled at 1 rev/s. If the string's length increases by 1 cm, what is its force constant?

26 Two 10 kg masses are suspended over a pulley by a rope with $k = 10^4$ N/m. By what length does the rope stretch?

27 When a used-car salesman sits down on the fender of a car, its springs sag vertically by 4 cm. If the spring constant is 2×10^4 N/m, how much does the salesman weigh?

28 You wish to purchase a fish scale that will be good up to 100 lbs. If the maximum compression of the spring is to be 5 cm, what spring constant should you look for?

Section 8.3 Stress and Strain

29 A force of 30 N is applied to a square block 15 cm on a side. What is the stress on the block?

30 What force will be needed to create a stress of 1 N/m^2 over a circular surface having radius 1 m?

31 A pillar of cross sectional area 1 m^2 supports a 10^4 kg mass. What are the stress and strain if the column is made of concrete?

32 What are stress and strain in Problem 31 if the column is made of brick?

33 If the column in Problem 31 is made of concrete, how many kg will it support?

34 If the column in Problem 31 is made of concrete and is 10 m tall, how much will it be compressed?

35 What is the strain experienced by an aluminum rod subjected to 9×10^8 N/m^2 stress?

36 How much larger in cross-sectional area should an iron rod be made to support the same weight under tension as a steel rod?

37 What is the maximum tension that can be supported by a brass rod with 2 cm^2 cross section?

PROBLEMS

- **38** What force is needed to keep a 1 mm radius steel piano wire stretched to 1.001 times its normal length?
- **39** A wood meterstick is compressed 1 mm. What are the stress and strain?
- **40¹** Estimate Young's modulus for an object whose stress-strain curve is shown in Fig. 8.20.

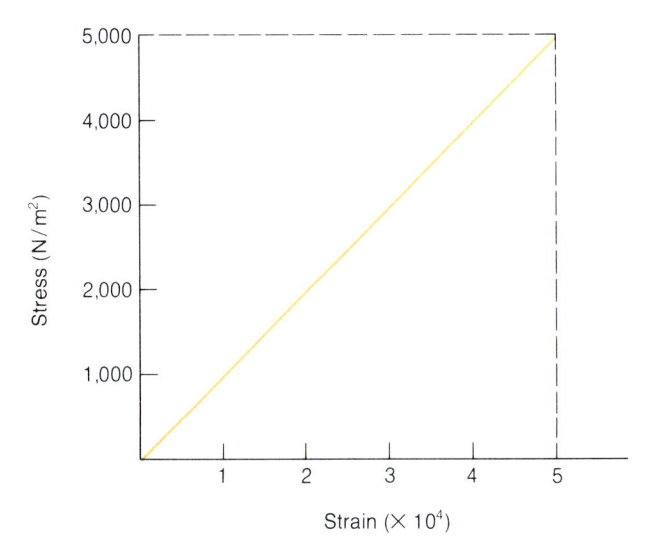

Figure 8.20

- **41** Suppose the weightlifter in Example 8.1 were Superman, the "man of steel." How much weight could he lift before crushing his legs if they were indeed made of steel?
- **42** Sketch the stress-strain curve for an object having $Y = 8 \times 10^{10}$ N/m².
- **43** A one-meter rod stretches by 0.01 m when subjected to a stress of 9×10^9 N/m². What is its elastic modulus Y?
- **44** An object subjected to a stress of 10^7 N/m² exhibits a strain of 10^{-4}. What is its Young's modulus?
- **45** A cubical block 1 m on a side is compressed by a 10^9 N force. The faces are found to change by 0.1 m on each side. What is the elastic modulus of the cube?
- **46** Assuming that the elastic modulus and ultimate compression strength for an elephant's bones are the same as for humans, estimate the maximum weight an elephant can support if its leg bones are 1 m long and 15 cm in diameter.
- **47** What is the decrease in bone length of the elephant in Problem 46 if it is lifting the maximum weight?
- **48** Find the angle through which a deformation takes place for the top cover of a 6 cm thick book displaced 1 cm relative to the bottom cover.
- **49** The top cover of a book 10 cm thick is displaced 2 cm relative to its bottom cover. What is its shear strain?
- **50** If the area over which a shear force acts is suddenly tripled, what happens to the shear stress?
- **51** How much shear strain will produce an angle of deformation of 60°?
- **52** One book has a shear modulus twice that of a second book. Compare the angle of deformation resulting from the same shear stress applied to each book.
- **53** Estimate the shear stress per atom if a force of 1 N is applied to the top of a layer with 1 m² area and one atom deep.
- **54** If the critical height of one cylinder is ten times that of a second cylinder of identical composition, find the ratio of the cylinders' radii.
- **55** A skier's anklebone is sheared by a force exerted at the end of his ski, one meter from his leg. If his bone breaks because of shear, estimate the force. Assume cross sectional area 10^{-3} m² and ultimate shear strength 150×10^6 N/m².
- **56** A steel cable supports an elevator whose total mass, including riders, is not to exceed 3,000 kg. The elevator is to have a maximum acceleration of $0.1g$. What radius cable should be used to have a factor of five safety margin?
- **57** A cubical object has a shear modulus of 2×10^{10} N/m² and a face area 0.5 m². Find the angle of deformation under 1,000 N force.
- **58** The top cover of a dictionary is pushed by $\Delta L = 2$ cm relative to the bottom cover, a distance $L = 5$ cm away. If the force was 10 N and the area of the dictionary cover 0.03 m², what is the shear modulus of the book?
- **59** A certain bone has an average cross section of 10^{-4} m² and a length 0.1 m. Find the force constant k under compression for the bone.
- **60** Show that Eq. 8–9 and Eq. 8–10 are equivalent.
- **61** A certain species of tree has an average trunk diameter three times that of a second species, when both are fully grown. How would the average heights of fully grown trees compare?
- **62** How would the average bone radius compare for two species of animals which differ in average height by a factor of two?

•• 63 If the baseball in Example 8.5 is dropped into the ocean, at what pressure would the ball's radius change by 0.001 m?

••• 64 The atmospheric pressure at sea level is about 10^5 N/m². How much larger or smaller will a cube-shaped object be in outer space if its edge is 2 cm and bulk modulus is 2×10^8 N/m²?

• 65 A cubical crystal exhibits a 10% change in volume when subjected to a volume strain. What is that strain?

•• 66 What increase in pressure is needed to change the volume of water by one part in 10^4?

• 67 A cubic meter of fluid changes its volume by 0.05% when subjected to change in a pressure of 5×10^6 N/m². Find the bulk modulus of the fluid.

•• 68 Find how much a 5 cm cube of iron is compressed by 10^3 atmospheres pressure. One atmosphere is approximately 10^5 N/m².

•• 69 As it descends to the surface of Planet X, a water droplet decreases in volume by 1%. What is the pressure change?

•• 70 The oil in a hydraulic system occupies a volume of 0.25 m³ and has bulk modulus 10^{10} N/m². If a 1,500 kg car is supported by a circular area of 5 cm radius, calculate the volume change of the fluid.

•• 71 The pressure of water increases by about 1 atmosphere (10^5 N/m²) for each 10 m depth. At what depth will an aluminum sphere decrease in volume by 10%?

••• 72 Using the data in Problem 71, estimate the depth at which a steel diving bell of 2 m radius and 1 cm thick plate is crushed by water pressure. Assume that the bell is crushed when the change in radius is approximately 2% the thickness of the plate.

SPECIAL PROBLEMS

1. The Depths of the Ocean

The deepest portion of the ocean is 10.9 km, in the Marianas Trench off the coast of the Philippines. One can show that the pressure at depth h equals the weight of the water above a 1 m² column.

(a) Show that the pressure at depth h may be written $P = \bar{\rho}gh$, where g is the gravitational acceleration and $\bar{\rho}$ is the average density of water above.

(b) Assuming that ρ is constant and approximately equal to its value at the surface, 1,000 kg/m³, estimate the pressure (N/m²) at the bottom of the trench.

(c) How many times atmospheric pressure is this? Use the fact that the earth's atmosphere is also a fluid, with density 1.2 kg/m³ and effective height 8.4 km.

(d) Using the bulk modulus of water, $B = 2 \times 10^9$ N/m², find the change in volume of 1 m³ of surface water lowered to the bottom of the trench. What is the corresponding density at the bottom? Is the assumption that $\bar{\rho}$ = constant justified?

2. Suspension Bridge

Consider one span of a suspension bridge (Fig. 8.21), supported by steel cables spaced every 25 m. Engineers wish to estimate the radius required for these cables to provide safe support for the bridge.

(a) Estimate the mass of the bridge roadway from the dimensions shown. Assume the roadway has a density 3,000 kg/m³.

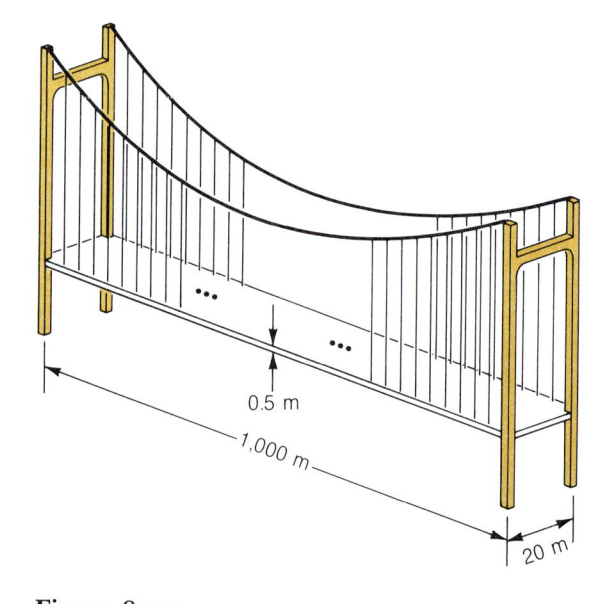

Figure 8.21

(b) Estimate the total number and total mass of 1,500 kg cars that could fit into four lanes of bumper-to-bumper traffic (allow 10 m per car). Will cars be a factor in total bridge weight?

(c) How many cables support the span? If the engineer allows a factor of ten safety margin below the tensile strength of steel, how thick should the cables be? By what fraction would their length increase under maximum load?

3. Mountaineer's Ropes

The standard nylon ropes used in climbing have 11 mm diameter and 150 ft length. In a test called working elongation a rope supporting 80 kg stretches in length by 8%.
(a) What is Young's modulus for the rope?
(b) Assuming a linear stress-strain relation, show that the rope's restoring force may be expressed as $F = kx$, where $x = (l - l_o)$ is the distance the rope is stretched beyond its standard length l_o. What is k?
(c) Plot F versus x. Show graphically from the area under the curve that the work done against the force in stretching the rope by x is $W = \frac{1}{2}kx^2$.
(d) A climber of mass m jumps off a cliff, supported by one end of the rope held fixed at the top. He falls the rope's length l_o and the rope stretches a further distance x_m before stopping him. From energy conservation, show that $x_m = (2mg/AY)^{1/2} l_o$, where A is the rope's cross sectional area and Y is its Young's modulus. Evaluate x_m/l_o for the above rope and $m = 80$ kg. (In practice most ropes become much stiffer for $x/l_o > 0.1$, and cease to obey the linear stress-strain relation. A typical maximum stretch is 20%.)

SUGGESTED READINGS

"Success in raquetball is enhanced by knowing the physics of the collision of ball with wall." J. Walker, *Scientific American,* Sep 84, p 215, **251**(3)

"Structural analysis of Gothic cathedrals." R. Mark, *Scientific American,* Nov 72, p 90, **227**(5)

"The Architecture of Christopher Wren." H. Dorn, R. Mark, *Scientific American,* Jul 81, p 160, **245**(1)

"The Forging of Metals." S. L. Semiatin, G. D. Lahoti, *Scientific American,* Aug 81, p 98, **245**(2)

"Gothic structural experimentation." R. Mark, W. W. Clark, *Scientific American,* Nov 84, p 176, **251**(5)

The flying circus of physics with answers. Jearl Walker, Wiley, New York, 1977

Newton at the bat: The science in sports. Eric W. Schrier, William F. Allman, eds, Scribners, New York, 1984

Sport Science. Peter J. Brancazio, Simon and Schuster, New York, 1984

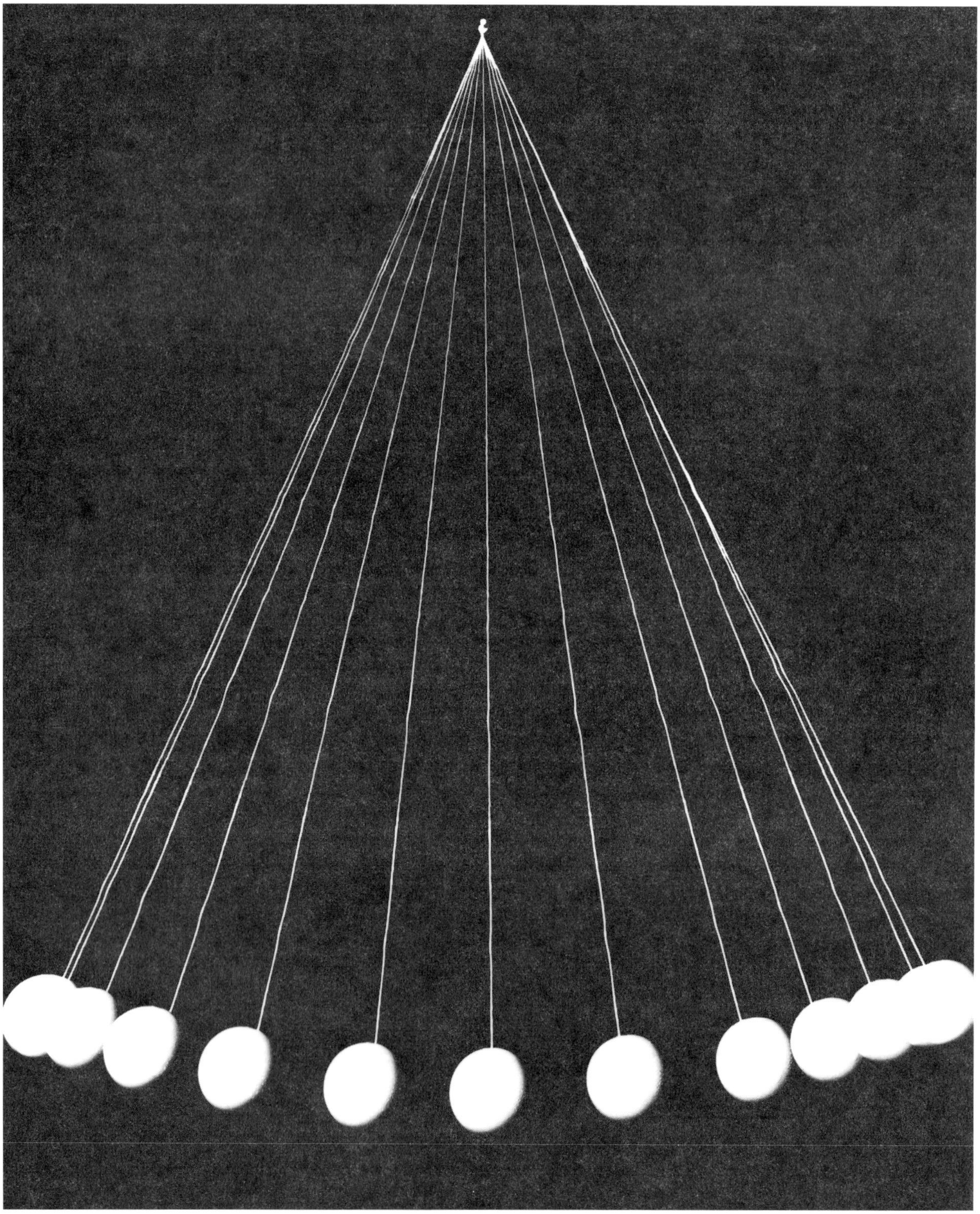

Chapter 9

Oscillations

One of the consequences of the fact that all materials are elastic is that many materials can vibrate. An object subjected to an impulse may vibrate thereafter. This may be obvious, as when a bell is struck by its clapper, or when a taut string is disturbed (Fig. 9.1). Some mechanical systems can vibrate as well, independent of elastic forces. For example, if a pendulum is disturbed from equilibrium, it oscillates.

In either case, the oscillations or vibrations are caused by a force acting to restore equilibrium when it has been disturbed. The elastic force acts to restore an object to equilibrium when it has been deformed, and gravity causes a pendulum to swing back toward vertical when it has been displaced. There are more complex kinds of vibrations or oscillations as well: a bouncing ball surely is oscillating, as is a fingernail being scratched across a blackboard.

A large class of vibrations are both common and simple to analyze, and we will devote most of our effort in this chapter to it. We begin by defining this class. In the following sections, we will discuss it quantitatively and then analyze some special applications before briefly treating oscillations of other types.

9.1 The Restoring Force and Simple Harmonic Motion

Let us examine an elastic body being deformed. A force causes the deformation, and when this force is released, internal forces cause the body to return to its equilibrium shape. As it does so, its momentum causes it to pass through its equilibrium shape; the object overshoots, and becomes deformed again, but with a displacement in the opposite direction. Again, the restoring force acts

256   CHAPTER 9   OSCILLATIONS

Figure 9.1

Vibrating strings. The bow creates vibrations in the strings of a violin. The vibrations in turn create sound waves.

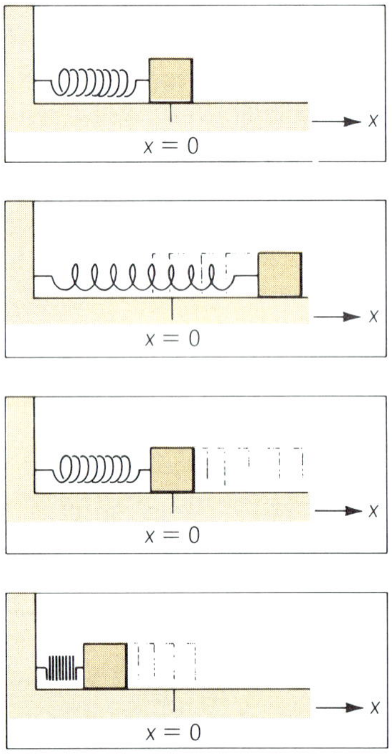

Figure 9.2

Simple harmonic oscillation. The mass oscillates about the equilibrium position because the spring exerts a restoring force when it is displaced. The restoring force obeys Hooke's law.

to return the object to equilibrium, overshoot happens again, and a new displacement in the original direction occurs. Thus oscillations continue for some time after the initial deformation. The oscillations would actually continue indefinitely if there were no dissipation of energy; that is, if there were no friction or other nonconservative forces. Of course, in most real systems such forces are present and the vibrations eventually stop, but we can profitably discuss vibrations by ignoring this for a while.

Oscillations will occur whenever there is a force that will act to restore equilibrium. The simplest oscillations involve a restoring force that is symmetric and continuous; that is, one that acts with equal strength for displacements in either direction from equilibrium, and one that varies smoothly as a function of displacement. It happens that many common situations involve restoring forces meeting these criteria. The most common of these are cases where the restoring force is proportional to the displacement (that is, Hooke's law applies). This is true of many kinds of elastic deformations, and it is approximately true for certain simple mechanical systems such as pendulums. Hence the term **simple harmonic motion (SHM)** is applied when the restoring force is of this type, and a system which exhibits this behavior is called a **simple harmonic oscillator (SHO)**.

Examples of SHOs include many kinds of springs and other objects subjected to length deformations, in which the vibrations are linear (Fig. 9.2) and pendulums and some types of springs, in which the vibrations are angular (Fig.

9.1 THE RESTORING FORCE AND SIMPLE HARMONIC MOTION

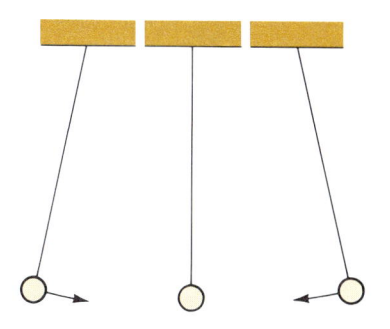

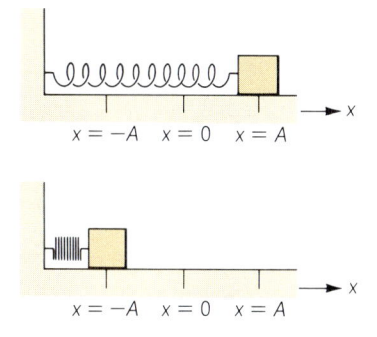

Figure 9.3

A pendulum. In this case, the oscillations are angular. (A pendulum such as this is approximately a simple harmonic oscillator if the angular displacement is small.)

Figure 9.4

Amplitude. The maximum displacement from the equilibrium position is the amplitude A.

9.3). Although there is displacement at any point within a deformed body, we normally speak of the displacement at the extremities. For example, when discussing vibrations in a spring, we will focus on the displacement of the end of the spring. For angular displacements we will speak of angular measures primarily, although it is sometimes of interest to calculate the corresponding linear displacement, for example at the bottom of a pendulum.

We must introduce some new terminology in order to discuss vibrations and SHM. Since Hooke's law applies to SHM and the restoring force is proportional to the displacement from equilibrium, we have a **force constant** k, defined by

$$k = \frac{-F}{x} \tag{9-1}$$

where F is the magnitude of the restoring force and x is the magnitude of the displacement from the equilibrium position. The minus sign shows that the restoring force is always opposite to the displacement.

The maximum displacement from equilibrium is called the **amplitude** (Fig. 9.4), denoted A. The motion continuously repeats itself over the full range of displacement, so that one full cycle or vibration is the complete motion from displacement $x = A$ to $x = -A$ to $x = A$ (or from any intermediate displacement to the same displacement one cycle later). The period T is the time it takes for a full cycle, and the frequency ν is the number of cycles per second. Hence

$$T = \frac{1}{\nu}. \tag{9-2}$$

9.2 Analyzing Simple Harmonic Motion

A number of powerful relationships among the quantities in SHM can be derived, and the problem-solving capability that we develop will be impressive. We first consider the energy in a simple harmonic oscillator, then we derive equations for the period and the displacement as a function of time.

Energy in a Simple Harmonic Oscillator

We can use the work-energy principle to calculate the potential energy in an oscillator when it is displaced from its equilibrium position. The work required to create displacement x is

$$W = \overline{F}x \qquad (9\text{-}3)$$

where $\overline{F}$ is the average force exerted on the oscillator over the displacement. We also know that the force at displacement x is

$$F = kx. \qquad (9\text{-}4)$$

(Here the minus sign from Eq. 9–1 has been dropped, because F refers to the force acting on the oscillator, which is directed against the restoring force). The average force $\overline{F}$ is half this value, because the force increases linearly from zero at equilibrium to kx at displacement x. Therefore the work put into the system to create the displacement is

$$W = \text{PE} = \frac{1}{2}kx^2. \qquad (9\text{-}5)$$

The kinetic energy, as always, is given by

$$\text{KE} = \frac{1}{2}mv^2. \qquad (9\text{-}6)$$

For an extended system that is deformed and set vibrating, the kinetic energy is the sum of the values $\frac{1}{2}mv^2$ for each point within the system. This can be complex to calculate, and normally requires either calculus or numerical integration. The calculation is simpler when the mass is effectively concentrated at the extremity of the oscillator and we can neglect the mass of the spring, such as in the case of a mass attached to a spring or the end of a string. For the remainder of this discussion we will assume that the mass in Eq. 9–6 is the attached mass.

The maximum potential energy occurs when the displacement reaches its maximum value, which is the amplitude A. The oscillator momentarily comes to rest at this position during every cycle, so the kinetic energy is zero at that point. Conversely, when the oscillator passes through its equilibrium position ($x = 0$), the potential energy is zero. If we neglect friction, the total energy

9.2 ANALYZING SIMPLE HARMONIC MOTION

must be constant. Therefore the kinetic energy is maximized when the oscillator passes through its equilibrium position. Thus a pendulum moves fastest when it is vertical.

Since the total energy is constant, the maximum potential energy must equal the maximum kinetic energy. Hence

$$\frac{1}{2}mv_0^2 = \frac{1}{2}kA^2$$

where v_0 is the maximum velocity. Therefore we have

$$v_0 = A\sqrt{\frac{k}{m}}. \tag{9-7}$$

We can similarly develop a general expression for the velocity of an oscillator at any displacement. The total energy is constant, so we can set the energy at arbitrary displacement x equal to the energy at maximum displacement A.

$$\frac{1}{2}mv^2 + \frac{1}{2}kx^2 = \frac{1}{2}kA^2$$

this leads to

$$v^2 = \left(\frac{k}{m}\right)(A^2 - x^2). \tag{9-8}$$

We then use Eq. 9–7 and take the square root.

$$v = v_0\sqrt{1 - \left(\frac{x}{A}\right)^2} \tag{9-9}$$

Example 9.1 On a frictionless surface a mass of 2 kg is attached to a spring with force constant $k = 1{,}500$ kg/s² (Fig. 9.5). If the mass is moved 1 cm to the left and released, what is its velocity when it passes through the equilibrium position? What is the maximum force exerted on the mass?

The velocity at the equilibrium position is the maximum velocity given by Eq. 9–7.

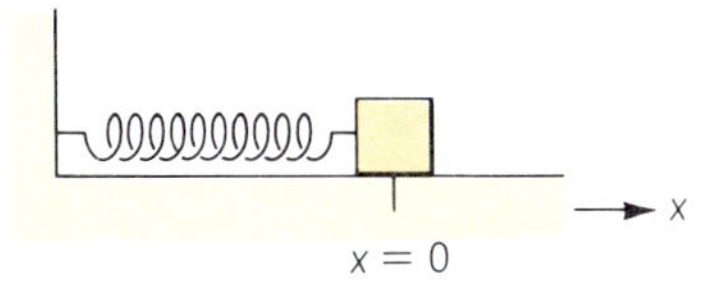

$$v_0 = A\sqrt{\frac{k}{m}} = (0.01 \text{ m})\sqrt{\frac{1{,}500 \text{ kg/s}^2}{2 \text{ kg}}} = 0.27 \text{ m/s}$$

Figure 9.5

CHAPTER 9 OSCILLATIONS

The maximum force F_0 can be found using Eq. 9–4, with $x = A = 1$ cm.

$$F_0 = kA = (1{,}500 \text{ kg/s}^2)(0.01 \text{ m}) = 15 \text{ N}$$

When we consider a mass hanging from a spring, the situation is a bit more complicated, because there is a downward gravitational force in addition to the restoring force. It is easy to show, however, that for small displacements, the motion is still SHM (if the spring's elastic limit is not exceeded, so that Hooke's law still applies). To see this, consider a spring with a weight attached, so that the spring is stretched to displacement X (Fig. 9.6). When the weight is at rest, equilibrium conditions dictate that the upward and downward forces acting on the weight must be equal. Therefore the net force F is zero.

$$F = kX - mg = 0$$

Now if the weight is displaced an additional distance x, the net force on it is

$$F = ma = k(X + x) - mg.$$

But $(kX - mg) = 0$, so

$$F = kx.$$

This is Hooke's law for the restoring force exerted by a spring, so we know that SHM will still occur, despite the presence of the gravitational force.

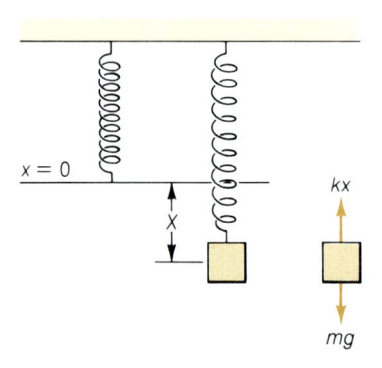

Figure 9.6

A mass hanging from a spring. At left there is no force on the spring (neglecting its own weight) and no displacement. At center, a weight is attached, and the spring stretches to a new equilibrium at X. At right the forces acting on the mass at equilibrium are shown.

Example 9.2 — Some scales consist of springs from which the weight to be measured is hung. Because the spring constant is known, the scale is calibrated to display the weight although it actually measures the displacement of the spring. Suppose such a scale stretches by 3 cm when a 20 N weight is attached (Fig. 9.7). How far will it stretch when the attached weight is 35 N? If the spring is stretched an additional 2 cm after the 35 N weight has been attached, what is the maximum velocity of the weight during its oscillation? What is its maximum acceleration? What is the velocity of the weight when it passes through a point 1 cm from the equilibrium position?

The first question can be answered very simply, since we know that Hooke's law applies and the displacement is proportional to the force. If the scale stretches 3 cm when a 20 N weight is attached, then it stretches by a factor of 35/20 more when the weight is 35 N. The displacement for the 35 N weight is therefore 5.25 cm.

To answer the other questions, we need to derive the value of k. From Eq. 9–1 we have

$$k = \frac{F}{x} = \frac{(20 \text{ N})}{(0.03 \text{ m})} = 667 \text{ kg/s}^2.$$

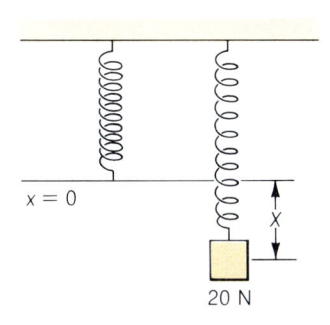

Figure 9.7

To find the maximum velocity reached by the 35 N weight after it has been stretched an additional 2 cm from its equilibrium position, we use Eq. 9–7, noting that the maximum displacement A is 2 cm, and that the mass of a 35 N weight is $m = W/g = (35 \text{ N}/9.8 \text{ m/s}^2) = 3.57$ kg.

$$v_0 = A\sqrt{\frac{k}{m}} = (0.02 \text{ m})\sqrt{\frac{667 \text{ kg/s}^2}{3.57 \text{ kg}}} = 0.27 \text{ m/s}$$

The maximum acceleration occurs when the force is greatest, which occurs when the displacement is greatest. The force when the displacement is $A = 2$ cm follows from Hooke's law.

$$F_0 = kA = (667 \text{ kg/s}^2)(0.02 \text{ m}) = 13.3 \text{ N}$$

Thus the maximum acceleration is

$$a_0 = \frac{F_0}{m} = \frac{13.3 \text{ N}}{3.57 \text{ kg}} = 3.73 \text{ m/s}^2.$$

The velocity at a displacement of 1 cm is found from Eq. 9–9.

$$v = v_0\sqrt{1 - \left(\frac{x}{A}\right)^2} = (0.27 \text{ m/s})\sqrt{1 - \left(\frac{0.01 \text{ m}}{0.02 \text{ m}}\right)^2} = 0.23 \text{ m/s}$$

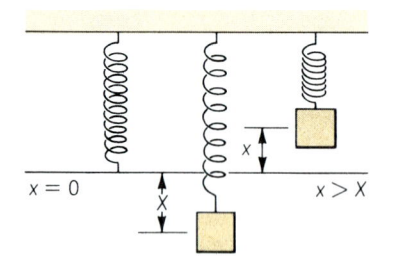

Figure 9.8

Nonsimple harmonic motion. At right, the mass has risen above the spring's unweighted equilibrium position, so that the spring is under compression. The restoring force is not symmetric about the equilibrium position x = X.

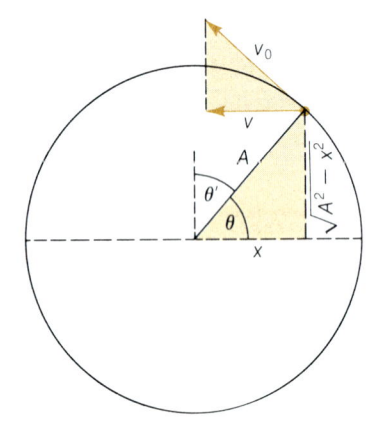

Figure 9.9

The reference circle. If the circle is viewed on edge (in the plane of the page), an object in uniform circular motion will appear to oscillate under SHM with displacements ±x and projected velocity v. The shaded triangles are similar.

CHAPTER 9 OSCILLATIONS

Things really do get more complex, and SHM is not valid, if the displacement x exceeds the equilibrium displacement X. In that case, the weight will rise above the position of the unweighted spring (Fig. 9.8), so that the elastic force caused by compression of the spring now adds to the downward force acting on the weight. Then the restoring force is not symmetric for displacements in opposite directions, and the simple treatment of SHM does not apply.

The Period of Simple Harmonic Motion

One characteristic of any system that undergoes SHM is its period. This is a function only of the mass and the spring constant; the amplitude has no effect. This is why SHM can be used for clocks: once the system is calibrated, its frequency will stay constant even if the amplitude varies.

To derive an expression for the period, we refer to the motion of an object moving in a circle, which is called the **reference circle**. If we assume that a mass m moves at constant velocity v_0 in a circle, and if we view the motion of the mass from the side, so that we see the mass moving back and forth in one dimension (Fig. 9.9), then we find that this projected motion simulates SHM, with amplitude equal to the radius of the circle. From the similar triangles shown, we see that the projected velocity v is related to v_0 by

$$\frac{v}{v_0} = \frac{1}{A}\sqrt{A^2 - x^2}$$

or

$$v = v_0 \sqrt{1 - \left(\frac{x}{A}\right)^2}.$$

This is identical to Eq. 9–9 for the velocity of a simple harmonic oscillator, so we conclude that our reference circle projection simulates SHM.

The period of a mass moving in a circle of radius A is

$$T = \frac{2\pi A}{v_0}.$$

From Eq. 9–7 we substitute for v_0 to find

$$T = 2\pi \sqrt{\frac{m}{k}}. \qquad (9\text{–}10)$$

As we noted earlier, the period depends on k and m, but not on A. The form of the dependence on k and m is not surprising because the period grows with increasing mass and decreases with increasing values of k.

The frequency is

$$\nu = \frac{1}{T} = \frac{1}{2\pi}\sqrt{\frac{k}{m}}. \qquad (9\text{–}11)$$

Example 9.3 At what frequency will the spring considered in Example 9.2 oscillate?

To answer this we must neglect the effects of gravity and assume that the spring is undergoing SHM. The spring scale in Example 9.2 has a spring constant $k = 667$ kg/s^2 and an attached mass of 3.57 kg. Hence its frequency is

$$\nu = \frac{1}{2\pi}\sqrt{\frac{k}{m}} = \frac{1}{2\pi}\sqrt{\frac{(667 \text{ kg/s}^2)}{(3.57 \text{ kg})}} = 2.2 \text{ Hz}.$$

The period of oscillation for the mass, $T = 1/\nu$, is just less than half a second.

The Equation of Motion for SHM

It is often useful to calculate the position of an oscillator at any point (or time) in its cycle, and to do this we resort again to the reference circle. From Fig. 9.9 we see that the projected position of the mass moving around the circle is

$$x = A \cos\theta$$

where θ, as are all other angular quantities in what follows, is measured in radians. From our discussion of circular motion in Chapter 6, specifically from Eq. 6–2, we know that $\theta = \omega t$, so we have

$$x = A \cos(\omega t). \tag{9–12}$$

We can substitute for the angular velocity ω using Eq. 6–6, and we find

$$x = A \cos(2\pi\nu t) \tag{9–13}$$

where ν is the frequency. In terms of the period T, the position is

$$x = A \cos\left(\frac{2\pi t}{T}\right). \tag{9–14}$$

If we had defined the angle θ differently, for example, as θ' in Fig. 9.9, we would find a different set of equations corresponding to Eq. 9–12 and Eq. 9–13. All the cosine terms would be sine terms (with the same arguments). Thus, in terms of the period, the position at time t would be

$$x = A \sin\left(\frac{2\pi t}{T}\right). \tag{9–15}$$

This illustrates an important point about SHM: we must always specify some initial condition which we call the **phase** of the motion. This means that

CHAPTER 9 OSCILLATIONS

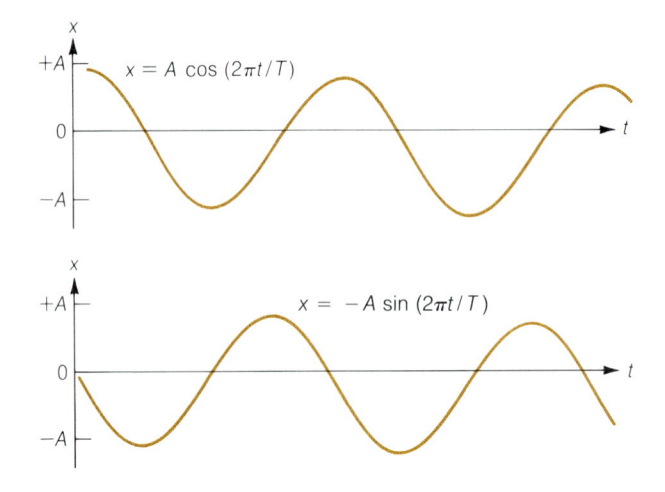

Figure 9.10

Sinusoidal motion. An object undergoing SHM follows a sinusoidal path, regardless of phase.

we must define where the starting, or zero, point is. When we use the cosine form of the equation of motion, we are saying that the zero point occurs when the oscillator is at its extreme displacement A. When we use the sine form, we are saying that the zero point is the equilibrium position $x = 0$. The form of the equation is the same for both. Plots of the motion represented in Eq. 9–14 and Eq. 9–15 are identical (Fig. 9.10), but out of phase with each other by one quarter-cycle.

Whether we use Eq. 9–14 or Eq. 9–15, we say that the motion of a simple harmonic oscillator is **sinusoidal**. The velocity and acceleration are also sinusoidal. To see this, we substitute Eq. 9–14 into Eq. 9–9.

$$v = \pm v_0 \sqrt{1 - \cos^2\left(\frac{2\pi t}{T}\right)}$$

We have explicitly written the square root as having either a positive or negative value. Now we consider the initial conditions, noting that when an oscillator is displaced a distance A at time $t = 0$, the velocity immediately thereafter is in the opposite direction from the displacement. Hence we choose a minus sign in order to express the phase relationship between displacement and velocity correctly. Since $(1 - \cos^2\theta) = \sin^2\theta$, we have

$$v = -v_0 \sin\left(\frac{2\pi t}{T}\right). \tag{9–16}$$

Similarly, we can write an expression for the acceleration by dividing Eq. 9–4 by the mass m, and substituting Eq. 9–14 into the resulting expression:

9.2 ANALYZING SIMPLE HARMONIC MOTION

$$a = -\frac{kx}{m} = -\left(\frac{kA}{m}\right)\cos\left(\frac{2\pi t}{T}\right) \qquad (9\text{-}17)$$

which simplifies to

$$a = -a_0 \cos\left(\frac{2\pi t}{T}\right) \qquad (9\text{-}18)$$

where $a_o = kA/m$ is the maximum acceleration. The minus sign indicates that the acceleration is in the opposite direction from the displacement at all times.

We note that the velocity and displacement are out of phase by one quarter-cycle (90° phase angle), while the acceleration and the displacement are out of phase by one half-cycle (180° phase angle). Thus the maximum velocity occurs one quarter-cycle before the maximum displacement, and maximum acceleration occurs one half-cycle before and after the maximum displacement.

Example 9.4 A mass of 100 kg is suspended from a spring, causing an equilibrium displacement of 3 cm. The mass is displaced an additional 2 cm and released. Find its displacement, its velocity, and its acceleration 5 seconds later.

First we need the period, so that we will know where the vibrating mass is in its cycle at t = 5 s. To calculate the period, we need to find the value of k. From Eq. 9–1 we use the equilibrium displacement to find

$$k = \frac{F}{x} = \frac{mg}{x} = \frac{(100 \text{ kg})(9.8 \text{ m/s}^2)}{(0.03 \text{ m})} = 3.27 \times 10^4 \text{ kg/s}^2.$$

The period, from Eq. 9–10, is

$$T = 2\pi\sqrt{\frac{m}{k}} = 2\pi\sqrt{\frac{(100 \text{ kg})}{(3.27 \times 10^4 \text{ kg/s}^2)}} = 0.347 \text{ s}.$$

Now we can answer the questions. Since we have set $t = 0$ when the mass is at its greatest displacement A, we use Eq. 9–14 for the displacement at time t.

$$x = A\cos\left(\frac{2\pi t}{T}\right) = (0.02 \text{ m})\cos\left(\frac{2\pi(5\text{s})}{0.347\text{s}}\right) = -0.0168 \text{ m}$$

The minus sign indicates that the displacement at time $t = 5$ s is in the opposite direction from the initial displacement. If the initial displacement was downward, then at time $t = 5$ s, the mass is above the equilibrium position. Note that the angle is in radians; its value is 90.54 rad. Note also that the oscillator is in its fifteenth cycle (since 90.54/2π = 14.41) at time $t = 5$ s, but that the cosine term is the same as if it had been in its first cycle at the same phase. The phase, the fraction of a cycle that the oscillator has completed, is (5 s/0.347 s) − 14 = 0.41. From this alone, we could have concluded that the mass was moving upward and was above the equilibrium position because it is between one quarter and one half phase.

The velocity at time $t = 5$ s is found from Eq. 9–16 with the value of v_o from Eq. 9–7.

$$v = -A\sqrt{\frac{k}{m}}\sin\left(\frac{2\pi t}{T}\right)$$

$$= -(0.02 \text{ m})\sqrt{\frac{3.27 \times 10^4 \text{ kg/s}^2}{100 \text{ kg}}}\sin(90.54 \text{ rad}) = -0.19 \text{ m/s}$$

This is a negative number, showing that the velocity is in the opposite direction from the initial displacement at this time.

Finally, the acceleration is given by Eq. 9–17.

$$a = -\left(\frac{kA}{m}\right)\cos\left(\frac{2\pi t}{T}\right)$$

$$= -\left[\frac{(3.27 \times 10^4 \text{ kg/s}^2)(0.02 \text{ m})}{100 \text{ kg}}\right]\cos(90.54 \text{ rad}) = 5.52 \text{ m/s}^2$$

The positive value indicates that the acceleration is in the same direction as the initial displacement, that is, downward. The mass is moving upward and slowing as it approaches the top of its range of motion where the displacement will be $-A$.

9.3 The Simple Pendulum

A mass at the end of a string will oscillate back and forth if disturbed, and under certain conditions the motion is approximately SHM. We need to check whether the restoring force is proportional to the displacement in order to verify this. From Fig. 9.11 and our discussions of angular quantities in Chapter 6, we see that if the angle θ (in radians) is small, the displacement x is

$$x \simeq \theta L \tag{9-19}$$

where L is the length of the pendulum. The restoring force is the component of gravity that acts tangent to the arc described by the mass, bringing the mass back to the equilibrium position. This force is given by

9.3 THE SIMPLE PENDULUM

$$F = -mg\sin\theta \qquad (9\text{-}20)$$

where the minus sign indicates that the direction of the force is opposite to the direction of θ. But $\sin\theta = x/L$, and we have just seen that for small θ, $x/L \simeq \theta$. Therefore the restoring force in this case becomes

$$F \simeq -mg\theta. \qquad (9\text{-}21)$$

Again, we note that x is proportional to θ, so that if the restoring force is proportional to θ, it must also be proportional to x. Hence for small θ the displacement and the restoring force approximately conform to Hooke's law, and we have SHM. The angle θ can be as large as 0.25 radians (14.3°) before the departure from Hooke's law becomes as great as 1 percent (that is, before $\sin\theta$ differs from θ by more than 1 percent).

We have already seen that in SHM the period is independent of the mass of the oscillator and the amplitude of the oscillations. This is therefore also true of a pendulum (as long as the amplitude remains small), and this is why pendulums are used for timekeeping. Galileo noticed this independence in the early seventeenth century and took advantage of it to invent the pendulum clock, which became the standard of timekeeping for centuries.

We can write an expression for the period of a pendulum, by substituting Eq. 9–19 into Eq. 9–21.

$$F = -\left(\frac{mg}{L}\right)x \qquad (9\text{-}22)$$

(We now use an equality sign, but keep in mind that this and subsequent expressions are approximations that are valid only for small θ). From this expression, we see that the force constant for a pendulum has the form

$$k = \frac{mg}{L} \qquad (9\text{-}23)$$

so we can adapt Eq. 9–10 to write an expression for the period of a pendulum.

$$T = 2\pi\sqrt{\frac{L}{g}} \qquad (9\text{-}24)$$

This expression quantifies what we have already said, that the period is independent of the mass and the amplitude (for small θ). It depends only on the length of the pendulum, since g is assumed constant (accurate measurements of the periods of pendulums are a useful way of measuring g and especially its small variations with altitude or proximity to large masses).

A pendulum set swinging will eventually come to a stop because of frictional forces. Therefore a clock has a drive mechanism that gives the pendulum a small push every cycle. Because of minor departures from SHM due to the use of the small-angle approximation, the period actually depends slightly on

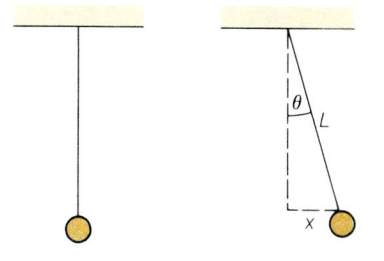

Figure 9.11

The simple pendulum. When the displacement θ is small, $x = \theta L$ and the restoring force is $F = mg\theta$ which is proportional to the angular displacement. Therefore when θ is small, a pendulum approximates SHM.

268 CHAPTER 9 OSCILLATIONS

the amplitude, so a clock works best if the amplitude is kept constant. This is also accomplished by the drive mechanism. Many early clocks used hanging weights to provide the drive, but in the past 150 years or so, coiled springs have been more common. In either case, the potential energy put into the mechanism when the clock is wound is released into the pendulum in small amounts each cycle to maintain the approximation to SHM.

Example 9.5 A pendulum clock is running a little slow, losing 2 minutes each day. If its pendulum is 0.2 m long, how much should the length be adjusted so that it will keep accurate time?

Since we do not know what the period is, we must solve this problem by finding the needed fractional change in period. A lag of 2 minutes per day is a fractional error of 0.00139. Therefore the ratio of the desired new period to the present period is

$$\frac{T'}{T} = \frac{1}{1.00139} = 0.9986.$$

From Eq. 9–24 we see that the period is proportional to the square root of L, so the fractional change in L required is found from

$$\frac{(L + \Delta L)}{L} = \left(\frac{T'}{T}\right)^2 = 0.9972.$$

Solving for ΔL yields

$$\Delta L = L(0.9972 - 1) = -0.00280(0.2\text{m}) = -0.000560 \text{ m} = -0.56 \text{ mm}.$$

The correction can be accomplished by shortening the pendulum just over half a millimeter. Pendulum clocks generally have a small screw for adjusting the position of the weight on the string or rod.

Example 9.6 By how much will the period of a pendulum vary when it is transported from the earth to the moon?

The value of g is the variable in this case, while L is held constant. In Chapter 7 we learned that

$$g = \frac{GM}{R^2}$$

where G is the gravitational constant, M is the mass of the earth, and R is the earth's radius. Substituting this into Eq. 9–24 yields

$$T = 2\pi R \sqrt{\frac{L}{GM}}.$$

9.3 THE SIMPLE PENDULUM

If T' is the period of the pendulum when it is on the moon, while T is its period on the earth, then

$$\frac{T'}{T} = \left(\frac{R'}{R}\right)\sqrt{\frac{M}{M'}}$$

where R' and M' refer to the moon, and R and M to the earth. From Appendix 4 we see that $R'/R = 0.272$, and $M/M' = 81.3$. Thus we have

$$\frac{T'}{T} = 2.45.$$

The period is therefore longer by a factor of nearly 2.5. It should be apparent from this example that it is possible to measure the value of g by observing the period of a pendulum. This technique can be used to measure very small variations in g with altitude and location on the earth's surface, but to do so requires very difficult and sensitive measurements.

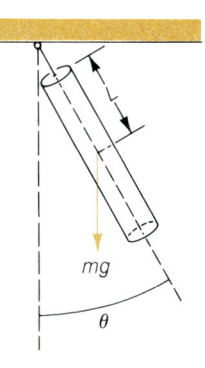

Figure 9.12

The physical pendulum. Gravity creates a torque $\tau \approx -mgL\theta$ on a pendulum of mass m. The result is SHM. The period depends on the moment of inertia of the pendulum.

So far when we have discussed simple pendulums, we considered only a point mass suspended at the end of a string whose mass was ignored. This is adequate for pendulums used in clocks, but, for many systems that hang from pivots and undergo oscillatory motion, these assumptions do not apply. The term **physical pendulum** is often used to describe such systems.

To analyze the motion of a physical pendulum, we use concepts developed in our discussions of rotational motion. We recall that torque is force times the lever arm. If the displacement is small so that the restoring force is proportional to the displacement, then it is easy to see that the restoring torque is proportional to the angular displacement θ. We simply multiply both sides of Eq. 9–21 by L.

$$\tau = -mgL\theta \tag{9–25}$$

L now represents the distance from the pivot point to the center of gravity (Fig. 9.12). This expression tells us that the requirement for SHM can be restated in terms of torque and angular displacement: SHM will occur if the restoring torque is proportional to the angular displacement. The condition for angular SHM can be written

$$\tau = -k'\theta \tag{9–26}$$

where k' is the effective torque constant, given by

$$k' = mgL. \tag{9–27}$$

For a physical pendulum, with an extended distribution of mass, deriving the period requires calculus, and will not be done here. The result is quite

FOCUS ON PHYSICS

Oscillations and Clocks

Timekeeping is a science that extends back into antiquity. There are many references to sundials and water clocks dated earlier than the time of Christ, for example. A sundial is a device that makes use of the position of the shadow of a pointer to estimate the time of day. A water clock is a mechanical wheel that rotates by a fixed amount each time one of several attached cups is filled with water whose flow rate is constant (Fig. 9.1.1). The steady source of water was usually a stream flowing out through a small hole in a container. Neither of these types of clocks was very accurate, and better techniques were eventually needed. The first mechanical clocks were developed during medieval times. A clock of that era used a weight to rotate a gear. Its rate of rotation was governed by a projection that moved in and out of the gear teeth, alternately blocking the rotation of the gear and then releasing it. These mechanical clocks, while better than sundials or water clocks, were still quite inaccurate because of friction and often lost or gained several minutes each day.

Vast improvements in timekeeping became possible with the development of devices that used steady oscillators to regulate the rotation of the main gear. The first steady oscillator used in this manner was the pendulum. The credit for recognizing this application of pendulums belongs to Galileo, who is said to have been the first to notice that the period of oscillation is constant (for small angles). Legend has it that when Galileo was a medical student in Pisa in 1582, while sitting in church he observed the regularity of oscillations of a chandelier that had been pulled to one side to be lighted. He quickly made practical use of this, and his first invention based on it was a medical instrument called a **pulsilogium**. This was a simple pendulum hanging from a board so that the string

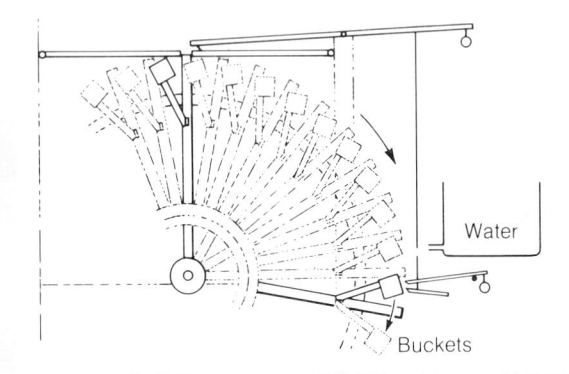

Figure 9.1.1

could be held against the board by the thumb, fixing the free length of the string and hence the rate of oscillation of the pendulum. The device was used to measure the pulse rate of a patient by adjusting the free length of the string until the pendulum's frequency matched the pulse. Galileo even calibrated the pulsilogium with markings on the board to indicate whether the pulse was normal or too fast or slow.

Galileo later used pendulums to keep time in recording astronomical observations, particularly of cyclical phenomena such as the orbital motions of Jupiter's moons. He did not attempt to build a general-purpose clock until his later years, but blindness prevented him from completing the project. He did design a clock, with the help of his son, but a working model was never completed (much later Galileo's plans were used to make working clocks, showing that his design was sound). Finally, in 1656 (13 years after Galileo's death), the great Dutch scientist Christiaan Huygens succeeded in building a very accurate clock using a pendulum.

9.3 THE SIMPLE PENDULUM

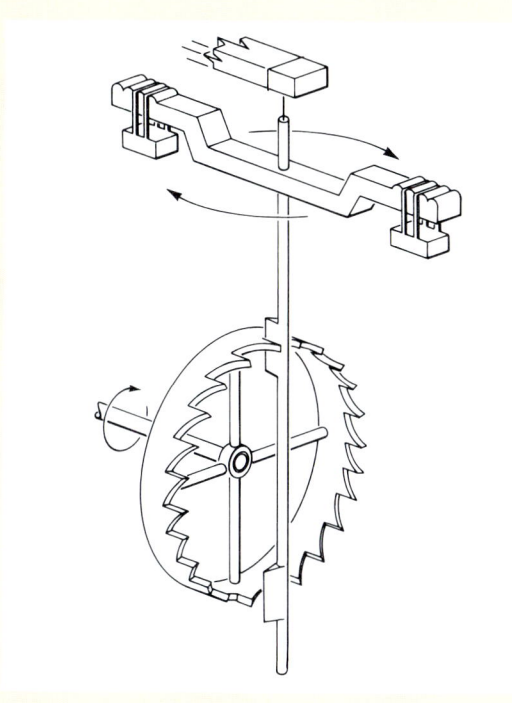

Figure 9.1.2

The role of the pendulum in a clock is to regulate the rate of rotation of a gear, which is linked to the hands that display the time. Huygens used a device much like the earlier mechanical clocks, a bar that swung back and forth, alternately stopping and releasing the gear (Fig. 9.1.2). The pendulum regulated the rate at which the bar swung back and forth, thereby regulating the rate at which the gear turned. The torque required to rotate the gear was supplied by a weight, as in the mechanical clocks. Huygens built clocks that lost or gained no more than 10 seconds per day, an unprecedented level of accuracy in that era.

Within a few years, another improvement in regulating the rotation of the gear was devised. The device that alternately stopped and released the gear was a specially shaped cog attached directly to the pivot axis of the pendulum and interfering less with the action of the pendulum (Fig. 9.1.3).

Clock technology changed very little for more than two centuries following the inventions of

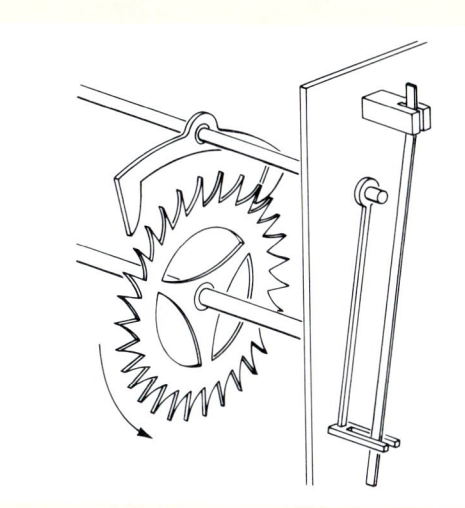

Figure 9.1.3

Huygens and his contemporaries. No better oscillator than a pendulum was found until electricity came into common use in this century. Today household electricity is alternating current (Chapter 17), meaning that the direction of current flow alternates regularly (with a frequency of 60 cycles per second in standard U.S. power systems). The frequency of oscillation, which is governed by the rotation rate of the generators at power plants, is very accurate and can be used to operate clocks. Thus, electric clocks that operate on household current use the frequency of the current to regulate the movement of the hands.

The most accurate clocks today use oscillations that are even more precisely regular than those of alternating current. Radio waves are forms of electromagnetic radiation (Chapter 18), and they obey a very precise relationship between wavelength and frequency, where the frequency may be viewed as the rate at which waves pass a fixed point. Radio waves are emitted at precisely fixed wavelengths by certain types of atoms; that is, the atom will always emit radio waves with exactly the same wavelength and the same frequency. Thus the radio emission from an atom can regulate the vibration frequency that regulates a clock. The major advantage of this, in addition to the precision of the frequency, is that scientists

(continued)

anywhere can produce the same frequency by using the radio waves from the same kind of atom. Since 1967, the official world's standard for atomic timekeeping has been an isotope of cesium, which emits at a particularly convenient frequency (9,192,631,770 cycles per second, in a portion of the radio spectrum that is easy to receive and use).

Regardless of how clocks are regulated, our timekeeping system is ultimately linked to the rotation of the earth. The length of the day, of the hour, the minute and the second are all based on the earth's rotation. The earth's rotation rate is not precisely constant, however; there are very small variations caused by a number of forces that create torques on the earth. For scientific purposes, when very high precision might be needed, measurements of time are based on atomic clocks. It is interesting that the second, derived originally from the earth's rotation, is now defined instead by the frequency of radio waves from cesium atoms.

simple, however, and could have been guessed from the analogy between rotational SHM and linear SHM. We simply substitute for restoring force, displacement, and mass the equivalent rotational quantities: torque, angular displacement, and moment of inertia I.

$$T = 2\pi \sqrt{\frac{I}{k'}} = 2\pi \sqrt{\frac{I}{mgL}} \qquad (9\text{--}28)$$

Note that the value of I is the moment of inertia about the pivot axis, which is never through the center of mass. Therefore caution must be used in solving problems; it may not be correct to simply find I in a standard table. Normally I must be calculated by carrying out an integration using calculus or numerical methods, or it may be measured experimentally.

Example 9.7 What is the period of oscillation for an earring that consists of a thin rod suspended from one end, if the length of the rod is 0.015 m and its mass is 0.5 g?

This is one of the few cases where the moment of inertia about an off-center pivot point can be found in a simple table, such as the one presented in Chapter 6. For a thin rod of length r pivoted about one end, the moment of inertia is

$$I = \frac{mr^2}{3}.$$

The value of L in Eq. 9–28 is the distance from the pivot point to the center of gravity, which is one half the length of the rod. Thus the expression for the period becomes

$$T = 2\pi \sqrt{\frac{2r}{3g}} = 0.201 \text{ s}.$$

9.4 Oscillations that are not SHM

Much of what we have said so far about SHM has been idealized. In the first place, the requirement for SHM given by Eq. 9–1 is almost never truly valid, because there are usually friction forces that are not simply proportional to the displacement. In the second place, there are many kinds of oscillatory motion that do not even arise from forces of the form given in Eq. 9–1, with or without friction. In this section we discuss some of these departures from SHM, but only in a qualitative way.

Damped SHM

When an oscillator loses energy to its surroundings through friction or some other mechanism such as air resistance, it must inevitably come to a stop. We say that the oscillations are **damped**. If the damping is weak, so that oscillations occur for many cycles before stopping, then it is approximately true that the description of SHM is valid. Although the amplitude gradually decreases, the period stays fairly constant. We can use the formulas of SHM to calculate the period, although they are not as good for calculating the displacement, velocity, or acceleration.

If the damping is strong, then the behavior is much less like SHM. Nevertheless, such cases cannot be ignored; strong damping is often deliberately introduced into mechanical systems so that oscillations will be suppressed. The role of shock absorbers in a car is precisely that, for example.

The three general cases are shown in Fig. 9.13. Curve (a) represents **overdamping**, where the oscillation is essentially stopped before it starts. The oscillator takes a long time to return to the equilibrium position following a displacement. Curve (c) represents **underdamping**, where many oscillations

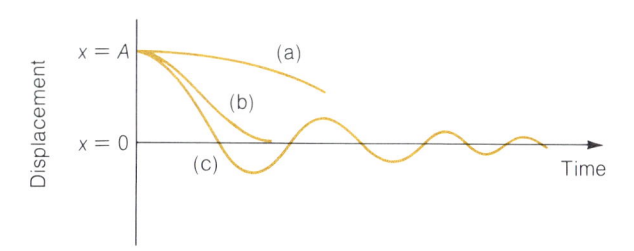

Figure 9.13

Damped SHM. The oscillator in (a) never reached equilibrium because of overdamping. In (b) the motion stopped at the equilibrium position on the first cycle. The oscillator was critically damped. In (c) the oscillator was underdamped, repeating several cycles before coming to rest at the equilibrium position.

274 CHAPTER 9 OSCILLATIONS

occur before the oscillator comes to rest. Any real pendulum (without a drive mechanism as in a clock) fits this description; as we have just said, some of the equations of SHM apply. Finally, curve (b) in the figure represents **critical damping,** where the system returns to equilibrium quickly, without oscillation, after the initial displacement. A door with a hydraulic cylinder adjusted so that the motion stops just as the door closes is an example of critical damping.

Relaxation Oscillations and Combined Oscillations

There are many kinds of systems that vibrate or oscillate without adhering to Eq. 9–1 at all. An entire class of vibrations, called **relaxation oscillations,** has been ignored so far, but is very important to us. This class includes all kinds of vibrations of membranes and strings, such as in human vocal cords and the strings of some musical instruments. It also includes bouncing balls, and the screech of two surfaces rubbed across one another. The only common characteristic of all these vibrations is that the acceleration of the oscillator is discontinuous, changing abruptly at some point in the cycle. When a bouncing ball strikes the pavement, for example, its compression causes an instantaneous reversal of the motion; when a violin bow is drawn across the strings, it alternately slips and sticks, so that the motion consists of a series of tiny jerks and stops. It is generally the displacing force, rather than the restoring force, that is discontinuous in relaxation oscillations.

Many systems vibrate with a combination of frequencies and their oscillations cannot be simply described. Even if the individual oscillations can be approximated by SHM, the net effect is not SHM at all.

There usually is no simple analytical technique for dealing with such oscillators; to write the equations of motion for non-SHM vibrators is a difficult task. There is, however, a very powerful empirical technique for analyzing non-SHM vibrations. In many cases a vibration, even if it is does not seem periodic, can be represented by the sum of two or more sinusoidal vibrations with different frequencies and amplitudes. This was first shown in the nineteenth century by the French mathematician J. Fourier, and the mathematical framework is still called **Fourier analysis.** When sinusoidal functions of the form of Eqs. 9–14 or 9–15 are added, the vibrations may add or subtract, depending on the relative phases and amplitudes. Any pattern of oscillations or variations can be reproduced to a good approximation by the sum of a properly constructed set of sinusoidal functions. If the motion being analyzed has a clear periodicity, then usually only a few such functions are required. If, on the other hand, the observed motion is random or nearly so, then many functions with a very broad range of frequencies is required. To carry out a Fourier analysis requires calculus.

9.5 Forced Vibrations and Resonance

In our discussions of SHM, we have developed various expressions for the period of an oscillating system. The period was always found to depend only on intrinsic characteristics of the oscillator, and to be independent of the

9.5 FORCED VIBRATIONS AND RESONANCE

Figure 9.14

Catastrophic resonance. Vibrations in this bridge caused by gusts of wind grew to large amplitude because they occurred at the bridge's natural frequency.

Figure 9.15

Instruments designed to be launched into space must be tested for resonances that might be destructive during launch, and are therefore shaken by equipment that simulates launch vibrations.

amplitude. This means that any system has a **natural frequency** at which it will tend to vibrate if disturbed.

If an external force is applied to an oscillator with a frequency other than its natural frequency, then the oscillator will vibrate with the applied frequency if the applied force is much greater than the internal restoring force of the oscillator. For example, a pendulum may be forced to oscillate at some arbitrary frequency by pushing on it at that frequency. If the applied force is comparable to the natural restoring force of the oscillator, then the result will be a complicated vibration in which the amplitude varies according to whether the two oscillations are in or out of phase.

If the natural frequency and the applied external force are of equal period and in phase, then the applied force acts in concert with the internal restoring force, and the amplitude is maximized. Anyone who has ever pushed a swing understands this; the best result occurs when the push is applied at the same point every cycle, in the direction of the swing's motion. This is called **resonance.**

Normally we speak of the energy transferred from a forcing oscillator to a natural oscillator. This energy transfer is characterized by the amplitude of the resultant vibration, so we see that the maximum energy transfer occurs when the applied force has the same frequency and is in phase with the natural oscillations of the system; that is, when resonance occurs.

There are many practical reasons to avoid resonance. It is best, for example, to design buildings so that their natural frequencies are far from any applied frequencies that might occur, such as those due to wind gusts. The famous collapse of the Tacoma Narrows Bridge in 1940 (Fig. 9.14) occurred because of a resonance between the bridge's natural frequency and the frequency at which strong winds made the bridge vibrate. Cars are designed so that their

Perspective

In this chapter we learned about one of the most basic phenomena in nature. Vibrations and oscillations occur in all materials and objects, although not always in an obvious fashion. We have focused on systems such as springs and pendulums, in which the oscillations are prominent, but we have also learned that natural vibrations of systems normally considered rigid are also important.

In the next chapter we will discuss a phenomenon closely related to vibrations and oscillations, namely, waves. We will see that natural frequencies play a big role in generating waves, and that resonance maintains them.

natural frequencies are far from the rotational frequencies of major moving parts such as the wheels or the engine (but sometimes minor parts have natural frequencies that are in resonance with these frequencies, and the car has a rattle at certain speeds). Instruments designed to be launched into space must be tested for resonances that might be destructive during launch, and are therefore shaken by equipment that simulates launch vibrations (Fig. 9.15).

PROBLEM SOLVING

Frequencies, Displacements, and Sinusoidal Motion

Most oscillators have either springs or pendulums. Problems in simple harmonic motion therefore require only a few basic ideas: (1) finding natural frequencies or periods, (2) finding displacements and velocities from an equilibrium point, and (3) analyzing phase and sinusoidal motion.

The natural angular frequency ω (rad/s) of a harmonic oscillator is given by

$$\omega = \sqrt{\frac{k}{m}} \quad \text{(spring)}$$

$$\omega = \sqrt{\frac{g}{L}} \quad \text{(pendulum)}$$

where k and m are the force constant and mass loading the spring and g and L are gravitational acceleration and length of the pendulum. The frequency in Hertz is $\nu = \omega/2\pi$, and the period is $T = 1/\nu$. Notice that the pendulum's frequency does not depend on the mass, and the spring's frequency does not depend on its length.

Finding the natural frequency is usually the first step in a problem. A spring's constant k is often found by measuring the displacement x for a known force F, say the weight mg loading it. Alternately, one may use a measured oscillation frequency to derive $k = m\omega^2$ (if m is known) or to find the gravitational acceleration $g = L\omega^2$ (if L is known).

Example 1 A 1 kg mass hung on a vertical spring has a 10 cm displacement. Find the force constant and the natural period of oscillation.

The loading weight $F = mg = 9.8$ N, so $k = F/x = 9.8$ N$/0.1$ m $= 98$ N/m. The natural frequency $\omega = \sqrt{(98 \text{ N/m})/1 \text{ kg}} = 9.90$ rad/s, so $\nu = \omega/2\pi = 1.58$ Hz. The natural period $T = 1/\nu = 0.63$ s. Notice that the units of k are N/m $=$ kg/s^2 so that $\omega = \sqrt{k/m}$ has the correct units of rad/s (a radian is a dimensionless angle).

A spring in harmonic motion conserves total energy, alternately changing kinetic energy, KE $= \frac{1}{2}mv^2$, to potential energy, PE $= \frac{1}{2}kx^2$. At maximum displacement the energy is entirely PE $= \frac{1}{2}kA^2$, while at zero displacement the energy is entirely KE $= \frac{1}{2}mv_0^2$, where A and v_0 are the amplitude and maximum velocity. If we know either A or v_0 we can find the other from energy conservation,

$$E = \frac{1}{2}kA^2 = \frac{1}{2}mv_0^2$$

provided that we know k and m (or equivalently, the natural frequency $\omega = \sqrt{k/m}$).

Example 2 An oscillating spring has 1 J total energy. If its natural frequency is 2 Hz and its mass is 50 g, what are its amplitude and maximum velocity? What is the displacement when $v = v_0/2$?

We wish to find the amplitude given the total energy E. Since $E = \frac{1}{2}kA^2$, we must first find the force constant $k = m\omega^2 = m(2\pi\nu)^2$.

$$k = (0.05 \text{ kg})[2\pi(2 \text{ Hz})]^2 = 7.90 \text{ kg/s}^2$$

Therefore,

$$A = \sqrt{\frac{2E}{k}} = \sqrt{\frac{2 \text{ J}}{7.9 \text{ kg/s}^2}} = 0.50 \text{ m}.$$

Similarly, we find the maximum velocity $v_0 = \sqrt{2E/m} = 6.32$ m/s. At a displacement x intermediate between zero and A, the velocity is given by

$$v = v_0\sqrt{1 - \left(\frac{x}{A}\right)^2}.$$

When $v = v_0/2$, we square this equation to find $1 - (x^2/A^2) = 1/4$. Thus, $x^2/A^2 = 3/4$, and $x = (\sqrt{3}/2)A = 0.866A = 0.433$ m.

Once the frequency ν and either A or v_0 are known for an oscillator, the resulting sinusoidal motion for displacement x, velocity v, or acceleration a can be written in terms of the functions $\sin(2\pi\nu t)$ or $\cos(2\pi\nu t)$. The argument of these functions is called the phase,

$$\phi = 2\pi\nu t = \frac{2\pi t}{T} = \omega t$$

where T is the period of oscillation. Phase ϕ is an angle in radians, and one complete cycle corresponds to 2π radians. Both $\sin\phi$ and $\cos\phi$ repeat in multiples of 2π, so oscillations at phase $\phi + 2\pi n$ ($n = 0, 1, 2, \ldots$) are equivalent.

A helpful hint in doing oscillator problems is to remember when $\sin\phi$ and $\cos\phi$ pass through 0 and ± 1. Then, you can use one of the above relations to find the corresponding time t, given the frequency ν or period T. For ϕ in radians, $\sin\phi$ equals 0 when $\phi = n\pi$, $\sin\phi = +1$ when $\phi = \pi/2 + 2\pi n$, and $\sin\phi = -1$ when $\phi = -\pi/2 + 2\pi n$ ($n = 0,1,2\ldots$). Similar expressions hold for $\cos\phi$, which is shifted in phase by one quarter-cycle ($\pi/2$ radians).

Example 3 An oscillator's displacement and velocity are given by $x = 5\sin(10t)$ and $v = 50\cos(10t)$. When are $x = 0$ and $v = 50$?

The phase $\phi = 10t$, and $\sin\phi = 0$ when $\phi = n\pi$. Thus, $x = 0$ when $\phi = 10t = n\pi$, or $t = (n\pi/10)$, $n = 0,1,2,\ldots$. Velocity reaches its maximum of $+50$ when $\cos\phi = 1$. Since $\cos\phi$ reaches its maximum at $t = 0$, it does so at multiples of 2π thereafter. So velocity equals 50 when $\phi = 10t = 2n\pi$, or when $t = n\pi/5$ ($n = 0,1,2,\ldots$). The period of the oscillator follows from the relation $2\pi\nu t = t/T = 10t$. Thus, $\nu = 10/2\pi$ Hz and $T = 1/\nu = 2\pi/10$ s.

SUMMARY TABLE OF FORMULAS

Force constant:
$$k = \frac{-F}{x} \quad (9\text{–}1)$$

Period and frequency:
$$T = \frac{1}{\nu} \quad (9\text{–}2)$$

Work and displacement for a SHO:
$$W = \bar{F}x \quad (9\text{–}3)$$

Force and displacement:
$$F = kx \quad (9\text{–}4)$$

Potential energy of a SHO:
$$PE = \frac{1}{2}kx^2 \quad (9\text{–}5)$$

Kinetic energy of a SHO:
$$KE = \frac{1}{2}mv^2 \quad (9\text{–}6)$$

Maximum velocity of a SHO:
$$v_0 = A\sqrt{\frac{k}{m}} \quad (9\text{–}7)$$

Instantaneous velocity of a SHO:
$$v^2 = \left(\frac{k}{m}\right)(A^2 - x^2) \quad (9\text{–}8)$$
$$v = v_0\sqrt{1 - \left(\frac{x}{A}\right)^2} \quad (9\text{–}9)$$

Period of a SHO:
$$T = 2\pi\sqrt{\frac{m}{k}} \quad (9\text{–}10)$$

Frequency of a SHO:
$$\nu = \frac{1}{2\pi}\sqrt{\frac{k}{m}} \quad (9\text{–}11)$$

Equation of motion for a SHO:
$$x = A\cos(\omega t) \quad (9\text{–}12)$$
$$x = A\cos(2\pi\nu t) \quad (9\text{–}13)$$

$$x = A\cos\left(\frac{2\pi t}{T}\right) \quad (9\text{–}14)$$

$$x = A\sin\left(\frac{2\pi t}{T}\right) \quad (9\text{–}15)$$

Velocity for a SHO:
$$v = -v_0 \sin(2\pi t/T) \quad (9\text{–}16)$$

Acceleration for a SHO:
$$a = -\left(\frac{kA}{m}\right)\cos\left(\frac{2\pi t}{T}\right) \quad (9\text{–}17)$$

$$a = -a_0 \cos\left(\frac{2\pi t}{T}\right) \quad (9\text{–}18)$$

Displacement of a pendulum:
$$x \simeq \theta L \quad (9\text{–}19)$$

Restoring force on a pendulum:
$$F = -mg\sin\theta \quad (9\text{–}20)$$
$$F \simeq -mg\theta \quad (9\text{–}21)$$
$$F = -\left(\frac{mg}{L}\right)x \quad (9\text{–}22)$$

Force constant for a pendulum:
$$k = \frac{mg}{L} \quad (9\text{–}23)$$

Period of a pendulum:
$$T = 2\pi\sqrt{\frac{L}{g}} \quad (9\text{–}24)$$

Restoring torque on a physical pendulum:
$$\tau = -mgL\theta \quad (9\text{–}25)$$
$$\tau = -k'\theta \quad (9\text{–}26)$$

Effective torque constant for a physical pendulum:
$$k' = mgL \quad (9\text{–}27)$$

Period of a physical pendulum:
$$T = 2\pi\sqrt{\frac{I}{mgL}} \quad (9\text{–}28)$$

THOUGHT QUESTIONS

1. Describe an example of oscillatory motion not given in the text which is produced by an elastic force.
2. Give an example of oscillatory motion not given in the text which is *not* produced by an elastic force.
3. Describe two examples of oscillatory motion which can be classified as simple harmonic motion and two which cannot.
4. Is a planet orbiting the sun an example of simple harmonic motion? Explain.
5. Sketch plots of v versus x, v versus A, and T versus k for a simple harmonic oscillator.
6. What happens to the motion of a harmonic oscillator if friction cannot be neglected?
7. List several practical applications of harmonic oscillators.
8. If a pendulum is to be treated as a simple harmonic oscillator, why is it necessary for the angle θ to be small?
9. Verify that the angle θ for a pendulum can be as large as 14° before the departure from Eq. 9–19 reaches 1%.
10. How might you use a pendulum to measure the acceleration of earth's gravity?
11. How might you use a pendulum to keep time?
12. Are the shock absorbers on an automobile an example of simple harmonic motion? Explain your reasoning.
13. Sketch a graph of PE and KE versus time for a simple pendulum.
14. How would you experimentally determine the spring constant of a spider web?
15. Give everyday examples of overdamping, underdamping, and critical damping.
16. What is resonance? Describe several examples not given in the text.
17. Will an object vibrate at frequencies other than v_o, its natural frequency?
18. How might you use a pendulum to determine the mass of the earth?
19. What is meant by a damped oscillator?
20. What might happen if cars were not designed so that their natural frequencies were far from the rotational frequencies of their moving parts?

PROBLEMS

Section 9.1 Restoring Forces, SHM

1. A force of 10 N produces a 10 cm extension of a spring. What is the force constant?
2. Show that the force constant k of a spring has units kg/s^2, as well as the units N/m.
3. A 20 cm spring has force constant 1,500 kg/s^2. If the spring is loaded by a mass of 0.1 kg, how much will it stretch when hung vertically?
4. The earth takes 365.24 days to orbit the sun at a distance of 1.5×10^8 km. Find the mean velocity of the earth in orbit. Find its frequency in Hz.
5. The sun orbits the center of the Milky Way galaxy at a rate of about 225 km/s at a distance of about 3×10^{17} km. Find the orbital period and its frequency.
6. What is the frequency of an oscillation with a one-hour period?
7. What is the period of an oscillatory motion of frequency 20 Hz?
8. What is the frequency (Hz) of the second hand on a clock?
9. Astronomers have discovered rapidly spinning neutron stars, called pulsars, whose frequencies are as large as 642 Hz. If these stars are spheres 10 km in radius, what is their shortest period and fastest surface velocity at the equator?
10. Musicians tune to concert A, 440 Hz. What is the period of vibration? How many vibrations would a tuning fork make in the course of a one-hour symphony?
11. Electric house current has a frequency of 60 Hz. How long do a million vibrations take?
12. An AM radio station has a frequency of 1,200 kHz. How many oscillations of the radio waves occur in one second?

- 13 What is the frequency of the orbits of Jupiter and Pluto (consult Appendix 4)?
- 14 The H_2 molecule vibrates with frequency 1.319×10^{14} Hz. Calculate how many vibrations it makes in the time it takes to fall from a height of 1 m under the force of gravity.
- 15 When a 60 kg person enters a car, its springs compress vertically by 1 cm. If the springs remain elastic, how far will they compress when two more people of the same mass enter?

Section 9.2 Harmonic Oscillators

- 16 Find the amplitude of a 2 kg harmonic oscillator which is moving with $v_0 = 10$ m/s and a force constant 8 kg/s².
- 17 A 5 kg mass is placed on a vertical spring with $k = 4.9 \times 10^3$ N/m. Find the potential energy of the spring if the displacement is 15 cm.
- 18 The mass in Problem 17 is now released. Describe the period, amplitude, and maximum velocity of the mass.
- 19 A 2 kg mass is hung on a vertical spring and produces an equilibrium displacement of 0.1 m. Then the mass is extended another 0.1 m and released. Find the displacement from equilibrium x, velocity v, and acceleration a of the mass after 5 s.
- 20 The planet Jupiter orbits the sun once every 11.86 yr with an orbital velocity of 13 km/s. Find the amplitude and frequency of Jupiter as viewed from the plane of Jupiter's orbit, far from the sun.
- 21 A 3 kg mass is placed on a vertical spring with constant k. Then the spring is replaced with a second spring having force constant $k/2$. How much more (or less) mass should be placed on the new spring so that the period of oscillation remains the same?
- 22 At what fractions of the period T of a simple harmonic oscillator will the displacement x have a value of (a) zero, (b) A, the initial displacement?
- 23 When a 80 kg salesman enters a 1,000 kg car, its springs compress vertically 1.2 cm. What will be the car's oscillation frequency, assuming no damping, if the salesman leans on the fender while standing outside?
- 24 A massless spring vibrates at 3 Hz with 3 kg attached to one end. What will be its frequency when 5 kg is attached?
- 25 A 120 lb woman steps into a 50 lb canoe, which sinks 6 cm deeper into the water, then oscillates when she jumps out. What is the period?
- 26 A 5 kg mass oscillates with amplitude $x = A \sin(2\pi\nu t)$, where $A = 1$ m and $\nu = 1$ Hz. What are the maximum velocity and maximum force?
- 27 A 0.5 kg mass is attached to a spring that vibrates at 10 Hz. What is the maximum kinetic energy of the mass, if the amplitude is 10 cm?
- 28 A tuning fork vibrates at 440 Hz with amplitude 1 mm. What is the maximum velocity of one prong?
- 29 A 5 kg mass is attached to a spring with constant 300 N/m. What is the maximum velocity of the mass if the amplitude is 0.2 m?
- 30 An oscillator's period is 10 s, and its maximum velocity 10 m/s. What is its amplitude?
- 31 The displacement of a harmonic oscillator is observed to be 30 cm maximum. What is the displacement at the moment the oscillator is moving at exactly half its maximum velocity?
- 32 An oscillator is moving at 0.8 of its maximum velocity when it has a displacement of 15 cm. What is the maximum amplitude of motion?
- 33 Two springs have identical masses attached. If the spring constants have a ratio of 2/1, what ratio of initial displacements is required to keep the maximum velocities of masses the same?
- 34 A harmonic oscillator obeys the equation $x = 4\cos(10t)$. Find the amplitude, frequency, and period of motion.
- 35 A harmonic oscillator vibrates with acceleration $a = 6\cos(5t)$. Find the maximum acceleration, period, and frequency of motion.
- 36 A harmonic oscillator vibrates with velocity $v = 8\sin(8t)$. Find the maximum velocity, period, and frequency of motion.
- 37 What is the frequency of a harmonic oscillator if its maximum acceleration equals g? Assume its amplitude is A and the mass is m.
- 38 What would be the period T of the harmonic oscillator in Problem 37?
- 39 A harmonic oscillator moves according to $x = 10\cos(3t)$. At what time(s) does the displacement x equal its maximum value?
- 40 A harmonic oscillator moves according to $v = 2\sin(5t)$. At what time(s) does the velocity v reach its largest value?

PROBLEMS

••• 41 A harmonic oscillator moves with acceleration $a = 7\sin(2t)$. At what time(s) does the acceleration reach 0.4 times its maximum value?

••• 42 A certain harmonic oscillator has a period of 10 s and reaches a maximum displacement at $t = 5$ s. At what time(s) do the acceleration and velocity reach maximum values?

••• 43 A certain harmonic oscillator has a period of 8 s and reaches its maximum acceleration at $t = 5$ s. At what time(s) do the displacement and velocity reach maximum values?

•• 44 A 50 kg mass is suspended from a spring, causing the spring to stretch 5 cm. The mass is then displaced an additional 3 cm and released. What are the period, amplitude, and frequency of the resulting motion?

•• 45 How much mass should be added to or subtracted from the system in Problem 44 to make the period 1 s?

••• 46 Find the displacement from equilibrium, velocity, and acceleration of the system in Problem 44 ten seconds after the mass is released.

•• 47 What are the maximum KE and PE of the system in Problem 44?

Section 9.3 Pendulums

• 48 A pendulum has a mass 1 kg and a length 3 m. What is its period?

• 49 A pendulum has a mass 2 kg and period 2 s. What is its length?

• 50 A simple pendulum has a period of 5 s. What is its length?

• 51 Find the frequency of a pendulum having a length of 2 meters.

• 52 Suppose you wish to make a pendulum clock with period 1 s. How long should the pendulum be?

••• 53 Which pendulum would have the higher frequency? One consists of a uniform one-meter rod with mass 1 kg; a second consists of a 1 kg mass at the end of a one-meter massless string.

••• 54 A 20 kg child plays on a massless swing 3 meters long with an amplitude of $\theta = 10°$ and a period of 2 s. What is the maximum acceleration to which the child will be subjected? Compare to g.

• 55 The acceleration of gravity on the moon's surface is about $g/6$. How much longer or shorter would one have to make a pendulum so that its period would be the same as on earth?

• 56 A simple pendulum has a period of 1 s on earth. Transported to an asteroid, the pendulum has a period 5 s. What is the gravity on the asteroid?

• 57 A pendulum has a length 1 m and a period of 5 s on the surface of an alien planet. What is the value of the surface gravity?

•• 58 Suppose a pendulum having period 5 s is taken halfway to the moon (192,000 km from the earth). What would be its period? Assume that the earth's radius is 6,378 km and neglect the gravity of the moon.

•• 59 How far above the earth's surface should a pendulum be moved to change its period by a factor of three?

•• 60 Derive an expression for the angular acceleration of a simple pendulum in which the displacement angle θ is not small.

•• 61 A 55 kg woman jumps from a ledge to a trampoline 10 m below, which sags 0.5 m. Assuming the net acts like a spring, estimate its force constant.

••• 62 Two spring oscillators, A and B, have a ratio of spring constants $k_A/k_B = C_1$ and a ratio of total energies, $E_A/E_B = C_2$, where C_1 and C_2 are constants. If the frequencies are the same, find the ratio of amplitudes and of maximum velocities.

•• 63 A 40 kg child stands on a trampoline, which sags 15 cm. Assuming elastic, springlike behavior, find the force constant and natural frequency. Neglect the mass of the trampoline material.

•• 64 If the child in Problem 63 jumps and lands on the trampoline, depressing it 40 cm, compute the PE stored in the trampoline and the maximum height above the equilibrium point she will reach upon elastic rebound.

•• 65 On a hot day, the length of a pendulum increases by 1% due to thermal expansion. Calculate the fractional period change $\Delta T/T$.

•• 66 A prisoner lies at the bottom of a deep pit, watching a pendulum blade slowly descending. The end of the 20 m pendulum is initially 10 mm from his chest, and is lowered at a rate of 1 mm/h. How many times will the blade pass his chest before he must be rescued?

•• 67 Assume that the earring in Example 9.7 was a sphere with radius 0.015 m and mass 0.5g, which was attached right at the surface. What would be the period of oscillation?

••• 68 Compute the period of a physical pendulum of mass M, consisting of a uniform rectangular board of cross sectional area A and length L, suspended about one end. (Consult Chapter 6 for a table of moments of inertia.)

•• 69 Estimate the natural period of a golf club, treated as a pendulum. For simplicity, neglect the mass of the 1-meter shank and approximate the club at the end by a circular disk of radius 4 cm, thickness 2 cm, and density 2.98 g/cm³. How does this period compare to a typical golf swing?

••• 70 What would be the period in Problem 69 if the cylindrical 1 meter shank had a mass of 300 grams? The club acts as a physical pendulum.

Section 9.4–9.5 Resonance

• 71 Calculate the resonant frequency of a spring with $k = 10^3$ N/m and $m = 2$ kg.

• 72 Calculate the resonant frequency of a pendulum with length 2 m.

•• 73 A 150 foot nylon climbing rope has diameter 11 mm and Young's modulus 10^8 N/m². What is its force constant k? What is its resonant frequency when supporting an 80 kg climber?

••• 74 A sphere of radius R is composed of material with density ρ and bulk modulus B. If it is compressed by a small amount in radius, ΔR, show that $(\Delta V/V) = 3(\Delta R)/R$. From the equations describing bulk stress (Chapter 8), find the force with which the sphere resists compression, and show that it acts like that in a spring. Evaluate the force constant and resonant frequency of the sphere.

••• 75 Fluid in a bent tube (Fig. 9.16) is displaced a distance Δx from equilibrium (the level on the other side falls by Δx). The fluid has density ρ, and the tube has cross sectional area A and length L. Assume that L is much longer than the tube's radius.

(a) Evaluate the gravitational PE stored when the fluid is displaced Δx.
(b) Show that the PE has the form $\frac{1}{2}k(\Delta x)^2$, and evaluate the effective spring constant and resonant frequency of the oscillation. Note that all the fluid oscillates.

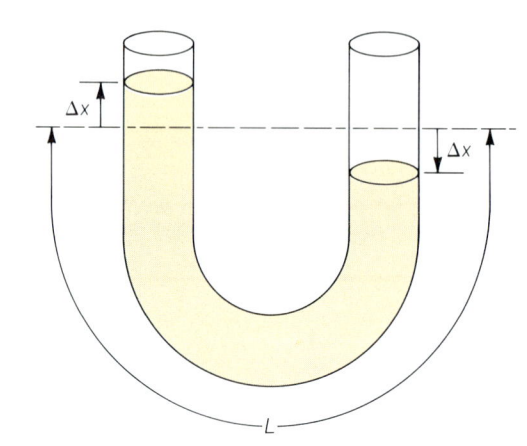

Figure 9.16

••• 76 Estimate one resonant vibrational frequency of a car's chassis. Assume the main steel beams are rectangular solids, 5 cm × 5 cm × 1 m. The density of steel is 7,900 kg/m³ and its Young's modulus 2×10^{11} N/m². Consider compression along the length of the beam. How does this frequency compare to those encountered on the road or by tire motion?

SPECIAL PROBLEMS

1. A Tunnel to China

In the future, engineers dig a tunnel through the earth, passing through its center. They construct a tube into which they release a capsule of mass m from rest. It falls toward the center, then passes through to the other side. Newton's law states that the gravitational force on the capsule is $F = -GM_r m/r^2$, where r is the distance from the center to the capsule and $M_r = (4\pi r^3 \rho/3)$ is the mass interior to radius r. We have assumed here that the earth is homogeneous with density ρ. (Geological experiments show that it is not.)

(a) Find the force on the capsule at radius r. Show that this force acts like a spring's restoring force, $F = -kr$, and evaluate k in terms of G, ρ, and other quantities.
(b) What is the period of oscillation of the capsule (the round trip tunnel time)? Assume that $\rho = 5,500$ kg/m³. Will this time depend on the capsule mass m?
(c) Explain why the tunnel time is also the period for the Earth to oscillate after an earthquake. Using data in the Appendix 4, estimate this period for the planets Jupiter and Mars.

2. Leg Motions

The legs of humans and animals act like physical pendulums, whose pivot is at the hip and whose period $T = 2\pi(I/mgL)^{1/2}$ (see Section 9.3).
  (a) Estimate the mass m and moment of inertia I of a human leg, assuming it to be a cylinder of length $l = 1$ m and average radius 7 cm, with average density 1 gram/cm^3.
  (b) Using $L = 0.4\,l$, calculate the natural period of the leg.
  (c) If a person walks by taking steps of angular amplitude $\theta = 30°$, how long will he or she take to walk a mile?
  (d) What is the natural period of the leg for an astronaut on the moon? Neglect the added mass of the spacesuit.

3. Molecular Vibrations

Diatomic (two-atom) molecules may be represented by two masses m_1 and m_2 connected by a spring (the chemical bond) with constant k. The effective mass loading the spring is given by $m = m_1 m_2/(m_1 + m_2)$.
  (a) For the carbon monoxide molecule (CO), find $m_1 = m(C)$, $m_2 = m(O)$, and m.
  (b) If the natural frequency of vibration of CO is measured as 6.5×10^{13} Hz, find the force constant k.
  (c) We will see when we study quantum mechanics that the CO vibration energies must be multiples of the fundamental energy 0.269 eV. Using the classical formulas for kinetic energy and potential energy of springs, estimate the amplitude x, velocity v, and restoring force of the molecular spring for this fundamental oscillation.

4. Sloshing the Water Trough

In the old days, cowboys watered their horses in troughs like the one in Fig. 9.17. We wish to estimate the resonant frequency with which a horse could create large waves by pushing on the trough periodically. Suppose the water sloshes to a height (Δx) on one side and falls the same amount on the other side (assume the displacement is symmetric for the length of the trough). Neglect friction.
  (a) Show that the mass displaced upward is $(WL\rho/2)(\Delta x)$ where ρ is the water density.
  (b) Evaluate the gravitational energy stored in the displacement. Then estimate the spring constant and resonant frequency $\nu_0 = (k/m)^{1/2}/2\pi$. Assume that m represents the mass of all the water in the trough (this is only an approximation).
  (c) Evaluate the period of sloshing for $D = W = 50$ cm, $L = 2$ m, and $\rho = 1{,}000$ kg/m^3. Does the period depend on any other variables?

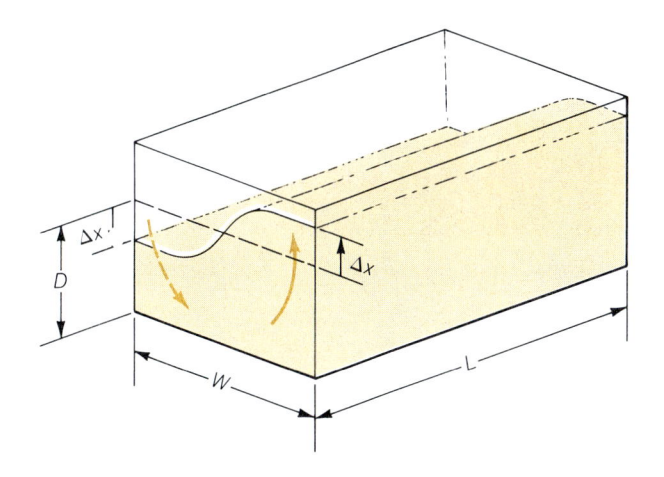

Figure 9.17

SUGGESTED READINGS

"Bridges." D. B. Steinman, *Scientific American,* Nov 54, p 60, **191**(5)

The flying circus of physics with answers. Jearl Walker, Wiley, New York, 1977

A source book in physics. William F. Magie, Harvard University Press, Cambridge, 1965

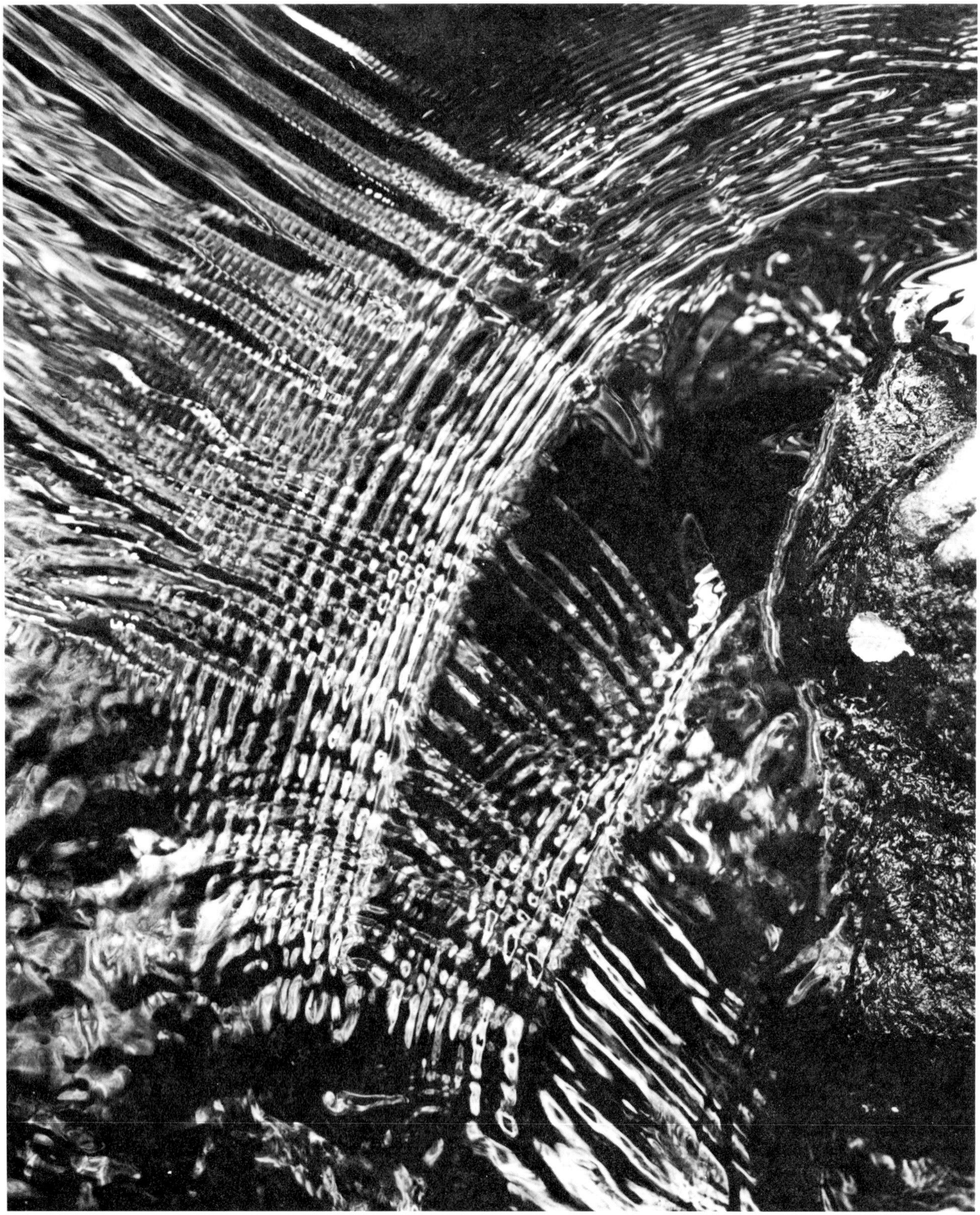

Chapter 10

Wave Phenomena

Wave motions are a fundamental characteristic of the physical universe. In its broadest context, the term "wave" can be applied to any transmission of energy through any medium. We will not be quite so liberal in this chapter, however, but even so, we will find that waves play a major role in a very broad range of common phenomena. Our most informative senses, sight and hearing, rely on waves to transmit information about our environment to our brains. Waves are the basis of all modern communication systems, and provide us with virtually all the information we have or perhaps ever will have on the universe beyond the solar system. Matter itself, as we will learn in Part VI of this text, exhibits wave behavior.

We begin with a discussion of the general properties of waves, followed by analyses of their physical characteristics and the ways in which they interact with their environment and each other. We will finish by discussing sound waves, while the treatment of the other major sensory input, light waves, will be deferred until we have discussed electricity and magnetism.

10.1 The Nature of Waves

The most easily visualized waves are water waves, because they are more tangible than sound or light waves, and because we see them in everyday life. Thus we begin our description of wave phenomena with water waves.

The Propagation of Disturbances

One of the first aspects of waves that should be noticed is the distinction between the wave and the medium through which it travels. A piece of wood floating on the surface of a pond (Fig. 10.1) does not travel across the pond

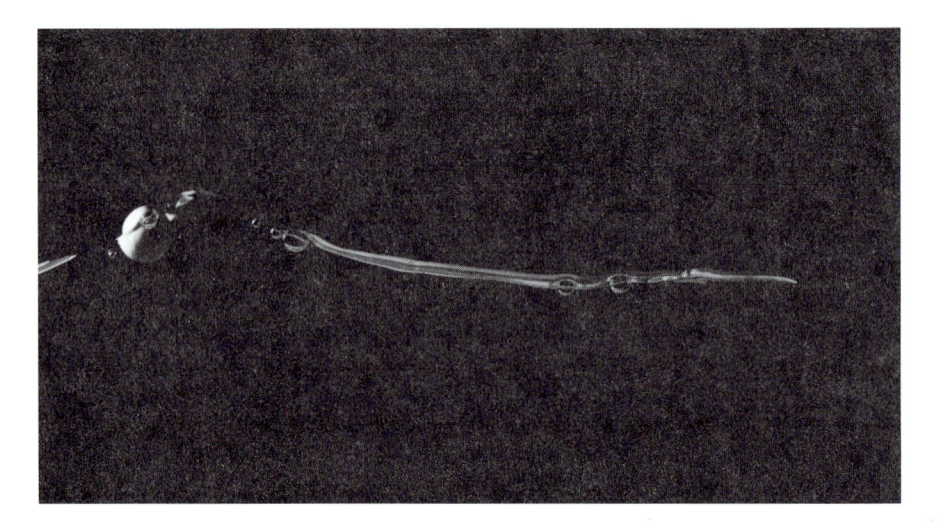

Figure 10.1

The distinction between a wave and the medium in which it travels. The ball oscillates in place as a wave passes by.

when a wave passes by; instead, it bobs up and down in place (with a slight horizontal oscillation as well). The wave, then, does not consist of a physical transport of matter. We notice, however, that the floating piece of wood, which was initially at rest, is put into motion by the wave and must therefore gain energy from the wave. We see that waves do transmit energy, even though they do not transport matter from one place to another. As we will see, they can also transmit information.

A wave may begin with a **pulse**; a momentary input of energy which creates a disturbance in a medium. It is this disturbance that somehow travels through the medium, and has the capability of releasing energy to another object. The way in which the disturbance is propagated depends on the nature of the pulse and of the medium. For waves on the surface of a body of water, the disturbance is transmitted by the pressure of displaced fluid elements on adjacent elements; for a wave travelling along a taut string, the disturbance is transmitted by tension in the string which acts to keep the string stretched between its fixed ends.

When we discuss light waves, we will find a significant contrast with what we will say about waves here. Unlike waves which travel through a medium, light can propagate through a vacuum. Thus light waves are distinctly different from the waves we discuss here, and would not fit in easily with the present discussion.

Types of Waves

A wave can be produced by a single pulse, so that a solitary disturbance is created, or a train of waves can be produced by a repeating pulse or a continuous oscillation. In this chapter we will concentrate on series of waves

10.1 THE NATURE OF WAVES

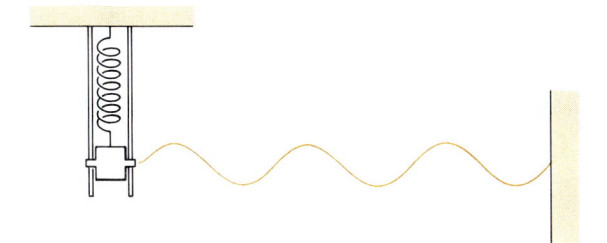

Figure 10.2

Continuous wave. Here a simple harmonic oscillator is attached to one end of a taut string, creating a continuous (periodic) wave.

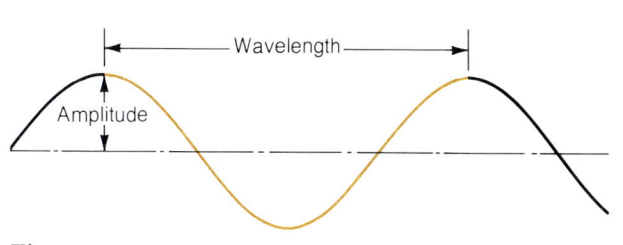

Figure 10.3

Properties of a continuous wave.

created by oscillations; such waves are called **periodic** or **continuous** waves (Fig. 10.2). These waves are particularly effective for transmitting information, because the pattern of disturbances as a series of pulses passes by a fixed point is constant, and reproduces the pattern of pulses that created the disturbances. As a simple example, imagine that a person repeatedly slaps the surface of a pool in a definite rhythm. The waves created by the slaps will travel to the other side of the pool, arriving there in the same rhythm. Thus, a person on the other side could reconstruct the sequence and derive information from it.

There are several terms we need to discuss periodic waves. The **wavelength** λ is the distance from one wave crest or trough to the next (Fig. 10.3). The **period** is the time required for a wave to travel a distance equal to one wavelength. If the wave velocity is v and the period is T, then we see that

$$v = \lambda/T. \tag{10-1}$$

The **frequency** ν (sometimes f is used instead of ν) is the number of waves per second that pass a fixed point, and is simply the reciprocal of the period.

$$\nu = \frac{1}{T}$$

For example, if a wave has a period of 0.5 s, then the frequency is 2 waves per second.

Combining our expression relating frequency and period with Eq. 10–1 leads to

$$v = \lambda\nu. \tag{10-2}$$

We will discuss the velocities of waves later. For now, we simply point out that the velocity is determined entirely by properties of the medium, and is independent of the nature of the initial pulse or oscillation.

Periodic waves occur in three types, each of which is characterized by the nature of the vibrations created in the medium as the waves travel (Fig. 10.4). The waves created by wiggling a taut string are called **transverse** or **shear**

(a) Transverse Wave

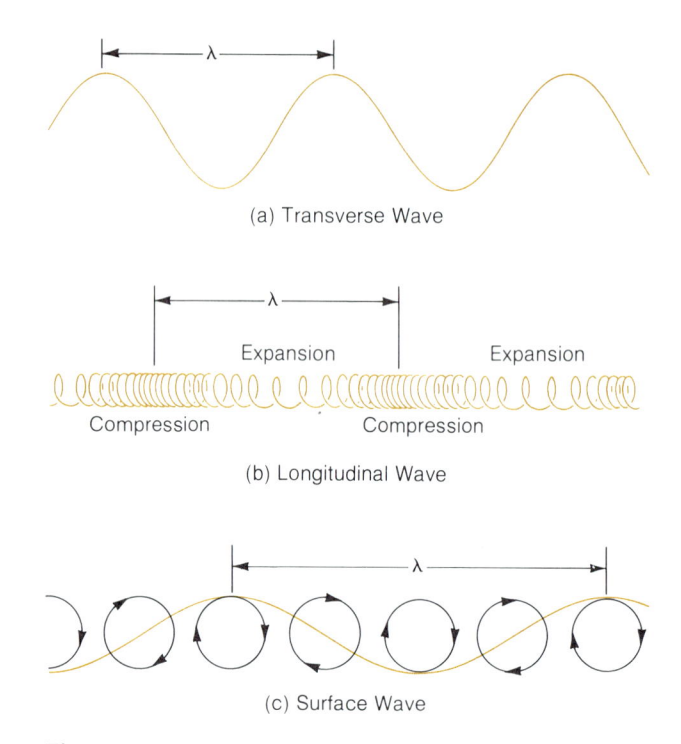

(b) Longitudinal Wave

(c) Surface Wave

Figure 10.4

Types of waves.

waves, because the medium is displaced at right angles to the direction of wave propagation. These waves require an elastic medium, therefore a solid material, in order to be transmitted, because they rely on the restoring force characteristic of solids. Sound waves and all others in which the vibrations are parallel to the direction of motion are called **compressional** or **longitudinal** waves. These are transmitted by pressure forces, and consist of a series of density enhancements and rarefactions. Because they do not depend on the elastic force, these waves can be transmitted through liquids and gases as well as solids. The third type of waves are **surface** waves such as the water waves so familiar to all of us. In these waves, the vibrations are circular, combining both transverse and longitudinal motions. We will devote most of our attention in this chapter to transverse and compressional waves.

Waves can occur on scales that we seldom visualize. For example, when an earthquake occurs, a localized pulse creates waves that travel literally around the world. Some of these **seismic** waves fall into each of the categories just described. The first waves to reach a position remote from the earthquake site, the **primary** or **P** waves, are compressional waves. The **secondary** or **S** waves arrive next; these are transverse waves. Finally, there are surface waves (usually called **L** waves) that travel around the earth rather than through it and arrive last. Geophysicists analyze seismic waves because their speed provides information on internal densities in the earth. Their distribution shows where liquid zones are in the earth's core, because the S waves cannot travel through such regions.

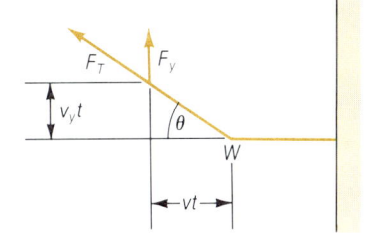

Figure 10.5

Velocity of a transverse wave. An idealized wave starts with a transverse displacement of one end of a taut string.

Wave Velocities

We have said that the velocity of a wave depends only on the nature of the medium. The relevant factors are the force that binds atoms or molecules together, expressed by the tension or the bulk modulus, and the density of the medium.

It is straightforward matter to derive an expression for the velocity of a transverse wave. We consider a string under tension, with the tension force designated F_T. If we define a coordinate system as in Fig. 10.5, we assume that the string is displaced vertically at one end by a force F_y which moves that end sideways with speed v_y. At some time after the initial disturbance, there is a point (W in the figure) beyond which the wave has not yet travelled. We want to find the velocity v of this point along the string.

In a time interval t, W moves along the string a distance vt, while the end of the string moves sideways a distance $v_y t$. If the segment of string between the end and W is a straight segment making an angular displacement θ then the two displaements are related to each other by

$$\sin\theta = \frac{v_y t}{vt} = \frac{v_y}{v}$$

and the sideways component of the tension, F_y, is related to the tension by

$$\sin\theta = \frac{F_y}{F_T}.$$

Therefore

$$F_y = F_T\left(\frac{v_y}{v}\right).$$

Recall from Chapter 5 that the impulse, which is the product of force times the time interval over which it is applied, is equal to the change in momentum that results. The impulse given to the segment of string that has been displaced is $F_y t$, while the change in momentum is the mass of the moving segment of string times its velocity v_y. Substituting for F_y from the above expression, we have

$$F_T\left(\frac{v_y}{v}\right)t = m v_y.$$

The mass m of a small string segment can be expressed as the mass per unit length $\mu = M/L$ (where M and L represent the total mass and length of the string) times the length of the segment which is displaced, vt. Making this substitution for m leads to

$$F_T\left(\frac{v_y}{v}\right)t = \left(\frac{M}{L}\right)vt v_y = \mu vt v_y$$

CHAPTER 10 WAVE PHENOMENA

which after solving for v becomes

$$v = \sqrt{F_T/\mu}. \tag{10-3}$$

Although this expression was derived for a taut string, it is valid for all transverse waves.

Example 10.1 A cave explorer being lowered into a deep pit reaches the end of his rope. He wishes to communicate this to his companions at the surface, so he creates a wave in the rope by jerking it. He weighs 185 lbs, and the rope is 150 ft long and weighs 40 lbs. How long does it take the wave created by the speleologist to reach the surface?

We need to know the tension in the rope, which varies with position in this example, because the amount that the rope's own weight contributes to the tension at any point depends on how much rope is below. We make the approximation that the rope's contribution to the tension is constant and equal to half its total weight. Thus, the tension in the rope equals 20 lbs plus the weight of the speleologist, so $T = 205$ lb $= 912$ N. The mass of the rope is 18.1 kg, and the length is 45.7 m. Therefore the wave velocity is

$$v = \sqrt{F_T/\mu} = \sqrt{F_T L/M} =$$
$$= \sqrt{\frac{(912 \text{ N})(45.7 \text{ m})}{(18.1 \text{ kg})}} = 48.0 \text{ m/s}.$$

The wave travels the full length of the rope in time

$$t = \frac{L}{v} = \frac{45.7 \text{ m}}{48.0 \text{ m/s}} = 0.95 \text{ s}.$$

As we will see, the dangling explorer could have signaled to the surface more quickly by shouting.

An expression for the velocity of a compressional wave can be derived in a manner similar to that used for transverse waves. Let us consider a metal rod whose density is ρ and whose cross-sectional area is A. One end of the rod is fixed and a force F applied to the other end causes a compression ΔL in time t (Fig. 10.6). During time t a segment L of the rod is compressed. The wave velocity we seek is the speed at which this compression travels along the rod. We apply the work-energy principle to set the work done on the rod equal to the energy it gains.

$$W = F\Delta L = \text{PE} + \text{KE}$$

10.1 THE NATURE OF WAVES

The potential energy is the force times $\frac{1}{2}\Delta L$, the average displacement during the interval t. The kinetic energy is $\frac{1}{2}mv^2$, where the mass can be expressed as $A\rho L$ and the velocity is the speed v_c of the compression (the speed of the end of the rod, not the same as the wave speed). Thus we have

$$F\Delta L = \frac{1}{2}F\Delta L + \frac{1}{2}A\rho L v_c^2.$$

Since the compression ΔL occurs in the same time the wave travels a distance L, the wave velocity v and the compression velocity v_c are related by

$$\frac{L}{v} = \frac{\Delta L}{v_c}$$

Solving this for v_c and substituting into the work-energy expression above leads, after solving for v, to

$$v = \sqrt{\frac{1}{\rho}\left(\frac{F}{A}\right)\left(\frac{L}{\Delta L}\right)}.$$

We recall from Chapter 8 that the stress F/A is related to the strain $\Delta L/L$ by

$$\frac{F}{A} = Y\left(\frac{\Delta L}{L}\right)$$

where Y is the elastic Young's modulus. Hence we have, finally,

$$v = \sqrt{\frac{Y}{\rho}}. \tag{10-4}$$

This expression, which gives the wave velocity entirely in terms of intrinsic properties of the medium, is the one we have been seeking. Like our expression for the velocity of a transverse wave, this expression applies to any compressional waves travelling in a solid medium such as a rod. For compressional waves travelling through a gas or liquid medium, the velocity is

$$v = \sqrt{\frac{B}{\rho}} \tag{10-5}$$

where B is the bulk modulus for rapid deformations in which little energy is exchanged with the surroundings.

Notice that, in general, the wave speed depends both on the elastic properties and the density of the medium. This is because the restoring force of the disturbance depends on the elastic modulus, while the acceleration or inertial

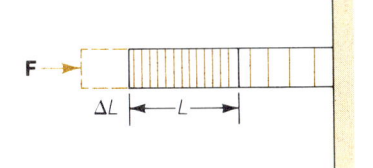

Figure 10.6

Velocity of a compressional wave. To calculate the velocity, we start by assuming a rod is compressed at its end.

Table 10.1
Summary of wave speeds

Wave type	Wave speed	Elastic modulus	Mass density	Equation
Transverse wave in rod	$\sqrt{\dfrac{F_T}{\mu}}$	Tension	μ (kg/m)	(10–3)
Compression wave in rod	$\sqrt{\dfrac{Y}{\rho}}$	Young's modulus	ρ (kg/m^3)	(10–4)
Compression wave in gas, liquid	$\sqrt{\dfrac{B}{\rho}}$	Bulk modulus	ρ (kg/m^3)	(10–5)
Compression wave in solid (*P*-wave)	$\sqrt{\dfrac{B + \tfrac{4}{3}S}{\rho}}$	Young's and Shear modulus	ρ (kg/m^3)	—
Transverse shear wave in solid (*S*-wave)	$\sqrt{\dfrac{S}{\rho}}$	Shear modulus	ρ (kg/m^3)	—

response depends on the density. Comparing Eqs. 10–3, 10–4, and 10–5, we see that the wave speed always has the form,

$$v = \sqrt{\frac{\text{elastic property}}{\text{mass density}}}$$

The various waves speeds of compressional and shear waves in gases, liquids, and solids are summarized in Table 10.1.

Example 10.2   The bulk modulus for air is 1.4×10^5 N/m^2, and the density in SI units is about 1.2 kg/m^3. If a person sees a distant lightning bolt, and hears the thunder 6 seconds later, how far is he from the bolt? (Light travels much faster than sound.)

We first calculate the speed of sound, which is a compressional wave travelling through a three-dimensional medium, so Eq. 10–5 applies:

$$v = \sqrt{\frac{B}{\rho}} = \sqrt{\frac{1.4 \times 10^5 \text{ N/m}^2}{1.2 \text{ kg/m}^3}} = 342 \text{ m/s}$$

The distance travelled in 6 s is

$$d = vt = (342 \text{ m/s})(6 \text{ s}) = 2{,}052 \text{ m} = 1.3 \text{ miles.}$$

10.1 THE NATURE OF WAVES

Energy Transport by Waves

We have shown that waves transmit energy, because they can create motion in the medium through which they travel. Now we consider the amount of energy transmitted. We make the simplifying assumption that the waves are transverse and sinusoidal; that is, they are produced by a sinusoidal oscillator. Then the vibrations created by the waves at any point in the medium are also sinusoidal. (This is easy to see if you imagine a string with one of its ends undergoing simple harmonic motion, so that the end is moving back and forth in a sinusoidal fashion.)

We learned in Chapter 9 that the energy of simple harmonic motion is $\frac{1}{2}kx_0^2$, where x_0 is the amplitude of the motion and k is the elastic constant. In that chapter we also developed an expression (Eq. 9–11) relating k to the frequency ν.

$$\nu = \frac{1}{2\pi}\sqrt{\frac{k}{m}}$$

Here m is the mass of the oscillator, a small element of the medium through which the wave is passing. Solving for k and substituting into the energy expression for SHM leads to

$$E = \frac{1}{2}m(2\pi\nu)^2 x_0^2 = 2\pi^2 m\nu^2 x_0^2.$$

We can express m as the volume density ρ times the volume of the small element in question, which is its cross-sectional area A times the length segment vt, where v is the wave velocity and t is the time required for the wave to pass through the element. Now substituting for m yields

$$E = 2\pi^2 \rho A v t \nu^2 x_0^2. \tag{10–6}$$

The power, the rate of energy transfer, is E/t, or

$$P = 2\pi^2 \rho A v \nu^2 x_0^2. \tag{10–7}$$

Note that both of these expressions vary as the square of the amplitude of the wave motion.

We define a new quantity for the transfer of energy in waves. This is the **intensity**, the power per unit area perpendicular to the direction of motion, or

$$I = \frac{P}{A} = 2\pi^2 \rho v \nu^2 x_0^2. \tag{10–8}$$

It is often the intensity that we measure or sense when receiving waves. For sound, this is the perceived volume, since the area of one's eardrum is fixed; for light, it is the the brightness of a source, again because the area of our eye is approximately constant (although for seeing and hearing the translation by

FOCUS ON PHYSICS

Waves in the Earth

Like any solid, the earth can transmit waves through its interior. The form and velocity of the waves are governed by the density and elasticity of the earth's material, just as waves in any medium would be. The earth's interior is complex, but the careful study of data from several decades of observation has allowed geophysicists to deduce a great deal about the internal structure of our planet.

As mentioned in the text, waves in the earth are called **seismic waves**. These waves are created by disturbances of various kinds, and usually die out after one passage through the earth. The most common disturbances that create seismic waves are earthquakes, although in principle any sudden deposition of energy in the earth would create them (in studying the moon, for example, astronauts created seismic waves with devices that thumped the surface).

A device for detecting seismic waves is called a **seismograph**. The first seismographs capable of detecting the waves created by distant earthquakes were developed in the late 1800s. A seismograph consists of three basic elements: an inertial member, a transducer, and a recorder (Fig. 10.1.1). The inertial member is a pendulum or a weight attached to a spring so that it is allowed to move freely in one dimension. When a seismic wave passes by the location of the seismograph, the weight tends to remain in place as the earth (and the rest of the instrument) oscillates. The transducer is a device, either mechanical or electromagnetic, that senses the relative motion of the weight and transmits the information to the recorder. A mechanical transducer might be a simple lever attached to the weight so that it pivots when a wave passes by. An electromagnetic transducer might consist of a coil of wire that moves back and forth in a magnetic field, so that an electric current is created by the wave oscillations (electromagnetic induction will

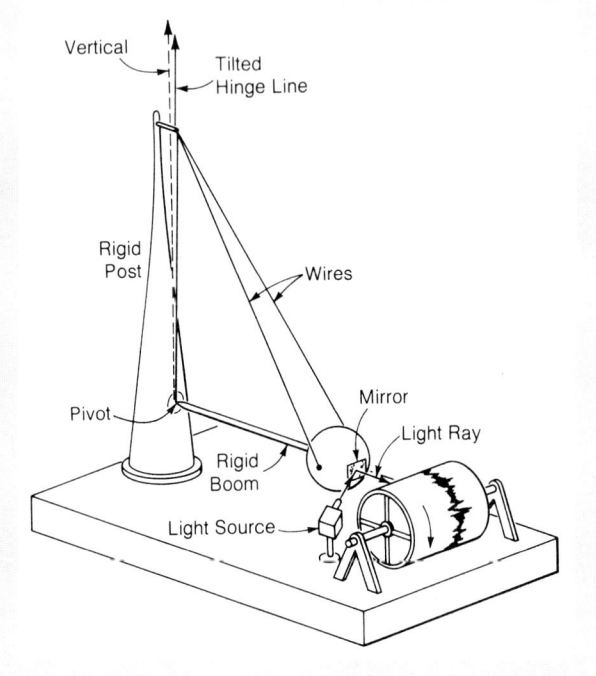

Figure 10.1.1

be discussed in Chapter 16). The recorder in a seismograph is a device that converts the signal from the transducer into a record which may be a simple chart or a computer memory. To measure wave motions in all three dimensions, three seismographs are needed.

The three types of waves described in this chapter are all created by earthquakes and recorded by seismographs. It always happens that the compressional waves, called P waves by geophysicists, arrive first, because they have the highest velocity. The transverse, or S waves arrive next, and the surface waves (L waves) arrive last. The speeds of all three types of waves are consistent,

10.1 THE NATURE OF WAVES

so that the distance to a faraway earthquake can be found quite accurately if the travel times of the waves are known (see Special Problem 2 and Fig. 10.32). Even if the time of the earthquake is not known initially, it can be deduced from the differences in arrival times of the different types of waves. (Imagine that two cars, one travelling at 40 mi/h and the other at 30 mi/h, leave the same point at the same time. You know that one outpaces the other at 10 mi/h, so if you note when they arrive at some distant point, you can deduce how long they have been travelling and how far they have come.)

The speeds of all three types of waves are constant as a function of distance travelled, but only up to a limit of about 11,000 miles. Then the *P* waves begin to be slowed, and the *S* waves stop arriving altogether. It was this observation that led geophysicists to realize that the earth has a liquid zone deep in its interior. In a liquid, compressional waves (*P* waves) travel more slowly than in a solid, and transverse waves (*S* waves) cannot propagate at all.

As records have accumulated from thousands of earthquakes, a detailed picture of the earth's interior structure has built up. Several distinct zones, distinguished by density, have been located in addition to the liquid zone (Fig. 10.1.2). There are three major zones, the **crust**, the **mantle**, and the **core**, which are themselves subdivided.

The crust varies in thickness from a few kilometers (under the oceans) to over 50 km (the continents). It consists primarily of silicate rocks, with an average density of roughly 3.0 g/cm^3, and with the uppermost part of the mantle it forms a rigid layer called the **lithosphere**. The lithosphere is broken into segments called plates, which float on the denser mantle. As the plates move about, they carry the continental landmasses. The surface of the earth is continually changing; during 100 to 200 million years the continents are rearranged.

At the bottom of the crust is an abrupt discontinuity where the density increases suddenly (seismic waves are partially reflected from this layer, called the **Mohorovicic discontinuity**). Below this is the mantle, which extends nearly halfway to the earth's center (its lower boundary is about 2900 km below the surface). Its upper portion (below the lithosphere), called the **asthenosphere**, is somewhat plastic and can be deformed. Slow flows of material in the asthenosphere, perhaps driven by convection (Chapter 12), are thought to be responsible for continental drift. The lower portion of the mantle, the **mesosphere**, is more rigid.

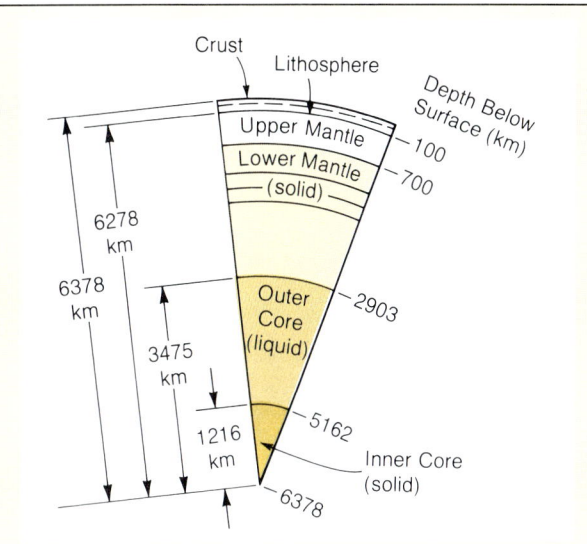

Figure 10.1.2

Below the mantle is the earth's core, consisting of an outer core extending to a depth of 5,100 km and an inner core reaching the earth's center, 6,378 km below the surface. The outer core, primarily iron and related metallic elements such as nickel and cobalt, is liquid. The major source of heat to melt the rock is natural radioactivity (Chapter 25). The earth's magnetic field is thought to result from electric currents in this liquid layer (see the *Focus on Physics* article in Chapter 16). The inner core has greater density than the outer core, and seismic waves travel more rapidly there, indicating that this region is solid. The density of the core is about 15 g/cm^3, giving the earth as a whole its average density of 5.5 g/cm^3.

Using seismic waves to deduce the internal structure of the earth is just one of a vast assortment of applications of the information discussed in this chapter. Waves of various kinds are involved in almost every aspect of our lives and our technology.

$A_1 = 4\pi r_1^2$

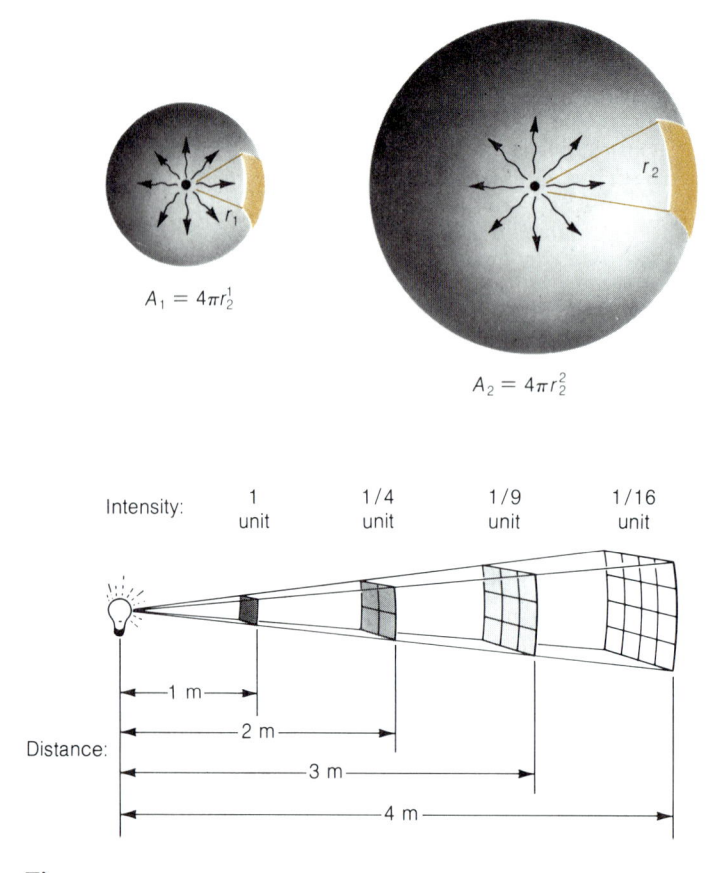

$A_2 = 4\pi r_2^2$

Figure 10.7

Intensity decrease with distance. For a source of waves propagating outward in all directions, the intensity at the surface of a sphere equals the energy emitted per second divided by the surface area of the sphere, $4\pi r^2$. Thus the intensity is proportional to $1/r^2$, where r is the radius of the sphere. This inverse square law is true for any source emitting radiation equally in all directions.

our nervous system into our brains is nonlinear, so that perceived loudness or brightness is not simply proportional to I).

It is interesting to note how the intensity varies with distance from a source of waves under different conditions. If a wave is three-dimensional, propagating outwards in all directions from some source, then the area over which the wave acts increases as the square of the distance from the source. Hence the intensity must decrease inversely as the distance squared (Fig. 10.7). The term in Eq. 10–8 that decreases is actually the amplitude x_o of the waves; the other quantities, such as the velocity, density, and frequency, are all constant. (Recall that the velocity has to be constant for fixed density; we also know that the frequency must be constant because in the last chapter we saw that it depends only on the displaced mass and the elastic constant, which are both intrinsic properties of the medium.) This inverse-square dependence on distance applies to light waves as well as to sound and other waves in this chapter.

Example 10.3 Suppose it is found that the minimum light intensity needed for reading without eye strain is 0.1 W/m². How far can a person sit from a 100 W light bulb and read without straining his eyes?

The intensity is the power divided by the surface area, which for a spherical surface is $4\pi r^2$. Therefore we have

$$I = \frac{P}{A} = \frac{P}{4\pi r^2}.$$

We solve for r, the distance from the source:

$$r = \sqrt{\frac{P}{4\pi I}}.$$

Substituting values for P, the power of the bulb, and I, the intensity, we have

$$r = \sqrt{\frac{100 \text{ W}}{4\pi(0.1 \text{ W/m}^2)}} = 8.9 \text{ m}.$$

10.2 Interactions of Waves

Waves can interact with their surroundings and with each other in many ways. In this section we explore some of them, setting the context for much of our later discussion of sound and light.

The Doppler Effect

We have discussed the wavelength and frequency of waves as though they were fixed quantities, dependent only on the nature of the source of waves. This is certainly true in the frame of reference of the source, but what an observer measures depends on the relative motion between source and observer. Many of us are familiar with the drop in pitch when a siren on a police car passes by. This shift in frequency is called a Doppler shift, and the entire phenomenon is called the **Doppler effect,** after the Austrian scientist who first explored its properties. The effect occurs for all waves, but for now we will discuss mostly sound waves. The Doppler effect for light and other forms of electromagnetic radiation differs from the form discussed here, and will be described when we analyze light waves.

In deriving the relationship between observed and emitted wavelengths and frequencies, we know that the medium through which the waves travel

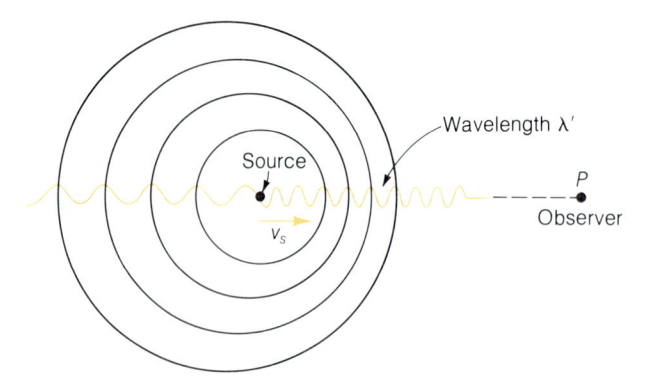

Figure 10.8

Doppler effect. A sound source moving to the right produces sound waves that are shifted in wavelength and frequency for a stationary observer. Here λ' is the observed wavelength for a listener at P.

provides a frame of reference that we can regard as fixed. The velocity of the waves in that frame of reference is determined only by properties of the medium (and is given by Eqs. 10–3, 10–4, and 10–5). We will treat two cases: (1) sources in motion with respect to the medium while the observer is at rest; and (2) sources at rest while the observer is in motion.

First we treat the case where the source of waves is moving, but the observer is at rest. The wavecrests become bunched up ahead of the source, and stretched out behind it (Fig. 10.8). Let us consider an observer ahead of the source. If λ is the rest wavelength of the waves, then the observed wavelength λ' is reduced by the distance the source travels between the emission of consecutive wavecrests. If the source velocity is v_s, then the observed wavelength is

$$\lambda' = \lambda - v_s T = \lambda - \frac{v_s}{\nu}$$

where T is the period and ν the frequency of the waves in the frame of the source. If we use Eq. 10–2 to substitute v/ν for λ and v/ν' for λ', then we have

$$\frac{v}{\nu'} = \frac{(v - v_s)}{\nu}$$

which can be rearranged to give us

$$\nu' = \left(\frac{\nu}{1 - \frac{v_s}{v}} \right) \qquad (10\text{–}9a)$$

Note that the expression is meaningless if the velocity of the source v_s equals

10.2 INTERACTIONS OF WAVES

or exceeds the sound velocity v. The expression depends in a fairly simple fashion on the relative velocity of the source and the waves. Suppose a siren has a rest frequency of 350 Hz, and is moving toward an observer at 70 mi/h = 31.3 m/s. The speed of sound in air is about 330 m/s, so the term v_s/v = 31.3/330 = 0.095. The observed frequency is then

$$\nu' = \frac{350 \text{ Hz}}{1 - 0.095} = 387 \text{ Hz}.$$

If we repeated this derivation for a source moving away from the observer, then the observed wavelength is

$$\lambda' = \lambda + \frac{v_s}{v}$$

and we would arrive at

$$\nu' = \left(\frac{\nu}{1 + \frac{v_s}{v}}\right). \qquad (10\text{–}9b)$$

For a police car moving at 70 mi/h, the observed frequency for a listener at rest behind the police car would be

$$\nu' = \frac{350 \text{ Hz}}{1 + 0.095} = 320 \text{ Hz}.$$

When the observer is in motion, but the source is at rest, there is a Doppler shift, but with a different form. The wavelength is not affected by the motion, since the source is at rest, but the wave velocity measured by the observer is affected, because the wave velocity is always fixed in the frame of the medium through which the waves travel. Solving Eq. 10–2 for ν and substituting $v' = v + v_o$ (where v_o is the velocity of the observer) yields

$$\nu' = \frac{v'}{\lambda} = \frac{(v + v_o)}{\lambda}$$

or

$$\nu' = \nu\left(1 + \frac{v_o}{v}\right) \qquad (10\text{–}10a)$$

if the observer approaches the source, and

$$\nu' = \nu\left(1 - \frac{v_o}{v}\right) \qquad (10\text{–}10b)$$

if the observer recedes from the source.

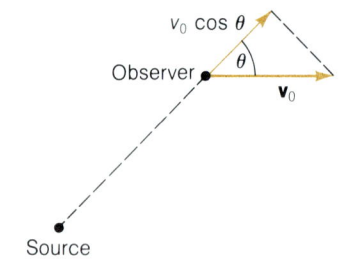

Figure 10.9

Doppler shift for an observer moving away from the source at angle θ to the line between source and observer. The observer's velocity in the Doppler shift calculation is $v_o \cos\theta$, the component along this line.

To see the effect quantitatively, we now assume the same police siren is at rest, and the observer travels first toward it, then away from it, at 70 mi/h. For the approaching observer, the observed frequency is

$$v' = (350 \text{ Hz})(1 + 0.095) = 383 \text{ Hz}$$

and for the receding observer, it is

$$v' = (350 \text{ Hz})(1 - 0.095) = 317 \text{ Hz}.$$

Note that the relevant velocity for a moving source or a moving observer is always the relative velocity along the line connecting source and observer. If the actual velocity is at some angle θ to the line connecting the two (Fig. 10.9), then we must substitute $v_s \cos\theta$ for v_s in Eq. 10–9a and Eq. 10–9b, and $v_o \cos\theta$ for v_o in Eq. 10–10a and Eq. 10–10b. If the motion is perpendicular to the line connecting the two, there is no Doppler shift at all. In most real situations, the angle between source and observer changes with time, so that the Doppler shift occurs over some time interval. When a police car passes a fixed observer, for example, most of the shift in frequency occurs during the interval when the $\cos\theta$ term is changing rapidly (that is, just as the car passes), but in fact there is a small shift occuring during the entire time the car is moving, because the angle θ is gradually changing. The Doppler shifts for source and observer in motion are summarized in Table 10.2.

Table 10.2

Summary of Doppler frequency shifts*

Source approaching	$v' = \dfrac{v}{\left(1 - \dfrac{v_s}{v}\right)}$	Observer approaching	$v' = v\left(1 + \dfrac{v_o}{v}\right)$
Source receding	$v' = \dfrac{v}{\left(1 + \dfrac{v_s}{v}\right)}$	Observer receding	$v' = v\left(1 - \dfrac{v_o}{v}\right)$

*Here, v_s and v_o are velocities of source and observer. The frequency emitted by source is v, and frequency received by observer is v'. The wave speed in the medium is v.

Example 10.4

A police car is travelling north at 55 mi/hr, siren blaring. The bad guys have just taken a fork in the road, so that they are travelling at 50 mi/h in a direction 35° east of north (Fig. 10.10). If the police siren has a frequency of 350 Hz, what frequency do the bad guys hear when the police car is at the fork in the road?

Here we have both kinds of Doppler effects at work: the police car is moving with respect to the medium, and so is the observer. Furthermore, the relative velocities are not directly along the line connecting the source and the observer. We first calculate the shift due to the motion of the source, which

10.2 INTERACTIONS OF WAVES

will tell us the frequency of the waves in the fixed frame of reference; then we will calculate the additional shift due to the motion of the observer. The component of the police car's velocity that is directed along the line towards the bad guys is (24.6 m/s)cos(35°) = 20.2 m/s, so the shifted frequency due to the motion of the source is given by Eq. 10–9a

$$\nu' = \frac{350 \text{ Hz}}{\left(1 - \frac{20.2 \text{ m/s}}{330 \text{ m/s}}\right)} = 373 \text{ Hz}.$$

This is the frequency the bad guys would observe if they were at rest. They are not, however; they are moving at 50 mi/h, so the component of their velocity along the line connecting them to the police car is (22.4 m/s)cos(35°) = 18.3 m/s. The Doppler-shifted frequency the bad guys observe is found by applying Eq. 10–10b (since they are receding with respect to the rest frame) to the frequency in the rest frame.

$$\nu'' = \nu'\left(1 - \frac{v_o}{v}\right) = (373 \text{ Hz})\left(1 - \frac{18.3 \text{ m/s}}{330 \text{ m/s}}\right) = 352 \text{ Hz}$$

As we might have expected, the result is a very small shift in frequency, because the bad guys and the police are moving at nearly the same velocity. Note that the bad guys could tell whether or not the police took the fork by carefully measuring the frequency of the siren to see whether the small Doppler shift when the police car changed direction did occur. Note also that there would be no net Doppler shift if the police and the bad guys were travelling at the same speed and in the same line because their relative velocity is zero (as can be seen by solving the problem for $v_s = v_o$ and $\theta = 0$).

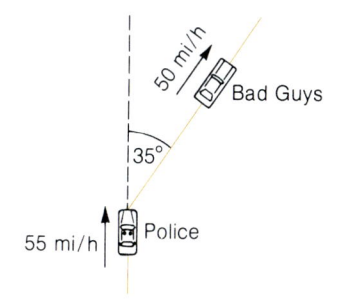

Figure 10.10

In many applications we write an expression for the amount of Doppler shift in frequency that occurs, which is defined as

$$\Delta\nu = \nu' - \nu. \tag{10–11}$$

When Eq. 10–9a or Eq. 10–9b is inserted for ν' (assuming that $\Delta\nu$ is much smaller than ν), we find

$$\frac{\Delta\nu}{\nu} = \frac{\pm v_s}{v} \tag{10–12}$$

with the sign determined by whether the source approaches or recedes from the observer (the Doppler shift in frequency, as defined by Eq. 10–11, is positive for approach and negative for recession). Similarly, Eqs. 10–10a or 10–10b, when substituted into Eq. 10–11, yield

$$\frac{\Delta\nu}{\nu} = \frac{\pm v_o}{v} \tag{10–13}$$

with no need to assume $\Delta \nu$ small. We have already seen from our examples that the expressions for $\Delta\nu/\nu$ are very similar when v_s or v_o is small compared to v. The shift is almost the same whether it is the source or the observer that moves.

For light, no medium is required, and there is no preferred frame of reference (this is one of the fundamental postulates of relativity theory). The simple expression for the Doppler shift of light (that is, for the case where the relative velocity v is much less than the speed of light $c = 3 \times 10^8$ m/s) is

$$\frac{\Delta \nu}{\nu} = \frac{\pm v}{c}. \tag{10-14}$$

This will be discussed and applied in Chapter 22.

Waves and Boundaries

Waves can interact with boundaries or surfaces that they meet. When a wave strikes a boundary between one medium and another, some of it is reflected and some of it enters the new medium (Fig. 10.11). If the boundary has an edge, part of the wave will bend around that edge (Fig. 10.12). We speak therefore of **reflection, refraction,** and **diffraction** of waves.

To analyze what happens in all three cases, it is helpful to imagine a periodic wave as a series of wavefronts; that is, parallel lines representing the wavecrests, much like ocean waves approaching the shore (Fig. 10.13). The **angle of incidence** is the angle of the incoming wavefront with respect to a boundary, and the **angle of reflection** is the angle of the outgoing wavefront with respect to the boundary. In both cases the standard convention is to define these angles as the angles between a perpendicular to the front (called a **ray**) and a line normal to the boundary. Similarly, the **angle of refraction**, the angle of the transmitted wavefront with respect to the boundary, is defined as the angle of the transmitted ray to the normal to the boundary.

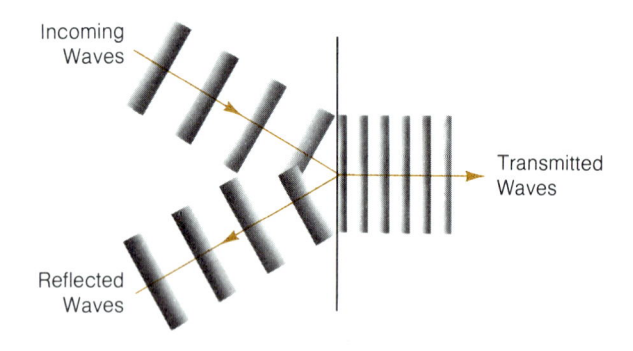

Figure 10.11

Reflection and transmission of waves. Here sound waves strike a boundary between two media. The waves are partially reflected and partially transmitted.

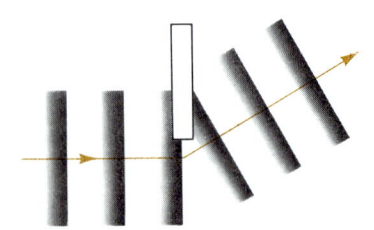

Figure 10.12

Diffraction. Waves will bend (diffract) around the edge of an obstruction.

10.2 INTERACTIONS OF WAVES

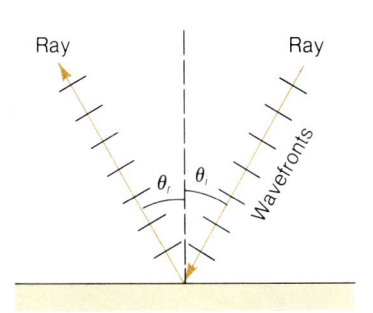

Figure 10.13

Wave fronts and reflection. The directions of motion of the incident and reflected waves are indicated by rays, straight lines perpendicular to the wavefronts. The angles of incidence and reflection are θ_i and θ_r.

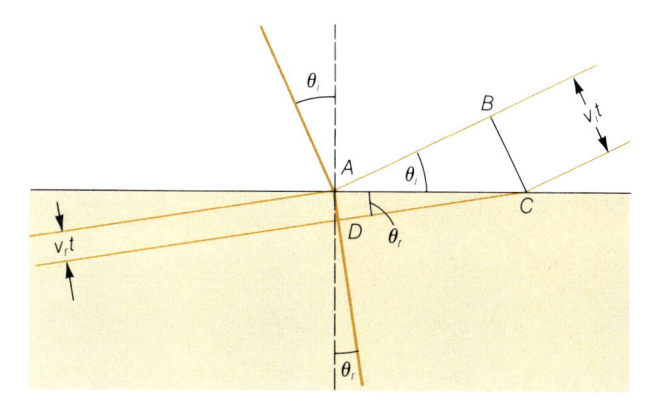

Figure 10.14

Refraction. Here a single wavefront is shown at two times as it passes from one medium into another. Refraction (bending) occurs because of the difference in the index of refraction.

There is a very simple relationship between the angle of incidence and the angle of reflection. If θ_i is the former and θ_r the latter, then

$$\theta_i = \theta_r. \tag{10-15}$$

This is sometimes known as **Snell's law,** and will be discussed further in Chapter 19. The wavelength and frequency of the waves are unaffected by the reflection, except when the reflecting boundary is moving. When the boundary moves there is a double Doppler effect: the boundary receives a shifted frequency, because it is in motion (so Eq. 10–10a or 10–10b applies) and the re-emitted wave is shifted further because the boundary acts as a moving source (so Eq. 10–9b or 10–9a must also be applied).

Refraction is more complicated than reflection, because the two media will in general have different properties, so that the wave velocity on one side of the boundary will differ from that on the other side. If we visualize a series of wavefronts in a medium where the wave velocity is v_i entering a medium where the velocity is v_r, then we can relate the angle of incidence θ_i to the angle of refraction θ_r. Fig. 10.14 shows a single wavefront at two times separated by t seconds. In triangles ABC and CDA we see that

$$\frac{\sin\theta_r}{v_r t} = \frac{\sin\theta_i}{v_i t}$$

or

$$\frac{\sin\theta_r}{\sin\theta_i} = \frac{v_r}{v_i}. \tag{10-16}$$

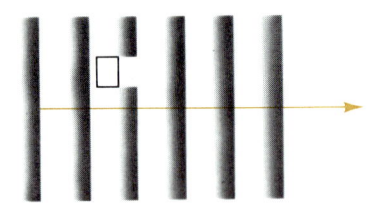

Figure 10.15

Diffraction when the obstacle is small compared to λ.

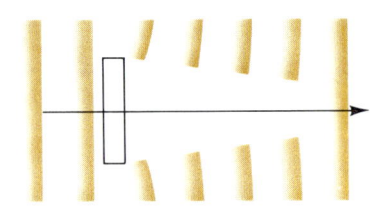

Figure 10.16

Diffraction when the obstacle is large compared to λ.

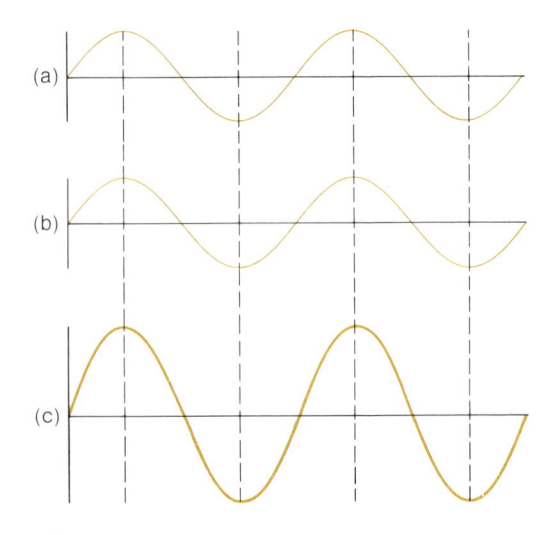

Figure 10.17

Constructive interference. When two waves (a) and (b) are in phase their amplitudes add (c).

The most common examples of refraction have to do with light; we are all familiar, for example, with the distortion of direction that occurs when we view an object that is underwater. We defer most of our treatment of refraction to Chapter 19.

Diffraction occurs when waves encounter an obstacle. This is a simple geometric effect; it is clear that wavefronts emanating from a point must be circular, and that those emitted by any other finite source will also spread as they propagate. It is also true that if parallel wavefronts are interrupted at some point by an obstacle, those that pass by bend around it, as though the obstacle defined a finite source (Fig. 10.12). The amount of bending depends on the relative size of the obstacle and the wavelength of the waves. If the obstacle is much smaller than the wavelength, then there is little effect (Fig. 10.15). If the obstacle is much larger than the wavelength, then there is a large effect (Fig. 10.16).

As in refraction, most of the practical applications of diffraction involve light, particularly light waves interacting after passing by some obstacle. For now we note that diffraction is unique to waves; that a stream of particles would not bend around a corner like a series of waves. The fact that light exhibits diffraction therefore had a great bearing on the early controversy over whether or not light consists of waves.

Interference of Waves

When two waves encounter each other while travelling through a medium, the result is a wave whose amplitude at the point of encounter is the algebraic sum of the amplitudes of the two waves. This is known as the **superposition**

10.2 INTERACTIONS OF WAVES

Figure 10.18

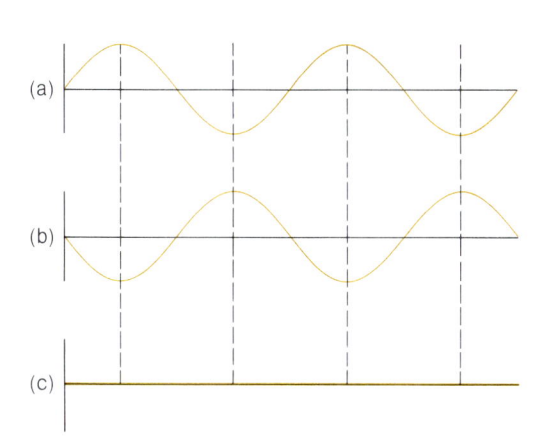

Destructive interference. Two waves of equal amplitude (a) and (b) are perfectly out of phase. The result is zero amplitude (c).

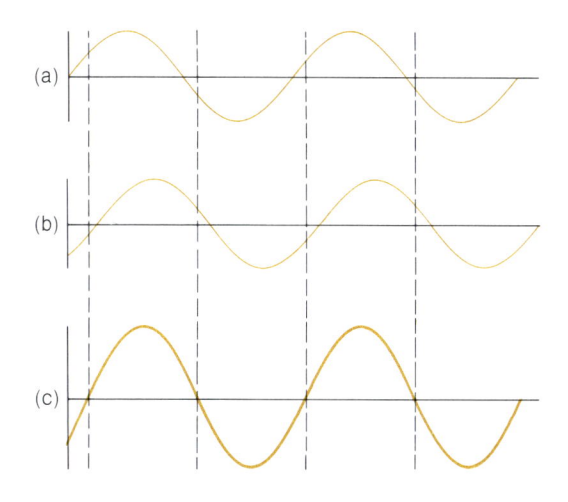

Figure 10.19

Partial interference. Waves (a) and (b) are out of phase. Their superposition (c) has the same frequency but its amplitude depends on the phase difference.

theorem. If, for example, two waves are travelling along a string so that the displacement of one of them at point x is $+0.002$ m, and the displacement of the other at the same point is -0.004 m, the actual displacement of the string at x will be -0.002 m. When the displacement is enhanced by superposition of two waves, we say that the waves undergo **constructive interference**. When the displacement is reduced, it is **destructive interference**.

If we consider two waves of equal amplitude and frequency moving in the same direction, we see that the nature of the interference depends on the relative positions of their crests and troughs when they are superposed. If the crests coincide, then we have purely constructive interference, and the result is a wave with the original frequency but twice the original amplitude (Fig. 10.17). In this case, we say that the waves are in phase. If the crests of one wave coincide with the troughs of the other, then we have purely destructive interference, and the sum is zero (Fig. 10.18). In this case the waves are perfectly out of phase; that is, out of phase by one half-cycle. If they are out of phase by a some other fraction of a cycle, then the result is a wave with the same frequency as the original waves, but with an amplitude smaller than the sum of the original individual amplitudes (Fig. 10.19).

It is clear that, if we have two continuous waves with different wavelengths, the effect of the superposition will vary with position. At some locations, the two waves will both have peak positive displacements, so that the sum is a very large positive displacement; at other places, both will have peak negative displacements, so that the superposition results in a very large negative displacement. At still other places, the positive displacement of one wave will coincide with the negative displacement of the other, and the net displacement will be small or zero.

If we consider a fixed point and plot the vibrations at that point due to

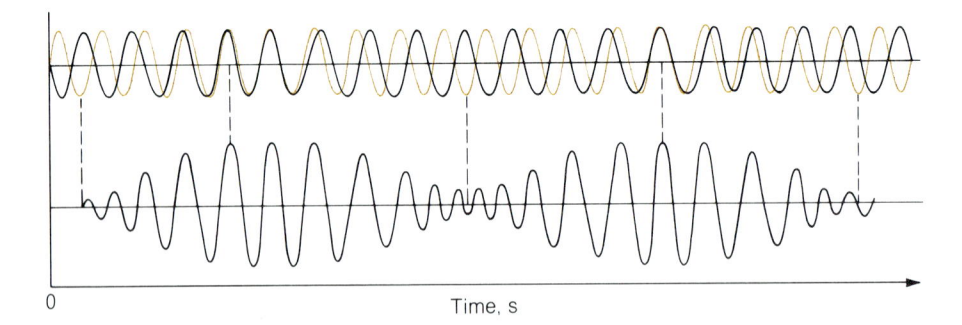

Figure 10.20

Superposition of unmatched frequencies. A 20 Hz wave and an 18 Hz wave interfere. The result is a wave with a 2 Hz beat frequency.

the passage of the two continuous waves, we see that the vibrations have a maximum amplitude at regular intervals. The frequency at which these maxima occur is simply the difference in frequency of the two waves. For example, if one wave has a frequency of 20 Hz, and the other a frequency of 18 Hz, the two waves will combine constructively to produce a maximum summed amplitude at a rate of 2 Hz (Fig. 10.20). If we are considering sound waves, then we see that a listener subjected to two slightly different frequencies will hear a regular modulation of the sound intensity (the loudness), at a frequency given by the difference of the two frequencies. Such a modulation is called a **beat**, and it can occur with any wave. The beat frequency in sound waves plays a role in determining the quality of music.

10.3 Stationary Waves

Interference occurs whenever waves are reflected from a boundary. In perfect reflection, the frequencies and amplitudes of incoming and outgoing waves are equal. The phase relationship of the incoming and outgoing waves determines the nature of the superposition. Ordinarily there will be a complex mixture of destructive and constructive interference, and no simple pattern appears.

There are certain wavelengths, however, for which a fixed pattern develops. These are wavelengths for which the number of waves between the source and the reflecting boundary is an integral (or half-integral) number, so that the locations where the displacement is zero in the wave and the reflected wave coincide (Fig. 10.21). There is no vibration at these positions, which are called **nodes**. The nodes remain fixed, and the wave motion consists simply of the oscillation of points between them, which are called **antinodes**. A fixed wave pattern such as this is called a **standing wave**, or **stationary wave**.

It is perhaps easiest to visualize standing waves by considering a string stretched between two fixed ends (Fig. 10.22). We note first that whenever a wave travelling along the string is reflected at the end, there is a phase shift

10.3 STATIONARY WAVES

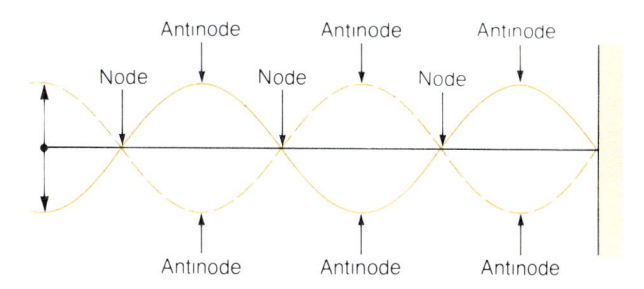

Figure 10.21

A standing wave. The incident and reflected waves are one half-cycle out of phase, so that a standing wave is created with nodes as shown.

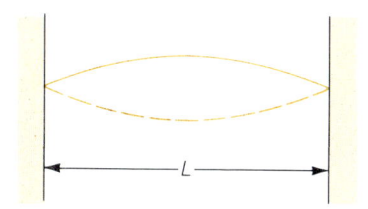

Figure 10.22

A standing wave with $\lambda = 2L$. This is the fundamental mode.

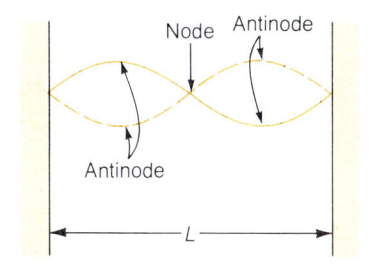

Figure 10.23

The first overtone is a standing wave with $\lambda = L$.

of one half-cycle as the wave is reflected; that is, the reflected wave is identical to the incident wave, but out of phase by one half-cycle. (This is sometimes called an inversion, because the the wave form is simply turned upside down upon reflection.) If the incident wave is exactly at a half or full cycle when it is reflected, then the reflected wave is in phase with the incident wave. This is the condition for a standing wave.

Now consider a string of length L. If a wave of wavelength $2L$ is established, then the resulting standing wave has only a single antinode, with nodes at either end (Fig. 10.22). This is the standing wave of longest wavelength, or lowest frequency, that can be established for a given L, and is referred to as the **fundamental mode** of vibration of the string. If the wavelength is now reduced, the next value for which a standing wave will be established has wavelength L. In this case, there is a node in the middle of the string, and two antinodes are between the nodes and the ends (Fig. 10.23). This is referred to as the **first overtone mode**. As we decrease the wavelength further, we find other standing wave modes when L is equal to an integral number of half-wavelengths. The condition for standing waves is

$$L = \frac{n\lambda_n}{2} \tag{10-17}$$

where $n = 1, 2, 3, \ldots$ is an integer that corresponds to the mode, and λ_n is the wavelength of the standing wave in that mode. The number of nodes is always equal to $(n + 1)$; recall that the two ends of the string are always nodes because they are fixed. The frequency of each mode, from Eq. 10–2, Eq. 10–3, and Eq. 10–17, is

$$\nu_n = \frac{nv}{2L} = \frac{n}{2L}\sqrt{F_T\left(\frac{L}{M}\right)}. \tag{10-18}$$

Although we discussed standing waves in terms of vibrating strings, similar equations apply to any kind of waves reflecting back and forth in a confined space (the exact treatment depends on the geometry of the situation; our discussion is confined to one-dimensional situations). There will always be standing waves possible at a variety of wavelengths given by Eq. 10–17 or frequencies given by Eq. 10–18.

The relationship between frequency and wavelength depends on the velocity of the waves, which in turn is determined by the properties of the medium, so the disturbing frequency required to create standing waves depends on the medium as well as the size of the confined region. Thus there is a natural wavelength (or frequency) at which waves can propagate easily in any finite medium, as well as many other possible overtone frequencies.

When a standing wave is established with one of the natural frequencies of the medium, it has a larger amplitude because the wave constructively interferes with its reflections. Waves of other frequencies will not have large amplitudes in the medium, because there is little constructive interference. If a string is plucked, waves of many frequencies are created. All the waves except for the ones which obey Eq. 10–17 and Eq. 10–18 will quickly die out. This is an example of **resonance,** and the frequencies given by Eq. 10–18 are called **resonant frequencies.** Waves with resonant frequencies are easy to create; to see this, image drawing a bow across the strings of a violin. Because of the jerky motion as the bow alternately sticks and slips, vibrations with a broad range of frequencies are created in the string. The only waves that persist in the strings, however, are those whose frequencies match the natural frequencies of the strings. Therefore the pitch of the sound produced by a string is determined by its natural, or resonant, frequencies.

We say that a resonance is **sharp** if the vibrating system responds strongly only to frequencies very close to the natural frequency. Nearly all energy of vibration is contained in vibrations very close to the resonant frequency. The sharper the resonance, the longer the vibration will persist before dying away. We know that a piano string will vibrate for a considerable time after being struck, so we infer that its resonance is a sharp one. In the next section, we will discuss at greater length the principles of musical instruments.

Example 10.5   Suppose a violin string has a fundamental frequency of 659 Hz. What is the frequency of its fourth overtone?

The fourth overtone means that the mode is $n = 5$ (recall that the first overtone is the first mode above the fundamental frequency, for which $n = 1$). We know that for $n = 1$, the frequency is $\nu = 659$ Hz. We recognize by inspection of Eq. 10–18 that the frequency for $n = 5$ is simply five times that for $n = 1$, so we conclude quickly that the frequency of the fourth overtone is

$$\nu_5 = 5\nu_1 = 5(659 \text{ Hz}) = 3{,}295 \text{ Hz}.$$

10.4 Sound Waves

We have mentioned sound waves here and there throughout this chapter, but have deferred discussions of several important aspects. Here we will describe not only the waves, but also the mechanisms, human and otherwise, that produce and receive them.

The Nature of Sound Waves

As we said earlier, sound waves are compressional waves, consisting of longitudinal vibrations of molecules. The waves can be described in terms of molecular motion, the to-and-fro oscillation of molecules as a disturbance travels through a medium, or they may be described as pressure enhancements and rarefactions that travel through a medium (Fig. 10.24). In either case, we speak of a wavelength, a wave speed, and a frequency, all defined in the usual sense. There is a phase difference between the two types of waves, however. The pressure is at its maximum value at a given position when the displacement of the molecules is zero, and vice versa. The two quantities are shifted by a quarter cycle with respect to each other. We will speak of the displacement wave when discussing resonance, but will refer to the pressure variation when we discuss energy transfer and intensity.

Sound waves are often thought of as occurring in air, but the term is used to describe compressional waves in any medium. As we mentioned earlier, only compressional waves can propagate in a liquid or a gas, while in a solid it is also possible to have transverse waves. In general when a solid is subjected to an impulse, both compressional and transverse waves are created, but the compressional waves travel faster. We will include only compressional waves in this treatment of sound.

The speed of sound, as we have already seen, is a function of the bulk modulus and the density of the medium. The expression for sound speed is

$$v_s = \sqrt{\frac{B}{\rho}} \qquad (10\text{-}19)$$

where B is the bulk modulus and ρ is the density. Intuitively this expression makes sense, because the bulk modulus is a measure of a material's resistance to compression, so a large bulk modulus implies that the material is stiff and returns quickly to its equilibrium state when deformed. The density, on the other hand, is a measure of the inertia of the material that must be deformed or displaced, so it follows that the sound speed should decrease with increasing density. The bulk modulus in this expression is called the **adiabatic bulk modulus,** which for some materials is significantly different from the **isothermal bulk modulus** that is often tabulated. The term "adiabatic" means that no energy is transferred to the surroundings as a fluid element is compressed; that is, the compression and subsequent rarefaction occur too quickly for any significant heat exchange to take place with the surroundings. The temperature

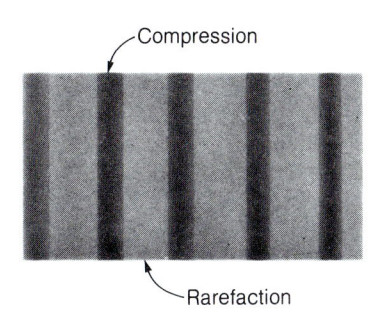

Figure 10.24

Sound waves are compressional waves consisting of alternating compressed and rarefied regions.

Table 10.3
The speed of sound*

Substance	Velocity (m/sec)
Wood (along grain)	3300–4700
O_2	316
†Air (0°C)	331
†Air (20°C)	343
N_2	334
Glass (flint)	3980
Glass (pyrex)	5640
Iron or Steel	5000–6000
Aluminum	6420
Helium	969
Rubber, hard	1450
Water (distilled)	1497
Water (sea)	1531
Lucite	2680
Copper	5010
Diamond	18100

*Longitudinal compression waves, 25°C unless noted
†$v_s = (331 + 0.6T)$ m/s, T in °C

at each point in a medium rises and falls as sound waves travel through. The isothermal bulk modulus, on the other hand, is valid when there is sufficient time for heat exchange to occur, so that the temperature of the medium stays uniform.

The speed of sound for a few media is tabulated in Table 10.3. The speed given for air is for a temperature of 0°C. Because the density of air varies with temperature, the sound speed does also. As a rule of thumb, the speed in air increases by 0.6 m/s for every degree above 0°C: $v = (331 + 0.6T)$ m/s for T in °C. The dependence of the sound speed on temperature is smaller for solids than for gases.

While we have said that all compressional waves are considered sound waves, for most practical purposes, when we speak of sound we mean waves that result in audible sensations in the human ear. This defines a restricted class of sound waves: those in the frequency range detectable to the ear, which is about 20 Hz to 20,000 Hz for most people. The pitch of a sound is a measure of its frequency; we say that a sound has high pitch if its frequency is high, and low pitch if the frequency is low. Sound waves with frequencies higher than those detectable to the ear are called **ultrasound** or **ultrasonic waves.** These waves are reflected easily at boundaries where the density is near that of living tissue; hence ultrasonic waves are useful to examine the internal structure of the human body. Ultrasonic waves can be used to establish standing waves, hence resonant vibrations, in hard solids such as gemstones and false teeth, so ultrasonic cleaning chambers are effective to shake dirt loose from such objects. At the other end of the spectrum of sound wave frequencies, those below the threshold for human detection, are **infrasonic waves.** These can fall into the frequency range of natural vibration frequencies of large

10.4 SOUND WAVES

Table 10.4
Intensities of sounds

	dB	W/m^2
Jet at 30m	140	100
Indoor rock concert	120	1
Threshold of pain	120	1
Siren at 30m	100	1.0×10^{-2}
Riveter	95	3.2×10^{-3}
Auto interior (90 km/h)	75	3.2×10^{-5}
Busy street traffic	70	1.0×10^{-5}
Ordinary conversation	65	3.2×10^{-6}
Quiet radio	40	1.0×10^{-8}
Whisper	20	1.0×10^{-10}
Rustle of leaves	10	1.0×10^{-11}
Threshold of hearing	0	1.0×10^{-12}

structures such as buildings, and therefore must be taken into account in building design, so that destructive resonances are prevented.

We noted earlier that the energy carried by waves of any kind is proportional to the square of the amplitude. We have defined the energy transmitted by waves, and the power and intensity. When dealing with sound, it is most often the intensity that we are concerned with, because this is the quantity that determines how loud something is. The intensity is a measure of the energy transmitted per unit time (that is, the power), and also takes into account the distance from the source. If we define sound waves in terms of their pressure variations, so that ΔP is the amplitude, then the intensity is

$$I = \frac{(\Delta P)^2}{2\rho v_s}. \qquad (10\text{–}20)$$

Typical sounds that are comfortable to the human ear have pressure amplitudes up to about 30 N/m^2 (in contrast with the atmospheric pressure at sea level, which is just over 10^5 N/m^2), hence intensities up to about 1 W/m^2. Because of the logarithmic response of the human ear, sound intensities are often measured in **decibels,** which are logarithmic units defined so that

$$\beta = 10 \log \frac{I}{I_0} \qquad (10\text{–}21)$$

where β is the intensity in decibels, and I/I_0 is the relative sound intensity with $I_0 = 10^{-12}$ W/m^2. This reference value is arbitrary, but corresponds roughly to the lower limit detectable to the human ear (Table 10.4). The ear has a remarkable range of sensitivity, covering a full 12 orders of magnitude, or a factor of a trillion, from the faintest sounds detectable to the loudest tolerable. In decibels, this range is 0 to about 120 dB.

Example 10.6 Suppose a speaker at a rock concert has an acoustic power output of 1,000 W. What intensity, in W/m² and in dB, does a concertgoer seated 50 m from this speaker hear? What are the limits of the comfort zone where the intensity is less than the upper limit tolerable to the human ear, but greater than the detection threshold?

At distance r from the speaker, the intensity is the speaker power divided by $4\pi r^2$, since the intensity varies according the the inverse square law. Thus, at $r = 50$ m, the intensity is

$$I = \frac{P}{4\pi r^2} = \frac{1{,}000 \text{ W}}{4\pi(50 \text{ m})^2} = 0.032 \text{ W/m}^2.$$

In decibels, this is

$$\beta = 10 \log(I/I_0) = 10 \log\left(\frac{3.2 \times 10^{-2}}{10^{-12}}\right) = 105 \text{ dB}.$$

The near boundary of the comfort zone is the distance from the speaker where the intensity is at the upper limit of human tolerance, which we have taken as about 1 W/m² (or 120 dB). The far boundary is the distance where the intensity has dropped to the limit of detectability, which is an intensity of about 10^{-12} W/m² (or 0 dB). If we solve our expression above for r, we find

$$r = \sqrt{\frac{P}{4\pi I}}.$$

For $I = 1$ W/m², this leads to

$$r = \sqrt{\frac{1{,}000 \text{ W}}{4\pi(1 \text{ W/m}^2)}} = 8.9 \text{ m}.$$

For $I = 10^{-12}$ W/m², we find

$$r = \sqrt{\frac{1{,}000 \text{ W}}{4\pi(10^{-12} \text{ W/m}^2)}} = 8.9 \times 10^6 \text{ m} = 8{,}900 \text{ km}.$$

Of course, it is impossible to hear a rock concert at such a great distance, because the normal noise background has much higher intensity than the threshhold for detection.

Resonances and Cavities

We might expect that in a finite medium sound waves can become standing waves if certain conditions are met. If we consider the longitudinal oscillations of air molecules, we see that at a boundary there can be no oscillation. Therefore

10.4 SOUND WAVES

any solid boundary must be a node. Hence a completely enclosed pipe would have nodes at both ends, identical to the situation for a string fixed at both ends. We usually do not consider sound waves within completely enclosed pipes, however.

Let us consider what happens at the open end of a pipe. Molecules are most free to oscillate longitudinally at this point because there is no restraint; hence the open end of a pipe is an antinode. That is, in a pipe with an open end, the natural wavelengths for standing sound waves are those with antinodes at the open end. If we have a pipe that is open at both ends, natural frequencies resonance occur when the length of the pipe is a multiple of one half-wavelength. This is exactly the same statement we made for a vibrating string fixed at both ends except that for sound waves in an open pipe, there is a phase shift of one quarter-cycle compared to the string (Fig. 10.25). The natural frequencies for an open pipe are

$$v_n = \frac{nv_s}{2L}, \quad n = 1, 2, 3, \ldots \tag{10-22}$$

where n is the mode number as before, and v_s is the speed of sound. This is identical in form to Eq. 10–18, except that v_s has been substituted for the expression for the wave speed in a string, $\sqrt{F_T(L/M)}$.

If a pipe is closed at one end and open at the other, however, there is a node at one end and an antinode at the other, so that standing waves will occur only when the length of the pipe is equal to an odd number of quarter-wavelengths. Thus, the fundamental frequency corresponds to a wavelength equal to four times the length of the pipe, and the overtone frequencies are odd multiples of this. Therefore the naural frequencies for a pipe that is open at one end are

$$v_n = \frac{nv_s}{4L}, \quad n = 1, 3, 5, \ldots \tag{10-23}$$

where only odd values of n are allowed.

When subjected to impulses with a range of frequencies, a pipe will resonate only at its natural frequencies. Hence by controlling the length of a pipe, it is possible to tune it so that it resonates only at certain frequencies or pitches. This is one technique used to construct musical instruments.

Fundamental

First Overtone

Second Overtone

Figure 10.25

Standing waves in a tube open at both ends.

Example 10.7 In the good old days, carbonated beverages were sold in small glass bottles. Back then, any kid knew that if you blew into the open mouth of the bottle, it would resonate, producing a really nifty sound (the driving vibration in this case is turbulence created by the flow of air across the mouth of the bottle). If we approximate one of those bottles by a pipe that is closed at one end and open at the other, and whose length is 20 cm, at what fundamental frequency will it resonate if empty? If half-full? In each case, what is the first overtone frequency? Assume $v_s = 330$ m/s.

If the bottle is empty, its effective length is $L = 20$ cm $= 0.2$ m. From Eq. 10–23 we find the fundamental frequency ($n = 1$) to be

$$\nu_1 = \frac{v_s}{4L} = \frac{330 \text{ m/s}}{4(0.2 \text{ m})} = 413 \text{ Hz}.$$

When the bottle is half full, its effective length is 10 cm $= 0.1$ m, so the fundamental frequency is

$$\nu_1 = \frac{330 \text{ m/s}}{4(0.1 \text{ m})} = 825 \text{ Hz}.$$

In each case, we find the first overtone by substituting $n = 3$ into Eq. 10–23, since this is a pipe with one open end, so that even values of n are not resonant frequencies. Hence the first overtones are

$$\nu_3 = \frac{3v_s}{4L} = 1{,}238 \text{ Hz}$$

for the empty bottle, and

$$\nu_3 = 2{,}475 \text{ Hz}$$

for the half-full bottle.

The Music of Sound

While much of our definition and enjoyment of music is subjective, there are some basic principles that can be analyzed objectively. Physics can say nothing about why some music is good and other music is not, because that distinction is largely cultural. We can, however, describe some of the characteristics of certain musical standards and of musical instruments.

Most instruments operate by exciting resonances, so most music consists of combinations of frequencies that are overtones of a few fundamental frequencies. A piano, for example, normally has 88 keys, each of which is connected to a hammer that strikes a taut string. Each string has a fundamental frequency and a series of overtone frequencies, so that when a number of keys are depressed at once, the result is a complex sound consisting of the fundamental frequencies and their overtones. Certain relative frequencies from key to key have become recognized as harmonious, and by long-standing convention, a piano is turned so that these relative key-to-key frequencies are maintained.

Certain frequency ratios have been given names. An **octave** is defined so that if two frequencies differ by a factor of two, they are one octave apart. The keys of a piano are arranged into seven groups of notes (plus four extra notes), each group being one octave away from the adjacent one. The notes themselves are arranged in frequency ratios according to a **scale**, a standard set of relative frequencies. The note C is usually taken as the representative note for each group; thus, the lowest C on a piano is a factor of 2^7 lower in frequency than the highest C. Middle C, the C note of the third group, has by convention a frequency of 264 Hz (then the note A in the same octave has the frequency 440 Hz; tuning is normally done by calibrating from middle A). A piano is tuned by adjusting the tension on the string corresponding to middle A to match its pitch with that of a standard emitter of sound, such as a tuning fork, at a frequency of 440 Hz. Even though most people may not be able to distinguish between sounds with small differences in pitch when played separately, it is easy to hear the beats when the piano and the tuning source are played together. Once middle A has been calibrated, the other notes in its group and the other As can be tuned by eliminating beats between middle A and overtones of the other keys. In an orchestra, the oboe is usually taken as the tuning standard for all the other instruments. Each musician matches the frequency of his instrument to that of an A played on the oboe.

In addition to the octave, there are other frequency intervals that have been recognized as harmonious and given special names. A **fifth**, for example, is a pair of tones whose frequencies have the ratio 3:2. A **major third** is a pair whose ratio is 5:4, and a **minor third** has the ratio 6:5. The notes within each grouping on a piano keyboard are calibrated to have these and other pleasing ratios of frequencies. In general, pleasant intervals are combinations of frequencies whose ratios are the ratios of two small integers.

The timbre or quality of a musical note is another characteristic that can be discussed objectively. Two different musical instruments can have the same fundamental frequency, but will sound very different when this frequency is played. Anybody can distinguish between a piano and a saxophone, for example, even though both may be playing the same note. The contrast lies in the number of overtones that are excited at the same time, and their amplitudes relative to that of the fundamental.

Most musical instruments consist of a **generator** and a **resonator**. The generator creates sound waves, usually with a range of frequencies, and the resonator amplifies only selected frequencies. Generators may consist of strings, as in a piano or violin, or openings where turbulence is created by blowing air across the opening, as in a flute, or by vibrating reeds or plates, as in a clarinet. Resonators can consist of cavities in which standing waves are created at resonant and overtone frequencies, as in a xylophone. In some instruments, both the generator and the resonator undergo resonances that govern the sound. In a violin, for example, the strings are tuned to certain resonant frequencies by adjusting the tension in the strings, while the body of the violin is sized and shaped so that certain frequencies and their harmonics are emphasized.

The human voice has as generator the vocal cords, a remarkable assembly of elastic membranes whose tension is variable. They vibrate when air passes through them, and their frequency is governed by changes in their tension. The tonal quality is controlled by the vibrating air column in the throat and mouth, whose resonant frequencies are also highly variable with modifications in the size and shape of the mouth opening.

FOCUS ON PHYSICS

Musical Harmony and Scales

Harmony in music is based on ratios of frequencies. People tend to find notes harmonious if the ratio of their frequencies equals the ratio of two small integers. For example, the ratios 2/1 (octave), 3/2 (fifth), 5/4 (major third), and 6/5 (minor third) are basic constituents of classical harmony. The reasons for the pleasing effect have their origins in the philosophy of the Pythagoreans, but more importantly in the physiological effect that beats between overtones are minimized when the basic frequencies have these ratios. The human ear perceives beats as dissonant. But when a note, its relative major third, and its relative fifth are played together, a pleasing chord known as a **major triad** is heard. Frequency ratios of somewhat larger integers, such as 9/8 or 10/9 are perceived as dissonant.

A scale is a collection of notes (frequencies) spanning the octave ratio of 2. The best-known scale in Western music is the **diatonic scale,** which consists of 8 notes (do-re-mi-fa-sol-la-ti-do). The second *do* completes the octave, having twice the frequency of the lower *do*. Each note is given a letter (A, B, C, D, E, F, G). Table 10.2.1 shows the diatonic C-major scale. The difference in pitch between each note is either a *whole interval* (frequency ratio 9/8 or 10/9) or a *half-interval* (frequency ratio 16/15). These intervals represent a compromise that maintains the pleasing frequency ratios across an octave. Here, notes C-E form a major third, C-G form a fifth, and C-E-G form a major triad.

There are also diatonic scales beginning on other notes. Musicians refer to these scales as **keys,** based on the beginning note. The spacing of intervals maintains the same frequency ratios, so that some notes must be raised or lowered a half-interval from the letters given in the C major scale. When a note is raised a half-interval it is called a

Table 10.2.1
Diatonic C major scale

Note	Letter	Frequency (Hz)	Frequency Ratio	Interval
do	C	264.0		
re	D	297.0	9/8	Whole
mi	E	330.0	10/9	Whole
fa	F	352.0	16/15	Half
sol	G	396.0	9/8	Whole
la	A	440.0	10/9	Whole
ti	B	495.0	9/8	Whole
do	C	528.0	16/15	Half

Table 10.2.2
Diatonic A major scale

Note	Letter	Frequency (Hz)	Frequency Ratio	Interval
do	A	220.0		
re	B	247.5	9/8	Whole
mi	$C^{\#}$	275.0	10/9	Whole
fa	D	293.3	16/15	Half
sol	E	330.0	9/8	Whole
la	$F^{\#}$	366.7	10/9	Whole
ti	$G^{\#}$	412.5	9/8	Whole
do	A	440.0	16/15	Half

sharp ($C^{\#}$); when it is lowered a half-interval it is called a **flat** (B^{b}). In the A major scale (Table 10.2.2) the notes C, F, and G are raised a half-interval to $C^{\#}$, $F^{\#}$, and $G^{\#}$. The notes A, B, and E have the same frequencies as they did in the C major scale (or a factor of two lower, since A and B are listed in the preceding octave). However, the D in the A major scale has a slightly lower frequency than the D in the C major scale. This is a

Table 10.2.3
Equally tempered scale

Letter	Frequency (Hz)	Frequency Ratio	Interval
C	261.6		
$C^\# = D^b$	277.2	$\sqrt[12]{2}$	Half
D	293.7	$\sqrt[12]{2}$	Half
$D^\# = E^b$	311.1	$\sqrt[12]{2}$	Half
E	329.6	$\sqrt[12]{2}$	Half
F	349.2	$\sqrt[12]{2}$	Half
$F^\# = G^b$	370.0	$\sqrt[12]{2}$	Half
G	392.0	$\sqrt[12]{2}$	Half
$G^\# = A^b$	415.3	$\sqrt[12]{2}$	Half
A	440.0	$\sqrt[12]{2}$	Half
$A^\# = B^b$	466.2	$\sqrt[12]{2}$	Half
B	493.9	$\sqrt[12]{2}$	Half
C	523.3	$\sqrt[12]{2}$	Half

consequence of maintaining the harmonic frequency ratios. If all the diatonic scales in all the keys were considered, many additional notes (frequencies) would be needed, which presents a definite problem on fixed-note instruments such as the piano.

A way around this complexity is the **equally tempered scale**. In this chromatic scale (a scale containing all the sharps and flats), there are 12 equally spaced half-interval notes (Table 10.2.3). Each note has a frequency 1.05946 (the twelfth root of 2) higher than the preceding one. Thus, after 12 half-intervals, the frequency has increased by $(1.05946)^{12} = 2$, or an octave. In this scale, $C^\#$ and D^b (and other pairs of sharps and flats between adjacent letters) are taken to have the same frequency. The total number of notes is greatly reduced, but at the expense of compromising some of the pleasing frequency ratios. For example, the C-E or C-G ratios are not precisely those of a major triad.

In addition to the 12 major scales (each starting on a different note in the equally-tempered scale), there are 12 minor scales in which the whole- and half-interval spacing is slightly different. Johann Sebastian Bach wrote 24 preludes and fugues based on the 12 major and 12 minor keys in his composition *The Well-Tempered Klavier*. Further discussion of musical harmony may be found in any good book on music theory. Other books discuss alternate scales in Oriental music and in modern music.

10.5 Shock Waves and Sonic Booms

In the course of our earlier discussion of the Doppler effect, we described the decrease of wavelength in the direction ahead of a moving source of waves. This is due to the motion of the source; each wavefront is emitted from a position closer to the preceding wavefront. Now consider what happens as the velocity of a source of sound approaches the speed of sound. The spacing between wavefronts will become shorter and shorter, and the waves will coincide when the speed of the source equals the speed of sound (Fig. 10.26). Constructive interference occurs, and the result is a single wavefront of very large amplitude. This wavefront carries a large amount of energy, and the large amplitude is heard as a loud boom when the wave strikes our eardrums. This wave is called a **shock wave,** and the noise associated with it is a **sonic boom.**

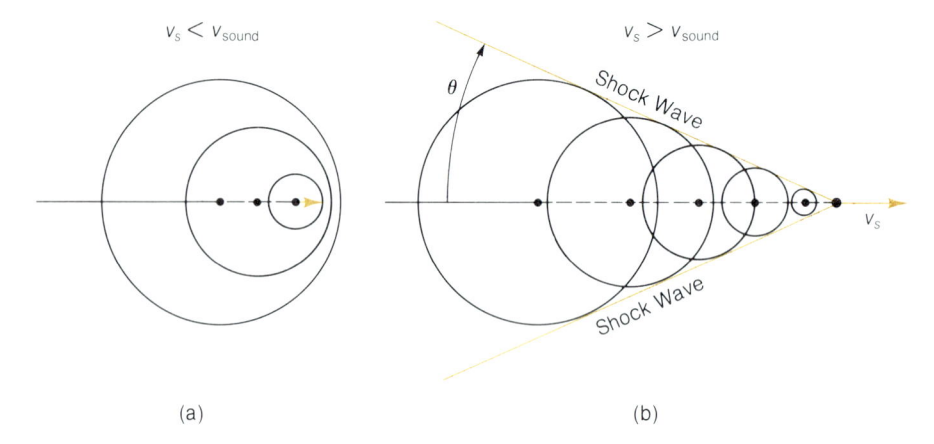

Figure 10.26

Shock wave formation. When a source moves slower than the sound speed, its wavefronts are compressed in front and rarefied behind (a). However, when v_s is larger than the sound speed (b), the source "outruns" the wave fronts. These fronts pile up along a conical shock wave (sonic boom), whose opening angle is given by $\sin\theta = (v_{sound}/v_s)$.

Modern aircraft are capable of exceeding the speed of sound, in some cases by large factors. The ratio of air speed to sound speed is called the **Mach number;** when this number exceeds 1, the aircraft surpasses the speed of sound. Because of the low temperature at very high altitudes, the speed of sound may be considerably less than at sea level, and Mach numbers are high as 6 or greater have been achieved by experimental aircraft such as the X-15. The large-amplitude wave that builds up in front of an aircraft as it approaches Mach 1 creates a physical barrier, since the wave is a compression wave, with high density and pressure. Because of this, the aircraft experiences difficulty in reaching the speed of sound, and the term "breaking the sound barrier" is often applied.

The shock wave created by an aircraft travelling at supersonic speed is actually double, because two waves develop (Fig. 10.27). One arises just in front of the aircraft, and the other to the rear. Hence a sonic boom is actually a double boom, with a small time interval between. The wave travels with the aircraft, so that the sonic boom follows the plane's path.

Mach numbers in excess of 1 occur in other situations as well, and for that reason a sophisticated mathematical treatment of shock waves has been developed. To a good approximation, it is reasonable to represent a shock as a discontinuity, with instantaneous changes in pressure, density, and temperature as the shock passes through a gas. In most treatments, a frame of reference is chosen in which the shock is stationary, so that the gas flows through the shock instead of the other way around. Shocks occur in many situations, particularly in astrophysical environments. Explosions of stars, for example, create shock waves that travel through interstellar space and form density and temperature discontinuities in interstellar gas clouds.

(a) (b)

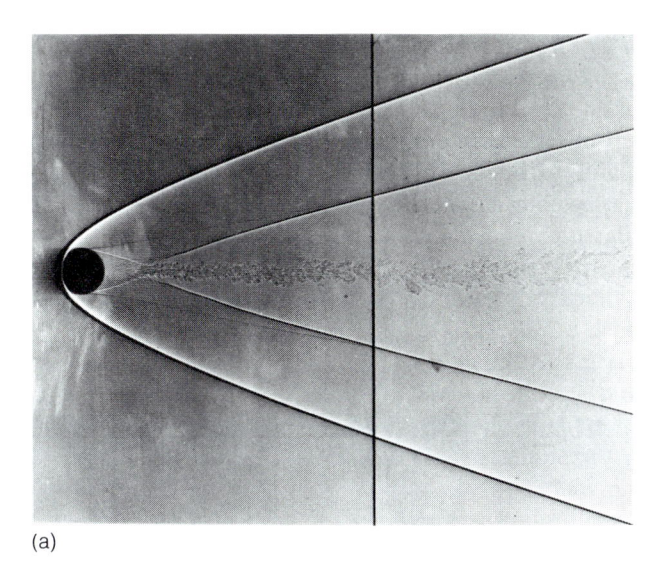

Figure 10.27

Formation of a sonic boom. An aircraft flying at supersonic speeds creates a double shock wave that people on the ground hear as a sonic boom. In these laboratory demonstrations, the double shock wave shows more clearly in (a) where a spherical object was used, than in (b) where an aircraft was used.

Perspective

While we have ignored many complicated manifestations of waves, we have succeeded in laying a groundwork of basic principles that should prepare us for more specialized treatments. Thus we have reached a natural endpoint of our overall discussion of deformations, elasticity, vibrations, and waves.

Next we will undertake an analysis of fluids and their properties. In the last section, we addressed some of these properties, but much more remains to be said.

SUMMARY TABLE OF FORMULAS

Velocity, wavelength, and period of a wave:
$$v = \lambda/T \tag{10-1}$$

Velocity, wavelength, and frequency of a wave:
$$v = \lambda \nu \tag{10-2}$$

Velocity of a transverse wave:
$$v = \sqrt{\frac{F_T}{\mu}} \tag{10-3}$$

Velocity of a compressional wave in a solid:
$$v = \sqrt{\frac{Y}{\rho}} \tag{10-4}$$

Velocity of a compressional wave in a fluid or gas:
$$v = \sqrt{\frac{B}{\rho}} \tag{10-5}$$

Energy of a transverse sinusoidal wave:
$$E = 2\pi^2 \rho A v t v^2 x_0^2 \quad (10\text{–}6)$$

Power of a transverse sinusoidal wave:
$$P = 2\pi^2 \rho A v v^2 x_0^2 \quad (10\text{–}7)$$

Intensity of a transverse sinusoidal wave:
$$I = \frac{P}{A} = 2\pi^2 \rho v v^2 x_0^2 \quad (10\text{–}8)$$

Doppler shift for moving source:
$$v' = \frac{v}{\left(1 - \dfrac{v_s}{v}\right)} \quad (10\text{–}9a)$$

$$v' = \frac{v}{\left(1 + \dfrac{v_s}{v}\right)} \quad (10\text{–}9b)$$

Doppler shift for moving observer:
$$v = v\left(1 + \frac{v_o}{v}\right) \quad (10\text{–}10a)$$

$$v' = v\left(1 - \frac{v_o}{v}\right) \quad (10\text{–}10b)$$

Amount of Doppler shift of frequency:
$$\Delta v = v' - v \quad (10\text{–}11)$$

Doppler shift for moving source:
$$\frac{\Delta v}{v} = \frac{\pm v_s}{v} \quad (10\text{–}12)$$

$$\frac{\Delta v}{v} = \frac{\pm v_o}{v} \quad (10\text{–}13)$$

Doppler shift for light:
$$\frac{\Delta v}{v} = \frac{\pm v}{c} \quad (10\text{–}14)$$

Angles of incidence and reflection:
$$\theta_i = \theta_r \quad (10\text{–}15)$$

Angles of incidence and refraction:
$$\frac{\sin\theta_r}{\sin\theta_i} = \frac{v_r}{v_i} \quad (10\text{–}16)$$

Modes for standing waves in string with fixed ends:
$$L = \frac{n\lambda_n}{2} \quad (10\text{–}17)$$

Frequency of mode n for string with fixed ends:
$$v_n = \frac{nv}{2L} = \frac{n}{2L}\sqrt{F_T\left(\frac{L}{M}\right)} \quad (10\text{–}18)$$

Speed of sound:
$$v_s = \sqrt{\frac{B}{\rho}} \quad (10\text{–}19)$$

Intensity of sound waves:
$$I = \frac{(\Delta P)^2}{2\rho v_s} \quad (10\text{–}20)$$

Sound intensity in decibels:
$$\beta = 10\log\frac{I}{I_0} \quad (10\text{–}21)$$

Natural frequencies for pipe open at both ends:
$$v_n = \frac{nv_s}{2L}, \quad n = 1, 2, 3, \ldots \quad (10\text{–}22)$$

Natural frequencies for a pipe open at one end:
$$v_n = \frac{nv_s}{4L}, \quad n = 1, 3, 5, \ldots \quad (10\text{–}23)$$

THOUGHT QUESTIONS

1. Can you think of another wave besides light which can travel through a vacuum?
2. Give an everyday example of a transverse, a longitudinal, and a surface wave.
3. Describe how you would deduce the structure of the earth's interior from a study of seismic waves.
4. Would sound travel carry better on a mountain peak or in the lowlands? Explain.
5. Why is Eq. 10–5 appropriate for waves travelling through a gas or liquid medium, while Eq. 10–4 is not?
6. Would it be easier to break the sound barrier in an aircraft at high altitude or low altitude? Explain.
7. A depth charge is set off under water at a distance of 100 m from a submarine. If the submarine were on the surface and the blast occurred in air, would the sub suffer more damage? Explain.

PROBLEMS

8 Why does the intensity of a three-dimensional wave decrease as the inverse square of distance from the source?

9 Which of the human senses detects waves? Explain.

10 Would you expect all three types of waves (transverse, longitudinal, surface) to exhibit a Doppler effect? Explain.

11 Describe how the frequency of the sound of a siren or train whistle appears to change as the source moves by. Are your observations consistent with the Doppler effect?

12 Explain how a young child's telephone made of two tin cans and a taut wire works.

13 When a person inhales helium, the pitch of his voice rises. Why?

14 Do ocean waves exhibit reflection, refraction, and diffraction?

15 Why is there a phase shift of a half-cycle when a wave moving along a string is reflected at a rigid boundary?

16 Suppose a vibrating string with fundamental frequency ν_o is immersed in water. How will the value of ν_o be affected?

17 Would sound travel faster in air, water, or iron? Explain why.

18 How could you demonstrate the existence of sound waves with frequencies outside the detectable range of the human ear?

19 Why are the natural frequencies for an open pipe different from those of a pipe closed at one end?

20 Choose a string or wind instrument and identify the generator and resonator.

21 Describe how a piano is tuned.

22 How is music generated by an electric guitar different from that produced by an acoustic guitar?

23 Can you think of a practical use for sounds that cannot be heard by the human ear?

24 Why can't transverse waves propagate in a liquid or gas?

25 Give an everyday example of a sharp resonance.

PROBLEMS

Section 10.1 Wavelength, Frequency, Velocity

• 1 Ocean waves pass a ship at 10 waves/min. If waves are 100 m apart, at what velocity do the waves pass the ship?

• 2 A wave moves at 300 m/s with frequency 20 Hz. What is its wavelength?

• 3 A wave has a wavelength of 10 cm and a speed of 330 m/s. What is its frequency?

• 4 A wave has a speed 330 m/s and a wavelength 1 cm. What is its period?

• 5 A wave has a frequency of 6×10^{14} Hz and a wavelength 5×10^{-7} m. What is its speed?

• 6 Ocean waves strike a beach once every 30 s. What are the frequency and period of these waves? If their speed is 10 km/h, what is their wavelength?

• 7 A string under 100 N tension transmits waves at 5 m/s. Find the mass per unit length of the string.

• 8 A rod transmits a compressional wave at 3,000 m/s. If the Young's modulus of the rod is 5×10^{10} N/m² what is its density?

•• 9 While being lowered into the sea, a diver exerts a tension of 900 N on a line with mass density 1 kg/m. The diver wishes to signal the surface by jerking on the line. How long will his signal take at a depth of 100 m?

•• 10 If the diver in Problem 9 wishes to double the wave speed what must be done?

•• 11 A submarine captain observes the flash from a depth charge that explodes directly overhead. He feels the concussion 0.3s later. How far above the submarine did the explosion occur?

• 12 A wave is observed to travel at 5,000 m/s through a solid medium having density 2,000 kg/m³. Find the elastic modulus for the medium.

• 13 Porpoises are sensitive to sound frequencies up to 200,000 Hz. What wavelength does this correspond to in (a) air, (b) water?

•• 14 People can hear sounds with frequencies between about 20 Hz and 20,000 Hz. Compute the corresponding wavelength range (a) at 0°C, (b) at 30 °C.

- 15 What is the wavelength of the pulse shown in Fig. 10.28?

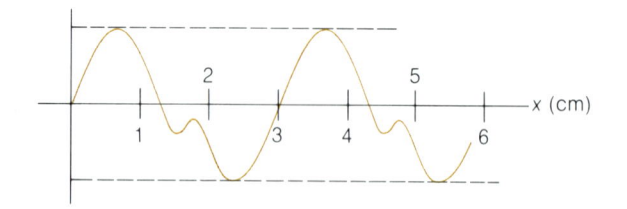

Figure 10.28

- 16 A bat sends out a sound and receives a reflected echo 0.05 s later. How far away was the object that reflected the sound if T = 20 °C?

- 17 Compute the speed of sound on a hot summer night (30 °C). If lightning flashes and the thunder is heard T seconds later, find a formula for the distance of the lightning in terms of T. Assume the speed of light is essentially infinite.

- 18 Compare the power of a wave whose amplitude is ⅓ that of a second wave of identical properties.

- 19 Calculate the intensity of sunlight on the earth's surface if the sun's power is 4×10^{26} watts and the earth-sun distance is 1.5×10^8 km. Using the criterion in Example 10.3, is this sufficient to read without eyestrain?

- 20 The light intensity from the full moon is about 10^{-6} that of the sun. How much closer to the earth would the moon need to be for a person to read without eyestrain?

- 21 The intensity of a wave 100 m from its source is 500 watts/m². What is the total power output from the source if it radiates isotropically?

- 22 What intensity would an observer measure 75 m from the source in Problem 21?

- 23 The total power of the sun is 4×10^{26} watts. What is the intensity of sunlight at the distance of Pluto (see Appendix 4)? How much fainter is the sun at Pluto than at earth?

- 24 If you hear a train coming when you place your ear to a steel track, 2 s before you hear a gunshot from a train robber, how far away is the train?

- 25 What is the wavelength of the radio waves from an AM station with frequency 1,200 kHz? Of an FM statio at 102 MHz? What are the intensities of the broadcast signals 15 km away if their powers are 50,000 watts? Light travels at 3×10^8 m/s.

- 26 You shout across a canyon and hear an echo 6 s later. How wide is the canyon? Assume that sound travels at velocity $v = (331 + 0.6T)$ m/s, where T is the temperature in degrees Celsius. Here, take $T = 30$ °C.

- 27 A piano is tuned during the day ($T = 32$ °C). By what fraction have its notes slipped in pitch during the evening concert, when $T = 25$ °C?

- 28 The amplitude of a wave is doubled. By what factor does its intensity increase?

- 29 Steel has twice the Young's modulus of iron but about the same density. How much faster will sound travel in steel than in iron?

- 30 Two waves have the same amplitude and travel through identical media with the same velocity. Wave 2 has twice the frequency of Wave 1. What is the ratio of their intensities?

Section 10.2 Doppler Shift, Refraction, Beats

- 31 If the speed of sound in air is 330 m/s, how fast does an airplane fly toward an observer if the observed frequency of its engine is 1.5 times its actual frequency?

- 32 A state trooper observes that the frequency of the sound from an approaching car is 160 Hz. The officer knows that the engine of that model car makes a sound at 150 Hz, as observed by a listener inside the car. Should he arrest the driver for exceeding the 55 mi/h speed limit? Explain.

- 33 Express Eq. 10–9a in terms of the wavelengths λ and λ'.

- 34 A spectral line with rest frequency 6.0×10^{14} Hz is seen in the spectrum of a star at 5.82×10^{14} Hz. Is the star approaching or receding from the earth? At what speed?

- 35 A typical Indianapolis 500 racecar can travel at speeds of 200 mi/h. What frequency shift would a spectator detect if the sound of the car's engine has a frequency of about 400 Hz and the car is approaching?

- 37 At what speed must an ambulance move for a siren with normal rest frequency 350 Hz to sound like it has a frequency of 370 Hz? Would the car need to move toward or away from the observer?

- 38 A concert pianist with perfect pitch stands at a platform, listening to a soprano singing abroad a train passing the station. He hears the singer's pitch drop one whole note, from A (440 Hz) to G (392 Hz). How fast was the train moving, if the singer was actually sustaining a single pitch? What was the frequency of that note as heard by passengers aboard the train?

PROBLEMS

··39 A sound from a jet has rest frequency 1,000 Hz. How fast must the jet fly in order that the frequency is shifted out of the range of the human ear (20,000 Hz)?

··40 What would be the frequency of the jet's sound in Problem 39 when the jet recedes?

··41 Derive Eq. 10–14 in terms of wavelength instead of frequency.

··42 Scientists at Loch Ness, Scotland, receive reflected signals from sound waves of rest frequency 1,000 Hz. The echoes return at 1,010 Hz after 1 s. How far away is Nessie, the Loch Ness monster, and how fast is she swimming?

···43 Two ambulances with sirens operating at 600 Hz are racing in opposite directions, at speeds of 20 m/s (A) and 30 m/s (B) respectively (Fig. 10.29).
(a) With what frequency does driver A hear the siren of ambulance B?
(b) With what frequency does driver B hear the siren of ambulance A?
(c) Do these frequencies differ? Explain why or why not.

Figure 10.29

···44 In music, two pitches whose frequencies are in ratio 5:4 are perceived to be harmonious. This pair of notes is called a third. Two musicians play the same note, but one is at rest while the other moves away at velocity v_s. What is v_s if the stationary musician hears a third? Does the moving musician then hear a third? Express v_s in units of v, the sound speed.

···45 Two sirens move at 30 m/s along streets making a 90° angle. If the pitch of each siren, as perceived by the other, is 500 Hz, what is their rest frequency? Assume that they are equidistant from the street intersection.

·46 A sound wave travels across a material interface and its velocity is halved. If the densities are the same, compare the bulk modulus ratio of the two materials.

··47 A sound wave travels from a steel medium into an iron medium. What is the ratio of sound speeds? If the angle of incidence is 20°, what are the angles of reflected and transmitted (refracted) waves?

··48 A sound wave impinges on a water surface, making an angle of incidence $\theta_i = 10°$ with the perpendicular to that surface. What are the angles of reflection and refraction?

···49 A mosquito buzzing on a post 50 cm above and 10 cm in from the edge of a lake cannot be heard by a frog 1 m below the water surface (Fig. 10.30) because of refraction of the sound waves in water. Find the minimum distance d offshore.

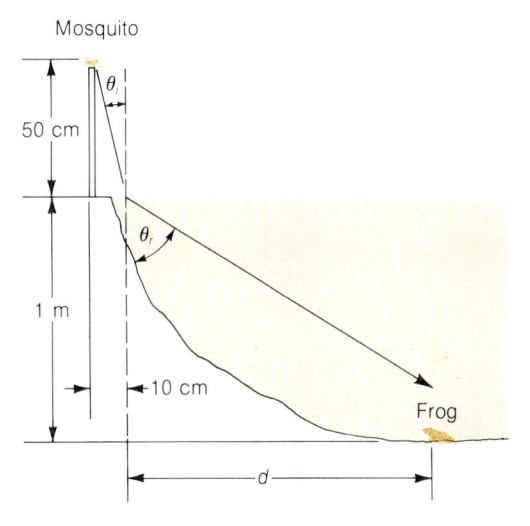

Figure 10.30

···50 A submarine rests on the bottom of the ocean, 1,000 m below the surface (Fig. 10.31). A reconnaissance plane equipped with sonar detects the sub at B because of the refraction of the sound waves, but it is actually at A. If the plane is 1,000 m above the surface, what is A's position, measured horizontally from a point on the surface directly below the plane?

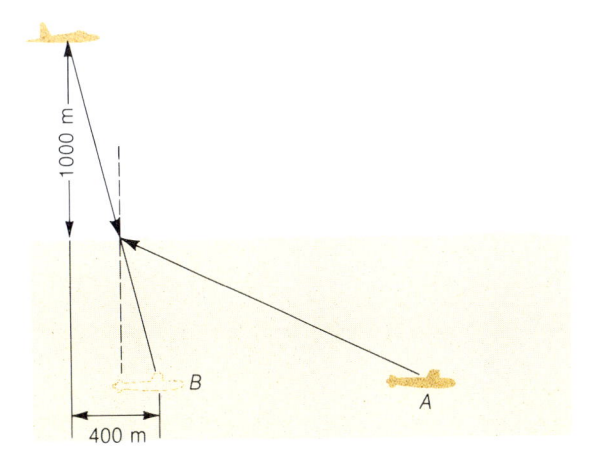

Figure 10.31

324 CHAPTER 10 WAVE PHENOMENA

••• 51 Solve Eq. 10–16 for $\sin\theta_r$ in terms of $\sin\theta_i$ and the ratio of sound speeds, v_r/v_i. Now suppose that v_r is less than v_i. Find the largest angle of incidence θ_i for which a realistic value of the reflection angle θ_r is possible (Recall that the sine of an angle can never be greater than 1). What happens to the transmitted (refracted) wave when θ_i is larger than this critical angle?

•• 52 What ratio of sound speeds, v_r/v_i, in two materials will result in a refracted angle twice that of the incident angle $\theta_i = 10°$, $20°$, and $30°$?

• 53 Two waves have amplitudes $x_1 = 3 \sin(3t)$ and $x_2 = 5 \sin(3t)$. What will be the maximum amplitude of their superposed wave?

•• 54 Two waves have amplitudes $x_1 = \sin(2t)$ and $x_2 = \sin(3t)$. What is the period of the regular modulation (beat)?

• 55 Two French horn notes of frequencies 660 Hz and 664 Hz produce beats of what frequency?

• 56 When two piano notes are struck together, a tuner hears beats at 4 Hz. How far apart are the notes in frequency? If one note is middle A (440 Hz), what are the possible frequencies for the other note?

•• 57 Two violins are tuned to concert A (440 Hz). If the tension in one violin's string decreases by 1%, how many beats per second are heard when the violins play that note together?

•• 58 In the equally tempered scale, a half-step corresponds to a ratio of frequencies 1.05946. How many beats per second are heard between the half-step from middle C (261.6 Hz) to C sharp (277.2 Hz)? How many beats per second are between the C and C sharp one octave higher in pitch?

•• 59 A superposition wave is made of two waves, one with frequency 100 Hz and amplitude 1 and a second with frequency 200 Hz and amplitude 2. Make a sketch of the total waveform. What is the beat frequency? What is the maximum amplitude of the superposed waveform?

••• 60 Sketch the superposition of the following waves: $x_1 = \cos(t)$ and $x_2 = \sin(t)$.

Section 10.3 Modes and Harmonics

• 61 A 1 m string under 100 N tension has a linear mass density 0.1 kg/m. Find the fundamental frequency and period of oscillation.

•• 62 What are the periods of the first three overtones of the string in Problem 61?

• 63 A violin string has a fundamental frequency of 600 Hz. What are the frequencies of its first four overtones?

•• 64 A piano maker places a 1 m steel string of mass density 0.1 g/cm under tension 500 N. What is its fundamental frequency of oscillation? What are its first three overtones?

• 65 A violin string vibrates at 440 Hz. How many overtones lie within the audible range to a person?

•• 66 A violinist wishes to produce a note at 700 Hz from a string whose fundamental frequency is 500 Hz. Where should the violinist place his fingers on a string of length L?

•• 67 A violin string has a fifth overtone at 3,400 Hz. What is its fundamental frequency?

Section 10.4 Sound Waves

• 68 The speed of sound in water is about 1,500 m/s. If the density of water is 1,000 kg/m^3, find its bulk modulus.

• 69 An underwater volcano explodes, sending debris into the air and sound waves into both air and water. If the volcano is 2,000 km away across the ocean, what is the time difference between sound waves in air and water?

• 70 An earthquake produces compressional (P) and shear (S) disturbances which propagate at 9 km/s and 4 km/s respectively. If the density of the material is constant, what is the ratio of the effective elastic (or shear) constants of the two waves?

• 71 What is the sound intensity (decibels) of a wave with intensity 10^{-7} W/m^2?

• 72 At 1,000 Hz the human ear can detect intensities from 0 to 120 decibels. What is the actual ratio of intensities (in W/m^2)?

• 73 Find the intensity of a 60 dB sound.

• 74 A rock band doubles its intensity. How many decibels did the sound increase?

• 75 What is the ratio in intensity between 120 dB and 110 dB?

• 76 If the human ear can detect a relative difference in sound of 1 dB, what is the actual ratio of intensities?

• 77 A stereo manufacturer claims a separation of 60 dB between channels. What ratio of intensities does this correspond to?

•• 78 Good tape recorders quote a signal-to-noise ratio of 50 dB. If the recording corresponds to an intensity

SPECIAL PROBLEMS

of 10^{-5} W/m², what is its intensity in dB? What is the intensity of the background noise on the tape?

•• 79 A 120 dB sound from a jet strikes the eardrum, a circular area of radius 4 mm. How much energy is absorbed per second?

• 80 A jet produces sound levels of 120 dB at a distance of 100 m. What is the decibel level at 1 km?

•• 81 Use Eq. 10–20 to estimate the pressure in a 120 dB sound wave travelling through air at 0° C. Assume air has density 1.2 kg/m³. What fraction of atmospheric pressure (10^5 N/m²) is this wave?

• 82 A jet engine at a distance of 50 m puts out 150 dB of sound. What is its total power (watts) in sound waves?

• 83 If the speed of sound is 330 m/s, how long must an organ pipe be to produce a fundamental mode at 120 Hz? One end is closed, one open.

• 84 What is the frequency ratio between the highest and lowest C's that can be struck on a piano? They are separated by seven octaves.

•• 85 Suppose that the bottle in Example 10.7 had a length of 30 cm. Find the fundamental frequencies, both empty and one-third filled.

•• 86 A clarinet is 60 cm long. What are the first three frequencies of vibration when all holes are closed? Treat the instrument as a cylinder open at one end.

•• 87 Compute the natural vibrational frequency of the Eisenhower tunnel through the Rocky Mountains. The tunnel is about 1.7 mile long (open at both ends). What harmonics would fall in the subaudible range 10–20 Hz?

•• 88 An oscillating pipe has frequencies of 100 Hz, 200 Hz, 300 Hz, and 400 Hz. No lower frequencies are found. What is the fundamental frequency? Is the pipe open or closed?

•• 89 The ear cavity may be approximated as a cylinder of depth about 2.5 cm, closed at the eardrum. What is the fundamental frequency of oscillation? What harmonics would fall in the audible range?

Section 10.5 Shock Waves

• 90 What is the difference in sound speed in an arctic region with mean temperature −60° C and in a tropical region (mean temperature 40 °C)?

• 91 What are the Mach numbers of a 660 mi/h plane flying in the regions of Problem 90?

•• 92 In the troposphere, the atmospheric temperature decreases with altitude up to about 25 km. At what temperature must a 640 mi/h jet fly to break the sound barrier?

•• 93 A plane flies at Mach 2.0 when the temperature is −50 °C. What is its speed? What would be its Mach number at sea level ($T = 25$ °C)?

•• 94 The world speed record for aircraft was set in 1967 by the *X-15*, which flew at 4,520 mi/h at 99,000 ft. If the temperature at that altitude was −50 °C, what was its Mach number?

••• 95 In interstellar space the sound speed in some hot ionized regions can be as large as 100 km/s. If a star orbits a large mass M at a distance of 10^{18} m, what is the largest M can be to prevent shock waves around the star from passing through the interstellar gas?

SPECIAL PROBLEMS

1. Musical Harmony

Refer to *Focus on Physics*, Musical Harmony and Scales (Chapter 10). These questions concern both the diatonic and equally tempered scales.

(a) Using the frequency ratios of the diatonic scale, show that over the seven whole or half steps (an octave), the frequency doubles.

(b) In the diatonic C major scale, the C-E major third has a frequency difference of $(9/8) \times (10/9) = 5/4$. Find the frequency ratios of the other intervals beginning with C: C-F (fourth), C-G (fifth), C-A (sixth), and C-B (seventh). Which are likely to be pleasing?

(c) In the diatonic C major scale, the notes in the C major triad (C-E-G) are in the frequency ratio 4:5:6. Find the other two major triads in this scale. Include the D in the next octave.

(d) In the equally tempered scale, compute the frequency ratios of the major third, fifth, sixth, and seventh found in part b. How close are these ratios to the harmonic ratios in the diatonic scale?

2. Earth's Interior

Seismologists have identified three major types of waves from earthquakes: surface waves and the somewhat faster P (compressional) and S (shear) waves that penetrate the earth's interior. The P and S wave velocities are given by

$$v_P = \sqrt{\frac{\left(B + \frac{4S}{3}\right)}{\rho}} \quad \text{and} \quad v_S = \sqrt{\frac{S}{\rho}}$$

where ρ is the mass density, B is the bulk modulus, and S is the shear modulus of the interior. The shear waves cannot penetrate the liquid core. The difference between arrival times of the P and S waves can be used to determine the distance of the earthquake's epicenter. There is a simple relationship between distance on the surface and arrival time (Fig. 10.32a) because the waves are bent (refracted) as they pass into the interior (Fig. 10.32b).

(a) If $v_P = 11$ km/s and $v_S = 6$ km/s in the upper mantle (1,000 km depth), where $\rho = 3,300$ kg/m³, find the moduli B and S.

(b) Both v_P and v_S increase with depth in the upper mantle. Explain why waves would refract when passing through this medium. Why do they bend away from the core? (This is the reason for the shape of the paths in Fig. 10.32b.)

(c) During an earthquake P waves are detected at 10:00 AM and S waves at 10:06 AM. Use Fig. 10.32a to estimate the distance to the quake's epicenter.

(d) Is information from just one groundsite sufficient to determine the actual epicenter? How many other groundsites are needed?

3. Doppler Velocity in Baseball and Medicine

The speed of a moving object may be measured by comparing the frequency of outgoing waves from a stationary transmitter with the Doppler-shifted waves reflected by the object. Let the outgoing waves have frequency ν, let the object approach at velocity v_s, and let c be the wave speed (either sound speed or light speed).

(a) Show that in the frame of the object the outgoing waves have frequency

$$\nu' = \frac{\nu}{[1 - v_s/c]}.$$

(b) The reflected waves also have frequency ν' in the frame of the moving object. Find the frequency of the reflected waves in the frame of the transmitter, and thus determine the beat frequency of the two.

(c) If a 1 MHz radar gun ($c = 3 \times 10^8$ m/s) is used to measure a baseball pitcher's 95 mi/h fastball by the above technique, what beat frequency is expected?

(d) If 2 MHz ultrasound waves are used to monitor bloodflow in the heart, and a beat frequency of 800 Hz is found, what is the velocity of the blood cells? The relevant wave speed is the speed of sound in human tissue, about 1,450 m/s.

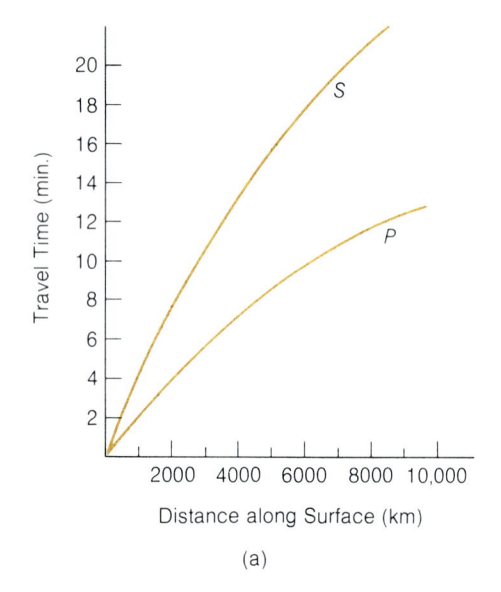

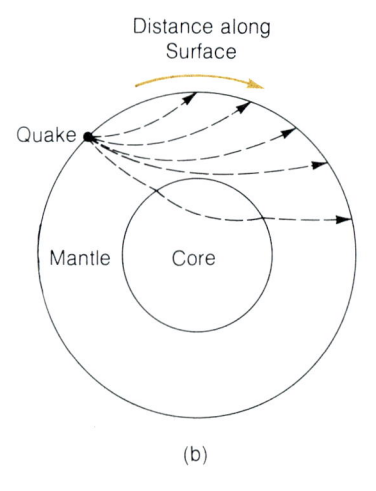

Figure 10.32

SUGGESTED READINGS

"What makes you sound so good when you sing in the shower?" J. Walker, *Scientific American*, May 82, p 170, **246**(5)

"How to analyze a city traffic-light system from the outside looking in." J. Walker, *Scientific American*, Mar 83, p 138, **248**(3)

"People listening to a bell can perceive sounds the bell does not really make." J. Walker, *Scientific American*, Jul 84, p 132, **251**(1)

"The physics of the piano." E. D. Blackham, *Scientific American*, Dec 65, p 88, **213**(6)

"Physics of brasses." A. H. Benade, *Scientific American*, Jul 73, p 24, **229**(1)

"Physics of the bowed string." J. C. Schelleng, *Scientific American*, Jan 74, p 87, **230**(1)

"Thunder." A. A. Few, *Scientific American*, Jul 75, p 80, **233**(1)

"Motion of the ground in earthquakes." D. M. Boore, *Scientific American*, Dec 77, p 68, **237**(6)

"Acoustics of the singing voice." J. Sundberg, *Scientific American*, Mar 77, p 82, **236**(3)

"The coupled motions of piano strings." G. Weinreich, *Scientific American*, Jan 79, p 118, **240**(1)

"The physics of kettledrums." T. D. Rossing, *Scientific American*, Nov 82, p 172, **247**(5)

"The physics of organ pipes." N. H. Fletcher, S. Thwaites, *Scientific American*, Jan 83, p 94, **248**(1)

"Seismic tomography." D. L. Anderson, A. M. Dziewonski, *Scientific American*, Oct 84, p 60, **251**(4)

"Musical String Vibrations." R. Johns, *The Physics Teacher*, Mar 77, p 145, **15**(3)

The flying circus of physics with answers. Jearl Walker, Wiley, New York, 1977

Science from your airplane window. Elizabeth Wood, Dover, New York, 1975

PART THREE

FLUIDS AND GASES

Chapter 11
Fluids 330

Chapter 12
Temperature and Heat 364

Chapter 13
Gas Kinetics, Thermodynamics and Heat Engines 408

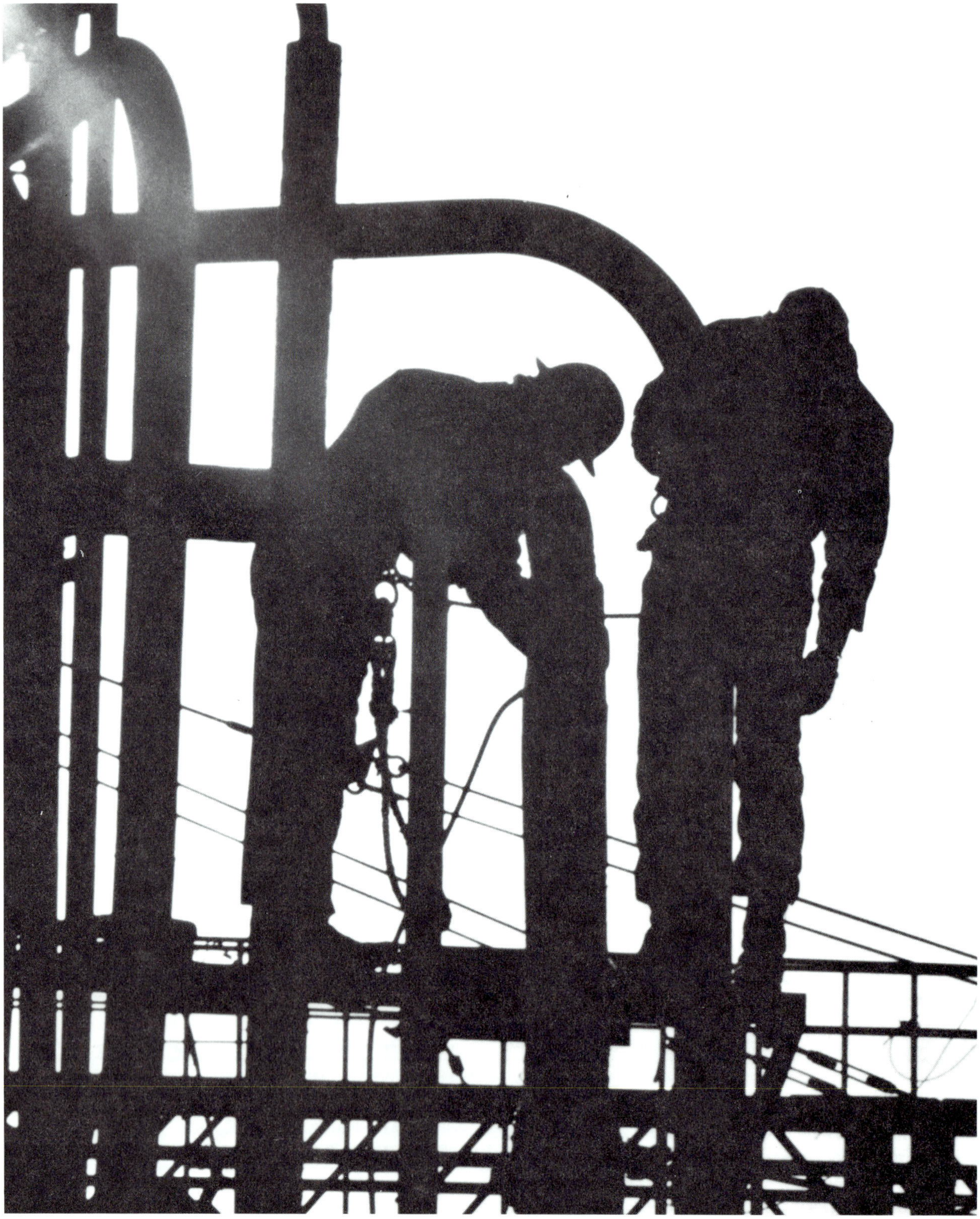

Chapter 11

Fluids

We are familiar with the notion that there are three states of matter: solid, liquid, and gas. So far in our study of physics we have dealt mostly with solids, except for the transmission of waves in air and in liquids. Now it is time to address the nature of gases and liquids and their interaction with each other and their environment.

A fluid is any substance that has no rigidity, and thus has no fixed shape. Both gases and liquids fit this description, and both are included in the treatment we are commencing here. The difference between gases and liquids is that liquids have a definite volume; the binding forces between molecules in a liquid are sufficiently strong that the molecules tend to maintain a certain fixed spacing. Hence when a quantity of liquid is placed in a container, it fills it only to a certain level. A gas, on the other hand, will expand to fill the container uniformly, since its molecules are independent of each other. Gases are also much more easily compressed than liquids.

In this chapter we will deal first with the properties of fluids at rest, and then with fluids in motion. Along the way we will develop an understanding of many general properties that will become relevant in the subsequent chapters on kinetic properties of fluids and heat transfer.

11.1 Pressure and Buoyancy in Stationary Fluids

Since a fluid has an indefinite shape, it is difficult to speak of forces exerted on or by a fluid. Such forces are always distributed over some area, so we normally speak of the **pressure,** which is force per unit area (with the force acting perpendicular to the area). In SI units, we define the **pascal** (after the French mathematician B. Pascal) as one Newton per square meter; that is, 1 Pa = 1 N/m^2.

332   CHAPTER 11   FLUIDS

Table 11.1
Conversion factors for pressure

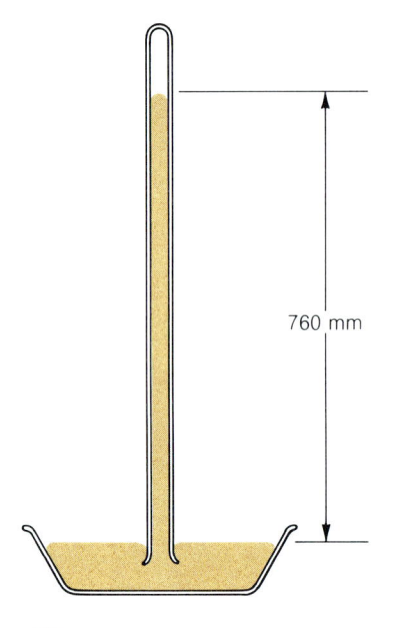

1 atm = 1.013×10^5 N/m^2	1 atm = 1.013×10^5 N/m^2
1 bar = 1.000×10^5 N/m^2	1 atm = 1.013 bar
1 dyne/cm^2 = 0.1 N/m^2	1 atm = 1.013×10^6 dyn/cm^2
1 kg/cm^2 = 9.85×10^4 N/m^2	1 atm = 1.03 kg/cm^2
1 lb/in^2 = 6.90×10^3 N/m^2	1 atm = 14.7 lb/in^2
1 mm Hg = 133 N/m^2	1 atm = 760 mm Hg
1 mm H$_2$O (4°C) = 9.81 N/m^2	1 atm = 1.03×10^4 mm H$_2$O (4°C)
1 torr = 133 N/m^2	1 atm = 760 torr
1 Pa = 1 N/m^2	1 atm = 1.013×10^5 Pa

Figure 11.1

The mercury barometer. The height of mercury in an evacuated tube is determined by the atmospheric pressure acting on the mercury reservoir. At normal sea-level pressure (1 atmosphere), the height of the mercury column is 760 mm.

760 mm

In many practical applications, pressure is expressed in terms of the earth's atmospheric pressure at sea level, which is reasonably constant (and whose definition has been standardized to mean the same thing to everybody). One **atmosphere** of pressure is equal to 1.013×10^5 Pa. Other units often used include the **bar,** which is equal to 10^5 Pa, so that 1 atm = 1.013 bar (or 1013 **millibars**). Other units are standardized to the height to which a column of mercury is raised, because measurement of such columns is a standard technique for measuring atmospheric pressure. If a column of mercury is placed in a tube that is closed at the top (the volume above the mercury is a vacuum) and opened into a reservoir of mercury at the bottom (Fig. 11.1), the column will rise to a height of about 29.9 in or 760 mm. English-speaking weather forecasters speak of the pressure in inches, but the standard definition is in millimeters; by definition, 1 atm = 760 mm Hg = 760 **torr.** The latter is a special unit named after E. Torricelli, the inventor of the barometer. Meteorologists typically use millibars in discussing atmospheric pressure, while most laboratory vacuum pumps are calibrated in torr. The units of pressure are summarized in Table 11.1.

Example 11.1

Severe weather is usually associated with regions of low atmospheric pressure, which create intense wind patterns and good conditions for the transport of moisture by air. The variation in pressure even in severe storms is relatively small, usually no more than 5%. Suppose a storm system has an atmospheric pressure of 29.13 in Hg. Express this pressure in Pa, atm, bar, mbar, and torr.

It is easiest to standardize everything in terms of atmospheres. The pressure of 29.13 in Hg corresponds to 740 mm Hg, or 740 torr. This is 740/760 or 0.974 atm, which corresponds to 0.986 bar or 986 mbar. Finally, since 1 bar = 10^5 Pa, this pressure corresponds to 9.86×10^4 Pa.

Different devices are used to measure pressure under different circumstances. Atmospheric pressure is measured by a **barometer,** which can be either a vertical tube of liquid (always mercury) or a vacuum chamber with an elastic

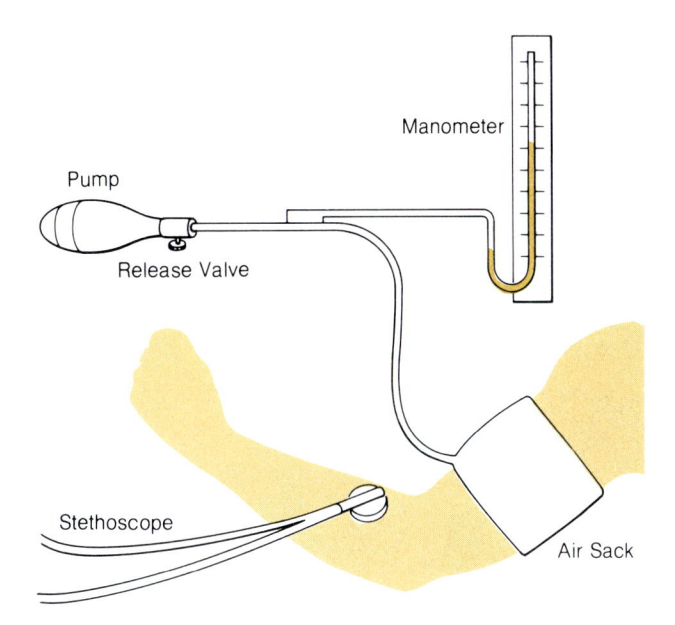

11.1 PRESSURE AND BUOYANCY IN STATIONARY FLUIDS

Figure 11.2

A sphygmomanometer.

membrane whose expansion or contraction with pressure changes is measured. Either barometer must compensate for the effects of temperature and moisture variations which can affect the readings. Human blood pressure is measured by a **sphygmomanometer,** which utilizes two columns of mercury. The air pressure in an inflatable sleeve can be adjusted to balance the pressure in the arteries of the arm (Fig. 11.2). Blood pressure is expressed in mm Hg, and is measured for the time when the heart contracts (when the pressure is maximized) and when it is at rest (when the pressure is minimized) between compressions. These readings correspond to the **systolic** and **diastolic** pressures.) For pressures very much lower than atmospheric pressure, completely different devices are used, because it becomes impractical to devise either liquid columns or elastic membranes that will survive very large pressure differences. Flow meters, which measure the rate of flow between two regions of different pressure, are used at low pressures, as are instruments that electrically charge gas particles and then sense the amount of charge, hence the amount of gas, that is present.

Density, Specific Gravity, and Pressure

The average density of any object or material is its mass divided by its volume. For solids and liquids, the density is fairly constant, since these materials have large bulk moduli (Chapter 8) and are therefore not easily compressed. Even so, the densities of solids and liquids do vary a little with such conditions as temperature, which affects the volume occupied by a given mass (thermal

Table 11.2
Densities of materials

Substance	Density, g/cm^3
Solids	
Platinum	21.4
Gold	19.3
Uranium	19.0
Lead	11.3
Copper	8.9
Iron and steel	7.9
Aluminum	2.7
Granite	2.7
Glass	2.6
Concrete	2.3
Bone	1.7
Ice	0.917
Wood (oak)	0.8
Wood (pine)	0.4
Liquids	
Mercury	13.6
Whole blood	1.05
Blood plasma	1.03
Sea water	1.025
Fresh water	1.000
Ethyl alcohol	0.79
Gasoline	0.68
Gases	
Carbon dioxide (0°C, 1 atm)	1.98×10^{-3}
Air (0°C, 1 atm)	1.29×10^{-3}
Helium (0°C, 1 atm)	0.79×10^{-3}
Steam (100°C, 1 atm)	0.60×10^{-3}

Densities in kg/m^3 are 1000 times densities in g/cm^3.

expansion is discussed in Chapter 12). For gases, which are easily compressed, density may vary significantly with conditions, and therefore statements of gas densities usually include the temperature and pressure.

In SI units, densities are given in kg/m^3. As we just saw for pressure, however, other units often are used. For density, the most common unit is g/cm^3 (Table 11.2). The conversion is simple: 1000 kg/m^3 = 1 g/cm^3. The historical reason for the use of g/cm^3 to measure density is that the density of water is very nearly 1 g/cm^3 (the original definition of the gram was the mass of one cubic centimeter of water at a specified temperature). Hence it is easy to compare densities in g/cm^3 with the density of water. The **specific gravity** (SG) is defined as the ratio of the density of a substance to that of water. Since water has a density of 1 g/cm^3, the value of the specific gravity is numerically equal to the density in g/cm^3. Any material whose specific gravity is greater than 1 will sink in water; if the SG is less than one, the material will float.

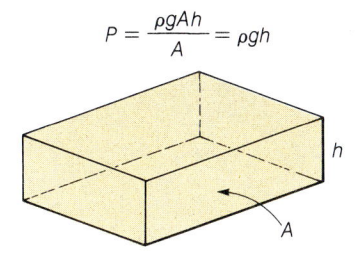

11.1 PRESSURE AND BUOYANCY IN STATIONARY FLUIDS

$$P = \frac{\rho g A h}{A} = \rho g h$$

Figure 11.3

Pressure due to a volume of fluid. The pressure at the bottom of a container of fluid with vertical sides is equal to the total weight of the fluid divided by the bottom surface area.

The density of air at 300 K and a pressure of 1 atm is about 1.2 kg/m^3, or 1.2×10^{-3} g/cm^3. The average density of the earth, on the other hand, is 5.5 g/cm^3, and that of a typical rock at the earth's surface is about 3 g/cm^3 (the earth is composed of iron or nickel at the center, so that its average density is higher than that of surface rock). The density of the human body is nearly 1 g/cm^3; hence most people just float in water.

The calculation of pressure exerted by a solid or a liquid due to its weight is quite straightforward, given the density of the material and the dimensions of the volume it fills (Fig. 11.3). The pressure at the bottom of a 50 m swimming pool of width 25 m and constant depth 2 m is simply the total weight of the water above, which is its density of 10^3 kg/m^3 times its volume of 2,500 m^3 times the acceleration of gravity, divided by its bottom surface area, which is 1,250 m^2. The result is a pressure of 1.96×10^4 Pa (this neglects atmospheric pressure, which should be added to correctly assess the total pressure at the bottom of the pool). Note that this result could have been obtained more directly, by simply calculating the weight of a 1 m^2 column of water of height 2 m; that is, the weight of 2 m^3 of water. In equation form, we can write

$$P = \rho g h \qquad (11\text{--}1)$$

where ρ is the density and h is the water depth. This simplification applies to liquids or solids with perfectly vertical sides.

Example 11.2 Calculate the pressure of seawater at an ocean depth of 5,000 m. Neglect the contribution of atmospheric pressure.

Here we assume that water is not compressed, so that its density is uniformly 10^3 kg/m^3 from the surface to this depth. The pressure at 5,000 m is proportional to the weight of a column of water 5,000 m high, having a cross section of 1 m^2; that is, the weight of a 5,000 m^3 volume of water. This is

$$P = \frac{W}{1 \text{ m}^2} = (5{,}000 \text{ m}^3)(10^3 \text{ kg/m}^3)\frac{(9.8 \text{ m/s}^2)}{1 \text{ m}^2} = 4.9 \times 10^7 \text{ Pa}.$$

We could have written this as

$$P = \rho g h = (1{,}000 \text{ kg/m}^3)(9.8 \text{ m/s}^2)(5{,}000 \text{ m}) = 4.9 \times 10^7 \text{ Pa}.$$

This is 484 times the atmospheric pressure at sea level.

Pascal's Principle

We mentioned in the preceding section that the pressure at the bottom of a swimming pool is the sum of the pressure due to the weight of the water and the pressure of the atmosphere. Yet the atmosphere is not acting directly on

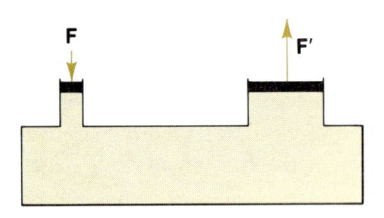

Figure 11.4

Hydraulic pistons. A downward force F on the small piston is converted into a larger upward force F' on the larger piston because the pressure is transmitted undiminished. The force is multiplied by the ratio of the areas of the pistons but acts over a displacement that is decreased by the same ratio.

the bottom of the pool; its pressure is transmitted by the water to the bottom. This phenomenon was analyzed in the early seventeenth century by B. Pascal, who also found that the transmission of pressure is uniform in all directions. At any depth in the pool, the pressure is due to the weight of the water above that depth, plus the atmospheric pressure acting on the water's surface. Furthermore, the pressure at any depth acts equally in all directions. Pascal's principle states

The pressure applied to a confined fluid is transmitted undiminished by the fluid in all directions.

Pascal's principle is the basis for a class of machines that use hydraulic fluids to convey pressure, hence force. Imagine liquid in a container with two movable (but sealed) pistons of different area (Fig. 11.4). If the small piston is pushed in, the pressure exerted on all sides of the confined volume is equal the pushing force divided by the area of the small piston. Therefore a force acts to push out the large piston. This force is equal to the pressure times the area of the large piston. Hence the force exerted on the small piston has been multiplied by the ratio of the piston areas. In this way, a relatively modest force can be used to create a large force. Recall from our discussion of machines (Chapter 5) that the output work cannot exceed the input work, so the force exerted on the small piston must be exerted over a relatively large displacement. Hydraulic systems are quite efficient, so that the actual mechanical advantage (the ratio of output force to input force is not much smaller than the ideal mechanical advantage (the ratio of input displacement to output displacement).

Example 11.3

A hydraulic lift in a garage is driven by a rectangular piston measuring 2 m by 0.5 m. The lift is activated by pushing down on a small piston, which is circular and has a radius of 0.1 m. What force is required on the small piston for the large one to support a car weighing 2 tons? How far must the small piston be displaced to lift the car to a height of 1 m?

The ratio of the piston areas is $(2 \text{ m})(0.5 \text{ m})/\pi(0.1 \text{ m})^2 = 31.8$. Assuming 100% efficiency, a force on the small piston of $(4,000 \text{ lbs})/31.8 = 126 \text{ lb}$ is required in order to support the car on the large piston. The ratio of piston displacements is also 31.8, so the small piston must be displaced a distance of 31.8 m to raise the car to a height of 1 m. But this is impractical in an ordinary garage, so the piston area ratio is actually much smaller than 31.8. Therefore the input force required is much larger than a person can exert directly, so another machine is required to displace the input piston.

Buoyancy and Archimedes' Principle

We learned in our preceding discussion that the pressure within a volume of liquid increases with depth, because the weight of the column of fluid increases with depth. We also learned that the pressure at any depth acts equally in all directions. It follows from these two statements that there is an upward pressure acting on any submerged object that counteracts its weight. We can see this by considering a cylindrical solid submerged in water (Fig. 11.5). The pressure due to the water acting on its upper surface is

$$P_u = \rho_w g h_u$$

where ρ_w is the density of the water and h_u is the depth of the water above the upper surface. The pressure exerted on the bottom of the cylinder is

$$P_l = \rho_w g h_l$$

where h_l represents the depth of the bottom of the cylinder. The buoyant force F_B is the difference between the force due to pressure acting on the lower surface of the cylinder and that due to pressure acting on the upper surface.

$$F_B = P_l A - P_u A.$$

A is the area of the top and bottom surfaces. Substituting for P_l and P_u yields

$$F_B = \rho_w g A (h_l - h_u).$$

The volume of the cylinder is

$$V = A(h_l - h_u)$$

so we have for the buoyant force

$$F_B = \rho_w V g. \tag{11-2}$$

Thus the upward force caused by the difference in pressure between the top and bottom of a submerged object is equal to the weight of the water in a volume equal to that of the object. We say that the buoyant force is equal to the weight of the fluid displaced by the submerged object. This statement is known as **Archimedes' principle,** who discovered it in the third century B.C.

We used a cylindrical object to rigorously demonstrate the principle because it was easy to calculate the pressure difference between its upper and lower surfaces. We could do this for other shapes, but there is a simpler argument to show that the principle holds for any shape. Consider an irregularly shaped object with volume V, submerged in a fluid with no forces acting on it except its own weight and the buoyant force (hence in general it will be falling, if its density is greater than that of the fluid). Now consider an equal volume of fluid at rest at the same depth and having the same shape. We know that for this volume of fluid, the weight and the buoyant force must be equal

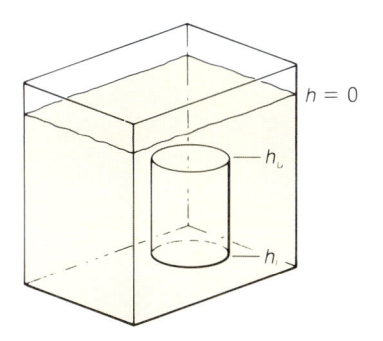

Figure 11.5

Buoyant force. The pressure exerted on the bottom of the submerged cylinder is greater than on the top. The difference creates an upward force on the cylinder.

because the fluid is in equilibrium. The buoyant force acting on the object must be equal to that acting on the identical fluid volume, because the conditions to which they are subjected are identical. Thus the buoyant force on the submerged object must be equal to the weight of an equal volume of water, as stated by Archimedes' principle.

When the density of the object is less than that of the fluid, the object will rise because the buoyant force exceeds the weight of the object. When the object reaches the surface, it will float. The buoyant force is then equal to the weight of the volume of fluid displaced by the submerged portion of the object. An equilibrium is reached in which the buoyant force is equal to the weight of the object; the level at which the object floats is dictated by this equilibrium condition. If the object sinks a little too low in the fluid, the buoyant force will become greater than the weight, and the object rises. If it rises above the equilibrium position, the buoyant force becomes less than the weight, and the object sinks to a lower level.

Example 11.4

A boulder weighing 3.6 tons on land is submerged in a lake (Fig. 11.6). Its density is 3 g/cm^3. What is its effective weight underwater?

We need to know the volume of the boulder to use Eq. 11–2. We are given the weight in air, W_a (3.6 tons = 7,200 lbs = 3.2×10^4 N), so we can find the mass.

$$m = \frac{W_a}{g} = \frac{3.2 \times 10^4 \text{ N}}{9.8 \text{ m/s}^2} = 3.3 \times 10^3 \text{ kg}$$

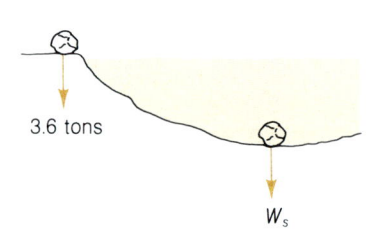

Figure 11.6

If ρ_r is the density of the rock, then the volume is

$$V = \frac{m}{\rho_r} = \frac{3.3 \times 10^3 \text{ kg}}{3,000 \text{ kg/m}^3} = 1.1 \text{ m}^3.$$

If ρ_w is the density of water, then the buoyant force is

$$F_B = \rho_w V g = (1,000 \text{ kg/m}^3)(1.1 \text{ m}^3)(9.8 \text{ m/s}^2) = 1.1 \times 10^4 \text{ N} = 1.2 \text{ tons}.$$

Finally, the effective weight W_s of the submerged boulder is equal to its dry weight less the buoyant force, or

$$W_s = W_a - F_B = 3.6 \text{ tons} - 1.2 \text{ tons} = 2.4 \text{ tons}.$$

It is easy to show, by using the above expressions for V and m, that $F_B = W_a(\rho_w/\rho_r)$. Then

$$W_s = W_a\left(1 - \frac{\rho_w}{\rho_r}\right).$$

Substituting values into this formula reproduces the result just found.

11.1 PRESSURE AND BUOYANCY IN STATIONARY FLUIDS

Example 11.5   A cubical crate with average density 0.65 g/cm³ floats in water. If the crate measures 0.8 m on a side, how far does it sink into the water as it floats?

We know that the crate will float at the equilibrium level where its weight is balanced by the buoyant force. Thus we need to find the level at which the displaced volume of water has the same weight as the crate. The equilibrium condition is

$$W = F_B$$

or

$$\rho_c V_c g = \rho_w V_d g$$

where ρ_c is the density of the crate, V_c is its total volume, ρ_w is the density of water, and V_d is the displaced volume when the crate is floating at equilibrium. We solve for this volume.

$$V_d = V_c \left(\frac{\rho_c}{\rho_w}\right) = \frac{(0.8 \text{ m})^3 (0.65 \text{ g/cm}^3)}{1 \text{ g/cm}^3} = 0.33 \text{ m}^3$$

The submerged volume V_d is given by

$$V_d = hA$$

where A is the cross-sectional area and h the depth of the crate's bottom surface below the water. We know that $A = (0.8 \text{ m})^2 = 0.64 \text{ m}^2$, so we have

$$h = \frac{V_d}{A} = \frac{0.33 \text{ m}^3}{0.64 \text{ m}^2} = 0.52 \text{ m}.$$

The crate will float with its bottom surface 0.52 m below the surface of the water.

Note that the density of an object of irregular shape can be determined by taking advantage of Archimedes' principle. A comparison of its dry weight and its effective weight under water yields a value for the buoyant force, which allows the volume of the object to be found (see Eq. 11–2). Given the volume and the mass (which is known from the dry weight), the density follows. It is said that Archimedes discovered his principle when seeking a method for determining the density of a crown, to see whether or not it was made of pure gold (it wasn't).

Example 11.6   Find the specific gravity of a rock whose weight out of water is 28.1 N, and whose effective weight under water is 19.8 N.

The buoyant force F_B is the difference between the dry weight and the underwater weight, or F_B = 8.3 N. From Eq. 11–2 we solve for the volume of the rock.

$$V = \frac{F_B}{\rho_w g}$$

The mass of the rock is

$$m = \frac{W}{g}$$

so the density is

$$\rho = \frac{m}{V} = \frac{W\rho_w}{F_B} = \frac{(28.1 \text{ N})(1{,}000 \text{ kg/m}^3)}{8.3 \text{ N}} = 3{,}400 \text{ kg/m}^3 = 3.4 \text{ g/cm}^3.$$

Thus the specific gravity is 3.4.

11.2 Surface Tension and Capillary Action

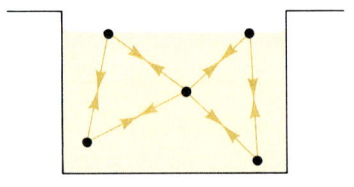

Figure 11.7

Surface tension. Molecules in a liquid attract each other. Those on the surface experience an unbalanced downward force that creates a slight compression of the surface. They resist being moved further apart or closer together, and this resistance allows the surface to support small objects that are denser than the fluid.

It is commonly observed that certain insects can walk on water, and that it is possible, with care, to make a needle float, even though its density is surely greater than that of water. These things are possible because the surface of a fluid acts as a weak membrane due to an effective elastic force acting to hold the molecules at the surface in place. We say that a fluid surface has **surface tension**.

If we consider the forces acting on molecules at the surface of a liquid, we see that there is a net downward force because there are no molecules above, which can exert binding forces to counteract those due to molecules below (Fig. 11.7). The surface becomes slightly compressed; this compressive force will always act to minimize the surface area of a liquid. Energy is required to stretch the surface area, so we say that the surface has a certain amount of potential energy, sometimes called the **surface energy**, that increases as the surface area is increased. Quantitatively, a **coefficient of surface tension** is defined as the potential energy per unit of area, and in the SI system it has units of J/m^2, or

$$\gamma = \frac{\text{PE}}{A}. \tag{11–3}$$

Measurement of the coefficient of surface tension (sometimes called simply the surface tension) in these units is not simple. It is done by measuring the amount of work needed to increase the surface area by a given amount, and applying the work-energy principle (Fig. 11.8). Suppose we have a rectangle of thin wire, across which a thin membrane of liquid forms, much like the

11.2 SURFACE TENSION AND CAPILLARY ACTIONS

membrane a child forms as he blows soap bubbles. The rectangular wire frame is adjustable, so that its area can be increased by moving one side outward. The work performed, which is equal to the increase in potential energy of the liquid surface, can be measured, as can the increase in surface area. The coefficient of surface tension is then

$$\gamma = \frac{W}{\Delta A} \qquad (11\text{--}4)$$

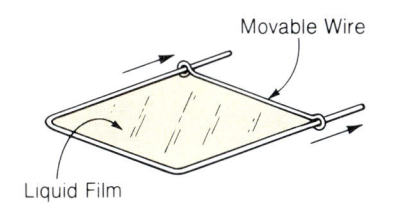

Figure 11.8

Measuring the coefficient of surface tension. The work needed to increase the liquid film's surface area is equal to the force times the displacement of the movable wire, and is equated with the change in the surface potential energy.

where W is the work performed, and ΔA is the increase in area.

This description of a hypothetical measurement of γ shows that the surface tension can be defined in terms of force per unit length, rather than energy per unit area. Since the work performed as the rectangle was expanded was equal to the force required to pull it apart times the displacement, and the increase of area is equal to the width l of the rectangle times the same displacement, we could write

$$\gamma = \frac{W}{\Delta A} = \frac{F \Delta s}{l \Delta s} = \frac{F}{l} \qquad (11\text{--}5)$$

where F is the force required to overcome the resistance of the surface tension, and l is the length of a line (imagine it as a fine wire) that is pulled across the surface.

In practice, surface tension is often measured by pulling a closed wire out of the surface and measuring the force at the point where the stretched surface is about to break (Fig. 11.9). The length l in Eq. 11–5 is then twice the circumference of the wire, because the sheet of liquid that is pulled out has two surfaces. Typical values of γ are given in Table 11.3.

Surface tension creates another well-known phenomenon, called **capillary action** or **capillarity**. The molecules in a confined liquid not only have cohesive forces for each other, which create the surface tension, but also **adhesive** forces,

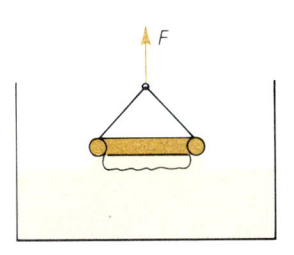

Figure 11.9

Alternative method for measuring the surface tension. As the thin ring is pulled from the liquid, surface tension resists. The total surface tension around the inner and outer diameters of the ring is equal to the measured force F when equilibrium is achieved.

Table 11.3
Surface tension

Liquid in contact with air	T(°C)	dynes/cm	N/m
Mercury	20	465	.465
Water	0	75.6	.0756
Water	20	72.8	.0728
Water	60	66.2	.0662
Water	100	58.9	.0589
Blood (whole)	37	58	.058
Glycerol	20	63.4	.0634
Carbon tetrachloride	20	26.95	.02695
Ethyl alcohol	20	22.75	.02275
Oxygen	−183	13.2	.0132
Neon	−248	5.50	.0055
Soap solution	20	2.50	.0025
Helium	−269	.12	.00012

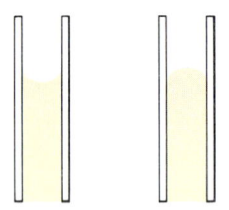

Figure 11.10

Meniscus formation. The surface of a liquid in a tube will curve upward or downward at the edges depending on whether the force binding molecules to the wall is greater or less than the cohesive forces of the molecules for each other.

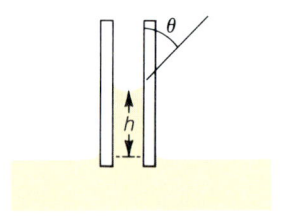

Figure 11.11

The rise of liquid in a capillary tube. This illustrates the geometry assumed in calculating h, the height to which the column rises above the reservoir level.

which tend to bind the molecules to the container walls. The relative strength of the cohesive and adhesive forces determines whether the **meniscus,** the surface of liquid in a small tube, curves upward or downward (Fig. 11.10). If the adhesive force is greater than the cohesive force, as it is for water in glass, then the meniscus curves upward. The liquid will actually rise upward until the upward force due to adhesion equals the weight of the column of liquid. It is this rise in a tube that is called capillary action or capillarity. When a fabric is dipped in water, the wet area will spread by capillary action if the adhesive force between the liquid and the fabric fibers exceeds the cohesive force, as for cotton in water. If, on the other hand, the adhesive force is less than the cohesive force, as for wool in water, then the wet area will not spread. For this reason, wool clothing is much better for wet weather or wet environments, such as hiking along a wet trail. The same factor determines whether a fabric is absorbent or not.

We can calculate how high a column of liquid will rise in a small tube whose end is submerged in a reservoir of the liquid (Fig. 11.11). We want to write an expression setting the upward force equal to the downward force, since this will be the equilibrium condition when the liquid has risen to its stopping point. The downward force is simply the weight of the vertical column, which is $\rho \pi r^2 h g$, where ρ is the density of the liquid and $\pi r^2 h$ is the volume of the cylindrical column. If θ is the angle between the direction of the surface tension and the vertical, then the upward force is equal to the cosine of θ times the surface tension given by Eq. 11–5, where l is the circumference of the tube (which is $2\pi r$). Putting this together yields

$$2\pi r \gamma \cos\theta = \pi r^2 \rho h g$$

or

$$h = \frac{2\gamma \cos\theta}{\rho g r}. \tag{11–6}$$

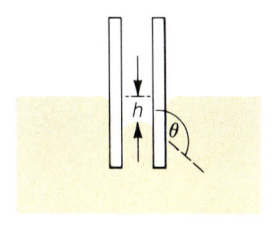

Figure 11.12

The fall of liquid in a capillary tube. If the cohesive forces between molecules exceed the adhesive forces attracting liquid molecules to the walls, then the liquid in the tube falls below the reservoir level. Equation 11–6 still yields h, if θ is defined as shown.

In many cases, the surface tension is sufficiently strong to pull the meniscus up so that its edges are almost vertical, so that θ is small and cosθ is nearly 1 (the value of θ is actually determined by the relative strengths of the cohesive and adhesive forces). Then the expression for h becomes

$$h = \frac{2\gamma}{\rho g r} \qquad (11\text{-}7)$$

which is much easier to use because it is not necessary to measure θ. The value of h can be quite large for sufficiently narrow tubes (that is, for small values of r).

Interestingly, capillary action can also cause the level of liquid in a tube to drop below the level of the surrounding reservoir (Fig. 11.12). This will happen whenever the adhesion force is less than the cohesion force, as for mercury in a glass tube. Thus, not only does water fail to spread through a wool fabric, it actually tries to compress itself.

Example 11.7 Suppose an ordinary drinking straw, a tube of radius 0.2 cm, stands in a glass of water. How high will the water inside the straw rise above the surface of the rest of the water in the glass at 0° C?

The value of γ for water in an environment of air is γ = 0.076 N/m. We assume that the straw material has a strong enough adhesion for water that Eq. 11–7 is valid. Then we have

$$h = \frac{2\gamma}{\rho g r} = \frac{2(0.076 \text{ N/m})}{(1{,}000 \text{ kg/m}^3)(9.8 \text{ m/s}^2)(0.002 \text{ m})} = 0.008 \text{ m}.$$

The water rises to a height of 8 mm above the surrounding liquid.

11.3 Fluids in Motion

Moving fluids play important roles in many situations. Water flows through pipes in buildings, providing water and carrying away sewage. Heated gases ejected through nozzles provide the thrust required to launch a rocket. Blood flowing through our veins and arteries provides the cells of our bodies with nutrients and carries away waste products.

A formal study of flowing fluids can be a very large topic, and fluid dynamics is the basis for entire courses. Here we will not be so thorough, but provide only the basic framework. Initially we will ignore some complexities such as turbulence and internal friction (viscosity), concentrating instead on smooth flows (called **laminar flows**). We will address the neglected complications in a later section.

The Equation of Continuity

If a fluid flows through a pipe, the quantity that comes out at one end is equal to the quantity that goes in at the other. This seems quite obvious, but is a basic concept in fluid dynamics. If we have a steadystate, where fluid flows continuously into one end of a pipe, then the rate of outflow at the other end must equal the rate of inflow. The **flow rate**, the mass per second passing a point in the pipe, is constant as a function of position in the pipe.

In mathematical terms, we may write an expression for the flow rate.

$$\frac{m}{t} = \frac{\rho A l}{t} = \rho A v \qquad (11\text{--}8)$$

Here l is the distance a fluid particle flows in time t and A is the cross-sectional area, so that $\rho A l$ is the mass of fluid that flows past a fixed point in this time. The flow velocity v is then equal to l/t.

Now suppose that the diameter of the pipe varies; that it has cross-sectional area A_1 at one point and A_2 at another (Fig. 11.13). The flow rate must be the same at both points, so we write

$$\rho_1 A_1 v_1 = \rho_2 A_2 v_2.$$

Most liquids are very difficult to compress, so that $\rho_1 = \rho_2$, and we can write

$$A_1 v_1 = A_2 v_2. \qquad (11\text{--}9)$$

This is the **equation of continuity** for a liquid. This equation tells us that the flow velocity is high where the cross-sectional area is small, and vice versa. Rivers obey the continuity equation, and consequently run rapidly where they are narrow, and slowly where they are wide.

The equation of continuity also comes into play when a fluid flowing through one pipe is split into two or more pipes. Then the cross-sectional area in Eq. 11–9 is the total cross-section of the pipes at any point. This happens in the human circulation system, when the blood flowing outward from the heart through a single major artery (called the aorta) eventually diverges into myriad tiny vessels (called capillaries). The flow rate in the aorta is equalled by the total flow rate in all the capillaries. The total cross section of the capillaries is much greater than that of the aorta, so the flow velocity in the capillaries is lower than it is in the aorta.

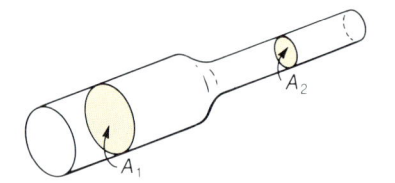

Figure 11.13

Flow in a pipe with varying cross section. Liquid must flow at a higher speed in the narrow portion, according to the equation of continuity (Eq. 11–9).

Example 11.8   A field is irrigated by a network of pipes of diameter 4 in, which is fed by a single large pipe 48 inches in diameter. If the large pipe supplies water to 50 of the smaller pipes, and the flow velocity in one of the small pipes is found to be 2 m/s, what is the flow velocity in the large pipe?

If v_1 and A_1 represent the single large pipe, we have

$$v_1 = v_2\left(\frac{A_2}{A_1}\right)$$

where A_2 represents the total cross-sectional area of the 50 small pipes. The ratio of cross-sectional areas is equal to the square of the ratio of the diameters, times the relative number of pipes of each diameter. Thus we have

$$v_1 = v_2\left[50\left(\frac{d_2}{d_1}\right)^2\right] = [2\text{ m/s}]\left[50\left(\frac{4\text{ in}}{48\text{ in}}\right)^2\right] = 0.7\text{ m/s}.$$

The total cross-sectional area of the 50 small pipes is smaller than that of the single large pipe, so the flow velocity in the small pipes is larger than in the large pipe.

Bernoulli's Equation

An important relationship can be derived for the pressure, velocity, and density of a fluid that is flowing. We will use some principles from our earlier discussion of mechanics: the work-energy principle and the conservation of energy.

An external force is required to make a fluid flow through a pipe. This force does work on the fluid, and work is done by the fluid on other fluid elements along the pipe. If we consider the work done on the fluid in a pipe at point 1, and that done by the fluid at point 2 (Fig. 11.14), then the net work done in time t is

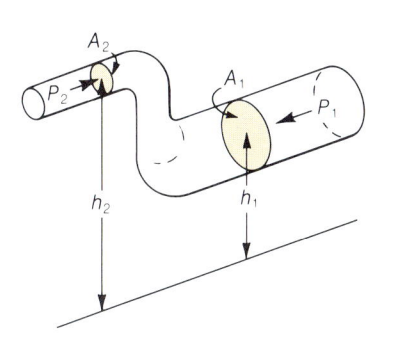

$$W = P_1 A_1 v_1 t - P_2 A_2 v_2 t$$

Figure 11.14

Diagram for Bernoulli's equation.

because the force acting at each point is the pressure times the cross-sectional area, and the displacement over which the force acts is the product of the velocity of flow times the time interval.

There is a net change in the energy of the fluid as a result of the work that is done. The kinetic energy change is the difference between the kinetic energy $\frac{1}{2}mv^2$ of a fluid element of mass m at the two points 1 and 2, where m is the mass contained in the volume V of fluid that flows past each point in time interval t. This mass is

$$m = \rho V = \rho A v t.$$

Thus the change in kinetic energy is

$$\Delta(\text{KE}) = \frac{1}{2}(\rho A_2 v_2 t)v_2^2 - \frac{1}{2}(\rho A_1 v_1 t)v_1^2$$

where the density is taken as constant on the assumption that the liquid is incompressible.

If the fluid changes height because the pipe curves up or down, then there is a change in potential energy as well. This is simply the difference between the potential energy at point 2 and that at point 1, the energy in each case being given by $mgh = \rho Avtgh$. Hence the change in potential energy is

$$\Delta(\text{PE}) = \rho A_2 v_2 t g h_2 - \rho A_1 v_1 t g h_1.$$

Now we can apply the work-energy principle and set the net work done equal to the total change in energy $\Delta(\text{KE} + \text{PE})$

$$P_1 A_1 v_1 t - P_2 A_2 v_2 t = \frac{1}{2}(\rho A_2 v_2 t) v_2^2 - \frac{1}{2}(\rho A_1 v_1 t) v_1^2 + \rho A_2 v_2 t g h_2 - \rho A_1 v_1 t g h_1$$

This simplifies, with the help of Eq. 11-9, to

$$\frac{1}{2}\rho v_1^2 + \rho g h_1 + P_1 = \frac{1}{2}\rho v_2^2 + \rho g h_2 + P_2. \qquad (11\text{-}10)$$

This is **Bernoulli's equation**, and it is a very powerful expression, linking the fundamental characteristics P, v, and ρ at different points in a flowing fluid. It has many useful applications, even though it contains a number of parameters, because the values or relative values of these parameters are often known.

One variable whose value may be easily known is the pressure difference $P_1 - P_2$. If we consider any system exposed to the atmosphere at both ends, for example, this difference is zero, and the terms P_1 and P_2 can be dropped from Eq. 11-10 (we explore this further in Example 11.9). Also, if the height does not change, the ρgh terms cancel. This is particularly interesting, because it tells us why an airplane flies or a sailboat moves upwind. If we set $h_1 = h_2$ in Eq. 11-10 so that the potential energy terms cancel, then we have

$$\frac{1}{2}\rho v_1^2 + P_1 = \frac{1}{2}\rho v_2^2 + P_2. \qquad (11\text{-}11)$$

If we now consider a pipe with a variable cross-section, we see from this expression that where the flow velocity is high because the cross section is small, the pressure is also small. We can rearrange this expression to find the pressure difference between two points where the velocities are v_1 and v_2.

$$P_1 - P_2 = \frac{1}{2}\rho(v_2^2 - v_1^2). \qquad (11\text{-}12)$$

Thus in a pipe that narrows as fluid flows along it, the pressure exerted on the walls decreases as the pipe narrows and the flow velocity increases. The underlying principle is the same for an airplane wing or a sailboat.

A fluid flowing over a surface, like fluid flowing through a pipe, exerts less pressure when the flow velocity is high than when it is low. An airplane wing is designed so that the air flowing over the top has to travel farther per unit time than the air flowing underneath (Fig. 11.15). Therefore the flow

11.3 FLUIDS IN MOTION 347

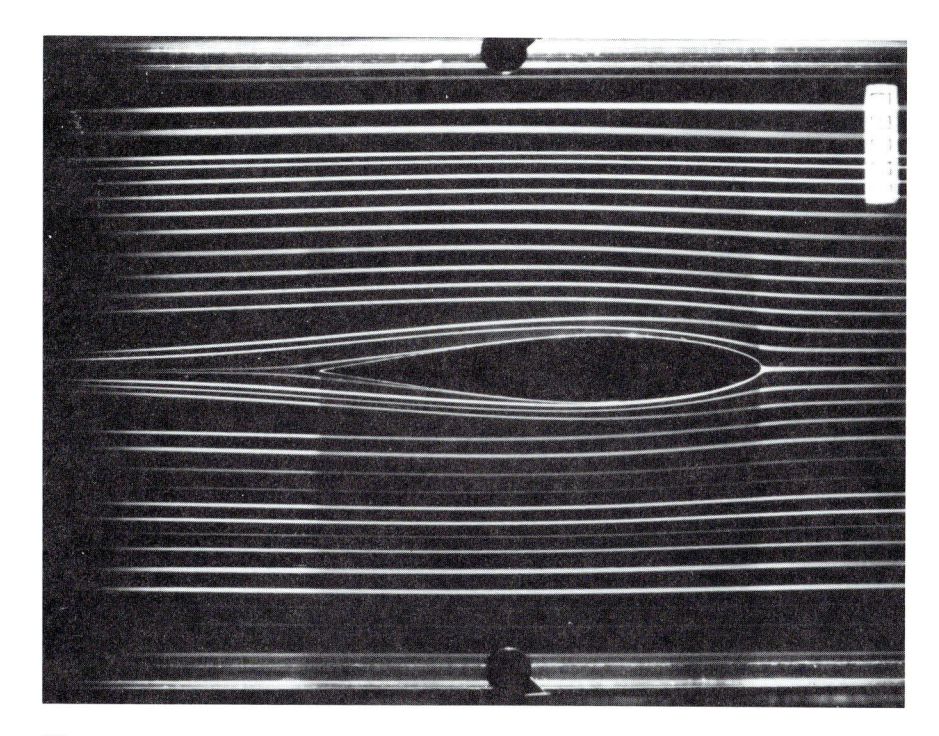

Figure 11.15

Cross section of an airplane wing.

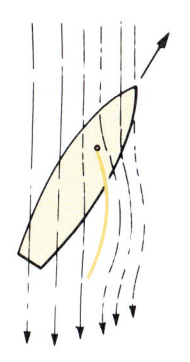

Figure 11.16

Curvature of the sail allows a sailboat to move upwind.

velocity over the top of the wing is greater than under the wing, and the air pressure acting on the top of the wing is less than on the bottom. The resultant upward force is called the **lift**. If the surface area of the wings is large enough and the air speed great enough, the lift exceeds the weight of the airplane, and it flies. A sailboat can actually move upwind (though not directly) because its sails curve, creating higher air speed on the front than on the back, and hence a pressure difference acting to push the boat forward (Fig. 11.16). In all these cases, we are ignoring (for the time being) other factors which influence the motion. One of these factors is turbulence, which can actually overwhelm the Bernoulli effect. A baseball, for example, curves in the direction opposite that predicted on the basis of the Bernoulli effect (Fig. 11.17).

Interestingly, Bernoulli's principle, as the statement of Eq. 11–10 is sometimes called, can be used to design an effective pump. Suppose that a chamber containing fluid has two openings, with air flowing rapidly past one of them (Fig. 11.18). The air velocity creates a reduction in pressure at that opening, so that the fluid is forced out by atmospheric pressure acting at the other one. This principle is the basis of the **Venturi tube**, which is used in automobile carburetors to pull fuel out of a reservoir and mix it with air for combustion. Such tubes are also used in a device called a **Venturi meter**, which measures the flow rate of a fluid by sensing the pressure at the end of a tube exposed to the flow.

CHAPTER 11 FLUIDS

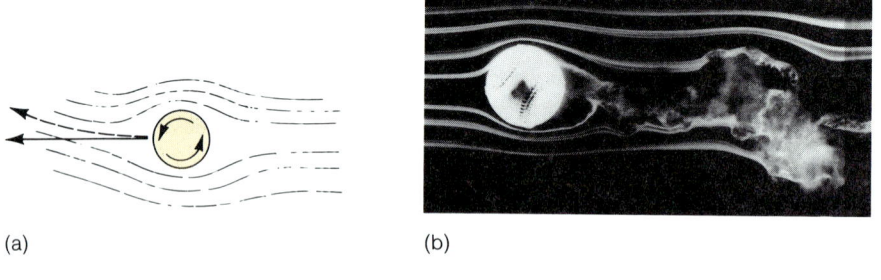

(a)   (b)

Figure 11.17

A curve ball. In (a), an idealization indicates that the spinning ball should curve upward, since the pressure on the top of the ball should be lower than on the bottom. In (b) we see that in reality turbulence causes the ball, which is moving and spinning in the same direction as in (a), to move downward.

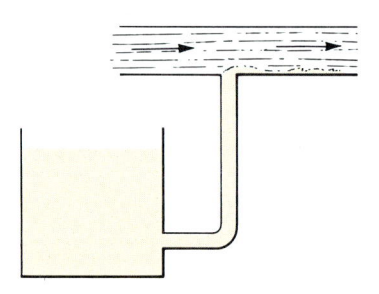

Figure 11.18

A Venturi tube. Air pressure in the upper pipe is reduced according to Bernoulli's equation, so fluid flows into it from the reservoir, which is exposed to atmospheric pressure.

Example 11.9   A lake with a surface of 2×10^8 m^2 is drained through a pipe in its bottom. The pipe has a one-meter diameter. What is the flow velocity in the pipe if the lake is 30 m deep?

Here $P_1 = P_2$, since both the lake's surface and the open end of the pipe are exposed to the same pressure, that of the atmosphere. The situation may be viewed as that of a pipe with a variable cross-sectional area, the area at point 1 being the area of the lake's surface, and at point 2 being the cross-section of the outlet pipe. Setting $P_1 = P_2$ in Eq. 11–10 and using Eq. 11–9 to substitute for v_1 leads to

$$\frac{1}{2}\rho v_2^2 \left[1 - \left(\frac{A_2}{A_1}\right)^2 \right] = \rho g(h_1 - h_2)$$

where A_2/A_1 is the ratio of the cross-sectional area of the pipe to the surface

11.3 FLUIDS IN MOTION

area of the lake, a ratio whose value is negligible. Hence we have

$$v_2 = \sqrt{2g(h_1 - h_2)}$$

which is sometimes known as **Torricelli's equation,** because it was discovered experimentally by E. Torricelli several decades before the work of Bernoulli.

Now, substituting for h_1 and h_2, we find

$$v = \sqrt{2(9.8 \text{ m/s}^2)(30 \text{ m})} = 24.2 \text{ m/s}.$$

Notice that the expression for the velocity is the same as for a freely-falling object dropping from rest through a height $h_1 - h_2$. Notice also that the flow velocity in the outlet pipe depends on the depth of the lake; that is, if a different lake had exactly the same volume of water, but were deeper (and therefore had a smaller surface area), the outflow velocity would be greater.

Example 11.10   Suppose radioactive coolant water flows through a 1-m diameter pipe in the basement of a reactor building, then is pumped through a series of 500 smaller pipes, each with a diameter of 0.1 m, that flow horizontally across the building at a level 10 m above the basement. If the flow velocity in the large pipe is 0.5 m/s and the pressure there is 6 atm (or 6.08×10^5 Pa), what is the flow velocity and pressure in the pipes in the upper level?

We find the velocity first, using Eq. 11–9. Following exactly the same procedure as in Example 11.8, we find that the velocity in the small pipes is

$$v_2 = \frac{v_1 A_1}{A_2} = \frac{(0.5 \text{ m/s})(1 \text{ m})^2}{(500)(0.1 \text{ m})^2} = 0.1 \text{ m/s}.$$

We now solve Eq. 11–10 for the pressure P_2.

$$P_2 = P_1 + \frac{1}{2}\rho(v_1^2 - v_2^2) + \rho g(h_1 - h_2)$$

Substituting for P_1, v_1, v_2, h_1 (which we set equal to zero), and h_2 yields

$$\begin{aligned}P_2 &= 6.08 \times 10^5 \text{ Pa} + (0.5)(1{,}000 \text{ kg/m}^3)[(0.5 \text{ m/s})^2 - (0.1 \text{ m/s})^2] \\ &\quad - (1{,}000 \text{ kg/m}^3)(9.8 \text{ m/s}^2)(10 \text{ m}) \\ &= 5.10 \times 10^5 \text{ Pa} = 5.04 \text{ atm}.\end{aligned}$$

Note that the velocity term played a very small role in determining the pressure in the small pipes; the major influence is their height above the large pipe in the basement. If the building were substantially taller (about a factor of six in this example), the pressure of 6 atm in the large pipe in the basement would have been inadequate to raise the water to the upper level. (Note that the value of P_2 would have been negative if h_2 had been much larger than 60 m).

Viscosity

So far we have discussed flowing fluids with laminar flow, meaning that the flow is smooth, without disruption or turbulence. We have also ignored resistance to flow caused by the interaction of molecules within the fluid. Real fluids are subject to such forces, which create a kind of internal friction called **viscosity**. In an intuitive sense, the viscosity of a fluid is a measure of its stickiness and its ability to flow. Syrup is more viscous than water, and grease is more viscous than syrup.

Viscosity in liquids is caused by cohesion of the molecules for each other, and by adhesion of the molecules for the walls of the containers through which the flow occurs. For gases, collisions cause viscosity between gas particles. So long as a flow is laminar, the resistive force caused by viscosity is proportional to the flow velocity. There is a point where the resistance suddenly increases much more rapidly; this is the point where turbulence sets in. We will discuss viscosity in laminar flows first, then we will tackle the subject of turbulence in the next section.

To see how viscosity is defined, we consider two plates of area A separated by a layer of fluid (Fig. 11.19). The viscosity governs how much force is required to move the plates with respect to each other. Experiments show that the viscous force acting to resist this motion is proportional to the flow velocity and the surface area of the plates, and inversely proportional to the separation of the plates. The proportionality constant is called the **coefficient of viscosity**, usually denoted η. Thus, in equation form, we write

$$F_v = \frac{\eta v A}{l}, \tag{11-13}$$

where F_v is the viscous force, η is the coefficient of viscosity, v is the relative velocity of the two plates, A is their surface area, and l is their separation.

We see from Eq. 11-13 that the units of η in the SI system are $N \cdot s/m^2$, which is the same as $Pa \cdot s$. In the cgs system we have $dyne \cdot s/cm^2$, which is called **poise** (P), after the French physicist J. Poiseuille (pronounced "pwah-say"), who developed much of the mathematical framework for the study of viscosity. Some representative values of η are given in Table 11.4. Note that temperatures are given also; the viscosity is a fairly sensitive function of temperature (consider the difference between warm and cold syrup).

The characterization of viscosity given by Eq. 11-13, while intuitively sensible, is not very practical because the movement of plates over a fluid layer is not a very common problem. It is probably more useful to consider the effect of viscosity on the flow of fluids through pipes. Experiments show that viscosity causes pressure in the fluid to drop along the length of a uniform pipe (we could say that a flow occurs only when there is a pressure difference between the ends of a pipe; otherwise the viscosity would inhibit flow). The amount of pressure drop depends on the length of the pipe, the viscosity, the volume rate of flow (V/t; the volume passing a fixed point in time t), and, very sensitively, on the radius of the pipe. The relationship among all these quantities, developed by Poiseuille, is

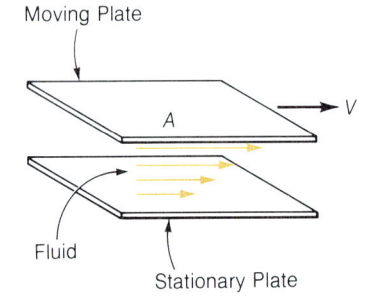

Figure 11.19

Measurement of viscosity. The coefficient of viscosity is determined by measuring the force required to move the upper plate at velocity v with respect to the lower plate.

11.3 FLUIDS IN MOTION

Table 11.4
Viscosity

	Temperature (°C)	Viscosity (η) (cp)	($N \cdot s/m^2$)
Glycerin	20	1490.	1.49
Castor oil	20	986.	0.986
Engine oil (SAE 10)	30	200.	0.200
Olive oil	20	84.	0.084
Sulfuric acid	20	25.4	0.0254
Ethylene glycol	20	19.9	0.0199
Mercury	20	1.554	1.55×10^{-3}
Water	0	1.79	1.79×10^{-3}
Water	20	1.00	1.00×10^{-3}
Water	100	0.282	2.82×10^{-4}
Air	0	0.0171	1.71×10^{-5}
Air	20	0.0181	1.81×10^{-5}
Air	100	0.0218	2.18×10^{-5}

1 poise = 0.1 N · s/m² = 1 dyne · s/cm²
1 centipoise (cp) = 0.01 poise
1 Pa · s = 1 N · s/m² = 10^3 cp

$$\frac{V}{t} = \pi r^4 \frac{(P_1 - P_2)}{8\eta L} \tag{11-14}$$

where r is the pipe radius, $(P_1 - P_2)$ is the pressure difference over the pipe length L, and η is the coefficient of viscosity, as usual. This equation is used to design pipe systems because it specifies the pressure difference required to achieve a certain flow rate over a specified distance. It is easy to rewrite Eq. 11–14 in terms of the flow velocity, rather than the volume flow rate, since the volume flow rate is just the pipe cross-sectional area times the flow velocity. This form is generally more practical, however.

Example 11.11

Suppose the Alaska pipeline (Fig. 11.20) flows horizontally, has a constant diameter of 2 m, and carries a volume flow rate of 3 m³/s. If it employs pumps capable of creating a pressure difference of 0.5 atm, how closely spaced must the pumps be? Assume a temperature of 0°C, for which oil has coefficient of viscosity $\eta = 1$ Pa · s.

We need to find the distance L over which the pressure in the pipe will drop by 0.5 atm = 5.1×10^4 Pa. Solving Eq. 11–14 for L yields

$$L = \frac{\pi r^4 (P_1 - P_2)}{8\eta (V/t)} = \frac{\pi (1 \text{ m})^2 (5.1 \times 10^4 \text{ Pa})}{8(1 \text{ Pa} \cdot \text{s})(3 \text{ m}^3/\text{s})}$$
$$= 6{,}700 \text{ m} = 6.7 \text{ km}.$$

Thus pumps should be no farther than 6.7 km apart to maintain the required volume flow rate.

Figure 11.20

So far we have concentrated on a fluid flowing through a pipe. Viscosity also plays a role when an object moves through a stationary fluid. Then we call the viscous force that resists motion **drag** or, in the earth's atmosphere, **air resistance** (already briefly described in Chapter 3). When the flow of fluid around the moving object is laminar, then the drag or viscous force is proportional to the velocity; that is,

$$F_v = kv \qquad (11\text{–}15)$$

where k is the **drag coefficient**. The units of k are N · s/m, which is equivalent to a length times the coefficient of viscosity. The actual value of k depends on the shape of the object and the viscosity of the fluid through which it moves. For a sphere of radius r, it is found that $k = 6\pi r\eta$, so the drag is

$$F_v = 6\pi r\eta v. \qquad (11\text{–}16)$$

(This expression is sometimes called **Stokes' law**).

An object falling through a fluid under the acceleration of gravity is subject to the downward force of gravity and two upward forces: drag and buoyancy. As the object accelerates, the drag force increases, eventually reaching equality with the force of gravity. After that time, the net force is zero, and no further acceleration occurs. The object falls at a constant velocity called the **terminal velocity**. The drag force is given by Eq. 11–15; and the buoyant force by Eq. 11–2. If we write the gravitational force mg in terms of the volume and density of the object, then set the gravitational force equal to the sum of the drag and buoyant forces, we find for the equilibrium case

$$\rho_o V g = k v_T + \rho_f V g$$

where ρ_o is the density of the object, V is its volume, ρ_f is the density of the fluid, and v_T is the terminal velocity. Rearranging, we find

$$v_T = \frac{(\rho_o - \rho_f) V g}{k}. \qquad (11\text{–}17)$$

For fluids other than air, the terminal velocity is often called the **sedimentation velocity,** since sedimentation is the process by which particles sink to the bottom of a liquid. For very small particles, the sedimentation velocity is rather small, but can be enhanced by using a centrifuge to increase the acceleration. Then the term g is replaced by the rotational acceleration $\omega^2 r$, where ω is the angular velocity of the centrifuge and r is the distance from the axis of rotation.

It turns out that laminar flow is an accurate approximation for falling objects only when the object is very small or the viscosity of the fluid is very large. For any object falling through the air or another gas, laminar flow does not occur, and Eq. 11–15 cannot be used to find the terminal velocity. We must, therefore, discuss turbulence.

11.3 FLUIDS IN MOTION   353

(a)

(b)

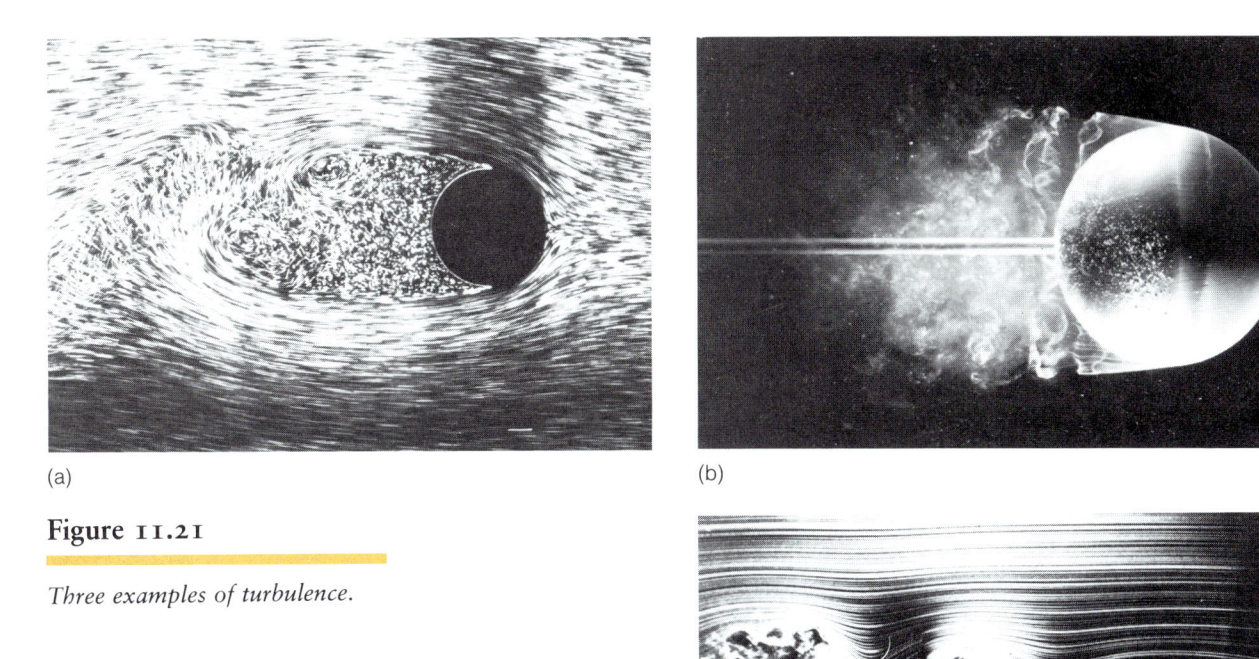

(c)

Figure 11.21

Three examples of turbulence.

Turbulence

The common intuitive view of turbulence is that of a flow accompanied by eddies and chaotic motions of the fluid (Fig. 11.21), rather than the smooth motion of laminar flow. Turbulence is very difficult to analyze, but we can discuss its effects. It is found that the viscous force depends much more sensitively on velocity when turbulence begins than in the case of laminar flow. Whereas the viscous force for laminar flow is proportional to the velocity, when turbulence sets in, the force depends on a higher power of the velocity, such as the square or the cube. This is the reason a great deal of effort goes into designing streamlined vehicles for ground or air travel.

A characteristic condition for the onset of turbulence has been found experimentally. We define the **Reynolds number (Re)** as

$$Re = \frac{2vr\rho}{\eta}, \qquad (11\text{--}18)$$

where v is the average flow velocity, ρ is the fluid density, r is the pipe radius, and η is the coefficient of viscosity. It has been found that turbulence will generally occur when $Re \geq 2,000$. The dependencies in Eq. 11–18 make some intuitive sense. We would expect that turbulence would occur most easily for high flow velocities and small viscosities, for example. On the other hand, the dependence on pipe radius and fluid density might not be so obvious.

Example 11.12 Heated air flows through ducts in the walls of a house. If the ducts are cylindrical with a radius of 0.2 m and the furnace provides a volume flow rate of 0.2 m³/s, is the flow turbulent or not? The coefficient of viscosity for air at room temperature is $\eta = 1.8 \times 10^{-5}$ Pa·s, and the density of air is 1.3 kg/m³.

We need to find the flow velocity, which is related to the volume flow rate by $v = V/A$, where A is the cross-sectional area of the duct. Therefore we have $v = (0.2 \text{ m}^3/\text{s})/\pi(0.2 \text{ m})^2 = 1.6$ m/s.

Now we can calculate the Reynolds number.

$$Re = \frac{2v r \rho}{\eta} = \frac{2(1.6 \text{ m/s})(0.2 \text{ m})(1.3 \text{ kg/m}^3)}{(1.8 \times 10^{-5} \text{ Pa} \cdot \text{s})} = 46,200$$

The flow is definitely turbulent.

The effect of turbulence for an object moving through a fluid is similar qualitatively to that for a fluid flowing in a pipe, in that turbulence causes the drag force to depend on a higher power of the velocity. Once turbulence sets in, the drag force is roughly proportional to the square of the velocity. For very high velocities, the dependence becomes even more sensitive.

A new Reynolds number can be defined for an object moving through a fluid.

$$Re' = \frac{vL\rho}{\eta}, \qquad (11\text{--}19)$$

where L is the characteristic size of the object (in one dimension), and ρ and η refer to the fluid, as before. Turbulence occurs when $Re' \geq 20$, which is usually the case for objects moving at ordinary speeds through air, as we have already mentioned.

Example 11.13 A 0.5 m diameter child's balloon drops from the ceiling to the floor slowly, taking about 5 seconds to travel 8 feet. Is its motion through the air laminar, or is it turbulent? The coefficient of viscosity for air at room temperature is $\eta = 1.8 \times 10^{-5}$ Pa·s, and its density is 1.3 kg/m³.

This is a simple matter to calculate. We just substitute a velocity of 1.6

ft/s = 0.5 m/s into Eq. 11–19, along with a characteristic size $L = 0.5$ m, and find

$$Re' = \frac{vL\rho}{\eta} = \frac{(0.5 \text{ m/s})(0.5 \text{ m})(1.3 \text{ kg/m}^3)}{(1.8 \times 10^{-5} \text{ Pa} \cdot \text{s})} = 18{,}100.$$

This is well into the turbulent range.

FOCUS ON PHYSICS

Drag and Fuel Economy in Cars*

In this chapter we have discussed turbulence and laminar flows, and mentioned wind resistance (also discussed in Chapter 3). One of the areas of greatest practical application of these studies is the design of road vehicles, because overcoming drag is a major factor in fuel consumption.

For cars travelling at relatively low speeds, as in city driving, most of the engine power is used to overcome the car's inertia, and aerodynamic drag is not significant. For suburban driving, however, speeds are in the 30 to 40 mph range, and about 25% of the power goes into overcoming drag. At highway speeds, more than half the power, that is, more than half of the fuel consumed, is required to overcome drag. Clearly there are great advantages to be gained by minimizing drag.

Aerodynamic drag is a force that always opposes the motion of the vehicle, much like friction. Mathematically, the drag force F_D can be represented by the equation

$$F_D = \tfrac{1}{2}\rho C_D A v^2$$

*Information for this article was taken largely from "The Aerodynamic Drag of Road Vehicles," by W. H. Bettes, in *Engineering and Science* (California Institute of Technology), 45(3):4.

where ρ is the air density, C_D is the **drag coefficient**, A is the cross-sectional (frontal) area of the vehicle, and v is the vehicle speed. Thus, in designing a vehicle for minimum drag, the goal is to reduce the factors in this equation to the smallest reasonable values. The density of air, of course, cannot be reduced, and vehicle speed must be varied for driving. That leaves the frontal area A and the dimensionless drag coefficient C_D as factors for improvement.

It is difficult to reduce the frontal area of cars much below present values, which are about 24 ft^2 (2.2 m^2) for full-size cars, because of the need for passenger space. (Compact and subcompact cars do have relatively small frontal areas, however, and economize by minimizing inertia as well, through reduced mass.)

Aerodynamic studies in wind tunnels show that the best hope for better fuel economy lies in reducing the drag coefficient C_D. For a full-size car, a 10% reduction in C_D would result in a 5% savings in fuel consumption. It would require greater than 15% reduction in car weight to achieve the same saving.

To minimize the value of C_D, the critical factors are turbulence and airstream geometry. It is

(continued)

particularly important to keep airstreams (flows of air over the surface of the vehicle) from separating from the vehicle if possible, and to control how and where they separate if they must. There is little or no separation of airstreams over a well-designed airplane wing, but of course it is impractical (except for specialized racing vehicles) to design cars shaped like airplane wings. When an airstream separates from the vehicle surface, vortices and eddies form, and these are the principal sources of drag.

As car designs have evolved since the first mass-production passenger cars of the early 1900s, the drag coefficients have steadily shrunk (Fig. 11.1.1). Typical values of C_D today are around 0.5, whereas for the squarish-looking cars of the 1920s, C_D was usually closer to 0.7. Even so, most of this reduction in C_D was more or less accidental; sleek-looking, lower body profiles became popular. There is still substantial room for improvement, according to wind tunnel tests, and modern automobile manufacturers are including drag reduction in their design criteria to a greater extent.

There are many contributions to the overall drag coefficient for a car. There is airstream separation and turbulence along the sides and underneath the car, and especially at the rear. But the most important area where turbulence and drag develop is at the front, where the full force of the incoming air confronts the car. Many modern cars have sharp corners rather than smoothly rounded ones, which could reduce C_D by as much as 15% (Fig. 11.1.2). Further reductions in drag could be achieved by making the front lower and more angled, although progress is limited by the need to have room for the engine (or luggage compartment, in rear-engine cars). Front-engine cars invariably use incoming air as a coolant, which means that air is allowed to flow in through the grill. This creates pressure against the back wall (the firewall) of the engine compartment that adds to the drag of the car. Some improvement could be made by providing better paths for this air to flow through and out of the engine compartment. Other improvements have been made in the front-end design of cars, primarily by making smooth

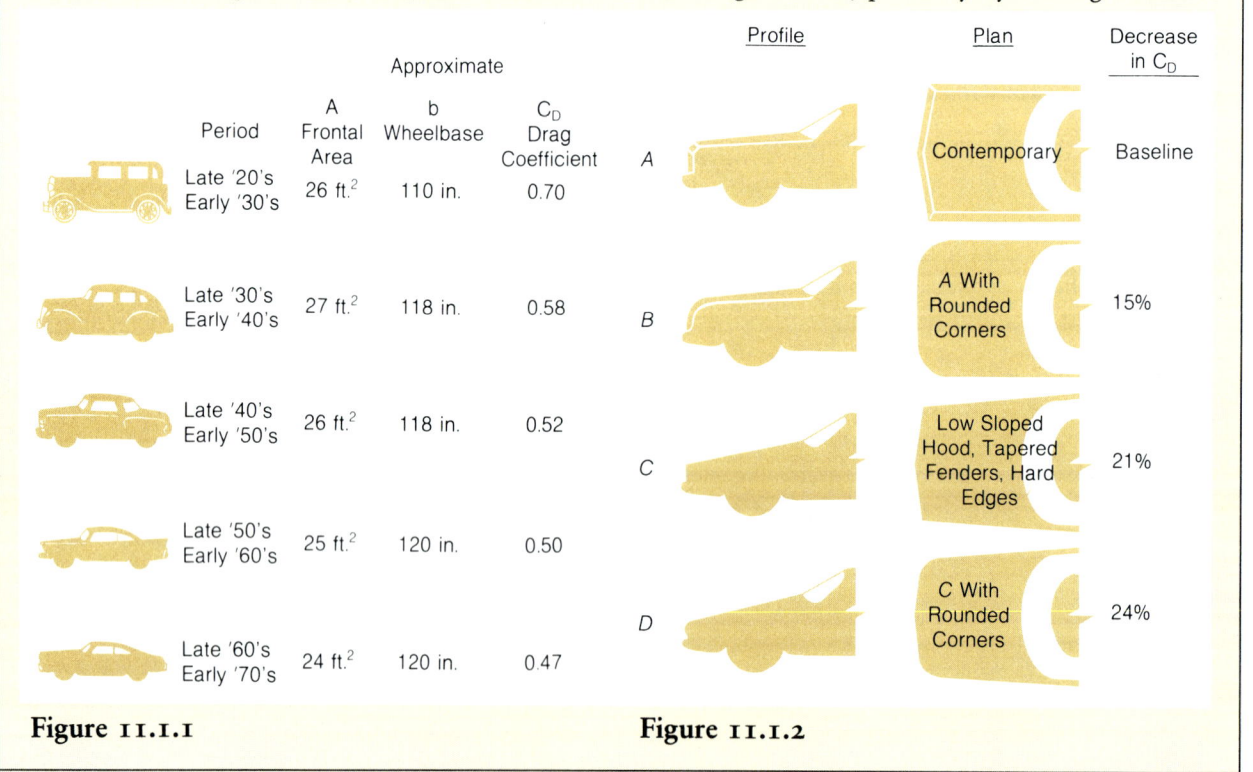

Period	A Frontal Area	b Wheelbase	C_D Drag Coefficient
Late '20s Early '30s	26 ft.^{2}	110 in.	0.70
Late '30s Early '40s	27 ft.^{2}	118 in.	0.58
Late '40s Early '50s	26 ft.^{2}	118 in.	0.52
Late '50s Early '60s	25 ft.^{2}	120 in.	0.50
Late '60s Early '70s	24 ft.^{2}	120 in.	0.47

Figure 11.1.1

	Profile	Plan	Decrease in C_D
A		Contemporary	Baseline
B		A With Rounded Corners	15%
C		Low Sloped Hood, Tapered Fenders, Hard Edges	21%
D		C With Rounded Corners	24%

Figure 11.1.2

fittings for headlights and windshields, so that the entire front has few projections.

At the rear of a car, there must necessarily be airstream separation, or else cars would have lengthy, tapering tails that would be very impractical. In contrast to the best strategy for the front of a car, at the rear it is best to have sharp edges rather than rounded corners. The reason is that if the corners are rounded, the separation points for the airstreams will wander about, and this causes greater turbulence and drag. It is better to keep the airstream separation points fixed by building the car with sharp edges at the rear.

The angle of the rear is also important. Here there are competing processes at work. Some angles create substantial turbulence due to air flowing around the sides of the car, but minimize the effects of air coming over the top, so that the best design is a compromise. Cars with rather steep rear ends, such as hatchbacks and especially stationwagons, can actually gain due to vortices created behind the car that exert a forward pressure and help the car along.

Another somewhat complicated area to design is the undercarriage of a car, which normally has many projections due to axles, shock absorbers, exhaust systems, fuel tanks, and other components, all of which create substantial drag. It is impractical and dangerous to cover these, but significant reductions in drag could be achieved, for example, with a smooth bottom pan that covers the entire underside of the car. While this is not feasible, it has been shown that a partial pan, extending from the front bumper to the front axle, could reduce C_D significantly.

Pressure on auto makers to economize in the fuel consumption of their products has built greatly in the past several years as gasoline prices have risen dramatically. We can expect cars of the future to take greater advantage of the existing opportunities to reduce aerodynamic drag.

Perspective

In this chapter we have been introduced to two states of matter that have until now been largely overlooked in this text. We have devloped some understanding of the properties of fluids, both at rest and in motion, but have stopped short of developing the dynamical treatment of fluids as fully as we did that of solid objects. We have emphasized practical applications, and have gained some useful knowledge for design and engineering. We have also laid the groundwork for more detailed studies of the transport of fluids in biological systems.

We turn now from the problem of flows to a more extensive analysis of static fluids and solids. Our next task is to analyze the internal energy of solids and fluids, and to learn the nature of heat and its effects on molecular motions.

SUMMARY TABLE OF FORMULAS

Pressure exerted by a fluid due to gravity:
$$P = \rho g h \quad (11\text{–}1)$$

Buoyant force:
$$F_B = \rho_w V g \quad (11\text{–}2)$$

Coefficient of surface tension:
$$\gamma = \frac{\text{PE}}{A} \quad (11\text{–}3)$$

$$\gamma = \frac{W}{\Delta A} \quad (11\text{–}4)$$

$$\gamma = \frac{F}{l} \tag{11-5}$$

Rise of fluid in a column:
$$h = \frac{2\gamma \cos\theta}{\rho g r} \tag{11-6}$$

Rise of fluid for strong surface tension:
$$h = \frac{2\gamma}{\rho g r} \tag{11-7}$$

Flow rate through a pipe:
$$\frac{m}{t} = \frac{\rho A l}{t} = \rho A v \tag{11-8}$$

Equation of continuity for fluid flow:
$$A_1 v_1 = A_2 v_2 \tag{11-9}$$

Bernoulli's equation:
$$\tfrac{1}{2}\rho v_1^2 + \rho g h_1 + P_1 = \tfrac{1}{2}\rho v_2^2 + \rho g h_2 + P_2 \tag{11-10}$$

Bernoulli's equation for constant gravitational potential:
$$\tfrac{1}{2}\rho v_1^2 + P_1 = \tfrac{1}{2}\rho v_2^2 + P_2 \tag{11-11}$$

Pressure difference due to varying velocity:
$$P_1 - P_2 = \tfrac{1}{2}\rho(v_2^2 - v_1^2) \tag{11-12}$$

Viscous force:
$$F_v = \frac{\eta v A}{l} \tag{11-13}$$

Volume rate of flow through a pipe:
$$\frac{V}{t} = \pi r^4 \frac{(P_1 - P_2)}{8\eta L} \tag{11-14}$$

Drag or viscous force for a body moving in a fluid:
$$F_v = kv \tag{11-15}$$

Drag or viscous force for a sphere moving in a fluid:
$$F_v = 6\pi r \eta v \tag{11-16}$$

Terminal velocity:
$$v_T = \frac{(\rho_o - \rho_f) V g}{k} \tag{11-17}$$

Reynolds number for fluid flow:
$$Re = \frac{2 v r \rho}{\eta} \tag{11-18}$$

Reynolds number for a body moving in a fluid:
$$Re' = \frac{v L \rho}{\eta} \tag{11-19}$$

THOUGHT QUESTIONS

1. Compare and contrast a gas, liquid, and solid.
2. What would happen if the volume above the mercury column in Fig. 11.1 were not a vacuum? Why is the liquid usually mercury?
3. Why is severe weather usually associated with regions of low atmospheric pressure and fair weather with high pressure?
4. Typical systolic/diastolic blood pressure readings for a quiet healthy adult are 120/80. What units are used?
5. Does Pascal's principle hold for solids? Why or why not?
6. In a hydraulic lift a modest force can be used to create a large force. Is this a violation of conservation of energy? Explain.
7. How is an iron ship able to float?
8. Is there a limit to the depth to which a ship will sink in the ocean?
9. When he discovered his principle, Archimedes was trying to determine whether or not a crown for the king of Syracuse contained the correct percentage of gold. How was Archimedes' principle used?
10. Would a fat person or a lean person float better in water? Explain.
11. Give an everyday example of capillary action.
12. Give an example of a fluid that is not difficult to compress.
13. Give a practical application of the Bernoulli effect.
14. Give examples of fluids with high and low viscosities.
15. How does a siphon work?
16. How does a chimney work?
17. Why are smokestacks constructed long and tall?

PROBLEMS

18. Why should the viscosity of a substance such as syrup be temperature dependent? Are there any substances whose viscosity decreases with increasing temperature?

19. Would you use a high or low viscosity fluid to reduce friction in a mechanical system? Explain why.

20. Where would you expect the largest flow velocity in a pipe?

21. Give an everyday example of viscous drag.

22. The atmosphere of Mars has a much lower density than that of Earth. Would a falling object reach a higher or lower terminal velocity on Mars?

23. How would fluid viscosity affect the terminal velocity reached by an object falling through that liquid?

24. Give an everyday example of (a) laminar flow (b) turbulent flow.

25. Would a fluid with high viscosity be more likely to exhibit turbulent flow? Explain.

PROBLEMS

Section 11.1 Pressure and Buoyancy

1. A high pressure system has an atmospheric pressure of 30.52 inches of mercury. Express this pressure in Pa, atm, bar, mbar, and torr.

2. The atmospheric pressure at the surface of the planet Venus is about 90 atm. Express this pressure in Pa, bar, mbar, torr, and inches of Hg.

3. The density of air is 1.3 kg/m^3. What is the density in g/cm^3? How much does 1 liter (10^3 cm^3) of air weigh?

4. The density of platinum is 21.45 g/cm^3. Compute the masses of 5-cm cubes of platinum and gold. If gold sells for $300 per ounce, how much is the gold bar worth? One troy ounce has a mass of 31.1 g.

5. If the ocean in Example 11.2 were composed of mercury, what pressure would be observed?

6. What is the pressure on the eardrums at the bottom of a 12 foot swimming pool? Express your answer in Pa and atm.

7. To what depth must one descend in the ocean to reach a pressure of 500 atm? Is any spot in the ocean that deep?

8. A mechanic wishes to lift a 5 ton truck 50 cm with a circular piston 1 m in diameter. The truck is to be supported by a 500 lb force exerted on a second piston. What is the diameter of the second piston? How far must it be displaced?

9. Calculate the buoyant force exerted on a fully submerged log of pine wood (density 480 kg/m^3) 1 m in diameter and 10 m long.

10. Suppose the radius of the log in Problem 9 could be shrunk without changing the log's mass. At what radius would the log sink?

11. A cube of material weighs 10% less when submerged in water than in air. What is its density?

12. A cylinder 1 m long and 1 m diameter has a top half composed of balsa wood (density 130 kg/m^3) and a bottom half of iron. Will this object float in water?

13. A cubical block of lead one meter on a side is dropped into a pool of mercury. To what depth will the block sink?

14. A crown, presumably made of gold, has a weight of 78.4 N in air and 74.4 N when submerged in water. Is the crown made of gold?

15. Calculate the number of pascals in a torr.

16. The total mass of the earth is about 6×10^{24} kg and its mean radius is 6,378 km. Find its mean density.

17. If the mean density of air is 1.3 kg/m^3, find the difference in atmospheric pressure exerted on the top of a 2 m person's head compared to that at his feet.

18. A 2-liter cylindrical bottle with 10 cm radius is filled with cola having density 1 g/cm^3. What is the pressure at the bottom of the bottle?

19. A hydraulic lift is required to raise 2,000 kg on a piston of area 4 m^2. If the lifting piston moves a distance of 2 m and the lifting force is 100 N, to what height can the 2,000 kg mass be raised?

20. A cube of copper 0.1 m on a side is placed in water. Find the buoyant force.

21. Calculate the buoyant force on the copper in Problem 20 if it is placed in a pan of mercury.

22. If the density of ice is about 0.92 g/cm^3, what fraction of an iceberg will be underwater? Assume that seawater has a density 1.02 g/cm^3.

- 23 An ore sample has a dry-land weight of 20 N and a weight of 15 N when submerged in water. If 1 liter of water was displaced when the sample was submerged, find the volume and density of the ore sample.
- 24 What is the difference in air pressure over a 1,000-foot rise in elevation? Assume that the density remains constant.
- 25 How high must an intravenous feeding tube be raised to create a pressure of 0.2 atm?

Section 11.2 Surface Tension and Capillary Action

- 26 A circular wire frame 1 cm in diameter holds a thin film of liquid. A total of 10^{-5} J work is required to increase the frame diameter to 2 cm. What is the coefficient of surface tension for this substance?
- 27 If the coefficient of surface tension in Problem 26 were suddenly doubled, how much work would be required to do the same task?
- 28 Compare the heights to which water and ethyl alcohol will rise in a tube.
- 29 How should the relative diameters of the tubes in Problem 28 be adjusted so that the heights remain the same?
- 30 A certain liquid in a glass tube 0.2 mm in diameter has a contact angle of 30° and rises to a height of 10 cm. If the density of the liquid is 0.6 g/cm^3, what is the coefficient of surface tension?
- 31 Tree sap, mostly water, has a density of about 1 g/cm^3 and rises in capillaries of radius 0.03 mm. If the contact angle between sap and capillary wall is zero, how high can the sap rise? Is capillary action a viable explanation for the growth of trees?
- 32 Consider a spherical membrane of radius r, having inside pressure P_i and outside pressure P_o. Show that the pressure difference $(P_i - P_o) = 4\gamma/r$. Consider the work done by pressure forces in expanding the bubble a small amount Δr in radius, and recall the binomial expansion $(1 + x)^2 \approx 1 + 2x$ for small x. Remember also that the membrane has two sides.
- 33 A soap bubble is blown to radius 1 cm. If soapy water has surface tension $\gamma = 0.03$ N/m, find the pressure difference between inside and outside.
- 34 Small capillaries in a plant have diameter 0.006 mm. Compute the heights to which water will rise at $T = 20°C$ and $T = 30°C$. You will need to interpolate between tabulated values of γ in Table 11.3.
- 35 A flat puddle of fluid of surface area A has depth d. Because of surface tension, the fluid beads up into hemispherical droplets of radius r.
  (a) How many droplets form?
  (b) Calculate the change in surface area.
  (c) If the total area covered on the ground decreases by a factor of 2, compute the ratio (r/d) and find the change in total surface energy associated with the surface tension γ.

Section 11.3 Fluid Dynamics

- 36 Water flows through a pipe whose radius suddenly doubles. What happens to the flow velocity?
- 37 How should the diameter of a certain pipe be constricted to speed the water flow by a factor of 10?
- 38 Water flows in a pipe of 1 m diameter at 0.5 m/s. Find the mass flow rate.
- 39 A pipe has 0.4 m diameter and water flow rate 1,000 kg/h. What is the velocity of the flow?
- 40 A river is 1 mile wide and 30 feet deep. If water flows with 8 mi/h current, find the flow rate. How much water passes by a given point each day?
- 41 A river varies in width from 10 m to 500 m. What is the ratio of highest and lowest flow velocities if depth remains constant?
- 42 The aorta leading from the heart has a radius of 0.9 cm. If blood leaves the heart with velocity 0.5 m/s, and blood velocity in the capillaries has been estimated at 0.75 mm/s, estimate the total cross sectional area of the capillaries.
- 43 A barrel filled with water has vertical sides 1 m high and a volume of 3 m^3. Where in the side of the barrel must a hole be drilled if water flows out at 3 m/s?
- 44 How tall should a water tank be to generate a flow velocity at its base of 10 m/s?
- 45 A vertical nozzle is opened in a tank containing water 2 m deep. To what height will water squirt out?
- 46 Water flows through a pipe whose diameter suddenly triples in size. What happens to the pressure on the walls? The initial velocity is v_1 and the density is ρ.
- 47 An engineer wishes to decrease the water pressure in a pipe by $\frac{1}{2}\rho v_1^2$, where v_1 is its flow speed. What ratio of pipe diameters is required?
- 48 If a flow pipe could somehow be attached to the bottom of the ocean at depth 5 km, what flow velocity would result?
- 49 A pipe is attached to the bottom of a cylinder 5 meters in diameter. If the water depth is 5 m, what is the flow velocity?
- 50 A water pipe of 1 m diameter lies along a 5° incline. The upper end is driven by atmospheric pressure. Compute the pressure P at a distance 100 m down the pipe. The initial flow velocity $v_1 = 5$ m/s.

51 Suppose the pipe in Problem 50 constricts in diameter at the rate of 1 m per 1,000 m. Compute the pressure at a point 200 m down the pipe.

52 The wind atop Mt. Washington blows at 200 mi/h across a hut with roof area 200 m^2. What is the force of lift?

53 The wind velocities above and below a plane's wing are 150 mi/h and 100 mi/h. If the air density is 1 kg/m^3 and the plane mass is 30,000 kg, what area of wing is needed to provide a lift force greater than gravity?

54 What is the lift on an airfoil if the air velocities above and below the foil are 200 m/s and 170 m/s, respectively? The foil has length 5 m and width 2 m.

55 Derive an expression for the power required to maintain the flow rate $Q = V/t$ through a pipe of area $A = \pi r^2$ and length L. Assume the liquid has a coefficient of viscosity η.

56 The units of viscosity in the SI system are N · s/m^2 and those in the cgs system are dyne · s/cm^2. Find the conversion factor between the two.

57 A force of 3 N is required to move the upper plate of a pair separated by 0.01 m. If the relative velocity between plates is 0.2 m/s and the plate area is 0.5 m^2, what is the viscosity of the fluid between?

58 A doctor changes needles on a hypodermic syringe, so that the diameter is smaller by a factor of 2. If her thumb pressure on the syringe remains the same, how does the flow rate change?

59 An artery in a certain animal has a radius 0.006 m and a pressure gradient (rate of change of pressure per length) of 8 Pa/m. If the average blood flow rate is measured at 0.01 m/s, find the blood's viscosity.

60 Water is pumped at 20°C through pipes of 1 m diameter at a flow rate of 10^6 kg/s. Compute the difference in pressure $P_1 - P_2$ needed over a length of 100 m.

61 For the water flow rate in Problem 60, estimate the average spacing needed between pumps capable of providing a pressure differential of 0.5 atm.

62 The viscosity of blood is about 0.004 N · s/m^2. If blood flows at 0.5 m/s through a 1 cm radius aorta, compute the required difference in pressure along its 40 cm length. Express your answer in Pa and in mm of mercury.

63 Derive an expression for the terminal velocity of a sphere of radius r and density ρ_s falling through a fluid of density ρ_f and viscosity η.

64 What will be the terminal velocity of an iron sphere 0.01 cm in radius which is released at the surface of a lake?

65 Calculate the Reynolds number for a cylindrical jet engine 1 m in diameter moving at 900 km/h. Assume that the air density is 1.1 kg/m^3 and that the air viscosity is 1.8×10^{-5} Pa · s. Is the flow laminar or turbulent?

66 At what velocity must blood flow through an artery 4×10^{-3} m in diameter to become turbulent? Assume that blood has viscosity 0.004 N · s/m^2 and that its density is 1,060 kg/m^3.

67 Using Stoke's law, compute the terminal velocity of a spherical raindrop of radius r. Evaluate your expression for $r = 0.5$ mm.

SPECIAL PROBLEMS

1. Fluid Velocity Gauges

The velocity in a moving fluid of density ρ_f can be measured using the pressure changes created by the Bernoulli effect. Such velocity gauges are used as airspeed indicators in planes and as flow meters in pipes.

(a) Consider a pipe (Fig. 11.22) whose cross-sectional area decreases from A_1 to A_2. The pressures and velocities are P_1, P_2, v_1, and v_2. Calculate the difference $h_1 - h_2$ in heights of small columns of fluid that rise in small tubes inserted into the flow. The upper ends of these tubes are exposed to atmospheric pressure P_a.

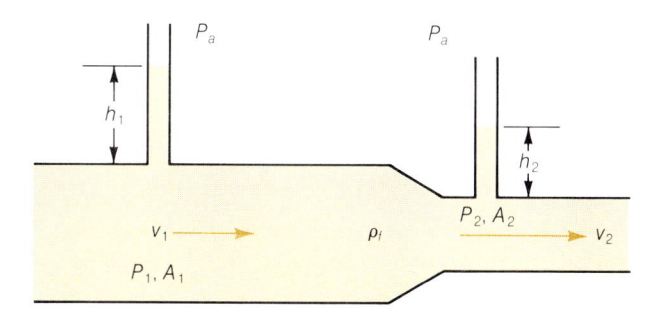

Figure 11.22

(b) Solve for the flow velocity v_1, using the equation of continuity in the flow. Explain how this device can be used as a speed indicator.

2. The Heart and Circulatory System

The heart acts as a pump which forces blood from the left ventricle into the aorta, smaller arteries (arterioles), and tiny capillaries. The aorta has a radius about 1 cm, arterioles about 10^{-3} cm, and capillaries about 3×10^{-4} cm. The pressure at the heart ranges from 120 to 80 mm Hg, and the volume flow rate is 80 cm^3/s out of the heart. The viscosity of blood is 0.004 N · s/m^2.

(a) Estimate the average flow velocity at the aorta. If the aorta were contracted in radius by a factor of 2, with no change in pressure, by what factor would the flow rate drop?

(b) If the total cross section of the arterioles is 1,000 cm^2, estimate the average flow speed and the pressure drop across their average 1 cm length. How many are there?

(c) If there are 7×10^9 capillaries, estimate their total cross sectional area, the average flow speed, and the pressure drop across their average 0.1 cm length.

(d) Estimate the Reynolds numbers for flow in the aorta, arterioles, and capillaries. Are the flows laminar or turbulent?

3. Geysers

Geysers are jets of hot water shot into the air from channels extending deep into the earth. One geyser at Yellowstone National Park shoots water 30 m vertically through a channel 1 cm in radius and 1 km deep.

(a) From the vertical height, estimate the velocity of the stream at the earth's surface. What is the mass flow rate (kg/s)?

(b) Water has a viscosity of 3×10^{-4} N · s/m^2 at 90°C. Estimate the pressure change, ΔP, over the 1 km channel if the geyser were horizontal. What additional pressure must be present to push the water 1 km upward against gravity? What is the total pressure at the bottom of the geyser, 1 km deep?

(c) From the properties of the fountain, estimate the power of the geyser in watts and in hp. How would your answer change if the channel had been only 1 mm in radius?

SUGGESTED READINGS

"The charm of hydraulic jumps, starting with those observed in the kitchen sink." J. Walker, *Scientific American,* Apr 81, p 176, **244**(4)

"The physics and chemistry of the lemon meringue pie." J. Walker, *Scientific American,* Jun 81, p 194, **244**(6)

"Why do honey and syrup form a coil when they are poured?" J. Walker, *Scientific American,* Sep 81, p 216 **245**(3)

"Reflections on the rising bubbles in a bottle of beer." J. Walker, *Scientific American,* Dec 81, p 172, **245**(6)

"What causes the 'tears' that form inside a glass of wine?" J. Walker, *Scientific American,* May 83, p 162, **248**(5)

"The troublesome teapot effect, or why a poured liquid clings to the container." J. Walker, *Scientific American,* Oct 84, p 144, **251**(4)

"Experiments with the external-combustion fluidyne engine, which has liquid pistons." J. Walker, *Scientific American,* Apr 85, p 140, **252**(4)

"Geometry of soap films and soap bubbles." F. J. Almgren, J. E. Taylor, *Scientific American,* Jul 76, p 82, **235**(1)

The flying circus of physics with answers. Jearl Walker, Wiley, New York, 1977

A source book in physics. William F. Magie, Harvard University Press, Cambridge, 1965

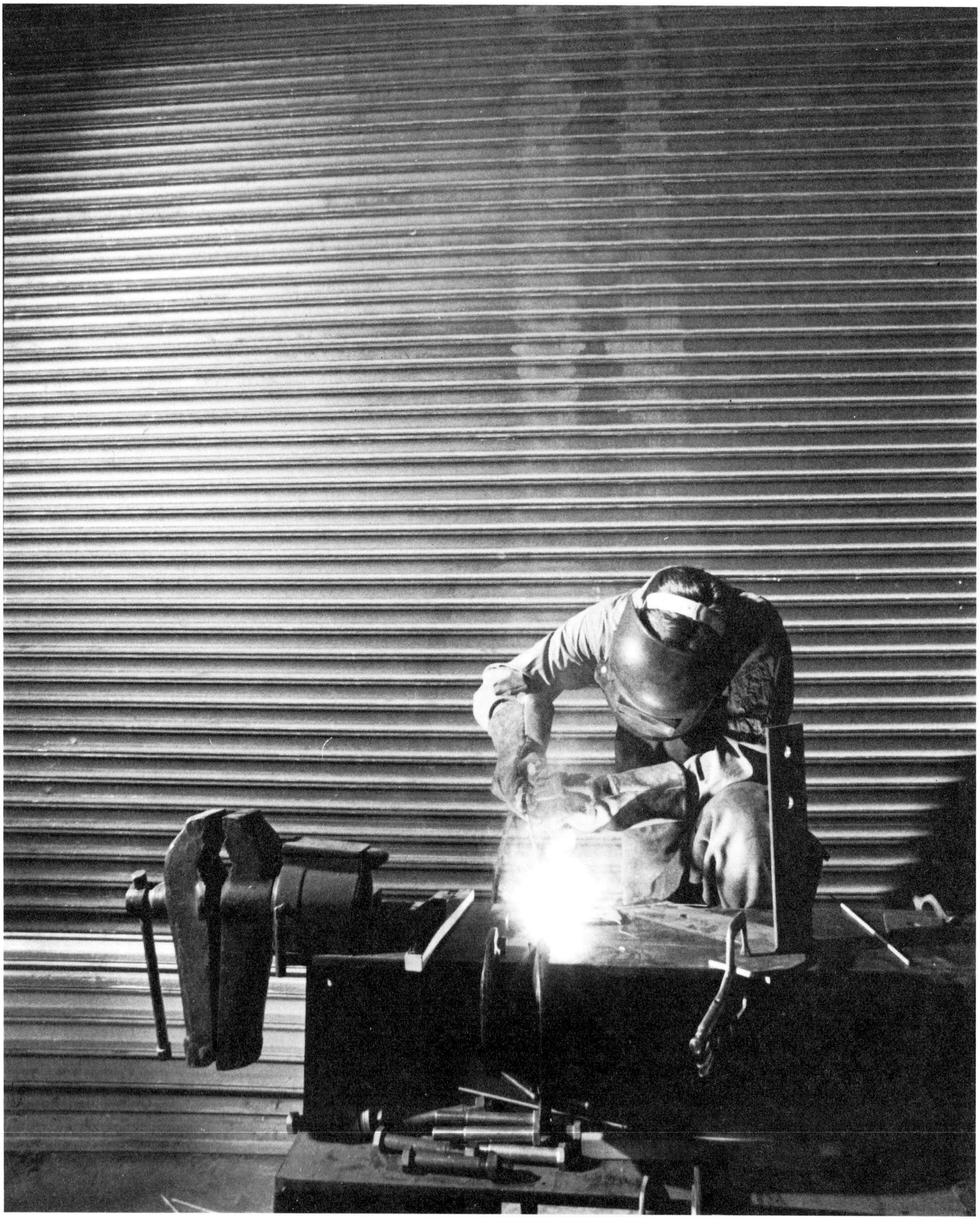

Chapter 12

Temperature and Heat

We have from time to time mentioned temperature in the preceding chapters. We assumed that the concept is well known and meaningful to anybody. Now it is time to examine more closely what we mean by temperature and the related concept of heat. In this chapter we will discuss these concepts and how they relate to solids and the large-scale behavior of fluids. The discussion of temperature and heat at the molecular level will be reserved for the next chapter.

12.1 Temperature

In the most intuitive terms, **temperature** refers to how hot or cold something is. Later we will see how this is related to the motions of individual molecules or atoms in a solid or fluid, but there is much to be gained by taking a macroscopic viewpoint first. We begin by discussing the scales and instruments with which temperature is measured.

Temperature Scales

There are three temperature scales in widespread use (Fig. 12.1). Two of them, the **Fahrenheit** scale and the **centigrade**, or **Celsius**, scale, are linked to certain properties of water. The third, the **absolute**, or **Kelvin** scale, is perhaps more fundamental, but is still defined in terms of the properties of water.

The most commonly used scale in the U.S. (but virtually nowhere else in the world) is the Fahrenheit scale. Devised by G. D. Fahrenheit, a physicst living in Holland, this scale has a zero point intended to be low enough so that negative temperatures would never occur. The value of 32° was chosen

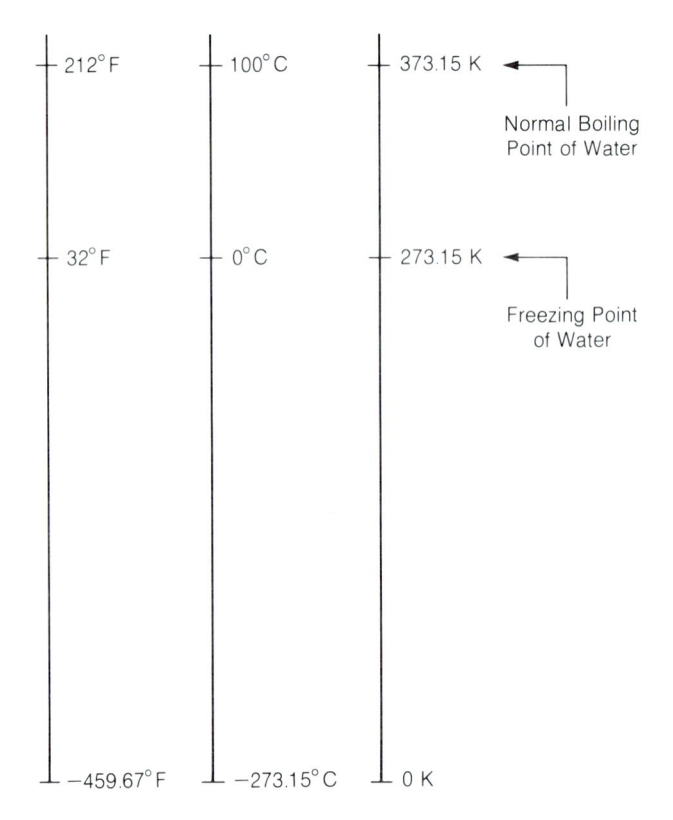

Figure 12.1

The relationship of the F, C, and K temperature scales.

for the freezing point of water, and 212° for the boiling point (this value was selected so that the human body temperature would be around 100°). (Both the freezing and the boiling point refer to water at sea-level atmospheric pressure). Comfortable temperatures for people without excessive clothing are near 70°F (although this is highly subjective), and most homes and offices are maintained at this temperature.

The centigrade, or Celsius scale is based more directly on the freezing and boiling points of water. The freezing point is defined as 0°C, and the boiling point as 100°C. Thus one centigrade degree is (212 − 32)/100 = 9/5 larger than one Fahrenheit degree. The equation relating the two is

$$T_C = \frac{5(T_F - 32)}{9} \tag{12-1}$$

or

$$T_F = 32 + \frac{9T_C}{5} \tag{12-2}$$

12.1 TEMPERATURE

where T_C represents the Celsius temperature and T_F the Fahrenheit temperature. From Eq. 12–1 we see that comfortable room temperature, 70°F, corresponds to a centigrade temperature of 21.1°C. Normal human body temperature (98.6°F) is 37.0°C.

The third scale, used most commonly by scientists, has the same degree size as the centigrade scale, but sets its zero at the lowest possible temperature, which is called **absolute zero.** We will be in position to understand the meaning of absolute zero only after our discussion in the next chapter of molecular motions and temperature; for now we simply point out that this lowest possible temperature corresponds to $-273.15°C$ (or $-459.67°F$). Usually this is rounded off to $-273°C$. Because the degree on the absolute scale is the same size as the centigrade degree, conversion between these two scales is very simple.

$$T_K = T_C + 273. \qquad (12-3)$$

Here T_K represents the temperature on the absolute scale (the K stands for the British scientist Lord Kelvin, who was a pioneer in understanding temperature and thermal properties of matter. In modern usage a degree on this scale is called a **Kelvin,** and the degree symbol (°) is dropped). The conversion between the absolute and Fahrenheit scales is accomplished by combining Eq. 12–1 or Eq. 12–2 with Eq. 12–3.

In this book we will use the centigrade and Kelvin scales for the most part, so it is advisable for the student to become familiar with them and with making conversions between scales. Some of the problems at the end of the chapter will require doing so.

Thermometers

A device that measures temperature is called a **thermometer.** There are many types, and they come in a wide variety of sizes and appearances. The majority of common thermometers rely on the fact that most materials expand when heated. The expansion is calibrated in terms of temperature. In the ordinary bulb thermometer, a glass tube contains a liquid (usually either mercury or alcohol) whose level in the tube rises or falls as the temperature varies. Another common type of thermometer utilizes the expansion of metals with temperature increases. Metals do not expand very much, but when a thin coiled strip of two metals bonded together is heated, the end of the strip will bend through a large angle (Fig. 12.2) if the two metals expand at different rates. Thus a pointer can be made to move back and forth along a calibrated scale. Thermometers of this type are widely used in thermostats and air temperature thermometers. Another, very simple kind of thermometer makes use of a **thermistor,** a small solid-state device whose ability to conduct electrical current depends on its temperature.

All of the simple thermometers just described are subject to minor irregularities, and are not sufficiently accurate for scientific purposes. Standard temperature calibrations are provided by a device called a **constant volume thermometer** (Fig. 12.3). This thermometer relies on the fact that the pressure in a gas varies quite uniformly with temperature. The pressure of a fixed volume

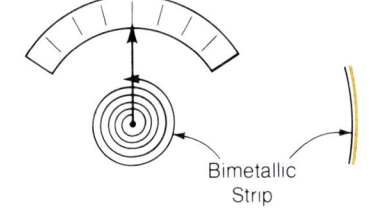

Figure 12.2

Bimetallic strip thermometer. A strip of two layers of metal with different coefficients of thermal expansion bends as the temperature changes. The strip is attached to a pointer on a scale.

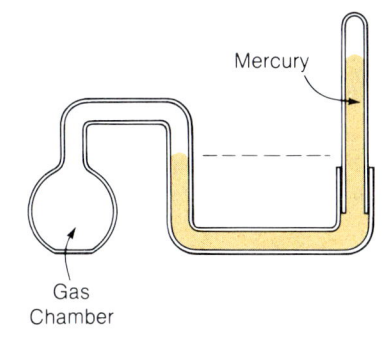

Figure 12.3

Constant volume thermometer. The tube at right is raised or lowered so that the height of the mercury at left is kept fixed, thereby keeping the volume of gas constant. Then the height of mercury at right indicates the pressure, hence the temperature, of the gas.

of gas is monitored by a pair of mercury-filled tubes, one of which is kept at constant level to maintain constant volume in the gas-filled chamber. The height of mercury in the other tube is a measure of the pressure in the gas chamber, and therefore measures the temperature of the gas chamber, which is equal to the ambient temperature (so long as the temperature is not changing rapidly).

All types of thermometers can be calibrated directly by subjecting them to the temperature of freezing water and boiling water, then setting their scales.

12.2 Thermal Expansion

We have already mentioned that most materials expand and contract when they are heated or cooled. This phenomenon is familiar to anyone who has seen a glass break when subjected to a sudden change in temperature; the expansion or contraction creates stress which may deform a material beyond its elastic limit. Thermal expansion includes not only the simple change of length in solid objects, but also the change of volume in solids and fluids.

Linear Expansion

It has been found experimentally that the change in length of an object subjected to a temperature change is proportional both to the amount of change in temperature and to the original length. In equation form, we have

$$\frac{\Delta L}{L} = \alpha \Delta T \tag{12-4}$$

where ΔL is the change in length, L is the initial length, ΔT is the temperature change, and α is the **coefficient of linear expansion**. This coefficient may be viewed as the fractional change in length per degree of temperature change, and is normally evaluated for the centigrade scale, so its units are $(C°)^{-1}$. The value of α depends on the material, and can vary quite significantly from one substance to another. Some typical values are presented in Table 12.1.

Thermal expansion is important in the design of any structure, from massive buildings to sophisticated scientific instruments. The expansion and contraction of girders with the weather must be allowed for in the construction of buildings and bridges, and the minute changes in precisely shaped and aligned components of technical instruments must also be taken into account. Long bridges typically have expansion joints (Fig. 12.4), where the bridge can expand and contract without creating stress. Sidewalks usually are laid in sections, with small gaps between to allow for thermal expansion. A number of materials with very small values of α have been developed for use in scientific instruments (and in cooking: there are now casserole dishes and even dinnerware made of substances that do not expand and contract much, and can therefore be used in the oven.).

12.2 THERMAL EXPANSION

Table 12.1
Coefficients of expansion (20 °C)

Material	Linear α (C°)$^{-1}$	Volume β (C°)$^{-1}$
Solids		
Lead	2.9×10^{-5}	8.7×10^{-5}
Aluminum	2.5×10^{-5}	7.2×10^{-5}
Brass	1.9×10^{-5}	5.6×10^{-5}
Copper	1.7×10^{-5}	5.1×10^{-5}
Iron or steel	1.2×10^{-5}	3.5×10^{-5}
Marble	$1.4\text{-}3.5 \times 10^{-6}$	$4\text{-}10 \times 10^{-6}$
Concrete	1.2×10^{-5}	3.6×10^{-5}
Glass (ordinary)	0.9×10^{-5}	2.7×10^{-5}
Glass (Pyrex)	0.3×10^{-5}	0.9×10^{-5}
Invar	0.1×10^{-5}	0.27×10^{-5}
Quartz	0.06×10^{-5}	0.12×10^{-5}
Liquids		
Mercury		18×10^{-5}
Water		21×10^{-5}
Glycerin		54×10^{-5}
Gasoline		95×10^{-5}
Ethyl alcohol		110×10^{-5}
Carbon tetrachloride		130×10^{-5}
Gases		
Air (and most gases at atmospheric pressure)		340×10^{-5}

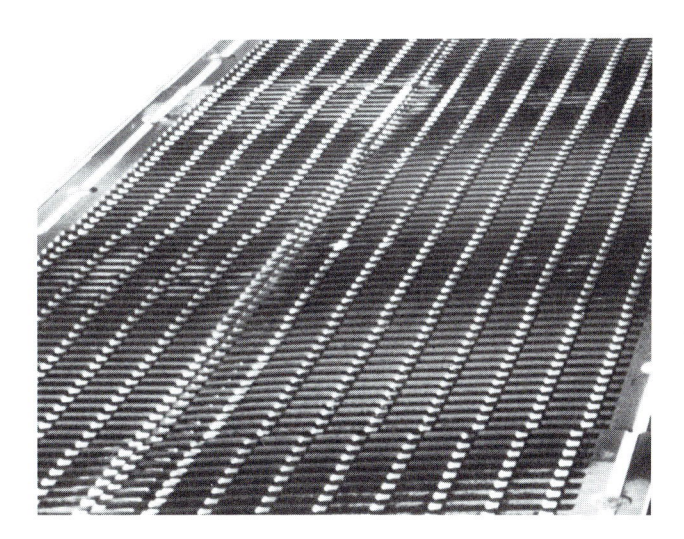

Figure 12.4

Expansion joint in a bridge. A joint of this type allows the bridge to expand and contract with temperature changes.

Example 12.1   A bridge is being designed and gaps must be planned near each end to allow for thermal expansion and contraction. The segment of bridge between gaps will have a length of 500 m at the mean temperature, and will be made of steel and concrete. If the mean temperature at the site is 45°F, with possible variations as much as 50F°, how large should the gaps be? Ignore the thermal expansion and contraction of the roadway to which the bridge connects.

The values of the linear expansion coefficients for steel and concrete are about the same: $\alpha = 12 \times 10^{-6}(C°)^{-1}$. A change in temperature of 50F° corresponds to $\Delta T = 27.8C°$. (Note that ΔT is a temperature change, so we must use the conversion of $1F° = 5/9C°$.) Therefore from Eq. 12–4 we have

$$\Delta L = \alpha L \Delta T = (1.2 \times 10^{-5} \, C°^{-1})(500 \, m)(27.8 C°) = 0.17 \, m.$$

The bridge will expand and contract by 0.17 m. Since there are two gaps, each must allow an expansion and contraction of half of this amount, or 0.085 m. At the mean temperature, each end of the bridge should rest midway in a gap 0.17 m wide.

Thermal Stress

If an object is fixed, then a stress is created when the object expands or contracts with temperature changes. Stresses are also created when different components of a structure expand or contract at different rates. This is what causes a glass to break when suddenly heated or cooled; the bottom, which is relatively thick, cannot respond as rapidly as the sides, which may change rather significantly. Stresses of this type are the basis of the bimetallic strip thermometers described above.

To calculate the **thermal stress,** as it is called when thermal expansion or contraction causes stress, we recall from Chapter 8 that

$$\frac{F}{A} = \frac{Y \Delta L}{L}$$

where F/A is the stress, Y is the elastic (Young's) modulus, and $\Delta L/L$ is the fractional change in length. Substituting for $\Delta L/L$ from Eq. 12–4 leads to

$$\frac{F}{A} = \alpha \Delta T Y \qquad (12\text{–}5)$$

where F/A is the stress that results from a temperature change of ΔT applied to an object of cross-sectional area A whose elastic modulus and coefficient of linear expansion are Y and α. Finding the value of F/A is often an important consideration in the design of structures, because it is generally advantageous to know whether thermal stresses will cause fractures or permanent deformations. Note that the dependence on L and ΔL has cancelled out, so that

12.2 THERMAL EXPANSION

the stress depends only on the cross-sectional area, the temperature change, and the intrisic properties of the material. Once the stress *F/A* has been calculated for a structure, its value can be compared with the known elastic limit and ultimate tensile strength or compression strength of the material to see what the effect will be.

Example 12.2 Suppose no allowance were made for thermal expansion in designing the bridge described in Example 12.1. The bridge is assembled at the mean temperature of 45°F. If the temperature then rises to 95°F, will the bridge fracture?

The temperature change is 50F° or 27.8C° (remember, ΔT is a temperature change, not an actual temperature). From Table 8.1, we see that the Young's modulus for concrete is $Y = 2 \times 10^{10}$ N/m², and from Table 8.2 the ultimate compression strength is 2×10^7 N/m². The stress created when the bridge is heated by 27.8C° is

$$\frac{F}{A} = \alpha \Delta T Y = (1.2 \times 10^{-5} \, (C°)^{-1})(27.8 C°)(2 \times 10^{10} \, N/m^2)$$
$$= 6.7 \times 10^6 \, N/m^2.$$

This stress is less than the ultimate compression strength of concrete. The bridge will not crumble.

Volume Expansion

Not only does the length of an object change when it is heated or cooled, but so does its volume. Furthermore, a fluid may expand or contract when subjected to a temperature change, so that its volume is altered. We refer therefore to the volume expansion as well as the linear expansion due to temperature changes.

We have an expression for volume expansion that is very similar to Eq. 12–4 for linear expansion,

$$\frac{\Delta V}{V} = \beta \Delta T \qquad (12-6)$$

where ΔV is the change in volume, V is the initial volume, ΔT is the change in temperature, and β is the **coefficient of volume expansion**. The units of this coefficient, like those of the linear coefficient α, are $(C°)^{-1}$. Typical values are given in Table 12.1.

Calculation of ΔV is usually not so critical for structural design as is the calculation of ΔL, simply because the important factor is usually the amount by which a component will expand or contract in one dimension. For solids that expand or contract uniformly in all three dimensions, the coefficient of volume expansion is approximately equal to three times the coefficient of linear expansion.

Example 12.3   Suppose a balloon is inflated so that its radius is 0.2 m, then heated 50C°. What will its new radius be?

The initial volume is $V = 0.034$ m^3. The coefficient of volume expansion for air is $\beta = 3.4 \times 10^{-3}$ (C°)$^{-1}$. Therefore from Eq. 12.6 we have

$$\Delta V = \beta V \Delta T = (3.4 \times 10^{-3} \text{ (C°)}^{-1})(0.034 \text{ m}^3)(50\text{C°}) = 0.0058 \text{ m}^3.$$

The new volume of the balloon is $V + \Delta V = 0.040$ m^3, so the new radius is

$$r = \left(\frac{3V}{4\pi}\right)^{1/3} = \left[\frac{3(0.040 \text{ m}^3)}{4\pi}\right]^{1/3} = 0.21 \text{ m}.$$

12.3 Equations of State

Although the thermal expansion of a gas was calculated in the last example, this can be done only in the few circumstances where the pressure on the gas is kept constant. It happens that this is approximately true of a balloon, since the pressure on the air inside is equal to the sum of the atmospheric pressure outside plus the pressure created by the elasticity of the balloon, which is roughly constant as the balloon expands by small amounts. The volume of a gas is generally a function of both the temperature and the pressure, because gases are highly compressible. Therefore we need a more general relationship, in terms of pressure, volume, and temperature. Such a relationship is called an **equation of state**. It is often a major task of the theoretical physicist to find the correct equation of state to describe the behavior of a gas under certain conditions. There are approximations, however, that are applicable to a wide variety of conditions.

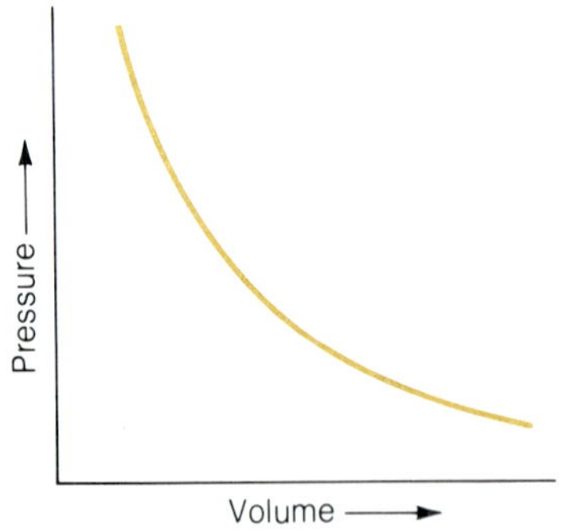

The Ideal Gas Law

Experiments carried out in the seventeenth century by R. Boyle and in the eighteenth century by J. Charles revealed several simple relations: the volume of a gas kept at constant temperature is inversely proportional to the pressure (sometimes known as **Boyle's law**; Fig. 12.5); and the volume of a gas is directly proportional to the temperature when the pressure is kept constant (**Charles's law**; Fig. 12.6). A third relation, that the pressure is proportional to the temperature if the volume is constant, was found by J. Gay-Lussac (and is known as **Gay-Lussac's law**; Fig. 12.7). Combining all three of these simple relationships leads to

$$PV \propto T. \tag{12-7}$$

This expression is still incomplete, however, because it does not take into account the quantity of gas that is present. All of the experiments mentioned

Figure 12.5

Boyle's law. In gas kept at constant temperature, the pressure and the volume are inversely proportional to each other.

12.3 EQUATIONS OF STATE

above used sealed containers of gas, so that the amount of gas remained constant and the dependence of the various quantities on the amount was obscured. It is clear, however, that there must be a direct relationship between pressure and quantity of gas; we know that the pressure in a tire increases as more air is pumped into it, even though the volume and the temperature may be constant.

The relevant parameter describing the quantity of gas is the number of molecules that are present. The effect of gas quantity on pressure is a simple function of the number of molecules, regardless of what kind of gas it is (the reason for this will become more clear in the next chapter). Hence two gases of different composition and mass density will have the same pressure if their volumes and temperatures are equal and the number of molecules of each is the same.

We therefore define a new unit for the quantity of a gas, a unit directly related to the number of molecules present. We recall from Chapter 8 that each element is characterized by its atomic mass number, which is the number of protons and neutrons (that is, nucleons) in its nucleus. The mass of a given nucleus is approximately equal to its mass number multiplied by the atomic mass unit, 1.6606×10^{-24} gram, the representative mass of an individual nucleon. The number of atoms present in a given amount of an element is therefore equal to the mass divided by the mass of a single nucleus, and the nuclear mass is equal to the product of the mass number times the atomic mass unit. If we consider two elements with mass numbers i and j, then i grams of the first will contain the same number of atoms as j grams of the second. This is a standard number that is the same for all elements. It is called **Avogadro's number,** and its value is

$$N_A = 6.02 \times 10^{23}.$$

Thus the number of atoms in 1 gram of hydrogen is 6.02×10^{23}, and the number in 12 grams of carbon is identical.

Now we extend the concept of mass number to molecules, defining the **molecular mass** as the total number of nucleons in a molecule. Then a quantity of gas whose mass is one gram times its molecular weight will contain Avogadro's number of molecules. Thus water vapor, with a molecular weight of 18, has 6.02×10^{23} molecules in a quantity whose mass is 18 g. We define one **mole** as the quantity of gas containing Avogadro's number of molecules. More specifically, one mole (mol) is the quantity of gas whose mass in grams is equal to the molecular weight.

If we include the dependence of pressure on the number of molecules in the relationship in Eq. 12–7, we have

$$PV \propto nT \qquad (12\text{--}8)$$

where n is the number of moles of gas present. We now need only a constant of proportionality, which is experimentally determined, and is denoted R. Hence we have the **ideal gas law**

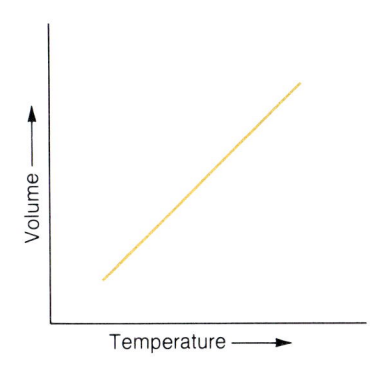

$$PV = nRT. \qquad (12\text{--}9)$$

Figure 12.6

Charles' law. When the pressure of a gas is constant, volume and temperature are directly proportional.

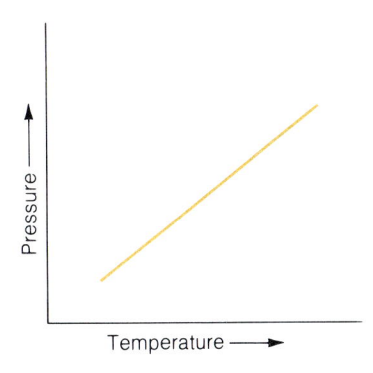

Figure 12.7

Guy-Lussac's law. When the volume of a gas is constant, pressure and temperature are directly proportional.

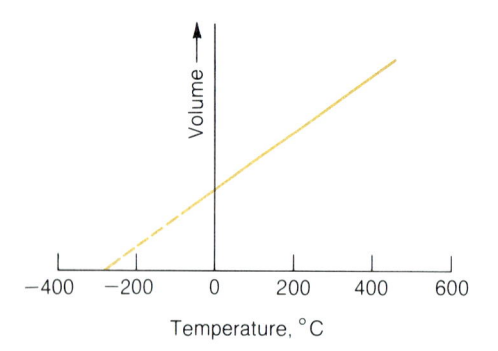

Figure 12.8

Absolute zero. If a plot is made of the relationship between volume and temperature at constant pressure, a linear relationship is found (Charles' law). If the graph is extrapolated, it reaches zero volume at −273.15°C.

The value of R, which is called the **universal gas constant,** is $R = 8.314$ J/(mol · K). Note that the units of R are energy per mole per degree absolute. We will discuss the connection of gas temperature with energy shortly. Note also that there is a mixture of SI and cgs units in this value of R, because the joule incorporates the standard unit of mass, the kilogram, whereas the mole is based on the cgs unit of mass, the gram. Nevertheless, for historical and practical reasons, this is the way R is usually specified, and the mole is the official SI unit of gas measures.

The reason that the absolute temperature is relevant for gases is that the proportionalities between temperature and volume and between temperature and pressure both are valid only for temperatures measured from absolute zero. If Charles' law is plotted and extrapolated to zero volume, the line crosses the temperature axis at a point corresponding to a value of −273.15°C (Fig. 12.8; this was the original basis for the notion of absolute zero). There would be no simple proportionality if some other, arbitrary zero point were used.

Because the state of a gas depends on temperature and pressure, it is always important to specify the values of these quantities. Generally a standard combination, referred to as **standard temperature and pressure** or **STP**, is used. The standard temperature is the freezing point of water (273.15K), and the standard pressure is one atmosphere (1.013×10^5 Pa).

Example 12.4   If a sealed chamber is evacuated so that the pressure inside is 0.001 atm, what mass of nitrogen gas (molecular weight 28) is contained inside if its volume is 2 m³ and the temperature is 50°F?

We need to solve Eq. 12–9 for n, and then convert to grams. We first compute that 50°F = 283K. We then find

$$n = \frac{PV}{RT} = \frac{(0.001)(1.013 \times 10^5 \text{ Pa})(2 \text{ m}^3)}{(8.314 \text{ J/mol} \cdot \text{K})(283\text{K})} = 0.086 \text{ mol.}$$

12.3 EQUATIONS OF STATE

The mass contained in this quantity of gas is

$$m = (0.086 \text{ mol})(28 \text{ g/mol}) = 2.4 \text{ g} = 0.0024 \text{ kg}.$$

It is useful to note that one mole of any gas always has the same volume at STP. We see this by solving Eq. 12–9 for V:

$$V = \frac{nRT}{P} \qquad (12\text{--}10)$$

and setting $n = 1$ mole, $T = 273.15$ K, and $P = 1$ atm $= 1.013 \times 10^5$ Pa. The resulting value of V is

$$V = 0.0224 \text{ m}^3 = 22.4 \text{ l}.$$

(Recall that a liter is a volume of 1000 cm^3, or 10^{-3} m^3.) It is often helpful to keep in mind that a mole of gas will always have this volume at STP. It is sometimes useful to define the value of R for volumes expressed in liters and pressures in atmospheres: $R = 0.0821$ (l · atm)/(mol · K).

Example 12.5 A helium balloon is filled with 20 lbs of helium and released (Fig. 12.9). What is its volume at sea level with $T = 50°F$, and at an altitude where the pressure is 0.1 atm and the temperature is $-50°F$? Neglect the pressure due to the elasticity of the balloon, so that the pressure inside the balloon is assumed equal to the ambient pressure.

Figure 12.9

First we convert 20 lbs to moles: 20 lbs corresponds to a mass of 9.1 kg = 9.1×10^3 g. The molecular weight of helium is 4, so the number of moles is $n = (9.1 \times 10^3 \text{ g})/(4 \text{ g/mol}) = 2{,}280$ mol. Now we solve for the volume at 1 atm when T = 50°F = 10°C = 283K.

$$V = \frac{nRT}{P} = \frac{(2{,}280 \text{ mol})(0.0821 \text{ l} \cdot \text{atm/mol} \cdot \text{K})(283\text{K})}{1 \text{ atm}} = 5.3 \times 10^4 \text{ l}$$

When the balloon reaches the altitude where the pressure is 0.1 atm and the temperature is −50°F (or 227°K), the volume is

$$V = \frac{(2{,}280 \text{ mol})(0.0821 \text{ l} \cdot \text{atm/mol} \cdot \text{K})(227°\text{K})}{(0.1 \text{ atm})} = 4.25 \times 10^5 \text{ l}.$$

The balloon has expanded by more than a factor of 8.

There is one more version of the ideal gas law that is sometimes useful. The number of molecules present in a volume of gas can be expressed directly in terms of Avogadro's number. If N is the number of molecules, then we have

$$N = nN_A$$

where N_A is Avogadro's number and n is the number of moles. If we write the ideal gas law in terms of N, then a new constant of proportionality must be used. Substituting for n yields

$$PV = \left(\frac{N}{N_A}\right)RT$$

so the new constant must be equal to R/N_A, and is defined as

$$k = \frac{R}{N_A} = 1.381 \times 10^{-23} \text{ J/K}.$$

This constant is called the **Boltzmann constant**. Now the ideal gas law is

$$PV = NkT. \tag{12-11}$$

This is fully equivalent to the other forms in which the ideal gas law has been expressed, but is sometimes more convenient.

Partial Pressure and Vapor Pressure

If a container of gas includes different molecules, then the total pressure on the container walls is the sum of the individual pressures exerted by the different gases. We can define the **partial pressure** as the pressure that each gas would

12.3 EQUATIONS OF STATE

exert if it were the only gas in the container. The **law of partial pressures,** first discovered experimentally, says that the total pressure is the sum of the partial pressures of the different gases that are present. If a gas contains two types of molecules, the total number is $N = N_1 + N_2$, and the ideal gas law (as stated in Eq. 12–11) becomes

$$PV = (P_1 + P_2)V = (N_1 + N_2)kT$$

and the partial pressures of gases 1 and 2 are

$$P_1 = \frac{N_1 kT}{V}, \qquad P_2 = \frac{N_2 kT}{V}.$$

The partial pressure of each constituent gas obeys the ideal gas law for its molecular concentration.

Example 12.6 What are the partial pressures of nitrogen and oxygen in the earth's atmosphere, if their relative concentrations are 4:1? (Assume that these are the only gases present.)

We are given that $N_N/N_O = 4$, so we conclude that $P_N/P_O = 4$ also, since T and V are constant. We also know that $P_N + P_O = 1$ atm. Solving, we find that

$$P_N = 1 \text{ atm} - P_O = 1 \text{ atm} - \frac{P_N}{4}$$

or

$$P_N = \frac{4 \text{ atm}}{5} = 0.8 \text{ atm} = 8.10 \times 10^4 \text{ Pa}.$$

Similarly,

$$P_O = \frac{1 \text{ atm}}{5} = 0.2 \text{ atm} = 2.03 \times 10^4 \text{ Pa}.$$

The relation of partial pressure to total pressure may seem obvious, but it has important consequences because it is often the partial pressure that governs how one constituent of a gas mixture will interact with its surroundings. In many biological systems, for example, the rate of penetration of certain gases through membranes depends on the partial pressure of those gases.

A common constituent of air is water vapor. Its partial pressure in the atmosphere at sea level is in the range 0.001 to 0.050 atm. Water vapor enters the atmosphere from the surfaces of oceans and rivers, where individual water molecules escape due to their own kinetic energy, which can exceed the energy

Table 12.2
Vapor pressure of water vs. temperature

Temperature (°C)	Saturated Vapor Pressure* 10^5 N/m²	Atmospheres	mm Hg	Saturated Vapor density (g/m³)
0	0.0061	0.0060	4.579	4.8
5	0.0087	0.0086	6.543	6.8
10	0.0123	0.0121	9.209	9.4
15	0.0170	0.0168	12.788	12.8
20	0.0234	0.0231	17.535	17.3
25	0.0317	0.0313	23.756	23.0
30	0.0424	0.0419	31.824	30.3
40	0.0734	0.0728	55.324	50.8
60	0.199	0.197	149.38	129.4
80	0.473	0.467	355.1	290.1
100	1.01	1.00	760.0	586.3
120	1.98	1.96	1489.1	1090.9
140	3.61	3.57	2710.9	1892.6
160	6.18	6.10	4636.0	3090.9
180	10.0	9.98	7520.2	4779.7
200	15.5	15.34	11659.	7095.3

*(760 mm Hg = 760 torr = 1.013×10^5 N/m² = 1 atmosphere)

binding them to the other molecules in the water. At the same time, water vapor molecules in the air strike the surface of the body of water and stick. The process by which they leave is called **evaporation,** and the return process is called **condensation.**

The evaporation rate depends on temperature, while the condensation rate depends largely on the partial pressure of water vapor. As evaporation occurs at some temperature, the partial pressure of water vapor will increase, and as it does so, the condensation rate will also increase. Eventually (as long as the water supply is not exhausted) the condensation rate will equal the evaporation rate, and an equilibrium will be established in which molecules leave the surface at a rate that is balanced by the return of other molecules to the surface. When that occurs, there is no further evaporation, and the air is said to be **saturated.** The partial pressure of water vapor at saturation is called the **saturated vapor pressure,** or simply the **vapor pressure.**

The vapor pressure is a function of temperature alone. It does not even depend on what gas, if any, the water vapor mixes with, because the processes of evaporation and condensation are unaffected by other gases that may be present. Even in a closed chamber, the vapor pressure will have the same dependence on temperature as it does in air.

Values of the water vapor pressure are given in Table 12.2 for a range of temperatures, as well as the densities of water vapor molecules, in g/m³, corresponding to saturation at each temperature.

Under most conditions, the partial pressure of water vapor in the atmosphere is less than the saturated vapor pressure. The **relative humidity** is defined as the ratio of partial pressure of water vapor to the vapor pressure. When

12.3 EQUATIONS OF STATE

the relative humidity is 100%, the air holds all the water vapor it can. High relative humidities are uncomfortable (particularly if the temperature is also high) because evaporation is inhibited, so the normal process by which the human body cools itself, sweating, is slowed down.

As the temperature drops, the atmosphere cannot hold as much water vapor. The temperature finally reaches a point at which the water vapor is saturated (the partial pressure of water equals the vapor pressure). At this temperature, called the **dew point,** moisture begins to form on grass, metal, and other exposed cool surfaces. Water vapor has reached saturation. Because of this fact, we can use the dew point to measure the relative humidity. Suppose the air temperature is 20°C and a metereologist reports that the dew point is 10°C. Then, according to Table 12.2, the partial pressure of water in the 20°C air is 9.21 mm Hg (the saturated vapor pressure at 10°C), whereas the vapor pressure at 20°C is 17.54 mm Hg. The relative humidity is then 9.21/17.54 = 0.53 or 53%.

Example 12.7   What is the partial pressure of water when the relative humidity is 75% and the temperature is 15°C? How much water is contained at these conditions in a room whose volume is 800 m^3?

From Table 12.2 we see that the saturated vapor pressure at 15°C is 12.8 torr, so a relative humidity of 75% implies that the partial pressure of water vapor is (0.75)(12.8 torr) = 9.6 torr = 0.0126 atm. From the table, we see that the saturated density of water at this temperature is 12.8 g/m^3, so the mass of water vapor in a room whose volume is 800 m^3 is

$$m = \rho V = (0.75)(0.0128 \text{ kg/m}^3)(800 \text{ m}^3) = 7.7 \text{ kg}.$$

This is a volume of 7.7 l of water, or about 2 gallons.

Changes of Phase

Evaporation and condensation are processes by which a substance changes its phase; that is, a liquid becomes a gas in one process, and a gas is converted to a liquid in another. Other changes of phase involve the conversion from liquid to solid (freezing), the change from solid to liquid (melting), the conversion from gas directly to solid (usually called condensation but sometimes referred to as deposition), and the reverse process (called sublimation). For each process there is an equilibrium state in which the two phases coexist without net conversion of one to the other because saturation has occurred. Thus we could construct more tables like Table 12.2, listing the saturated vapor pressure (which is always defined as the pressure at which equilibrium exists between the two phases) versus temperature. It is much more instructive to make a plot instead, showing the equilibrium pressure versus temperature. Such a plot for water is shown in Fig. 12.10. This is called a **phase diagram,**

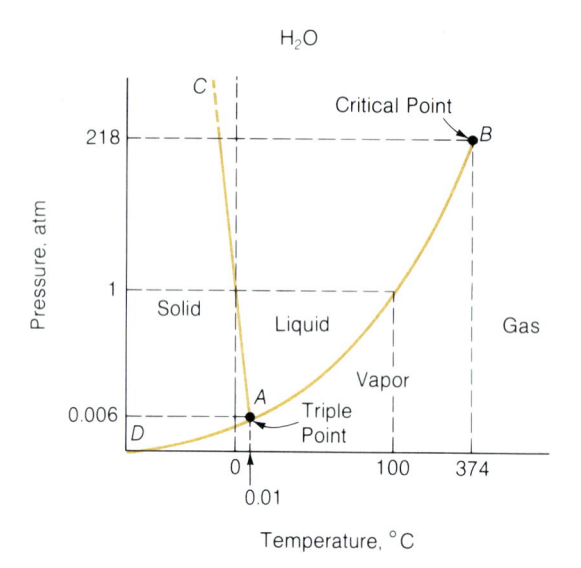

Figure 12.10

Phase diagram for water.

because it shows the conditions under which equilibrium exists between phases, and therefore it shows the conditions for which transitions between the phases will occur.

The solid curves in Fig. 12.10 show the equilibrium conditions under which two phases can coexist. Curve AB, for example, representing the data from Table 12.2, shows the combinations of pressure and temperature for which liquid water and water vapor are in equilibrium. Curve AC, representing higher pressures than curve AD for a given temperature, shows the conditions under which liquid water and ice can coexist in balance. The lower left-hand segment, curve AD, represents the conditions for equilibrium between ice and water vapor. Hence at low temperatures and pressures, it is possible for water to exist in solid and gaseous phase, with no liquid present. It is important to stress that different phases can exist together for combinations of temperature and pressure not lying on one of the solid curves; they are not in equilibrium, however, so the material in one of the phases is gradually changing to the other.

At point A in Fig. 12.10, where the three curves meet, water can exist in all three phases and be in equilibrium. This is called a **triple point,** and for water it occurs at $T = 0.01°C$ and $P = 0.00603$ atm $= 4.58$ torr. Triple points for other substances are given in Table 12.3. Another important combination of pressure and temperature is represented by point B, which is called the **critical point.** The curve does not extend any farther to the right because liquid water cannot exist for any temperatures higher than that at the critical point; water is gaseous for these temperatures, regardless of the pressure. Thus it is possible for water vapor to have higher density than liquid water, yet remain gaseous. Similarly, the central density in the sun is much higher than

12.3 EQUATIONS OF STATE

Table 12.3
Triple points

	Temperature (K)	Pressure N/m^2	Atm
Water	273.16	6.10×10^2	0.00603
Carbon dioxide	216.55	5.16×10^5	5.10
Ammonia	195.40	6.06×10^3	0.060
Nitrogen	63.18	1.25×10^4	0.124
Oxygen	54.36	1.52×10^2	0.00150
Hydrogen	13.84	7.03×10^3	0.0695

Table 12.4
Critical temperatures and pressures

	Temperature (°C)	Pressure 10^5 N/m^2	Atm
Iodine	512	117.5	116
Water	374	221.1	218.3
Carbon dioxide	31	73.8	72.9
Methane	−82	46.4	45.8
Oxygen	−119	50.8	50.1
Nitrogen	−147	33.9	33.5
Hydrogen	−240	13.0	12.8
Helium	−267	2.29	2.26

the density of lead, yet the sun is gaseous throughout. Critical points of other substances are given in Table 12.4.

When material in one phase is subjected to changing temperature and crosses one of the unbroken lines on the phase diagram, an abrupt change of phase will occur. We can see this by considering a change in temperature at constant pressure, corresponding to following a horizontal line across Fig. 12.10. For example, if ice, initially in the far left-hand section of the diagram in Fig. 12.10, is heated so that its temperature rises to reach line AC, the ice will begin to melt. If more heat is added, the temperature will remain constant until all the ice is melted, and begin to rise again only after that. Similarly, if the pressure is low enough to lie below point A, then the ice will suddenly start subliming when the temperature rises to meet curve AD, and the temperature will then remain constant until all the ice is converted to water vapor. The process of boiling occurs when liquid water is heated until it reaches curve AB; that is, when the vapor pressure of water equals the surrounding atmospheric pressure. The boiling water will remain at the boiling temperature while additional heat is added; thus the setting on the stove is not very important once foods are boiling, because the cooking temperature is constant, regardless of the rate of heat input. Note that the temperature of the boiling

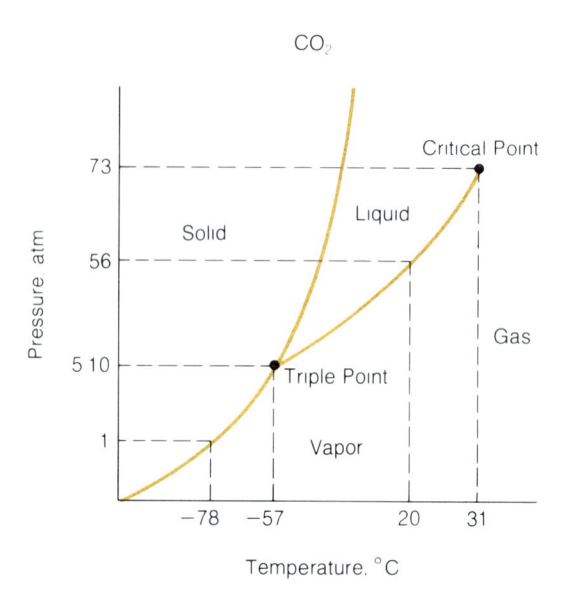

Figure 12.11

Phase diagram for carbon dioxide.

point decreases with decreasing atmospheric pressure, so that the boiling point is lower at high elevations than at sea level.

Any substance can undergo changes of phase and can therefore be represented by a phase diagram. Fig. 12.11 shows a phase diagram for carbon dioxide, in which it is seen that the triple point lies well above atmospheric pressure. Therefore liquid CO_2 is never stable at atmospheric pressure, whereas solid and gaseous CO_2 can be, depending on the temperature. At room temperatures, solid CO_2 will sublime, rather than melting, and this is why we refer to it as dry ice.

12.4 Heat and Energy

We have spoken of temperature at some length, and have mentioned the concepts of heat and heat transfer, particularly in the preceding section, where we discussed phase changes. To carry these discussions further, we must say what we mean by heat and how it is transferred.

Internal Energy, Heat, and the First Law of Thermodynamics

Any substance at a temperature above absolute zero has a certain amount of internal energy, in the form of kinetic energy of its constituent particles. Molecules in a gas move about freely, atoms or molecules in a liquid roam around (while retaining more or less fixed separations), and atoms or molecules in a

12.4 HEAT AND ENERGY

solid vibrate about their equilibrium positions. We can always define an average internal energy per particle, and it is this average that defines the **temperature.** A pot of hot soup has a higher temperature than a glass of cold water because the molecules in the soup are moving more rapidly than those in the water. In a solid or liquid, the average molecular energy determines the spacing between molecules; this is why most substances undergo thermal expansion when the temperature increases. A gas expands with increasing temperature because it exerts greater pressure on its surroundings as its individual molecules move more rapidly. Recall that most of the thermometers we described early in this chapter depend on the thermal expansion of liquids or solids, or the dependence of pressure on temperature for gases.

Absolute zero now can be defined more fundamentally than in our earlier discussion. There is a temperature at which all molecular motion comes to a stop, a temperature where the average molecular energy is zero. This is the temperature at which a confined gas will have zero pressure at constant volume, or zero volume at constant pressure. The temperature measured from absolute zero is therefore the only one that is proportional to internal energy, and is therefore the temperature we use in the ideal gas law. We will comment further on the meaning of absolute zero in the next chapter.

Heat refers to the total quantity of internal energy. The amount of heat contained in a large quantity of cold water may be much greater than the amount of heat contained in a pot of hot soup, even though the temperature of the soup is much higher. Heat is certainly related to temperature, in that the addition of heat increases the average kinetic energy of particles in a substance and therefore increases the temperature. In most contexts, we will deal with heat as a means of transferring internal energy from one body to another. In that sense, heat can be viewed as a substance that can flow, and it follows certain specific rules in doing so (an early theory of heat actually viewed it as a material fluid called **caloric**).

We have discussed and used the work-energy principle, which states that the increase of energy of an object is equal to the work done on it. Now we are defining heat as a form of energy that can be transferred to an object, so the concept of heat must be closely related to that of work. It is valid to consider that the quantity of heat transferred to an object represents the work done on the object. The units of work, energy, and heat are all the same (although for historical reasons, there are a couple of other units defined for heat, as defined below).

The law of conservation of energy applies to the transfer of heat, just as it does when energy is transferred by mechanical work. If heat is added to an object, its total internal energy increases by the amount of heat that is added. If the addition of heat to an object or a system such as a gas causes that object or system to perform work, such as when a gas is heated so that it expands and pushes a piston, then the increased internal energy plus the work done by the system must equal the quantity of heat that was added. In other words,

All the heat that is added to a system must be converted into internal energy or work done by the system.

384 CHAPTER 12 TEMPERATURE AND HEAT

This statement of the conservation of energy is known as the **first law of thermodynamics**. In equation form, we may write

$$\Delta U = Q - W \qquad (12\text{–}12)$$

where ΔU is the change of internal energy of the system that gains a quantity of heat Q and performs an amount of work W. We will see that the first law of thermodynamics has many important and useful applications.

While the joule is the standard SI unit for heat, other units are frequently used. These other units were developed before the relationship between energy and heat was established (which happened in the mid-nineteenth century), and are based on the amount of heat needed to accomplish a given change in temperature for a given quantity of some substance such as water. The **calorie** (**cal**) is defined as the quantity of heat needed to raise the temperature of a gram of water one degree centigrade*, while the **British thermal unit** (**BTU**) is the quantity of heat required to raise the temperature of one pound of water one degree Fahrenheit. One calorie is equal to 4.18 J, and one BTU is equivalent to 1,054 J. Because of the historical precedent, and because it is occasionally convenient to express heat quantities in terms of the temperature changes they induce, we will use calories as much as joules in our problems and discussion. We are not prepared to tackle problems yet, however; we still need to know a bit more about heat transfer.

Specific Heat

The change of temperature of a material that is caused by the loss or gain of a certain amount of heat depends strongly on the nature of the material. Thus we must take into account the rate at which the temperature of a substance changes with changes in heat content. We define the **specific heat** as the amount of heat absorbed or lost per unit mass per degree of temperature change. Then the specific heat of water is 1 cal/g · C°, because, by definition, the calorie is the amount of heat required to raise the temperature of a gram of water one degree centigrade (actually, the specific heat of water varies slightly as a function of temperature). Since the BTU is the amount of heat required to raise one pound of water one degree Fahrenheit, the specific heat of water may also be expressed as 1 BTU/lb · F°. It is generally true that specific heats in cal/g · C° are numerically equal to those expressed in BTU/lb · F°.

Many substances have much smaller specific heats than water; metals, for example, experience a 1C° change of temperature for heat gains or losses that are only a small fraction of a calorie. Specific heats are measured for gases as well as for solids and liquids, but for gases we must be careful because the specific heat is affected by changes in pressure or volume that may accompany changes in temperature. Normally, two distinct specific heats are specified for

---

*Because the specific heat of water (and other substances) varies slightly with temperature, the strict definition of the calorie refers to water at a specific temperature, 15°C, being raised one degree, to 16°C. Also, the calorie defined here is not the same as the Calorie referred to by dieters (sometimes called the nutritional calorie). The Calorie is equal to 1,000 calories (that is, one kcal), and is the amount of heat required to raise the temperature of a kilogram of water one degree centigrade.

12.4 HEAT AND ENERGY

gases: one which refers to a heat input or loss when the pressure is kept constant; and another which refers to a heat transfer when the volume is kept constant. Table 12.5 lists specific heats for some common substances. The table is valid only for the indicated temperature range, because for most substances the specific heat varies with temperature.

To calculate how much heat is required to bring about a given change in temperature, we can write the expression

$$Q = mc\Delta T \qquad (12\text{–}13)$$

where Q is the quantity of heat lost or gained to create a change in temperature ΔT for a substance with mass m and specific heat c.

Example 12.8

How much heat must be lost by a person whose body temperature drops from the normal value of 98.6°F to the hypothermic value of 94°F? Assume the person weighs 150 lb and that the human body has the same specific heat as water.

To make this calculation in British units, we substitute the weight in pounds for m in Eq. 12–13. We find

$$Q = mc\Delta T = (150\text{ lb})(1\text{ BTU/lb} \cdot \text{F}°)(4.6\text{F}°) = 690\text{ BTU} = 1.74 \times 10^5\text{ cal} = 7.27 \times 10^5\text{ J}.$$

Such a large heat loss can occur in a human body only when the metabolic processes by which heat is produced internally cannot keep up with the heat loss rate, as when a person is exposed without protection to extremely low temperatures.

Table 12.5

Specific heats (at 25°C, 1 atm)

Substance	Joule / kg · C°	kcal / kg · C°
Mercury	138	0.033
Lead	159	0.038
Gold	129	0.0308
Silver	237	0.0566
Copper	385	0.092
Iron	443	0.106
Glass	840	0.20
Aluminum	900	0.215
Wood	1700	0.40
Beryllium	1820	0.436
Steam (110°C)	2010	0.48
Ice (−5°C)	2100	0.50
Liquid water (15°C)	4180	1.00

We can combine the principle of conservation of energy with what we have just learned about specific heats to solve more complex problems. We know that energy is not created or destroyed when it is transferred, so we can write an equation setting the total heat lost within a system of bodies equal to the total heat gained. If, for example, a freshly boiled egg is dropped into a pot of cold water, heat will be transferred from the egg to the water and to the pot as the entire system reaches a uniform temperature (a uniform temperature will be reached in time because of the second law of thermodynamics, which we will treat in Chapter 13). If we neglect heat lost to the surroundings, we could write an equation setting the heat lost by the egg equal to the sum of the heat gained by the water and by the pot. If we know the weights or masses of the egg, the water, and the pot, and the initial temperatures of all three, we would have only one unknown in our equation, namely, the final temperature, and we could solve for it. Conversely, if we measured the initial and final temperatures, but did not know the value of one of the specific heats (that of the egg, for example), we could solve for the unknown specific heat. This is one method for measuring specific heats.

Example 12.9 Suppose a 1 kg mass of lead is heated to 100°C by immersion in boiling water, then transferred to 1 l of icewater whose temperature is 0°C, contained in a glass jar whose mass is 0.2 kg. If the final temperature reached when the lead, the glass, and the water have come to equilibrium is 2.9°C, what is the specific heat of lead?

We write an equation setting the heat lost by the lead equal to that gained by the water and the glass jar.

$$m_L c_L (\Delta T)_L = m_W c_W (\Delta T)_W + m_G c_G (\Delta T)_G$$

The L, W, and G subscripts refer to the lead, the water, and the glass, respectively. Solving for c_L and substituting the known values for the other quantities (Table 12.5) leads to

$$\begin{aligned} c_L &= \frac{(m_W c_W + m_G c_G)(\Delta T)_{W,\,G}}{m_L (\Delta T)_L} \\ &= \frac{[(1{,}000\text{ g})(1\text{ cal/g}\cdot\text{C}°) + (200\text{ g})(0.20\text{ cal/g}\cdot\text{C}°)](2.9\text{C}°)}{(1{,}000\text{ g})(97.1\text{C}°)} \\ &= 0.031\text{ cal/g}\cdot\text{C}°. \end{aligned}$$

This value is indeed similar to the one tabulated for lead in Table 12.5.

Latent Heats and Phase Changes

We have mentioned that when a phase change occurs as heat is lost or gained by a substance, the temperature will not change until all the substance has changed phase. This implies that a certain amount of heat is required for a change of phase; that the change of phase is a change of internal structure that either absorbs or releases heat. The heat required to bring about a change of phase without any change of temperature is called the **latent heat** (latent means hidden; the heat may be considered hidden because it does not cause a change of temperature). When the change of phase is melting or freezing, it is the **latent heat of fusion.** When a substance changes from liquid to gas or back, then we refer to the **latent heat of vaporization.** Latent heats depend on the nature of the substance, and are tabulated in units of cal/g, kcal/kg, or J/kg. A few examples are given in Table 12.6.

Now if we wish to write an equation setting heat losses within a system equal to heat gains, we can take into account the heat lost or gained during changes of phase. The method is identical to that already used, except that we must now add terms reflecting the latent heats. For example, if 1 kg of ice is added to a container of hot water, and the system is then allowed to come to a uniform temperature, the ice will melt, and the temperature of the released water will rise higher. Therefore there are two terms in the equation representing heat gains by the ice: one for the latent heat of fusion (equal to the mass of the ice times the latent heat of fusion for water, which is 80 cal/g),

12.4 HEAT AND ENERGY

Table 12.6
Heats of fusion and vaporization (1 atm)

Substance	Melting Point °C	Heat of fusion J/kg	kcal/kg	Boiling Point °C	Heat of Vaporization J/kg	kcal/kg
Helium	−269.5	5.23×10^3	1.25	−268.8	2.093×10^5	5.0
Hydrogen	−259.0	5.86×10^4	14.0	−252.6	4.605×10^5	110.0
Nitrogen	−209.8	1.38×10^4	3.30	−182.8	2.132×10^5	50.9
Chlorine	−100.8	1.83×10^5	43.7	−33.9	5.834×10^5	139.4
Mercury	−38.87	1.15×10^4	2.75	356.6	2.959×10^5	70.7
Water	0	3.33×10^5	80.	100	2.260×10^6	539
Lead	327.4	2.31×10^4	5.52	1749	8.671×10^5	207.1
Aluminum	660	3.95×10^5	94.4	2467	1.220×10^7	2914.7
Copper	1083	2.07×10^5	49.5	2582	4.837×10^6	1155.5
Gold	1063	6.28×10^4	15.0	2967	1.648×10^6	393.7
Carbon	3620	8.7×10^3	2.08	4200	5.930×10^7	14166

and one representing the heat gained thereafter by the water released as the ice melted. (If the ice were initially colder than the freezing point, then there would be an additional term for the heat gained by the ice as its temperature rose to the melting point.)

Example 12.10 Suppose 1 kg of ice at 0°C is added to 5 l of water initially at 50°C. Neglecting heat lost or gained by the container, what is the final temperature of the water after the ice has melted and reached a uniform temperature?

Our equation setting heat gains equal to heat losses is

$$l_i m_i + m_i c_w (T - 0°C) = m_w c_w (50°C - T)$$

where l_i represents the latent heat of fusion for ice, m_i is the mass of the ice, and m_w and c_w represent the mass and the specific heat of water. Thus, the term $l_i m_i$ represents the heat required to melt the ice, the term $m_i c_w (T - 0°C)$ is the heat needed to raise the temperature of the resulting water from 0°C to its final value T, and the term $m_w c_w (50°C - T)$ represents the heat lost by the initial quantity of water to the ice as it is melted and warmed up. The mass of 5 l of water at 4°C would be exactly 5 kg; at 50°C we can still take $m_w = 5$ kg as a good approximation. Solving for T and substituting the given values leads to

$$T = \frac{m_w c_w (50°C) - l_i m_i}{m_i c_w + m_w c_w}$$
$$= \frac{(5{,}000 \text{ g})(1 \text{ cal/g} \cdot \text{C°})(50°C) - (80 \text{ cal/g})(1{,}000 \text{ g})}{(1{,}000 \text{ g})(1 \text{ cal/g} \cdot \text{C°}) + (5{,}000 \text{ g})(1 \text{ cal/g} \cdot \text{C°})}$$
$$= 28.3°C.$$

388   CHAPTER 12   TEMPERATURE AND HEAT

Table 12.7
Heat of combustion

Substance	Heat of Combustion MJ/kg	kcal/kg
Methane	55.5	13300
Propane	50.4	12000
Gasoline	47.5	11300
Natural gas	56.0	13400
Diesel oil	44.7	10700
Fats	39.0	9300
Coal (bituminous)	30.2	7200
Carbohydrates	16.5	3900
Wood	15.0	3600
TNT	6.78	960

(1 MJ = 10^6 J)

In addition to changes of phase, other processes involve the internal absorption or release of energy, and they must be included in the conservation of heat energy. All chemical reactions, for example, either release energy or require energy; that is, they are either **exothermic reactions** or **endothermic reactions**. Thus, when a reaction takes place in a mixture of substances, the energy absorbed or released must be represented by a term in the energy conservation equation. In general form, an exothermic reaction could be represented by

$$A + B \rightarrow C + \text{heat}$$

and an endothermic reaction by

$$A + B + \text{heat} \rightarrow C.$$

One common type of exothermic reaction is **combustion**, or burning, the very rapid oxidation that can occur in many materials under the proper conditions of temperature and oxygen supply. We refer therefore to the **heat of combustion**, which is the amount of energy released per unit mass when combustion occurs. Highly **volatile** materials, those that vaporize easily, are often good fuels because they ignite easily and have high heats of combustion. Values for some materials are given in Table 12.7.

Example 12.11   Suppose a house requires 2,000 kW · hrs/month if it is heated by electric heaters. How much wood is required to heat the same house with a wood-burning stove instead? Assume that the electricity is converted to heat with a 75% efficiency, and the wood is converted with a 40% efficiency.

Here we need to consider only the net energy needed to heat the house, without concern for the process of conversion. First we must convert 75% of

2,000 kW · hrs/month to its equivalent in units more useful to us. We find that 1 kW · hr = (1,000 J/s)(3,600 s) = 3.6×10^6 J = 3.6 MJ. The amount of energy needed to heat the house for a month is 75% of 2,000 kW · hr, or 5,400 MJ.

The heat of combustion of wood is about 15 MJ/kg, so the amount of wood needed to produce 5,400 MJ is

$$m_w = \frac{5,400 \text{ MJ}}{(15 \text{ MJ/kg})(.40)} = 900 \text{ kg}.$$

The factor (.40) in the denominator represents the 40% efficiency of conversion. If we know the density of wood, we can estimate what 900 kg represents in more common units, such as cords, where 1 cord = 128 ft^3 = 3.6 m^3 of stacked wood. If the wood is dry, so that its density is low, say 0.6 g/cm^3 or 600 kg/m^3, we estimate the density of the stacked wood (allowing for air gaps between logs) to be 300 kg/m^3, so the mass of dry wood in a cord is about (300 kg/m^3)(3.6 m^3) = 1,080 kg. Finally, the estimated quantity of wood needed to heat the house for a month is (900 kg/1,080 kg/cord) = 0.83 cord.

12.5 Heat Transport

To complete our discussion of the macroscopic properties of heat, we need to consider how it is transferred. We have analyzed the conversion of energy as heat is transferred, but we have said nothing about how the transfer actually takes place. There are three basic heat transport processes whose relative importance varies depending on conditions.

Conduction

When two substances at different temperatures come into direct contact, heat **conduction** is the principal process by which heat energy is transferred. It is also the mechanism by which an object that is hotter at one point than elsewhere gradually becomes uniformly heated. In conduction, kinetic energy is transferred at the molecular level, as adjacent molecules or atoms collide with each other (Fig. 12.12). If one particular molecule has more kinetic energy than its neighbor, some of that energy will be transferred when the two collide. Eventually an equilibrium will be reached where the average kinetic energy per molecule is uniform throughout the object.

It follows that conduction of heat will occur only when there is a temperature difference between adjacent points. On the average, energy is transferred from one molecule to its neighbor only if there is an imbalance to begin with. Thus, the amount of heat transferred per unit time is proportional to the **temperature gradient,** the temperature difference per unit length between

Figure 12.12

Heat conduction. Molecules in the hot end of the bar move rapidly. Their kinetic energy is transferred by collisions to other molecules along the bar; hence, the temperature rises along the bar.

Table 12.8
Thermal conductivity

Substance	J/s·m·(C°)	cal/s·cm·(C°)
Metals		
Silver	429	1.03
Copper	401	0.96
Aluminum	237	0.57
Brass	130	0.31
Steel	50.2	0.12
Lead	35.3	0.084
Mercury	8.3	0.020
Solids		
Ice	1.6	0.004
Concrete	0.8	0.002
Glass	0.8	0.002
Red brick	0.6	0.0015
Wood	0.04–0.12	0.0001–0.0003
Felt	0.04	0.0001
Styrofoam	0.01	0.00002
Human tissue	0.2	0.0005
Gases (at room temperature)		
Hydrogen	0.19	0.00045
Helium	0.15	0.00036
Oxygen	0.027	0.000065
Nitrogen	0.026	0.000062
Air	0.026	0.000062
Liquids		
Water (27°C)	0.606	0.00145

two points. If Q is the quantity of heat transferred in time t, we write

$$\frac{Q}{t} = \frac{kA(T_1 - T_2)}{d} \qquad (12\text{–}14)$$

where $(T_1 - T_2)/d$ is the temperature gradient between points where the temperatures are T_1 and T_2 that are separated by distance d, A is the cross-sectional area of the object, and k is the **thermal conductivity** of the material. This constant, dependent on the material, is expressed in cal/cm · s C° or in J/m · s C°.

Thermal conductivities vary enormously. Some sample values are given in Table 12.8. Metals tend to have the highest values, and are generally good heat conductors. Amorphous solids such as asbestos tend to have low thermal conductivities, as do most gases. Such materials are good insulators. This is why the walls of houses are filled with such materials, and modern windows are often made of two layers of glass, separated by an air gap.

Example 12.12 Suppose a copper rod of 0.5 cm² cross section and length 25 cm is brought into contact at one end with a cubical block of ice whose mass is 5 kg. The other end of the rod is kept at a temperature of 100°C. How long will it take to melt the block of ice, if it is perfectly insulated everywhere except where the rod touches it?

First we must find the amount of heat needed, which is the mass of the ice times the latent heat of fusion, which is 80 cal/g. Hence the total heat energy needed is

$$Q = m_i l_i = (5{,}000 \text{ g})(80 \text{ cal/g}) = 4 \times 10^5 \text{ cal}.$$

The end of the rod in contact with the ice has a temperature of 0°C, so the temperature difference from one end to the other is 100°C. Copper has a thermal conductivity $k = 0.96$ cal/cm·s·C°. If we assume that the temperature gradient is uniform over the length of the rod (not always the case), we can solve Eq. 12–14 for t.

$$\begin{aligned} t &= \frac{Qd}{kA(T_1 - T_2)} \\ &= \frac{(4 \times 10^5 \text{ cal})(25 \text{ cm})}{(0.96 \text{ cal/cm} \cdot \text{s} \cdot \text{C}°)(0.5 \text{ cm}^2)(100\text{C}°)} \\ &= 2.1 \times 10^5 \text{ s} = 58 \text{ h} \end{aligned}$$

Note that the block of ice would melt much more slowly if there were no contact with a good conductor of heat such as the copper rod. For example, suppose that the entire 2,000 cm² surface of the ice were exposed to air inside a compartment which maintained an air temperature of 15°C at a distance from the ice of 5 cm. Again assuming that the temperature gradient is uniform, we have

$$\begin{aligned} t &= \frac{(4 \times 10^5 \text{ cal})(5 \text{ cm})}{(6.2 \times 10^{-5} \text{ cal/cm} \cdot \text{s} \cdot \text{C}°)(2{,}000 \text{ cm}^2)(15\text{C}°)} \\ &= 1.1 \times 10^6 \text{ s} = 300 \text{ hrs} = 12.5 \text{ days}. \end{aligned}$$

A block of ice will normally melt much faster than this because the air around it is not static, so other heat transport mechanisms come into play.

Convection

Currents and flows can occur in a fluid or a gas, and therefore heat energy can be carried from one place to another. This process is called **convection,** and it is a very common and important form of heat transport. If it were not for convection, we would find ourselves quite comfortable in many situations where we otherwise would be cold. Indeed, the major reason clothing keeps us warm is that it traps air and inhibits convection.

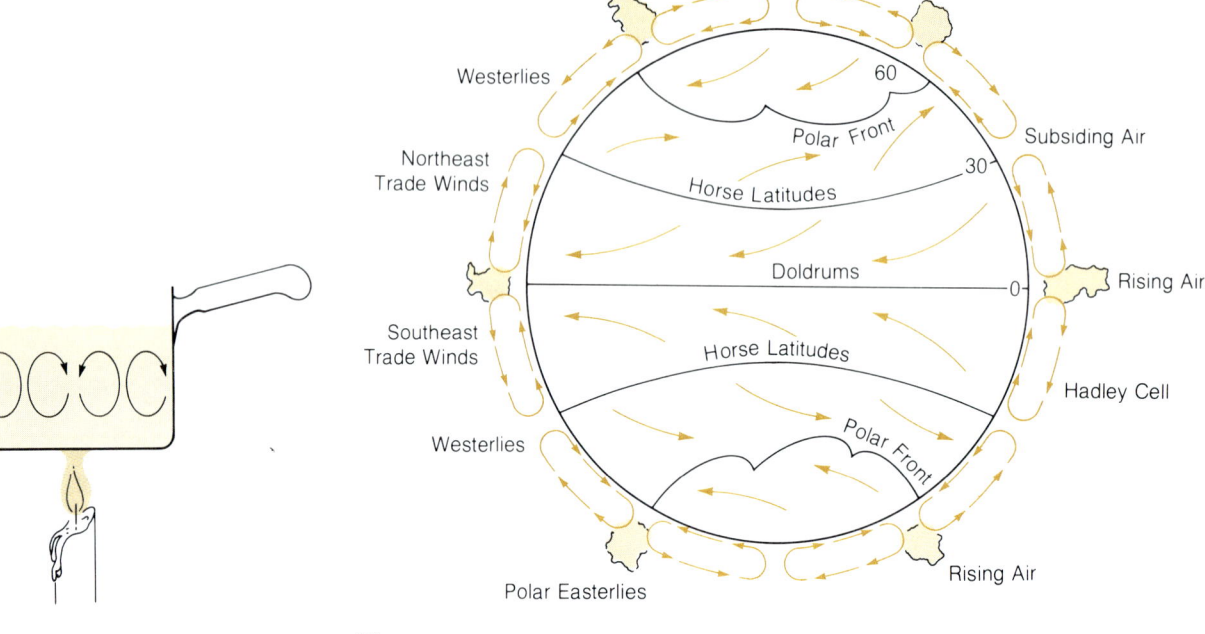

Figure 12.13

Convection. As a pan filled with liquid is heated from the bottom, warm liquid rises, and cooler liquid descends. This is an example of convection driven by the buoyant force, rather than by an applied force.

Figure 12.14

The pattern of circulation in the earth's atmosphere.

While convection may be induced by anything that causes a fluid to move, it may occur naturally when there is no external force to create fluid movement (Fig. 12.13). The reason is that a gas expands when heated, so its density is lowered. Therefore there is a buoyant force causing the warmed gas to rise (whether this actually happens depends on how rapidly the expansion occurs compared to the rate of heat loss to the surroundings). Thus, the air next to a warm body will be heated by conduction, and then will rise, starting the convection process. As the warm air rises, cooler air from the surroundings fills in. At equilibrium, cool air descends at a sufficient rate to compensate for the warm air that is rising. The earth's weather is controlled to a large extent by convection (Fig. 12.14), as is the movement of continental land masses over its surface.

Within the human body, convection due to flowing blood is the principal mode of heat transport. Many furnaces rely on forced-air convection to distribute heat throughout buildings, and even other forms of heaters, such as fireplaces and electric heaters, rely on natural convection for heat transport.

It is very difficult to generalize the quantitative treatment of heat transport by convection, because it depends to such a great extent on the particular situation. An equation much like Eq. 12–14 is appropriate, because convection is proportional to the temperature gradient and the exposed surface area, but the equivalent of the thermal conductivity constant is difficult to define. This

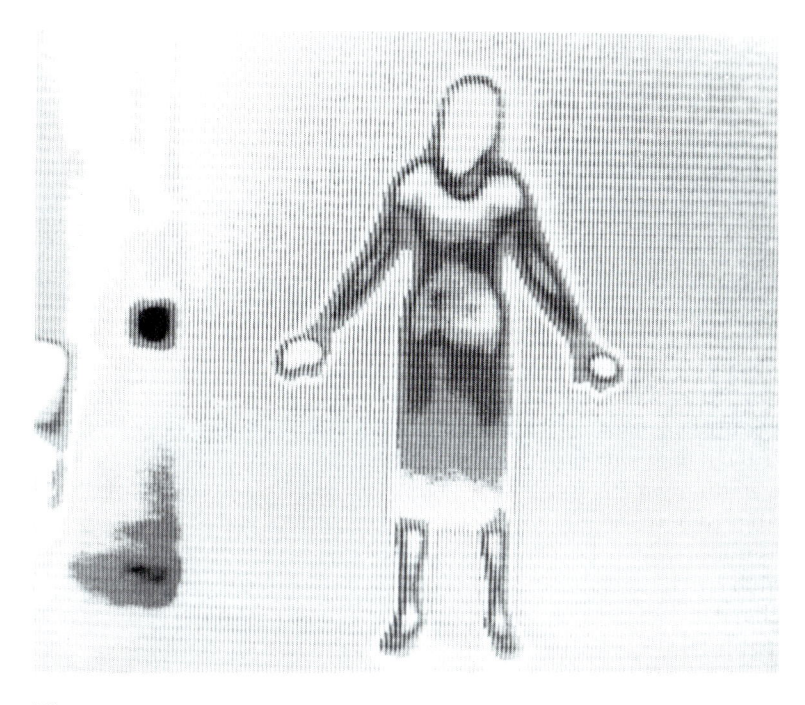

Figure 12.15

Infrared glow from a human. At normal body temperature, a human radiates infrared light. This image was used to find diseased areas, which are warmer than their surroundings, and therefore glow more brightly in infrared light.

term depends on the velocity with which fluid elements or bubbles move, and on the average distance they travel (this is usually referred to as the mixing length), both of which are difficult to evaluate.

Radiation

The third type of energy transfer does not require a material medium, but instead is the result of **radiation**. To fully discuss the process of energy transfer by radiation requires an understanding of the properties of electromagnetic waves, which we will not attempt until later chapters. It is feasible, however, to analyze the rate of heat transfer without detailed knowledge of the waves themselves.

Any object with a temperature above absolute zero emits radiation. This is not obvious, because the wavelength of the radiation depends on the temperature of the object, and objects at ordinary temperatures radiate primarily in the infrared, which is invisible to the human eye. Thus, a human body or the walls of a room are glowing (Fig. 12.15), but the glow can only be seen by specialized sensors. Hotter objects, such as a stove burner or a star, are visible because they are hot enough to emit radiation in wavelengths that we know as visible light.

It has been found that the rate of energy emission by a glowing object is proportional to its surface area and to the fourth power of its temperature. This represents a remarkable sensitivity to temperature; if one object is only twice as hot as another, for example, it will radiate $2^4 = 16$ times more energy per unit of surface area.

The rate at which emitted energy Q_e leaves an object is given by

$$\frac{Q_e}{t} = e\sigma A T^4 \tag{12-15}$$

where e is called the **emissivity**, a dimensionless number between 0 and 1 which represents the efficiency of the object as an emitter of energy; σ is the **Stefan-Boltzmann constant**, A is the surface area of the object, and T is its temperature on the absolute scale. The dependence on the specific properties of the material, which is represented by the emissivity, has been separated from a coefficient, σ, that is universal. In standard SI terms the value of σ is 5.67×10^{-8} J/s · m² · K⁴, or, in the more colloquial units we have been using in this chapter, $\sigma = 1.36 \times 10^{-12}$ cal/s · cm² · K⁴.

The emissivity varies with apparent shininess; that is, a shiny material such as polished metal has a nearly zero value of e, while a very black substance such as coal has a value near one. Thus a material that is a good absorber of radiation (that is, a black material) is also a good emitter, whereas one that absorbs little also emits little.

The amount of energy lost by an object is easy to calculate from Eq. 12-15, but the amount gained by absorption is more difficult to assess. This depends on the relative positions of the object and the source of the radiation it is absorbing, and on the surroundings (that is, whether both objects are in a room whose walls radiate and reflect, for example, or in open space where radiant energy not striking one object directly is lost). There is one situation which is fairly easy to treat; namely, an object at temperature T_1 in an environment of uniform temperature T_2. Then the radiation impinging everywhere on the surface of the object is that of an object of temperature T_2, and the net heat flow from the object is

$$\frac{Q}{t} = \frac{Q_e}{t} - \frac{Q_a}{t} = e\sigma A(T_1^4 - T_2^4) \tag{12-16}$$

where Q_a is the heat absorbed. The proportionality constant e is the same for both absorption and emission because we know that if $T_1 = T_2$ there must be zero net energy transfer.

Usually, calculating the absorbed energy is more difficult because the radiation of the emitting body diminishes with distance. If the emitting body sends out radiation in all directions, then the intensity of its radiation at any point is inversely proportional to the square of the distance from the emitter. Thus, if L is the **luminosity** (that is, the total energy emitted per unit time, given by Eq. 12-15) of the emitting object, then the intensity of the radiation at distance d from the emitting object is $L/4\pi d^2$. Hence the rate of energy absorption by a body of cross-sectional area a, which is a distance d from an emitter whose luminosity is L, is

12.5 HEAT TRANSPORT

$$\frac{Q_a}{t} = L\left(\frac{a}{4\pi d^2}\right). \tag{12-17}$$

In practice, the rate of energy absorption is less than this, because some fraction of the incident radiation is usually reflected.

Example 12.13 Calculate the equilibrium temperature of the earth, if its only heat source is radiant heat from the sun, and its only heat loss is its own radiation into space. Assume $e = 0.6$ and that .6 of the light incident on the earth is absorbed.

We can write expressions for the rate of heat loss and rate of heat gain, then set them equal and solve for T. The rate of heat loss is given in Eq. 12–15. The earth's surface area is $A = 4\pi R^2$ and its cross-sectional area $a = \pi R^2$, where $R = 6,378$ km.

$$\frac{Q_e}{t} = e\sigma A T^4 = (0.6)(5.67 \times 10^{-8} \text{ J/s} \cdot \text{m}^2 \cdot \text{K}^4) 4\pi (6.38 \times 10^6 \text{ m})^2 T^4$$
$$= (1.7 \times 10^7 \text{ J/s} \cdot \text{K}^4) T^4$$

The rate of heat absorption by the earth comes from Eq 12–17, where L is the sun's luminosity, $L = 4 \times 10^{26}$ J/s. (This could have easily been calculated from the sun's temperature and radius, using Eq. 12–15.) The distance from the sun to the earth is $d = 1.5 \times 10^{11}$ m. Since .6 of the incoming light is absorbed by the earth, we have

$$\frac{Q_a}{t} = (.6) L \left(\frac{a}{4\pi d^2}\right)$$
$$= \frac{(.6)(4 \times 10^{26} \text{ J/s}) \pi (6.38 \times 10^6 \text{ m})^2}{4\pi (1.5 \times 10^{11} \text{ m})^2}$$
$$= 1.1 \times 10^{17} \text{ J/s}.$$

Now we set the rate of heat loss equal to the rate of heat gain, since this is the condition for equilibrium.

$$(1.7 \times 10^7 \text{ J/s} \cdot \text{K}^4) T^4 = 1.1 \times 10^{17} \text{ J/s}$$

Solving for T yields

$$T = \left[\frac{1.1 \times 10^{17} \text{ J/s}}{1.7 \times 10^7 \text{ J/s} \cdot \text{K}^4}\right]^{1/4} = 284 \text{K}$$

This oversimplified energy conservation analysis gives us a temperature for the solid earth that is a bit low; around 11°C or just about 52°F. The effects of the atmosphere can be significant in trapping or radiating energy.

FOCUS ON PHYSICS

Convection and the Earth

In this chapter we discussed the three basic forms of energy transport. One of the three, conduction, is usually of lesser importance in nature (although it is important in some astrophysical situations), but the other two are quite common. Radiation is, of course, the primary means by which solar energy reaches the earth, and solar energy is the ultimate source of most forms of energy with which we are familiar. Convection is very important as well, particularly in transporting energy from place to place, on or within the earth.

Convection requires a source of heat and a gravitational force. The source of heat produces a **temperature gradient;** that is, temperature changes steadily with height. Gravity creates a buoyant force whenever a fluid element or bubble has lower density than its surroundings (Chapter 11). Convection will occur spontaneously if a fluid element that is displaced slightly from its equilibrium position experiences a buoyant force that keeps it moving. Whether the buoyant force acts depends on the temperature gradient and the thermal properties of the fluid. For example, if a bubble of fluid that is displaced upward retains its internal heat and stays warmer than the fluid at its new level, then it will expand and decrease its density, so that the buoyant force will cause it to continue to rise. Under this condition we say that the fluid is unstable against convection (meteorologists refer to air in this condition simply as "unstable air"). Note that the source of heat must be at the bottom of the fluid for convective instability to occur.

Convection occurs in the earth's atmosphere when sunlight sufficiently heats the surface. If the earth were not rotating, the atmosphere would overturn steadily in a very simple pattern consisting of two giant cells, one on each side of the equator (Fig. 12.1.1a). Because of rotation, how-

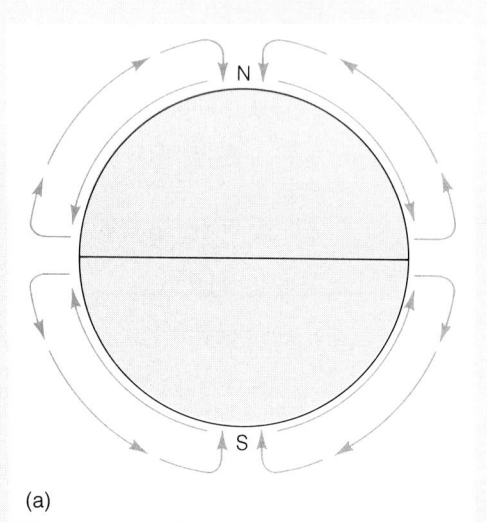

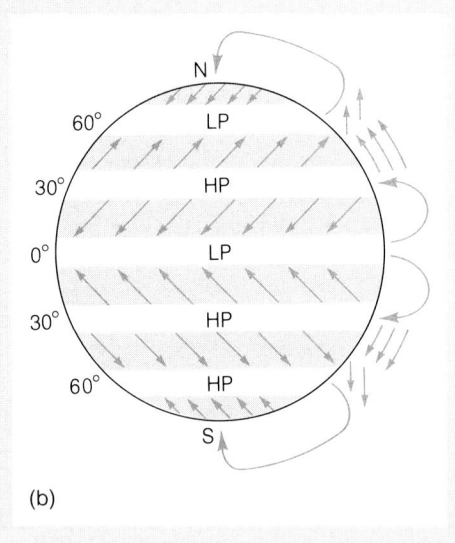

Figure 12.1.1

ever, a force called the **Coriolis force** acts perpendicular to any motion along the earth's surface. Thus winds do not flow directly north or south from the equator, but are instead diverted into circular flow patterns, so that the general circulation of the atmosphere is quite complex (Fig. 12.1.1b). There are zones of very little wind at the equator (called the "doldrums" in earth science), where air rises and there is little horizontal flow; and near 30° N or S latitude (the "horse latitudes"), where air descends. Between these zones there are steady lateral winds, toward the west in the northern or southern tropics, and toward the east in the northern and southern temperate zones. Thus, in the latitudes of the continental U.S., the prevailing winds are from west to east.

Apart from the overall circulation of the atmosphere, convection governs more localized motions. High- and low-pressure regions are places where cool air descends (creating high pressure at the surface) or warm air rises (creating low pressure at the surface). Again, the earth's rotation modifies the flows, causing air to move in one direction around low-pressure centers, and in the opposite direction around high-pressure regions. The flow around a low is called a **cyclone**; in the northern hemisphere, the direction of flow is counterclockwise. The flow around a high is called an **anticyclone,** which is in the clockwise direction in the northern hemisphere (the directions of cyclones and anticyclones are reversed in the southern hemisphere).

Severe storms are usually associated with cyclonic flows. Often the magnitude of these storms is enhanced by a special effect that causes very strong convection. If the warm, rising air in a convective flow contains a large quantity of water vapor, the water vapor may begin to condense as the air rises and cools. As the vapor condenses, the latent heat of vaporization is released and the air is warmed. This increases the buoyant force and makes the upward flow more rapid and energetic. This runaway process increases the intensity of cyclones so that they become tropical storms such as hurricanes, and it also acts on very localized disturbances to produce such storms as tornados and thunderstorms. Severe thunderstorms are characterized by towering cumulo-nimbus clouds that develop because of this enhanced convection and contain very strong updrafts (and counterflowing downdrafts).

Whereas the role of convection in atmospheric motions is quite readily observed and understood, it was not suspected until rather recently that the interior of the earth might also undergo convection. Even there the convection is not believed to occur in the liquid outer core, where it might at first be expected (for a summary of the earth's internal structure, see the *Focus on Physics* article on seismic waves in Chapter 10). The reason convection probably does not occur in the outer core is that the radioactivity that creates the major heat input to that zone is thought to lie primarily above it (but this point is controversial). If this is so, one of the requirements for convection, that the heat source be below the fluid, is not met.

It is now thought that convection occurs in the outer portion of the earth's mantle, in a zone called the **asthenosphere.** The main evidence for such convection is that the earth's crustal plates are known to move about slowly but steadily. A variety of evidence (including direct satellite measurements of the motions) shows that continental drift is taking place, causing the earth's landmasses to rearrange themselves over hundreds of millions of years. It has been found, from measurements of seafloor temperatures and fossil magnetic fields in seafloor rocks, that the ocean beds are slowly spreading away from undersea ridges in several locations.

The best explanation for this **tectonic activity** is that convection in the asthenosphere, driven by radioactive heating from below, is responsible (Fig. 12.1.2). The undersea ridges are thought to be places where warm material rises and spreads, and deep trenches called **subduction zones** are the places where the plates return to the interior. The crustal plates themselves, which carry the continents, float on the denser mantle material, and are carried along by the convective motions.

(continued)

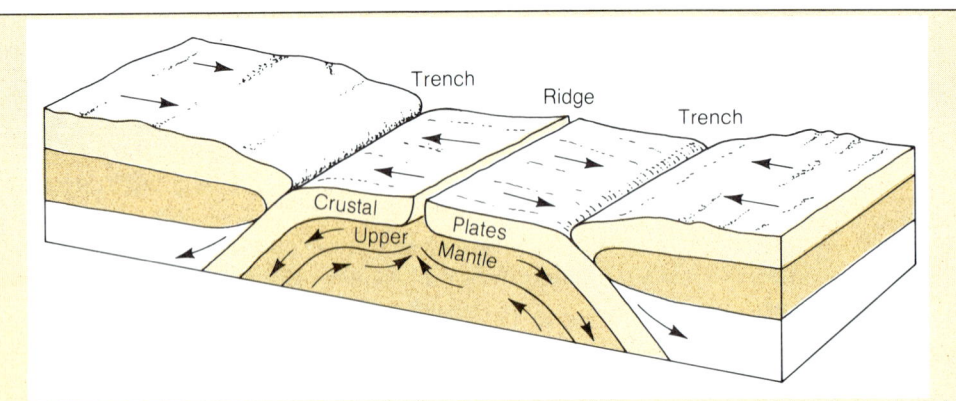

Figure 12.1.2

For convection in the mantle, the mantle material must not be rigid. Tidal distortions of the earth (Chapter 7) and other evidence indicate that it is somewhat pliable, like some plastics.

Convection is a universal phenomenon. The sun and stars have convective zones, for example, which play important roles in their energy distributions, just as they do in the earth. In addition, the atmospheres of other planets have flow patterns governed partially by convection.

Perspective

In this chapter we have covered quite a lot of ground, treating all macroscopic aspects of heat and heat transfer. We have seen how temperature changes cause thermal expansion of solids and liquids, and we have analyzed the complex relationship among temperature, volume, and pressure for gases. In addition, we have learned about changes of phase and have applied the principle of conservation of energy or, its equivalent, heat, to many examples of heat transfer.

To complete our general treatment of fluids and heat, we still need to analyze the activity of individual molecules in gases. This will be the subject of our next chapter.

PROBLEM SOLVING

Problems with Ideal Gases and Heat Transport

In many problems, only changes in temperature ΔT are important. For example thermal expansion depends only on ΔT,

$$\Delta L/L = \alpha \Delta T, \quad \Delta V/V = \beta \Delta T$$

where T is measured in degrees Centigrade or

Kelvin. However, the ideal gas law defines an absolute temperature scale (Kelvins),

$$P = \left(\frac{N}{V}\right)kT = \left(\frac{n}{V}\right)RT$$

where N/V or n/V is the gas density in particles/m^3 or moles/m^3. We use Boltzmann's constant k when we deal with numbers of particles and the universal gas constant R when we deal with moles (1 mole equals 6.022×10^{23} particles). Moles are useful in problems because, by definition, 1 mole of particles has a mass (in grams) equal to the atomic mass of the particle. For example, one mole of water vapor (H$_2$O) has a mass of 18 g (oxygen is 16 g per mole and each of the hydrogen moles is 1 g).

If we regard N/V or n/V as a single variable (density), then the ideal gas law is a relation between three thermal variables—pressure P, temperature T, and density. If we know two of these variables we can find the third.

Example 1 A mass 88 g of carbon dioxide gas (CO$_2$) fills a 2 liter container at 25 °C. What is its pressure?

The atomic mass of CO$_2$ is 44 g (carbon is 12 g and each of the oxygens is 16 g). Thus we have 2 moles of CO$_2$. We therefore know two thermal variables, the temperature T = 25 °C = 298 K and the density n/V = 2 moles/2 l. The pressure is

$$P = \left(\frac{n}{V}\right)RT$$
$$= \left(\frac{2 \text{ mol}}{2 \times 10^{-3} \text{ m}^3}\right)(8.314 \text{ J/mol K})(298 \text{ K})$$
$$= 2.48 \times 10^6 \text{ J/m}^3.$$

Since 1 J = 1 N · m, we see that the units of pressure are N/m^2, which is correct. One atmosphere equals 1.013×10^5 N/m^2, so this is 24.5 atm.

A good application of the ideal gas law is to problems about vapor pressure and relative humidity. Table 12.2 gives the saturated vapor pressure of water for various temperatures. As T rises, the vapor pressure (P_v) rises and the atmosphere can hold more water. The density ρ can be computed from the ideal gas law,

$$\rho = m(N/V) = m(P_v/kT)$$

where m = 18 u = 2.99×10^{-23} g is the mass of a water molecule. The relative humidity is the ratio of the actual (partial) pressure of water vapor to the saturated vapor pressure. By the gas law, it is also the ratio of the actual density to the saturated density.

Example 2 What is the saturated density of water vapor at 20 °C? If a 300 m^3 room contains 2 l of water vapor, what is the relative humidity?

We begin by using the gas law to find the saturated vapor density ρ. From Table 12.2, the saturated vapor pressure at 20 °C is 0.0234×10^5 N/m^2. Thus,

$$\rho = \frac{(2.99 \times 10^{-23} \text{ g})(2.34 \times 10^3 \text{ N/m}^2)}{(1.38 \times 10^{-23} \text{ J/K})(293 \text{ K})}$$
$$= 17.3 \text{ g/m}^3$$

which agrees with the last column of the table. The room contains 2 l of water, or a mass of 2 kg. The actual density is then 2000 g/300 m^3 = 6.67 g/m^3, and the relative humidity is 6.67/17.3 = 0.386 or 38.6%.

Heat transport involves thermodynamics and the first law, $\Delta U = Q - W$. To apply this law to problems, one must be careful about the signs of the heat transfer Q and the work W. Just remember that when heat flows into a system or work is done on the system, the internal energy increases ($\Delta U > 0$). That means that in the first law of thermodynamics, Q is positive for heat flow into the system

(continued)

and W is positive for work done by the system on the surroundings. If work is done on the system by the surroundings, W is negative.

Heat transfer may occur by conduction, convection, or radiation. Equations 12–14 and 12–15 describe the transfer rate by conduction and radiation; convection is more difficult to analyze and we will not discuss it further. To compute Q/t for conduction, one must know 3 quantities, the coefficient of conduction k, the ratio $(T_1 - T_2)/d$, which we call the temperature gradient, and the surface area A. To compute Q/t by radiation, we must know the surface temperature T, the surface area A, and the emissivity e, a number between 0 and 1. Conductive heat transport is proportional to the temperature gradient, that is, to the difference in temperatures on the inside and outside of the object. Radiative heat loss, on the other hand, depends on the fourth power of the surface temperature. The radiative heat loss from hot objects is enormously greater than from warm objects.

The heat transferred to an object is directly related to its temperature change through the specific heat, $Q = mc\Delta T$. If a phase change occurs, then an extra Q is required even though $\Delta T = 0$. In problems involving a solid changing to a liquid (melting) or a liquid changing to a gas (evaporation), be sure to include the appropriate latent heat in finding Q.

Example 3 One kg of ice at -5 °C is heated to $+110$ °C steam, first melting to water and then vaporizing. How much heat is required?

The ice increased in temperature by $\Delta T_i = 5$ C°, the water by $\Delta T_w = 100$ C° and the vapor by $\Delta T_v = 10$ C°. We must add to this the latent heats of fusion L_f and vaporization L_v. Referring to Tables 12.5 and 12.6, we find

$$Q = m(C_i \Delta T_i + C_w \Delta T_w + C_v \Delta T_v + L_f + L_v)$$
$$= (1 \text{ kg})[(2100 \text{ J/kg} \cdot °\text{C})(5 °\text{C}) +$$
$$(4180 \text{ J/kg} \cdot °\text{C})(100 °\text{C}) +$$
$$(2010 \text{ J/kg} \cdot °\text{C})(10 °\text{C}) +$$
$$3.33 \times 10^5 \text{ J/kg} + 2.26 \times 10^6 \text{ J/kg}]$$
$$= 3.04 \times 10^6 \text{ J}$$

Since 1 kcal equals 4180 J, this is 727 kcal.

SUMMARY TABLE OF FORMULAS

Conversion between centigrade and Fahrenheit scales:
$$T_C = \frac{5(T_F - 32)}{9} \tag{12-1}$$

$$T_F = 32 + \frac{9}{5}T_C \tag{12-2}$$

Relationship between absolute and centigrade temperature:
$$T_K = T_C + 273 \tag{12-3}$$

Linear thermal strain:
$$\frac{\Delta L}{L} = \alpha \Delta T \tag{12-4}$$

Linear thermal stress:
$$\frac{F}{A} = \alpha \Delta T Y \tag{12-5}$$

Volume thermal strain:
$$\frac{\Delta V}{V} = \beta \Delta T \tag{12-6}$$

Ideal gas law:
$$PV \propto T \tag{12-7}$$
$$PV \propto nT \tag{12-8}$$
$$PV = nRT \tag{12-9}$$
$$V = \frac{nRT}{P} \tag{12-10}$$
$$PV = NkT \tag{12-11}$$

First law of thermodynamics:
$$\Delta U = Q - W \tag{12-12}$$

Heat and temperature change:
$$Q = mc\Delta T \quad (12\text{–}13)$$

Conduction of heat:
$$\frac{Q}{t} = \frac{kA(T_1 - T_2)}{d} \quad (12\text{–}14)$$

Emitted energy:
$$\frac{Q_e}{t} = e\sigma A T^4 \quad (12\text{–}15)$$

Heat loss for an object that absorbs and emits:
$$\frac{Q}{t} = \frac{Q_e}{t} - \frac{Q_a}{t} = e\sigma A(T_1^4 - T_2^4) \quad (12\text{–}16)$$

Absorption of energy from an object of luminosity L;
$$\frac{Q_a}{t} = L\left(\frac{a}{4\pi d^2}\right) \quad (12\text{–}17)$$

THOUGHT QUESTIONS

1. Why do scientists often prefer the Kelvin temperature scale? How is it related to the properties of water?

2. Can you devise a way, not mentioned in the text, to construct a thermometer?

3. Suppose you wished to make a mirror or lens for a telescope. Would you choose a glass with high or low coefficient of volume expansion? Explain.

4. How would you construct a temperature scale based on the properties of a substance other than water? What properties of the substance would be relevant?

5. Why is it often possible to remove a stubborn cap from a bottle by running hot water over the top of the bottle?

6. Why is the coefficient of volume expansion equal to three times the coefficient of linear expansion for isotropic substances?

7. Discuss how a hot air balloon works.

8. How would you experimentally determine the value of Avogadro's number and the universal gas constant?

9. Can you think of a biological process in which the law of partial pressures plays an important role? Explain the role.

10. Explain the phenomenon of dew. Can dew occur on a clear night?

11. Compare and contrast heat and temperature.

12. How would you measure the specific heat of (a) a solid, (b) a liquid, (c) a gas?

13. How does the specific heat of a typical gas compare to that of a typical solid? Explain the difference.

14. Why is a certain amount of (latent) heat required to produce a phase change?

15. Under what circumstances, if any, will carbon dioxide exist as a liquid?

16. If a block of ice is placed under pressure, will the heat of fusion or melting point be altered? Explain the relevance for ice skating.

17. Give an everyday example of heat transfer by (a) conduction, (b) convection, (c) radiation. Which method is most important in determining overall weather patterns on the earth?

18. Why do metals tend to have high values of thermal conductivity?

19. Explain what is meant by wind chill factor. Is there an equivalent effect at hot temperature? Explain why or why not.

20. Why can't there be a emissivity outside the range 0–1?

21. What, if anything, can a high or low emissivity for a planet's surface tell us about that planet?

22. Describe the factors that determine whether energy will be transported by conduction, convection, or radiation.

23. Can more than one type of energy transport (see Question 22) occur at once? Explain.

24. Would you expect human skin to have a high or low thermal conductivity? Explain.

25. Would a rough surface have a high or low emissivity? Explain.

PROBLEMS

Section 12.1 Temperature Scales

- 1. Show that Eq. (12–2) follows from Eq. (12–1).
- 2. Verify that absolute zero on the Fahrenheit scale is $-459.67°$ if 0 K corresponds to $-273.15°C$.
- 3. The surface temperature of the planet Venus is about 700 K. What is this temperature in Centigrade and Fahrenheit?
- 4. The photospheric temperature of the sun is about 5800 K. What is this on the Fahrenheit scale?
- 5. At what temperature will a Fahrenheit thermometer have the same reading as a Celsius thermometer?
- 6. Calculate the Centigrade temperatures of (a) a cold day in Canada ($-30°F$), (b) a hot day in Death Valley ($120°F$), (c) a summer day in Colorado ($85°F$).
- 7. What is body temperature in Centigrade degrees?
- 8. Ethanol freezes at $-114°C$. At what Kelvin and Fahrenheit temperatures does it freeze?
- 9. Write an expression for converting from T_F (Fahrenheit temperature) to T_K (Kelvin temperature).
- 10. What are the record temperature extremes on earth ($+136°F$ in Libya and $-127°F$ in Antarctica) in degrees Kelvin?
- 11. The gas temperature in the center of some dark interstellar gas clouds has been measured as 10 K. How many degrees below zero Fahrenheit is this?

Section 12.2 Thermal Expansion

- 12. A 2 m rod of metal is heated from $-100°C$ to $+500°C$ and expands by 1 cm. Find the coefficient of linear expansion.
- 13. How much would a steel meterstick have to be heated (or cooled) so that its length changes by 1 mm?
- 14. A rod of a certain substance measures 1.50 m at 10°C. At 40°C its length is measured at 1.52 m. What is its coefficient of linear expansion?
- 15. A sidewalk is to be made of 1 m concrete slabs. The temperature range in that locale is $-20°F$ to $+100°F$. How large a gap should be left between slabs if they are laid at 60°F?
- 16. The steel rails on a railroad track are 4 m long and have spacings of 1 cm. Over what temperature range, ΔT, will the track be able to withstand buckling?
- 17. Suppose you could assemble the bridge in Example 12.2 at a temperature of your choice within the stated range. At what minimum temperature would it be safe to build the bridge without gaps?
- 18. A square sheet of aluminum 10 cm on a side has a hole 1 mm in diameter punched through its center. If the aluminum is heated from 0°C to 200°C, how large is the hole?
- 19. A 1 m iron rail can expand only 1 mm before structural problems occur. If the rail was placed in the structure at 20°C, to what temperature can the structure rise without difficulty?
- 20. A certain volume of mercury occupies 0.5 m^3 at 50°C. What volume will it occupy at 0°C?
- 21. To what temperature must the mercury in Problem 20 be heated to increase the volume to 0.501 m^3?
- 22. A cube 0.5 m on a side expands its volume by 0.01% when its temperature rises 20°C. Find the coefficients of linear and volume expansion.
- 23. An object with volume 1 cm^3 has a density 1.1 g/cm^3 and is immersed in water at 10°C. As the water and object are heated together, the object begins to float at 80°C. What are the coefficients of linear and volume expansion for the object?
- 24. A brass pendulum has length 1 m. If the temperature range on a given day is from 60° to 100°F, what is the fractional accuracy of the pendulum's period?
- 25. Carbonated beverages (mostly water) often come in aluminum cans designed to hold 354 ml. If the cans are made and filled at 5°C and the beverage is consumed at room temperature (25°C), how much volume should be left as empty space in the can?
- 26. A mercury thermometer consists of a spherical glass bulb of radius 5 mm connected to a long tube of 0.1 mm radius. Calibrate the thermometer. How many cm does the mercury column rise in the tube for a 1C° increase in temperature? Ignore expansion of the glass.
- 27. The U.S. is crossed by many interstate highways, which have an average width of 20 m and average depth of 0.3 m of concrete pavement. If there are three main east-west highways (5,000 km) and five main north-south (2,000 km) highways, estimate the extra volume of concrete created by thermal expansion between winter and summer. Assume mean temperatures of 20°F and 80°F.

PROBLEMS

Section 12.3 Equations of State

- **28** What are the molecular masses of water (H_2O), carbon dioxide (CO_2), and ethyl alcohol (C_2H_6)?

- **29** What is the molecular mass of a uranium 238 atom?

- **30** If an atom has a mass of 1.99×10^{-26} kg, what atom is it likely to be?

- **31** How many water molecules are in 1 liter of liquid? How many moles?

- **32** Ten grams of a gas occupy 15 liters at 1 atm and 20°C. Find the molecular mass of the gas.

- **33** A gas at 0°C and constant pressure has a volume of 10 liters. What will be the volume at 10 K (Kelvin)?

- **34** Show that the ideal gas law can be written $P = (\text{constant})\rho T$, where ρ is the mass density and T the Kelvin temperature. What is the constant?

- **35** The atmospheric pressure at 100,000 feet is about 0.01 atm. To what volume will the gas in a scientific balloon expand if it begins at 3 m³ at sea level? Assume that the sea level temperature is 20°C and drops to −55°C at 100,000 feet.

- **36** The atmosphere of Venus is composed almost entirely of carbon dioxide (CO_2), to a height of 90 miles. If the mean temperature is 700 K and the mean pressure 50 atm, estimate the number of moles of CO_2 in the atmosphere. Venus has a surface radius of 6,050 km.

- **37** Assuming that an average atom has a radius of 10^{-10} m, find the volume that an Avogadro's number of atoms will occupy when packed together.

- **38** Using the value of the universal gas constant R, and assuming that $P = 1$ atm and $n = 1$ mole, plot the volume V versus the Kelvin temperature T. Find the temperature at which $V = 0$.

- **39** A room 3 m × 10 m × 10 m is at temperature 20°C and 1 atm pressure. How many grams of oxygen are present? Assume the earth's atmosphere is 21% oxygen (O_2) by number.

- **40** The average density of atomic hydrogen gas between the stars is about 1 atom per cm³. Find the mass density (g/cm³) and the gas pressure, if the mean temperature is 10^4 K.

- **41** Verify the value of Boltzmann's constant k from the values of R and N_A given in the text.

- **42** Calculate the number of atoms in a 10 m³ container at 2 atm pressure and 20°C.

- **43** By mass, the earth's atmosphere is 21% oxygen (O_2) and 79% nitrogen (N_2). We neglect other gases, mostly argon. For 1 liter of atmosphere at STP, calculate (a) the mean molecular mass, (b) the number of N_2 and O_2 molecules.

- **44** Suppose three gases are in a container. Write the law of partial pressures for this system.

- **45** The relative concentrations of carbon dioxide (CO_2) and nitrogen (N_2) in Venus's atmosphere are 27:1 by number. If the total atmospheric pressure at the surface is 90 atm, find the partial pressure of each gas.

- **46** The temperature of Mars ranges from 130 K to 290 K, and its surface pressure is 0.007 atm. Using Fig. 12–10 and Fig. 12–11, discuss the phase behavior of water and CO_2 on Mars.

- **47** What is the partial pressure of water when the relative humidity is 90% and the temperature 90°F? How many grams of water are contained in a room of volume 300 m³?

- **48** A planetary atmosphere contains 30% argon (Ar), 40% CO_2, and 30% N_2 by number. If the total pressure of the atmosphere is 4 atm, calculate the partial pressures of each gas.

- **49** Suppose the average temperature and relative humidity over the earth are 20°C and 50%. If the atmosphere has a mean thickness of 8.4 km, estimate the mass of water molecules in the atmosphere. The earth's radius is 6,378 km.

- **50** Normal car tire pressure is 24 lb/sq in at 20°C. What pressure will the tire pressure gauge read after a long day's drive at temperature 40°C?

- **51** A diver fills his lungs to 2 m³ capacity at a depth of 30 m. If he were so foolish as to hold his breath and rise to the surface, what would be his lungs' volume?

- **52** What are the phases of water under the following circumstances:
  (a) 1 atm and 90°C, (b) 100 atm and 0.01°C, (c) 0.005 atm and 50°C.

- **53** Two labor negotiators are confined to a 5 m × 5 m × 3 m room at 30°C. If the relative humidity is 50%, how many grams of water can evaporate from each person's skin?

- **54** What is the dew point on a day when the relative humidity is 70% and the temperature 10°C? The dew point is the temperature at which vapor is saturated.

- **55** Find the relative humidity if the dew point is 20°C and the air temperature is 35°C.

- **56** Suppose we wish water to boil in a pressure cooker at 120°C. At what pressure should the cooker operate?

- 57 Describe the changes, if any, that would occur if water at 10 atm pressure were heated from −100°C to 400°C.

- 58 The average atmospheric pressure on Mars is about 0.01 atm, and its temperature ranges from −140°C to 20°C. Describe the phase(s) in which water can exist on the planet's surface.

- 59 Find the value of the triple point of water in degrees Kelvin and mm Hg.

Section 12.4 Heat, Energy, and Phase Changes

- 60 Verify by direct calculation from definitions of the Joule and BTU that one BTU equals 1,054 J. Remember that 1 lb has a mass of 2.205 kg.

- 61 Derive the conversion between the specific heat expressed in BTU/lb · F° and in cal/g · C°.

- 62 A typical adult diet includes 1,500–3,000 Cal of food (1 Cal equals 1000 cal) per day. What is the energy content in Joules?

- 63 A system gains 40 J of heat and produces 20 J of work. What is the change in the internal energy of the system? How efficient is the system at performing work?

- 64 How much work will an engine perform if it gains 100 J of heat and 30 J of internal energy?

- 65 An object is heated with 10 J of energy, and the temperature rises 15°C. If the object has a mass of 5 g, find its specific heat.

- 66 Calculate the heat needed to change the temperature of an ice cube 5 cm on a side from −30°C to −20°C.

- 67 How much heat is needed to change the temperature of the ice cube in Problem 66 from −30°C to liquid at 10°C?

- 68 A copper cube 2 cm on a side has an unknown temperature. It is plunged into a 1 kg bath of water at 5°C, and comes to equilibrium. If the temperature of the water rises 2°C, what was the cube's original temperature?

- 69 How much energy is needed to melt a 5-kg block of ice?

- 70 A refrigerator removes 10^5 J of heat from a pan containing 1 kg of water initially at 10°C. Will the water freeze? If not, how much more heat must be removed?

- 71 Compute the energy required to (a) boil 1 liter of water at 100°C, (b) melt 1 kg of ice at 0°C.

- 72 Power generators typically generate waste heat, which must be dumped to the atmosphere or a water supply. If one power plant generates 500 megawatts of heat, compute (a) the amount of water that must be evaporated per second in a cooling tower, (b) the temperature increase of a lake of 10^7 m^3 volume if one day's heat output is dumped and none radiates away.

- 73 A substance has heat capacity 0.05 cal/g · C° as a solid and twice that value as a liquid. The substance is heated from 15C° below its melting point to 5C° above. If the total energy needed is 80 cal and the mass is 2 grams, find the heat of fusion.

- 74 The specific heat of nitrogen (N$_2$) gas is about 1,000 J/kg/K. How much heat will be transferred to 10^3 m^3 of N$_2$ gas at STP if the temperature is increased 10C°?

- 75 Five kg of water at 0°C are mixed with 10 kg of water at 25°C. Find the temperature of the final mixture.

- 76 A 1 kg of mass of an unknown sample is heated to 90°C, then immersed in a 2 l water bath at 10°C. If the final temperature of the bath after immersion is 15°C, what is the specific heat of the sample?

- 77 The heat of combustion of natural gas (methane CH$_4$) is about 13,000 cal/g. Express this in J/kg, and estimate (a) the energy contained in 1 liter of methane at STP, (b) the energy released in burning one CH$_4$ molecule.

- 78 Gasoline has a heat of combustion of about 11,000 cal/g. If its density is 0.68 g/cm^3, estimate (a) the energy contained in one gallon (3.786 l), (b) the energy in a barrel of oil containing 42 gallons, (c) the number of barrels of oil needed to meet the United States' annual oil-based energy consumption of 4 × 10^{19} J. Assume that the energy contents of oil and gasoline are comparable, and that energy generation is 34% efficient (part c).

Section 12.5 Heat Transport

- 79 A steel fireplace poker has length 1 m and radius 0.5 cm. One end is in the flame at 250°C. How much heat is transferred to the cool end (35°C) in 10 minutes?

- 80 Calculate the rate of heat conduction through a 10-inch concrete wall, 15 m^2 in area. Assume the temperature of the outside is 0°C and the inside 20°C.

- 81 Calculate the heat loss (watts) through a glass window, 1 m^2 in area and 0.5 cm thick, separating temperatures of 10°C and 20°C. Repeat the calculation for a pane twice as thick, but with twice the area.

SPECIAL PROBLEMS

82 Calculate the rate of heat conduction from the core of a human body (37°C) to the surface. The surface area is 1.7 m² and its temperature is 34°C. Assume the mean thickness of tissue is 2 inches.

83 Estimate the conductive heat loss from a wooden house constructed of 2-inch thick boards. The four walls have area 40 m² each, and the flat roof, 100 m². The temperature difference between inside and outside is 20°C, and the conductivity is 0.1 J/s/m/°C. Would insulation be a good idea?

84 Compare the amount of heat conducted per unit time along two rods of length 1 m and 2 m. The temperature difference between ends is the same, but their areas $A_1/A_2 = 4$.

85 A sphere of temperature T_1 is immersed in a medium of temperature $T_2 > T_1$. If the temperature difference, $(T_2 - T_1)$ is doubled, how should the radius of the sphere R be changed to maintain the same heat loss rate?

86 The tungsten filament in a 100 watt light bulb has an emissivity $e = 0.3$ and burns at 3,000°C. What is its radiating area?

87 A tree's leaves have 4 cm² area per side and emissivity $e = 0.7$. If sunlight through the atmosphere has a flux of 600 watts/m², estimate (a) the rate of heat absorption, (b) the equilibrium temperature of the leaf if radiative losses are the only means of cooling.

88 Estimate the net heat output (watts) of a human, with surface area 1.8 m², surface temperature 34°C, and emissivity $e = 0.7$. Assume the effective temperature of the environment is 80°F (see Eq. 12–16).

89 Two spherical bodies generate heat at rates W_1 and W_2 (watts). Derive an expression for the ratio of their radii in terms of their surface temperatures and emissivities.

90 A heating filament in a toaster has $e = 0.5$ and a surface area of 10^{-5} m². If its melting point is 3,000°C, what is the maximum power that can be run through the filament?

91 A radiating object has an intensity I (watts/m²) when viewed at a distance of 1 m. If the surface temperature is doubled, at what distance will the intensity equal I?

92 The solar constant is defined as the amount of energy reaching the earth per unit time per unit area. If this constant is 1,400 watts/m² and the earth is 1.5×10^{11} m from the sun, what is the luminosity (watts) of the sun? If the sun's surface temperature is 5,800 K and $e = 0.95$, what is its radius?

93 Suppose the sun's luminosity suddenly tripled, as a result of new nuclear energy generation in the core. What effect would this ultimately have on the surface temperature? If the surface temperature remained constant, what would happen to the sun's radius?

SPECIAL PROBLEMS

1. Planetary Temperatures

The surface temperature of planets in the solar system can be estimated by equating the solar heating with their radiative cooling. The sun's luminosity $L = 3.83 \times 10^{26}$ watts. Distances are often expressed in terms of the sun-earth distance, 1 AU $= 1.496 \times 10^{11}$ m, and mean radii are given in units of the earth's radius, $R_e = 6378$ km.

(a) Show that the heat absorption rate by a planet of mean radius R, a distance r from the sun, is given by $H = (L/4\pi r^2)(\pi R^2)(1 - a)$, where a is the fraction of light reflected (called the albedo).

(b) If the planet has a surface emissivity e (which we will take equal to 1), find its equilibrium temperature T in terms of L, r, R, and a.

(c) Evaluate T for the planets whose properties are given below. Is this the value of T you would expect for the earth?

Planet	r (AU)	R/R_e	a
Mercury	0.387	0.382	0.06
Venus	0.723	0.949	0.75
Earth	1.000	1.000	0.31
Mars	1.524	0.531	0.15
Jupiter	5.203	10.85	0.34
Saturn	9.555	9.01	0.34
Uranus	19.218	4.05	0.34
Neptune	30.110	3.91	0.33
Pluto	39.44	0.227	0.4

2. Solar Heating

The sun's intensity through the atmosphere is about 600 watts/m² on a clear day. A solar house has two sets of panels on its roof, a 100 m² water heating system (50% efficient) and a 200 m² photovoltaic system for generating electricity (10% efficient).

(a) Calculate the energy delivered to water heating and to electricity in one 10-hour day of full sunlight.
(b) How much water per day can this system heat from 50°F to 130°F? One gallon of water has a mass of 3.79 kg.
(c) How much electricity is produced (in Joules and kilowatt-hours)?
(d) Compare these figures to the average consumption in your home or apartment.

3. The Human Energy Budget

An average human diet provides about 2,400 Cal/day (1 Cal = 4,180 J). The table gives the mean metabolic rate for a 70 kg human engaged in various activities. The body metabolizes the food into energy, which can perform work or generate heat. The remainder is stored in fat, used for growth of cells, and eliminated as waste. The body is less than 40% efficient in converting Calories into useful work and heat.

Activity	Calories/h
Sleeping	60
Sitting at rest	100
Light activity (desk work)	200
Moderate activity (walking)	400
Moderate exercise (jogging)	600
Heavy exercise (running)	800

(a) Estimate the Caloric requirements for one day in an active person's life for sleep (8 h), sitting (9 h), walking (2 h), light activity at work (4 h), moderate exercise (1 h).
(b) Show that 1 Cal/h is equal to an energy consumption rate of 1.16 watts.
(c) What is the Caloric difference between running 5 miles at 8 minutes/mi and walking the same distance at half that speed?
(d) Estimate the water evaporation rate (g/s) needed to cool the body during running. Assume that 50% of the calories go into heat. What weight of water is lost in a one-hour run?
(e) What is the heat loss rate through radiative cooling? Assume the body has a surface area 1.8 m², surface temperature 34°C, and emissivity 0.7. The net cooling is the difference between heat radiated and heat absorbed from the environment at 80°F.

4. Thermal Pane Windows

Three layers of substances with widths d_1, d_2, d_3, and conductivities k_1, k_2, k_3 separate regions of temperatures T_0 and T_3 (Fig. 12.16). In a steady state, a temperature profile is set up, with intermediate temperatures T_1 and T_2 at the two interfaces.

(a) Explain why the heat flow Q/t must be the same through each of the layers.
(b) Use Eq. (12–14) to evaluate the temperature differences, $(T_0 - T_1)$, $(T_1 - T_2)$, and $(T_2 - T_3)$ in terms of Q/t, the area A, and the constants k_1, k_2, k_3.
(c) Show that $Q/t = A(T_0 - T_3)/D$, where D is a constant. What is D in terms of other parameters?
(d) Apply your results in part c to a sandwich pane window, in which two planes of glass 0.25 cm thick are separated by 0.10 cm of air. If the temperature difference between inside and outside is 20°C, compute the heat flow through a 1 m² window. What would be the heat flow through a solid window 0.6 cm thick?

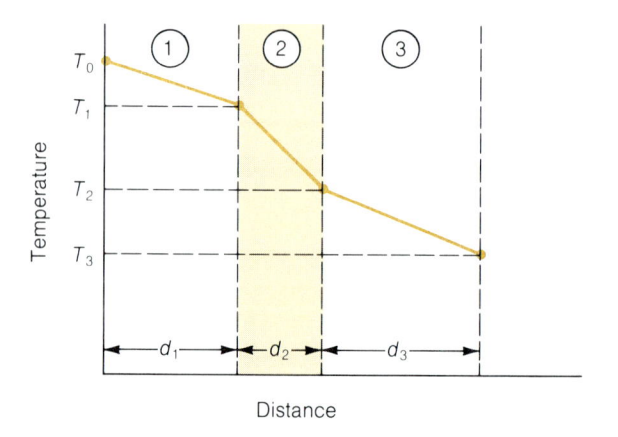

Figure 12.16

SUGGESTED READINGS

"Drops of water dance on a hot skillet and the experimenter walks on hot coals." J. Walker, *Scientific American*, Aug 77, p 126, **237**(2)

"Hot water freezes faster than cold water. Why does it do so?" J. Walker, *Scientific American*, Sep 77, p 246, **237**(3)

"Wonders of physics that can be found in a cup of coffee or tea." J. Walker, *Scientific American*, Nov 77, p 152, **237**(5)

"What happens when water boils is a lot more complicated than you might think." J. Walker, *Scientific American*, Dec 82, p 162, **247**(6)

"The physics of grandmother's peerless homemade ice cream." J. Walker, *Scientific American*, Apr 84, p 150, **250**(4)

"Convection currents in the earth's mantle." D. P. McKenzie, F. Richter, *Scientific American*, Nov 76, p 72, **235**(5)

"The thermostat of vertebrate animals." H. C. Heller, L. I. Crawshaw, H. T. Hammel, *Scientific American*, Aug 78, p 88, **239**(2)

"Convection." M. G. Velarde, C. Normand, *Scientific American*, Jul 80, p 92, **243**(1)

"The invention of the balloon and the birth of modern chemistry." A. F. Scott, *Scientific American*, Jan 84, p 126, **250**(1)

"Towards the absolute zero." O. V. Lounasmaa, *Physics Today*, Dec 79, p 32, **32**(12)

The flying circus of physics with answers. Jearl Walker, Wiley, New York, 1977.

A source book in physics. William F. Magie, Harvard University Press, Cambridge, 1965

Bioenergetics—Molecular Basis of Biological Energy Transformations. Albert L. Lehninger, Benjamin Press, Menlo Park, 1971

Concepts in Bioenergetics. Leonard Peusner, Prentice-Hall, Englewood Cliffs, 1974

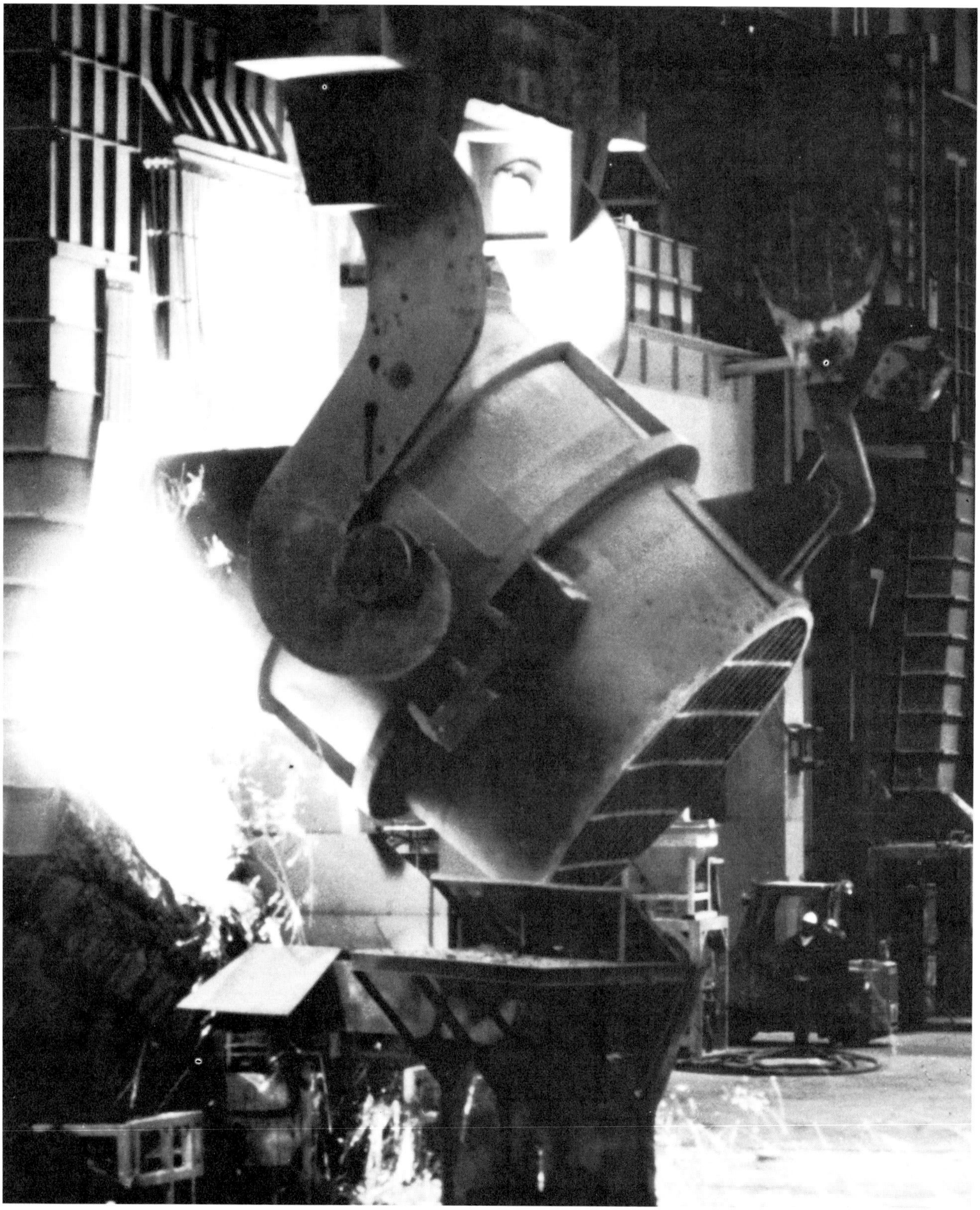

Chapter 13

Gas Kinetics, Thermodynamics, and Heat Engines

We have learned much about the behavior of gases by treating them as fluids and examining their macroscopic properties, with little consideration of what happens to individual particles. It has been clear, however, that the macroscopic behavior of a gas is the result of the aggregate motions of the individual particles of which it is composed. In this chapter we will analyze the properties of gases from this point of view.

While it is impractical to treat the motion of each individual atom or molecule in a gas, it is both practical and useful to consider average or probable motions, because it is possible to represent the overall behavior of a gas in those terms. Thus we will begin this chapter by developing relationships between particle properties and overall characteristics of gases. This will include a reconsideration of some of the behavior of gases discussed in the preceding chapter, and will also allow us to describe other behavior that can only be treated in this way. We will then consider the thermodynamic properties of gases in more detail than we could before, and we will study some practical applications.

13.1 Kinetic Theory

The mathematical framework which treats gases as collections of individual particles is called **kinetic theory.** This framework was developed in the late nineteenth century, primarily by Ludwig Boltzmann in Vienna and his Scottish contemporary, James Clerk Maxwell.

In kinetic theory, it is assumed that the general properties of a collection of particles can be represented by summing the behavior of all the individual particles. For this to be a manageable mathematical problem, certain simplifying assumptions are necessary. The first is that the total number of particles is large enough that statistical analyses are valid. This means that there is some representative behavior of the collection as a whole, and that the random variations of a few particles do not dominate. This is comparable to defining, for example, the average height of people in a room and comparing it to the average for the entire population; if the number of people in the room is small, then by chance their average height could be quite different from that of the general population, but if the number in the room is large, then their average will likely represent the overall population quite well.

We also assume that the distribution of individual particle velocities is **isotropic**, meaning that there is no preferred direction. Another important assumption is that the average distance between particles in a gas is large compared to the size of the particles, and that the particles exert forces on each other only when they collide (this is essentially true so long as the particles are not electrically charged). Finally, it is assumed that all collisions between pairs of particles or between particles and surfaces are perfectly elastic, so that no kinetic energy is converted into potential energy or radiation that is lost, nor is any of the kinetic energy of the particles transferred to the walls of a container.

Pressure, Temperature, and Particle Kinetic Energies

In Chapter 12 we made the assertion that the temperature of a gas is proportional to the average kinetic energy of its particles, but we said little more than that, and we did not define the constant of proportionality. Now we will do so.

We can begin with an intuitive discussion of the ideal gas law. The pressure exerted on the walls of a container of gas is the force due to collisions of gas particles with the wall. It is clear that if the volume of the container is increased, the average distance between particles must increase, so that the frequency of collisions with the walls must decrease (if the average particle speed remains constant). Hence, the pressure must decrease when the volume is expanded. On the other hand, if we keep the volume fixed, but increase the temperature, then the average speed of the particles increases (recall the discussion of heat and temperature in Chapter 12). The particles strike the walls with greater frequency and impact, so the pressure increases as the temperature is increased. Thus we can understand the general form of the ideal gas law.

Let us now consider the relationship between pressure and particle speed in more detail. Imagine a cubical box with sides L. Consider one wall of this box, lying in the yz plane (Fig. 13.1). A gas particle in this box will have a velocity vector **v** with components v_x, v_y, and v_z, but only the v_x component can exert a force on the wall. Therefore we begin with this component.

We learned in our discussion of mechanics that the force exerted by an object that strikes a surface is equal to its change of momentum per unit time. The momentum of a single particle in the x-direction is simply mv_x, where m is the particle's mass. Upon striking the wall in a perfectly elastic collision, the particle rebounds. There is no change in the y- and z-components of its velocity, and the x-component is simply reversed in direction. The momentum of the particle after the collision is $-mv_x$, so the change in momentum is $2mv_x$.

13.1 KINETIC THEORY

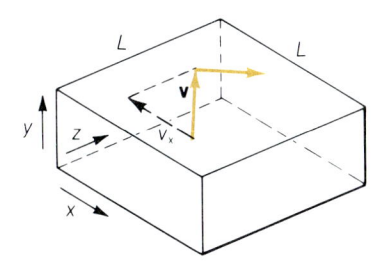

Figure 13.1

*A particle in a box. A gas particle with velocity **v** has a component of motion v_x perpendicular to wall y-z. The change in the particle's momentum when it strikes wall y-z and rebounds elastically is $2mv_x$, where m is the particle mass.*

It is impractical to assess the duration of the impact, and we are interested in the average of many impacts over time anyway, so we find the average force exerted on the wall by this individual particle by calculating how often it hits the wall as it bounces back and forth in the container. The particle must travel a distance $2L$ following a collision with the wall before it strikes that wall again, and this takes a time

$$t = \frac{2L}{v_x}.$$

The average force exerted on the wall by this particle is therefore

$$\overline{F} = \frac{\Delta p}{t} = \frac{2mv_x}{t} = \frac{mv_x^2}{L}.$$

Now we introduce the notion of an average value for the square of the velocity, so that the total force exerted on the wall is

$$F = \frac{Nm\overline{v_x^2}}{L} \qquad (13\text{--}1)$$

where $\overline{v_x^2}$ is the mean square velocity in the x-direction, and N is the number of particles in the container.

Our assumption that the motions are isotropic means that the average velocity components in all three dimensions must be equal, so

$$\overline{v}_x = \overline{v}_y = \overline{v}_z$$

and the resultant velocity for a particle is related to its components by

$$\overline{v^2} = \overline{v_x^2} + \overline{v_y^2} + \overline{v_z^2} = 3\overline{v_x^2}.$$

Now we can substitute for $\overline{v_x^2}$ in Eq. 13–1 for the total force exerted on a single wall due to all the particles in the gas.

$$F = \frac{Nm\overline{v^2}}{3L} \qquad (13\text{--}2)$$

We note carefully that $\overline{v^2}$ is the average squared velocity, which is not the same as the square of the average velocity (to see this, write a few numbers at random, then compare the average of their squares with the square of their average).

Finally, we divide the force by the area of the wall, which is L^2, and find

$$P = \frac{F}{L^2} = \frac{Nm\overline{v^2}}{3L^3} = \frac{Nm\overline{v^2}}{3V} \qquad (13\text{--}3)$$

where V is the volume of the container. We note that Pascal's theorem, which states that the pressure in a gas is transmitted equally in all directions, insures that this will be the pressure on all the walls, and that the shape of the container has no effect, so long as its volume is V. Thus, Eq. 13-3 is a general statement of the relationship among pressure, temperature, and particle velocity.

We can now find how temperature is related to the other quantities by invoking the ideal gas law. First we rewrite Eq. 13-3,

$$PV = \frac{1}{3}Nm\overline{v^2} \tag{13-4}$$

and then recall the ideal gas law,

$$PV = NkT \tag{13-5}$$

where k is the Boltzmann constant. Comparing these two expressions and noting that the average kinetic energy per particle is $\frac{1}{2}m\overline{v^2}$ leads to

$$\frac{1}{2}m\overline{v^2} = \frac{3}{2}kT. \tag{13-6}$$

Now we have a quantitative relationship showing that the temperature of a gas is proportional to the average kinetic energy per particle, consistent with our definition of temperature in the preceding chapter.

We may carry this analysis one step farther, and solve for the velocity,

$$v_{rms} = \sqrt{\frac{3kT}{m}} \tag{13-7}$$

where v_{rms} is the **root mean square velocity**, since it represents the square root of the mean square velocity $\overline{v^2}$. As we noted earlier, this is not the same as the square of the mean or average velocity $(\bar{v})^2$.

We are now in position to solve a variety of practical problems. For example, we can find the rms speed of molecules in any gas whose temperature we know. In a room at normal temperature, say 300 K, the oxygen molecules (O_2, which has a molecular weight of 32, so the particle mass is 32 u = 5.31×10^{-26} kg) move at velocity $v_{rms} = 484$ m/s, while the slightly less massive nitrogen molecules (N_2, with molecular weight 28) have $v_{rms} = 517$ m/s.

Example 13.1   It has been found that a species of gas will eventually escape a planetary atmosphere entirely if its rms velocity exceeds one sixth of the planet's escape velocity. If the temperature in the upper levels of the atmosphere of Venus is 240 K (Fig. 13.2), will hydrogen (H_2) escape? Will CO_2? The escape velocity for Venus is 10.3 km/s.

We need to calculate v_{rms} for H_2 and CO_2. For H_2 the molecular mass is

$$m = 2 \text{ u} = 2(1.66 \times 10^{-27} \text{ kg}) = 3.32 \times 10^{-27} \text{ kg}$$

13.1 KINETIC THEORY

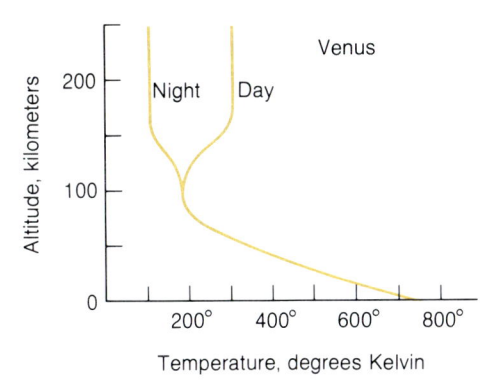

Figure 13.2

and the rms velocity is

$$v_{rms} = \sqrt{\frac{3kT}{m}} = \sqrt{\frac{3(1.38 \times 10^{-23} \text{ J/K})(240 \text{ K})}{(3.32 \times 10^{-27} \text{ kg})}}$$
$$= 1{,}730 \text{ m/s} = 1.73 \text{ km/s}.$$

This is just over one sixth of the escape velocity for Venus, so we conclude that hydrogen would eventually escape from the atmosphere of Venus.

For CO_2, the molecular weight is 44, so the particle mass is

$$m = 44 \text{ u} = 44(1.66 \times 10^{-27} \text{ kg}) = 7.30 \times 10^{-26} \text{ kg}$$

and the rms velocity is

$$v_{rms} = \sqrt{\frac{3(1.38 \times 10^{-23} \text{ J/K})(240 \text{ K})}{(7.30 \times 10^{-26} \text{ kg})}}$$
$$= 369 \text{ m/s} = 0.37 \text{ km/s}.$$

Therefore CO_2 will be retained. It has been found that this molecule is the dominant constituent of the atmosphere of Venus, while hydrogen is present only in molecular combination with other elements such as sulfuric acid (H_2SO_4).

The Distribution of Speeds

We have treated gases as though all the particles moved at the same velocity, but of course they do not actually do so. There is instead a range of velocities, and we can speak of a **velocity distribution,** which is the relative number of particles at each velocity. More useful is the **distribution of speeds,** which refers to the range of particle speeds in a gas, independent of direction.

In a gas at a uniform temperature, that is, an **isothermal** gas, it has proven possible to derive expressions for the velocity and speed distributions. This was first done by Maxwell, and the result is referred to as a **Maxwell distribution**. The principal assumption upon which this derivation rests is that the gas is in **thermodynamic equilibrium**, which means that the speeds of the particles are due entirely to the temperature of the gas. If any outside influence, such as the insertion of particles at a different temperature, has taken place, collisions among the particles will soon restore thermodynamic equilibrium, so this assumption is reasonable.

In general form, the Maxwell distribution of speeds is

$$n(v) = 4\pi \left(\frac{m}{2\pi kT}\right)^{3/2} v^2 e^{-mv^2/2kT} \qquad (13\text{–}8)$$

where $n(v)$ is the relative number of particles with speed v (actually the number in a narrow velocity interval centered at v). The key factor in this is the product of v^2 and the exponential term $e^{-mv^2/2kT}$, because the other terms simply represent a proportionality constant. As we consider increasing speeds starting from zero, at first the growth of the v^2 term dominates the decreasing exponential factor, and the number of particles rises (Fig. 13.3). Later the exponential decline dominates the v^2 term, and the number of particles begins to decrease. The value of v for which the maximum number of particles is found is

$$v_p = \sqrt{\frac{2kT}{m}} \qquad (13\text{–}9)$$

which is called the **most probable velocity**, because more particles will have this speed than any other. The entire distribution may be viewed as a probability distribution, because it represents the probability that any particle will have a particular speed. The mean velocity for a Maxwell distribution is

$$v_m = \sqrt{\frac{8kT}{\pi m}}. \qquad (13\text{–}10)$$

The most probable velocity v_p, the rms velocity v_{rms}, and the mean velocity v_m are indicated in Fig. 13.3 The most probable speed is not equal to either the mean velocity or the rms velocity. This is because the distribution is not symmetric about the peak; there are more particles at larger velocities than there are at smaller velocities. It is technically true that the distribution continues indefinitely towards high velocities (although with a vanishingly small number of particles as the velocity becomes very great), whereas there is a strict low-velocity cutoff at $v = 0$.

The Maxwell distribution is important in any process that depends on particle speeds. Most chemical activity, for example, including that associated with biological functions, depends on the relative velocities of particles that

13.1 KINETIC THEORY

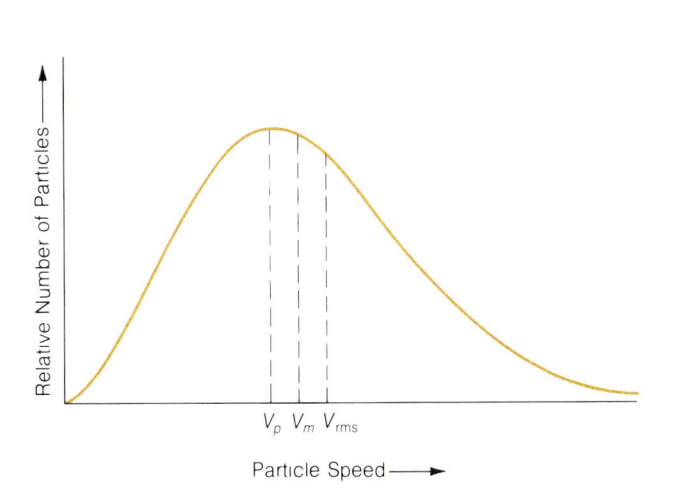

Figure 13.3

The Maxwell velocity distribution.

Figure 13.4

Maxwell distributions for different temperatures. Particle velocities have a greater range for the higher temperature. If v_{min} is the minimum particle speed for particles to react if they collide, in the hotter gas (T_2) many more particles can undergo reactions.

collide. Reactions will generally not occur unless a certain minimum amount of energy is available, and this energy usually comes from the kinetic energy of the two particles that collide. Thus many chemical reactions are much more effective at high temperatures than at low temperatures. Fig. 13.4 shows two Maxwellian distributions for different temperatures, with a vertical line indicating a minimum energy for some reaction to occur. We see that the number of particles exceeding this minimum is much greater for the hotter gas, so that the reaction will occur much more readily in this gas.

Although it is true that not many particles have velocities very much higher than the most probable velocity, the fact that there are some particles in the high-velocity tail of the Maxwell distribution can have very important consequences. In Example 13.1, we mentioned that a planetary atmosphere will lose a species of molecule completely if the rms velocity is only one sixth of the escape velocity. Now we can see how this happens. Even if the rms velocity is well below the escape velocity, there will be some molecules in the high-velocity tail of the distribution, whose velocities exceed the escape velocity (Fig. 13.5). These molecules will escape in time. So long as the temperature of the atmosphere is constant, however, thermodynamic equilibrium will ensure that a constant proportion of the particles will always be in the high-velocity tail. In effect, collisions among particles will boost new ones up to the portion of the velocity distribution that was vacated by the particles that escaped. The new members of the high-velocity tail will escape in their turn, to be replaced by more. This process continues until essentially all the molecules of the species are gone.

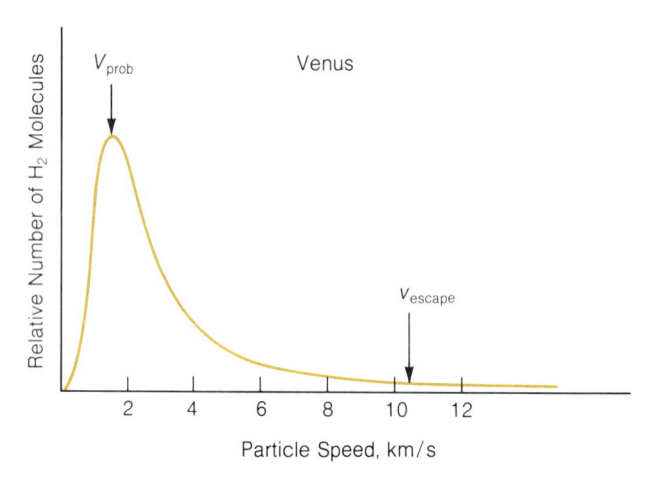

Figure 13.5

The escape of a gas from the atmosphere of Venus. For H_2 molecules at $T = 240$ K, the rms velocity is 1.73 km/sec. Only a tiny fraction of the H_2 molecules exceed the escape velocity of 10.3 km/sec, but all the H_2 eventually escapes.

The length of time required for all molecules of a particular gas to escape depends on how far from the rms velocity the escape velocity is. If the escape velocity is only a little greater than the rms velocity, then the gas will leave rather quickly; whereas if the rms velocity is much less than the escape velocity, the process will take longer. It happens that when the rms velocity is less than about one-sixth of the escape velocity, the time for all of the gas to escape becomes so great that it effectively never happens.

The high-velocity tail of the Maxwell distribution is also important in nuclear reactions, such as occur inside stars and in reactors. For two positively-charged atomic nuclei to react requires that they collide with enormous relative velocities, much higher than the rms velocities, even at the high temperatures inside stars. Therefore reactions occur only in the small fraction of collisions in which one or both particles are in the high-velocity tail of the distribution. The process of nuclear reactions will be discussed more completely in Chapter 25, but its dependence on the Maxwell distribution will be a major consideration.

To solve quantitative problems using the Maxwell distribution often requires methods of calculus or numerical integration (which, as always, can be done graphically, by summing the areas of narrow rectangular strips under the curve). If, for example, we wanted to know what fraction of the molecules in a gas had kinetic energies or velocities greater than some fixed value, we would add up the area under the curve that lies beyond that velocity, and compare it with the area under the curve for velocities less than that value. There are some simple problems that can be solved algebraically, however, particularly where what is needed is a comparison of relative numbers of particles at specific velocities, rather than over some range in velocity.

13.1 KINETIC THEORY

Example 13.2 Suppose a container of gaseous helium has a temperature of 300 K. Calculate the rms velocity, the mean velocity, and the most probable velocity, as well as the relative number of molecules at each of these velocities.

Helium remains in atomic form, so the molecular weight is nearly equal to the atomic number for helium, 4. Hence the particle mass is

$$m = 4\text{ u} = 4(1.66 \times 10^{-27}\text{ kg}) = 6.64 \times 10^{-27}\text{ kg}.$$

The rms velocity, from Eq. 13–7, is

$$v_{rms} = \sqrt{\frac{3kT}{m}} = \sqrt{\frac{3(1.38 \times 10^{-23}\text{ J/K})(300\text{ K})}{6.64 \times 10^{-27}\text{ kg}}}$$
$$= 1{,}370\text{ m/s}.$$

The mean velocity, from Eq. 13–10, is

$$v_m = \sqrt{\frac{8kT}{\pi m}} = \sqrt{\frac{8(1.38 \times 10^{-23}\text{ J/K})(300\text{ K})}{\pi(6.64 \times 10^{-27}\text{ kg})}}$$
$$= 1{,}260\text{ m/s}.$$

The most probable velocity, from Eq. 13–9, is

$$v_p = \sqrt{\frac{2kT}{m}} = \sqrt{\frac{2}{3}}\, v_{rms} = 1{,}120\text{ m/s}.$$

To find the relative numbers of molecules at each of these velocities, we can write the Maxwellian, $n(v)$, for each and then divide.

$$\frac{n_{rms}}{n_p} = \left(\frac{v_{rms}}{v_p}\right)^2 e^{-m(v_{rms}^2 - v_p^2)/2kT}$$

But $(v_{rms}/v_p)^2 = 3/2$, and $v_{rms}^2 - v_p^2 = kT/m$, so we have

$$\frac{n_{rms}}{n_p} = \left(\frac{3}{2}\right) e^{-1/2} = 0.91.$$

Thus the number of particles having the rms velocity is a little more than 90% of the number having the most probable velocity.

The number of particles at the mean speed, relative to the number at the most probable speed, is

CHAPTER 13 GAS KINETICS, THERMODYNAMICS

$$\frac{n_m}{n_p} = \left(\frac{v_m}{v_p}\right)^2 e^{-m(v_m^2 - v_p^2)/2kT} = \left(\frac{4}{\pi}\right) e^{-\left(\frac{4}{\pi} - 1\right)} = 0.97.$$

Hence about 97% as many particles are at the mean speed as are at the most probable speed.

13.2 Diffusion and Osmosis

The random motions of individual particles cause a gas to fill its container uniformly or a liquid to have a level surface. When a gas enters an empty container and fills it, or when one species intermixes with another, this process is called **diffusion**. Liquids and gases alike undergo diffusion, although it happens more rapidly for gases. Common examples of diffusion are the steady rise of pressure in a vacuum chamber when air enters, the spread of a dye through a liquid, or the permeation of an entire house with aromas from the kitchen (although convection is likely to play a role also).

It is easy to see why diffusion occurs. Consider a double container, with gas in one chamber, and vacuum in the other chamber (Fig. 13.6a). We see that the frequency with which molecules strike the walls of the chamber containing gas depends on the density of particles in that chamber. At the wall separating the two chambers, there are frequent impacts of molecules on one side, and none on the other. Now imagine that the wall between the chambers is removed. The molecules that would have hit the wall will now enter the other chamber. Hence molecules will initially pass through in one direction only, beginning the process of filling the double chamber uniformly. A short time later, when there is still a higher concentration of molecules on one side than the other, there will be molecules passing through in both directions, but more are passing in the direction away from the original concentration than towards it (Fig. 13.6b). At an even later time, the concentrations on both sides will be more nearly equal, and the difference in flow rates in the two directions will be smaller. Eventually the diffusion process will ensure an even concentration of molecules throughout the container, after which an equilibrium is maintained and the flow rates in opposite directions are equal.

The rate of diffusion, that is, the mass flow rate per unit time through a perpendicular surface between two points separated by distance d, is given by **Fick's law**,

$$J = \frac{DA(\rho_1 - \rho_2)}{d} \tag{13-11}$$

where J is the mass flow rate (in kg/s or, as is often the convention for gases, in mol/s), A is the cross-sectional area through which diffusion occurs, ρ_1 and ρ_2 are the densities on either side of the plane through which diffusion is taking place (in kg/m^3 or mol/m^3); and D is a constant known as the **diffusion constant** (in m^2/s).

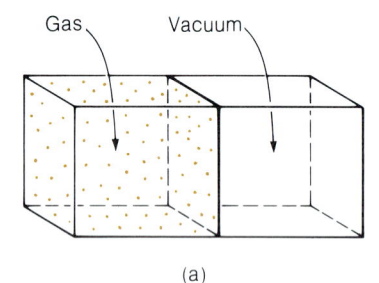

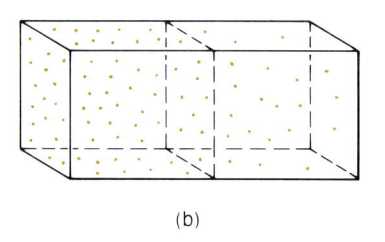

Figure 13.6

Diffusion. The barrier initially keeps all the gas particles in one half of the box (a). When the barrier is removed, there is a net flow of particles from the region of higher concentration toward the region of lower concentration (b) until the concentrations are equal.

13.2 DIFFUSION AND OSMOSIS

The value of the diffusion constant depends on a number of factors, including not only the molecular mass of the diffusing gas or liquid, but also on the temperature, the pressure, and the properties of the fluid into which the diffusion occurs. Obviously, the diffusion will be more rapid when a gas fills a previously evacuated chamber than in the case where the chamber is already filled with another gas, so that the two types must mix. In practice, diffusion does involve mixing with a fluid already present so the value of D is correspondingly small.

It is sometimes useful to know the time required for the diffusion, that is, the time needed for some initial **concentration gradient,** the quantity $(\rho_1 - \rho_2)/d$, to be reduced to some smaller value. This is useful, for example, when it is desirable to know how quickly a mixture of gases or liquids will reach some required degree of uniformity after the addition of one of the fluids to the other. If we express the mass flow rate J as m/t, then we can rewrite Eq. 13–11 and solve for t.

$$t = \frac{md}{DA(\rho_1 - \rho_2)}$$

Next we multiply the numerator and the denominator by d, and note that $dA = V$, where V is the volume of fluid between the points where the densities are ρ_1 and ρ_2.

$$t = \frac{md^2}{DV(\rho_1 - \rho_2)}$$

Finally, we substitute for m/V the average density $\bar{\rho}$ in the region between the points where the densities are ρ_1 and ρ_2, ending up with

$$t = \frac{d^2\bar{\rho}}{D(\rho_1 - \rho_2)}. \qquad (13\text{–}12)$$

This can be written in a different form by substituting $(\rho_1 + \rho_2)/2$ for $\bar{\rho}$.

$$t = \frac{d^2(\rho_1 + \rho_2)}{2D(\rho_1 - \rho_2)} \qquad (13\text{–}13)$$

The equations have the same form whether we express the concentrations ρ_1, ρ_2, and $\bar{\rho}$ in units of mol/m^3, kg/m^3, or any units proportional to these.

The time t may be viewed as the average time for an individual molecule to cross the distance d. It is interesting that this relationship shows that the distance travelled is proportional to the square root of the time, rather than being directly proportional to time. The reason for this is that the individual molecules proceed by a process called **random walk.** They bounce off of other molecules repeatedly, with a random change of direction in each collision (Fig. 13.7). Therefore a molecule does not simply travel in a straight line at a fixed velocity. Statistical methods (see Special Problem 1) are required to prove that the dependence has the square-root form but this conclusion is fully supported by experiment.

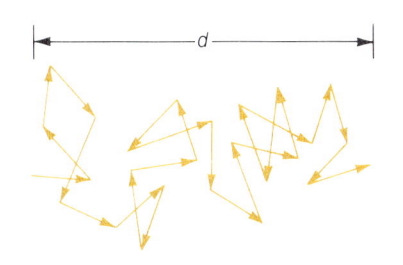

Figure 13.7

Random walk. Each time a particle collides, its direction is randomly changed. The linear distance d *is proportional to the square root of the travel time.*

Example 13.3 Suppose red dye is added to a test tube containing water, so that initially the concentration of dye at one end of the tube is 5 parts per million (ppm). If the tube has a radius of 1 cm and a length of 15 cm, what is the average time for a dye molecule to reach the other end of the tube? Assume that the diffusion constant is $D = 5 \times 10^{-11}$ m²/s.

When the initial concentration at one end is zero ($\rho_2 = 0$), as in this example, then the solution (Eq. 13–13) is very simple.

$$t = \frac{d^2}{2D} = \frac{(0.15 \text{ m})^2}{2(5 \times 10^{-11} \text{ m}^2/\text{s})} = 2.25 \times 10^8 \text{ s} = 7.1 \text{ yr}$$

This is a surprising result, because we normally find dyes spreading much more quickly than this. The reason is that ordinarily liquid flows transport the added substance much more quickly than diffusion.

Diffusion is a very slow process when a substance must diffuse into a liquid, because the diffusion constant D typically has values of the order of magnitude 10^{-11} m²/s. For gases the values are much larger, about 10^{-5} m²/s. Thus, if Example 13.3 had involved a gas diffusing into another gas, rather than a liquid into a liquid, the time would have been much shorter, about 20 minutes.

Despite the slowness of diffusion into liquids, this process is very important in biological functions. The only way diffusion can take place rapidly enough is if the distances are very small. In the human circulatory system for example, oxygen has to diffuse only a very short distance into the blood in the lungs, where capillaries are very close to the tiny chambers (called alveoli) that fill with air each time a person inhales (Fig. 13.8). The circulation of blood through the arteries then provides rapid transport for the oxygen.

Figure 13.8

Air sacs (alveoli) in human lung. Oxygen from these tiny chambers reaches the blood stream by diffusion.

13.2 DIFFUSION AND OSMOSIS

In most biological systems, diffusion must occur through some kind of membrane, since the two fluids involved are not in direct contact. Oxygen molecules in the lungs must pass through the walls of the alveoli and capillaries as they enter the blood stream, for example. This situation can be treated by the same mathematical representation as where there is no membrane, but with a different value of D. The presence of a membrane inhibits diffusion to a degree that depends on the size of the molecules and the properties of the membrane, as well as other factors. Thus, the value of D depends on these factors in addition to all of its usual dependencies on the nature of the two fluids and physical conditions such as temperature and pressure. Often the diffusion rate is expressed in terms of the **permeability** P, which is the diffusion constant D divided by the thickness of the membrane d. The units of permeability therefore are m/s, and Fick's law becomes

$$J = PA(\rho_1 - \rho_2) \qquad (13\text{--}14)$$

and from Eq. 13–13 we see that the diffusion time if ρ_2 is zero becomes

$$t = \frac{d}{2P}. \qquad (13\text{--}15)$$

Example 13.4

If the permeability of the membrane separating the blood in a capillary from the air in an alveolus is $P = 2 \times 10^{-3}$ m/s, and the thickness of the membrane is 10^{-8} m, what is the diffusion time for oxygen to pass from the alveolus into the blood?

The time is given by Eq. 13–15.

$$t = \frac{d}{2P} = \frac{(1 \times 10^{-8}\ \text{m})}{2(2 \times 10^{-3}\ \text{m/s})} = 2.5 \times 10^{-6}\ \text{s}$$

This is very rapid, and is the reason oxygen exchange in the lungs is efficient.

When a membrane is permeable to one substance but not to another, it is said to be **semipermeable**. The reason this can happen is that the molecules of one substance are small enough to fit through the pores in the membrane, while those of the other substance are too large (sometimes molecules are blocked for reasons of their chemical structure). The presence of a semipermeable membrane can create some strange effects. For example, if a container is divided into two sections by a semipermeable membrane, and one section is filled with a liquid that is a mixture of substances, only one of which can pass through the membrane, then at equilibrium the liquid in the two sections will not be at the same height (Fig. 13.9). There is an apparent pressure difference between the two sections. The process by which a substance passes through a semipermeable membrane is called **osmosis**, and the pressure difference that is created is called the **osmotic pressure**.

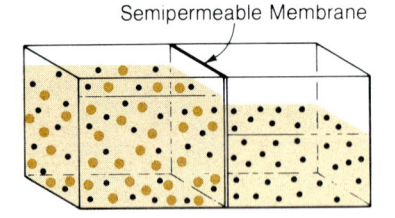

Figure 13.9

Osmosis. Some particles can pass through the semipermeable membrane, while others cannot. At equilibrium, the partial pressure of the free-flowing substance is the same on both sides of the membrane, but the additional partial pressure of the confined substance makes the pressure on the left greater.

To understand the source of this pressure difference, we must consider the partial pressures in the fluids on opposite sides of the membrane. After equilibrium has been reached, one fluid consists of a mixture of two substances, each of which has a partial pressure. On the other side of the membrane is a single substance whose pressure equals the partial pressure due to that substance alone. Therefore the side with the mixture has a greater total pressure than the side with only one substance, and the difference, the osmotic pressure, is equal to the partial pressure of the substance that is on only one side of the membrane. The seemingly perplexing aspect of osmosis is that, starting with equal total pressures on both sides of a membrane, fluid can flow through and establish a pressure difference between the two sides. This happens if there is initially a difference between the two sides in the concentration of some impurity to which the membrane is impermeable.

A semipermeable membrane acts as a pump because of its ability to create a pressure difference. It also acts as a filter, because it allows some substances to pass but not others. Both functions are important in biological systems. This pressure enhancement is responsible, for example, for the stiffness of a plant or leaf that is adequately watered. The filtering function of osmosis is important in waste removal from the bloodstream in humans, because the kidneys have semipermeable membranes that allow certain small waste molecules to pass through, but not larger molecules that are needed by blood cells.

13.3 Heat Capacity and the First Law of Thermodynamics

Just as we have used the kinetic theory to discuss molecular transport mechanisms, we now use it to continue our analysis of heat and heat transport, the science we call **thermodynamics**. We have already begun to discuss the first law of thermodynamics, because it can be stated readily in terms of the macroscopic properties of fluids, which were the subject of the previous chapter. In this section we will explore further the consequences for heat capacity of the fact that heat energy obeys the principle of conservation of energy.

The heat capacity of a gas depends on whether it is allowed to expand as it is heated, or instead is confined so that its pressure increases. To understand this, we must examine at the molecular level what heat capacity means. First, suppose the volume is kept constant. All the added heat goes into increased kinetic energy of the molecules in the gas, hence the temperature of the gas rises readily. If the gas is allowed to expand, however, some of the input energy goes into work as the gas expands against the external pressure. To raise the temperature of the gas by a given amount, more heat input is required than if the volume is kept constant.

We can easily assess the difference in the amount of heat required for both constant pressure and constant volume. Suppose two equal quantities of the same gas are heated so that their temperatures both increase by the same amount. One of the gases is kept at constant volume, and the other is allowed to expand at constant pressure. More heat is required for the latter gas, because

13.3 FIRST LAW OF THERMODYNAMICS

of the work it does in expansion. The quantity of extra heat needed is equal to the amount of work done, which is

$$W = F\Delta d$$

where F is the force exerted by the gas and Δd the distance over which the force is exerted. Since we normally deal with pressure rather than force, and pressure, force, and area A are related by

$$P = \frac{F}{A}$$

we can write

$$W = PA\Delta d = P\Delta V \qquad (13\text{--}16)$$

since the total surface area A times the displacement Δd equals the total change in volume ΔV. Hence the amount of heat needed for the same change in temperature is greater by the amount $P\Delta V$ if pressure is constant than if volume is constant.

It is possible to express this in terms of specific heat. Recall that the specific heat is the amount of heat required to raise the temperature of a unit quantity of a substance by one degree. Since more heat is required to do this for a gas at constant pressure than for one at constant volume, the specific heat at constant pressure is larger than the specific heat at constant volume. We can calculate the difference between c_p and c_v, the specific heats for constant pressure and constant volume. This is simplest if we define a new specific heat, the **molar heat capacity**, which is the amount of heat required to raise the temperature of a mole of gas by 1C°. Then C_p and C_v are the molar heat capacities for constant pressure and constant volume. The mass in kg of a mole of gas is numerically equal to the molecular weight m, so we have

$$C_p = mc_p$$

and

$$C_v = mc_v$$

where the specific heats c_p and c_v are expressed in terms of heat per kg · K.

If n moles of gas are heated at constant pressure or at constant volume so that the temperature changes by ΔT, then the amounts of heat required are $nC_p\Delta T$ and $nC_v\Delta T$. The difference between these two quantities, as we have just argued, is equal to $P\Delta V$, the work done by expansion under constant pressure. Hence we have

$$nC_p\Delta T - nC_v\Delta T = P\Delta V.$$

Now we can assess this, with the help of the ideal gas law,

$$PV = nRT$$

so if P is constant, then

$$P\Delta V = nR\Delta T$$

and

$$nC_p\Delta T - nC_v\Delta T = nR\Delta T$$

which leads, finally, to

$$C_p - C_v = R. \qquad (13\text{--}17)$$

Now we can evaluate the molar heat capacities, using the first law of thermodynamics. This law dictates that when heat is added to a gas that is kept at constant volume, all of that heat must go into increased internal energy in the gas. In our earlier discussion of the relationship between internal energy and temperature, we considered only the kinetic energy per particle, which is the quantity that defines the temperature of a gas. Now, however, when we consider heat gained or lost by a gas, we must consider other forms of internal energy. The kinetic energy of a single atom is due only to its translational motion in three dimensions, whereas molecules can have rotational and vibrational energy as well as kinetic energy of motion. Hence when heat is added to a **monatomic gas** (one consisting of single atoms) all of that heat goes into increased kinetic energy of the atoms, and hence into an increase in temperature. We treat this simple case first.

From Eq. 13–6 we know that the kinetic energy per particle is $3kT/2$, so the total internal energy in a gas containing N particles is

$$U = \frac{3NkT}{2}.$$

The Boltzmann constant k is equal to R/N_A or nR/N, so we can write

$$U = \frac{3nRT}{2}.$$

If the temperature of the gas is changed by the amount ΔT, then the internal energy changes by ΔU, where

$$\Delta U = \frac{3nR\Delta T}{2} = nC_v\Delta T$$

which reduces to

$$C_v = \frac{3}{2}R. \qquad (13\text{--}18)$$

The value of R is 8.314 J/mol · K, which is equivalent to 1.99 cal/mol · K, so we find that the value of the molar heat capacity for constant volume is $C_v =$

13.3 FIRST LAW OF THERMODYNAMICS

$3(1.99 \text{ cal/mol} \cdot \text{K})/2 = 2.99 \text{ cal/mol} \cdot \text{K}$ for a monatomic gas. It follows from Eq. 13–17 that the molar heat capacity for constant pressure is $C_p = 5R/2 = 4.98 \text{ cal/mol} \cdot \text{K}$. Experiment bears out both of these theoretical values.

For gases which are not monatomic, the values of C_p and C_v are greater. Molecules can rotate and the individual atoms within molecules can vibrate about their equilibrium positions, and both types of motion represent internal energy. Therefore, when heat is added to a molecular gas, some of the heat goes into these forms of internal energy, and this increases the amount of heat needed to raise the temperature one degree.

Each available form of energy that a molecule can have is called a **degree of freedom**, because each represents a way in which the molecule is free to act. Measurements of specific heats for various gases led to the realization that the energy tends to be equally divided among the possible forms. The principle of **equipartition of energy** states that the internal energy in a gas is distributed uniformly among the available degrees of freedom. A monatomic gas has three degrees of freedom, because the individual atoms are free to move in each of three dimensions. We have seen that the internal energy in a monatomic gas is $3kT/2$, so we conclude that the energy per degree of freedom is $kT/2$. A diatomic gas under normal conditions has five degrees of freedom, because the molecules can rotate about either of two axes (Fig. 13.10), so the internal energy per molecule is $5kT/2$. At high temperatures the atoms in a diatomic molecule can vibrate, and there is both kinetic and potential energy associated with these vibrations, so there are a total of seven degrees of freedom, and the internal energy is $7kT/2$. For gases consisting of more complex molecules the number of degrees of freedom can be even larger. In all cases the molar specific heat at constant volume is equal to $iR/2$, where i is the number of degrees of freedom. Thus the value of C_v for a hot diatomic gas is $C_v = 7R/2 = 6.97$ cal/mol · K, for example.

The fact that some degrees of freedom are not available below a certain temperature is quite significant, and was not understood for a long time. The explanation is that there is a certain minimum amount of energy that a molecule can have in each degree of freedom, and if the temperature is too low, there may not be sufficient internal energy in the gas to activate a particular degree of freedom. Thus, a molecule cannot have an arbitrarily small amount of rotational energy, but instead has no rotational energy at all until it has enough to surpass a certain minimum value. This is due to a general principle, that all forms of energy can exist only in certain discrete amounts called **quanta**. In most of our discussion so far, this has been unimportant, because we have dealt with large-scale properties of matter, but we will see later that the quantum nature of energy is fundamentally important on the atomic and molecular level.

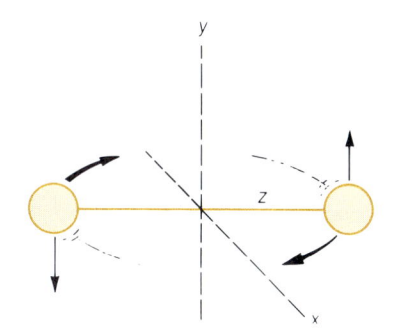

Figure 13.10

Degrees of freedom for a diatomic molecule. A diatomic molecule can rotate about the x-axis and the y-axis, in addition to moving in the x, y, and z directions. Thus it has 5 degrees of freedom. If it has sufficient internal energy to vibrate, the kinetic and potential energy of vibration create two additional degrees of freedom.

Example 13.5 Air is heated in a furnace so that its temperature is increased 5C°. The volume is kept constant during the heating. If the volume is 1 m³ and the initial temperature is 15°C, how much heat is required?

First we need to know how many moles of gas are present. We assume the air is at atmospheric pressure initially, so from the ideal gas law we find

$$n = PV/RT = \frac{(1.013 \times 10^5 \text{ Pa})(1 \text{ m}^3)}{(8.314 \text{ J/mol} \cdot \text{K})(288 \text{ K})} = 42.3 \text{ mol}.$$

Air consists almost entirely of diatomic gases (N_2 and O_2), so the number of degrees of freedom is 5 (at room temperature, there is insufficient internal energy for vibration). Then the molar specific heat for constant volume is $C_v = 5R/2 = 20.79$ J/mol · K (or 4.98 cal/mol · K, but we prefer to work in joules for this problem). The heat energy required to raise the temperature of the gas 5 C° is

$$Q = nC_v \Delta T = (42.3 \text{ mol})(20.79 \text{ J/mol} \cdot \text{K})(5 \text{ K}) = 4{,}400 \text{ J}.$$

If this heating is done in a short time, say in one minute, as in most furnaces, then the power requirement for the heater is

$$P = \frac{Q}{t} = \frac{4{,}400 \text{ J}}{60 \text{ s}} = 73.3 \text{ W}.$$

13.4 Heat Engines and the Second Law of Thermodynamics

We have discussed transfers of heat and the relation between heat and work when a gas is allowed to expand. In doing so, we have relied on the first law of thermodynamics, the statement that heat and other forms of energy are equivalent and are conserved. To say that heat and other forms of energy are equivalent is not the same as saying that they can always be interchanged, however.

Entropy and the Second Law

It is generally easy to convert mechanical energy into heat. When an object is rubbed along a surface, for example, both the object and the surface are heated by friction. When an object falls, its initial potential energy ends up being converted into heat. When mechanical energy is converted into heat, the internal energy is increased; in such a process, kinetic energy of the large-scale motions of objects is transferred to individual molecules and increases their random motion.

Once a quantity of energy is converted into internal energy, that is, heat, it cannot be fully recovered. Thus in any process that produces heat, some energy must be irrevocably lost. The random motions of individual molecules will never spontaneously be converted back into some form of organized mechanical energy. The heat energy of an object that has been pushed across a surface will not spontaneously be converted back into kinetic energy and push

13.4 SECOND LAW OF THERMODYNAMICS

the object back along its path, and the object that has fallen will not suddenly jump back up into the air as the molecules act in concert to restore its potential energy. There is something fundamental about this; in all mechanical interactions, some energy gets lost as heat, never to be recovered. To be a bit more general, rather than saying that energy is always converted into heat, we say that the randomness of molecular motions, that is, the disorder of matter, either stays the same or increases. We define a quantity called **entropy**, which is a measure of the disorder.

The general statement that the disorder of matter always stays the same or increases is one version of the **second law of thermodynamics**:

The entropy of the universe never decreases, but always remains constant or increases.

This statement has truly cosmic implications. Taken to its logical conclusion, it means that the universe will eventually run down as all forms of energy become randomized. We refer to the "heat death" of the universe, an era far in the future, when the universe consists of nothing but elementary particles at a constant temperature. Estimates based on the present distribution of matter and energy in the universe set the heat death as far as to 10^{50} years in the future. (the current age of the universe is of order 10^{10} years).

The irreversible trend towards ever-increasing entropy helps us to demonstrate that time can flow only one way. Many processes in the universe are cyclical, and those who ponder the meaning of time speculate that, since many cycles are reversible, time, in effect, could run backwards. The second law of thermodynamics gives us a fundamental reference direction for time, however; if time ran backwards, the second law would be violated because entropy would decrease instead of increase. Thus, the direction of always-increasing entropy is referred to as "time's arrow". As the philosopher knows, you can never go back.

On small scales, entropy can be reversed. A living organism, for example, consists of well-organized molecules in cells, and was formed from disorganized molecules. Any machine or process that manufactures something from raw materials reverses entropy. The point of the second law is that in all such processes, there is a net increase in disorder. Thus the living organism releases heat to its surroundings as it grows, as do the machines that produce goods. An ideal machine would just break even; the decrease of disorder that it produces would be just balanced by the increase of disorder of its surroundings.

Heat Engines

Let us return to the consideration of heat and internal energy. The second law implies that disorder will always stay the same or increase. If we have an object that is hotter in one place than others, there is some order present which would be decreased if the heat energy were distributed randomly throughout the object. Heat will flow from the high-temperature region to the lower-temperature regions around it until the temperature is uniform throughout. This is the state of maximum disorder. It would be a violation of the second law, however, if the random motions of the molecules in an object acted in concert to heat one portion to a temperature higher than the rest. This leads to an alternative statement of the second law:

Heat will always flow from regions of high temperature to regions of lower temperature, but will never flow spontaneously from regions of low temperature to regions of higher temperature.

Thus all the temperature differences in the universe tend to eliminate themselves as heat flows from regions of high to low temperature. Whenever a temperature difference is eliminated, some energy becomes unavailable to do work, because heat can no longer be made to flow between the two regions. Thus, when the universe reaches its heat death, there will be no energy available to do any work.

This statement of the second law forms the basis for a class of machines called **heat engines,** in which heat energy is transformed into mechanical energy. Such engines were first analyzed in the early nineteenth century by the French scientist N. L. S. Carnot, who founded the science of thermodynamics. A heat engine does work when heat flows from a region of high temperature to a region of lower temperature, and work is performed. (Usually the heat input creates expansion in a gas). The source of high temperature is usually provided by burning some kind of fuel, whose heat of combustion heats the gas. Steam engines, turbines, diesel engines, and gasoline-powered motors such as those used in most cars are all heat engines.

If we reduce a heat engine to its basic components, then we have just three essential parts: a heat reservoir at high temperature, another reservoir at low temperature, and a device that converts heat into work (Fig. 13.11). When the engine operates, heat flows from the first to the second of these reservoirs, and a portion of it is converted to work on the way. The most useful heat engines are those that operate cyclically, allowing heat to flow between the reservoirs repeatedly, extracting work during each cycle.

Carnot was able to demonstrate several general properties of heat engines. One is that the maximum possible efficiency is completely determined by the temperatures T_1 and T_2 of the two reservoirs; these two temperatures are known as the **operating temperatures** of the engine. He showed further that this maximum efficiency is reached only by reversible engines, those that can be operated in either direction. This would imply that no energy is converted into random motion; that no heat is lost to the material of which the engine was constructed, and that there are no losses due to friction. Entropy would remain constant instead of increasing, but this never quite happens. There is

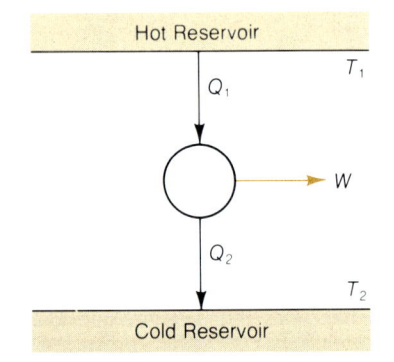

Figure 13.11

Heat engine. Heat flows from a region of high temperature (T_1) *to a region of low temperature* (T_2). *Some heat is used to do work* W.

13.4 SECOND LAW OF THERMODYNAMICS

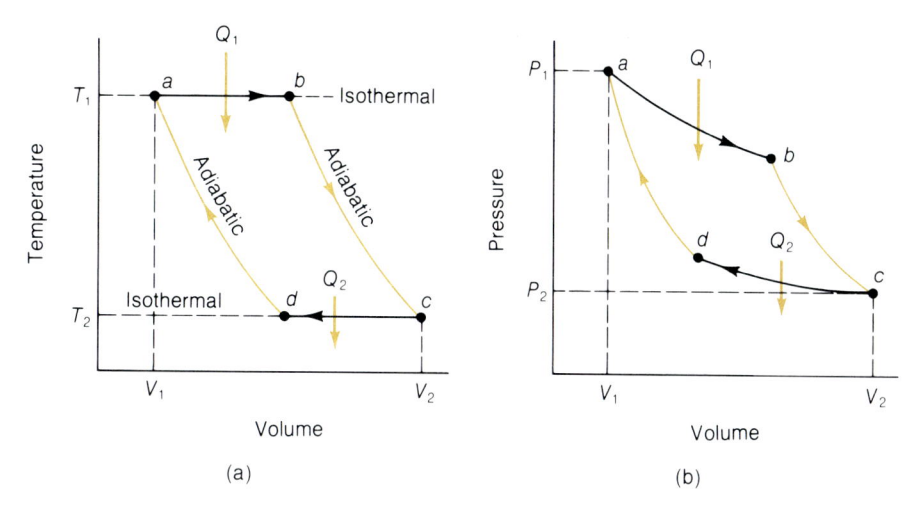

Figure 13.12

The Carnot cycle. The temperature and pressure of an ideal gas change as it expands and contracts in one cycle (abcd). Heat Q_1 flows in, heat Q_2 flows out, and the net work $W = Q_1 - Q_2$.

no real engine that is perfectly reversible, and the theoretical maximum efficiency is never quite attained. Nevertheless the analysis of the properties of such engines is fruitful.

Carnot analyzed the properties of a particularly simple, idealized engine now known as a **Carnot engine.** There are four phases to the cycle of a Carnot engine (Fig. 13.12). In the first part of the cycle (section ab), heat is added to a quantity of an ideal gas at initial pressure P_1 and temperature T_1, and, as the gas expands, heat continues to be added so that the temperature is constant. The total amount of heat added during this stage is Q_1. In the second stage (section bc), the gas is isolated from its surroundings so that no heat is added or lost (this is called an **adiabatic** process). The gas is allowed to expand during this phase, so it cools, eventually reaching temperature T_2. At that point it is compressed and heat is allowed to flow out of it (section cd), so that again the temperature remains constant. The heat lost from the gas during this stage is Q_2. Finally, it is thermally isolated and compressed further adiabatically (section da), and the temperature rises until T_1 is reached again. The net input of heat during the entire cycle is $Q_1 - Q_2$. The net work done by the gas during the four cycles, by the work-energy principle, is

$$W = \Delta Q = Q_1 - Q_2.$$

Carnot showed that the efficiency of such an engine is

$$\text{Efficiency} = \frac{W}{Q_1} = \frac{(Q_1 - Q_2)}{Q_1} = \frac{T_1 - T_2}{T_1}. \qquad (13\text{–}19)$$

Thus the theoretical limit on engine efficiency is a simple function of its two operating temperatures.

An ideal engine can reach 100% efficiency only if $T_1 - T_2 = T_1$, a condition that would be met if $T_2 = 0$. Thus, if it were possible to have a heat engine with absolute zero as its lower operating temperature, it would have 100% efficiency.

One way to define absolute zero is to say that it is the temperature at which a Carnot engine ejects no heat; that is, the temperature T_2 is such that $Q_2 = 0$. This may seem like an odd way to define absolute zero, but, as with many other concepts in physics, it is most useful to define it according to its effect. To say that absolute zero is the temperature for which all molecular motion stops, as we did in the previous chapter, is a bit misleading because even if molecular kinetic energies are zero, there can still be some energy present (in subatomic energy states). If there is any energy present, then in principle it is possible to drive a heat engine by having the energy flow from regions of higher concentration to regions of lower concentration.

Any real engine is not perfectly reversible, because heat flows into the engine parts or the surroundings, and there are usually friction losses as well. The true efficiency of an engine is the ratio of the input heat to the output work (times 100 to convert the fraction to a percentage). The best efficiencies that have actually been reached are about 60 to 80% of the Carnot efficiency, which itself is usually 60 to 70 percent.

Example 13.6   Suppose an autmobile engine has operating temperatures $T_1 = 2,500°F$ and $T_2 = 500°F$ (these are the temperature of the gas in a cylinder at the moment of ignition and the temperature of the exhaust gas). Measurements show that a gallon of gasoline produces a total work of 5.3×10^8 J. What is the Carnot efficiency of this engine, and what is its true efficiency? The density of gasoline is 0.68 g/cm^3 and its heat of combustion of 4.8×10^7 J/kg.

The Carnot efficiency can be calculated directly from Eq. 13–19, once we have converted T_1 and T_2 to the absolute scale. We find

$$\text{Efficiency} = \frac{1{,}644 \text{ K} - 533 \text{ K}}{1{,}644 \text{ K}} = 0.68.$$

To find the true efficiency, we must compare the input energy with the output energy, which is the quantity of work. The input energy is equal to the heat of combustion of 1 gal of gasoline (we assume that the gasoline is completely burned, which is never actually true). To calculate this, we must convert 1 gal into kg: 1 gal = 4 quarts = 3.785 l = 0.003785 m^3. Therefore the mass of 1 gal of gasoline is (680 kg/m^3)(0.003785 m^3) = 2.57 kg, and the heat released by its combustion is

$$Q = (2.57 \text{ kg})(4.8 \times 10^7 \text{ J/kg}) = 1.23 \times 10^8 \text{ J}.$$

The efficiency is therefore

13.4 SECOND LAW OF THERMODYNAMICS

$$\text{Efficiency} = \frac{Q_{input}}{Q_{output}} = \frac{(4.8 \times 10^7 \text{ J})}{(1.23 \times 10^8 \text{ J})} = 0.39.$$

Thus the true efficiency is 39%, or 57% of the Carnot efficiency. This is higher than normally achieved in real automobile engines.

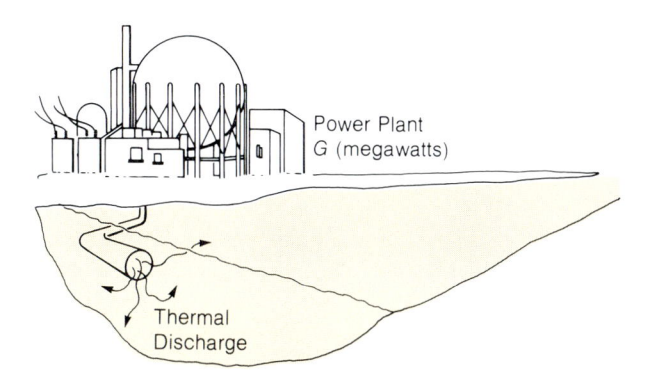

Figure 13.13

Excess heat from a power plant can increase the river's temperature by several degrees.

There is always a considerable amount of heat lost to the surroundings during the operation of any heat engine, even one that operates at the Carnot efficiency (recall that the quantity Q_2 is released as part of the Carnot cycle, even though no other losses were assumed). Only an engine operating with 100% efficiency would produce no excess heat. We do not usually think of the effect of this waste heat when we consider the pollution caused by engines in our environment, but thermal pollution can be a significant problem. A typical large power plant that operates by boiling water (either by burning coal or using nuclear power) and using the steam to drive generator turbines produces enough excess heat to raise the temperature of a fair-sized river by one or two degrees (Fig. 13.13). Even this small an effect can be harmful to some aquatic organisms, and if the same river serves as coolant for several power plants, the cumulative effect can be disastrous.

Heat Pumps

We have discussed heat engines that perform work using the natural flow of heat from regions of high temperature to regions of low temperature. Another class of engines, called **heat pumps,** force heat to flow in the opposite direction, from cool places to hotter ones. The most common are refrigerators, air conditioners, and reversible heat pumps that are used for heating and cooling houses in regions of moderate seasonal temperature variation.

Since heat will never flow spontaneously from a cool region toward a region of higher temperature, this requires work. Thus, a refrigerator or air conditioner has a motor to perform the work. The effectiveness of a heat pump is therefore represented by the ratio of heat transferred to the amount of work required to transfer it.

A heat pump is simply a heat engine in reverse (Fig. 13.14). There are two operating temperatures, and a device that transfers heat from a reservoir at one temperature to another reservoir at the other temperature. This is identical to the schematic operation of a heat engine, except that the flow of heat is in the opposite direction, and work is put in, rather than extracted from it.

Most heat pumps make use of a coolant fluid, which changes phase from liquid to vapor at temperature near the temperature of the cold region in the cycle. A common refrigerant is freon, although others, such as ammonia, methyl chloride, and sulfur dioxide are used as well. The fluid, in gaseous form, is pressurized and heated by a compressor, a motor-driven device with a piston (Fig. 13.15). The hot gas is then circulated outside the refrigerator (or air conditioner) through a series of tubes called a condenser; as heat is lost to the outside, the refrigerant cools and liquifies. Then an expansion valve forces the liquid to expand and cool further before it flows through the evaporator, a series of tubes in the region that is being cooled. There heat is absorbed from the cold region, causing the refrigerant to vaporize and the vapor to heat further. Finally, the gas enters the compressor again, and the cycle repeats.

The figure normally used to evaluate the effectiveness of a heat pump is called the **coefficient of performance** (**CP**), whose definition depends on the purpose of the heat pump. If it is being used as a cooler, then the goal is to remove heat from the cool reservoir, and the effectiveness is the ratio of Q_2, the heat removed, to W, the work required to remove it. Hence the coefficient of performance for a cooler is given by

$$CP = \frac{Q_2}{W}. \qquad (13\text{--}20)$$

If the pump is designed to bring heat from a region of low temperature to one at a higher temperature, as for heating a house in the winter, then the effectiveness depends on how much heat is brought to the warmer reservoir for a given amount of work. The coefficient of performance in that case is

$$CP = \frac{Q_1}{W} \qquad (13\text{--}21)$$

where Q_1 is the quantity of heat added to the warm reservoir.

For a device that has the Carnot efficiency if run as a heat engine, the coefficient of performance is maximized, and is given by

$$CP_{max} = \frac{T_2}{(T_1 - T_2)} \qquad (13\text{--}22)$$

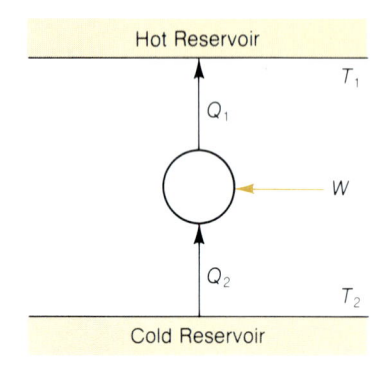

Figure 13.14

A heat pump. Work W is put in, causing heat to flow from a low-temperature region (T_2) to a region at higher temperature (T_1).

13.4 SECOND LAW OF THERMODYNAMICS

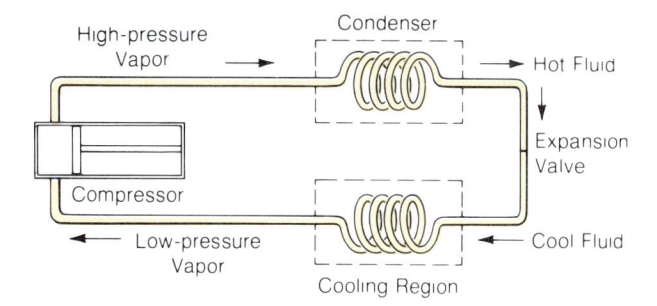

Figure 13.15

The refrigerator.

for a cooler, and by

$$CP_{max} = \frac{T_1}{(T_1 - T_2)} \quad (13\text{-}23)$$

for a heater. In both cases T_1 is the higher of the two operating temperatures, and T_2 is the lower. This theoretical maximum value of the CP is always greater than one, but actual values are always less than this.

Example 13.7 A household refrigerator operates between temperatures $T_1 = 38°C$ and $T_2 = -8°C$. It is operated by an electric motor whose power is 300 W, and in one hour it removes 2.0×10^6 J of heat energy. What is the maximum coefficient of performance for a refrigerator operating between these two temperatures, and what is the actual coefficient?

The maximum coefficient of performance is given by Eq. 13–22.

$$CP_{max} = \frac{T_2}{T_1 - T_2} = \frac{265 \text{ K}}{311 \text{ K} - 265 \text{ K}} = 5.76$$

To calculate the actual coefficient of performance we need to know the work performed. We assume that the electric motor is 100% efficient; therefore the work performed in an hour is

$$W = Pt = (300 \text{ W})(3{,}600 \text{ s}) = 1.08 \times 10^6 \text{ J}.$$

The actual coefficient of performance is given by Eq. 13–20.

$$CP = \frac{Q_2}{W} = \frac{(2.0 \times 10^6 \text{ J})}{(1.08 \times 10^6 \text{ J})} = 1.85$$

This is a typical value for refrigerators.

FOCUS ON PHYSICS

Throwing Away Heat

As we learned in this chapter, a perfectly efficient engine, one whose output temperature is absolute zero, is a practical impossibility. In reality, all engines have substantial exhaust temperatures, and the heat added to our environment in this way can have significant consequences.

Today power plants are by far the most important source of waste heat. A typical steam-operated power plant is about 40% efficient, that is, for every kilowatt-hour of energy produced, 1.5 kWh is lost as heat. (The kilowatt-hour is the standard unit of energy measure in power generation; 1 kWh = (1,000 J/s)(3,600 s) = 3.6×10^6 J.) Some of this heat goes into the power plant structure and is radiated away, and some goes into heating air that escapes into the atmosphere, but most of it (about 75%) needs to be disposed of. These numbers represent a typical fossil fuel-burning plant; nuclear reactors operate at lower input steam temperatures (T_2 in our efficiency equation) and therefore are even less efficient (about 33%), producing proportionally more waste heat per kWh of power generated.

By far the best way to carry away the waste heat from power plants is to use water. The only other possibility would be to use air, but the specific heat of air is very much lower than that of water, and no practical way to use air-cooling has been devised. Water is plentiful and has excellent properties for heat disposal, but of course it is needed for many other purposes. Therefore the impact of using water to carry away waste heat must be carefully assessed.

The easiest way to use water for cooling is to pump it from a lake, ocean, or river through the power plant and then release it into the source. This **once-through cooling** method requires about 8 gallons of water per second per 1,000 kWh of waste heat energy, assuming that the water temperature is allowed to rise 15 F° (8.3 C°) in the process (the student can check this by calculating the heat needed to raise the temperature of 8 gallons, or about 30 liters, of water 8.3 C°). Thus, a power plant that produces 1,000 kW of power, hence 1,500 kW of waste heat, must use about 8 gallons of water per second for cooling. This is a lot of water, and a 1,000 kW plant is only moderate in size. It is estimated that the total amount of water needed by all of the power plants in the U.S. today represents roughly one sixth of the total water available from runoff in streams and rivers! Of course, water can be reused many times for cooling, if sufficient time is allowed between uses, so this does not mean that actually one sixth of our water supply is being devoted to eliminating waste heat. Furthermore, water that has been used for cooling is still available for other uses. Nevertheless, the demands placed on our water supply by the need to dispose of waste heat are enormous.

If the source of cooling water for a power plant is a river, then the heated water ejected into the river after circulation through the power plant will return to normal ambient temperatures a few miles downstream. Still, there can be significant impact on aquatic life forms which are often unable to tolerate these temperature increases. In addition, to kill algae and reduce corrosion, water is often treated with chemicals before it is circulated through a power plant, so there are chemical dangers introduced by water cooling as well. Several major fish kills in American rivers, particularly in the East, have resulted from the use of rivers for disposing of waste heat.

If a lake is used, then the heat dissipation is less efficient, occurring primarily through evaporation at the surface. Again, major impacts on aquatic life can occur. Power plants built on seashores have a very large reservoir of water, and in areas where the water close to shore is deep enough, little environmental impact results (in fact, pumping cold bottom water to the surface can have beneficial effects by providing enriched water for fish living near the surface). In shallow areas and marine bays, however, there has been extensive thermal damage to aquatic life. Corrosion is especially difficult to prevent when salt water is used, so the danger of chemical pollution is worsened because the water is treated to minimize corrosion.

Cooling ponds are used by some power plants. These are large, shallow ponds devoted entirely to the disposal of waste heat, and therefore they have little negative impact on the environment. Cooling occurs by evaporation into the atmosphere, so a large surface area is desirable. A surface area of about 2 acres is needed for a 1,000 kW plant, so for a very large plant (1 megawatt or larger), many acres of ponds would be needed. The size of the ponds can be reduced greatly if the hot water is sprayed into the air as it is returned to the ponds, so that the surface area available for evaporation is temporarily increased while the water is in the form of droplets. Loss of water through evaporation must be balanced by a source of incoming water, usually a small stream.

A promising method for disposing of waste heat is the use of **cooling towers**. These may circulate water through the tower, or create a rising column of air that carries away heat. In the so-called **wet towers**, water is either sprayed into the air or allowed to flow in a thin film over a lattice, to provide an enlarged surface area for evaporation. Wet towers create localized regions of warm, humid air, and chemicals that may be in the water (for algae control, for example) can be dispersed into the air, creating environmental hazards. Also, water lost to evaporation must be replenished.

In a **dry tower**, cool air flows over the surface of the hot water, is heated by conduction and convection, and rises (by convection or by force, with the use of fans) through an open chimney, carrying away the heat. Dry towers have the smallest environmental impact of all the types of waste heat disposal discussed here, but are the most expensive. Nevertheless, the distinctive shapes of these immense towers are becoming ever more familiar landmarks on the countryside. Fig. 13.1.1 shows the natural draft cooling tower for the Portland General Electric Company's Trojan Plant.

Figure 13.1.1

It is interesting to contemplate the enormous quantity of energy that is represented by the waste heat from power plants. For every kWh of usable electrical energy, 1.5 kWh are lost. Perhaps some day a way will be found to use some of this lost energy.

Perspective

While we have left many sidestreets unexplored, we have now travelled most of the main roads in our study of thermal properties of matter, fluid mechanics, and thermodynamics. We have developed more than a passing acquaintance with the principles, and have learned how to tackle a variety of practical problems.

We are ready to move on to an entirely different area of physics, one that has only begun to be explored in the past century or so. Our next major task is the study of electricity and magnetism.

SUMMARY TABLE OF FORMULAS

Force on box wall due to gas particles moving in one dimension:

$$F = \frac{Nm\overline{v_x^2}}{L} \tag{13-1}$$

Force due to particles moving in three dimensions:

$$F = \frac{Nm\overline{v^2}}{3L} \tag{13-2}$$

Gas pressure:

$$P = \frac{F}{L^2} = \frac{Nm\overline{v^2}}{3V} \tag{13-3}$$

$$PV = \frac{1}{3}Nm\overline{v^2} \tag{13-4}$$

Ideal gas law:
$$PV = NkT \tag{13-5}$$

Relationship between average particle KE and temperature:

$$\frac{1}{2}m\overline{v^2} = \frac{3}{2}kT \tag{13-6}$$

Root mean square velocity of gas particles:

$$v_{rms} = \sqrt{\frac{3kT}{m}} \tag{13-7}$$

Maxwell velocity distribution:

$$n(v) = 4\pi \left(\frac{m}{2\pi kT}\right)^{3/2} v^2 e^{-mv^2/2kT} \tag{13-8}$$

Most probable velocity:

$$v_p = \sqrt{\frac{2kT}{m}} \tag{13-9}$$

Mean velocity:

$$v_m = \sqrt{\frac{8kT}{\pi m}} \tag{13-10}$$

Diffusion rate (Fick's law):

$$J = \frac{DA(\rho_1 - \rho_2)}{d} \tag{13-11}$$

Diffusion time:

$$t = \frac{d^2\overline{\rho}}{D(\rho_1 - \rho_2)} \tag{13-12}$$

$$t = \frac{d^2(\rho_1 + \rho_2)}{2D(\rho_1 - \rho_2)} \tag{13-13}$$

Fick's law for permeable membrane:
$$J = PA(\rho_1 - \rho_2) \tag{13-14}$$

Diffusion time for permeable membrane:

$$t = \frac{d}{2P} \tag{13-15}$$

Work done by expanding gas:
$$W = P\Delta V \tag{13-16}$$

Molar heat capacities
$$C_P - C_V = R \tag{13-17}$$

$$C_V = \frac{3}{2}R \tag{13–18}$$

Efficiency of a Carnot engine:
$$\text{Efficiency} = \frac{W}{Q_1} = \frac{T_1 - T_2}{T_1} \tag{13–19}$$

Coefficient of performance for a cooler:
$$\text{CP} = \frac{Q_2}{W} \tag{13–20}$$

Coefficient of performance for a heater:
$$\text{CP} = \frac{Q_1}{W} \tag{13–21}$$

Coefficients of performance for Carnot engines:
$$\text{CP}_{max} = \frac{T_2}{(T_1 - T_2)} \text{ (cooler)} \tag{13–22}$$

$$\text{CP}_{max} = \frac{T_1}{(T_1 - T_2)} \text{ (heater)} \tag{13–23}$$

THOUGHT QUESTIONS

1. What is meant by macroscopic. Explain the statement, "The macroscopic behavior of a gas is the result of the aggregate motions of the individual particles of which it is composed."

2. What does the kinetic theory assume about the structure of individual atoms?

3. What is the cause of the random movement of dust particles in air (called Brownian motion)? How does it relate to the kinetic theory of gases?

4. Explain how kinetic theory accounts for the following: boiling, pressure, evaporation, diffusion, and osmosis.

5. Does the law of partial pressures follow from the kinetic theory? Explain.

6. Show that the mean squared velocity $\langle v^2 \rangle$ is not in general equal to the square of the mean velocity $\langle v \rangle^2$. Under what circumstances would they be equal?

7. Describe an instance in which the motion of gas particles might not be isotropic.

8. What evidence do we have that particles in a gas move with a distribution of speeds, rather than all with the same speed?

9. How do the principles in this chapter allow us to deduce the possible composition of a planet's atmosphere by knowing its mean temperature?

10. Why do most chemical reactions occur more efficiently at high temperatures? Is it possible for a reaction to occur more rapidly at low temperature? Explain.

11. How can all the liquid in a dish evaporate without boiling?

12. Hydrogen and oxygen are placed in identical containers at identical temperatures. Which container has the higher pressure? Why?

13. Will diffusion occur between two gases having different molecular masses but the same density? Explain why or why not.

14. Why are diffusion rates for gases larger than those for liquids? Can you think of an exception to this?

15. Describe a device that makes use of diffusion. Describe several processes in the human body that depend on diffusion.

16. Suppose you wished to create a system with a long diffusion time. Would you choose a substance with a high or low permeability for a membrane? Why?

17. Is it possible for two solids, placed face to face, to diffuse at their boundary? Explain.

18. Give an example of a semipermeable membrane and its role in the human body.

19. Why is there a difference between the heat capacities of a gas at constant pressure and a gas at constant volume?

20. Give an example in which a semipermeable membrane functions as (a) a pump, (b) a filter.

21. Describe how you would measure experimentally the values of C_V and C_P for a gas.

22. Give two everyday examples of the first law and the second law of thermodynamics at work.

23. Scientists often refer to the "entropy monster." What do you think is meant by this?

24. Pick out a heat engine and identify the high temperature and low temperature reservoirs. What is the mechanism that converts heat into work?

25. Why is the true efficiency of an engine less than the theoretical Carnot efficiency?

PROBLEMS

Section 13.1 Kinetic Theory

- **1** Calculate the number of atoms in 1 m³ of an ideal gas at STP. Do you think this is a large enough number for valid statistical analyses?

- **2** A container is filled with helium gas at 350 K. Find the rms velocity of the gas.

- **3** Find the rms velocity of argon atoms in a gas at −10 °C.

- **4** A certain gas has a rms velocity of 600 m/s at 300 K. What is the molecular mass of the gas particles?

- **5** An atom moving with velocity $v_x = 5$ m/s strikes the walls of a container every 10^{-3} s. What is the length of the side of the box? Neglect collisions with other atoms.

- •• **6** Would we expect to see hydrogen gas (H_2) in the upper atmosphere of the planet Jupiter, which has a temperature of 160 K and an escape velocity of 60 km/s?

- •• **7** Saturn's satellite Titan has a mean temperature of about 100 K, a radius 2,575 km, and a mass 1.35×10^{23} kg. What is the highest molecular mass of gases that could easily exist in Titan's atmosphere? Could nitrogen (N_2) gas exist?

- •• **8** To what temperature would the earth's atmosphere have to be heated to lose its oxygen (O_2)?

- **9** Find the rms velocity of radon gas at 500 K.

- **10** Calculate the ratio between the rms velocity and the most probable velocity.

- •• **11** Avogadro's number of atoms produces a pressure of 10 N/m² in a container of volume 20 m³. If the rms velocity is 10^2 m/s, what is the mass of one of the atoms?

- •• **12** Using Section 13.1, calculate the range of molecular masses which could, in theory, be retained by the planet Mercury, with escape velocity 4.3 km/s and daytime temperature 620 K. Would you expect Mercury to have an atmosphere?

- ••• **13** A gas at 350 K has a ratio of $n(v_{rms})/n(100$ m/s$) = 3.0$. What is the molecular mass of the gas?

- ••• **14** Describe how you would derive Eq. 13–9 from Eq. 13–8.

- •• **15** A quantity of N_2 gas has temperature 300 K. What is the ratio of molecules moving at the speed of sound (330 m/s) to those moving at v_{rms}?

- •• **16** Construct a graph of the most probable velocity v_p versus temperature T.

- •• **17** A mixture of helium and carbon dioxide gases is heated to 400 K. Find and compare the most probable velocities of each.

- •• **18** Estimate the relative number of hydrogen gas ions (H^+) at 10^{10} K that have a velocity $0.1c$, compared to those with $v = v_p$.

- **19** Estimate the rms speed of hydrogen ions on the surface of (a) the sun (5,800 K), (b) a very hot star (50,000 K), (c) a neutron star (10^8 K).

- •• **20** At what temperature will the rms velocity of a hydrogen ion approach the speed of light?

- •• **21** Consider a cubical room, 5 m on a side, filled with oxygen (O_2) molecules at 25 °C and 10^{-10} atm pressure. (a) What is the rms velocity of molecules in the x-direction and the total rms velocity? (b) On average, how long does a molecule take to cross the room? Assume a molecule rarely collides with other molecules.

- •• **22** In Problem 21, what is the number density of O_2 molecules? If a man with body surface area 1.8 m² sits at the center of the room, how many molecules strike him each second?

Section 13.2 Molecular Transport

- **23** How long will it take a hydrogen molecule (H_2) to diffuse a distance of 1 km in the atmosphere if the value of D is 6.4×10^{-5} m²/s?

- **24** How long will it take a perfume molecule to diffuse 10 m across the room if its diffusion constant is 2×10^{-4} m²/s?

- •• **25** If the diffusion constant in Example 13.3 had been 10^{-5} m²/s (like a gas), how would the results have been affected?

- •• **26** The pathlength between collisions for a photon leaving the center of a star is very small. If the photon diffuses outward towards the surface, taking 10^6 years to reach the surface radius of 5×10^8 m, estimate the diffusion constant.

- •• **27** A certain membrane 10^{-7} m thick has a diffusion time of 2×10^{-5} s. What is the permeability of the membrane?

- **28** Find the permeability of the system in Example 13.3 if the dye had been separated from the water by a

PROBLEMS

membrane 0.1 mm thick with the same diffusion constant.

- **29** Derive Eq. 13–13.
- **30** A membrane with permeability 10^{-2} m/s and thickness 10^{-6} m separates two solutions with concentrations of impurities 1 part per billion (ppb) and 5 ppb. How long do the impurities take to diffuse through the membrane?
- **31** Nutrients diffuse through a liquid solution at a rate of 10^{-5} moles/m²/s when their concentrations are 10 moles/m³ and 100 moles/m³ at either end of its 0.1 mm length. What is the liquid's permeability and diffusion constant?
- **32** A leaky cork is placed in one end of wine bottle, and begins to dry out by diffusion. The cork is 1 cm in radius, 2 cm long, has a density of 0.8 g/cm³, and absorbs 10% of its mass in wine (mostly water). The air outside the bottle is 20°C with relative humidity 50%.
  - **(a)** Estimate the water concentration (molecules/cm³) inside and outside the cork.
  - **(b)** If the diffusion constant of water through the cork is 10^{-8} m²/s, estimate how long the cork takes to dry out.
- **33** How thick must a membrane of permeability 2×10^{-3} m/s be in order for the diffusion time to be 1 s?
- **34** A membrane with $P = 10^{-2}$ m/s separates two fluids. If the membrane is 10^{-5} m thick, what is the diffusion time?
- **35** A membrane 10^{-4} m thick separates two fluids and has a diffusion time of 10^{-4} s. What is its permeability?

Section 13.3 Heat Capacities and the First Law

- **36** Calculate the total internal energy of Avogadro's number of molecules at 300 K.
- **37** Suppose the Kelvin temperature of the system in Problem 36 was increased by a factor of three. What effect would this have on the total internal energy?
- **38** Find the internal energy of 1 mole of helium gas at 400 K.
- **39** A box of gas is at 300 K and has 10 J internal energy. How many atoms are in the box?
- **40** Find the heat capacity C_V for monatomic helium gas in J/kg · K.
- **41** A monatomic gas has heat capacity 10 cal/K. How many moles are present?
- **42** How much heat is required to raise the temperature of 3 moles of helium gas 5 °C? Assume volume is kept constant.
- **43** How much heat is required to heat 2 kg of argon gas from 10 °C to 20 °C? Assume pressure is kept constant.
- **44** A gas at 1 atm pressure changes its volume from 2 l to 1 l. How much work was done?
- **45** Three moles of a gas are heated with 3,000 J of energy, while volume is kept constant. If the temperature rises from 10 °C to 40 °C, find C_P and C_V.
- **46** Find the ratio C_P/C_V for a diatomic gas at room temperature. Neglect vibrations.
- **47** How many degrees of freedom does helium gas have?
- **48** How many degrees of freedom does O_2 gas have at room temperature? Neglect vibrations.
- **49** How many degrees of freedom does nitrogen gas (N_2) have when it is shocked to a high temperature? Include rotational and vibrational excitations. What is the ratio C_P/C_V?
- **50** How many degrees of freedom does a methane (CH_4) molecule have? Assume it can rotate about three axes and vibrate in several modes.
- **51** Find the energy per degree of freedom of argon gas at 350 K. Argon is a noble gas and does not form molecules.
- **52** In equilibrium, how much rotational energy does an O_2 molecule have at 300 K?
- **53** How much energy is required to heat a 300 m³ room from 10 °C to 30 °C? Assume the air is 20% O_2 and 80% N_2 by number, at 1 atm pressure.

Section 13.4 The Second Law and Heat Engines

- **54** A heat engine is operated between room temperature of 25 °C and a liquid helium reservoir at 4 K. Find the theoretical efficiency.
- **55** How hot must the upper reservoir of a Carnot engine be to have 90% efficiency if the lower reservoir is at 300 K?
- **56** The input of heat to an engine is 400 J, while the work done is 250 J. If the temperature change during the engine cycle is 500 K, what is the efficiency? What is the temperature of the upper reservoir?
- **57** A refrigerator with 30% efficiency is run by an engine which removes heat and dumps it outside the box. If the exhaust temperature is 30 °C, what is the temperature inside?

•• 58 Derive a relation between the maximum coefficient of performance of a heat pump cooler and its Carnot efficiency.

• 59 Suppose that we wished the refrigerator in Example 13.7 to operate between 40 °C and −20 °C. What are the maximum and actual coefficients of performance?

••• 60 A power plant generates 5×10^9 watts of power at 50% efficiency. The water flow rate into the cooling system is 5×10^5 kg/s from a nearby river. How much will the temperature of this water rise as a result of the power plant? Does the result depend on the river water's initial temperature?

• 61 The exhaust of a heat engine is 300 K. What must be the upper temperature to achieve 40% efficiency?

••• 62 Burning 1 gallon of gasoline releases about 1.3×10^8 J. Estimate the efficiency of a car's engine if the car gets 30 mi/gal at 60 mi/h and 30 hp.

••• 63 A 300 m³ room contains 10 people, whose basic metabolic rates each generate 100 Cal/h while sitting. How much does the temperature rise from 20 °C in a 1 h meeting if no heat escapes and no cool air is brought in? If a heat engine is used to cool the room, dumping the heat at a lower temperature of 20 °C, estimate how much work can be done with the metabolic heat.

••• 64 A refrigerator is run by an engine with 500 W power and 100% efficiency and can remove 2×10^6 J/h energy. If the outside of the refrigerator is at 30 °C, to what theoretical temperature can the refrigerator cool the inside?

SPECIAL PROBLEMS

1. Random Walk Diffusion

A particle moves randomly away from the origin along the x-axis (one dimension). In every interval τ it takes a step of length $\Delta x = \pm \lambda$ to the right or left with equal probability. After time t it has taken N steps. Fig. 13.16 shows a random walk in two dimensions.

(a) Show that $t = N\tau$.
(b) Show that the mean distance moved from the origin is $\langle x \rangle = 0$, but the mean square distance moved is $\langle x^2 \rangle = N\lambda^2$.
(c) If the particle takes steps at the rate $1/\tau$ (steps/unit time), show that $\langle x^2 \rangle = \lambda^2 t/\tau$. That is, show that the rms distance diffused from the origin increases with time, proportional to $t^{1/2}$.

(d) Find the time needed to diffuse an rms distance d from the origin. What is the diffusion constant D in terms of λ and τ? Compare to Eq. 13–13 with $\rho_2 = 0$.
(e) For rapid diffusion, which is more important, a large step size λ or a short time interval between steps τ?

2. Atmospheric Scale Height

According to the principle of equipartition, the internal energy of a gas is distributed uniformly among the available degrees of freedom. Each degree of freedom has $\frac{1}{2}kT$ energy. Suppose a gas is in equilibrium with the gravitational field $g = GM/R^2$ of a planet whose mass is M and whose radius is R.

(a) Show that the potential energy per particle at altitude h is mgh, where m is the mean molecular mass.
(b) From the equipartition principle, estimate the mean altitude, h_m, known as the **scale height** of the atmosphere.
(c) The Earth's atmosphere is about 20% O_2 and 80% N_2 by number, and $T = 290$ K. Estimate the atmosphere's scale height in km.
(d) What is the scale height of Jupiter's atmosphere, if $T = 160$ K and methane (CH_4) is the predominant constituent? See Appendix 4 for mass and radius.
(e) What is the scale height of the sun's atmosphere, with $T = 5800$ K and hydrogen atoms the dominant constituent?

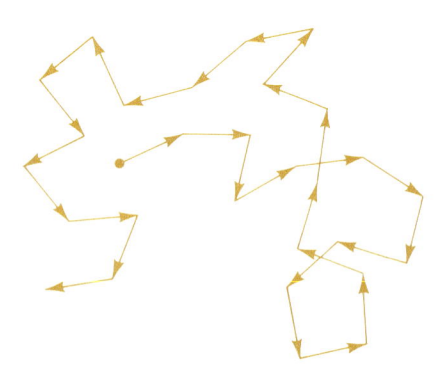

Figure 13.16

3. Osmotic Pressure

Osmotic pressure arises at the interface of two solutions with different dissolved particle concentrations, separated by a semipermeable membrane. The pressure excess is given by the law of partial pressure for the solute (dissolved particles).

(a) Show that the excess pressure of the solute is $P_s = n_s RT/V$, where n_s is the number of moles of particles dissolved in volume V at temperature T.

(b) What is the excess height h of the column of fluid on the right hand side of the U-shaped tube in Fig. 13.17? The solute on the left has n_s moles of particles dissolved in a solution of volume V, and the tube has cross sectional area A. The fluid density is ρ kg/m^3.

(c) If 10 g of table salt (NaCl) are dissolved in 1 liter of water at 300 K on the left of the U-shaped tube, what is the excess height on the right?

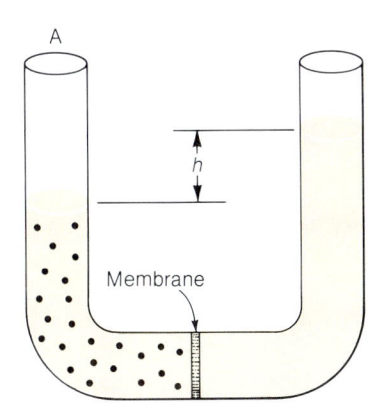

Figure 13.17

4. Entropy and Statistics

Entropy is a measure of the state of disorder in a system. However, it is best understood by a statistical or probabilistic analysis, first made by Ludwig Boltzmann in the nineteenth century. The entropy of a system (more specifically of a **macrostate** which specifies the overall properties of the system) is related to the number of different ways (**microstates**) of achieving that macrostate. Quantitatively, Boltzmann showed that the entropy of a macrostate is $S = 2.303\, k \log(\Omega)$, where Ω is the number of microstates that result in a given macrostate. The factor 2.303 arises because the equation is usually written $S = k \ln(\Omega)$, where $\ln(\Omega) = 2.303 \log(\Omega)$ is the natural logarithm, whose base is $e = 2.71828$.

We illustrate the concepts of macrostate, microstate, and entropy with a simple example of rolling two dice. Suppose the dice can each land with six equally probable numbers, 1 through 6. The total score is the macrostate, which can be 2, 3, 4, ..., 12. The individual die scores are the microstates. For example, the macrostate score of 3 can be obtained in two ways (two microstates): die A rolls 1, die B rolls 2, or else die A rolls 2 and die B rolls 1.

(a) Show that the total number of possible microstates is 36.

(b) Make a list of the total scores, 2 through 12, and give the number of microstates that correspond to each score. Do the microstates total 36?

(c) The probability of rolling a given score is just the number of microstates divided by 36. Which score is most likely?

(d) Explain why maximizing entropy is equivalent to maximizing the probability of obtaining a given macrostate.

SUGGESTED READINGS

"Geothermal power." J. Barnea, *Scientific American*, Jan 72, p 70, **226**(1)

"The Wankel engine." D. E. Cole, *Scientific American*, Aug 72, p 14, **227**(2)

"The arrow of time." D. Layzer, *Scientific American*, Dec 75, p 56, **233**(6)

"Alternative automobile engines." D. G. Wilson, *Scientific American*, Jul 78, p 39, **239**(1)

"The origins of the water turbine." N. Smith, *Scientific American*, Jan 80, p 114, **242**(1)

"Sadi Carnot." S. S. Wilson, *Scientific American*, Aug 81, p 134, **245**(2)

"Brownian motion." B. H. Lavenda, *Scientific American*, Feb 85, p 70, **252**(2)

The flying circus of physics with answers. Jearl Walker, Wiley, New York, 1977

A source book in physics. William F. Magie, Harvard University Press, Cambridge, 1965

The second law. P. W. Atkins, Freeman, NY, 1984

PART FOUR

ELECTRICITY AND MAGNETISM

Chapter 14
Electrical Charge and Field: Electrostatics 444

Chapter 15
Current and Circuits 476

Chapter 16
Magnetism and Electricity 512

Chapter 17
Applied Electricity and Magnetism 550

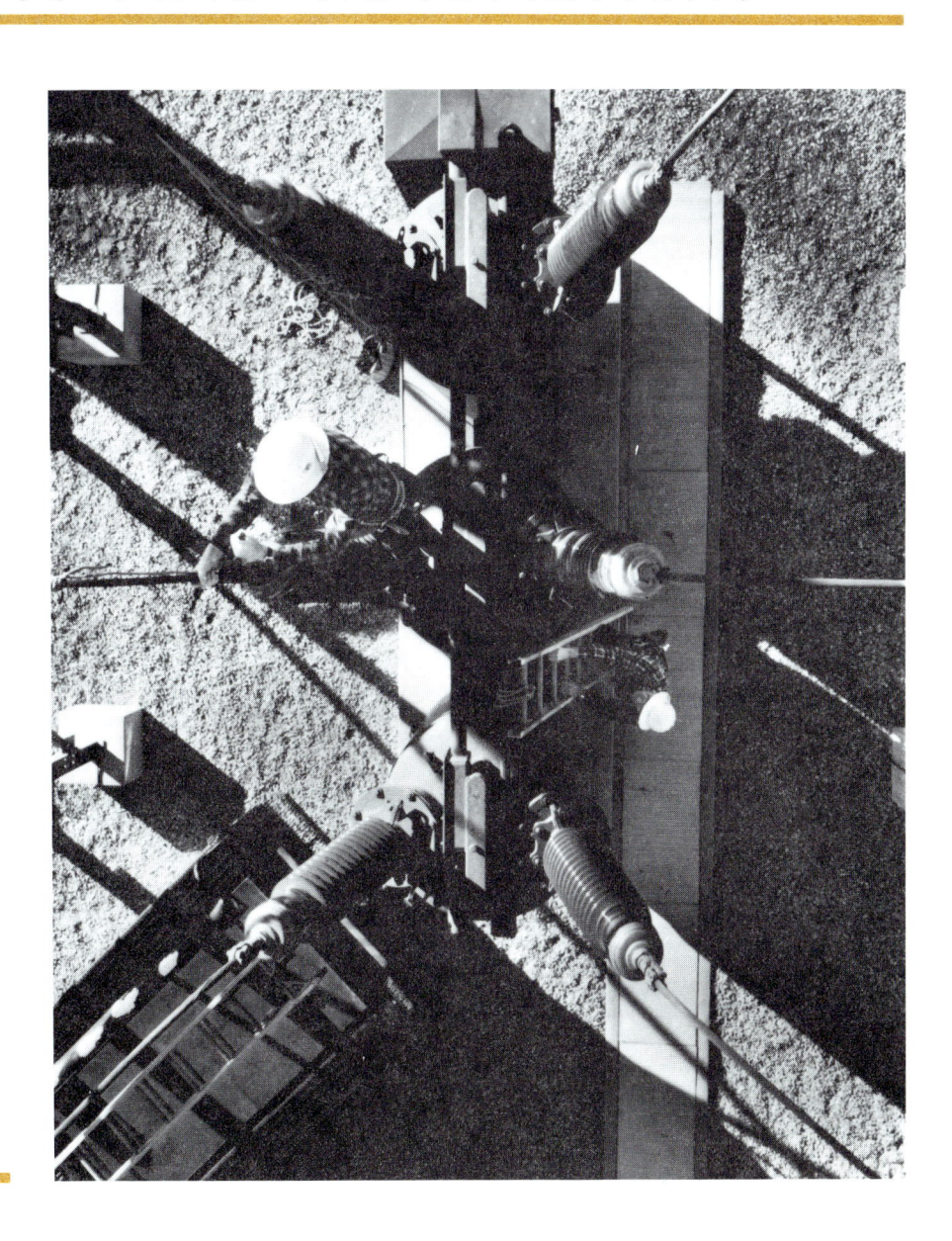

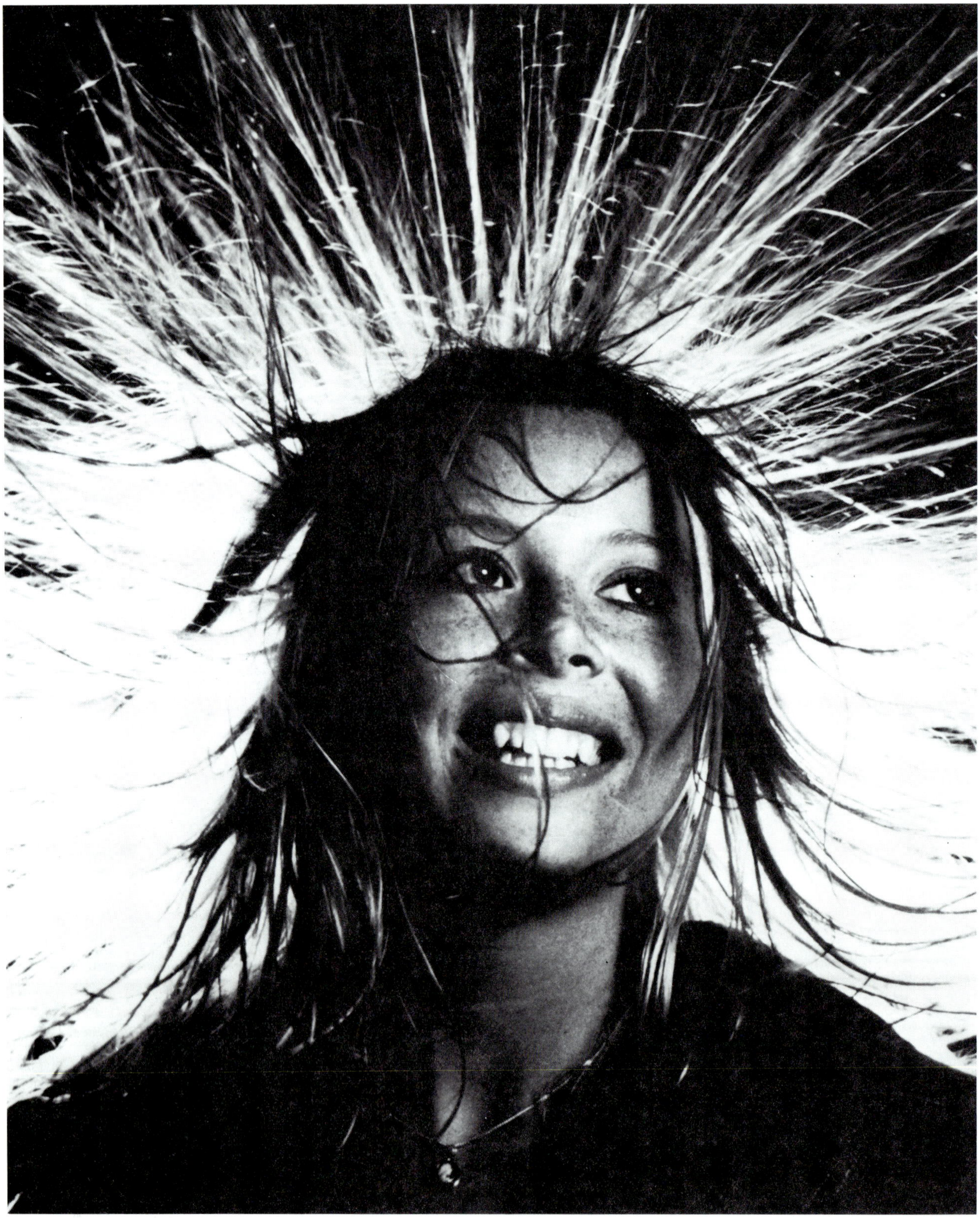

Chapter 14

Electrical Charge and Field: Electrostatics

So much of our technology depends on the properties of electrical charge and current that society would come to a standstill if for some reason the use of electricity became impossible. Even the most rugged backpacker relies on battery-operated devices such as flashlights, and clearly none of the conveniences of modern city life would be possible without electricity. The study of this phenomenon therefore holds the key to understanding most of today's technology.

In this chapter we will first discuss the nature of electrical charge, then its relationship to energy and work. In later chapters we will treat the flow of charge and the use of currents in technology.

14.1 Charge

Some phenomena related to electrical charge have been observed for centuries. The word "electricity" itself is derived from the Greek term for amber, a substance easily charged by friction so that it attracts bits of debris. The first modern systematic investigations were made in the eighteenth century. Two kinds of electrical charge were discovered, and the American statesman and scientist Benjamin Franklin named them "positive" and "negative". (He thought that there was really only one kind of charge, which could either be present or absent.)

Even today we often think of charge as some kind of physical entity, but all we really know about it is how to describe its effects. Mathematical treatment of charge deals with it as a quantity that obeys certain relationships (such as conservation laws), which supports the notion that it is a physical substance, but we should keep in mind that this view is artificial. Scientists have developed this formalism for charge because it helps describe its effects, not because we think of it as a fundamental description of what charge really is.

CHAPTER 14 ELECTRICAL CHARGE AND FIELD: ELECTROSTATICS

Figure 14.1

Static cling results from attraction of unlike charges.

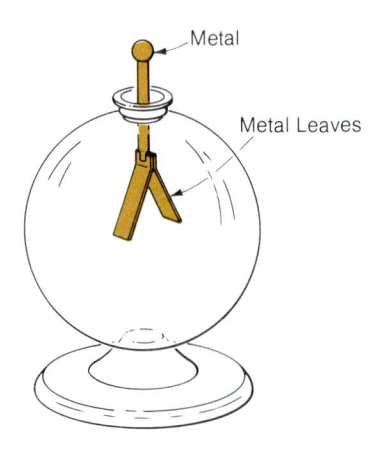

Figure 14.2

The electroscope. Charge is redistributed between the leaves and the metal knob at the top. Like charges make the leaves separate due to the repulsive force.

Experimental Observations

Several properties of electrical charge were revealed by the experiments of Franklin and others. It was found, for example, that charges can repel each other or attract each other (Fig. 14.1). It was this property that led to the discovery that there are two kinds of charge; different substances were found to take on opposite charges when rubbed with a cloth. Another early discovery was that some materials conduct charge much more readily than others. A simple device called an **electroscope** can demonstrate this. An electroscope consists of two foil leaves joined at the top and sealed inside a chamber that is insulated from contact with the outside except through a metal connector attached to the leaves (Fig. 14.2). When an electrical charge enters the electroscope through the metal connector, the leaves gain like charges and separate due to the repulsive force. To compare how well materials conduct electricity, a remote object (perhaps a piece of glass or plastic) can be charged by rubbing it, and then connected to the electroscope by different materials. For some materials, such as metals, the leaves will separate quickly, showing that the charge has been transported readily; for other materials, such as glass, the leaves separate only very slowly, if at all.

Early experiments also showed that electrical charges can act at a distance; the attractive or repulsive force between two charged objects is observed even when they are separated. Thus the electrical force became the second one known to act remotely, after Newton's studies of gravity. Now we know that the electrical force is a field force, one of only four known to exist (for a discussion of field forces, review Section 3.1). Another manifestation of the action of electrical charge at a distance is the observation that a charge can be **induced** in an object by the nearby presence of a charged body. If we hold a charged object near an electroscope, for example, the leaves will separate. Since no contact is made with the electroscope, charge is not added to it, but is redistributed within it. Charge of the same sign as the nearby object flees from the electroscope terminal down to the leaves, and unlike charge flows towards the terminal (Fig. 14.3).

Today we understand the nature of electrical charge in terms of atomic structure, which we have already discussed in Chapter 8. An atom consists of a positively charged nucleus and some negatively charged electrons, which normally contain a total charge equal to that of the nucleus, so that the net charge is zero. We say that an object is electrically charged if its positive and negative particles are out of balance. We have mentioned that such an imbalance can be created by friction. For example, when certain substances are rubbed, electrons tend to be transferred, so that the object being rubbed and the material doing the rubbing both become charged, with opposite signs. It is common for a person to build up an electrical charge (particularly in very dry weather) by sliding his feet across a carpet. This is the **static electricity** that sometimes causes minor shocks when we touch a metal doorway or another person.

The capability of a material for conducting electricity depends on how tightly its electrons are fixed in place. In discussing solids in Chapter 8, we generally assumed that the atoms within a solid are fixed quite rigidly in place, and are not free to move about. This is true, but does not necessarily apply

14.2 ELECTRICAL FORCE AND FIELD

Figure 14.3

Induced charge. A positive charge is held near, but not touching, the terminal on an electroscope. The total charge on the electroscope is unchanged but it is redistributed.

to the electrons. In metals especially, a certain fraction of the electrons are free to move about, and are not tied to any particular atom. Thus, in a metallic conductor, the flow of electrical charge always consists of the movement of electrons which redistribute the overall charge. When a charge on an object is induced by the presence of another charged body, for example, it is these free electrons that either flow towards or away from the charged body, depending on the sign of its charge. It is ironic that the active agent in transporting charge corresponds to Franklin's "absence" of charge; his picture was correct, but he had the sign wrong.

The Conservation of Charge

From our description of the induced charge created by the presence of a charged body, it is clear that the total charge remains constant, even though it may be redistributed. The flow of electrons creates a net negative charge in one portion of a body, but creates an equal positive charge in other regions which the electrons have vacated.

Other simple experiments using an electroscope verify that the total charge is constant. If a piece of plastic is rubbed with a cloth, the plastic gains electrons (hence a net negative charge) while the cloth loses them (and develops a net positive charge). Exposure of the plastic and then the cloth to the electroscope shows that the amount of charge on each is the same.

Other, more elaborate experiments always confirm the law of conservation of electric charge:

Electrical charge can be neither created nor destroyed in any process.

This is a formal statement of the experimental finding that a body can gain electrical charge only when the charge is lost by some other body. The total quantity of charge is always constant.

Just as we found in our discussions of the conservation of momentum and energy, this conservation law has many useful applications. We will see this more clearly after further investigation of electrical fields and forces.

14.2 Electrical Force and Field

We have mentioned action at a distance and the induction of charge, both of which imply that there is some intangible connection between electric charges that are separated from each other. The exact nature of this connection is not fully understood, but a great deal is known about its properties. Just as in many other instances, we are better able to define this phenomenon by its effects than by its ultimate cause. As in the treatment of electrical charge, the

Figure 14.4

Electrical force inside a uniformly charged sphere. Experiments show that the net force acting on a charge at any point P inside a hollow charged sphere is zero. The force contributions from opposite directions always cancel each other. The wall area contributing to the force at P within a conical angle is proportional to the square of the distance from P to the wall (because the area subtended by the conical angle is proportional to the square of the distance). For all contributions at P to cancel out the force must be inversely proportional to the square of the distance.

concept of electrical field is really just our mathematical representation of the effects of isolated charges on each other, rather than being a fundamental description of something tangible that exists in the space between charges.

Coulomb's Law

Many investigators studied the force of attraction or repulsion between electrical charges during the century following Newton's great work on mechanics and gravitation. It was known that the gravitational force decreases as the square of the distance between masses, and this led to a general expectation that the same might be true of the electrical force. The first real demonstration of this followed Franklin's experimental finding that the electrical force exerted at all points inside a charged, hollow container was zero. Soon the English physicist Joseph Priestley showed that this is expected if the force decreases as the inverse square of the distance because, inside a sphere, such a force will balance at all points. This can be seen intuitively by imagining the force due to opposite walls acting on some interior point and considering the force from each wall within a given conical angle (Fig. 14.4). Then we see that the area of wall acting on a given interior point increases as the square of the distance to the wall; the force must decrease as the square of the distance for all the contributions due to the walls to cancel. Priestley's publication of his finding was little noted, and others took up the task. The French engineer C. Coulomb, in particular, refined the techniques for measuring small electrical forces and directly determined to a high degree of accuracy that the force decreases as the inverse square of the distance. Today this result is known as **Coulomb's law.**

In more explicit terms, the law states that the force between two electric charges at rest is proportional to the product of the charges and inversely proportional to the square of the distance between them. In equation form, we write the force F as

$$F = \frac{kQ_1Q_2}{r^2} \qquad (14\text{--}1)$$

where Q_1 and Q_2 are the two charges, r is the distance between them, and k is a proportionality constant. Note the emphasis on charges at rest; the situation is modified when charges are in motion, and this will not be discussed until Chapter 16. Coulomb's law applies only to the **electrostatic** force.

There is another form of Coulomb's law, in which the constant k is replaced by another, called the **permittivity of free space** ϵ_o, which is defined so that

$$k = \frac{1}{4\pi\epsilon_o}. \qquad (14\text{--}2)$$

Whether to use ϵ_o or k is purely a matter of preference; most find it easier to use k, since it is a single number, with no factor of 4π to carry through calculations. The reason ϵ_o is sometimes used is that it is in a sense more fundamental, having to do with the manner in which the force actually is

14.2 ELECTRICAL FORCE AND FIELD

transmitted over a distance. We will generally apply Coulomb's law in the form of Eq. 14–1, with k as the constant of proportionality.

To evaluate k, we need a unit of measure for the charge Q. This could be done in any number of ways, but we shall adhere to the modern SI units, in which charge is measured in a unit called the **coulomb** (**C**). Ironically, the definition of the coulomb is based not on some atomic parameter such as the charge on the electron, but is instead based on measurements of **current**, the flow of charge through a conducting medium. The basic SI unit of current is the **ampere** (**A**), and the coulomb is defined as the amount of charge that flows past a fixed point in one second in a current of one ampere. The reason the unit of charge is defined in this way is that current can be measured far more precisely than any natural unit of charge (such as the charge on an electron), and it is more consistent with other SI standards to maintain as high a degree of accuracy as possible.

In practice, however, we often use the charge on the electron as a unit of charge, because this quantity of charge is relevant for many applications. Several experiments have shown that, within the accuracy of measurement, the charge on the electron is constant. The measured value of the charge on an electron is $-e = -1.602 \times 10^{-19}$ C. (We define the fundamental charge e to be a positive number.)

Now that we have a unit of electrical charge, we can find the value of the proportionality constant in Eq. 14–1. Experiments have determined $k = 9.0 \times 10^9$ N·m²/C² (hence the value of ϵ_o, from Eq. 14–2, is $\epsilon_o = 8.85 \times 10^{-12}$ C²/N·m²). We can now calculate the magnitude of actual electrical forces in various situations. For example, we use the average distance between the nucleus and the electron in a hydrogen atom (Fig. 14.5), 5.3×10^{-11} m, to find the force between the electron and the nucleus (a single proton, whose charge is equal in magnitude but opposite in sign to that of the electron).

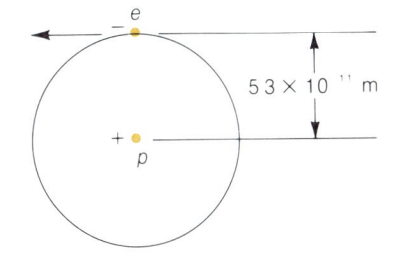

Figure 14.5

The hydrogen atom. The nucleus consists of a single proton (positive charge), and the electron (negative charge) orbits it at an average distance of 5.3×10^{-11} m. The electric force of attraction is far stronger than the gravitational force between the proton and the electron.

$$F = \frac{kQ_1Q_2}{r^2}$$
$$= \frac{-(9.0 \times 10^9 \text{ N·m}^2/\text{C}^2)(1.6 \times 10^{-19} \text{ C})^2}{(5.3 \times 10^{-11} \text{ m})^2}$$
$$= -8.2 \times 10^{-8} \text{ N}$$

The negative sign indicates that the charges are unlike, so the force is attractive. This is a very small force, but far more than the gravitational force between the proton and the electron, which is $F_g = 3.6 \times 10^{-47}$ N. Thus the structure of the atom, at least on the scale of orbital radii of electrons, is dominated by the electrical force (the details are complicated, but this general conclusion is valid).

Example 14.1

Suppose a table tennis ball has a mass of 1 g. If an object containing a charge of 10μC (where 1μC = 10^{-6} C) is held 0.1 m above such a ball, how much charge must the ball carry to be lifted by the electrostatic force (Fig. 14.6)? Treat the ball as a point charge.

We are asked to find the charge that would create an electrostatic force equal to the weight of the ball. The weight is

$$W = mg = (0.001 \text{ kg})(9.8 \text{ m/s}^2) = 0.0098 \text{ N}.$$

We substitute W for the force in Eq. 14–1 and solve for the charge Q_1 on the ball, where Q_2 is the charge on the second object.

$$Q_1 = \frac{r^2 W}{k Q_2} = \frac{(0.1 \text{ m})^2 (0.0098 \text{ N})}{(9.0 \times 10^9 \text{ N} \cdot \text{m}^2/\text{C}^2)(10 \times 10^{-6} \text{ C})}$$
$$= 1.1 \times 10^{-9} \text{ C} = 0.0011 \text{ μC}$$

This is a small amount of charge, the equivalent of about 7 billion electrons, which can easily be added to a table tennis ball by friction.

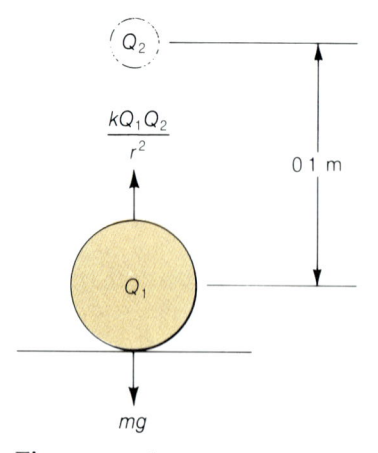

Figure 14.6

Just as Newton's law of gravitation is universal, so is the electrostatic force. Every charge in the universe exerts a force on every other charge. Normally we can neglect the force due to faraway charges when we are considering a close pair, but there certainly can be more than two charges near each other. Then the net force acting on each charge is the vector sum of the forces due to the others.

Example 14.2 Suppose we have three point charges, with the magnitudes, signs, and positions shown in Fig. 14.7. What is the magnitude, sign, and direction of the net force on charge A?

The vectors in this example are quite simple. There is a downward force F_1 due to the -1μC charge, and a force F_2 to the right due to the -4μC charge. The magnitudes are

$$F_1 = \frac{(9.0 \times 10^9 \text{ N} \cdot \text{m}^2/\text{C}^2)(2 \times 10^{-6} \text{ C})(-1 \times 10^{-6} \text{ C})}{(2 \text{ m})^2} = -0.0045 \text{ N}$$

and

$$F_2 = \frac{(9.0 \times 10^9 \text{ N} \cdot \text{m}^2/\text{C}^2)(2 \times 10^{-6} \text{ C})(-4 \times 10^{-6} \text{ C})}{(3 \text{ m})^2} = -0.008 \text{ N}.$$

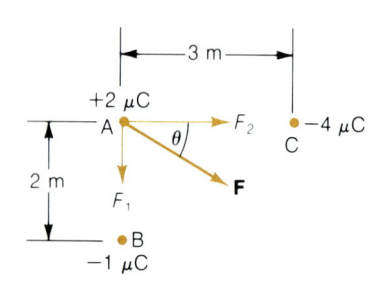

Figure 14.7

The net force due to multiple charges. Charge A experiences a force F_1 due to charge B, and a force F_2 due to charge C. The resultant force on A is F.

The resultant vector is the hypotenuse of a right triangle with F_1 and F_2 as the legs, so its magnitude is

$$F = \sqrt{F_1^2 + F_2^2} = \sqrt{(0.0045 \text{ N})^2 + (0.008 \text{ N})^2} = 0.0092 \text{ N}$$

14.2 ELECTRICAL FORCE AND FIELD

and its direction below horizontal is the angle whose tangent is equal to the ratio F_1/F_2.

$$\theta = \arctan\left(\frac{F_1}{F_2}\right) = 29.4°$$

Without explicitly saying so, we have been assuming that the charges leading to a force given by Coulomb's law are separated only by a vacuum or some medium that has no effect on the magnitude of the force. This is not generally true, however; the charges can be immersed in a medium that effectively reduces the force. There is still an inverse-square dependence on distance, but a scale factor must be introduced to reduce the force. This scale factor is called the **dielectric constant**, and is usually denoted K (Table 14.1). Now Coulomb's law is written

$$F = \left(\frac{k}{K}\right)\left(\frac{Q_1 Q_2}{r^2}\right). \tag{14-3}$$

A dielectric reduces the effective electric force because of the electrical properties of certain molecules. A molecule consists of a number of atoms bonded together rigidly, so that individual atoms are held in fixed positions relative to each other. In certain molecules, the atoms are effectively charged, so that the rigid structure produces a separation of positive and negative charges. Such molecules are said to be **polarized**. When a substance containing such molecules is placed between two unlike external electric charges, the molecules are forced to align their positively-charged portions toward the

Table 14.1
Dielectric constants (20°C)

Material	Dielectric constant K
Vacuum	1.00000
Air (1 atm)	1.00059
Air (100 atm)	1.0548
Paraffin	2.2
Amber	2.65
Rubber, hard	2.8
Mica	3–6
Mylar	3.1
Paper	3–7
Quartz	4.3
Glass	4–7
Methyl alcohol	33.1
Water	80.4

external negative charge, and negative charges toward the external positive charge. Thus, the electric force between the separated charges within each molecule opposes the force between the external charges, and a dielectric material placed between two charges reduces the electric force between them.

The reduction of the electrostatic force by the introduction of a dielectric medium can have important effects. For example, water has a relatively large value of K (approximately 80). Some molecular compounds dissolve in water because the molecular binding force is reduced so far that the individual atoms can break free. For air and other gases, the value of K is generally near 1 (which is its precise value by definition for a vacuum), so we can ignore the effect of the dielectric constant when charges are separated by a gaseous medium.

Example 14.3 Repeat the calculation in Example 14.1 for a table tennis ball immersed in a liquid whose dielectric constant is $K = 5$.

Because of the dielectric, the force is diminished by a factor of five now, so the charge required for the same force must be five times greater. Therefore a charge of 5.5×10^{-9} C is required for the electric force on the ball to equal its weight.

The Electric Field

When we discussed the gravitational force, we spoke casually of the gravitational "field", meaning in some vague way the presence of an intangible medium that somehow gives rise to a force acting on any mass. We used the term "field forces" in Chapter 3, when we discussed the fundamental nature of forces, but even then we did not say what we mean by the term "field".

There is no detectable physical medium that transmits forces, yet somehow forces are communicated over distances. The nature of fields is one of the major concerns of modern particle physics. As usual, we will have to be content to say what a field does, rather than what it is. For the electrical field, we say that it is a field that will produce a force on a charge. It permeates the space surrounding any electrical charge, and its intensity is measured by the magnitude of the force it exerts on a known charge (Fig. 14.8). The direction of the field is always radial; that is, straight toward or away from the charge that creates the field. By convention, we say that the direction of the field is the same as the force it exerts on a positive charge. We introduce the notion of a **test charge** q, which has a positive sign but such a small magnitude that it does not affect the distribution of other charges near it. Then we write

$$\mathbf{E} = \frac{\mathbf{F}}{q} \qquad (14\text{--}4)$$

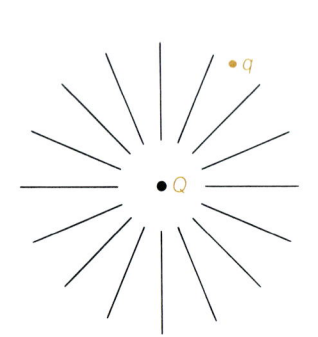

Figure 14.8

The electrical field of a point charge. Surrounding any electrical charge Q is an electric field, which is manifested by the force exerted on a test charge q.

14.2 ELECTRICAL FORCE AND FIELD

where **E** is the vector representing the electric field, and **F** is the vector representing the force exerted on a positive test charge q. Here it is convenient to use vector notation, since we will be concerned with the direction of the field and with the net effect of fields due to multiple charges.

The role of the test charge is seen clearly if we calculate the electric field for a simple example, a single point charge Q. The force exerted on a test charge at distance r from the charge Q is

$$F = \frac{kqQ}{r^2}.$$

and the magnitude of the electric field given by Eq. 14–4 is

$$E = \frac{F}{q} = \frac{kQ}{r^2}.$$

We see that the magnitude of the test charge has cancelled, so that the electric field is defined entirely by the charge Q. The field therefore represents the force per unit charge that would be exerted on a charge placed in the field. The SI units of E are N/C.

The field due to an assemblage of charges is the vector sum of the fields due to all the individual charges. This is true even if, rather than a group of point charges, we have a continuous charge distribution, as for an extended charged object. The calculation becomes more difficult in that case, however, and normally calculus techniques are required to sum the contributions from all points on the charged body. We will deal only with situations where a simple vector sum of fields will be sufficient.

Example 14.4 Suppose we have a square formed by three point charges and a fourth point that has no charge (Fig. 14.9). What is the electric field at the fourth point due to the three charges, if the charges are each $+3\mu C$, and the sides of the square are 1 m in length?

We establish a coordinate system as shown, and calculate the x- and y-components at the fourth point for each of the three charges. For two of the three this is simple: the field due to charge Q_1, for example, is entirely in the x-direction, while that due to charge Q_3 is entirely in the y-direction. The field due to charge Q_1 is

$$E_{x1} = \frac{kQ}{r^2} = \frac{(9.0 \times 10^9 \text{ N} \cdot \text{m}^2/\text{C}^2)(3 \times 10^{-6} \text{ C})}{(1 \text{ m})^2} = 2.7 \times 10^4 \text{ N/C}$$

in the negative x direction, since the repulsive force on a positive charge at the lower left corner of the square would be toward the left. The notation E_{x1} refers to the x-component of the field due to charge 1.

The distance to charge Q_2 is $\sqrt{2}$ m = 1.41 m and the direction is 45° from the x-axis, so the x-component of force due to charge 2 is

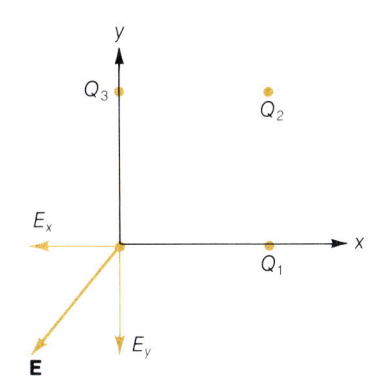

Figure 14.9

$$E_{x2} = \frac{(\cos 45°)(9.0 \times 10^9 \text{ N} \cdot \text{m}^2/\text{C}^2)(3 \times 10^{-6} \text{ C})}{(1.41 \text{ m})^2}$$
$$= 9.5 \times 10^3 \text{ N/C}$$

again directed towards the left (negative *x*-direction). There is no *x*-component due to charge 3, so the total *x*-component of force is

$$E_x = E_{x1} + E_{x2} = -2.7 \times 10^4 \text{ N/C} - 9.5 \times 10^3 \text{ N/C}$$
$$= -3.7 \times 10^4 \text{ N/C}.$$

Similarly, the total *y*-component is

$$E_y = E_{y2} + E_{y3} = -9.5 \times 10^3 \text{ N/C} - 2.7 \times 10^4 \text{ N/C}$$
$$= -3.7 \times 10^4 \text{ N/C}$$

where the minus sign indicates that the *y*-component is downward, toward the negative *y* direction.

Finally, the resultant field has the magnitude

$$E = \sqrt{E_x^2 + E_y^2}$$
$$= \sqrt{(-3.7 \times 10^4 \text{ N/C})^2 + (-3.7 \times 10^4 \text{ N/C})^2}$$
$$= 5.2 \times 10^4 \text{ N/C}$$

and its direction is towards the angle 225° as shown. (In this problem we know the angle because of symmetry; normally, we would have to derive it from the ratio of the two components, which equals the tangent of the angle.)

It is sometimes helpful when visualizing an electric field to draw **lines of force** to indicate the direction of the force on a positive electric charge (Fig. 14.10). The spacing of the lines is a measure of the magnitude of the field. Near a point charge, for example, the radial lines are close together, and they diverge with distance from the charge. (If this were drawn in three dimensions, we would see that the density of lines varies as $1/r^2$, where *r* is the distance

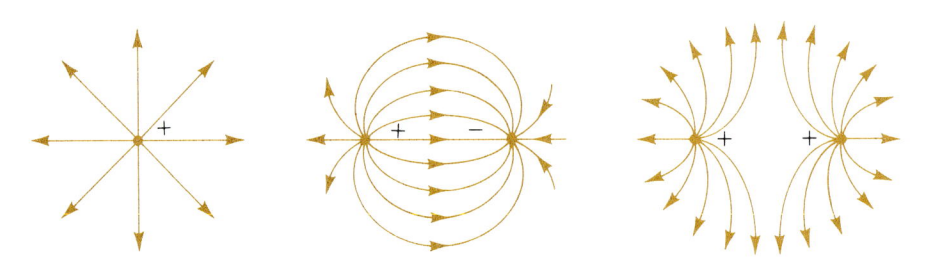

Figure 14.10

Lines of force.

from the charge. The density of lines is proportional to the field intensity.) The number of lines actually drawn to depict a field is arbitrary, since a line could be drawn through any point. The normal practice is to draw enough lines to show the structure of the field, which means a higher density of lines is needed in regions where there is a strong field or complex structure due to a complex charge distribution.

We emphasize that lines of force, sometimes called field lines, are not physical entities, but only represent the elusive concept of a field. While there were once theories that invoked a real physical medium that constituted a field, we now think of a field only in terms of the forces that will act on particles at various positions.

14.3 Electric Energy and Potential

In our discussion of gravitation we said that a system of bodies has potential energy due to the relative positions of the bodies and the gravitational forces they exert on each other. We were able to solve problems about orbital motions by including this potential energy in the total energy. Now we will find that an electric potential energy can be defined in a similar manner, and that it has a wide range of applications.

The Electric Potential

If a charged particle is moved against the force exerted on it by an electric field, then work is done; and the total energy of the particle must be changed by an amount equal to the work. We say that the particle has electric potential energy, and that the change in this potential energy must equal the work done; that is, the work-energy principle is valid for electrical potential energy, just as for other forms of potential energy.

When we discussed gravitational potential energy, we said that the potential energy was zero for an infinite separation of two masses. We will in some cases define a zero point for electric potential in a similar manner, but we can begin our discussion by considering only changes in potential energy, with no absolute reference. Whether potential energy is gained or lost when a charge is moved within an electric field depends on the sign of the charge and the direction of the field, complications that did not arise when we discussed gravitational potential energy. We see that work is required to bring two like charges together or to move unlike charges apart; both result in a gain of potential energy (Fig. 14.11). On the other hand, the electric potential energy decreases as like charges move apart or as unlike charges are brought together, so there is a loss of potential energy.

Just as it proved convenient to define a force per unit charge (the electric field), now we define the **electric potential** as the electric potential energy per unit charge. Therefore a change in electric potential energy is a **potential difference,** which is designated V. By this definition we have

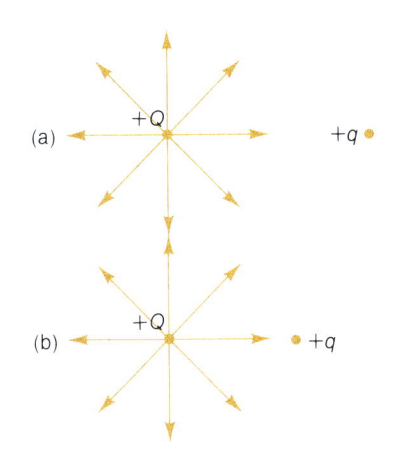

Figure 14.11

Electric potential. Work is required to move a test charge q with respect to a charge Q. Here the potential is higher in (b) because the like charges are closer together.

$$V = \frac{PE}{q} \qquad (14\text{--}5)$$

where q is the charge on a particle that undergoes a change in electric potential energy PE. Potential energy is gained when work is required to move a charge; that is, when the charge is moved against the electric force exerted on it by the field. Conversely, potential energy is lost when a charge is allowed to move in the direction of the force exerted by the field. Thus we say that the potential difference is positive in the first case, and negative in the second.

We are careful to refer to V as a potential **difference,** although there are conventions for establishing a zero point (such as the potential of the ground), so that sometimes people speak simply of the potential, with the understanding that they mean the potential with respect to the zero point.

The dimensions of potential and potential difference are energy per unit charge, so the SI units should be J/C. A special name is used instead: the **volt** (V) is defined as equal to one J/C (the volt is named in honor of A. Volta, the Italian scientist who invented the battery). Thus we often refer to a potential difference as a **voltage,** a term familiar to us all. Because electrical appliances operate by extracting work from electrons accelerated by potential differences, we are used to the idea that an appliance requires a certain voltage. We will not be in position to analyze the operation of appliances until we have discussed current and circuits, but there are practical problems that we can already solve.

Example 14.5   Suppose electrons in a particle accelerator are accelerated from rest by a potential difference of 2,000 V. How much potential energy do they lose, and what is their velocity after being accelerated?

The total energy is constant, so the gain in kinetic energy must equal the loss in potential energy; therefore

$$PE = qV = (-1.6 \times 10^{-19} \text{ C})(2{,}000 \text{ V}) = -3.2 \times 10^{-16} \text{ J}.$$

We note that the change in potential energy is negative, since the potential difference is allowed to accelerate the charged particle.

The electron was initially at rest, so there was zero kinetic energy. The total energy of the electron remains constant as it is accelerated, so the kinetic energy after the acceleration equals the loss of potential energy. Hence we have

$$\frac{1}{2}mv^2 = -PE$$

or

$$v = \sqrt{\frac{-2PE}{m}} = \sqrt{\frac{-2(-3.2 \times 10^{-16} \text{ J})}{(9.1 \times 10^{-31} \text{ kg})}} = 2.7 \times 10^7 \text{ m/s}.$$

14.3 ELECTRICAL ENERGY AND POTENTIAL

This is nearly one tenth of the speed of light; it is not uncommon for electrons to reach even higher speeds in particle accelerators.

Because potential differences often accelerate electrons, a special energy unit is commonly used. This is the **electron-volt,** defined as the potential energy change of an electron accelerated by a 1-volt potential difference. Thus 1 eV = $(1.6 \times 10^{-19}$ C$) \times (1$ V$) = 1.6 \times 10^{-19}$ J. This unit will be especially useful when we discuss atomic structure and electron energy levels in a later chapter, because the energies involved have typical magnitudes of a few electron-volts.

Electric Potential and Electric Field

It should be obvious that there is an intimate relationship between the potential difference and the electric field. The work done, as a charged particle is moved against an electric field, changes its potential energy. Since we have defined the potential difference in terms of a change in potential energy, an electric field must somehow be equivalent to a potential difference. The relationship is easy to see for a uniform electric field: the work done as a particle is moved through a distance d by force F is simply

$$W = Fd = -qEd$$

where the minus sign arises because when q, E, and d are all positive, the work is negative. (The direction of **E** is the direction toward which a positive charge will move. Hence when the positive charge is moved in the direction of positive d, the particle is being moved in the direction of the force due to the field, and is losing potential energy.) Hence from Eq. 14-5 we see that the potential difference must be

$$V = \frac{W}{q} = -Ed. \qquad (14\text{-}6)$$

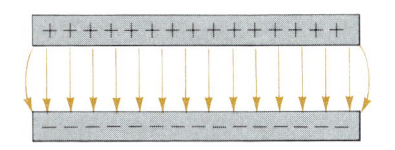

Figure 14.12

Uniform electric field between parallel charged plates. Note that the field is not quite uniform at the edges.

We see from this expression that an electric field can be expressed in terms of the potential; that is,

$$E = \frac{-V}{d} \qquad (14\text{-}7)$$

where $-V/d$ is the rate of change of the potential with displacement. The units therefore are V/m, yet earlier we expressed electric fields in units of N/C. It is worthwhile to demonstrate to yourself that the two are equivalent.

Uniform electric fields are not typical, but they sometimes exist. If two oppositely-charged plates are parallel to each other, for example, then the field lines between them are all perpendicular to their surface (Fig. 14.12), and the

CHAPTER 14 ELECTRICAL CHARGE AND FIELD: ELECTROSTATICS

lines are uniformly spaced. Except near the edges, where the lines curve outward, the field is uniform between the two charged plates. Devices consisting of parallel metal plates that contain opposing charges are discussed later in the chapter. Another example is a conducting wire, where the field is essentially confined within the wire, so that over long distances the field lines are essentially parallel to each other.

Example 14.6

Suppose a wire half a meter long is attached to opposite terminals of a 12 V battery. Assuming that the resulting electric field inside the wire is uniform, what is its strength?

This is a straightforward application of Eq. 14–7.

$$E = \frac{-V}{d} = \frac{-12 \text{ V}}{0.5 \text{ m}} = -24 \text{ V/m}$$

The direction of the field is along the wire, towards the negative terminal of the battery, since a positive charge would move in this direction.

For many geometric configurations, electric charges produce fields that are not uniform, and the calculation of the potential is more complex than in the examples we have just cited. Even for a single point charge, for example, the field diminishes with distance. Even though the field may not be uniform, we can always express it in terms of the rate of change of the potential. By considering the limit as the displacement d gets smaller and smaller, so that Eq. 14–7 becomes more and more valid, we see that the local electric field is equal to the negative of the rate of change with distance of the potential. If we plot V versus d, then the field at some displacement d is equal to the negative slope of the plot at that point. Unless the changes all lie in a plane, we must treat the three components of the vectors **d** and **E** independently.

For some simple charge configurations, the methods of calculus can be used to find an explicit expression for the potential. One of these is the single point charge, for which

$$V = \frac{kQ}{r} \tag{14–8}$$

where r is the distance from the charge. Here V is the potential, assuming its value is zero at an infinite distance. (At infinite separation, the force between a pair of charges would be zero, and the potential is defined to be zero there. As the charges are brought closer together, work must be performed on them, and the potential becomes nonzero. This definition of the zero for electrical potential is analogous to the zero for gravitational potential energy due to two point masses, discussed in Chapter 7.)

The potential due to a collection of point charges is simply the algebraic sum of the individual potentials, since the potential is a scalar and has no direction.

14.3 ELECTRICAL ENERGY AND POTENTIAL

Example 14.7 What is the potential at a position midway between two point charges, separated by 0.2 m, if the charges are $+15\mu C$ and $-30\mu C$?

The potential due to the $+15\mu C$ charge is

$$V_1 = \frac{kQ}{r} = \frac{(9.0 \times 10^9 \text{ N} \cdot \text{m}^2/\text{C}^2)(15 \times 10^{-6} \text{ C})}{0.2 \text{ m}} = 6.75 \times 10^5 \text{ V}$$

and that due to the $-30\mu C$ charge is

$$V_2 = \frac{(9.0 \times 10^9 \text{ N} \cdot \text{m}^2/\text{C}^2)(-30 \times 10^{-6} \text{ C})}{0.2 \text{ m}} = -1.35 \times 10^6 \text{ V}.$$

The sum is

$$V = V_1 + V_2 = -6.75 \times 10^5 \text{ V}.$$

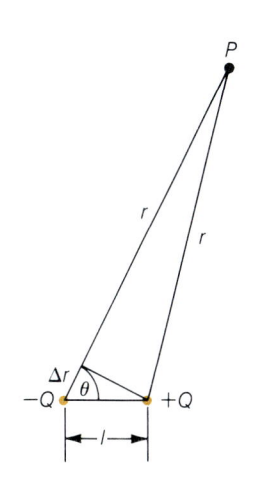

Figure 14.13

The electric dipole.

A charge configuration that often occurs naturally is the **electric dipole**, which consists of two equal but opposite charges separated by distance l (Fig. 14.13). Let us find the potential due to this pair of charges. If we consider an arbitrary point a distance r from one of the charges (and a distance $r + \Delta r$ from the other), then the potential is

$$V = \frac{kQ}{r} + \frac{k(-Q)}{(r + \Delta r)} = \frac{kQ\Delta r}{r(r + \Delta r)}.$$

If our arbitrary point is far from the dipole, so that r is much greater than l, then the sum $r + \Delta r$ in the denominator is approximately equal to r, and the expression becomes

$$V = \frac{kQ\Delta r}{r^2}.$$

If we define θ as the angle formed by the line from our point to one of the charges and the dipole axis, then for large r we see that Δr is approximately equal to $l \cos\theta$, and we have

$$V = \frac{kQl \cos\theta}{r^2}. \tag{14-9}$$

This is a general expression for the potential due to a dipole. The quantity Ql is called the **dipole moment**, p. Hence an equivalent expression for the potential is

$$V = \frac{kp \cos\theta}{r^2}. \tag{14-10}$$

Table 14.2
Dipole moments for molecules

Molecule	Dipole moment (C·m)
HNO_3	7.2×10^{-30}
H_2O	6.2×10^{-30}
NH_3	4.9×10^{-30}
HCl	3.6×10^{-30}
CO	0.37×10^{-30}
CO_2	0.0

Figure 14.14b shows the configuration of the potential due to a dipole. Note that the potential due to a dipole falls off more rapidly with distance (that is, it varies as $1/r^2$) than does the potential due to a point charge (which varies as $1/r$).

Dipoles occur naturally in many diatomic molecules (Table 14.2), which may be regarded as pairs of equal but opposite electrical charges separated by a distance l (recall our earlier description of polarized molecules and dielectric materials). Because chemical reactions are governed by electrical forces, the dipole moment of a molecule, an intrinsic quantity, influences its likelihood of reacting.

It is sometimes useful to visualize the potential due to an electric field by drawing lines indicating surfaces of constant potential. The field created by a single point charge, for example, diminishes uniformly in all directions, so that the potential is constant over the surface of a sphere centered on the point charge. Such a surface is called an **equipotential surface**. Because there is no potential difference from place to place on an equipotential surface, no work is required to move a charged particle along such a surface. Thus a particle moving on an equipotential surface can have no component of motion that is parallel to the electric field, for that would require work; therefore, we see

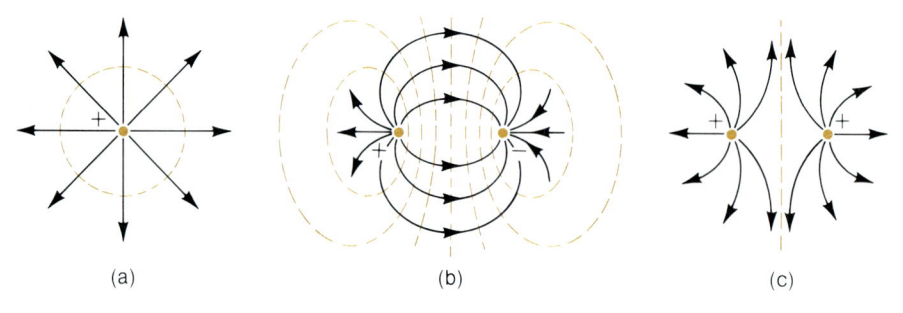

Figure 14.14

Some equipotential surfaces. The dashed lines indicate surfaces of constant potential for each charge configuration shown. Figure (b) is an electric dipole.

14.4 Storage of Electrical Energy

that an equipotential surface is everywhere perpendicular to the electric field. Hence we can draw the equipotential surface easily when the electric field lines are already drawn. Such drawings can help us understand the energies and motions of charged particles in complex electric fields. Fig. 14.14 shows equipotential surfaces for a few common charge distributions.

It can be shown that the surface of any statically charged conducting body is an equipotential surface. In a conductor, charge is free to flow, so if there were any potential differences from place to place, the charge would quickly redistribute itself so that the potential differences are nullified. Hence the entire conductor must be at the same potential, and therefore the surface is an equipotential surface.

14.4 Storage of Electrical Energy

We have mentioned that electricity can be made to do work, and we have studied in some detail the changes in potential energy that can come about when charges are moved in an electric field. Energy is required to create an electric field, and this energy can be recovered when the field is destroyed. Any source of electric charge is a source of energy. In this section we discuss the storage and sources of electric energy.

Capacitors

One way to store electric energy is to create an electric field, since, as we have seen, doing so requires a potential difference which can be created only by supplying energy. A useful device for doing this is a **capacitor**, which consists of two conducting bodies, often in the form of parallel metal plates, placed near each other (Fig. 14.15). When opposing charges are placed on such a pair of conductors, an electric field is established in the space between them. Work is required to put the charges on the conductors and create the electric field; this work becomes potential energy that is stored in the field. The charge is placed on the conductors by some source of energy such as a battery; batteries are described later in this chapter.

If a potential difference exists between the two plates of a capacitor, then they will acquire equal but opposite charges whose magnitude is proportional to the potential difference. In equation form, we write

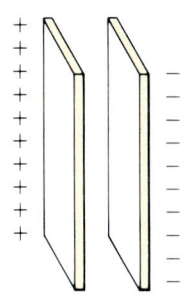

$$Q = CV \qquad (14\text{--}11)$$

Figure 14.15

Parallel plate capacitor. The work required to create the charge separation goes into electric potential energy that is stored in the electrical field between the plates.

where Q is the charge on the capacitor, V is the potential difference between the plates, and C is a proportionality constant called the **capacitance**. The value of C depends on the size and separation of the conducting plates, and is fixed for each capacitor. Its units are C/V, where 1 C/V is called a **farad** (F), after Michael Faraday, a pioneer in the understanding of electric fields. Typical values of C for capacitors used in electronics circuits are in the range of 10^{-12} F (**picofarads**, pF) to 10^{-6} F (**microfarads**, μF).

For a parallel-plate capacitor, the potential between the plates is

$$V = \frac{Qd}{A\epsilon_o}$$

where Q is the charge on the capacitor, d is the plate separation, A is the area of each plate, and ϵ_o is the permittivity of free space, a quantity we defined in our first discussion of Coulomb's law. The capacitance is therefore

$$C = \frac{Q}{V} = \frac{\epsilon_o A}{d}. \qquad (14\text{--}12)$$

If a dielectric is between the plates of a capacitor, which is often the case, then the force between the plates is diminished by a factor K, the dielectric constant. This reduction of the force means that the potential is also reduced by a dimensionless factor K, which means in turn that the capacitance is increased by a factor K:

$$C = \frac{K\epsilon_o A}{d} = \frac{\epsilon A}{d} \qquad (14\text{--}13)$$

where $\epsilon = K\epsilon_o$. As we learned in an earlier section of this chapter, the dielectric constant has the value 1 for a vacuum, is near 1 for most gases, but is much larger for solids and liquids. Normally, paper, for which K is in the range 3–7, is used in capacitors. Paper helps isolate the plates from each other and reduces the chance of accidental discharge (if the charge can jump the gap between plates, then all the energy stored in the capacitor is lost). Furthermore, the dielectric increases the capacitance, so a given plate size and separation yields a greater capacity for storing electric energy.

The energy stored in a capacitor is equal to the work required to place the charge on it. The work required to add charge to a capacitor plate that already has some charge on it is equal to the gain in potential energy when the charge is added. In a circuit that contains a capacitor, a voltage is applied across the gap between the plates; the effect of this voltage is to move charge from one plate to the other. The increase in potential energy when a charge increment Δq is transferred across the gap is equal to the work done; that is,

$$\Delta PE = \Delta q V_i$$

where V_i is the potential difference at the time the charge is transferred. The total work done, equal to the total increase in potential energy, is the product of the total charge Q that is transferred times the average potential difference as the charge is being transferred. We start with zero charge and zero potential, and let V be the final potential difference on the capacitor when the total charge Q has been placed on it. Then the average potential is $V/2$, and the total energy stored in the capacitor is

$$PE = \frac{1}{2}QV. \qquad (14\text{--}14)$$

14.4 STORAGE OF ELECTRICAL ENERGY

Substituting for Q from Eq. 14–11 yields

$$\text{PE} = \frac{1}{2}CV^2 \qquad (14\text{–}15)$$

and replacing C with the expression in Eq. 14–13 gives us

$$\text{PE} = \frac{K\epsilon_o A V^2}{2d}. \qquad (14\text{–}16)$$

Finally, we recall that for a uniform electric field, which is a good approximation for a parallel-plate capacitor, the magnitude of the electric field is related to the potential difference by

$$V = Ed$$

so we can express the stored energy in terms of the field.

$$\text{PE} = \frac{1}{2}K\epsilon_o E^2 A d \qquad (14\text{–}17)$$

This last expression allows us to calculate the energy density u of the field, which is the total energy divided by the volume Ad.

$$u = \frac{1}{2}K\epsilon_o E^2 \qquad (14\text{–}18)$$

Although we have derived this expression explicitly for a parallel-plate capacitor, it is actually a general equation for the energy stored in an electric field. We emphasize, however, that many electric fields vary with position, so that the energy density does also, and is not easy to calculate.

Example 14.8 A 6 V battery is attached to a 20 pF capacitor. What is the resulting charge on each plate of the capacitor, and what is the total energy stored in it? If the space between the plates is filled by a layer of paper 0.1 mm thick whose dielectric constant is $K = 5$, what is the energy density in the gap between plates?

First, the charge on the capacitor is given by Eq. 14–11.

$$Q = CV = (20 \times 10^{-12}\text{ F})(6\text{ V}) = 1.2 \times 10^{-10}\text{ C}$$

The stored energy, from Eq. 14–15, is

$$\text{PE} = \frac{1}{2}CV^2 = (0.5)(20 \times 10^{-12}\text{ F})(6\text{ V})^2 = 3.6 \times 10^{-10}\text{ J}.$$

464 CHAPTER 14 ELECTRICAL CHARGE AND FIELD: ELECTROSTATICS

To calculate the energy density, we use Eq. 14–18, with V/d substituted for E.

$$u = \frac{1}{2}K\epsilon_o\left(\frac{V}{d}\right)^2 = \frac{(2.5)(8.85 \times 10^{-12} \text{ C}^2/\text{N}\cdot\text{m}^2)(6\text{ V})^2}{(1 \times 10^{-4}\text{ m})^2}$$
$$= 0.08 \text{ J/m}^3$$

Batteries

A battery is not, strictly speaking, a device for storing electric energy, but instead is one that produces electric charge by converting other forms of energy (usually chemical) into electricity. It might be more accurate to say that a battery creates charge separation, since we know that charge cannot actually be created. The separated charges from a battery create a potential difference that can be used to do work.

The earliest batteries, as well as most modern ones, take advantage of certain chemical reactions which release free **ions,** which are charged atomic or molecular particles such as electrons (negative ions) or atoms which have lost one or more electrons (which are therefore positive ions). When two materials capable of mutual interaction come into contact, one will gain a net positive charge, and the other a net negative charge. This was discovered in the late eighteenth century by A. Volta, who built the first battery.

A typical battery consists of two **electrodes** imbedded in an **electrolyte** (Fig. 14.16). The electrodes are made of different substances, and the electrolyte is a fluid whose chemical reactions with the electrodes create a net positive charge on one, the **anode,** and a net negative charge on the other, the **cathode** (an **electrolyte** is a substance that separates into free ions when dissolved in water). Thus charges are separated and a potential difference is created between the external projections of the electrodes, which are called the **terminals** of the battery. After a certain amount of charge separation has occurred, the natural repulsion of like charges for each other prevents a further buildup of charge on the two electrodes. Thus each battery has a fixed potential difference. As long as there is some electrolyte left, and the electrodes are not excessively contaminated with products of the reactions, a battery will maintain its normal potential difference because the reactions will always proceed until the normal charge separation is produced. When the two terminals of a battery are connected through an external circuit, charge flows between them, tending to neutralize the potential difference. This allows the chemical reactions to resume, however, so that the potential difference is maintained. Thus a continuous potential difference created by chemical reactions makes the battery a source of electrical energy.

In a typical car battery (Fig. 14.17), the cathodes are made of lead dioxide (PbO_2), the anodes are pure lead, and the electrolyte is sulfuric acid (H_2SO_4), which in water dissolves to form H^+ and $(SO_4)^-$ ions. In an ordinary flashlight battery, the anode is zinc, the cathode carbon, and the electrolyte is a sulfuric

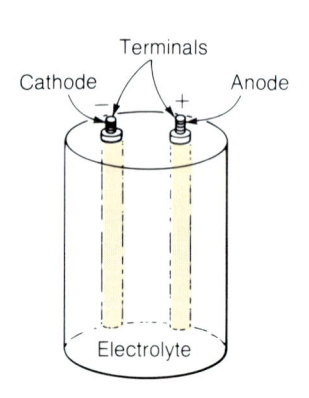

Figure 14.16

A simple battery.

14.4 STORAGE OF ELECTRICAL ENERGY

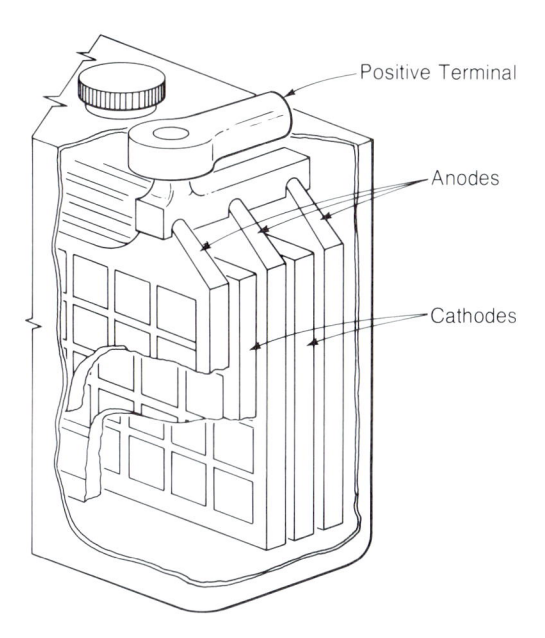

Figure 14.17

An automobile battery.

acid paste. When the electrodes are nickel and cadmium and the electrolyte is potassium hydroxide, the batteries last longer.

Eventually a battery becomes depleted because all the molecules in the electrolyte have undergone reactions, or because the electrodes have become coated with such a thick layer of ions that the interior material is no longer available for further reactions. Many types of batteries can be recharged by allowing current to flow through them in the opposite direction from the flow during normal operation. The reverse chemical reaction restores the electrolyte or the electrodes to their original state, and creates anew the chemical potential that allows charges to be built up on the electrodes. Thus, a battery may be viewed as a storage device for electric energy, since electricity is converted into chemical potential energy to be released later as electric energy.

We note in passing that it is sometimes desirable to cause a substance to be deposited on an electrode by applying a voltage to an electrolyte solution; that is, by the reverse of the battery function. In a process called **electrolysis** (sometimes **electroplating**), metal ions are deposited on an electrode which is a piece of jewelry or tableware, coating it with that metal. If the electrolyte is silver nitrate ($AgNO_3$), for example, the current releases silver ions (Ag^+) which are then attracted to the cathode where they combine with electrons to form silver atoms. If the cathode is a piece of tableware, then the result is a silver-coated piece of tableware, made of some ordinary metal but covered with a thin layer of silver atoms. In other cases the products formed at the electrodes may be free gas molecules. If a dilute acid is the electrolyte, for example, the products of electrolysis are free oxygen and hydrogen molecules.

FOCUS ON PHYSICS

Lightning and Thunder

Nothing is quite so startling as a flash of lightning or a clap of thunder on a warm summer night. These two atmospheric phenomena have been explained by such colorful events as angry war gods or mythological rams butting heads. In fact, both thunder and lightning are the result of the accumulation of large charges and powerful electric currents.

Thunderstorms develop when large updrafts of warm moist air rise to altitudes over 50,000 feet (about 15 km). Most thunderstorms form over land, and scientists estimate that between 1,500 and 3,000 thunderstorms are active around the earth at any time. An important fact about the land-based storms is that they tend to involve cold air. At an altitude of about 5 km, water droplets begin to freeze, when the temperature is around -10 °C. Lightning usually originates near the base of the clouds when the droplets freeze into ice crystals. Through a combination of air currents and droplet-ice interactions, a charge separation develops in the cloud. Positively charged particles are convected to the tops of the clouds, 10–15 km in altitude, while the cloud bases hold a negative charge (Fig. 14.1.1). The cloud acts like a capacitor or a battery, which stores opposite charges on two "plates" separated by 5 to 10 km. Lightning represents the current which flows between these charge separations, transferring anywhere between 1 and 100 Coulombs.

A potential difference of up to 300,000 V can exist between the cloud base and the ground, 5 km below. Different portions of the cloud also have large potential differences. As large as these voltages are, however, they are insufficient to produce a single current discharge, or arc, across the entire 5 km to the ground. An arc will jump across a space of dry air, as in the *Jacob's Ladder* demonstration done in many laboratories, when the

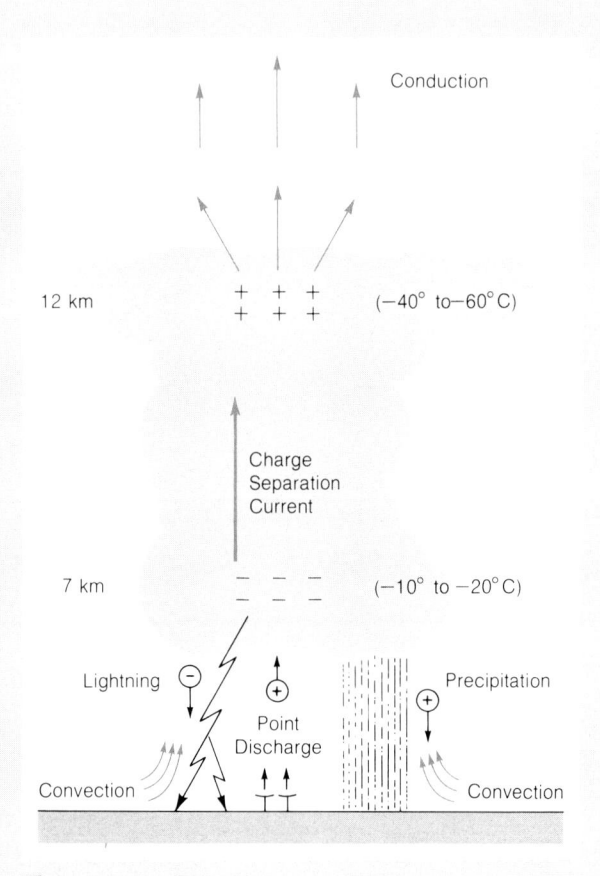

Figure 14.1.1

electric field is 3×10^6 V/m (somewhat less if the air is moist). The average electric field near the Earth's surface is only 100 V/m in fair weather. Instead, the lightning takes many small steps.

The trigger to lightning is believed to be a cosmic ray or radioactivity from gases that leak into the atmosphere from the soil. These ionize an air molecule (O_2 or N_2), producing an electron and a positively charged molecular ion. The electron is accelerated to high speed by the electric fields over small regions in the cloud, which are of the order 10^6 V/m. The electron soon collides with another air molecule, liberating another electron. The process repeats, and quickly an avalanche of 10^7 electrons is produced. The entire process takes about 1 μs and produces a small discharge called a **stepped leader.**

The path of the leader to the ground requires hundreds of these small steps, with pauses of perhaps 50 μs between steps. A channel to the ground is therefore erratic and may take 20 ms to develop. As the leader approaches the ground, the electric field increases, and sparks can jump up from tall objects, such as buildings, trees, or even humans (Fig. 14.1.2). A frightening experience can be to find one's hair standing on end during a thunderstorm. For safety, it is best to avoid standing beneath trees or in high points during these storms.

Once a channel is opened to ground, the lowered resistance allows a large current to flow. Currents of 500,000 amperes are possible, transferring over 100 C during a period of 200 μs. The power in these bolts can be as great as 10^{12} W for a short instant. The energy from this discharge heats the atmosphere in the channel to 30,000 °C and produces pressures of 100 to 1,000 atmospheres. The

Figure 14.1.2

(continued)

channel expands explosively as a shock wave, much like a sonic boom, and we hear the acoustic sound waves as thunder. The more powerful the lightning stroke, the wider the channel and the deeper the pitch of the thunder. Channel widths of an inch or more are common, although it is difficult to see the actual size because of the blinding light.

The sound from the lightning travels at about 340 m/s (1,100 ft/s) and has a range of between 7 and 15 miles, depending on atmospheric conditions and the inclination of the bolt. A good rule for estimating the distance of the lightning is to count the seconds between the flash and when you hear the thunder. The distance is approximately one mile for each 5 seconds delay. Thus, a lightning strike 10 miles away will be heard as thunder 50 seconds later.

Atmospheric scientists are learning more about lightning and the role it plays in the global electric circuit. Many of the models of the ionosphere rely heavily on the charge storage and current discharges that occur in thunderstorms across the large continental land masses.

Perspective

In this chapter we have developed the rudiments of electrostatics, including the concepts of charge, electrostatic force, electric field, and the very important notion of potential difference. We have an inkling of how potential difference can be made to do work, but we will not fully understand this until we have discussed the flow of charge. Thus, the next chapter is devoted to conductors and circuits.

SUMMARY TABLE OF FORMULAS

Electrostatic force (Coulomb's law):
$$F = \frac{kQ_1Q_2}{r^2} \tag{14-1}$$

Permittivity of free space:
$$k = \frac{1}{4\pi\epsilon_0} \tag{14-2}$$

Coulomb's law for a dielectric medium:
$$F = \left(\frac{k}{K}\right)\left(\frac{Q_1Q_2}{r^2}\right) \tag{14-3}$$

Electric field:
$$\mathbf{E} = \frac{\mathbf{F}}{q} \tag{14-4}$$

Potential difference:
$$V = \frac{PE}{q} \tag{14-5}$$

$$V = \frac{W}{q} = -Ed \tag{14-6}$$

Electric field:
$$E = \frac{-V}{d} \tag{14-7}$$

Potential due to a point charge:
$$V = \frac{kQ}{r} \tag{14-8}$$

Potential due to a dipole:
$$V = \frac{kQl\cos\theta}{r^2} \qquad (14\text{–}9)$$

$$V = \frac{kp\cos\theta}{r^2} \qquad (14\text{–}10)$$

Charge on a capacitor:
$$Q = CV \qquad (14\text{–}11)$$

Capacitance:
$$C = \frac{\epsilon_0 A}{d} \qquad (14\text{–}12)$$

Capacitance for a dielectric medium:
$$C = \frac{K\epsilon_0 A}{d} = \frac{\epsilon A}{d} \qquad (14\text{–}13)$$

Energy stored in a capacitor:
$$PE = \frac{1}{2}QV \qquad (14\text{–}14)$$

$$PE = \frac{1}{2}CV^2 \qquad (14\text{–}15)$$

$$PE = \frac{K\epsilon_0 AV^2}{2d} \qquad (14\text{–}16)$$

$$PE = \frac{1}{2}K\epsilon_0 E^2 Ad \qquad (14\text{–}17)$$

Density of energy stored in a capacitor:
$$u = \frac{1}{2}K\epsilon_0 E^2 \qquad (14\text{–}18)$$

THOUGHT QUESTIONS

1. The field of mechanics was first systematically studied in the seventeenth century, but electrostatics was not investigated until about a century later. Why do you think this was so?

2. What evidence do we have that charge is conserved in nature?

3. What is the difference between an electric force and electric field? Is the electric force a vector quantity? What are lines of force?

4. How would you show that there is a basic, indivisible charge in nature?

5. Why does the ability of a material to conduct electricity depend on how rigidly its electrons are held in place? Would you expect a solid, liquid, or gas to be the best conductor?

6. Why is there no electric force inside a conducting sphere?

7. Suppose we lived in an "antimatter universe" in which all charges were reversed. Atoms would have negatively charged nuclei and positively charged electrons (positrons). Would Coulomb's law hold?

8. What is the gravitational equivalent of the electric field E? Compare and contrast the behavior of the electric field with a gravitational field. What is the gravitational equivalent of the electric potential?

9. Why is work done when a charged particle is moved in the presence of an electric field?

10. How might you demonstrate that atoms consist of positively charged nuclei surrounded by negatively charged electrons?

11. Why do you suppose that uniform electric fields are not typical? Can you think of an example, other than charged capacitor plates?

12. Can you think of several processes in the human body that make use of electrical charge? Explain.

13. How can we be sure there is not some equivalent of a gravitational dielectric medium, which exists between the solar system and other stars and which affects gravity the way a dielectric medium affects the electric force between two charges?

14. Can you think of a phenomenon in nature which uses capacitance? Explain.

15. Why are equipotential surfaces interesting and important in electrostatics?

16. How would you demonstrate that the surface of any conducting body is an equipotential surface?

17. What shape capacitor would be most practical: two parallel plates, two concentric spheres, or two colinear cylinders? Explain why, or describe an even better shape.

18. Can you think of another way to create a battery besides chemical reactions?

19. Suppose you wished to construct a series of batteries from the same substance, all having different voltages. What parameters would you vary?

20. Explain how lightning works. Where is the charge stored?
21. Explain why you receive a shock when walking on wool carpet in the winter.
22. Why does a record attract dust more readily after it has been played?
23. Why is most matter electrically neutral to a very good approximation?
24. Explain why the electric potential can be zero, and yet an electric force and electric field exist. If the potential is constant in a region of space, what is the value of the electric field?
25. In electronic circuits, the voltages or potentials are usually measured with respect to "ground." Why does connecting a ground wire to a large body (such as the earth) function as a good reference point?

PROBLEMS

Sections 14.1–14.2 Electric Charge and Forces

1. What is the force between charges of 5 C and -8 C, separated by 10 m? Is the force attractive or repulsive?
2. In Problem 1, where would you place a third charge, of 7 C, so that the net force on the -8 C charge is zero?
3. How much charge would be needed to balance the gravity of a proton if the charge is held at a distance of 1 cm?
4. Find the magnitude of the electrostatic force on a 15 C charge at one corner of an equilateral triangle having sides 2 m and charges 20 C and -20 C located at the other corners.
5. Calculate the electric field at the 15 C charge in Problem 4.
6. Find the magnitude and direction of the net force on a 5 C charge arising from three other charges of -10 C placed at the other three corners of a square 1 m on a side.
7. Calculate the electric field on the 5 C charge in Problem 6.
8. Find the magnitude of the electric field 1 m from a proton.
9. If the radius of a proton is about 10^{-15} m, find the electric charge density.
10. An electric field has polar coordinates $E_r = E$ and $E_\theta = 0$. Find the rectangular components E_x and E_y for this field.
11. Find the magnitude and direction of the electric field due to two 10 C charges at a point 0.5 m from each charge. The two charges are separated by 0.5 m.
12. Two charges exert an attractive force of 10 N on each other. How must the distance be changed to double the force?
13. If the distance between charges in Problem 12 is tripled, what will the force be?
14. A charge of 8 C is placed 50 cm from a second charge of 8 C. What is the magnitude and direction of the force?
15. A body with 20 C charge is held 40 cm above the floor. A second body of mass 0.5 kg is placed so that it is in equilibrium, 30 cm directly below the first charge. What charge does it carry?
16. Calculate directly the value of ϵ_0 from k.
17. If the separation of a proton and electron in an atom is 3×10^{-11} m, find the electrostatic force between them.
18. Calculate the charge on 1 mole of electrons.
19. Calculate the charge on 1 mole of singly ionized hydrogen ions, H^+.
20. A person walks across a thick carpet and picks up an excess charge of 10^{10} electrons. How much mass can the person lift electrostatically, if that mass picks up an equal but opposite charge and is 1 cm away?
21. Find the force between two spheres having a charge of 100 C and separated by the distance between the earth and moon (384,000 km). The force on two spheres acts as though the charge was concentrated at the centers.
22. Three 10 C charges are placed at corners of an equilateral triangle, 50 cm on a side. What is the direction and magnitude of the force exerted on a -10 C charge placed in the plane of the triangle, 50 cm from two of the charges, so the array is now a parallelogram (Fig. 14.18)?
23. Two charges are placed 1.5 m apart and have charges 10 C and 15 C. Where must a 5 C charge be placed so that the net force on it is zero?
24. Two charges of 20 C and -10 C are placed 80 cm apart. What is the magnitude and direction of the

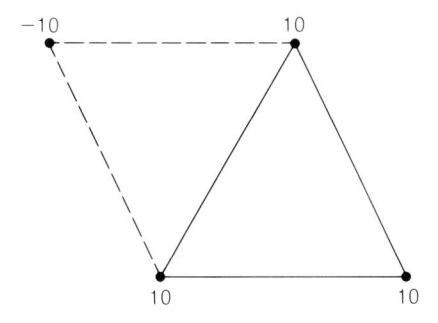

Figure 14.18

(Charges: −10 and 10 on top, 10 and 10 on bottom)

net force exerted on a 5 C charge placed 20 cm beyond the −10 C charge on a line extended from the first two?

•• 25 Four 5 C charges are placed in a plane square array, 0.5 m on a side. Where must a −5 C charge be placed if the net force on it is zero?

•• 26 How would your answer in Problem 25 change if the test charge had been +5 C instead of −5 C? Explain.

••• 27 To a good approximation, most matter is electrically neutral. Suppose two 60 kg people 1 m apart had excess charges amounting to 1 part in 10^{12} of their protons. Calculate the net force between the two individuals and compare it to their weights.

••• 28 If the earth and moon each had an excess electric charge, of one part in 10^x of their protons, find the value of x at which their electrostatic force would equal their attractive gravitational force.

• 29 A 15 C charge is placed 80 cm from a −20 C charge. What force is exerted if a substance having dielectric constant $K = 20$ is inserted between them?

• 30 Two 5 C charges separated by 40 cm are suddenly immersed in a gas with dielectric constant 5. To what distance must the charges be separated so that the electric force between them has the same initial value?

• 31 A pair of charges is placed in a gas having unknown dielectric constant K. If the charges must be moved from 1 m to 0.2 m apart to regain the initial value of the electrostatic force, what is K?

• 32 Find the electric field due to a 3 C charge at a distance of 2 m.

• 33 How far away must a 3 C charge be so that its field is 1 N/C?

•• 34 How will the results of Problems 32 and 33 be altered if the charges are placed in a medium with dielectric constant 25?

• 35 Find the electric field due to two 8 C charges 80 cm apart at a location halfway between.

•• 36 Two charges of 4 C and −8 C are placed 80 cm apart. What is the electric field at a point 20 cm from the −8 C charge on the side opposite the 4 C charge?

•• 37 How would the results of Problems 35 and 36 be affected if the charges were placed in a medium of dielectric constant 15?

• 38 The electric field of a certain charge decreases by a factor of eight when it is placed in a certain gas. What is the dielectric constant?

•• 39 In an old model of the hydrogen atom, the electron moves in a circular orbit of radius 0.529×10^{-10} m about the proton. The electron has charge $-e$ and the proton $+e$. The masses are $m_p = 1.67 \times 10^{-27}$ kg and $m_e = 9.11 \times 10^{-31}$ kg. Find the ratio of the electrostatic force and gravitational force between proton and electron. Does gravity play a role in atomic physics?

•• 40 For the electron orbit in Problem 39, calculate the velocity needed for equilibrium between electrostatic attraction and centripetal acceleration.

Section 14.3 Electric Energy and Potential

• 41 Show that the units of V/m are equivalent to N/C.

• 42 The potential difference across a 0.22 mm slab is 1,000 volts. What is the electric field?

• 43 How much voltage is required to slow an electron moving at $0.5c$ to rest?

• 44 Find the magnitude and direction of the electric field inside a 9 cm long wire connected to the terminals of a 1.5 volt dry-cell battery.

• 45 The electric field in a wire 30 cm long is 20 N/C. What is the terminal voltage of the battery to which the wire is attached?

• 46 Find the electric potential of a 5 C charge which is 20 cm distant.

•• 47 Find the electric potential at a point midway between two charges, 8 C and −4 C, separated by 80 cm.

• 48 How would the results in Problem 47 be altered if the charges were immersed in a dielectric gas with $K = 4$?

•• 49 Two charges of 10 C and −15 C are placed 60 cm apart. Is there any location along the line between the two where the electric potential is zero?

• 50 A scientist wishes to have a field of 30 N/C within a wire connected to a 12 volt power supply. How long should the wire be?

- 51 What potential difference will accelerate an electron to $0.1c$?
- 52 Over what distance will an electric field of 1 N/C produce a potential difference of 5 volts?
- 53 What is the ratio of the electric potential of a charge Q viewed at distances of 2 m and 18 m?
- ••• 54 Sketch a diagram of the force field lines and the equipotential surfaces of two charges, $+Q$ and $-Q$, separated by distance d. Why are the surfaces perpendicular to the field lines?
- ••• 55 Suppose the earth was held in orbit by electrostatic forces rather than by gravity. If the sun has charge $+Q$ and the earth $-Q$, what is Q? Assume the earth's orbital speed is 30 km/s at a distance 1 AU $= 1.5 \times 10^{11}$ m.
- ••• 56 In the ground state of the hydrogen atom, an electron orbits a proton at a distance 0.529×10^{-10} m. What is the electron's velocity? What are its electrostatic PE and its orbital KE? Show that PE $= -2$ KE, and $E = -$PE$/2$ where E is the total energy. How much energy is needed to ionize the atom; that is, how much energy will remove the electron to infinity?
- •• 57 In a television tube, electrons are accelerated across plates with 25 kV potential difference (1 kV = 1,000 V). Find the final velocity of the electron in m/s and as a fraction of $c = 3 \times 10^8$ m/s.
- • 58 What is the speed of a 100 eV electron?
- • 59 What is the speed of a 1 MeV proton?
- •• 60 A lightning bolt transfers 200 C of charge to earth in a 1 km long channel. If the electric field in the cloud is 10^4 volts/cm, what is the total voltage? If the stroke lasts 200 microseconds, what is the power?

Section 14.4 Capacitors and Batteries

- • 61 How much charge can be stored on one plate of a 1 μF capacitor connected to a 12 volt battery?
- • 62 Two plates of a capacitor store $\pm 10^{-3}$ C when connected to a power supply of 1 kilovolt. What is the capacitance of the plates?
- • 63 Two parallel plates with 10 pF capacitance are separated by 1 mm of dielectric with $K = 20$. What area must the plates have?
- •• 64 The electric field between two capacitor plates is 10^6 V/m. Their area is 10^{-3} m^2 and the dielectric material has $K = 5$. What is the charge stored on one plate?
- • 65 A 100 pF capacitor has area 100 cm^2 and separation 1 mm. What is the dielectric constant of the material separating the plates?
- • 66 A 100 μF capacitor is connected to a 100 volt power supply. How much energy is stored by the plates?
- • 67 A 10 pF capacitor stores a charge of 10^{-8} C. How much energy is stored?
- •• 68 Two parallel plates of 1 cm^2 area are separated by 1 mm of air. How much energy is stored if the electric field is 10^5 V/m?
- • 69 Dry air will break down in an electrostatic discharge when the electric field exceeds about 3×10^6 V/m. What is the maximum amount of charge that can be stored on capacitor plates of 10^{-2} m^2 area separated by air?
- ••• 70 In a capacitor connected to a 100 volt battery, you wish to store 10^{-4} C charge. What size capacitance is needed? If the plates are separated by air, what minimum area is required to avoid breakdown of the air? (See Problem 69.)
- • 71 An air capacitor with square plates 1 cm on a side has a capacitance 2 μF. If paper with $K = 5$ is inserted between the plates, by what factor must the area be decreased to keep the capacitance the same?
- •• 72 Find the energy density between the plates of a capacitor with $K = 5$, hooked up to a 12 volt battery. The plates are separated by 1 mm.
- • 73 A capacitor is made from two 100 m square plates, separated by 1 km on the moon's surface (a very good vacuum). Find the capacitance in F.
- ••• 74 What is the electric field between the plates of a capacitor which can store 10^{-6} J when connected to a 24 volt battery? The plates have 0.1 m^2 area, and the dielectric constant is $K = 5$.
- ••• 75 A ⌣ volt battery is attached to a 20 μF capacitor which has a dielectric constant $K = 5$. The dielectric is suddenly pulled out. How much work is required? Assume the voltage, plate area, and plate separation are fixed.
- ••• 76 Two colinear, concentric conducting cylinders have length L, radii a and b, and surface charges $+Q$ and $-Q$, respectively. If the dielectric constant of the material between them is K, find the capacitance. Assume that the length L is much greater than either a or b and that a and b are nearly the same $(b > a)$.

SPECIAL PROBLEMS

1. Electroscope

One method of measuring charge is to allow like charges to accumulate on adjacent leaves of an electroscope (Fig. 14.19). Suppose charges of $+Q$ are deposited on two balls of mass m, suspended at the end of thin, massless rods of length L. The Coulomb repulsion holds the balls apart at an angle θ.

(a) Write the equations of force balance on the balls for the apparatus as shown.
(b) Show that tension in the rod must balance the forces of gravity and electrostatic repulsion in the direction along the rod.
(c) From force balance in a direction perpendicular to the rod, find an expression for Q in terms of m, L, g, and θ.
(d) If the balls have mass 1 g each, $L = 10$ cm, and $\theta = 20°$, what is Q?

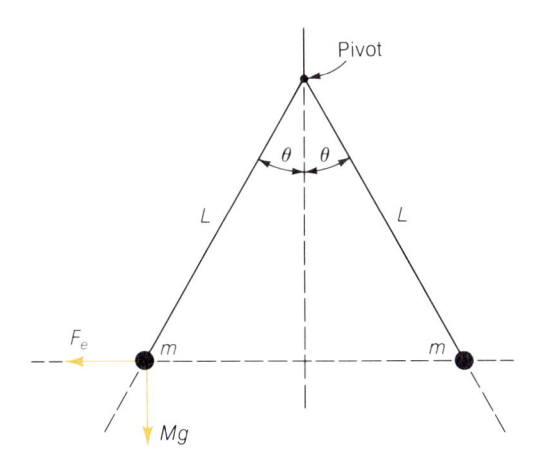

Figure 14.19

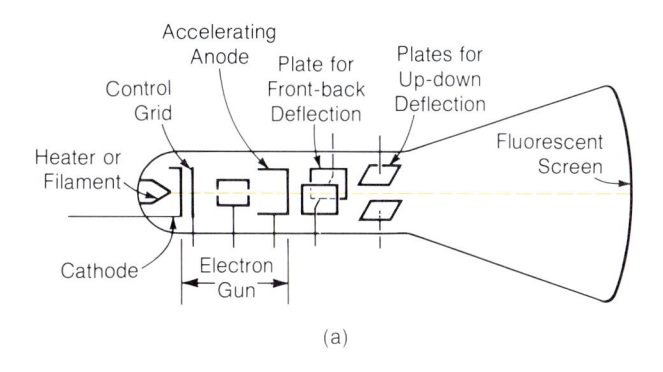

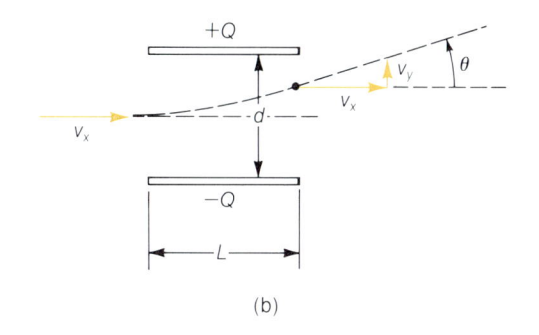

Figure 14.20

2. Cathode Ray Tube

Figure 14.20a is a diagram of a cathode ray tube (CRT) similar to those used in oscilloscopes and computer terminals, and related to the television picture tube. Electrons are accelerated in the "electron gun" to voltages V_1 between 20 and 30 kV, and then deflected horizontally or vertically by small plates before striking the fluorescent screen.

(a) What is the horizontal velocity v_x of an electron after passing through the electron gun with $V_1 = 25$ kV?
(b) What is the vertical force F_y and vertical acceleration a_y (Fig. 14.20b) imparted by the deflection plate of voltage $V_2 = 5$ kV? If the deflection plates have dimensions $L = 3$ cm and $d = 1$ cm, what is the vertical velocity v_y acquired by the electron as it passes through?
(c) What is the deflection angle θ?

3. Millikan Oil Drop Experiment

About 1913, R. A. Millikan measured the charge on the electron and showed that charge was always a multiple of $e = 1.6021892 \times 10^{-19}$ C (positive or negative). His experiment consisted of balancing the electrostatic force on charged oil drops with gravity and with viscous drag. The experiment has since been repeated with latex balls and even with superconducting niobium balls, as part of a search for fractionally charged quarks (theory predicts quarks to have charges $\pm e/3, \pm 2e/3$). Figure 14.21 shows Millikan's apparatus. Oil is sprayed in drops into the region between charged capacitor plates (voltage V, separation d). An oil drop of radius r and density ρ_0 is brought to rest between the plates, balanced between the forces of gravity $\mathbf{F}_g$ and electrostatic attraction $\mathbf{F}_e$. The charge on the drop is q.

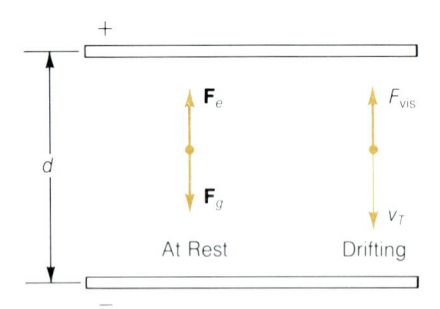

Figure 14.21

(a) When the drop is at rest, show that $q = mgd/V$, where m is the mass of the drop. Write an expression for q in terms of ρ_0, V, g, d, and r. All of these parameters are easy to measure except r.

(b) Now, the oil drop is allowed to drift through air (viscosity η) at a terminal velocity v_T given by Stokes law (Section 11.3). Derive an expression for r in terms of η, v_T, ρ_0, and g. Would buoyancy forces be important?

(c) Combine your results of parts a and b to find q in terms of the known parameters.

(d) An oil drop of density 0.8 g/cm³ can be held at rest between plates with electric field $V/d = 6.46 \times 10^5$ V/m. The drop then drifts through air at terminal velocity 10^{-4} m/s. What is the drop's radius? How many excess electrons are on the drop? Air has viscosity $\eta = 1.8 \times 10^{-5}$ N·s/m² at 20 °C. What does your result signify?

4. Penetrating the Nucleus

When a positively charged particle, such as a proton or alpha particle, is fired into an atom, it easily passes through the electron cloud but is repelled by the positively charged nu-

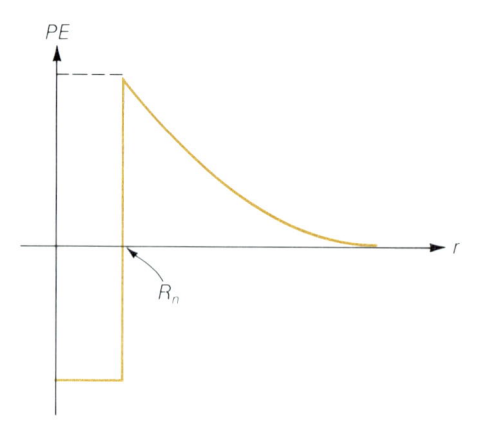

Figure 14.22

cleus. To penetrate the nucleus, the projectile must have sufficient energy to overcome the electrostatic repulsion, known as the "Coulomb barrier." Figure 14.22 shows the potential energy curve for a positively charged particle approaching the nucleus. The electrostatic potential energy, kQ_1Q_2/r, increases as the particle approaches the nuclear radius, $R_n = (1.2 \times 10^{-15} \text{ m})A^{1/3}$, where A is the atomic mass number. The potential energy then plummets as the particle falls into the nucleus, which has negative PE because of the strong nuclear binding forces.

(a) Estimate the energy (in MeV) of the Coulomb barrier of the $^{12}_{6}C$ nucleus, which has charge $6e$ and mass number $A = 12$, and a colliding proton.

(b) What velocity must be given to a proton to penetrate the carbon nucleus? What velocity must be given to an alpha particle (a $^{4}_{2}He$ nucleus)?

(c) Estimate the energy required for a proton to penetrate the uranium nucleus ($^{235}_{92}U$). The enormous size of the barrier explains why nuclei are better penetrated by uncharged neutrons.

SUGGESTED READINGS

"Electrostatics." A. D. Moore, *Scientific American*, Mar 72, p 47, **226**(3)

"Thunder." A. A. Few, *Scientific American*, Jul 75, p 80, **233**(1)

"Robert A. Millikan." D. J. Kevles, *Scientific American*, Jan 79, p 142, **240**(1)

"Franklin's physics." J. Heilbron, *Physics Today*, Jul 76, p 32, **29**(7)

"My work with Millikan on the oil drop experiment." H. Fletcher, *Physics Today*, Jun 82, p 43, **35**(6)

A source book in physics. William F. Magie, Harvard University Press, Cambridge, 1965

The discovery of subatomic particles. Steven Weinberg, Freeman, NY, 1984

The flying circus of physics with answers. Jearl Walker, Wiley, New York, 1977

Great scientific experiments. Rom Harre, Phaidon Press, Oxford, 1981 (J. J. Thomson and the discovery of the electron)

"The Mechanism of Lightning", L. B. Loeb, in *Atmospheric Phenomena: Readings from Scientific American*, Freeman, San Francisco, p 99, 1980

"Thunder", A. A. Few, in *Atmospheric Phenomena: Readings from Scientific American*, Freeman, San Francisco, p 111, 1980.

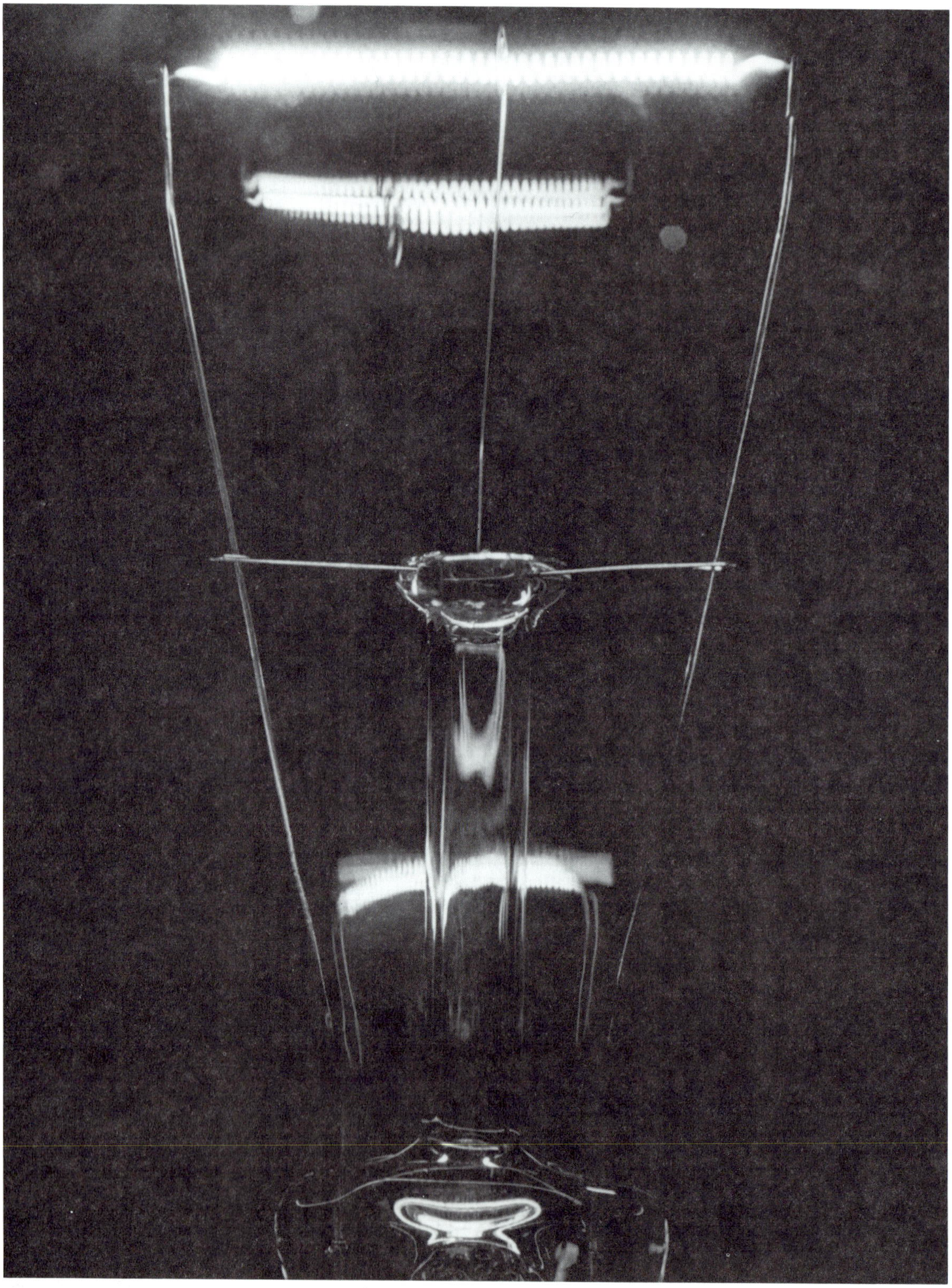

Chapter 15

Current and Circuits

We made several references in the previous chapter to the conduction of electricity, the flow of electric charge that can occur when there is a potential difference between two points connected by a conducting medium. Such a flow is called a **current,** and in this chapter we will discuss how currents interact with components in electric circuits.

Our discussion here will focus entirely on **direct-current** (DC) circuits, which are circuits in which the current flows steadily in one direction. In a later chapter we will discuss **alternating-current** (AC) circuits, in which the flow reverses periodically. Alternating currents are commonly used in households and businesses, but the reasons for this have little to do with the principles of how circuits operate. The information gained in this chapter on direct-current circuits will prove useful when we analyze modern electric machinery.

We will begin with sources of electric potential, which are required in order to make current flow, and then we will discuss several important rules and components of circuits before using all this information to analyze applications.

15.1 Electromotive Force

We have seen that no current will flow in a conductor that is left to its own devices. Since it is free to move about, any charge in the conductor will arrange itself so that there is no potential difference between any two points, and then there will be no further flow. If there were a potential difference, for example, and some charge were added at some localized point, the force exerted by the electric field due to the potential difference would quickly move the charge to negate the potential difference.

To make a current flow, therefore, we must have a source of potential difference. This would be some device that could create and maintain a voltage, so that as charge flowed, the potential difference between points on a conductor would not be neutralized. We have already described one such device, the battery. Others, to be discussed after we understand electromagnetic induction, include generators and alternators. For now we will restrict ourselves to currents driven by batteries. The exact nature of the source of potential is not important for understanding how circuits work.

The potential difference created by a battery or other device is called an **electromotive force**, abbreviated **emf** and usually designated $\mathscr{E}$. Its units are those of potential difference, that is, volts. A source of emf is a device that converts some other form of energy into electric potential difference. In a battery, as we have seen, chemical potential energy is converted; in a generator, it is mechanical energy; and other devices convert other forms of energy such as heat and sunlight into electric potential difference.

15.2 Current

Under the proper conditions, charge will flow along a conductor, creating a current. A source of emf is a minimum requirement if a current is to flow in a conductor. Another requirement is that there be a path along which charge can flow between the points of unequal voltage at the source of emf. This path is called a **circuit** (Fig. 15.1).

Since current is the flow of charge along a conductor, its dimensions are charge per unit time. The SI unit of charge is the coulomb, which was defined in the previous chapter, so the units of current are coulombs per second. The special name **ampere** is given to these units. One ampere (A) is equal to one coulomb per sec. This is not the technical definition of the ampere, however. We recall, in fact, that the coulomb was defined in terms of the ampere, which is considered the more fundamental unit. The ampere is actually defined in terms of the magnetic force exerted by two current-carrying wires upon each other, something we will not discuss until the next chapter. For now we must continue to be content to know only the numerical magnitude of the unit.

Since currents usually consist of flowing electrons, it is interesting to see how many electrons are involved in typical currents. The number of electron charges in one coulomb is 6.2422×10^{18}, so a current of one ampere is equivalent to a flow of 6.2422×10^{18} electrons per second.

We have stated the magnitude of our unit of current, but have said nothing about the sign. Charge can be either positive or negative, so we must adopt a convention for assigning a sign to a current. Here the rather arbitrary historical adoption of the electron charge as negative leads to a minor inconsistency: the actual flow of particles in most currents is said to be a **negative** current, because the particles carry negative charges. We usually define the direction in which the current flows as the direction a **positive** charge would flow (Fig. 15.1).

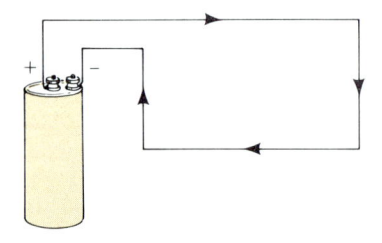

Figure 15.1

A simple circuit. The potential difference between the terminals of a battery causes a current to flow. The arrows indicate the direction of flow for positive charges, the conventional definition of current direction.

15.3 RESISTANCE

Hence in an electric circuit, where electrons are flowing in one direction, we say that the current is in the opposite direction.

Sometimes the flow actually does consist of positive charges or positive and negative charges, as in particle accelerators, in electrolyte solutions such as we discussed in the previous chapter, and in **plasmas,** which are hot gases where electrons and positive ions are not bound together in atoms, but move independently.

When we defined electric potential energy in the previous chapter, we established the sign by saying that the potential increases when a test charge approaches a positive charge. When there is a potential difference between two points, work is required to move a positive charge from the region of lower potential to the region of higher potential. Thus, a positive charge will flow away from the higher potential toward the lower potential, so a positive current flows from the positive terminal on a battery to the negative one. Thus, our convention for defining current fits in nicely with our intuitive sense of current flow when a battery is the source of emf in a circuit.

15.3 Resistance

We have described current as though it were perfectly conservative, that is, as though no energy were lost as current flowed through a circuit. Although occasionally this is very nearly true, in most real circuits it is not. Devices are commonly inserted into circuits to convert electric energy into other forms, but even if this is not done, some energy is lost. It is an intrinsic property of currents.

Heat and Current: Joule's Law

When electrons flow through a conductor, they are impeded to some extent by forces exerted by the atoms of the conducting material. The exact mechanism by which these forces are exerted is a quantum-mechanical effect, and its description is beyond the scope of this chapter. The result, however, is something we are prepared to understand: energy is transferred to the atoms of the conductor, increasing its internal energy. We know from our discussions in Chapters 12 and 13 that an increase in internal energy corresponds to an increase of temperature. Thus, the flow of current through a conductor causes heating of the conductor (Fig. 15.2).

A conductor that resists the flow of current is called a **resistor;** as we have said, any conductor is a resistor. The degree to which a conductor acts as a resistor depends on its atomic or molecular properties. We say that a material is a good conductor if it does not strongly resist the flow of current. Certain metals, especially copper and silver, are good conductors.

The energy converted into heat when a current flows cannot be converted back into electric energy (except by a machine that requires even more energy

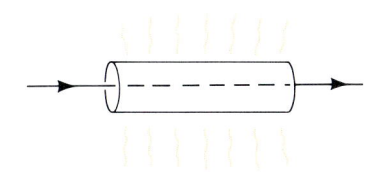

Figure 15.2

Resistance. Here current flows through a resistor and converts electrical energy into heat.

for the operation), so we say that the energy has been dissipated. This is in accord with the second law of thermodynamics (recall our discussion in Chapter 13). The first person to evaluate the amount of energy dissipated was James Joule, who measured the heat produced by various currents. He found that the heat produced per unit time, which is equal to the energy dissipated per unit time (which in turn is equal to the power), is proportional to the square of the current. In mathematical terms,

$$P = I^2 R \tag{15-1}$$

where P is the power, I is the current, and R is a proportionality constant called the **resistance**. In the SI system, the unit of power is the watt (1 J/s), so the resistance is measured in W/A^2, which are given the special name **ohms** (Ω). Thus $1\Omega = (1 \text{ W})/(1 \text{ A})^2$.

The relationship among power, current, and resistance given by Eq. 15–1 is known as **Joule's law,** and is the first of several simple relationships that we will use to analyze circuits. Joule's law is an empirical law, not a fundamental relationship that is always precisely correct; nevertheless it is quite accurate for practical purposes.

Example 15.1

A small electric coil is marketed as a heating device for water (Fig. 15.3). Its resistance is 15Ω, and it draws a current of 8 A. How much power is required to operate it? How long does it take the coil to increase the temperature of a half-liter of water by 50C°?

The answer to the first question requires a straightforward application of Eq. 15–1.

$$P = I^2 R = (8 \text{ A})^2 (15\Omega) = 960 \text{ W}$$

The second question requires that we know how much heat is needed to reach the higher temperature. We recall our application of the first law of thermodynamics (Chapter 12), where we set the heat required, Q, equal to the heat gained by the water when its temperature rises,

$$Q = m_W c_W (\Delta T)$$

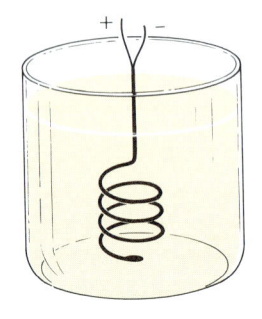

Figure 15.3

where m_W is the mass of water, c_W is the specific heat of water, and ΔT is the temperature change. Thus we have

$$Q = (500 \text{ g})(1 \text{ cal/g} \cdot \text{C°})(50 \text{ C°}) = 25{,}000 \text{ cal} = 1.05 \times 10^5 \text{ J}.$$

Finally, the time required to produce this quantity of heat is

$$t = \frac{Q}{P} = \frac{(1.05 \times 10^5 \text{ J})}{(960 \text{ W})} = 109 \text{ s} = 1.8 \text{ min}.$$

15.3 RESISTANCE

Resistance and Potential Difference: Ohm's Law

Since we have characterized resistance as the tendency of a conductor to inhibit the flow of current, we might expect that the amount of current that flows through a conductor subjected to a given potential difference should depend on the resistance of the conductor. We would expect that the amount of current would be less for a greater resistance.

To see what the dependence of current on resistance is, we consider the work required to move a charge Q across a potential difference V. We learned in the previous chapter that the work required is

$$W = QV$$

when there is a source of emf, so that the potential is constant. We can consider the work required per unit time,

$$\frac{W}{t} = \frac{QV}{t}$$

which is equivalent to

$$P = VI \qquad (15\text{--}2)$$

since the work per unit time is the power P, and the charge per unit time is the current I. Next we take advantage of Joule's law (Eq. 15–1) to substitute for P, then

$$I^2 R = IV$$

or

$$I = \frac{V}{R}. \qquad (15\text{--}3)$$

Thus there is indeed a simple relationship between current and resistance of the sort we expected; for a given potential difference, the current is inversely proportional to the resistance. This expression is known as **Ohm's law,** and is named after G. S. Ohm, the German scientist who discovered it experimentally in the early nineteenth century. This relation is often written in the form

$$V = IR$$

which says that voltage is linearly proportional to current. Like Joule's law, Ohm's law is an idealized expression, and is highly accurate only for metals. Even so, it is very useful, because nearly all circuits of interest in practical applications have metal components.

We add Eqs. 15–2 and 15–3 to our repertoire of expressions that can be used in analyzing circuits. We know from Eq. 15–3 that whenever a current flows through a circuit or any segment of one, the current is given by the ratio V/R, where V is the potential difference across that segment and R is the resistance. If the potential difference is maintained by a source of emf, then $V = \mathscr{E}$, and the current is equal to the ratio $\mathscr{E}/R$. From Eq. 15–2 we have a convenient relationship between power and current for a given potential difference (or emf). Now we can use these relationships to solve problems.

Example 15.2 How much current flows through a 100 W light bulb that operates on ordinary household voltage, 120 V? Assume for the sake of the example that the current is constant (DC), not alternating (AC).

We are given the power P and the potential difference V, so we need only solve Eq. 15–2 for I.

$$I = \frac{P}{V} = \frac{100 \text{ W}}{120 \text{ V}} = 0.83 \text{ A}.$$

Example 15.3 The starter motor in a car draws a current of 100 A during the short time when the car is being started, and this motor drives the engine. Cars operate on 12 V electrical systems; that is, the emf of the battery is 12 V. What is the resistance of the circuit from the battery through the starter motor? How much energy is dissipated, if the motor operates for 5 s as it starts the engine?

We use Eq. 15–3 here to find

$$R = \frac{\mathscr{E}}{I} = \frac{12 \text{ V}}{100 \text{ A}} = 0.12 \Omega$$

The power dissipated is calculated from Joule's law (Eq. 15–1).

$$P = I^2 R = (100 \text{ A})^2 (0.12 \Omega) = 1{,}200 \text{ W}$$

The energy consumed in 5 s is

$$Q = Pt = (1{,}200 \text{ W})(5 \text{ s}) = 6{,}000 \text{ J}.$$

This example is unrealistic, because it assumes that all the power in the current goes into heat dissipation. This would actually be true only at the first instant as the motor is turned on, before it begins to rotate. After that, more and more of the power goes into rotating the motor. This will be discussed later, in Chapter 17.

Resistivity and Resistors

Resistors with accurately known values of R (Fig. 15.4) are available. Furthermore, it is often necessary, when designing circuits, to know the resistance of each component, even those that consist only of wire leads. This is important

15.3 RESISTANCE

Figure 15.4

Standard resistors. Calibrated resistors are inexpensive and can be purchased in large quantities. Color-coded stripes indicate the values of their resistances.

for estimating the heat that will be dissipated and the equilibrium operating temperature. Also, the potential difference across each component in a circuit driven by a source of emf is affected by the resistance of other components. For these reasons, it is useful to measure resistances.

In principle it is possible to determine R for a resistor as Joule did, by measuring the power dissipated when a known current is applied. This is not very practical, however, and normally it is much easier to use Ohm's law (Eq. 15–3) and measure the current for a known potential difference. Devices for measuring current, called **ammeters,** are described later in this chapter.

The resistance of any resistor depends not only on the intrinsic properties of the material from which it is made, but also on its shape. If we maintain our intuitive view of resistance as the tendency of a conductor to inhibit the flow of current, we see that the resistance should be proportional to the length of the resistor, and inversely proportional to its cross-sectional area. Resistance should be proportional to the length because electrons that travel a greater distance will be more effectively inhibited. Resistance should be universely proportional to cross-sectional area because the number of paths available to an electron increases as the area increases. Experiment bears out these expectations, and we have the relationship

$$R = \frac{\rho L}{A} \qquad (15-4)$$

where L is the length of the resistor, A is the cross-sectional area, and the constant of proportionality ρ is called the **resistivity**. The resistivity is a fundamental property of the conducting material and can therefore be measured and tabulated for various substances, so that the resistance can be calculated

Table 15.1
Coefficients of resistivity

Substance	α (C°)$^{-1}$	ρ(20°C) (Ω·m)	ρ$_0$(0°C) (Ω·m)	Melting point (C°)
Conductors				
Gold	0.0083	2.24 × 10^{-8}	1.92 × 10^{-8}	1064
Copper	0.0068	1.68 × 10^{-8}	1.48 × 10^{-8}	1083
Iron	0.00651	9.71 × 10^{-8}	8.59 × 10^{-8}	1535
Silver	0.0061	1.59 × 10^{-8}	1.42 × 10^{-8}	962
Tungsten	0.0045	5.47 × 10^{-8}	5.02 × 10^{-8}	3400
Platinum	0.003927	10.6 × 10^{-8}	9.83 × 10^{-8}	1772
Aluminum	0.00429	2.65 × 10^{-8}	2.44 × 10^{-8}	660
Mercury	0.00089	95.8 × 10^{-8}	94.1 × 10^{-8}	−38.8
Nichrome (Fe, Cr, Ni alloy)	0.0004	100. × 10^{-8}	99.2 × 10^{-8}	1500
Semiconductors				
Silicon	−0.07	(0.1–2000)		1410
Germanium	−0.05	(1–500) × 10^{-3}		937
Carbon (graphite)	−0.0005	(1–60) × 10^{-5}		3550
Insulators				
Glass		10^9 − 10^{14}		
Rubber		10^{13} − 10^{15}		
Wood		10^8 − 10^{11}		
Quartz (fused)		7.0 × 10^{14}		

for a resistor of known size and shape. Some example data are given in Table 15.1, where we see that the units of resistivity are ohm-meters (Ω · m). We see that resistivities have an enormous range, approaching infinity for the kinds of materials known commonly as electrical **insulators**. Just as no conductor has zero resistivity, no insulator has infinite resistivity, however, so there will always be some current flowing when a potential difference exists between different parts of an object. The role of an insulator is to keep the current so small that it is negligible.

As in many other cases of intrinsic properties of materials, the resistivity varies with temperature. For most substances, the variation is regular and can be represented in similar fashion to the thermal expansion of solids (Chapter 12). Hence we have

$$\frac{\Delta \rho}{\rho} = \alpha \Delta T \tag{15-5}$$

where Δρ/ρ is the fractional change in the resistivity for temperature change ΔT, and α is a proportionality constant called the **temperature coefficient of resistance**. If we define a standard temperature (usually 0°C) so that the resis-

15.3 RESISTANCE

tivity at this temperature is ρ_0, and denote as ρ_T the resistivity for temperature T (in C°), then simple algebraic manipulation of Eq. 15–5 leads to

$$\rho_T = \rho_0(1 + \alpha T). \qquad (15\text{–}6)$$

This form is most convenient to use, since values of ρ_0 and α are available in Table 15.1.

Since the resistance of a conductor is proportional to the resistivity, we could have derived Eqs. 15–5 and 15–6 in terms of R instead of ρ, which would have yielded

$$R_T = R_0(1 + \alpha T). \qquad (15\text{–}7)$$

In Chapter 12 we mentioned a type of thermometer called a **thermistor,** which provides information on temperature by measuring how resistance varies with temperature. Now we see how such a device works. What is needed is a calibrated scale (as is true of all thermometers), and a material that survives the temperature extremes and other environmental effects to which it will be exposed. One great advantage of thermistors is their compactness (they are only small segments of wire); another advantage is that they can easily be used remotely, since the only data measurement, of the current in the thermistor, can be carried out by rather simple measurement devices. Thermistors are often placed in equipment that is used in remote places (such as in space; Fig. 15.5).

It is interesting to see what happens to the resistivities of metals for temperatures approaching absolute zero. The resistivities of many (but not all) metals abruptly become very nearly zero when the temperature falls below a certain value called the **transition temperature.** Transition temperatures for metals are usually below 20 K, and may be as small as a few tenths of a kelvin. Materials whose resistivities have been lowered in this fashion are called **superconductors,** and have many potentially important applications. Extremely large currents can flow in superconductors without dissipating much heat, so they can profitably be used to carry very large currents (such as in very large electromagnets), or to transmit current over very large distances. Research in superconductor applications is a very active field today.

Semiconductors are a class of materials that have moderate resistivities and negative temperature coefficients of resistance (that is, the resistance decreases with increasing temperature). Such materials are vital to electronics technology (Chapter 17).

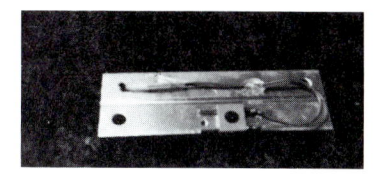

Figure 15.5

A thermistor in place on a portion of a space experiment. This is part of an instrument flown into space on a sounding rocket. The thermistor is the tiny dark segment at the end of the wire.

Example 15.4 A 12 V battery operates a circuit containing 3 m of copper wire, which has a 1 mm radius. How much power is dissipated, if the copper wire is the only source of resistance in the circuit, and the temperature is 20°C?

The resistivity, from Eq. 15–6, is

$$\rho_T = \rho_0(1 + \alpha T)$$

CHAPTER 15 CURRENT AND CIRCUITS

where we see from Table 15.1 that $\rho_0 = 1.48 \times 10^{-8}\ \Omega \cdot m$ and $\alpha = 0.0068(C°)^{-1}$. Therefore we find, for $T = 20°C$,

$$\rho_{20} = [1.48 \times 10^{-8}\ \Omega \cdot m][1 + (0.0068 C°^{-1})(20 C°)]$$
$$= 1.68 \times 10^{-8}\ \Omega \cdot m.$$

The length of the wire, L, is 3 m, and the cross-sectional area is

$$A = \pi r^2 = \pi(0.001\ m)^2 = 3.14 \times 10^{-6}\ m^2$$

so the resistance is

$$R = \frac{\rho L}{A} = \frac{(1.68 \times 10^{-8}\ \Omega \cdot m)(3\ m)}{(3.14 \times 10^{-6}\ m^2)} = 0.0161\ \Omega.$$

Now we can calculate the power dissipated, by using Eq. 15-3 to find the current,

$$I = \frac{V}{R} = \frac{12\ V}{0.0161\ \Omega} = 745\ A$$

and then, using Eq. 15-1,

$$P = I^2 R = (745\ A)^2(0.0161\ \Omega) = 9{,}000\ W.$$

This is an enormous amount of power, and the wire will quickly heat beyond its melting point (Table 15.1). Fuses operate on this principle, consisting of wire segments that will melt and break if too much current is allowed to pass through (Fig. 15.6). Ordinarily this happens only when some component in the circuit loses its resistance because it fails or is accidentally circumvented by a misplaced conductor. When that happens, we say that a **short circuit** has occurred. Without the protection of a fuse (or a circuit breaker), a short circuit can be very dangerous because of the high temperatures created by the enormous current, as this example shows.

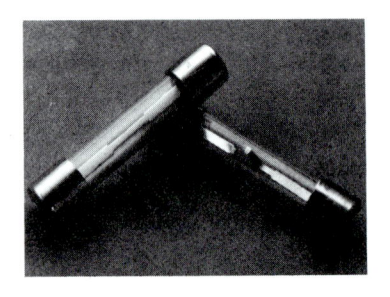

Figure 15.6

Fuses. Here are two fuses, one intact, and one that has "blown" due to a high current.

Example 15.5 A spacecraft in orbit has a thermistor consisting of a 2 cm segment of platinum wire whose cross-sectional area is $2.6 \times 10^{-9}\ m^2$. What is the temperature of the thermistor when the current through it is 18 A? The spacecraft electrical system operates on 12 V.

We use Eq. 15-3 to find R_T, where T is the unknown temperature.

$$R_T = \frac{V}{I} = \frac{12\ V}{18\ A} = 0.667\ \Omega$$

Now we use Eq. 15-4 to find the resistivity ρ_T.

15.4 ELEMENTS OF CIRCUITS

$$\rho_T = \frac{R_T A}{L} = \frac{(0.667\,\Omega)(2.6 \times 10^{-9}\,\text{m}^2)}{(0.02\,\text{m})} = 8.67 \times 10^{-8}\,\Omega \cdot \text{m}$$

Finally, we solve Eq. 15–6 for T.

$$\begin{aligned}
T &= \frac{\rho_T - \rho_0}{\rho_0 \alpha} \\
&= \frac{(8.67 \times 10^{-8}\,\Omega \cdot \text{m} - 9.83 \times 10^{-8}\,\Omega \cdot \text{m})}{(9.83 \times 10^{-8}\,\Omega \cdot \text{m})(0.003927(\text{C}°)^{-1})} \\
&= -30°\text{C}
\end{aligned}$$

15.4 Elements of Circuits

Electrical circuits can be designed to perform a broad range of useful tasks. In every case, electrical energy is converted into some other form by some device, perhaps as simple as a light bulb filament or as complex as a motor (an electric motor is actually quite simple, as are all components). The role of the circuit is to provide the electric energy in the necessary quantity at the proper time. This requires resistors, capacitors, batteries and other circuit elements, along with a few simple rules. Here we will not discuss the energy-conversion devices, but will instead focus on the circuit elements used to deliver the required current to them.

It is useful to introduce some standard notation before we proceed. It is common practice to use symbols to represent the components in circuit diagrams. A source of emf is symbolized by one or more pairs of parallel lines perpendicular to the line representing the circuit, with the longer of the parallel lines representing the positive side of the potential difference (Fig. 15.7). A capacitor is indicated by a very similar symbol, except that the two parallel lines are of equal length, and one of them may be curved (Fig. 15.8). A resistor is shown by a simple jagged line (Fig. 15.9), and a variable resistor (called a **rheostat**) is indicated by a resistor symbol with an arrow crossing or pointing to it (Fig. 15.10). It is usually assumed that connecting wires have negligible resistance. Junctions between wires, where electrical contact is made and cur-

Figure 15.7

Standard symbol for a source of emf.

Figure 15.8

Standard symbols for a capacitor.

Figure 15.9

Standard symbol for a resistor.

Figure 15.10

Symbols for rheostats.

Figure 15.11

A heavy dot indicates a circuit junction.

Figure 15.12

Standard symbols for a connection to ground.

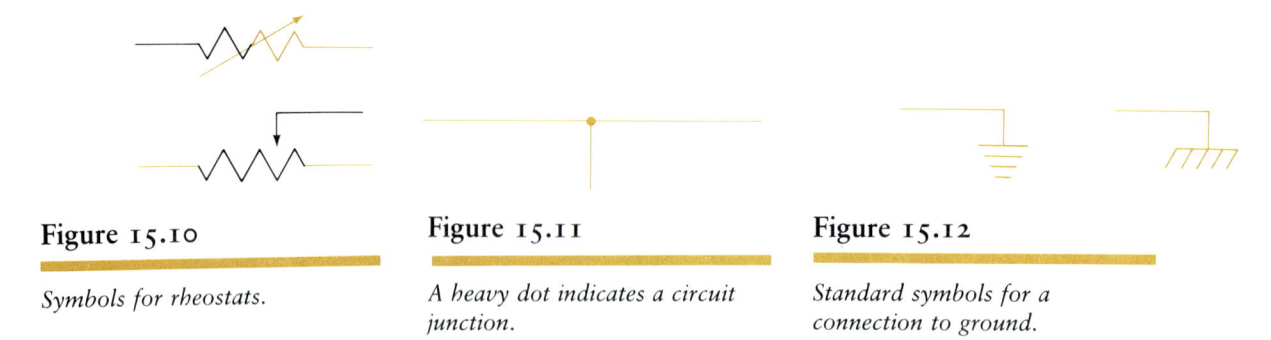

rent is allowed to flow, are indicated by heavy dots (Fig. 15.11). Finally, if a circuit is "grounded", which is symbolized by either of two forms of symbol (Fig. 15.12), the potential is defined to be zero where the circuit is grounded.

In a schematic drawing the components are drawn in the sequence in which they are encountered by the current. The actual physical position and spacing, however, is not reproduced to scale, and usually does not matter, so long as the relationships are properly represented.

Kirchhoff's Rules

In the previous chapter we learned that certain conservation laws apply to electricity: the conservation of charge, and the conservation of energy. Now we will see how these laws translate into principles of operation for circuits. In the process we will see strong analogies with the behavior of fluids.

Gustav Kirchhoff, a German scientist working in the mid-nineteenth century, recognized a pair of important applications of conservation principles to the flow of current in circuits. (Kirchhoff's name will come up again in our studies. He was a pioneer in understanding the emission and absorption processes for light.) The first of Kirchhoff's rules is based on the principle of conservation of charge:

The algebraic sum of the currents entering any junction point in a circuit is zero.

This states that charge cannot accumulate at any junction in a circuit and is analogous to the equation of continuity that we discussed in our study of fluids (Chapter 11). In equation form we write

$$\Sigma I = 0. \tag{15-8}$$

where the Σ symbol means "the sum of". As we will see, this equation is useful in finding the currents flowing through junctions in circuits.

15.4 ELEMENTS OF CIRCUITS

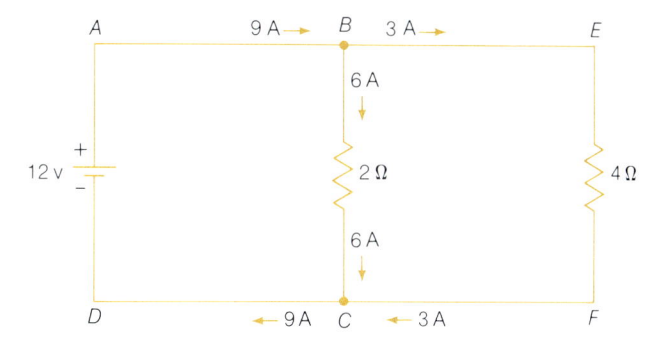

Figure 15.13

Kirchhoff's first rule. The sum of the currents at a junction is always zero. For example, at B and C the total current entering is equal to the total current leaving.

A simple circuit (Fig. 15.13) can be used to illustrate Kirchhoff's first rule. It is easy to see that the total current into the junction points B and C is zero, since the currents entering and leaving each of these points add to zero. It may not be so easy to see why the currents have the values indicated. Somehow the current flowing in a circuit adjusts itself so that Kirchhoff's first rule is satisfied; Kirchhoff's second rule tells how this is so.

Kirchhoff's second rule deals with the total potential change as current flows around a circuit:

The algebraic sum of the changes in potential around any closed loop is zero.

This is a statement of the conservation of energy, since a change in potential is a change in potential energy. We recall that work is required to move a charge against a potential difference; if it were possible to move a charge around a circuit without doing work, then it would be possible to obtain more work from a circuit than was put in. This would violate the principle of conservation of energy. Again, we can write an equation representing this law:

$$\Sigma \mathscr{E} - \Sigma IR = 0 \tag{15-9}$$

where the first term on the left represents the sum of the gains in potential due to sources of emf in the circuit, and the second term represents the sum of the drops in potential due to the passage of current through resistances.

Changes of potential in a circuit are created by sources of emf and by elements that have resistance, which change the potential by the quantity $\Delta V = IR$. The current will always adjust itself so that the total change in

490   CHAPTER 15   CURRENT AND CIRCUITS

potential is zero; we saw in an earlier example in this chapter that this can lead to a massive current if there are no substantial resistances in a circuit. To determine the current in each segment of a circuit, we first draw the circuit and the components that can cause changes in the potential, then use Kirchhoff's second rule to find each current. We trace the circuit in one direction, accounting for every rise or fall in potential.

We can see how this works by referring again to Fig. 15.13, this time imagining that only the emf and the resistances are known. There are three closed loops in this circuit. Let us first consider loop ABCD. In the segment BC the potential drop across the resistor must equal the gain in the segment DA, which is equal to the emf, 12 V. Hence the product IR for the resistor in segment BC must equal -12 V. The resistance is 2Ω, so the current must therefore be 6 A. The product IR is negative because the resistor always causes a drop in potential if we follow the circuit in the direction of the current. Similarly, in loop AEFD, the product IR for the resistor in segment EF must equal -12 V. Since the resistance is 4Ω, the current in that segment must be 3 A. Finally, we consider loop BEFC, and find the fall in potential in segment EF must be matched by a gain in potential in segment CB. We see that this condition is met because as we follow the loop in the clockwise direction, we move against the current in segment CB, so that the potential rises as we encounter the resistor there. Hence the total change in potential around this loop is zero, as it should be. We could have analyzed this circuit just as well by tracing the loops in the opposite direction; it might be a good exercise for the student to do so.

Note that the source of emf determines the direction of all changes in potential in a circuit, because it governs the direction of flow of the current. Resistors have no intrinsic preferred direction; they always act to decrease the potential in the direction of current flow, or, equivalently, to increase the potential in the direction opposite to the current flow.

Sometimes in complex circuits it is necessary to write algebraic equations for the various loops, and to solve them simultaneously. This happens when some or all of the closed loops contain more than one resistor, so that we cannot find the current independently by examination of a single loop. We must instead rely on the interaction of the loops with each other. It is always possible to write as many independent equations as there are unknowns, so long as the resistances and emfs are given. There is some art to deciding which equations are simplest to write and solve, but this can best be learned by experience.

Example 15.6   Consider the circuit shown in Fig. 15.14a. We want to find the current at all points in the circuit. All resistances (R_1, R_2, R_3, and R_4) are given, and the currents I_{BE}, I_{CF}, I_{DE}, and I_{GH} are unknown.

We will analyze this circuit by brute force. We will find in the next section that some simplifications can be made, but it is useful to go through the analysis without these simplifications first, because in some circuits such simplifications are not possible.

15.4 ELEMENTS OF CIRCUITS

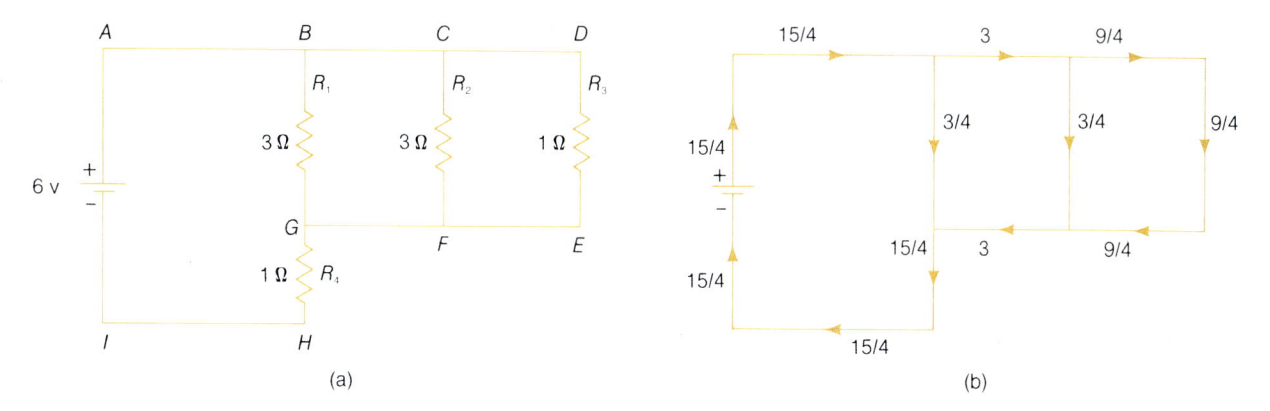

Figure 15.14

We need to find a set of independent equations such that the number of equations equals the number of unknown quantities. We can use both of Kirchhoff's rules to find such equations, but we must avoid the pitfall that not all of the possible equations are independent. We can write equations for loops ABGHI, BCFG, and CDEF, but no new information would be added by also writing equations for loops ABCDEFGHI, ABCFGHI, or BCDEFG. These equations would contain no new relationships among the unknowns (this is generally true when loops are nested inside larger loops; it is usually most straightforward just to write equations for the innermost loops).

We start by writing three independent equations implied by Kirchhoff's second rule. For the loop ABGHI, we have the equation

$$\mathcal{E} - I_{BG}R_1 - I_{GH}R_4 = 0$$

where I_{BG} and I_{GH} represent the currents in segments BG and GH. Substituting the known values for $\mathcal{E}$, R_1, and R_4, we have

$$6 - 3I_{BG} - 1I_{GH} = 0. \tag{a}$$

Similarly, for loops BCFG, and CDEF we have

$$-3I_{CF} + 3I_{BG} = 0 \tag{b}$$

and

$$-1I_{DE} + 3I_{CF} = 0. \tag{c}$$

We need an additional equation in the same four unknowns, which we can find from Kirchhoff's first rule.

$$I_{BG} + I_{CF} + I_{DE} = I_{GH} \qquad (d)$$

(If you have trouble seeing this one, note that the circuit would be exactly equivalent if junction points E, F, and G were all merged into a single junction.)

Now we have four independent equations in four unknowns, so there is sufficient information available to find the four unknown currents. Ordinarily we assume that the student can solve such a set of equations, but we provide some guidance in this case.

A good first step is to use equations (b) and (c) to express I_{CF} and I_{DE} in terms of I_{BG}, and then to substitute the results into equation (d), which then becomes an expression for I_{GH} in terms of I_{BG} alone: $I_{GH} = 5I_{BG}$. We then substitute for I_{GH} in equation (a), which can then be solved for the value of $I_{BG} = 3/4$ A. From this, a value for I_{GH} is immediately found: $I_{GH} = 5I_{BG} = 15/4$ A. Then equation (b) can be solved for I_{CF}, yielding $I_{CF} = 3/4$ A. Finally, equation (c) yields $I_{DE} = 9/4$ A.

We can find the current in each of the remaining segments of the circuit (those, such as AB, BC, etc., that contain no resistors) by using Kirchhoff's first rule. Without showing the details (left as an exercise), we find $I_{HI} = I_{IA} = I_{AB} = I_{GH} = 15/4$ A; $I_{BC} = I_{FG} = 3$ A, and $I_{CD} = I_{EF} = 9/4$ A. The current flow is shown schematically in Fig. 15–14b. Note how the current is divided among segments to keep the total flow constant at 15/4 A.

Resistors in Combination

As we have just seen, the analysis of a circuit containing many resistors can be quite burdensome. We now show that sometimes a combination of resistors can be represented by a single equivalent resistor, simplifying the analysis.

When resistors are connected in series, with no junctions that can draw current away (Fig. 15.15), then the current through each of the resistors is the same, and the total drop in potential of the series is equal to the sum of the individual drops at each resistor. Hence we have

$$V = IR_1 + IR_2 + IR_3 + \cdots = I(R_1 + R_2 + R_3 + \cdots).$$

That is, the net effect of a series of resistors is equivalent to a single resistor whose resistance is equal to the sum of the resistances. Hence the equivalent resistance R_s for a series of resistors is

$$R_s = R_1 + R_2 + R_3 + \cdots. \qquad (15\text{--}10)$$

In Figure 15.16, a number of resistors are connected in parallel, meaning that they are connected in a loop so that the potential drop across each is the same. In this case the current is divided among the circuit segments according to the resistances. If the total current is I, then

$$I = I_1 + I_2 + I_3 + \cdots = \frac{V}{R_p}$$

15.4 ELEMENTS OF CIRCUITS

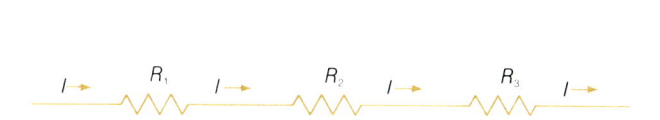

Figure 15.15

Resistors in series. The net resistance is equal to the sum of the individual resistances.

Figure 15.16

Resistances in parallel. The reciprocal of the net resistance is equal to the sum of the reciprocals of the individual resistances.

where R_p represents the equivalent resistance of the entire assemblage of parallel resistors. Solving for $1/R_p$ yields

$$\frac{1}{R_p} = \frac{I_1 + I_2 + I_3 + \cdots}{V} = \frac{I_1}{V} + \frac{I_2}{V} + \frac{I_3}{V} + \cdots \qquad (15\text{--}11)$$
$$= \frac{1}{R_1} + \frac{1}{R_2} + \frac{1}{R_3} + \cdots$$

The equivalent resistance is found from its reciprocal, which is equal to the sum of the reciprocals of the individual resistances.

Example 15.7 Analyze the circuit in Example 15.6 (Fig. 15.14) again, this time using the formulas for resistors in combination.

We note that resistors 1, 2, and 3 in the circuit are in parallel, so we may use Eq. 15–11 to find the equivalent resistance.

$$\frac{1}{R_p} = \frac{1}{R_1} + \frac{1}{R_2} + \frac{1}{R_3} = \frac{1}{3\Omega} + \frac{1}{3\Omega} + \frac{1}{1\Omega} = 1.67\Omega^{-1}$$

Therefore the equivalent resistance is

$$R_p = 0.6\Omega.$$

Now we have a circuit that consists of a single loop, with two resistors, one the equivalent of the three in parallel and, in series with that, the resistor R_4. Fig. 15.17 shows a new schematic diagram. The equivalent of these two resistances in series is

$$R_s = R_p + R_4 = 0.6\Omega + 1\Omega = 1.6\Omega.$$

Now we find the current in the loop.

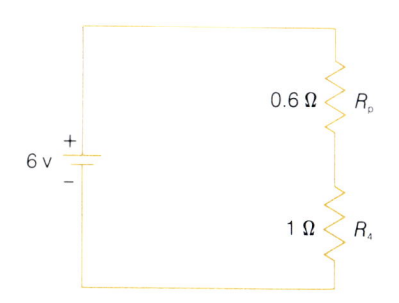

Figure 15.17

$$I = \frac{V}{R_s} = \frac{6\text{ V}}{1.6\Omega} = 3.75\text{ A}$$

This is the current in loop ABGHI, except for segment BG, where this current is divided over the three parallel resistors R_1, R_2, and R_3. Note that this answer matches our result in Example 15.6, obtained much more laboriously.

We can also easily find the current through each of the three parallel resistors, since each is subjected to an equal potential difference, which is the product IR_p, where I is the current that is divided over the three parallel resistors and R_p is the equivalent resistance. The value of R_p is 0.6Ω, and the value of I is 3.75 A, so the potential difference across the three resistors is 2.25 V. Now we solve for the current through each resistor.

$$I_1 = \frac{V}{R_1} = \frac{2.25 \text{ V}}{3\Omega} = 0.75 \text{ A} \qquad \text{(segment BG)}$$

$$I_2 = \frac{V}{R_2} = \frac{2.25 \text{ V}}{3\Omega} = 0.75 \text{ A} \qquad \text{(segment CF)}$$

$$I_3 = \frac{V}{R_3} = \frac{2.25 \text{ V}}{1\Omega} = 2.25 \text{ A} \qquad \text{(segment DE)}$$

The same values of I were found in Example 15.6.

Internal Resistance and Sources of emf in Combination

It can happen that a circuit contains more than one source of emf. Most batteries consist of series of cells, each of which is by itself a source of emf, so even if there is only a single battery, we are implicitly dealing with sources of emf in combination.

When the sources are in series, then the result is quite straightforward: the net potential difference across the series is equal to the sum of the potential differences of the individual sources (Fig. 15.18). This is seen by considering the work that must be done to transport a charge through the series of potential differences, which is analogous to the work done when an object is lifted against the earth's gravitational field. Whether the object is lifted in steps or

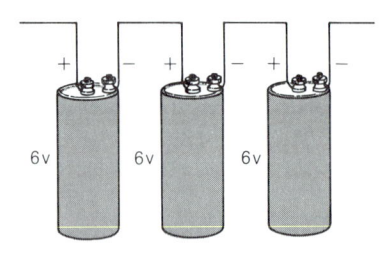

Figure 15.18

Emfs in series. The total emf, 18 V, is equal to sum of the individual values.

15.4 ELEMENTS OF CIRCUITS

all at once, the potential energy gained is the same, so long as the total height increase is the same.

The situation is more complicated when sources of emf are used in parallel. The only easy case is when the sources have equal potential differences, because then the potential difference across the set of parallel sources of emf is equal to that of any of the individual sources.

It may not be clear why sources of emf would be used in parallel, since the net potential difference is the same as if only one source were used. The reason is that batteries always have **internal resistance,** which decreases the actual potential difference V between the terminals, according to

$$V = \mathcal{E} - Ir \qquad (15\text{--}12)$$

where r is the internal resistance. Normally r is quite small, so that the voltage across the terminals is close to the emf of the battery, but as batteries age or if very large currents are drawn, the product Ir may become significant.

When batteries are connected in parallel, then the current is divided among them, so that the product Ir for each is reduced. Thus, a set of batteries in parallel is better able to maintain a potential difference equal (or nearly so) to the emf of the batteries. This technique is especially useful when currents are very large.

Example 15.8 An electric motor, powered by a set of parallel 12 V batteries, requires a minimum of 10.5 V to operate properly. It draws a current of 200 A. If the internal resistance of each battery is 0.04Ω, how many are needed?

We substitute $V = 10.5$ V, $\mathcal{E} = 12.0$ V, and $r = 0.04Ω$ in Eq. 15–12, and solve for I.

$$I = \frac{\mathcal{E} - V}{r} = \frac{12.0 \text{ V} - 10.5 \text{ V}}{0.04Ω} = 37.5 \text{ A}$$

Thus, each battery will have a terminal voltage below the required 10.5 V if the current drawn from it exceeds 37.5 A. Therefore the current drawn from each battery must be less than 37.5 A. Since the required total current is 200 A, and (200 A)/(37.5 A) = 5.33, at least six batteries are needed.

Capacitors in Combination

We know that capacitors can be used to store charge, although we will not see the practical reasons for doing so until we discuss time-dependent circuits such as alternating circuits (after all, when a continuous current flows through a circuit containing a capacitor, it is not at equilibrium because the charge on the capacitor keeps building up). Nevertheless it is appropriate to discuss capacitors in combination here, because the treatment is similar to that of resistors and sources of emf in combination.

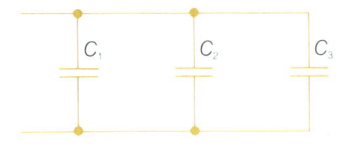

Figure 15.19

Capacitors in parallel. The net capacitance is equal to the sum of the individual capacitances.

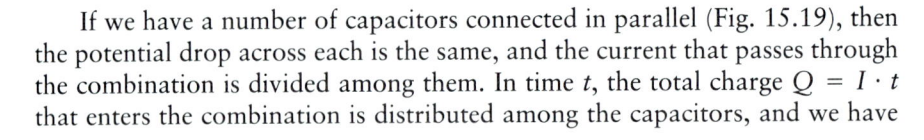

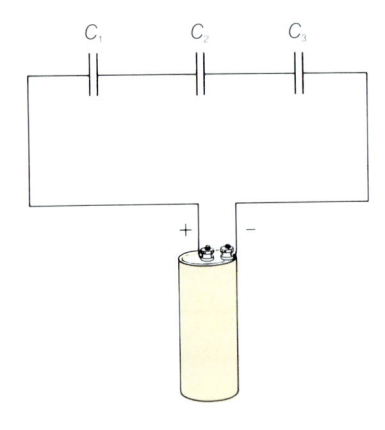

Figure 15.20

Capacitors in series. The reciprocal of the net capacitance is equal to the sum of the reciprocals of the individual capacitances.

If we have a number of capacitors connected in parallel (Fig. 15.19), then the potential drop across each is the same, and the current that passes through the combination is divided among them. In time t, the total charge $Q = I \cdot t$ that enters the combination is distributed among the capacitors, and we have

$$Q = Q_1 + Q_2 + Q_3 + \cdots$$

where $Q_1, Q_2, Q_3, \ldots$ are the charges stored on the individual capacitors. From Eq. 14–11, we know that for each capacitor $Q = CV$, where C is the capacitance. Then we have

$$C_p V = C_1 V + C_2 V + C_3 V + \cdots$$

where C_p is the equivalent capacitance of the set of parallel capacitors. This leads immediately to

$$C_p = C_1 + C_2 + C_3 + \cdots . \qquad (15\text{–}13)$$

When capacitors are connected in series, the conservation of charge requires that the total charge entering the series equal the charge on each individual capacitor. To see this, imagine a series of capacitors connected between the terminals of a battery (Fig. 15.20). The charge leaving the positive terminal must equal the charge entering the negative terminal. On each capacitor, positive charge accumulates on the plate nearest the positive terminal of the battery, and an equal but negative charge is induced on the other plate. This in turn creates an equal but positive charge on the near plate of the next capacitor in the series. At the last capacitor in line, an equal but negative charge is induced, which flows back to the battery.

The total drop in potential across a series of capacitors equals the sum of the individual potential drops, or

$$V = V_1 + V_2 + V_3 + \cdots$$

From Eq. 14–11, we therefore have the relation

$$\frac{Q}{C_s} = \frac{Q}{C_1} + \frac{Q}{C_2} + \frac{Q}{C_3} + \cdots$$

where C_s is the equivalent capacitance of the series. Since the charges on the individual capacitors are each equal to the total charge entering the series, we now divide by this charge Q.

$$\frac{1}{C_s} = \frac{1}{C_1} + \frac{1}{C_2} + \frac{1}{C_3} + \cdots \qquad (15\text{–}14)$$

When a source of emf is applied to a capacitor, charge will accumulate on the capacitor until the potential across it is equal to the emf. Thus, current will not flow after the capacitor is charged. In most practical applications, capacitors are used to regulate a variable current or voltage. The analysis of such circuits is beyond the scope of the present discussion.

15.5 Experimental Analysis of Circuits

We have seen how circuits may be constructed from components, and we have learned some of the rudiments of analyzing them. By working the examples, and especially by solving the problems at the end of the chapter, the student will acquire proficiency in mathematically analyzing circuits. It remains to be seen, however, how the numerical values for such analyses are measured.

Measuring Current: The Ammeter

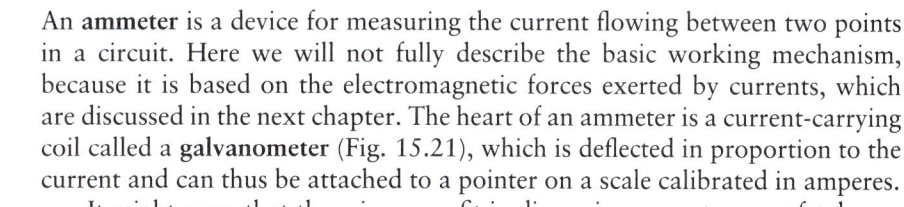

An **ammeter** is a device for measuring the current flowing between two points in a circuit. Here we will not fully describe the basic working mechanism, because it is based on the electromagnetic forces exerted by currents, which are discussed in the next chapter. The heart of an ammeter is a current-carrying coil called a **galvanometer** (Fig. 15.21), which is deflected in proportion to the current and can thus be attached to a pointer on a scale calibrated in amperes.

It might seem that there is no profit in discussing ammeters any further at this point, but we can discuss how some of the principles learned in this chapter apply. A practical ammeter should measure currents over a wide range, and we see that this could be difficult on a single scale, since we may encounter currents as small as a few microamperes (μA) all the way up to many amperes, a factor of more than a million.

In order to be sensitive to currents over a large range of values, an ammeter employs resistors in parallel with the galvanometer to reduce the current. The galvanometer is designed so that it can measure very small currents directly, but cannot respond properly to large currents. To measure larger currents, a parallel resistor (called a **shunt resistor**) is switched into the circuit so that a known fraction of the current is diverted through it. The galvanometer itself has a resistance, so the division of current between the two parallel circuit segments (see Fig. 15.22) is determined by the relative values of r, the galvanometer resistance, and R, the shunt resistance.

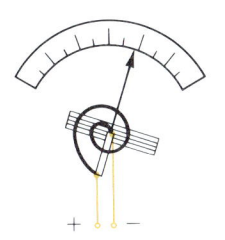

Figure 15.21

The galvanometer. This is a coil whose deflection is proportional to the current.

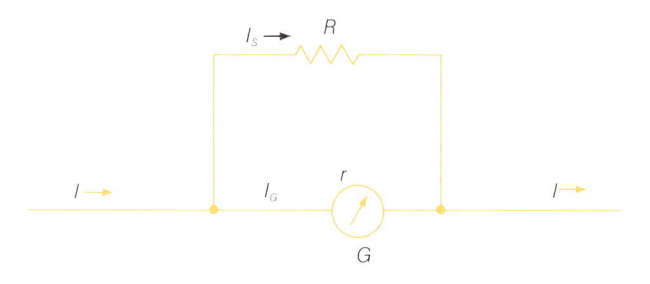

Figure 15.22

The ammeter. The shunt resistor R is used in parallel with the galvanometer G to divide the current so I_G does not exceed the galvanometer's range.

The potential difference is the same for both the galvanometer and the shunt, so we have

$$I_G r = I_S R \qquad (15\text{--}15)$$

where I_G is the current through the galvanometer, and I_S is the current through the shunt. The value of R must be chosen so that I_G falls within the range where the galvanometer responds properly. An ammeter usually has a rotary switch to select the shunt resistance R according to the range of current being measured. When reading the dial, the scale for which the switch is set must be kept in mind.

Example 15.9 When set on its most sensitive scale (with no shunt resistor), an ammeter reads 1 μA at its maximum value. On another setting, it reads 1 A at the same position. If the internal resistance of the galvanometer is $r = 100\Omega$, what must the value of the shunt resistance be when the switch is set for the 1 A scale?

If a current of 1 μA flows through the galvanometer, the rest of the 1 A current must flow through the shunt. Hence $I_G = 1$ μA, and $I_S = 1$ A $-$ 1 μA $= 0.999999$ A. We solve Eq. 15–15 for R.

$$R = r\left(\frac{I_G}{I_S}\right) = \frac{(100\Omega)(1 \times 10^{-6}\text{ A})}{0.999999\text{ A}} = 1.0 \times 10^{-4}\Omega$$

Measuring Potential Difference: The Voltmeter

The **voltmeter** is a device for measuring potential difference. Unlike an ammeter, which is inserted into the circuit in series, so that all the current flows through it, a voltmeter is used in parallel with the circuit. Some of the current is diverted into the meter, but it is usually desirable to minimize this effect, so voltmeters are designed to have very large resistances.

When two points in a circuit are connected through a voltmeter, current flows through the meter (Fig. 15.23). According to Eq. 15–3, the current is equal to V/R', where V is the potential difference between the two points, and R' is the resistance in the meter. The current is measured by a galvanometer, whose deflection drives a meter calibrated in volts. To minimize the current flowing through the meter, the meter is designed so that R' is large. In order to achieve this, a resistor is connected in series with the galvanometer. If its resistance is R, and that of the galvanometer is r, then we have

$$V = IR'$$

or

$$V = I(R + r). \qquad (15\text{--}16)$$

15.5 EXPERIMENTAL ANALYSIS OF CIRCUITS

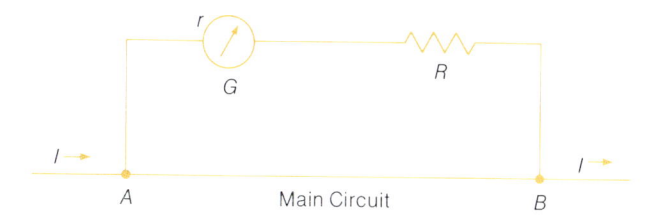

Figure 15.23

The voltmeter. Resistance R is inserted in series with the galvanometer G so little current flows through the meter. The potential between A and B in the main circuit is not significantly reduced by the voltmeter.

Usually it is desirable to make the voltmeter capable of measuring a wide range of voltages, so a switch allows any of several resistors to be selected.

Example 15.10 A voltmeter contains a galvanometer designed to give full deflection when the current passing through it is 5 μA. The internal resistance of the galvanometer is 100Ω. What resistance should be connected in series with the galvanometer to have full deflection of the galvanometer for a potential difference of 100 V?

We solve Eq. 15–16 for R.

$$R = \frac{V}{I} - r = \frac{100 \text{ V}}{5 \times 10^{-6} \text{ A}} - 100\Omega = 2.0 \times 10^7 \Omega$$

Measuring Resistance: The Wheatstone Bridge

Resistance can be measured quite directly by a device called an **ohmmeter**, which applies a known emf to a circuit containing the unknown resistance and, in series with the resistor, an ammeter (Fig. 15.24). The ammeter measures the current flowing through the resistor; since the voltage is also known, the ratio V/I yields the resistance R. Often circuits to do this, using a battery as the source of emf, are built into a combined meter called a **volt-ohm-meter** *(VOM),* which measures voltage, current, or resistance according to the setting of the switch.

A more precise method of measuring resistance uses a circuit called a **Wheatstone bridge,** which balances an unknown resistance against a known one, so that the value of the unknown one can be found. The advantage of such devices is that no current is drawn when balance is achieved, so there is no impact on the circuit being measured.

Figure 15.24

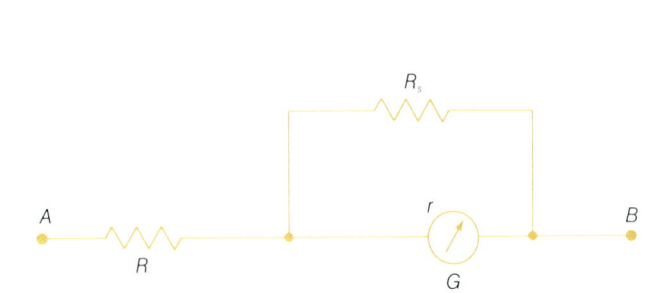

The ohmmeter. The unknown resistance R is connected in series with an ammeter in a circuit with known potential difference between A and B.

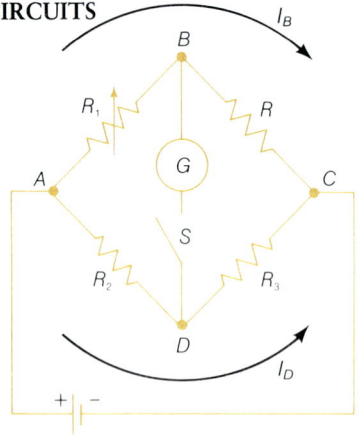

Figure 15.25

The Wheatstone bridge. The unknown resistance R is found by adjusting R_1 until the current through BD is zero, indicating that the potential difference between B and D is zero. R can then be found from R_1, R_2, and R_3.

The circuit for a Wheatstone bridge is shown in Fig. 15.25. When switch S is opened, the current created by the applied emf flows through two parallel paths, one containing two known resistances, and the other containing the unknown resistance and a variable resistance. The variable resistance is adjusted so that the potential difference between points B and D is zero; this is checked by momentarily closing switch S and reading an ammeter in the circuit segment BD. The current in the ammeter is zero when the potential difference between B and D is zero, and the bridge is said to be balanced.

Now we know that the potential difference between points A and B must be equal to that between points A and D, so that

$$I_B R_1 = I_D R_2$$

where I_B and I_D are the currents flowing through the two parallel paths. Similarly, we have

$$I_B R = I_D R_3$$

where R is the unknown resistance. If we divide these two equations and solve for R, we find

$$R = \frac{(R_1)(R_3)}{R_2}.$$

Thus the unknown resistance is expressed in terms of the three known ones, and can be measured as accurately as they are known. This is better than using an ohm-meter, because an ohm-meter depends on the terminal voltage in a battery, which varies with its internal resistance.

15.5 EXPERIMENTAL ANALYSIS OF CIRCUITS

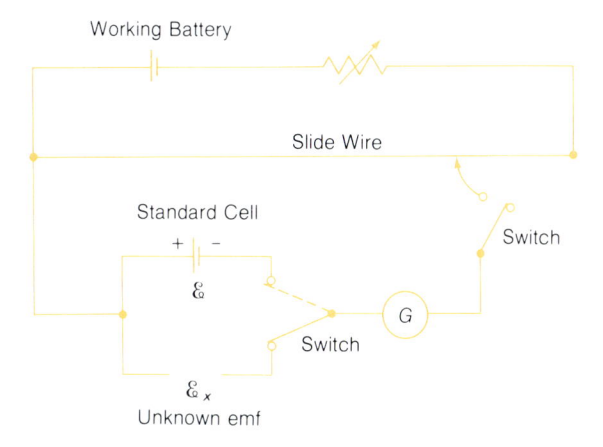

Figure 15.26

The potentiometer. A sliding contact with the circuit loop containing the working battery is adjusted until the current through the galvanometer is zero and the unknown emf is balanced by the emf of the working battery. The emf of the working battery is calibrated against a standard cell.

Measuring emf: The Potentiometer

We have seen that a voltmeter draws current, which alters the characteristics of the circuit being measured. This is especially serious when the emf of a battery is measured, because current will flow, so that the terminal voltage is less than the emf by the quantity Ir, where r is the internal resistance (recall Eq. 15–12). Hence it is desirable to measure potential differences without drawing current.

A device called a **potentiometer** accomplishes this, and is more accurate than a voltmeter. A potentiometer allows a known, variable emf to be applied to a circuit containing the unknown emf, so that the two can be balanced, yielding a value for the unknown emf. A circuit (Fig. 15.26) is constructed so that the unknown emf drives a current through a loop, one segment of which is a "slide wire," which acts as a variable resistor. A contact can be moved along the slide wire, varying the potential supplied to the circuit by a battery called the "working battery." This is done until the current in the loop is zero (as determined by an ammeter). When the current is zero, then there is zero potential difference around the loop, and the potential drop along the slide wire must equal the unknown potential difference. Hence the emf supplied by the working battery has been balanced against the unknown emf. The position of the sliding switch therefore indicates the unknown potential.

It is difficult to calibrate the working battery precisely, so normally the potentiometer is calibrated by first placing a standard source of emf in the circuit, so that the position of the sliding switch for a known emf is established. Then the position of the switch when the circuit is balanced with the unknown source in the circuit indicates the value of the unknown emf relative to the standard source. To switch back and forth easily between the standard and unknown sources of emf, parallel circuit segments are usually constructed with a switch so that the two can be alternately connected into the circuit.

Potentiometers are useful whenever very small voltages are to be measured, which is common in modern electronics. They are also quite useful in biological applications, because they are sensitive to small voltages and are unaffected by internal resistance. For example, potentiometers can measure varying potentials created in nerve cells by the transmission of electric impulses.

FOCUS ON PHYSICS

Household Circuits

Electric circuits are part of our everyday lives, not only because every machine or appliance that runs on electricity contains circuits, but also because the source of electricity for those devices is brought into our houses by circuits. Whenever an appliance is plugged into a wall socket, we complete a circuit, supplying electricity to the appliance (of course, a switch usually has to be closed as well).

Household circuitry in the United States (as in most other countries) has become standardized throughout the country. In the U.S. the standard emf is 120 V, and 60 Hz alternating current is used. In much of the rest of the world, the standard emf is 240 V, although the frequency is still 60 Hz in most places (in parts of Europe, the standard frequency is 50 Hz).

To supply electric current to a circuit, only two wires are needed. One (the hot wire) carries the voltage, while the other is grounded (zero potential) so that current flows from the hot side through the appliance to the ground. In practice, however, more than two wires are used. Modern electrical outlets have two slots for the blades on a plug that carry the hot and ground sides of the circuit and a third hole for the rounded prong that acts as a shield against short circuits. This third prong is part of a circuit that connects the outer metal case or frame of the appliance to ground. It normally carries no current, but in the event of a short circuit, when the hot wire accidentally touches the frame or case of the appliance, this wire will divert the current (usually very large) to ground so that the operator of the appliance is not shocked. (Usually a circuit breaker or fuse will stop the current altogether. These devices are described below.)

The circuit leading to a house normally has an emf of 240 V. This large emf is required by a few major appliances, such as clothes dryers and water heaters. In the house, the 240 V is reduced to 120 V for most of the outlets, but not for the ones that operate the heavy-duty appliances (the outlets for these have different slots, so that it is impossible to accidentally plug a 120 V appliance into a 240 V outlet, and vice versa). This means that the usual cable leading into a house has four wires: two hot wires that have potentials of +120 V and −120 V relative to ground, a ground wire, and a fourth conductor to act as the shield against short circuits, as described above. The potential difference of 240 V needed for some appliances is obtained by connecting these appliances between the two hot wires that have voltages of ±120 V. The normal 120 V outlets are connected between either of the hot wires and the ground wire. Whether an outlet is 240 V or 120 V, the fourth wire protects against short circuits.

A short circuit means that, because of accidental direct contact between the hot and ground wires, current flows with little or no resistance. The current can be immense and can cause resistive

(continued)

heating to temperatures sufficient to start a fire. It is important therefore to install some protection against the possibility of a short circuit.

Protection against short circuits is provided by fuses, or, more often today, circuit-breakers. Both are devices that will interrupt the circuit if the current exceeds some predetermined safe value. A fuse is a wire that will melt if the current is too high, thereby breaking the circuit and stopping the dangerously high current. Once a fuse has been "blown," it must be replaced before the circuit can be operated again. A circuit-breaker, on the other hand, can be reset (by flipping a switch) if it has broken the circuit. A circuit-breaker uses electromagnetic forces (Chapter 16) to physically open a switch to interrupt the current if it gets too high. Most houses built today have a "breaker box," where circuit breakers are located for all the household circuits. Circuit breakers are more convenient to use than fuses, but are considered less reliable.

We have discussed the reduction of potential ($V = IR$) caused by resistance in the wire carrying a current. We also learned that the power dissipated by a conductor as heat is proportional to the resistance and to the square of the current ($P = I^2R$). For these reasons it is important that the resistance of a wire carrying current by minimized; on the one hand, this minimizes the drop in potential, and on the other hand, it also minimizes the heat. The resistance of a wire is inversely proportional to the cross-sectional area, so it is always best to have a wire of large cross section, so that resistance is minimized. This is expensive, however (copper is usually the conductor, and very large copper wires are costly). Therefore, the diameters of wires are selected so that the potential drop and the resistive heating, while not the lowest possible, are acceptable. For most 120 V appliances, a 12-gauge wire (diameter 2.05 mm), which can carry currents up to 20 A without danger, is suitable. Usually an 8-gauge wire (diameter 3.26 mm) or larger is required for heavy-duty appliances, and a much larger wire (2-gauge, or 6.54 mm) is used for the main lines bringing power to a house.

Contractors in the U.S. are normally required to employ licensed electricians and to follow established building codes when installing electric circuits. These measures are taken to guarantee that the wiring is both safe and adequate to meet the needs of a normal household. The codes usually describe exactly how circuit connections are to be made (usually within isolated, nonconducting boxes), and specify the diameters of the wires that can be used in various applications.

We have described here the distribution of power in a single house. Imagine how complex the circuits are that bring electric power across the countryside from the generating station to the typical household. That our country, and many others as well, have been able to develop systems to distribute power safely to all its citizens is a remarkable achievement.

Perspective

In this chapter we have covered quite a large body of material, all of it involving charge flowing in circuits. We have learned about sources of emf, which are required to bring about the flow, and about the fundamental relationships among potential, current, and resistance. We have examined the various components of circuits, and in the process we have learned how to measure and analyze them.

We now must fill in our picture of electrical processes by considering magnetic phenomena, which are interesting for their own right, but also crucial to understanding the full range of uses to which electricity can be put.

PROBLEM SOLVING

Strategies for Analyzing Circuits

The many types of circuits in this chapter are somewhat bewildering, but there are only a few basic concepts to remember. These are summarized by the formulas

$$V = IR \qquad V = Q/C \qquad P = IV = I^2R = V^2/R.$$

These express the voltage V across a resistor R or capacitor C and the power dissipated by Joule heating in a resistor R. Notice that Ohm's law, $V = IR$ has been used to write the power in three alternate forms.

In analyzing a circuit, then, one must find the total or effective resistance R_{eff} and capacitance C_{eff}. The first step is to reduce all series and parallel circuit components to a single component, using the rules discussed in the chapter. In series, resistors add directly and capacitors add in reciprocal. In parallel, the reverse is true. Of course, one must be able to recognize series and parallel components, which is not always easy if the circuit is not drawn in ideal form. Therefore, redraw the circuit to highlight the series and parallel components. Start with series components, since they are easier to identify. Combine the series elements into a single component, then look for junctions between parts of the circuit. A junction tells you that current is being divided into several parallel channels. Sometimes you will need to move certain junctions together into a single multiple junction to see the parallel channels better. If there are no circuit elements between two points in the circuit, there is no potential drop and you are allowed to eliminate that section, almost as though you had a pair of scissors.

When you have the total resistance, you can find the total current $I_{tot} = V/R_{eff}$. The currents and potential differences across individual components can be found by techniques illustrated in Example 1.

Example 1 Find the effective resistance of the circuit in Fig. 15.2.1a. What are the currents and potential drops across each resistor? What is the total power?

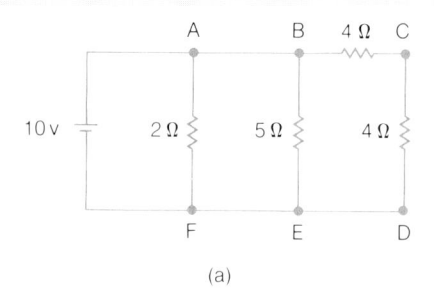

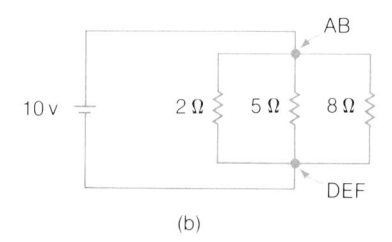

Figure 15.2.1

We start with the series components. The two 4 Ω resistors are in series, and we can combine them into a single 8 Ω component. Next we look at the junctions. Since there are no components between AB, and DE or EF, we can combine them

PROBLEM SOLVING

into single junctions as shown in Fig. 15.2.1b. In redrawing the circuit, we have eliminated the portions between A and B, between E and F, and between D and E. We now recognize that we have three parallel channels, whose effective resistance is given by

$$\frac{1}{R_{eff}} = \frac{1}{2\,\Omega} + \frac{1}{5\,\Omega} + \frac{1}{8\,\Omega} = 0.825\,\Omega^{-1}.$$

So $R_{eff} = 1.21\,\Omega$. Notice that R_{eff} is less than any individual resistance, since there are several channels to carry the current.

The total current $I_{tot} = 10\,V/1.21\,\Omega = 8.25\,A$ and the total power dissipated is $P = VI = (10\,V)(8.25\,A) = 82.5\,W$. Since the voltage across each of the parallel channels must be the same (a junction point means a single potential), the product of IR across each channel must equal 10 V. Therefore, the 2 Ω resistor carries $10\,V/2\,\Omega = 5\,A$, the 5 Ω resistor carries 2 A, and the 8 Ω resistor carries 5/4 A. Actually, since the 8 Ω resistor is composed of two 4 Ω resistors in series, each carries 5/4 A (the electrons don't pile up, so the current through these series components is the same).

A final strategy in circuit analysis is to check that the currents through all parallel channels add up to the total current $I_{tot} = V/R_{eff}$. If they do not, you made an error somewhere. In this circuit, the three parallel channels carry 5 A, 2 A and 5/4 A, for a total of 8.25 A, which agrees with the value for I_{tot}.

Capacitor problems are similar, except for the addition rules. Also, instead of current dividing between parallel resistors, we have charge divided between parallel capacitors.

Example 2 Find the effective capacitance of the circuit in Fig. 15.2.2a. What are the charge and voltage drop on each capacitor?

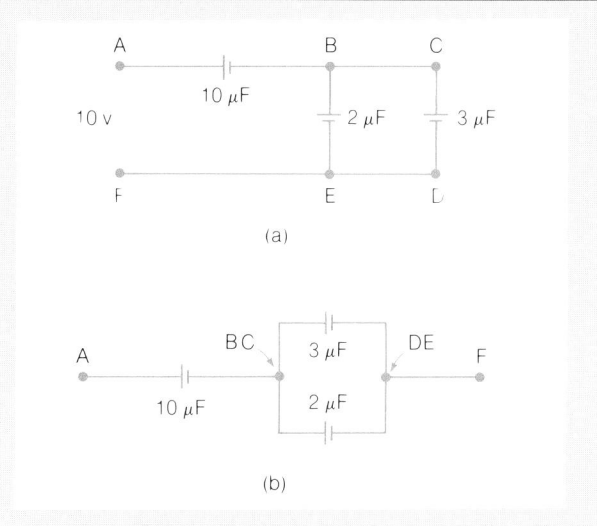

Figure 15.2.2

The 2 μF and 3 μF capacitors are in parallel, since BC and DE may be moved together into single junctions (Fig. 15.2.2b). Capacitances in parallel add directly (reverse of resistances), so these are equivalent to 5 μF. The 10 μF capacitor is then in series with 5 μF, and the reciprocal of the total capacitance is the sum of reciprocals, $C_{eff} = 3.33$ μF. The total charge is

$$Q_{tot} = C_{eff} V = (3.33 \times 10^{-6}\,F)(10\,V)$$
$$= 3.33 \times 10^{-5}\,C$$

which must be the same on each component in series (otherwise there would be a potential difference and current would flow to equalize the charges). The voltage across the 10 μF capacitor is $V = Q/C = 3.33 \times 10^{-5}\,C/10 \times 10^{-6}\,F = 3.33\,V$, and the voltage across the parallel capacitors is $3.33 \times 10^{-5}\,C/5 \times 10^{-6}\,F = 6.67\,V$. Notice that the voltage drops add to 10 V (a good check). The charge is divided between the two parallel capacitors in proportion to their capacitances. The 2 μF capacitor holds 2/5 and the 3 μF holds 3/5 of the 3.33×10^{-5} C.

SUMMARY TABLE OF FORMULAS

Power dissipation by a current (Joule's law):
$P = I^2R$ (15–1)

Power, potential difference, and current:
$P = VI$ (15–2)

Ohm's law:
$I = \dfrac{V}{R}$ (15–3)

Resistance of a conductor:
$R = \dfrac{\rho L}{A}$ (15–4)

Thermal dependence of resistivity:
$\dfrac{\Delta \rho}{\rho} = \alpha \Delta T$ (15–5)
$\rho_T = \rho_0(1 + \alpha T)$ (15–6)

Thermal dependence of resistance:
$R_T = R_0(1 + \alpha T)$ (15–7)

Kirchhoff's first rule:
$\Sigma I = 0$ (15–8)

Kirchhoff's second rule:
$\Sigma \mathscr{E} - \Sigma IR = 0$ (15–9)

Resistors in series:
$R_s = R_1 + R_2 + R_3 + \ldots$ (15–10)

Resistors in parallel:
$\dfrac{1}{R_p} = \dfrac{1}{R_1} + \dfrac{1}{R_2} + \dfrac{1}{R_3} + \ldots$ (15–11)

Internal resistance of a battery:
$V = \mathscr{E} - Ir$ (15–12)

Capacitors in parallel:
$C_p = C_1 + C_2 + C_3 + \ldots$ (15–13)

Capacitors in series:
$\dfrac{1}{C_s} = \dfrac{1}{C_1} + \dfrac{1}{C_2} + \dfrac{1}{C_3} + \ldots$ (15–14)

Shunt current in a galvanometer:
$I_G r = I_S R$ (15–15)

Potential difference in a voltmeter:
$V = I(R + r)$ (15–16)

THOUGHT QUESTIONS

1. Compare and contrast electric current with river current.
2. Are there any sources of emf in nature? Describe them.
3. Describe a conductor which might not have a path along which charge can flow.
4. Why is the ampere considered a more fundamental unit than the coulomb?
5. Would the flow of "holes" described in the last chapter be an example of positive or negative current?
6. Give examples of materials which are very poor conductors. What electrical use would they have?
7. Would you expect materials with high thermal conductivity to have high electrical conductivity also? Why or why not?
8. Explain why a resistor being heated by a current is an example of the second law of thermodynamics.
9. List several devices which make use of Joule's law.
10. Is wood a good insulator for electrical wires in a house? Why or why not?
11. Why are we advised against replacing a fuse with a penny?
12. Discuss the possible uses of a material which is superconducting at room temperature.
13. Some thermometers measure temperatures by gauging the variation of electrical resistance in a conductor. Are there any practical limits to this device?
14. Would the thermal expansion of a wire affect the validity of Eq. 15–6 and Eq. 15–7? Explain.
15. Why can't charge pile up at a junction the way cars can at an intersection?
16. Design a circuit that would measure the emf of a battery without drawing any current from it.

17 Why are the expressions for combining capacitors in series and parallel the opposite of the expressions for combining resistors in parallel or series?

18 Are there any circumstances in which it would be better to measure the value of a resistor by an ohmmeter rather than a Wheatstone bridge? Explain.

19 Compare and contrast a voltmeter and a potentiometer.

20 Are there any limits to how much charge a capacitor can hold?

21 Batteries are sometimes rated according to "ampere hours." What does this mean?

22 What is the meaning of electrical "ground"?

23 What happens to electrons after they traverse a circuit?

24 Discuss the analogy between emf in a circuit and water running downhill. Are emf and gravitational potential energy similar?

25 Why do electrons flow in the opposite direction to the positive convention of current? Which way do electrons flow in lightning?

PROBLEMS

Sections 15.1–15.2 EMF and Current

- 1 How many coulombs are in one ampere-hour?
- 2 Use the charge on the electron to show that 6.24×10^{18} electrons flow past a point each second if the current is one ampere.
- 3 If you charge a battery for 5 hours at 6 A, how many coulombs flow?
- 4 In a lightning strike, 200 C flow to the ground. How many electrons flow?
- 5 How long will it take 180 C to pass if the current is 0.2 A?
- 6 What current represents one electron passing per second?
- •• 7 How many doubly ionized helium particles will pass per second if the current is 0.25 A?
- 8 A current of 10 A flows through a wire. How many electrons pass in 1 hour?
- 9 If an Avogadro's number of electrons pass per second, what is the current?
- ••• 10 Suppose the density of electrons per m^3 in a conductor is n_e. If the electrons flow at velocity v, and the conductor has cross sectional area A, what is the current?
- ••• 11 Copper has density 9 g/cm^3. If a copper wire has a radius of 1 mm and carries 1 A current, what is the mean velocity of electrons? Assume that each atom of copper donates one free electron.
- ••• 12 In a plasma device for controlled fusion, the ionized hydrogen gas (plasma) has an ion density of 10^{14} cm^{-3}, a current of 500 A, and a cross sectional area of 1 m^2. What is the mean velocity of the ions?

Section 15.3 Resistance

- 13 The filament of a 100 W light bulb draws a current of 0.7 A. What is its resistance?
- 14 What voltage is required to operate the lightbulb in Problem 13?
- 15 Using the definitions of the ohm and volt, verify that V^2/Ω does indeed have units of power.
- 16 What resistance is required for the filament of a 60 W light bulb if it operates at a current of 1.5 A?
- 17 A 1,000 Ω resistor is connected across the terminals of a 15 volt battery. How much current flows? How much power is dissipated?
- •• 18 Suppose we wished to heat the water in Example 15.1 for 1 min with 8 A. What resistance would be required for the electric coil?
- 19 A person with resistance 10^5 Ω accidentally touches the terminals of a 115 V household outlet. If the outlet were DC, what current would flow through the person? What would be the power dissipated in the person?
- •• 20 What resistance is required for a coil to heat a 3 liter bucket of water from 10 °C to 80 °C in 10 min? The coil is connected to a 120 V line.
- 21 How much power is generated by a 20 Ω resistance operating at 115 volts?
- 22 The heating elements of an electric oven generate 5,000 W of heat when connected to the 220 V power source. What is the resistance of the elements?
- •• 23 An electric hair dryer generates 1,500 W and is connected to a household outlet of 115 V. What current flows? What is the resistance?

- 24. What is the maximum voltage that can be connected to a 1,000 Ω resistor rated at 1 W?
- 25. An industrial power supply is connected to a 100 A fusebox. If the voltage is 440 V, what is the maximum power that can be drawn from the circuit?
- 26. A 1 m long wire has 5 Ω resistance. If the wire is cut in half, how should its diameter be altered to maintain the same resistance?
- 27. A wire is connected to a 3 V battery, and 2 A current flows. A second wire of identical dimensions is connected to the same battery, and 3 A flows. What is the ratio of resistances?
- 28. A wire 2 mm in diameter and 40 cm long has a current of 0.1 A when connected to a 3 V dry cell. What is its resistivity?
- 29. Calculate the total resistance of a pine log, 0.5 m in radius and 10 m long. The resistivity of pine wood is 10^{10} Ω · m.
- 30. In a 1 km lightning strike, the current is 500,000 A and the voltage is 10^9 V. What is the effective resistance of air? What power is generated?
- 31. Number 12 gauge copper wire at 20 °C is designed to produce no more than 2 W per m length, to avoid fires in house wiring. If its diameter is 2 mm, what is the maximum current that can be carried?
- 32. An engineer wishes to make a cylindrical resistor with the same resistance as one in a circuit, but would like it half as long. How should this be done?
- 33. Find the resistance along its length of a 2cm × 4cm slab of silicon, 5 cm long. Assume its resistivity is 2,000 Ω · m.
- 34. Show that Eq. 15–6 follows from Eq. 15–5, and then derive Eq. 15–7.
- 35. A 1 l container of water changes its temperature from 20 °C to 22 °C in 5 min when a 10 Ω heating coil is inserted. What current flows through the coil?
- 36. A resistor has 34 Ω resistance at 300 K and 42 Ω at 400 K. What are R_0 and α?
- 37. A resistor has $R_0 = 5$ Ω and α = 0.002 °C^{-1}. To what temperature must the resistor be heated to achieve 10 Ω?
- 38. What resistance would be observed in Problem 37 if $T = 350$ K?
- 39. A resistor has a temperature coefficient α. To what temperature must this resistor be heated to increase its resistivity from ρ_0 to $2\rho_0$?
- 40. If the wire in Example 15.4 had a radius of 2 mm, how much power would be dissipated?
- 41. Find the thermal coefficient α for a 1 mm (radius) wire 10 m long which has resistance 0.05 Ω at 0 °C and 0.07 Ω at 60 °C.
- 42. To what temperature must we heat a copper wire, initially at −10 °C, for its resistance to double?
- 43. A certain substance 5 cm long and 1 cm in radius has 4 Ω resistance at 20 °C and 8 Ω at 50 °C. Find ρ_0, ρ(20 °C); and α for this material. Is this substance a conductor or insulator?
- 44. For a given circuit, at constant voltage, is more power dissipated at high or low resistance? Why?
- 45. Which draws more current, a 100 W bulb or a 60 W bulb? Assume household voltage is 115 V.
- 46. A 12 V car battery is rated at 100 A · h. If the two headlights require 35 W each, how long will the battery take to drain its charge?
- 47. Estimate the cost of a monthly household electric bill. Assume each of 4 rooms has two 100 W bulbs used 5 hours per day. The house has a 5 kW electric stove used 2 h a day, a 1.5 kW hairdryer used 30 min/day, and a refrigerator (200 W average). Electricity costs about $0.10/kWh.
- 48. Using the resistivity of copper at 20 °C, estimate how much power is dissipated in 20 km of 1 cm radius copper wires from a 100,000 W power plant which transports its power over 10 kV lines.
- 49. How much power would be dissipated in Problem 48 if the voltage were 30 kV?
- 50. Copper wire designed to carry 100 A maximum current at 20 °C must not produce more than 1 W of heat per meter length. What is the minimum radius of the wire?
- 51. How many watts must an electric space heater have to heat a room of 50 m^3? Assume that fresh air at 10 °C is circulated into the room once each hour, after heating it at constant pressure to 20 °C. Heat loss through windows, doors, and walls is about 500 W.

Sections 15.4–15.5 Current Elements

- 52. Find the equivalent resistance of the system of resistors in Fig. 15.27.
- 53. Find the equivalent resistance of the system in Fig. 15.28. All resistors are 2 Ω.
- 54. Find the current flowing through each of the resistors in Problem 52 if the voltage across AB is 15 V.
- 55. In Problem 53 what voltage should be applied across AB for the current in the diagonal resistor to be 1 A?

PROBLEMS

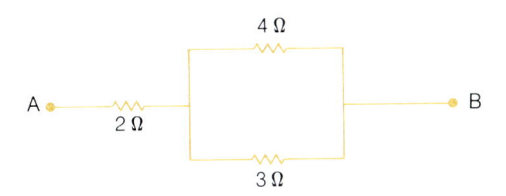

Figure 15.27

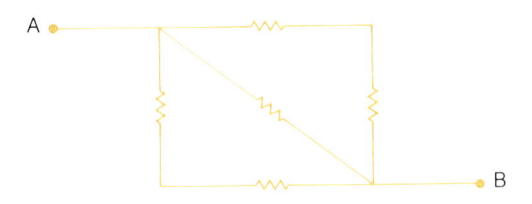

Figure 15.28

··56 Find the current flowing through each of the 5 Ω resistors in Fig. 15.29 if the emf is 15 V.

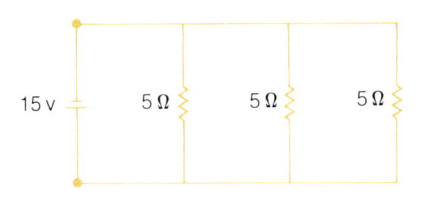

Figure 15.29

··57 For a net current of 4 A in the system of resistors in Fig. 15.30, what resistance should R be if the potential across AB is 15 V?

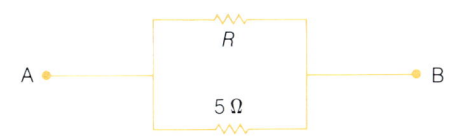

Figure 15.30

··58 A voltage of 100 V is applied across AB in Fig. 15.31. Find the value of resistance R for it to dissipate 20 W.

Figure 15.31

···59 A voltage of 1.5 V is applied across AB in Fig. 15.32. Find the resistance R needed to develop a total power of 3 W in the circuit.

Figure 15.32

···60 Find the current at all points in the circuit of Fig. 15.33.

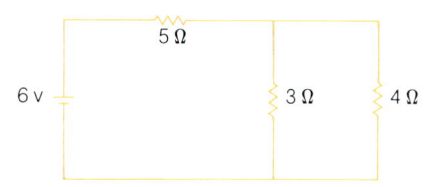

Figure 15.33

··61 Find the current at all points in the circuit of Fig. 15.34.

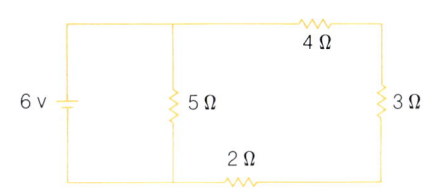

Figure 15.34

···62 Find the change in potential across each element of the circuit of Fig. 15.35.

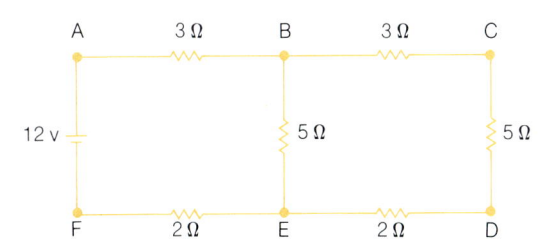

Figure 15.35

···63 Find the change in potential across each element of the circuit in Fig. 15.36.

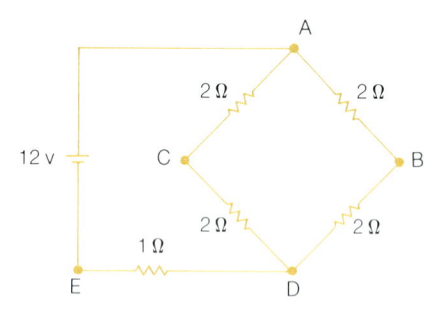

Figure 15.36

- 64 Find the equivalent resistance of the following set of resistors (Fig. 15.37).

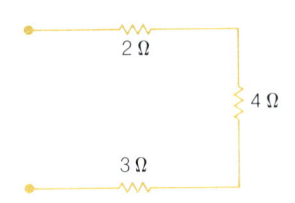

Figure 15.37

- 65 Find the equivalent resistance of the following set of resistors (Fig. 15.38).

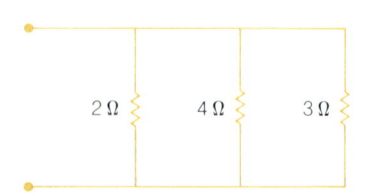

Figure 15.38

- 66 Find the equivalent resistance of the following set of resistors (Fig. 15.39).

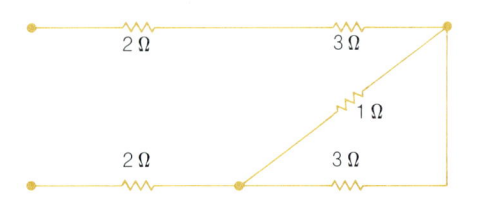

Figure 15.39

- 67 Find the equivalent resistance of the following set of resistors (Fig. 15.40).

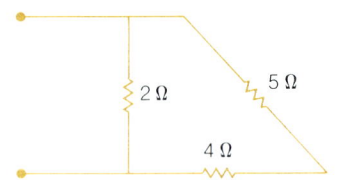

Figure 15.40

•• 68 Find the equivalent resistance of the set of resistors in Fig. 15.41.

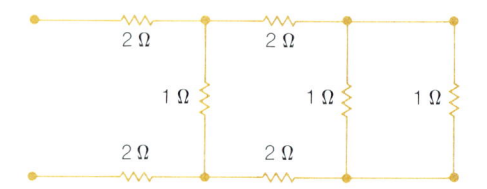

Figure 15.41

- 69 Find the equivalent capacitance of the set of capacitors in Fig. 15.42.

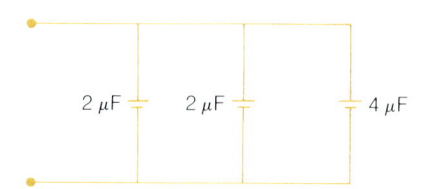

Figure 15.42

- 70 Find the equivalent capacitance of the set of capacitors shown in Fig. 15.43.

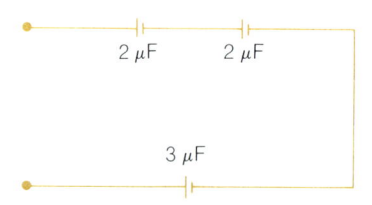

Figure 15.43

•• 71 Find the equivalent capacitance of the set of capacitors shown in Fig. 15.44.

•• 72 Find the equivalent capacitance of the set of capacitors shown in Fig. 15.45.

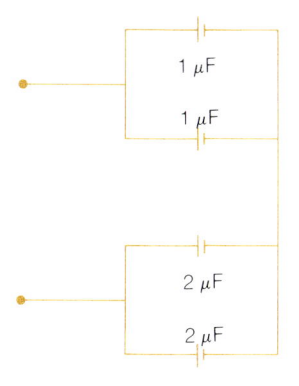

Figure 15.44

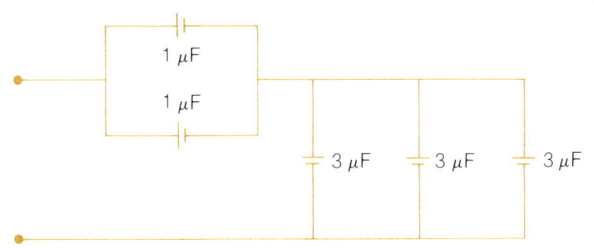

Figure 15.45

73 Find the equivalent capacitance of the set of capacitors in Fig. 15.46.

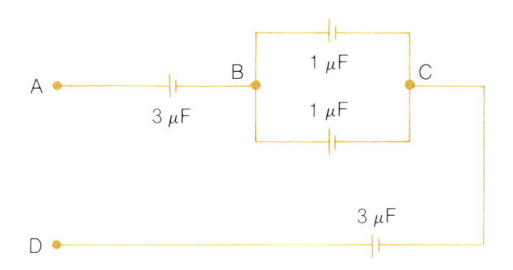

Figure 15.46

74 Find the charges and potential differences for each capacitor in Problem 73 if 12 V is applied across AD.

75 Find the equivalent capacitance of the system in Fig. 15.47. If the voltage across AB is 15 V, determine the charge and potential difference across each capacitor.

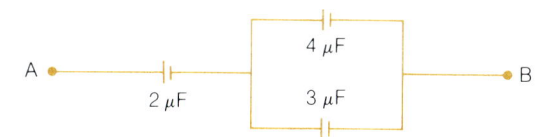

Figure 15.47

76 A 20 pF capacitor is charged by a 15 V battery and then disconnected. The capacitor is then attached to a 2 Ω resistor. How much energy will be dissipated in the resistor?

77 What shunt resistance should be used in Example 15.9 if the internal resistance of the galvanometer is 50 Ω?

78 The Wheatstone bridge in Fig. 15.48 is to be used to measure an unknown resistance R. If the value of R_V is found to be 5 Ω when the ammeter reads zero, what is R?

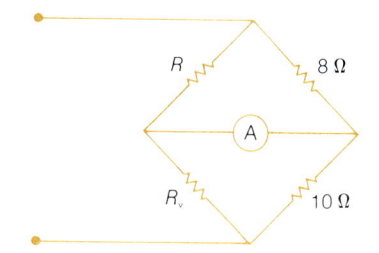

Figure 15.48

79 A 12 V battery draws 10 A of current if the actual operating voltage is 11 V. What is the battery's internal resistance?

80 In Fig. 15.49, $R_1 = 15$ Ω and $R_2 = 30$ Ω. If the galvanometer at the center registers zero current when R_V is set at 3 Ω, find the value of R_x.

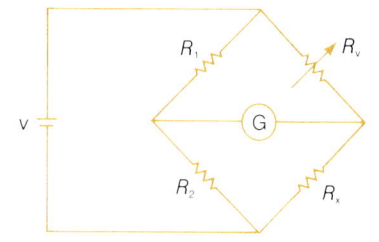

Figure 15.49

SPECIAL PROBLEMS

1. Lightning Bolts

Lightning is caused by the accumulation of static electricity on charged water droplets or ice crystals in convective thunderclouds (Fig. 15.50). When the electric fields in the cloud become sufficiently large (3×10^4 V/cm in dry air, less when droplets are present), a spark discharge is triggered by a stray ion. An avalanche of secondary ionizations follows, opening a "streamer channel" for the lightning bolt.

(a) Estimate the energy given to one ion by the initial spark discharge over 1 cm. If it requires about 35 eV to create one new ion pair (a molecular ion plus free electron), how many secondary ion pairs are created?

(b) If a single bolt transfers 100 C charge, what is the current in amperes, the total energy, and power during the 100 microsecond flash? Assume the bolt is 1 km long across an electric field 10^4 V/cm.

(c) What mass and volume of air can this energy heat to 30,000 K? Assume $C_P = 6.99$ cal/mole/K.

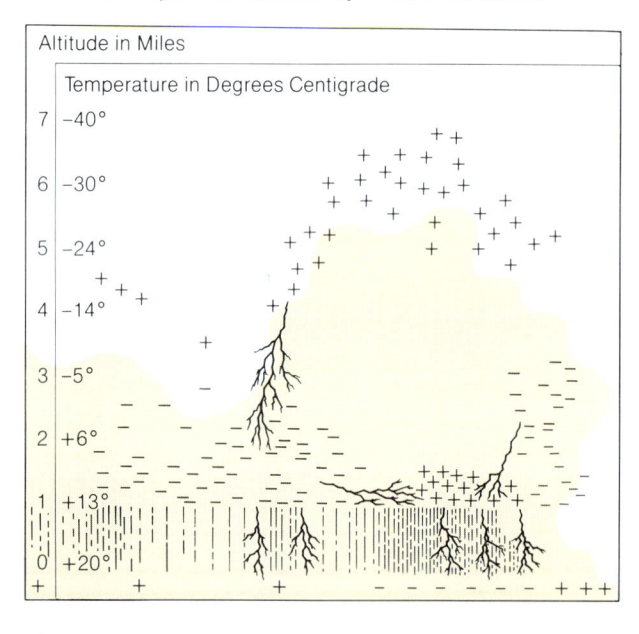

Figure 15.50

2. Electricity and Human Physiology

Because it contains fluids and electrolytic ions (Na^+, K^+, Mg^{+2}), internal human tissue is a far better conductor than dry skin, which is a good insulator. Dry skin has a resistance of $10^4 \, \Omega - 10^6 \, \Omega$, while the resistivity of internal tissue is about $5 \, \Omega \cdot m$. If the skin becomes wet, its resistance decreases rapidly. The physiological effects of electric currents range from mild tingling to severe burns or even death (Fig. 15.51). Currents from about 0.1 A to about 3 A send the heart into spasmodic contraction, called ventricular fibrillation, which endangers the blood supply and can result in death.

(a) Estimate the resistance of the internal human tissue between one hand and foot. Consider the arm to be a cylinder 1 m long and 4 cm in radius, the trunk to be a cylinder 0.6 m long and 13 cm in radius, and the leg to be a cylinder 0.7 m long and 8 cm in radius.

(b) What would be the current and effects in a person touching a 120 V electric receptacle with dry skin (20,000 Ω skin resistance) and with damp skin (negligible skin resistance)?

(c) What current and physiological effects would result if the person in part a touched a 20 kV power line, while standing on the ground with bare feet? Assume a skin resistance of 100,000 Ω.

3. Hair Dryer

An electric hair dryer uses Joule heating of Nichrome wire ($\rho = 10^{-6} \, \Omega \cdot m$, radius 0.4 mm). The dryer has a circular opening of radius 4 cm and blows hot air at 3 m/s.

(a) Estimate the volume and mass of air blown each second.

(b) Estimate the power needed to heat this air from 20 °C to 70 °C. Show that the kinetic energy of moving air is small compared to this heat. Assume that the dryer is 50% efficient in heating the air.

(c) What resistance is required to generate this power by Joule heating if the voltage is 120 V? What current is drawn?

(d) What length of Nichrome wire is required?

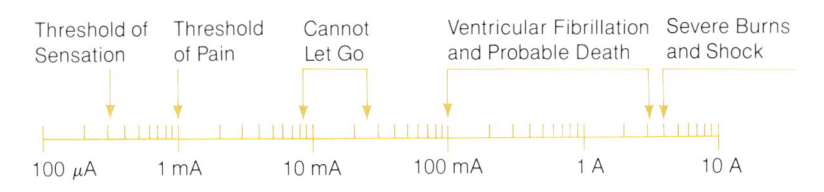

Figure 15.51

4. Model for Resistivity

In a simple model for electron conduction in metals, the current is carried by outer electrons donated by atoms in the metal. These free electrons are accelerated by an applied electric field E, but suffer collisions with the stationary ions and lose their newly acquired energy as heat. In between collisions, an electron is accelerated at $a = eE/m_e$, where e and m_e are the charge and mass of the electron. The average distance between collisions is λ, the mean free path. Consider a metal with length L, cross sectional area A, and electron density, $n_e(m^{-3})$.

(a) Show that the mean time between collisions is approximately $t = \lambda/v_t$, where v_t is the random thermal velocity of the electrons.

(b) Calculate the final velocity along the wire acquired by the accelerated electron during time t. The mean drift velocity v_d is half this value.

(c) From the definition of current, $I = n_e e v_d A$, find the resistance of the metal in response to a voltage $V = E/L$. Show that the resistivity is $\rho = (2m_e v_t / n_e e^2 \lambda)$.

(d) For copper, in which each atom of atomic mass 63 u donates one free electron, estimate the resistivity at 0 °C. Assume $v_t = (3kT/m_e)^{1/2}$ and $\lambda = 5 \times 10^{-9}$ m. Copper has a density 8,900 kg/m³.

5. Hot Tubs

Hot tubs are popular, but they can add to your electric bill. Consider a 300 gallon hot tub, warmed by Joule heating.

(a) Estimate the energy required to heat the water from 60 °F to 100 °F. Give your answer both in Joules and kWh.

(b) An uncovered tub will lose 15 °F per hour. What power (watts) is needed to maintain the water temperature at 100 °F?

(c) If electricity costs $0.10 per kWh, estimate the cost of heating the tub once and the hourly cost of maintaining the temperature.

(d) If you plan on using the tub in the morning for one hour and in the evening for one hour, should you leave the heat on in between? What is the monthly cost of operating the tub for 15 days, for a single one-hour sitting each day?

SUGGESTED READINGS

"Photovoltaic generation of electricity." B. Chalmers, *Scientific American*, Aug 76, p 34, **235**(3)

"The hair cells of the inner ear." A. J. Hudspeth, *Scientific American*, Jan 83, p 54, **248**(1)

Basic electronics for scientists. J. J. Brophy, McGraw-Hill, New York, 1977

The flying circus of physics with answers. Jearl Walker, Wiley, New York, 1977

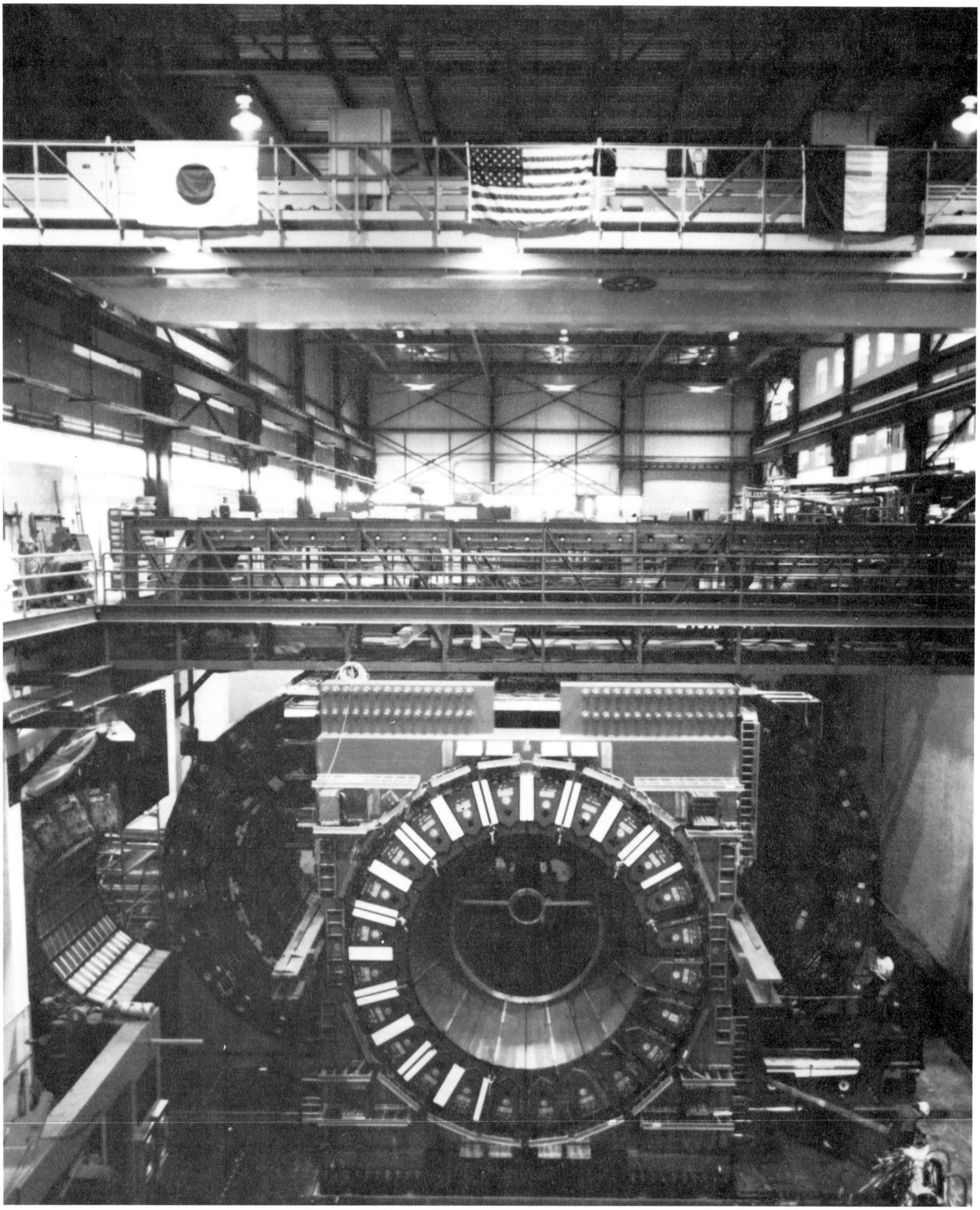

Chapter 16

Magnetism and Electricity

Most of us become acquainted as children with magnets, and develop some concept of magnetic fields and forces. We discover that a magnet will be attracted to certain materials and not others, and we find that a magnet has two poles, so that like poles of two magnets repel each other, and unlike poles attract each other.

Our ancient predecessors in science knew little more than the modern child learns by playing with toy magnets. Certain stones called **lodestones** or **magnets** (after Magnesia, a region of modern Turkey where they can be found) were discovered to have properties we now recognize as magnetism. Little more than that was learned until the nineteenth century when, in parallel with the many pioneering investigations of electricity that took place, there were great advances in the understanding of magnetism.

In this chapter we begin by describing what magnetism is, then how it is related to electricity, which will in turn require us to discuss magnetic forces on electrical currents. We will conclude by discussing the creation of currents by magnetic forces, which is the basis for many important applications of circuits in the next chapter.

16.1 Magnetism

We have already described some of the empirical facts about magnetism: certain natural substances have it; those that do always have two poles; the orientation of the two poles determines the direction of the forces that magnets always exert on each other (Fig. 16.1). Other observations show that the natural materials that exhibit magnetism at ordinary temperatures are few, and are metallic, most notably iron, nickel, and cobalt. These substances are said to be **ferromagnetic**.

515

Figure 16.1

Magnetic poles and forces. Every magnet has a pair of poles, designated north (N) and south (S). Unlike poles attract each other; like poles repel each other.

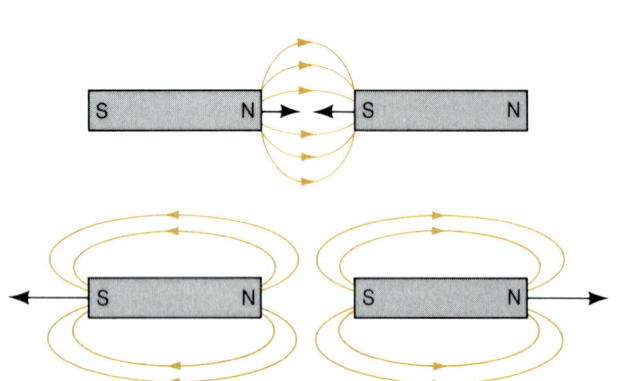

Figure 16.2

The earth's magnetic field. The lines indicate magnetic field lines. This is idealized. Note that the magnetic pole near the north geographic pole is a *south* magnetic pole.

A magnet creates a **magnetic field,** a region where a force is exerted on another magnet (or, as we shall see, on a conductor carrying an electric current). Just as in the case of electric fields, we speak of magnetic field lines and a field direction (defined later), yet we understand that there is no material medium.

The earth itself has a magnetic field (Fig. 16.2). This is evident because a force is always exerted on a magnet at the earth's surface, which can be seen most easily if the magnet is elongated as in a compass and allowed to rotate in response to the force. One end will always point towards the north, and by convention we say this is the north pole of the magnet. Therefore the magnetic pole of the earth that lies near the north pole is the *south* magnetic pole, since it attracts the north pole of a magnet. A compass always provides an indication of north, but not precisely true north, because the earth's magnetic poles are not exactly located at the poles of its rotation axis. (Furthermore, geological evidence shows that the magnetic north and south poles move about, and have exchanged places many times in the earth's history, doing so at sporadic intervals, usually of a few hundred thousand years).

Magnets always have two poles; if one is cut in half, each half then has both a north and a south pole. At the microscopic level, a ferromagnetic substance is found to consist of tiny magnetic regions called **domains,** each of which has a north and a south pole (Fig. 16.3). When an object made of such a material is subjected to a magnetic field, the domains become aligned, so that the entire object becomes polarized (that is, develops poles) and a force is exerted on the object by the magnetic field. When magnetism is induced in this way, the orientation of the poles is always opposite to that of the applied external field, and the force is therefore always attractive; thus, a magnet always attracts iron filings or paper clips, and does not repel them. A **permanent**

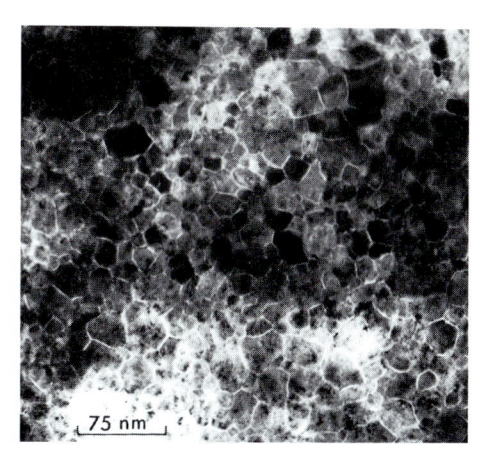

16.2 MOVING CHARGES AND MAGNETISM   517

Figure 16.3

Magnetic domains. These microscopic regions in ferromagnetic materials act as tiny magnets, each with a north and a south magnetic pole. The domains become aligned if an external magnetic field is applied, and are permanently aligned in permanent magnets.

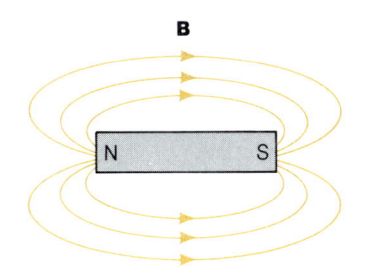

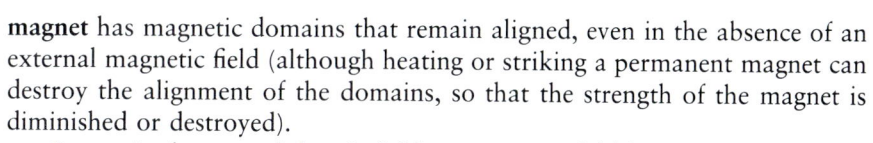

Figure 16.4

*The magnetic field vector **B** is parallel to the magnetic field lines and proportional to their density.*

magnet has magnetic domains that remain aligned, even in the absence of an external magnetic field (although heating or striking a permanent magnet can destroy the alignment of the domains, so that the strength of the magnet is diminished or destroyed).

Just as in the case of electric fields, a magnetic field has both a magnitude and a direction associated with it, and is therefore properly represented as a vector. Usually denoted **B**, this vector is closely related to the concept of magnetic lines of force, since its direction is that of the lines of force, and its strength is related to the density of the lines (Fig. 16.4). This is not a very practical definition, however, given the tenuous notion of lines of force. In practice, we may define the magnitude in terms of the torque exerted on a pivoted magnet, and we do so in Section 16.3.

16.2 Moving Charges and Magnetism

It was discovered in 1820 by H. C. Oersted, a Danish physicist seeking a connection between electricity and magnetism, that an electric current creates a magnetic field. This was the first step to a comprehensive understanding, later in the century, of the intimate relationship between the two phenomena. Today we refer to electricity and magnetism as two manifestations of the same phenomenon, which we call simply **electromagnetism.**

If we have a straight segment of wire which carries a current, and we observe the direction of the magnetic field it produces, we find that the field is everywhere tangent to a circle centered on the wire (Fig. 16.5). Thus the

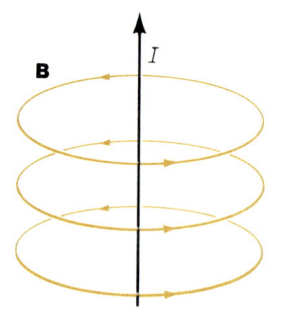

Figure 16.5

Magnetic field due to a straight wire carrying a current.

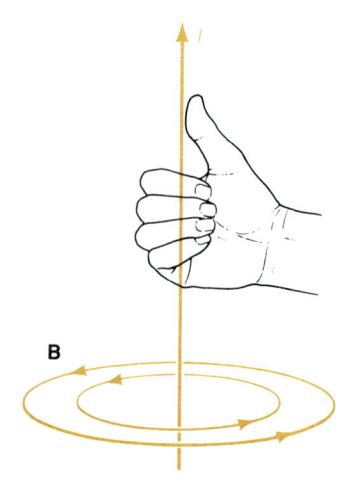

Figure 16.6

The right-hand rule. When the right-hand thumb points in the direction of the conventional current, the fingers curl in the direction of the magnetic field lines.

direction of the magnetic field is perpendicular to that of the current. The positive direction of a magnetic field is defined as the direction toward which the north pole of a compass points. When a field is created by a current, the direction of the field relative to that of the current is given by the **right-hand rule** (Chapter 4): if we use the curled fingers of our right hand to indicate the direction of the field, then the right thumb points in the direction of the current (Fig. 16.6). Note that the magnetic field direction is perpendicular to the direction of the current.

A moving electric charge will always create a magnetic field, even if it is an isolated moving charge instead of the continuous flow of a current. Thus, a free electron moving through space creates a magnetic field in the reference frame through which the electron moves. Even a charge's rotational motion can create a magnetic field; thus, the spin of an electron creates a field. This spin is thought to be responsible for the properties of ferromagnetic materials; it has been hypothesized that the electrons within a ferromagnetic domain somehow align their spins, so there is a net magnetic field due to the ensemble of electrons. (In nonferromagnetic materials, the electron spins are random, so the sum of all the individual magnetic fields they create is zero.)

It is not entirely clear that all magnetic fields are created by currents, but this possibility is certainly consistent with the observation that magnets always have two poles. The two poles are the result of flowing electric charge, and even the smallest possible unit of charge creates a bipolar field. If a magnetic field that has only one pole is ever found, our understanding of how magnetic fields are created will have to be revised. This would hardly be worth mentioning, except that some modern unified field theories, as well as one experimental result not yet confirmed, call for the existence of **magnetic monopoles,** subatomic particles containing a single magnetic pole, rather than a pair. This possibility will be considered again in our discussions of modern particle theory and cosmology, near the end of this text. Meanwhile, we assume that magnetic fields are bipolar, created only by moving electric charges.

Let us consider the magnetic field due to a loop of current, (Fig. 16.7). If we apply the right-hand rule at various points around the loop, we find that the field lines are all parallel within the loop and act in the same direction. Outside the current loop, the field lines curve around and join, forming closed loops themselves.

If we have a number of current loops together, with the current flowing in the same direction in each, then the fields they create are aligned and in the same direction, and they add together. This effect can be created by coiling a single continuous wire and allowing current to flow through it. Such a coil is called a **solenoid,** and it has an overall magnetic field oriented as shown in Fig. 16.8.

Solenoids have several practical applications. If an iron rod is inserted in the coil, then the domains in the iron become aligned by the solenoid's field, so that the iron itself becomes magnetic. Then the overall magnetic field is the sum of that due to the solenoid and that due to the iron, and it can be immense. **Electromagnets,** as such devices are called, are used widely to create magnetic fields for such diverse uses as lifting metallic objects (Fig. 16.9) and confining charged particles in accelerators. Mechanical switches often use solenoids, in

16.2 MOVING CHARGES AND MAGNETISM 519

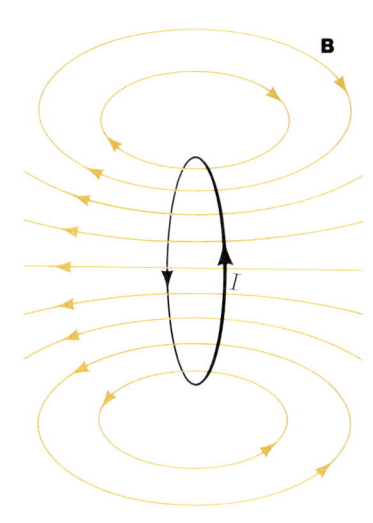

Figure 16.7

Magnetic field around a current loop. Inside the loop, the field lines from all segments of the current point in the same general direction.

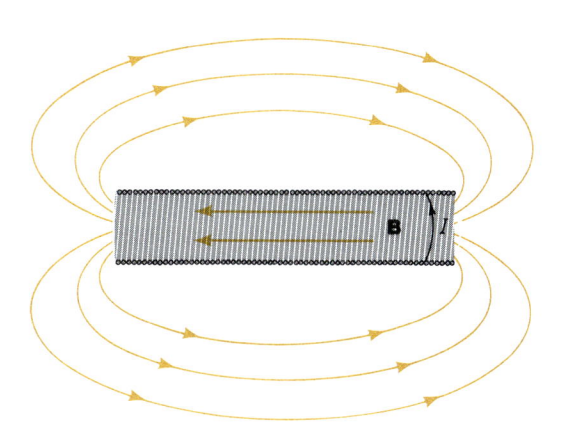

Figure 16.8

A solenoid. Magnetic field lines are parallel inside a solenoid, which is a tube consisting of many circular current loops created by coiling a single wire.

Figure 16.9

Powerful electromagnets. A strong magnetic field is created by wrapping current-carrying wires around a ferromagnetic core.

which a magnetized central iron rod moves along the axis of the solenoid (due to the magnetic force) when the current is turned on. The solenoid in a car's starter moves the starter motor into position so that its gear engages with the gear of the engine, for example, and solenoids are used in many other applications involving electrical motors.

16.3 Magnetic Forces on Moving Charges

We need to explore further the creation of magnetic fields by moving charges, and will do so later in this chapter. Before we do that, however, we must continue to analyze the forces between moving charges and magnetic fields, which we do in this section.

A moving charge creates a magnetic field, which exerts a force on a magnet. Therefore, Newton's third law dictates that a magnetic field must exert a force on a moving charge. Such forces, also discovered by Oersted, have many important consequences, and will be discussed extensively in the rest of this chapter.

The Force on a Current-Carrying Wire

We first consider a current-carrying wire in a uniform magnetic field (Fig. 16.10). We can assess both the direction and the magnitude of the force exerted on the wire. The force exerted on the wire by the magnetic field is perpendicular to both the current and the applied field. The right-hand rule is useful again: first the fingers are aligned with the current direction, then rotated to align with the direction of the magnetic field; the thumb then points in the direction of the force. If the field is not perpendicular to the current, then only the perpendicular component contributes to the force.

The strength of the force is used to define the magnitude of the magnetic field. The force is found from experiment to be proportional to the magnitude of the current and to the length of the wire that is subjected to the field. By definition, the magnetic field is also proportional to the magnitude of the force, so we have

$$F \propto IlB \sin\theta$$

where B is the strength of the magnetic field, I is the current, l is the length of the wire that is immersed in the field, and θ is the angle between **I** and **B**. By convention, the proportionality constant in this expression is defined to be precisely equal to 1, so we have the equation

$$F = IlB \sin\theta \qquad (16-1)$$

which then defines the units of B. In the SI system the unit of magnetic field strength is the **tesla** (**T**), which has dimensions F/Il, so that 1 T = 1 N/A · m. The tesla used to be called the weber per square meter, and this terminology may still be found (we will define the weber later in this chapter).

16.3 MAGNETIC FORCES ON MOVING CHARGES

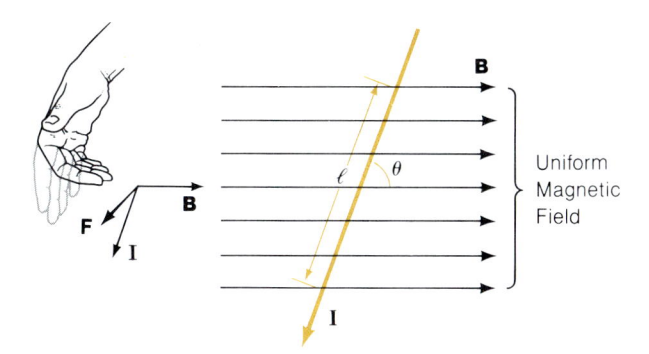

Figure 16.10

*The force on a current-carrying wire in a uniform magnetic field. The force is proportional to **B**, l, and the component of **I** that is perpendicular to **B**. Here the right-hand rule tells us that the direction of the force is out of the page.*

Another common unit for magnetic field strength is the **gauss** (**G**), which is the cgs unit: $1 \text{ G} = 10^{-4}$ T. The magnetic field strength at the earth's surface is about 0.5 G, or 5×10^{-5} T.

Example 16.1   A power line is suspended nearly horizontally between two utility poles 100 m apart, and aligned so that the direction of the current it carries is 65° east of north. If the current is 10,000 A, what is the magnitude and direction of the force exerted on the wire between the two poles by the earth's magnetic field, which we take as 0.5 G due north?

The current and the magnetic field lie in the horizontal plane, with the direction of the field pointed towards the north and that of the current 65° east of north. Hence the strength of the force is

$$F = IlB \sin\theta = (1 \times 10^4 \text{ A})(100 \text{ m})(5 \times 10^{-5} \text{ T})\sin(65°) = 45.3 \text{ N}.$$

The right-hand rule tells us that the force is upward.

The Force on a Free Charge

Just as a magnetic field exerts a force on a current, it also exerts a force on a single moving charge. If a charge q is moving at speed v, it is equivalent to a current I:

$$I = \frac{q}{t} = \frac{qv}{l}$$

where l is the distance travelled in time t. If we substitute this into Eq. 16–1, we find

$$F = qvB \sin\theta \qquad (16\text{–}2)$$

CHAPTER 16 MAGNETISM AND ELECTRICITY

where θ is the angle between the directions of the vectors **v** and **B**, whose magnitudes are v and B. The direction of **F** is given by the right-hand rule, as before, but now we must be careful to specify the sign of the charge. The right-hand rule gives the direction of **F** for a positive charge, whereas the force on a negative charge is in the opposite direction.

Example 16.2 An electron is moving horizontally at 3×10^6 m/s in a horizontal magnetic field of strength $B = 1 \times 10^{-3}$ T that is oriented perpendicular to the electron motion (Fig. 16.11). Find both the magnitude and direction of the force on the electron. How far away from a straight path will it be deflected as it traverses a distance of 1 cm?

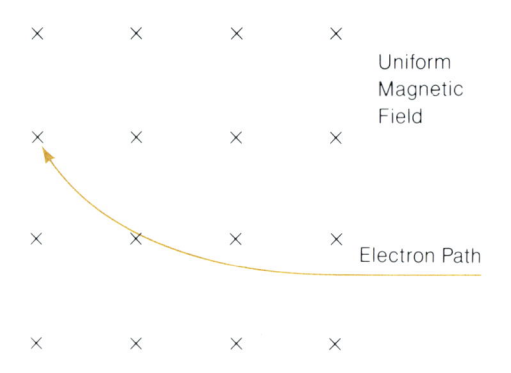

Figure 16.11

We have specified that the motion is perpendicular to the field, so $\sin\theta = 1$, and the strength of the force is

$$F = qvB = (1.6 \times 10^{-19} \text{ C})(3 \times 10^6 \text{ m/s})(1 \times 10^{-3} \text{ T}) = 4.8 \times 10^{-16} \text{ N}.$$

The electron moves right to left in Fig. 16.11 and the direction of the field is into the page. The right-hand rule therefore indicates that a positive charge would experience a force directed downward; hence the force on an electron is upward.

To find the displacement of the electron from a straight path requires calculating its acceleration due to the force.

$$a = \frac{F}{m} = \frac{4.8 \times 10^{-16} \text{ N}}{9.11 \times 10^{-31} \text{ kg}} = 5.3 \times 10^{14} \text{ m/s}^2$$

This is far in excess of the downward gravitational acceleration, which can therefore be ignored. This acceleration acts on the electron for a time $t = (0.01 \text{ m})/(3 \times 10^6 \text{ m/s}) = 3.3 \times 10^{-9}$ s. Initially there is no motion in the upward direction, so the displacement is simply

$$d = \frac{1}{2}at^2 = (0.5)(5.3 \times 10^{14} \text{ m/s}^2)(3.3 \times 10^{-9} \text{ s})^2 = 0.0029 \text{ m} = 0.29 \text{ cm}.$$

16.3 MAGNETIC FORCES ON MOVING CHARGES

We have neglected the change in acceleration as the direction of the electron's motion changes. The upward deflection that we calculated is nearly one third of the horizontal distance covered by the electron, so the direction of electron motion is no longer perpendicular to the force exerted on it by the magnetic field. Hence we have found only an approximate answer to the last question.

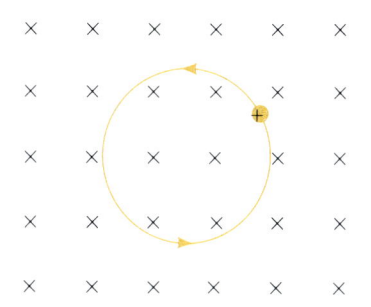

Figure 16.12

Motion of a charged particle in a magnetic field. The field direction is into the page, and the positive charge moves in a circular path because the force on it is always perpendicular to its direction of motion.

If a freely moving charged particle travels through a uniform magnetic field, then its direction of motion will constantly change due to the force exerted by the field. The force is always perpendicular to the field and to the direction of motion, so the acceleration is also perpendicular to both, and there is no change of the particle's speed. There is only a change in direction, resulting in a circular or spiral motion. If the charged particle is initially moving perpendicular to the direction of the field, then the resulting motion will be a circle in a plane perpendicular to the field (Fig. 16.12). If the charged particle has a velocity component parallel to the field, then it will continue to move along the field, but will also execute a circular motion in the perpendicular plane, the result being a spiral path (Fig. 16.13).

We can use simple considerations from our earlier discussions of mechanics to find the radius of the circular motion and its frequency. The force exerted on the particle by the field is always perpendicular to the particle's velocity, and is given by Eq. 16–2. We set this equal to the centripetal force:

$$qvB = \frac{mv^2}{r}$$

where we have dropped the $\sin\theta$ term because the angle is always 90°. Solving for r leads to

$$r = \frac{mv}{qB}. \qquad (16\text{--}3)$$

The period of the circular motion is the time it takes for the charged particle to complete a circle of circumference $2\pi r$, which is

$$T = \frac{2\pi r}{v} = \frac{2\pi m}{qB} \qquad (16\text{--}4)$$

and the frequency ν is

$$\nu = \frac{1}{T} = \frac{qB}{2\pi m}. \qquad (16\text{--}5)$$

These relations are used in the design of some particle accelerators call **cyclotrons** to keep charged particles moving in circular paths. The radius in Eq. 16–3 is called the **cyclotron radius,** and the frequency in Eq. 16–5 is the **cyclotron frequency.**

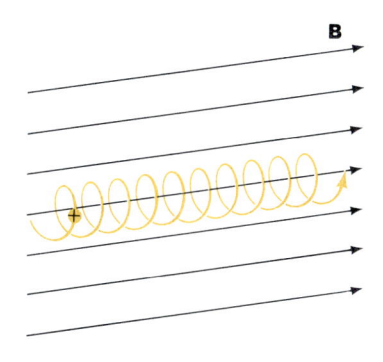

Figure 16.13

Spiral path of a positively charged particle moving along magnetic field lines. The field is directed toward the right.

Example 16.3 What are the cyclotron radius and frequency for electrons moving at one tenth the speed of light, in a magnetic field where $B = 5$ T? What are the radius and frequency for protons under the same conditions?

The radius for the electron is

$$r = \frac{mv}{qB}$$
$$= \frac{(9.1 \times 10^{-31} \text{ kg})(3 \times 10^7 \text{ m/s})}{(1.6 \times 10^{-19} \text{ C})(5 \text{ T})} = 3.4 \times 10^{-5} \text{ m}.$$

The frequency is

$$\nu = \frac{qB}{2\pi m} = \frac{(1.6 \times 10^{-19} \text{ C})(5 \text{ T})}{2\pi(9.1 \times 10^{-31} \text{ kg})} = 1.4 \times 10^{11} \text{ Hz}.$$

For protons, the calculations differ only by the ratio of the proton mass to that of the electron, a ratio of $(1.672 \times 10^{-27} \text{ kg})/(9.109 \times 10^{-31} \text{ kg})$ or 1,836, so the radius for the proton is

$$r = (1,836)(3.4 \times 10^{-5} \text{ m}) = 0.062 \text{ m} = 6.2 \text{ cm}.$$

The frequency for the proton is

$$\nu = \frac{1.4 \times 10^{11} \text{ Hz}}{1,836} = 7.6 \times 10^7 \text{ Hz}.$$

Charge Separation in Conductors: The Hall Effect

We have spoken of electrical currents as though they consisted of strictly one-dimensional flows of charged particles, whereas the charged particles are actually able to move in three dimensions. Consider a current in a wire. In the absence of an external magnetic or electric field, the charged particles will indeed move in one dimension, in a straight line dictated by the direction of the potential difference that is driving the current. When there is an external field, however, the motion is more complex.

Consider a wire carrying a current through a magnetic field. The field exerts a force on the charged particles in the current, so that the particles are deflected to one side of the conductor, depending on the direction of the force. If electrons move through the wire, then they will travel along the side of the wire opposite to that given by the right-hand rule (Fig. 16.14), since they carry negative charges. The tendency of the electrons to travel along one side of the wire creates an internal charge separation that in turn creates an internal potential difference perpendicular to the wire. This potential difference acts against the effect of the magnetic field, tending to accelerate the electrons

16.3 MAGNETIC FORCES ON MOVING CHARGES

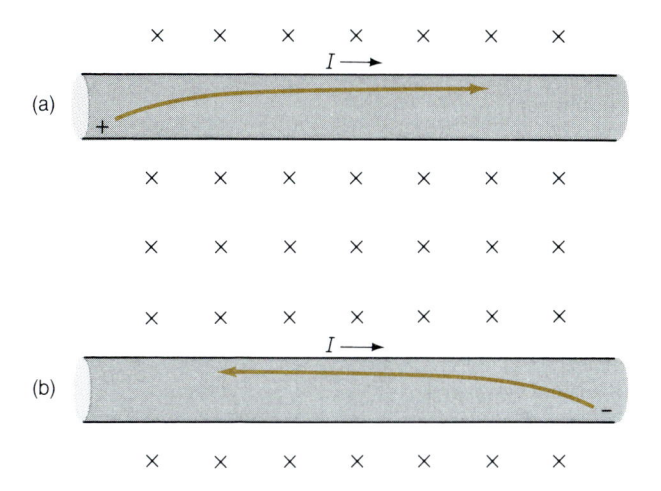

Figure 16.14

The Hall effect. The magnetic field is directed into the page, and the conventional current is to the right in both (a) and (b). In (a), positive charges moving to the right are deflected upward. In (b), negative charges moving to the left are also deflected upward. Thus the direction of the emf due to charge separation is opposite in the two cases.

toward the other side of the wire. At equilibrium, the charge separation due to the magnetic field is balanced by this potential difference.

This separation of charge in a conductor within a magnetic field allows us to determine whether a current consists of negative or positive particles. In Fig. 16.14, because the direction of the magnetic field is into the page, positive charges moving to the right or negative charges moving to the left will be deflected toward the top of the figure. The net current in both cases may be the same, but the direction of the internal charge separation, hence of the internal potential difference across the conductor, is not (see Special Problem 1 at the end of this chapter). The sign of the potential difference that is created in the conductor reveals the sign of the charge on the moving particles. Thus this effect, called the **Hall effect** after its discoverer E. H. Hall, was used to establish that most currents consist of moving electrons.

The Hall effect also provides a convenient method for measuring magnetic field strengths, because the potential difference created within a conductor is proportional to the field strength. A conductor, usually consisting of a flat cable carrying a small current, is placed in the magnetic field. The potential difference across the conductor (in the direction perpendicular to both the current and the field) is measured with a potentiometer or voltmeter. Regardless of sign, the potential difference is proportional to the strength of the field. The device, called variously a **Hall probe** or a **gaussmeter**, must be calibrated against a standard field, a procedure we discuss later in the chapter.

CHAPTER 16 MAGNETISM AND ELECTRICITY

The Force on a Coil: The Galvanometer

Consider a loop of wire that carries a current in a magnetic field aligned parallel to the plane of the loop (Fig. 16.15). The field exerts a force on the components of the current that are perpendicular to the field. If the field is horizontal as shown, then a force is directed into the page on the right-hand side of the loop, and out of the page on the left-hand side (there is no force on the top and bottom segments, which are parallel to the field). The result is a torque that acts to rotate the loop about a vertical axis. The magnitude of this torque is increased if we have a coil of many loops because the magnetic field acts on more charges. Both the galvanometer and the electric motor take advantage of these torques. The motor normally requires an AC current, and will be described in the next chapter, but the galvanometer represents a straightforward application that we can describe here.

The galvanometer is a small coil (usually in a rectangular shape) with a current I in each of N loops, as shown in Fig. 16.16. The coil is suspended between opposite poles of a permanent magnet, so that the magnetic field is uniform and parallel to the plane of the coil. The coil is mounted in a pivot, with one end of the axis attached to a spring. The coil rotates against the restoring force produced by the spring, reaching equilibrium when the torque produced by the magnetic field is balanced by the opposing torque due to the restoring force of the spring. Since the magnetic torque is proportional to the current in the coil, and the restoring torque is proportional to the angular displacement, the displacement is proportional to the current. Thus a pointer attached to the pivot axis of the coil indicates the strength of the current. After a galvanometer is calibrated against a standard current, it can be used to measure unknown currents, as we have already discussed (Chapter 15).

We can derive an expression for the torque exerted on a coil by a magnetic field. For simplicity, let us consider a rectangular coil of dimensions X and Y. Then we need consider only the torque due to the two coil segments of length Y that are perpendicular to the field. Recall that torque is the product of force and the perpendicular component of the lever arm, the displacement from the pivot. Thus the torque on one side of the coil is

$$\tau = \frac{FX}{2} = \frac{NIYBX}{2}$$

where the force F has been set equal to $NIYB$, following Eq. 16–1, and the net current is NI, where N is the number of loops of wire in the coil. The contribution to the torque by the other vertical segment of the coil is equal to this torque, and since both torques act in the same direction, the total torque is

$$\tau = NIAB \qquad (16\text{--}6)$$

where $A = XY$ is the area of the coil. It can be shown, using methods of calculus, that Eq. 16–6 holds for any shape of coil. It is this torque that is opposed by the restoring force of the spring in the galvanometer.

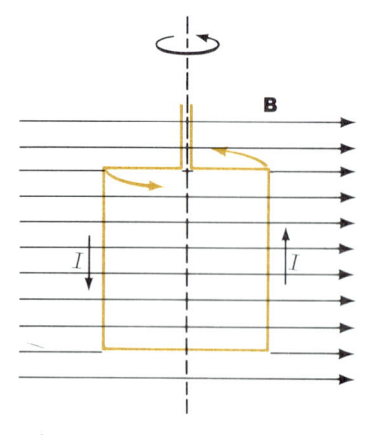

Figure 16.15

The torque on a current loop in a magnetic field. The components of the current perpendicular to the magnetic field create forces which apply a net torque about the vertical axis.

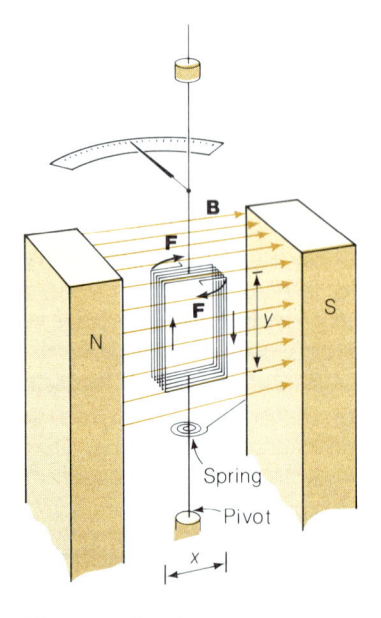

Figure 16.16

The galvanometer. Current passing through multiple loops creates a force due to the magnetic field. The resulting torque rotates the current loop against a spring; the equilibrium deflection is proportional to the current.

16.4 ANALYZING MAGNETIC FIELDS

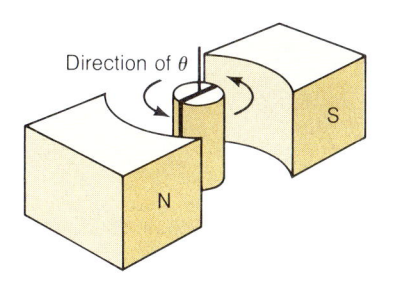

Figure 16.17

Cylindrical geometry in a galvanometer. The galvanometer coil is wrapped around an iron cylinder in a radial magnetic field, to ensure that the torque exerted on the current is independent of the angle of rotation.

The restoring torque for a coil spring is

$$\tau = k\phi$$

where k is the spring constant and ϕ is the angle of the spring's displacement. Thus, when equilibrium has been reached in a galvanometer, the torque from Eq. 16–6 is balanced by the restoring torque. Setting the two equal and solving for the displacement angle yields

$$\phi = \frac{NIAB}{k}. \qquad (16-7)$$

It is the angle ϕ that is proportional to the current in the galvanometer coil.

There is still a complexity that we need to address. Our calculation of the torque is valid only when the magnetic field is aligned with the plane of the coil, and this is not true once the coil has rotated. Hence the torque varies. The variable factor in Eq. 16–6 is the area A which, strictly speaking, is the component of the area that is parallel to the magnetic field. Thus the full expression for the torque is

$$\tau = NIAB \cos\theta \qquad (16-8)$$

where θ is the angle between the field and the plane of the coil.

The purpose of a galvanometer is to provide a measure that is proportional to the current, so the $\cos\theta$ term must be eliminated. This is normally done by using a magnet whose poles have curved faces, creating a radially-directed magnetic field (Fig. 16.17). Then the field is always parallel to the plane of the coil, even though the coil rotates through large angles.

16.4 Analyzing Magnetic Fields

In an earlier section of this chapter we discussed magnetism, and we found that magnetic fields are created by moving electric charges. We did not go much beyond that, however, because until we discussed the forces created by magnetic fields, we could not quantitatively discuss the fields themselves. Recall that a field is characterized by the forces it exerts; now that we know something about these forces, we can return to our discussion of magnetic fields.

Ampere's Law

There is a general relationship between current and magnetic field, which we will develop by first considering the case of a straight wire in which a current flows. We already know that the field is cylindrical, with field lines that are circles centered on the wire (Fig. 16.18). Experiments show that the strength

CHAPTER 16 MAGNETISM AND ELECTRICITY

of the field at a distance r from the wire is proportional to the current I and inversely proportional to r:

$$B \propto \frac{I}{r}.$$

The proportionality constant has the form $\mu_0/2\pi$, where μ_0 is a constant called the **permeability of free space**, having the value $\mu_0 = 4\pi \times 10^{-7}$ T · m/A. The expression for the magnetic field therefore becomes

$$B = \frac{\mu_0 I}{2\pi r}. \qquad (16\text{–}9)$$

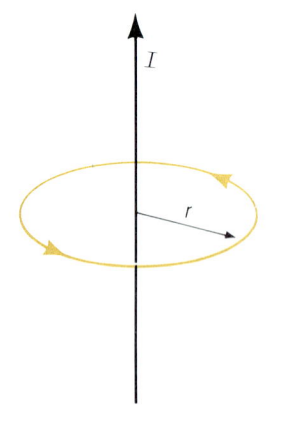

Figure 16.18

Magnetic field due to a straight current-carrying wire.

This expression is valid for the special case of a long straight wire. The French scientist A. M. Ampère developed a more general expression which gives the magnetic field for any shape conductor. Consider a current-carrying conductor of arbitrary shape, and imagine a closed path around it (Fig. 16.19). The value of B along the closed path will vary, depending on the shape of the conductor and of the path. The general expression for B at any point on the path is not simple, but Ampere did find a general relation between the field along the entire path and the current in the conductor. Ampere found that the sum of the product $B_\parallel \Delta l$ for all segments is related directly to the current I:

$$\Sigma B_\parallel \Delta l = \mu_0 I \qquad (16\text{–}10)$$

where $B_\parallel$ represents the component of **B** that is parallel to the segment Δl, and I is the total current passing through the enclosed path. This expression, known as **Ampère's law**, is difficult to evaluate for arbitrary shapes without calculus or numerical integration.

In special cases it is possible to choose an enclosed path for which B is constant, however, and B can be evaluated with only simple algebra. To show this, we reconsider the case of a long straight wire carrying current I. By symmetry, we know that B will be the same at any r from the wire, so we choose as our enclosed path a circle of radius r centered on the wire. Then the sum on the left-hand side of Eq. 16–10 can be rewritten as

$$B\Sigma \Delta l = \mu_0 I.$$

The sum of all the segments around a circle is the circumference of the circle, which is $2\pi r$, so we have

$$B(2\pi r) = \mu_0 I$$

which easily reduces to the form of Eq. 16–9. We see why the proportionality constant in Eq. 16–9 is usually written in terms of $\mu_0/2\pi$; this reflects the more fundamental expression in Eq. 16–10, where the proportionality constant is simply μ_0.

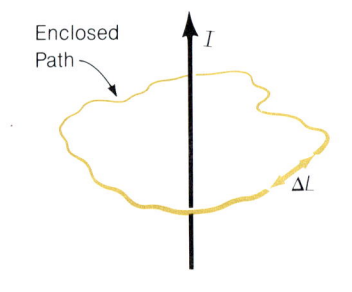

Figure 16.19

Ampere's law. Here an enclosed path of arbitrary shape surrounds a conductor carrying a current. Ampere found a relationship between the current and the sum of all products $B_\parallel \Delta l$ around the enclosed path, where $B_\parallel$ is the component of the magnetic field parallel to each path segment Δl.

16.4 ANALYZING MAGNETIC FIELDS

Example 16.4   We know that a compass locates the direction of the earth's north pole, which means that it can detect and align itself with a field whose strength is 5×10^{-5} T. Suppose a wire in the wall of a house carries a current of 0.9 A (roughly the current drawn by a 100 W light bulb). Could an ordinary compass be used to locate the wire, if the wire is 10 cm inside the wall? In which direction would the needle point, if the current flows vertically downward? (Household current is actually not a steady current in one direction, but assume it is for the time being.)

To answer the first part of the question, we need to calculate the strength of the magnetic field due to the wire at 10 cm. From Eq. 16–9, we have

$$B = \frac{\mu_0 I}{2\pi r} = \frac{(4\pi \times 10^{-7} \text{ T} \cdot \text{m/A})(0.9 \text{A})}{2\pi(0.1 \text{ m})} = 1.8 \times 10^{-6} \text{ T}.$$

Thus the field is only about 3.6% as strong as the earth's field, and the compass may not be sensitive enough to be deflected by it.

We can still answer the second part of the question, however, on the assumption that the needle would be deflected. The direction of the field is given by the right-hand rule. We see that the field is in the clockwise direction (as seen from above), so that the needle will point horizontally and to the left as we face the wall where the wire is located.

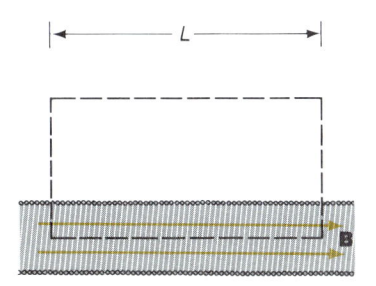

Figure 16.20

Magnetic field inside a solenoid. Only the side of the rectangular path inside the solenoid parallel to the field contributes to the summation $\Sigma B_\parallel \Delta l$ in Ampere's law.

Another configuration of current for which Eq. 16–10 gives B is the solenoid. Recall that the solenoid is a coil that carries a current, so that inside the coil the field created by the current is parallel to the axis of the coil (Fig. 16.8). Except near the ends, where the field lines diverge, the field inside a solenoid is uniform along its length. We can also show that it has the same value at all points inside the solenoid.

Consider an enclosed rectangular path (Fig. 16.20). One side of the rectangle is inside the solenoid, and the opposite side is outside. We note first that the current passing through this enclosed path is the same, regardless of where the side of the rectangle that is inside the solenoid lies. Furthermore, the value of B along this side of the rectangle is constant, so the product $B_\parallel \Delta l$ for that segment is simply BL. We note also that there is no component of B that is parallel to the two ends of the rectangle, so the product $B_\parallel \Delta l$ for them is zero. Finally, the magnetic field outside a solenoid is very small, so we can neglect the product $B_\parallel \Delta l$ for the fourth side of the rectangle. Therefore the left-hand side of Eq. 16–10 reduces to BL. On the right-hand side, we need to write the value of the current flowing through the rectangle. If there are N loops of wire within the length L, then the total current passing through the rectangle is NI, where I is the current in the wire. Putting these expressions together, we have

$$BL = \mu_0 NI$$

or

$$B = \left(\frac{N}{L}\right)\mu_0 I = n\mu_0 I \tag{16-11}$$

where n is the number of current loops per unit length. Because this expression does not depend on position within the solenoid, we see that solenoids are practical devices for creating uniform magnetic fields. The magnetic field at the center of a single loop of radius R is $B = \mu_0 I/2R$.

Example 16.5

Suppose we have a solenoid with 200 wire loops per cm. What current is needed to produce a magnetic field equal in strength to that of the earth?

The required field strength is 5×10^{-5} T, so we solve for I:

$$I = \frac{B}{n\mu_0} = \frac{5 \times 10^{-5} \text{ T}}{(2 \times 10^4/\text{m})(4\pi \times 10^{-7} \text{ T} \cdot \text{m/A})} = 2.0 \times 10^{-3} \text{ A}$$

We find that a rather small current can match the magnetic field of the earth.

The equations considered so far assume that only air fills the loops or solenoid. If iron or other ferromagnetic material is inserted, the total magnetic field will be enhanced by a large factor, often 100 – 10,000, because the magnetic domains in the iron become aligned by the external field. The total field can be written

$$\mathbf{B} = \mathbf{B}_0 + \mathbf{B}_M$$

where $\mathbf{B}_0$ is the applied field due to the current loops and $\mathbf{B}_M$ is the additional field due to the aligned domains. We can account for the extra field simply by changing the constant μ_0 to μ, the **magnetic permeability** of the material inside the loops. Thus, the field inside an iron-core solenoid becomes

$$B = n\mu I$$

For free space (air), $\mu = \mu_0$. For iron, steel, and other ferromagnetic alloys, μ can be much greater, although the precise value depends on the strength of B_0. For example, for $B_0 = 20$ Gauss, pure iron has a permeability 5,000 times that of free space, and some alloys reach permeabilities 100,000 times μ_0. The magnetic properties of alloys and superconductors are a research field of vast importance for developing large electromagnets and related devices.

The Ampere

Until now we have used the ampere as a unit of current without knowing much more about it than its value. Now that we have discussed forces exerted

16.5 MAGNETIC INDUCTION

by magnetic fields, and the fields created by moving charges, we are finally in position to understand the basic definition of the ampere.

Consider two parallel wires of length l, separated by distance R, carrying currents I_1 and I_2 (Fig. 16.21). Each wire's magnetic field exerts a force on the other. The field at the position of wire 2 due to the current I_1 is

$$B_1 = \frac{\mu_0 I_1}{2\pi R}$$

and the force it exerts on wire 2 is

$$F = I_2 l B_1 = \frac{I_2 l \mu_0 I_1}{2\pi R}.$$

If we divide by the length l to find the force per unit length, we have

$$\frac{F}{l} = \frac{\mu_0 I_1 I_2}{2\pi R}. \tag{16–12}$$

If the two currents are in the same direction, we find that the force is attractive (to see this, apply the right-hand rule first to find the direction of $\mathbf{B}_1$, then again to find the direction of $\mathbf{F}$). If the currents are in opposite directions, then the force is repulsive.

If each wire carries 1 A, and they are separated by 1 m, then the force per unit length is

$$\frac{F}{l} = \frac{(4\pi \times 10^{-7}\text{ T}\cdot\text{m/A})(1\text{ A})(1\text{ A})}{2\pi(1\text{ m})} = 2 \times 10^{-7}\text{ N/m}.$$

(recall that 1 T = 1 N/A · m). At last we have the definition of the ampere: the ampere is the current flowing in each of two long parallel wires one meter apart which produces a force on a unit length of either wire of exactly 2×10^{-7} N/m. This may seem like an odd way to define one of the fundamental quantities in physics, but it has the great advantage that the force between two wires can be precisely measured. Earlier (in Chapter 14) we discussed the coulomb, noting that it is not so easy to precisely measure a quantity of charge. Hence the coulomb is considered a derived unit, and, as we already have stated, is defined so that 1 C = 1 A · s.

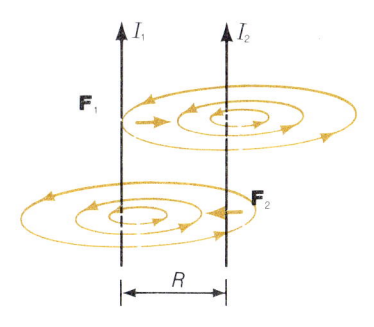

Figure 16.21

Definition of the ampere. The two parallel wires exert attractive forces on each other. One ampere is the current needed in each wire to provide a force of 2×10^{-7} N/m when the wires are separated by 1 m.

16.5 Magnetic Induction

To close our discussion of magnetism and electricity, we need to consider the creation of electric current by a magnetic field. We might have suspected that this is possible, since the reverse process, the creation of a magnetic field by a current, has already been discussed. The process of inducing a current (or, equivalently, an electromotive force) is crucial to modern technology, for this

CHAPTER 16   MAGNETISM AND ELECTRICITY

is how most of our electrical power is generated. We discuss practical aspects of generators in the next chapter, but here we can analyze their principles.

The American Joseph Henry and the Englishman Michael Faraday performed experiments in the early nineteenth century that revealed the relationship between magnetic field and induced current. They found that no induced current is created when the conductor does not move in the magnetic field, and the field strength does not vary. It was found, however, that a current is produced when the conductor moves or when the field strength varies. We consider the moving conductor first.

Induced emf and Faraday's Law

We already know that a magnetic field exerts a force on a moving charge. When a conductor moves in a magnetic field, a force is exerted on the free electrons in the conductor, creating an induced emf. We can find the value of this emf by considering the straight wire of length L moving in a magnetic field (Fig. 16.22). A charged particle in the wire is subjected to a force whose strength, from Eq. 16–2, is

$$F = qvB \sin\theta$$

where q is the charge on the particle, v is the velocity of the wire, B is the field strength, and θ is the angle between the direction of motion and the field direction. The direction of **F**, by the right-hand rule, is parallel to the wire, so the charged particle moves along the rod. As it does so, work is done, given by

$$W = FL = qvBL \sin\theta$$

where L is the distance travelled by the charge. The work done creates a potential difference, which is the induced emf $\mathscr{E}$, and is equal to the work done per unit charge:

$$\mathscr{E} = \frac{W}{q} = BvL \sin\theta. \qquad (16\text{–}13)$$

If the velocity **v** is perpendicular to the direction of **B**, then the maximum work is done, because $\sin\theta = 1$, and we have

$$\mathscr{E} = BvL. \qquad (16\text{–}14)$$

The current induced in the wire is

$$I = \frac{\mathscr{E}}{R} = \frac{BvL}{R} \qquad (16\text{–}15)$$

where R is the resistance. The work done, which creates electric potential, has to be supplied by some outside agent as the wire is moved.

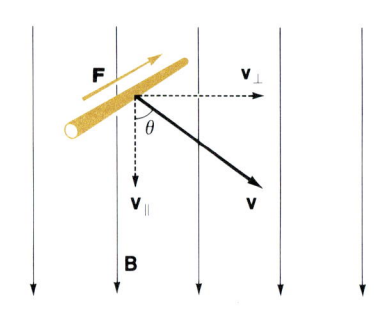

Figure 16.22

Induced emf. As a conductor moves through a magnetic field, forces are exerted on electrical charges in the conductor, causing charge separation and hence an induced emf.

16.5 MAGNETIC INDUCTION

The conservation of energy requires that the direction of the induced current is such that the force on that current acts to oppose the motion of the wire. This ensures that work must be done to move the wire through the magnetic field; if the current flowed the other way, it would create a force that acted to move the wire in the same direction as it is already being moved, and this would mean that the total energy of the system would be greater than the amount of work put in. This statement of conservation of energy was first formulated by H. Lenz, and is known as **Lenz's law**:

The induced emf creates a current whose action opposes the change that induced it.

Thus, in Fig. 16.22, the induced emf in the wire creates a current flowing into the page. From the right-hand rule, we see that this is the direction that creates a force (given by Eq. 16–1) toward the left, acting against the motion of the wire, which is to the right.

In a generator, wires arranged in a circuit loop called an **armature** are rotated through a magnetic field (Fig. 16.23). For maximum efficiency, the circuit loop is perpendicular to both its direction of motion and to the magnetic field (so the field is radial, as in galvanometer). The total current produced is enhanced by an armature of many wires, so that the total length of wire moving through the magnetic field can be very large. The work done must be supplied by whatever agent rotates the armature; in many generating plants, steam-powered turbines are used.

It is instructive to consider the direction of the current in the armature. When the loop is at the position pictured in Fig. 16.23 the force created by the current in the upper segment of the armature must act toward the left. Since the direction of **B** is upward, the right-hand rule tells us that the current in this segment must be flowing out of the page. At the bottom, similar considerations tell us that the current must be flowing into the page. Half a revolution later, however, the two segments of the armature have exchanged places, but the direction of the currents in the upper and lower portions must still be the same. Hence the direction of flow in the armature loop must reverse itself every half revolution. Thus a simple generator such as this one produces an alternating current. If a direct current is required, then the armature must use a switching device called a **commutator**. Generators and related devices will be discussed more fully in the next chapter.

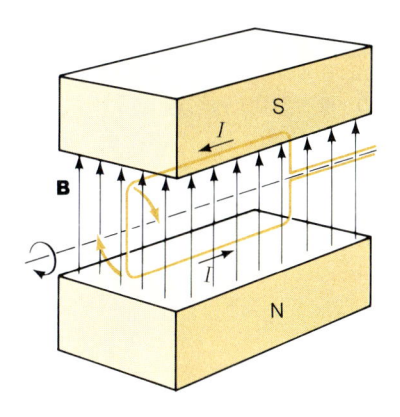

Figure 16.23

A simple generator. A conducting loop is rotated in a magnetic field, inducing a current. B is upward, and the rotation of the loop is clockwise. The induced current flows out of the page in the upper segment, and into the page in the lower segment.

Example 16.6 Suppose a generator has 1,000 rectangular circuit loops 50 cm long wound around a cylindrical armature of radius 10 cm (Fig. 16.24) which rotates at 100 Hz in a radial magnetic field of strength 0.1 T. What is the induced emf in this generator and what is the current produced if the total resistance in the wire is 20Ω?

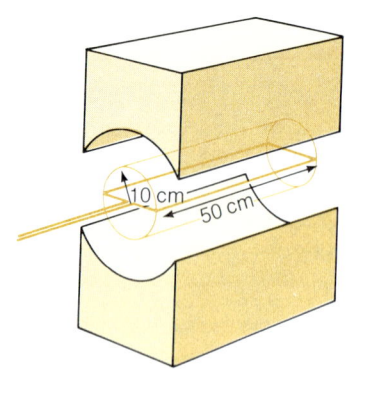

Figure 16.24

This really asks what the current will be in a moving wire (0.5 m)(10³) = 500 m long, where the velocity is the tangential velocity at radius 10 cm for a circular motion whose period is $T = 1/\nu = 1/100$ Hz = 0.01 s. The velocity v is

$$v = \frac{2\pi r}{T} = \frac{2\pi (0.1 \text{ m})}{(0.01 \text{ s})} = 62.8 \text{ m/s}.$$

The wires are perpendicular to the field and to their own motion, so the emf is

$$\mathcal{E} = BvL = (0.1 \text{ T})(62.8 \text{ m/s})(500 \text{ m}) = 3.1 \text{ kV}.$$

The current is

$$I = \frac{\mathcal{E}}{R} = \frac{(3.1 \times 10^3 \text{ V})}{20\Omega} = 160 \text{ A}.$$

Thus, this rather modest generator produces a substantial current.

Induction by a Variable Magnetic Field

The experiments of Faraday showed that a current is induced in a conductor subjected to a varying magnetic field even if there is no physical motion. Careful observation showed that the magnitude of the induced emf does not depend precisely on the strength of the magnetic field, but instead is somehow related to the total field encompassed by the circuit loop in which the current is induced. The induced emf is related to the **magnetic flux,** which is defined as

$$\Phi = B_\perp A = BA \cos\phi \qquad (16-16)$$

where Φ is the magnetic flux and $B_\perp$ is the component of the magnetic field that is perpendicular to the plane of the circuit loop whose area is A. We see that $B_\perp$ is equal to $B \cos\phi$, where ϕ is the angle between the direction of **B** and a normal to the plane of the circuit loop (Fig. 16.25). The magnetic flux can be thought of as the total intensity of the magnetic field over the area A. In terms of magnetic field lines, if the density of lines indicates the intensity of the field, then the magnetic flux is proportional to the total number of lines enclosed by the loop.

The units of magnetic flux are those of magnetic field intensity B (sometimes called **magnetic induction**) times units of area. A special name, the **weber** (Wb), is used for this; 1 Wb = 1 T · m². (This explains why, until the tesla was adopted as the unit of magnetic field intensity, it used to be expressed in Wb/m², as mentioned earlier in this chapter.)

To see how a change in magnetic flux creates an induced emf, we consider Fig. 16.26, where we have a rectangular current loop in a magnetic field. One

Figure 16.25

*Magnetic flux. The area of the circuit loop is A, and ϕ is the angle between the directions of **B** and a normal to the plane of the loop. The flux is $\Phi = BA\cos\phi$.*

16.5 MAGNETIC INDUCTION

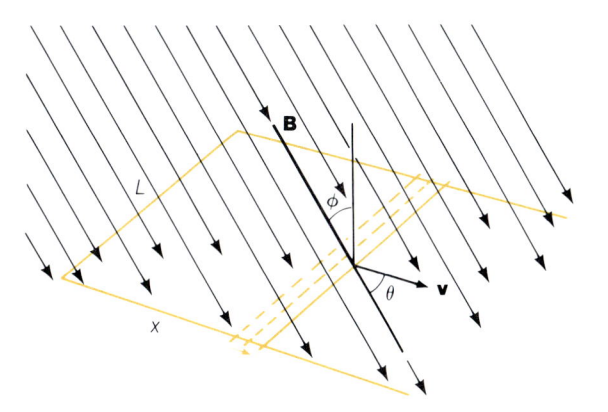

Figure 16.26

Faraday's law. A rectangular circuit with one movable side encloses a variable area. Thus the magnetic flux changes with time and creates an induced emf.

side of the loop is movable. We have already seen that, when the wire moves, its induced emf is given by Eq. 16–13:

$$\mathscr{E} = BvL \sin\theta$$

where θ is the angle between the direction of motion and the direction of **B**. This angle is the complement of the angle ϕ between the field direction and the perpendicular to the plane of the current loop, so $\sin\theta = \cos\phi$. The velocity v can be expressed in terms of the rate of change of x, the dimension of the rectangle that is perpendicular to L:

$$v = \frac{\Delta x}{\Delta t}.$$

Then the expression for the induced emf becomes

$$\mathscr{E} = BL\left(\frac{\Delta x}{\Delta t}\right)\cos\phi.$$

The area of the current loop is $A = Lx$, so the quantity $L(\Delta x/\Delta t)$ is equal to the rate of change of the area. We therefore have

$$\mathscr{E} = B\left(\frac{\Delta A}{\Delta t}\right)\cos\phi$$

and this is equivalent to

$$\mathscr{E} = -\frac{\Delta \Phi}{\Delta t}. \tag{16–17}$$

Thus, we have an expression for the induced emf in terms of the rate of change of the magnetic flux. This expression is known as **Faraday's law** (it is usually written with a minus sign to indicate that the direction of the induced emf acts to oppose the change in Φ). If the circuit loop is a coil of N turns, then Faraday's law becomes

$$\mathcal{E} = -N\left(\frac{\Delta \Phi}{\Delta t}\right). \tag{16–18}$$

Faraday's law (Eq. 16–17 or Eq. 16–18) is considered a more general expression for induction than our Eq. 16–13 through Eq. 16–15, because emf is induced for any change in Φ, whether it is due to a changing area, as in our derivation here, or to a changing magnetic field intensity. In the next chapter we will discuss how to produce a variable magnetic field.

Example 16.7   One way to produce a momentary, time-variable magnetic field is to turn a current on or off. Suppose the circuit described in Example 16.4 is turned on, so that the current in the wire in the wall rises from zero to 0.9 A in 10 μs. What would be the maximum induced emf during that time in a circuit loop of area 0.1 m², 10 cm from the wire? How should the circuit loop be oriented so maximum emf is induced? If the resistance in the circuit loop is 0.005Ω, what is the induced current?

We answer the second question first. We found in Example 16.4 that the direction of the magnetic field 10 cm from the wire is horizontal and parallel to the wall. Therefore the maximum induced emf will occur when the plane of the circuit loop is held vertical and perpendicular to the wall.

We found in Example 16.4 that the intensity of the magnetic field at this point is $B = 1.8 \times 10^{-6}$ T when the current is flowing down the wire in the wall. We want to find the maximum induced emf, which occurs when the plane of the circuit loop is perpendicular to **B**, so the magnetic flux is simply

$$\Phi = AB.$$

Therefore the induced emf, from Eq. 16–17, is

$$\mathcal{E} = \frac{\Delta \Phi}{\Delta t} = A\left(\frac{\Delta B}{\Delta t}\right) = \frac{(0.1 \text{ m}^2)(1.8 \times 10^{-6} \text{ T})}{10 \times 10^{-6} \text{ s}} = 0.018 \text{ V}.$$

The current is

$$I = \frac{\mathcal{E}}{R} = \frac{(0.018 \text{ V})}{(0.005 \Omega)} = 3.6 \text{ A}.$$

This is substantial, but recall that it flows only briefly.

FOCUS ON PHYSICS

The Earth's Magnetic Dynamo

Simple experiments with a compass, or even a needle floating on a glass of water, show that the earth has a magnetic field. The strength of the field is small, less than one percent of the strength of a typical child's horseshoe magnet. Nevertheless, the earth's magnetic field has had a profound impact on life forms on the earth's surface. Furthermore, the existence and properties of the field can tell us a great deal about the deep interior of our planet.

The earth's magnetic field has two opposing poles with field lines connecting them (see Fig. 16.2). A compass needle will always align itself so that it points toward the south magnetic pole, which is near (but not precisely at) the geographic north pole of the earth. It is found that the magnetic poles move about steadily, covering some 70 miles every ten years (Fig. 16.1.1 shows the location of the north magnetic pole at several times since 1900). In addition, there are often rapid fluctuations that can result in temporary excursions of fifty miles or more in less than a day.

Perhaps most surprising is that the earth's magnetic field occasionally reverses its poles, so that the north magnetic pole moves to the opposite side of the planet. These reversals occur sporadically, but on the average about 300,000 years apart, and none has been recorded directly. We know about the reversals through the study of rocks that record the magnetic field direction at the time of their formation. About half the rocks analyzed for magnetic alignment show one orientation of the earth's field, and the rest show the opposite orientation, indicating that neither alignment has been favored over the lifetime of the earth. The best chronometer for dating ancient field reversals is found under the oceans, where the seafloor spreads slowly away from ridges where material from the earth's interior rises (Fig. 16.1.2). The magnetic alignment of seafloor rocks forms a pattern of alternating directions, showing that the earth's field has alternated steadily (but irregularly) over hundreds of millions of years while the seafloor has been spreading.

We know from the discussions in this chapter that a magnetic field is created by a changing electric field or by an electric current. We can guess that the earth's magnetic field must be somehow created by electric currents in its interior. The location and cause of these currents has been very difficult for geophysicists to determine, however. It is known from studies of seismic waves that the earth has a liquid zone in its outer core. It is suspected that flowing motions in this liquid zone create the currents needed to produce the earth's magnetic field, in a process called a magnetic dynamo. Electric currents are induced when molten material, which has a high metallic content and can therefore conduct electricity, flows across the earth's magnetic field lines. These currents, in turn, create a magnetic field. Thus, the creation of a field by the dynamo requires the prior presence of a field. It is thought that the earth had a field when it formed, and that the dynamo sustains it, but could not have initiated it.

Even though there is general agreement that the dynamo mechanism is responsible for maintaining the earth's magnetic field, a very important question remains. The dynamo depends on flows in the earth's interior, but it is not yet known with certainty what causes these flows. The possibilities include radioactive heating (which is known to be important, but which was thought until recently to occur primarily in the earth's crust, rather than in the outer core), chemical reactions, and tidal stresses. Another hypothesis is that the currents are driven by released gravitational energy, as heavy elements gradually sink through the molten core toward the earth's center. This process, called dif-

(continued)

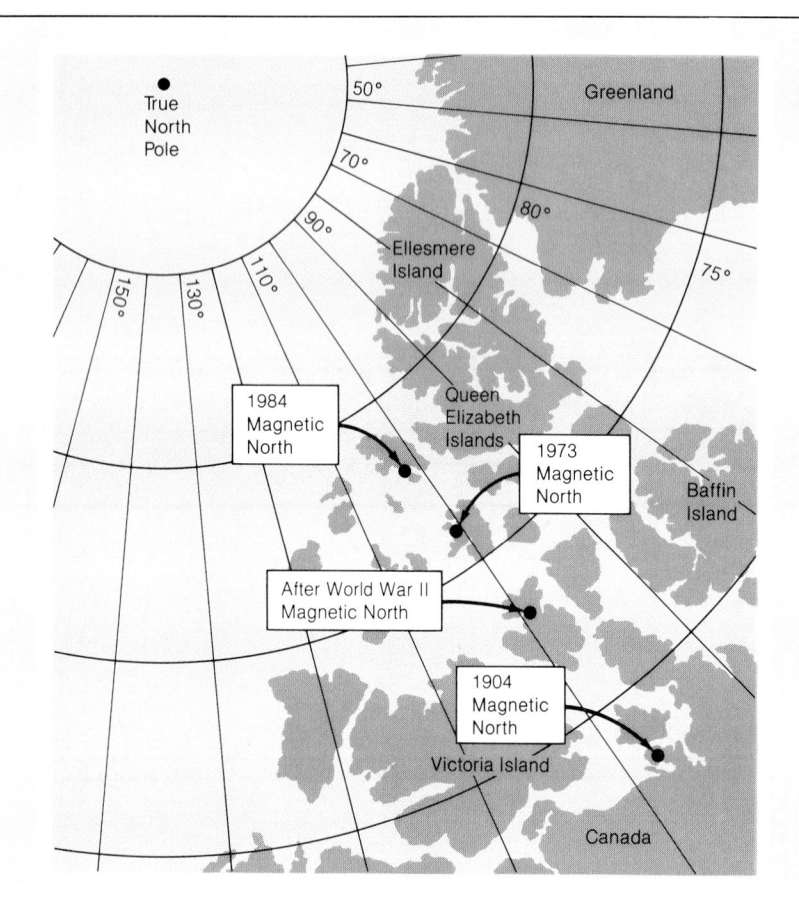

Figure 16.1.1

ferentiation, causes an overturning of the earth's core, which leads to the fluid motions required for the magnetic dynamo to operate. A recent suggestion, favored by many geophysicists, involves a separation of elements at the outer boundary of the solid inner core. Heavy elements such as iron "freeze" into solid form there, while lighter species remain fluid and rise through the liquid outer core. The overturning motion created in this way is called compositional convection.

If the question of a driving force for the dynamo is unanswered, then the problem of explaining reversals of the earth's magnetic field is even further from being solved. The dynamo favors neither orientation of the poles over the other, so it is easy to see that either alignment is possible. But there is no good explanation yet for the reversals or for the steady wandering of the magnetic poles. The

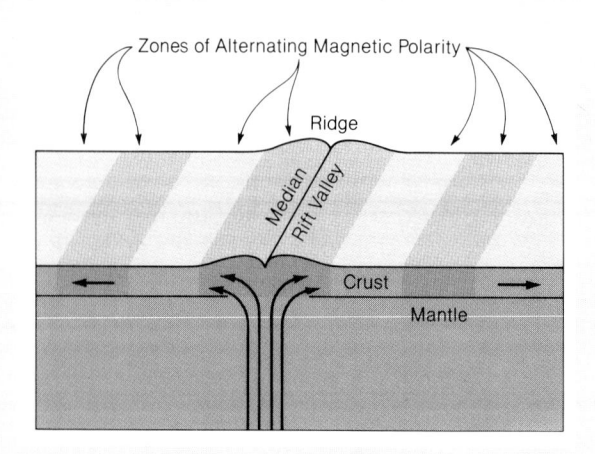

Figure 16.1.2

16.5 MAGNETIC INDUCTION

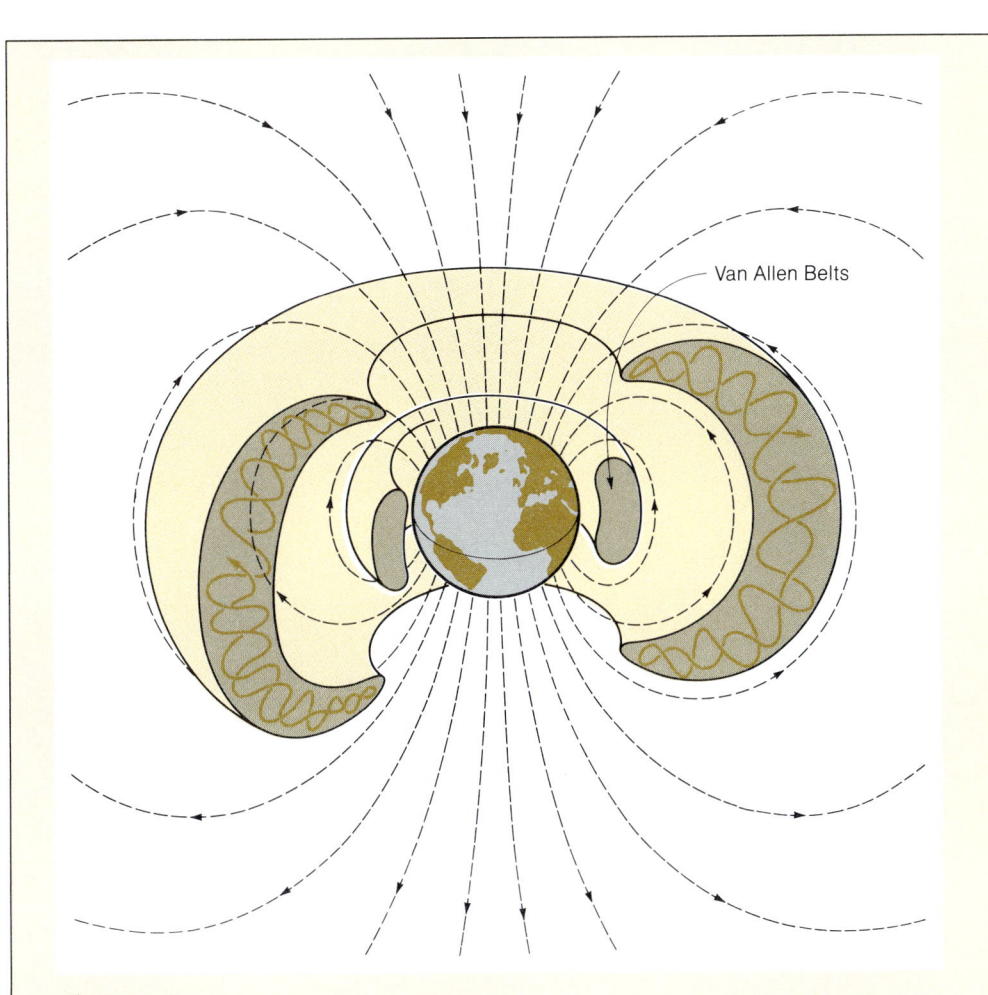

Figure 16.1.3

wild daily fluctuations in the locations of the poles are known to be due to an external influence. A vast quantity of charged particles are trapped above the earth's atmosphere by the magnetic field, and occasionally the concentration of these particles is altered dramatically but temporarily by the arrival of bursts of particles from the sun. When this happens, electric currents in the particle zone above the atmosphere disturb the earth's magnetic field, causing the temporary movements of the poles.

The charged particle belts (Fig. 16.1.3; these are called the Van Allen belts, after the scientist who discovered them) are created by the earth's magnetic field. The magnetic field exerts forces on charged particles that reach the earth from space, causing the trapped particles to move along the field lines, rather than penetrating to the ground.

Thus, the earth's magnetic field acts as a shield, protecting the surface from particles that might otherwise have harmful effects on life. Interestingly, when the earth's field reverses direction, there is a brief interval when the field is disorganized and this protective shield is weakened. Thus, every 300,000 years or so, the earth's surface undergoes a bombardment of charged particles. It has been suggested that these episodes might play a role in the evolution of life, because the incoming particles might cause an unusually high rate of genetic mutations.

Study of the earth's magnetic field, both past and present, is a major discipline in the field of geophysics. Understanding electricity and magnetism is important, as is knowledge of fluids and thermodynamics.

Perspective

In this chapter we have tackled a new phenomenon, magnetism, and found that it is closely linked to another that we have already discussed, electric currents. The mutual interaction of the two is complex, and we can never have one without the other. We ignored magnetism in the preceding chapters in electricity, but now we know that it was there all along; that every circuit we analyzed created a magnetic field, which in turn exerted a force on other circuits, and so on.

In the next chapter we will take advantage of our wider understanding of electricity and magnetism to discuss devices that are used in modern technology, either to create electrical energy or to use it. Although we will leave vast quantities of material uncovered, we will be able to comprehend the workings of most modern circuitry.

PROBLEM SOLVING

The Cross Product and EM Forces

The magnetic force on a moving charge or current is perpendicular to both the magnetic field direction and to the charge's velocity. This type of force is an example of the vector *cross product*. Two vectors **A** and **B** form a plane (Fig. 16.2.1). Their cross product is another vector **C** = **A** × **B**, with magnitude

$$C = |\mathbf{A} \times \mathbf{B}| = AB \sin\theta$$

where θ is the angle between **A** and **B** (this angle is less than 180°) and where vertical bars symbolize the vector's magnitude. The direction of **C** is perpendicular to both **A** and **B** (that is, perpendicular to the plane defined by **A** and **B**) and oriented in the direction given by the right hand rule. As shown in Fig. 16.2.1, if you place your right hand so that its fingers curl in the direction from **A** to **B**, then your thumb points in the direction of **C**. It is helpful to picture your fingers curling so as to rotate the vector **A** into **B**.

We can now use the cross product to express

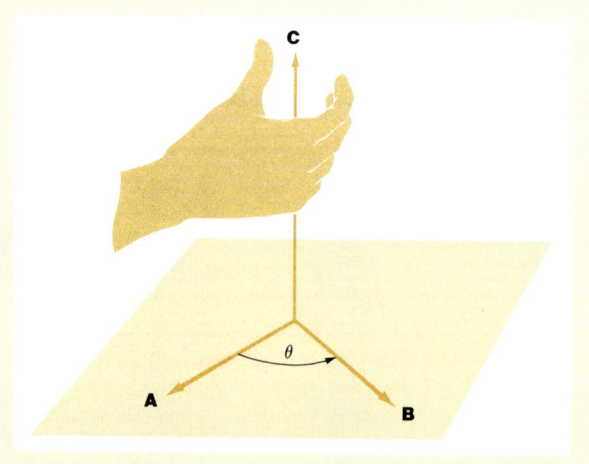

Figure 16.2.1

both the magnitude and direction of the magnetic force on a charge q moving at velocity **v** in a magnetic field **B**,

$$\mathbf{F} = q\,(\mathbf{v} \times \mathbf{B}).$$

Both the direction (perpendicular to **v** and **B**) and magnitude ($qvB\sin\theta$) are properly given by this relation. Likewise, we can write the magnetic force on a wire of length l carrying a current I,

$$\mathbf{F} = I\,(\mathbf{l} \times \mathbf{B})$$

where the length vector **l** lies along the wire in the direction of positive current. The magnitude of **F** equals $IlB\sin\theta$ (Eq. 16–1), and its direction is perpendicular to both **l** and **B**.

When both an electric field **E** and a magnetic field **B** are present, the force can be written

$$\mathbf{F} = q\,(\mathbf{E} + \mathbf{v} \times \mathbf{B}).$$

This relation is known as the Lorentz force. If only **E** is present, the acceleration of a particle of mass m, $\mathbf{a} = \mathbf{F}/m = e\mathbf{E}/m$, is along the direction of **E** and the particle picks up speed. However, if only **B** is present, the speed remains constant, but the direction of velocity **v** changes. Because **F** is perpendicular to **v**, the charge is bent into a circular orbit of radius mv/qB.

In most problems, the first step should be to identify the magnitude and direction of the force, using the cross product and right-hand rule. Remember that the force on a negative charge is opposite to that on a positive charge. Electrons bend in the opposite direction from protons in a magnetic field (this is the cause of the Hall Effect).

Example 1 A solenoid with current I produces a magnetic field **B** in a direction given by the right-hand rule (Fig. 16.2.2). A proton and electron, each with speed v, move in cyclotron orbits. What is the radius and direction of these orbits?

We know that for a circular orbit of radius r and velocity v, the magnetic force qvB must be directed radially inward and have a magnitude equal to the centripetal force, mv^2/r. Thus,

$$\frac{mv^2}{r} = qvB$$

and

$$r = \frac{mv}{qB}.$$

Since q, v, and B have the same magnitude for protons and electrons, r is proportional to mass m, and the radius of the proton's orbit is 1836 times larger than the electron's. However, the particles move in opposite directions. According to the right-hand rule, the proton's velocity **v** must point opposite to the direction of the solenoid current I in order for $\mathbf{v} \times \mathbf{B}$ to point radially inward (Fig. 16.2.2). The negatively charged electron will move in the same direction as I. Now, the small current produced by these protons and electrons will produce a magnetic field in the opposite direction to **B** of the solenoid. This effect is known as **Lenz's law**. The motion of the protons and electrons is the result of the induced emf, and their current is in the direction opposite to that which produced it (here, the solenoid current).

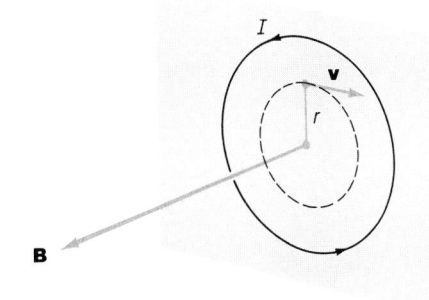

Figure 16.2.2

SUMMARY TABLE OF FORMULAS

Force on a current-carrying wire:
$$F = IlB \sin\theta \quad (16\text{-}1)$$

Force on a free charge:
$$F = qvB \sin\theta \quad (16\text{-}2)$$

Cyclotron radius:
$$r = \frac{mv}{qB} \quad (16\text{-}3)$$

Cyclotron period:
$$T = \frac{2\pi m}{qB} \quad (16\text{-}4)$$

Cyclotron frequency:
$$\nu = \frac{qB}{2\pi m} \quad (16\text{-}5)$$

Torque on a current loop for normal magnetic field:
$$\tau = NIAB \quad (16\text{-}6)$$

Displacement angle in a galvanometer:
$$\phi = \frac{NIAB}{k} \quad (16\text{-}7)$$

Torque on a current loop:
$$\tau = NIAB \cos\theta \quad (16\text{-}8)$$

Magnetic field due to a current in a straight wire:
$$B = \frac{\mu_0 I}{2\pi r} \quad (16\text{-}9)$$

Ampere's law:
$$\Sigma B_\parallel \Delta l = \mu_0 I \quad (16\text{-}10)$$

Magnetic field in a solenoid:
$$B = n\mu_0 I \quad (16\text{-}11)$$

Force per unit length on a wire due to a second wire:
$$\frac{F}{l} = \frac{\mu_0 I_1 I_2}{2\pi R} \quad (16\text{-}12)$$

Induced emf:
$$\mathcal{E} = BvL \sin\theta \quad (16\text{-}13)$$

Induced emf when v is perpendicular to B:
$$\mathcal{E} = BvL \quad (16\text{-}14)$$

Induced current:
$$I = \frac{BvL}{R} \quad (16\text{-}15)$$

Magnetic flux:
$$\Phi = B_\perp A = BA \cos\phi \quad (16\text{-}16)$$

Faraday's law:
$$\mathcal{E} = -\frac{\Delta\Phi}{\Delta t} \quad (16\text{-}17)$$

Faraday's law for a coil:
$$\mathcal{E} = -N\left(\frac{\Delta\Phi}{\Delta t}\right) \quad (16\text{-}18)$$

THOUGHT QUESTIONS

1. How would you look for lodestones? How would you determine the location of the magnetic poles of an irregularly shaped lodestone?

2. What might be some examples of geological evidence that the earth's magnetic poles have moved?

3. Suppose you cut a magnetic domain in half. What would the magnetic poles and field of this half-domain look like?

4. Why can heating or striking a permanent magnet cause a loss of its magnetism?

5. Compare and contrast a magnetic field with an electric field. Do the same for magnetic forces and gravitational forces.

6. If the earth's magnetic field were generated by current loops, which way (east or west) would the current have to flow?

7. Are there processes in the human body which depend on magnetism? Explain.

8. One explanation of the earth's magnetic field is the "dynamo theory," in which a liquid ferromagnetic core ro-

PROBLEMS

tates and produces currents, which then generate the magnetism. Discuss how well this theory predicts certain aspects of the magnetic field, including location of the magnetic poles and pole reversal. Could the sun's magnetic field be generated in a similar way?

9   Identical currents are running in opposite directions in two parallel wires. Draw the magnetic field lines and describe the forces on the wires.

10 Will a permanent magnet attract any object, or just ferromagnetic metals like iron? Explain.

11 The *Aurora Borealis* (Northern Lights) are the result of charged particles striking the earth's atmosphere. Why are these effects more common near the north pole? Are there similar effects in the southern hemisphere?

12 Describe several practical uses for solenoids.

13 How could you measure the earth's magnetic field? How might you measure the field of a distant planet using a space probe?

14 Is there a magnetic equivalent of a dielectric medium? Explain.

15 Sketch the magnetic field of a current loop. Show how the field varies with distance from the center of the loop in a direction
    (a) perpendicular to the loop,
    (b) coplanar with the loop.

16 What is an induced emf?

17 What is an electromagnet? How might you use superconductivity to create large magnetic fields?

18 Sketch the magnetic field lines of a horseshoe magnet. If you bring a wire carrying a current near the ends of this magnet, which way will the force point?

19 What is meant by a magnetic fluid? What would be its useful properties?

20 What are the differences between magnetic flux, magnetic intensity, and magnetic field?

21 Compare and contrast Faraday's Law, Ampere's Law, and Lenz's Law.

22 Would the migration of a "hole" along a conducting wire produce a magnetic field? How could the Hall effect be used in answering this question?

23 Compare and contrast the electric and magnetic fields arising from an electric charge at rest with those arising when the charge moves.

24 A physics professor observes a conducting wire at rest in the laboratory. He measures no current and no magnetic field. However, relative to some students racing around the classroom, the electrons in the wire do have a velocity and therefore a current. Do the students measure a magnetic field? If so, explain why they and the professor disagree about the magnetism.

25 Will a magnetic compass work in the southern hemisphere? What will happen to an American compass when it is taken to Australia?

26 Are there magnetic equivalents to a conductor, a semiconductor, and an insulator? Explain.

27 Antimatter consists of atoms with negatively charged nuclei and positively charged electrons. Can you think of an experimental method, involving magnetism, by which you could detect antimatter atoms? Explain.

28 Discuss the idea that magnetic fields and field reversals may have played an important role in the evolution of life on earth.

29 Are there any instances of Lenz's Law and induced emf in the functioning of the human body?

30 Discuss the future of magnetic fields and superconductivity for power generation. You may rely on reading about magnetohydrodynamic (MHD) generators, superconducting magnets, and magnetic plasma fusion devices.

PROBLEMS

Sections 16.1–16.3 Moving Charges and Magnetism

- 1 Calculate the force on a 30 cm wire with 1 A current aligned perpendicular to the earth's magnetic field.

- 2 Using the right-hand rule, find the direction of the magnetic force on a positively charged particle moving at velocity **v** for each case in Fig. 16.27. (The symbol $\otimes$ indicates a vector into the paper and $\odot$ indicates outward.)

- 3 The force on a current-carrying wire is zero when the wire points 60° west of geographic north. In what direction(s) could magnetic north lie?

- 4 What magnetic field is required to exert a force of 10^{-10} N on an electron moving at $0.9c$?

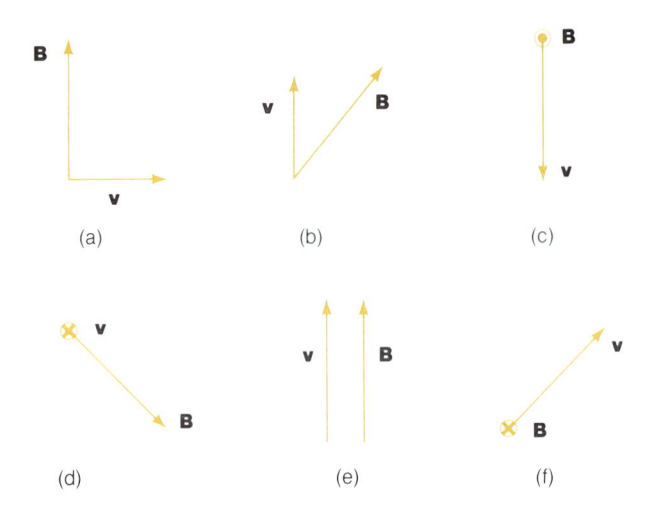

Figure 16.27

- 5 What mass particle would suffer no deflection in Example 16.2?
- 6 How much current flows in a 1 m long wire in a 0.05 T magnetic field if the force is 1 N?
- 7 What is the force per unit length on a high voltage wire carrying 50 A, aligned perpendicular to the earth's field of 0.5 G?
- 8 A cable carrying 100 A is aligned at 60° to the earth's magnetic field of 5×10^{-5} T. What is the force per unit length?
- 9 A charged particle approaches the earth from the sun, perpendicular to the earth's magnetic axis. If the particle is a 400 m/s proton, calculate the direction and magnitude of the magnetic force.
- 10 A proton moves through interstellar space at 400 m/s, perpendicular to a magnetic field of 3×10^{-6} G. By how much has the proton deflected after travelling 1 km?
- 11 A pendulum of mass 1 kg and length 2 m swings with amplitude 10° in a magnetic field of 1 G (Fig. 16.28). If the mass has charge 10^{-1} C, estimate the magnetic force on the ball at mid-swing. What is the force at maximum θ?

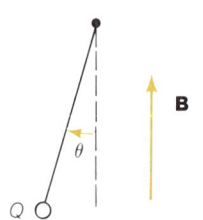

Figure 16.28

- 12 For each of the situations in Fig. 16.29, find the direction of magnetic field that produces force **F** on an electron of velocity **v**.

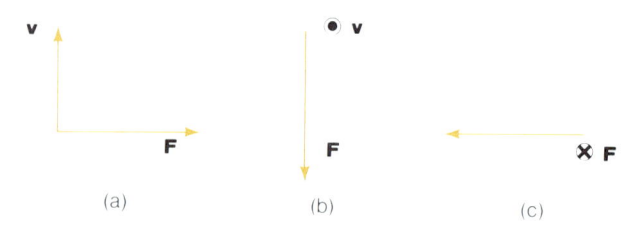

Figure 16.29

- 13 A proton moves in a circle, perpendicular to a magnetic field of 1 T. If the radius is 5 cm, what is the proton's kinetic energy?
- 14 In Problem 13, what would be the radius for an electron of 1 keV energy?
- 15 A particle of mass m and velocity v moves in a circular path of radius r in a magnetic field B. Find the particle's momentum p in terms of its charge q and other parameters. If all one knows about a particle is its radius in the field, can one determine its mass alone? Can one determine its velocity alone?
- 16 Suppose we wished the electron and proton in Example 16.3 to move at $0.8c$, but retain the same cyclotron radius. What value of B would be required? What change, if any, would occur in the cyclotron frequency for each?
- 17 An electron moves through space at 500 km/s. If the magnetic field of the galaxy is about 3×10^{-6} G between the stars, find the cyclotron frequency, radius, and period for this electron.
- 18 Cosmic rays are high velocity particles, mostly protons, which travel through interstellar space. What is the cyclotron radius of a cosmic ray with kinetic energy E, in terms of the mean interstellar magnetic field B_g? If $B_g = 3 \times 10^{-6}$ G, what is the cyclotron radius for a cosmic ray proton with $E = 2$ MeV?
- 19 An electron accelerated to 30 keV by a cathode ray tube (CRT) encounters a magnetic field of 10^2 G over a horizontal distance of 10 cm. How much does the electron deflect, if its velocity is perpendicular to the magnetic field?
- 20 A galvanometer coil has 100 turns and a total area 0.01 m². Through what angle will the galvanometer spring be turned if its torque obeys the law $\tau = k\phi$, where $k = 10^{-6}$ N·m, the current is 10^{-3} A, and $B = 1$ G?

PROBLEMS

Section 16.4 Ampere's Law

- **21** Find the force per unit length between two wires separated by 50 cm, each carrying 0.5 A current.

- **22** Two wires are separated by 2 m and experience a force per unit length of 5×10^{-6} N/m. If the currents are equal, what is their value?

- **23** Two wires are separated by 0.5 m. The current in the first wire is three times that in the second, and the force per unit length is 4×10^{-7} N/m. What are the currents?

- **24** Suppose the compass in Example 16.4 passes the wire at a distance 10 m. What is the magnetic field? What fraction of the earth's 0.5 G field is this?

- **25** Two parallel wires 50 cm apart each carry 10 A current in the same direction. Where should a third parallel wire carrying 20 A be placed for the net magnetic field on the central 10 A wire to be zero? Assume all wires are in the same plane.

- **26** How large would the current in Example 16.4 need to be for the compass to detect the wire in the wall? Assume that a field 30% of the earth's is detectable.

- **27** What is the magnetic field 1 m away from a long straight wire carrying 10 A? What would be its force per length on a similar parallel wire, with 10 A current in the opposite direction, located 5 m away?

- **28** Two parallel wires 1 cm apart have identical currents running in opposite directions. If the linear mass density of the wires is 1 g/m, how much current is needed to lift one wire above the other?

- **29** Suppose we wish to make a solenoid with 10^4 G magnetic field, using 10^4 wire loops per cm. What current is required?

- **30** Find the magnetic field generated by a solenoid with 15 loops/mm operating with 15 A current.

- **31** How many turns will be needed for a solenoid 0.5 m long if it must produce a magnetic field of 5 T operating on a current 5 A?

- **32** The magnetic field at the center of a circular loop of radius R is $B = \mu_0 I/2R$. Find B at the center of the circuit loop shown in Fig. 16.30, with $R = 0.2$ m.

- **33** Find the magnetic field at the center of the current loop in Problem 32 if the 4 μF capacitor is removed.

- **34** A satellite is placed in orbit just above the earth's surface in a plane perpendicular to the earth's magnetic field. What is its velocity? In what direction must the satellite orbit, and what positive charge must it carry to neutralize the magnetic field of the earth?

- **35** A lightning bolt strikes the ground from a cloud at altitude 1 km. If the electric field is 10^4 V/cm, the power is 10^{15} W and the discharge lasts 0.1 s, find the magnetic field generated by this flow of charge at a distance 1 m from the channel of the bolt.

- **36** A solenoid has 150 turns/cm and a resistance of 50 Ω. If the ends of the solenoid are connected to 30 V, find the magnetic field.

- **37** Suppose the solenoid in Problem 36 is cooled until it becomes superconducting and exhibits a resistance of 0.01 Ω. What is the magnetic field now?

Section 16.5 Magnetic Induction

- **38** Find the magnetic flux Φ for a circular wire loop of radius 20 cm, inserted at 30° to a magnetic field of 1 T.

- **39** A square loop of wire, 30 cm on a side, is inserted in a 5 T magnetic field. What are the maximum and minimum possible values for magnetic flux as the loop rotates by an angle between ±20° from the field **B**?

- **40** A rectangular loop of wire, 20 cm × 40 cm, is inserted in a magnetic field of 1.5 T. At what angle must the loop be oriented to produce a flux of 10^{-2} T·m²?

- **41** A circular loop of wire, 5 cm in radius, is immersed in a magnetic field **B** = 1 T. The loop rotates about an axis perpendicular to **B** at 1 rev/min. Plot the magnetic flux as a function of time if the initial angle φ between **B** and the perpendicular to the plane of the loop is 0°.

- **42** If the circular loop in Problem 41 had an initial value φ = 90°, how would the plot of flux versus time change?

- **43** A long solenoid with 100 turns/cm carries a current of 10 A. The radius of the loops is 1 cm, and the current is increasing at the rate 100 A/s. What is the instantaneous magnetic flux, and what is the induced emf?

- **44** Suppose a power line 5 m overhead, carrying 10^4 A, is suddenly shut off in 0.01 s. Find the maximum

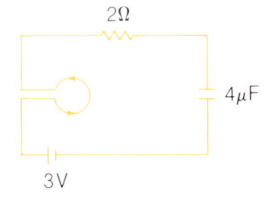

Figure 16.30

possible induced emf in a circuit loop of 1 m² area. How should the loop be oriented to attain the maximum emf? What current will flow in the loop if its resistance is 0.01 Ω?

•• 45 At what rate would a 5 A wire current need to be changed to produce a maximum induced emf of 0.001 V in a current loop of area 0.2 m², 15 cm away?

•• 46 A 1 m² wire loop is oriented with its normal parallel to the earth's rotation axis, but is located at the north magnetic pole (74° latitude). What is the magnetic flux through the loop? Does it change as the earth rotates?

•• 47 A large bar magnet of strength 5 Wb/m² is moved into the plane of a 0.4 cm × 0.4 cm square loop at a rate of 2 m/s. The loop is much smaller than the pole plate of the magnet, so that the field may be considered uniform. What is the current in the loop if its resistance is 0.1 Ω?

•• 48 A conducting bar is moved to the left on a U-shaped conductor in a uniform magnetic field **B** (Fig. 16.31). If the width of the bar is L, find the induced emf in the conductor if the bar's velocity is v.

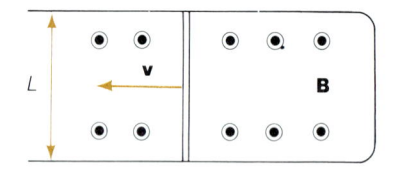

Figure 16.31

••• 49 What is the force F needed to maintain the bar's velocity v in Problem 48? Show that the rate at which work is done on the bar equals the power dissipated in the conductor, which has resistance R.

••• 50 A 10 cm bar moves to the left on a U-shaped conductor as in Problem 48. If the magnetic field is 1 T, the velocity $v = 5$ m/s, and the resistance of the bar is 0.1 Ω, find the induced emf, the current, and power dissipated.

••• 51 A 20 turn coil, 2 m × 2 m, is rotated at 100 rev/s in a magnetic field of 2 T. Find the maximum induced emf and current, if the resistance is 1 Ω.

• 52 The magnetic flux inside a coil changes from +10 Wb to −5 Wb in 1.5 s. What is the induced emf?

••• 53 A special pendulum of length L consists of a loop of mass m and radius r, insulated from the massless pendulum by support rods (Fig. 16.32). The pendulum swings with maximum amplitude θ_0 through a vertical magnetic field **B**. What is the magnetic flux Φ at time t, if $\theta = \theta_0$ at $t = 0$? Make a plot of the induced emf versus t. You need not compute the absolute value. (The rate of change of $\cos \theta$ equals $-\sin \theta$ times the rate of change of θ.)

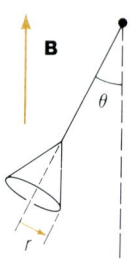

Figure 16.32

••• 54 When an ionized gaseous star collapses under its own gravity, after it stops burning nuclear fuel, its magnetic field collapses too (Fig. 16.33). The total magnetic flux contained within its circumference is conserved. Suppose stars like the sun have fields of 100 G and initial radius 7×10^8 m.
(a) Estimate the magnetic field of a white dwarf star, which has radius 10^6 m.
(b) Estimate the magnetic field of a neutron star of radius 10 km.

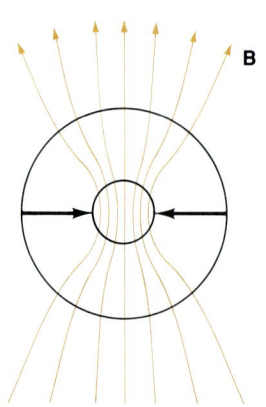

Figure 16.33

SPECIAL PROBLEMS

1. Hall Effect

When a conducting slab of cross-sectional area A and width d is placed in a uniform magnetic field **B**, the magnetic force **F** deflects charges to one side (Fig. 16.34). If the current I is to the right, then by convention positive charges move to the right and negative charges move to the left with velocity **v**. The charge separation produces an electric field **E**, whose direction depends on the sign of the charges which carry the current. Suppose these carriers have charge q and number density n.

(a) Show that in equilibrium, the magnetic force qvB must balance the electric force qE.
(b) From the definition of current, I/A, find an expression for nq in terms of I, A, B, and E.
(c) How can one use measurements of I, B, A, and E to determine the sign of the charge carriers? Can one also determine the number density n?
(d) If a voltage V is measured across the width d of the slab, and electrons are the charge carriers, what is their velocity v?

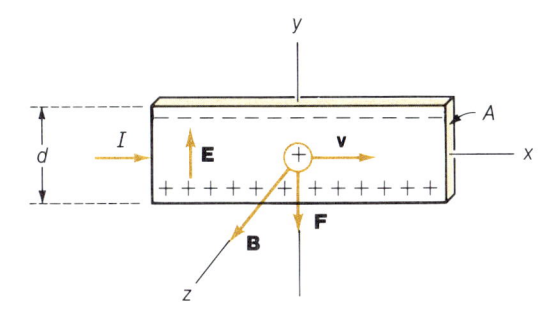

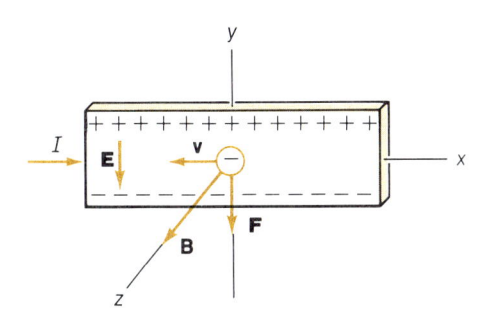

Figure 16.34

2. Mass Spectrometer

In a mass spectrometer, positive ions of charge q, mass m, and velocity v are injected into a region of uniform magnetic field **B** (Fig. 16.35). They move in a semicircle and strike a detector at a distance d from the injection point.

(a) From force balance, show that $d = (2mv/qB)$.
(b) If a voltage V accelerates the ions to their injection velocity v, find a formula for the mass m in terms of V and other variables.
(c) Explain why the mass spectrometer can measure only the particle's charge-to-mass ratio, q/m, not q and m separately.
(d) Suppose a voltage of 10 kV is used to inject bare carbon nuclei into a mass spectrometer with $B = 10^4$ G. How far apart on the detector will the two stable isotopes, $^{12}_6\text{C}$ and $^{13}_6\text{C}$, land?

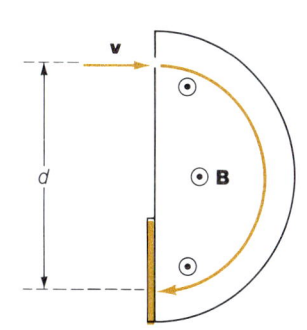

Figure 16.35

3. Magnetic Mirrors

When a positively charged ion of mass m, charge q, and velocity **v** enters a magnetic field **B** at an angle θ, it moves in a helix (Fig. 16.36a). The helical motion is a combination of a constant velocity $v_\parallel$ parallel to **B** plus a circular motion of radius $r_L = (mv_\perp/qB)$, where $v_\perp$ is the component of **v** perpendicular to **B**. The radius r_L is often called the particle's **Larmor radius**, and the angle θ is called the **pitch angle** of the helix. In a **magnetic mirror**, the magnetic field strength increases toward the outside, and the field lines converge (Fig. 16.36b). The helical motion becomes tighter, the pitch angle θ increases until finally $\theta = 90°$ and the particle reflects the other way at the mirroring point.

(a)

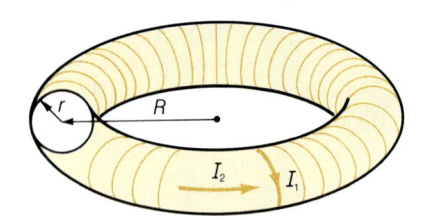

Figure 16.37

(b)

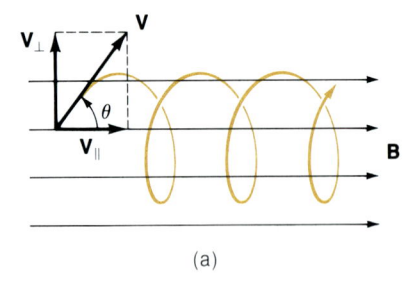

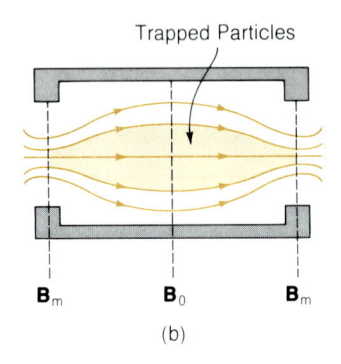

Figure 16.36

(a) One can show that the flux of magnetic field contained within the particle's Larmor radius must remain constant. Show that $\pi r_L^2 B$ is therefore constant, and that $v_\perp^2/B$ and $(\sin^2\theta)/B$ must also remain constant for the helix.

(b) A particle whose initial pitch angle $\theta = \theta_0$ when $B = B_0$ will reach a mirroring point when the field $B = B_m$. Since $\sin^2\theta$ can never be larger than 1, find B_m in terms of B_0 and θ_0.

(c) Describe the motion of a charged particle between the two points at which $B = B_m$. Explain how such a field configuration could be used to contain a plasma of hot ions. (Magnetic mirror machines were used in early attempts to obtain controlled nuclear fusion reactions in hot plasmas.)

(d) Show that particles with small initial pitch angles ($\sin \theta_0 < (B_0/B_m)^{1/2}$) will not reflect. Discuss how these particles could hurt the effectiveness of a magnetic mirror machine for controlled fusion reactions.

4. Magnetic fields in toroids

A solenoid of radius r may be bent into a ring or toroid (Fig. 16.37) of major radius R. A current I_1 flows in N windings.

(a) Use Ampere's law for a circular path inside the ring to show that the magnetic field inside the toroid is $B_1 = (\mu_0 N I_1 / 2\pi R)$. This field is called the **toroidal field**.

(b) Suppose the iron core of the solenoid is replaced by a conducting ion plasma, with a current I_2 flowing around the ring. What is the magnetic field B_2 that is produced? Show that this field is in the same direction as the windings. This field is called the **poloidal field**. Plasma devices using this combination of toroidal and poloidal fields are called **tokamaks** and have been used in experimental test reactors for controlled nuclear fusion.

(c) What ratios of B_2/B_1 will result in field lines that close on each other? That is, what ratios result in field lines which meet again after going around the torus many times? Such ratios turn out to produce plasma instabilities which spoil the plasma confinement.

(d) In a recent version of a tokamak, $R = 130$ cm, $r = 45$ cm, $N = 1{,}000$, $B_1 = 35$ kG, and $B_2 = 3$ kG. What are I_1 and I_2?

5. Dipole Magnetic Fields

The magnetic field of a small current loop or a magnetized sphere is described by the **magnetic dipole field** (Fig. 16.38). Inside the sphere, the field is constant and lies along the z-axis, $B_0 = (2m/R^3)$, where R is the sphere's radius and m is the **magnetic dipole moment**. (For a small current loop, m equals the current times the area of the loop.) Outside the

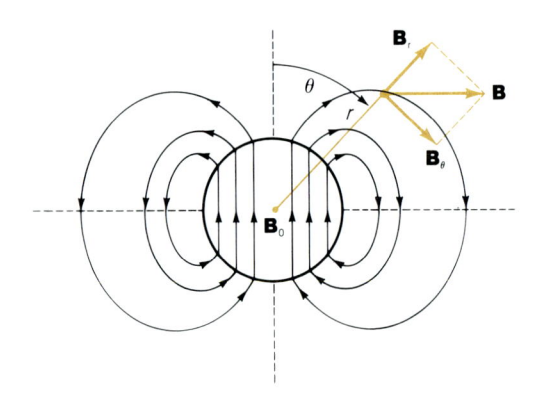

Figure 16.38

sphere, the magnetic field at a point with radius r and polar angle θ has components in the r and θ directions:

$$B_r = (2m/r^3)\cos\theta \quad \text{and} \quad B_\theta = (m/r^3)\sin\theta$$

Notice that the dipole field decreases as the inverse cube of the radius.

(a) Write the field components B_r and B_θ outside the sphere in terms of B_0, r, and θ. Then find the magnitude of **B** as a function of arbitrary r and θ. (Use the Pythagorean theorem.)

(b) What is the field magnitude in the equatorial plane ($\theta = 90°$) for arbitrary r? What is the field at the surface ($r = R$) for arbitrary θ? What is the ratio of the field at the poles to that at the equator?

(c) The earth has a polar magnetic field $B_0 = 0.5$ G at its surface ($R = 6{,}378$ km). What is the field at the distance of the moon ($r = 384{,}000$ km) at $\theta = 90°$?

(d) A satellite flies by a planet in the equatorial plane, at a distance 100,000 km. If it detects a field of 10^{-3} G, what is the planet's surface polar field B_0 at $R = 10^4$ km?

SUGGESTED READINGS

"Reversals of the Earth's magnetic field." A. Cox, G. B. Dalrymple, R. R. Doell, *Scientific American*, Feb 67, p 44, **216**(2)

"Gauss." I. Stewart, *Scientific American*, Jul 77, p 122, **237**(1)

"The source of the Earth's magnetic field." C. R. Carrigan, D. Gubbins, *Scientific American*, Feb 79, p 118, **240**(2)

"Magnetic navigation in bacteria." R. P. Blakemore, R. B. Frankel, *Scientific American*, Dec 81, p 58, **245**(6)

"The active solar corona." R. Wolfson, *Scientific American*, Feb 83, p 104, **248**(2)

"Magnetic fields in the cosmos." E. N. Parker, *Scientific American*, Aug 83, p 44, **249**(2)

"Magnetic fields of the human body." D. Cohen, *Physics Today*, Aug 75, p 34, **28**(8)

"The Earth's magnetosphere." S. Akasofu, L. Lanzerotti, *Physics Today*, Dec 75, p 28, **28**(12)

"Rowland's physics." J. D. Miller, *Physics Today*, Jul 76, p 39, **29**(7)

"The anomalous Hall effect." G. Bergmann, *Physics Today*, Aug 79, p 25, **32**(8)

"High magnetic fields for physics." L. G. Rubin, P. A. Wolff, *Physics Today*, Aug 84, p 24, **37**(8)

"Searches for magnetic monopoles and fractional electric charge." S. B. Felch, *The Physics Teacher*, Mar 84, p 142, **22**(3)

The flying circus of physics with answers. Jearl Walker, Wiley, New York, 1977

A source book in physics. William F. Magie, Harvard University Press, Cambridge, 1965

Great scientific experiments. Rom Harre, Phaidon Press, Oxford, 1981, (Michael Faraday and electromagnetic induction)

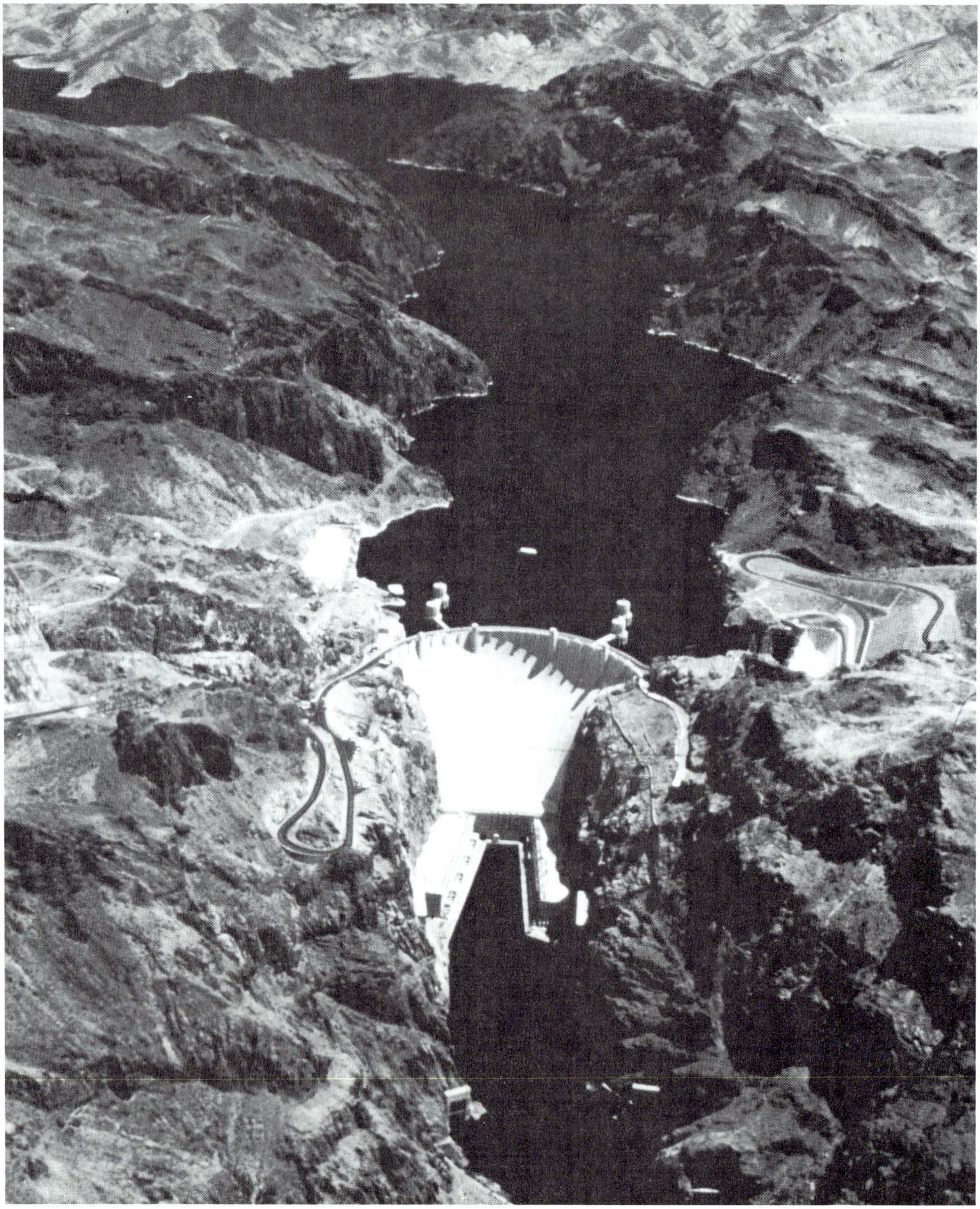

Chapter 17

Applied Electricity and Magnetism

We have learned enough to understand in principle the workings of most modern machinery and technology, but in practice there is still much that we have not discussed. In this chapter we complete our discussions of electricity and magnetism by learning how the principles we have digested in the previous three chapters are applied.

We will begin with the generation and transmission of electrical power; then we will discuss various devices for deriving work from the flow of electricity. Our discussion of generators will lead naturally to alternating-current (AC) circuits, which will be our topic for most of the rest of the chapter.

17.1 Generators and Motors

Generators and motors are devices that take advantage of inverse processes either to create current induced by an applied torque, or to create torque through magnetic forces acting on an applied current. We will find that these processes are reversible, so that a generator can act as a motor and vice versa.

Generators

As we learned in the previous chapter, a generator works on the principle of inductance, in which a changing magnetic flux Φ produces an electrical current. The change in Φ is brought about by rotating an armature, an iron core with a conducting coil wrapped around it (Fig. 17.1). The magnetic field can be held constant, while rotating the armature varies the magnetic flux through the coil. We already have seen that the sign of the current produced in a generator alternates as the armature revolves, because the direction of the

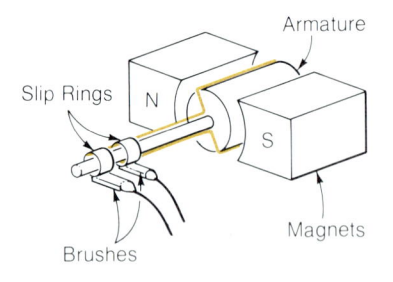

Figure 17.1

A simple AC generator. An alternating current is induced in coils on a rotating armature as the magnetic flux changes. The current is conducted from the rotating shaft to fixed external wires by slip rings and brushes. An external energy source rotates the armature.

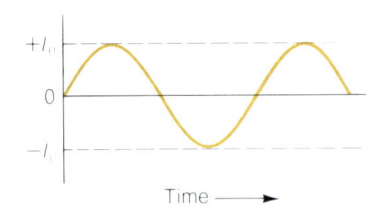

Figure 17.2

Alternating current. The current produced by an armature rotating at constant frequency in a steady magnetic field has a sinusoidal form.

magnetic flux through the coil reverses twice each revolution. Recall that the magnetic flux is related to the component of the magnetic field that is perpendicular to the plane of the coil; hence, the magnitude of the flux varies as the cosine of the angle between the direction of the field and the normal to the plane of the coil. Because the armature is rotated at a constant frequency, the current produced has a sinusoidal time dependence (Fig. 17.2). Such a current is called an **alternating current**.

Alternating current is widely used in modern technology, not only because of the ease of generating it but also because of many other practical advantages, for example, in transmitting power and transforming voltages (described in the next section). The standard household current in the United States is alternating current with a frequency of 60 Hz and an emf of 120 V.

Because the armature rotates, the current produced in it must be connected to the external circuit by means of some kind of sliding contact; otherwise, the wires would become tangled and wound up. Normally there is a **slip-ring** mounted on the shaft, which makes contact with metallic wires called **brushes.** In most generators, the magnetic field is produced by electromagnets, which were mentioned in the previous chapter. In a few generators, permanent magnets are used. The generator may then be called a **magneto.**

As we have seen, a generator requires a source of work, because some agent must be available to rotate the armature. Today two sources are in general use for large-scale power generation: steam-operated turbines, and those powered by falling water (these are known as **hydroelectric generators**). In a steam-powered generating plant (Fig. 17.3), water is boiled and the steam is heated further, so that its expansion forces turbines to rotate. The steam is then allowed to recondense, so that the water can be recycled.

The fuel for a steam-powered plant is most commonly coal or a thermonuclear reactor. Each has its own disadvantages (coal pollutes the atmosphere, thermonuclear reactors produce radioactive waste and may accidentally release radiation). The plants have a common disadvantage: the problem of dissipating heat. This is usually done by circulating cool water through the plant, then releasing it into a nearby river, where it can have serious effects on aquatic organisms. (Review the discussion of heat engines in Chapter 13 to understand why heat must be dissipated.)

In a hydroelectric plant (Fig. 17.4), water falls through turbines, so that gravitational potential energy is converted into kinetic energy, some of which is transferred to the turbines. This is environmentally cleaner than the steam-operated plants, but can be used only in mountainous regions with plentiful streams. Hydroelectric plants are generally located at dams. The water controlled by the dam is a source of gravitational potential energy which does not vary with seasonal fluctuations in the water level of rivers and streams.

Despite the advantages of AC currents, direct currents still are widely used in modern technology. Generating DC current is more complicated than AC current, but the basic principle is the same. The armature in a DC generator supplies alternating current to a special switch called a **commutator** (Fig. 17.5), which replaces the slip-ring used in AC generators. The commutator has two fixed contacts that act as sliding switches arranged so that as the armature rotates, the coil comes into contact with them alternately. The contacts are positioned so that one of them always contacts the side of the armature coil

17.1 GENERATORS AND MOTORS

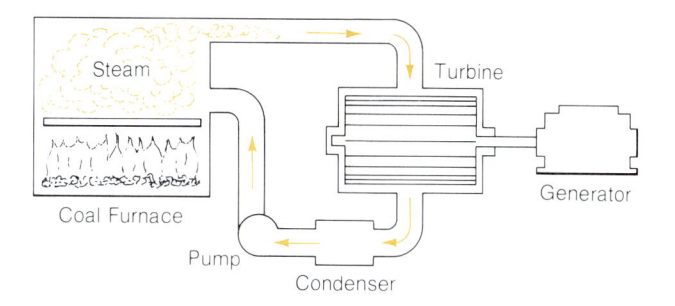

Figure 17.3

A steam-powered generating plant.

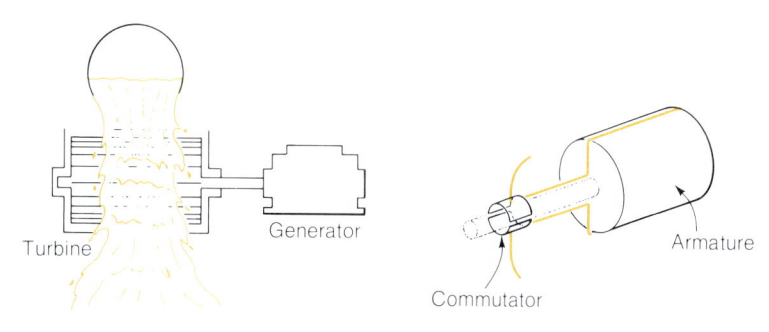

Figure 17.4

A hydroelectric plant. The force of falling water rotates turbines attached to the generators.

Figure 17.5

A commutator. The pair of half-rings is fixed and alternately contacts opposite sides of the coil carrying current from the armature. Hence the direction of the current in each wire does not change.

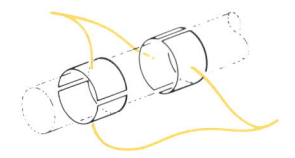

Figure 17.6

A pair of commutators. When a pair of commutators, arranged at right angles, is used, a constant DC current is produced.

that carries positive current; the other contacts only the side carrying negative current. As the armature rotates, there is a "dead spot" twice each cycle, as the sides of the coil are exchanged. Furthermore, the magnitude of the current varies sinusoidally over the portion of the cycle when the contact is made, so the current produced, while DC, is not steady. To produce a steady DC current, two coils can be mounted on the armature at right angles to each other (Fig. 17.6), so that one produces maximum current just as the other reaches its minimum.

Most modern DC generators use another technique, rather than the commutator. A special device called a **diode** or, in this application, a **rectifier,** is inserted into the circuit. These devices have the property that current can flow through them in only one direction, and they filter out the opposing current in an AC circuit. Thus an AC current is converted into a periodic DC current. The combination of AC generator and rectifier is called an **alternator,** and is commonly used in cars. Diodes can be used in pairs to produce a reasonably steady DC current from an AC source. Diodes and rectifiers are discussed later in this chapter.

In some alternators, particularly at high voltages, the roles of the armature and the magnetic field coils are reversed. Current is supplied to the armature coil, so that as it rotates it produces a rotating magnetic field which induces alternating current in the fixed coil surrounding the armature. The advantage is that the current is induced in a fixed circuit, and sliding contacts are not needed to carry the current to external circuits. This is especially important for high voltages, because it reduces the danger of arcing.

Electric Motors

The operation of an electric motor is very similar to that of the galvanometer (Chapter 16). A magnetic field is applied to a current-carrying coil wound on an armature, so that the force exerted on the coil causes it to rotate. In a galvanometer, the rotation is limited to a small fraction of a revolution, but in a motor, the rotation is continuous.

The motor is also similar to the generator, where allowance had to be made for continuous rotation. The coil must be connected to the source of emf by a sliding contact to prevent tangling the wires. Furthermore, as the armature turns, its orientation with respect to the magnetic field varies, so that the force varies sinusoidally (if the field is uniform, with parallel field lines) and changes sign every half-revolution. If the current in the coil were DC, the coil would not rotate continuously because the direction of the torque acting to rotate it would reverse every half-cycle. AC current is a natural solution to this problem, so long as the current reverses direction just as the coil passes through the position where the torque would reverse.

It is possible to construct a DC motor with a commutator, as described in our discussion of generators. As the shaft rotates, the brushes alternately contact the two sides of the commutator (Fig. 17.7), so that the DC current is effectively converted into AC. The commutator contacts are arranged so that the alternating current will be in phase with the rotation of the armature.

Back emf and Back Torque

A continuous torque is applied to the armature in an electric motor, and it might seem that the armature would therefore undergo continuous angular acceleration. This is not the case, however, because a counteracting torque develops, so that an equilibrium is reached where there is no angular acceleration (but quite a high angular velocity). As the armature rotates, magnetic induction creates a **back emf** that reduces the current in the coil. As stated by Lenz's law, the induced emf acts to counter the motion that creates it. To help see this, consider Fig. 17.8. The direction of the magnetic field is vertical, with the positive direction upward. At the moment pictured, the induced current in the upper segment of the coil is flowing out of the page, because the area of the coil perpendicular to the magnetic field is decreasing. The right-hand rule tells us that the force in this segment is to the left, which produces a

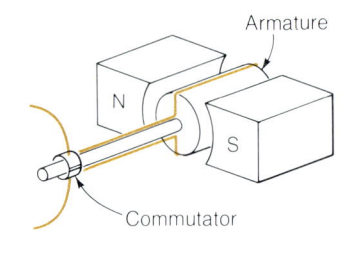

Figure 17.7

A DC motor. A commutator, in effect, converts direct current into alternating current so that the torque acting on the armature is always in the same direction.

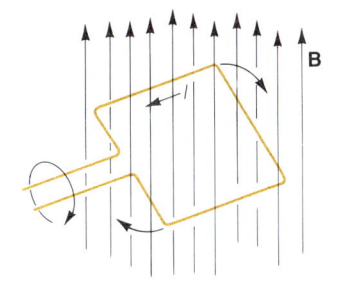

Figure 17.8

Back torque. As the coil rotates, the changing magnetic flux creates an induced current, hence an induced torque, in the direction opposite to the applied current. This induced counter-torque (back torque) increases as the armature accelerates, until it equals the torque due to the applied current. Thereafter, the angular velocity of the armature is constant.

counterclockwise torque, opposite to the sense of rotation. During the next quarter-cycle, the flux will increase again, creating an induced current in the upper segment of the coil that is directed into the page. The rate of change of flux increases as the armature's angular velocity increases, so this counter-current increases also, until the angular acceleration of the armature becomes zero.

The back emf develops over a period of time as a motor starts up. Therefore the motor initially draws a much higher current than it does after it reaches equilibrium. The momentary dimming of house lights when a refrigerator motor starts is a consequence of this; while the motor draws very high current during start-up, the effective voltage in the household circuit drops by the product IR, where R is the resistance of the refrigerator.

Back emf shows that a motor acts simultaneously as a generator. In the same way, a generator always acts simultaneously as a motor, because the magnetic field exerts a force on the current induced in the coil, and, according to Lenz's law, this force will always act to counter the change that created it. For a generator, this means that a **back torque** is created. As the armature of a generator undergoes angular acceleration when it is started up, the back torque builds up, until eventually an equilibrium is reached in which the angular acceleration is zero. This equilibrium is altered if the current drawn from the generator varies because the force exerted on the coil by the magnetic field will vary. Thus, if increased current is drawn, more work is required to turn the armature.

Example 17.1

Suppose the starter motor in a car has armature windings whose resistance is 0.6Ω. The motor runs on a 12 V battery. If the back emf when the motor is running is 10.5 V, how much current does the motor draw when starting, and how much when it is running at full speed?

At first, as the motor is just beginning to turn, the induced back emf is zero. The current at that time is

$$I = \frac{V}{R} = \frac{12 \text{ V}}{0.6\Omega} = 20 \text{ A}.$$

When the motor has reached full speed, the back emf opposes the applied emf, so that the net emf driving the motor is 12 V − 10.5 V = 1.5 V. Now the current is

$$I = \frac{V}{R} = \frac{1.5 \text{ V}}{0.6\Omega} = 2.5 \text{ A}.$$

Eddy Currents

Whenever a conductor moves in a magnetic field, currents will be induced. If the conductor does not provide a well-defined path for these currents, then they will swirl around as dictated by the induced emf and the need for a return

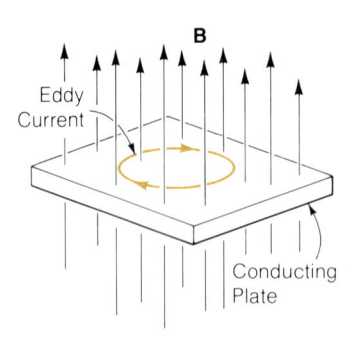

Figure 17.9

Eddy currents. If the magnetic field is increasing with time, eddy currents will be induced.

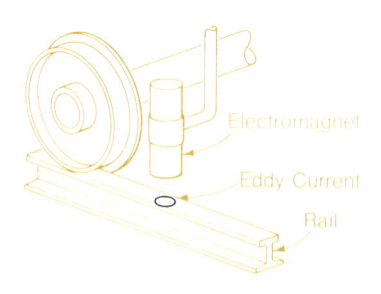

Figure 17.10

Magnetic brake. As the field of the electromagnet increases, eddy currents in the metal rail create a force opposing the motion of the train.

flow for any current that is created (Fig. 17.9). Thus internal circuit loops called **eddy currents** are created in extended conductors subjected to a changing magnetic flux.

Eddy currents can be very useful, because, following Lenz's law, they always act to oppose the change that creates them. Among the applications of eddy currents are the electromagnetic brakes used in some trains, and electromagnetic vibration dampers. An electromagnetic brake permits smooth braking when a magnetic field is applied either to the metal wheels of a train car or to the rails underneath (Fig. 17.10). In a vibration damper a conductor oscillates in a magnetic field. The eddy currents, which reverse sign as the conductor moves back and forth, act to counter the motion, bringing the conductor smoothly to a halt.

According to Joule's law, some energy is dissipated by eddy currents, so it is usually helpful to minimize them. For this reason, the core of a motor armature is usually not solid iron, but is laminated with layers of a nonconducting substance. This reduces the eddy currents, and minimizes the power dissipation.

17.2 Transmission and Transformation of Power

Once current has been produced by a generator or alternator, it often must be transported to some remote location where it is used to produce work. Along the way there will be losses due to resistive heating, and there may be a need to transform the current to a different voltage. In small-scale, specialized applications such as in a car engine, these are not serious problems, but producing power in a centralized plant and distributing it far and wide for household and industrial use does present problems.

17.2 TRANSMISSION AND TRANSFORMATION OF POWER

Transmission of Power

When current flows in a resistor, power is dissipated according to

$$P = VI = I^2R$$

as we learned in Chapter 15. The dissipated power is lost as heat, and therefore must be minimized for efficient transmission over long distances. Even with highly conductive materials such as copper these losses can be significant.

We find that the loss is minimized if the power is transmitted at the highest practical voltage. Since $P = IV$, the higher the voltage, the lower the current. We also know that the power loss is proportional to the square of the current. We see this quantitatively by calculating the loss for a power plant that produces power P which is then transmitted over wires with total resistance R. The current is

$$I = \frac{P}{V}$$

where V is the potential difference. The power dissipated, P_D, is equal to

$$P_D = I^2R = \left(\frac{P}{V}\right)^2 R \qquad (17\text{–}1)$$

which tells us that the power loss is inversely proportional to the square of the potential difference.

Example 17.2 Suppose a generating plant produces 50 kW of power, and transmits this power over lines having a total resistance of 0.2Ω. Compare the power lost if the potential difference is 120 V (ready for household use) to the power lost if $V = 50$ kV.

We simply solve Eq. 17–1 for each case. For $V = 120$ V, we find

$$P_D = \left(\frac{P}{V}\right)^2 R = \left(\frac{5 \times 10^4 \text{ W}}{120 \text{ V}}\right)^2 (0.2\Omega) = 3.47 \times 10^4 \text{ W}.$$

nearly 70% of the power is lost!

When the voltage is 50 kV, then the loss is

$$P_D = \left(\frac{5 \times 10^4 \text{ W}}{5 \times 10^4 \text{ V}}\right)^2 (0.2\Omega) = 0.2 \text{ W}$$

which is insignificant.

The strategy is quite straightforward: power should be transmitted at the highest possible voltage. Therefore cross-country transmission lines have potential differences as high as a few hundred kilovolts, and the power lines through urban neighborhoods have potentials of a few kilovolts.

Such high voltages are impractical for household use and are potentially very dangerous because they can create lethal currents in human bodies, and because their strong electric fields may spontaneously arc. At the other end, it is impractical to produce current at such high voltages, so clearly devices are needed to raise the voltage at the power plant and to reduce it at the household.

Transformers

The **transformer** is a device that alters the voltage in a circuit. It does so by using the principle of inductance in two steps: the incoming current is allowed to create a changing magnetic flux, which in turn creates an outgoing current. The change in voltage occurs because the length of the circuits exposed to the magnetic flux is different for the two circuits.

A typical transformer consists of two coils that are either wrapped around different parts of the same ferromagnetic core (Fig. 17.11) or are interwoven. One coil carries the primary, or incoming current, and the other the secondary, or outgoing current. If the primary current were steady, as in DC current, then there would be no variation in the magnetic flux it creates, and no current would be induced in the secondary coil. If, however, the current in the primary coil varies, as in AC current, then the magnetic flux through the secondary coil varies (at the same frequency), so that a varying current is created in the secondary coil. The role of the ferromagnetic core is to enhance the magnetic flux as its domains keep themselves aligned with the varying magnetic field produced by the primary coil.

From Faraday's law (Eq. 16–18), we see that the voltage induced in the secondary coil is

$$V_s = N_s \left(\frac{\Delta \Phi}{\Delta t} \right)$$

where V_s is the induced voltage, N_s is the number of loops in the secondary coil, and $\Delta \Phi / \Delta t$ is the rate of change of the magnetic flux. In the absence of internal resistance in the coils, the primary voltage is given by the same equation:

$$V_p = N_p \left(\frac{\Delta \Phi}{\Delta t} \right)$$

where it is assumed that the two coils are identical except for the number of loops. Hence we have the relationship

$$\frac{V_s}{V_p} = \frac{N_s}{N_p} \tag{17–2}$$

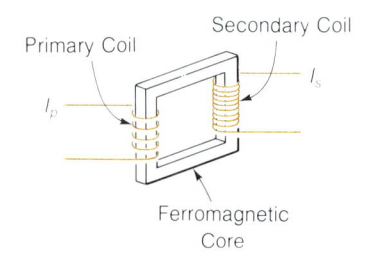

Figure 17.11

A transformer. The alternating voltage created in the primary coil by the current I_p creates an alternating voltage in the secondary coil, hence an alternating current I_s. The ratio of the voltages in the two coils, hence of the amplitudes of the two currents, is equal to the ratio of the number of loops in the two coils.

17.3 TIME-DEPENDENT CIRCUITS

which is sometimes known as the **transformer equation,** because it relates the input voltage to the output voltage.

A transformer is therefore a very simple device for either raising or lowering the voltage in an alternating current. It is very difficult to do this for a direct current, and the ease of doing this (as explained in the previous section) has played a major role in the nearly universal adoption of AC current as the standard.

Of course, any real transformer will lose some power due to dissipation in the internal circuits. This can be minimized so that it is nearly negligible, however, and efficiencies of 99% are common. If we neglect losses, then the currents in the two coils may be easily compared. We assume we have equal power in each, so

$$V_p I_p = V_s I_s$$

which, with Eq. 17–2, leads to

$$\frac{I_s}{I_p} = \frac{N_p}{N_s}. \qquad (17\text{–}3)$$

While we have said that transformers are not useful for DC circuits, there are exceptions. For a brief moment when the current in a circuit is turned on or off, there is a changing current and therefore a changing magnetic flux. Therefore at that moment a secondary current, at some other voltage, can be induced, so we can have a transformer that acts essentially instantaneously. For example, in car ignition systems the rotating distributor momentarily touches contacts (the "points"), creating low-voltage currents that are transformed by the ignition coil into the brief high-voltage currents that fire the spark plugs.

Figure 17.12

The standard symbol for an oscillating source of emf.

17.3 Time-Dependent Circuits

Not only is AC current easier to generate and transmit, but it also has numerous unique applications in circuits. In this section we will discuss several of these applications. In drawing a diagram for an AC circuit, the source of sinusoidal emf is designated by the special symbol shown in Fig. 17.12.

RC Circuits

In an **RC circuit** containing a resistor and a capacitor in series, the charge on the capacitor, hence the current in the circuit, changes with time. Suppose such a circuit (Fig. 17.13) has a source of emf that oscillates or can be turned on and off (RC circuits have DC as well as AC applications). When the source of emf is turned on, current flows in the circuit and charge accumulates on the capacitor, creating a potential difference across it that opposes the source of emf. The charge accumulating on the capacitor tends to inhibit further

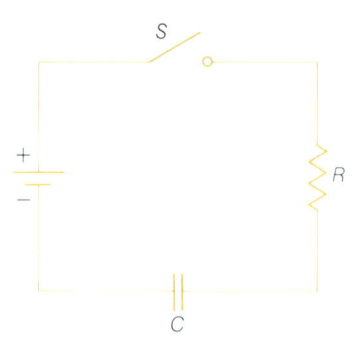

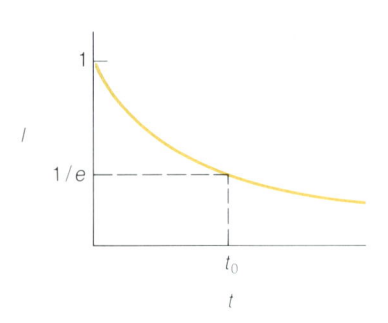

Figure 17.13

An RC circuit. When the switch S is closed, current flows and charge builds up on the capacitor. A potential difference develops across the capacitor, opposing the source of emf and causing an exponential decrease in the current.

Figure 17.14

Exponential decay. The current I decreases by a factor of 1/e = 0.368 in every time interval t_0. For an RC circuit, the time constant is t_0 = RC.

charge from being added, so the charging of the capacitor slows, and then stops when the potential difference across the capacitor equals the emf. The slowing of the charging, hence the reduction of the current in the circuit, is an example of **exponential decay.** It has a mathematical and graphical form (Fig. 17.14) that is encountered quite often in physics (see, for example, the discussion of radioactive half-lives in Chapter 25).

The product *RC* is called the **time constant** of an *RC* circuit, and it represents the time required for the current to drop by a factor of 1/e, where e = 2.718 . . . is the base for natural logarithms. (To derive this value and the form of exponential decay curves requires the use of calculus.) Thus a decay by a factor of 1/e means that the current has dropped to 0.368 of its initial value. The *RC* delay time is discussed further in Special Problem 5 at the end of the Chapter.

The time constant can be adjusted by varying R and C. This is commonly done to provide protection against sudden changes in voltages or currents in the event of short-circuits, for example, and has many specialized applications as well.

Inductance

Suppose we have two coils situated near each other (Fig. 17.15). If an AC current passes through one of them, it creates a changing magnetic flux, which induces a current in the other and in itself. First let us consider the current induced in the second coil. From Faraday's law we know that the emf induced in the second coil is proportional to the rate of change of flux, which is in turn proportional to the rate of change of the current in the first coil. Hence we have

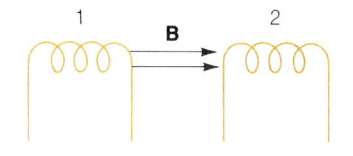

Figure 17.15

Mutual inductance. An alternating current in coil 1 creates a changing magnetic flux, which in turn induces an alternating current in coil 2.

$$\mathcal{E}_2 = -M\left(\frac{\Delta I_1}{\Delta t}\right) \qquad (17\text{--}4)$$

where $\mathcal{E}_2$ is the induced emf in coil 2, $\Delta I_1/\Delta t$ is the rate of change of current in coil 1, and M is a proportionality constant called the **mutual inductance**. The dimensions of M are $V \cdot s/A = \Omega \cdot s$, and this unit is given the special name the **henry** (H), after Joseph Henry. The value of M depends on a number of factors, including the shape and relative sizes of the two coils, and is usually determined experimentally.

Mutual inductance transfers a current from one circuit to another without physical contact. This is how a heart pacemaker works, for example; a battery-driven coil outside the body induces an AC current in the pacemaker, which is implanted in the body. But mutual inductance can also be a nuisance, because it creates AC currents where they may not be wanted. One of the major tasks in "debugging" complex circuits is finding and minimizing all the unwanted induced currents.

Not only does a coil carrying an AC current induce an emf in a second coil, it also induces an emf in itself. This **self-inductance,** occurs because the changing magnetic flux created by the changing current produces an induced emf. This emf obeys Lenz's law, and acts to oppose the changing current in the coil. The induced emf is given by

$$\mathcal{E} = -L\left(\frac{\Delta I}{\Delta t}\right) \qquad (17\text{--}5)$$

where the minus sign indicates that the emf is always directed against the current, and L is a proportionality constant called the **self-inductance**. The value of L, in henries, is usually determined experimentally.

Self-inductance is not always useful, and in many cases efforts are made to minimize it. The reasons for this are made clearer in the next section.

Example 17.3

What is the self-inductance of a solenoid consisting of n wire loops per unit length and having cross-sectional area A?

For this simple solenoid, we can determine L analytically, without need for measurement. In Chapter 16 (Eq. 16–11) we learned that the magnetic field in a solenoid is

$$B = n\mu_0 I.$$

The magnetic flux Φ is

$$\Phi = BA = n\mu_0 IA.$$

The induced emf, from Faraday's law, is

$$\mathscr{E} = -N\left(\frac{\Delta\Phi}{\Delta t}\right) = -Nn\mu_0 A\left(\frac{\Delta I}{\Delta t}\right)$$

where N is the total number of loops in the coil. Finally, Eq. 17–5 can be set equal to this expression for the induced emf, yielding

$$-L\left(\frac{\Delta I}{\Delta t}\right) = -Nn\mu_0 A\left(\frac{\Delta I}{\Delta t}\right)$$

This can be solved for L:

$$L = \mu_0 NnA = \frac{\mu_0 N^2 A}{l}$$

where l is the length of the solenoid and the substitution $n = N/l$ has been made.

Reactance and Impedance

Components in AC circuits often impede the current, even though they may have little or no resistance. The reason for this is that a time-varying current produces induced emfs that oppose the current. This opposition results in a drop of potential, and thus plays a similar role and is measured in the same units as resistance. The tendency of an AC circuit component to resist current is called the **reactance,** and it is measured in ohms. We did not discuss this in connection with transformers, but it does play an important role. If it were not for reactance, a transformer would draw such a large current that it would be damaged. The sum of the reactance and the resistance of a component is the **impedance.** Often the resistance is so small that the impedance is practically equal to the reactance.

If an **inductor**, that is, a coil with large self-inductance, has zero resistance, then the potential drop through it must equal the applied emf, or

$$V = L\left(\frac{\Delta I}{\Delta t}\right). \qquad (17\text{–}6)$$

The reactance creates a voltage drop in accordance with Ohm's law; that is,

$$V = IX_L \qquad (17\text{–}7)$$

where X_L is the **inductive reactance**. In this expression V and I represent the maximum values (or the average values) of the potential difference and the current, which is not the same as saying they represent the simultaneous values at a particular instant; we explain this below. We see by combining these two expressions that the reactance has the form

17.3 TIME-DEPENDENT CIRCUITS

$$X_L = \left(\frac{L}{I}\right)\left(\frac{\Delta I}{\Delta t}\right).$$

For an AC circuit the sinusoidal applied emf is $\mathcal{E} = \mathcal{E}_0\cos(2\pi\nu t)$, where ν is the frequency. By using calculus to find the rate of change of the sinusoidal current, we can express the reactance as

$$X_L = 2\pi\nu L. \qquad (17\text{-}8)$$

Note that the induced reactance is proportional to the frequency; hence, the reactance becomes very high if the frequency of the current is high. For this reason, inductors can be used as filters to screen out unwanted high frequencies in circuits (see Special Problem 1 at the end of the chapter).

The relationship between the voltage and the current is shown graphically in Fig. 17.16. We note that the current and the voltage across an inductor are out of phase by a quarter-cycle, since the voltage is proportional to the rate of change of current, and the rate of change of the current is greatest during the part of the cycle between peaks. The net effect is that energy is temporarily stored in the magnetic field of the inductor and then restored to the form of current. No energy is lost in the process; there is no dissipation of power in an inductor, quite unlike a resistor.

The fact that V and I are out of phase means that Eq. 17–7 is valid only for the peak values of V and I, not at any particular instant. If we speak of average values (usually the rms average), then the expression is also valid.

A capacitor in an AC circuit also has an impedance, again because the current is out of phase with the applied emf. Unlike DC current, AC current flows across a capacitor. This happens because the charge that builds up on one side of the capacitor as the current rises creates an induced opposite charge on the other side; as the current oscillates, the signs of the two charges change. The charge on the capacitor is in phase with the applied emf, but the current is not. This is due to the fact that charge builds on the capacitor while the voltage is building, but the current stops when the voltage reaches its peak. Thus the current is one quarter-cycle ahead of the voltage, as shown in Fig. 17.17. Compare this with the inductor, where the current is one quarter-cycle behind the voltage.

As in the coil, there is no power dissipation in a capacitor. The energy that flows through it is alternately in the form of stored electric field energy (instead of magnetic field energy) and flowing electric charge.

The relationship between voltage and reactance for a capacitor can be derived from Ohm's law:

$$V = IX_C \qquad (17\text{-}9)$$

where X_C is the reactance of the capacitor. The relationship between potential difference and capacitance, along with the application of calculus to a sinusoidal emf, leads to a **capacitive reactance**,

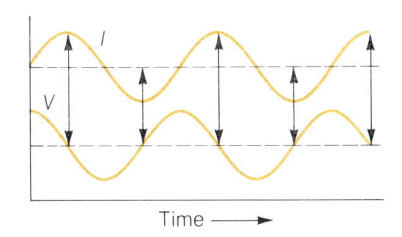

$$X_C = \frac{1}{2\pi\nu C} \qquad (17\text{-}10)$$

Figure 17.16

The phase relationship between current and voltage in an inductor in an AC circuit. The current lags the voltage by a quarter-cycle (90° in phase).

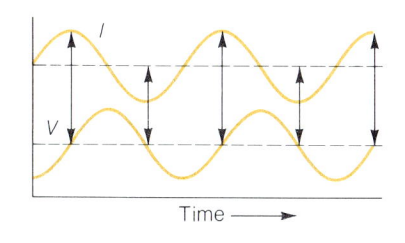

Figure 17.17

The phase relationship between current and voltage for a capacitor in an AC circuit. The current leads the voltage by a quarter-cycle.

where C is the capacitance. Currents of low frequency produce high reactances, and therefore a capacitor can filter out low frequencies.

The impedance of circuit components such as inductors and capacitors must be taken into account when analyzing AC circuits. This is important when applying Ohm's law to find unknown voltages or currents, and furthermore the interplay between the impedances of circuit components must be considered. It can be shown that, if the output of one circuit provides the emf for a second circuit, then the maximum transfer of power to the second circuit is achieved when the impedances of the two are matched. Thus efficient sound systems are designed to have the impedance of the amplifier circuit match that of the speaker system. We derive an expression for the total impedance of a circuit in the next section.

LRC Circuits and Phase Angles

We can best appreciate the utility of various circuit components by seeing how they interact with each other. We can do this by discussing circuits containing an inductor, a resistor, and a capacitor; such circuits are often called **LRC circuits**. We will find that the phase differences in the various components between voltage and current can play important roles in the behavior of such a circuit.

At any moment, the current in all parts of the circuit must be the same. We have seen, however, that for a capacitor, the peak current occurs one quarter-cycle before the peak voltage, and for an inductor, the peak current occurs one quarter-cycle after the peak voltage. Thus, when a sinusoidal emf is applied in an AC circuit, the voltage drops across the individual components are out of phase with the current and with each other. The net potential difference around the circuit at any instant is zero, but the sum of the average voltages is not.

To assess an LRC circuit, a **phasor diagram** is helpful (Fig. 17.18). Arrows called **phasors** represent the peak voltages V_{Lo}, V_{Ro}, and V_{Co} across the inductor, the resistor, and the capacitor. The length of each arrow represents the magnitude of the peak voltage, and the angle ϕ it makes with the x-axis represents its phase with respect to the phase angle of the current. It is convenient to let the phase angle of the current be the zero-point, since we know that the current is in phase throughout the circuit. Therefore, let $I(t) = I_0 \cos(2\pi\nu t)$ represent the current of frequency ν and peak value I_0. In Fig. 17.18(a), at $t = 0$, the arrow representing V_{Ro} lies along the x-axis, since the voltage in the resistor stays in phase with the current, while the arrows for V_{Lo} and V_{Co} are aligned in the positive and negative y directions, respectively. This corresponds to inductor voltage peaking one quarter-cycle *ahead* of the current (a leading phase $\phi = 90°$) and to capacitor voltage peaking one quarter-cycle *behind* the current (a lagging phase $\phi = -90°$). Since V_{Lo} and V_{Co} both lie along the y-axis, they combine to produce a net y-component.

When the circuit is activated by a sinusoidal source of emf, the phasors representing the peak voltages rotate about the origin of the diagram with the frequency, ν, of the sinusoidal emf. Fig. 17.18(b) shows the phasors at a time t shortly after $t = 0$. The phasors have rotated through an angle $2\pi\nu t$. At time

17.3 TIME-DEPENDENT CIRCUITS

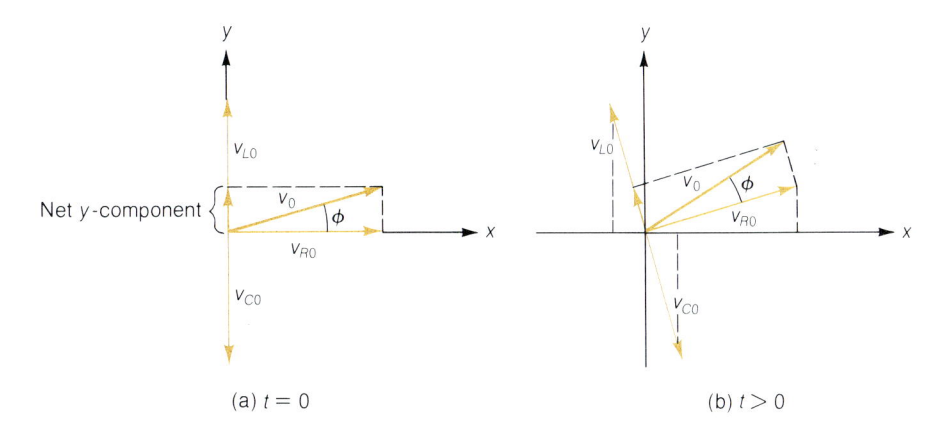

Figure 17.18

Phasor diagrams. Vectors called phasors represent the phase differences between the voltages on the resistor, the inductor, and the capacitor.

t, the voltages across each circuit element are represented by the projections of the phasor arrows onto the x-axis, as indicated by the dashed lines in Fig. 17.18(b). The sum of these projected voltages is the net instantaneous voltage across the three circuit elements, which must equal the source voltage at time t. Hence the phasor diagram can be used to find the voltages anywhere in the circuit at any time.

The net projection of the three phasors onto the x-axis can be viewed as the projection of a new phasor representing the vector sum of the peak voltages (Fig. 17.18). This phasor, V_0, in a sense represents the net behavior of the circuit, and is therefore useful in characterizing the circuit as a whole. We note that, as the phasors representing the individual components rotate, V_0 rotates with them. If we let ϕ be the angle V_0 makes with the x-axis at $t = 0$, then ϕ represents the phase difference between the net voltage in the circuit and the current. Hence ϕ is called the **phase angle** of the circuit, and the net voltage across the terminals of the circuit at t is

$$V = V_0 \cos(2\pi \nu t + \phi). \tag{17-11}$$

We can assess V_0 in terms of the individual voltages. From Fig. 17.18(a) we see from the Pythagorean theorem that

$$V_0^2 = V_{Ro}^2 + (V_{Lo} - V_{Co})^2$$

into which we can substitute (from Eq. 17-7 and Eq. 17-9) expressions for the peak voltages of each component in terms of their impedances and the peak current I_0:

$$V_0^2 = I_0^2 R^2 + (I_0 X_L - I_0 X_C)^2$$

Thus we have

$$V_0 = I_0\sqrt{R^2 + (X_L - X_C)^2} = I_0 Z \qquad (17\text{–}12)$$

where Z is the total impedance of the circuit, defined as

$$Z = \sqrt{R^2 + (X_L - X_C)^2} = \sqrt{R^2 + \left(2\pi\nu L - \frac{1}{2\pi\nu C}\right)^2} \qquad (17\text{–}13)$$

where Eq. 17–8 and Eq. 17–10 have been used.

We can also express the phase angle ϕ in terms of the total impedance of the circuit. In Fig. 17.18(a) we see that the phase angle is given by

$$\cos\phi = \frac{V_{R_o}}{V_0}.$$

We know that $V_{R_o} = I_0 R$, and we have just found that $V_0 = I_0 Z$, so we can substitute for V_{R_o} and V_0, and find

$$\cos\phi = \frac{R}{Z}. \qquad (17\text{–}14)$$

We now have a series of expressions that characterize the overall behavior of a circuit. Although derived for a circuit with an inductor, a resistor, and a capacitor in series, these expressions are actually quite general, because many circuits can be reduced to an equivalent of a simple LRC circuit (recall the expressions for combining resistances and capacitors derived in Chapter 15, for example). If a circuit lacks one of the three types of components, the same expressions are valid with the substitution $R = 0$, $L = 0$, or $C = \infty$ (the latter because, from Eq. 17–10, an infinitely large capacitance corresponds to zero induced reactance).

Example 17.4

Suppose we have a simple LRC circuit, as shown in Fig. 17.19, where $R = 50\,\Omega$, $L = 0.1$ H, and $C = 100\,\mu$F. The emf is provided by ordinary household power, with $V_0 = 120$ V and $\nu = 60$ Hz. Find the peak current, the total impedance, and the phase angle of this circuit.

First we find the individual impedances, from Eq. 17–8 and Eq. 17–10:

$$X_L = 2\pi\nu L = 2\pi(60 \text{ Hz})(0.1 \text{ H}) = 37.7\,\Omega$$

$$X_c = \frac{1}{2\pi\nu C} = \frac{1}{2\pi(60 \text{ Hz})(100 \times 10^{-6} \text{ F})} = 26.5\,\Omega$$

The total impedance (Eq. 17–13) is therefore

$$Z = \sqrt{R^2 + (X_L - X_C)^2} = \sqrt{(50\,\Omega)^2 + (37.7\,\Omega - 26.5\,\Omega)^2} = 51.2\,\Omega.$$

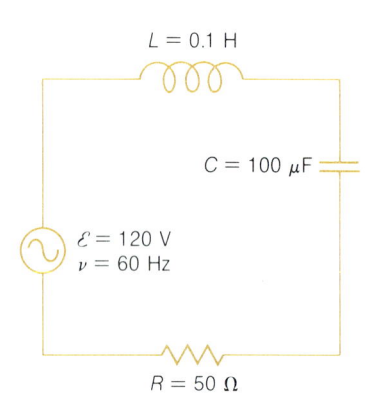

Figure 17.19

$L = 0.1$ H
$C = 100\,\mu$F
$\mathcal{E} = 120$ V
$\nu = 60$ Hz
$R = 50\,\Omega$

The phase angle, from Eq. 17–14, is

17.3 TIME-DEPENDENT CIRCUITS

$$\cos\phi = \frac{R}{Z} = \frac{(50\,\Omega)}{(51.2\,\Omega)} = 0.977$$

so that

$$\phi = 12.3°.$$

Finally, the peak current is

$$I_0 = \frac{V_0}{Z} = \frac{120\text{ V}}{51.2\,\Omega} = 2.34\text{ A}.$$

Power in an AC Circuit

We know that the power in a DC circuit is simply I^2R, where I is the current and R the resistance. Clearly, this formula must be modified for an AC circuit because the potential varies with time. We first consider a circuit containing only resistance.

In an AC circuit with sinusoidal potential of the form

$$V = V_0 \cos(2\pi\nu t)$$

the current across the resistor is in phase and has the form

$$I = I_0 \cos(2\pi\nu t).$$

At time t, the power is

$$P = I^2 R = R I_0^2 \cos^2(2\pi\nu t). \tag{17-15}$$

The average value of the square of the cosine function is ½, so the average power is

$$\overline{P} = \frac{1}{2} R I_0^2 = R (I_{\text{rms}})^2 \tag{17-16}$$

where I_{rms} is the root mean square current, given by

$$I_{\text{rms}} = \frac{I_0}{\sqrt{2}}. \tag{17-17}$$

The rms current is the equivalent DC current that would produce the same power as the AC circuit. A voltage equivalent to that in an AC circuit may also be defined as

$$V_{\text{rms}} = \frac{V_0}{\sqrt{2}}.$$

The scales of ammeters and voltmeters are normally set to display the rms values of current and voltage when applied to AC circuits.

The power in an AC circuit containing impedances is a bit more complicated to derive because of the phase angle. We start with $P = VI$ and then substitute Eq. 17–11 for V and a sinusoidal term for I.

$$P = VI = [V_0 \cos(2\pi\nu t + \phi)][(I_0 \cos(2\pi\nu t)]$$

If we apply the rule of trigonometry for the cosine of the sum of two angles and rearrange, we get

$$P = V_0 I_0 \cos^2(2\pi\nu t) \cos\phi - \frac{1}{2} V_0 I_0 \sin(4\pi\nu t) \sin\phi.$$

This gives the instantaneous power in the circuit. It is often more useful to speak of the average power. The average of the $\cos^2(2\pi\nu t)$ term is ½, and the average of the $\sin(4\pi\nu t)$ term is zero, so we have

$$\overline{P} = \frac{1}{2} V_0 I_0 \cos\phi.$$

We have already seen that $V_{\text{rms}} = V_0/\sqrt{2}$ and $I_{\text{rms}} = I_0/\sqrt{2}$, so we can express the average power in terms of the rms voltage and current.

$$\overline{P} = V_{\text{rms}} I_{\text{rms}} \cos\phi \qquad (17\text{–}18)$$

We also have

$$\overline{P} = I_{\text{rms}}^2 Z \cos\phi \qquad (17\text{–}19)$$

and

$$\overline{P} = \left(\frac{V_{\text{rms}}^2}{Z}\right) \cos\phi. \qquad (17\text{–}20)$$

Each of these is analogous to a similar expression for DC circuits, except for the $\cos\phi$ factor, which is called the **power factor** and accounts for the fact that voltage and current are out of phase by an angle ϕ. Note that for a circuit with only a resistance R, the phase is zero so the power factor is $\cos\phi = 1$, whereas for a circuit containing only a capacitance or an inductance, the power factor is $\cos\phi = 0$, and no power is dissipated.

Example 17.5   Find the rms voltages and currents in the three circuit components described in Example 17.4, as well as the average power in the circuit.
We need to find the rms current first.

17.3 TIME-DEPENDENT CIRCUITS

$$I_{rms} = \frac{I_0}{\sqrt{2}} = \frac{2.34 \text{ A}}{\sqrt{2}} = 1.65 \text{ A}$$

Now we can find the individual rms voltages.

$$(V_R)_{rms} = I_{rms}R = (1.65 \text{ A})(50\Omega) = 82.5 \text{ V}$$
$$(V_L)_{rms} = I_{rms}X_L = (1.65 \text{ A})(37.7\Omega) = 62.2 \text{ V}$$
$$(V_C)_{rms} = I_{rms}X_C = (1.65 \text{ A})(26.5\Omega) = 43.7 \text{ V}$$

These would be the values indicated by an AC voltmeter.
The average power dissipated in the circuit is

$$\overline{P} = I_{rms}^2 Z \cos\phi = (1.65 \text{ A})^2 (51.2\Omega)(0.977) = 136 \text{ W}.$$

Note that this is the same as $I_{rms}^2 R$, as expected from Eq. 17–16.

Resonances in an AC Circuit

The current in an AC circuit is determined by the source of emf and by the impedances of the circuit components. These impedances, in turn, depend on the source frequency, through Eq. 17–8 and Eq. 17–10. Therefore the current in the circuit depends on the source frequency. The rms voltage and current are related by

$$I_{rms} = \frac{V_{rms}}{Z}.$$

into which Eq. 17–13 can be substituted

$$I_{rms} = \frac{V_{rms}}{\sqrt{R^2 + \left(2\pi\nu L - \frac{1}{2\pi\nu C}\right)^2}} \qquad (17\text{–}21)$$

We see that the value of I_{rms} will be maximized when the denominator is smallest, which occurs when

$$2\pi\nu L - \frac{1}{2\pi\nu C} = 0$$

Then the frequency is

$$\nu_0 = \frac{1}{2\pi\sqrt{LC}}. \qquad (17\text{–}22)$$

This expression represents the frequency at which maximum current flows, and therefore maximum power is carried by the circuit. This reminds us of our discussion of resonances in Chapter 9, and we refer to this as the **resonant frequency** of the circuit.

When Eq. 17–22 is satisfied, the reactances X_L and X_C are equal, and the impedance of the circuit is entirely due to the resistance R. The phase angle is zero, so $\cos\phi = 1$. Fig. 17.20 shows how the rms current or the power varies with circuit frequency. A sharp peak is centered on the resonant frequency v_0.

The sharpness of the peak around v_0 is determined by the ratio of the voltage across the capacitor to the voltage across the resistor. The larger the ratio, the sharper the peak (Fig. 17.20) and the better the circuit is tuned to the resonant frequency. The sharpness or quality of the resonance is usually expressed in terms of the **Q-factor,** defined as

$$Q = \frac{V_C}{V_R} \qquad (17\text{–}23)$$

where V_C and V_R are the voltages across the capacitor and resistor at resonance ($v = v_0$). It is often desirable to maximize the Q-factor in designing a circuit, so the relative values of V_C and V_R are planned accordingly. One can show that $Q = \sqrt{LC}/RC$ (see Special Problem 2 at the end of the chapter).

There are many practical uses of resonance in circuits, and there are pitfalls as well. Unwanted resonances can create large drains of power into undesired circuit loops. On the other hand, resonances can be used as filters or tuners, allowing significant current to flow only at some desired frequency. In a radio or television tuner, for example, either the inductance or the capacitance is adjusted (although usually it is the capacitance) so that the circuit frequency matches the desired station's broadcast frequency. When a match is achieved, maximum current flows in the tuner circuit in response to the received signal, while signals received at other frequencies do not create significant current. Thus the unwanted frequencies are effectively screened out.

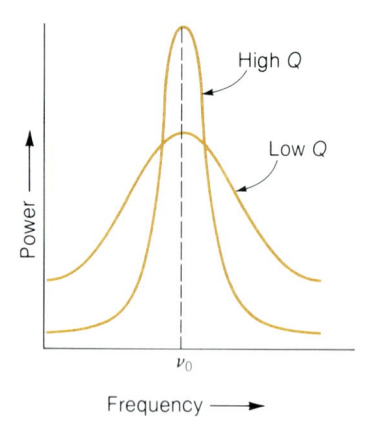

Figure 17.20

Resonance. The power (and rms current) in a circuit peak sharply at the resonance frequency, when $X_L = X_C$. The sharpness of the peak is determined by the Q-factor, where $Q = V_C/V_R$ at $v = v_0$.

Example 17.6   Suppose you want to listen to a radio station whose broadcast frequency is 1,200 kHz. The tuner circuit is an LRC circuit with a fixed capacitance of 50 pF and a variable inductance L, which is controlled by turning a knob. When the radio is properly tuned, what is the inductance?

We are given v_0 and C, and want to solve Eq. 17–22 for L.

$$L = \frac{1}{4\pi^2 v_0^2 C} = \frac{1}{4\pi^2 (1.2 \times 10^6 \text{ Hz})^2 (50 \times 10^{-12} \text{ F})} = 3.5 \times 10^{-4} \text{ H}$$

17.4 Vacuum Tubes and Semiconductors

We have previously referred to devices called diodes, particularly in reference to rectifiers, which allow current to pass through them in only one direction. Diodes were originally constructed from vacuum tubes, as were related devices

17.4 VACUUM TUBES AND SEMICONDUCTORS

called **triodes,** which can be used to amplify an AC voltage. In modern circuitry, both functions are performed by a different class of devices, which use semiconductor materials. In this section we briefly discuss vacuum tubes, then dwell at more length on semiconductor devices.

Vacuum Tubes

The basis for the vacuum tube is **thermionic emission** (Fig. 17.21). When a source of electrons is placed near an exposed conducting plate which has a positive voltage (with respect to the electron source), electrons are attracted to the plate, creating a current in it. If the conductor has a negative voltage applied to it, it repels electrons, and no current flows. If the source of electrons, called the **cathode,** is in series with the plate in a circuit with an alternating source of emf, the result is a DC current, because current flows only when the potential difference between plate (also called the anode) and cathode is positive. This is the basis of the vacuum-tube diode.

This process does not operate effectively at normal atmospheric pressure, because the electrons collide with gas particles, and no current flows in the plate. Hence a diode of this type must be operated in a vacuum. The term thermionic emission refers to the source of electrons, which is a heated filament or plate. Electrons within the material of this plate can acquire sufficient energy to escape as free particles, if the temperature is high enough. As in gases, the energy of the individual electrons within a solid depends on temperature; therefore, the higher its temperature, the more electrons are released by the cathode. The temperature of the filament is raised by a resistive heater, operated by a separate circuit (Fig. 17.22), which is close to the cathode in the tube.

A **bias voltage** is a negative potential difference between the cathode and the plate that repels electrons (Fig. 17.23). In a vacuum tube, the bias voltage is usually supplied by a wire-mesh grid between the cathode and the plate, so that electrons are not physically blocked. The electrons have a spread of kinetic energies, so the bias voltage, if set at an intermediate value, can screen out the low-kinetic energy electrons. Thus the bias voltage controls the current that reaches the plate.

If the bias voltage is adjusted to oscillate about a level that is negative with respect to the voltage on the plate, then an oscillating DC current flows to the plate. If the relative potentials are adjusted properly, small oscillations in the bias voltage can produce very large current oscillations in the plate. In that sense such a device, called a **triode,** is an amplifier, because the current amplitude, hence the power, is enhanced (Fig. 17.23).

Vacuum tubes have many shortcomings. They are bulky, fragile, require time to heat the cathode, and they dissipate a lot of power, which makes them inefficient and creates excess heat. In modern devices, semiconductors have largely replaced vacuum tubes, but vacuum tubes still have some applications. With very large voltages, for example, semiconductor devices tend to break down, so vacuum tubes are used instead. The picture tube in a television set, like all modern video display screens, is a large vacuum tube called a **cathode ray tube** (CRT) in which the electrons flow through a hole in the anode plate to spots that are controlled by varying electric or magnetic fields (Fig. 17.24). An image is created by sweeping the electron beam across the front of the

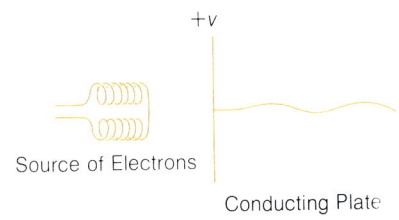

Figure 17.21

Thermionic emission. When a source of electrons (such as a heated filament) is placed near a conductor with positive potential, a current flows in the conductor. No current flows if the voltage on the conductor is negative with respect to the electron source.

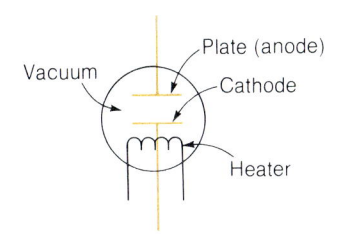

Figure 17.22

A vacuum-tube diode. Electrons from the heated cathode are attracted to the anode when the anode has a positive relative potential. Hence an AC current is converted to an intermittent DC current.

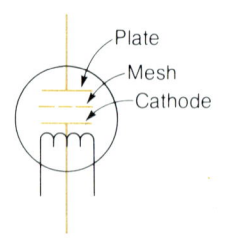

Figure 17.23

A vacuum-tube triode. A negative bias voltage on the mesh is adjusted to repel electrons having low kinetic energies. By adjusting the bias voltage, the current to the plate is adjusted. An oscillating bias voltage can be used to create very large current oscillations in the plate to amplify an input current.

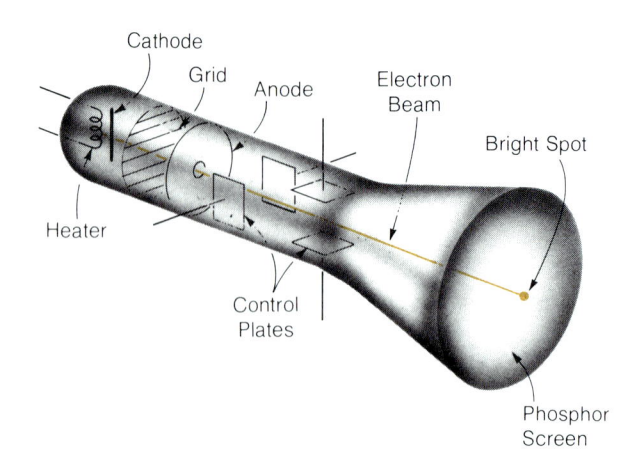

Figure 17.24

A cathode-ray tube.

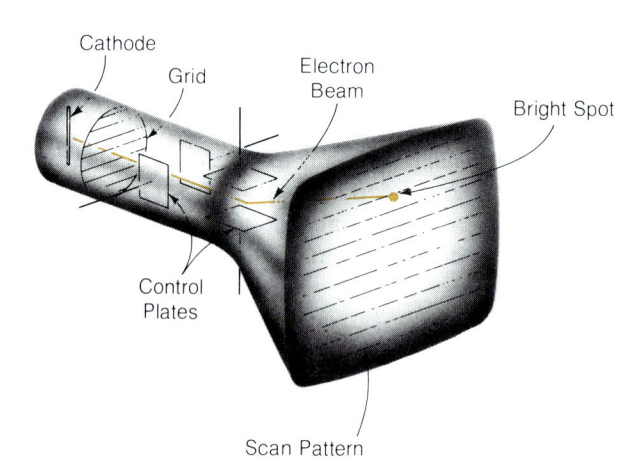

Figure 17.25

A television picture tube. The intensity of the electron beam is controlled by the voltage on the grid; the location of the bright spot on the screen is controlled by the voltages on the control plates. The beam is swept across the screen in horizontal rows, in phase with the intensity variations.

tube, which is coated with a phosphor, so that the spot where the electrons strike it glows briefly. An image on a television screen consists of a large number of horizontal rows, each of which has variations in intensity that are created by time variations in the bias voltage supplied by the grid (Fig. 17.25). The entire sequence of rows is swept very rapidly (in 1/60 of a second), so that the image is renewed frequently. The glow of the phosphor lingers, as does the

17.4 VACUUM TUBES AND SEMICONDUCTORS

image in the human eye, so that the image appears to be continuously present on the screen. The rate at which the beam is swept across the tube face by the deflection fields must be synchronized with the variations in the bias voltage, so that the correct intensity fluctuations appear in the correct places on the screen. The beam on the screen must be placed precisely. In color television, even more precision is required, because the phosphor coating consists of tiny spots with different color responses, and the electron beam must vary properly as it excites the different colors to glow. Sometimes separate electron beams are used for each of the three primary colors.

Semiconductors

Most semiconductors are crystals of silicon or germanium with a trace of an impurity such as arsenic or boron. A pure crystal of either silicon or germanium has a regular cubic structure because these elements have four outer electrons which create bonds between atoms (Fig. 17.26). A pure crystal of either of these elements does not conduct significantly, because the electrons are bound rather strongly in their lattice positions, and are therefore not free to roam throughout the solid.

When an impurity is introduced, isolated lattice positions here and there in the crystal are taken by another kind of atom, which may have some number of outer electrons other than four. Adding an impurity to a crystal is called **doping.** If the atom has extra electrons, as in the case of arsenic (which has five outer electrons), then these extras are not rigidly bound, and can move about if a modest potential difference is applied. If, on the other hand, the impurity element has fewer than four outer electrons, as in the case of boron (which has three), then there are vacancies here and there in the lattice where electrons would normally be. An electron can move from a nearby lattice position into one of these "holes", as the vacancies are called, which results in the hole moving to a new position. The electron holes become free to move through the lattice if a potential difference is applied. When there are extra

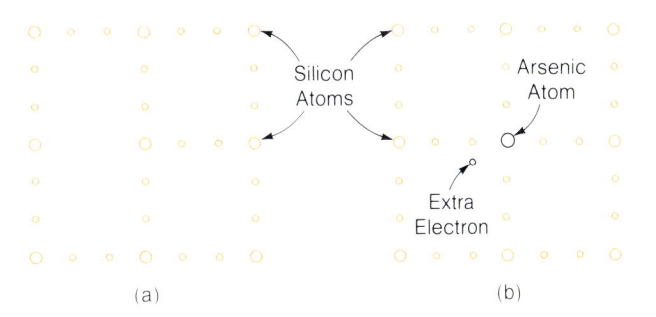

Figure 17.26

Crystal lattices. (a) A pure silicon crystal, showing the cubic arrangement of silicon atoms. (b) A doped silicon crystal, with an arsenic atom replacing one of the silicon atoms, and providing an extra electron (n-type).

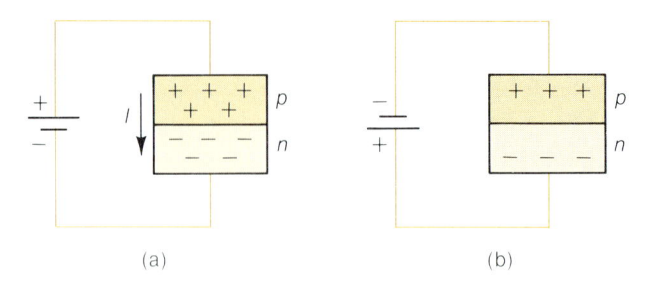

Figure 17.27

A semiconductor diode. When the voltage is forward-biased (a), current flows. When it is reverse-biased (b), no current flows.

electrons, the semiconductor is said to be an **n-type** semiconductor, because negative charge flows; when there are electron holes, we have a **p-type** semiconductor, because, in effect, positive charge flows. The type of semiconductor is determined by the impurity.

A minimum amount of energy is required for an electron or an electron hole to move freely under an applied potential difference. This is because the electrons or the electron holes are weakly bound in the crystal lattice, and require a certain amount of energy to break the bond. The energy of electrons increases with temperature, so the bonds become easier to break if the temperature is high. Thus the conductivity of a semiconductor, unlike that of an ordinary conductor or resistor, increases very rapidly with temperature. Such materials are called semiconductors because of the energy barrier (called the **band gap**) that must be overcome for charge to flow; there is no such energy barrier in a conductor, where the electrons are free to move.

A semiconductor diode is a pair of adjacent *n*-type and *p*-type semiconductors (Fig. 17.27). When such a device is inserted into a circuit so that the positive terminal of the source of emf is connected to the *p* side, a strong current flows. Electrons flow through the *n* side towards the junction with the *p* side. Meanwhile, holes flow through the *p* side towards the junction. At the junction, electrons cross over to fill the holes (a process called recombination), and the net result is a current flowing in the conventional sense. (In this configuration the diode is said to be **forward-biased**.)

If the direction of the emf is now reversed, the electrons and holes on opposing sides of the junction will move away from each other, and no current will flow. (In this configuration the diode is **reverse-biased**.) Thus a semiconductor diode converts AC current into DC current, just as a vacuum-tube diode does.

The advantages of the semiconductor device are that it is very compact and durable, it requires very little power, it has no warm-up time, and it does not require any heating. One disadvantage, though it can be put to practical use in certain circumstances, is that a sufficiently high applied voltage can cause a semiconductor diode to break down. This means that lattice electrons are pulled free by the large voltage and allowed to flow in the reverse-biased

17.4 VACUUM TUBES AND SEMICONDUCTORS

Figure 17.28

Output voltage from a half-wave rectifier.

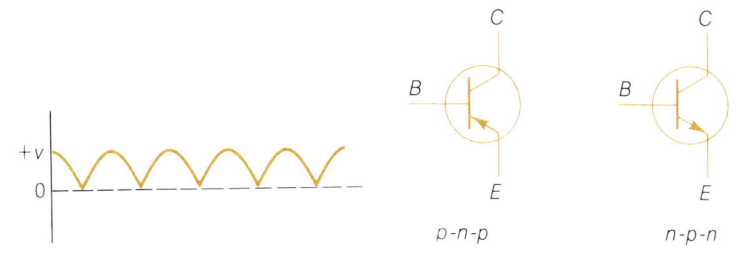

Figure 17.29

Output voltage from a full-wave rectifier.

Figure 17.30

Standard symbols for transistors. The direction of conventional current, indicated by the arrows, depends on whether the transistor is p-n-p *or* n-p-n.

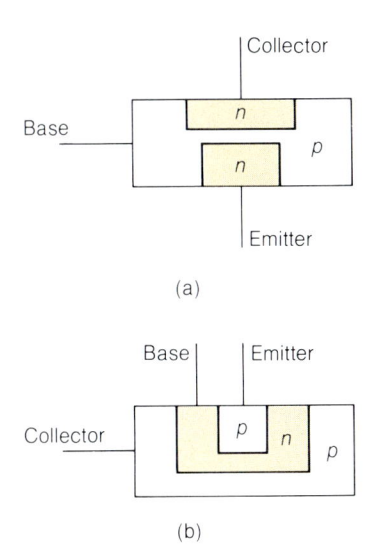

Figure 17.31

Transistors. An n-p-n *transistor is shown in (a), and a* p-n-p *transistor is shown in (b).*

direction normally forbidden by the diode. Hence, as we noted in the previous section, vacuum-tube diodes must be used with very large voltages.

When a high enough voltage is applied to a semiconductor diode so that it breaks down, current flows as electrons are freed from the crystal lattice. Any further increase in voltage is compensated for by additional electrons and additional current, so the voltage remains constant over a wide range of current levels. Such a device, called a **zener diode,** can therefore be used to regulate the voltage level in a circuit. It is possible to construct zener diodes for any voltage.

When used to convert AC current into DC current, a diode is called a rectifier, as we have already seen. We have noted also that the DC current is not steady, but is instead periodic. A diode truncates one half the sinusoidal form of AC current, producing half-wave rectification with time-dependence as shown in Fig. 17.28. Two diodes used in parallel produce full-wave rectification (Fig. 17.29). A capacitor inserted into the circuit can help smooth the current because of the time delay while charge builds on the capacitor.

Just as the *p-n* junction semiconductor is the analog of a vacuum-tube diode, the **transistor** is the analog of the vacuum-tube triode. The symbols for a transistor in a circuit are shown in Fig. 17.30. A transistor consists of three adjacent layers of semiconductor material, which can be arranged in a *p-n-p* or *n-p-n* sequence (Fig. 17.31). To see how a transistor works, let us consider a *p-n-p* example in a circuit (Fig. 17.32). The three layers are called the **emitter,** the **base,** and the **collector.** If a source of emf is connected so that conventional current flows into the emitter, then the *p-n* junction between the emitter and the base is forward-biased, and positive holes would ordinarily flow into the base. The base is very thin, however, and is not very strongly doped, so that it has few extra electrons. Furthermore, it tends to lose its extra electrons if they are not resupplied, because they recombine with holes or flow through the emitter toward the positive side of the source of emf. These losses of electrons can leave the base with a slight positive charge, which inhibits the flow of positive holes into the base from the emitter. Little current flows through the base into the collector in that case. On the other hand, if electrons are supplied to the base by a current, then the positive charge is overcome, and

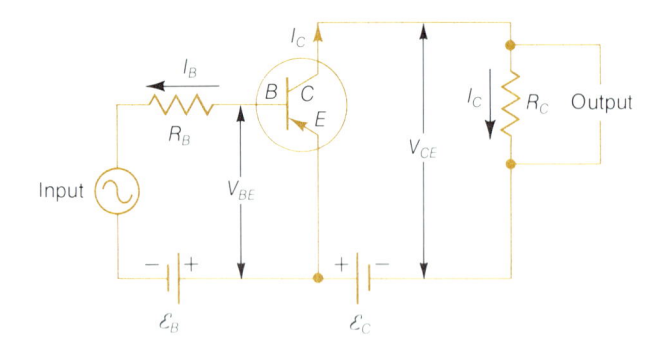

Figure 17.32

A p-n-p transistor used as an amplifier. The input current supplies electrons to the base, controlling the flow of current in the circuit. A small change in input voltage produces a large change in collector current I_C and thus a large amplified voltage across the output resistor R_C. The voltages V_{BE} and V_{CE} are called bias voltages and are maintained by small batteries.

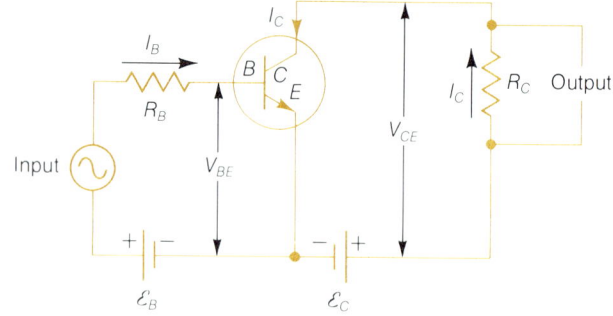

Figure 17.33

An n-p-n transistor used as an amplifier. The input current supplies electrons to the base, controlling the flow of current in the circuit. A small change in input voltage produces a large change in collector current I_C and thus a large amplified voltage across the output resistor R_C. The voltages V_{BE} and V_{CE} are called bias voltages and are maintained by small batteries.

holes flow easily from the emitter into the base. Most of the holes flow right through the base into the collector and out of the transistor, because the base is so thin that recombinations between holes and extra electrons are rare.

The base current, which supplies electrons to the base, controls the ability of emitter current to flow through to the collector and out of the transistor. In a sense, the transistor acts as a valve for current driven by the source of emf, with the base current controlling how widely open the valve is. Because a very small change in the base current can create a large change in the emitter current, a transistor is a very effective amplifier of AC current. If, for example, the base current is the tiny current from a brain wave, this current can be amplified by the transistor so that it can be measured and recorded in an electroencephalograph.

An *n-p-n* transistor works in just the same way as a *p-n-p* transistor, except that it is electrons rather than holes that flow from the emitter through the base to the collector. The emitter, base, and collector in an *n-p-n* transistor correspond in function to the cathode, grid, and anode in a vacuum-tube triode. Fig. 17.33 shows an *n-p-n* transistor used as an amplifier in a circuit.

The performance of transistors is normally represented by graphs called **characteristic curves** (Fig. 17.34), which show the current through the collector (I_C) as a function of the voltage across the transistor (V_{CE}). Separate curves show I_C versus V_{CE} for different base currents (I_B). From such a graph it is possible to determine the **gain**, the ratio of the amplitude of the collector current to that of the base current. The gain is the factor by which the amplitude of a current is multiplied. Suppose, for example, that the base current oscillates between 2 and 6 μA, and the voltage V_{CE} is constant at 5 V. From Fig. 17.34 we see that the collector current varies between 1.0 and 2.8 mA. The gain β is the ratio of the amplitude, or range, of the collector current to that of the base current, or

17.4 VACUUM TUBES AND SEMICONDUCTORS

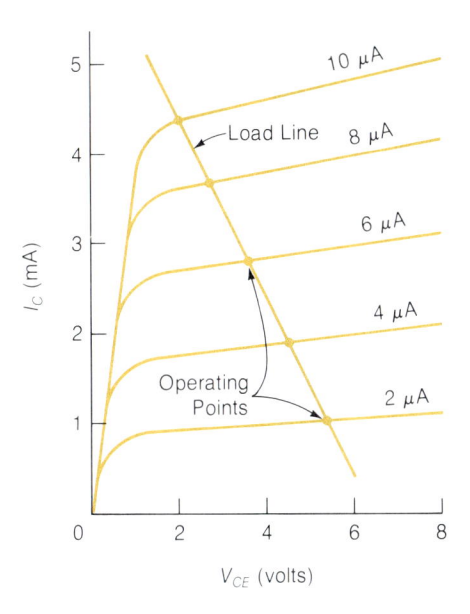

Figure 17.34

Characteristic curves. For each value of the base current (here labelled in units of μA), a characteristic curve shows the relationship between collector current I_C and transistor voltage V_{CE}. The load line represents the allowed combinations of I_C and V_{CE}, and intersects the characteristic curves at operating points.

$$\beta = \frac{(2.8 \times 10^{-3}\text{ A} - 1.0 \times 10^{-3}\text{ A})}{(6.0 \times 10^{-6}\text{ A} - 2.0 \times 10^{-6}\text{ A})}$$
$$= \frac{(1.8 \times 10^{-3}\text{ A})}{(4.0 \times 10^{-6}\text{ A})} = 450.$$

Thus, in this example the input current (I_B) is amplified by a factor of 450.

It is not quite correct to assume, as we did in this example, that the voltage (V_{CE}) across the transistor is steady. It actually varies with the oscillating current such that

$$V_{CE} = \mathscr{E} - I_C R_C,$$

where R_C is the resistance in the circuit that contains the source of emf $\mathscr{E}$. A line representing this relationship is called the **load line,** which cuts diagonally across the graph of characteristic curves (Fig. 17.34). The intersection points are called **operating points** because a transistor can operate only when its combination of I_C and V_{CE} satisfy both the characteristic curve and the load line for any value of the base current I_B.

Since semiconductors are solid crystals, devices made of them are often called **solid-state** devices. These are the foundation of modern electronics, much of which would be impossible with vacuum-tube technology. Not only can

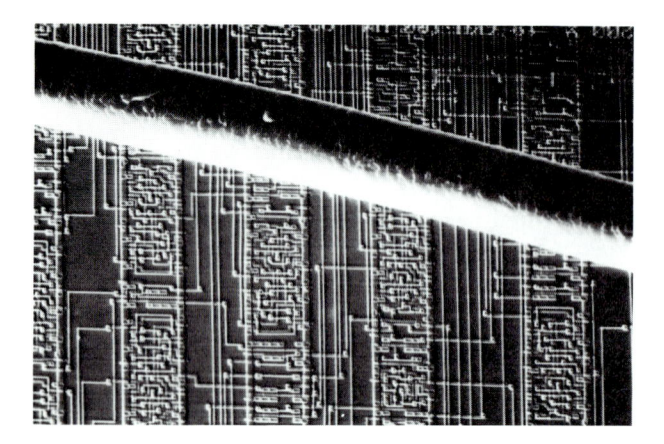

Figure 17.35

A compact solid state circuit chip. This is a section of an electronic circuit on a silicon chip. Individual conducting segments can be clearly seen. Lying across the chip is a human hair, to illustrate the scale.

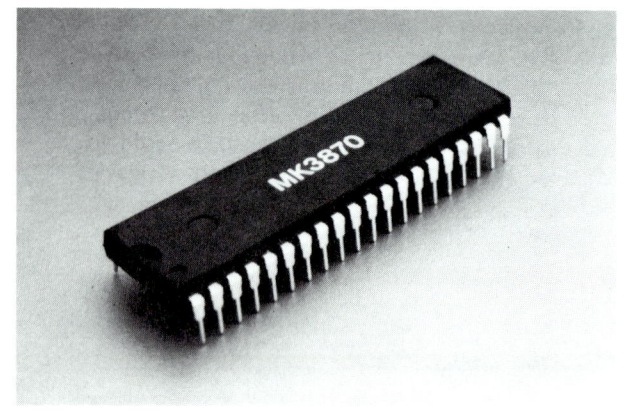

Figure 17.36

An integrated circuit. This device, perhaps 1 cm in length, contains hundreds or thousands of individual circuit elements.

currents be amplified, but so can voltages. One of the greatest efficiencies in solid-state electronics is the incredible compactness that is possible (Fig. 17.35). A single piece of silicon can effectively contain a myriad of electronic components if different impurities are added in different places. We have seen that arsenic and boron create *n*-type and *p*-type semiconductors when added to a silicon crystal; it is also possible to attach conducting metal layers; or to deposit an insulator, such as silicon dioxide. With the proper treatment of microscopic areas of a silicon crystal wafer (commonly known as a circuit chip), a single tiny piece can contain thousands or even hundreds of thousands of diodes, transistors, amplifiers, and other circuit components, and can perform a very complex set of functions. Such a device is called an **integrated circuit** (Fig. 17.36).

FOCUS ON PHYSICS

Computer Memory and Data Storage*

At the heart of the information revolution is memory, the ability to store large amounts of data in one place. In the most primitive computers, data were stored on punch cards. Today, magnetic tape and disks have largely superseded cards, and integrated circuit memory chips have led to an amazing number of devices undreamed of twenty-five years ago: rockets and satellites with on-board computers, personal computers, pocket calculators, and even robots!

The breakthrough in miniaturization came when electrical engineers were able to place large numbers of circuit components on a single semiconductor chip (usually pure silicon). Thin layers of silicon dioxide and p- or n-type silicon are deposited on the pure silicon, together with silicon dioxide (an insulator) and conducting channels. Circuits that use this technology are called **integrated circuits**. A large semiconductor industry based on these chips rapidly grew up in California's Santa Clara Valley (often called "Silicon Valley"), in the Boston area, and in Japan.

The process of laying down circuit elements on a silicon wafer and ensuring that they are connected electrically is called **integration**. Medium-scale integration involves placing hundreds of circuit components on a single chip, and large-scale integration places several thousand per chip. The industrial state of the art, called **Very Large Scale Integration**, integrates up to 400,000 components on a chip. The complexities of designing these circuits exceed ordinary humans' capabilities, and computers are generally used in the design of such chips.

Semiconductor memories can be classified into two main types: **RAMs** (random-access memories) and **ROMs** (read-only memories). The term *random access* means that any one of their storage locations can be accessed quickly, either to be read from or written to. Similarly, a *read only* memory is designed so that it can only be written (programmed) once by the manufacturer, to be read many times later by the user. This type of memory is useful for storing instructions for computers, word-processors, video games, and robots. RAMs lose their data when the power supply is turned off, whereas ROMs retain the information programmed into them, even when the power is off. Thus, ROMs are ideal for the start-up and control routines that must be pre-programmed into personal computers, and the RAMs are useful for reading and writing data and programs in personal computers. Another distinction is that static RAMs

*Information provided by Scott Shull, Intel Corp. Photographs courtesy of Lattice Semiconductor, Inc.

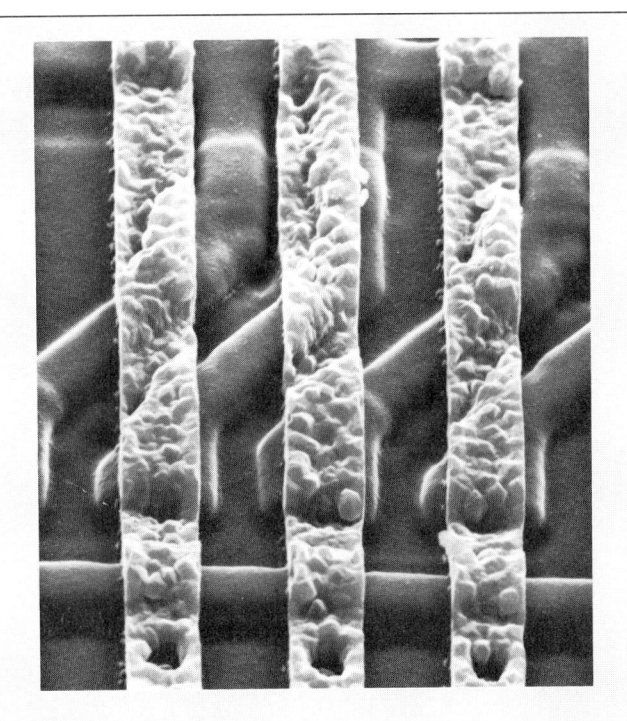

Figure 17.1.1

Microscopic view of 64K RAM chip, showing one storage cell and three bit lines.

(SRAMs) can be accessed much faster than either RAMs or ROMs, so that the fast computer memories tend to use SRAMs chips, even though they use somewhat more power and do not have as much storage capacity.

All memories, SRAM, RAM or ROM, consist of an array of memory cells, which are made of transistors, resistors, and capacitors. Each cell stores a digital bit of information, either a 1 or a 0, in the form of the presence or absence of an electrical charge or current. The storage array of cells is placed on a chip and surrounded by connectors and devices which control the input and output of information. The storage capacity is measured in **bits**, the number of independent cells placed on the chip. The current popular model of static RAM is known as the 64K RAM chip, which stores over 64 kilobits of data (the exact number is 65,536, which is 2^{16} bits). Figure 17.1.1 shows a microscopic view of one such chip, and Figure

(continued)

17.1.2 shows the chip with its peripheral connections.

The typical size of these RAM chips is about 0.5 cm square. Since each cell on the chip usually requires about 6 circuit devices (4 transistors and 2 voltage-regulating resistors), a 64K RAM chip integrates almost 400,000 components. A 256K RAM chip is now on the market (this stores 2^{18} or 262,144 bits), and plans are being made for RAM chips with 1 megabit (2^{20} or 1,048,576 bits) and 4 megabits (2^{22} or 4,194,304 bits)! The 1 megabit chip will integrate over 2 million devices onto an area of 60 mm^2, using a line spacing of 1.2 microns (a human hair has a diameter of about 25 microns. The 4 megabit chip will use line spacings of 0.8 microns.

Other than their density, perhaps the most amazing thing about these chips is their small cost. A ROM of typical density costs about $1. A RAM of the same density costs about $2, and prices of 256K RAM chips will soon fall to about $3. Because the cost of memory has been decreasing at the same time as the storage capacities increase, most computer designers predict that supercomputers will produce another revolution in the next decade.

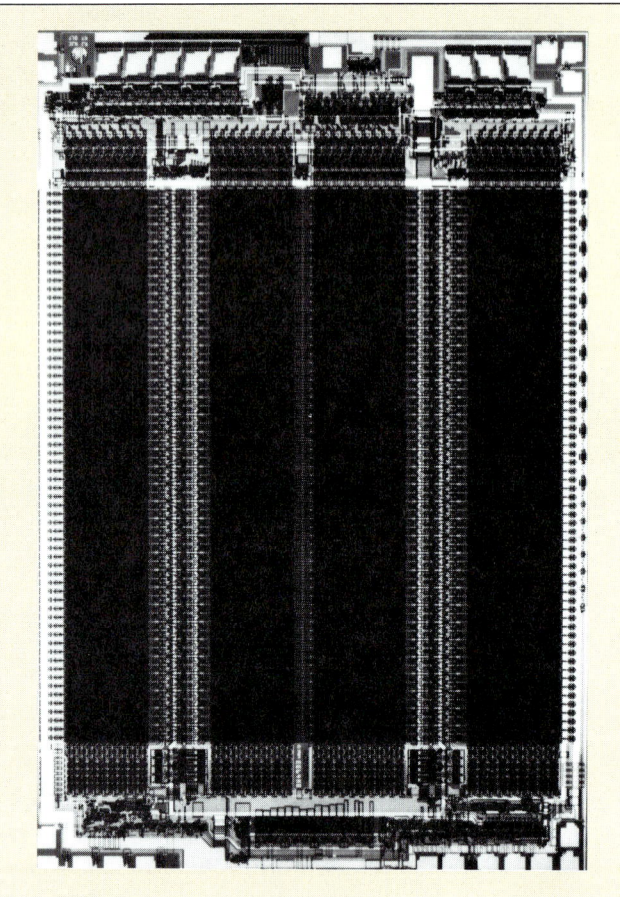

Figure 17.1.2

A 64K RAM chip with connectors.

Perspective

With this discussion of various practical applications of electricity and magnetism, we have completed our treatment of charge and currents. As in many other topics, we have developed an understanding of the principles so that further investigation will be comprehensible, but we are forced to turn our attention elsewhere. Electronics is not only the subject of entire courses, but entire degree programs at the undergraduate and graduate levels.

We are not yet finished with our discussions of electric and magnetic fields, however, for very important phenomena remain. In Part Five, we will examine electromagnetic radiation.

PROBLEM SOLVING

LRC Circuit Problems

Problems with AC circuits have many similarities to DC circuits discussed in Chapter 15, except that we must use impedances and reactances rather than resistances. Impedance depends on frequency for inductors (L) and capacitors (C). However, the voltage-current relations are similar to Ohm's law.

$$V_L = IX_L = I(2\pi\nu L) \quad \text{Inductor}$$
$$V_C = IX_C = I/(2\pi\nu C) \quad \text{Capacitor}$$
$$V_R = IR \quad \text{Resistor}$$

For a circuit containing L, R, and C components, the total impedance is not the direct sum of the resistances and reactances, but instead involves the Pythagorean sum (the phasor diagram involves triangles),

$$Z = \sqrt{R^2 + (X_L - X_C)^2}$$

and the phase angle between voltage and current is given by

$$\cos\phi = \frac{R}{Z}.$$

Joule's law for average power also has an analog for AC circuits,

$$\overline{P} = VI = I^2 Z \cos\phi = \left(\frac{V^2}{Z}\right)\cos\phi$$

where I and V are either the peak or the rms values of current and voltage. Peak values are a factor $\sqrt{2}$ larger than rms values, so take care not to mix them.

To analyze an LRC circuit, one is generally interested in three quantities: (1) the total impedance Z, (2) the phase factor $\cos\phi$, and (3) the resonant frequency $\nu_0 = 1/(2\pi\sqrt{LC})$. Given these three quantities, other parameters such as current or power can be derived.

Example 1 A circuit has $R = 5\ \Omega$, $L = 1$ mH, and $C = 5\ \mu$F. What is the resonant frequency? Find the current, I_{rms}, the phase angle ϕ, and the average power dissipated in the circuit if a 10 V (peak) voltage is applied at 2,000 Hz.

The resonant frequency is

$$\nu_0 = \frac{1}{2\pi\ [(10^{-3}\ \text{H})(5\times 10^{-6}\ \text{F})]^{1/2}} = 2{,}250\ \text{Hz}.$$

The reactances at frequency 2,000 Hz are

$$X_L = (2\pi)(2{,}000\ \text{Hz})(10^{-3}\ \text{H}) = 12.6\ \Omega$$
$$X_C = 1/[(2\pi)(2{,}000\ \text{Hz})(5\times 10^{-6}\ \text{F})] = 15.9\ \Omega$$

so the total impedance is

$$Z = \sqrt{(5\ \Omega)^2 + (12.6\ \Omega - 15.9\ \Omega)^2}$$
$$= 5.99\ \Omega.$$

The phase factor is

$$\cos\phi = \frac{R}{Z} = 5\ \Omega/5.99\ \Omega = 0.835$$

or $\phi = 33.4°$. To find the current I_{rms} we need the rms voltage, $V_{rms} = V_0/\sqrt{2} = 7.07$ V. Thus,

$$I_{rms} = \frac{V_{rms}}{Z} = \frac{7.07\ \text{V}}{5.99\ \Omega} = 1.18\ \text{A}$$

and the average power is

$$\overline{P} = V_{rms}\, I_{rms}\, \cos\phi$$
$$= (7.07\ \text{V})(1.18\ \text{A})(0.835) = 6.97\ \text{W}.$$

SUMMARY TABLE OF FORMULAS

Power dissipated by a current:
$$P_D = I^2R = \left(\frac{P}{V}\right)^2 R \qquad (17\text{-}1)$$

Transformer equation:
$$\frac{V_s}{V_p} = \frac{N_s}{N_p} \qquad (17\text{-}2)$$

Current in a transformer:
$$\frac{I_s}{I_p} = \frac{N_p}{N_s} \qquad (17\text{-}3)$$

Mutual inductance:
$$\mathscr{E}_2 = -M\left(\frac{\Delta I_1}{\Delta t}\right) \qquad (17\text{-}4)$$

Self inductance:
$$\mathscr{E} = -L\left(\frac{\Delta I}{\Delta t}\right) \qquad (17\text{-}5)$$

Potential drop through an inductor:
$$V = L\left(\frac{\Delta I}{\Delta t}\right) \qquad (17\text{-}6)$$
$$V = IX_L \qquad (17\text{-}7)$$

Inductive reactance for a sinusoidal current:
$$X_L = 2\pi\nu L \qquad (17\text{-}8)$$

Voltage and reactance in a capacitor:
$$V = IX_C \qquad (17\text{-}9)$$

Reactance in a capacitor:
$$X_C = \frac{1}{2\pi\nu C} \qquad (17\text{-}10)$$

Voltage across an LRC circuit:
$$V = V_0 \cos(2\pi\nu t + \phi) \qquad (17\text{-}11)$$

Magnitude of the phasor in an LRC circuit:
$$V_0 = I_0\sqrt{R^2 + (X_L - X_C)^2} = I_0 Z \qquad (17\text{-}12)$$

Total impedance:
$$Z = \sqrt{R^2 + \left(2\pi\nu L - \frac{1}{2\pi\nu C}\right)^2} \qquad (17\text{-}13)$$

Phase angle:
$$\cos\phi = \frac{R}{Z} \qquad (17\text{-}14)$$

Power in an AC circuit:
$$P = RI_0^2 \cos^2(2\pi\nu t) \qquad (17\text{-}15)$$

Average power:
$$\overline{P} = \frac{1}{2}RI_0^2 = R(I_{rms})^2 \qquad (17\text{-}16)$$

Root mean square current:
$$I_{rms} = \frac{I_0}{\sqrt{2}} \qquad (17\text{-}17)$$

Average power in terms of I_{rms}:
$$\overline{P} = V_{rms} I_{rms} \cos\phi \qquad (17\text{-}18)$$
$$\overline{P} = I_{rms}^2 Z \cos\phi \qquad (17\text{-}19)$$
$$\overline{P} = \left(\frac{V_{rms}^2}{Z}\right)\cos\phi \qquad (17\text{-}20)$$

Root mean square current:
$$I_{rms} = \frac{V_{rms}}{\sqrt{R^2 + \left(2\pi\nu L - \frac{1}{2\pi\nu C}\right)^2}} \qquad (17\text{-}21)$$

Resonant frequency in an AC circuit:
$$\nu_0 = \frac{1}{2\pi\sqrt{LC}} \qquad (17\text{-}22)$$

Quality factor for a resonant circuit:
$$Q = \frac{V_C}{V_R} \qquad (17\text{-}23)$$

THOUGHT QUESTIONS

1. How would you convert a motor into an electrical generator?
2. Why are most electric devices used in modern technology based on alternating current (AC) rather than on direct current (DC)?
3. Explain how you would convert an AC generator into a DC generator.
4. Which countries would be most likely to generate significant amounts of hydroelectric power? Discuss why most hydroelectric power plants are built in narrow canyons.
5. Could the ocean's tides be used to run an electric generator? Explain the ultimate source of energy.
6. Why would we expect a continuous torque applied to an armature to cause continuous angular acceleration? What modifies this?
7. Compare and contrast an armature and a commutator.
8. How would you prevent the momentary decrease in line power in a laboratory during the start-up of a motor?
9. Is there a fluid equivalent to electrical eddy currents? Explain.
10. The wall voltages in foreign countries often differ from those in the United States. If you wish to take an electric appliance overseas, what would you do to use it?
11. What are the advantages and disadvantages of using superconducting power lines? What are the advantages and disadvantages of transmitting power at megavolt potential differences?
12. For what uses would you want
    (a) high voltages,
    (b) low voltages?
13. Compare and contrast
    (a) reactance and resistance,
    (b) impedance and resistance.
14. Why are the current and voltage across an inductor out of phase?
15. Compare and contrast a phasor and a vector. What are the limits on the phase angle ϕ? What does $\phi = 0$ tell you about a circuit?
16. What reading will register on a voltmeter if its leads are connected to a standard 120 V household outlet? Explain.
17. Compare the concept of resonance in an LC circuit with that of mechanical resonance (Chapter 9). In the electrical circuit, what oscillates? What do inductance L and capacitance C correspond to in a mechanical oscillator?
18. Can you think of an example of inductance in the human body? In airport metal detectors?
19. What is the impedance in a LRC circuit? For what values of L, R, or C does the circuit resonate? For what values is the impedance a minimum?
20. Compare and contrast the operation of
    (a) a semiconductor diode and a vacuum tube diode,
    (b) a transistor amplifier and a vacuum tube triode amplifier.
21. Give an example of an everyday device that uses transistors. What is the function of the transistors?
22. Is there a limit to the compactness with which a transistor may be manufactured?
23. Will a transistor operate on DC voltage? Explain.
24. Explain how a transistor operates as
    (a) an amplifier,
    (b) a switch.
25. Does the power enhancement produced by a triode or transistor violate conservation of energy? If not, why not?

PROBLEMS

Section 17.1 Generators and Motors

1. What current would you feed into the generator of Example 16.6 so that it will run as a motor at 2,000 rev/min?
2. Find the rate of change of magnetic flux required to induce a voltage of 120 V in a coil.
3. A starter motor has armature windings with 10 Ω resistance and runs at 120 V. If the back emf is 12 V when the motor is running, how much current is drawn during start up and how much at full speed?
4. The back emf in a motor is 50 V at 1,000 rev/min. What would be the back emf at 2,000 rev/min if the magnetic field remains the same?

- 5 The electric generator for a car produces 12 V at 500 rev/min. What is its output at 750 rev/min if the magnetic field remains constant?

•• 6 An armature coil, 10 cm × 10 cm, rotates at 10^3 rev/min and produces 120 V (peak). If it has 500 windings, what is the magnetic field?

•• 7 A generator has 100 turns, an area of 0.05 m², and rotates in a 500 G field. What must its frequency be to produce a peak voltage of 120 V?

•• 8 A shipwrecked sailor constructs a coil 1m × 1m, with 5,000 windings. How fast must he rotate the coil in the earth's 0.5 G field to produce 15 V (peak)?

Section 17.2 Power Transmission and Transformation

- 9 The high voltage neon tubes used in signs require 12 kV. What ratio of turns is required for the transformer?

- 10 If the neon tube in Problem 9 requires 10 W, what was the input current for the transformer?

•• 11 A transformer draws 20 A and puts out 100 W at 15 V. What is the input voltage? What is the ratio of secondary to primary turns?

- 12 A 100 kW power plant transmits its power over 0.5 Ω lines. How much more power is lost by transmitting at 10^3 V than at 100 kV?

•• 13 At what voltage should the power in Problem 12 be transmitted to keep power losses in the lines at 0.01%?

- 14 A town's 10 MW power supply is transmitted via a 50 kV line with 2 Ω resistance. What is the power lost in the line?

••• 15 A town's power supply is carried by one high voltage cable, rated at 200 A. Each of the town's 10,000 inhabitants uses 300 kWh of electricity per month.
    (a) What total power must be carried by the line? Assume a factor of 2 over the town's average power use.
    (b) What line voltage is needed at 200 A current?
    (c) What ratio of turns should the transformers use to step down to 120 V?

•• 16 How would you alter Eq. 17–3 to take into account internal loss of power?

- 17 A transformer having 100 turns in a 12 V primary coil has a secondary coil voltage of 3 V. How many turns does the secondary have?

•• 18 A power plant produces 100 kW and transmits this power over 0.5 Ω lines. At what voltage should the line be operated to keep losses at 1%?

- 19 A transformer is designed to convert 120 V power into 50 kV power. What ratio of turns should be used? What will be the current ratio?

- 20 The secondary coil of a transformer is designed to operate at 120 V. If the primary and secondary coils have 100 turns and 50 turns, at what input voltage must the primary be operated?

- 21 If no internal power losses occur, what is the current ratio for the transformer in Problem 20?

Section 17.3 AC and LRC Circuits

- 22 Starting with Eq. 17–8 and Eq. 17–10, show that the units of X_L and X_C are ohms.

- 23 How long should a 0.2 m² solenoid with 1,000 turns be to have an inductance of 1 Henry?

- 24 A coil with cross section 0.1 m² has a length of 50 cm and 10 turns per cm. What is its inductance?

- 25 At what rate should the current be increased in the coil of Problem 24 to produce an induced emf of −5 V?

- 26 An inductor 2 cm long with 100 turns runs on 1 A current. What is the coil's area if it produces an inductance of 1 Henry?

••• 27 Derive an expression for the energy that can be stored in an inductor.

- 28 At what frequency will a 100 mH inductor have an impedance of 500 Ω?

- 29 At what frequency will a 1 μF capacitor have an impedance of 2,000 Ω?

•• 30 Plot a graph of the impedances of a 10^{-2} F capacitor and a 10 μH inductor between frequencies 10 to 10,000 Hz. You may want to use logarithmic graph paper.

- 31 At what frequency are the impedances of the two elements in Problem 30 equal?

- 32 What is the inductance of a coil which draws 1 A current at 120 V and 60 Hz?

- 33 What is the capacitance of a circuit which draws 1 A current at 120 V and 60 Hz?

•• 34 What is the self-inductance of a coil, 10 cm long, 1 cm in radius, with 50 windings? If $\Delta I/\Delta t = 100$ A/s, what is the self-induced emf?

- 35 The current passing through a 0.01 H coil changes from 1.0 A to 0.1 A in 1 s. What voltage is induced?

•• 36 What is the self-inductance of a solenoid, 5 cm long and 3 mm in radius, with 500 windings? Assume the

PROBLEMS

core is air rather than the usual iron. If the current changes by 10 A/s, what emf is induced?

•• 37 Inductances add, in series or in parallel, according to rules similar to resistors. Derive an expression for the equivalent inductance of the circuit shown in Fig. 17.37. Assume no coupling (mutual inductance).

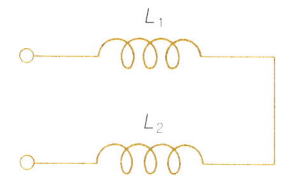

Figure 17.37

••• 38 Derive an expression for the equivalent inductance for the circuit shown in Fig. 17.38. Assume no mutual inductance.

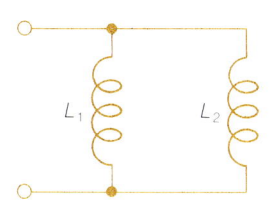

Figure 17.38

• 39 What is the total impedance at 1,000 Hz of the circuit shown in Fig. 17.39?

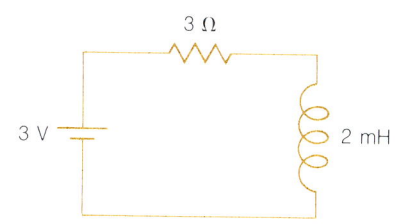

Figure 17.39

• 40 What is the total impedance at 1,000 Hz of the circuit shown in Fig. 17.40?

•• 41 What is the total impedance at 1,000 Hz of the circuit shown in Fig. 17.41?

••• 42 What is the total impedance at 1,000 Hz of the circuit shown in Fig. 17.42?

•• 43 Find the phase angle for each of the circuits in Problems 39–42.

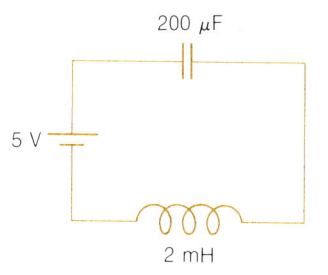

Figure 17.40

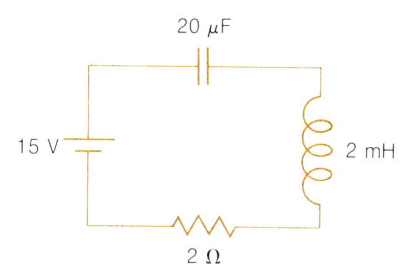

Figure 17.41

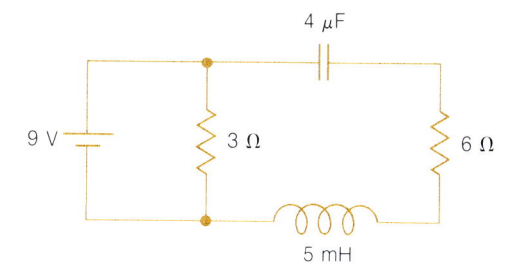

Figure 17.42

•• 44 Find the total impedance and phase angle for the circuit in Fig. 17.43 at 60 Hz.

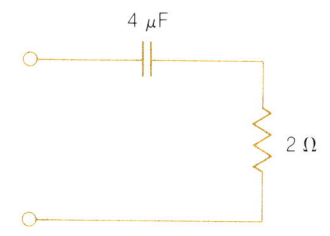

Figure 17.43

45 Find the total impedance and phase angle for the circuit in Fig. 17.44 at 60 Hz.

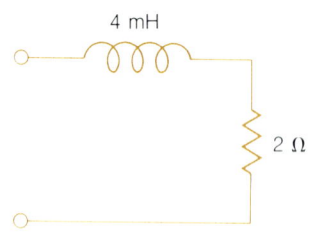

Figure 17.44

46 Find the total impedance and phase angle for the circuit in Fig. 17.45 at 60 Hz.

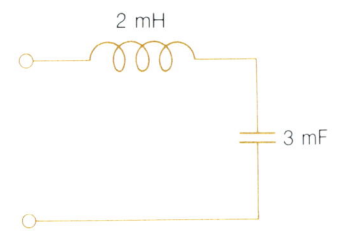

Figure 17.45

47 Find the total impedance at 60 Hz of the circuit in Fig. 17.46. You must first show that the reciprocal of the equivalent impedance for circuit elements in parallel equals the sum of the reciprocals of the individual impedances.

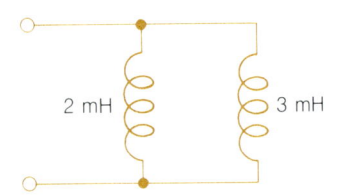

Figure 17.46

48 Suppose 2 mH and 3 mH inductors are placed in series with a 1 Ω resistor. Find the total impedance at 60 Hz. If the applied voltage is 10 volts, find the phase angle, peak current, and average power dissipated.

49 What are the resonant frequencies of the circuits in Probs. 44–46? Which resonant frequency is closest to the frequency of the input voltage?

50 A 1 Ω resistor is in series with a 3 V dry cell and a switch. Sketch a graph of the current versus time after the switch is closed.

51 A 2 Ω resistor is in series with a 3 V battery, a 30 mF capacitor, and a switch. Sketch a graph of the current versus time after the switch is closed.

52 A 3 Ω resistor is in series with a 3 V battery, a 2 µF capacitor, and a switch. Sketch a graph of the current versus time after the switch is closed.

53 Sketch graphs of current versus time for the circuits of Problems 50 and 51 if the 3 V battery is replaced by a 3 V alternating 60 Hz source.

54 Sketch a graph of current versus voltage for the circuit of Problem 52 if the battery is replaced by a 3 V source of alternating 60 Hz voltage.

55 Find the peak current, total impedance, and phase angle for an RLC circuit with peak voltage $V_0 = 120$ V, frequency 60 Hz, $R = 1$ Ω, $L = 1$ H, and $C = 1$ F.

56 What is the average power of the circuit in Problem 55?

57 What is the resonant frequency of the circuit in Problem 55?

58 An astronomer wishes to observe the 21-cm line of atomic hydrogen, using a radio telescope of inductance 10 mH. What must the telescope's capacitance be to properly tune to this line's frequency?

59 An LR circuit has $R = 2$ Ω and $L = 0.1$ H. Find the total impedance, the peak current, the phase angle ϕ, and the average power $\overline{P}$ when an AC voltage of 115 V is applied at 60 Hz.

60 Find the rms current in an RLC circuit with $R = 1{,}000$ Ω, $L = 10$ mH, and $C = 5$ µF, when a voltage $V = (100 \text{ V})\sin(250t)$ is applied. What is the phase angle and average power dissipated?

61 In the winter it is easy to accumulate a static charge from a rug. If the resistance through your shoes to the floor is 10^5 Ω and the charge bleeds off in 2 min when you stand still, estimate the capacitance of your body.

62 A capacitor of 1 µF is charged to 10 V and then connected to an inductor with $L = 100$ mH. Find the frequency of oscillation in the circuit and the maximum potential energy stored in the inductor.

63 To what capacitance should the capacitor in an AM radio be tuned to listen to a station at 600 kHz? The inductance of the tuning circuit is 10^{-4} H.

64 An RLC circuit has $L = 1$ mH and $R = 100$ Ω. What capacitance should the circuit have to produce a resonance at $\nu_0 = 100$ kHz?

65 What will be the peak current in the circuit of Problem 64 at resonance if $V = (10 \text{ V})\cos(2\pi\nu_0 t)$?

SPECIAL PROBLEMS

1. Frequency Filters

It is often useful to design circuits that filter out unwanted high or low frequencies. For example, high frequency noise can cause hiss on stereo systems, and low frequency noise can create instabilities in some devices. In the RC circuit shown in Fig. 17.47, an AC voltage $V_0 \sin(2\pi\nu t)$ is applied and the output voltages V_1 and V_2 are filtered of the unwanted frequencies.

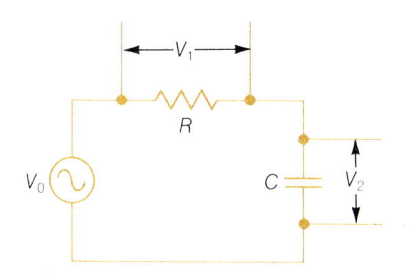

Figure 17.47

(a) Find the impedance of the circuit and derive the peak output voltage V_1 across the resistor as a function of R, C, and frequency ν.
(b) Find the peak output voltage V_2 across the capacitor as a function of R, C, and ν.
(c) Show that $V_1/V_0 = 2\pi\nu RC/[1 + (2\pi\nu RC)^2]^{1/2}$ and $V_2/V_0 = 1/[1 + (2\pi\nu RC)^2]^{1/2}$.
(d) Plot the ratios V_1/V_0 and V_2/V_0 versus frequency ν. Which output corresponds to a "high pass filter" (low frequencies attenuated) and which corresponds to a "low pass filter" (high frequencies attenuated)? At what frequency are the outputs equal: $V_1 = V_2$?

2. Resonant RLC Circuits

The current in a resonant circuit never goes to infinity at the resonant frequency $\nu_0 = 1/(2\pi\sqrt{LC})$, but is limited by the resistance R. The rms current in the circuit is $I_{rms} = V_{rms}/Z$ and the average power is $\overline{P} = V_{rms} I_{rms} \cos\phi$ where Z is the impedance and ϕ the phase angle.

(a) Use Eq. 17–13 and Eq. 17–14 to show that

$$\overline{P} = \frac{V_{rms}^2 R}{Z^2} = \frac{V_{rms}^2 R}{[R^2 + \{2\pi\nu L - 1/2\pi\nu C\}^2]}.$$

(b) Using the definition of the resonant frequency ν_0, show that the power may be written

$$\overline{P} = \frac{V_{rms}^2/R}{\left[1 + \left\{\dfrac{(\nu^2 - \nu_0^2)}{2\pi RC\nu\nu_0^2}\right\}^2\right]}.$$

(c) Plot $\overline{P}$ versus frequency ν and discuss the effect of high and low resistance R on the width of the resonance. Ignore frequencies much lower than the resonant frequency ν_0.
(d) The "Q-factor" of a resonance is approximately equal to the ratio $\nu_0/\Delta\nu$ of the resonant frequency to the frequency width $\Delta\nu$ of the resonance. For the RLC circuit, Q is defined as the ratio of voltage across the capacitor (or inductor) to the voltage across the resistor at resonance ($\nu = \nu_0$). Show that $Q = \sqrt{LC}/RC$ and derive the frequency width of the resonance.
(e) If a circuit has $L = 1$ mH, $C = 1$ μF, and $R = 1$ Ω, what are ν_0, Q, and $\Delta\nu$?

3. Metal Detectors

Metal detectors at airports operate on the principle of inductance. An AC voltage is applied to a large coil (Fig. 17.48). Any metal carried through the coil increases its inductance, the current changes, and a monitor activates a buzzer. The detector works because the magnetic susceptibility of metal is far greater than that of the human body.

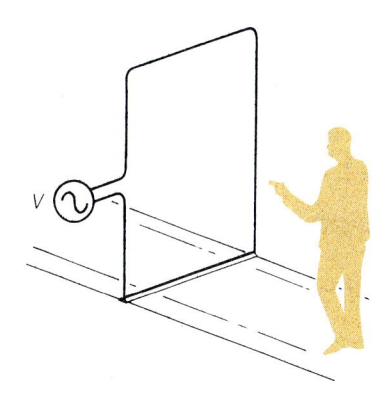

Figure 17.48

(a) Suppose the inductance of the coil is 0.05 H, the resistance 1 Ω, and the capacitance 10^{-4} F. What is the resonant frequency ν_0?
(b) If $V = 115$ V at 60 Hz, what peak current flows in the coil?

(c) Calculate the fractional change in peak current if a person carries through pieces of metal which change the coil's inductance by 1% and by 5%. What determines the change in inductance? What sort of objects might trigger the alarm?

4. Energy Storage

The electrical potential energy (PE) stored in a capacitor is $\frac{1}{2}CV^2$ and that in an inductor is $\frac{1}{2}LI^2$, where V and I are the voltage and current, and C and L are the capacitance and inductance. This energy is stored in the electric field E (capacitor) and magnetic field B (inductor).

(a) Repeat the arguments in Section 14.3 (Eq. 14–14 through Eq. 14–17) to show that the PE of a capacitor of plate area A and plate separation d may be written

$$\text{PE} = \frac{1}{2}QV = \frac{1}{2}Q^2/C = \frac{1}{2}(\epsilon_0 A d)E^2.$$

Show also that the energy density in the electric field is $u_E = \frac{1}{2}\epsilon_0 E^2$.

(b) Consider an inductor coil (solenoid) with area A, length l, and N windings, which carries current I. Use Eq. 16–11 to derive the magnetic field B and magnetic flux Φ in the coil. Then, from Faraday's law, derive the coil's inductance L.

(c) Derive an expression for the energy stored in the inductor, in terms of B^2, A, μ_0, and l. Show that the energy density in the magnetic field is $u_B = \frac{1}{2} B^2/\mu_0$.

(d) Show that the ratio of field strengths in these two components is $E/B = (u_E/u_B)^{1/2}(1/\sqrt{\epsilon_0\mu_0})$. What is the constant $1/\sqrt{\epsilon_0\mu_0}$ numerically? Is its value familiar? (It has the units of velocity.)

5. Delay Times

Special Problem 4 described energy storage in capacitors or inductors. When these elements are discharged into a circuit with resistance R, a transient state exists which approaches a steady state after a time τ, known as the circuit **time constant.** For an RC circuit $\tau = RC$, while for an RL circuit $\tau = L/R$.

(a) Show from the definitions of R, L, and C that RC and L/R have units of time.

(b) If a capacitor holding initial charge Q_0 is connected to a circuit with resistance R at time $t = 0$, techniques of calculus show that the charge decays according to the law $Q(t) = Q_0 e^{-t/RC} = e^{-t/\tau}$, where we have used the exponential function (see Appendix 2) with $e = 2.71828$. Plot $Q(t)$ in units of t/τ. What fraction of the initial charge remains after $t = 5\tau$? Derive the current $I(t)$ in the circuit.

(c) An inductor L is connected to a battery with voltage V and carries current I_0. At $t = 0$ the battery shorts out. Calculus techniques show that the current decays according to $I(t) = I_0 e^{-tR/L} = I_0 e^{-t/\tau}$. Plot $I(t)$ and derive the current flowing after $t = 3\tau$.

(d) Explain the role of RC or L/R time constants in the following: delay between flashes in a camera, discharge time for the high voltage in a television set after it is shut off, transient time for the field in an electromagnet.

SUGGESTED READINGS

[Issue on microelectronics]. *Scientific American,* Sep 77, **237**(3)

"Microcircuits in the nervous system." G. M. Shepherd, *Scientific American,* Feb 78, p 92, **238**(2)

"NMR imaging in medicine." I. L. Pykett, *Scientific American,* May 82, p 78, **246**(5)

"NMR spectroscopy of living cells." R. G. Shulman, *Scientific American,* Jan 83, p 86, **248**(1)

"The electric lamp: 100 years of applied physics." J. M. Anderson, J. S. Saby, *Physics Today,* Oct 79, p 32, **32**(10)

"Superconducting electronics." D. G. McDonald, *Physics Today,* Feb 81, p 36, **34**(2)

"Superconducting materials." M. R. Beasley, T. H. Geballe, *Physics Today,* Oct 84, p 60, **37**(10)

"To a solid state." J. Bardeen, *Science 84,* Nov 84, p 143, **5**(9)(the invention of the transistor)

"From Edison's wastebasket." L. S. Reich, *Science 84,* Nov 84, p 73, **5**(9)(the vacuum tube)

"The tube." D. G. Fink, *Science 84,* Nov 84, p 121, **5**(9)(television)

The flying circus of physics with answers. Jearl Walker, Wiley, New York, 1977

Basic electronics for scientists. J. J. Brophy, McGraw-Hill, New York, 1977

PART FIVE

ELECTROMAGNETIC RADIATION

Chapter 18
Electromagnetic Waves 590

Chapter 19
Geometric Optics 614

Chapter 20
Wave Phenomena in Light 660

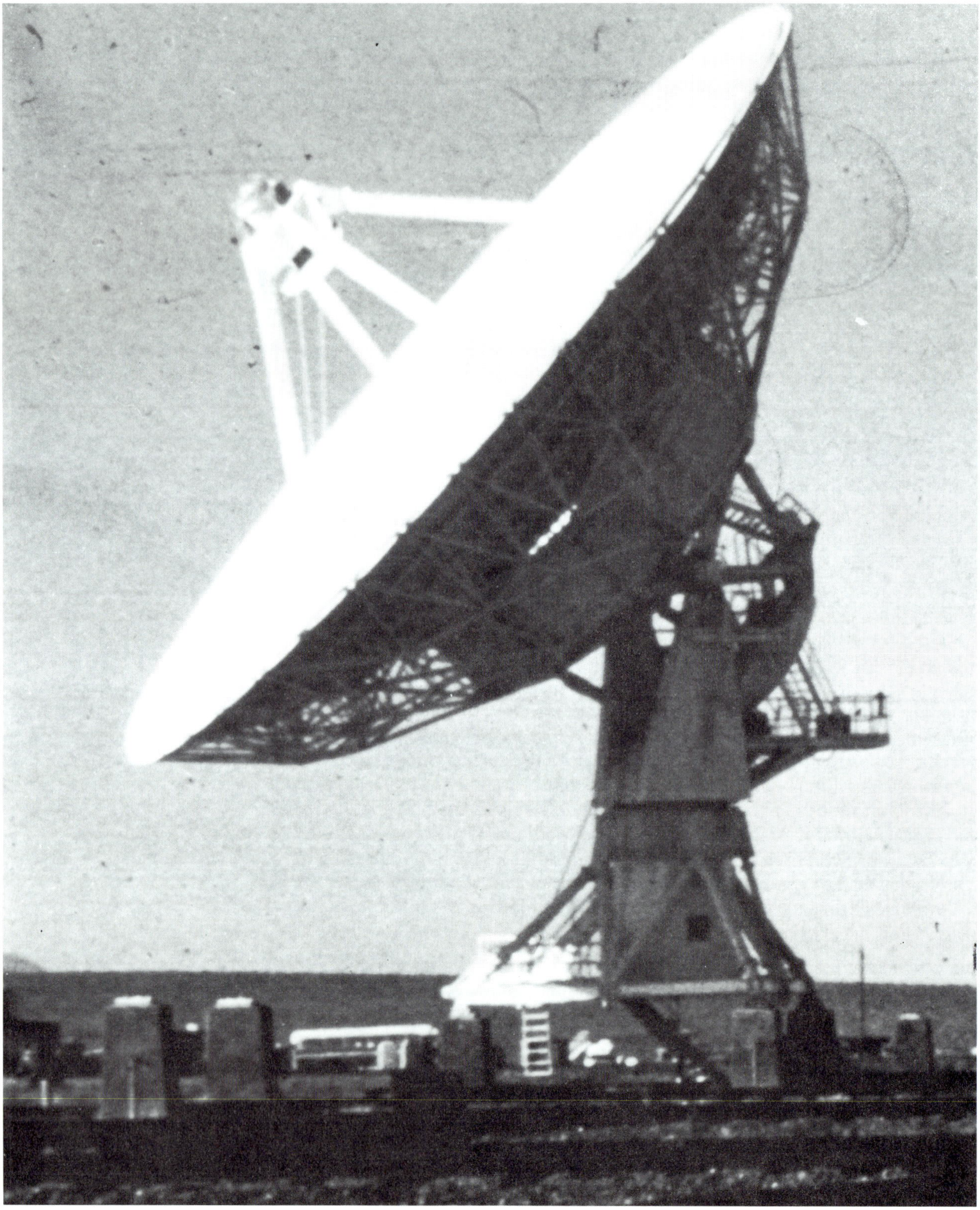

Chapter 18

Electromagnetic Waves

When we studied wave phenomena earlier in this text, we pointed out that waves do not transport matter, but instead propagate disturbances which can transmit energy and information. In those discussions, we focused our attention on waves that propagate in various kinds of physical media, and we devoted some effort to understanding the vibrations and resonances that cause the disturbances.

Electromagnetic waves have many qualities in common with mechanical waves, as well as some important differences. These waves are crucial to all aspects of our lives. They transmit nearly all forms of energy on the earth (except nuclear and hydroelectric), and they carry information to our primary sense, sight, and make most long-distance communication possible (Fig. 18.1).

In this chapter we discuss the formation of electromagnetic waves, their nature, and how they transport energy. In later chapters we will explore the interaction of these waves with matter and with each other.

18.1 The Origin of Electromagnetic Waves

That light has wave characteristics was known by the early nineteenth century, but little was known about the nature of the waves. Later in the century, the existence of electromagnetic waves was predicted, but their relation to light was not immediately demonstrated. In this section we discuss first the theory and then the formation of electromagnetic waves.

Maxwell's Prediction

In our discussion of electricity and magnetism we invoked the concept of fields, but used it only with coulomb forces and energy storage in capacitors and inductors. Indeed, the early pioneers in electricity and magnetism did not think

Figure 18.1

Communication using electromagnetic waves. The signal from this automobile telephone is transmitted by EM waves, as are most long-range communications.

in terms of fields as they carried out their largely empirical studies. It was later shown by the Scottish physicist James Clerk Maxwell, however, that all electromagnetic phenomena could be understood in terms of electric and magnetic fields.

Maxwell, who carried out his principal work in the latter half of the nineteenth century, developed four equations describing the relationships of electricity and magnetism that neatly summarized all the previously known behavior and made predictions about behavior not yet observed. These equations, known today as the **Maxwell equations,** are as fundamental as Newton's laws of mechanics and gravitation. Since they are expressed quantitatively in calculus, we will only state them qualitatively and explore one of their predictions.

The first of Maxwell's equations presents a general relationship between electric charge and the electric field it creates, allowing for all possible charge distributions (we discussed only the point charge). The second equation describes magnetic fields, stating that they always consist of closed field lines; thus, we observe no magnetic equivalent of the electric charge on which magnetic field lines could start or stop. The third relation states that an electric field can be created by a changing magnetic field (we already know this as Faraday's law, although we have spoken of electric forces rather than fields). The fourth Maxwell equation, which broke entirely new ground, *predicted* that a magnetic field can be created by a changing electric field as well as by an electric current. We now focus on this fourth equation.

18.1 THE ORIGIN OF ELECTROMAGNETIC WAVES

We know already (from Oersted and Ampere's mathematical formulation) that an electric *current* produces a magnetic field. This is incorporated in the fourth Maxwell equation, but the equation is more general. Maxwell realized that a changing electric field is equivalent to an electric current, and he developed the notion of a **displacement current**. To understand this idea, consider a capacitor in a circuit (Fig. 18.2). We know from Ampere's law that if we choose any surface normal to a current-carrying wire, there is a magnetic field around the surface whose strength is proportional to the total current passing through the surface.

Now suppose that our imaginary surface intersects the circuit through the gap between the plates in a capacitor (Fig. 18.2), through which no physical current passes. We know that if the current in the circuit is changing with time, as for an AC source of emf, current, in effect, passes through the capacitor; this happens because the changing charge on one side of the gap induces a changing charge of opposite sign on the other side. This charge is induced by the changing electric field in the gap between the capacitor plates. Maxwell postulated that Ampere's law should hold for this surface; that is, a changing electric field should create a magnetic field, just as a physical current would. For mathematical simplicity, Maxwell defined the **displacement current** as the equivalent current required to produce a given magnetic field. Including the displacement current, Ampere's law became

$$\Sigma B_{\parallel} \Delta l = \mu_0 (I_C + I_D) \qquad (18\text{--}1)$$

where the left-hand side represents the sum of the product of the length of each small segment of perimeter around the surface and the component of the magnetic field parallel to that segment, and on the right-hand side, μ_0 is the permeability of free space, I_C is the physical current, now called the **conduction current,** and the new term I_D is the displacement current.

We can use this form of Ampere's law to find an expression for the displacement current. We learned in Chapter 14 that the potential difference V between the plates of a capacitor is related to the electric field E by

$$V = Ed$$

where d is the separation between the plates. We also learned that

$$Q = CV$$

where Q is the charge on the capacitor and C is the capacitance of the two plates, which is given by

$$C = \frac{\epsilon_0 A}{d}$$

where ϵ_0 is the permittivity of free space, and A is the area of the capacitor plates. Combining these three expressions leads to

$$Q = \epsilon_0 A E. \qquad (18\text{--}2)$$

This expression relates the charge on the plates to the field created by the charge. When the charge on the plates is changing with time, the charge is

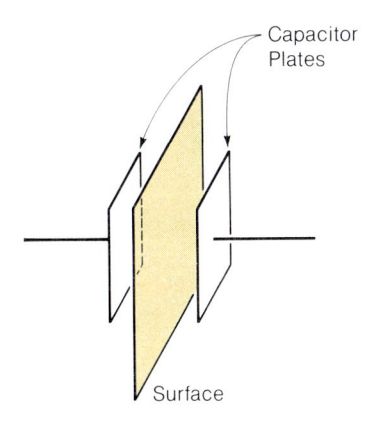

Figure 18.2

Displacement current. An alternating current can be transmitted across the gap in a capacitor by the changing electric field in the gap. Maxwell postulated that this displacement current obeys Ampere's law, and creates a magnetic field around the surface.

transferred (by induction) across the gap by the changing electric field, which Maxwell expressed in terms of the displacement current. From Eq. 18–2, we see that the rate of change of the charge is

$$\frac{\Delta Q}{\Delta t} = \epsilon_0 A \left(\frac{\Delta E}{\Delta t} \right)$$

which is equal to the current that flows through the capacitor; that is, the displacement current. Hence we have

$$I_D = \epsilon_0 A \left(\frac{\Delta E}{\Delta t} \right). \tag{18–3}$$

We see that the term on the right bears a resemblence to our expression (from Chapter 16) for a changing magnetic flux, and we define an **electric flux** Φ_E, such that

$$\Phi_E = EA. \tag{18–4}$$

In terms of the electric flux, the displacement current is

$$I_D = \epsilon_0 \left(\frac{\Delta \Phi_E}{\Delta t} \right) \tag{18–5}$$

and Ampere's law becomes

$$\Sigma B_\parallel \Delta l = \mu_0 I_C + \mu_0 \epsilon_0 \left(\frac{\Delta \Phi_E}{\Delta t} \right). \tag{18–6}$$

This expression is the complete form of Maxwell's fourth equation except for an additional term on the right that allows for a magnetic field created by a magnetized body, which need not concern us at present.

The magnetic field created by a changing electric field is itself changing with time. We already know that a changing magnetic field creates an electric field, and we find that we have a self-propagating situation: a changing electric field gives rise to a changing magnetic field, which in turn creates a changing electric field, and so on. Maxwell realized that the result would be a train of **electromagnetic waves** that could propagate through space. Such waves are commonly called **EM waves**.

The Formation of Electromagnetic Waves

We used the capacitor to demonstrate that changing electric fields give rise to magnetic fields, and then to state that electromagnetic waves are the consequence. We have not really shown how EM waves are created, however.

We note that, like a steady current, a charge moving at constant velocity

18.1 THE ORIGIN OF ELECTROMAGNETIC WAVES

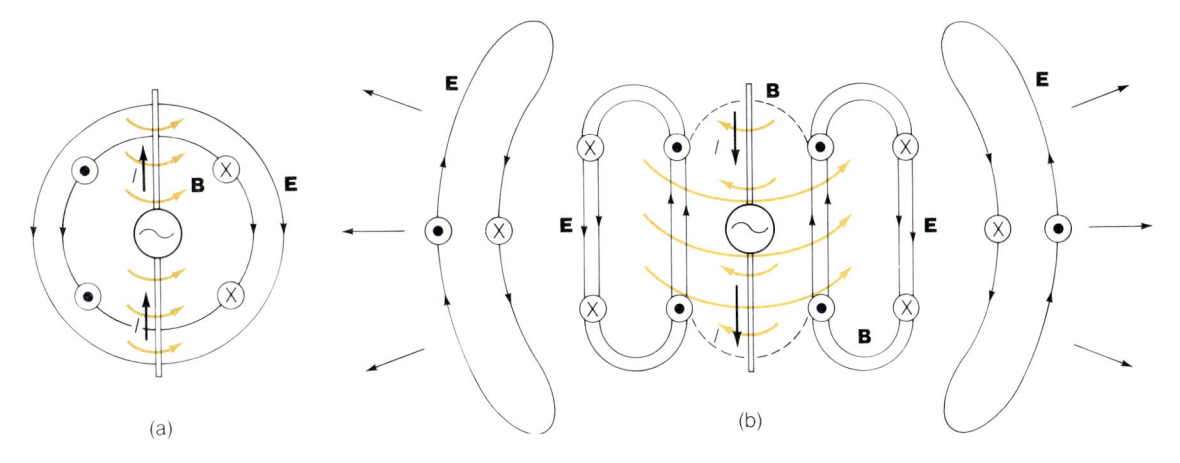

Figure 18.3

(a) Creation of an electromagnetic wave. An alternating source of emf drives an oscillating current in a conducting rod. The current is upward, creating a counterclockwise cylindrical magnetic field pointing into the paper (⊗) on the right side and out of the paper (⊙) on the left. The electric field is donut-shaped, from top to bottom. (b) A short time later, the current has reversed, and the counterclockwise magnetic field has moved outward. Now electric and magnetic fields are created in the opposite direction. As the current oscillates, electromagnetic waves propagate away from the conductor.

creates a steady magnetic field. A steady magnetic field, however, does not give rise to an electric field, so the condition for creating EM waves is not satisfied. If, however, an electric charge is accelerating, then it not only creates a magnetic field but one that is changing. A changing magnetic field creates a changing electric field, the changing electric field creates a changing magnetic field, and so on. Hence we conclude that an accelerating charge is required to create EM waves.

When we discussed waves previously, we distinguished between an isolated disturbance, which can propagate as a single wave, and an oscillating disturbance, which creates continuous waves, so that the wave pattern repeats. We make the same distinction now; an electric charge that undergoes a steady acceleration creates a single EM wave, whereas one that oscillates creates continuous EM waves. To visualize this, imagine a conducting rod with a quantity of electric charge that is oscillating because of an AC source of emf (Fig. 18.3a). During the part of the oscillation cycle when positive charge is building in the upper part of the rod (and negative charge in the lower part), a current flows upward in the figure. An electric field, directed downward, is building up, and this changing electric field creates a cylindrical magnetic field that is directed into the page to the right of the rod, and out of the page on the left. The fields are at right angles to each other, according to the right-hand rule.

After the positive charge in the upper part of the rod passes its maximum, positive charge flows toward the bottom of the rod. Now the direction of the

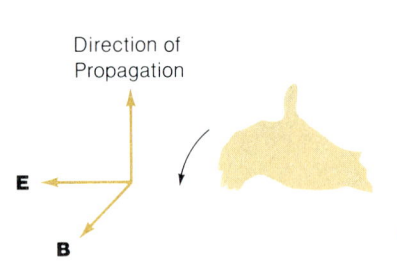

Figure 18.4

The right-hand rule. When the fingers of the right hand are rotated from the direction of the electric field to that of the magnetic field, the thumb points in the direction of propagation.

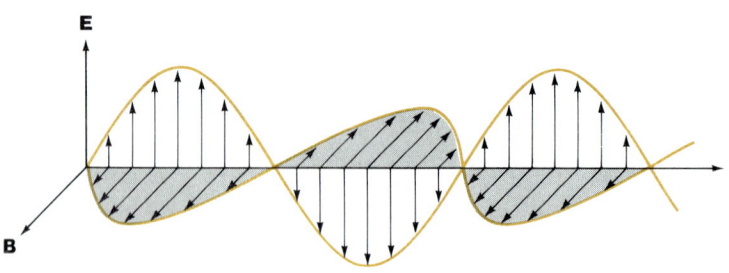

Figure 18.5

A sinusoidal EM wave. The electric and magnetic fields are in phase and vary sinusoidally.

current is reversed, the electric field is downward, and the cylindrical magnetic field is directed out of the page on the right, into it on the left (Fig. 18.3b). The fields created during the previous part of the cycle have moved away from the rod, as do the new fields now being created. To a fixed observer at some distance from the rod, an alternating pattern of electric and magnetic fields passes by at the speed of light; he detects a continuous electromagnetic wave. The direction of propagation of the wave is at right angles to both the electric and magnetic fields; the propagation is in the direction of the right-hand thumb when the fingers are rotated from the electric field into the direction of the magnetic field (Fig. 18.4).

If the source of emf varies sinusoidally, then the strengths of the alternating electric and magnetic fields do also (Fig. 18.5). Furthermore, the two fields are in phase; that is, their maximum and minimum values are at the same points in the cycle. Electromagnetic waves resemble transverse waves in form, but there is a very important distinction: these are waves consisting entirely of electric and magnetic fields, not disturbances in some physical medium. Electromagnetic waves can propagate in a vacuum.

18.2 The Nature of EM waves

We have already described the alternating fields that comprise an EM wave. We can describe other characteristics a bit more quantitatively, such as the wavelengths of the waves, and the speed of propagation.

The Electromagnetic Spectrum

After it was realized that visible light is composed of waves, experiments with sunlight and prisms demonstrated a connection between wavelength and color. Techniques (discussed in Chapter 20) were developed for measuring the wavelengths, and it was found that the wavelengths of visible light are about

18.2 THE NATURE OF EM WAVES

Table 18.1

Wavelengths of Visible Light

Color	Wavelength Band (nm)
Red	650–700 nm
Orange	600–650
Yellow	550–600
Green	500–550
Blue	450–500
Indigo	400–450
Violet	350–400

3.5×10^{-7} m to 7.5×10^{-7} m (Table 18.1). At the low end of this range lies violet light; at the high end, red light. A common unit for expressing visible wavelengths is the **Ångstrom** (Å), which is defined so that $1 \text{ Å} = 10^{-10}$ m. In these units, visible light lies between 3,500Å and 7,500Å. In modern physics convention, the **nanometer** (equal to 10^{-9} m) is often used instead, so that visible light falls between 350 nm and 750 nm.

The fact that the sun emits light over this broad range of wavelengths led to the suspicion that perhaps it also emits at wavelengths not perceptible to the human eye. Simple experiments, carried out in the mid-nineteenth century, verified this. Beyond the red end of the visible range lies **infrared** light, and below the violet end lies **ultraviolet** light. Quite independently, **x-rays** and **γ-rays** (gamma rays) were discovered later, and eventually proved to be electromagnetic waves of very short wavelength. By this time, electromagnetic waves were known alternately as electromagnetic radiation. Again, through independent experimentation, radiation at wavelengths even longer than infrared was discovered, and became known as radio waves. H. Hertz made the first deliberate attempt to produce EM waves in 1887 (after Maxwell had died). Using a spark-gap apparatus, Hertz generated radio waves with a wavelength of about 3 mm. He was able to show that these waves, apart from their much longer wavelength, exhibited all the properties of light waves, thus confirming Maxwell's hypothesis that light consists of EM waves.

The full range of the **electromagnetic spectrum** is shown in Fig. 18.6 (**spectrum** is a general term for the spread of waves according to wavelength or color). It is remarkable that the wavelengths vary by more than a factor of 10^{23}. In principle, EM radiation can have any wavelength, but there are few physical processes which generate waves outside the limits in the figure.

The Speed of Light

It was a straightforward matter for Maxwell to calculate the speed with which his predicted EM waves should travel. He was able to show that the speed is equal to the ratio of the magnetic and electric field strengths at any point in space (this was done by considering a fixed surface at some point and analyzing the relationship between the two fields and the velocity as a wave passes by). In equation form, we have

$$c = \frac{E}{B}. \tag{18-7}$$

Ampere's law gives an independent expression for the ratio of the two field strengths which can be solved for the velocity c. The result is

$$c = \sqrt{\frac{1}{\epsilon_0 \mu_0}} \tag{18-8}$$

where $\epsilon_0 = 8.85 \times 10^{-12}$ C^2/Nm^2 is the permittivity of free space, and $\mu_0 = 4\pi \times 10^{-7}$ Ns^2/C^2 is the permeability of free space. Evaluation of this expres-

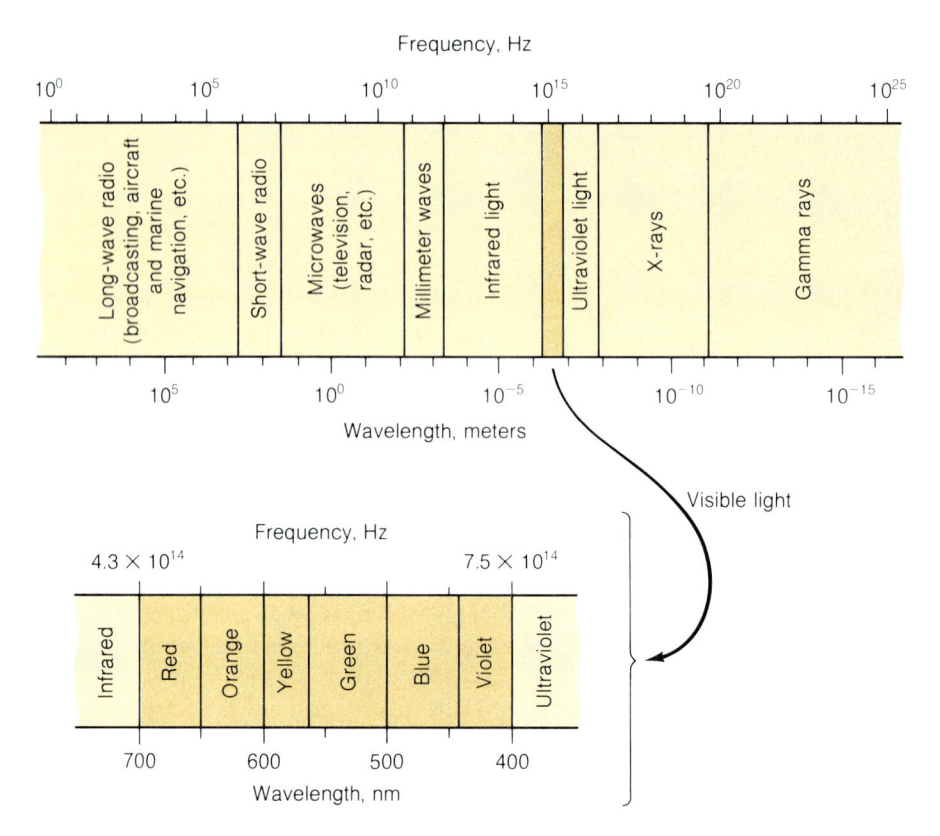

Figure 18.6

The electromagnetic spectrum.

sion leads to $c = 2.9979 \times 10^8$ m/s. It is significant that this speed does not depend on the wavelength.

Experiments to measure the speed of light had been attempted long before it was even suspected that light consists of waves (it is reported, for example, that Galileo reached the conclusion that light was either very fast or instantaneous; much more sophisticated measurements were made later). The first accurate measurements (see the *Focus on Physics* article in this chapter) took place in the last century, and verified Maxwell's predicted value.

Once the speed of EM radiation is known, it is a simple matter to find the frequency of oscillation of an EM wave. From Chapter 10 we have the simple relation $\nu = v/\lambda$, where ν is the frequency, v the velocity, and λ the wavelength. Inserting c for the velocity of EM radiation, we have

$$\nu = \frac{c}{\lambda}. \tag{18-9}$$

Visible light has frequencies in the interval 4×10^{14} Hz to 8.5×10^{14} Hz, whereas radio frequencies cover a broad range, about 10^3 Hz to perhaps 10^{11} Hz. The frequencies of γ-rays go as high as 10^{26} Hz (in principal there is no upper limit), while x-rays lie in the range 10^{16} Hz to 10^{18} Hz.

FOCUS ON PHYSICS

Measuring the Speed of Light

In this chapter we have described how Maxwell was able to derive the speed of light from theoretical grounds, and we have shown that experiments have verified the predicted speed. We said little, however, about how the speed of light was measured.

One of the first scientists to consider the problem was Galileo, who, in his *Dialogue on the Two Chief World Systems,* published in 1638, described a thought experiment, a hypothetical method for making the measurement. He proposed that two people with lanterns could stand at large distances from each other, and measure the light travel time for a round trip from one person to the other and back again. The first person would open the shutter on his lantern, and when the second person saw the light, he would open the shutter on his lantern. The first person would then see the light from the second lantern, and the time since he had opened his shutter in the first place would be the elapsed time for light to travel from him to the other person and back. Of course, we now know that this method would not work, because the light travel time over any reasonable distance would be very much shorter than the human reflex time required to open the shutters, and because very precise time measurement would be required, something surely not possible in Galileo's era.

What is needed to measure the speed of light is either a very long pathlength over which to make the measurement (Galileo's technique would have a chance of working, if, for example, one person were on the earth and the other on the moon), or very precise methods for measuring time. Both possibilities have been exploited in successful experiments.

Two methods making use of astronomical phenomena were among the first to accomplish the task of measuring the speed of light. In 1675, the

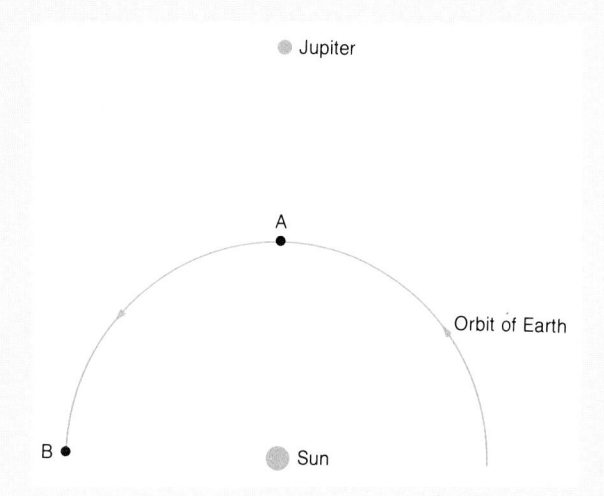

Figure 18.1.1

Danish astronomer O. Roemer used observations of the moons of Jupiter to derive a value of $c = 2 \times 10^8$ m/s. He carefully determined the orbital periods of the moons when the earth was at its nearest approach to Jupiter (point A in Fig. 18.1.1), and then predicted the moons' positions for a time some three months later, when the earth was at point B. He found that the moons all lagged behind their predicted positions, and attributed the lag to light travel time, since the earth was now farther from Jupiter. Knowledge of the extra distance the light travelled to reach the earth at B, along with the length of the delay, allowed Roemer to calculate the speed of light.

In 1727, the Englishman J. Bradley measured the **aberration of starlight,** the apparent shift of a star's position due to the motion of the earth. If a star is observed in the direction perpendicular to the earth's motion, the light from that star actually approaches the earth's surface at a slight angle

(continued)

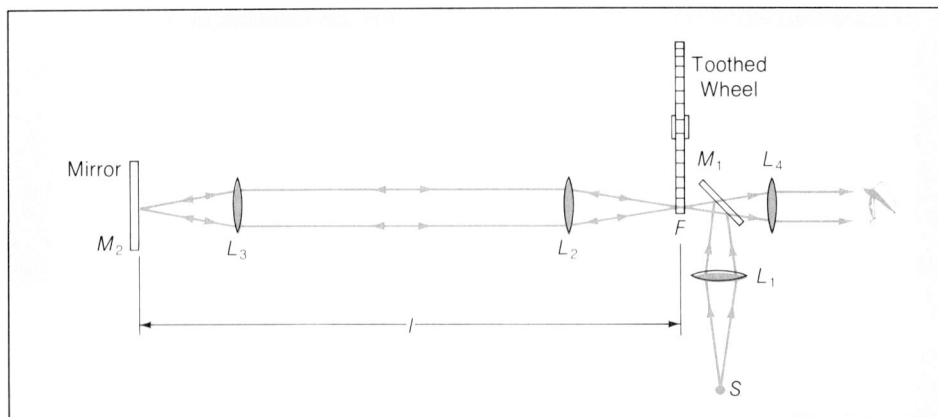

Figure 18.1.2

from the perpendicular, just as raindrops appear to fall at an angle to a person in motion. The angle of deflection from the perpendicular is determined by the relative velocities of the earth and of light; Bradley's measurement of this angle (about 20.5 arcseconds) allowed the speed of light to be quite accurately estimated as 3.0×10^8 m/s. (Of perhaps even greater historical significance, Bradley's observation is considered the first direct proof that the earth moves through space, orbiting the sun, rather than being stationary, as the ancient philosophers believed. Of course, by Bradley's time, the works of Copernicus, Kepler, Galileo, and especially Newton had already convinced scientists that the earth does orbit the sun.)

The first technique to make use of accurate timing, rather than astronomical distances, to measure the speed of light was employed by H. L. Fizeau in 1849. Employing a system of lenses and mirrors, he created a beam of light that passed back and forth along the same path before reaching the observer's eye (Fig. 18.1.2). Into this beam was inserted a toothed wheel that could be rotated at variable speeds, alternately blocking and passing the beam. If we consider first the outgoing beam from the wheel to mirror M_2, we see that the rotation of the wheel creates bursts of light, rather than a steady beam. In general, a burst of light will reach mirror M_2 and return along the path to the wheel, where it can either pass through a gap between teeth, or be blocked by another tooth. If the wheel has turned just far enough so that the burst from one gap in the teeth is blocked by the next tooth, then the time required for light to travel from the wheel to mirror M_2 and back is known, if the rotation rate of the toothed wheel is known. The experiment consisted of starting from a slow wheel rotation rate and increasing the rate until the observer saw no light coming back through the toothed wheel. When the light stopped passing through, then the rotation rate of wheel provided the light travel time between the wheel and mirror M_2. Fizeau measured a value of 3.13×10^8 m/s using this technique.

In 1862, another French physicist, J. B. L. Foucault, developed a variation on the experiment of Fizeau, using a rotating mirror instead of a toothed wheel. Later the American A. A. Michelson carried out several experiments using this technique, arriving at the most accurate value of c then known. In Michelson's experiments, a rotating octagonal mirror was used (Fig. 18.1.3) to deflect light from a source to a distant mirror and then to an observer. The first reflection off the rotating mirror turned the steady beam from the source into a series of beams that swept across the distant flat mirror. Each time a beam swept across the flat mirror, a burst of light was reflected back to the rotating mirror. If the rate of rotation was adjusted correctly, the return bursts would strike one of the faces of the rotating mirror at the correct angle to be reflected to the observer. The experimental procedure was to adjust the rotation rate until the

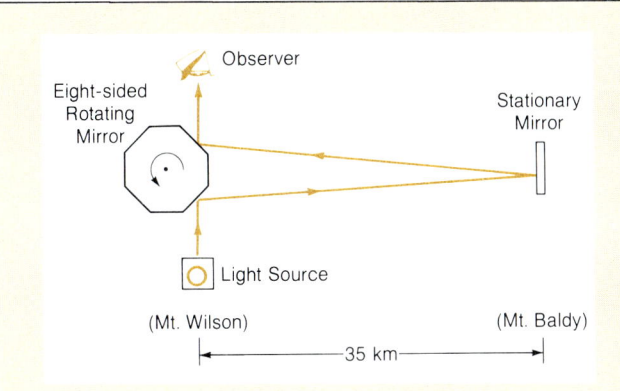

Figure 18.1.3

observer saw the light from the source, which only occurred for a specific relationship between mirror rotation rate and light travel time. Then knowledge of the distance from the rotating mirror to the flat mirror led to a value for the speed of light. In one of Michelson's most famous experiments, the flat mirror was placed 35 km from the rotating mirror, which was on Mt. Wilson, overlooking Pasadena, California. Michelson's best value for c was 2.99796×10^8 m/s, with an uncertainty of only 0.00004×10^8 m/s (that is, 4 km/s uncertainty in a measured value near 300,000 km/s).

Modern measurements of the speed of light have generally used radio waves in cavities. In analogy with the creation of standing sound waves in a closed pipe (Chapter 10), radio waves in an enclosed metal cavity will interfere with each other to create standing waves whose wavelengths are precisely related to the dimensions of the cavity. The size of the cavity is measured accurately, as is the frequency of the radio waves when standing waves occur. Then the speed of light is determined from the relation $c = \lambda \nu$.

The best modern value for c is 2.99792458×10^8 m/s. In fact, because scientists can measure time more accurately than distance, the speed of light is now defined to be *exactly* this value, and the meter is defined in terms of this speed. This value of c is the speed of light in a vacuum, but it is only slightly different (slower) in air. For most practical purposes, it is sufficient to round this off to 3.00×10^8 m/s. In the next chapter we will see how the speed of light depends on the medium in which it travels.

Example 18.1   What would be the wavelength of EM waves produced by charges oscillating at the frequency of ordinary household current?

The frequency is 60 Hz, so the wavelength would be

$$\lambda = \frac{c}{\nu} = \left(\frac{3 \times 10^8 \text{ m/s}}{60 \text{ Hz}} \right) = 5 \times 10^6 \text{ m, or } 5{,}000 \text{ km.}$$

This radiation no doubt exists, but cannot easily be detected.

18.3 Transmission of Energy

We have pointed out that EM waves transmit energy, and that this is the ultimate source of most of the energy we use. This referred to fossil fuels which are derived from the remains of organisms that depended on sunlight for their existence.

The energy carried by EM waves is stored in their alternating electric and magnetic fields. We have already developed expressions for the energy density (the energy per unit volume) in an electric field and in a magnetic field:

$$u_E = \frac{1}{2}\epsilon_0 E^2$$

and

$$u_B = \frac{1}{2}\frac{B^2}{\mu_0}$$

where u_E and u_B represent the energy densities for electric and magnetic fields, respectively, and E and B are the field strengths. The total energy stored in an EM wave at any moment is therefore

$$u = \frac{1}{2}\epsilon_0 E^2 + \frac{1}{2}\frac{B^2}{\mu_0}. \tag{18-10}$$

We learned in the previous section that $B = E/c$ and that $c = 1/\sqrt{\epsilon_0\mu_0}$; substituting these into Eq. 18-10 and manipulating algebraically leads to the following three expressions:

$$u = \epsilon_0 E^2 \tag{18-11}$$

$$u = \frac{B^2}{\mu_0} \tag{18-12}$$

$$u = EB\sqrt{\frac{\epsilon_0}{\mu_0}} \tag{18-13}$$

We can now use these expressions to find the rate at which energy is transported by an EM wave. Consider a surface of area A, perpendicular to the direction of travel of an EM wave (Fig. 18.7). Let S be the energy transported through the surface per unit of surface area per unit time. During time interval Δt, the amount of energy Δu that passes through this surface is the amount that fills the volume $Ac\Delta t$, where $c\Delta t$ is the distance travelled by the wave in time Δt. The amount of energy in the volume $Ac\Delta t$ is $uAc\Delta t$, so $S = (\Delta u/A\Delta t) = uc$, which we can evaluate using each of our expressions for u from Eq. 18-11 through Eq. 18-13 and the relation $c = 1/\sqrt{\epsilon_0\mu_0}$. We get the following three expressions for S:

$$S = \epsilon_0 c E^2 \tag{18-14}$$

$$S = \frac{cB^2}{\mu_0} \tag{18-15}$$

$$S = \frac{EB}{\mu_0} \tag{18-16}$$

18.3 TRANSMISSION OF ENERGY

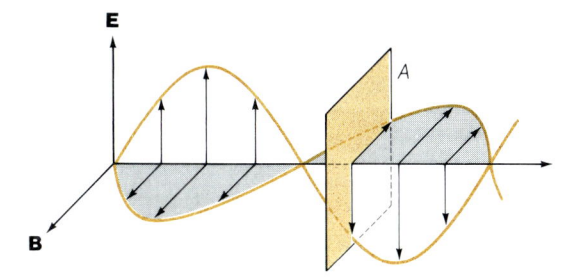

Figure 18.7

Energy transport by an EM wave. The energy contained in the alternating electric and magnetic fields passes through the surface at the speed of light.

To fully specify the rate of energy transport requires designating the direction as well as the magnitude, which leads us to define the vector **S**, whose magnitude is given by Eq. 18–14, Eq. 18–15, or Eq. 18–16 and whose direction is given by the right-hand rule (the direction of wave propagation). The vector **S** is called the **Poynting vector**. The dimensions of S are energy per unit time per unit area, or W/m².

The expressions for energy transport given by Eq. 18–14 through Eq. 18–16 are for the instantaneous rate of energy transport, which varies as E and B vary. When an EM wave passes through a point in space, the energy passing through varies periodically with the frequency of the wave. It is often more useful to consider the average energy transport over some time interval. If E and B are sinusoidal with maximum values E_0 and B_0, then their average values are the rms values $E_0/\sqrt{2}$ and $B_0/\sqrt{2}$, respectively. Therefore the expressions for the average rate of energy transport derived from Eq. 18–14 through Eq. 18–16 are

$$\overline{S} = \frac{1}{2} \epsilon_0 c E_0^2 \tag{18-17}$$

$$\overline{S} = \frac{1}{2} \frac{cB_0^2}{\mu_0} \tag{18-18}$$

$$\overline{S} = \frac{1}{2} \frac{E_0 B_0}{\mu_0}. \tag{18-19}$$

These expressions for $\overline{S}$ are equivalent to the intensity of energy transport by waves, as discussed in Chapter 10. Thus, if radiation is emitted uniformly in all directions by a source of EM waves, then the intensity decreases as the inverse square of the distance from the source. The apparent brightness of an object emitting visible light is a function of the intensity reaching the eye times the area of the eye's opening, the pupil. The total energy received by a radio antenna, a telescope, or any other device for collecting EM waves is the product of the intensity of the radiation times the collecting area of the receiver; hence receivers must be built in large sizes for the detection of weak signals.

Example 18.2

The sun's luminosity is 3.83×10^{27} Watts, so that the intensity of sunlight reaching the Earth (above the atmosphere) is 1360 W/m² (this quantity is called the **solar constant**). If this energy were transmitted by a single EM wave, what are its maximum values of E and B? What is the intensity of sunlight on the surface of Mercury, which is 0.39 as far from the sun as the Earth, and on Pluto, which is 39.4 times farther from the sun than the Earth? What are the values of E_0 and B_0 at the surfaces of these two planets?

We find the values for the Earth first. The value of E_0, from Eq. 18–17, is

$$E_0 = \sqrt{\frac{2\overline{S}}{\epsilon_0 c}} = \sqrt{\frac{2(1{,}360 \text{ W/m}^2)}{(8.85 \times 10^{-12} \text{ C}^2/\text{N} \cdot \text{m}^2)(3 \times 10^8 \text{ m/s})}}$$
$$= 1.01 \times 10^3 \text{ V/m}.$$

The value of B_0, from Eq. 18–18, is

$$B_0 = \sqrt{\frac{2\overline{S}\mu_0}{c}} = \sqrt{\frac{2(1{,}360 \text{ W/m}^2)(4\pi \times 10^{-7} \text{ T} \cdot \text{m/A})}{(3 \times 10^8 \text{ m/s})}}$$
$$= 3.4 \times 10^{-6} \text{ T}.$$

The intensity of sunlight on Mercury is scaled from the solar constant by the square of the ratio of the distances of the Earth and Mercury from the sun:

$$I_M = I_E \left(\frac{r_E}{r_M}\right)^2$$

where I_M and I_E represent the intensities reaching Mercury and the Earth, respectively, and r_E and r_M are the distances of the Earth and Mercury from the sun. Substituting values, we find

$$I_M = (1{,}360 \text{ W/m}^2)\left(\frac{1}{0.39}\right)^2 = 8.94 \times 10^3 \text{ W/m}^2.$$

For Pluto, the same reasoning yields

$$I_P = I_E \left(\frac{r_E}{r_P}\right)^2 = (1{,}360 \text{ W/m}^2)\left(\frac{1}{39.4}\right)^2 = 0.88 \text{ W/m}^2.$$

Since the intensity is proportional to the square of E_0 and B_0, their values at Mercury and Pluto are simply proportional to the distances of the planets from the sun. Therefore we have for Mercury

$$E_{0M} = E_{0E}\left(\frac{r_E}{r_M}\right) = (1.01 \times 10^3 \text{ V/m})\left(\frac{1}{0.39}\right) = 2.6 \times 10^3 \text{ V/m}$$

and

$$B_{0M} = B_{0E}\left(\frac{r_E}{r_M}\right) = (3.4 \times 10^{-6} \text{ T})\left(\frac{1}{0.39}\right) = 8.7 \times 10^{-6} \text{ T}.$$

Similarly, for Pluto we find

$$E_{0P} = 25.6 \text{ V/m}$$

and

$$B_{0P} = 8.6 \times 10^{-8} \text{ T}.$$

18.4 Information Transport

Just as most plant species have developed the ability to extract energy directly from EM waves, so have many animal species developed the ability to derive information from EM waves. For both, the primary type of EM radiation utilized is visible light, which is understandable because the sun emits most of its radiant energy in this wavelength band, and the earth's atmosphere is transparent to this band. Today modern technology allows us to use not only visible light, but also other wavelengths of EM radiation, for the transmission of information. Usually, visible light transfers information by forming images. This process requires understanding of optics, the subject of the next chapter. Information can also be transmitted as wavepatterns, which can transmit messages over vast distances. The radio portion of the EM spectrum is most widely used in this way, so our discussion here will center on this wavelength band.

For a message to be sent by EM waves, it first must be converted into a specific pattern of electric charge oscillations, which in turn produces a specific pattern of waves when the charge oscillations take place in an antenna (Fig. 18.8). When the pattern of waves is received at another antenna, their electric and magnetic fields induce a pattern of charges there that can be reconverted to the original form of the message. In the first applications of this concept, beginning with the experiments of G. Marconi in the late nineteenth century, the pattern of waves consisted of a series of "on" and "off" states, such as the dots and dashes of the Morse code. In modern transmission the amplitudes or the frequencies of the waves themselves are arranged into a pattern that codes the message. This allows information to be transmitted at a much greater rate, because many more variables can be used to create the pattern of waves. Sufficient information can be carried by EM waves for the receiver to reproduce the full range of frequencies of the original sounds and images.

A radio message is transmitted in several distinct steps. First, sounds are converted into a pattern of electrical currents by a device called a **transducer**,

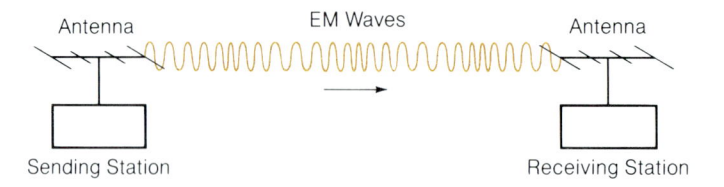

Figure 18.8

Transmission of information. The radio signal is converted into a pattern of alternating charges at the sending station, and the alternating charge pattern in the antenna creates EM waves with a corresponding pattern. These waves are received by another antenna, and converted into a pattern of alternating charges from which the original information can be extracted.

which produces a current when vibrations caused by sound waves create either a varying electric field or a varying magnetic field. For example, in a **condenser microphone,** the plates of a capacitor are allowed to vibrate with the frequency of the incoming sound waves; varying the plate separation causes variations in the capacitance, which in turn create variations in the charge on the plates (through the relation $Q = CV$). In a **magnetic microphone,** an inducting coil near a small permanent magnet is attached to a membrane that vibrates with the sound waves; moving the coil in the magnetic field induces an emf in the coil with the same frequency as the sound waves.

The output of the microphone is an electrical signal which varies with the sound-wave frequency. The next step is to mix this signal with an oscillating electrical signal whose frequency is equal to that of the EM waves to be transmitted, which is typically a factor of 100 to 1,000 times higher than that of sound waves. The radio-frequency oscillations are governed by an LC circuit whose resonant frequency depends on the inductance L and capacitance C (recall our discussion of electrical resonance in Chapter 17). The frequency of this circuit is called the **carrier frequency** of the transmission. The mixing can be done in either of two ways: the sound-frequency oscillations can be used to modify the amplitude of the radio-frequency oscillations, or they can be used to create variations in the frequency of the radio-frequency oscillations. In the first case, the amplitudes of the audio-frequency and radio-frequency signals are added; in the second case, their frequencies are added. Therefore we speak of **amplitude-modulation,** or **AM** transmission; and **frequency-modulation,** or **FM** transmission. These two possibilities are illustrated in Fig. 18.9. In either case, the result is an electric oscillation that can create EM waves in an antenna.

Because the electrical signals in the audio-frequency and radio-frequency circuits are normally quite weak, they need amplification by triodes (as we saw for vacuum-tube radios) or transistors.

When the radio wave is received at a distant antenna, it first must be filtered so that only the desired radio-frequency message is retained. This is done through the use of a tuner circuit whose resonant frequency is controlled (by a variable capacitance or a variable inductance) so that it matches the

18.4 INFORMATION TRANSPORT

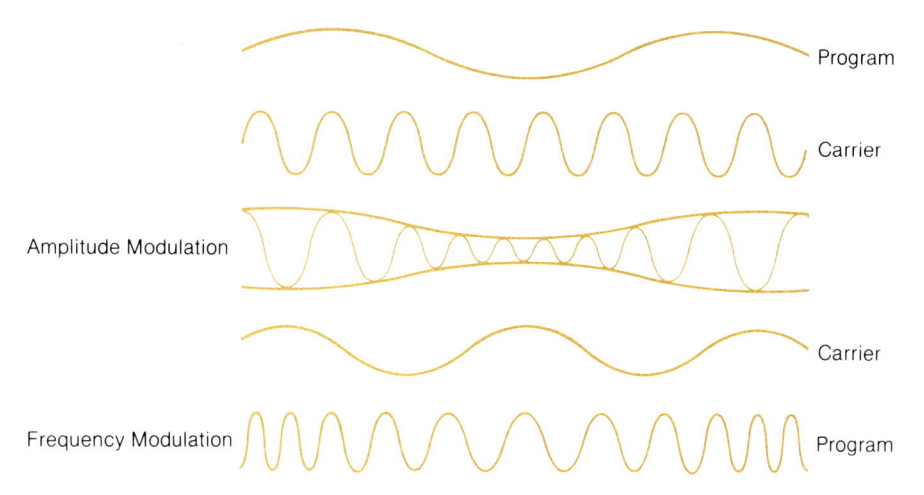

Figure 18.9

AM and FM transmission. This illustrates schematically how the program and carrier signals are combined.

carrier frequency of the signal. The signal is demodulated; that is, the audio-frequency signal is separated from the summed signal that is received. The demodulation is just the inverse process of the mixing that took place at the transmitter. One simple way to demodulate the signal is to connect the receiver circuit to ground through a capacitor. Recall (from Chapter 17) that the induced reactance in a capacitor is inversely proportional to the current frequency; hence, high-frequency current oscillations will pass through it, whereas low-frequency oscillations will meet a large impedance and flow instead through the receiver circuit. In modern radio receivers, diodes or transistors are used for the demodulation.

The transmission and reception of television signals is done in just the same way as radio signals, except that a video signal is mixed in as well as an audio signal. The video signal is generated by a device that converts an image into a pattern of electrical signals corresponding to the brightness variations as the image is scanned; this pattern is demodulated by the receiver and then it recreates the image in a cathode ray tube, the picture tube discussed in Chapter 17.

Perspective

We have gained an inkling of the workings of what is, from a human point of view, one of the most important phenomena in the universe. We now understand how light and other EM waves are created and received, and how these waves transport energy and information.

We still have much to discuss concerning these waves, however. In the next two chapters, we will complete the classical picture of EM radiation, but then we will have to learn a great deal more about atoms before we can develop the full modern understanding of EM waves.

SUMMARY TABLE OF FORMULAS

Ampere's law, including displacement current:
$$\Sigma B_{\parallel} \Delta l = \mu_0 (I_C + I_D) \quad (18\text{–}1)$$

Charge on a capacitor:
$$Q = \epsilon_0 A E \quad (18\text{–}2)$$

Displacement current:
$$I_D = \epsilon_0 A \left(\frac{\Delta E}{\Delta t} \right) \quad (18\text{–}3)$$

Electric flux:
$$\Phi_E = EA \quad (18\text{–}4)$$

Displacement current:
$$I_D = \epsilon_0 \left(\frac{\Delta \Phi_E}{\Delta t} \right) \quad (18\text{–}5)$$

Ampere's law:
$$\Sigma B_{\parallel} \Delta l = \mu_0 I_C + \mu_0 \epsilon_0 \left(\frac{\Delta \Phi_E}{\Delta t} \right) \quad (18\text{–}6)$$

Speed of light:
$$c = \frac{E}{B} \quad (18\text{–}7)$$

$$c = \sqrt{\frac{1}{\epsilon_0 \mu_0}} \quad (18\text{–}8)$$

Frequency, wavelength, and the speed of light:
$$\nu = \frac{c}{\lambda} \quad (18\text{–}9)$$

Energy stored in an EM wave:
$$u = \frac{1}{2} \epsilon_0 E^2 + \frac{1}{2} \frac{B^2}{\mu_0} \quad (18\text{–}10)$$

$$u = \epsilon_0 E^2 \quad (18\text{–}11)$$

$$u = \frac{B^2}{\mu_0} \quad (18\text{–}12)$$

$$u = EB \sqrt{\frac{\epsilon_0}{\mu_0}} \quad (18\text{–}13)$$

Instantaneous rate of energy transport:
$$S = \epsilon_0 c E^2 \quad (18\text{–}14)$$

$$S = \frac{cB^2}{\mu_0} \quad (18\text{–}15)$$

$$S = \frac{EB}{\mu_0} \quad (18\text{–}16)$$

Average rate of energy transport:
$$\bar{S} = \frac{1}{2} \epsilon_0 c E_0^2 \quad (18\text{–}17)$$

$$\bar{S} = \frac{1}{2} \frac{cB_0^2}{\mu_0} \quad (18\text{–}18)$$

$$\bar{S} = \frac{1}{2} \frac{E_0 B_0}{\mu_0} \quad (18\text{–}19)$$

THOUGHT QUESTIONS

1. Compare and contrast electromagnetic (EM) waves with sound waves.
2. Give an example of how the human body uses EM waves.
3. Can you give an example of long distance communication that does not depend on EM waves?
4. Which of the human senses employ EM waves?
5. Describe a practical use of each of the following EM waves: gamma rays, x-rays, ultraviolet rays, infrared rays, and radio waves.
6. Why are gamma rays and x-rays better able to penetrate matter than infrared or radio waves?
7. Is it possible for longitudinal EM waves to exist? Explain why or why not.
8. Why is a changing electric field equivalent to an electric current? Why is displacement current a useful concept in electromagnetic theory?
9. Why won't a steady magnetic field create an electric field?
10. Is it possible for EM waves to have their electric and magnetic fields out of phase with each other? Explain.
11. How would you verify that the sun emits EM waves at wavelengths beyond the ability of the human eye to see them?

PROBLEMS

12 Why is the Poynting flux **S** a vector?

13 Why is the radio portion of the spectrum commonly used to transmit information as wave patterns?

14 Can you think of another reason for humans and animals making the most use of visible light, other than that the sun emits most of its energy in this wavelength range?

15 Compare and contrast the methods by which sound is converted into a variable electric signal by a condenser microphone and a magnetic microphone.

16 Compare and contrast amplitude modulation and frequency modulation.

17 How might you jam a radio broadcast?

18 How does a radio antenna work? Can you use an antenna to broadcast as well as to receive signals?

19 Does the average rate of energy transport as expressed by Eq. 18–19 depend on the frequency of **E** and **B**? Explain.

20 Why can EM waves propagate through a vacuum, whereas sound and other mechanical waves require a physical medium? Is the vacuum a physical medium?

21 What effects, if any, does a nonvacuous medium have on EM waves?

22 How would the equations in this chapter be affected if **E** and **B** were not sinusoidal in nature?

23 Why does the intensity of an EM wave vary with the inverse square of the distance to a source, while the values of E_0 and B_0 vary with the inverse of the distance?

24 How might you measure the speed of light using celestial objects?

25 Can you give any everyday evidence that the speed of light is finite?

PROBLEMS

Section 18.1 Origin of EM Waves

1 A capacitor consists of two parallel plates, 30 cm square. At what rate must the electric field be changed to produce a displacement current of 10^{-3} A?

2 If we wish to generate the displacement current in Problem 1 by altering the voltage on the capacitor, while keeping the plate separation constant at 0.1 mm, at what rate must the voltage increase?

3 If we wish to generate the displacement current in Problem 1 by changing the plate separation, while keeping the voltage constant at 100 V, at what rate must the plate separation change? (The rate of change of a product of two variables is $\Delta(a \cdot b)/\Delta t = a(\Delta b/\Delta t) + b(\Delta a/\Delta t)$.)

4 Describe how Eq. 18–2 through Eq. 18–4 would change if the capacitor was filled with a dielectric of constant K and the dimensions and voltage remained constant.

5 A capacitor consists of two circular plates of radius R, separated by distance d. Compute the electrostatic potential energy stored and the rate at which it changes when the voltage increases at a rate $\Delta V/\Delta t$. (The rate of change of V^2 equals $2V$ times the rate of change of V.)

6 For the capacitor in Problem 5, show that the rate of increase in electrostatic PE equals the rate at which the displacement current does work in charging the plates against a voltage V.

7 Two parallel circular plates of radius R and separation d are being charged by a voltage which increases at rate $\Delta V/\Delta t$. What is the magnetic field at a radial distance $r > R$ from the center of the plates? Neglect edge effects.

8 Compute the magnetic field inside the plates in Problem 7 at a radial distance $r < R$ from plate center.

9 Find the electric flux Φ_E between the plates of a 10 pF capacitor operating at 12 V with plate separation 0.3 mm.

10 Find the electric flux between the plates of a 6 μF paper ($K = 5$) capacitor operating at 3 V.

11 What capacitance would be required for the electric flux to be 10^2 N · m²/C between the plates of an air capacitor operating at 24 V?

12 A straight 100 kV, 60 Hz power line is 10 km long and 1 cm in radius, and made of copper with resistivity 1.7×10^{-8} Ω · m. Calculate the resistance of the line and the conduction current I_C. Calculate the

peak displacement current I_D. (The rate of change of $\sin(2\pi\nu_0 t)$ is $(2\pi\nu_0)\cos(2\pi\nu_0 t)$.) What is the ratio I_C/I_D? Is the displacement current important in determining the magnetic field surrounding the line?

Section 18.2 EM Spectrum and Light Travel

- 13 The distance between the Earth and Sun is 1.496×10^{11} m. What is this distance in Å? In nanometers?
- 14 What is the wavelength of the hydrogen 21-cm line in Å? In nm?
- 15 A typical atom has a diameter of about 10^{-10} m. What is this is Å and in nm?
- 16 Find the wavelength of a beam of electromagnetic radiation with frequency 1 Hz.
- 17 Find the frequency ranges of
  (a) visible light,
  (b) radio waves,
  (c) x-rays.
- 18 The shortest wavelengths of EM radiation thus far observed with laboratory detectors are about 10^{-4} Å. What is the frequency of such a γ-ray?
- 19 Astrophysicists have recently discovered γ-rays of frequency 2×10^{26} Hz by their effects in the Earth's atmosphere. What is the γ-ray's wavelength? Compare this to the size of the nucleus.
- 20 What is the wavelength of the EM radiation given off by AC currents in household wiring?
- 21 Evaluate the quantity $1/\sqrt{\epsilon_0 \mu_0}$ and verify that it is equal to the speed of light with the correct units.
- 22 Galileo once tried to measure the speed of light by measuring the round trip travel time to a hill 2 km away. How long would the light take?
- 23 If the radius of the Earth's orbit is 1.496×10^{11} m, how long will it take light to reach the Earth from the Sun?
- 24 Communication with deep space probes becomes difficult because of their distance. Calculate the minimum times for two-way communications between Earth and the Voyager probe as it passes Uranus and then Neptune (see Appendix 4).
- 25 As the Voyager II spacecraft flew past Saturn, it transmitted photographs back to Earth. If Saturn and Earth were aligned on the same side of the Sun, estimate the delay in receiving the photos.
- 26 Find the velocity of light in a medium with dielectric constant $K = 3$.
- 27 Find the velocity of a wave (not EM) which has a frequency 10^9 Hz and a wavelength 0.1 cm.
- 28 Calculate the wavelength range of sound waves which are audible to the human ear. Assume the velocity of sound is 340 m/s.
- 29 Calculate the frequency with which an electron would have to oscillate between two energy levels separated by 2 Å in an atom to produce a spectral line (EM wave) with wavelength 6,563 Å.
- 30 The electron beam in a television tube moves across the screen at 10^{-4} the speed of light. If a ghost image is displaced by 3 cm from the main image, estimate the delay time for the signal that reflects off some nearby large body. How far away is the body, if the signals take the same time to you reach you and the body?

Section 18.3 Transmission of Energy

- 31 Find the energy density in the Earth's magnetic field of 0.5 G.
- 32 How much total magnetic energy is stored in the Earth's interior? Assume the field is constant at 0.5 G, over a sphere of radius 6,378 km.
- 33 Show from the units of ϵ_0, E, μ_0, and B that the expressions given in the text for energy densities u_E and u_B have units J/m^3.
- 34 What value of E will produce an energy density 1 J/m^3?
- 35 What value of B will produce an energy density 1 J/m^3?
- 36 Find the energy densities resulting from an electric field of 100 V/m and from a magnetic field of 10 T.
- 37 Suppose a wave has $u_E = u_B$. Find the ratio of E_0 to B_0.
- 38 If the radius of the Sun is 6.96×10^8 m, find the values of E_0 and B_0 for EM waves leaving the Sun's surface.
- 39 Find the values of E_0 and B_0 a human eye would receive from a 100 W bulb 2 m away.
- 40 Astronomers define the light-year as the distance light travels in 1 yr. What is this distance in m? The nearest star is Alpha Centauri, about 4.3 light years away. Find the intensity of light from our sun on an imaginary planet orbiting close to Alpha Centauri.
- 41 Light coming from a certain star is observed to have $E_0 = 10^{-4}$ V/m and $B_0 = 3.3 \times 10^{-13}$ T. If the star is similar in properties to the Sun, estimate how far away it is.

- 42 Find the values of E_0, B_0, and $\bar{S}$ for the full moon. Assume the moon reflects 50% of the sunlight incident on one hemisphere, and see Appendix 4 for radii and distances.
- 43 Calculate the energy stored in the magnetic field of the Sun if its interior magnetic field is constant at 100 G and its radius is 6.96×10^8 m.
- 44 Find the magnitude of the Poynting vector S for an EM wave which has $E = 150,000 \sin(60t)$ and $B = (0.0005) \sin(60t)$. What are the SI units for E and B as given?
- 45 What are E_0 and B_0 for the radio waves from a 50,000 W station 10 km away? Assume the station broadcasts uniformly in all directions.
- 46 If microwaves were to transmit power equivalent to power lines with 100 kV and 1,000 A, what values of E_0 and B_0 would be required for a straight cylindrical beam 1 m in diameter?
- 47 A light source has an intensity of 1 W/m². Find E_0 and B_0 for the EM waves.
- 48 An EM wave has $E_0 = 10^{-5}$ V/m and $B_0 = 3.3 \times 10^{-14}$ T. Find the magnitude of the Poynting vector.
- 49 Calculate the magnitude of the Poynting vector as EM waves leave the surface of the Sun if the solar flux on the earth is 1,360 W/m².
- 50 Using your result from Problem 49, estimate the total power output of the Sun, assuming it radiates isotropically.
- 51 When EM waves strike a surface, they may either reflect or be absorbed. In either case, they do work on the surface and create radiation pressure. Calculate the radiation pressure of sunlight on a panel deployed in earth orbit if the Sun's intensity is 1,360 W/m² and the panel is aligned face-on. Assume the light is fully absorbed.

Section 18.4 Radio Stations

- 52 Radio dials show frequencies in kHz for AM stations and in MHz for FM stations. What are the frequencies and wavelengths of radio waves from stations at AM 1200 and FM 103?
- 53 The oscillating tuner circuit for an AM station at 1,000 kHz has a capacitance of 10^{-9} F. What is its inductance? What must the capacitance be for a station at 1,600 kHz if the inductance remains fixed?
- 54 The owners of a 10,000 W radio station at 600 kHz wish to reach a listener 10 km away. If the listener erects a vertical antenna, of length equal to one quarter-wavelength, what voltage will develop from the electric field of the wave? Assume the wave's electric field is parallel to the antenna and that the station broadcasts power equally in all directions.
- 55 What is the intensity of a 50,000 W radio station at 100 km? Would it make sense to use such power for FM stations (frequencies 100 MHz)? Compare the wavelength of the FM station with that of a typical AM station at 1,000 kHz and think about diffraction and the earth's curvature.

SPECIAL PROBLEMS

1. Communication with Other Planets

In an ideal alignment, the planets would have positions shown in Fig. 18.10, with initial angles θ_0 measured from the horizontal axis. The earth has $\theta_0 = 65°$. A deep space probe is launched at a constant velocity 10,000 m/s and maintains a constant direction of 65°. (In reality, the gravity of Jupiter or Saturn would deflect it.) Planetary orbits obey Kepler's law, $T^2 = a^3$, where T is the period in years and a is the distance from the Sun in AU (1 AU = 1.496×10^{11} m).

(a) Show that a planet's angular position at a time after launch is $\theta = \theta_0 + \omega t$, where $\omega = 2\pi/T$ is the angular velocity in rad/s.

(b) Find the encounter times, t_e, for the outer planets Mars, Jupiter, Saturn, Uranus, and Neptune (Pluto's orbit is eccentric and out of the plane of the others). What are the changes in angle $\Delta \theta$ for each planet at t_e?

(c) Compute the initial angles θ_0 for the 5 planets so that the probe encounters them as it flies by their orbits. Consult Appendix 4 for periods and orbital distances.

(d) Calculate the time needed for two-way radio communication at the time of encounter. Remember that the earth has moved and convert from polar to rectangular coordinates.

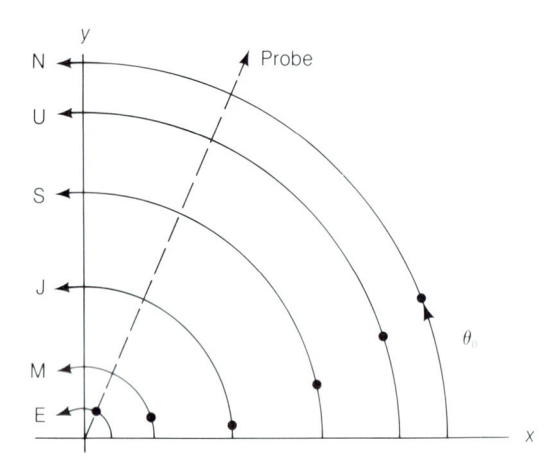

Figure 18.10

2. Radio and TV Frequency Bands

Certain radio wave frequency bands have been allocated by the Federal Communications Commission to television, AM, and FM radio. These are 535–1605 kHz for AM, 88–108 MHz for FM, and 460–470 MHz for citizen's band. The television channels are listed in the table.

Channel	Frequencies (MHz)
2	54–60
3	60–66
4	66–72
5	76–82
6	82–88
7	174–180
8	180–186
9	186–192
10	192–198
11	198–204
12	204–210
13	210–216

(a) For television channels 2–13, which are each 6 MHz wide, compute the equivalent wavelength bands in meters. Which channel has the largest wavelength band?

(b) Compute the wavelengths of the AM, FM, and citizen's band.

(c) Why does channel 6 often get interference? Are harmonics a problem for television?

(d) Channels 14–83 are in the UHF (ultrahigh frequency) band, and range from 470 MHz to 890 MHz. What are the corresponding wavelengths? How much shorter is the wavelength of channel 35 (596–602 MHz) than channel 4?

3. Standing Waves

Electromagnetic waves may be trapped in cavities to form standing waves (Chapter 10). These standing waves play a major role in the operation of lasers. Suppose a cavity has length L.

(a) Show that standing waves form when light has wavelength $\lambda_n = 2L/n$, $n = 1, 2, 3 \ldots$.

(b) Where are the positions of the nodes (points of zero electric field)? If $L = 30$ cm and $\lambda = 500$ nm, how many nodes are in the cavity? What is n? What is the round trip light travel time from one side to the other and back?

(c) If the cavity's length is slowly changed by one part in 10^5, how many nodes pass by a fixed position? Can you devise a way to use this as a precision technique for measuring lengths?

4. Blackbody Spectrum

When a hot objective radiates energy, it emits EM waves over a broad range of frequencies known as a spectrum. In an ideal case, when the emissivity $e = 1$ (Section 12.5), the spectrum is that of a **blackbody,** whose frequency distribution is given by the Planck distribution,

$$B(\nu) = \frac{(2h\nu^3/c^2)}{(e^{h\nu/kT} - 1)}$$

where $h = 6.626 \times 10^{-34}$ J/Hz is Planck's constant, $k = 1.381 \times 10^{-23}$ J/K is Boltzmann's constant, and T is absolute temperature in K. The Planck distribution for EM radiation frequencies is analogous to the Maxwellian distribution for particle energies (Chapter 13) in thermal equilibrium.

(a) Define the dimensionless variable $x = h\nu/kT$ and show that the Planck distribution may be written as $B(x) = C\, x^3/(e^x - 1)$, where C is a constant. Make a schematic graph of $B(x)$ from $x = 0$ to $x = 6$.

(b) Using trial and error, show that $B(x)$ reaches a maximum value when $x = x_p = 2.82$. Show that the most probable frequency is $\nu_p = (2.82k/h)T$. What wavelength does this correspond to?

(c) If the Sun has $T = 5{,}700$ K, at what wavelength does its blackbody spectrum peak? What is the intensity of blue light ($\lambda = 475$ nm) relative to red light ($\lambda = 675$ nm)?

(d) If an object is hotter, its blackbody spectrum will peak at shorter wavelengths. Three stars have surface temperatures of 2,500 K, 10,000 K, and 50,000 K. Find the peak wavelengths, and identify which star appears blue, which infrared, and which ultraviolet.

SUGGESTED READINGS

"James Clerk Maxwell." J. R. Newman, *Scientific American,* Jun 55, p 58 **192**(6)

"Light wave communications." W. S. Boyle, *Scientific American,* Aug 77, p 40, **237**(2)

"Phased-array radars." E. Brookner, *Scientific American,* Feb 85, p 94, **252**(2)

"Lightwave communication." T. Li, *Physics Today,* May 85, p 24, **38**(5)

The flying circus of physics with answers. Jearl Walker, Wiley, New York, 1977

A source book in physics. William F. Magie, Harvard University Press, Cambridge, 1965

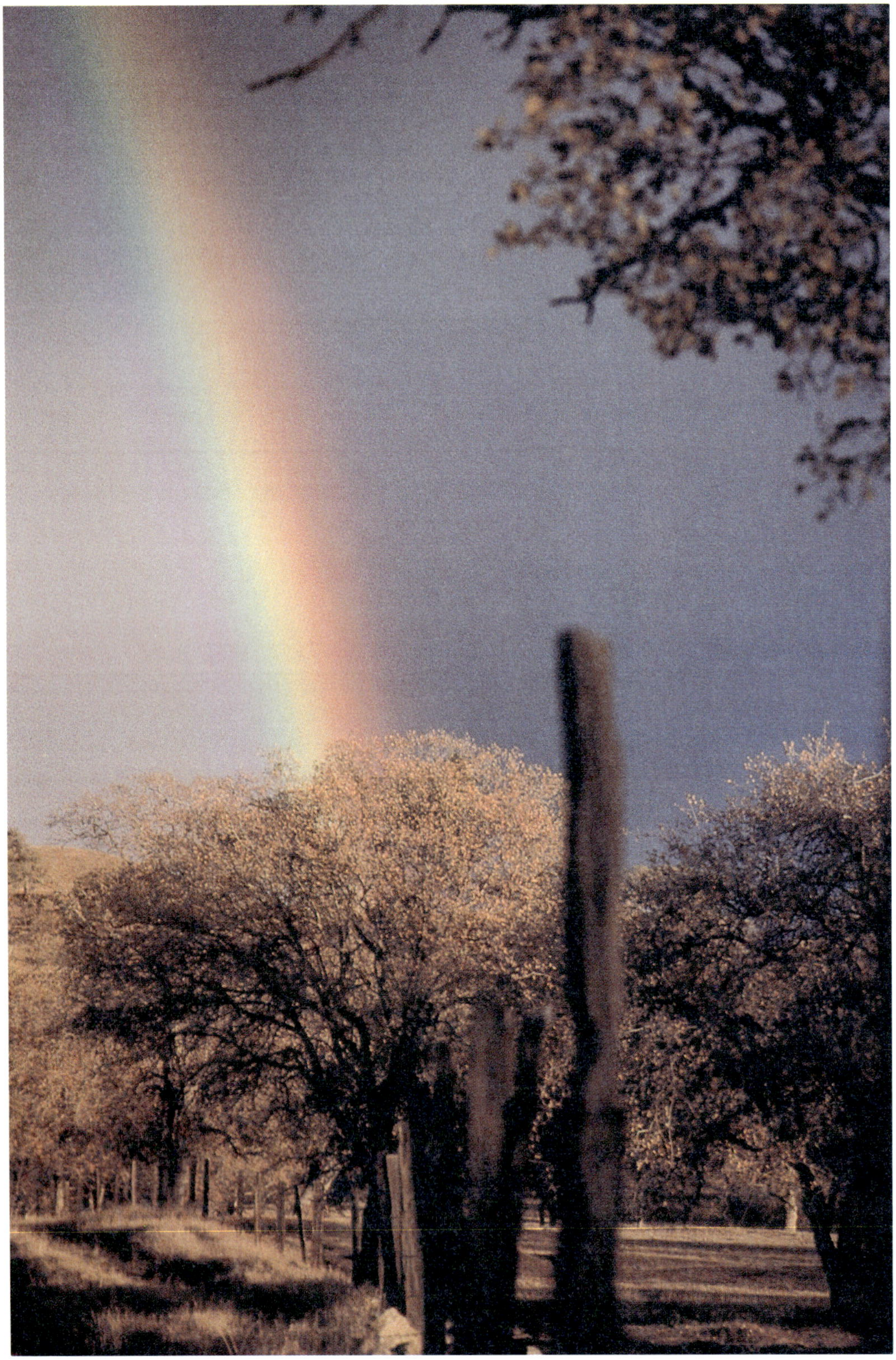

Chapter 19

Geometric Optics

Figure 19.1

Christiaan Huygens (1629–1695).

In the previous chapter we alluded to images as one form of information transported by EM waves. When the human eye sees an object, the light from it forms an image on an interior surface (the retina) of the eye. This image is brought to a focus by the optical action of the lens of the eye. Although radiation of any EM wavelength can form an image, we shall confine ourselves to the visible wavelengths in our discussion of optics.

The foundations of optical science were laid in the early part of the seventeenth century with the invention of the telescope and Galileo's well-known application of it to astronomy. A prominent figure in developments immediately thereafter was the Dutch physicist C. Huygens (Fig. 19.1), who was active in the latter half of the seventeenth century. Huygens discovered evidence for the wave nature of light (which was never accepted by his great contemporary Newton, who thought of light as consisting of minute particles which he called corpuscles). Huygens was able to explain many aspects of geometric optics with his wave model. Some of the ideas of Huygens have already been mentioned in this text (Chapter 10).

In this chapter we will first describe some general principles of optics, then discuss in some detail the formation of images by mirrors and lenses. The latter part of the chapter will be devoted to several important kinds of optical instruments.

19.1 General Principles of Optics

Before we can profitably discuss the formation of images, we must establish a few ground rules. These have to do with the nature of light propagation in a vacuum and in media such as air or glass, and with a few simple laws governing the reflection and refraction of light.

615

CHAPTER 19 GEOMETRIC OPTICS

The Ray Model and Ray-Tracing

It seems intuitively obvious that light travels in a straight line, and indeed this is so as long as it passes through a vacuum or a uniform medium. Experiment verifies this; we find, for example, that light passing through an opening strikes a distant surface at the point where a straight line from the source through the opening would (Fig. 19.2). Huygens was able to demonstrate, on the basis of his wave model of light, that its travel should be in a straight line. He portrayed light as consisting of wave fronts that emanate spherically from a point; in this view, it is straightforward to show that the propagation away from any point is always in a straight line. As we mentioned in Chapter 10 and will see again in this chapter, the concept of propagating wavefronts also successfully explains how light is deflected by mirrors and lenses.

Since light travels in straight lines, it is convenient to visualize light **rays,** which are represented as arrows pointing in the direction of propagation (Fig. 19.3). In this model, we see that an image is formed when rays from each point on an object come together; the principal goal of this chapter will be to analyze how this occurs. We will find it useful to trace the paths of rays from key points on the object to the surface where the image forms; this **ray tracing** is an invaluable tool, as we shall see.

One simple geometrical effect is readily explained by the ray model. The apparent size of an object corresponds to its **angular size,** which is the angular spread of the rays from it (Fig. 19.4). The more distant an object is, the smaller its angular size becomes, so the smaller its apparent size. This explains **perspective;** objects appear smaller as their distance increases (Fig. 19.5).

In Chapter 10 we learned about some wave effects, such as diffraction, in which straight-line propagation does not occur. Hence the ray model is apparently oversimplified. We will find in our discussions in the next chapter that the important factor in creating departures from straight-line travel when light passes an obstacle is the relative size of the wavelength and the obstacle. For visible light, the wavelength is far smaller than most obstacles or openings that rays may pass, so for now we can ignore departures from straight-line travel, and adhere to the ray model. We would not be able to do so if we discussed longer-wavelength radiation such as radio waves, for example, and we will find that sometimes we cannot ignore the departures even for visible light.

The Index of Refraction

It is well known that light rays are bent when they pass through a boundary between two substances. In Chapter 10 we learned that this effect is called **refraction.** Refraction occurs because the speed of light depends on the properties of the medium through which it passes. (The property of a substance that slows the speed of light is called the **optical density,** and it is generally related to the mass density.) The standard value for c given in the previous chapter is the speed of light in a vacuum; it is slower in any physical medium.

In Chapter 10 we showed why refraction occurs, using the wavefront model (Fig. 19.6). As a wavefront approaches a boundary between two media

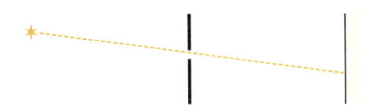

Figure 19.2

The path of light. In a vacuum or uniform medium, light travels in a straight line.

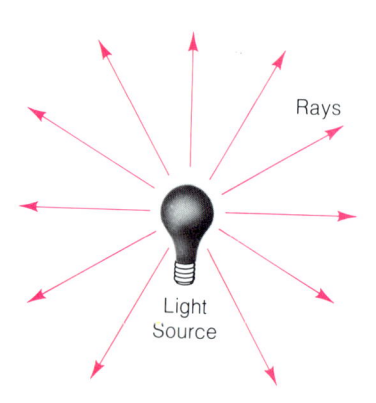

Figure 19.3

Rays. The direction of propagation of light can be represented by straight arrows known as rays.

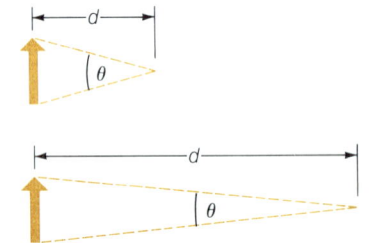

Figure 19.4

Angular size. The angle θ represents the angular size of the arrow, as viewed from two different distances. The angular size is inversely proportional to the distance.

19.1 GENERAL PRINCIPLES OF OPTICS

Figure 19.5

Perspective. Distant objects appear smaller than nearby ones, because their angular size is smaller.

at an oblique angle, the portion of the front that enters the boundary first is slowed (for example, as light passes into a region of higher density), while the other portions are still travelling at higher speed. Thus the wavefront bends. In the ray model, the ray is bent closer to the normal vector, when the light passes from a low-density medium into a higher density one (Fig. 19.7). If the light passes from a high-density medium to a lower-density one, then the ray is bent away from the normal vector.

The speed of light in a medium, and hence the degree to which it is bent, depends on the wavelength of the light. This is the reason that white light can be spread into a colorful spectrum by a glass prism or by raindrops. Within the visible light band, the different wavelengths have nearly the same speed, so that when we characterize a medium according to the speed of light in it, we treat visible light as if it had only one speed.

If c is the speed of light in a vacuum, and c_n the speed in another medium, then we define the **index of refraction** n for that medium as

$$n = \frac{c}{c_n}. \tag{19-1}$$

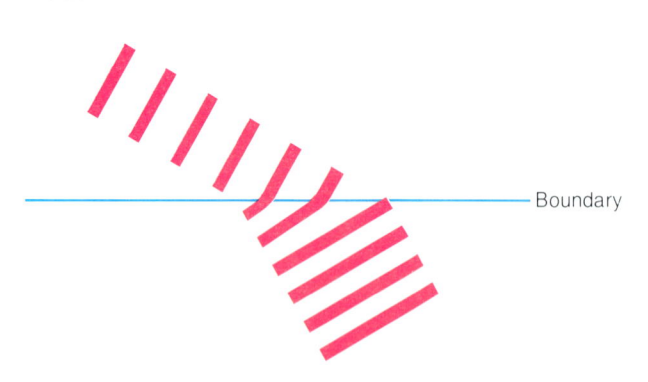

Figure 19.6

Refraction. Parallel wavefronts are refracted (bent) upon entering a medium of higher (or lower) optical density, because the speed of light is reduced (or increased).

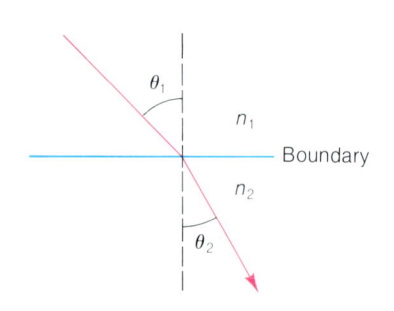

Figure 19.7

Refraction. Here a ray passes through a boundary into a medium of increased index of refraction, so the angle of refraction θ_2 is smaller than the angle of incidence θ_1. The angles are related by $n_1\sin\theta_1 = n_2\sin\theta_2$, where n_1 and n_2 are the indices of refraction.

Table 19.1

Indices of Refraction ($\lambda = 589$ nm)

Material	n
Diamond	2.42
Zircon	1.92
Dense flint glass	1.66
Light flint glass	1.58
Rock salt	1.54
Quartz	1.54
Crown glass	1.52
Benzene	1.50
Glycerine	1.47
Fused quartz	1.46
Carbon tetrachloride	1.46
Ethyl Alcohol	1.36
Water	1.33
Ice	1.31
Air	1.00

Typical values for common substances such as water and glass are in the range 1.3 to 1.6, but can be considerably higher for some transparent substances, such as certain gemstones. For air the value of n is very nearly 1. Some example indices of refraction are listed in Table 19.1.

The frequency of light waves remains constant when light passes through a boundary between media of differing index of refraction. The waves get crowded closer together when entering a denser medium, but the frequency of the oscillations is not changed. Therefore the wavelength must be changed. The relation

$$\lambda = \frac{c}{\nu} \quad (19\text{--}2)$$

leads to

$$\lambda_n = \frac{c_n}{\nu} = \frac{\lambda}{n} \quad (19\text{--}3)$$

where λ_n is the wavelength in the medium with index of refraction n, and λ is the wavelength in a vacuum, where the speed of light is c.

Example 19.1 Calculate the speed of light in water. What is the wavelength in water of light that has a wavelength of 500 nm in vacuum?

The index of refraction is $n = 1.333$, so the speed of light is

$$c_n = \frac{c}{n} = \frac{3.00 \times 10^8 \text{ m/s}}{1.333} = 2.25 \times 10^8 \text{ m/s}.$$

19.1 GENERAL PRINCIPLES OF OPTICS

The wavelength in water of light whose vacuum wavelength is 500 nm is

$$\lambda_n = \frac{\lambda}{n} = \frac{500 \text{ nm}}{1.333} = 375 \text{ nm}.$$

As we mentioned, the index of refraction varies slightly with wavelength for many materials (Fig. 19.8). A beam of white light passing through a prism separates into a rainbow of colors, because the refraction depends on the wavelength. For example, the index of refraction of a glass prism is larger for blue light than for red light. Therefore, blue light is bent more by a prism (Fig. 19.9). This phenomenon is called **dispersion** and is responsible for many physical phenomena, including the rainbows produced by dispersion of light by water droplets (see Special Problems 1, 2, and 3 at the end of the chapter).

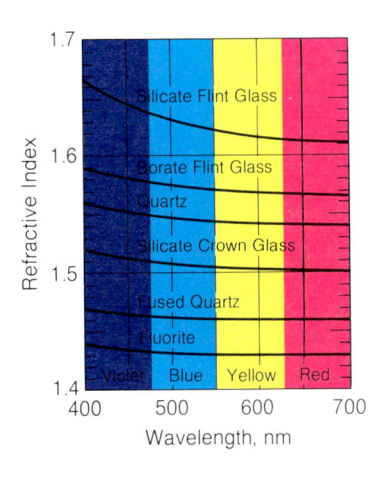

Figure 19.8

Index of refraction versus wavelength. For most substances, the index of refraction varies with wavelength, resulting in dispersion of white light into a rainbow of colors.

Figure 19.9

Dispersion by a prism. A beam of white light is dispersed into the colors of the visible spectrum when it passes through a material whose index of refraction varies with wavelength.

Laws of Refraction and Reflection

We learned in Chapter 10 that the angle of refraction is related to the angle of incidence (where both are defined as the angle from the perpendicular to the surface; see Fig. 19.7) by a relationship called **Snell's law.** If we define θ_1 and θ_2 as the angles between the incoming and outgoing rays and the perpendicular to a surface between media whose indices of refraction are n_1 and n_2, then we have

$$n_1 \sin\theta_1 = n_2 \sin\theta_2. \tag{19-4}$$

CHAPTER 19 GEOMETRIC OPTICS

Figure 19.10

Passage of light through a windowpane. The ray is bent toward the normal as it passes from air into glass, then bent away from the normal, back to its original direction, as it passes back into air.

Figure 19.11

Reflection. When light strikes a reflecting surface, the angle of reflection θ_2 is equal to the angle of incidence θ_1.

This relationship is valid for light travelling in either direction across a boundary between media with different indices of refraction. We need only remember that the angle is smaller in the medium with higher optical density. Hence, light passing through a windowpane is actually bent twice, once upon entering the glass and once upon leaving it, and, as it passes through, its direction is more nearly perpendicular to the plane of the pane than it is in the air on either side (Fig. 19.10). The outgoing ray leaving the glass after undergoing both refractions is parallel to the incoming ray, but displaced.

When light strikes a reflecting surface, the law is equally simple, again as discussed in Chapter 10. The angle of reflection is always equal to the angle of incidence, where both are measured with respect to the perpendicular to the reflecting surface (Fig. 19.11). Thus we have

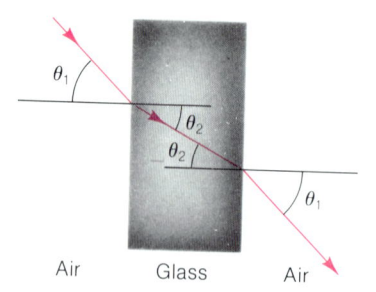

$$\theta_1 = \theta_2 \qquad (19\text{--}5)$$

which is known as the **law of reflection.**

The two simple expressions, Eqs. 19–4 and 19–5, form the basis for the analysis of optical systems. The major complexities that come into play arise when curved surfaces, rather than plane surfaces, are used for reflection and refraction. We will address these complexities in Sections 19.2 and 19.3.

Example 19.2 A four-foot-tall child (eye level 3 ft 7 in) wishes to see a coin at the bottom of a pool whose sides are raised so that the surface of the water, which is even with the top of the side, is 3 ft 3 in above the surface on which the child stands (Fig. 19.12). If the coin is 3 ft from the base of the wall, how far on the other side of the wall must the child stand to see it?

We need to find the point where a ray from the coin that just clears the wall intersects the child's eye level. We know that, in the water, the angle of the ray from the vertical is the angle given by

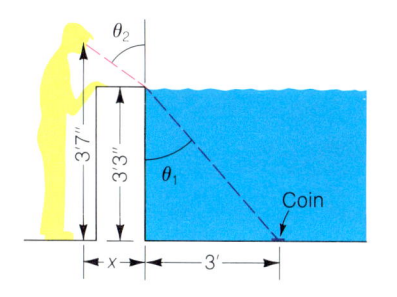

$$\theta_2 = \arctan \frac{3 \text{ ft}}{3.25 \text{ ft}} = 42.7°.$$

If we designate the air as medium 1 and the water as medium 2, then from Eq. 19–4 we can find the angle the emergent ray makes with the vertical.

$$\theta_1 = \arcsin\left[\left(\frac{n_2}{n_1}\right)\sin\theta_2\right] = \arcsin\left[\left(\frac{1.333}{1.000}\right)\sin(42.7°)\right]$$
$$= \arcsin(0.904) = 64.7°$$

Now we need to find the horizontal distance required for the ray emerging at this angle from the vertical to reach a level 4 in higher than the surface of the pool. From Fig. 19.12 we see that this distance x is

$$x = 4 \tan(64.7°) = 8.5 \text{ in.}$$

If the child stands so that his eye is 8.5 in horizontally from the edge of the water, he will just see the coin.

Total Internal Reflection

We have not discussed the efficiency of refraction or reflection, but this can be important. It may seem obvious that mirrors in general are not perfect; that they absorb some light so that the reflected ray has slightly less intensity than the incident ray. This is true, although with a very smooth, well-polished and clean surface, the loss can be less than 1 percent. It is also true that refraction is ordinarily not a perfectly efficient process, because normally some reflection occurs at any boundary between media with different indices of refraction. In one process, no losses occur at all, however.

As we have seen, when a ray travels from one medium to another with a lower index of refraction, the ray is bent away from the perpendicular to the boundary. If we use Snell's law to find the angle of refraction for larger and larger angles of incidence, we find that the angle of refraction approaches 90° as its sine (as calculated from Eq. 19–4) approaches 1. Snell's law fails to describe what happens for larger incident angles, because there is no angle whose sine is greater than 1. It is found experimentally that for larger angles of incidence, the ray does not pass through the boundary at all, but instead is reflected back into the medium of higher index of refraction (Fig. 19.13). This is called **total internal reflection,** and is one of the few processes in physics which is 100% efficient. The incident angle for which the angle of refraction equals 90° is called the **critical angle.**

The critical angle is a function of the indices of refraction. For example,

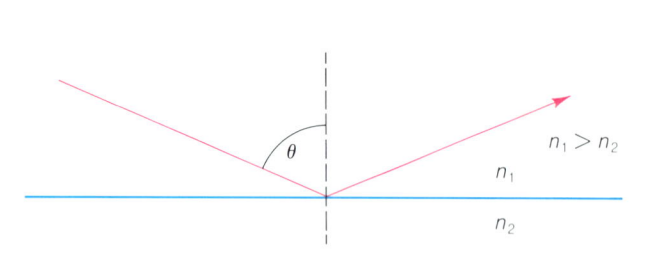

Figure 19.13

Total internal reflection. If the angle of incidence exceeds the critical value, then refraction does not occur, and the ray is totally reflected at the boundary. This can occur when the index of refraction in the initial medium is greater.

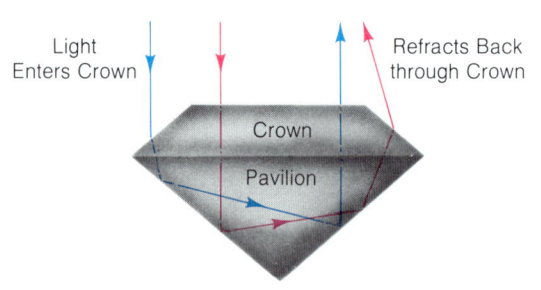

Figure 19.14

The brilliant cut for diamonds. The angles are calculated so that light entering through the top of the stone is reflected back out through the top, because of total internal reflection.

Figure 19.15

Optical fibers. The ends of the thin fibers appear to glow, because the light transmitted through them escapes only at the ends.

Figure 19.16

Optical fiber. A ray entering one end of an optical fiber is repeatedly reflected internally, eventually emerging from the other end with little loss of energy.

light passing through water or glass is reflected from a boundary with air. The critical angle is easily calculated from Eq. 19–4, with the refracted angle set equal to 90°. For water the result is $\theta_c = 49°$, and for a typical glass it is 42° (air is assumed to be the lower-density medium in each case).

Because of its perfect efficiency, total internal reflection can be very useful. For example, in cameras and binoculars, glass prisms are employed to invert images without any loss of brightness. The angles of the facets on a transparent gemstone are often cut so that entering light rays are totally reflected back out of the top of the stone, enhancing its glitter (the "brilliant" cut for diamonds is illustrated in Fig. 19.14). Another very important example of total internal

19.1 GENERAL PRINCIPLES OF OPTICS

reflection occurs in **optical fibers,** which are flexible glass fibers that can be made as small as a few micrometers in diameter (Fig. 19.15). Light entering one end of such a fiber is totally reflected each time it encounters the inside wall, and so it travels the length of the fiber without any loss (Fig. 19.16). A bundle of fibers, sometimes called a **light pipe,** can transmit an image if the fibers are in the same arrangement at both ends of the bundle. Fiber optics are used in medicine to examine internal body cavities, and in astronomical and other optical imaging systems.

Example 19.3 Calculate the critical angle for total internal reflection inside a diamond, whose index of refraction is 2.42.

If we let air be medium 1 and diamond medium 2, we solve for θ_c by setting $\sin\theta_1 = 1$.

$$\theta_2 = \theta_c = \arcsin\left[\left(\frac{n_1}{n_2}\right)\sin\theta_1\right] = \arcsin\left(\frac{n_1}{n_2}\right) = \arcsin\left(\frac{1.00}{2.42}\right) = 24.4°$$

Hence for the brilliant cut the angles of incidence in the diamond are always at least 24.4°.

FOCUS ON PHYSICS

Optical Fiber Communications

Many people are becoming aware of optical fibers and some of their more novel features (see Fig. 19.1.1, for example). The many practical uses of these devices are not so widely recognized, however.

Optical fibers are capable of transmitting light over very long distances with little loss of energy (recall that total internal reflection can actually be perfectly efficient, with no loss at all). In a long fiber, because of very many reflections, a small amount of energy is lost, primarily because the material of the fiber has unavoidable slight impurities which scatter light. After several years of intense development, however, fibers with energy losses as low as 25 percent per kilometer of fiber length have been manufactured. The best efficiencies are possible in the infrared portion of the spectrum, where scattering losses are minimized.

Optical fibers can transmit images or pulses of light in patterns that convey information. To transmit an image requires a bundle of fibers, because each single fiber carries information on the inten-

(continued)

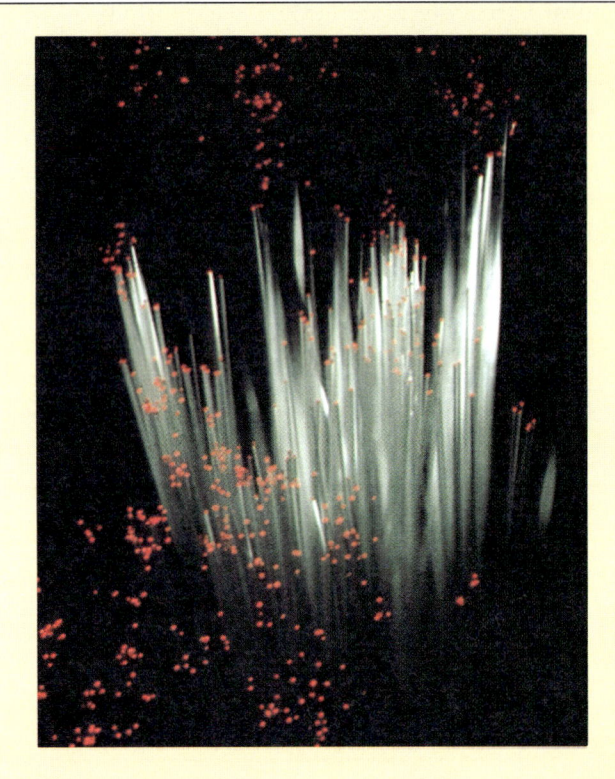

Figure 19.1.1

sity of light from only a small area on the object being viewed. Image transmission with optical fibers is used in medicine and astronomy, among other fields. Modern surgical techniques using optical fibers make it possible for doctors to see inside small parts of the body (the inside of a knee joint, for example, or even the interiors of the heart and major blood vessels), so that operations can be performed with minimum intrusion into the body. In astronomy, often a single fiber suffices, because the goal is simply to transport a point image from the focal point of the telescope to a remote instrument where the light is analyzed.

By far the most widespread potential use of optical fibers in technology is in communications systems, rather than image transmission. In communications, infrared light is most commonly used, because it is more efficient than visible light. Pulses of light are transmitted over long distances through optical fibers in an invariant pattern that can convey information. In practice, the method is very similar to using radio waves to send signals: a carrier frequency of pulses is sent along the fiber, and is modulated by the signal frequency (see the discussion of AM and FM radio communications in Section 18.4). The pulses are created by a laser or some other light-emitting device that produces significant power and can be controlled very precisely in response to an electrical input signal. This input could be the pattern of electrical pulses from a microphone, for example, or it could be a video signal produced by a television camera. The frequency or the amplitude of the carrier signal can be modulated, as AM or FM radio transmissions are.

Optical fibers have been used successfully in several pilot projects in telephone systems for major cities, and are expected eventually to supersede conventional electrical cables in many other applications (such as cable TV systems). An undersea telecommunications cable is even being designed to cross the Atlantic Ocean, with installation expected in the late 1980s. In an optical fiber system spanning such great distances, repeaters along the fiber receive the signal, convert it to electrical form, amplify it, and then retransmit it.

The advantages of optical fibers over electrical wires include lower loss of energy, particularly at high frequencies; greater flexibility and lower weight; and immunity to electromagnetic disturbances such as nearby wires or lightning strikes. Many of us have used optical fiber communication systems without being aware of it, and in the future we may expect our telephone and television messages to rely on this technology to an increasing extent.

19.2 Image Formation by Mirrors

We are ready now to study how optical systems form images, and we begin with reflection. The simplest example, which we treat first, is a plane mirror. We will then analyze image formation by curved mirrors.

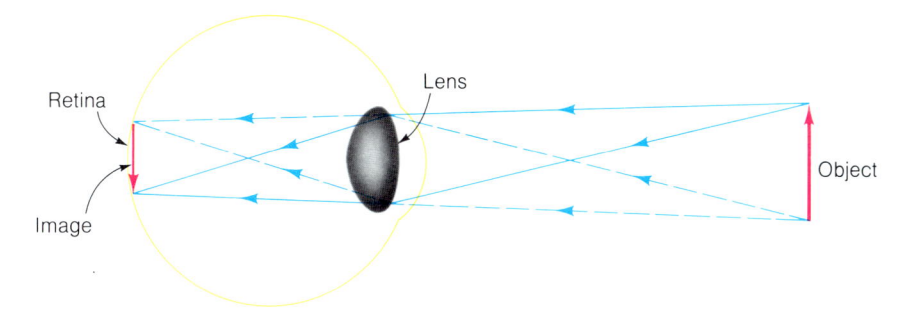

Figure 19.17

Formation of an image in the eye. Rays from each point on the object (arrow) are brought to a focus on the retina, the light-sensing back surface in the eye. The image is inverted by the eye.

Plane Mirrors

When a person looks at an object, the rays of light reaching his eye from various points on the object form an image (Fig. 19.17). From each point on the object, a divergent bundle of rays enters the eye and is brought to a focus on the retina. The size of the image is determined by the angular size of the object, as we mentioned earlier in this chapter. The angular size, in turn, depends on the actual size of the object and the distance between it and the eye.

Suppose we now look at the reflection of the same object in a plane mirror. The image is formed in the same manner, but the light rays reach the eye by an indirect route (Fig. 19.18). Because of the symmetry of reflection (Eq. 19–5), the apparent distance of the object behind the mirror is equal to its actual distance in front of it. The angular size of the object is now determined by the total distance from the eye to the mirror and from there to the object.

The image of the object appears to be behind the mirror. Light rays are not actually focused at the image; a camera positioned where the image appears to be would not record the image. Because of this, we say that the image is a **virtual image**, because it does not physically exist at its apparent location. Shortly we will see how **real images** can be formed by other mirrors and by refraction.

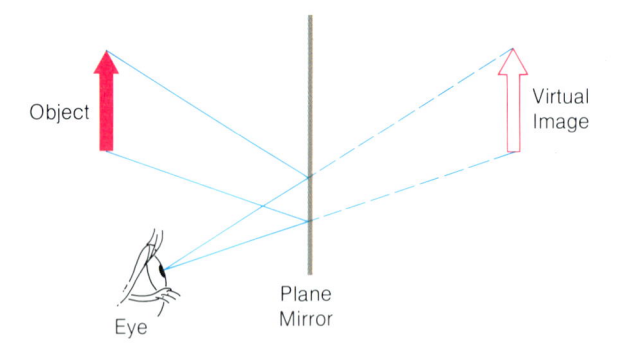

Figure 19.18

Formation of a virtual image. To the eye, the image of the object appears to be behind the mirror.

Spherical Mirrors

Mirrors can be curved, rather than flat, and curved mirrors can form images, too (Fig. 19.19). We will discuss primarily **concave** mirrors that consist of sections of spheres (although later we will see that other shapes are sometimes used). **Convex** mirrors also form images, which is useful especially when a wide field of view is desired.

First let us consider the formation of an image by a spherical concave mirror when the object being viewed is very distant, so that the rays from each point on it are essentially parallel when they reach the mirror (Fig. 19.20). If point C is the center of curvature of the sphere, we can find the focus, the point where the rays from the distant object meet. Consider the ray that strikes the mirror at A. The line CA, a radius of the sphere, is perpendicular to the mirror at A. Therefore, according to the law of reflection, the angle θ_1 is equal to the angle θ_2. These angles are also equal to angle θ_3 between the line CA and the axis of the mirror, because the incoming ray is parallel to the axis. Hence triangle CAF has two equal angles, which means that the sides opposite these two angles must also be equal. Thus, the lines CF and FA are equal.

For any point A on the spherical mirror, distances CF and FA would be equal, but the position of F would vary as A was moved around. We see that if A does not stray far from the axis of the mirror, so that the angles θ_1, θ_2, and θ_3 are small, then F is nearly fixed in position, just halfway between the center of curvature C and the point where the axis meets the mirror (O). Thus, for a mirror which is small compared to its radius of curvature, all rays parallel to the axis of the mirror converge at F, which is called the **focal point** of the mirror. The distance FO is the **focal distance,** often denoted simply f. We have shown that

$$f = \frac{r}{2} \tag{19-6}$$

where r is the radius of curvature of the mirror.

19.2 IMAGE FORMATIONS BY MIRRORS

Figure 19.19

Image formation by a concave mirror. This is the primary mirror for the Space Telescope, *which will be launched into Earth's orbit in 1986. It's diameter is 2.4 m.*

Although the rays are not precisely focused at *F*, the focus is adequate for many applications. For a mirror made from a parabolic surface, however, the rays are focused precisely at a common point. Thus, when high precision is required, as in astronomical telescopes, parabolic mirrors are often used.

The image of a point object forms where the rays from it converge, so a spherical mirror forms an image of a very distant point object at F, where parallel rays meet. For rays that are not parallel, as from a nearby extended object, we trace rays from various parts of the object (Fig. 19.21). In principle this could be a tedious task, since we could trace rays from many positions on the object, but in practice it is made simpler by a judicious choice of rays.

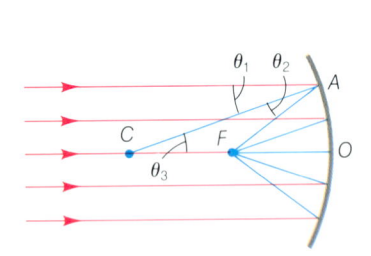

Figure 19.20

Image formation by a spherical mirror. Parallel rays (from a distant object) are brought to a focus at point F. Simple geometric arguments show that F is approximately fixed for all rays parallel to the axis.

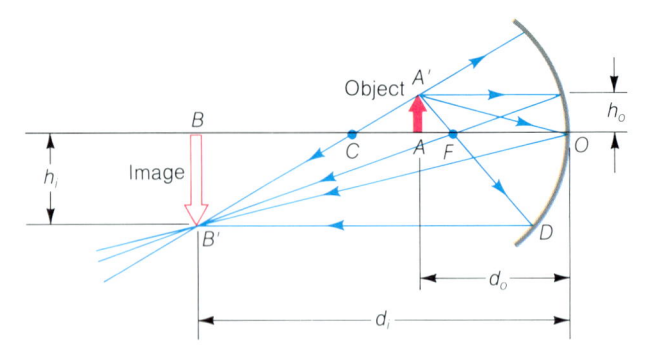

Figure 19.21

Image formation by a spherical mirror. Here an object is located between the focal point and the center of curvature.

It is often convenient to start at the end of the object (such as the tip of the arrow shown in Fig. 19.21), and trace three or four easy rays, one through the center of curvature, one parallel to the axis of the mirror, one through the mirror center, and one through the focal point. A ray through the center of curvature strikes the mirror in the normal direction, and is thus reflected back along itself. A ray parallel to the axis reflects back through the focal point, one through the mirror center reflects at an equal angle from the axis, and one through the focal point is reflected parallel to the axis. All meet at a point beyond the center of curvature, if the object is between there and the focal point (Fig. 19.21), and in between the focal point and the center of curvature if the object is farther from the mirror than the center of curvature. Where the rays converge, an image of the point on the object where they originate is formed. This is a **real image,** since light physically passes through it. If we trace rays for any other point on the object, we find that the rays converge at the same distance from the mirror, so that an image of the entire object forms there. The image is inverted because it has been reflected about the axis.

We can derive an expression to find where the image forms for any object location. If d_o and d_i represent the distances of the object and its image from the center of the mirror, and h_o and h_i the sizes of the object and the image, then Fig. 19.21 shows that

$$\frac{h_o}{h_i} = \frac{d_o}{d_i}$$

because the triangles $A'OA$ and $B'OB$ are similar. Furthermore, keeping in mind the assumption that the extent of the mirror is small, we see that triangles $A'FA$ and DOF are approximately similar, so that

$$\frac{h_o}{h_i} = \frac{d_o - f}{f}.$$

Equating these two expressions for h_o/h_i and rearranging leads to

19.2 IMAGE FORMATIONS BY MIRRORS

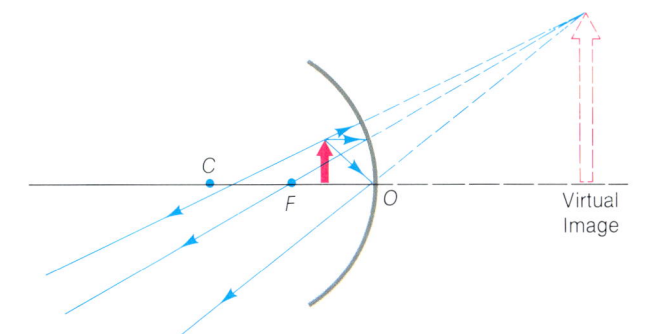

Figure 19.22

Image formation by a spherical mirror when the object is inside the focus. The reflected rays diverge, so there is no real image, but there is a magnified virtual image.

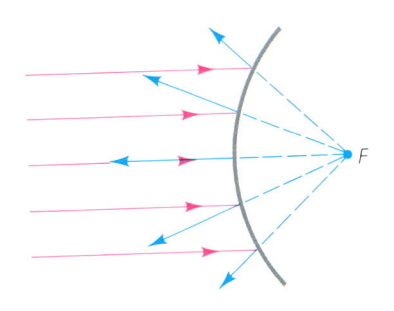

Figure 19.23

Image formation by a convex spherical mirror for a distant object. Incident rays are parallel. The reflected rays diverge, and there is no real image. A virtual image is formed at the distance F behind the mirror.

$$\frac{1}{d_o} + \frac{1}{d_i} = \frac{1}{f} \qquad (19\text{--}7)$$

which is called the **mirror equation**.

The ratio of image size h_i to object size h_o is called the **magnification** of the image, which we designate m. Normally a minus sign is used to indicate an inverted image, so from our similar triangles $A'OA$ and $B'OB$ we have

$$m = \frac{-h_i}{h_o} = \frac{-d_i}{d_o}. \qquad (19\text{--}8)$$

Eqs. 19–7 and 19–8 are valid for any object position that is farther from the mirror than the focal point, because for all such positions, the rays converge after reflection from the mirror. For an object closer to the mirror than the focal point, however, the reflected rays diverge (Fig. 19.22). The situation is similar to the case of a plane mirror in that the rays appear to be diverging from a point behind the mirror. The image is therefore upright and virtual. It is also magnified, because the rays diverge at a wider angle than for a plane mirror. The mirror equation is still valid, but the image distance d_i is negative because the image is behind the mirror. Slightly concave mirrors are sometimes used for magnification of objects placed within the focal point; an example is a make-up or shaving mirror, designed so that people can scrutinize themselves more closely than with a plane mirror.

Note that the mirror equation also applies to plane mirrors if we set $f = \infty$. Then $d_i = -d_o$, consistent with our discussion in the previous section. The magnification for a plane mirror, derived through Eq. 19–8, is $m = 1$, also as expected.

For a convex spherical mirror (Fig. 19.23), the reflected rays appear to diverge from focal point behind the mirror. If the focal distance is taken as negative, then the mirror equation is valid for such a mirror. The magnification is always less than 1 for a convex mirror, so that the image is reduced. Such mirrors are commonly used to provide a wide field of view, as in a rear-view mirror on a car or truck.

It is worth noting that whenever an image is magnified, the intensity of light in the image is altered. If the image is larger than the object, the intensity of light in the image must be reduced, and when the image is reduced in size, the intensity is increased. This becomes important in some applications, such as photography.

Example 19.4   Suppose we have a concave spherical mirror with radius of curvature 20 cm and an object 4 cm in height. Find the position and magnification of the image if the object is placed at distances of 5 cm and 15 cm from the mirror.

The focal length of this mirror, from Eq. 19–6, is $f = 10$ cm. Solving the mirror equation for the image distance d_i yields

$$d_i = \frac{f d_o}{d_o - f}.$$

When the object distance is $d_o = 5$ cm, then the image distance is

$$d_i = \frac{(10 \text{ cm})(5 \text{ cm})}{5 \text{ cm} - 10 \text{ cm}} = -10 \text{ cm}.$$

The image is 10 cm behind the mirror (and is virtual and upright). The magnification is

$$m = \frac{-d_i}{d_o} = -\frac{(-10 \text{ cm})}{5 \text{ cm}} = 2.$$

Thus, the image is twice the size of the object.

When the object distance is $d_o = 15$ cm, then the image distance is

$$d_i = \frac{(10 \text{ cm})(15 \text{ cm})}{15 \text{ cm} - 10 \text{ cm}} = 30 \text{ cm}$$

and the magnification is

$$m = -\frac{(30 \text{ cm})}{15 \text{ cm}} = -2.$$

The image is real, is formed 30 cm in front of the mirror, is inverted, and is twice the size of the object.

19.3 Image Formation with Lenses

No doubt the most widely used device for image formation is the lens, in which light is focused by refraction. The lens is the basic optical element in the human eye (as well as those of all sighted animals), and also proves the most versatile for most human-made optical systems. (An exception is the telescope, for which mirrors are generally used; see Section 19.4). The first eyeglasses and telescopes, made in the fifteenth century, used lenses; Newton's invention of the reflecting telescope in the late seventeenth century was the first use of concave mirrors to form images.

In constructing lenses, there are many factors to consider: the curvature, the index of refraction of the material used, and the thickness are all independently variable. Many modern optical systems, such as in cameras and microscopes, use compound lenses (multiple lenses used together). We start our discussion at a much more basic level, however, and discuss the more sophisticated applications in Section 19.4.

The Thin Lens

If we construct a lens with spherical surfaces, simple ray-tracing shows that the light passing through it (parallel to the axis) converges to a focus at nearly a single point (Fig. 19.24). If the lens is thin (which is equivalent to saying that the radius of curvature of the surfaces is large compared to the diameter of the lens), the image is focused nearly at a single point.

Rays that pass through a thin spherical lens parallel to each other, but not parallel to the axis of the lens, will also come to a focus at the same distance from the lens as rays parallel to the axis do (Fig. 19.25). The point on the axis where focus occurs is called the **principal focus,** and the plane in which off-

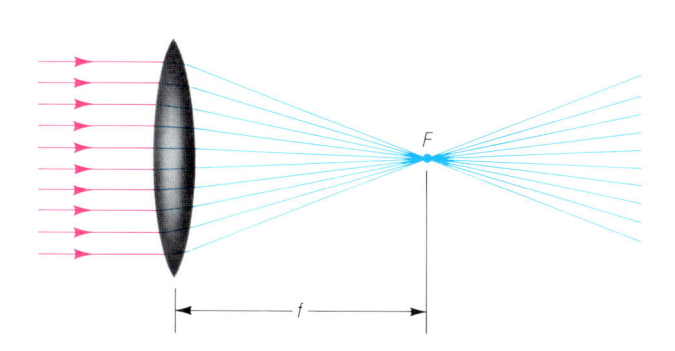

Figure 19.24

Focus by a lens. Parallel rays from a distant object, along the axis of the lens, are brought to a focus a distance f *behind a thin spherical lens.*

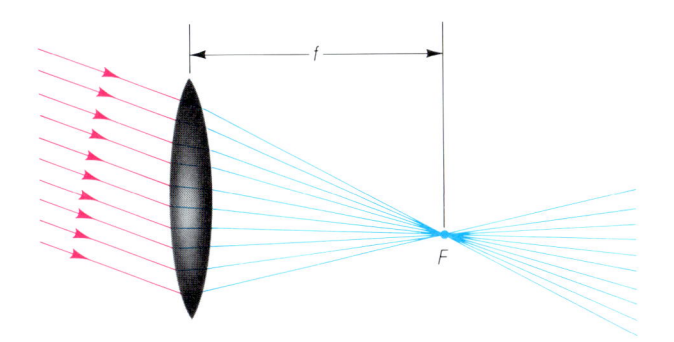

Figure 19.25

Focus of off-axis rays. Parallel rays are focused at the same distance f *behind the lens as on-axis rays.*

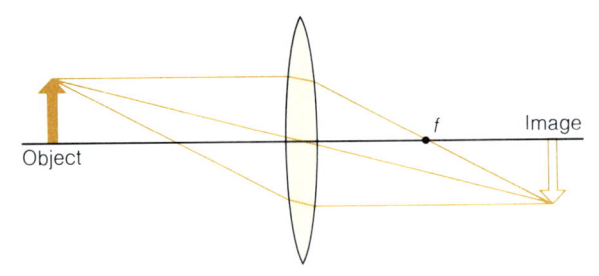

Figure 19.26

Image formation by a thin spherical lens. The image is real and inverted.

axis rays converge is called the **focal plane.** The **focal length** is the distance from the lens (technically, from its center) to the focal point. The image formed by a lens, such as those pictured, with two convex surfaces is a real image. If we trace the rays from various points on an extended object (Fig. 19.26), we see that the image is inverted as well.

We can derive an expression relating the distances and sizes of an object and its image, just as we did for a concave spherical mirror. Again, we note that certain rays are easier to trace than others. First, a ray that enters the lens parallel to its axis will pass through the focal point on the other side. Second, a ray through the center of the lens will pass through without any net refraction because the two surfaces of the lens are parallel to each other at this point (see Fig. 19.27). Third, a ray that passes through the focal point on one side of the lens before entering the lens will come out the other side parallel to the axis of the lens. Keeping these rays in mind, we find in Fig. 19.27 that triangles $FB'B$ and FOD are similar, so that

$$\frac{h_i}{h_o} = \frac{d_i - f}{f}$$

where h_i and h_o represent the object and image heights, respectively, and d_i is the image distance from the lens. Triangles $A'OA$ and $B'OB$ are also similar, so that

$$\frac{h_i}{h_o} = \frac{d_i}{d_o}$$

where d_o is the object distance from the lens. Equating these two expressions and rearranging yields

$$\frac{1}{d_o} + \frac{1}{d_i} = \frac{1}{f} \tag{19–9}$$

which is known as the **lens equation,** and is identical to Eq. 19–7, the mirror equation.

We can check the lens equation for an object at infinite distance. The rays

19.3 IMAGE FORMATION WITH LENSES

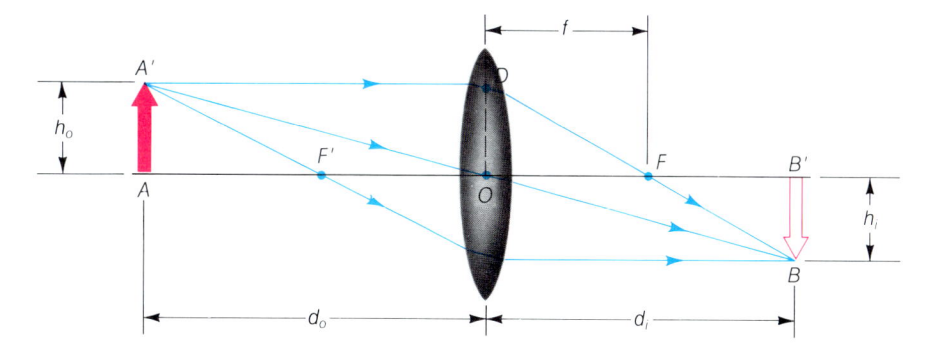

Figure 19.27

Derivation of the lens equation. The triangles FB'B and FOD are similar, as are triangles A'OA and B'OB. Use of these similarities leads to a relationship among object distance, image distance, and focal length known as the lens equation.

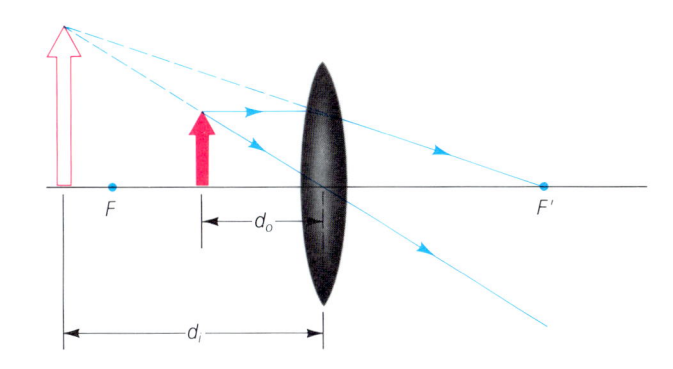

Figure 19.28

Image formation when d_o is less than the focal length. The image is virtual and upright, and located on the same side of the lens as the object. The lens equation is valid, with d_i negative.

are parallel when they enter the lens, $1/d_o = 0$, and we find that $d_i = f$, as it should.

As before, the magnification is the ratio of the image size to the object size. Similar triangles $A'OA$ and $B'OB$ show that

$$m = \frac{-h_i}{h_o} = \frac{-d_i}{d_o} \qquad (19\text{–}10)$$

just as for a spherical mirror (Eq. 19–8). The minus sign indicates that the image is inverted.

The same equations show that when an object is closer to the lens than the focal plane, a virtual, upright image is formed by the diverging rays (Fig. 19.28). The image appears to be on the same side as the object (that is, the

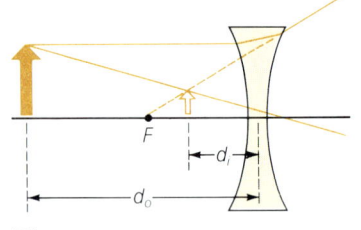

Figure 19.29

A diverging lens. A lens constructed of concave spherical surfaces causes rays to diverge. The image is virtual and upright, and formed on the same side of the lens as the object.

image is viewed by looking through the lens toward the object), so the sign of d_i is negative. Thus, the magnification is positive, as it should be for an upright image.

Identical considerations can be applied to a lens constructed of two concave spherical surfaces (Fig. 19.29). The rays diverge upon passing through the lens, and the result is a virtual, upright image on the same side as the object. The expressions for magnification and for the relationship between image distance, object distance, and focal length are the same as for a converging lens, except that the signs of d_i and f are taken as negative. The lens equation is therefore

$$\frac{1}{d_o} - \frac{1}{d_i} = \frac{-1}{f}. \tag{19-11}$$

It may be a useful exercise for the reader to derive this by tracing rays.

Example 19.5 A thin lens is spherical and convex on both sides. If its focal length is 50 mm, where and how big is the image of a 5 mm-high object if it is placed 70 mm from the lens, and if it is 30 mm from the lens?

When the object is 70 mm from the lens, it will form a real, inverted image on the other side. The lens equation can be solved for the image distance d_i.

$$d_i = \frac{d_o f}{d_o - f} = \frac{(70 \text{ mm})(50 \text{ mm})}{70 \text{ mm} - 50 \text{ mm}} = 175 \text{ mm}$$

The magnification is

$$m = \frac{-d_i}{d_o} = \frac{-175 \text{ mm}}{70 \text{ mm}} = -2.5.$$

The image is 12.5 mm high, is real, inverted, and forms 175 mm from the lens.

If the object is placed 30 mm from the lens, then we find

$$d_i = \frac{(30 \text{ mm})(50 \text{ mm})}{30 \text{ mm} - 50 \text{ mm}} = -75 \text{ mm}.$$

The negative sign indicates that the image appears on the same side of the lens as the object. The magnification is

$$m = -\frac{(-75 \text{ mm})}{30 \text{ mm}} = 2.5.$$

This time the image is upright, virtual, 12.5 mm high, and forms 75 mm from the lens, on the same side as the object.

19.3 IMAGE FORMATION WITH LENSES

Designing and Analyzing Lenses

Because lenses are very widely used in many kinds of technology, methods for designing lenses for specific applications are required. Furthermore, there are standard conventions for characterizing the properties of lenses. To explore either the design or analysis of lenses in any detail would be an unnecessary diversion from the goals of this textbook, but it is useful to include a brief description.

Most thin lenses have two spherical surfaces. To design a lens with a certain focal length and diameter, it is necessary to find the correct radii of curvature for the two surfaces. A careful analysis of the paths travelled by rays as they pass through different portions of a lens leads to the **lens-maker's equation**:

$$\frac{1}{f} = (n - 1)\left(\frac{1}{R_1} + \frac{1}{R_2}\right) \qquad (19\text{--}12)$$

where f is the focal length, n is the index of refraction, and R_1 and R_2 are the radii of curvature of the two surfaces. For many lenses R_1 and R_2 are equal, but that need not be true for all. The values of R_1 and R_2 are defined as positive for convex surfaces, and negative for concave surfaces.

In characterizing lenses, it is sometimes convenient to think in terms of wavefronts. A wavefront that is diverging as it travels away from a point has a certain radius of curvature, depending on how far it is from its point of origin (Fig. 19.30). The **curvature** is defined as the inverse of the radius. The curvature is said to be negative if the wavefront is diverging (so that the curvature is decreasing as the wavefront propagates), and positive if it is converging. Thus, a wavefront reaching the earth from the sun has curvature $-1/(1.5 \times 10^{11} \text{ m}) = -6.7 \times 10^{-12} \text{ m}^{-1}$, which is nearly zero, the value for a plane wave having no curvature.

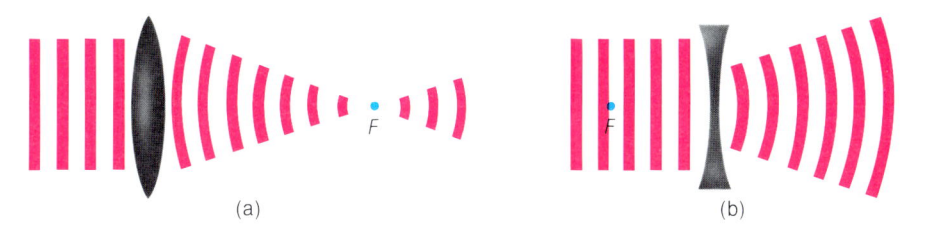

Figure 19.30

Curvature of lenses. Parallel incoming wavefronts are converted to curved wavefronts by a lens. (a) A converging lens is said to be a positive lens, because it creates positive curvature of the wavefronts. (b) A diverging lens is a negative lens.

A lens alters the curvature of a wavefront. For example, parallel light having zero curvature is converted into converging light having positive curvature by a typical lens with two convex surfaces (Fig. 19.30). Hence a converging lens is called a **positive lens**. A diverging lens, on the other hand, creates negative curvature, and is called a **negative lens**. The degree to which a lens alters the curvature depends on its focal length; the shorter the focal length, the greater its effect. We often speak therefore of the **power** of a lens, defined as the reciprocal of its focal length. If the focal length is expressed in meters, then the unit in which power is expressed is m^{-1}, which is given the special name **diopter**.

It is possible and, in some circumstances, advantageous to use curvatures of wavefronts and lenses to solve problems. If we re-examine the lens equation (Eq. 19–9), we see that the terms correspond to the curvatures of the wavefronts approaching and leaving the lens, and the power of the lens. The wavefront of light from the object approaching the lens is diverging, so the curvature $1/d$ is negative. Substituting into Eq. 19–9 and rearranging, we have

$$\text{Initial curvature} + \text{Power of lens} = \text{Final curvature}$$

Solving practical problems now becomes a matter of calculating the curvature of the incoming wavefront and the power of the lens to find the curvature of the outgoing wavefront (hence the image distance), or using another pair of known quantities to find a different unknown. This may seem not to have gained us anything over ray tracing, and actually the two are equivalent, but the method of curvatures has the advantage that it eliminates much of the confusion over the signs. It is generally easier to decide whether the power of a lens or the curvature of a wavefront is positive or negative than it is to remember all the conventions for positive and negative distances in ray tracing.

Example 19.6   Suppose a convex camera lens is to be designed with a focal length of 500 mm. The material to be used is glass with an index of refraction $n = 1.5$. What are the radii of curvature of the two surfaces, if the lens is symmetric, and what is the power of the lens in diopters?

We are given that R_1 and R_2 in the lens-maker's equation are equal, and we know that their values are positive for a convex lens, so the equation becomes

$$\frac{1}{f} = (n - 1)\frac{2}{R}$$

which can be solved for the radius of curvature.

$$R = 2f(n - 1) = 2(500 \text{ mm})(1.5 - 1) = 500 \text{ mm} = 0.5 \text{ m}$$

19.4 OPTICAL INSTRUMENTS

The power of this lens is

$$P = \frac{1}{f} = \frac{1}{0.5 \text{ m}} = 2 \text{ diopters}.$$

Example 19.7 A lens consisting of two concave surfaces has focal length -25 cm (Fig. 19.31). Locate the image of an object placed 100 cm from the lens.

This is a negative lens, since it creates a diverging beam, and its power is $1/f = 1/(-0.25 \text{ m}) = -4$ diopters. The initial curvature (the curvature of the incoming light from the object to the lens) is also negative, because the light is diverging, and is $-1/(1 \text{ m}) = -1 \text{ m}^{-1}$. Adding the initial curvature and the lens power yields the final curvature, which is -5 m^{-1}. The image distance is the reciprocal of this, or -20 cm. The minus sign indicates that the image is virtual, appearing on the same side of the lens as the object.

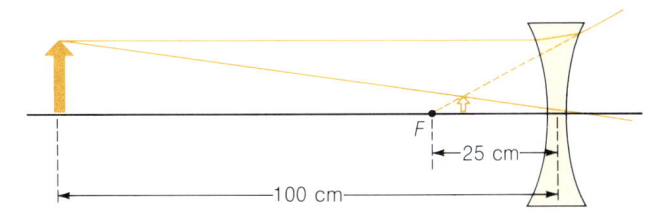

Figure 19.31

19.4 Optical Instruments

To complete our discussion of geometric optics, we now explore the workings of several kinds of optical systems, beginning with the human eye. We will see how the simple principles we have already developed are applied to each.

The Human Eye

The eye is a remarkable mechanism, beautifully designed to take best advantage of the lighting conditions that prevail on the earth. We already learned (in Chapter 18) that the eye responds to wavelengths in the range 350 nm–750 nm, which is just the wavelengths that are most intense in natural light. The reasons for this are twofold: the sun emits most strongly at a wavelength near 550 nm, in the middle of the visible band, and the earth's atmosphere happens to be transparent to this band, whereas it is opaque throughout the ultraviolet and shorter wavelengths and partially or wholly opaque over much of the

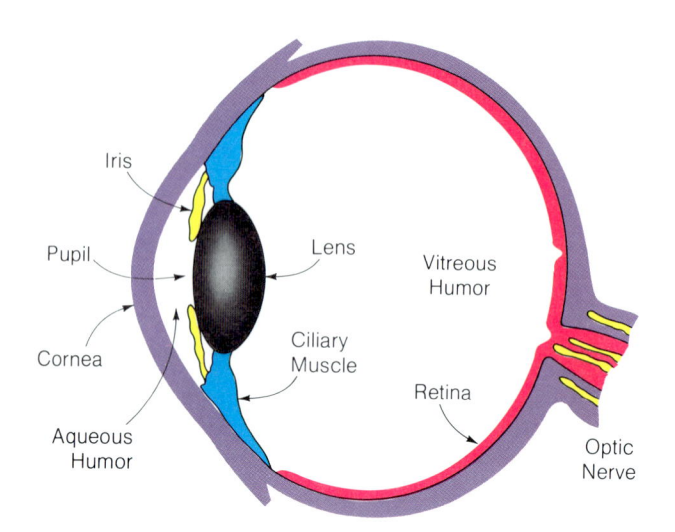

Figure 19.32

The human eye.

infrared. No doubt the human eye would have evolved for maximum sensitivity in other wavelengths if the sun or the atmosphere had different light-emitting and light-transmitting properties.

Optically, the eye consists of a simple lens and a front surface known as the **cornea** which create images on a curved focal plane known as the **retina** (Fig. 19.32). The cornea performs most of the refraction, and the lens provides variable focusing for objects at different distances. The images are inverted, but the brain, which interprets the image from the tiny electrical impulses transmitted to it from the retina, perceives an upright image. These electrical impulses are created by receptors called **rods** and **cones,** the former being sensitive to light intensity gradations, the latter being responsible for color perception.

The intensity of light that reaches the retina is governed by the pupil, a small opening into the eyeball whose size varies in response to the light level. The focus is controlled by the lens, which varies its radius of curvature. In a normal eye, good focus is possible between a **near point,** typically about 25 cm away, to a **far point** at infinity (that is, the curvature can be varied sufficiently to focus diverging rays coming from as close as 25 cm and parallel rays coming from an infinite distance). Various eye defects may prevent this full range from being achieved.

Myopia or **nearsightedness** occurs when the eye cannot focus on distant objects. It usually results because the eyeball is too long (or the corneal curvature is too strong), so that the image of a distant object is focused in front of the retina. The near and far points of a nearsighted eye might be 10 cm and 100 cm, for example. A concave diverging lens can correct for this by bringing the light to a focus farther back on the retina. **Farsightedness** or

19.4 OPTICAL INSTRUMENTS

hyperopia results when the eye cannot focus on nearby objects because the eyeball (usually) is too short. The near point might be 1 m instead of 25 cm. This effect can be corrected by a convex converging lens, which makes the focus shorter. **Astigmatism** is usually a result of improper curvature of the cornea, which causes pointlike objects to focus into lines. It can be corrected by a combination of spherical and cylindrical lenses.

We commonly think of the eye as continuously receiving images, but the nervous system acts as a shuttered camera, recording about 30 images per second. Movie and television images are renewed on the screen at a comparable rate, so the eye sees as continuous movement what is actually a discontinuous series of images.

In the next section, we will see many parallels between the design of a camera and the function of the eye, as well as some contrasts.

Example 19.8   A nearsighted person finds that his eyes cannot focus on an object that is more than 5 m away. What power of eyeglass lens is required to enable him to focus on an object at an infinite distance? If the lens is made of two equally curved surfaces, what is their radius of curvature?

We assume for the sake of simplicity that the eyeglass lens is very close to the person's eye. We are given the object distance and the distance (5 m) at which the image must be for the eye to focus well. The image must be virtual, since it must appear on the same side of the lens as the object. Solving the lens equation, (Eq. 19–11, for a diverging lens, in which f and d are both negative) we find

$$\frac{-1}{f} = \frac{1}{d_o} - \frac{1}{d_i} = 0 - \frac{1}{(-5\,\text{m})} = 0.2\ \text{m}^{-1}.$$

The power of the lens is -0.2 diopters, the minus sign indicating that it is a diverging lens, so the focal length is -5 m. Therefore the eyeglasses consist of two concave surfaces (this is why extremely nearsighted people have glasses whose lenses are very thick around the edges), or perhaps one flat and one concave. If both are concave with equal radii of curvature, the lens-maker's equation can be solved for R:

$$R = 2f(n - 1) = 2(-5\ \text{m})(1.5 - 1) = -5\ \text{m}$$

The radius of curvature is -5 m. The minus sign indicates that the lens is concave.

FOCUS ON PHYSICS

Optical Defects of the Eye

The human eye is a marvelously complex and adaptable optical instrument, but it can develop defects which limit its ability to focus properly. In addition, there is a steady change of the focusing ability that occurs with age in most people. For these reasons many people wear corrective lenses, either in the form of glasses or contact lenses.

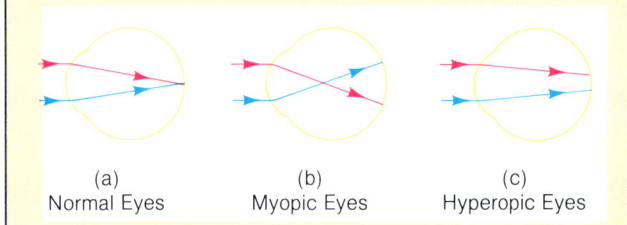

(a) Normal Eyes (b) Myopic Eyes (c) Hyperopic Eyes

Figure 19.2.1

As described in the text, the eye focuses by changing the focal length of the lens. The change in focal length is accomplished by the ciliary muscles, which change the curvature of the lens (see Fig. 19.32). In a relaxed state, the lens is thin, and objects at a large distance are properly focused. The lens is thickened to increase the curvature for focusing the light from nearby objects. Young children can usually focus on objects as close as 7 cm; that is, the near point can be as close as that. In a process called **presbyopia**, the near point moves steadily outward as a person ages and the lens becomes less flexible. The normal near point of 25 cm referred to in the text is really just a representative value for adults. The increase of the near point continues into old age for most people, and can become as large as 200 cm. This is why many

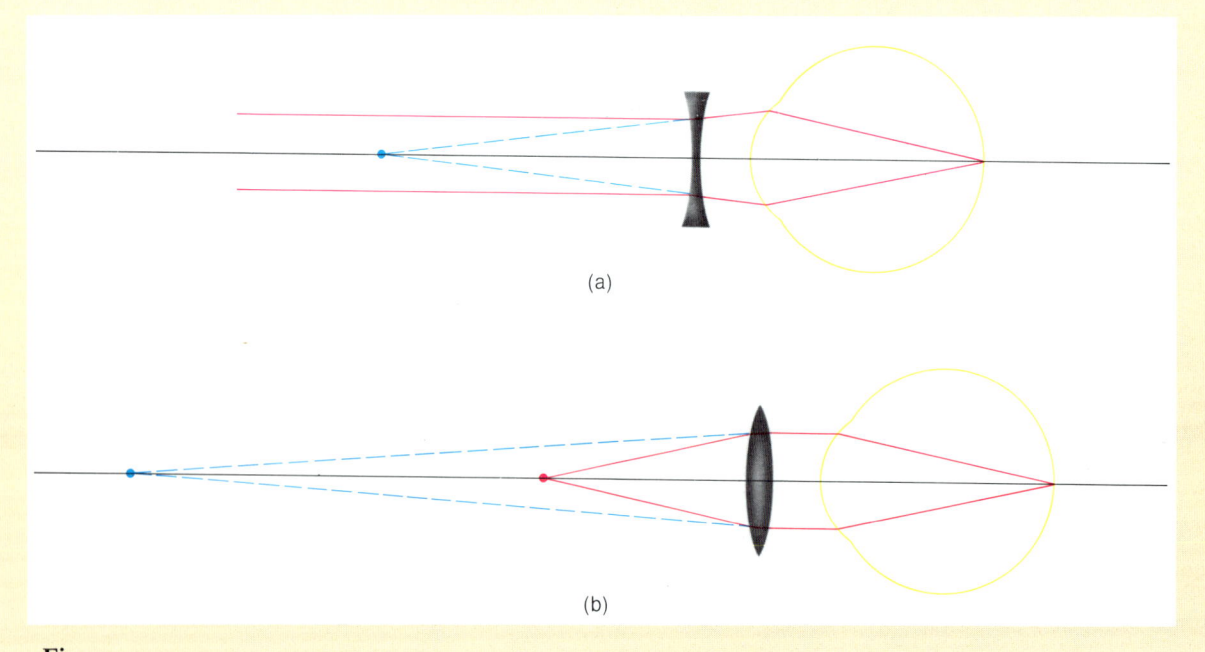

Figure 19.2.2

19.4 OPTICAL INSTRUMENTS

people become far-sighted as they age, even if their eyes are normal.

The eye can, at any age, lose its ability to focus on objects at certain distances. This can occur due to a defect in the shape of the cornea, or, more commonly, because the eyeball itself is either too long or too short, so that the focused image does not fall on the retina. If the eyeball is too long, which is far more common, the condition is called **myopia** or nearsightedness. If it is too short, the condition is **hyperopia** or farsightedness. The diagrams in Fig. 19.2.1 illustrate both of these conditions.

Treatment for either myopia or hyperopia consists of designing eyeglass lenses with the appropriate focal length so that images form properly on the retina (a new surgical treatment is now being performed in some cases to treat myopia: several small radial slits are made in the lens, allowing it to relax and thin, increasing its focal length). A diverging lens is needed to correct for myopia (Fig. 19.2.2a), and a converging lens is needed to shorten the focal length to correct for hyperopia (Fig. 19.2.2b). Example 19.8 in the text illustrates how to calculate the power of the lens needed in a typical case of myopia. For hyperopia the method is similar. Suppose, for example, that a person's near point is 200 cm from the eye, and we want to design a lens that allows this person to focus on an object 25 cm away. The image of an object that is 25 cm away must form at a distance of 200 cm. Hence the object distance d_o is 25 cm and the image distance d_i is -200 cm (remember, the image must be virtual, since it must form on the same side of the lens as the object; therefore d_i is negative). If we solve the lens equation for the focal length f, we find that $f = 28.6$ cm $= 0.286$ m. The power of this converging lens is $P = 1/f = +3.5$ diopters.

Strictly speaking, the lens power found here applies to contact lenses, because we assumed that the lens was at the eye, whereas glasses are placed a small distance (about 2 cm) in front of the eye. For glasses, the calculation would be the same, except that the image distance assumed allows for the small displacement of the lenses from the eye.

For hyperopia correction, where the image distance is large (200 cm in the example above), the correction is negligible, but for myopia, where the image distance is much smaller, it is important to allow for the displacement. This is done simply by subtracting the displacement (2 cm) from the image distance that is desired. Thus, if a myopic eye has a far point (greatest object distance for which the eye focuses) of 20 cm, the lens for glasses would be designed to form an image of a distant object 18 cm in front of the lens, which is 20 cm from the eye.

For eyes that cannot properly accomodate different object distances, it is sometimes necessary to prescribe **bifocals** or even **trifocals**. These are glasses whose lenses consist of sections with differing powers. This is most often needed by hyperoptic people, who need little or no correction for distant viewing, but must have a correction for close-up work such as reading. Bifocals are sometimes prescribed for myopia as well, particularly if the correction to extend the far point results in a near point that is too far from the eye to allow focusing on nearby objects such as printed matter.

Sometimes the eye has a different form of defect called **astigmatism.** This can occur even in an eye that has normal vision, although it is more common in eyes that are myopic. Astigmatism is the result of an asymmetry in the cornea or the lens of the eye, so that images are not focused properly in all dimensions. A point object may have an image that is elongated, for example. To correct for astigmatism, an asymmetric lens must be used to offset the asymmetry of the image. If an eye is astigmatic, the correcting lens must have a radius of curvature that varies from place to place. Because the lens is not symmetric, it must be placed in front of the eye with a specific orientation. For glasses, this is easy to accomplish, since the lenses are fixed in place by the frames. For contact lenses, however, it is more difficult. For hard contact lenses, it is possible to weight one side of the lens so that the weighted side is always down, and the lens maintains constant orientation. For soft contact lenses, no successful technique has yet been developed for correcting astigmatism (this is due not

(continued)

only to the difficulty in maintaining a fixed orientation, but also to the difficulty of shaping the soft lens material with sufficient precision).

The basic optical principles of correcting eyesight have been known for a long time, but many advances are being made in the implementation of these principles. In addition to the surgical thinning of lenses to correct myopia, techniques are also under development for the surgical implantation of artificial lenses. In cases of blindness, techniques such as corneal transplants and removal of cataracts have been in use for some time. The day may come when everybody can have unimpaired vision without contact lenses or glasses.

The Camera

A camera is very much like the human eye. There are an enclosed volume, a lens, and a focal plane where film records the image (Fig. 19.33). The volume must be enclosed so that only focused light reaches the film; similarly, the eye would become hopelessly confused if unfocused light reached the retina from all directions and distances.

The image that reaches the focal plane of a camera is real and inverted. Furthermore, its magnification is usually much less than 1 (except when very close-up photos of small objects are made), because the image must be small enough to fit onto conveniently-sized film, and the distance from the lens must be rather small.

Unlike the eye, the camera is not equipped with a sophisticated nervous system and pliable materials, so that some of the adjustments needed to record sharp, well-illuminated images must be made by other techniques. Focus in a

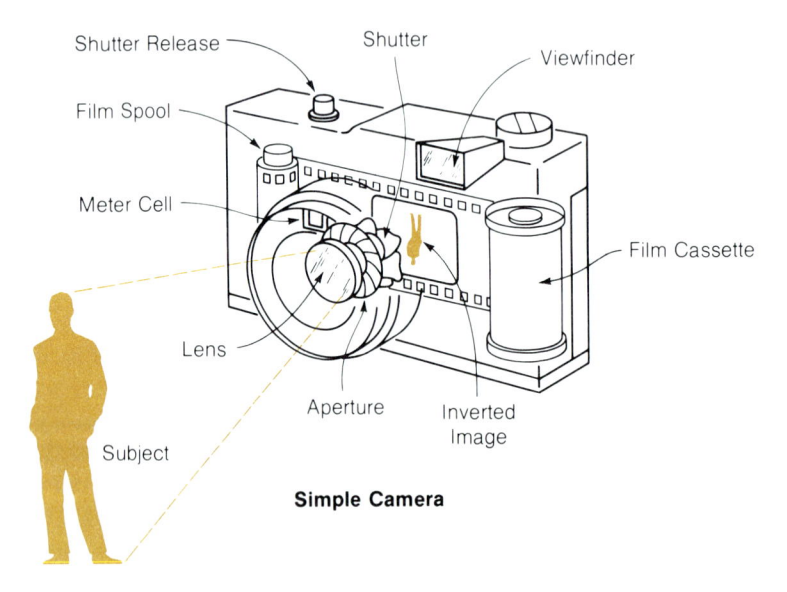

Figure 19.33

A simple camera.

19.4 OPTICAL INSTRUMENTS

Figure 19.34

A single-lens reflex camera.

camera, for example, is controlled by moving the lens so that the image position coincides with the location of the film, whereas the eye adjusts the focal length of the lens by changing its radius of curvature. In a *rangefinder* camera, the lens is moved according to the estimated object distance (hence these cameras often have focus adjustments on which a scale of feet or meters is displayed). It is possible to design lenses to achieve a large range in depth of field (the distance over which the focus is good) so that small errors in estimating the object distance can be tolerated. In a *single-lens reflex* (SLR) camera, the focus is accomplished by directly viewing the focal plane, so that the camera operator can actually see when proper focus is achieved. The converging light behind the lens is diverted to a viewfinder, through the use of a prism which utilizes total internal reflection (Fig. 19.34). The specially designed prism re-inverts the inverted image, so that it appears upright. The optical path from the lens through the prism to the viewfinder is carefully designed to be equal in length to the path from the lens to the film, so that when the light is diverted to the film during exposure, the light is focused on the film.

The light intensity that reaches the film is governed by the aperture size, which is variable through a mechanical setting on the camera (or may be automatically controlled in some cameras). Normally the aperture setting is expressed in terms of the **focal ratio, f-ratio,** or **f-stop,** which in any case is the ratio of the focal length to the diameter of the aperture. For example, a typical camera lens may have a focal length of 50 mm. An f-stop of f/2 would mean that the aperture was set for a diameter of 25 mm. The smaller the f-stop, the larger the aperture, and the "faster" the camera is, that is, it is better able to obtain sufficient light under dim conditions. A typical camera may have f-stop settings ranging from f/1.4 to f/32. The intermediate values are usually designed to vary the light intensity by a factor of two for each f-stop, hence adjacent f-stops are in the ratio 1.41 : 1. (The square root of 2 is about 1.41. Recall that the intensity of light entering the camera is proportional to the *area* of the aperture, which in turn varies as the square of its diameter.)

Another important element of a camera is the **shutter,** which governs the total amount of light reaching the film by controlling the length of time light is allowed to enter the camera. A shutter may consist of a set of leaves that are operated by springs (Fig. 19.35a), or it may consist of a pair of blinds that move in front of the film, so that the size of the space between them controls the amount of time that light is allowed to reach the film (Fig. 19.35b). It may seem that we have some redundancy, in that the intensity of light reaching the film is controlled by the aperture and by the shutter, but both are needed. Each kind of film responds well to a limited range of light intensities, and keeping the intensity within this range cannot always be achieved by adjusting the shutter speed alone. Therefore, controlling the aperture size is also needed to ensure that the light level on the film will be within its acceptable range. Furthermore, the depth of field is governed by the aperture size, because for large apertures off-axis rays are allowed to enter the camera, and for these rays the focal length varies with object distance more than for rays entering near the lens axis. Hence for artistic or technical reasons, it may be desirable to set a certain aperture size, and then adjust the shutter speed accordingly. Many semi-automatic cameras allow the user to set the aperture, and then

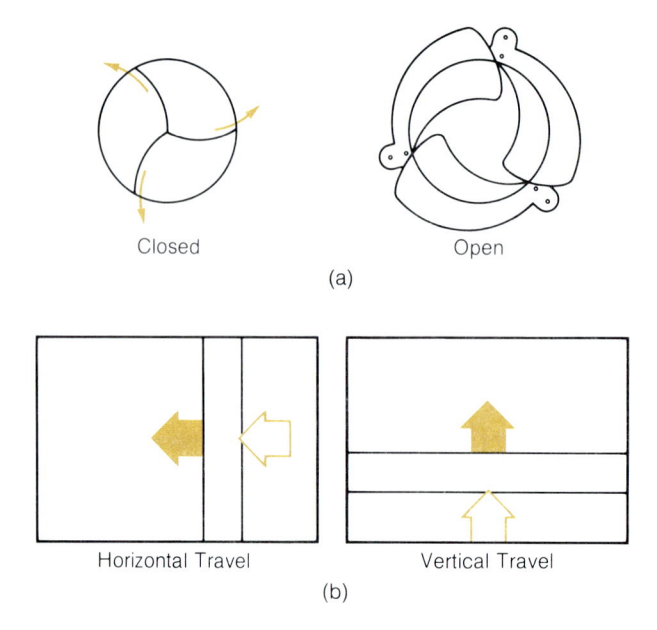

Figure 19.35

Typical shutter mechanisms. (a) Leaf-type shutter, (b) Movable blind shutter.

automatically set the shutter speed appropriate for the film being used, or allow the user to set the exposure time and then automatically adjust the aperture so that the proper intensity will reach the film.

Another variable available with single-lens reflex cameras is the capability of changing lenses. A telephoto lens, for example, has a long focal length and therefore a high f-ratio, so that its field of view is relatively small (because the angle of convergence of the rays reaching the focal plane is small). This creates relatively large images at the focal plane, because less demagnification occurs than in the case where the focal length is short (to see this, consider the effect in Eqs. 19–9 and 19–10 of increasing the value of f). Interchangeable lenses are normally available only for single-lens reflex cameras because it is still possible to focus them by directly viewing the focal plane in the viewfinder, whereas in other cameras it is difficult or impossible to achieve focus for lenses other than the standard built-in one.

Finally, another variable in cameras is the film. Film is coated with an **emulsion** which consists of chemicals that undergo changes when light strikes them. These chemical changes are then converted into intensity and color gradations when the film is developed. Different chemicals can be treated differently and applied in various densities to achieve different degrees of response to light. A very "fast" film, for example can form a good image under dim conditions or with a very short exposure time because it is highly sensitive to low light levels. Practical considerations often dictate the type of film that is to be used, so that typically the photographer has only to choose the lens and then adjust the f-stop and shutter speed.

19.4 OPTICAL INSTRUMENTS

Example 19.9 Suppose a camera has a 50 mm lens. How far must the lens be from the film to obtain sharp focus on an object that is very far away, and for an object that is only 2 m away?

The answer to the first part is straightforward, because we know that the point where parallel rays are focused defines the focal length of a lens: the lens must be 50 mm from the film.

For the second part, we use the lens equation. The object distance d_o is 2 m, and the focal length is $f = 0.05$ m. Solving for the image distance d_i yields

$$d_i = \frac{f d_o}{d_o - f} = \frac{(0.05 \text{ m})(2 \text{ m})}{2 \text{ m} - 0.05 \text{ m}} = 0.0513 \text{ m}.$$

Hence, even for a relatively nearby object, the lens needs to be moved only 1.3 mm to focus.

The Magnifying Glass

The key factor determining the amount of detail that can be perceived by the human eye is the angular size (that is, the angle subtended; see Fig. 19.4) of the image formed on the retina. Up to a point, the angular size of the image can be increased by moving the object very close to the eye, but the normal eye cannot focus on objects closer than about 25 cm. The purpose of a magnifying glass is to allow the eye to focus on a very nearby object, taking advantage of the large angle subtended by the object when it is very close.

We have already seen that when an object is inside the focal length of a thin convex lens, the image formed is virtual and upright, and on the same side of the lens as the object (Fig. 19.28). To view the image requires looking through the lens toward the object. The angle subtended by the image is the same as the angle subtended by the object, but the image distance is greater than the object distance. For the eye to focus comfortably, the image distance must be at least 25 cm. If the object is exactly at the focal point of the lens, then the lens equation tells us that the image distance is infinity. For most people, this is the most comfortable distance at which to focus, because it corresponds to relaxation of the muscles that govern the eye's focal length. If the eye is focused at some closer point, then proper focus is achieved when the object is just inside the focal point of the magnifying glass.

Because it is the angle subtended by the image that governs the amount of detail that can be seen, the term "magnification" is commonly thought to mean enhancement of the subtended angle. This is in contrast to the magnification we have discussed so far in this chapter, which is the ratio of actual image size to object size. We can now define the **angular magnification** as the ratio of the angle subtended by the image as seen through the lens to the angle subtended by the object if it were at the image distance (Fig. 19.36):

$$M = \frac{\theta'}{\theta} \qquad (19\text{-}13)$$

where θ' is the angular size of the image seen through the lens, and θ is the angular size of the image formed in the eye without the lens.

The **magnifying power** of the lens is the angular magnification when the image is as close to the eye as the eye can accomodate. We can find an expression for the magnifying power by writing the lens equation (Eq. 19–9) for an image distance of -25 cm, and solving for the object distance:

$$d_o = \frac{d_i f}{(d_i - f)} = \frac{(25 \text{ cm})f}{(f + 25 \text{ cm})}$$

The angle subtended by the image seen through the lens is the angle whose tangent is h/d_o, where h is the object height and d_o the object distance. If the angle is small, which is often a reasonable approximation, then the angle itself (in radians) is approximately equal to its tangent, so we have

$$\theta' \approx \frac{h}{d_o} = \frac{h(f + 25 \text{ cm})}{(25 \text{ cm})f} \text{ rad.}$$

The angle subtended by the object when it is at the image distance but seen without the lens is

$$\theta = \frac{h}{25 \text{ cm}} \text{ rad.}$$

Substituting these expressions for θ' and θ into Eq. 19–13 and cancelling yields

$$M = \frac{\theta'}{\theta} = 1 + \frac{25 \text{ cm}}{f}. \qquad (19\text{-}14)$$

This is the angular magnification when the eye is focused as close as possible, a distance we have been taking to be 25 cm (there are minor variations from person to person). This shows that the magnifying power is enhanced by choosing a lens with a short focal length.

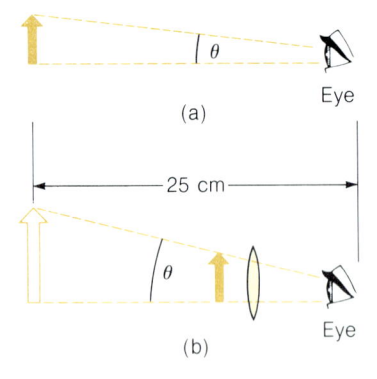

Figure 19.36

The magnifying glass. The eye can focus comfortably no closer than about 25 cm. (a) The angular size θ of an object viewed from 25 cm distance. (b) The object is much closer to the eye, and a magnifying glass forms an enlarged image at a distance of 25 cm. The angular size θ is larger.

Example 19.10

What is the magnifying power of a lens whose focal length is 5 cm? What is its angular magnification when the eye is relaxed, so that the image distance is infinity?

We simply substitute $f = 5$ cm into Eq. 19–14 to answer the first question.

$$M = 1 + \frac{25 \text{ cm}}{5 \text{ cm}} = 6.$$

The maximum angular magnification possible with this lens is 6.

When the eye is relaxed and the image distance is infinity, then we have $\theta' = h/f$ (recall that the angular size of the image is equal to the angular size of the object at its position, which is at the focal point when the image is at infinity). The angular magnification is

$$M = \frac{\theta'}{\theta} = \frac{h/f}{h/25 \text{ cm}} = \frac{25 \text{ cm}}{f} = \frac{25 \text{ cm}}{5 \text{ cm}} = 5.$$

The magnification is only slightly less than the magnifying power, so not much is lost by allowing the eye to relax and focus at infinity.

The Telescope

Telescopes form images of very distant objects. They may be designed for viewing objects on the earth's surface, or they may be designed for astronomical observations. The major design criteria are different for each purpose.

A telescope designed for terrestrial viewing usually requires substantial magnification because usually the purpose is to allow the viewer to make out details of faraway objects. By contrast, an astronomical telescope is designed to have the largest possible aperture because most astronomical light sources are quite dim. Stars are essentially point sources of light, so that no structure can be seen under any possible magnification; therefore, magnification is not normally important in astronomical telescopes (except for some observations of relatively nearby objects, such as the Sun and planets).

Telescopes can be **refractors,** in which lenses are used to form an image, or **reflectors,** in which mirrors are used. The simplest form of refractor is sketched in Fig. 19.37. An **objective lens** brings light rays to a focus. The incoming rays are usually essentially parallel, because generally only distant objects are viewed, so the focus occurs at the focal point of the objective lens. The image formed there is real (and inverted), so it can be the object for the

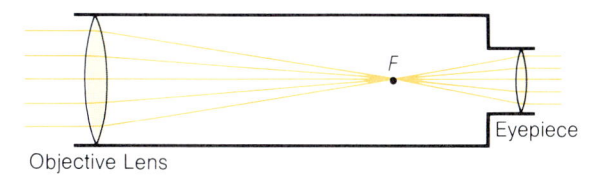

Figure 19.37

The refracting telescope. Parallel rays from a distant object are focused at the focal point F of the objective lens. The real and inverted image is then magnified by the eyepiece. The viewer sees an enlarged, virtual image.

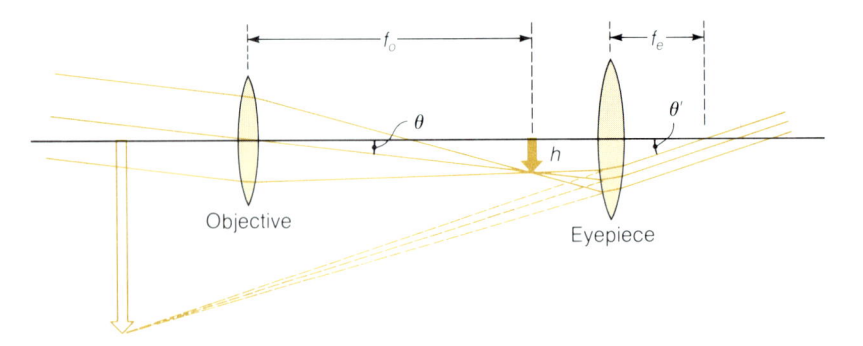

Figure 19.38

Magnification in a refracting telescope.

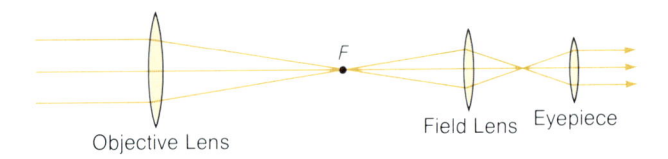

Figure 19.39

Reinverting the image. The addition of a field lens to a refracting telescope reinverts the image. The viewer sees an erect enlargement of the distant object.

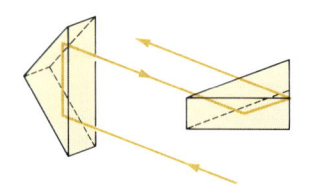

Figure 19.40

Image inversion by prisms. Internal reflection in a 45° prism inverts an image in one plane. In binoculars, two prisms, oriented at right angles to each other, allow the inverted image from the objective lens to be reinverted in both planes, in a compact space.

eyepiece, a second lens whose role is to magnify the image. The eyepiece acts just as a magnifying glass does, providing angular magnification. The image seen by the eye is inverted and virtual.

The magnification is simple to calculate (see Fig. 19.38): the angle subtended by the image that would be seen by the unaided eye is simply the angle subtended by the image at the objective, so that $\theta = h/f_o$ (in radians), where h is the image diameter, f_o is the focal length of the objective, and the small-angle approximation that $\tan \theta \approx \theta$ was assumed. The angle subtended by the magnified image is h/f_e, where f_e is the focal length of the eyepiece lens. Thus the angular magnification of the telescope is

$$M = \frac{\theta'}{\theta} = \frac{-f_o}{f_e} \tag{19-15}$$

where the minus sign indicates that the image is inverted.

In astronomical applications, the fact that the image is inverted is usually of little concern, but for terrestrial uses of telescopes, re-inversion of the image is desirable. This is accomplished by an additional optical element, either another lens (Fig. 19.39), or, as in most binoculars, prisms (Fig. 19.40). A third method, actually used by Galileo, is to replace the eyepiece with a di-

19.4 OPTICAL INSTRUMENTS

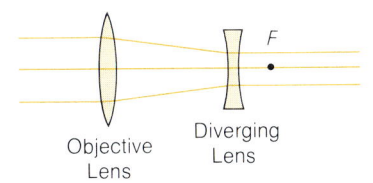

Figure 19.41

The Galilean telescope. The diverging lens is placed before the focal point F of the objective lens, enlarging the image and leaving it upright. The image is virtual.

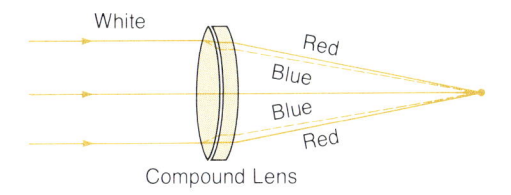

Figure 19.42

Correction for chromatic aberration. A lens made of carefully shaped layers of glass with differing indices of refraction can compensate for the color separation that would occur in a lens made of a single material.

verging lens, which intercepts the converging light from the objective before it reaches a focus (Fig. 19.41). The diverging lens creates a virtual, upright image.

As noted earlier, most astronomical telescopes are reflectors, primarily because mirrors can be made much larger than lenses. A mirror can be supported from behind, whereas a lens can only be supported at its edges, and may therefore be deformed by gravitational forces. Any such deformation of the lens causes poorly-focused images. Other advantages of mirrors are that only one optical surface need be formed accurately, and that **chromatic aberration** is avoided. Recall that the refractive index of most materials varies slightly with wavelength of light; this is, after all, the reason that a prism creates a rainbow of colors. A lens therefore has slightly different focal lengths for different wavelengths, so that the different colors come to a focus at slightly different distances from the lens. This can be corrected by the careful design of layered lenses, in which glasses of differing indices of refraction are shaped so that the chromatic aberrations offset one another (Fig. 19.42), but mirrors simply do not have the problem in the first place.

Some typical focal arrangements for reflecting telescopes are shown in Fig. 19.43. In each, the large light-collecting mirror is called the **primary mirror.** This is a concave mirror (usually parabolic) that brings incoming light to a focal point in front of the mirror. It is usually not practical to record the image at this position, because to do so requires placing a person or a camera inside the telescope (some telescopes are actually so large, however, that it is possible to place a person at this position, called the **prime focus,** for astronomical observations; see Fig. 19.44). Therefore, a **secondary mirror** is needed to deflect the image outside the telescope. This secondary mirror must block some of the incoming light on its way to the primary mirror, but usually the fraction that is blocked is very small.

For most practical applications, astronomical telescopes must produce real images, so that the light can be recorded. Therefore, the secondary mirror must form a real image, which generally means that the beam from the secondary must be converging. The secondary mirror always intercepts the converging beam from the primary before it reaches a focus, and reflects this converging beam out of the telescope. If a flat secondary mirror is used, then

650   CHAPTER 19   GEOMETRIC OPTICS

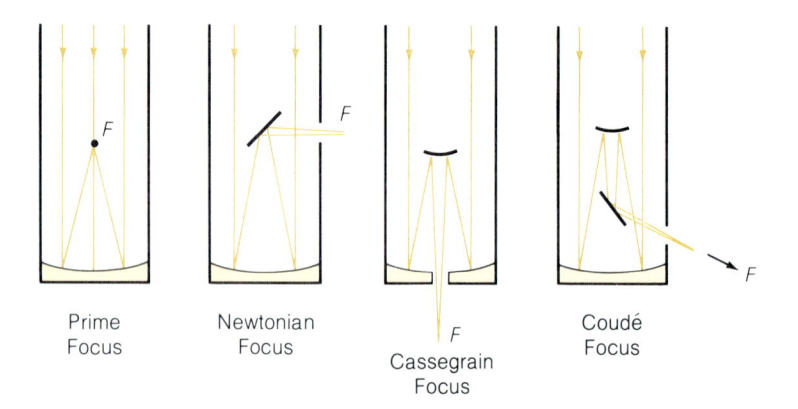

Prime Focus   Newtonian Focus   Cassegrain Focus   Coudé Focus

Figure 19.43

Reflecting telescopes. These sketches illustrate several commonly-used focus arrangements for reflecting telescopes.

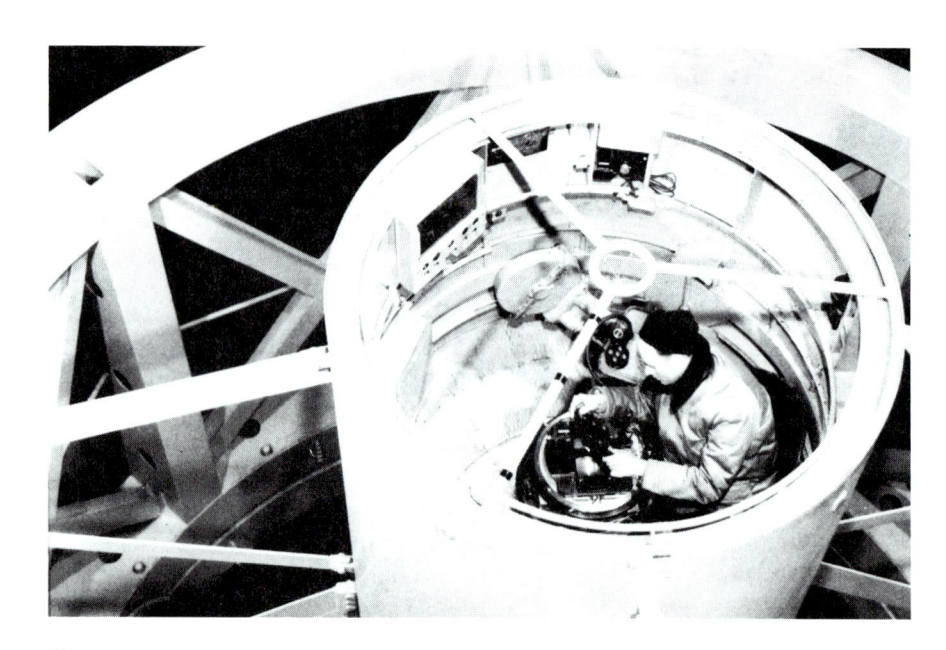

Figure 19.44

An observer at the prime focus of the Palomar 200-inch telescope.

the focal plane is simply displaced, with no effect on the net focal length. If a convex secondary mirror is used, the angle of convergence is reduced and the focal length is extended. Then the image may be formed just behind the primary mirror (through a hole in it), or may be diverted through a series of flat mirrors to a remote location.

19.4 OPTICAL INSTRUMENTS

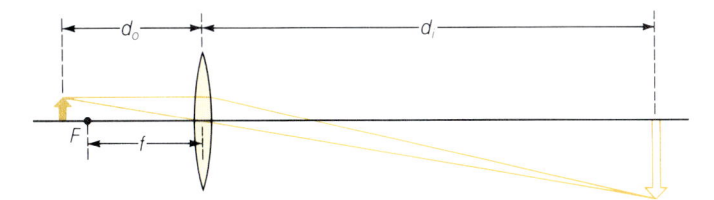

Figure 19.45

High magnification by a single lens. When the object is placed just beyond the focal point of a thin lens, the image distance d_i is large, hence the magnification is high.

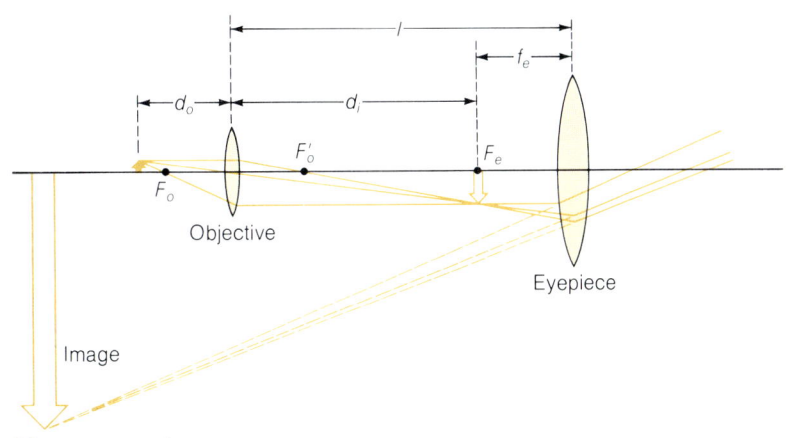

Figure 19.46

The compound microscope. An eyepiece magnifies the already highly magnified real inverted image formed by the objective.

The Microscope

A typical microscope, called a **compound microscope,** is very similar in general design to a simple refracting telescope. The principal difference is that the microscope is designed for maximum magnification of objects located very close to the objective lens, rather than for magnification of faraway objects.

We see from Eq. 19–9 and Eq. 19–10 that the magnification is very large when an object is placed just beyond the focal point of a thin double convex lens (Fig. 19.45). That is, when the object distance d_o is only slightly greater than the focal length f_o, the image distance d_i is large. This leads to a large ratio d_i/d_o and hence a large magnification. A compound microscope consists of an objective lens and an eyepiece (Fig. 19.46). The object to be viewed is placed just beyond the focal point of the objective. The image produced is real, inverted, and highly magnified. Further magnification is created by the eyepiece, which produces a virtual image (which remains inverted). Ideally, the image formed by the objective lies at the focal point of the eyepiece, so that parallel rays emerge, and the eye can relax and focus at infinity to view the image.

The total magnification is the product of the magnifications created by the objective and by the eyepiece. If the distance of the object from the objective

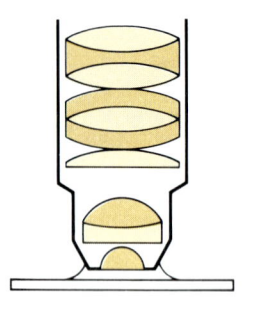

Figure 19.47

A typical microscope lens. The thin-lens approximation is not valid, and multiple lenses are used to offset aberrations introduced by thick lenses.

is d_o and the image formed by the objective lies at the focal point of the eyepiece, then the image distance is $d_i = l - f_e$, where f_e is the focal length of the eyepiece, and l is the total distance between the objective and the eyepiece. The magnification produced by the objective is therefore

$$M_o = \frac{-d_i}{d_o} = \frac{-(l - f_e)}{d_o}.$$

The eyepiece is a simple magnifier, whose magnification when the object is placed at the focal point is

$$M_e = \frac{25 \text{ cm}}{f_e}$$

where 25 cm is taken to be the closest point at which the eye can focus comfortably (this expression was derived in Example 19.10). The total magnification of the microscope is therefore

$$M = M_o M_e = \frac{-(25 \text{ cm})(l - f_e)}{f_e d_o}. \tag{19-16}$$

The magnification is enhanced when the focal lengths of the lenses are very small. Lenses with small focal lengths have very small radii of curvature, but they have serious aberrations because they are not thin lenses. Thus, microscope lenses usually consist of several layers, each designed to rectify aberrations while creating an overall short focal length (Fig. 19.47).

Example 19.11   Suppose a microscope has $f_o = 1$ cm and $f_e = 2$ cm. If the image is clearly focused when the object is 1.1 cm from the objective, what is the length of the microscope tube (that is, the separation of the two lenses) and what is the overall magnifying power?

To answer the first part, we must find the distance from the objective to the image it forms. Since we know that $d_i = l - f_e$, then $l = d_i + f_e$. From the thin lens equation, we have

$$d_i = \frac{f_o d_o}{d_o - f_o} = \frac{(1 \text{ cm})(1.1 \text{ cm})}{1.1 \text{ cm} - 1 \text{ cm}} = 11 \text{ cm}.$$

Hence the length of the tube is

$$l = d_i + f_e = 11 \text{ cm} + 2 \text{ cm} = 13 \text{ cm}.$$

The magnification is

$$M = \frac{-(25 \text{ cm})(l - f_e)}{f_e d_o} = \frac{-(25 \text{ cm})(11 \text{ cm})}{(2 \text{ cm})(1.1 \text{ cm})} = -125.$$

Perspective

A variety of specialized microscopes are used to provide both higher magnifications and higher contrast than are possible with a compound microscope, since there is often little difference in color or intensity of light from a specimen. Some of these specialized microscopes use light wave interference, and others such as electron microscopes, use particle beams. The interference of light waves is discussed in the next chapter, along with diffraction, which has important impact on the images formed with microscopes and telescopes. Electron microscopes are discussed in Chapter 22.

Perspective

In this chapter we have explored the behavior of light from a very simplistic point of view, taking into account only how rays are bent as they encounter reflecting or refracting surfaces. Even so, we have been able to understand the principles of many kinds of optical systems, and can now look a camera, microscope, or telescope in the eye with a sense of comprehension.

While the geometric optics studied in this chapter are derived from the wave properties of light, there are still more wave properties that we have not yet applied to light. These include diffraction and interference, which are discussed in the next chapter.

SUMMARY TABLE OF FORMULAS

Index of refraction:
$$n = \frac{c}{c_n} \tag{19-1}$$

Wavelength:
$$\lambda = \frac{c}{\nu} \tag{19-2}$$

Wavelength in a medium with index of refraction n:
$$\lambda_n = \frac{c_n}{\nu} = \frac{\lambda}{n} \tag{19-3}$$

Snell's law for refraction:
$$n_1 \sin\theta_1 = n_2 \sin\theta_2 \tag{19-4}$$

Law of reflection:
$$\theta_1 = \theta_2 \tag{19-5}$$

Focal length of a spherical concave mirror:
$$f = \frac{r}{2} \tag{19-6}$$

Mirror equation:
$$\frac{1}{d_o} + \frac{1}{d_i} = \frac{1}{f} \tag{19-7}$$

Magnification by a mirror:
$$m = -\frac{h_i}{h_o} = -\frac{d_i}{d_o} \tag{19-8}$$

Lens equation:
$$\frac{1}{d_o} + \frac{1}{d_i} = \frac{1}{f} \tag{19-9}$$

Magnification by a lens:
$$m = -\frac{h_i}{h_o} = -\frac{d_i}{d_o} \tag{19-10}$$

Lens equation for doubly convex lens:
$$\frac{1}{d_o} - \frac{1}{d_i} = -\frac{1}{f} \tag{19-11}$$

Lens-maker's equation:
$$\frac{1}{f} = (n-1)\left(\frac{1}{R_1} + \frac{1}{R_2}\right) \tag{19-12}$$

Angular magnification:
$$M = \frac{\theta'}{\theta} \quad (19\text{–}13)$$

Angular magnification for a magnifying glass:
$$M = 1 + \frac{(25 \text{ cm})}{f} \quad (19\text{–}14)$$

Angular magnification for a telescope:
$$M = -\frac{f_o}{f_e} \quad (19\text{–}15)$$

Total magnification of a microscope:
$$M = M_o M_e = \frac{-(25 \text{ cm})(l - f_e)}{f_e d_o} \quad (19\text{–}16)$$

THOUGHT QUESTIONS

1. What are the units of indices of refraction n? Are there any limits to n? Can n be less than 1.0?

2. How would you experimentally measure the index of refraction?

3. Is there an equivalent of the index of refraction for sound waves?

4. Would a ray model work for sound waves? Explain.

5. Why does the frequency of a light wave not change as it passes through a boundary between media of different indices of refraction?

6. Example 19.1 predicts that the color of an object will change if it is viewed under water. Does this really occur? Explain.

7. What physical characteristics would a substance with high index of refraction possess? Can you list several practical uses for such substances?

8. Give everyday examples in nature of the phenomena of reflection and refraction.

9. What effects will refraction have on the sunset or sunrise? Discuss the shape of the sun and the time of sunset or sunrise.

10. If you hold a sign up to a mirror, why are the letters reversed left to right, but not upside down?

11. How would you distinguish between a real and a virtual image in an optical system?

12. Can you think of an optical instrument that uses a virtual image? Explain.

13. Can you think of a practical use of optical fibers not mentioned in the text?

14. Why do the length of a person submerged in water and the depth of a swimming pool appear shorter than they really are? Sketch a ray diagram.

15. Why is the index of refraction of a gem, say a diamond, related to its brilliance?

16. Why is it easier to look out of a lighted room through a glass window during the day than it is during the night? What is responsible for the one, and often two, reflected images seen at night?

17. Can you think of an optical device other than a telescope that uses a mirror or mirrors?

18. How would the lens equation be altered if you used a "thick" lens—one whose surface curvature radius is not large compared to the lens diameter?

19. Why is the lens equation independent of the index of refraction of the lens?

20. How would you experimentally measure the radius of curvature for a thin lens?

21. What changes, if any, would be made to the lens-maker's equation if the lens were immersed in a medium different from air?

22. What is the difference between power and magnification of a lens?

23. Are there animals or insects whose wavelength band of vision lies outside that of the human eye? Explain.

24. Compare and contrast the human eye with a camera.

25. Compare and contrast a microscope and a telescope.

26. Why does the image distance in astronomical reflecting telescopes always equal the focal length of the mirror?

27. Why does the image intensity decrease with larger magnification and increase with smaller magnification in a lens?

28. Is the image of the moon viewed through a refracting telescope inverted?

29. Explain the phenomena of nearsightedness, farsightedness, and astigmatism in terms of lenses and focal lengths. Why are bifocals often needed as one gets older?

30. How does a zoom lens work?

PROBLEMS

Section 19.1 Reflection and Refraction

- 1 A light ray moves through a substance at 2.4×10^8 m/s. What is the index of refraction?
- 2 Find the speed of light in a medium with index of refraction 1.5. What would be the wavelength of a 400 nm light wave in this medium?
- 3 A beam of light has a wavelength 600 nm in a vacuum and 300 nm in a medium. Find the index of refraction and the speed of light in this medium.
- 4 A beam of light travels from water into crown glass. What must the incident angle be if the refracted angle is 45°?
- 5 What index of refraction is required to change the apparent color of an object from red (675 nm) in air to blue (475 nm) in the medium?
- •• 6 A light beam travels through a window pane 2 mm thick, with index of refraction 1.6. How much is the light beam displaced vertically from its original path if its angle of incidence is 45°?
- •• 7 Calculate the displacement of the beam in Problem 6 if the angle of incidence had been 30°. Generalize the effect to an arbitrary angle of incidence.
- ••• 8 A beam of light is displaced 1 mm vertically as it passes through a 1 cm thick slab. If the angle of incidence was 45°, what is the index of refraction of the slab?
- 9 Calculate the critical angle of a strong alcoholic beverage (ethyl alcohol).
- 10 A diamond is immersed in a medium and has a critical angle of 45°. What is the index of refraction of the medium?
- 11 What index of refraction is required for a critical angle of 30° into air?
- 12 An astronomer wishes to reflect the sun's image down a vertical shaft into a laboratory for analysis. If the sun is 30° from the vertical, at what angle should the astronomer tilt a mirror at the top of the shaft?
- •• 13 A beam of light is broken into a reflected portion and a refracted portion at an interface from air to crown glass. At what angle of incidence will the reflected angle be twice the refracted angle?
- •• 14 If the index of refraction of air is 1.000277 for visible light, but varies by 1 part in 10^6 for two different colors, how far apart will waves of these colors be after the beam has passed through 30 km of atmosphere?
- •• 15 At an interface between air and water, what angle of incidence will produce an angle of refraction equal to 0.7 that amount?
- •• 16 Two parallel rays encounter a prism with apex angle ϕ (Fig. 19.48). What is the angle of divergence of the rays after reflection?

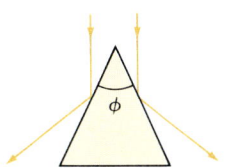

Figure 19.48

- ••• 17 A mirror is placed symmetrically at the corner of a right-angled hall (Fig. 19.49). What portion of the near side of the hall around the corner is not visible to a person standing in the middle of the hall?

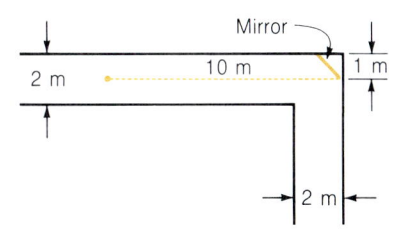

Figure 19.49

- ••• 18 A light ray encounters a 1 cm glass slab ($n = 1.52$) at an angle of incidence 30° and refracts (Fig. 19.50). At every interface inside the slab, a fraction 0.20 is

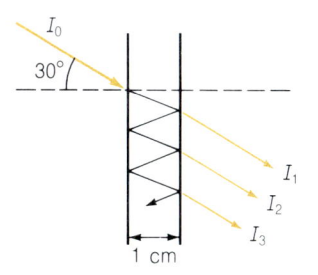

Figure 19.50

656 CHAPTER 19 GEOMETRIC OPTICS

reflected and a fraction 0.80 is transmitted. Describe the beams of light exiting the slab. Give their lateral displacements and intensities relative to the incident intensity I_0.

••• 19 Two parallel light rays encounter a quartz prism with base angles θ (Fig. 19.51). What is the angle φ between them after emergence?

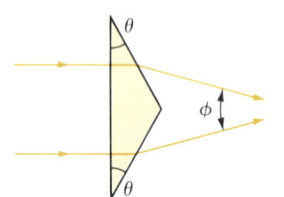

Figure 19.51

Section 19.2 Mirrors

• 20 A concave mirror produces a magnification of −5 for an image distance of 1 m. Find the focal length and radius of curvature for the mirror.

•• 21 A spherical mirror with 30 cm radius of curvature is used to observe an object 4 cm tall. Find the position and magnitude of the image if the object is placed at distances of 10 cm and then 20 cm from the mirror.

• 22 With the aid of a compass, draw a mirror surface with 16 cm radius of curvature. Using ray-tracing, determine the size and location of the image of an arrow 2 cm long, perpendicular to the mirror's optical axis and located 20 cm from the mirror along the optical axis.

• 23 Repeat Problem 22 for a similar arrow 5 cm from the mirror.

• 24 Calculate the theoretical magnification for the images in Problems 22 and 23. Compare the results to those measured directly from your diagram.

•• 25 Show by means of ray diagrams why parallel rays off the optical axis do not all converge to the focus at half the radius of curvature.

•• 26 Show by ray diagrams that the focal distance approximately equals $r/2$ for a convex mirror of radius of curvature r.

• 27 A mirror has a focal distance of 1 m. What is its radius of curvature?

• 28 A concave mirror has a focal length 1 m. At what distance will an image be formed of an object 3 m from the mirror?

• 29 What will be the magnification of the image of the object in Problem 28?

• 30 What focal length should a mirror have to produce a magnification of 10 for an object 3 m away?

••• 31 Show that the mirror or lens equation may be written in the alternate form, stated by Newton, $x_o x_i = f^2$, where x_o is the distance of the object from the focus on the object's side and x_i is the distance of the image from the focus on the opposite side.

Section 19.3 Lenses

• 32 A lens with focal length 20 cm is used to observe an object 5 cm tall. Find the position and magnification of the image if the object is placed at distances of 10 cm and 50 cm from the lens.

• 33 An optician wishes to have a lens form an image 20 cm from a lens of focal length 8 cm. How far from the lens must the object be?

• 34 An optician wishes to have an object 1 m from a lens imaged at a distance 15 cm. What focal length should the lens have?

• 35 What are the magnifications of the lenses in Problems 33 and 34?

• 36 How would the focal length of a diamond lens compare with that of a lens made of fused quartz, assuming all dimensions are equal?

••• 37 Two lenses are constructed from fused quartz and from crown glass, with their focal lengths and front side radii of curvature all equal to the same value f. What is the ratio of $R_2(\text{glass})/R_2(\text{quartz})$?

• 38 Using ray-tracing, determine the size and location of an image formed by a lens of focal length 10 cm of a 2 cm arrow perpendicular to the optical axis and 20 cm away along that axis.

• 39 Repeat Problem 38 for a similar arrow 5 cm from the lens.

•• 40 Derive Eq. 19–11 directly by using ray tracing.

• 41 Calculate the theoretical magnifications for the images in Problems 38 and 39. Compare the results with those measured from your ray diagrams.

•• 42 A lens has a focal length 40 mm. Indicate the regions of object distance for which the image formed is
(a) real,
(b) virtual.

•• 43 What focal length is required to produce a 10 cm image of an object 2 cm tall and located 2 m away?

• 44 Find the curvature of a wavefront arriving from the moon, if the mean distance is 384,000 km.

• 45 A lens has a radius of curvature 10 cm for one surface and 5 cm for the other. What index of refraction would be needed to have a focal length 6 cm?

PROBLEMS

- **46** What index of refraction is required to make a 5 diopter lens with radii of curvature $R_1 = R_2 = 0.5$ m?

- **47** Sketch a plot of index of refraction versus focal length for a lens having $R_1 = R_2 = 5$ cm.

- **48** A lens with focal length 20 cm consists of convex and concave surfaces. Locate the image of an object placed 50 cm from the convex side.

- **49** A convergent lens produces an image at 25 cm of an object at 60 cm. What are the focal length, power, and magnification?

- **50** Show that if two thin lenses with focal lengths f_1 and f_2 are placed together, the total focal length of the pair is given by $1/f_T = 1/f_1 + 1/f_2$.

- **51** Use the result of Problem 50 to find the focal length of the combination of a diverging lens and converging lens (Fig. 19.52) of focal lengths -30 cm and 15 cm.

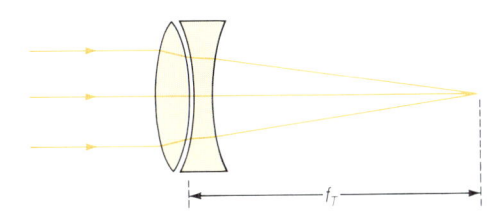

Figure 19.52

- **52** In a slide projector, a 35 mm wide slide is projected onto a screen through a lens of 12 cm focal length. If the screen is 5 m away, how far from the lens should the slide be placed? How wide will the image be on the screen?

Section 19.4 Optical Instruments

- **53** A camera iris has an area of 200 mm² when the f-stop is f/1.4. What area will the iris have at f/32?

- **54** What focal length lens is required to produce a magnification of 2 for an image distance of 10 cm?

- **55** Find the f-ratio for a lens with focal length 20 cm and a 0.625 m diameter aperture.

- **56** Find the power of the lenses in Problems 32–34.

- **57** A nearsighted person cannot focus on objects beyond a far point of 0.5 m. What power lens will enable him to focus clearly on objects at infinity? What is the focal length of that contact lens?

- **58** A nearsighted eye has a lens that can accomodate between near and far points of 10 cm and 20 cm. What contact lens power will allow this eye to see distant objects clearly? What will be the new near and far points?

- **59** A farsighted person cannot focus on objects nearer than 50 cm. What contact lens power will enable the person to focus on an object at 25 cm, the normal near point?

- **60** Corrective lenses of 8 diopters or more were once considered to be a definition for effective blindness. What focal length lens is this? If these lenses corrected a nearsighted person's vision, what is the maximum distance to which the uncorrected eyes could focus?

- **61** If regular eye glasses were used to correct the nearsightedness in Problem 57, the lens would be 2 cm from the eye instead of in direct contact. What power lens would then be required?

- **62** What power of lens would be needed to correct the farsightedness in Problem 59 if the eyeglasses were located 2 cm from the eye?

- **63** What is the focal length of a magnifying glass with 3X magnification for an eye focused at the near point of 25 cm?

- **64** A magnifying glass has a focal length of 5 cm. How far from the eye should the lens be placed for maximum magnification? What is this magnification? How far from the lens is the object in this configuration? Assume the eye has a normal near point of 25 cm.

- **65** What are the focal length and maximum magnification of a 20 diopter lens?

- **66** Explain why a farsighted person's eyes look enlarged through their glasses. If the lenses are 6 diopters and are located 2 cm from the person's eyes, what is the magnification of the eyes as seen by a nearby observer?

- **67** A microscope with an objective focal length 1.8 cm is designed so that the object's inverted image is 2 cm from the objective. What focal length should the eyepiece have to produce a magnification of 200?

- **68** A 1,000X compound microscope has an objective focal length of 0.5 cm. If the total distance between eyepiece and objective is 15 cm, what is the focal length of the eyepiece? Assume the human eye's near point is 25 cm.

- **69** A compound microscope has an eyepiece of 20X magnification and an objective of 50X. If they are separated by 20 cm and the human eye has a near point of 25 cm, what is the total magnification? What are the focal lengths of each lens?

- **70** An optician wishes to construct a compound microscope from an eyepiece with $f_e = 0.5$ cm and an objective with $f_o = 0.94$ cm. If an overall magnification of 900X is desired, how far apart should the two lenses be? If the sample to be observed is placed 1 cm from the objective lens, how far away is its image? Assume the human eye is relaxed and has a near point of 25 cm.

• 71 What is the total magnification of a telescope whose objective and eyepiece have focal lengths of 1 m and 5 cm?

•• 72 What is the total magnification of an astronomical reflecting telescope whose primary mirror has a radius of curvature 5 m and whose eyepiece has a power of 50 diopters?

••• 73 Mars has a diameter of 6,768 km and, at its closest approach to earth, is 78 million km away. If Mars is observed through a telescope with objective and eyepiece focal lengths 1 m and 0.5 cm, what is the magnification of the image? How far away does Mars appear from the eyepiece? Assume that the objective and eyepiece are 1.005 m apart.

SPECIAL PROBLEMS

1. Refraction in Air

The index of refraction of light in air is nearly 1, but it varies from color to color. To a good approximation, n in dry air deviates from 1.0 by the formula

$$(n - 1) = 10^{-7} \times (2{,}726.43 + 12.888/\lambda^2 + 0.3555/\lambda^4)$$

where λ is the wavelength measured in μm (1 μm = 10^{-6} m = 1,000 nm).

(a) A spectral line is measured to have a wavelength of 500.000 nm in dry air. What is the true wavelength in a vacuum? What percentage error was made?

(b) Derive the speeds of light at the wavelengths corresponding to the centers of the 7 rainbow color bands:

Red	650–700 nm	Blue	450–500 nm
Orange	600–650 nm	Indigo	400–450 nm
Yellow	550–600 nm	Violet	350–400 nm
Green	500–550 nm		

(c) If a beam of white light passes through a 20 km layer of air, at an angle of incidence $\theta_i = 1°$ below the horizon, what will be the vertical deviation between red light (675 nm) and blue light (475 nm) after 20 km (Fig. 19.53)? This effect is responsible for the "green flash" when the sun sets at sea.

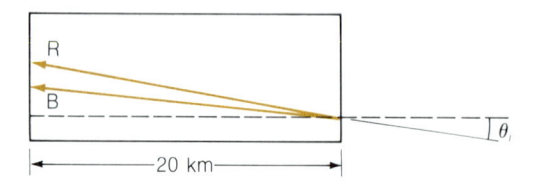

Figure 19.53

2. Prism Dispersion

When light passes symmetrically through an isosceles prism (Fig. 19.54) the angle of deviation δ can be used to determine the index of refraction n.

(a) Using geometry, show that $\theta_{in} = \theta_{out} = \alpha + \theta_2$, that $\delta = 2\alpha$, and that $\theta_2 = \phi/2$.

(b) Use these relations with Snell's law to show
$$\sin\left(\frac{\phi + \delta}{2}\right) = n \sin(\phi/2).$$
Explain what measurements must be performed to measure n.

(c) Estimate the angular width of the rainbow produced when white light passes symmetrically through a prism with $\phi = 60°$. Assume that the index of refraction of crown glass is 1.539 for violet light (361 nm) and 1.514 for red light (656 nm).

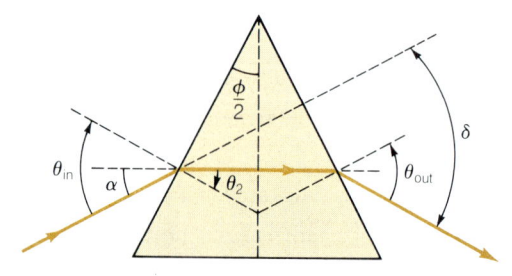

Figure 19.54

3. The Rainbow

The French mathematician René Descartes and the English physicist Isaac Newton first explained the rainbow in terms of refraction and internal reflection in water droplets illuminated by the sun's nearly parallel rays (Fig. 19.55). The rays enter the drop over a range of angles θ_1 from 0° to 90°. After one internal reflection, the ray emerges at angle γ_1.

arise because the rainbow angle depends on the index of refraction of water, which varies with wavelength. A secondary rainbow resulting from two internal reflections can sometimes be seen at $\gamma_2 = 50°$.

(a) From geometry show that the internal angles $\theta_2 = \theta_3 = \theta_4 = \theta_5$ and that $\theta_6 = \theta_1$.

(b) Show that the emergent angle $\gamma_1 = 4\theta_2 - 2\theta_1$, where $\sin\theta_1 = n \sin\theta_2$.

(c) The index of refraction for water at 20°C is 1.3435 for violet light (397 nm) and 1.3320 for red light (620 nm). Make tables of θ_1, θ_2, and γ_1 for θ_1 varying from 58° to 60°. Find the maximum of γ_1, and find the rainbow angle for each color.

(d) Estimate the width of the rainbow in arc minutes. Remember to add 30 arc min for the sun's angular diameter (the sun's rays are not exactly parallel).

4. Fish-eye View

A fish is 1 m under the smooth surface of a pond. Its view of the surface world is affected by refraction and internal reflection with $n = 1.34$.

(a) Describe the fish's view of the surface world. What angles are visible? Are some portions hidden? Are dimensions accurate?

(b) The fish is 1 m from shore and watches a fisherman standing on land, 1 m in from the shore. What apparent height does the fish perceive for the 1.8 m man? You will need to solve this by trial and error using Snell's law.

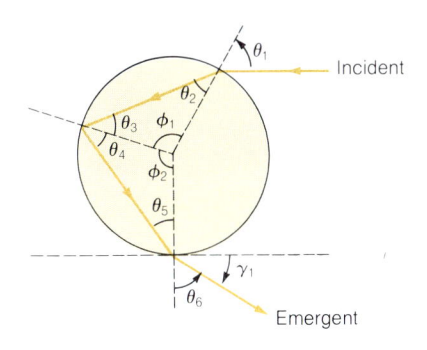

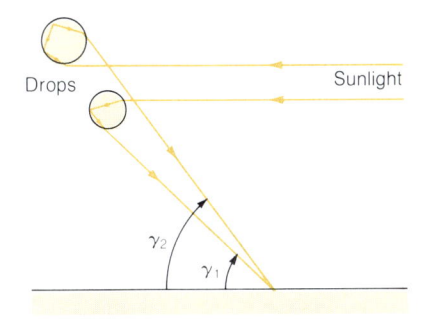

Figure 19.55

Over this range γ_1 reaches a maximum of about $\gamma_1 \approx 42°$ (the **rainbow angle**), and the concentration of emergent rays near that value leads to an enhanced light intensity which we see as a rainbow arc with opening angle 42°. The colors

---

SUGGESTED READINGS

"Mysteries of rainbows, notably their rare supernumerary arcs." J. Walker, *Scientific American*, Jun 80, p 146, **242**(6)

"What is a fish's view of a fisherman and the fly he has cast on the water?" J. Walker, *Scientific American*, Mar 84, p 138, **250**(3)

"Mirages." A. B. Fraser, W. H. Mach, *Scientific American*, Jan 76, p 102, **234**(1)

"Photographic lens." W. H. Price, *Scientific American*, Aug 76, p 72, **235**(2)

"The theory of the rainbow." H. M. Nussenzveig, *Scientific American*, Apr 77, p 116, **236**(4)

"Compound eye of insects." G. A. Horridge, *Scientific American*, Jul 77, p 108, **237**(1)

"Atmospheric halos." D. K. Lynch, *Scientific American*, Apr 78, p 144, **238**(4)

"The interpretation of visual illusions." D. D. Hoffman, *Scientific American*, Dec 83, p 154, **249**(6)

"The topology of mirages." W. Tape, *Scientific American*, Jun 85, p 120, **252**(6)

"The renaissance of x-ray optics." J. H. Underwood, D. T. Atwood, *Physics Today*, Apr 84, p 44, **37**(4)

"Experiment and mathematics in Newton's theory of color." A. Shapiro, *Physics Today*, Sep 84, p 34, **37**(9)

The flying circus of physics with answers. Jearl Walker, Wiley, New York, 1977

A source book in physics. William F. Magie, Harvard University Press, Cambridge, 1965

Great scientific experiments. Rom Harre, Phaidon Press, Oxford, 1981 (Isaac Newton's experiments with light)

Atmospheric Phenomena: Readings from Scientific American. Freeman, San Francisco, 1980

Rainbows, Halos, and Glories, Robert Greenlee, Cambridge University Press, Cambridge, 1980

Sunsets, Twilights, and Evening Skies, Aden and Marjorie Meinel, Cambridge University Press, Cambridge, 1983

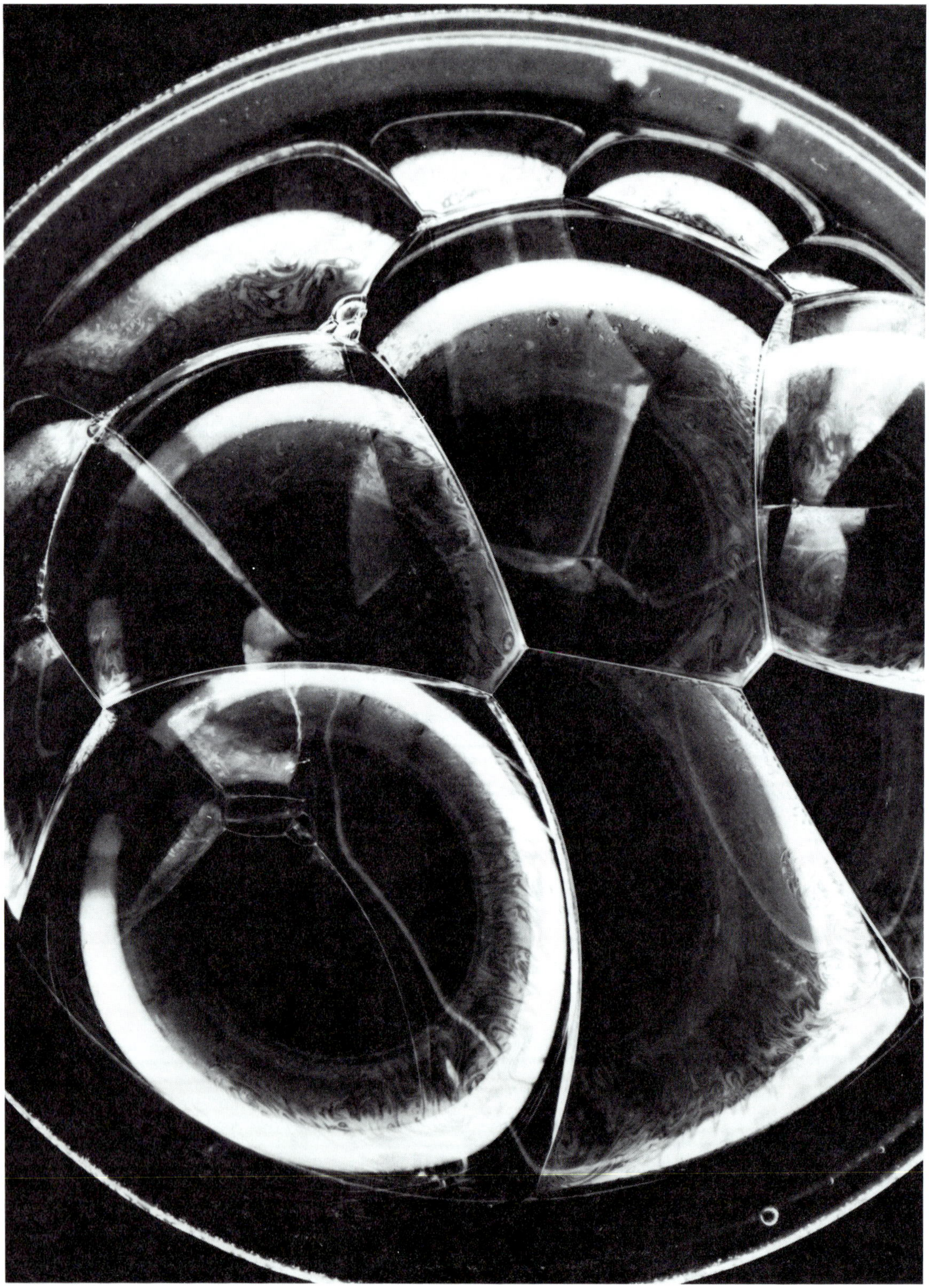

Chapter 20

Wave Phenomena in Light

From the time of Newton and Huygens until the early twentieth century, the apparent dual nature of light was an important unsettled issue. We have alluded to this conflict: we have discussed Huygens' experimental evidence demonstrating wave properties, and we have mentioned Newton's belief that light consists of particles.

In later chapters the full story will unfold as we discuss the modern atomic theory and light emission and absorption by atoms. For now, we continue to confine ourselves to the classical point of view, deducing the nature of light by analyzing its overall behavior. In this chapter we explore properties of light that can be understood only with the wave hypothesis, and we will find that these properties have a variety of applications.

20.1 The Evidence for Light Waves

It is instructive to review the basis for the controversy over the nature of light. It helps demonstrate how science makes progress as hypotheses are made and tested, and it helps set the stage for our later discussions of modern atomic theory. We begin by discussing the conflicting optical evidence that confronted scientists following Newton and Huygens.

Ambiguous Evidence: Reflection and Refraction

In the previous chapter we were able to analyze the optical properties of mirrors and lenses by assuming only two simple geometric laws: the law of reflection and Snell's law for refraction. It is easy to see that the law of reflection holds for waves or particles (so long as the collisions between light particles and

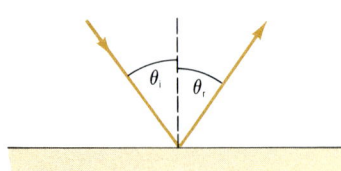

Figure 20.1

Reflection. Because momentum and energy are conserved in a reflection, the angle of reflection equals the angle of incidence.

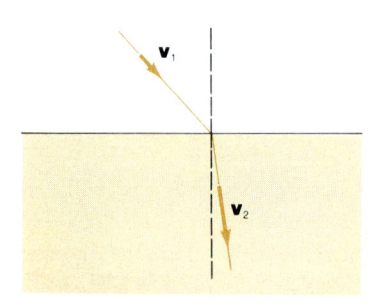

Figure 20.2

Newton's explanation of refraction. Newton hypothesized that light entering a medium such as glass or water is accelerated at the boundary by a force acting perpendicular to it.

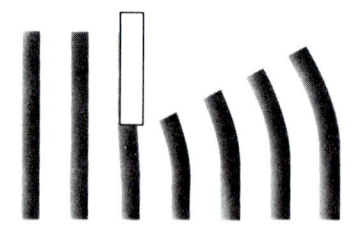

Figure 20.3

Diffraction. Unlike particles, light waves will bend around an obstacle, just as water waves do.

reflecting surfaces are perfectly elastic). If momentum and energy are conserved by a particle as it collides with a surface, then there is no change in the component of velocity parallel to the surface, and a change in direction but not magnitude in the component perpendicular to the surface (Fig. 20.1). Thus, the reflected angle equals the incident angle.

Refraction is another matter, however. In chapters 10 and 19, we found that refraction in a medium such as water or glass can be explained by the wave hypothesis, because of the slowing of the wavefronts as they enter the medium of higher optical density. To explain refraction under his corpuscular theory, Newton was forced to conclude that light travels faster in water or glass than in air, because the component of particle motion into the water or glass had to increase to produce a net deflection toward the perpendicular (Fig. 20.2). Thus, Newton hypothesized that a force is exerted on light particles at the boundary between the two media, and that the force acts perpendicular to the boundary, so that the component of velocity into the medium is increased. Thus, the wave and particle hypotheses led to opposing predictions regarding the speed of light in a medium, and in principle the issue could have been resolved experimentally. Unfortunately, no one at the time knew how to measure the speed of light in a medium such as water or glass, so the two hypotheses continued to compete, with Newton's theory dominant for a long time because of the great success of his mechanics.

In Chapter 10 we discussed other wave properties in addition to reflection and refraction, however, and these properties could not be successfully explained by Newton's particle theory of light. In particular, if diffraction and interference could be attributed to light, then a very strong case could be made for the wave hypothesis.

Proof of Wave Properties: Young's Experiment

In Chapter 10 we described diffraction as the bending of waves when they pass by an obstruction (Fig. 20.3). We noted that waves from a point source spread as they propagate, and that they do so at any point where there is an obstruction. This picture was applied to light by Huygens, whose wavefront model enjoyed great success. This was mentioned in Chapter 19; Huygens found that the behavior of a wavefront could be explained by assuming that it consists of tiny spherical waves propagating from any point along the front (Fig. 20.4). For an unimpeded wavefront, this results in straight-line travel, as we have already seen. When there is an obstruction, however, then the small spherical waves propagate into the shadow region behind it (Fig. 20.4). Water wave experiments showed that the amount of bending, which is called diffraction, depends on the relative sizes of the wavelength and the obstacle. Maximum diffraction occurs when the two are comparable.

Newton was aware of earlier experiments showing that light passing through a small opening diverges, but he attributed this to the interaction between the light particles and the obstruction (recall that Newton envisioned a force acting between matter and light particles). The amount of bending was not very large, so the effect of the obstruction on light particles did not need to be very strong. We now recognize that the diffraction was slight because

20.1 THE EVIDENCE FOR LIGHT WAVES

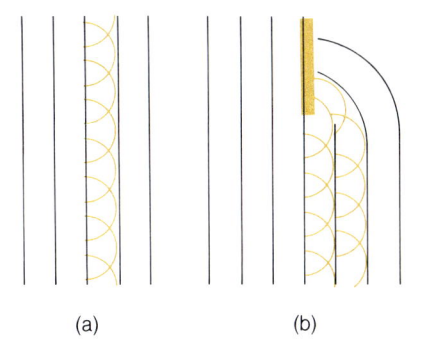

(a) (b)

Figure 20.4

Huygens' principle. Huygens found that the behavior of a wavefront could be explained by imagining tiny spherical waves propagating from any point on the wavefront. (a). Plane wavefronts travelling in a uniform medium. (b). Diffraction around an obstacle.

the wavelengths of light are very short compared to the size of any obstruction or hole that they might normally encounter. Hence, even though the diffraction of light is sufficient to establish its wave nature, during Newton's time it was not well accepted that light really diffracts.

The definitive demonstration of the wave character of light did not occur until 74 years after Newton's death. This demonstration was related to how waves are superposed. The wave crests and troughs can either match, producing constructive interference; or they can cancel each other, creating destructive interference (or partial interference can occur in intermediate cases). We explored this property in Chapter 10, particularly in discussing sound waves.

Ordinarily, light is **incoherent,** meaning that it consists of a mixture of phase differences; that is, the crests and troughs of the individual waves are randomly arranged. When incoherent waves are superposed, no interference pattern can be easily perceived. Interference certainly occurs in incoherent light, but it is randomly constructive and destructive, and has no net effect. If light from some source has only one wavelength and the rays are parallel, however, then the phase difference between waves is constant, and the light is said to be **coherent.** If the waves in coherent light are in phase, so that the troughs and crests are superimposed on one another, then the interference is constructive.

Interference in light was demonstrated in 1801 by the English scientist Thomas Young, who ingeniously used diffraction of light to produce beams of light that were in phase. Young allowed light to pass through a single slit in an otherwise opaque screen, then to pass through two parallel slits in a second screen (Fig. 20.5). If light consisted of particles, he would have seen two bright strips on a third screen, where the particles passing through the slits would strike. Instead, Young observed a series of alternating bright and dark strips on the third screen, and was able to interpret them as due to a combination of diffraction and interference.

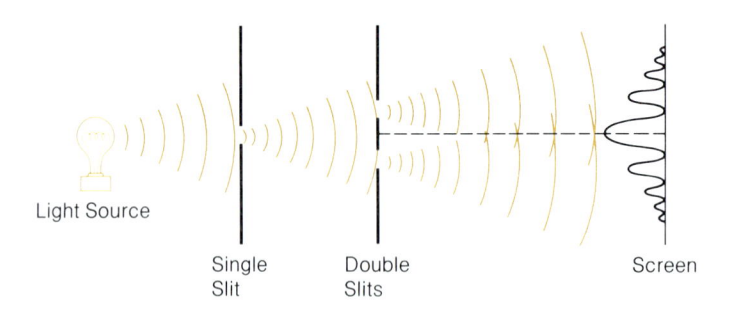

Figure 20.5

Young's double-slit experiment. The wavefronts emerging from the double slits interfere with each other to produce a pattern of bright and dark strips on the screen.

It is easiest to understand this by imagining that the light consists of only one wavelength. Light passing through the first single slit is diffracted, so that a curved wavefront enters the two slits in the second screen. Diffraction occurs also at this pair of slits, so that curved wavefronts emerge. Because they are parts of the same single wavefront from the first slit, these two wavefronts are in phase with each other but are separated as they emerge from the pair of slits in the second screen (Fig. 20.5). The pair of slits therefore acts as a pair of coherent light sources. As the fronts diverge and overlap, they interfere. Whether the interference is constructive or destructive at the point where the waves strike the third screen depends entirely on the distances the fronts have travelled from the pair of slits. If they have travelled the same distance, then they will be in the same phase, and the interference is constructive. Hence there is a bright strip on the third screen directly behind the pair of slits. At other positions on the third screen, the two wavefronts have travelled different distances. If one wavefront has travelled an integral number of wavelengths farther than the other, then again there is constructive interference. Therefore there are many other positions on the third screen where bright strips appear. These bright strips are called **interference fringes.** If, on the other hand, one wavefront has travelled ½, 1½, 2½, etc. wavelengths farther than the other, then they are out of phase when they reach the screen, so the interference is destructive, and dark strips occur at these positions.

If the distance between the two slits is very small compared to the distance from the slits to the third screen, then the light rays from the two slits that meet at a single point on the screen are nearly parallel (Fig. 20.6). If we represent their direction by the angle θ that they make with the horizontal, then we see that the difference in pathlength from the two slits to a single point on the third screen is $d \sin\theta$. Thus, the condition for constructive interference is

$$d \sin\theta = m\lambda, \quad m = 0, 1, 2, \ldots \quad (20\text{--}1)$$

where m is an integer called the **order** of the interference fringe. When the pathlength difference is an odd number of half wavelengths, so that destructive interference occurs, then

20.1 THE EVIDENCE FOR LIGHT WAVES

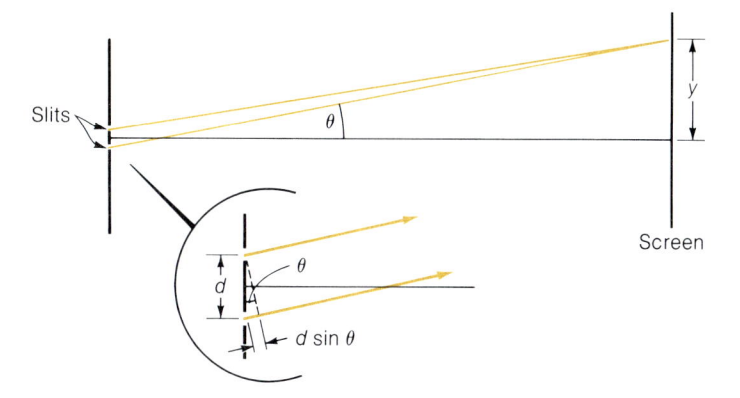

Figure 20.6

The geometry of the double-slit experiment. If the slit separation d is much smaller than the distance to the screen, then the two rays are essentially parallel, and the pathlength difference between them is dsinθ. Constructive interference occurs at the screen when this difference is an integral multiple of the wavelength.

$$d\sin\theta = \left[m + \frac{1}{2}\right]\lambda, \quad m = 0, 1, 2, \ldots \quad (20\text{--}2)$$

The intensity of each wavefront decreases with distance from the slits, so the maximum brightness in the interference fringes occurs where the combined distances that the two fronts have travelled is smallest, which corresponds to the central fringe (zero order). The brightness of the other fringes drops off as m increases.

Young's demonstration of interference in light proved that light has wave properties, although the question of a particle nature for light was to arise again, as we shall see in the next chapters of this text.

Example 20.1 Suppose that, in Young's experiment, the slits are separated by 0.1 mm, the wavelength of light being observed is 500 nm, and the distance from the slits to the screen is 1 m. How far apart are the zero-order and first-order interference fringes on the screen?

We solve Eq. 20–1 for the angle θ corresponding to first order ($m = 1$):

$$\theta = \arcsin\left(\frac{m\lambda}{d}\right) = \arcsin\left[\frac{(1)(500 \times 10^{-9}\text{ m})}{1 \times 10^{-4}\text{ m}}\right] = 0.005\text{ rad} = 0.286°$$

(Actually, for such a small angle, we could have assumed that $\sin\theta = \theta$.)

The separation of the zeroth and first orders on the screen, which we designate y, is given by

$$y = l \tan\theta$$

where l is the distance from the slits to the screen. We use the small angle approximation $\tan\theta \approx \theta$ with θ in radians, so we find

$$y = l\theta = (1 \text{ m})(0.005 \text{ rad}) = 0.005 \text{ m} = 5 \text{ mm}.$$

We note that the second-order fringe occurs at twice the angle, hence at twice the distance, from the zeroth order fringe. Therefore, so long as the small-angle approximation is valid, the fringes are spaced regularly on the screen.

20.2 Diffraction and Interference

Now that we understand how interference occurs, we can delve into it a little deeper. In many situations, usually involving diffracted light, interference effects can be important. In this section we explore some of them.

Measuring Wavelengths

In Chapter 18, when we began our discussions of light, we avoided describing the methods used to measure its wavelength, merely stating that this began to be done in the nineteenth century. Now we are in a position to see how.

In discussing Young's experiment, we assumed for simplicity that the light is **monochromatic**, that is, only one wavelength is present. If we stick with that assumption a little longer, we see readily that it should be possible to determine the wavelength of the light by measuring the distance between interference fringes so that the angle θ is known as a function of m, the order number. Then it is a simple matter to solve Eq. 20–1 for λ, the wavelength. This is straightforward, but in Young's time, sources of monochromatic light were not available.

Now let us consider white light, which consists of a continuous range of wavelengths. When white light is used in Young's experiment (as it originally was), each wavelength has its own value of θ for each order m. Thus, the first-order fringe for red light (wavelength 700 nm), for example, will appear at a slightly larger angle than blue light (wavelength 450 nm). Since the light consists of a continuous range of wavelengths, each interference fringe is broadened, with the shortest wavelengths appearing on the side closer to the zeroth order, and the longest wavelengths appearing on the side of the fringe that is away from the zeroth order. Each fringe becomes a tiny rainbow (except for the zeroth-order fringe, which is narrow and white, since $\theta = 0$ equally for all wavelengths). This is what Young actually saw, and what is found in laboratory

20.2 DIFFRACTION AND INTERFERENCE

re-enactments of his experiment. The pattern of interference fringes is not as sharp as we described in the previous section, but instead consists of a series of broadened, color-separated bands that may actually overlap a bit. Spreading light according to wavelength is called **dispersion,** whether it is done by a slit or any other device (such as a prism, where the index of refraction varies with wavelength).

Measuring wavelengths is still quite straightforward in principle. What is needed is to find the separation on the screen of the fringes for a particular wavelength, and solve for λ in Eq. 20–1. Young was able to do this, and was the first to measure wavelengths of visible light.

Example 20.2 Consider a double-slit experiment where the slits are separated by 0.05 mm, and the distance from the slits to the back screen is 1.0 m. If the zeroth-order and first-order fringes for the same color (that is, the same wavelength) of light are separated by 12 mm, what is the wavelength that is being observed?

First, we use the small-angle approximation, noting (from Example 20.1) that

$$\theta \approx \frac{y}{l} = \frac{1.2 \times 10^{-2} \text{ m}}{1.0 \text{ m}} = 0.012 \text{ rad.}$$

From Eq. 20–1, we find the wavelength λ.

$$\lambda = \frac{d}{m} \sin\theta = \frac{(5 \times 10^{-5} \text{ m})(0.012)}{1} = 6.0 \times 10^{-7} \text{ m} = 600 \text{ nm}$$

This is in the orange portion of the visible spectrum.

Example 20.3 Suppose in Young's experiment the slits are separated by 0.1 mm and the distance from them to the screen is 1 m, as in Example 20.1. White light is allowed to enter the front slit. Find the locations of the first- and second-order fringes for red light (700 nm) and blue light (450 nm).

This is a straightforward duplication of Example 20.1, except that we are asked to do the calculation for two different wavelengths, and for the first and second orders. If we use the small angle approximation $y = l\theta$ and Eq. 20–1, we find the expression

$$y = l\theta = \frac{m\lambda l}{d}$$

Solving this for all the cases, we find

first-order red light: $y = \dfrac{(1)(700 \times 10^{-9} \text{ m})(1 \text{ m})}{(1 \times 10^{-4} \text{ m})} = 7$ mm

second-order red light: $y = 14$ mm

first-order blue light: $y = 4.5$ mm
second-order blue light: $y = 9$ mm.
We see that the first-order red light almost overlaps the second-order blue.

Gratings

Suppose that we perform Young's experiment, but instead of a pair of slits in the second screen, we have a large number of closely-spaced parallel slits (Fig. 20.7). When parallel, coherent light enters all of these slits, interference occurs just as for two slits. Diffraction occurs at each slit, so that a series of curved wavefronts emerge on the other side, and these wavefronts all overlap. Constructive interference at the screen will occur when the pathlengths of rays from separate slits differ by integral multiples of the wavelength.

Consider a ray of light from one slit. At some angle this ray and the one from the adjacent slit will differ in pathlength by one wavelength, so that the two rays are in phase and a first-order interference fringe appears on the screen. At the same angle, the ray from the next slit, two spacings away from our initial one, will differ in pathlength by twice the wavelength, so this ray also interferes constructively, adding to the first-order fringe. The ray from the next slit has a pathlength difference of three wavelengths, so it also contributes to the fringe, and so on. The result is a first-order interference fringe just as for two slits, except that now we are dealing with an incoming beam of light that illuminates a whole series of slits (Fig. 20.7). The fringe occurs at the same angle as for two slits, as given by Eq. 20–1 for $m = 1$. Fringes for other orders are formed in the same way, each one consisting of parallel rays of light leaving the grating at an angle θ given by Eq. 20–1 for the order m.

If white light is used, then each order consists of dispersed light, just as for two slits. The requirement for seeing the fringes on a screen is that the distance to the screen be large compared to the slit separation, so that the rays are essentially parallel and Eq. 20–1 correctly describes the condition for constructive interference. It is also possible to use a lens to focus the parallel rays, so that a large distance is not required (Fig. 20.7).

There is a significant advantage in using a grating of many slits instead of a single pair. It is easiest to see this by again considering monochromatic light. When we discussed this before, we tacitly described an idealized situation where the fringes are sharp because constructive interference was assumed to occur only when the two rays are precisely in phase. In reality, the waves from the two slits become out of phase with each other rather gradually as the angle departs from the correct angle for an order. Therefore, the fringes are somewhat broadened, being brightest at their centers where θ exactly satisfies Eq. 20–1, but having substantial extent to either side of that. When there are many slits involved, however, then for even slight displacements from the exact value of θ for an order, there is destructive interference from some distant slit. Consider again a particular ray from a single slit. At an angle departing only slightly from the value of θ at the center of one of the fringes, the phase difference between this ray and the one from the adjacent slit is small, so that the interference is partially constructive. The pathlength difference between our par-

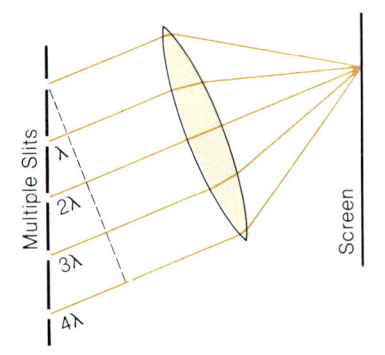

Figure 20.7

Multiple slits. If coherent light enters a series of slits, all the emerging wavefronts will be in phase, so that they can interfere constructively at any angle θ for which the pathlength difference is an integral multiple of the wavelength. Here, the pathlength difference increases by one wavelength from one slit to the next. A lens focuses the parallel diffracted rays onto a screen.

20.2 DIFFRACTION AND INTERFERENCE

ticular ray and the one from a faraway slit is much larger, however, and if there are enough slits, there will be one whose ray has sufficiently different pathlength to be perfectly out of phase with the one under consideration, so that there is highly destructive interference. Thus, the interference fringes for a grating are much sharper than for a pair of slits. The total energy remains constant, however, so that in effect the light is better concentrated into the fringes, which are now brighter. The greater the number of slits, the greater the sharpness and brightness of the fringes.

A grating typically consists of a glass plate on which closely-spaced grooves have been ruled with a diamond stylus, or of a photographic replica of such a grating. The sharpness and brightness of the fringes is governed by the density of the grooves, because this effectively determines how many slits act to produce the interference pattern for a beam of light, thereby controlling the sharpness and intensity of the interference fringes. Gratings with as many as a few thousand grooves per millimeter are now produced fairly routinely.

In many applications, **reflection gratings** are used instead of the type just described, which are often called **transmission gratings** because light is transmitted through them. In a reflection grating, grooves are ruled in a reflecting surface, so that the reflected light consists of many rays whose phase differences are governed by their relative pathlengths (Fig. 20.8). The effect is analogous to that of a transmission grating, and the analysis is similar, leading to identical results. Any grooved surface can act as a reflection grating; for example, a phonograph record held at just the right angle reflects a color spectrum.

A device using a grating (or a prism) to disperse light according to wavelength is called a **spectroscope**. Normally a lens is used to bring the parallel rays to a focus. The angle at which a certain wavelength is found is usually measured and calibrated to give a measurement of the wavelength. If the dispersed light is recorded, for example, with photographic film, then the instrument is usually called a **spectrometer** or a **spectrograph**. Such instruments have far-reaching applications, because a great deal of information about a source of light can be derived from its spectrum. The reasons for this are complex, and only understood in the context of quantum theory (discussed progressively in the next three chapters). For now, one simple statement should suffice to show the potential of the science called **spectroscopy**: each chemical element absorbs and emits light only at certain specific wavelengths, so the spectrum of light from a source can tell us what the source is made of. Many other characteristics of the source can be derived as well.

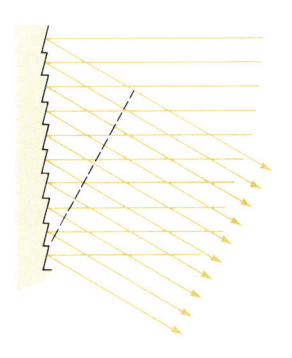

Figure 20.8

Reflection grating. Incident light is reflected from a series of angled grooves, creating an interference pattern because of the pathlength differences between rays.

Example 20.4

Find the locations of the blue (λ = 450 nm) and red (λ = 700 nm) light in the first-, second-, and third-order fringes from a transmission grating with 500 grooves/mm on a screen placed 2 m away.

The slit spacing is 1/500 mm or 2×10^{-6} m. As shown in Eq. 20–1, the interference fringes occur at angles

$$\theta = \arcsin\left(\frac{m\lambda}{d}\right)$$

and the separations of various orders on the screen are given by $y = l \tan\theta$. Notice that we cannot use the small angle approximation, $y \simeq m\lambda l/d$ as in Example 20.3 since θ will turn out to be much larger than 10°. Substituting $m = 1, 2,$ and 3, we find that the blue fringes in the first three orders occur at the following angles and positions relative to the zeroth-order fringe:

$$\theta_1 = \arcsin\left[\frac{(1)(4.5 \times 10^{-7} \text{ m})}{2 \times 10^{-6} \text{ m}}\right] = 13.0°$$
$$y_1 = (2 \text{ m}) \tan(13.0°) = 0.46 \text{ m}$$
$$\theta_2 = 26.7°$$
$$y_2 = 1.01 \text{ m}$$
$$\theta_3 = 42.5°$$
$$y_3 = 1.83 \text{ m}$$

The red fringes occur at

$$\theta_1 = 20.5° \quad y_1 = 0.75 \text{ m}$$
$$\theta_2 = 44.4° \quad y_2 = 1.96 \text{ m}$$

There is no third order red fringe, since $\sin(m\lambda/d)$ cannot be greater than 1.0. There are a couple of significant points to be made about this result: (1) given a fine grating and a substantial distance to the screen on which the fringes form, rather large physical separations of the orders and large extents of the orders can be achieved, which is useful if measuring instruments are to be placed there to analyze the fringes, and (2) the orders begin to overlap more and more as the order number increases. In this example, the first and second orders do not overlap, but the second and third begin to (third-order blue lies within second-order red). Often filters (described later in this chapter) are used to screen out unwanted orders in measuring instruments.

Diffraction by a Single Slit or a Disk

We have seen that diffraction occurs whenever light passes an obstruction. The light waves spread out behind the obstruction, and waves that pass it at different distances traverse different pathlengths as they travel to a distant location (Fig. 20.9). If the incoming light waves are in phase, as from a point source or a slit, then there will be noticeable interference effects (interference always happens, but it is disorganized and unrecognizable if the light is not parallel and in phase when it encounters the obstruction).

A solid disk is a simple example. If coherent monochromatic light illuminates such a disk, the shadow of the disk is not a clean circular shadow, but instead has dark rings around it, corresponding to positions where the waves which diffracted at the edge of the disk have travelled odd multiples of a half-wavelength and therefore interfere destructively (Fig. 20.10). Furthermore, there is a bright spot in the middle of the shadow, because light waves diffracted to that spot from around the edge of the disk have all travelled equal

20.2 DIFFRACTION AND INTERFERENCE

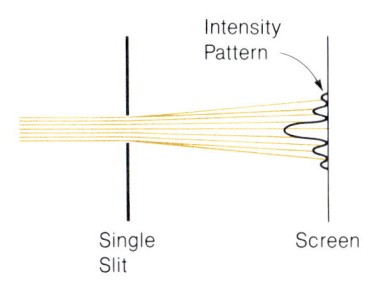

Figure 20.9

Single slit diffraction. Light passing through an opening such as a slit diffracts, so that some light strikes the screen at displaced positions. If the incoming light is coherent, then the intensity pattern on the screen shows interference fringes.

Figure 20.10

The diffraction pattern of a solid disk. Fringes appear around the edges, and a bright spot lies at the center of the shadow, where the interference is constructive.

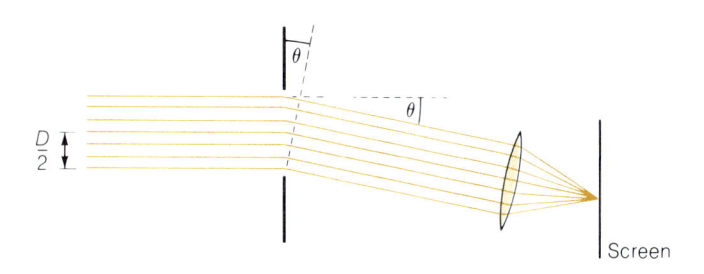

Figure 20.11

Interference in single slit diffraction. A dark strip occurs in the interference pattern when the angle θ is such that the pathlength difference between the ray through the center of the slit and one at the edge is a multiple of λ/2. Bright fringes occur when the difference is a multiple of λ.

distances, and are therefore in phase, so that they interfere constructively. If the light is not monochromatic, but still parallel, then interference occurs, but the rings around the shadow are more diffuse because their location varies with wavelength. There is still a bright central spot, however, because all wavelengths interfere constructively at the center. This situation is similar to that of a double slit or a grating, where the zeroth order fringe consists of white light, but the higher-order fringes consist of dispersed light.

When coherent light passes through a single slit, interference also occurs. Because the slit has some width, waves that pass through different parts of it and then travel to a distant location will have phase differences determined by their separation at the slit and the angle of diffraction (Fig. 20.11). Again, it is simplest to consider monochromatic light first. If the slit width is D, then we see that for some diffraction angle θ a wave from the edge of the slit will interfere destructively with one from the center of the slit, because the two waves will be out of phase. The path difference $(D/2)\sin\theta$ (see Fig. 20.6) must equal a half-wavelength, so the angle must satisfy the condition

$$\frac{\lambda}{2} = \frac{D}{2}\sin\theta$$

or

$$\sin\theta = \frac{\lambda/2}{D/2} = \frac{\lambda}{D}$$

because at this angle, the wave from the edge of the slit has a pathlength exactly one half-wavelength longer than the wave from the center. For any pair of waves with separation ½D at the slit, the same condition prevails. Therefore, there is a dark band at this angle. Similarly, there is an angle such that

$$\sin\theta = \frac{\lambda/2}{D/4} = \frac{2\lambda}{D}$$

for which destructive interference also occurs because the pathlength difference between waves separated by one quarter the slit width are out of phase by one half of the wavelength. Thus, the diffracted light has a dark band at each angle given by

$$D\sin\theta = m\lambda, \quad m = 1, 2, 3, \ldots \quad (20\text{--}3)$$

where m is again the order number. Since this interference depends on the assumption that the diffracted rays are parallel to each other for any angle, the interference pattern shows up well only if the diffracted light is allowed to travel a large distance to a screen, or if a lens is used to focus the parallel rays (Fig. 20.11). There is not a dark band at $\theta = 0$, because at this angle all the waves interfere constructively, and there is a bright band. Lesser bright bands occur between the other dark bands. As for two or more slits, when white light is used instead of monochromatic light, interference still occurs, but the diffraction bands are less distinct, because the light of different wavelengths interferes destructively at different locations.

Note that the angular size θ of the diffraction pattern from a single slit is proportional to λ/D, where D is the slit width. This is a quantitative statement of something we have postulated before: the diffraction of waves is greatest when the obstruction or opening is small compared to the wavelength.

Example 20.5 Monochromatic light of wavelength 400 nm passes through a 0.01 mm-slit, and a 1 mm-slit. Find the location of the first dark band for each, if the light is allowed to fall on a screen 2 m behind the slit.

The linear distance from the central bright fringe is

$$y = l\tan\theta = l\theta$$

20.2 DIFFRACTION AND INTERFERENCE

if we use the small-angle approximation. Letting $m = 1$ in Eq. 20–3 and substituting $m\lambda/D$ for θ yields

$$y = \frac{l\lambda}{D}.$$

For the 0.01 mm slit, the value of y is

$$y = \frac{(2 \text{ m})(4 \times 10^{-7} \text{ m})}{1 \times 10^{-5} \text{ m}} = 0.08 \text{ m}.$$

For the larger slit, the distance is

$$y = \frac{(2 \text{ m})(4 \times 10^{-7} \text{ m})}{1 \times 10^{-3} \text{ m}} = 8 \times 10^{-4} \text{ m} = 0.8 \text{ mm}.$$

Hence the bright bands are much wider for a small slit.

Diffraction Limits in Optical Instruments

If we consider a circular opening instead of a linear slit, the analysis is a bit more complicated. But if we assume that the circular opening approximates a slit, we can use our result for the single slit. Numerical and experimental analysis shows that a circular opening of diameter D is equivalent to a slit of width 1.22 D. Therefore the angle at which the first dark fringe occurs (that is, the case where $m = 1$) is given by

$$\theta = \frac{1.22\lambda}{D} \tag{20-4}$$

where θ now represents the angular deviation in radians from the direction of the incoming light (Fig. 20.11), and the small angle approximation has been applied.

Diffraction in light passing through circular openings is very important in such optical instruments as telescopes and microscopes. The **resolution** of such an instrument, its ability to distinguish closely-separated objects, is limited by diffraction. Two closely-spaced objects, such as a pair of stars, will not appear as separate objects if their bright central images overlap. The size of the bright spot at the center of the diffraction pattern is governed by the location of the first dark fringe that bounds it, and this location is given by Eq. 20–4. The generally-accepted criterion (called the **Rayleigh criterion**) is that two objects are resolved, that is, recognized as separate objects, if the center of the diffraction pattern of one is just at the location of the first dark fringe from the other. Thus, the angle θ given by Eq. 20–4 is called the **diffraction limit** of a telescope or microscope. Because of the inverse dependence of this limit on the diameter of the opening, the resolution of such an instrument is enhanced if the instrument is large. This provides a second reason (in addition to maximizing light-gathering power) for building telescopes as large as possible.

For microscopes, the ability to resolve detail is often expressed in terms of the linear separation of two objects that are just resolved, rather than stating the angle at which resolution barely occurs, as we have just done. As we learned in Chapter 19, the object viewed with a microscope is typically placed very near the focal point of the objective lens; that is, at the focal distance f. If we use the small-angle approximation, then the distance s between two objects that are separated by an angle θ as given in Eq. 20–4 is

$$s = f\theta = \frac{1.22\lambda f}{D} \tag{20-5}$$

where D is the diameter of the microscope aperture. This distance s is often called the **resolving power** of a microscope.

Example 20.6 — What is the diffraction limit of the human eye (assume typical visible light, wavelength 550 nm, and a pupil diameter of 1 mm), of a 5-m telescope (assuming the same wavelength), and of a radio telescope of 30 m diameter, used to observe radiation of wavelength 0.21 m?

For each, we simply apply Eq. 20–4. For the human eye, we have

$$\theta = \frac{1.22\lambda}{D} = \frac{(1.22)(5.5 \times 10^{-7}\text{ m})}{0.001\text{ m}} = 6.7 \times 10^{-4}\text{ rad} = 2.3'.$$

For a 5-m telescope, the limit is

$$\theta = \frac{(1.22)(5.5 \times 10^{-7}\text{ m})}{5\text{ m}} = 1.34 \times 10^{-7}\text{ rad} = 0.03''.$$

Finally, for the radio telescope receiving 21-cm radiation, the diffraction limit is

$$\theta = \frac{(1.22)(0.21\text{ m})}{30\text{ m}} = 0.0085\text{ rad} = 29'.$$

We see that radio telescopes generally have poor resolution compared to visible-wavelength telescopes.

20.3 Reflection and Interference

We have seen that diffracted light, if made coherent by passage through a slit or if it is coherent because it is emitted by a point source, produces important interference effects. There can also be important interference effects when

20.3 REFLECTION AND INTERFERENCE

systematic phase differences arise in reflected beams of light. In this section we explore such situations.

We first point out that reflection can change the phase: if a light wave encounters a boundary with a medium of higher index of refraction than the one it is in, reflection changes the phase of the wave by 180° (this is analogous to waves travelling along a taut string whose end is fixed, as discussed in Chapter 10). If the second medium has lower index of refraction than the medium the light wave is in, then there is no change of phase in the reflected wave. We will keep these points in mind in the following discussion.

Thin Films and Newton's Rings

It is commonly observed that a film of oil floating on water creates a colorful display (Fig. 20.12). This is an interference effect; light waves reflected from the upper and lower surfaces of the oil film interfere with each other. There is a phase shift at both reflecting surfaces, because the index of refraction increases progressively from air to oil to water. For light perpendicular to the surface, the difference in pathlength for the rays reflected from the top and bottom surfaces of the oil is $2d$, where d is the thickness of the oil layer (Fig. 20.13); the ray traverses an extra pathlength of d in each direction. (The same effect occurs for light striking the surface at other angles, but the expression for the pathlength difference is much more complex because of the lateral displacement between the two reflected rays.) The wave reflected from the

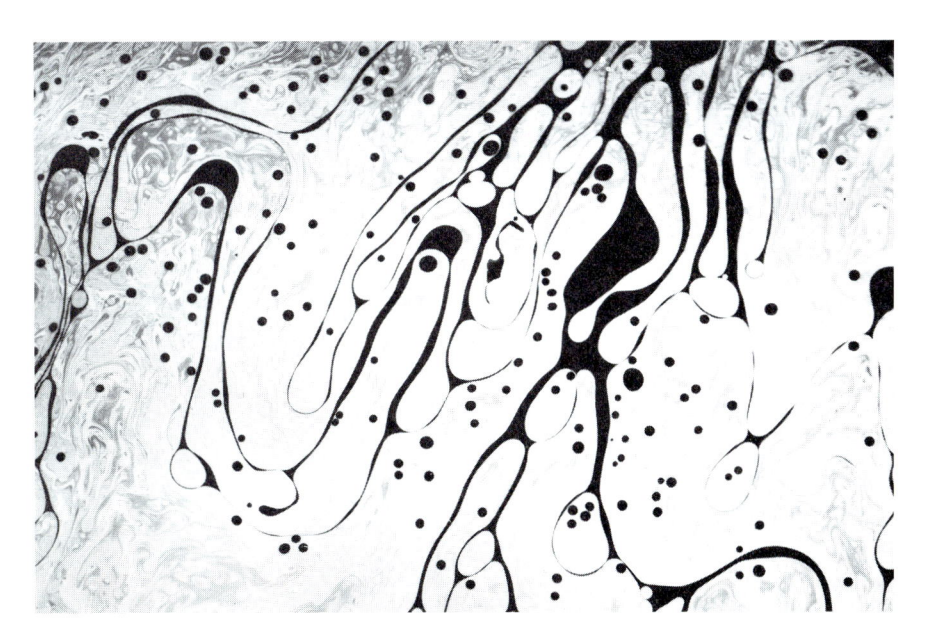

Figure 20.12

Interference in a thin film. Here is a photograph of a multicolored pattern produced by interference in a thin film of oil floating on water.

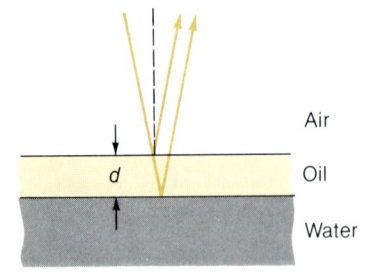

Figure 20.13

Interference in a thin film. The rays reflected from the upper and lower surfaces of the oil layer have a pathlength difference equal to 2d. When the pathlength difference is a multiple of the wavelength, constructive interference occurs. Rays are drawn at an angle for clarity.

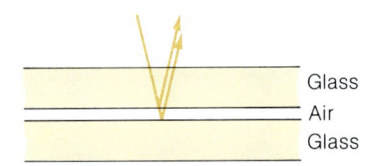

Figure 20.14

Interference in a cavity. Light reflected from the upper and lower surfaces of a thin air gap between layers of glass interferes. Unlike in a thin film, the light changes phase at one reflection but not the other, so the pathlength difference required for constructive interference is now a multiple of $\lambda/2$.

upper surface is in phase with that from the lower surface if this pathlength difference is a multiple of the wavelength. The relevant quantity is the wavelength in the film, which is affected by the index of refraction, as we learned in the previous chapter. For an index of refraction n, the wavelength λ' becomes

$$\lambda' = \frac{\lambda}{n}$$

where λ is the wavelength in a vacuum. Since the pathlength difference must be a multiple of the wavelength, the criterion for constructive interference is

$$d = \frac{1}{2}m\lambda' = \frac{m\lambda}{2n}, \qquad m = 1, 2, 3, \ldots \qquad (20\text{–}6)$$

where m is analogous to the order number we discussed previously. When the incident ray is not perpendicular to the surface, light of different wavelengths will interfere constructively for different angles. For a given angle of incidence, a certain wavelength will interfere constructively, while others interfere destructively. At a slightly different angle, a slightly different wavelength survives the interference, so that as we vary the angle of incidence (by moving our viewing angle) we see a continuous spread of color. Thus white light is dispersed by a thin film, much as it is by a pair or a series of slits.

Reflected light waves in a thin film are similar to standing sound waves in a cavity, including the overtones, because for a given angle of incidence, many wavelengths are evenly divisible into the pathlength through the film (this is the same as saying that the fringes from different orders overlap). Hence the colors produced by a thin film are not pure, but are mixtures of colors created by wavelengths that are simple multiples of each other. In practice, however, only one of the wavelengths falls within the range visible to the human eye, so the colors are actually quite vivid. Note that the film must be truly thin for the color effect. If the film is much thicker than the wavelength of visible light, then there are many wavelengths in the visible spectrum that are able to interfere constructively, and the reflected light will appear colorless.

A phenomenon similar to interference fringes in a thin film occurs when there is a thin cavity between two media, such as two layers of glass (Fig. 20.14). In this case the relative indices of refraction result in a phase shift at one reflecting surface, but not at the other, so the criterion for constructive interference is that the difference in pathlength between two waves be an odd multiple of half-wavelengths. Thus such a cavity forms color bands just as a thin film does, except that the relationship between the angle of incidence and the wavelength for constructive interference is

$$2d = \left(m + \frac{1}{2}\right)\lambda,$$

or

$$d = \frac{1}{2}\left(m + \frac{1}{2}\right)\lambda, \qquad m = 0, 1, 2, \ldots \qquad (20\text{–}7)$$

(The index of refraction does not appear in this expression because n is very close to 1 in air. If the gap were filled with a substance with an index of

20.3 REFLECTION AND INTERFERENCE 677

refraction significantly different than 1, the expression would have $\lambda' = \lambda/n$ instead of λ.)

Consider a cavity with variable thickness, as in Fig. 20.15. For a fixed λ (that is, for monochromatic light), there will be an alternating series of bright and dark bands as the pathlength through the cavity varies smoothly with position. For example, when a curved surface is placed on a flat one (Fig. 20.16) a series of concentric rings is formed at locations where the thickness of the gap creates alternately constructive and destructive interference. These are called **Newton's rings** (Fig. 20.17).

The alternating bright and dark rings (or straight lines, for two plane surfaces and a wedge-shaped cavity) will appear regular only if the two surfaces are very smooth, meaning that variations in their shape are small compared to the wavelength of light being used. Therefore placing surfaces in contact and observing the interference pattern is a useful method for determining how flat and smooth they are.

Thin films on glass can be used as filters to screen out certain wavelengths of light; such devices are called **interference filters**. These are usually designed for light perpendicular to the surfaces, so that Eq. 20–6 or Eq. 20–7 applies (which equation is appropriate depends on the relative index of refraction of the thin film coating compared to that of air or glass). The thickness of the film is designed so that destructive interference occurs for the unwanted wavelength. Destructive interference will occur when the waves reflected from the top and bottom surfaces of the thin film are out of phase. Therefore, if the index of refraction of the film is between the indices for air and glass, so that a phase shift occurs at both surfaces, then destructive interference occurs when

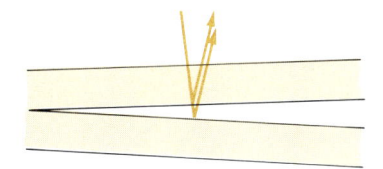

Figure 20.15

Interference in a cavity of variable thickness. The pathlength difference between the two reflected rays varies with position.

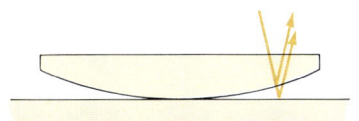

Figure 20.16

Formation of Newton's rings. The convex glass resting on a flat surface forms a circular cavity of variable thickness. The reflected light forms a circular interference pattern known as Newton's rings.

Figure 20.17

A photograph of Newton's rings.

the pathlength difference equals an odd integral number of half-wavelengths; that is, when

$$2d = \frac{\left(m + \frac{1}{2}\right)\lambda}{n}, \qquad m = 0, 1, 2, \ldots \qquad (20\text{--}8)$$

where n is the index of refraction of the film material. When the index of refraction of the film is greater than that of glass,

$$2d = \frac{m\lambda}{n}, \qquad m = 1, 2, 3, \ldots. \qquad (20\text{--}9)$$

Example 20.7 A thin film of oil on the surface of a pond has a thickness of 500 nm. What wavelengths will this film produce for an observer who views it from above? The index of refraction of the oil is $n = 1.25$.

We want to know what visible wavelengths this film will allow to interfere constructively at this angle. The index of refraction of the film is between that of air and water, so there is a 180° change of phase at both reflecting surfaces. Therefore, Eq. 20–6 is appropriate; solving for λ and setting $m = 1$ as an initial guess yields

$$\lambda_1 = \frac{2nd}{m} = \frac{2(1.25)(5 \times 10^{-7}\text{ m})}{(1)} = 1{,}250\text{ nm}.$$

This is in the infrared portion of the spectrum, and will not be visible. We try $m = 2$.

$$\lambda_2 = \frac{\lambda_1}{2} = 625\text{ nm}$$

This is squarely in the visible (orange) portion of the spectrum. The value for $m = 3$ is $\lambda_3 = 417$ nm, which is in the violet, where the eye is less sensitive. Hence the film will appear orange to the observer.

The Michelson Interferometer

A. A. Michelson developed an instrument for making accurate wavelength measurements which uses thin-film interference. Two beams of light from a common source are allowed to interfere. The pathlength of one beam is adjusted to find the pathlength differences corresponding to constructive and destructive interference, which yields the wavelength of the light from the source.

20.3 REFLECTION AND INTERFERENCE

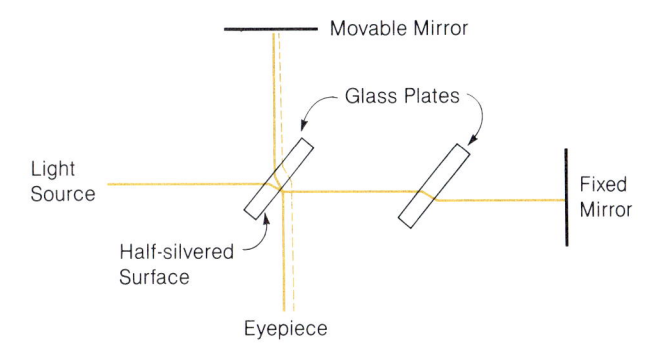

Figure 20.18

The Michelson interferometer. The dashed line illustrates the effect of slightly tilting the movable mirror, shifting the interference pattern at the eyepiece.

The interferometer is illustrated in Fig. 20.18. The beam of light from the source enters a plate of glass set at a 45° angle, which has its back surface half-silvered, so that some of the light is reflected and some passes through. The light that passes through reaches a fixed flat mirror and is returned to the half-silvered surface, from which a portion is reflected to an eyepiece. The part of the incoming beam that is reflected from the half-silvered surface is next reflected to another flat mirror and then back through the half-silvered mirror to the same eyepiece. Each beam passes through the same combination of optical elements (an extra glass plate is inserted in the path towards the fixed mirror to ensure this), and the pathlength difference between the two beams is controlled by moving one of the flat mirrors. When the two beams traverse equal pathlengths, they are in phase when they reach the eyepiece, and they interfere constructively. When the movable mirror is shifted by a quarter-wavelength, the pathlength difference becomes one half-wavelength (because the beam must traverse the extra one quarter-wavelength twice), so the interference at the eyepiece is destructive. If the source of light is monochromatic, then the viewer will see brightness in the first case, and darkness in the second, and will deduce that the wavelength of the light is four times the amount by which the movable mirror was shifted. If the source is white light, then moving the mirror in effect disperses the light, because different wavelengths will interfere constructively as the pathlength difference is varied. Therefore, devices similar to the Michelson interferometer are used in spectroscopy (see Special Problem 2 at the end of the chapter).

For monochromatic light, greater accuracy in measuring the wavelength can be achieved by slightly tilting the movable mirror, so that it is not quite perpendicular to the beam of light. This corresponds to a wedge-shaped cavity (discussed earlier in this section), and the viewer sees a series of bright and dark bands at positions where the pathlength difference between the two beams corresponds to a multiple of the wavelength or to an odd multiple of the half-wavelength. Now as the mirror is moved, the fringes move sideways in the field of view (Fig. 20.18). In effect, the very small motion of the mirror along

FOCUS ON PHYSICS

Space Geodesy and Satellite Ranging

Measuring distances accurately has been important to society for millenia. The architects of Egypt and Babylon were expert surveyors, and the Greeks invented geometry, which contains the mathematics of measurement. In today's world of satellites and radio waves, however, measurement has expanded to the scale of continental and interplanetary distances.

Geodesy is the study of the size and shape of the Earth, including the precise location of positions on its surface. With lasers and radio interferometry, distances can now be measured 10 to 100 times more accurately than with conventional techniques. And relative movements as small as 1–2 cm/yr along earthquake faults can be detected. Geologists and oceanographers are also using these ranging techniques to study the wobble of the Earth's poles, the height and speed of ocean currents, and the distances to the moon and planets.

The basis for the radio measurements is **Very Long Baseline Interferometry** or **VLBI**. This technique was first developed by radio astronomers to study the structure of very distant radio galaxies and quasars with unprecedented angular resolutions, $\Delta\theta \approx 10^{-3}$ arc sec. Because of the diffraction limit (Eq. 20–4), the only way to achieve such resolution for radio wavelengths (6–20 cm) is to use a large separation between antennas ($D = 1{,}000$ km or more). The antenna separation is known as the **baseline,** analogous to a surveyor's baseline. Hence the name VLBI (Fig. 20.1.1).

To measure distances, the signals from the distant sources are recorded on tape by the two antennas and compared to atomic clocks, which keep time to an accuracy of one part in 10^{10}. Because the radio waves arrive at different times, they are slightly out of phase. This phase difference can then

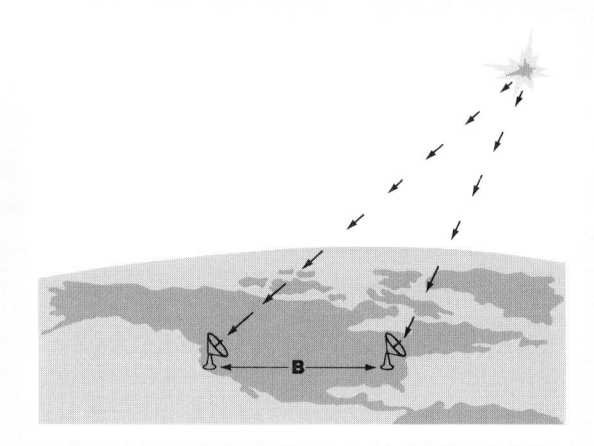

Figure 20.1.1

be used to determine the distance between antennas. Because of the extreme accuracy of the atomic clocks, distances can be measured to an accuracy comparable to the wavelength of the radio waves—that is, centimeters.

Distance ranging can also be done using lasers instead of radio interferometry. Laser ranging observatories on the ground transmit short pulses of light to satellites equipped with special reflectors. When the round-trip time of the light is recorded and compared with the known orbit of the satellite, the distances to the ground stations can be determined and monitored. The United States and other countries have launched over fourteen ranging satellites, and reflectors were left on the surface of the moon by the United States *Apollo* astronauts and by unmanned Soviet probes. Laser ranging to the moon has provided important information on its shape, internal structure, and orbit. Lasers have

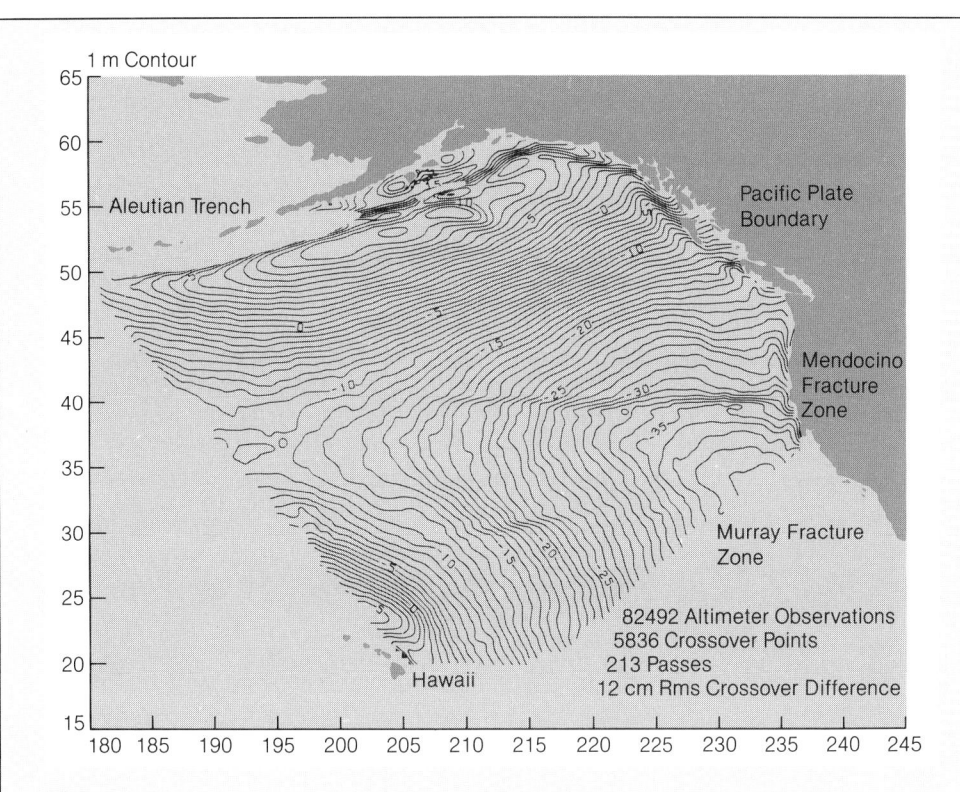

Figure 20.1.2

*Mean sea-surface topography at 1 m contours of eastern Pacific ocean, measured by NASA's **SEASAT** satellite with radar ranging. (From J. G. Marsh et al., EOS, Vol. 63, No. 9, 1982). Note the Aleutian trench and Mendocino fracture zone.*

also monitored the Earth's rate of rotation and the wobble of its pole.

One of the first scientific uses of VLBI and laser ranging was in the study of earthquakes. Geologists know that earthquakes are common along the boundaries between the approximately 20 rigid tectonic plates that move across the earth's surface (see *Focus on Physics* in Chapter 10). By monitoring the distance to the antenna stations on either side of a fault line, geologists can study plate movement and determine whether the fault is locked, suggesting that an earthquake might occur at some time in the future. For example, stations across the San Andreas Fault in Southern California show that the 900-mile baseline between stations is getting shorter by 8 cm/yr. The boundary between the Pacific and Nazca plates is moving at about 15 cm/yr! However VLBI measurements across the 4,000-mile baseline between California and Massachusetts show no detectable movement of the mid-continent (less than 1 cm/yr).

Satellite ranging has important uses in oceanography as well. Using radio measurements from NASA's **SEASAT** satellite, launched in 1978, scientists have measured the altitude of the ocean (Fig. 20.1.2) to an accuracy of 3–4 cm. Maps of sea-surface topography can then be used to compute ocean currents, which control long-range weather and climate forecasts and provide valuable information for understanding the migrations of fish,

(continued)

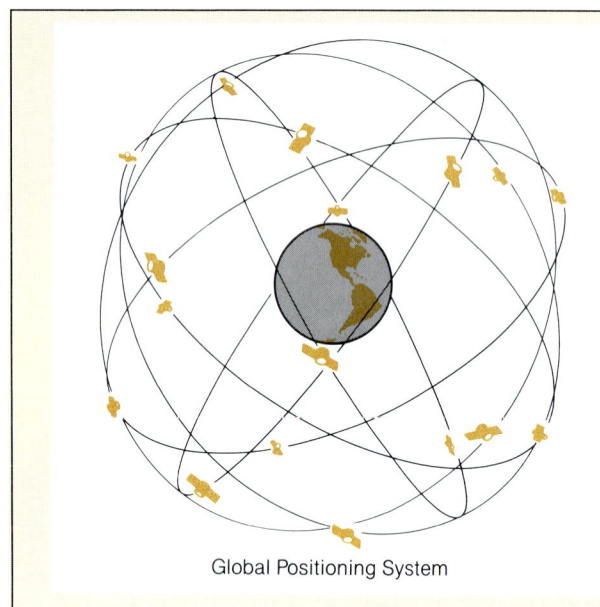

> marine nutrients, and pollution. In the coming decade, NASA plans to launch an even more accurate oceanographic topography satellite known as TOPEX.
>
> Perhaps the most exciting use of radio ranging is the **Global Positioning System** or **GPS**, an advanced navigation system of 18 satellites to be in place by 1988 (Fig. 20.1.3). With a single GPS radio receiver, a ground station, aircraft, or ship at sea can locate its position to better than 10 m relative to four satellites that will always be visible. After the satellite orbits are better known, accuracies of 1–2 cm are expected over distances of 200 km. The US Geodetic Survey expects GPS to provide its primary surveying tool in the coming decades.
>
> Global Positioning System
>
> **Figure 20.1.3**

the direction of the beam has been magnified into a much larger sideways motion of the interference fringes, which can therefore be measured more accurately.

The accurate wavelength measurements allowed by the Michelson interferometer led to a definition of the meter that was the standard until very recently. Previously, the meter had been defined as the length of a standard metal bar, but careful measurement with the interferometer showed that the bar is equal to 1,650,763.73 times the wavelength of light emitted by the isotope ^{89}Kr of the element krypton. This was adopted as the definition of the meter, because in principle any laboratory can use a krypton light source and establish its own accurate meter, whereas this was not so easily done when the standard was a particular metal bar located in France. (The meter is currently defined in terms of the well-established and constant speed of light and the frequency of vibration of certain atoms.)

20.4 Polarization and its Applications

In addition to diffraction, refraction, and interference, light displays another property that is associated with waves. This is polarization, and it has to do with the orientation of the electric and magnetic fields that constitute light waves. In this section we first explore the nature of polarization and how it occurs, then see some of its consequences.

20.4 POLARIZATION AND ITS APPLICATIONS

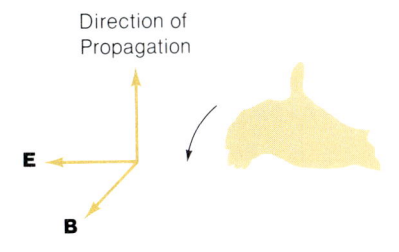

Figure 20.19

Polarization of light. The orientation of the electric field vector (which is in the plane of the page) determines the plane of polarization of light.

Plane and Circular Polarization

We learned in Chapter 18 that light consists of alternating transverse electric and magnetic fields. Because each of these fields operates in a single plane, a light wave has a certain orientation that can be defined by the orientation of these planes. That is, the plane in which the electric field oscillates remains fixed as the wave moves through space, and the same is true of the plane perpendicular to this one, in which the magnetic field oscillates. Thus, we can define a fixed orientation of an electromagnetic wave. This orientation is called the **polarization** of the wave, and by convention it is taken as the direction of the electric field vector (Fig. 20.19).

Under most circumstances, in a group of waves from a light source, the orientations of the individual waves are randomly distributed so that there is no overall polarization. This is true of the light from an ordinary light bulb or from the Sun, for example. There are, however, many circumstances in which the waves are preferentially aligned in a certain orientation, and the aggregate of waves is said to be polarized or **plane polarized.** This alignment can come about because of the manner in which the light is emitted by the source, but more commonly it is due to the way in which the light has been filtered or reflected.

Imagine a narrow slit with transverse waves of various orientations incident upon it (Fig. 20.20). Waves whose planes of vibration are parallel to the slit pass through unimpeded, whereas those oriented perpendicular to the slit may be blocked entirely. Thus, the slit acts as a filter to convert randomly-oriented waves into a polarized aggregate.

Certain natural crystals and manmade materials can perform the same function for light waves. A Polaroid filter, marketed commercially for many years, contains very long molecules that are aligned parallel to each other, so that light waves can pass through only if their electric vectors oscillate in the plane parallel to the axis of the filter, which is the direction of the long molecules. Ideally, waves oriented perpendicular to the filter axis are completely blocked, while those that are parallel to the axis pass through freely. For other waves, only the component of the electric field vector **E** parallel to the filter

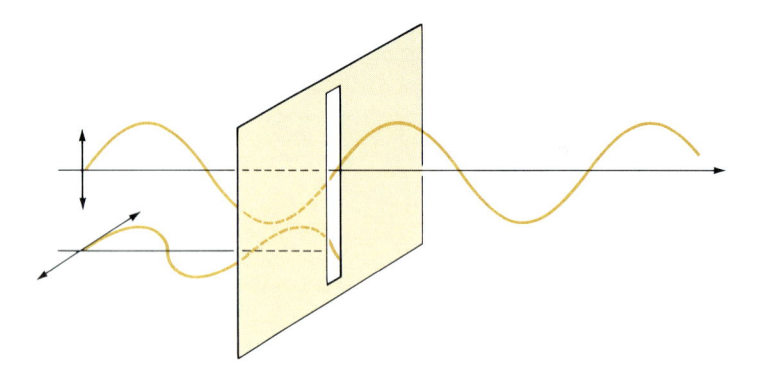

Figure 20.20

A polarizing filter. The effect of a polarizing filter is the same as that of a screen. The slit allows only waves whose polarization is parallel to the slit to pass through. Incident light containing random orientations emerges as polarized light of lower intensity.

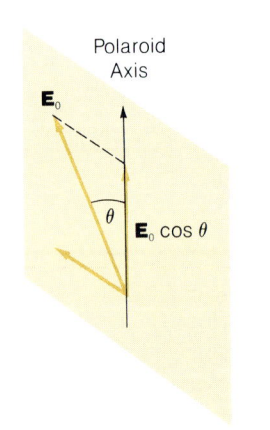

Figure 20.21

Transmitted intensity in a polaroid filter. If an EM wave has an electric field vector E_0 plane-polarized at angle θ with respect to the axis of a polaroid filter, only the parallel component $E_0 \cos\theta$ is transmitted. The transmitted intensity is $I_0 \cos^2\theta$.

axis gets through, so these waves are reduced in amplitude but polarized in the same plane as all the others that pass through the filter. Suppose a plane-polarized wave of amplitude E_0 and intensity I_0 (W/m^2) is incident on a polaroid filter whose axis makes an angle θ with the initial plane of polarization (Fig. 20.21). Only the parallel component of electric field, $E_0 \cos\theta$, is transmitted. Because the wave's intensity is proportional to the amplitude squared, the wave's emergent intensity is

$$I = I_0 \cos^2\theta. \qquad (20\text{--}10)$$

A polaroid filter can be used to determine if the incoming light is already polarized, and to find the angle of orientation of the electric vector. The filter is rotated until the maximum intensity of light passes through it; this corresponds to having the filter axis aligned with the polarization of the incoming light ($\cos\theta = 1$). The angle at which this occurs gives the direction of the electric vector of the light.

There are few natural polarizing filters, but sometimes polarization occurs naturally. When light is reflected at an oblique angle from a surface, waves whose electric vectors are parallel to the surface are most readily reflected (Fig. 20.22), so that the reflected light is at least partially polarized. For this reason, sunglasses made of polarizing filters effectively reduce glare. Light reflected from clouds also tends to be polarized, so the use of a polarizing filter on a camera makes the contrast between clouds and sky sharper.

The degree to which reflection polarizes light depends on the angle of reflection and on the relative indices of refraction of the medium in which the light travels and the medium from which it reflects (see also Special Problem 3). For a given pair of substances, the polarization increases as the angle of incidence θ_i goes from zero (the light strikes the surface at normal incidence)

20.4 POLARIZATION AND ITS APPLICATIONS

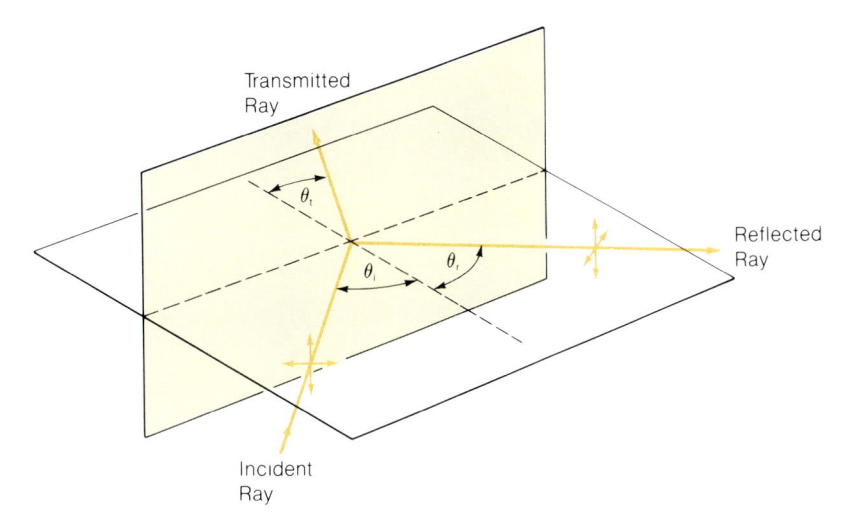

Figure 20.22

Polarization by reflection. Waves whose electric vector is parallel to the reflecting surface are more efficiently reflected. The emergent light tends to be polarized parallel to the surface. When $\theta_i = \theta_p$ the reflected ray is 100% polarized.

to some critical angle where the reflected light becomes completely polarized. This critical angle, called the **polarizing angle,** is given by

$$\tan\theta_p = \frac{n_2}{n_1} \qquad (20\text{–}11)$$

where n_1 is the index of refraction in the medium the light travels in, and n_2 is the index of refraction in the medium from which the light reflects. When the first medium is air, $n_1 = 1$, and the angle of polarization is called **Brewster's angle** ($\tan\theta_p = n_2$). When $\theta_i = \theta_p$, the 100% polarized reflected ray makes an angle of 90° with the transmitted ray (Fig. 20.22). That is, one can show that $\theta_r + \theta_t = 90°$ if $\theta_r = \theta_i = \theta_p$ (see Special Problem 3 at the end of the chapter).

Until now we have only discussed plane polarized light. We know, however, that in general an EM wave may be the superposition of two waves of angular frequency ω, $E_x \cos(\omega t)$ and $E_y \cos(\omega t)$, whose planes of polarization are at right angles to one another (Fig. 20.23). If these waves have equal amplitudes $E_x = E_y = E_0$, then the superposition electric vector is just another plane polarized wave at 45°. However if one wave is shifted in phase by one-quarter cycle, say $E_y = E_0 \cos(\omega t \pm \pi/2)$, the net electric field vector rotates to the right or the left as the wave propagates, tracing out a circle. Waves that rotate to the left are called **left circularly polarized** and those that rotate to the right are called **right circular polarized**; the sense of rotation depends on whether E_y leads or lags E_x by one-quarter cycle. If the amplitudes E_x and E_y

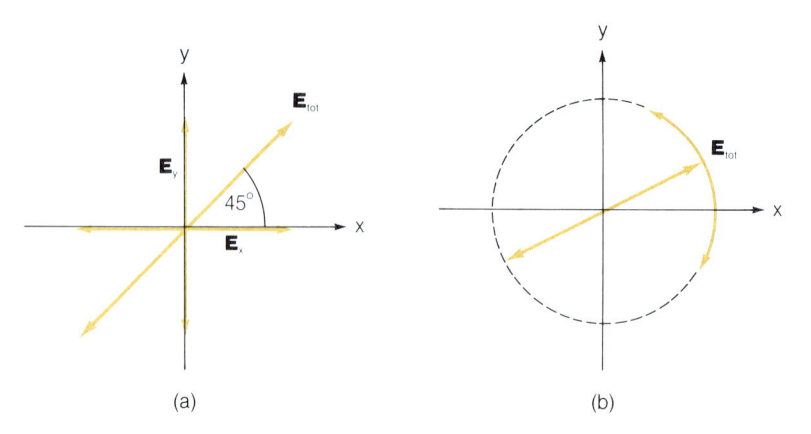

Figure 20.23

Circularly polarized light. An EM wave E_{tot} is the superposition of two waves, E_x and E_y, polarized at right angles. (a) If E_x and E_y are in phase and have equal amplitude, then E_{tot} is also plane polarized at 45°. (b) If $E_x = E_y$ but the waves are one-quarter cycle out of phase, E_{tot} is circularly polarized. The tip of the the electric vector traces out a circle and rotates to the left or right. If E_x and E_y are not equal, or if the phase shift is not one-quarter cycle, the wave is elliptically polarized.

are not equal or if the phase shift is other than one-quarter cycle, the electric vector traces out an ellipse and the wave is **elliptically polarized.**

Some crystals have the property that their index of refraction differs for left and right circular polarizations (these materials are called **birefringent**). If two waves with perpendicular polarization planes enter a birefringent crystal in phase, they propagate at different speeds because of the difference in index of refraction. When they emerge, they are therefore out of phase by an amount that depends on the thickness of the crystal. If the phase shift is exactly one-quarter cycle, the material will transform linearly polarized light into circularly polarized light (remember the definition above). Such crystals are called **quarter-wave plates** and are used to go back and forth between plane polarized and circularly polarized light waves.

Birefringence and Dichroism

We have seen that the speed of light in a medium is slower than in a vacuum. For most substances, the speed of light is the same in all directions, and we say that such a medium is **isotropic,** that is, no direction is preferred. There are **anisotropic** media, however, in which the speed of light does depend on direction of propagation or on the sense of circular polarization. Many crystals display this trait, which gives rise to some unusual phenomena. If the speed of light depends on direction, so does the index of refraction, and such crystals are said to be **doubly refracting,** or, equivalently, **birefringent.**

In certain materials, such as calcite, there is a preferred direction called the **optic axis,** along which light is refracted according to Snell's law. Light

20.4 POLARIZATION AND ITS APPLICATIONS

entering such a material at an angle to the optic axis, however, is split into two rays with polarization planes perpendicular to each other (Fig. 20.24). The **ordinary** ray (called simply the *o* ray), has its plane of polarization perpendicular to the optic axis of the crystal, and its speed is the same in all directions. This ray obeys Snell's law. The second ray, consisting of waves whose polarization plane has a component parallel to the optic axis, has a speed that is different from that of the *o* ray. In most crystals, the speed of this **extraordinary,** or *e* ray is greater than that of the *o* ray. The maximum speed of the *e* ray occurs when its plane of polarization is parallel to the optic axis. If the *e* ray has a speed different from that of the *o* ray, then its index of refraction must also be different, so the *e* ray is refracted differently. Two rays emerge from the crystal. The *o* ray passes through in the direction dictated by Snell's law, while the *e* ray is refracted according to its orientation with respect to the optic axis (the *e* ray actually obeys Snell's law also, but according to the index of refraction for its orientation). The *e* ray may be refracted away from the direction of the optic axis, as in most cases, or closer to its direction, depending on whether the *e* ray has a higher or lower speed than the *o* ray. Objects viewed through a birefringent crystal such as calcite appear doubled; many kinds of gemstones show the same effect.

A related effect is **dichroism.** If a birefringent crystal absorbs waves of one polarization more readily than those of the other, then the intensity of the emerging light varies according to polarization. In a polarizing filter of dichroic material, the pathlength is sufficiently great that one polarization component is essentially eliminated.

Optical Activity

In some materials, the orientation of plane-polarized light is rotated as the light passes through because right and left circular polarized light travel at different speeds. This occurs because the molecules of which the medium is composed are asymmetric; this is common in many complex organic substances such as amino acids, where very large molecules have a spiral structure. This rotation of the polarization plane is called **optical activity.**

The amount by which the polarization plane is rotated depends on the material and the pathlength. It is therefore possible to infer the pathlength (actually the concentration of optically active molecules along a given distance) by measuring the amount of rotation of plane-polarized light (see Special Problem 4). The degree of rotation may also vary with physical conditions that affect the structure of the molecules. Some techniques for analyzing stress in plastics or acidity in solutions, for example, actually measure the degree of rotation.

The direction of the rotation also depends on the substance. The plane of polarization may be rotated either to the left or to the right as viewed along the direction of light travel. Most sugars produce right-handed rotation. The complex protein molecules (RNA and DNA) that make up all amino acids in organisms always create rotation to the left, but it is possible for mirror images of these molecules to exist so that they would rotate the plane of polarization to the right. It is very interesting that some amino acids found a few years ago

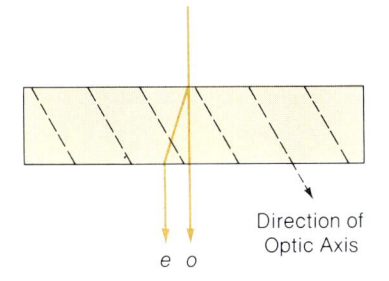

Figure 20.24

Birefringence. When light enters a birefringent crystal, the rays whose polarization is perpendicular to the optic axis obey Snell's law and form the o *ray. Rays with a component of polarization parallel to the optic axis emerge as the* e *ray, because the medium has a different index of refraction for these rays.*

embedded in a meteorite are of the type that produce rotation to the right; this helps support the interpretation that these amino acids might not be a product of life forms that developed on earth, but instead may have formed in some extraterrestrial environment.

Perspective

We have now explored the properties of light that can be understood in the classical picture of waves. We have examined them in terms of their alternating electric and magnetic fields and we have analyzed how they interact with matter in several ways, including reflection, refraction, and diffraction. We have also studied their interaction with each other, interference phenomena.

We have reached a state of physics sophistication comparable to that of a well-educated science student at the turn of this century. To advance our understanding of physical phenomena, we must now explore the vast new territory that began to be charted with atomic theory and quantum mechanics.

SUMMARY TABLE OF FORMULAS

Constructive interference condition, double-slit experiment:
$$d \sin\theta = m\lambda, \quad m = 0, 1, 2, \ldots \tag{20-1}$$

Destructive interference, double slit experiment:
$$d \sin\theta = \left(m + \frac{1}{2}\right)\lambda, \quad m = 0, 1, 2, \ldots \tag{20-2}$$

Destructive interference condition for a single slit:
$$D \sin\theta = m\lambda, \quad m = 1, 2, 3, \ldots \tag{20-3}$$

Destructive interference for a circular opening:
$$\theta = \frac{1.22\lambda}{D} \tag{20-4}$$

Resolving power of a microscope:
$$s = f\theta = \frac{1.22\lambda f}{D} \tag{20-5}$$

Constructive interference in a thin film:
$$d = \frac{m\lambda}{2n}, \quad m = 1, 2, 3, \ldots \tag{20-6}$$

Constructive interference in a cavity:
$$d = \frac{1}{2}\left(m + \frac{1}{2}\right)\lambda, \quad m = 0, 1, 2, \ldots \tag{20-7}$$

Destructive interference for a film on glass, $n < n_{\text{glass}}$:
$$2d = \frac{\left(m + \frac{1}{2}\right)\lambda}{n}, \quad m = 0, 1, 2, \ldots \tag{20-8}$$

Destructive interference for a film on glass, $n > n_{\text{glass}}$:
$$2d = \frac{m\lambda}{n}, \quad m = 1, 2, 3, \ldots \tag{20-9}$$

Intensity of a polarized EM wave:
$$I = I_0 \cos^2\theta \tag{20-10}$$

Critical angle for polarization by reflection:
$$\tan\theta_p = n_2/n_1 \tag{20-11}$$

THOUGHT QUESTIONS

1. Give an example in which light exhibits wave characteristics.
2. Give an example in which light behaves like a particle.
3. Why does the particle theory of light require that light travel faster in a physical medium than in a vacuum?
4. Give everyday examples of light interference and diffraction. Compare and contrast interference and diffraction.
5. What are the similarities and differences between the diffraction from a single slit and a double slit?
6. Why does destructive interference occur at half-integral wavelengths?
7. How might you use interference to measure the angular diameters of light sources of very small apparent size, such as stars?
8. What would the interference pattern look like if there were three slits instead of two slits?
9. Compare and contrast the dispersion produced by a prism and by a grating.
10. How are optical engineers able to produce gratings with several thousand grooves per millimeter?
11. Is a rainbow an example of a diffraction grating or a prism?
12. What would the diffraction pattern look like for a triangular shaped aperture?
13. Using the principles of this chapter, can you think of a method for measuring the thickness of a thin oil film on the ground?
14. How does a pinhole camera work?
15. Why do the individual waves in a group of waves from a light source exhibit no overall polarization?
16. Explain how you might use the phenomenon of polarization to determine the index of refraction of a substance.
17. Explain how the speed of light could depend on the direction of travel in a crystal or other anisotropic medium.
18. Explain how the concept of optical activity can be used to analyze the acidity of a solution or the stress in a plastic.
19. At which visible wavelength does an optical microscope have the best resolution?
20. If an interference experiment is done with white light, containing all frequencies, what effects are observed?
21. Why do polaroid sunglasses help reduce the glare off car windows and other shiny substances?
22. Explain the properties of wave behavior, polarization, interference, and energy in terms of the classical picture of light as an electromagnetic wave.
23. We can hear sounds around corners and receive radio broadcasts from stations over mountains, and yet the human eye can only see objects in a straight line. Explain the difference in terms of wave behavior.
24. Can sound waves interfere? Give an example of a water wave interference pattern.
25. Explain how standing waves in cavities are examples of wave interference.

PROBLEMS

Section 20.1 Young's Slit Experiment

1. A Young's experiment has slits separated by 0.2 mm. At what wavelength will the separation between zero and first order fringes be 8 mm on a screen 20 cm away?

2. What separation of slits is required for a Young's apparatus so that the zero and first order fringes are separated by 20 mm on a screen 50 cm away when light of wavelength 450 nm is used?

3. A Young's apparatus has slits separated by 0.05 mm and is used with light of wavelength 600 nm. At what distance should a screen be placed so that zero and first order fringes are separated by 4 mm?

4. A Young's apparatus has slits 0.3 mm apart and uses 500 nm light. What is the separation between third and fourth order fringes if the back screen is 75 cm from the slits?

5. A Young's apparatus has a screen 1 m from the slits.

690   CHAPTER 20   WAVE PHENOMENA IN LIGHT

If the distance between first order blue light and first order red light is 10 mm, what is the slit separation?

- 6 What slit separation is needed for a Young's apparatus in which the screen is 60 cm away and the first order dispersion, $\Delta\lambda/\Delta y$, along the screen is 50 nm/mm?

- 7 A Young's apparatus is set up with slits separated by 0.05 mm and a screen 50 cm away. If the light has wavelength 600 nm, where is the first dark fringe on the screen?

- 8 What fringe spacing would be observed for the apparatus in Problem 7?

- 9 Suppose the slit separation in Problem 7 was widened to 0.15 mm. Where should the screen be placed to obtain the same fringe separation?

- 10 At what angle relative to zero order would a second order bright fringe appear for the apparatus in Problem 7?

Section 20.2 Diffraction and Interference

- 11 A grating is ruled with 2,000 grooves/mm. What is the groove separation? Is this separation comparable to wavelengths of visible light?

- • 12 Find the second order dispersion in nm/mm of a Young's apparatus in which the back screen is 150 cm from the slits and the slit spacing is 0.05 mm.

- • 13 How far away should a screen be placed for a Young's apparatus with slit separation 0.2 mm to have a first order dispersion of 75 nm/mm?

- • 14 Find the locations of the red light in the first, second, and third order fringes from a transmission grating with 100 grooves/mm on a screen 150 cm away.

- • • 15 Find the dispersion near $\lambda = 500$ nm of the first, second, and third order fringes for a transmission grating with 200 grooves/mm on a screen 2 m away. You may not use the small-angle approximation.

- • 16 How many grooves/mm should a grating have to produce first order dispersion of 100 nm/mm on a screen 1 m away?

- • 17 Suppose you wish to construct a radio wave version of Young's apparatus, using 20 cm waves and a screen 5 m from the slits. What slit spacing will result in a separation of 20 cm between first and second order fringes?

- 18 What single slit width will produce a first order dispersion of 200 nm/mm on a screen 150 cm away?

- 19 Calculate the angle θ for a second order Young's apparatus fringe in blue light (475 nm) with a slit separation of 1 cm. Is the angle detectable?

- • 20 Plot the diffraction angle θ versus slit width D for a single slit transmitting light with 500 nm wavelength.

- 21 Compare the diffraction angle θ for a visible telescope 1 m in diameter observing 500 nm light with that of a 4 m infrared telescope operating at 2,000 nm.

- 22 How large would a radio telescope operating at a wavelength of 21 cm need to be to obtain the same angular resolution as a 2 m telescope operating at 500 nm?

- • 23 What diameter lens is required in an optical microscope to resolve an amoeba, 0.1 mm in diameter and 2 cm from the lens?

- 24 Can the human eye distinguish two points separated by 1 cm at a distance of 100 m? Assume the eye operates with light at 500 nm.

- 25 What is the smallest size object the human eye can resolve at a distance of 30 cm?

- 26 A telescope can resolve a quarter (2.5 cm diameter) at a distance of 10 km. What is the diameter of the telescope's lens? Assume 500 nm light and neglect effects of the earth's atmosphere.

- 27 If the quarter in Problem 26 were moved to 50 km, what aperture size would be required? Assume 500 nm light and neglect atmospheric effects.

- • • 28 What is the diffraction limited angular resolution for the 2.4 m diameter Space Telescope, operating at 200 nm? What is the closest distance between stars this telescope could resolve in the Andromeda galaxy, two million light-years away? One light-year equals 9.45×10^{15} m.

- 29 A biologist wishes to study an amoeba with diameter 5×10^{-5} m, using light of wavelength 450 nm. What fraction of the amoeba's diameter can be resolved if the microscope lens has a diameter of 0.5 cm and focal length 0.3 cm?

- 30 Find the resolving power of a microscope which operates with 500 nm light, and has a diameter 0.5 cm and a focal length of 0.5 cm.

- • • 31 Light of wavelength 500 nm passes through a circular aperture of 0.2 mm diameter. How large is the bright spot at the center of the diffraction pattern formed on a screen 1 m away?

- • • 32 What angles can be resolved by the aperture in Problem 31?

- • • • 33 An FM radio station at 103 MHz uses two antennas separated by 10 m. The antennas are identical and are operated in phase, so that the radiation has a maximum at an angle of 0° from the perpendicular

PROBLEMS

to the line between centers. Find the other angles at which there are maxima and minima in intensity.

••• 34 An AM radio station at 1,560 kHz uses two identical antennas separated by a distance d. If the line between antennas is aligned north-south and the station wishes to broadcast with an intensity maximum in a direction 30° north of east, what minimum separation d is needed?

•• 35 In a 35 mm camera, a lens has a focal length of 50 mm and an f-ratio of f/1.8. If the angular resolution of this lens is determined by the diffraction limit, what is the smallest object that it can resolve at a distance of 30 m in blue light (475 nm)?

••• 36 The resolution of a camera lens is often expressed in terms of the number of lines/mm in the image that are distinguishable. For an object beyond several meters from the camera in Problem 35, what is the size of a line just resolved according to the diffraction limit? What is the number of lines/mm at f/1.8? What is the number of lines/mm at f/16?

Section 20.3 Reflection and Interference

•• 37 Discuss the difference between Eq. 20–8 and Eq. 20–9. What is the number of wavelength shifts in a thin film whose index of refraction is greater than that of glass?

• 38 How would Eq. 20–7 be modified, if at all, if the index of refraction were significantly different from unity?

•• 39 A thin film has a reddish color (650 nm) when viewed from the vertical. If the film is 300 nm thick, what is its index of refraction?

• 40 Suppose the oil film in Example 20.7 had $n = 1.35$. What color would the observer see?

•• 41 An optician wishes to have a thin film with $n = 1.3$ appear red (650 nm). What minimum thickness film is required if the observer's angle is vertical?

••• 42 Suppose the index of refraction of the oil film in Problem 40 were 1.50 instead of 1.35. How would this change the results?

•• 43 What visible wavelengths will constructively interfere in a film 500 nm thick, observed from above? Assume the film has $n = 1.2$ and lies on glass.

•• 44 Calculate the wavelengths of the first three orders of light which would constructively interfere when reflecting from a film on glass 1,000 nm thick with $n = 1.3$, observed from above.

• 45 A glass prism with index of refraction 1.55 is made nonreflecting at 500 nm by a thin film coating. What is the wavelength of this light in the film of index 1.30? How thick must the film be so that light reflected at 0° from inner and outer surfaces is one half-wavelength out of phase?

• 46 Radio waves from a great distance reflect from a nearby rock formation and arrive at your antenna out of phase. If the reflection does not change the phase of the waves, how far must the rock be for (a) constructive interference, (b) destructive interference? Assume the station is AM 1,400 kHz.

• 47 What is the wavelength of a light in an interferometer if 1,000 fringes are counted as the movable mirror moves 0.25 mm?

• 48 How far must the mirror in a Michelson interferometer be moved if 500 fringes pass a reference point when 589 nm light is used?

• 49 The krypton-86 light previously used as the meter standard has 1,650,763.73 wavelengths per meter. What is its wavelength? How far must the movable mirror in a Michelson interferometer move to produce 600 fringes passing a reference point?

Section 20.4 Polarization

• 50 Calculate the polarizing angle for light in water, reflected from glass.

• 51 The polarizing angle for light in water, reflected from an immersed object, is 56°. What is the substance's index of refraction?

• 52 What is Brewster's angle for the substance in Problem 51?

•• 53 Find the polarizing angles for diamond-water interfaces and Brewster's angle for light reflecting from diamond.

•• 54 The polarizing angle for an interface between a substance with index n_1 and glass with $n_2 = 1.5$ is 60°. What is n_1? Are any substances likely candidates?

• 55 For air, with n approximately 1.00, calculate the limits on n_2 if Brewster's angle is less than 60°.

•• 56 Unpolarized light passes through two polarizing filters whose axes have an angle θ between them. After passing through the first filter, the waves have an electric field E_0 and an intensity I_0.
  (a) Show that the electric field after passing through the second filter is $E_0 \cos\theta$.
  (b) Show that the intensity of the emergent waves, after passing through both filters, is $I_0 \cos^2\theta$.

• 57 Use the results of Problem 56 to find the intensity of a wave emerging from two polarizing filters oriented at 30° to one another if the intensity after the first filter is 10^{-5} W/m².

- 58 A wave is reduced in intensity by a factor 0.25 by passing through a second polarizing filter of the type described in Problem 56. What is the angle between the filters?

••• 59 Unpolarized light passes through three identical polarizing filters, each of whose axes is rotated by 30° with respect to the previous one. After one filter the waves have electric field E_0 with intensity I_0.
   (a) What are the electric field strength and intensity after the second filter?
   (b) What are the electric field strength and intensity after the third filter?
   (c) Compare the intensity in part (b) to that which would emerge if the second filter were removed. Explain the difference in terms of the wave nature of light.

• 60 An undersea vessel sends a radio wave upward at angle θ from vertical. At what angle are the waves internally reflected? At what angle is the reflected wave most strongly polarized? Assume the index of refraction is constant at 1.34.

••• 61 Sunlight reflects from the windshields ($n = 1.5$) of cars headed directly toward or away from the sun. The windshields are inclined at an angle α < 90° from horizontal, and the sun makes an angle φ with the vertical.
   (a) What is the angle φ for which the reflected rays leave the windshields horizontally and reflect into the eyes of oncoming drivers? Express your answer in terms of α.
   (b) If α = 60°, what is φ in part (a)? Which driver receives the glare?
   (c) If α = 34°, what angle φ corresponds to reflections with the most polarization? Which driver receives the glare? At what angle from horizontal do these polarized rays reflect?

SPECIAL PROBLEMS

1. X-Ray Crystallography

In the early 20th century, physicists found regular diffraction patterns when x-rays were scattered off crystals. These patterns can be used to measure the interatomic spacings in crystals, and the technique has become known as x-ray crystallography. X-ray diffraction has also led to the determination of the structures of many biologically important molecules, including proteins and the DNA "double helix" (*Focus on Physics*, Chapter 24). The simplest crystal structure is the cubic lattice, as in table salt (NaCl).
   (a) Consider the beam of x-rays incident on a cubic lattice at angle θ (Fig. 20.25). Two x-rays are reflected from successive horizontal planes of atoms. Find the extra distance travelled by the second ray and show that these waves will constructively interfere if $2d \sin\theta = m\lambda$ ($m = 1, 2, 3 \ldots$), where d is the distance between planes (here d equals a, the interatomic spacing). This conditions is known as the **Bragg condition** and the angles θ are known as **Bragg angles**.
   (b) The Bragg angles correspond to x-ray reflections from successive planes of atoms. There are other planes besides horizontal that will lead to this effect (like trees lining up in an orchard). Draw a two-dimensional cubic lattice and find three other planes besides horizontal and vertical. What are their separations d?

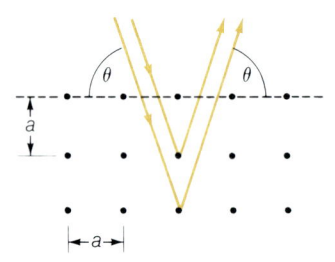

Figure 20.25

   (c) Suppose 50 keV x-rays are found to diffract at angles θ = 1.26°, 2.52°, and 3.78°. What interatomic spacing a would you assign to this cubic lattice?

2. Michelson Interferometer

The Michelson interferometer is used to measure wavelengths and distances very precisely through interference fringes viewed through an eyepiece (Fig. 20.18). Suppose one laser beam of the interferometer, with vacuum wavelength λ_0, passes through a cavity of length L and cross-sectional area A, initially a vacuum. As gas is slowly admitted into the cavity to a final pressure P at 25°C, N interference fringes pass a reference wire in the eyepiece.
   (a) Show that the change in the number N of wavelengths in the beam's path is $(2L/\lambda_0)(n - 1)$, where

n is the index of refraction of the gas at pressure P.

(b) Find a formula for $(n - 1)$ in terms of N, L, and λ_0. Suppose $L = 20$ cm, $A = 2$ cm^2, and $\lambda_0 = 633$ nm. If 20 fringes pass while the pressure is increased to 0.1 atm, what is the index of refraction at 0.1 atm?

(c) If $(n - 1)$ is proportional to the density of gas molecules (molecules/cm^3) in the cavity, what index of refraction would you expect at 1 atm pressure? How many fringes would pass the reference point? Assume the ideal gas law.

3. Reflection and Transmission

When a beam of light strikes an interface at angle of incidence θ_i, part is reflected at angle $\theta_r = \theta_i$ and part is refracted (transmitted) at angle θ_t satisfying Snell's law, $n_1 \sin\theta_i = n_2 \sin\theta_t$. Here n_1 and n_2 are the indexes of refraction, with $n_2 > n_1$. Electromagnetic theory gives the reflection coefficients, equal to the ratio I_r/I_t of reflected and transmitted intensities. For light polarized with electric field **E** perpendicular to the plane of incidence (plane of the paper), the coefficient is

$$R_\perp = \frac{\sin^2(\theta_i - \theta_t)}{\sin^2(\theta_i + \theta_t)}$$

while for **E** polarized parallel to the plane of incidence,

$$R_\parallel = \frac{\tan^2(\theta_i - \theta_t)}{\tan^2(\theta_i + \theta_t)}.$$

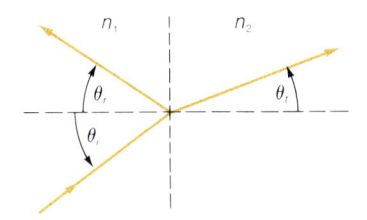

Figure 20.26

(a) Suppose light travelling in air ($n_1 = 1.00$) strikes a pane of glass ($n_2 = 1.52$) at angle $\theta_i = 30°$. Calculate the reflection coefficients for both polarizations. What fraction of light is transmitted in each case?

(b) Show that when $\theta_i = \theta_p$, the polarizing angle, light is reflected with 100% polarization perpendicular to the plane of incidence. That is, show that $R_\parallel = 0$ while $R_\perp > 0$. Note that $R_\parallel = 0$ when $\theta_i + \theta_t = 90°$. (Polaroid sunglasses take advantage of this fact, since reflected light is predominantly polarized.)

(c) Show that when light strikes the interface at normal incidence, that is, $\theta_i = 0°$, the reflection coefficients are the same for either polarization.

$$R_\perp = R_\parallel = \frac{(n_2 - n_1)^2}{(n_2 + n_1)^2}$$

To avoid the meaningless expression 0/0, use the small angle approximations for $\sin\theta$ and $\tan\theta$ and Snell's law, then let θ_i and θ_t go to zero.

(d) From the results in part (c), compute the fraction of light reflected at normal incidence from glass ($n_2 = 1.52$) and from water ($n_2 = 1.34$).

4. Optical Activity

When a plane polarized beam of light passes through some crystals and many solutions, the plane of polarization is rotated by an angle θ that depends on the path length L and, in the case of solutions, on the concentration of the dissolved optically active molecules. Quartz, sugars, proteins, and amino acids are optically active. Those which rotate the plane to the right are called **dextrorotatory** and those which rotate to the left are **levorotatory** (dextro and levo for short). Most sugars are dextro, while most amino acids and proteins are levo.

(a) For a solid with pathlength L, the rotation angle $\theta = \alpha_1 L$, where α_1 is a proportionality constant called the **specific rotation,** or sometimes the optical rotatory power. This constant depends on both wavelength and temperature. For quartz at 20°C, α_1 equals 17,318 degrees/m at 656 nm (red), 27,543 degrees/m at 527 nm (green), and 47,481 degrees/m at 410 nm (blue). If plane polarized white light passes through a 2 cm thick quartz crystal, what will be the emergent polarization angles between red, green, and blue light?

(b) For a solution the rotation angle $\theta = \alpha_2 nL$, where n (g/cm^3) is the density of dissolved molecules in the solution and L is the pathlength. The specific rotations α_2 (degrees cm^2/g) at 589 nm for three sugars are $+6.64$ (sucrose), $+5.25$ (dextrose), and -5.14 (L-glucose). Three samples are made with each sugar dissolved in 100 grams of water at 20°C, and plane polarized light at 589 nm is passed through a 50 cm column of solution. The rotation angles are $+3.32°$, $+6.56°$, and $-7.71°$ for sucrose, dextrose, and L-glucose. How many grams of each are in solution?

SUGGESTED READINGS

"Studying polarized light with quarter-wave and half-wave plates of one's own making." J. Walker, *Scientific American,* Dec 77, p 172, **237**(6)

"The bright colors in a soap film are a lesson in wave interference." J. Walker, *Scientific American,* Sep 78, p 232, **239**(3)

"Mysteries of rainbows, notably their rare supernumerary arcs." J. Walker, *Scientific American,* Jun 80, p 146, **242**(6)

"Dazzling laser displays that shed light on light." J. Walker, *Scientific American,* Aug 80, p 158, **243**(2)

"Interference patterns made by motes on dusty mirrors." J. Walker, *Scientific American,* Aug 81, p 146, **245**(2)

"Heinrich Hertz." P. Morrison, E. Morrison, *Scientific American,* Dec 57, p 98, **197**(6)

"Polarized-light navigation by insects." R. Wehner, *Scientific American,* Jul 76, p 106, **235**(1)

"Henry A. Rowland." A. D. Moore, *Scientific American,* Feb 82, p 150, **246**(2)

"Radio astronomy by very-long-baseline interferometry." A. C. S. Readhead, *Scientific American,* Jun 82, p 52, **246**(6)

"Michelson and his interferometer." R. S. Shankland, *Physics Today,* Apr 74, p 36, **27**(4)

The flying circus of physics with answers. Jearl Walker, Wiley, New York, 1977

A source book in physics. William F. Magie, Harvard University Press, Cambridge, 1965

Science from your airplane window. Elizabeth Wood, Dover, NY, 1975

PART SIX

ATOMIC STRUCTURE AND MODERN PHYSICS

Chapter 21
Relativity 696

Chapter 22
Particles and Waves 728

Chapter 23
Light and the Atom 756

Chapter 24
Essentials of Quantum Mechanics 786

Chapter 25
The Nucleus and Nuclear Energy 826

Chapter 26
Elementary Particles and Cosmology 874

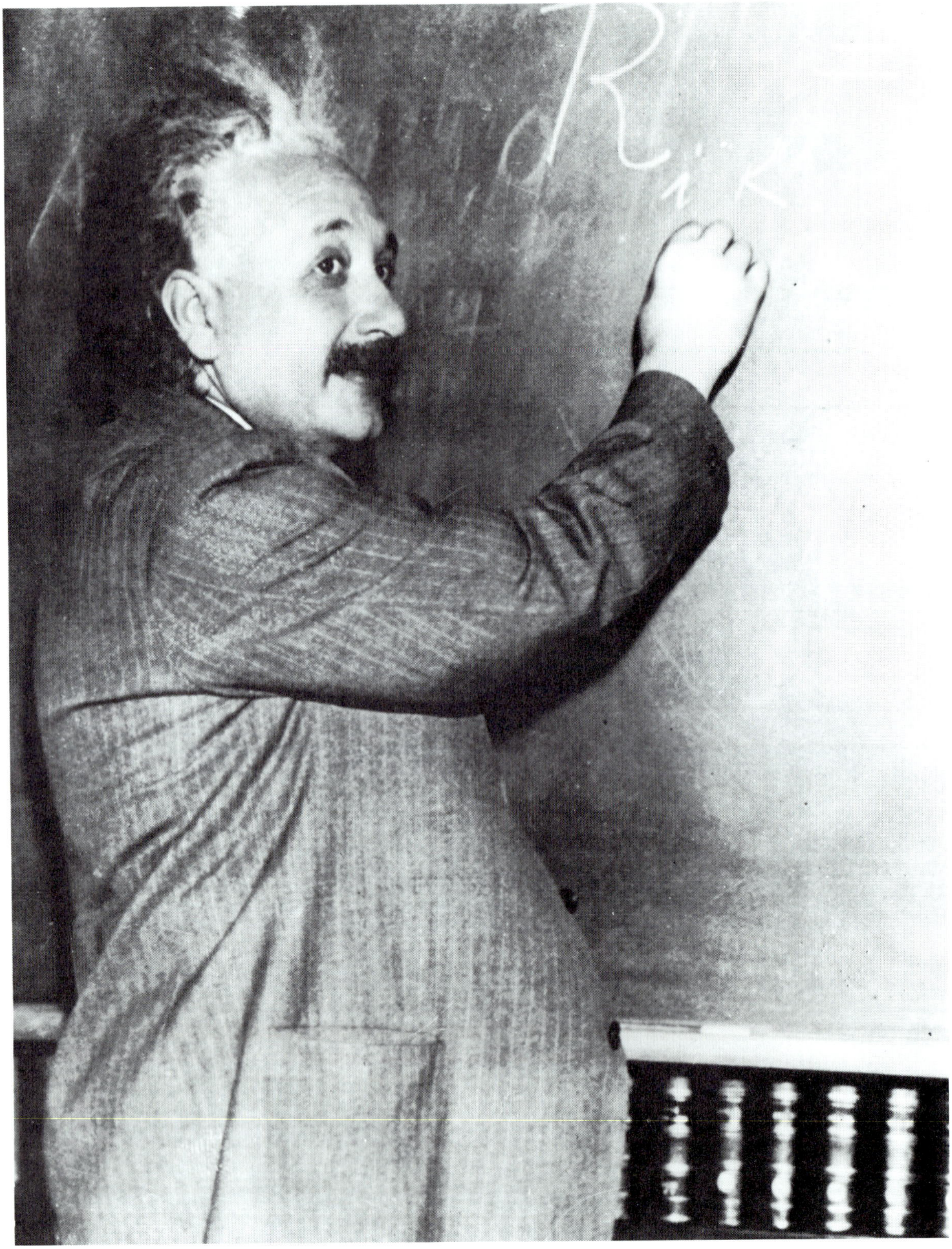

Chapter 21

Relativity

It is necessary at this point to return to the subject of mechanics, even though we devoted some seven chapters to it at the beginning of this text. The material included in those chapters was *classical* mechanics, which can be defined as the science of motion as described by Newton's laws. Classical mechanics served us well in our discussions of the properties of solids, thermodynamics and fluid dynamics, and played only a small role in our studies of electricity and magnetism and of electromagnetic radiation. As we embark on a course leading us through the essentials of modern physics, we will find classical mechanics inadequate to explain many important phenomena. We must therefore insert an interlude for a discussion of relativistic mechanics or relativity theory.

The word "relativity" itself has been applied classically to any situation where physical behavior is observed relative to different frames of reference. Thus, we may speak of Newtonian relativity (or Galilean relativity, since its principles were first expressed by Galileo) when Newton's laws are adequate, or we may discuss **special relativity** or **general relativity**, both developed by Albert Einstein during the first few years of this century, which must be used when Newtonian mechanics is inadequate. In this chapter we will briefly describe Galilean relativity before undertaking our more thorough treatment of special relativity. We also include a short qualitative discussion of general relativity.

21.1 Galilean Relativity

The concept of inertia can be extended to allow us to define an **inertial reference frame** as any frame of reference in which Newton's first law is valid. In an inertial frame, an object in a state of rest or of uniform motion remains in

698 CHAPTER 21 RELATIVITY

that state unless an outside force acts upon it. This is clear in a frame that is at rest, and it is also true in a frame that is moving at constant velocity. As an example, consider an advertisement used to promote air travel during the early days of commercial jets: the ride was said to be so smooth that a coin could be balanced on edge in a speeding plane. Assuming a lack of turbulence or other factors that might exert outside forces, this is possible, but only because the first law is valid in the frame of reference of the airplane moving at constant velocity. If a frame is accelerated, it is **noninertial**.

Having defined an inertial frame of reference, we can now state a principle postulated by both Galileo and Newton, and now known as the **Galilean relativity principle**:

The laws of mechanics are the same in all inertial reference frames.

This postulate may seem quite straightforward, but it has interesting consequences. For example, the coin standing on edge in a moving airplane has quite different states of motion to observers in different reference frames. To a person on board the plane, the coin is motionless, but to a person on the ground or in some other inertial frame, the coin is moving at constant velocity.

The behavior of the coin as seen from the reference frames of the plane and the earth is consistent with the relativity principle. The physical laws that describe the coin's motion in the two frames are the same; the only difference is that in one frame (that of the plane) there is zero initial velocity, whereas in the other frame (that of the earth) there is an initial lateral velocity. Thus, the laws of motion developed in Chapter 3 of this text can be applied directly to either frame.

If we now consider an object that is dropped on a moving airplane, similar effects appear (Fig. 21.1). Neglecting air resistance, the observer on the plane would see the object drop straight down, accelerating at 9.8 m/s^2 and hitting the floor directly beneath the point of release. With respect to the earth, however, the object follows a curving path, still accelerating downward at 9.8 m/s^2, but also moving laterally in the direction of the plane's motion.

If we have two inertial frames whose relative positions and velocities are known, it is easy to find the velocity of an object in one frame if it is known in the other (Fig. 21.2). If vector **v** represents the velocity of an object in one frame, and **v**$_{fr}$ the velocity of that frame with respect to a second frame, then the velocity of the object in the second frame is **v**′, the vector sum of **v** and **v**$_{fr}$ (note that the frame velocity is **v**$_{fr}$ = **v**′ − **v**). Displacement vectors can be added similarly to find the displacement of an object in one frame from its known displacement in the other.

Acceleration translates even more easily from one inertial frame to another. By definition, the acceleration of an inertial frame is zero, so there is no relative acceleration between frames. Thus, an object's acceleration in one inertial frame is the same as in any other. When an object was dropped inside a moving plane, the acceleration in the frame of the plane was the same as that in the frame of the Earth: 9.8 m/s^2 downward.

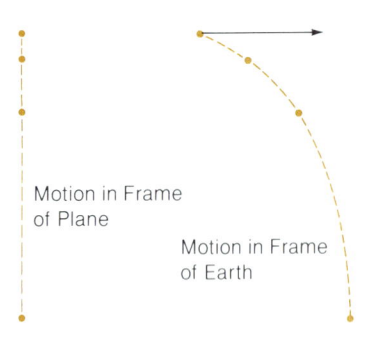

Figure 21.1

The path of an object dropped in a moving airplane. To a person on a moving airplane, a dropped object appears to fall straight down under the acceleration of gravity. To a person at rest in the earth's frame of reference, the downward acceleration is the same but there is lateral motion at the velocity of the plane.

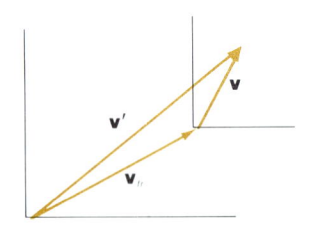

Figure 21.2

Relative velocity. If frame fr *is moving at velocity* **v**$_{fr}$ *with respect to some other reference frame, and an object has velocity* **v** *in frame* fr, *then the object's velocity in the other frame is* **v**′, *the vector sum of* **v** + **v**$_{fr}$.

21.2 SPECIAL RELATIVITY

Because the laws of motion operate in the same way in all inertial frames, there is no fundamental reference frame. It is not possible, or necessary, to define a frame that is absolutely at rest. This was implicit even in the early chapters of this book, where we ignored the fact that the Earth is moving through space at substantial velocity with respect to the Sun, that the Sun moves in the galaxy, and so on. Without saying so, we assumed that the Earth is an inertial frame, and developed the laws of motion as if it were at rest. (Actually, rotational effects create minor deviations, and a position on the surface of the Earth is technically not an inertial frame.)

Following Newton's work in the late seventeenth century, no shortcomings of the relativity principle were uncovered for nearly two centuries. Any complacency that physicists may have felt over this success began to erode in the late 1800s, however, as new developments revealed inconsistencies.

21.2 Special Relativity

The first hint that the relativity principle as expressed by Galileo and Newton might not fully describe all situations came with Maxwell's theory of electromagnetic radiation. As we learned in Chapter 18, Maxwell's wave theory gives a specific prediction of the speed of light, $c = 1/\sqrt{\epsilon_0 \mu_0}$. The dilemma was to define the frame of reference in which the theoretical speed of light is the speed that is actually measured. It appeared that if there were such a frame, its existence would violate the relativity principle, because it would be a preferred, fundamental frame of reference.

Related to this was the difficulty that mechanical waves always required a medium in which to propagate, yet light did not. It was therefore assumed that there was some sort of universal, nonmaterial medium for light propagation, called the **ether**, and that the ether established the fundamental frame of reference in which the speed of light is that predicted by Maxwell. In an important experiment carried out in 1887 by A. A. Michelson and E. W. Morley, direct evidence for the existence of the ether was sought. An interferometer was used to measure the speed of light in different directions (Fig. 21.3). No difference was found, even though the motion of the Earth with respect to the ether should have created one. The failure of the Michelson-Morley experiment to detect an ether created a serious problem for Galilean relativity.

The experiment further suggested that the speed of light was the same in all reference frames. This problem and its implications for electromagnetic radiation stimulated Einstein particularly. If light travels at a certain speed in some inertial frame, then the Galilean relativity principle implies that there should be an observer's frame, one travelling at the speed of light with respect to the fundamental frame, in which the speed of light is zero. It was difficult (although not logically inconsistent) to envision light waves with zero velocity, yet Maxwell's theory states that electromagnetic radiation is always emitted by accelerating electrical charges, which could exist in an inertial frame moving at any velocity. The difficulty of accepting electromagnetic waves with zero velocity and the results of the Michelson-Morley experiment prompted Einstein to modify the conventional view of the nature of space and time itself.

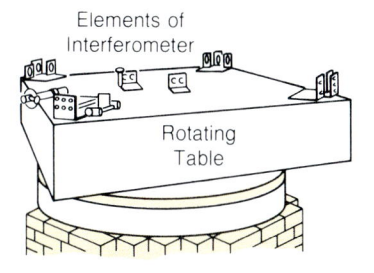

Figure 21.3

The Michelson-Morley experiment. A Michelson interferometer (Chapter 20) was placed on a rotating table, so that light from stars could be measured in different directions. They detected no shift in the interference fringes, hence in the speed of light, implying that there is no ether.

The Postulates of Special Relativity

The inconsistency created by the assumption that the velocity of light might vary with the inertial reference frame led Einstein to generalize the Galilean relativity principle. Replacing the word "mechanics" with "physics", Einstein stated

---

The laws of physics are the same in all inertial frames.

If we accept the assumption that Maxwell's equations represent laws of physics in the sense of this first postulate, then it is implied that Maxwell's equations are independent of reference frame, which led directly to Einstein's second postulate:

The measured speed of light is the same regardless of the relative velocity between source and observer.

One implication of the first postulate is that Maxwell's equations must have the same form in all inertial frames. Therefore, since these equations lead directly to a predicted value for the speed of light, this speed must be the same in all frames, as stated in the second postulate.

To accept the second postulate, however, required making assumptions that violate Newtonian concepts. Consider, for example, the apparent inconsistency presented by the often-cited example of a train with a headlight approaching an observer at some velocity v. Galilean relativity would tell us that the speed with which the light reaches the observer is $c + v$, yet Einstein's second postulate insists that it is c. Speed is the rate of change of displacement with time; Einstein resolved the apparent paradox by realizing that time passes at different rates for observers in different inertial frames. As a consequence, other quantities such as length and mass also vary with inertial frame. We explore these aspects of special relativity in the following sections.

Time Dilation

We can see how the rate of passage of time varies with inertial frame by considering a simple example. Suppose we have a pair of parallel mirrors that bounce a photon of light* back and forth between them, and suppose that we can record the arrival time of the photon every time it strikes one of the mirrors

---

*So far we have treated light as consisting of waves, but as we will discuss in Chapter 22, it also exhibits particle behavior, with the particles called **photons.**

21.2 SPECIAL RELATIVITY

(Fig. 21.4a). In the frame of reference of our pair of mirrors, the time between photon arrivals at this mirror is

$$t_0 = \frac{2d}{c} \qquad (21\text{-}1)$$

where d is the separation between mirrors. Here t_0 may be referred to as the **proper time**, meaning that it is measured at a fixed point within a single inertial frame. In the frame of the mirrors, the photon will reach the mirror regularly at intervals of t_0.

Now suppose the inertial frame containing the set of mirrors is moving at velocity **v** with respect to an observer's frame of reference (Fig. 21.4b). In the observer's frame, let t be the time it takes for the photon to make a round trip between mirrors; this is analogous to the interval we called t_0 in the reference frame of the mirrors. In the stationary frame, the photon travels a distance

$$d' = ct = 2\sqrt{d^2 + \left(\frac{vt}{2}\right)^2} \qquad (21\text{-}2)$$

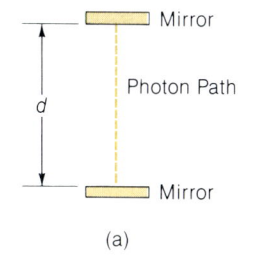

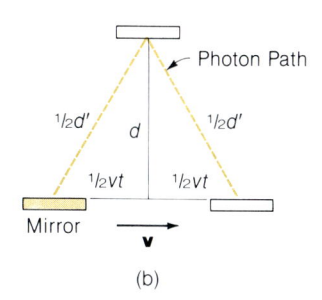

Figure 21.4

*Time dilation. A pair of parallel mirrors, moving at velocity **v** in an observer's reference frame. (a) The path, in the frame of the mirrors, of a photon bouncing between them. (b) The photon path in the frame where the mirrors are moving at velocity **v**.*

between reflections from one of the mirrors. By the second postulate, we know that c is invariant. Therefore we can solve Eq. 21–1 for c and substitute into Eq. 21–2:

$$ct = \left(\frac{2d}{t_0}\right)t = 2\sqrt{d^2 + \left(\frac{vt}{2}\right)^2}$$

or

$$\frac{4d^2 t^2}{t_0^2} = 4\left[d^2 + \left(\frac{vt}{2}\right)^2\right].$$

After some algebraic manipulation, we can solve for t.

$$t^2 = \frac{4d^2 t_0^2}{4d^2 - v^2 t_0^2} = \frac{t_0^2}{1 - \frac{v^2 t_0^2}{4d^2}}$$

Since $d = ct_0/2$,

$$t = \frac{t_0}{\sqrt{1 - \frac{v^2}{c^2}}} = \gamma t_0. \qquad (21\text{-}3)$$

CHAPTER 21 RELATIVITY

We see that when the observer and the mirrors are moving at different velocities, the time interval is longer than when the observer is at rest in the frame of the mirrors. This effect is called **time dilation.**

The amount by which time is expanded (dilated) is given by the **Lorentz factor:**

$$\gamma = \frac{1}{\sqrt{1 - \frac{v^2}{c^2}}} \qquad (21-4)$$

Notice that γ is always larger than 1; as viewed from a frame at rest, time in a moving frame appears to slow down by this factor. With a little algebra, we can solve Eq. 21–4 for the ratio v/c, which is often given the symbol β.

$$\beta = \frac{v}{c} = \sqrt{1 - \frac{1}{\gamma^2}} = \frac{\sqrt{\gamma^2 - 1}}{\gamma}. \qquad (21-5)$$

We stress that time dilation is not just a superficial phenomenon affecting the rate at which a clock runs. It is a literal slowing of the rate of passage of time, as measured by an observer in a different inertial frame. This means that biological processes, as well as any mechanical motions, occur at different rates as observed from different inertial frames. This has interesting consequences. A space traveller, for example, having travelled at high velocity for some interval of time, would find that he had aged less while away than his colleagues who remained on Earth.

Note in Eq. 21–3 that if $v = 0$, then $t = t_0$ as expected. Furthermore, the difference between t and t_0 is very small for values of v much less than the speed of light. Therefore time dilation is not an obvious effect under ordinary circumstances, and it can be ignored unless v becomes very large, as demonstrated by the success of Newtonian mechanics in so many applications. Note also, however, that the ratio t/t_0 approaches infinity as v approaches c; this implies substantial effects for objects travelling at speeds close to that of light. As we will see in later chapters, subatomic particles can travel at such speeds, as a result of acceleration by artificial or natural mechanisms. Therefore we will have to keep time dilation (and other relativistic effects) in mind when we discuss atomic structure and elementary particles.

Example 21.1 Muons are short-lived subatomic particles produced in nuclear collisions. At rest, a muon decays after 2.2×10^{-6} s. If a muon is moving at $v = 0.98c$ relative to an observer in the laboratory, how far from its source does it travel before decaying?

Because of time dilation, time appears to pass more slowly in the frame of the high-speed muon than it does in the laboratory. The time interval t in the laboratory is longer than the muon's (proper) lifetime $t_0 = 2.2 \times 10^{-6}$ s by a factor

21.2 SPECIAL RELATIVITY

$$\gamma = \frac{1}{\sqrt{1-(0.98)^2}} = 5.03.$$

This means that in the laboratory frame, the muon's lifetime is extended by a factor γ to

$$t = \gamma t_0 = (5.03)(2.2 \times 10^{-6} \text{ s}) = 1.1 \times 10^{-5} \text{ s}.$$

At $0.98c$, it therefore travels a distance

$$d = vt = (0.98)(3 \times 10^8 \text{ m/s})(1.1 \times 10^{-5} \text{ s}) = 3.2 \times 10^3 \text{ m}.$$

Example 21.2 An astronaut orbiting the Earth on the Space Shuttle travels at approximately 7.5 km/s. After six days in orbit, how much younger would he be than if he had stayed on the Earth?

We ignore the effects of acceleration, and assume that the astronaut moved at a constant speed of 75,000 m/s for six days (treatment of accelerating frames of reference requires general relativity theory, as discussed later in this chapter). Six days is 5.2×10^5 s or t, the time observed by a person on the Earth. What we are required to find is t_0, the time perceived by the astronaut. The ratio v/c is 2.5×10^{-5}, which is very much smaller than 1. Hence we use the binomial expansion $\sqrt{1-x} \approx 1 - x/2$ where $x = (v/c)^2$, which leads to

$$t - t_0 = t - t\sqrt{1 - \left(\frac{v}{c}\right)^2}$$

$$\approx t - t\left[1 - \frac{1}{2}\left(\frac{v}{c}\right)^2\right] = \left(\frac{v}{c}\right)^2 \frac{t}{2}$$

$$\frac{(2.5 \times 10^{-5})^2 (5.2 \times 10^5 \text{ s})}{2} = 1.63 \times 10^{-4} \text{ s}.$$

Hence the effect would not be noticeable. It is noteworthy, however, that in one of the experiments that verified the time dilation effect, a very accurate atomic clock was flown around the world (at ordinary airline speeds) and then compared with a similar clock that had been left in one place. The tiny difference between the time intervals on the two clocks was measured successfully.

Another important consequence of relativity is that time can no longer be regarded as an absolute quantity. Because a time interval depends on an observer's velocity and reference frame, two observers may disagree about the rate of time's passage and about whether two events are simultaneous.

Two events are said to be **simultaneous** (to an observer) if light from these two events reaches the observer at the same time. For example, in Fig. 21.5a, light rays from flashes at points 1 and 2 reach the observer at the same time.

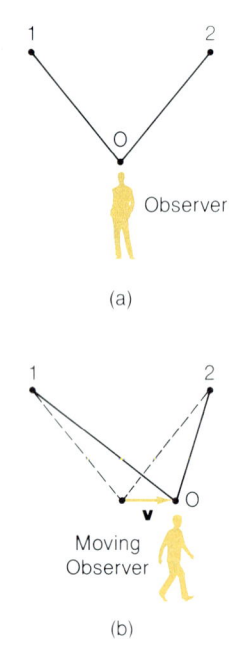

Figure 21.5

Simultaneity. Light flashes from points 1 and 2 in (a) reach the observer at rest simultaneously. In (b), however, the observer is moving, and sees the flash from point 2 before the flash from point 1, even though the flashes may have been simultaneous in the frame of the resting observer.

Because the distances O—1 and O—2 are equal, the light travel time is the same along each path and the flashes appear simultaneous.

But are the flashes simultaneous in all frames? We know that time intervals depend on the relative velocity of the observer. Thus, if a second observer is moving to the right at velocity **v** with respect to points 1 and 2 (Fig. 21.5b), then light from flash 2 reaches the observer *before* light from flash 1. During the time the light takes to reach the observer, the observer has moved to the right, so that the path from flash 2 is shorter. The observer sees flash 2 before flash 1. Similarly, if the observer were moving to the left, he would see flash 1 before flash 2. Thus, we find that two events which are simultaneous in one observer's frame are not necessarily simultaneous in a moving frame.

We may interpret this phenomenon by remembering that time is not absolute in special relativity. Because of time dilation, the interval between two events and the concepts of simultaneity and chronology ("before" and "after") depend on the frame of the observer. This appears to contradict everyday experience, but it is important to remember that such effects are noticeable only when the relative velocity approaches c. These speeds do occur in particle physics, however, and in principle they predict interesting phenomena in high-speed space travel as well.

Length Contraction

Time dilation has several important consequences for our understanding of physical quantities. One such quantity is length. Consider a spaceship moving at a high velocity v relative to frame 1 at rest with respect to the Earth and the planets. Suppose this spaceship passes by Mars and then by the Earth when the two planets are separated by a fixed distance L_0, in frame 1. We know that time dilation causes the time interval between the spaceship's encounters with Mars and the Earth to appear smaller for an observer in frame 2 on the ship than for an observer on the Earth or Mars (Fig. 21.6). The relative velocity between the ship and the planets is the same in both frames of reference, however, so we have the interesting dilemma that the relative velocity of the two objects is the same in each frame, yet it takes less time to pass each other in frame 2 than in frame 1. The solution to this seeming paradox is that the distance L between the planets in the ship's frame (2) is less than in the frame of the Earth and the planets themselves (1). This effect is called **length contraction**.

We can easily find an expression for length contraction. Since the relative velocity v is the same in both frames, we can write expressions for v in both frames and then set them equal. In the frame of the ship (frame 2), $v = L/t_0$, where L is the distance from Mars to the Earth as measured by a shipboard observer, and t_0 is the time between the ship's encounters with the two planets. This is a proper time, because it is measured in a single inertial frame at a single point. In this frame the ship is fixed at one location, and the planets pass that point at velocity v. In the frame of the Earth, the velocity is $v = L_0/t$. Here L_0 is, as mentioned above, the distance from Mars to the Earth as measured in the frame of the planets (so that L_0 can be called a **proper length**), and t is the time it takes the spaceship to pass from Mars to the Earth. Eq. 21–3 can be used for the relationship between proper time

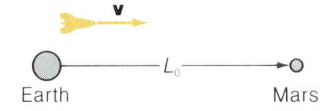

21.2 SPECIAL RELATIVITY

(the interval t_0 measured by a shipboard observer) and t, so that we can express the velocity v in the frame of the planets as $v = L_0/t = (L_0/t_0)\sqrt{1 - (v/c)^2}$. Now, setting the two expressions for v equal, we find

$$v = \frac{L}{t_0} = \frac{L_0\sqrt{1 - \left(\frac{v}{c}\right)^2}}{t_0}$$

which we solve for L:

$$L = L_0\sqrt{1 - \left(\frac{v}{c}\right)^2} = \frac{L_0}{\gamma} \qquad (21\text{-}6)$$

We see that the distance L between Earth and Mars seen by the astronauts is contracted by the same Lorentz factor γ that affects time dilation (Eq. 21–4). This contraction is a general phenomenon: as viewed by any observer (who by definition considers himself at rest), a moving object's length is contracted by a factor γ. Consider, for example, the measurement of the length of the ship itself. Suppose we first measure it while the ship is at rest on the Earth, and find its rest length to be d_0 (this corresponds to what an observer on the ship would measure, whether the ship was in motion with respect to the Earth or not). Now suppose the ship is in motion, and we measure its length from the Earth, finding it to be $d = d_0/\gamma$. Because d is less than d_0, from our point of view on the Earth, the ship has decreased in length because it is moving rapidly with respect to us.

Only the dimension of the ship parallel to the relative velocity between reference frames is affected; its diameter (the dimension perpendicular to its direction of travel) would be the same in both reference frames.

The length contraction given in Eq. 21–6 is sometimes called the **Lorentz contraction,** because it was derived before Einstein's work by H. A. Lorentz. Lorentz developed the expression given in Eq. 21–6 not from relativistic principles, but by realizing that such an expression would explain the results of the Michelson-Morley experiment and would rectify the inconsistency cited earlier in this chapter between Maxwell's equations and the classical expressions for wave velocity.

Figure 21.6

Length contraction. Here a spaceship passes by Earth and Mars. In the frame of the earth (a), the rocket has velocity v and the Earth-Mars distance is L_0. In the frame of the rocket (b), the Earth and Mars have velocity v to the left, and the distance between them is L. Because of time dilation, L is less than L_0.

Example 21.3

If the spaceship described in the text had length $L_0 = 100$ m in its own inertial frame, how long would it appear to an observer on Earth if its velocity were $0.5c$?

We need to find the length L in Eq. 21–6.

$$L = L_0\sqrt{1 - \left(\frac{v}{c}\right)^2} = (100 \text{ m})\sqrt{1 - (0.5)^2} = 87 \text{ m}$$

The Electromagnetic Force

In earlier chapters we discussed electric and magnetic forces, and, though we treated them separately, we sometimes used the term "electromagnetic" force. The reason for this can now be understood, as we show that the magnetic force is actually an electric force caused by relativistic effects.

Consider a pair of thin parallel wires, as shown in Fig. 21.7, each carrying a current with positive charges moving to the right, electrons to the left. In the reference frame of a positive charge in *either* wire, the electrons are streaming past it at some velocity v. Therefore the relativistic length contraction reduces the distance between electrons in each wire compared to the distance between positive charges.

Now consider an individual positive charge in *one* of the wires. It feels an attractive electric force due to all the electrons in both wires. The positive charge feels no net force due to electrons in its own wire, because these electrons are distributed uniformly in both directions along the wire. What our positive charge "sees" in the other wire, however, is a higher density of electrons than positive charges, that is, a net negative charge, because of the relativistic length contraction. The electrons appear to be packed in tighter. Therefore the positive charge in one wire feels an attractive force due to the higher density of electrons in the other wire. If the two currents flow in opposite directions, then the net force is repulsive, because the relativistic length contraction creates an excess of like charges in one wire, as seen by the charges in the other.

Recall from Ampere's law (Chapter 16) that wires carrying electrical currents exert magnetic forces on each other. Now we see that the force is actually the Coulomb force, created by relativistic effects. An electric field in one frame of reference becomes both an electric and a magnetic field in a moving frame. Because special relativity states that physical laws are the same in all inertial frames, we do not recognize the magnetic force as fundamentally different from the electrical force, and we often use the term "electromagnetic" force to refer to the entire class of forces caused by separated electric charges. Relativity has therefore modified our understanding of the electric and magnetic forces, unifying them into a single force.

Mass Increase and the Ultimate Velocity

It is possible, though somewhat complicated, to show that because of time dilation the apparent mass of an object depends on its velocity. Let us consider once again the spaceship travelling at a high velocity with respect to the Earth. If m_0, often called the **rest mass**, is the mass of the spaceship as measured in its own frame of reference (the rest frame), and m its apparent mass as measured by an observer on the Earth, then

$$m = \frac{m_0}{\sqrt{1 - \left(\frac{v}{c}\right)^2}} = \gamma m_0. \tag{21-7}$$

21.2 SPECIAL RELATIVITY

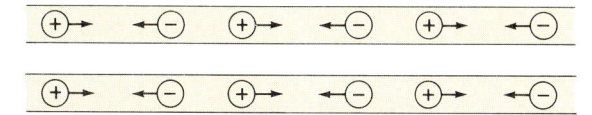

Figure 21.7

The electromagnetic force. Each wire carries a current with positive charges moving to the right and negative charges moving to the left. Because of length contraction, charges "see" a higher density of opposite charges than like charges in the other wire, so that the coulomb force creates a net attraction between the two wires.

This equation is derived by writing an expression for conservation of momentum, as in a collision between two bodies, and using the time dilation and length contraction expressions derived earlier. After all, it is the momentum p of a moving object that one measures in order to infer the mass. Whereas $p = mv$ in Newtonian mechanics, we now see that in Einstein's special relativity,

$$p = mv = \frac{m_0 v}{\sqrt{1 - \left(\frac{v}{c}\right)^2}} = \gamma m_0 v. \qquad (21\text{--}8)$$

As the velocity increases, a particle's momentum (and mass) increase more rapidly than Newton believed, by a factor γ. The total energy E and kinetic energy KE also increase, in a more complex way (see discussion later in this Section).

If an object moves relative to an observer, the observer will measure a larger mass in his frame of reference than would an observer in the rest frame of the object. This means, for example, that the gravitational force between the spaceship and the Earth would be enhanced by the relative motion between them.

Notice that as v approaches c, the mass observed in the frame of the observer approaches an infinite value. An infinite mass is a logical impossibility for a variety of reasons. For example, an infinite energy would be required to move an object of infinite mass (see how many other reasons you can think of), and therefore it is considered impossible to accelerate an object to a speed equal to or greater than the speed of light. The same result can be found by letting $v = c$ in the expressions we have derived for time dilation or length contraction.

Some theoretical physicists speak of particles called **tachyons** which have speeds greater than c, but which can never be decelerated below c. Such particles have never been detected, but it is difficult to see how they could be. Only photons and other massless particles can travel at speed c; indeed they can travel at no other speed.

Example 21.4 If an astronaut has a rest mass of 85 kg, what are his apparent mass and momentum (in the frame of the Earth) when travelling in a spaceship at a speed of 2×10^5 km/s relative to the Earth?

We calculate the Lorentz factor γ first.

$$\gamma = \frac{1}{\sqrt{1 - \left(\frac{v}{c}\right)^2}} = 1.34$$

Thus, from Eq. 21.7, the astronaut's apparent mass is

$$m = \gamma m_0 = (1.34)(85 \text{ kg}) = 114 \text{ kg}$$

and his momentum is

$$p = \gamma m_0 v = (1.34)(85 \text{ kg})(2 \times 10^5 \text{ km/s}) = 2.28 \times 10^7 \text{ kg} \cdot \text{m/s}.$$

In the frame of the Earth, the mass and momentum of the astronaut are 34% greater at this velocity than at rest. (In the frame of the spaceship, of course, the astronaut's mass is the rest mass, 85 kg.)

Mass and Energy

The phenomenon that mass increases with speed can be used to argue that energy and mass are equivalent, one of the better-known consequences of special relativity theory. As the velocity of an object increases, its kinetic energy increases as well, not only because of the increased velocity, but also because of the increased mass. When the velocity nears c, the body cannot be accelerated much further, but the energy can still grow substantially because of the mass increase. This implies that an increase in mass is the same as an increase in energy, so it follows that mass and energy must be equivalent. Einstein found that the **total energy** of an object is

$$E = mc^2 \tag{21-9}$$

where c is the speed of light, as usual. This equivalence has been verified experimentally many times. Mass is converted into energy in many processes, but this is particularly noticeable in nuclear reactions, where the amount of energy released can be very large (nuclear reactions are discussed in Chapter 25).

The total energy E given by Eq. 21–9 includes an object's kinetic energy KE, in addition to the energy in the form of rest mass. Thus we have

21.2 SPECIAL RELATIVITY

$$E = mc^2 = E_0 + \text{KE} = m_0c^2 + \text{KE} \qquad (21\text{--}10)$$

where the quantity $E_0 = m_0c^2$ is the **rest mass energy**, the total energy when the object has zero velocity and zero kinetic energy. If we solve this expression for the kinetic energy, we find

$$\text{KE} = mc^2 - m_0c^2.$$

If we use Eq. 21–7 to substitute for m, then we find

$$\text{KE} = \frac{m_0c^2}{\sqrt{1 - \left(\dfrac{v}{c}\right)^2}} - m_0c^2 = (\gamma - 1)m_0c^2. \qquad (21\text{--}11)$$

This is rather different from the classical expression $\text{KE} = \tfrac{1}{2}mv^2$. It can be seen, however, that if v is much less than c, Eq. 21–11 reduces to the familiar expression (to show this requires use of the binomial expansion $1/\sqrt{1-x} \simeq 1 + x/2$ when x is much smaller than 1). Thus the classical definition $\text{KE} = \tfrac{1}{2}mv^2$ is actually a low-speed approximation to the correct expression for kinetic energy.

Example 21.5

The Sun produces its energy by the conversion of mass into energy in nuclear reactions in its core. In one second, the quantity of energy produced is 3.8×10^{26} J. How much rest mass per second is converted into energy by the Sun? How long could the Sun produce energy in this way if all its mass (2×10^{30} kg) could be converted into energy?

We solve for m_0 in the expression for rest energy.

$$m_0 = \frac{E_0}{c^2} = \frac{3.8 \times 10^{26} \text{ J}}{(3 \times 10^8 \text{ m/s})^2} = 4.2 \times 10^9 \text{ kg}$$

The rate of mass consumption is 4.2×10^9 kg/s. If the total mass of the Sun is 2×10^{30} kg, then its lifetime at this rate is

$$t = \frac{2 \times 10^{30} \text{ kg}}{4.2 \times 10^9 \text{ kg/s}} = 4.8 \times 10^{20} \text{ s} = 1.5 \times 10^{13} \text{ yr}.$$

Because nuclear reactions are less than 1% efficient in converting mass to energy, and because the reactions can occur only in the innermost regions of the Sun (for reasons described in Chapter 25), the actual energy available to the Sun over its lifetime is considerably less than estimated here. The Sun's actual lifetime is expected to be around 10^{10} yr.

Example 21.6 Find the total energy, the rest energy, and the kinetic energy of an electron travelling at speed $v = 0.95c$.

The rest energy is

$$E_0 = m_0c^2 = (9.11 \times 10^{-31} \text{ kg})(3 \times 10^8 \text{ m/s})^2 = 8.2 \times 10^{-14} \text{ J}.$$

The kinetic energy, from Eq. 21–11, is therefore

$$KE = (\gamma - 1)m_0c^2 = \left[\frac{1}{\sqrt{1-(0.95)^2}} - 1\right](8.2 \times 10^{-14} \text{ J})$$
$$= 1.8 \times 10^{-13} \text{ J}.$$

Finally, the total energy is

$$E = E_0 + KE = 8.2 \times 10^{-14} \text{ J} + 1.8 \times 10^{-13} \text{ J} = 2.6 \times 10^{-13} \text{ J}.$$

It is noteworthy that at such a highly relativistic velocity (the Lorentz factor γ is 3.2), the kinetic energy of the electron is substantially greater than the rest energy.

Addition of Relativistic Velocities

Before completing our discussion of special relativity, we briefly consider the situation where we have three inertial frames which may have relative velocities large enough for relativistic effects to be important. For example, suppose an observer on earth wanted to know the velocity of a missile launched at relativistic velocity from a spaceship that was moving relativistically with respect to the earth. We already know that this cannot be as simple as the Newtonian case (where straightforward vector addition suffices), because we know that the sum of the velocities must never exceed c.

Consider two objects moving relative to an observer at rest (Fig. 21.8). If v is the velocity of object 1 relative to the observer, and u' is the velocity of object 2 relative to object 1, then Einstein showed that the velocity of object 2 relative to the observer is

$$u = \frac{v + u'}{1 + \dfrac{vu'}{c^2}} \qquad (21\text{--}12)$$

if u' and v are in the same direction. Notice that in Newtonian (Galilean) relativity the resultant velocity is $u = v + u'$, which could exceed c if v and u' were large. The term $(1 + vu'/c^2)$ in the denominator of Eq. 21–12 assures that this does not happen.

For example, suppose that a high-speed locomotive, with $v = 0.8c$ as seen by an observer at rest, turns on a headlight with $u' = c$ relative to the train.

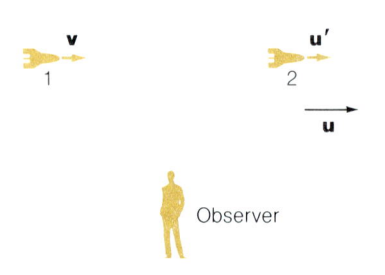

Figure 21.8

*Addition of relativistic velocities. Here **v** is the velocity of object 1 in the frame of the observer, **u** is the velocity of object 2 in the same frame, and **u'** is the velocity of object 2 in the frame of object 1. Einstein developed a relativistic formula for computing **u** in terms of **v** and **u'**.*

21.2 SPECIAL RELATIVITY

If we wish to solve for the speed of the light beam relative to the observer, Eq. 21–12 yields

$$u = \frac{0.8\,c + c}{1 + \frac{(0.8c)(c)}{c^2}} = \frac{1.8c}{1.8} = c.$$

We obtain the expected result that $u = c$; light always travels at c in all reference frames. We see also that if the product vu' is much smaller than c^2, then $u \approx v + u'$ as expected in classical mechanics.

Example 21.7 Suppose a spaceship, moving toward the Earth at a speed of 2.2×10^5 km/s with respect to the Earth, fires a missile toward the Earth at a speed (relative to the spaceship) of 1.5×10^5 km/s. What is the speed of the missile in the reference frame of the Earth?

Here $v = 2.2 \times 10^5$ km/s, and $u' = 1.5 \times 10^5$ km/s, and we want to find u. From Eq. 21–12, we have

$$u = \frac{v + u'}{1 + \frac{vu'}{c^2}}$$

$$= \frac{(2.2 \times 10^5 + 1.5 \times 10^5)\ \text{km/s}}{1 + \frac{(2.2 \times 10^5\ \text{km/s})(1.5 \times 10^5\ \text{km/s})}{(3.0 \times 10^5\ \text{km/s})^2}}$$

$$= 2.7 \times 10^5\ \text{km/s}.$$

Example 21.8 Suppose now that the spaceship in Example 21.7 is moving *away* from the Earth at $0.8c$ relative to the Earth, and fires a missile *toward* the Earth at $0.9c$ relative to the spaceship. What is the missile's speed relative to the Earth?

The velocities of the spaceship and the missile are in opposite directions, and we must be careful about signs. We choose the positive direction as away from the Earth, so that the spaceship's velocity is $v = 0.8c$, and the missile's velocity with respect to the spaceship is $u' = -0.9c$. Eq. 21–12 then yields

$$u = \frac{0.8c - 0.9c}{1 + \frac{(0.8c)(-0.9c)}{c^2}} = \frac{-0.1c}{1 - 0.72} = -0.36c.$$

The missile approaches the Earth at 36% of the speed of light, as seen from the Earth.

21.3 General Relativity

In discussing special relativity, we considered its effect on measurable quantities of uniform motion; that is, we confined ourselves to inertial reference frames. Einstein generalized his theory of special relativity to include reference frames which *accelerate* with respect to each other. His theory of general relativity, published in 1915, provided new understandings, not only of accelerated reference frames, but also of gravity. To properly discuss general relativity requires sophisticated mathematical techniques that are well beyond the scope of this text, but nevertheless we can develop a qualitative comprehension of the theory.

Spacetime

As a prelude to describing general relativity theory, we must introduce a notion that could actually have been discussed in our section on special relativity. Whereas in classical mechanics we are used to thinking of space as having three dimensions, special relativity shows us that a fourth dimension, time, plays an equally important role. We distinguished the notion of proper time, the time interval measured at the same point within a single inertial frame, from the time measured in one frame by an observer in another frame, or at two locations within a single frame. To fully specify the location of an event, we must say not only *where* it is in the three spatial dimensions, but *when* it is there.

In addition, Einstein's theory of special relativity demonstrated that our notions of space and time become entangled when one makes measurements of length and time at very high velocity. We have seen that when an object travels at very high speed in the reference frame of an observer, the observer finds that time itself slows for that object, while the length of the object shrinks and its mass increases. Absolute space and absolute time do not exist, since different observers cannot agree on lengths or time intervals. Space and time intervals are *relative* concepts, dependent on the velocity of the frame from which they are measured, but related in a fundamental way in the four-dimensional continuum known as **spacetime.** Instead of speaking separately of an object's three spatial coordinates (x,y,z) and its time coordinate (t), we now refer to four-dimensional spacetime (x,y,z,t). In a different reference frame, (x',y',z',t'), the coordinates would all differ, but in a calculable and interdependent fashion. This mixing of space and time destroys our conventional notions of simultaneity and of "before" and "after," since absolute time is not a relevant concept. In special relativity, space and time are inextricably bound together, and we will hereafter speak of spacetime routinely.

The Basis of General Relativity

In discussing Newton's laws of motion and gravitation, we found that the acceleration of an object falling at the surface of the Earth does not depend on its mass. It is simple to show algebraically why this is so in classical me-

21.3 GENERAL RELATIVITY

chanics, and Newton was satisfied that he had explained the phenomenon in a straightforward way.

Einstein, on the other hand, realized that there were far broader implications. Suppose that a person in a completely enclosed room with no windows is taken into space and accelerated in one direction at a rate of 9.8 m/s² (Fig. 21.9). Locally, the person will feel a force against the floor of the room that is equal to his mass times the acceleration. This is the same as the force he would feel if the room were at rest on the surface of the Earth. The person in the room has no experimental way to tell whether the room is accelerating in space or is at rest on the Earth's surface.

From this kind of thought experiment, Einstein developed the **principle of equivalence:**

An observer in an enclosed space cannot distinguish whether he is undergoing inertial acceleration or is at rest in a gravitational field.

By inertial acceleration, we mean acceleration of the reference frame of the observer, for example, the enclosed room being accelerated at 9.8 m/s². In other words, an upward acceleration is equivalent to a downward force of gravity.

The principle of equivalence leads immediately to other profound results. One of them is that light can be curved by a gravitational field. Again, consider a person in an enclosed room, and let the room be accelerating at 9.8 m/s². Suppose a photon of light in the reference frame of an external observer enters the room through a small slit, and passes across it in the direction perpendicular to the acceleration (Fig. 21.10). The photon must follow a straight path in the external reference frame, but to the observer inside the accelerating room, the photon will appear to follow a curving path (actually, in general relativity there is additional bending because the coordinate system also changes).

If the equivalence principle holds, then the room must appear to the observer inside as if it were at rest on the Earth's surface. Therefore, Einstein predicted that a gravitational field would cause a photon's path to curve. This should actually be stated a bit more precisely: what Einstein really said was that spacetime itself is curved in the presence of a gravitational field, and that a photon, constrained to follow the shortest possible path between any two points, must therefore follow a curved path. Hence modern physicists speak of curved spacetime.

The degree of curvature depends on the strength of the gravitational field; that is, on the concentration of mass per unit volume. The mass of the Earth creates very little curvature (imagine how little the path of a photon would bend while passing across a room that is being accelerated at 1 g), and it is difficult to detect the curvature of spacetime due to the Earth's mass. For larger astronomical bodies it is not so difficult, and in 1919 the phenomenon was confirmed experimentally (Fig. 21.11). The position of a star seen just past the edge of the Sun's disk was carefully measured, and was found to deviate by just the predicted amount (1.75″) from its measured position when the Sun

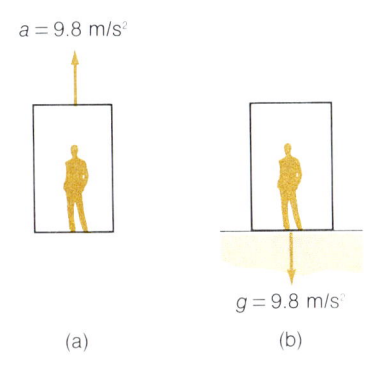

Figure 21.9

Inertial and gravitational acceleration. (a) A person stands in an enclosed room in space that is accelerating at 9.8 m/s². (b) The same person stands in the same enclosed room, at rest on the surface of the Earth, whose surface gravitational acceleration is 9.8 m/s². The principle of equivalence states that there is no observable difference to the person in the enclosure.

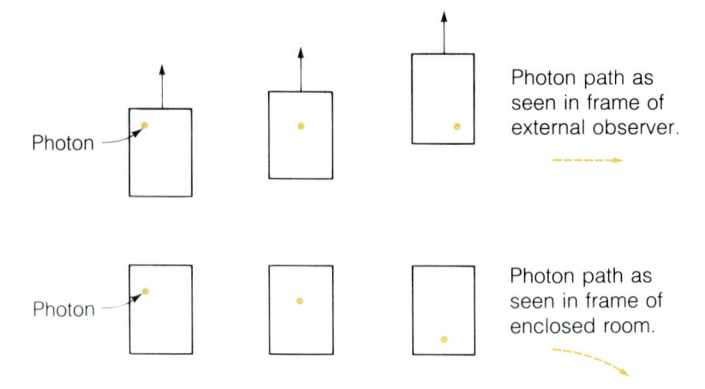

Figure 21.10

Curvature of light. To an observer in an accelerating reference frame, a photon follows a curving path, even though the photon path would appear straight to an unaccelerated outside observer. Because of the principle of equivalence, a photon follows a curved path in a gravitational field.

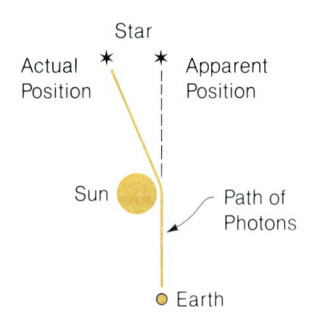

Figure 21.11

Confirmation of bending of light, a test of general relativity carried out in 1919. The apparent position of a star near the Sun (seen during an eclipse) was shifted due to the curved path that photons from the star followed as they passed through the Sun's gravitational field.

was not so near. (This measurement had to be made during a total solar eclipse, so that the star could be seen.) The light from the star was deflected by the Sun's gravitational field as it passed close to the Sun; in other words, the Sun creates local curvature of spacetime, so that light travels along curved paths in its vicinity.

Stars more massive than the Sun, and in particular collapsed stars with very large masses compressed into small volumes, create even more significant curvature of spacetime. The so-called **black hole** is an object whose gravitational field is so immense that light cannot escape. The radius of a black hole's surface, also known as its **Schwarzschild radius,** is given by

$$R_s = \frac{2GM}{c^2} \quad (21\text{--}13)$$

where M is the mass of the object and G is the universal gravitational constant. The value of R_s for the Sun is 3 km, and R_s is directly proportional to M, so we can simplify Eq. 21–13 by writing

$$R_s = (3 \text{ km})\left(\frac{M}{M_\odot}\right)$$

where $M/M_\odot$ is the object's mass relative to that of the Sun. One can easily show, using the classical formula for escape velocity (derived in Chapter 7), that the escape velocity from the surface of a black hole is the speed of light. Just outside the Schwarzschild radius, spacetime is strongly curved (Fig. 21.12), resulting in unusual effects such as the slowing of time (as perceived by a faraway observer), the red-shift of light waves, and enormous tidal forces which

21.3 GENERAL RELATIVITY

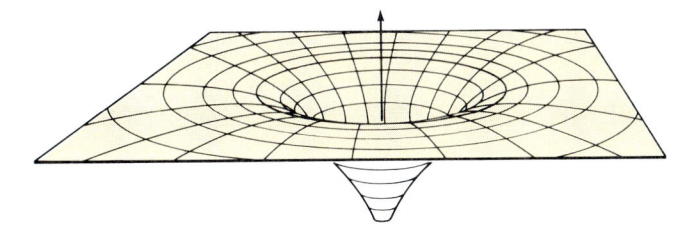

Figure 21.12

The curvature of spacetime near a black hole. This two-dimensional sketch of a three-dimensional figure illustrates how the spacetime curvature varies with distance from the black hole. Photons of light cannot escape if they are within the Schwarzschild radius.

can rip apart normal matter. Such an object would be nearly impossible to observe directly, although several suspected black holes are known to astronomers by their gravitational effects on nearby matter.

Because significant curvature of spacetime requires very large masses, it is not easy to test general relativity in the laboratory. Most of the verifications have used astronomical observations of large masses with significant general relativistic effects. As we will see in the next section, general relativity also provides the best framework known for describing the universe as a whole.

In the theory of general relativity, gravity is seen as somehow unique with respect to the other three natural forces, because only gravity is thought to have the property of bending spacetime. It remains to be seen whether this will be reconciled with the developing unified field theories (Chapter 26), in which it is hoped that all four forces can be shown to be the same, with different manifestations under different conditions.

General Relativity and Cosmology

As soon as Einstein published his theory of general relativity in 1915, he and others began to apply it to the universe as a whole. Because the universe contains a vast quantity of mass, according to general relativity theory it must have an overall curvature. Einstein developed a set of equations, called **field equations,** to describe the mass and energy of the universe and the curvature of spacetime. Cosmology, the study of the overall properties of the universe, became an exercise in finding solutions to the field equations.

The first solution found by Einstein had some unfortunate aspects. Einstein found that the universe must either be empty of mass (obviously not the case), or that it was unstable. To solve this dilemma, he arbitrarily added a new term to the field equations called the **cosmological constant** in order to allow his model universe to be static. Soon thereafter, the Dutch physicist W. de Sitter and the Belgian Abbé G. LeMaitre found solutions in which the universe was expanding. LeMaitre realized that such a universe must have begun from a compact, hot state, and the concept of a **big bang** was introduced. Seven years

Figure 21.13

Positive curvature. A sphere is a two-dimensional surface with positive curvature; it has no boundaries, but is finite in extent.

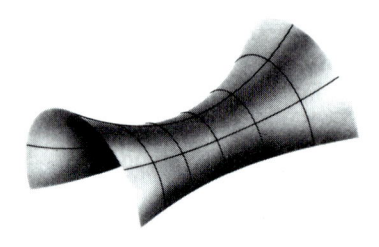

Figure 21.14

Negative curvature. A saddle surface is a two-dimensional surface with negative curvature; it has no boundaries (if extended to infinity) and is infinite in extent.

later, in 1929, the American astronomer Edwin Hubble deduced, from the Doppler shifts of light from distant galaxies, that the universe is indeed expanding. Today most cosmologists assume that in the present universe there is no cosmological constant, and concentrate instead on the details of the expansion.

Until very recently, it was generally assumed that there were three possible solutions to the field equations, representing three possible types of curvature of spacetime. In the **closed** universe, the curvature is said to be positive, that is, the universe has a finite extent but has no boundaries; a two-dimensional analog to this is the surface of a sphere (Fig. 21.13). In the **open** universe, the curvature is negative, meaning that the universe is infinite in extent and has no boundaries; a two-dimensional analog for this is a saddle surface (Fig. 21.14). The intermediate case (actually the boundary between the other two cases), where the curvature is said to be zero, corresponds to a **flat** universe. The major question in observational cosmology for the past four decades has been to determine which type of curvature describes our universe.

The essential unknown is the total mass density of the universe, because this determines how curved spacetime is. Therefore various techniques have been used for determining the mass content, with the result that the amount of mass contained in observable forms such as galaxies is considerably less than that needed to have a closed universe. There may be matter in invisible forms, and the observations are complicated by the vast distances involved, so this result is not considered conclusive. The most recent indications are that the universe is not far from the intermediate, flat case, which leads some to suspect that it is indeed flat. This may seem like an improbable coincidence, since a precise quantity of total mass is required for zero curvature, but we are only beginning to speculate about the laws that governed the initial mass of the universe.

If the universe is flat, then it will continue to expand forever, slowing its expansion rate to zero as its age becomes infinite. In the closed case, there is sufficient mass content (that is, sufficient curvature) to reverse the expansion eventually, so that the universe will contract to its original state of very high density and small size. Using a classical analogy, one can picture the expansion rate of the closed universe as being below "escape velocity," just as a rock thrown upward will return to earth under the influence of gravity. Some prefer to think, on largely philosophical grounds, that this would lead to a cyclical universe in which expansion and contraction phases would alternate. The third model, that the universe is open, implies that the expansion would keep slowing forever, but would not stop. In this case, we could say that the universe has a velocity greater than the escape velocity. In either the flat or open universe, the eventual final state is "heat death," in which the universe consists of a very low density distribution of elementary particles or dead stars with extremely low temperatures (recall our discussion of entropy in Chapter 13).

Not all cosmologists accept general relativity as the only correct description of gravity or the only adequate means to predict the fate of the universe. So far, however, all experimental evidence suggests that general relativity is correct (see *Focus on Physics*, this chapter). This still leaves open the possibility, however, that general relativity, like Newton's laws, is only a limited case of some more general theory. Some imaginative cosmological theories in which this is postulated have begun to be developed in recent years.

FOCUS ON PHYSICS

Testing Relativity

When Einstein announced his special theory of relativity in 1905, it came partially as a result of many experiments challenging the classical ideas of space and time. Various workers had demonstrated that light travelled at a constant speed, and Maxwell had shown theoretically that this speed in a vacuum was

$$c = \frac{1}{\sqrt{\epsilon_0 \mu_0}} = 2.99792458 \times 10^8 \text{ m/s.}$$

In 1887 the failure of the Michelson-Morley experiment to detect any motion of the earth through the hypothetical ether suggested that light travels at the same speed in all inertial reference frames. Thus, special relativity was developed to reconcile certain experimental facts with physical theory.

In contrast, the general theory of relativity, published in 1915, was foremost a result of Einstein's philosophical search for a geometrical theory that placed the gravitational field on an equivalent footing with an accelerated reference frame (the Equivalence Principle). Rather than arising from experiments, general relativity grew out of pure theory and 19th-century developments in abstract geometry and tensor calculus. However, any theory must ultimately be tested by experiment. Relativity is no exception, although the tests are exceedingly difficult, since the deviations from Newton's theory of gravity are small.

To test general relativity, Einstein and others proposed three "classical tests",

1. Mercury's perihelion precession
2. Deflection of starlight by the Sun
3. Gravitational redshift

A fourth test, the gravitational time delay of radar waves reflected from planets behind the Sun, was carried out in the 1960's. Each of these tests involves large masses (the Sun, planets, or other stars), since masses in the laboratory are usually too small to give a detectable effect.

Test 1 was an immediate success for general relativity. Since 1846, astronomers had puzzled over an unexplained irregularity in Mercury's orbit. Newton's laws of gravitation predict that a planet should orbit the Sun in an ellipse. The gravity of other planets, particularly Venus, Earth, and Jupiter, produces small changes to Mercury's orbit, causing its major axis to precess (Fig. 21.1.1) by an angle of about 1.56° per century. The observed precession was 43" (arc seconds) per century greater than Newton's theory predicted (one arc second is 1/3600 of a degree). However, the gravitational field is slightly larger near the sun in Einstein's theory, and leads to an extra precession of 0.1038" per revolution. Since Mercury makes 415 revolutions each century, that adds up to the observed 43" per century!

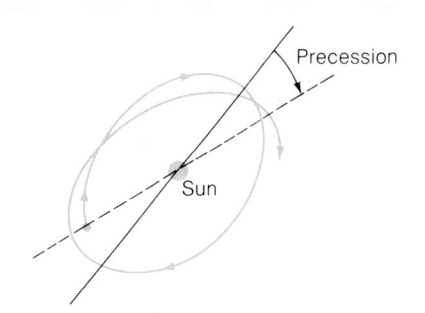

Figure 21.1.1

Precession of Mercury's orbit. Because of the gravitational effects of other planets, the major axis of Mercury's orbit rotates slowly by about 1.56° per century (exaggerated here). An extra 43" precession per century, unexplained by Newton's theory, was one of the first major predictions of Einstein's general relativity.

(continued)

The second test received worldwide publicity in 1919, when two expeditions to Africa and Brazil measured the deflection of the positions of stars near the Sun during a solar eclipse (Fig. 21.11). Einstein's theory predicted a shift by an angle of 1.75" at the edge of the Sun's disk, and the expeditions found angles of 1.68 ± 0.40" and 1.98 ± 0.16", in fair agreement with general relativity. Many eclipse expeditions since that time have found angles near Einstein's prediction, but the accuracy of their experiments was never better than about 10%. In 1974, however, radio astronomers measured the deflection of radio waves from distant sources called quasars as they passed near the Sun. Their measurement of 1.761 ± 0.016" confirmed Einstein's prediction to an accuracy better than 1%.

The third test measures time dilation, or the slowing of the passage of time near a strong source of gravity. In Einstein's theory, all clocks, including atomic clocks based on the frequency of atomic oscillations, should run slower in strong gravitational fields. For example, people living at sea level will age more slowly than people at 5,000 feet elevation, because the people at sea level are closer to the Earth's center. A slowing of time and frequency v of an atom results in a longer wavelength, $\lambda = c/v$, for the light emitted by that atom. This shift is called the **gravitational redshift,** and was confirmed in the early 1960's by measuring the change in frequency of γ-rays as they rose or fell in a tower. In 1976, an atomic clock carried to high altitude by a rocket confirmed the slowing of time to an accuracy of 0.02%. The time dilation near sources such as the Earth or Sun is so small that it is virtually unnoticed in everyday life. However, it leads to dramatic effects near strong sources of gravity, such as neutron stars or black holes.

In the past decade several new tests have been carried out or proposed. In one, physicists have measured the time delay of radio waves reflected off Mars or Venus as they pass behind the Sun. According to relativity a wave that passes near the Sun will have its travel time slowed by time dilation. The best results, based on active radio instruments left on the surface of Mars by the *Viking* mission, have confirmed the relativistic time delay to 0.1%. A new generation of experiments will be searching for more exotic predictions of Einstein's theory, including gravitational waves (the gravitational analogy to electromagnetic waves) and small effects of the Earth's rotation on the motion of gyroscopes. These experiments push the technology of precision measurements to their limits, and they have produced many valuable "spin-offs" for industrial use. With every new advance in our ability to measure the physical universe, Einstein's theory has stood the test. General relativity stands as one of the major achievements of 20th century theoretical physics, and its tests have provided some of the most challenging problems for experimental physics.

Additional Reading:
"Testing Relativity: 20 Years of Progress," Clifford Will, *Sky and Telescope,* Vol. 66, No. 4, p. 294 (October 1983).

"Testing Einstein's General Relativity During Eclipses of the Sun," Jack B. Zirker, *Mercury,* Vol. 14, No. 4, p 98 (July 1985).

Perspective

We have now developed the full picture of the laws of mechanics that we must have to proceed to modern physics. As we learned for the magnetic force, we will find as we go on that many common phenomena depend on relativistic effects. Even though velocities approaching the speed of light are rare in our macroscopic world, they are not unusual in the world of the atom and its constituent particles.

Nearly in parallel with the development of relativity theory in the first two decades of the twentieth century were the foundations of another revolution in physics. A new field was developed, which violates classical mechanics as severely as does relativity theory, dealing with the nature of matter and its interactions on the smallest scales. In the coming chapters we will build an understanding of quantum mechanics, which lies at the very heart of all modern physics.

PROBLEM SOLVING

Special Relativity

The results of special relativistic kinematics can be summarized by a few formulas. These formulas begin with the definition $\beta = v/c$, and are useful in solving problems about energy, rest mass, momentum, and kinetic energy:

Lorentz factor $\quad \gamma = \dfrac{1}{\sqrt{1-\beta^2}}$

Mass $\quad\quad\quad\quad m = \dfrac{m_0}{\sqrt{1-\beta^2}} = \gamma m_0$

Momentum $\quad\quad p = \dfrac{m_0 v}{\sqrt{1-\beta^2}} = \gamma m_0 v$

Total energy $\quad\quad E = mc^2 = \gamma m_0 c^2$

Kinetic energy $\quad \text{KE} = E - m_0 c^2 = (\gamma - 1) m_0 c^2$

With a little algebra, we may deduce the following related expressions:

$$\beta = \dfrac{\sqrt{\gamma^2 - 1}}{\gamma}$$

$$E^2 = p^2 c^2 + m_0^2 c^4$$

$$p = \dfrac{\sqrt{(\text{KE})^2 + 2(\text{KE})(m_0 c^2)}}{c}$$

Graphically, the relations between E, p, and m_0 obey the Pythagorean theorem for an imaginary triangle (Fig. 21.2.1) with sides of "length" $m_0 c^2$ and pc, and hypotenuse E. Thus,

$$E^2 = (pc)^2 + (m_0 c^2)^2$$

and the angle θ is given by

$$\sin\theta = \dfrac{pc}{E} = \dfrac{\gamma m_0 v c}{\gamma m_0 c^2} = \dfrac{v}{c} = \beta$$

and

$$\cos\theta = \dfrac{m_0 c^2}{E} = \dfrac{m_0 c^2}{\gamma m_0 c^2} = \dfrac{1}{\gamma}.$$

These relations are often useful in solving problems.

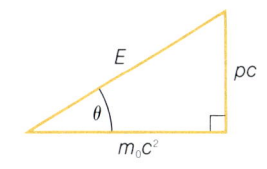

Figure 21.2.1

Example 1 A proton has rest mass $m_0 = 1.67 \times 10^{-27}$ kg and is accelerated to a total energy $E = 3 \times 10^9$ eV. What are its velocity v (or β), Lorentz factor, and kinetic energy?

(continued)

We first calculate the proton's rest energy,

$$m_0 c^2 = (1.67 \times 10^{-27} \text{ kg})(3 \times 10^8 \text{ m/s})^2$$
$$= 1.50 \times 10^{-10} \text{ J}$$

which corresponds to about 10^9 eV (1 eV = 1.6×10^{-19} J). Thus, the Lorentz factor is

$$\gamma = \frac{E}{m_0 c^2} = \frac{3 \times 10^9 \text{ eV}}{10^9 \text{ eV}} = 3.$$

then

$$\beta = \frac{\sqrt{\gamma^2 - 1}}{\gamma} = 0.94$$

and

$$\text{KE} = (\gamma - 1) m_0 c^2 = 2 \times 10^9 \text{ eV}.$$

Notice how easy this problem is after we find the Lorentz factor γ.

Example 2 An electron of rest energy $E_0 = 5 \times 10^5$ eV is accelerated to a kinetic energy of KE = 2×10^6 eV. What is its Lorentz factor γ? Find its velocity, momentum, and total energy.

We first find γ from the relation

$$\text{KE} = (\gamma - 1) m_0 c^2$$

which yields

$$\gamma - 1 = \frac{\text{KE}}{m_0 c^2} = \frac{2 \times 10^6 \text{ eV}}{5 \times 10^5 \text{ eV}} = 4$$

or

$$\gamma = 5.$$

The total energy is then

$$E = \gamma m_0 c^2 = (5)(5 \times 10^5 \text{ eV}) = 2.5 \times 10^6 \text{ eV}$$
$$= 4.0 \times 10^{-13} \text{ J}.$$

The velocity is

$$\beta = \frac{\sqrt{\gamma^2 - 1}}{\gamma} = 0.98$$

so that

$$v = 0.98c.$$

To calculate the electron's momentum, we must write an expression for its rest mass:

$$m_0 = \frac{E_0}{c^2}$$

where E_0 is the rest energy,

$$E_0 = 5 \times 10^5 \text{ eV} = 8 \times 10^{-14} \text{ J}.$$

Substituting into the expression for the momentum yields

$$p = \gamma m_0 v = \gamma (E_0 / c^2) v$$
$$= \frac{(5)(8 \times 10^{-14} \text{ J})(0.98)(3 \times 10^8 \text{ m/s})}{(3 \times 10^8 \text{ m/s})^2}$$
$$= 1.3 \times 10^{-21} \text{ kg} \cdot \text{m/s}.$$

Notice that we needed to convert the electron's rest energy into SI units (5×10^5 eV = 8×10^{-14} J) before dividing by c^2 to obtain its rest mass m_0 in kg. Alternatively, the momentum could have been found from the Pythagorean theorem and our imaginary triangle.

$$p = \frac{\sqrt{E^2 - (m_0 c^2)^2}}{c}$$
$$= \frac{\sqrt{(4.0 \times 10^{-13} \text{ J})^2 - (8.0 \times 10^{-14} \text{ J})^2}}{3 \times 10^8 \text{ m/s}}$$
$$= 1.3 \times 10^{-21} \text{ kg} \cdot \text{m/s}$$

SUMMARY TABLE OF FORMULAS

Photon travel time between parallel mirrors:
$$t_0 = \frac{2d}{c} \tag{21-1}$$

Photon travel distance between moving mirrors, in stationary frame:
$$d' = ct = 2\sqrt{d^2 + \left(\frac{vt}{2}\right)^2} \tag{21-2}$$

Time dilation:
$$t = \frac{t_0}{\sqrt{1 - \left(\frac{v}{c}\right)^2}} = \gamma t_0 \tag{21-3}$$

Lorentz factor:
$$\gamma = \frac{1}{\sqrt{1 - \left(\frac{v}{c}\right)^2}} \tag{21-4}$$

Relation between β and γ:
$$\beta = \frac{v}{c} = \sqrt{1 - \frac{1}{\gamma^2}} = \frac{\sqrt{\gamma^2 - 1}}{\gamma} \tag{21-5}$$

Length contraction:
$$L = L_0 \sqrt{1 - \left(\frac{v}{c}\right)^2} = \frac{L_0}{\gamma} \tag{21-6}$$

Mass increase:
$$m = \frac{m_0}{\sqrt{1 - \left(\frac{v}{c}\right)^2}} = \gamma m_0 \tag{21-7}$$

Relativistic momentum:
$$p = mv = \frac{m_0 v}{\sqrt{1 - \left(\frac{v}{c}\right)^2}} = \gamma m_0 v \tag{21-8}$$

Mass-energy equivalence:
$$E = mc^2 \tag{21-9}$$

Total energy:
$$E = mc^2 = E_0 + KE = m_0 c^2 + KE \tag{21-10}$$

Relativistic kinetic energy:
$$KE = \frac{m_0 c^2}{\sqrt{1 - \left(\frac{v}{c}\right)^2}} - m_0 c^2 = (\gamma - 1) m_0 c^2 \tag{21-11}$$

Addition of relativistic velocities:
$$u = \frac{v + u'}{1 + \frac{vu'}{c^2}} \tag{21-12}$$

Schwarzschild radius for a black hole:
$$R_s = \frac{2GM}{c^2} \tag{21-13}$$

QUESTIONS

1. What is meant by an inertial frame? Is constant velocity sufficient? Is the Earth an inertial frame?

2. Explain why an elevator's acceleration upward simulates downward gravity. If you were in a box in space, accelerated horizontally, would you think gravity was horizontal in the opposite direction?

3. Does time dilation affect the biological aging process as well as physical objects such as watches?

4. The Lorentz contraction foreshortens only lengths parallel to the velocity. How would this affect the appearance of a three-dimensional object (a cube, say)?

5. If a distant galaxy is receding from you at $0.5c$, how fast does its light travel toward you, according to special relativity? What happen to the colors of the light, according to the Doppler effect?

6 If you move vertically at 0.6c, describe what would happen to your mass, height, girth, and heartbeat as monitored by a doctor on earth through a remote television camera at rest relative to the Earth.

7 Why does special relativity do away with the concept of absolute simultaneity in time?

8 In one inertial frame, event 2 follows event 1. Can one always find another frame, moving at velocity $v < c$, in which event 2 precedes event 1? Does it matter how far apart the two events are spaced in distance?

9 If one wishes to observe significant relativistic effects on mass, length, or time (say 1% changes), how large must v/c be?

10 Does hot matter have greater mass than cold matter?

11 In classical mechanics, momentum $p = mv$. Since v cannot exceed c, does this place an upper bound on p?

12 Since matter can be destroyed in nuclear reactions, what physical law should replace conservation of mass?

13 In a particle accelerator, explain why it becomes increasingly difficult to accelerate a particle as its velocity approaches c.

14 When one increases a particle's energy, does most of the energy go into increasing its velocity or its mass?

15 Which has greater velocity, a proton with kinetic energy 1 MeV or an electron with the same kinetic energy?

PROBLEMS

Section 21.1 Galilean Relativity

- 1 While riding inside a train moving at 50 km/h in a straight line, you toss a coin straight up, and it falls 1 s later. Where does it land relative to its initial position, as viewed inside the train and outside on the ground?

- 2 What is the downward acceleration of the coin in Problem 1, as viewed inside and outside the train?

- 3 Two cars approach each other at 60 mi/h from opposite directions. What is their relative velocity?

•• 4 A child swims a distance L downstream at velocity p in a river with current velocity q, then returns upstream at the same speed with respect to the water. Compute the child's velocity upstream and downstream as viewed from shore, then show that the total time elapsed is $t = (2L/p)/(1 - q^2/p^2)$. Why is this time longer than $2L/p$, the time the trip would take in still water?

••• 5 While standing 1 km from the two sides and end of a dead-end canyon (Fig. 21.15), a lonesome cowboy shouts out. If a strong wind is blowing down the canyon at velocity $v = 80$ km/h, how long do the echoes take to reach him? Assume sound travels at velocity $v_s = 340$ m/s in still air.

••• 6 At closest approach, Jupiter is 4.2 AU from Earth. If the solar system were moving at 250 km/s through the "ether," what would be the difference in the round-trip travel times of radio waves bounced off Jupiter and returning to Earth
  (a) when the path is parallel to the velocity through the ether, and
  (b) when the path is perpendicular to the velocity?

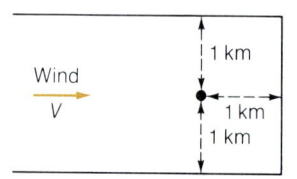

Figure 21.15

Section 21.2 Special Relativity

- 7 A spaceship streaks by the Earth at a speed $0.8c$, appearing on Earth to take 5.31×10^{-2} s to cross the Earth's diameter. How long did the trip take in the frame of the ship?

- 8 At rest, a pi meson (π^+) has a lifetime of 2.6×10^{-8} s. How long does it last in a lab if it travels at 2.8×10^8 m/s?

•• 9 When cosmic rays strike the Earth's upper atmosphere, they produce μ mesons, which travel downward at nearly the speed of light. During their rest lifetime of 2.2×10^{-6} s, they should travel only 660 m before decaying. Yet many are observed at the Earth's surface, 10 km from where they were produced.
  (a) What minimum time dilation factor γ is required to prolong their life?
  (b) At what minimum fraction of c do they travel?

- 10 If you could travel at $v = 0.9c$, how long would you perceive the following trips to take: to Mars, Pluto, and the star Betelgeuse (500 light years away)?

PROBLEMS

•• 11 Through ecological incentives, Arizona coyotes develop the ability to run at velocity $0.98c$. Unfortunately, roadrunners can run at $0.99c$.
  (a) Compute the time dilation factors γ for the coyote and roadrunner.
  (b) If the animals spend their entire lifetimes chasing one another, what are their apparent lifespans as observed by a park ranger at rest? Assume their normal (rest) lifetimes are 10 years each.

•• 12 A subatomic particle with $v = 0.999c$ travels 10 m before decaying. What is the particle's rest lifetime?

• 13 If space travel were possible at $v = 0.9c$, how long would a trip to the nearest star, Alpha Centauri (4.2 light years away) take, as perceived on board the ship?

• 14 Suppose a spacecraft journeyed to the Orion Nebula, 1500 light years away, at $v = 0.99c$. How long would the trip take according to the traveller and according to observers on Earth?

••• 15 The rotational velocity of a point on Earth varies from pole to equator. If normal human lifespans are 70 years, calculate the added lifetime of a resident of Ecuador compared to a resident of Antartica.

••• 16 A 5-year-old girl watches as her 30-year-old mother departs on a high speed space voyage. When the 40-year-old mother returns, she is greeted by a 75-year-old woman who claims to be her daughter. How fast did the spacecraft move?

• 17 A rocket, 100 m long at rest, passes by at $0.7c$. How long does it appear to you?

•• 18 If the star Sirius is 8.8 light years away, how fast must one travel so that the distance appears to be 5 light years?

• 19 While cruising at $0.65c$ in your 1999-model sportscar, passing billboards appear to be 10 m long. What is their rest length?

•• 20 Two art critics disagree about the dimensions of a painting. One says it is 1 m × 2 m; the other says 1 m × 1 m. What is the relative velocity between the critics?

••• 21 An alien observer of the solar system notes that the Earth travels around the Sun on an eccentric orbit, with $e = 0.5$. What is the velocity of the spacecraft relative to the solar system? (See Chapter 7 on orbits.)

•• 22 A cube is L_0 on a side and has a rest volume of L_0^3. What is its volume as measured by an observer moving parallel to one side at velocity v?

•• 23 Two sides of a right triangle make an angle θ (Fig. 21.16), such that $\tan \theta = y/x$. An observer moves at velocity v along the x-direction.

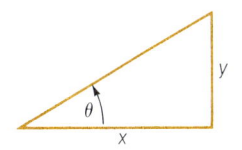

Figure 21.16

  (a) Show that in the observer's frame, $y' = y$ and $x' = x(1 - v^2/c^2)^{1/2}$.
  (b) Show that the new angle is perceived by this observer to have

$$\tan \theta' = \gamma \tan \theta$$

  where $\gamma = 1/(1 - v^2/c^2)^{1/2}$ is the Lorentz contraction factor.
  (c) Is θ' larger or smaller than θ?

•• 24 A beam of fast particles travels at $v = 0.999c$ in a circular ring of 1 km radius. In the lab frame (at rest), the particles make 10^4 circuits before slamming into a target. How far (km) and how long (s) does the trip last in the frame of the particles?

• 25 What is the apparent mass of a proton ($m_0 = 1.67 \times 10^{-27}$ kg) moving at $0.8c$?

• 26 An electron ($m_0 = 9.11 \times 10^{-31}$ kg) has a mass measured to be 8×10^{-30} kg. What fraction of c is its velocity?

• 27 At what speed will a mass be increased by 1% over its rest mass?

• 28 At what speed will a particle's mass be increased by a factor of 10?

• 29 In the hydrogen atom, the electron orbits the proton at velocity $v = 2187$ km/s. By what fraction is the electron's mass increased over its rest mass?

•• 30 A mass falls freely (no air resistance) toward the Earth from a height h much less than the Earth's radius. Show that just before the object strikes ground, its mass $m = m_0/(1 - 2gh/c^2)^{1/2}$.

••• 31 Halley's comet moves in an eccentric orbit with $e = 0.967$ and $a = 17.95$ AU (see Chapter 7). Find the velocities at perihelion and aphelion, and compute the fractional difference in mass $\Delta m/m$ between the two points.

••• 32 A mass m with charge q moves at velocity v perpendicular to a uniform magnetic field B.
  (a) From force balance (Chapter 16) show that the particle moves in a circular orbit of radius $r = (mv/qB)$ with angular frequency $\omega = (qB/m)$.

Notice that for $v \ll c$, the frequency ω does not depend on v.

(b) If v becomes large, show that the radius and frequency become $r = (\gamma m_0 v/qB)$ and $\omega = (qB/m_0\gamma)$, where $\gamma = 1/(1 - v^2/c^2)^{1/2}$ is the Lorentz factor and m_0 the rest mass.

(c) As a cyclotron accelerates a particle in a circular orbit to higher and higher velocities, what happens to the frequency ω?

··· 33 Using a mass spectrometer, physicists are able to find a particle's mass by measuring its velocity v and radius of curvature R in a magnetic field B (Fig. 21.17). Refer to Problem 32.

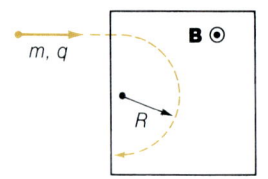

Figure 21.17

(a) Show that $R = (\gamma m_0 v/qB)$ where m_0 and q are the rest mass and charge of the particle and where γ is the Lorentz factor.

(b) For small velocities ($v \ll c$) what will be the ratio of radii R_2/R_1 for two particles with the same charge-to-rest-mass ratio q/m_0?

(c) If two particles have the same ratio q/m_0 but mass 1 moves at $0.98c$ while mass 2 moves at $0.99c$, what will be the ratio of their radii R_2/R_1? (Compute the Lorentz factors for each particle.)

··· 34 Starting at rest a very large distance from a planet of mass M and radius R, an astronaut falls freely toward the surface with no air resistance.

(a) From Newtonian energy conservation, find the velocity v as the astronaut strikes the surface.

(b) Show that, at that point, his mass is $m = m_0/(1 - 2GM/c^2R)^{1/2}$.

(c) If the object were a black hole (with radius $R = 2GM/c^2$) instead of a planet, what would be the astronaut's mass at radius R? Neglect the fact that he would be torn apart by tidal forces long before passing into the hole.

· 35 How much mass (in grams) is equivalent to 1 J of energy?

· 36 One ton of TNT explosive is equivalent to 5×10^9 J energy. What is the mass equivalent of this energy?

· 37 Radioactive nitrogen 12 decays into carbon 12, releasing 2.72×10^{-12} J of energy. What is the difference in masses (or mass equivalents)?

·· 38 When one uranium-235 nucleus is split, or fissioned, it releases about 3.2×10^{-11} J of energy. If the first atomic bomb released the equivalent of 12 kilotons of TNT (see Problem 36), how many U-235 nuclei were split? What was the total mass of fissionable U-235 in the bomb? Assume $m_U \cong 235\, m_p$.

· 39 When 1 kg of coal is burned it releases 3.3×10^7 J energy. What is the mass difference between reactants and products?

· 40 What is the rest mass energy equivalent of 1 kg of coal? Compare to the combustion energy release (3.3×10^7 J/kg).

· 41 One calorie (*not* nutritional Calorie) is the heat energy required to raise the temperature of 1 gram of water one degree Celsius (1 cal equals 4.18 J). What is the mass increase of 1 g of water heated 1 C°?

· 42 One gram of water vapor releases 540 calories of heat when it condenses into liquid. What is the decrease in mass?

·· 43 A very hot star, with mass 60 $M_\odot = 1.19 \times 10^{32}$ kg produces energy at the rate of $10^6 L_\odot = 3.83 \times 10^{32}$ J/s. The nuclear reactions that power it have an efficiency of 0.007 in converting mass into energy.

(a) How much mass is converted into energy each second?

(b) How long would such a star live, assuming that only 30% of the star's mass is available for nuclear burning?

·· 44 If one 10 g marshmallow were converted entirely into energy, how long could it supply the energy needs of a small community of 90,000? Assume each person uses 1 kilowatt of power.

·· 45 Calculate the total energy and kinetic energy of a proton ($m_p = 1.67 \times 10^{-27}$ kg) travelling at $0.9c$.

·· 46 A proton has KE = 500 MeV. What is its velocity? (Use 1 MeV = 10^6 eV and 1 eV = 1.6×10^{-19} J.)

·· 47 Use the relativistic forms of total energy E and momentum p to show that $E^2 = p^2c^2 + m_0^2c^4$.

·· 48 What is the velocity of an electron whose KE is 0.8 MeV?

·· 49 Use the definition of the Lorentz factor, $\gamma = 1/(1 - v^2/c^2)^{1/2}$, to show that the total energy $E = \gamma m_0 c^2$.

··· 50 If a particle has rest mass m_0, kinetic energy T, and total energy E, show that:

(a) $E = T + m_0 c^2$

(b) Momentum is given by $p = (T^2 + 2Tm_0c^2)^{1/2}/c$. (Use Prob. 47.)

•• 51 A proton has kinetic energy T = 400 MeV and rest energy m_pc^2 = 938 MeV. What is its momentum p? Use Problem 50.

••• 52 A neutral pi meson (π^0) of rest mass energy 134.96 MeV decays at rest into two photons.
(a) Show that the photons must fly off in opposite directions, to conserve linear momentum.
(b) What is the energy of each photon (in MeV)?

•• 53 Show that the kinetic energy is $(\gamma - 1)m_0c^2$, where γ is the Lorentz factor (see Problems 49 and 50).

•• 54 Show that for a particle with zero rest mass (photon or neutrino), $E = pc$.

•• 55 Two spaceships each move at $0.6c$ in opposite directions. When they pass, what is the relative velocity each perceives the other to have?

••• 56 As observed on ground, rocket 1 with velocity v_1 = $0.5c$ chases after rocket 2 with v_2 = $0.3c$. With what relative velocity does 1 overtake 2?

••• 57 As viewed by an observer moving at velocity v with respect to a fixed laboratory, two particles move apart in opposite directions, each at $0.8c$. In the laboratory frame, particle 1 is at rest, and particle 2 moves away at velocity v_2. What are v and v_2?

SPECIAL PROBLEMS

Special relativity has many implications that are often paradoxical. Most of the paradoxes, however, can be resolved when one thinks about space and time properly. For example, two events that are simultaneous in one frame are not simultaneous in another frame. Also, a situation that seems symmetric may not be, because one frame undergoes acceleration. These problems illustrate some of these paradoxes, and they may be used by students and instructors to see how time dilation and length contraction work.

1. The Racecar and the Trench

A racecar travelling at $0.5c$ encounters a trench with rest width 3.2 m. From the point of view of the driver, the trench appears Lorentz contracted by a factor 1.15 and has a length of 2.77 m. The driver thinks that his car's wheelbase (Fig. 21.18) is wider than the trench. However from the trench, the car is moving toward it at $0.5c$ and is Lorentz contracted by a factor 1.15 and has a length 2.61 m. In the trench's frame, the car appears shorter than the trench and would seem to be in danger of falling in. Does the car fall into the trench? How do you resolve the paradox?

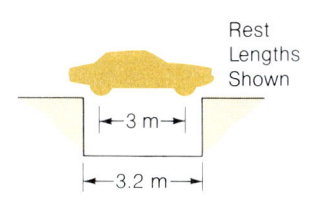

Figure 21.18

2. Carousel Rings

Two children, Michael and Jennifer, rotate about a common center, on rings that move with equal but opposite angular velocities (Fig. 21.19). At the moment when they pass, their watches agree. Michael notices that Jennifer's watch runs more slowly because of time dilation, and he therefore believes that his watch will be ahead the next time they meet. However, Jennifer perceives Michael's watch to be running more slowly, and believes her watch will be ahead next time. What really happens? Explain why they disagree about whose watch will be ahead.

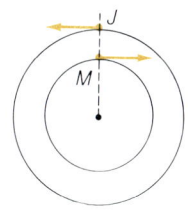

Figure 21.19

3. Pole in the Barn Paradox

A pole vaulter runs at $0.6c$ toward a barn 4.5 m long. Since his pole is 5 m long (at rest), he fears it will not fit. At the moment when the front end of the pole reaches the forward barn door, a farmer closes and opens both doors very quickly. Lorentz contraction makes the pole appear only 4 m to the farmer (it fits!). However, to the vaulter the barn is Lorentz contracted to 3.6 m, so it would appear it doesn't fit. What really happens? Does the pole fit, or is it squashed by the doors? (Think about simultaneity of the doors closing in the two frames.)

4. Pion Decay

A pi meson (π^-) decays at rest into a muon (μ^-) and an antineutrino ($\bar{\nu}$) as shown in Fig. 21.20. To conserve momentum, the μ^- and $\bar{\nu}$ fly off in opposite directions with equal momenta $p_\mu = p_\nu$. Let the total energies be E_μ and E_ν, and the kinetic energies be T_μ and T_ν. The rest mass energies are $m_\pi c^2 = 139.57$ MeV, $m_\mu c^2 = 105.66$ MeV, and $m_\nu = 0$ (massless).

(a) From energy and momentum conservation, show that

$$E_\nu^2 = E_\mu^2 - m_\mu^2 c^4 \quad \text{and} \quad E_\mu + E_\nu = m_\pi c^2.$$

(b) Use the relation $E = T + m_0 c^2$ to show that

$$T_\mu = (m_\pi - m_\mu)^2 c^2/2m_\pi$$
$$T_\nu = E_\nu = (m_\pi^2 - m_\nu^2) c^2/2m_\pi.$$

(c) What are T_μ and T_ν in MeV?

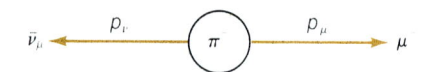

Figure 21.20

5. The Sprinter versus the Lightbeam

After winning the Olympics, a famous sprinter decides to race a lightbeam. Relative to spectators in the stands, he can run at $0.99c$. Use the five choices to answer the following questions.

Choices:

(1) There is no race; the sprinter is frozen in time along with light.

(2) The light beam is moving $0.01c$ faster than the sprinter.

(3) The light beam is moving the full speed of light faster than the sprinter.

(4) The sprinter is moving $0.01c$ faster than the light beam.

(5) The sprinter appears to move at zero velocity.

Questions:

(a) How does the race appear to spectators in the stands?

(b) How does the race appear in the sprinter's frame of reference?

(c) How does the race appear to the lightbeam?

SUGGESTED READINGS

"The Michelson-Morley experiment." R. S. Shankland, *Scientific American*, Nov 64, p 107, **211**(5)

"The discovery of a gravitational lens." F. H. Chaffee, *Scientific American*, Nov 80, p 60, **243**(5)

"Quantum gravity." B. S. DeWitt, *Scientific American*, Dec 83, p 112, **249**(6)

[Einstein centennial issue]. *Physics Today*, Mar 79, **32**(3)

"How I created the theory of relativity." A. Einstein, Y. Ono, tr, *Physics Today*, Aug 82, p 45, **35**(8)

"Intuition, time dilation and the twin paradox." D. E. Hall, *The Physics Teacher*, Apr 78, p 209, **16**(4)

"Einstein's wonderful year." T. Ferris, *Science 84*, Nov 84, p 61, **5**(9)

A source book in physics. William F. Magie, Harvard University Press, Cambridge, 1965

The 4th dimension. Rudy Rucker, Houghton-Mifflin, Boston, 1984

Mr. Tompkins in wonderland. George Gamow, Cambridge University Press, New York, 1940

Constructing the universe. David Layzer, Freeman, New York, 1984

Spacetime physics. Edwin F. Taylor, John A. Wheeler, Freeman, New York, 1966

Relativity: The special and general theory. Albert Einstein, Crown, New York, 1961

Great scientific experiments. Rom Harre, Phaidon Press, Oxford, 1981 (the Michelson-Morley Experiment)

"Testing Einstein's General Relativity during Eclipses of the sun", Jack B. Zirker, *Mercury*, Vol. 14, No. 4, July 1985, p 98

The Relativity Explosion. Martin Gardner, Viking Books, New York, 1976

Cosmic Frontiers of Relativity, William Kaufmann, Little-Brown, Boston, MA, 1977

"Testing General Relativity", Clifford Will, *Sky and Telescope*, Vol. 66, No. 4, p 294, Oct. 1983

General Relativity: An Einstein Centenary Survey, eds. S. W. Hawking and W. Israel, Cambridge, Cambridge University Press, 1979

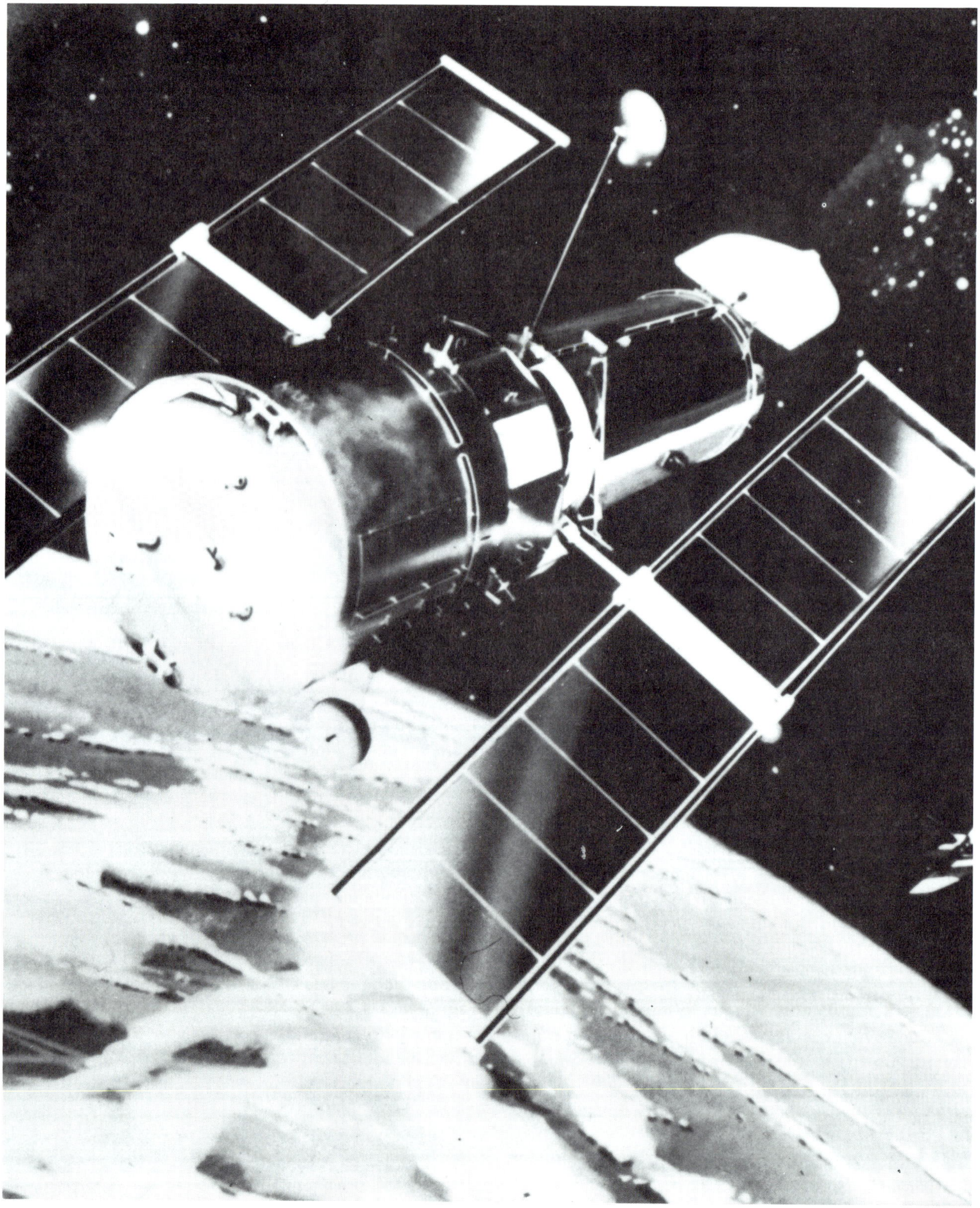

Chapter 22

Particles and Waves

A major revolution occurred in physics, starting about the beginning of the twentieth century. Before the revolution, it appeared that physics was a mature science, and that most natural phenomena were understandable in terms of the known laws. In Newton's time, mechanics had been developed, along with the requisite mathematical techniques. The behavior of fluids and the laws of thermodynamics were analyzed satisfactorily (for the most part) by the mid-nineteenth century, and the wave properties of light and the nature and behavior of electromagnetic phenomena in general seemed comprehensible in view of Maxwell's great work. After the revolution, every one of these disciplines had been vastly modified; the structure of the atom (a major remaining unknown in the late 1800s) had been explored; and even the mechanics of Newton had been altered in the wake of Einstein's theories of special and general relativity.

In this chapter we continue to explore the new picture that began to develop at the beginning of this century. Some of the first major cracks to appear in the established foundations of physical theory were in the study of the properties of light. We have seen (in Chapters 18 through 20) that many important phenomena connected with light can be explained in terms of waves (that a wave nature of light is *required* to explain them), but there were nagging inconsistencies with this picture. We begin this chapter by discussing some of these inconsistencies and the unanticipated directions in which they led.

22.1 The Particle Nature of Light

We have described Newton's preference for the view that light consists of tiny particles, but we have seen that several experiments during and since his time have established beyond doubt that light consists of waves. Now we are going to find that, in some circumstances, light unmistakably acts as though it consists of particles, though not in ways envisioned by Newton.

The Evidence for a Particle Nature

Throughout the nineteenth century, many scientists performed experiments designed to elucidate the properties of light emitted from various sources. In contrast with the experiments discussed in Chapter 20, where the nature of light itself was in question, other experiments were aimed at determining the relationship between a source of light and the light emitted. Many of these experiments produced results that were not immediately understood, and whose eventual explanation was to become part of the revolution in physics that we have mentioned.

One class of experiments, carried out by W. H. Wollaston and J. Fraunhofer and others in the first decades of the nineteenth century, and by R. Bunsen and G. Kirchhoff in the 1850s and 1860s, involved **spectral lines.** When light from a source such as a star or a flame is dispersed (by prisms in all the early experiments), dark or bright features appear in the spectrum at specific wavelengths, superposed on a continuous spectrum of smoothly varying intensity (Fig. 22.1). The work of Bunsen and Kirchhoff showed that these features are created by individual chemical elements, so that the composition of a distant source of light could be determined. But for a considerable time there was no explanation of why individual elements produced the lines only at specific wavelengths. The full understanding of spectra came well into the twentieth century, when great strides were made in understanding the structure of the atom. Therefore, we will not be in a position to discuss spectral lines until the next chapter, where we treat atomic structure.

Meanwhile, another class of experiments was carried out during the nineteenth century, having to do with the continuous radiation from light sources such as stars. It was found that any body whose temperature is above absolute zero emits radiation over a broad range of wavelengths (actually, it emits over all wavelengths). The spectrum of this continuous radiation was found generally to have a distinctive shape, with a peak at some particular wavelength, often designated λ_{max} (Fig. 22.2).

It was found by experiment (and later explained by theory) that the wavelength of maximum emission λ_{max} is inversely related to the temperature T (in Kelvins).

$$\lambda_{max} = \frac{2.90 \times 10^{-3} \text{ m} \cdot \text{K}}{T} \tag{22–1}$$

This relationship, known as **Wien's law,** enables us to predict that the Sun's greatest intensity of radiation is emitted at wavelength

$$\lambda_{max} = \frac{2.90 \times 10^{-3} \text{ m} \cdot \text{K}}{5700 \text{ K}} = 509 \text{ nm}$$

in the visible part of the spectrum.

22.1 THE PARTICLE NATURE OF LIGHT

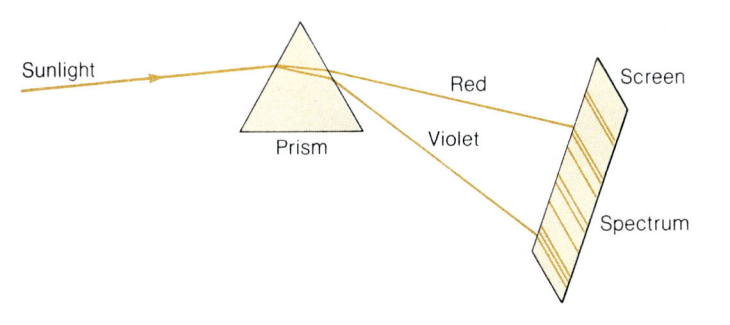

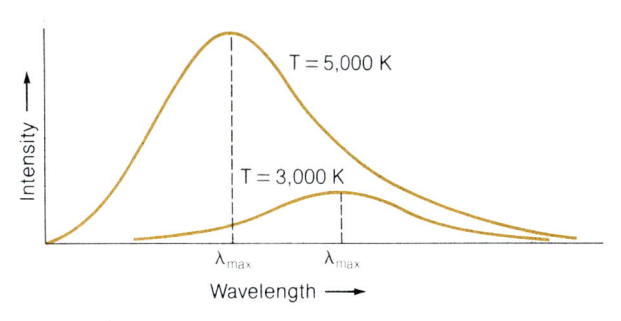

Figure 22.1

Spectral lines. Light from the Sun is dispersed by a prism and displayed on a screen. Dark lines appear at specific positions or wavelengths in the spectrum. Other light sources (such as a candle, for example) produce bright lines. In either case, these spectral lines are formed by specific chemical elements.

Figure 22.2

Blackbody radiation. An idealized object that absorbs all light incident on it, so that its emission is strictly due to its own temperature, emits at these wavelengths. Two curves are shown, for T = 5000 K and T = 3000 K, illustrating Wien's law.

In the ideal case, in which the surface emits and absorbs all radiation with 100% efficiency, no light would be reflected and no spectral lines would appear. Such an ideal emitter of light is called a **blackbody radiator,** and is both the best absorber and best emitter of light. As discussed in Chapter 12, the rate at which emitted energy leaves the surface of a glowing hot object (that is, the power radiated) is proportional to the surface area A of the object and to the fourth power of its temperature T:

$$\text{Power} = e\sigma A T^4 \qquad (22\text{--}2)$$

Here, e is called the **emissivity,** a dimensionless number between 0 and 1 representing the efficiency of the object as a radiator, and $\sigma = 5.67 \times 10^{-8}$ W/m² · K⁴, the **Stefan-Boltzmann constant.** For an ideal blackbody, $e = 1$. Some surfaces and natural sources of light come reasonably close to being blackbodies ($e > 0.9$). One example is a cavity with no contact with the outside; another is a special black paint used in *Space Shuttle* experiments with $e \approx 0.99$ (that is, it reflects less than 1%). The efficiency of a metal, such as copper, may be as low as $e = 0.3$.

Example 22.1   A square plate, 1 m on a side, is heated to 600°C. If its emissivity is $e = 1.0$, what is the wavelength of maximum radiation? How much energy (in Joules) is radiated in each minute?

First, we convert 600°C to 873 K, since absolute temperatures are required in Eq. 22–1 and Eq. 22–2. We find that the wavelength of maximum emission is

$$\lambda_{max} = \frac{2.9 \times 10^{-3} \text{ m} \cdot \text{K}}{T} = \frac{2.9 \times 10^{-3} \text{ m} \cdot \text{K}}{873 \text{ K}} = 3320 \text{ nm}$$

which is in the infrared portion of the spectrum.

The radiated power is

$$\text{Power} = e\sigma AT^4 = (1.0)(5.67 \times 10^{-8} \text{ W} \cdot \text{m}^{-2} \cdot \text{K}^{-4})(1 \text{ m}^2)(873 \text{ K})^4$$
$$= 3.29 \times 10^4 \text{ W}.$$

One W = 1 J/s, so in one minute, the surface radiates a total energy

$$E = (60 \text{ s})(3.29 \times 10^4 \text{ J/s}) = 2 \times 10^6 \text{ J}.$$

Neither Wien's law (Eq. 22–1) nor the Stefan-Boltzmann law (Eq. 22–2) could be explained by the classical laws of physics. Maxwell's theory of electromagnetism was used to describe the radiation from the oscillations of electric charges in the atoms of the material absorbing and emitting light. Maxwell's classical theory predicted, incorrectly, that the intensity of the observed spectrum of emitted light should increase toward the shorter wavelengths (higher frequencies) as shown in Fig. 22.3. Thus, a blackbody of any temperature would emit most of its radiation at very short wavelengths (the spectrum actually goes to infinite energy at small wavelengths), and this problem with classical theory was known as the "ultraviolet catastrophe." Classical theory was evidently wrong.

The German scientist Max Planck proposed a new theory in 1900 that explained the blackbody spectrum, including the turnover in intensity at short wavelengths. He had to postulate a remarkable new property of light: that light can be emitted only in discrete packets of energy called **quanta,** rather than in a continuous range of energies. Planck viewed this as a consequence of the emission process, which by this time was envisioned to be due to tiny electric oscillators (recall that Maxwell had shown that EM waves consist of oscillating electric and magnetic fields, and that these waves were emitted whenever electric charges oscillated). Einstein further extended Planck's quantum concept in 1905, postulating that light itself could only carry energy in discrete quantities. The energy carried by a single particle of light was found to be

$$E = h\nu = \frac{hc}{\lambda} \tag{22–3}$$

where ν is the frequency of the radiation, λ is the wavelength, and h is **Planck's constant,** whose value is 6.626×10^{-34} J · s. The success of this hypothesis now forced acceptance of the notion that light consists somehow of particles, tiny bullets, each carrying a definite amount of energy. The earlier evidence for the wave nature of light remained, however, and a new concept, the **photon,** was born. In this picture, light is said to have a dual nature; it acts as a wave in some circumstances, and as a particle in others.

The simple relationship (Eq. 22–3) between photon energy and frequency allows us to see immediately that there is a great range in photon energies. The highest-energy detected photons, those of γ-rays, have energies over 10^{23}

22.1 THE PARTICLE NATURE OF LIGHT

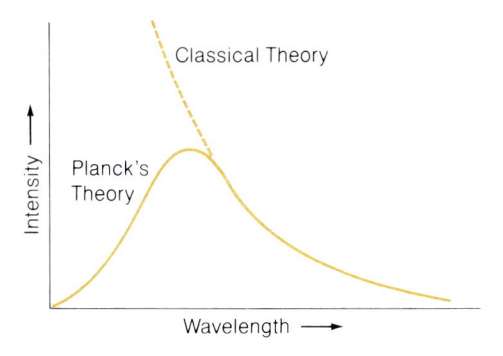

Figure 22.3

Comparison of classical theory of radiation with Planck's theory for a blackbody. In the classical theory, the intensity continues to rise toward short wavelengths, a result known as the "ultraviolet catastrophe."

times greater than the longest-wavelength radio photons. This implies that high-frequency photons can do more work than low-frequency ones, something that we will see applied in the next section.

Example 22.2 What is the energy of a photon of yellow light, whose wavelength is 600 nm? What is the energy of an x-ray photon, whose wavelength is 1 nm?

We recall that the relationship between wavelength and frequency is

$$\lambda = \frac{c}{\nu}$$

where λ is the wavelength, ν is the frequency, and c is the speed of light. Hence the energy of a photon can be expressed as

$$E = \frac{hc}{\lambda}.$$

For the yellow photon, we find

$$E = \frac{hc}{\lambda} = \frac{(6.63 \times 10^{-34} \text{ J} \cdot \text{s})(3 \times 10^8 \text{ m/s})}{6 \times 10^{-7} \text{ m}} = 3.3 \times 10^{-19} \text{ J}.$$

For the x-ray photon, the energy is

$$E = \frac{(6.63 \times 10^{-34} \text{ J} \cdot \text{s})(3 \times 10^8 \text{ m/s})}{1 \times 10^{-9} \text{ m}} = 2.0 \times 10^{-16} \text{ J}.$$

CHAPTER 22 PARTICLES AND WAVES

The Photoelectric Effect

Einstein's theories about the particle nature of light also explained another experimental mystery, known as the **photoelectric effect**. It had been known since 1887 that this effect occurred when light shone on a surface, especially if the surface were a metal, with loosely-bound electrons. Consider an apparatus in which a pair of metal plates inside a vacuum tube have a potential difference applied by a battery (Fig. 22.4). When light strikes the plate with negative potential, a current flows in the circuit because electrons acquire enough energy to be ejected from this plate and cross to the plate with positive potential. Such a tube is called a **photocell,** and today many practical uses for them have been found.

Detailed studies of the photoelectric effect revealed some properties that could not be understood by the classical wave theory of light. It was found that when light illuminated the photocell, no electrons were emitted unless the frequency of light was greater than some **threshold frequency** v_T, typically in the ultraviolet. In addition, the wave theory and photon theory made quite different predictions about the energies of the ejected **photoelectrons** as the light intensity varied.

Consider light of a single frequency v. In the **wave theory,** the light intensity is governed by the amplitude of the electric and magnetic fields in the waves. The light rays are made more intense by increasing the wave amplitude, while holding v constant. Thus, in this view the energies of the photoelectrons should depend on the intensity of the incident light, but not on its frequency. A higher intensity means a larger amplitude, so that the increased electric fields will accelerate the electrons to greater speeds.

The **quantum theory,** or photon theory, makes dramatically different predictions. This theory postulates that light consists of particles with discrete quantities of energy, hv. The light intensity for a given frequency is governed only by the number of incident photons, since the energy per photon is fixed strictly by the frequency. Hence the number of ejected photoelectrons should increase as the light intensity increases, but each electron's energy should depend only on the frequency and not on the intensity. In contrast with the wave theory, which predicts that light of any frequency can produce a current of photoelectrons if the intensity is high enough, the photon theory predicts that no current will be produced, regardless of intensity, unless the photons have frequencies above the threshold frequency. Careful experiment by R. A. Millikan in 1913 showed that Einstein's theory was correct, and the classical wave theory was wrong. Einstein was later awarded the Nobel Prize for his theory of the photoelectric effect and the particle nature of light.

The reasons for the success of Einstein's theory depend on the nature of electrons in solids. Electrons are bound to the metal surface so that a certain amount of energy, called the **work function,** is required for an electron to break free and be ejected. In the wave theory, this threshold would be crossed by increasing the intensity of light until the amplitude of the waves is sufficient, whereas in the photon theory, the threshold energy is obtained by increasing the frequency until the individual photons have sufficient energy. In the photon theory, the intensity can be increased indefinitely and still no electrons would be ejected if the frequency of the photons were too low.

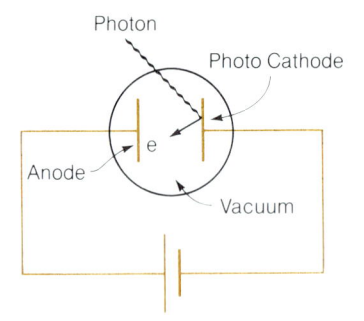

Figure 22.4

A photocell. When a photon of sufficient energy strikes the photocathode, an electron is ejected and accelerated by the potential difference to the anode. Thus, incident light creates a current in the circuit that is proportional to the intensity of the light.

22.1 THE PARTICLE NATURE OF LIGHT

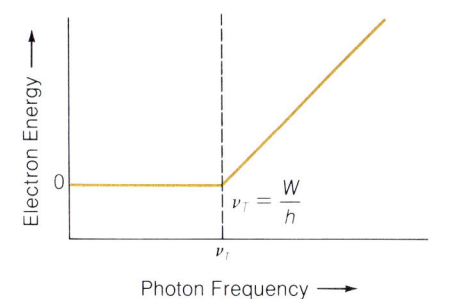

Figure 22.5

Photoelectric effect. When photons strike a surface, electrons are ejected only by photons with energy equaling or exceeding the work function W. Therefore the photons have a threshold frequency $v_T = W/h$, above which electrons are ejected with energy proportional to $(v - v_T)$.

The kinetic energy of an ejected electron is equal to the energy given it by the photon it absorbs, less the amount required to overcome the work function. Thus, in the quantum theory,

$$KE = hv - W \qquad (22\text{-}4)$$

where hv is the energy of the incident photon, and W is the work function of the surface. This equation predicts what should be seen as the frequency of monochromatic light on a photocell is varied. At low frequencies, such that hv is less than W, the electrons have negative energies and do not escape the surface. As the frequency is increased, there continues to be no electron ejection, until the photon energy hv exceeds the work function W. Then a current suddenly begins to flow. The threshold frequency is thus given by $hv_T = W$. The energy of the ejected electron then rises in proportion to the frequency. A plot of electron energy versus photon frequency is shown in Fig. 22.5. The electron energy can be measured by varying a known opposing voltage between the two plates in the photocell until the electrons just stop reaching the positive plate; this voltage corresponds to the kinetic energy $\frac{1}{2}m_e v_e^2$ of the fastest-moving electrons.

Example 22.3 What is the speed of the ejected electrons in a photocell, whose work function is 2×10^{-19} J = 1.25 eV, subjected to monochromatic red light of frequency 5×10^{14} Hz? What happens if the intensity of the light is doubled?

The kinetic energy of the ejected electrons is given by Eq. 22-4:

$$KE = hv - W$$
$$= (6.63 \times 10^{-34} \text{ J} \cdot \text{s})(5 \times 10^{14} \text{ Hz}) - 2 \times 10^{-19} \text{ J} = 1.3 \times 10^{-19} \text{ J}$$

The kinetic energy is equal to $\frac{1}{2}mv^2$, so solving for v yields

$$v = \sqrt{\frac{2\text{KE}}{m}} = \sqrt{\frac{2(1.3 \times 10^{-19}\,\text{J})}{9.1 \times 10^{-31}\,\text{kg}}} = 5.4 \times 10^5 \text{ m/s}.$$

If the intensity is doubled, the frequency remains the same. Thus, the kinetic energy and the speed of the electrons remain the same, but the number of ejected electrons is doubled. Therefore the current from the photocell is doubled. This simple proportionality makes photocells useful devices for measuring the intensity of light.

A concept related to the work function of metals is the **band gap** of a semiconductor or insulator. Every solid contains electrons, but they are not always free to move about the solid and conduct electricity. For quantum mechanical reasons beyond the scope of this text, the electrons in a crystalline solid are distributed in wide bands of energy separated by forbidden regions (Fig. 22.6). The electrons are normally in either the lower energy **valence band,** where they are tightly bound to individual atoms, or in the higher energy **conduction band,** where they are free to roam about the solid. No electrons are found in the forbidden region, which is separated by the band gap energy of between 1 and 5 eV (Table 22.1).

In insulators (Fig. 22.6a), the valence band is full, and the band gap is large enough to prevent electrons from becoming conductive. The resistivity is therefore large. In a conductor (Fig. 22.6b), the valence band and part of the conduction band is filled. Thus, electrons move quickly in response to an applied electric field and the resistivity is small. A semiconductor (Fig. 22.6c) is intermediate between an insulator and conductor. The electrons fill only the valence band, but the band gap is small enough (usually about 1 eV) that a

Table 22.1
Band gap energies*

Substance	Band gap, eV
Diamond (carbon)	5.4
Zinc Sulfide (ZnS)	3.54
Cadmium Sulfide (CdS)	2.4
Silicon Carbide (SiC)	2.3
Silicon (amorphous)	1.75
Cadmium Telluride (CdTe)	1.44
Gallium Arsenide (GaAs)	1.35
Indium Phosphide (InP)	1.27
Silicon (crystalline)	1.11
Copper Indium Selenide (CuInSe$_2$)	1.0
Germanium	0.67
Indium Antimonide (InSb)	0.165

*Band gaps at room temperature

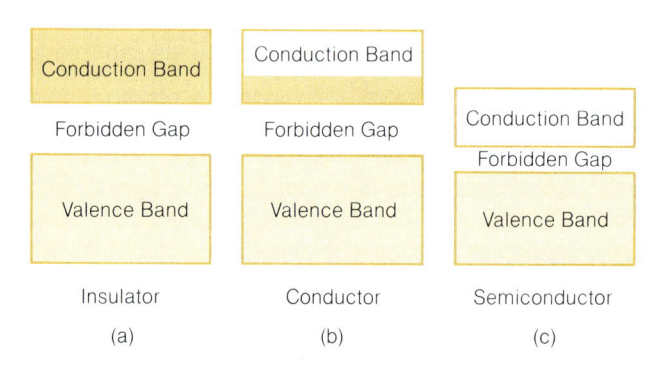

Figure 22.6

Energy bands in solids. (a) In an insulator, the valence band is full, and the conduction band is separated by a larger forbidden gap. (b) In a conductor, electrons fill the valence band and part of the conduction band. (c) In a semiconductor electrons fill only the valence band, but they may be thermally excited across the small band gap into the conduction band.

22.1 THE PARTICLE NATURE OF LIGHT

small number may be thermally excited across the forbidden gap into the conduction band. The resistivity of a semiconductor is therefore quite dependent on the temperature. When small impurities are added to the semiconductor, new energy levels appear just below the conduction band. This doping process (Chapter 17.4) makes the conductive properties of p-type and n-type semiconductors even more sensitive to small changes in temperature or applied voltage.

An electron may also jump the forbidden gap by absorbing a photon whose energy is greater than the band gap energy, analogous to the photoelectric effect in a metal. Thus, semiconductors such as silicon, gallium arsenide, or more complex compounds such as copper indium selenide may be used in solar cells to generate electricity from photons in the solar spectrum.

FOCUS ON PHYSICS

Solar Cells and Solar Energy*

Energy from the sun is widely used to heat homes and buildings. Nearly all such applications have converted sunlight into heat by absorption in collecting panels through which water flows. This technique has proven quite effective for heating buildings and water, but is not capable of supplying all the energy needed. Solar cells, which convert sunlight directly into electricity, have the potential of doing just that.

Solar cells have no moving parts, are silent, and use no fuel except sunlight. Semiconductor materials such as silicon crystals absorb photons and release electrons with sufficient energy to enter the conduction band. This is an excited level where electrons can flow freely throughout the crystalline structure of the material. The spaces vacated by the electrons act as positive charges, which can shift position as electrons move about. Charge separation occurs and current flows when sunlight strikes such a material.

Solar cells have been used to power satellites in space for over twenty years. The cells used in space cannot be used on earth to provide economical electricity, however, because space cells were designed for peak performance, and are ten times more expensive than conventional means (nuclear, coal, oil, or natural gas). The main thrusts of current research are to adapt the high-performance space cells to earth, and to develop a new generation of low-cost solar cells. Over the past decade, the latter strategy has begun to bear fruit in the technologies of thin films and concentrators.

Thin-film solar cells are made of special semiconductors that can transform the available sunlight in materials as thin as 10^{-6} m (about 1/25 the diameter of a human hair). These cells are inexpensive because relatively little material is used. Thin-film cells are designed to be used with inexpensive supporting substances such as glass, steel, or metal foil. The projected cost of these cells is $50/m^2, which is a great improvement over the $1,000/m^2 for space cells. Current thin-film cells transform sunlight into electricity only half as efficiently as space cells, however. More effort is required to improve their economy by raising their performance.

Thin-film solar cells of amorphous silicon (silicon layers that are not perfectly ordered in a crystalline structure) are being used, mostly by Japanese companies, to power small commercial

*Contributed by Ken Zweibel, Solar Energy Research Institute, Golden, Colorado.

(continued)

electronic products such as calculators, watches, and radios. Profits from the sale of these devices are funding improvements in amorphous silicon and other promising materials.

Another new solar cell technology uses concentrators in which lenses focus sunlight on small but highly efficient cells. A large area of low-cost lenses focused on a small cell reduces the need for a large area of expensive cells. The efficiencies of the best cells exceed 25% when illuminated by the equivalent of 100 suns. Concentrator cells have been built from single-crystal silicon or from gallium arsenide. The effort is to improve the high-performance cells while increasing the durability of the system.

As the new-generation solar cell technologies improve rapidly, they provide new paths to achieve economical solar electricity. One promising new device is the cascade cell, a hybrid structure in which two different cells are stacked on top of each other (Fig. 22.1.1). Each layer is "tuned" to photons in a different energy range, thus increasing the overall efficiency for conversion to current. The combination is about 50% more efficient than either cell alone. Before thin-films, this multicell option was not available. Cascade cells may provide the best path to ultimate success: efficient, stable, low-cost solar cells capable of producing a significant portion of our electric power economically.

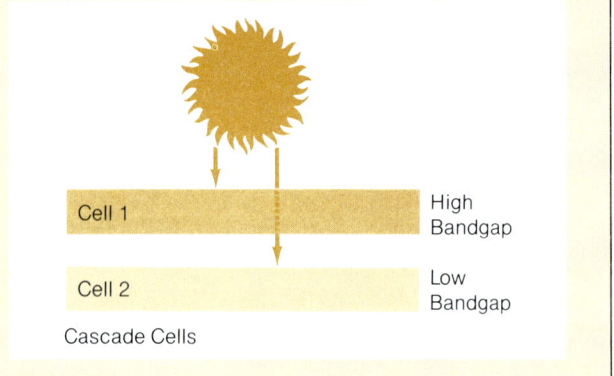

Figure 22.1.1

Compton Scattering

In 1922, A. H. Compton made another important demonstration of the photon theory. His experiments involved the interaction of electrons and x-ray photons, and could not have been performed until techniques for dispersing and measuring the wavelengths of x-rays were available. This capability was developed by W. H. Bragg and W. L. Bragg, who found that natural crystals have regular atomic spacings that can act as transmission gratings for x-rays (Fig. 22.7). This allowed the development of an x-ray spectrometer that operated on the same principle as grating spectrometers (Chapter 20) and created an entire field of research in which crystal structures are deduced from x-ray diffraction patterns (see Special Problem 1, Chapter 20).

With an x-ray spectrometer, Compton observed that the wavelength of an x-ray photon is shifted to longer wavelength when it scatters from an electron (Fig. 22.8). Compton interpreted this scattering as a collision between the photon and the electron, in which both energy and momentum are transferred to the electron. This was a daring interpretation, since it attributed particle-like properties (momentum) to the photon, although it was consistent with Einstein's interpretation of the photoelectric effect and Planck's analysis of the blackbody spectrum. To derive the momentum of a photon, one must use Einstein's theory of special relativity, which provides the relationship

$$E^2 = p^2 c^2 + m_0^2 c^4$$

22.1 THE PARTICLE NATURE OF LIGHT

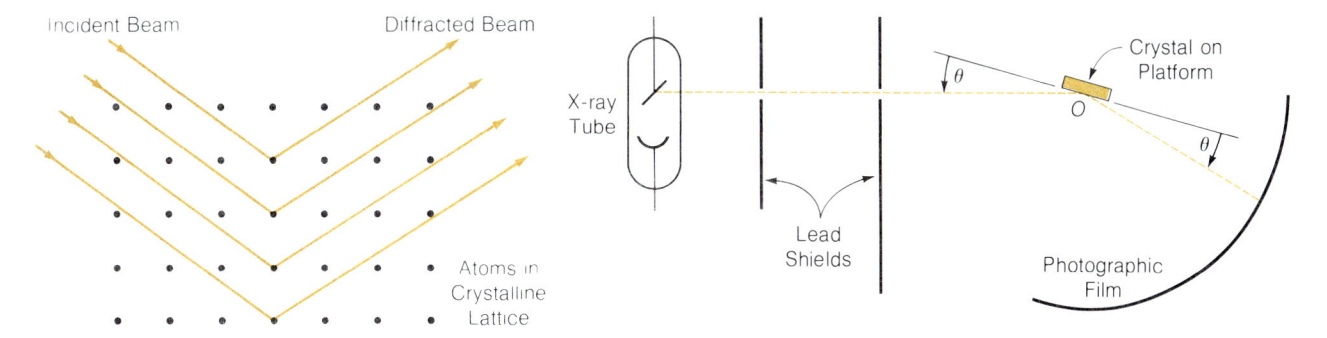

Figure 22.7

X-ray transmission grating. Many crystals have regular atomic spacings, which allow x-rays to be diffracted at certain angles. These crystals can also be used in a spectrometer to measure accurate wavelengths of x-rays.

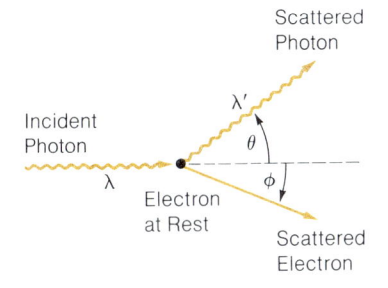

Figure 22.8

Compton scattering. This shows the incident photon (wavelength λ), the scattered photon (wavelength λ'), and the scattered electron. The scattered photon moves away at an angle θ from the direction of the incident photon.

between a particle's total energy E, its momentum p, and its rest mass m_0. A photon has zero rest mass, so it obeys the simpler relation

$$E^2 = p^2 c^2$$

or

$$E = pc.$$

Thus, since $E = h\nu$ for a photon of frequency $\nu = c/\lambda$, the photon's momentum is

$$p = \frac{E}{c} = \frac{h\nu}{c} = \frac{h}{\lambda}. \tag{22-5}$$

Compton was able to derive expressions for the conservation of energy and momentum for the photon-electron collision (Fig. 22.8). If λ and λ' represent the wavelengths of the incident and scattered photon and if θ is the scattering angle of the photon, then the **Compton formula** is

$$\lambda' - \lambda = \left(\frac{h}{m_e c}\right)(1 - \cos\theta) = \lambda_c(1 - \cos\theta). \tag{22-6}$$

Since momentum is transferred from the photon to the electron, the photon's wavelength is always increased. The amount of increase depends on the angle θ; the maximum shift of $2\lambda_c$ occurs when $\theta = 180°$, because the photon is reflected directly backward and the maximum amount of momentum is transferred to the electron. The quantity $\lambda_c = (h/m_e c) = 2.426 \times 10^{-12}$ m, known as the **Compton wavelength,** represents a very small shift for visible photons ($\lambda \simeq 5 \times 10^{-7}$ m). For x-rays or γ-rays, however, the Compton shift can be a substantial fraction of the photon's initial wavelength λ.

The significance of Compton's discovery merits further emphasis. The shift in wavelength of a photon upon scattering can be understood only if we assume that photons behave like particles and not like waves. A wave would scatter with no change in wavelength. A particle, however, can collide elastically with other particles, transferring momentum according to the laws of mechanics.

Example 22.4 An x-ray photon of energy 50 keV scatters off an electron at $\theta = 45°$. What are the energies of the scattered x-ray and electron?

We first find the x-ray's initial wavelength, using the formula $E = h\nu = hc/\lambda$ and converting eV to Joules.

$$\lambda = \frac{hc}{E} = \frac{(6.63 \times 10^{-34} \text{ J} \cdot \text{s})(3 \times 10^8 \text{ m/s})}{(5 \times 10^4 \text{ eV})(1.6 \times 10^{-19} \text{ J/eV})}$$
$$= 2.49 \times 10^{-11} \text{ m}$$

From the Compton scattering formula (Eq. 22–6), we find

$$\lambda' - \lambda = (2.426 \times 10^{-12} \text{ m})(1 - \cos(45°)) = 7.11 \times 10^{-13} \text{ m}.$$

The scattered x-ray thus has

$$\lambda' = 2.56 \times 10^{-11} \text{ m}$$

and energy

$$E' = \frac{hc}{\lambda'} = 7.77 \times 10^{-15} \text{ J} = 48.5 \text{ keV}.$$

The decrease in the photon's energy, 1.5 keV, was transferred to the electron, so we may compute its velocity from the relation $E_e = \frac{1}{2}m_e v_e^2$.

$$v_e = \sqrt{\frac{2E_e}{m_e}} = \sqrt{\frac{2(1.5 \times 10^3 \text{ eV})(1.6 \times 10^{-19} \text{ J/eV})}{9.11 \times 10^{-31} \text{ kg}}}$$
$$= 2.3 \times 10^7 \text{ m/s}$$

Note that we used the nonrelativistic formula for the electron's energy. This approximation is justified because the electron's velocity is only $0.077c$. In general, we would need to use the formulas of Chapter 21 for the electron's kinetic energy.

The Wave-Particle Duality

We are presented with an apparent dilemma: we saw in preceding chapters that light has unmistakable wave properties, yet now we find that it has equally unambiguous particle characteristics. The conclusion is clear: light has both wave and particle properties. This requires a modification of classical theory; a successful modification (quantum mechanics) exists, but because of its mathematical complexity, we cannot go into much detail. We can, however, describe the philosophy behind quantum mechanics.

We must discard our conventional way of visualizing things on all scales as analogs to familiar objects. We simply cannot view light as consisting of waves that look like transverse waves in a taut string, nor can we properly view it as consisting of tiny bullets. We must view it as having both types of characteristics, and recognize that its behavior is dependent on the conditions under which it is observed. When light interacts with matter, its particle characteristics dominate its behavior, but when it interacts with itself, its wave nature governs the result. Once again, we must be content to define something by how it acts, rather than by what it is. There is no analog to EM radiation that we can visualize; there is no physical object that corresponds to it.

It may be helpful to view light as consisting strictly of particles, but with the paths of the particles governed by nonmaterial waves. This allows the photons (the particles) to act as material objects when interacting with matter, but at the same time explains why their locations and paths are governed by wave characteristics. These waves represent the probability that the particle exists at a certain location (technically, the probability is the square of the wave amplitude, analogous to wave intensity). Interference then is dictated by how the wave amplitudes combine. The concept of probability waves will turn out to be useful throughout our discussion of quantum mechanics.

22.2 The Wave Nature of Particles

If photons have a dual nature, acting as particles in some situations and as waves in others, then perhaps particles of ordinary matter also can act as waves in some sense. This question began to be explored after the work of Compton.

The Hypothesis of de Broglie

In 1923, the French physicist L. de Broglie proposed that ordinary particles act as waves under certain circumstances. This bold assertion was based on an assumption that nature is symmetric, and was no doubt encouraged in part by the new climate in which traditional views in many areas of physics were being successfully questioned.

It was proposed by de Broglie that the locations of particles are governed by probability waves, just as those of photons are. He modified the relationship between wavelength and momentum of a photon (Eq. 22–5) by inserting for the momentum p the classical momentum mv of a material body, and solved for the wavelength.

$$\lambda = \frac{h}{mv} \quad (22\text{--}7)$$

CHAPTER 22 PARTICLES AND WAVES

This is called the **de Broglie wavelength.** For an ordinary particle, it is so small as to be unmeasurable (for example, for a mass of 1 kg moving at 100 m/s, the de Broglie wavelength is only 6.6×10^{-36} m). For a relativistic particle, $\lambda = h/p$, where p is the relativistic momentum, from the relation

$$E^2 = p^2c^2 + m_0^2 c^4$$

or from

$$p = \frac{m_0 v}{\sqrt{1 - \left(\frac{v}{c}\right)^2}} = \gamma m_0 v.$$

As we learned in Chapter 20, wave behavior such as diffraction becomes significant only when waves interact with openings or obstructions comparable in size to the wavelength; therefore, there is no easy way to detect wave behavior in ordinary objects.

Now consider an electron travelling at a typical nonrelativistic velocity such as might occur in the presence of an accelerating voltage. If, for example, the electron moves at a speed about 10^6 m/s (about what we found in Example 22.3), then the de Broglie wavelength is several times 10^{-10} m; that is, a fraction of a nanometer. This is comparable to the wavelengths of x-ray photons and to the typical spacings in crystal lattices. This opened the possibility of measuring diffraction effects in electrons scattered by crystal lattices. In 1927 C. J. Davisson and L. H. Germer and, independently, G. P. Thomson, found that electrons created interference fringes when scattered from a metal surface, and they were able to measure the wavelength of the electrons in a manner analogous to measuring wavelengths of photons with a grating. The result agreed with the calculated de Broglie wavelength for electrons travelling at the appropriate velocity, and therefore confirmed de Broglie's hypothesis. Later experiments verified the wave nature of other subatomic particles, such as the proton and the neutron.

Example 22.5 What is the de Broglie wavelength for a proton that has been accelerated from rest by a potential difference of 1000 V?

The proton gains an amount of kinetic energy equal to the potential energy lost during acceleration. The potential energy lost is $e \cdot V$, where e is the charge on the proton (equal but opposite in sign to the electron charge), and V is the potential difference. The proton is nonrelativistic, since its KE is much smaller than its rest energy, which is $m_p c^2 = 938$ MeV. Thus, the speed of the proton after acceleration is

$$v = \sqrt{\frac{2\text{KE}}{m}} = \sqrt{\frac{2eV}{m}} = \sqrt{\frac{2(1.6 \times 10^{-19} \text{ C})(1000 \text{ V})}{1.67 \times 10^{-27} \text{ kg}}}$$
$$= 4.4 \times 10^5 \text{ m/s}.$$

22.2 THE WAVE NATURE OF PARTICLES

Therefore the de Broglie wavelength is

$$\lambda = \frac{h}{mv}$$

$$= \frac{(6.63 \times 10^{-34} \text{ J} \cdot \text{s})}{(1.67 \times 10^{-27} \text{ kg})(4.4 \times 10^5 \text{ m/s})} = 9.0 \times 10^{-13} \text{ m}.$$

Example 22.6 What is the de Broglie wavelength of a relativistic electron whose kinetic energy is 2 MeV or 2×10^6 eV? The electron's rest mass energy is $m_e c^2 = 0.511$ MeV.

We must compute the electron's momentum using the relativistic formula from Chapter 21, $E^2 = p^2 c^2 + m_0^2 c^4$. Here E is the total energy, equal to the sum of the particle's rest energy and kinetic energy.

$$E = m_0 c^2 + KE = 0.511 \text{ MeV} + 2.0 \text{ MeV}$$
$$= (2.511 \times 10^6 \text{ eV})(1.6 \times 10^{-19} \text{ eV/J}) = 4.02 \times 10^{-13} \text{ J}.$$

Thus the momentum is

$$p = \frac{\sqrt{E^2 - m_0^2 c^4}}{c} = \frac{\sqrt{(4.02 \times 10^{-13} \text{ J})^2 - (8.18 \times 10^{-14} \text{ J})^2}}{3 \times 10^8 \text{ m/s}}$$
$$= 1.31 \times 10^{-21} \text{ kg} \cdot \text{m/s}$$

where we have converted the rest energy, 0.511 MeV, into SI units, 8.18×10^{-14} J.

Thus, the de Broglie wavelength is

$$\lambda = \frac{h}{p} = \frac{(6.63 \times 10^{-34} \text{ J} \cdot \text{s})}{1.31 \times 10^{-21} \text{ kg} \cdot \text{m/s}} = 5.1 \times 10^{-13} \text{ m}.$$

The Electron Microscope

If particles act as waves, then it follows that it might be possible to construct optical tools using particles instead of light. This is particularly advantageous when a very small wavelength is needed, for example, in microscopes. Recall that diffraction effects limit the resolving power of a microscope, which is the smallest linear dimension (or the smallest separation of points) that can be seen. For an ordinary compound microscope, the resolving power is typically a few hundred nm.

The wavelengths associated with high-energy electrons are typically a factor of 1000 or so smaller than those of visible light, so in principle it is possible that resolving powers of only a fraction of a nm are possible if electrons are used instead of visible light. In other words, the diffraction rings created by waves passing through a circular opening will be much closer together if the waves are those associated with electrons than if they are waves associated

744   CHAPTER 22   PARTICLES AND WAVES

with photons. Hence, if we can use a beam of electrons to form an image of an object, it should be possible to see much more detail than if we use visible light. This is the principle of the electron microscope.

There are at least two different ways in which electrons are used to form images. In the **transmission electron microscope** (Fig. 22.10), electrons are accelerated by a potential difference so that they have very short de Broglie wavelengths. Shaped magnetic fields are used first to create a parallel beam of electrons, and then to focus the beam after it has passed through the specimen. It is possible to bring the beam to a sharp focus, although not quite perfectly because of minor variations in electron velocities and other difficulties in shaping the focusing magnetic fields properly. Nevertheless, transmission electron microscopes routinely provide resolving powers of half a nm or less, about 1,000 times better than visible-light microscopes (Fig. 22.9).

In the **scanning electron microscope**, an electron beam is focused on a specimen. The incoming electrons cause the ejection of secondary electrons, which are accelerated by a small potential difference and collected on a cathode

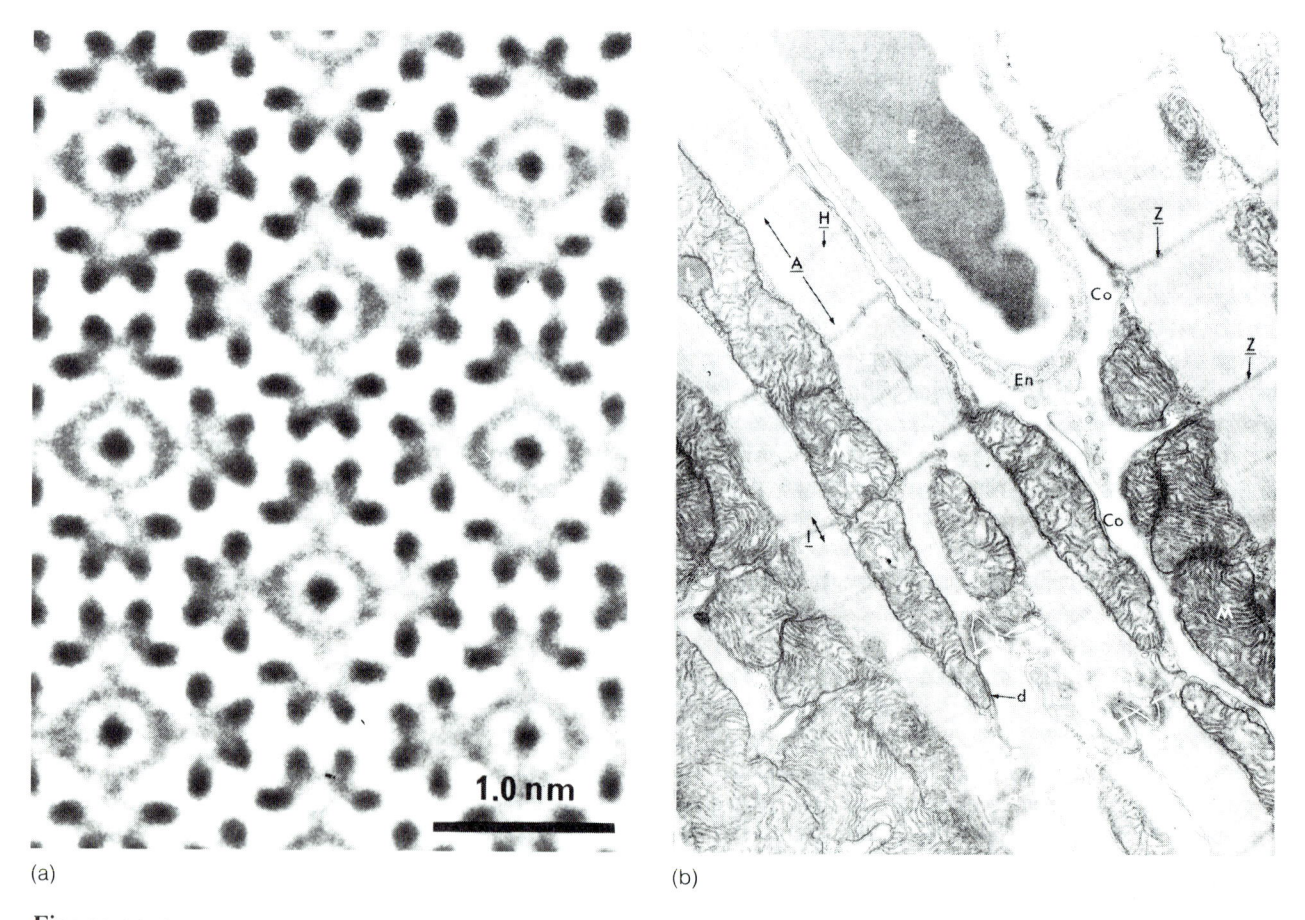

(a)   (b)

Figure 22.9

Images obtained with a transmission electron microscope. In (a) we see the remarkable resolving power of such an instrument. The dark spots are individual atoms in a complex, regular molecule. Note the scale indicated at lower right. In (b) we see an image of human cardiac muscle, magnified roughly 10,000 times in this reproduction.

22.2 THE WAVE NATURE OF PARTICLES

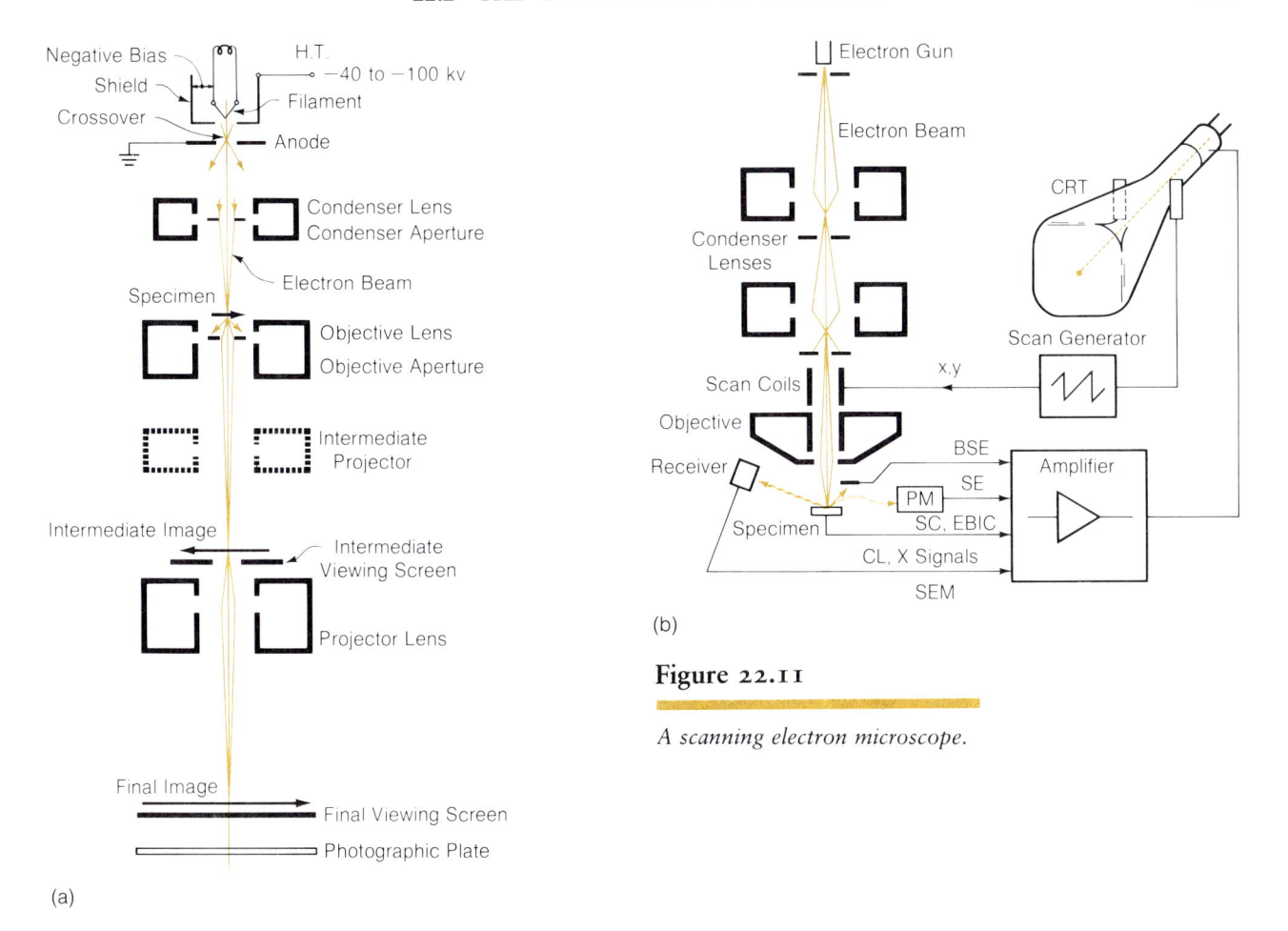

Figure 22.10

A transmission electron microscope.

Figure 22.11

A scanning electron microscope.

(Fig. 22.11). The beam of high-energy electrons is scanned across the specimen, and the current created by the secondary electrons is converted into a signal that is displayed on a cathode ray tube (CRT). The intensity of the signal is proportional to the number of secondary electrons ejected from the specimen, and the location of the signal on the CRT corresponds to the location of beam impact on the specimen. Thus, as the beam scans the specimen, an image of the specimen is built up on the CRT. A great advantage of the scanning microscope over the transmission electron microscope is that the number of secondary electrons ejected at any particular point on the specimen is a sensitive function of the angle of impact of the electron beam, so that variations in the elevation of the surface of the specimen show up on the CRT as substantial variations in intensity (Fig. 22.12). Thus, three-dimensional information can be derived from a high-contrast image. The disadvantage of the scanning electron microscope is that the resolutions achieved so far are not as great as with the transmission electron microscopes; typically, the difference in resolving power is about a factor of ten.

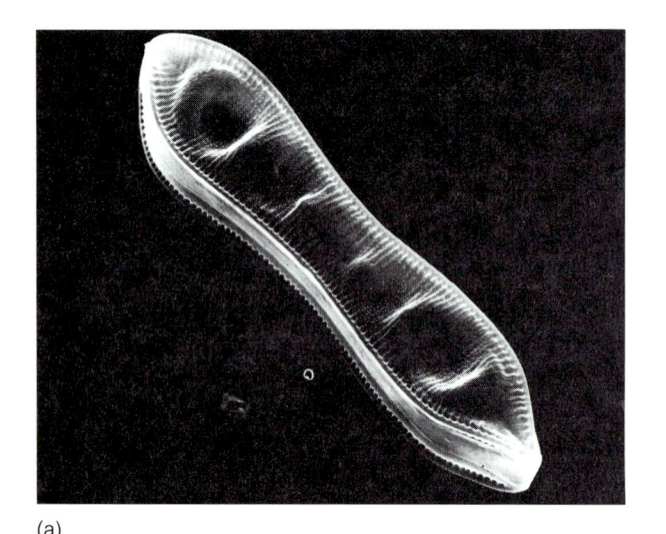

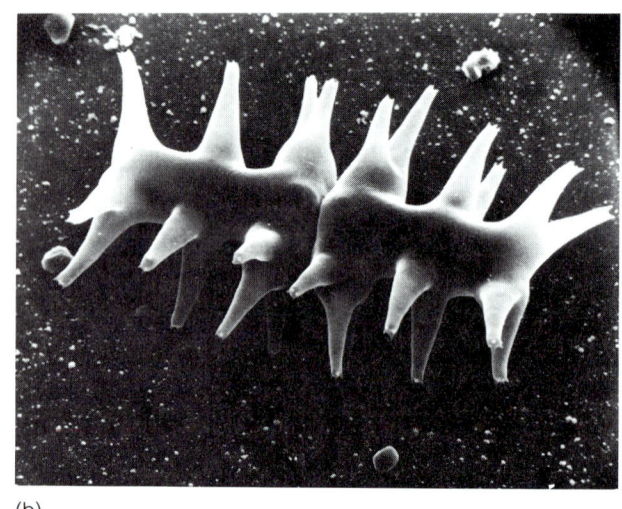

Figure 22.12

Images obtained with a scanning electronic microscope. In (a) we see a diatom, and in (b) an alga cell. Note the three-dimensional appearance in each case.

22.3 The Uncertainty Principle

In Chapter 20, when we discussed the diffraction of light in the double-slit experiment, we found that light that initially passed through a single slit, and then a pair of slits, created two beams that could interfere with each other. At that time we were considering light to consist of waves, and the wavefront model of Huygens was perfectly consistent with splitting a wave. Now, however, we have had to modify our view of the nature of light, and we realize that in the double-slit experiment we have discrete entities called photons, some of which go through one slit, and some through the other. In essence, two photons incident along identical paths are now seen to be capable of reaching different destinations. This implies a certain imprecision in the laws that govern the motion of photons. Photons are guided by probability waves, but the positions of photons are not precisely determined. This uncertainty violates the classical notion of causality, which assumes that, given the position and momentum of a body, its future position and momentum can be predicted exactly from the laws of motion.

A different consideration leads to a similar conclusion. Suppose that we wish to measure the precise position and momentum of a particle. To do this, we must detect it with some sort of probe, whether it is a beam of light that might be reflected to our eyes, or a physical tool with which we could "feel" the presence of the particle. In either case, the act of probing for the position of the particle must unavoidably affect the particle. If we view it with photons of light, the photons impart some momentum to the particle as they are reflected from it. If we sense the position of the particle by touching it with a physical

22.3 THE UNCERTAINTY PRINCIPLE

probe, we necessarily impart some momentum to it. In either case, we can accurately measure the position of the particle, but because we have affected it in the process, we cannot know its future position with perfect accuracy.

The idea that it is inherently impossible to simultaneously know with perfect accuracy both the position and the momentum of a particle is known as the **uncertainty principle**. It was formulated quantitatively by Werner Heisenberg in 1927. We can carry out a very simple analysis to demonstrate this principle. If we use light of wavelength λ to probe the position of a particle, then the uncertainty Δx in the measured position is approximately

$$\Delta x \simeq \lambda.$$

For minimum impact on the particle, only one photon is required to measure its position. The momentum imparted to the particle by a single photon is nonzero (or no scattering would occur) but no greater than the photon's total initial momentum, which is h/λ. Hence the uncertainty in the particle's momentum following its interaction with a single photon is approximately

$$\Delta p \simeq \frac{h}{\lambda}.$$

The product of the uncertainties in both position and momentum is therefore about

$$(\Delta x)(\Delta p) \simeq h.$$

Heisenberg's more detailed calculation gives a slightly modified form of this.

$$(\Delta x)(\Delta p) \approx \left(\frac{h}{2\pi}\right) = \hbar \qquad (22\text{–}8)$$

Thus, the constant term $h/2\pi$ (often designated by the symbol $\hbar$) represents the smallest possible value of the product of uncertainties in position and momentum. This is a fundamental limitation of nature, having nothing to do with the limitations of experimental apparatus or other practical constraints. It follows simply from the fact that a material object is not precisely at any single location, but instead has a probability distribution. If we reduce the uncertainty in position Δx, we accordingly increase the uncertainty in momentum Δp, and vice versa. For objects large enough to be perceptible to our senses, the uncertainty expressed by Eq. 22–8 is negligible because the value of $\hbar$ is so small. As a practical matter we could measure the position and momentum of a car or a baseball with any accuracy our instruments could provide.

For subatomic particles, however, the uncertainty becomes significant, and we can succeed in describing them only in a statistical sense. Whereas we cannot know both the position and momentum of an individual photon or electron, we can quite accurately describe the behavior of a large number. Thus, in the double slit experiment, the probabilities that an individual photon will pass through one slit or the other are equal, so we cannot predict which way an individual photon will go. But we can be quite confident that, given a large number of incident photons, half of them will go through each slit. If we consider the process of recording an image on film, we find that if only a few photons are used, the image is blurred because of the imprecision of the probability waves associated with individual photons. As more and more photons strike the film, the image becomes sharper because the total collection of incident photons conforms more and more closely with the probability distribution.

Example 22.7

An electron of mass m_e is confined to a one-dimensional box of width L. Use the Heisenberg uncertainty principle to estimate the particle's momentum, velocity, and lowest possible kinetic energy.

This is a tricky problem, but is straightforward if we make a certain approximation. The basic idea is that the electron may be viewed as a charge cloud spread over a distance $\Delta x \simeq L$. Thus, the uncertainty principle requires that the electron's momentum be spread out by

$$\Delta p \simeq \frac{\hbar}{\Delta x} = \frac{h}{2\pi L}.$$

Now, the key approximation to make is that Δp is of the same order of magnitude as p, the particle's momentum, just as Δx is comparable to L, the size of the box. With this assumption, the three quantities quickly follow:

$$p \simeq \Delta p = \frac{h}{2\pi L}$$
$$v = \frac{p}{m_e} = \frac{h}{2\pi m_e L}$$
$$\text{KE} = \frac{1}{2}m_e v^2 = \frac{h^2}{8\pi^2 m_e L^2} = \frac{\hbar^2}{2m_e L^2}$$

The uncertainty principle can also be stated in terms of the uncertainty in energy of a particle. To see how this is assessed, suppose we probe the position of a test particle with a photon. As we have seen, the uncertainty in the measured position is approximately equal to the wavelength λ of the photon. If the position of the test particle is uncertain by this amount, then the time when the particle is in the measured position is uncertain by the amount

$$\Delta t \simeq \frac{\lambda}{c}$$

where c is the velocity of the photon.

The photon has initial energy hc/λ, some fraction of which is transferred to the test particle. Hence, the uncertainty in the energy of the test particle when it interacts with the photon is

22.3 THE UNCERTAINTY PRINCIPLE

$$\Delta E \simeq \frac{hc}{\lambda}.$$

The product of these two uncertainties is

$$(\Delta t)(\Delta E) \simeq h.$$

Again, Heisenberg's more detailed analysis gives

$$(\Delta t)(\Delta E) \simeq \frac{h}{2\pi} = \hbar. \tag{22-9}$$

This form of the uncertainty principle has the interesting implication that the energy of a particle is not precisely conserved, but is instead uncertain by some amount ΔE for a short time $\Delta t \simeq h/2\pi\Delta E$. Again, this uncertainty becomes significant only on the scale of subatomic particles. For larger-scale objects, the uncertainty in energy is present, but is insignificant compared to the energies involved.

Example 22.8 An atom remains in an excited state for an average time $\tau = 10^{-8}$ s before making a transition to the ground state by emitting a photon of energy 5 eV. What is the uncertainty in energy of the upper level resulting from the uncertainty principle? What is the corresponding uncertainty of the photon's frequency?

From Eq. 22-9 we find, using $\Delta t \approx \tau$, that

$$\Delta E \approx \frac{h}{2\pi\Delta t} = \frac{6.63 \times 10^{-34} \text{ J} \cdot \text{s}}{(6.28)(10^{-8} \text{ s})} = 1.06 \times 10^{-26} \text{ J}.$$

The frequency uncertainty follows from the relation $E = h\nu$ for a photon,

$$\Delta\nu = \frac{\Delta E}{h} = \frac{1.06 \times 10^{-26} \text{ J}}{6.63 \times 10^{-34} \text{ J} \cdot \text{s}} = 1.6 \times 10^7 \text{ Hz}.$$

This frequency uncertainty represents a fundamental inability to know the frequency (or wavelength) of the photon. We will learn in Chapter 23 that when an atom makes a transition from an excited state to the ground state, it emits a photon with frequency distributed about a well-defined frequency. The width of this distribution that results from the uncertainty principle is called the **natural line width**. In general,

$$\Delta\nu = \frac{(h/2\pi\tau)}{h} = \frac{1}{2\pi\tau}.$$

We may ask what the uncertainty principle tells us about the physical universe. How are we to reconcile this basic inability to know something precisely with our traditional viewpoint that any quantity can be measured, and that the principle of cause and effect is capable of making perfect predictions, so long as the appropriate physical laws are known? Now we find that the laws themselves allow ambiguities that are inconsistent with our normal perceptions of the properties of matter.

The picture of reality brought to us by our senses is a picture of probability patterns. As we have noted, the wave nature of matter and the ambiguities recognized by the uncertainty principle become significant only on very small scales, much smaller than we can perceive directly. On the larger scales that our senses record, the numbers of quanta are so vast that their statistical adherence to the expected probability patterns is very accurate. Thus, Newton's mechanics is accurate for large-scale bodies, and a large number of photons create a sharp image as expected under the wave hypothesis. We can therefore continue to succeed in describing and predicting the behavior of objects by using the classical methods, but now we know that these methods are inadequate at a very fundamental level.

Perspective

The material in this chapter has served notice on us that nature is more subtle and complex than was previously believed. Light is found to have a dual nature, with properties that cannot be cleanly ascribed to either wave or particle phenomena. Furthermore, particles are subject to the same ambiguity, displaying both particle and wave characteristics when examined properly. Our traditional view of a precise, well-defined physical universe has been replaced with one which cannot be precisely known.

As we prepare to delve into the nature of atoms and molecules, we must adopt a new formalism for the behavior of particles on the subatomic scale, where the classical methods often fail. Thus, as we now embark on an analysis of the structure of the atom and how it absorbs and emits light, we will invoke the principles of quantum theory. In the next chapter we begin this process by discussing the atom in the light of the quantum theory.

SUMMARY TABLE OF FORMULAS

Wavelength of maximum emission from a blackbody:
$$\lambda_{max} = \frac{2.90 \times 10^{-3} \text{ m} \cdot K}{T} \tag{22-1}$$

Power emitted from the surface of a blackbody:
$$\text{Power} = e\sigma A T^4 \tag{22-2}$$

Energy of a photon:
$$E = h\nu = \frac{hc}{\lambda} \tag{22-3}$$

Kinetic energy of a photoelectron:
$$KE = h\nu - W \tag{22-4}$$

Momentum of a photon:
$$p = \frac{E}{c} = \frac{h\nu}{c} = \frac{h}{\lambda} \tag{22-5}$$

PROBLEMS

Compton formula:
$$\lambda' - \lambda = \left(\frac{h}{m_e c}\right)(1 - \cos\theta) = \lambda_c(1 - \cos\theta) \quad (22\text{--}6)$$

de Broglie wavelength:
$$\lambda = \frac{h}{mv} \quad (22\text{--}7)$$

Uncertainty principle:
$$(\Delta x)(\Delta p) \simeq h/2\pi = \hbar \quad (22\text{--}8)$$
$$(\Delta t)(\Delta E) \simeq h/2\pi = \hbar \quad (22\text{--}9)$$

THOUGHT QUESTIONS

1. Which light carries the most momentum per photon: red, blue, or yellow?

2. List several differences between x-ray photons and radio photons.

3. Explain briefly the "photon nature" of the following: photoelectric effect, semiconductor bandgap, Compton scattering, photographic film exposure.

4. List several differences between particle waves and photon waves.

5. What happens to a particle wave of a fixed energy as its rest mass goes to zero? Does the particle then resemble a photon?

6. As a light source gets faint, what happens to the interval between arrival of photons to the eye?

7. In the photoelectric effect, what happens to a photon's energy if $h\nu$ is less than the work function, W?

8. It is said that the Compton effect is evidence that photons scatter like particles, not waves. Explain the evidence and why it argues for particle behavior.

9. Why is the wavelength shift in Compton scattering noticeable only for x-rays or γ-rays and not for visible light?

10. Why are wave properties of particles not observable for large, everyday objects?

11. Would an electron microscope of given voltage have better resolution if it could accelerate particles less massive than electrons?

12. According to the Heisenberg uncertainty principle, it is more difficult to predict the future position of a light object then a heavy object. Why?

13. The uncertainty principle has been interpreted as saying that "energy is not always conserved." Is this true?

14. In what cases of particle or photon behavior is Newtonian classical mechanics wrong compared to quantum mechanics?

15. According to the uncertainty principle, there is a small probability that you could suddenly find yourself outside the room in which you are now sitting. Explain why.

PROBLEMS

Section 22.1 Energy and Momentum of Light

- 1. Use the Wien displacement law (Eq. 22–1) to find the peak wavelength of the continuous emission spectrum, λ_{max}, for a metal heated to 1,000 K.

- 2. A hot star has a surface temperature of 40,000 K. At what wavelength does it emit most of its energy? What type of light is this?

- 3. What is the energy (in both Joules and eV) of an ultraviolet photon (200 nm)?

- 4. What is the wavelength (in nm) of a 1 eV photon?

- 5. The sun has an effective surface temperature of about 5,700 K. What is the energy (in eV) of the photons at its peak of emission?

- 6. A local radio station broadcasts at a frequency of 660 kHz. What are the wavelength (m) and energy (eV) of the photons?

- 7. What is the energy (in eV) of an x-ray photon with wavelength 0.1 nm?

- 8. Which has more momentum: a green photon (500 nm) or an electron moving at 1 km/s?

- 9. Show that ultraviolet light (200 nm) carries three times the momentum of red light (600 nm).

- •• 10. How many photons per second does a 100 Watt sodium vapor light put out? Assume the light is orange ($\lambda = 589$ nm).

11 How many photons per second does the 50,000-Watt radio station KZAP put out, if it broadcasts at 1,100 kHz?

12 A 100-kilowatt laser shoots a burst of blue light (380 nm) lasting 0.01 s. How many photons were released? What was the total momentum in the burst?

13 Approximately how many visible photons per second are emitted by our Sun? Assume 1 $L_\odot$ = 3.83 × 10^{26} J/s and surface temperature T = 5,700 K.

14 A 50,000-Watt laser beam (λ = 400 nm) is aimed at a shiny 100 kg object at rest on a frictionless surface. Assume that the light reflects fully (180° off the object).
  (a) Calculate the number of photons striking per second.
  (b) Show that the momentum imparted by a single photon is $(2h/\lambda)$.
  (c) What is the acceleration of the object?

15 What is the wavelength of light which will just initiate a current from sodium metal (work function W = 2.46 eV)?

16 Sunlight falls on two semiconductor solar cells: silicon (bandgap 1.11 eV) and germanium (bandgap 0.67 eV). If the light has λ = 500 nm, find the kinetic energy of the photoelectrons in each case.

17 Suppose we lived on a planet orbiting a cool star, whose radiation peaks in the infrared (5,000 nm). What bandgap material (in eV) should one search for to use as a photocell?

18 The sun has a luminosity of 3.83 × 10^{26} Watts and is 1.5 × 10^{11} m away.
  (a) Find the flux of energy (Watts/m^2) passing through a 1 m^2 area each second at the distance of the earth.
  (b) Compute the power delivered (Watts) by a 2 m × 4 m solar panel constructed from silicon solar cells of 10% efficiency. Assume that 70% of the sunlight is usable, after accounting for photons of energies below the bandgap or for light absorbed and scattered in the atmosphere.

19 Blue sensitive film responds only to light with λ < 450 nm. Calculate the energy (eV) needed for a molecule of the emulsion to undergo the chemical reaction responsible for darkening the film.

20 When 270 nm ultraviolet light shines on a metal, a reverse voltage of 2.1 V stops the photoelectric current. What is the work function of the metal?

21 A 0.01 nm gamma ray strikes an electron at rest and scatters directly backward. From conservation of relativistic momentum and energy, compute the KE of the electron (which is knocked forward) and the wavelength of the scattered photon.

22 In a process known as pair production, the energy of a gamma ray is used to produce an electron (e^-) and its antiparticle, a positron (e^+), each having rest mass energy 511 keV (see Fig. 22.13). What minimum photon energy is required? (Assume that the momentum is taken up by a nearby nucleus.)

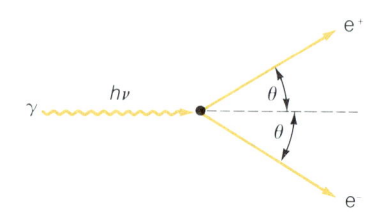

Figure 22.13

23 Suppose a 2.02 MeV gamma ray produces an e^+ – e^- pair, which move away at angles θ = 30° (Fig. 22.13).
  (a) Using conservation of energy, find the KE of each of the particles.
  (b) Using the relativistic formula $E^2 = p^2c^2 + m_0^2c^4$, find the momenta of the e^+ and e^-. Show that their sum does not equal the momentum of the photon, $h\nu/c$.
  (c) How could the presence of another body (say an atomic nucleus) help conserve momentum?

24 A 10^4 eV x-ray scatters off an electron, moving off an angle θ = 50°. Use the Compton scattering formula (Eq. 22–6) to find the energy E' of the scattered photon (eV).

25 An x-ray of 10.000 nm wavelength scatters off an electron at angle θ. If the scattered photon has wavelength 10.000 nm, what was θ?

26 A 100,000-volt power supply accelerates electrons, which then radiate x-rays. What are the shortest wavelength x-rays emitted?

Section 22.2 Particle Waves

27 Calculate the wavelength of a 100-gram baseball thrown at 96 mi/h.

28 What is the wavelength of a neutron (m_n = 1.67 × 10^{-27} kg) moving at 0.01c?

29 Calculate the wavelength of an electron moving with KE = 10 eV.

30 An electron and proton have the same KE (both are nonrelativistic). What is the ratio of their wavelengths, λ_p/λ_e?

31 What is the wavelength of a thermal O_2 molecule (mass = 32 m_p) in the atmosphere? Assume T =

PROBLEMS

70°F and refer to Chapter 13 for rms thermal velocities.

32 The wavelength of a massive particle is $\lambda = h/p$, where p is the momentum. Suppose that the particle's velocity is no longer small compared to c.
  (a) Show that the relativistically correct expression is $\lambda = hc/(E^2 - m_0^2c^4)^{1/2}$, where E is the particle's total energy.
  (b) If the particle's KE is given by T, show that $\lambda = hc/(T^2 + 2m_0c^2T)^{1/2}$.

33 What is the ratio of a particle's de Broglie (nonrelativistic) wavelength, $\lambda = h/mv$, to its Compton wavelength, λ_c?

34 What is the wavelength of a relativistic electron with KE = 1 MeV?

35 A Bragg crystal spectrometer has a crystal lattice spacing $d = 0.25$ nm. Estimate the energy of the x-rays and the KE of the neutrons ($m_n = 1.67 \times 10^{-27}$ kg) needed to produce a diffraction pattern. (Choose $\lambda \simeq d$.)

36 What approximate voltage is needed to produce electrons in a microscope with sufficient momentum to obtain 0.1 nm resolution?

37 What approximate resolution (nm) is possible in principle for an electron microscope with a 50 keV power source?

38 An electron gun shoots 10 keV electrons through 2 slits separated by a width $d \ll a$. (Fig 22.14). Use the wave nature of the electrons to find the separation l, from zero-order of the first interference maximum along the screen (see Chapter 20) if $a = 1$ m and $d = 1$ mm.

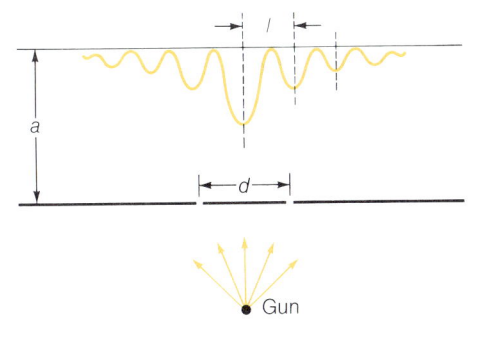

Figure 22.14

Section 22.3 Uncertainty Principle

39 A proton moves with velocity uncertainty $0.01c$. With what accuracy can its position be determined?

40 An electron is confined to a region 10 nm in size. How accurately can one know its velocity?

41 A proton is confined to the nucleus, 10^{-15} m in size. From the uncertainty principle, estimate the velocity and kinetic energy of the lowest energy level.

42 A rifle shoots a 40 g bullet at approximately 700 m/s toward a target 400 m away horizontally. If the bullet's position is uncertain by 5 mm (the barrel width), make the idealized assumption that the uncertainty principle governs the error in the bullet's trajectory and find
  (a) The uncertainty in the bullet's velocity, transverse to the horizontal.
  (b) The size of the error circle at the target.

43 A particle of mass m and velocity v is confined to a one-dimensional box of length L. From the wave nature of the particle ($\lambda = h/mv$) and the uncertainty principle,
  (a) Draw the possible standing wave modes (Chapter 10) and show that $n\lambda/2 = L$ ($n = 1, 2, 3, \ldots$).
  (b) Find the kinetic energy of the nth mode.
  (c) Show that the lowest mode ($n = 1$) satisfies the uncertainty principle: $(\Delta p)(\Delta x) \simeq \hbar$.

44 Estimate the momentum, and thus the kinetic energy, of the electron in the ground state of the hydrogen atom confined to a size 5.3×10^{-11} m.

45 Estimate the momentum and the kinetic energy (in MeV) of a relativistic electron confined to a region the size of a nucleus, 10^{-15} m.

46 An electron remains in the excited state of an atom for a time τ before dropping down to the ground state, emitting a photon at a discrete energy E. Show from the uncertainty principle that the resulting photon distribution has an energy uncertainty $\Delta E \simeq \hbar/\tau$ and a frequency uncertainty $\Delta \nu \simeq (2\pi\tau)^{-1}$.

47 The first excited state of hydrogen has a radiative lifetime $\tau = 1.6 \times 10^{-9}$ s. What is the frequency uncertainty of the emitted photon?

48 A pendulum has a natural oscillation frequency ω_0 (see Chapter 9). From the uncertainty principle, show that its minimum energy state has an energy $E_{min} \simeq \hbar\omega_0$. (Use the relationship between period and frequency.)

49 A 1 kg mass is attached to a spring with force constant $k = 1{,}500$ kg/s² (see Chapter 9).
  (a) Calculate the natural frequency of the spring, ω_0.
  (b) From the uncertainty principle, estimate the uncertainty in kinetic energy of the spring.
  (c) What is the uncertainty in position x and in velocity v of the mass? Recall that KE = $\frac{1}{2}mv^2$ and PE = $\frac{1}{2}kx^2$ for a spring.

50 What is the uncertainty in the velocity of a 50 kg angel dancing on the head of a pin, 10^{-4} m in radius?

SPECIAL PROBLEMS

1. Solar Sails

When a body is far from the Earth's gravitational field, one can use the momentum in sunlight to accelerate it outward between the planets. The Sun puts out 3.83×10^{26} J/s.

(a) Calculate the momentum per second delivered by sunlight to a 1 m² area located at distances of the orbits of Earth (1.00 AU) and Mars (1.52 AU). (Recall momentum $p = E/c$ for a photon.)

(b) Calculate the outward acceleration of a 100 m × 100 m sail, with mass 10^3 kg.

(c) How long would the sail take to move from the orbit of Earth to Mars? You may assume an acceleration equal to the average between the two planets.

2. Maximum Luminosity of Stars

Several effects determine how bright a star can become. The English astrophysicist Sir Arthur Eddington pointed out that when a star reaches a certain luminosity, known today as the "Eddington Luminosity," its starlight pushes on matter in its atmosphere harder than gravity holds it in. The result is an outflowing atmosphere or stellar wind.

Consider the atmosphere to consist of proton-electron pairs (Fig. 22.15). The effective area of the pair "seen" by a photon is 7×10^{-29} m². Suppose a star has a mass equal to that of the Sun (2×10^{30} kg) and puts out a luminosity L (Joules/s).

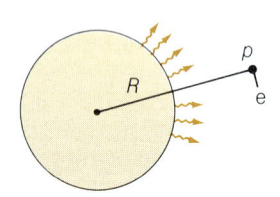

Figure 22.15

(a) Compute the momentum delivered to a proton-electron pair each second by absorbing photons if the pair is located a distance R from the center of the star. This is the outward force due to "radiation pressure."

(b) Compute the force of gravity on the electron-proton pair ($m_p = 1.67 \times 10^{-27}$ kg; the electron mass is negligible here). Assume, as before, the pair is a distance R from the star's center.

(c) Compute the critical luminosity L such that the two forces are equal (the Eddington luminosity). Notice that the R's cancel. What is this luminosity in terms of solar luminosities? $1\ L_\odot = 3.83 \times 10^{26}$ J/s, $G = 6.67 \times 10^{-11}$ N·m²/kg².

3. Photon Counting

As a source of light gets fainter, the interval between arrival of its photons gets longer. The following examples illustrate this effect.

(a) Calculate the time between photon arrival of a 100-Watt lightbulb 5 miles away from the human eye. Assume the light has wavelength 400 nm, and that the eye has an effective area of 0.008 cm² for "capturing" photons. (First work out the photon flux of the bulb in photons per m² per second at a distance of 5 miles.)

(b) A star like the Sun puts out 3.83×10^{26} J/s, with a typical wavelength of 500 nm. How many light years away would such a star have to be for its photons to strike a telescope of radius 2 cm at a rate of once a second? One light year is 9.46×10^{15} m.

(c) An extremely bright galaxy puts out 10^{10} times the luminosity of our Sun in 1 keV x-rays. A telescope with effective area 10 cm² observes these x-ray photons arriving at an average interval of 5 seconds. How far away is the galaxy?

4. Atmospheric Ozone and UV Radiation

The Earth's atmosphere is about 78% nitrogen (N_2) and 21% oxygen (O_2), but in the stratosphere (25–55 km altitude) trace amounts (1–6 parts per million) of ozone (O_3) are formed by ultraviolet sunlight. The ability of ozone to absorb the sun's ultraviolet and infrared light causes a stratospheric temperature inversion in which the atmospheric temperature rises with altitude (Fig. 22.16a). Ozone is also the dominant absorber of the sun's "tanning ultraviolet rays," between 200 and 300 nm (Fig. 22.16b). These rays can also cause genetic mutations, and for this reason the ozone layer is of key importance to life.

(a) Ozone is formed when atomic oxygen, formed by ultraviolet sunlight, attaches to O_2 in the presence of a catalyst (a third body called M).

$$O + O_2 + M \rightarrow O_3 + M$$

The atomic oxygen is produced by UV photons γ which break apart O_2.

$$O_2 + \gamma \to O + O$$

From Fig. 22.16b, which shows the altitude at which sunlight of various wavelengths is absorbed by O_2, O_3, etc., estimate the photon energies that will break apart O_2.

(b) Ozone is destroyed by UV photons, $O_3 + \gamma \to O + O_2$. Estimate the photon energies that will break apart O_3.

(c) If the average number density of ozone molecules in the stratosphere is 3×10^{18} m^{-3}, estimate the total number of O_3 molecules in a square column of 1 m^2 area that passes vertically through the 35 km-thick stratosphere.

(d) If the earth is a sphere of radius 6378 km, estimate the total number of ozone molecules in the atmosphere. What is their total mass?

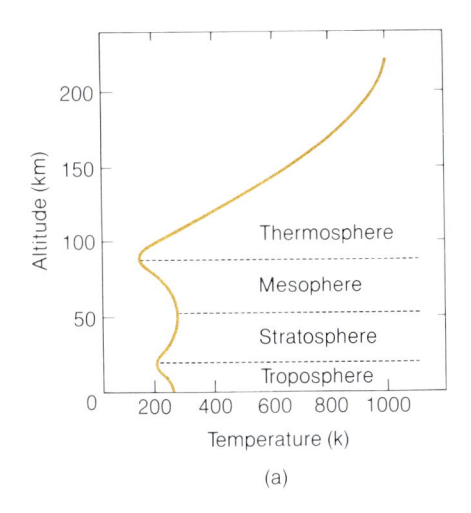

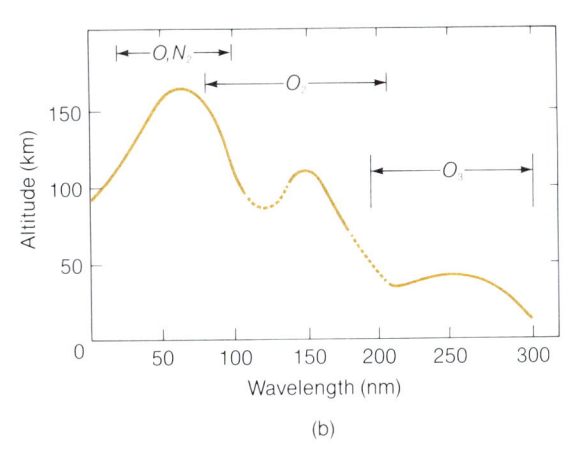

Figure 22.16

SUGGESTED READINGS

"The principle of uncertainty." G. Gamow, *Scientific American*, Mar 52, p 47, **186**(3)

"Heisenberg and the early days of quantum mechanics." F. Bloch, *Physics Today*, Dec 76, p 23, **29**(12)

"Heisenberg's first paper." D. C. Cassidy, *Physics Today*, Jul 78, p 23, **31**(7)

"Rayleigh scattering." A. T. Young, *Physics Today*, Jan 82, p 42, **35**(1)

"Is the moon there when nobody looks? Reality and the quantum theory." D. Mermin, *Physics Today*, Apr 85, p 38, **38**(4)

Mr. Tompkins in Wonderland. George Gamow, Cambridge University Press, New York, 1940

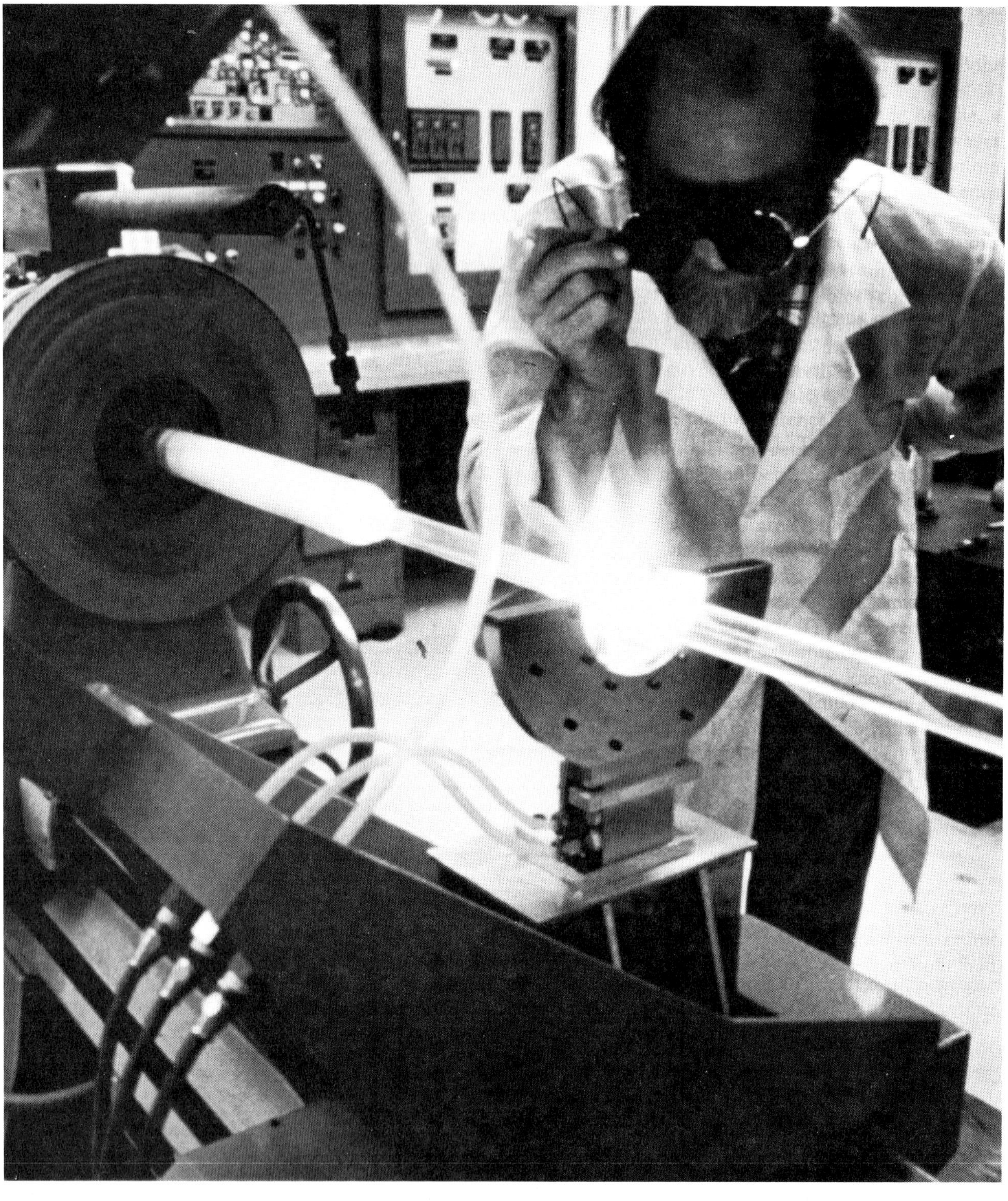

Chapter 23

Light and the Atom

The notion that matter consists of tiny, indivisible particles is a very old one, dating back to the early Greek philosophers. The word "atom" has an equally long history, and has always stood for the component particles of which material objects consist. The notion of what an atom is has evolved considerably over the ages, however.

In this chapter we will discuss the development of the modern theory of atomic structure, showing how this development was intertwined with the revolution in physics that began at the turn of the twentieth century. We will see that the concept of the quantum again plays a central role, setting the stage for the new science of quantum mechanics.

23.1 Experimental and Observational Clues

By the late 1800s many of the chemical elements had been isolated and the idea was generally accepted that each was composed of atoms of a unique type. Just before 1900 the electron was discovered by J. J. Thomson, who deduced that the discharge produced when a high voltage was applied to a rarefied gas in a sealed tube consisted of negatively charged particles. He was able to use magnetic and electric fields to bend their paths, and was able to accurately measure the ratio of their electrical charge to their mass. Based on this work, Thomson proposed a model for the atom consisting of a sphere of positively-charged material with the negatively-charged electrons embedded in it (Fig. 23.1). While this particular model was soon disproved, it was the first to envision atoms as being composed of subordinate particles. This set the stage for more refined models, which followed shortly.

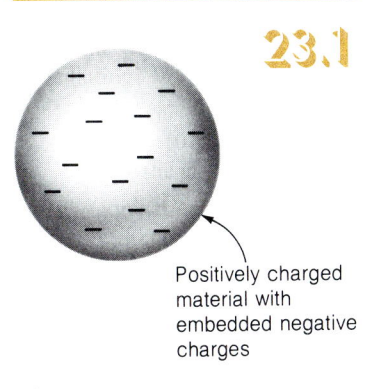

Figure 23.1

The Thomson model of the atom.

CHAPTER 23 LIGHT AND THE ATOM

The Rutherford Model

Shortly after 1910 Ernest Rutherford found that the positive charge in the atom is concentrated in a single location, which was assumed to be the center and which became known as the nucleus. It had been found that certain radioactive substances emit positively charged particles called **alpha rays** or **alpha particles** (these were soon identified as the nuclei of helium atoms). When these particles were allowed to pass through a thin sheet of metal foil, most passed straight through. This indicated that they had not closely encountered any positively charged material which would have deflected or slowed them. Those few α-particles that were scattered were often deflected by large angles, as though they had encountered a strong repulsive force. Rutherford concluded that the atoms of the foil sheet must contain isolated concentrations of positive electric charge, because a distributed positive charge would not have produced such a spotty pattern of deflection.

Rutherford envisioned the atom as a miniature planetary system, with electrons orbiting the central positive nucleus (Fig. 23.2). Rutherford was even able to estimate the radius of the nucleus at 10^{-15} to 10^{-14} m. Other evidence (primarily x-ray diffraction by crystals) indicated that the radius of an atom was typically some 10^4 to 10^5 times greater, so it appeared that atoms consisted mostly of empty space. It was clear that the electrons had to be in motion, orbiting the nucleus, because otherwise the attractive electrical force would compel the electrons and nuclei to come together. This led to a dilemma, however, for classical electromagnetic theory insists that an electrical charge that is undergoing acceleration must emit EM waves. Hence it was not clear how electrons could stay in orbit without quickly giving up energy in the form of radiation and falling into the nucleus.

Detailed analysis of the scattering of α-particles by metal foil enabled Rutherford to deduce the magnitude of the positive electrical charge on the nuclei of the metal atoms in the foil. By experimenting with different kinds of metallic foil, Rutherford derived the value of Z, the nuclear charge in units of the electron charge, for a few elements such as gold, silver, and copper. Since atoms are electrically neutral, the number of electrons orbiting any nucleus must be equal to the value of Z for that element, so that the total negative charge equals the positive charge on the nucleus. This number became known as the **atomic number**, the primary identifying characteristic of each element.

The Link with Atomic Spectra

There remained substantial missing information in the structure of the atom, and the key lay in the spectra of light absorbed and emitted by atoms. We have mentioned that experiments in the middle and late 1800s demonstrated that individual chemical elements emit and absorb light at specific wavelengths. If a rarefied gas is heated or subjected to a strong electric field, it emits light at certain wavelengths, forming an **emission-line spectrum** (Fig. 23.3). If the continuous spectrum of a heated solid is allowed to pass through a cooler cloud of gas, then light is removed at certain wavelengths, forming an **absorption-line spectrum** (Fig. 23.4). For each gas, the emission lines coincide in

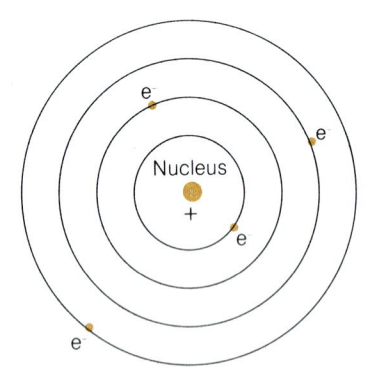

Figure 23.2

The Rutherford planetary model. Based on his scattering experiments, Rutherford envisioned the atom as consisting of a central, positively-charged nucleus, with electrons orbiting it like planets orbiting the sun.

23.1 EXPERIMENTAL AND OBSERVATIONAL CLUES

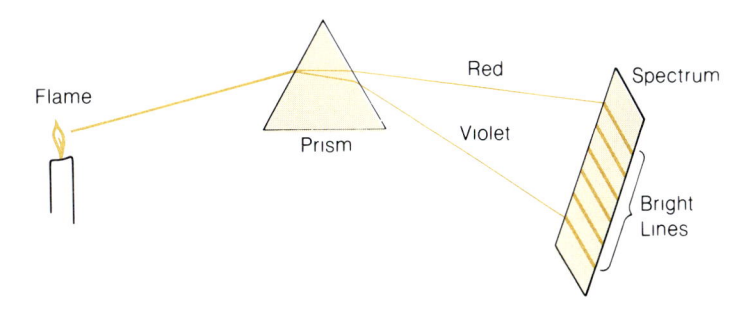

Figure 23.3

Emission lines. When a rarified gas is heated, it emits light only at specific wavelengths, and forms emission lines.

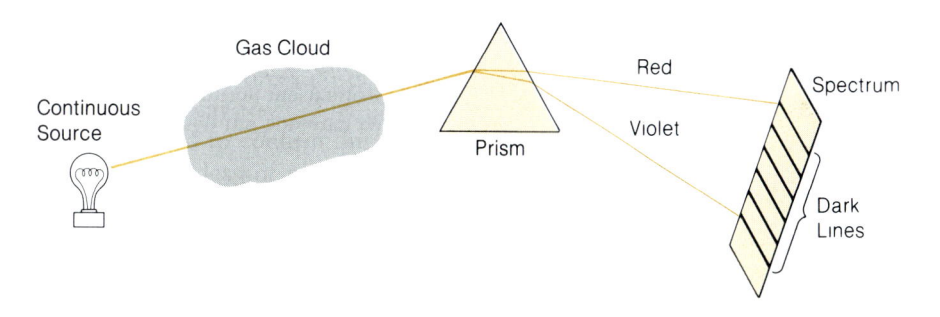

Figure 23.4

Absorption lines. When light from a continuous source passes through a cool gas cloud, the gas absorbs light at specific wavelengths, and produces absorption lines.

wavelength with the absorption lines. Thus, each chemical element has its own characteristic spectrum. This discovery (elucidated primarily by R. Bunsen and G. Kirchhoff) had enormous practical implications, for it meant that the chemical composition of a gas could be deduced from its spectrum. It also had far-ranging implications for understanding the structure of the atom.

The element with the simplest spectrum is hydrogen, whose visible-wavelength spectral lines appear in a regular series (Fig. 23.5). This is called the **Balmer series** after J. J. Balmer, who found a simple algebraic formula that matched the positions of the lines:

$$\frac{1}{\lambda} = R\left(\frac{1}{2^2} - \frac{1}{n^2}\right), \quad n = 3, 4, 5, \ldots \tag{23-1}$$

where λ is the wavelength of line n, and R is a constant known as the **Rydberg constant**. The value of R is 1.097×10^7 m^{-1}, so the first line in the series ($n = 3$) has the wavelength $\lambda = 656$ nm; the second ($n = 4$) has $\lambda = 486$ nm; and so on. If n increases to infinity, then the value of λ approaches the

CHAPTER 23 LIGHT AND THE ATOM

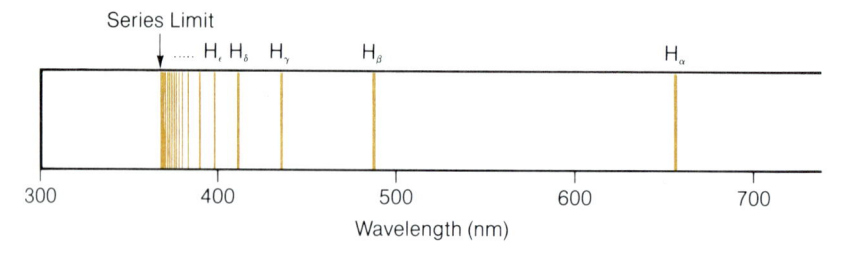

Figure 23.5

The Balmer series of spectral lines. Formed by hydrogen atoms, the wavelengths of the lines are predicted by a simple formula found by J. J. Balmer.

series limit, 365 nm. Indeed, ultraviolet spectroscopy soon showed that the Balmer series abruptly stops at this wavelength, with lines crowding closer and closer together at wavelengths just slightly longer (Fig. 23.5). In addition, other series of hydrogen lines were soon found, which fit similar formulas, with the exception that the fraction $1/2^2$ is replaced by other simple fractions, and the starting value for n is different. For example, in one series, $1/2^2$ is replaced by $1/1^2$ and the first value of n is 2 (this is the **Lyman series,** which is entirely in the far ultraviolet). In another, the fraction is $1/3^2$ and the first value of n is 4 (this is the **Paschen series,** in the near infrared). In another, the fraction is $1/4^2$ and n begins with 5 (this is the **Brackett series,** farther in the infrared), and so on.

These regular series of spectral lines appeared likely to require an interpretation in terms of atomic structure, but for some time the relationship was elusive. Finally, in 1912 and 1913, some 27 years after the work of Balmer but right on the heels of Rutherford's experiments, the Danish physicist Niels Bohr made the essential breakthrough, as described in the next section.

Example 23.1 Calculate the wavelengths of the Balmer lines for $n = 5$ and 100, and of the first line in the Lyman series.

For the Balmer series lines, we use Eq. 23–1 and substitute for n.

$$\frac{1}{\lambda} = R\left(\frac{1}{2^2} - \frac{1}{n^2}\right) = (1.097 \times 10^7 \text{ m}^{-1})\left(\frac{1}{2^2} - \frac{1}{5^2}\right) = 2.304 \times 10^6 \text{ m}^{-1}$$

therefore

$$\lambda = 434 \text{ nm}.$$

Similarly, for $n = 100$, we find

$$\frac{1}{\lambda} = (1.097 \times 10^7 \text{ m}^{-1})\left(\frac{1}{2^2} - \frac{1}{100^2}\right) = 2.741 \times 10^6 \text{ m}^{-1}$$

and

$$\lambda = 365 \text{ nm}$$

which is virtually equal to the Balmer series limit, the value for $n = \infty$.

For the Lyman series, we replace the fraction 1/2 by 1, so for the first line ($n = 2$), we find

$$\frac{1}{\lambda} = R\left(1 - \frac{1}{n^2}\right) = (1.097 \times 10^7 \text{ m}^{-1})\left(1 - \frac{1}{2^2}\right) = 8.228 \times 10^6 \text{ m}^{-1}$$

which yields the wavelength

$$\lambda = 122 \text{ nm}$$

which is in the far ultraviolet, as we noted.

23.2 The Bohr Model

By 1912 the assorted clues necessary for solving the riddle of atomic structure were in place. Planck and Einstein had shown that light is quantized in energy, and Rutherford had found the basic elements of atomic structure. Bohr in 1912 invoked the new quantum concept in attempting to explain the line spectra of atomic elements, and developed a mathematical model for the hydrogen atom that nicely fit the spectroscopic behavior of that element.

The Elements of the Model

Bohr accepted the Rutherford model of the atom, and found that its major shortcomings could be avoided if fixed energy states were associated with the electron orbits. This meant that as long as an electron stayed in its orbit, its energy was constant, so that it did not emit EM radiation and the orbit did not decay. Furthermore, in Bohr's hypothesis, an electron could change its orbit only by gaining or losing a discrete quantity of energy, by the absorption or emission of a single photon. Thus, the new picture of the atom consisted of a nucleus surrounded by electrons, each with a specific, fixed energy. An electron could jump to a higher energy state (corresponding to an orbit farther from the nucleus) if it absorbed a photon with precisely the correct amount of energy. An electron could drop to a lower energy state (an orbit closer to the nucleus) by emitting a photon whose energy equalled the difference between states. It is as though the atom has a predetermined set of possible orbits for the electrons, so that the electrons cannot exist in any other orbits in between. (We will see later that in the quantum-mechanical view of the atom, these orbits correspond to maxima in the probability distribution of the electron cloud.)

Therefore each element, with its own unique structure of electron energy states, could absorb or emit light only with specific, discrete energies; hence each element absorbs or emits only at a specific, characteristic set of wavelengths. We already know that the energy of a photon of frequency v is hv, where h is Planck's constant. Therefore, the relationship between electron energy states E_u and E_l (the upper and lower of two levels) and photon frequency or wavelength is given by

$$hv = \frac{hc}{\lambda} = E_u - E_l. \quad (23-2)$$

In this way the Bohr hypothesis neatly explained the discrete spectra of rarefied gases while avoiding the classical pitfalls of the Rutherford model.

Example 23.2   Find the frequency and the wavelength of the photon emitted when an electron drops from one level to a lower one if the energy difference between the levels is 4.83×10^{-17} J (302 eV).

We are given that $E_u - E_l = 4.83 \times 10^{-17}$ J. We use Eq. 23-2 to find the frequency of the emitted photon.

$$v = \frac{E_u - E_l}{h} = \frac{4.83 \times 10^{-17} \text{ J}}{6.63 \times 10^{-34} \text{ J} \cdot \text{s}} = 7.29 \times 10^{16} \text{ Hz}.$$

The wavelength of this photon is

$$\lambda = \frac{c}{v} = \frac{3 \times 10^8 \text{ m/s}}{7.29 \times 10^{16} \text{ Hz}} = 4.12 \text{ nm}$$

which is in the x-ray part of the EM spectrum.

Calculating the Energy Levels

Having made his quantum hypothesis, Bohr was able to find the energies of the electron orbits by assuming that the angular momentum of an orbiting electron was also quantized, having values equal to integral multiples of $\hbar = h/2\pi$. This hypothesis, like Bohr's original assumption that the electron energy states were discrete, was made on a purely empirical basis, that is, it provided mathematical consistency with the observed behavior of atoms, but was not derived from any known fundamental laws. In this way Bohr's work was akin to that of Kepler, who found laws of planetary motion by making mathematical fits to observational data, but was distinct from that of Newton, who found the underlying physical principles from which Kepler's laws could be derived. We will learn about the quantum mechanical equivalent of Newton's work in due course.

23.2 THE BOHR MODEL

The classical expression (Chapter 6) for the orbital angular momentum L of a body of mass m in circular orbit of radius r and orbital velocity v is

$$L = mvr.$$

Bohr simply set this expression equal to a multiple of $h/2\pi$, his postulated quantized angular momentum, and found

$$L_n = m_e v_n r_n = \frac{nh}{2\pi}, \quad n = 1, 2, 3, \ldots \qquad (23\text{--}3)$$

where n is an integer known as the **quantum number** of the orbit.

With this expression for the orbital angular momentum of the electron, Bohr was then able to calculate the energies associated with the electron orbits, and compare them with the observed spectral line wavelengths using Eq. 23–2. First he set the centripetal force required to keep an electron in orbit equal to the Coulomb force of attraction between the electron and the nucleus (Fig. 23.6). Thus,

$$\frac{ke^2}{r_n^2} = \frac{m_e v_n^2}{r_n} \qquad (23\text{--}4)$$

where the left-hand term is just Coulomb's law ($k = 8.9876 \times 10^9$ Nm²/C²) for the attraction between a single electron and an equal positive charge (the nucleus of the hydrogen atom consists of a single particle with a positive charge equal in magnitude to the negative charge of the electron), and r_n is the radius of the orbit for quantum number n. From Eq. 23–3 we find

$$v_n = \frac{nh}{2\pi m_e r_n}$$

which, when substituted for the centripetal force in Eq. 23–4 leads to

$$r_n = \frac{n^2 h^2}{4\pi^2 m_e ke^2} = \frac{n^2 \hbar^2}{m_e ke^2}. \qquad (23\text{--}5)$$

It is interesting to substitute numerical values and see what the radii of the orbits are. For $n = 1$ (the smallest orbit), the radius is $r_1 = 5.29 \times 10^{-11}$ m, a value sometimes called the **Bohr radius**. As n increases, subsequent orbits in sequence grow in radius as n^2, so we find $r_2 = 4r_1 = 2.12 \times 10^{-10}$ m, $r_3 = 9r_1 = 4.76 \times 10^{-10}$ m, and so on.

Now we are in position to calculate the energy associated with a particular electron orbit. We learned in Chapter 14 that the potential energy of a point charge in the electric field of another point charge is

$$\text{PE} = \frac{kq_1 q_2}{r}$$

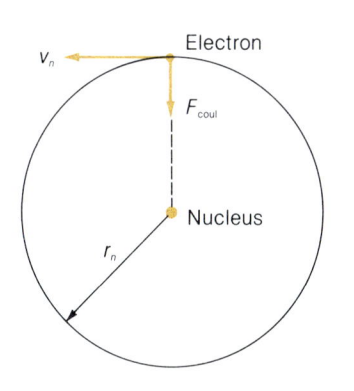

Figure 23.6

An electron in a circular orbit of radius r_n about the nucleus. The coulomb force $F_{coul} = \frac{ke^2}{r_n}$ equals the centripetal force $F_{cen} = \frac{mv_n^2}{r_n}$.

CHAPTER 23 LIGHT AND THE ATOM

where q_1 and q_2 are the two charges and r their separation. This expression is based on the convention that the potential energy is zero when r is infinite, so that for any bound pair of charges, the potential energy will always be negative. Thus, the potential energy of the single electron ($q = -e$) orbiting a hydrogen atom nucleus ($q = +e$) is

$$\text{PE} = \frac{-ke^2}{r_n}. \tag{23-6}$$

The kinetic energy of a particle with mass m and velocity v is $\frac{1}{2}mv^2$, so the total energy of the orbiting electron is

$$E_n = \text{KE} + \text{PE} = \frac{1}{2}m_e v_n^2 - \frac{ke^2}{r_n}.$$

When we substitute for v_n from Eq. 23-3 and then for r_n from Eq. 23-5, we find

$$E_n = -\frac{2\pi^2 e^4 m_e k^2}{h^2 n^2}$$
$$= -\frac{m_e k^2 e^4}{2n^2 \hbar^2}, \quad n = 1, 2, 3, \ldots. \tag{23-7}$$

Comparing Predictions with Observed Spectra

Now that we can calculate the energies associated with the electron orbits, we are in position to compute the wavelengths of the photons whose energies correspond to the differences between levels. Again, it is helpful first to find numerical values for a few cases. For the lowest orbit ($n = 1$), we find $E_1 = -2.17 \times 10^{-18}$ J. For $n = 2$ and $n = 3$ we find $E_2 = -5.43 \times 10^{-19}$ J and $E_3 = -2.41 \times 10^{-19}$ J. We notice that the difference between the energies of orbits 2 and 3 is

$$E_3 - E_2 = 3.02 \times 10^{-17} \text{ J}.$$

The wavelength of a photon with this energy is

$$\lambda = \frac{hc}{E} = 658 \text{ nm}$$

which is quite close to the measured wavelength of the first line in the Balmer series for hydrogen (a precise match is found when the calculation is made using very accurate values for the constants). Furthermore, the difference $E_4 - E_2$ corresponds to the wavelength of the next line in the series; the difference $E_5 - E_2$ corresponds to the next line, and so on. Thus, the Bohr model explains the Balmer series as a family of lines produced when electrons

23.2 THE BOHR MODEL

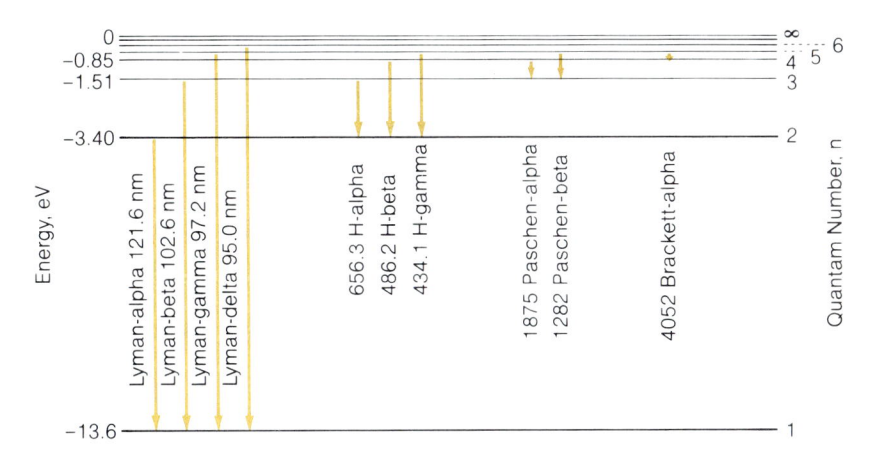

Figure 23.7

Energy levels and spectral lines in hydrogen line series. Balmer series is labelled H-alpha, H-beta, etc.

Figure 23.8

Absorption and emission lines. For a given element (hydrogen here), absorption lines form when an electron absorbs energy and moves to a higher state (a), and emission lines form when an electron spontaneously drops from a high energy state to a lower one (b). The wavelengths of the absorption and emission lines are the same.

in high-energy states drop to the $n = 2$ state (Fig. 23.7). The other observed series of hydrogen lines are due to electron transitions to other lower states: the Lyman series, for example, is the result of electrons from higher states dropping into the lowest orbit ($n = 1$); the Paschen series has $n = 3$ as the lower state; the Brackett series has $n = 4$ as the lower state, etc. Each can appear as a series of absorption lines when continuous radiation passes through a cloud of hydrogen atoms, so that photons with appropriate energies are absorbed, and electrons move from the lower level to higher levels (Fig. 23.8). The conditions under which absorption occurs instead of emission are described more fully later in this chapter.

In most atomic physics and spectroscopic applications, the energies of the electron orbits are expressed in electron volts. Thus, the first few energy levels of the hydrogen atom are $E_1 = -13.6$ eV, $E_2 = -3.4$ eV, $E_3 = -1.5$ eV, etc. It is often useful to depict the energy levels of an atom as horizontal lines in an **energy-level diagram** (Fig. 23.7). Then it is easy to visualize the electron jumps or transitions which produce the spectral lines of the atom.

The lowest energy state the electron can be in, corresponding to the lowest orbit, is called the **ground state,** and the higher states are called **excited states.** In general, the electron will stay in the ground state unless it gains the energy needed to reach an excited state through the absorption of a photon (it can also be excited by a collision with an electron or another atom, as discussed later in this chapter). Once in an excited state, the electron will quickly drop back to the ground state, emitting a photon in the process (the electron is actually most likely to drop back in a series of steps, emitting a photon each time it goes from one level to a lower one).

If we write Eq. 23–7 separately for levels n' and n, where n' represents the upper of two states and n the lower, then the difference in energy between the two states is

$$E_{n'} - E_n = \left(\frac{2\pi^2 e^4 m_e k^2}{h^2}\right)\left(\frac{1}{n^2} - \frac{1}{n'^2}\right) \tag{23-8}$$

which is always positive if n' is greater than n. From Eq. 23–1 we see that

$$\frac{1}{\lambda} = R\left(\frac{1}{n^2} - \frac{1}{n'^2}\right), \quad n = 1, 2, 3, \ldots, (n' - 1) \tag{23-9}$$

where

$$R = \frac{2\pi^2 e^4 m k^2}{h^3 c} = 1.096776 \times 10^7 \text{ m}^{-1} \tag{23-10}$$

is exactly the hydrogen Rydberg constant, whose numerical value was known from the earlier experiments of Balmer. We have used a corrected value of mass, $m = m_e/(1 + m_e/m_p)$, slightly less than m_e, in order to account for the small motion of the proton nucleus, of mass m_p. Thus, we have come full circle: by adopting Bohr's quantum hypothesis for the orbital angular momentum, we have now reproduced the earlier expression for the wavelengths of all the series of lines in the hydrogen spectrum.

Let us reconsider the energy associated with the ground state, -13.6 eV. Consider what happens if an electron in this state gains energy in an amount equal to or greater than 13.6 eV. The total energy of the electron will then be zero or positive, so it will no longer be bound to the nucleus. The electron is freed in a process called **ionization** (because the remaining nucleus, with a net positive charge, is now an **ion**). Whereas the electron in the ground state could gain an amount of energy less than 13.6 eV only if it corresponded exactly to a transition to some higher state, it can accept *any* quantity of energy greater than 13.6 eV. Thus, we say that there is a **continuum** of energy levels above the highest bound level. To find the energy required for ionization from any lower state E_n, simply let n' equal infinity, and solve Eq. 23–8 for E_n. The

23.2 THE BOHR MODEL

wavelength of a photon required to just ionize the atom is given by Eq. 23–9, again with n' equal to infinity. If we solve Eq. 23–9 when $n = 2$, we find that ionization from the second level occurs for $\lambda = 365$ nm, corresponding to the limit of the Balmer series. For high values of n', the wavelengths become very close together, corresponding to the crowding of lines that is observed at wavelengths just longer than the series limit.

We stress that the Bohr model was still an empirical one in that it was constructed on the basis of an assumption that allowed a fit to the observed facts, not on the basis of some underlying property of atoms that would explain why the angular momentum should be quantized. This was left for later workers, as we shall soon see.

Example 23.3 Calculate the wavelengths of the first two lines of the Lyman series of hydrogen.

The Lyman series corresponds to $n = 1$, and for the first two lines $n' = 2$ and $n' = 3$. Hence we find the inverse wavelengths from Eq. 23–9. For $n' = 2$, we find

$$\frac{1}{\lambda} = R\left(\frac{1}{n^2} - \frac{1}{n'^2}\right) = (1.096776 \times 10^7 \text{ m}^{-1})\left(1 - \frac{1}{2^2}\right)$$
$$= 8.2258 \times 10^6 \text{ m}^{-1}$$

so that the wavelength is

$$\lambda = 121.57 \text{ nm}.$$

(We are now justified in keeping five significant figures, since the Rydberg constant in Eq. 23–10 invokes rather precise values of the constants.)

The next line corresponds to $n' = 3$, for which we find

$$\frac{1}{\lambda} = (1.096776 \times 10^7 \text{ m}^{-1})\left(1 - \frac{1}{3^2}\right) = 9.7491 \times 10^6 \text{ m}^{-1}$$

and

$$\lambda = 102.57 \text{ nm}.$$

Application to Other Atoms

The Bohr model was originally constructed for the hydrogen atom, and does not generally apply to other atoms, for which the electron energy level structure is more complex. The model is correct for other atoms that have only a single electron orbiting the nucleus, however. There is no normal (that is, electrically neutral) atom such as this other than hydrogen, but there is a class of ions, having lost one or more electrons, that do satisfy this requirement. For example, helium, which normally has two electrons, becomes hydrogen-like if one of the electrons is lost. Then a nucleus with a positive charge twice the electron charge is orbited by a single electron. The next most complex element is lithium, which has a nuclear charge of $Z = 3$ and normally has three electrons. If two of its electrons are lost, the lithium ion is analogous to the hydrogen atom, except for a stronger attractive force. So that hydrogen-like ions can be included, the value of e^2 should be replaced by Ze^2 in the expression for the Coulomb force in Eq. 23–4, and in the potential energy term in Eq. 23–6. Then the results of the derivation are

$$E_n = \frac{-2\pi^2 Z^2 e^4 m k^2}{h^2 n^2}, \quad n = 1, 2, 3, \ldots \quad (23\text{–}11)$$

$$r_n = \frac{n^2 h^2}{4\pi^2 m_e k Z e^2}, \quad n = 1, 2, 3, \ldots \quad (23\text{–}12)$$

$$\frac{1}{\lambda} = RZ^2 \left(\frac{1}{n^2} - \frac{1}{n'^2}\right), \quad n = 1, 2, 3, \ldots, (n' - 1) \quad (23\text{–}13)$$

for the energy levels and the wavelengths of the spectral lines. Here, $m = m_e/(1 + m_e/m_N)$ to account for the small motion of the nucleus of mass m_N.

In the next chapter, when we discuss atomic structure in the context of the full quantum theory, we will find that there are classes of atoms whose electron energy levels are similar, and whose spectra can therefore be described by similar equations. For now we must be content with the great success of the Bohr model for the hydrogen atom and hydrogen-like ions.

Example 23.4 Find the energies of the ground state and the first excited state of ionized helium, and the wavelength of the line corresponding to the transition between these levels.

For helium, we have $Z = 2$. The ground state corresponds to $n = 1$, so we find from Eq. 23–11

$$E_1 = \frac{-2\pi^2 Z^2 e^4 m k^2}{h^2 n^2} = 4 E_{1H}$$

where E_{1H} refers to the ground state of hydrogen. Thus, the energy of the ground state of ionized helium is

$$E_1 = 4(-13.6 \text{ eV}) = -54.4 \text{ eV}.$$

The first excited state for ionized helium similarly has four times the (negative) energy of the first excited state of hydrogen, so we have

$$E_2 = 4(-3.4 \text{ eV}) = -13.6 \text{ eV}$$

just equal to the ground state energy for hydrogen.

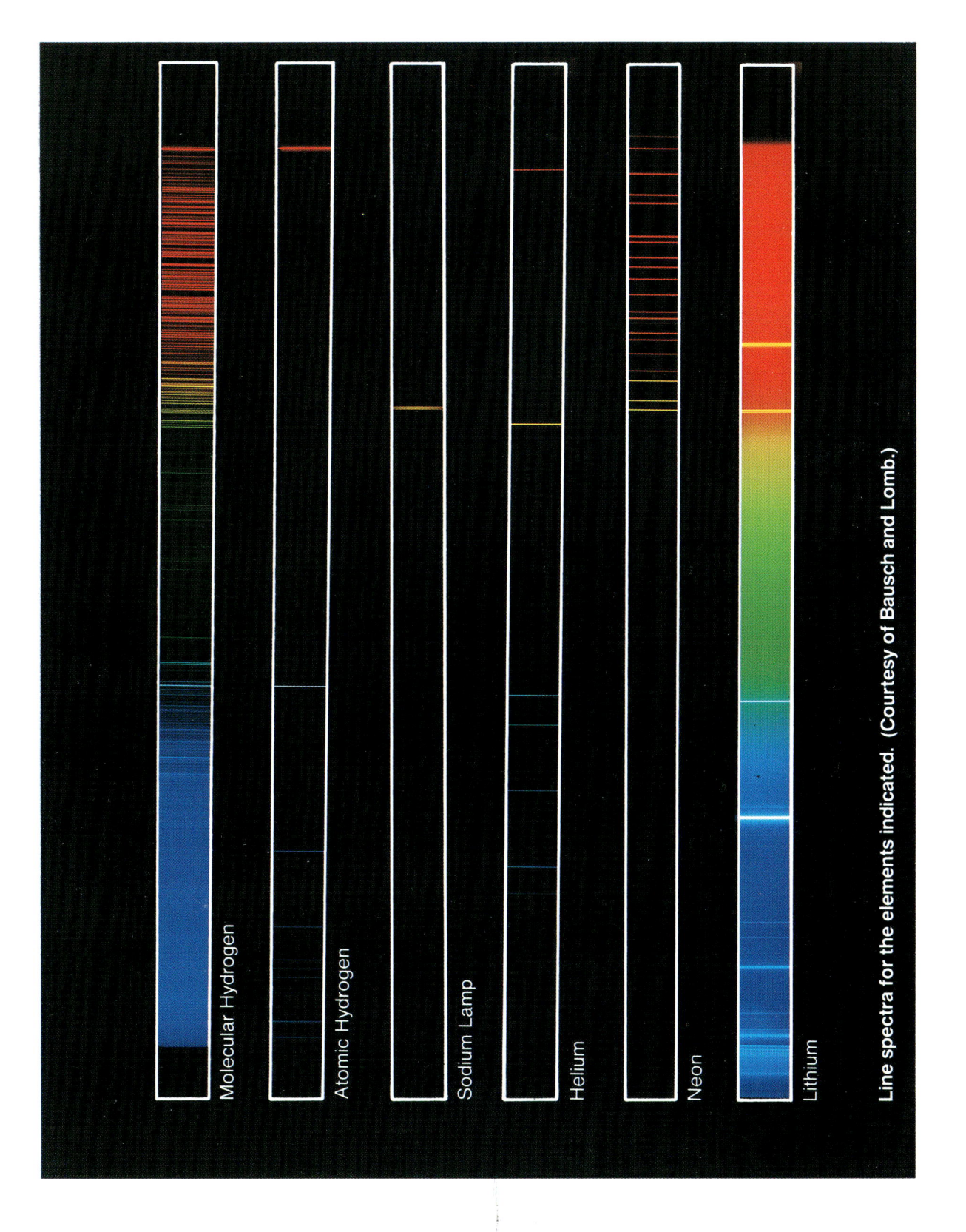

Line spectra for the elements indicated. (Courtesy of Bausch and Lomb.)

Mercury Lamp | Iron Arc | Barium | Calcium | Fraunhofer Lines | Tungsten Lamp | Fluorescent Lamp

The inverse wavelength for the transition between the first excited state and the ground state of ionized helium is

$$\frac{1}{\lambda} = RZ^2\left(\frac{1}{n^2} - \frac{1}{n'^2}\right)$$
$$= (1.097 \times 10^7 \text{ m}^{-1})(4)\left(1 - \frac{1}{2^2}\right) = 3.29 \times 10^7 \text{ m}^{-1}$$

and the wavelength of this line is

$$\lambda = 30.4 \text{ nm}.$$

This is in the far ultraviolet, in a spectral region sometimes called the extreme ultraviolet.

23.3 Practical Spectroscopy

We have followed the development of Bohr's successful model of the hydrogen atom, and we have a picture of the process by which photons are emitted or absorbed by atoms. To fully understand the implications of this model for atomic structure, we will have to explore the science of quantum mechanics, which we will do in the next chapter. Meanwhile, we can use the information already in hand to discuss some important aspects of atomic spectra from a practical point of view.

Absorption and Emission: Kirchhoff's Rules

We have seen that photons are emitted or absorbed when electrons make transitions between energy levels, but we have not said much about why emission occurs in some cases and absorption in others. We have also not explained why solids emit continuous spectra, and why rarefied gases have line spectra. The explanation of the type of spectrum observed lies in understanding the physical conditions in the source of light and in any intervening matter that may lie between the source and the observer.

Most of the early laboratory work, such as that of Balmer, Bunsen, Kirchhoff, and others, involved heated, rarefied gases that produced emission spectra. Typically a substance was subjected to high temperature or to a high voltage so that it became vaporized and glowed (Fig. 23.9). The result was always the emission of light at certain wavelengths; that is, emission lines. Now we understand that emission lines result when the electrons of an atom are allowed to drop from high energy levels to lower ones. We conclude therefore that the process of heating a substance and creating a glowing vapor causes electrons to reach high energy levels, from which they can then descend to lower ones.

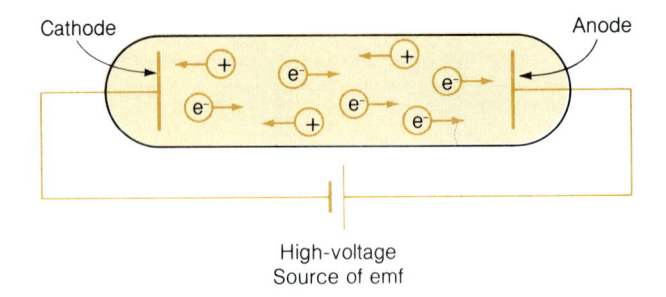

Figure 23.9

Gas discharge tube. A high voltage applied to a rarefied gas produces emission lines, as electrons are excited by collisions to high energy states from which they drop spontaneously and emit radiation.

Kirchhoff also found conditions under which absorption lines are formed. He discovered that a cool, rarefied gas would produce absorption lines if a continuous source were observed through it (Fig. 23.4). Evidently the electrons in the cool gas tend to be in low energy states, so that they can absorb photons with the correct energies to excite these electrons to higher energy states.

It was also known that a solid or a dense gas, heated sufficiently, would produce a continuous spectrum; this is the blackbody radiation we mentioned in the previous chapter. Without knowing anything about atomic structure, Kirchhoff formulated three simple empirical laws of spectroscopy:

1. A heated solid emits a continuous spectrum.
2. A heated, rarefied gas produces an emission-line spectrum.
3. Light passing from a continuous source through a cool, rarefied gas produces an absorption spectrum.

With these rules in mind, it is possible to say something about the physical conditions of a distant light source, in addition to identifying the elements of which it is composed. This kind of analysis, which becomes quite complex, is the basis of the science of **spectroscopy**, which has far-reaching applications.

Following the work of Bohr and the subsequent development of quantum mechanics, the basis for Kirchhoff's laws became better understood. We already have some idea that the second and third laws have to do with whether the electrons in a gas tend to be in high or low energy levels (see the more detailed discussion later in this section), but we have not explained why solids emit continuously. Individual photons are emitted and absorbed in a solid by electrons making discrete transitions between fixed energy levels, just as they are in a rarefied gas. In a solid, however, the energy levels in individual atoms are no longer dictated simply by the structure of the atoms themselves, but are

23.3 PRACTICAL SPECTROSCOPY

instead modified by neighboring atoms into energy bands (Chapter 22). Energy levels may be raised or lowered by electrical forces between atoms, and in a large collection of atoms the result is a continuous distribution of energy levels separated by forbidden gaps (Fig. 22.6). Hence a solid typically has a continuous spectrum. If it has some regular structure, as in a crystal, or a surface layer where some discrete, constant energy levels survive, a solid will have features in its spectrum. In this case we can no longer view the object as a perfect blackbody.

Excitation and Ionization: The Spectra of Gases

In Chapters 12 and 13 we discussed the behavior of gases, and in particular the relationship between average motions of atoms or molecules and macroscopic gas properties such as temperature. Now we can see how these concepts govern the spectra of gases.

In any gas whose temperature is above absolute zero, individual gas particles are in random motion, but in a large collection of particles, the *average* behavior is quite predictable. The concept of temperature is intimately linked to the average speed of the particles, as we have learned. In our earlier discussions of gas kinematics, we assumed that all collisions between particles are perfectly elastic, so that no exchange of energy occurs. In reality, there can be energy transfer between collision partners, and usually the result is conversion of kinetic energy into electron excitation. When an atom collides with another particle, it often happens that some of the kinetic energy of motion is imparted to one or more electrons. The electrons can then gain energy and move to higher states. This process is called **collisional excitation.** Once an electron is excited, usually it will immediately drop back to the ground state, emitting photons as it does so. Hence the ultimate result is the conversion of gas kinetic energy into radiated energy (if the radiation escapes from the gas without being re-absorbed, the energy is lost and the gas is cooled by the process).

Because electrons usually drop back to the ground state quickly, only a small fraction of the atoms in a gas have excited electrons at any moment. The fraction depends primarily on the density of the gas, because the density governs the frequency of collisions (the temperature of the gas also plays a role, because the collisions must be sufficiently energetic to excite the electrons). It is possible to infer the density if the fraction of atoms with excited electrons is known. This can be determined from the spectrum of the gas. If the gas is observed in absorption, that is, a background continuous source is observed through it, then the particular absorption lines that are seen depend on how highly excited the gas is. The Balmer series of hydrogen, for example, arises from the $n = 2$ electron energy level, so this series will be seen in the absorption spectrum of hydrogen only if sufficient excitation occurs to maintain a number of atoms with electrons in the second energy level.

If collisions between atoms and other particles are sufficiently energetic, then electrons may gain enough energy to be freed entirely, resulting in **collisional ionization.** The spectrum of a gas can tell us how highly ionized it is, because the spectrum of an atom that has lost an electron is completely different from its spectrum if it has not lost any electrons. Once an electron is freed, it may eventually find an ion with which it can join, so that the fraction of atoms that are ionized at any moment depends on how rapidly this **recombination** process occurs compared to the rate of ionization. In a very hot gas, ionizations may occur much more rapidly than recombinations, so that the gas is essentially fully ionized. At moderate temperatures, the rates may be comparable, so that comparable numbers of ions and atoms are present. Most elements have several electrons, so that there can be varying degrees of ionization, depending on how many electrons are lost. Iron, for example, normally has 26 electrons. At temperatures above about 12,000 K, little atomic iron exists. Between 12,000 K and about 20,000 K, most iron is in the state of ionization where one electron has been lost (thus it is singly ionized). Between 20,000 K and 40,000 K, iron is doubly ionized, that is, most of the atoms have lost two electrons. At higher temperatures, more electrons are lost. When T exceeds 10^8 K, iron has lost all 26 electrons. Iron in each state of ionization has a unique spectrum of lines, so that the state of ionization, hence the temperature of the gas, can be determined from the spectrum. The same is true of any element, but iron is particularly interesting because lines of very highly ionized iron were found many years ago in the spectrum of the Sun and were for a long time unidentified because it was not expected that the Sun would have such high temperatures (over 10^6 K!). When laboratory experiments showed that iron in highly ionized states produces lines matching those observed in the solar spectrum, it was realized that the Sun's corona consists of very hot, tenuous gas.

Historically, much of the impetus for identifying gases by analyzing their spectra came from astronomers, who have no other means of probing physical conditions in the remote objects they study. In modern times, however, the need for these techniques arises in laboratory situations, particularly in studies of gases at extreme temperatures, as in controlled nuclear fusion.

Transition Probabilities

We have said that an electron, once excited, usually drops quickly back to the ground state, emitting one or more photons on the way (depending on how many discrete steps are taken by the electron as it drops to lower levels). The time before a downward transition occurs depends on the quantum mechanical probability that it will do so. Usually, the probability is high and the lifetime of the electron in the excited state consequently short. A typical lifetime is 10^{-8} s. Sometimes, however, the lifetime is longer (between 10^{-3} s and 10^3 s) because the probability of the downward transition is lower. Energy levels such as this are called **metastable** levels, and when collisional excitations are taking place in a gas, electrons accumulate in these levels.

When Einstein studied the emission of light by atoms (in his work on the photoelectric effect), he expressed the likelihood that an electron will make a given transition by an **absorption** or **emission coefficient**. The coefficient for spontaneous emission, as just discussed, depends exclusively on atomic properties. Just as in classical electromagnetic theory, the orbiting electrons act as oscillating dipoles, emitting radiation in analogy with radio antennas. The difference is that in quantum mechanics, radiation occurs in discrete jumps or

23.3 PRACTICAL SPECTROSCOPY

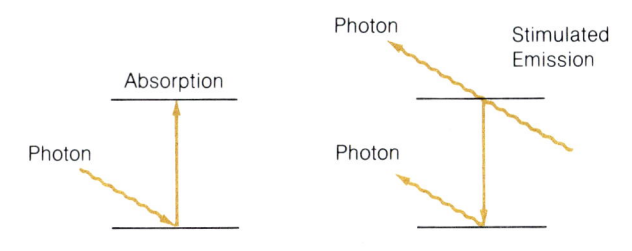

Figure 23.10

Absorption and stimulated emission. A photon of energy, corresponding to the difference between two levels in an atom, can be absorbed or can stimulate emission, depending on which level the electron is in initially.

transitions in energy between one bound state and a lower state. The *lifetime* of an electron in an excited state depends only on the structure of the atom or ion to which the electron is bound.

On the other hand, the *probability* that an electron will absorb a photon and move into a higher state depends not only on the intrinsic properties of the atom and its energy level structure but also on the intensity of the incident radiation (more specifically, on the intensity of photons with energies corresponding to transitions to existing higher levels). A third process is possible, in which an excited electron can be induced to drop to a lower level under stimulation of a nearby photon whose energy matches the energy level difference. In this process, called **stimulated emission** (Fig. 23.10), the result is two photons of identical frequency and phase: the initial incident photon and the new one produced by the downward transition of the electron. Both kinds of stimulated transitions (that is, absorption and stimulated emission) can be viewed as a resonance between the energy of an incoming photon and an energy-level difference in the atom or ion. The probability that an electron will make a stimulated downward transition depends both on the intrinsic properties of the atom and on the intensity of the radiation (specifically, on the intensity of incident photons with the correct energy).

To interpret spectra, it is necessary to know the absorption or emission probabilities for the processes observed. Therefore much of the science of spectroscopy is involved with determining these probabilities. In principle, they can be determined theoretically from quantum mechanical calculations, but in practice many atoms and ions are sufficiently complex that it is more efficient to measure the transition probabilities in laboratory experiments. Only when these probabilities are known is it possible to determine compositions and physical conditions in remote objects through spectral analysis.

The Laser

Normally, stimulated emission is an unimportant process, because electrons spend so little time in excited states that there is little chance that a photon of the correct frequency will happen to strike the atom while the electron is

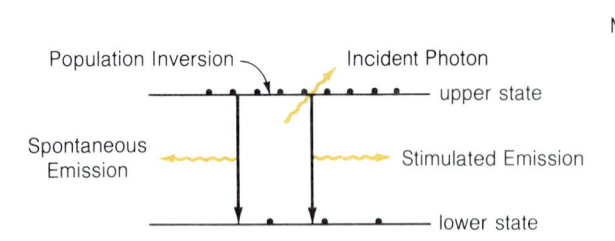

Figure 23.11

Energy levels in a laser. If an excess of atoms have their electrons in the upper, metastable state, then stimulated emission from this state produces a strong, coherent emission line.

Figure 23.12

Laser cavity. To increase the path length of the laser, the beam is reflected many times off the cavity walls. Stimulated emission enhances the intensity of the radiation at the laser frequency, and a small fraction of the emitted light is transmitted from one end.

in the excited state. If, however, the electron is in a metastable excited state, and has a substantial lifetime, then incident radiation can produce significant stimulated emission. If a collection of atoms with electrons in a metastable state is subjected to radiation containing photons with the correct energy, then stimulated emission creates new photons with the same energy. In this way, the number of photons with that energy is increased. This is the basic principle of the **laser** (the name stems from the original acronym: Light Amplification by Stimulated Emission of Radiation).

Consider, for simplicity, a particular kind of atom that has only two electron energy levels: the ground state and one excited state, which is metastable (Fig. 23.11). If photons with energy corresponding to the difference between these two levels are incident on a collection of these atoms, there will be no net gain in the number of photons of this energy. The reason is that the incident photons are just as likely to create upward as downward transitions. Most of the atoms at any time will have their electrons in the ground state, because spontaneous transitions to this state from the excited state are more probable.

Now suppose that for some reason most of the atoms have their electrons in the excited state. When photons interact with this **inverted** population, there is a net gain in the number of photons with the appropriate energy, because the stimulated downward transitions will outnumber the upward transitions due to photon absorption, and there is a net emission of photons. In other words, the incident light is amplified if electrons can be pumped into the excited state, creating an inverted population. In a laser the overpopulation of the upper state is accomplished either by subjecting the material to flashes of white light or electrical discharges.

Another necessity for a laser is to have a high probability for a given photon to stimulate the emission of another. This could be accomplished by having a very long pathlength for the light in the material, but this is impractical. Instead, surfaces reflect the light back and forth from opposite ends of a chamber (Fig. 23.12). The reflecting surfaces are actually partially transmitting, so that some photons escape and form the emergent beam from the

23.3 PRACTICAL SPECTROSCOPY

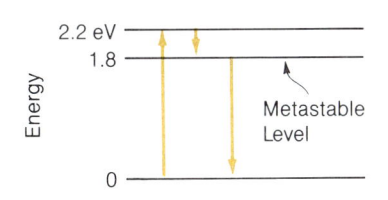

Figure 23.13

Energy levels in a ruby laser. Photons of wavelength 550 nm excite electrons to the 2.2 eV level, from which they drop to the 1.8 eV metastable state. Occasional spontaneous transitions back to the ground state produce photons at 694.3 nm, which create additional photons at that wavelength through stimulated emission.

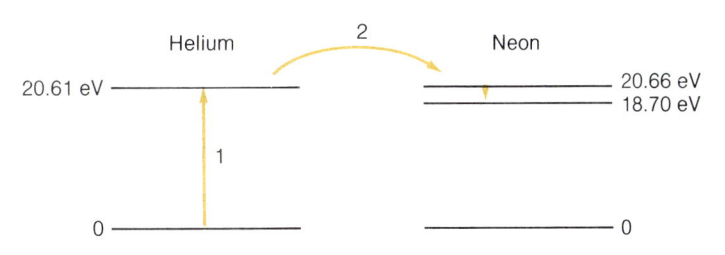

Figure 23.14

The helium-neon laser. After helium is excited by electrical discharge to a metastable level 20.61 eV above the ground state, a collision with a neon atom can excite the neon atom to a metastable state 20.66 eV above the ground state, from which it may drop back to the 18.70 eV state and emit laser radiation at 632.8 nm.

laser, while a large number remain inside the tube, stimulating further photon emission. This has the additional benefit that the light becomes highly **collimated**, meaning that the photons are all very nearly parallel, having essentially travelled a very long, narrow column. Any photons emitted in a direction not perfectly parallel to the long axis of the chamber eventually escape through the sides and do not contribute to the emergent beam. Thus, the divergence of the beam is due only to diffraction, and is therefore comparable to λ/D, where λ is the wavelength of the light and D the diameter of the tube (Chapter 20). The value of λ/D can be incredibly small.

Many different types of lasers have been developed using different materials as the source of the inverted electron population and different pumping techniques. In a ruby laser, one of the earliest types, chromium atoms at isolated lattice locations in synthetic ruby (also known as corundum, aluminum oxide in a crystalline structure) are excited by the absorption of photons of wavelength 550 nm (Fig. 23.13), then they quickly drop to an intermediate metastable state whose lifetime is about 3×10^{-3} s. Some spontaneous emission occurs as electrons drop from the metastable level to the ground state. The resulting photons then stimulate further photon emission, and the laser amplification is under way. The energy difference between the metastable level and the ground state is 1.8 eV, corresponding to a photon wavelength $\lambda = hc/E = 694.3$ nm, in the deep red part of the visible spectrum. Using light to create an inverted electron population is known as **optical pumping.**

In a helium-neon laser, a mixture of helium and neon atoms fills a cylindrical chamber. An electrical discharge excites helium atoms to a metastable state 20.61 eV above the ground state (Fig. 23.14). The lifetime of helium atoms in this state is sufficiently long that a collision with a neon atom will typically occur before the helium atom has time to spontaneously drop out of the excited state. Thus, when a collision occurs, there is 20.61 eV of excitation energy that can be transferred to the neon atom (along with some kinetic energy of the colliding particles) and the neon atom is excited to a metastable state of its own, 20.66 eV above the ground state. In this way, a population inversion in the neon atoms is created, and the neon begins to act as a laser

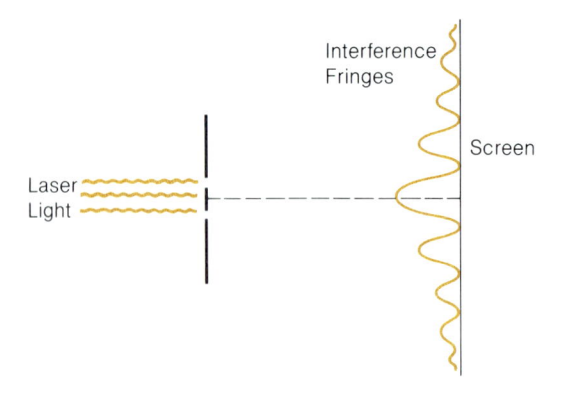

Figure 23.15

Double-slit interference using a laser. The double-slit interference experiment can be performed without an initial single slit, if the incident light is coherent light from a laser.

when transitions occur from this metastable level to an intermediate level 1.96 eV lower. The wavelength of the emergent beam is 632.8 nm, again in the red part of the visible spectrum.

The output of a laser is an intense, parallel beam of coherent light. The photons are all in phase and have identical wavelengths. If the pumping can be done continuously, as is possible for He-Ne lasers, then the laser beam output can be maintained continuously. Another possibility is to pump the electrons occasionally, as with flashes of light in optical pumping, so that the beam is emitted in pulses. It is usually possible to achieve higher intensities with pulsed lasers. In any case, a laser does not create energy; it converts supplied energy (from the pumping mechanism) into coherent light.

The uses of lasers are far-ranging, and increasing with time. The high intensity of pulsed lasers permits precise microscopic surgery and welding when very high temperatures are needed in very small places. Similarly, laser beams have been explored as heat sources for highly localized nuclear fusion reactions (Chapter 25). The narrow, parallel beams from lasers are useful for making precise alignments and measurements of direction. Lasers have been widely used in surveying and in aligning precision optical instruments and experiments. Furthermore, laser beams diverge very little and can be transmitted with little loss over large distances. Lasers are being used as carriers of information (through encoded pulses) and for long-distance ranging experiments. In one current program, laser beams from the Earth are reflected from a mirror left on the Moon during one of the manned landings, and then received at the sending station. Thus very precise measurements of the Earth-Moon distance can be made. Similarly, lasers on satellites are used to precisely measure distances on the Earth's surface. Motions of continents of only a few centimeters per year have actually been measured.

23.3 PRACTICAL SPECTROSCOPY

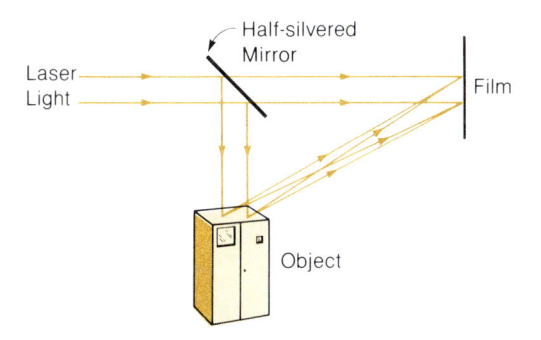

Figure 23.16

Holography. Light from a laser is split into two beams which are allowed to interfere with each other after one has been reflected by an object. The interference pattern, dependent at each point on the pathlength difference between the two beams, can be used to create a three-dimensional image of the object.

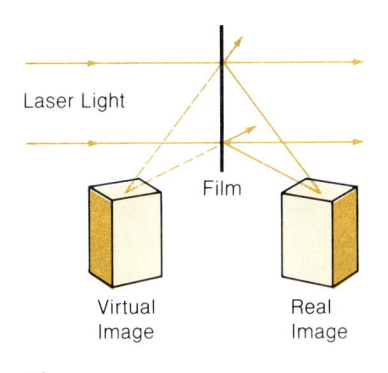

Figure 23.17

A holographic image. Both a real and a virtual image are formed when laser light is incident on a holographic film. Because the aspect of each image depends on the viewing angle, it creates a three-dimensional effect as the viewer moves.

The coherence of laser light is also very useful, particularly in applications of interference effects. The double-slit experiment can be performed without an initial single slit, for example, since the light is already in phase (Fig. 23.15). Another use of interference that is becoming widespread is **holography,** the use of interference in coherent light to create three-dimensional images (Fig. 23.16). In this technique, a laser beam is split by a half-silvered mirror so that part reaches the target object and is reflected before interfering with the other part as they reach a photographic film where they create interference fringes. At any location on the film, the fringe pattern depends on the phase relationship between the two laser beams, so the intensity of the fringes is governed by the distance between the film and the point on the object from which the reflected light comes. This distance, in turn, depends on both the shape of the object and the direction to the part of it that produces the reflected beam. Hence, the **hologram,** as the image of interference fringes on the film is called, is a three-dimensional representation of the object.

If the film is developed and laser light shined through it, the fringes act as a complex diffraction grating, creating images that the eye can see. Both a virtual image and a real image are formed (Fig. 23.17). Because the interference pattern is originally created by the three-dimensional relationship of the various parts of the target object, the images formed when a laser beam is shined through the hologram show the parts of the object in their proper three-dimensional relationship. When the viewing angle is varied, the appearance of the image varies also, just as if the object were being viewed directly.

The science of holography is being developed rapidly, and soon there may be holographic microscopes and moving pictures, among other things. Lasers that operate in ultraviolet, infrared, or x-ray wavelengths may prove useful for holographic applications.

FOCUS ON PHYSICS

Natural Lasers in Space

Lasers are finding more technological applications every day, and have virtually revolutionized several fields since they were first developed in the early 1960s. They are used in medicine, for precision surgery; they are used in many situations where precise geometrical measurements or alignments are needed; they provide calibrated, intense light sources for many experimental measurements in physics and chemistry; and they are even being discussed as potential weapons. While lasers are generally thought of as very new high-technology devices, they have roots here on earth that go farther back than the 1960s, and there are natural situations where the laser mechanism occurs as well.

The predecessor to the laser was the **maser**, which was developed in the years following World War II, as physicists pursued knowledge of radio and radar waves developed during the war. The term *maser* stands for **M**icrowave **A**mplification by **S**timulated **E**mission of **R**adiation, and it applies to a process in which microwaves (short-wavelength radio emission) are amplified in a cavity by stimulated emission of photons, in close analogy with the way photons of visible light are amplified in a laser. Masers have many practical applications, particularly in molecular spectroscopy.

Beginning in the mid-1960s, astronomers began to detect microwave emission from molecules in space. In dense interstellar clouds (Fig. 23.1.1), molecules form from free atoms. Collisions between molecules raise them occasionally to excited energy levels (these are rotational energy states, not quite the same as the electron energy states discussed in this chapter, but similar in that they are discrete and that photons are emitted or absorbed as molecules change energy states; see the discussion of molecular spectroscopy in Chapter 24). Once a molecule is in an excited state, it soon drops

Figure 23.1.1

A very young cluster with nebulosity and dark interstellar clouds.

to a lower state, emitting a photon in the process. The energy states are close together, so the emitted photons have low energies and long wavelengths. Whereas an atom typically emits visible light when its electron drops from one level to a lower one, a molecule in an interstellar cloud emits at radio (microwave) wavelengths. Since the first observations of radio emission from interstellar molecules more than twenty years ago, about 100 different molecular species have been detected in space. Most are simple molecules, consisting of only two

23.3 PRACTICAL SPECTROSCOPY

Figure 23.1.2

The Orion nebula.

or three atoms (although species with as many as 13 atoms have been found).

The measurement of radio emission lines from molecules is useful for astronomers not only because it reveals the abundances of various molecular species in space, but also because it provides information on physical conditions. Because collisions between molecules create the emission lines (by raising molecules to excited states), the relative strengths of the lines from different excited levels depend on the rate of collisions, which in turn depends on the density and temperature of the gas. Therefore astronomers carefully analyze the line strengths in order to measure the densities and temperatures of interstellar clouds (this information is used in understanding other processes such as the formation of new stars, which occurs in these dense clouds).

The first molecule detected by radio emission was the hydroxyl radical, OH. In the course of observing this species in several regions in space, astronomers soon found some very surprising line strengths. The line strengths from this molecule, in certain regions such as parts of the great nebula in the constellation Orion (Fig. 23.1.2), were far stronger than could be explained by the normal process of collisional excitation (unless a temperature of more than 10^{12} K was assumed, which is entirely unreasonable; typical temperatures for these clouds are less than 50 K). It was soon realized that a natural maser was at work.

The emission from OH arises from transitions among four energy levels, whose separations are indicated (but not drawn to scale) in Fig. 23.1.3. The line that was found to be overly strong is the 1,665 MHz line (wavelength 18.02 cm), whereas the 1,667 MHz line (18.00 cm) is normally the strongest. The upper level of the 1,665 MHz transition (level 3 in the diagram) has a relatively low probability of making the spontaneous transition down to the lowest level (level 1), so in order for this line to be so strong, there had to be some mechanism at work to enhance the population of level 3, or to enhance the rate of transitions from there to level 1, or both.

It appears that both things are occurring. Level 3 is overpopulated (compared to the process of collisional excitation) by molecules that are excited to higher levels and then drop down to level 3. The excitation or "pumping" is apparently due to the absorption of infrared light from nearby cool stars. Hence this overpopulation occurs only in regions where there are stars close by to provide the photons needed for the pumping. Given an overpopulation in level 3, then stimulated emission begins to occur as soon as any of the molecules make the 1,665 MHz transition. Even though this is a low-probability transition, it still does occur, so that 1,665 MHz photons are available to stimulate the emission of more 1,665 MHz photons,

(continued)

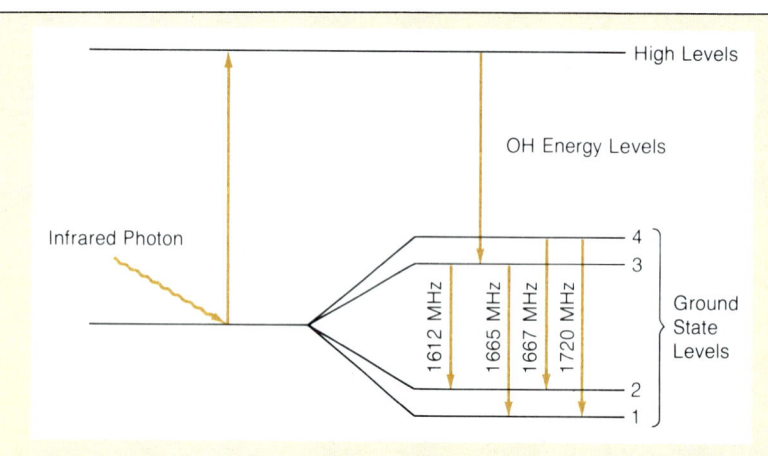

Figure 23.1.3

and so on. Thus a cloud near a source of pumping infrared photons emits a very strong line at 18.02 cm wavelength, accounting for the observed line intensity.

Other molecules have been observed as interstellar masers. Two of the most common are silicon monoxide (SiO) and water vapor (H_2O). It is interesting to speculate whether observations of interstellar masers would have led to the invention of the laser, if the technology of radio astronomy had progressed more rapidly and the strange emission line intensities had been discovered sooner.

Perspective

In this chapter we have begun to explore the properties of the atom, and we have found them to be closely intertwined with the nature of light and its emission and absorption. The quantum nature of light has now been revealed as intimately linked with the discrete nature of energy states in atoms, and we have begun to see how spectroscopy can be used to diagnose physical conditions in remote materials.

The quantum theme has been building through the last two chapters, as has the notion that the material universe is governed not by precisely knowable quantities, but by probabilities. The stage is now set for us to undertake a discussion of modern quantum mechanics, in which these themes will be seen as basic to the fundamental structure of matter.

SUMMARY TABLE OF FORMULAS

Balmer formula for hydrogen line wavelengths:

$$\frac{1}{\lambda} = R\left(\frac{1}{2^2} - \frac{1}{n^2}\right), \quad n = 3, 4, 5, \ldots \quad (23\text{--}1)$$

Photon energy and energy level difference:

$$h\nu = \frac{hc}{\lambda} = E_u - E_l \quad (23\text{--}2)$$

Quantization of orbital angular momentum:
$$L_n = m_e v_n r_n = \frac{nh}{2\pi}, \quad n = 1, 2, 3, \ldots \quad (23\text{--}3)$$

Centripetal force for circular electron orbit, hydrogen:
$$\frac{ke^2}{r_n^2} = \frac{m_e v_n^2}{r_n} \quad (23\text{--}4)$$

Orbital radii, hydrogen:
$$r_n = \frac{n^2 h^2}{4\pi^2 m_e k e^2} = \frac{n^2 \hbar^2}{m_e k e^2} \quad (23\text{--}5)$$

Electron potential energy, hydrogen:
$$PE = \frac{-ke^2}{r_n} \quad (23\text{--}6)$$

Total energy of state n, hydrogen
$$E_n = -\frac{2\pi^2 e^4 m_e k^2}{h^2 n^2} = -\frac{m_e k^2 e^4}{2n^2 \hbar^2}, \quad n = 1, 2, 3, \ldots \quad (23\text{--}7)$$

Energy difference between levels, hydrogen:
$$E_{n'} - E_n = \left(\frac{2\pi^2 e^4 m_e k^2}{h^2}\right)\left(\frac{1}{n^2} - \frac{1}{n'^2}\right) \quad (23\text{--}8)$$

Photon wavelengths, hydrogen ($n' > n$):
$$\frac{1}{\lambda} = R\left(\frac{1}{n^2} - \frac{1}{n'^2}\right), \quad n = 1, 2, 3, \ldots, (n'-1) \quad (23\text{--}9)$$

Rydberg constant:
$$R = \frac{2\pi^2 e^4 m k^2}{h^3 c} = 1.096776 \times 10^7 \text{ m}^{-1} \quad (23\text{--}10)$$

Energy levels for hydrogen-like ions, charge Z:
$$E_n = \frac{-2\pi^2 Z^2 e^4 m k^2}{h^2 n^2}, \quad n = 1, 2, 3, \ldots \quad (23\text{--}11)$$

Orbital radii for hydrogen-like ions:
$$r_n = \frac{n^2 h^2}{4\pi^2 m_e k Z e^2}, \quad n = 1, 2, 3, \ldots \quad (23\text{--}12)$$

Photon wavelengths, hydrogen-like ions ($n' > n$):
$$\frac{1}{\lambda} = RZ^2\left(\frac{1}{n^2} - \frac{1}{n'^2}\right), \quad n = 1, 2, 3, \ldots, (n'-1) \quad (23\text{--}13)$$

THOUGHT QUESTIONS

1. In Rutherford's model of the atom, what force keeps the electron in orbit about the nucleus? Why was this model unsatisfactory?

2. What is the experimental evidence that the positively charged matter in the atom is concentrated in a tiny nucleus?

3. Explain why the temperature and density of a gas and its location relative to any outside source of radiation determine whether it produces an emission or absorption spectrum.

4. Why do all hydrogen lines (Lyman series, Balmer series, etc.) have the same Rydberg constant?

5. In the Bohr model of the atom, why does quantizing orbital angular momentum ($L_n = n\hbar$) imply quantized energy levels E_n?

6. Explain how the Bohr model of the atom differs from the Rutherford model, and why it is inconsistent with the uncertainty principle.

7. In the Bohr model, when an atom emits a photon, an electron drops down from one radius $r_{n'}$ to a smaller radius r_n. Does the electron's orbital angular momentum change? If so, where does it go? Recall that in classical mechanics a change in angular momentum requires torque.

8. How many energy levels and possible transitions exist for the hydrogen atom?

9. Why is it impossible for an electron to orbit the hydrogen atom in the Bohr model any closer than r_1, the ground state radius?

10. Give several examples of how an atom may be excited. How does it decay from these energy levels?

11. What is the difference between continuum and discrete energy transitions in the hydrogen atom? Why are only negative energy states discrete (quantized)?

12. As the charge Z of a hydrogen-like ion increases, what happens to its Bohr radius, its energy levels, and the velocity of the electrons?

13. The Lyman lines of hydrogen lie in the ultraviolet (91.2 nm – 121.5 nm), while the corresponding lines of hydrogenic ions of oxygen, iron, and other heavy elements lie in the x-ray portion of the EM spectrum. Why?

CHAPTER 23 LIGHT AND THE ATOM

14 What is the difference between collisional excitation and collisional ionization? What happens to the bound electron in each case?

15 What is the relationship between the temperature of a gas and its degree of ionization? If a gas is extremely hot, can ions lose all their electrons?

16 Explain the difference between spontaneous and stimulated emission and describe their role in the laser.

17 Why is there no such thing as spontaneous absorption?

18 List several ways in which radiation (light) can interact with atoms.

PROBLEMS

Section 23.1 Experiments and Rutherford Model

- 1 Calculate the charge-to-mass ratio, q/m, of a particle which moves at velocity v in a magnetic field B with circular orbit of radius R.

- 2 With the experiment in Problem 1, how can one tell whether a particle has a positive or negative charge?

- 3 In Rutherford's scattering experiment, alpha particles (mass $4m_p$, KE = 5 MeV) were directed at a sheet of gold foil. If gold nuclei have a positive charge $Q = +79e$ and alpha particles have $q = +2e$, how close could the alpha particles get to the nucleus? (Consider the balance between initial KE and the coulomb PE at the point of closest approach.)

- 4 Consider a 1-gram sheet of gold foil, composed of gold nuclei (mass $197m_p$) and numerous electrons. The atomic number Z is 79.
    (a) For the gold to be electrically neutral, how many electrons are there per nucleus?
    (b) What is the ratio of the mass in electrons to mass in nuclei?
    (c) How many nuclei are in the foil?

- 5 Silver has an atomic number 47. How many protons and electrons are present in a neutral silver atom?

- 6 A new line of hydrogen at wavelength 434 nm is believed to belong to the Balmer series. What transition ($n' \rightarrow n$) does it correspond to?

- 7 What are the wavelengths of the first three lines in the Lyman series?

- 8 What is the wavelength of the hydrogen Lyman series limit (nm)?

- 9 What is the wavelength of the first line in the hydrogen Paschen series? What type of radiation (what color) is this?

- 10 What are the wavelength (cm) and frequency (Hz) of the hydrogen line from $n' = 110$ to $n = 109$? In what spectral region is the line?

- 11 Calculate the series limits (nm) for the Balmer, Paschen, and Brackett series.

- 12 A hydrogen spectrum has a line at 2626 nm. Identify its series and n' and n.

Section 23.2 The Bohr Model

- 13 Write expressions in symbols for the radii, r_n, of Bohr levels in hydrogen for $n = 1$ through 6. Draw a sketch of the orbits.

- 14 What is the radius of the $n = 100$ level of hydrogen?

- 15 For what quantum number n does the electron's orbital radius equal 0.1 cm?

- 16 A rarified gas of atomic hydrogen has a density of 1,000 atoms/cm^3.
    (a) What is the average distance L between atoms? (Consider the atoms to lie at the corners of a cubic lattice, a distance L on a side.)
    (b) For what value of quantum number n does the electron's orbit take it to the nearest atom ($r_n = L$)?

- 17 What is the ratio of the hydrogen atom's Bohr radius (r_1) to the radius of the nucleus (10^{-15} m)? What fraction of the volume of the atom is therefore not empty space? Assume the electron to be very small.

- 18 Derive an expression in symbols for the velocity v_n in orbit about hydrogen with quantum number n.

- 19 What is the velocity of the electron at the $n = 1$ Bohr radius of hydrogen? Give your answer in m/s and mi/h.

- 20 At what value of quantum number n does an electron's orbital velocity, v_n, equal the average thermal speed of a free electron, $v_{rms} = (3kT/m_e)^{1/2}$, at room temperature, $T = 300$ K?

- 21 Calculate the total energy E_n in eV for an electron of hydrogen in $n = 3$.

PROBLEMS

- **22** What is the energy difference (eV) between the $n' = 5$ and $n = 2$ levels of hydrogen? What wavelength transition is emitted?

- **23** What are the ionization energies (eV) of hydrogen from $n = 1$, 2, and 5?

- **24** For the Bohr model, show that the potential energy, kinetic energy, and total energy are related by

$$PE_n = -2(KE)_n; \quad E_n = -KE_n = \frac{1}{2}PE_n.$$

  What are the PE and KE (in eV) for an electron of hydrogen if $n = 1$?

- **25** Radiation of 2500 nm shines on hydrogen gas. From which levels n can atoms be ionized by this radiation?

- **26** A 12.2 eV electron moves through hydrogen gas in the ground state. Which levels n can it excite? What emission lines will be radiated?

- **27** Suppose the electron mass m_e was 207 times greater. Estimate, using the Bohr model, the resulting Bohr radius r_1 and ionization energy E_1 of hydrogen. (Such atoms form when a muon, μ^-, replaces an electron, e^-, in a heavy atom, often lead).

- **28** In Problem 27, estimate the wavelength of the Lyman transition from $n' = 2$ to $n = 1$ for the "muonic" atom of hydrogen.

- **29** For a hydrogen-like ion of an element with atomic number Z, and only one electron, write down symbolic expressions for:
  - (a) The orbital radius r_n
  - (b) The total energy E_n
  - (c) The velocity v_n.

- **30** Draw the energy level diagram for the hydrogen-like ion $He^+ (Z = 2)$. Show the first five levels and label them with n and E_n (in eV).

- **31** What are r_1 and E_1 for hydrogen-like iron ($Z = 26$)?

- **32** What is the ground state ionization energy (eV) of hydrogen-like carbon ($Z = 6$)?

- **33** What is the energy of the $n' = 2$ to $n = 1$ transition of hydrogen-like iron ($Z = 26$)?

- **34** For hydrogen-like ions with large Z, the velocity v_1 of the innermost orbit becomes relativistic. Write an expression for v_1 when the nuclear charge is Ze. For what Z does v_1 equal $0.1c$?

- **35** A Lyman line ($n' = 2 \to n = 1$) of a hydrogen-like ion has an energy of 1469 eV. What is the atomic number Z of the element? What is the element?

- **36** A 1 keV x-ray ionizes a hydrogen-like ion of oxygen ($Z = 8$) from its ground state ($n = 1$). How much extra KE does the newly freed electron carry off?

Section 23.3 Practical Spectroscopy

- **37** Fig. 23.18 shows the first five energy levels of the twice-ionized O^{+2} ion. Give the wavelengths of the three emission lines indicated by arrows.

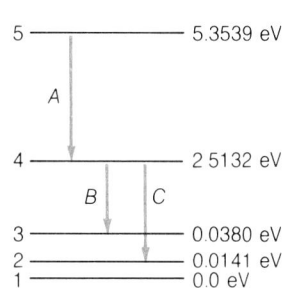

```
5 ─────────── 5.3539 eV
    │
    │ A
    ▼
4 ─────────── 2.5132 eV
    │   │
    │ B │ C
    ▼   ▼
3 ─────────── 0.0380 eV
2 ─────────── 0.0141 eV
1 ─────────── 0.0 eV
```

Figure 23.18

- **38** For energy level 4 of O^{+2} (Fig. 23.18), the radiative transition probabilities are: $A(4 \to 2) = 7.1 \times 10^{-3}$ s^{-1} and $A(4 \to 3) = 2.1 \times 10^{-2}$ s^{-1}. These are the probabilities per second that an ion in level four will spontaneously emit a photon in one of the lines $(4 \to 2)$ or $(4 \to 3)$.
  - (a) What is the total transition probability A_{tot} from level four?
  - (b) What is the average lifetime (s) for an excited ion in level four?

- **39** Each O^{+2} ion in Problem 38 is struck by electrons and excited from level 1 to level 4 at a rate $C_e \cdot n_e$ (excitations per second), where $C_e = 1.28 \times 10^{-9}$ cm^3/s at $T = 10^4$ K.
  - (a) Show that when excitations balance radiative decays back to the lower states, $n_e \cdot C_e \cdot n(1) = n(4) \cdot A_{tot}$, where $n(1)$ and $n(4)$ are number densities of O^{+2} ions in levels 1 and 4.
  - (b) What is the equilibrium ratio, $n(4)/n(1)$, of O^{+2} ions in the excited state 4 to the ground state 1 at 10^4 K if the electron density $n_e = 100$ electrons/cm^3?

- **40** The spectrum of a hot (10,000 K) gas of hydrogen atoms and electrons shows a set of emission lines.
  - (a) Compute the average KE of electrons in the gas.
  - (b) Based on this KE and your knowledge of hydrogen energy levels, what emission lines are likely to be excited from the ground state?

•• 41 Continuous ultraviolet light, ranging in wavelength from 91.2 nm to 330 nm, shines through a cool, low density gas of H^0 atoms. Calculate and list the wavelengths of three Lyman absorption lines that will appear to an observer.

•• 42 If the gas has high enough density in Problem 41, continuous visible light between 330 nm and 700 nm will be absorbed out at various Balmer lines.
(a) Why is high density required for Balmer line absorption?
(b) List two wavelengths likely to be seen.

SPECIAL PROBLEMS

1. Hydrogen Ionization Fraction

In a hot, low density gas, the fraction of hydrogen atoms is determined by a balance between collisional ionization ($H^0 + e^- \rightarrow H^+ + e^- + e^-$), which produces ions, and recombination ($H^+ + e^- \rightarrow H^0 + $ photon), which produces atoms. Suppose the particle densities of atoms, ions, and electrons are $n(H^0)$, $n(H^+)$, and $n(e)$ in particles/cm^3.

(a) Suppose that the average time between ionizing collisions is τ_i and the average time between recombinations is τ_r. Show that in a steady state, in which the rate of ionizations balances the rate of recombinations,

$$\frac{n(H^0)}{\tau_i} = \frac{n(H^+)}{\tau_r}.$$

Equivalently, one may write this relation as $n(H^0)/n(H^+) = (\tau_i/\tau_r)$.

(b) Suppose that when the electron density is $n(e) = 1$ electron/cm^3 and $T = 10^4$ K, the average times between ionizations or recombinations are $\tau_i = 1.23 \times 10^{15}$ s and $\tau_r = 3.85 \times 10^{12}$ s. What is the ratio of atoms to ions, $n(H^0)/n(H^+)$?

(c) Under these conditions, what fraction of the total hydrogen is ionized? (The total hydrogen density, $n(H_{tot})$ is $n(H^0) + n(H^+)$.)

2. Sodium Vapor Lights

While incandescent lightbulbs put out a continuous spectrum, many lights contain sodium or mercury vapor and emit light at discrete wavelengths, giving them a characteristic color. Figure 23.19 shows the first five energy levels of sodium. The two lines labeled C and D are responsible for the yellow color of sodium lights.

(a) What are the wavelengths (nm) of the four lines indicated by arrows?
(b) What resolution spectrometer, $\lambda/\Delta\lambda$, is required to separate the two lines C and D? You may see this in a lab experiment.

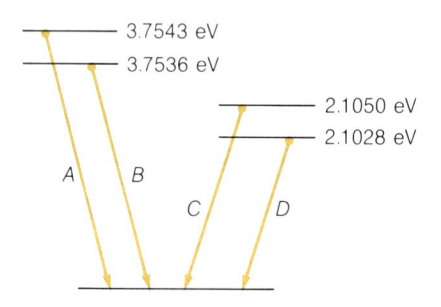

Figure 23.19

(c) Give two reasons why lines A and B would be difficult for humans to observe from a low-temperature sodium light.

3. Doppler Shift

When an object moves away from an observer at velocity v, its light is Doppler shifted to longer wavelengths or lower frequencies (see Chapter 10). If the velocity v is much less than c, the shift in wavelength $\Delta\lambda$ is given by the formula $z = \Delta\lambda/\lambda_0 \approx v/c$, where $\Delta\lambda = \lambda - \lambda_0$ (λ is the observed wavelength, λ_0 is the wavelength emitted at rest, and z is a dimensionless number known as the "redshift" factor).

(a) In a star's spectrum, the first line in the Balmer series (rest wavelength $\lambda_0 = 656.28$ nm) appears at $\lambda = 656.60$ nm. How fast is the star moving away (in km/s)?
(b) A galaxy in the Virgo Galaxy Cluster is moving away from our Sun at a velocity $v = 1,200$ km/s. At what wavelength would the Balmer line appear?
(c) Suppose an object was moving *toward* us at 500 km/s. At what wavelength would the Balmer line appear? (Think about what happens to frequencies and wavelengths when the source approaches the observer.)

4. Relativistic Doppler Shift and Quasars

When v becomes comparable to the speed of light c, the relativistic Doppler shift formula must be used:

$$z = \frac{\Delta \lambda}{\lambda_0} = \sqrt{\frac{1 + \frac{v}{c}}{1 - \frac{v}{c}}} - 1.$$

(a) Show that this leads to the solution for the velocity:

$$v = c \left[\frac{(1 + z)^2 - 1}{(1 + z)^2 + 1} \right].$$

(b) Some very distant objects in space, known as quasars, appear to move away at velocities approaching c. The fastest one has a redshift factor $z = 3.78$. What is its velocity v in units of c?

(c) At what wavelength would the Lyman-alpha line ($n' = 2 \rightarrow n = 1$) of hydrogen appear in the quasar's observed spectrum? What color would it have? (This line is normally emitted in the far ultraviolet at 121.5 nm.)

SUGGESTED READINGS

"How to make holograms." C. L. Strong, *Scientific American*, Feb 67, p 122, **216**(2)

"In which a Lifesaver lights up in the mouth." J. Walker, *Scientific American*, Jul 82, p 146, **247**(1)

"The spectra of streetlights illuminate basic principles of quantum mechanics." J. Walker, *Scientific American*, Jan 84, p 138, **250**(1)

"The spectrum of atomic hydrogen." T. W. Hänsch, A. L. Schawlow, G. W. Series, *Scientific American*, Mar 79, p 94, **240**(3)

"The causes of color." K. Nassau, *Scientific American*, Oct 80, p 124, **243**(4)

"The spectroscopy of supercooled gases." D. H. Levy, *Scientific American*, Feb 84, p 96, **250**(2)

"J. J. Thomson and the Bohr atom." J. L. Heilbron, *Physics Today*, Apr 77, p 23, **30**(4)

"Current trends in atomic spectroscopy." J. J. Wynne, *Physics Today*, Nov 83, p 52, **36**(11)

"Laser linewidth." A. Mooradian, *Physics Today*, May 85, p 42, **38**(5)

"The laser: A splendid light." A. A. Boraiko, photos by C. O'Rear, *National Geographic*, Mar 84, p 335, **165**(3)

"The wonder of holography." H. J. Caulfield, photos by C. O'Rear, *National Geographic*, Mar 84, p 364, **165**(3)

"Harnessing light." C. H. Townes, *Science 84*, Nov 84, p 153, **5**(9)(lasers)

Mr. Tompkins explores the atom. George Gamow, Cambridge University Press, New York, 1942

Science from your airplane window. Elizabeth Wood, Dover, NY, 1975

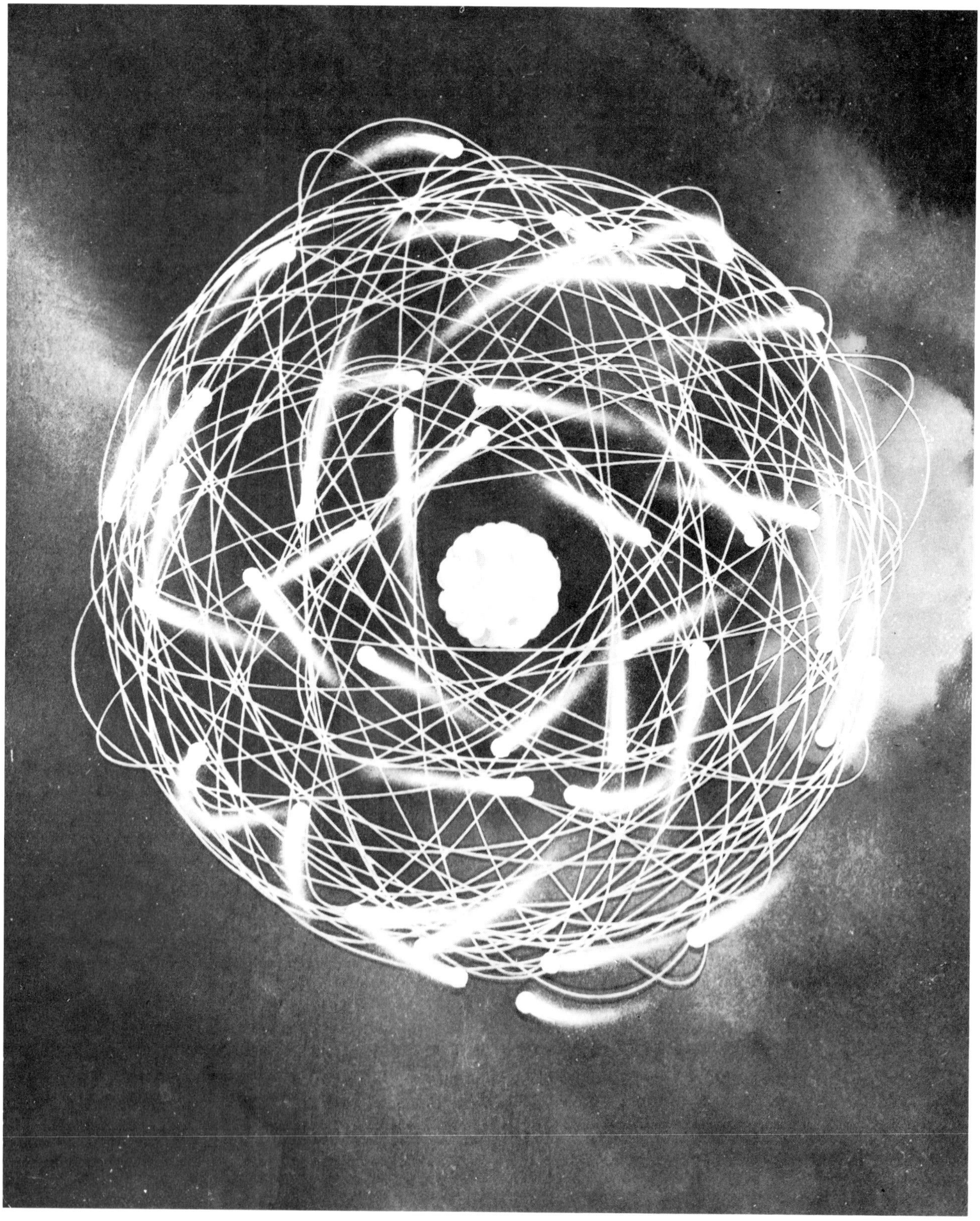

Chapter 24

Essentials of Quantum Mechanics

Despite the success of the Bohr model for hydrogen atoms and for ions with only a single electron, this model was not a physical theory of atomic structure. It did not offer any explanation of electron behavior—why electrons should occupy only discrete energy levels. In postulating well-defined energy levels, the Bohr model was also inconsistent with the uncertainty principle. Furthermore, the Bohr model could not account for the observed spectra of atoms with more than one electron.

The work of Bohr was nevertheless valuable, not only because of its great success with a certain class of atoms, but perhaps more importantly because its radical assumptions stimulated the development of a full theoretical framework for understanding atomic processes. Thus the science of quantum mechanics was in large part initiated by the Bohr model.

24.1 Wave Mechanics

We have discussed the fact that particles of matter have wave properties, and we have mentioned that it is more realistic to think of probability distributions than of precisely knowable positions, because of the uncertainty principle. Now we are ready to tie these ideas together with quantum mechanics (sometimes called wave mechanics). We will have to be content with description, since a mathematical treatment without calculus is difficult.

The Hypothesis of de Broglie

When de Broglie suggested that particles might have wave properties, the Bohr model had been proposed more than ten years earlier. (We have discussed some of the developments out of sequence.) One of de Broglie's arguments for

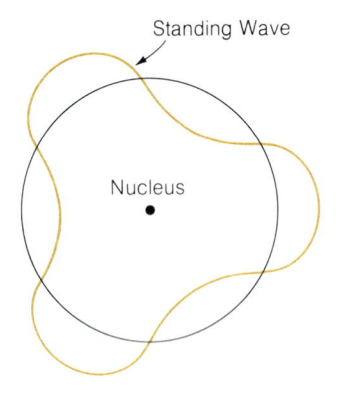

Figure 24.1

De Broglie's hypothesis. It was De Broglie's idea that electrons could exist only in orbits whose circumferences correspond to integral numbers of particle waves, so constructive interference would create standing waves.

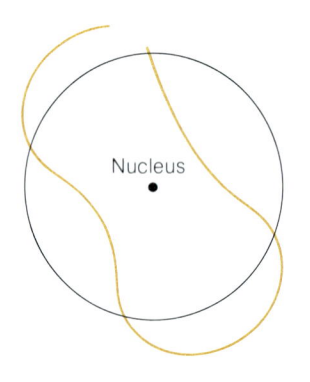

Figure 24.2

A non-allowed electron orbit. When the orbital circumference is not equal to an integral number of wavelengths, destructive interference occurs, and an electron cannot have such an orbit.

a dual nature of matter was that it provided some explanation of the Bohr model.

If particles are associated with waves, then, it occurred to de Broglie, a particle in a confined space might be subject to resonances and standing waves much as sound waves in a cavity are. We learned in Chapter 10 that within a cavity of fixed length, sound waves will persist if waves reflected at the ends of the cavity interfere constructively, but will quickly die out if they interfere destructively. Thus we found a definite relationship between the dimensions of a cavity and the wavelengths of its possible standing waves.

De Broglie reasoned that, if electrons are associated with waves, then perhaps the allowed electron orbits about an atomic nucleus correspond to standing waves. If the circumference of an electron orbit is an even multiple of the electron wavelength, then constructive interference occurs and an electron can have such an orbit (Fig. 24.1). If the circumference is not an even multiple of the electron wavelength, then the interference is destructive, and the electron cannot have such an orbit (Fig. 24.2). The de Broglie wavelength is

$$\lambda = \frac{h}{p} = \frac{h}{mv} \quad (24\text{–}1)$$

where h is the Planck constant, and p is the nonrelativistic particle momentum, the product $m_e v$ of the electron's mass and its velocity. Thus, the requirement for a standing wave, that the circumference of the orbit equal an integral number n of electron wavelengths, is

$$2\pi r_n = n\lambda = \frac{nh}{m_e v_n}, \quad n = 1, 2, 3, \ldots$$

Rearranging, we find

$$m_e v_n r_n = \frac{nh}{2\pi} = n\hbar \quad (24\text{–}2)$$

which is precisely Bohr's equation (Eq. 23–3 in our text) for the quantization of orbital angular momentum. It was from this assumption that Bohr had derived his model. Thus de Broglie's hypothesis provided theoretical underpinnings for the Bohr model. It also pointed the way for a complete theoretical framework in which the wave-particle duality plays the central role.

The Wave Function

Shortly after de Broglie published his innovative work, others began to enlarge upon the nature of matter waves. The de Broglie hypothesis dealt only with the wavelength and did not address other wave properties such as amplitude. When we discussed electromagnetic waves, we were able to characterize their amplitude in terms of the maximum value of the oscillating electric field. But what aspect of matter waves can we associate with amplitude?

24.1 WAVE MECHANICS

Without saying precisely what it means physically, E. Schrödinger in the mid-1920s adopted a formalism in which the amplitude is called the **wave function**, usually designated ψ. In this formalism, the task of mathematically describing the motion of a particle consisted primarily of writing and solving the **Schrödinger equation**, a differential equation involving the quantization of energy states and the uncertainty principle. Let us try to grasp the meaning of ψ.

In the particle interpretation of photons, the meaning of intensity of light is the number of photons arriving per unit time per unit surface area. We have said that the motion of an individual photon is not predictable, but that when many photons are considered, they will conform statistically to a predictable pattern. Thus, the intensity is a measure of the probability distribution for photons; individual photons may depart from the most probable path, but collectively they conform.

At the same time, in the wave interpretation of light, the intensity is proportional to E^2, the square of the amplitude of EM waves. Thus, in a picture of light invoking a wave-particle duality, E^2 is equivalent to a probability distribution for photons. Where E^2 is large, the probability of a photon's presence is high; where E^2 is small, the probability is low. Thus, in the wave-particle model for photons, the photons can be treated as particles whose precise positions are unknowable, but whose probable positions are governed by their wavelike behavior. It is as though photons are particles whose motions are somehow guided by waves.

With that analogy in mind, we can approach the question of the meaning of ψ. If ψ represents the amplitude of a particle wave, then ψ^2 must represent the probability distribution for a particle, if the wave-particle duality applies in the same manner as it does for photons. If ψ represents the amplitude of the wave associated with a single particle, then ψ^2 represents the probability of finding the particle at any given place. For a large collection of particles, ψ^2 gives the particle distribution, so that the behavior of a large number can be precisely predicted, just as can the behavior of a large number of photons.

The probability distribution we speak of here is a distribution in both space and time. When we derive ψ for a particle, we can find the probability that the particle will be in a certain place at a certain time. Thus, the process of solving the Schrödinger equation for ψ is equivalent to solving the equations of motion in classical mechanics. Furthermore, we may expect that for large objects or collections of particles, the quantum theory will agree with classical mechanics. Therefore, we may regard quantum mechanics as more fundamental, and regard classical mechanics as a special application to scales where quantum effects become unimportant.

There are certain aspects of the quantum theory that defy easy visualization. To simply say that only the probable behavior of a single particle can be predicted is not enough, because it implies that a particle may indeed be in a specific position, even though that position is unpredictable. On the contrary, we must view a particle as actually being distributed in space and time. To see this, reconsider the double slit experiment. If particles pass through the double slits at a very low rate, so that only one at a time passes through and reaches the screen (Fig. 24.3), the usual pattern of interference fringes will build up. This implies that individual particles "know" about the presence of

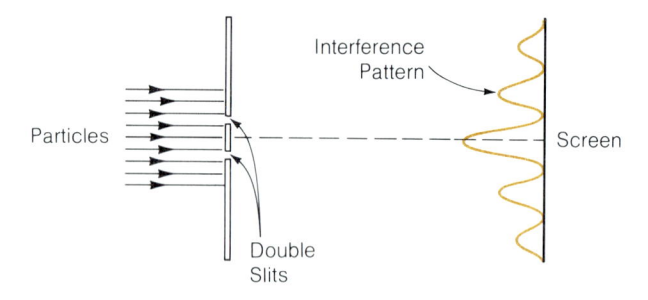

Figure 24.3

Particle interference. Particles passing through a pair of slits create an interference pattern. This demonstrates that particles have wave properties.

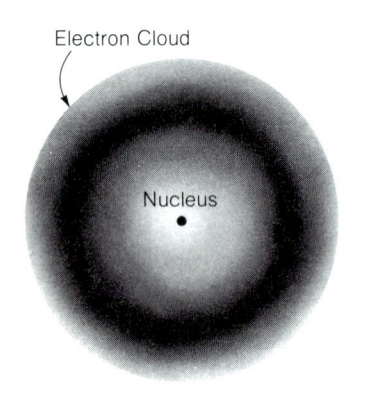

Figure 24.4

An electron cloud. Rather than viewing an electron orbit as a precise, discrete path, the modern view is that the electron has a probability distribution, so that we speak of an electron cloud rather than a single, point-like particle.

two slits, even though our conventional statement would be that each particle goes through one slit or the other. If one slit is blocked, then no interference pattern results; the particles simply go through the one open slit and straight on to the screen. Therefore we must conclude that when both slits are open an individual particle somehow passes through both of them at the same time. This is the only way to understand how individual particles can follow paths dictated by the interference of waves passing through two slits.

When we apply this to an electron orbiting an atomic nucleus, we realize that it is better to think of the electron cloud as a distribution of negative charge, rather than a discrete point charge orbiting in a discrete path (Fig. 24.4). The de Broglie/Bohr quantization condition, $m_e v_n r_n = n\hbar$, may also be viewed as an example of the uncertainty principle, $(\Delta p)(\Delta x) \gtrsim \hbar$. The electron's momentum and position are both uncertain ("spread out") by amounts comparable to their own values, so that

$$\Delta p \simeq p \simeq m_e v_n$$

and

$$\Delta x \simeq \frac{r_n}{n},$$

since in the *n*th energy level, the electron's wavefunction fluctuates *n* times going out from the nucleus. Thus, $(\Delta p)(r_n/n) \simeq \hbar$; that is, $m_e v_n r_n \simeq n\hbar$.

In the Bohr model, it is visualized that the electron physically and instantaneously jumps to a different orbit when a photon is emitted or absorbed; now, however, we must view the transition as a sudden but finite change in the electron's wave function or as a change in the shape of the electron cloud. We will find that there are quantum mechanical laws that govern which kinds of changes occur with high probability, and which cannot, so that it becomes possible to predict the wavelengths and relative strengths of the spectral lines produced by any kind of atom.

Example 24.1 An electron mass m_e is constrained to move in a one-dimensional box of width L. In quantum theory the possible modes of oscillation of the wave function ψ correspond to standing waves with nodes (points of zero amplitude) at the edges of the box. Show that the nth mode satisfies the condition $(n\lambda_n/2) = L$, where λ_n is the particle's wavelength. Then show that the kinetic energy of the particle in the nth mode has quantized values $E_n = n^2h^2/8m_eL^2$, where $n = 1, 2, 3, \ldots$.

Fig. 24.5 shows the first three modes of oscillation (standing waves). We see from the figure that $\lambda_n = 2L/n$. Using the de Broglie condition $\lambda_n = h/m_ev_n$, we see that

$$v_n = \frac{h}{m_e\lambda_n} = \frac{nh}{2m_eL}$$

and

$$KE = \frac{1}{2}m_ev_n^2 = n^2h^2/8m_eL^2.$$

Thus we see that discrete energy states (energy quantization) are a direct result of quantum mechanics.

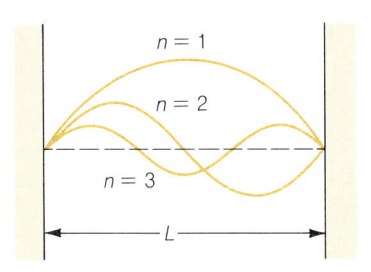

Figure 24.5

24.2 Atomic Structure

With the tools developed by Schrödinger and by his contemporaries, including most notably W. Heisenberg, who independently developed the concepts of quantum mechanics, it became possible mathematically to reproduce the structure and the behavior of atoms. The success of quantum mechanics in explaining the spectra of atoms, as well as many other atomic processes such as the formation of bonds and the historical problem presented by the specific heats of gases at low temperatures (described in Chapter 13), has verified the validity of the theory beyond any reasonable doubt. In this section we reexamine the structure of the atom in the light of quantum mechanics.

Quantum Numbers and Energy States

Let us consider the hydrogen atom at first, again because of its relative simplicity. Quantum mechanics reproduces the basic energy level structure yielded by the Bohr model; that is,

$$E_n = \frac{-2\pi^2e^4mk^2}{h^2n^2} = \frac{-13.6 \text{ eV}}{n^2}, \qquad n = 1, 2, 3, \ldots \qquad (24\text{--}3)$$

where e represents the electron charge and $m = m_e/(1 + m_e/m_p)$ is the effective electron mass (correcting for the small motion of the proton), k is the constant in Coulomb's law, h is Planck's constant, and n is the principal quantum number, which can have any positive integral value.

The Bohr model of the hydrogen atom needed only one quantum number (n) to predict the energy levels. Both experiment and theory showed the atom to be more complex than that, however. In the quantum mechanical theory of the hydrogen atom, we must specify four distinct quantum numbers (n, l, m_l, and m_s) to fully define the quantum state of the electron. At this point we must be more precise about what a **quantum state** really is. Quantum mechanics successfully describes all observable properties of a system from its wave function ψ. To know the state of a system is to know (or measure) ψ. Of course, the Heisenberg uncertainty principle states that it is impossible to know all of the properties of a quantum state with arbitrary accuracy. For example, we can measure a particle's position to an accuracy of Δx, but only at the expense of a large uncertainty in its momentum $\Delta p = h/(\Delta x)$. Similarly, we can measure the energy of an excited electron to precision ΔE, but at the cost of an indefinite lifetime $\Delta t = h/(\Delta E)$ for its state.

For the quantum state of an electron in a hydrogen atom, the wave function ψ depends on n, l, m_l, and m_s. These quantum numbers specify the energy, orbital angular momentum, and the components of the orbital and spin angular momentum along a reference axis. To a first approximation, all states of hydrogen with the same value of the **principal quantum number** n have the same energy $E_n = (-13.6 \text{ eV})/n^2$. This coincidence of energies, called **degeneracy**, no longer holds when we examine the hydrogen energy level structure in greater detail, however. We find that the spectral lines of hydrogen actually consist of two or more closely-spaced lines. Evidently the energy levels of hydrogen are split. This splitting is called **fine structure** and results from effects of the electron's angular momentum. Fine structure is the reason that a single quantum number is insufficient to specify fully the quantum state of an electron.

The second quantum number is the **orbital quantum number** l, which is associated with the orbital angular momentum of the electron about the nucleus. For each value of the principal quantum number n, there are n degenerate states (all of energy E_n) with $l = 0, 1, 2, \ldots, (n-1)$. For example, the ground state ($n = 1$) can have only $l = 0$; the next state ($n = 2$) can have $l = 0$ or $l = 1$; and so on. If $l = 0$, then the electron cloud is spherically symmetric, having no preferred plane. If $l > 0$, there is no spherical symmetry, and we may visualize the electron cloud as rotating symmetrically about some axis (Fig. 24.6). Angular momentum is a vector quantity **L**, and both its magnitude L and one component in a given direction (say the z-direction) are quantized. In terms of the fundamental unit of angular momentum, $\hbar = h/2\pi$, the quantization rules are

$$L = \sqrt{l(l+1)}\hbar, \quad l = 0, 1, 2, \ldots, (n-1) \quad (24\text{--}4)$$

and

$$L_z = m_l \hbar, \quad m_l = 0, \pm 1, \pm 2, \ldots \pm l. \quad (24\text{--}5)$$

Here L_z represents the z-component of **L**, and m_l is the third quantum number, the **magnetic quantum number,** which can take on $(2l + 1)$ integral values

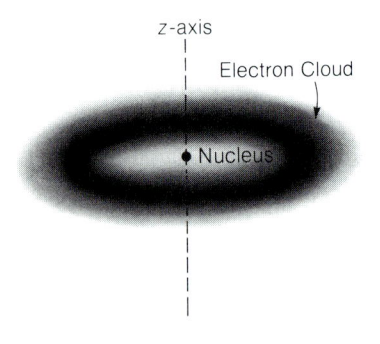

Figure 24.6

The electron cloud for $l = 1$. When the orbital quantum number $l > 0$, there may be axial symmetry but no spherical symmetry.

24.2 ATOMIC STRUCTURE

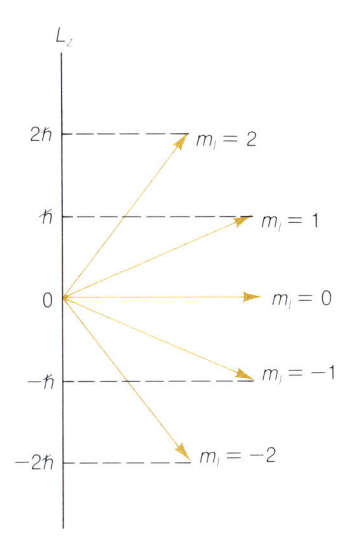

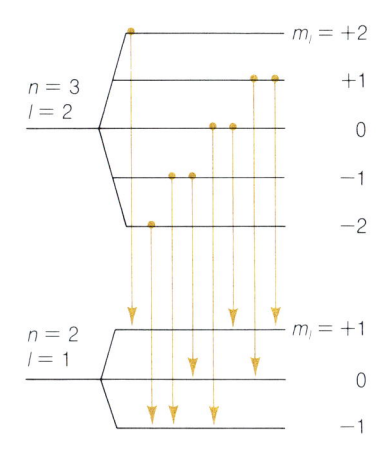

Figure 24.7

Magnetic quantum number and angular momentum. If the orbital quantum number is $l = 2$, the magnetic quantum number can take on 5 values, corresponding to the z-component $L_z = 0, \pm\hbar, \pm 2\hbar$. The magnitude of the total angular momentum is $L = \sqrt{l(l+1)}\,\hbar = 2.45\hbar$, which is greater than the maximum possible value of L_z.

Figure 24.8

Zeeman splitting. When the hydrogen atom is placed in a magnetic field, the energy levels corresponding to the $(2l + 1)$ different values of m_l are separated. Transitions between levels result in photons of several wavelengths.

from $-l$ to $+l$. For example, if $l = 2$, there are $(2l + 1) = 5$ values of m_l: $-2, -1, 0, +1, +2$. Notice that L_z is always less than L (unless $l = 0$). For example, if $l = 2$, $L = \sqrt{2(3)}\hbar = 2.45\hbar$, whereas $L_z = m_l\hbar$ can never be larger than $2\hbar$. Each of these m_l values corresponds to a different orientation of the angular momentum vector (Fig. 24.7). Ordinarily the orientation of the angular momentum vector is random in any external frame of reference, and the different values of m_l simply correspond to different relative orientations. When there is an external magnetic field, however, then the magnetic quantum number is related to the orientation of the angular momentum vector with respect to the direction of the field, usually defined to be the z-axis.

When a magnetic field is applied, there is a small potential energy due to the interaction between the electron's current and the field. Each energy level is split into $(2l + 1)$ closely-spaced levels, one corresponding to each allowed value of m_l. This splits the spectral lines into closely-spaced components (Fig. 24.8), a phenomenon called the **Zeeman effect**, that was discovered in 1896 by the Dutch physicist P. Zeeman. It is easy to see why there should be such an effect, since a moving electron represents an electric current, and we know that a magnetic field interacts with a current. In a collection of atoms, there will be a distribution of electron angular momentum orientations with respect to the magnetic field, so that a spectrum will show the separated lines. If there is no external field, the energy associated with each allowed value of m_l is precisely the same, and there is no splitting of levels. Further details are discussed in Special Problem 1 at the end of this chapter.

The fourth quantum number, m_s, is the **spin quantum number,** which bears some similarity to m_l, except that it can take on only the values $m_s = +½$ (spin up) and $-½$ (spin down) for the electron. The existence of spin was discovered from the fine structure splitting of hydrogen lines and from the relativistic modification of Schrödinger's quantum mechanics by P. A. M. Dirac. As an analogy, consider the motion of the earth, which rotates on its axis (spin) as it revolves around the sun (orbit). With each motion there is an associated angular momentum. Of course, the electron is not a localized spinning body. It is best to picture spin as an intrinsic property of the electron that has some similarities (but also some contrasts) to orbital angular momentum.

Like orbital angular momentum, spin angular momentum is a vector **S**, whose magnitude S and z-component S_z are quantized according to the usual rules:

$$S = \sqrt{s(s+1)}\hbar, \qquad s = \frac{1}{2} \qquad (24\text{-}6)$$

and

$$S_z = m_s\hbar, \qquad m_s = \pm\frac{1}{2}. \qquad (24\text{-}7)$$

Note that these quantization rules are the same as for L and L_z (Eq. 24–4 and Eq. 24–5), except that the electron has spin $s = ½$, so that $m_s = \pm½$. This is the primary difference between spin and orbital angular momentum: l can only be integral (0, 1, 2, . . .), whereas s is integral or half-integral ($s = ½$ for a single electron, but in general particles can have $s = ½$, 1, 3/2, 2, . . .).

Let us put these concepts together for the common hydrogen isotope, which consists of a single proton as the nucleus and an orbiting electron with spin. Recall that the basic energy levels are determined by n, which has n degenerate states of orbital angular momentum l ranging from $l = 0$ to $l = n - 1$. Each state (n,l) is split into $(2l + 1)$ quantum states (n,l,m_l) with m_l ranging from $-l$ to $+l$. These states can be further split by the effects of electron spin ($m_s = \pm½$) into quantum states specified by all four quantum numbers (n,l,m_l,m_s). Note that $s = ½$ for the electron is assumed, so we do not list it as an independent quantum number.

The proton nucleus also has spin $s = ½$, which may interact very weakly with the spin of the electron to produce the **hyperfine splitting** of spectral lines. In the simplest interpretation, the electron and the nucleus each have minute magnetic fields due to the tiny currents created by their spins. The energy state of the electron depends on whether its magnetic field is parallel ($m_s = +½$) or anti-parallel ($m_s = -½$) to that of the proton. The hyperfine splitting of energy levels is much smaller than the spacings of levels due to the different values of the principal quantum number or the fine-structure splitting, and very high-resolution spectrographs are needed to detect it.

The four quantum numbers and their allowed values are summarized in Table 24.1. It is noteworthy that in the nonrelativistic hydrogen atom the only

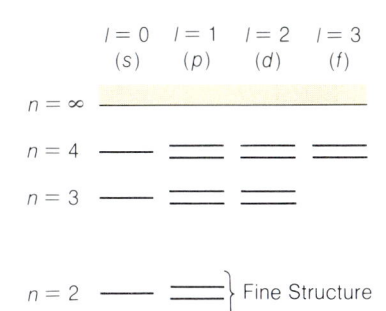

24.2 ATOMIC STRUCTURE

Table 24.1
Quantum numbers in atoms

Symbol	Quantum number	Possible values
n	Principal	$0, 1, 2, \ldots$
l	Orbital	$0, 1, 2, \ldots (n-1)$
m_l	Magnetic	$0, \pm 1, \pm 2, \ldots \pm l$
m_s	Spin	$\pm \frac{1}{2}$

quantum numbers that affect the energy of a given level are n and m_s (in the absence of an external magnetic field). That is, two states with the same n and m_s have the same energy, regardless of the values of l and m_l (Fig. 24.9). The electron wave functions for two such states are different, however, and so is the probability distribution. Therefore the electron clouds have different shapes for different values of l (Fig. 24.10).

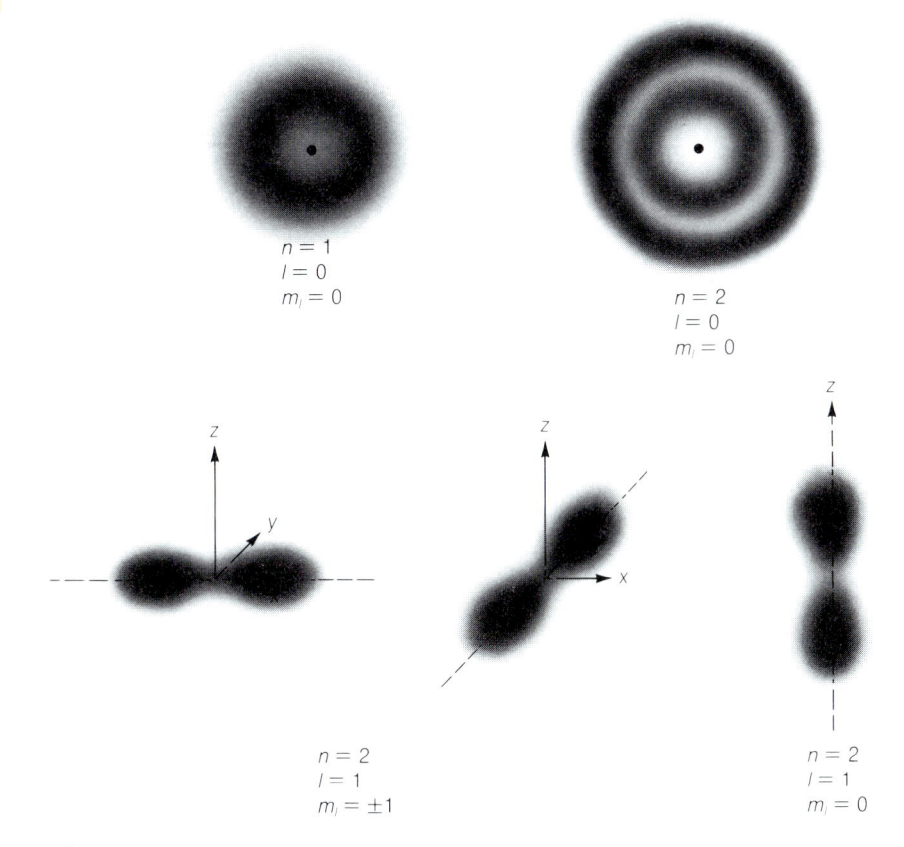

Figure 24.9

Energy levels and fine structure. In the hydrogen atom, each value of n has n values of l, all with approximately the same energy. The fine structure is evident in the doubling of the levels with l > 0.

Figure 24.10

Electron clouds for the n = 1, n = 2 states of hydrogen. The states with l = 0 are spherically symmetric, whereas those with l = 1 are not. The n = 2, l = 1, m_l = ±1 clouds lie in planes perpendicular to the z-axis, while the n = 2, l = 1, m_l = 0 cloud is symmetric about the z-axis.

Example 24.2 List the possible quantum states for an electron in the $n = 2$ level in a hydrogen atom.

There are two allowed values of the orbital quantum number l; they are $l = 0$ and $l = 1$. For $l = 0$, there is only one allowed value of m_l, which is $m_l = 0$. The value of m_s can be either $+\frac{1}{2}$ or $-\frac{1}{2}$. For $l = 1$, there are three possible values of m_l: $-1, 0, +1$. For each of these possibilities, m_s can be either $+\frac{1}{2}$ or $-\frac{1}{2}$. In all, there are eight possible states for the electron, summarized in the following table.

n	l	m_l	m_s
2	0	0	$+\frac{1}{2}$
2	0	0	$\frac{1}{2}$
2	1	-1	$+\frac{1}{2}$
2	1	-1	$-\frac{1}{2}$
2	1	0	$+\frac{1}{2}$
2	1	0	$-\frac{1}{2}$
2	1	$+1$	$+\frac{1}{2}$
2	1	$+1$	$-\frac{1}{2}$

The Exclusion Principle

In atoms with more than one electron, the possible energy states become much more numerous. In general, each value of n now has the potential to be split into several closely-spaced levels, even in the absence of a magnetic field. The reason for this is that interactions between several electrons are now important and the energy level now depends on the value of l, in contrast with hydrogen, in which the energy of a state for a given value of n is the same, regardless of the value of l.

The energy level structure for many complex atoms is determined experimentally from their spectra. From careful measurements of the spectral lines (including measurements of their energy-level splitting) it is possible to construct an energy-level diagram similar to those for hydrogen that were discussed in the previous chapter (see Fig. 24.11, for example).

In principle it is possible to write and solve the Schrödinger equation for all the electrons in a complex atom, but in practice it is complicated. Each electron now interacts not only with the nucleus, but also with the other electrons. Fortunately, there is a simple rule, first enunciated by W. Pauli, that governs the possible energy states in an atom, and which therefore greatly simplifies the mathematics. Called the **Pauli exclusion principle**, this rule states

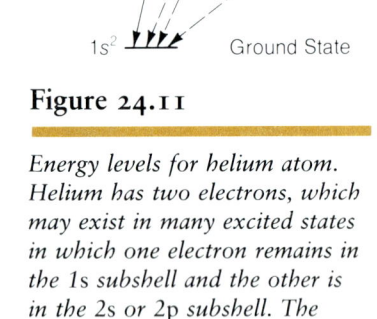

Figure 24.11

Energy levels for helium atom. Helium has two electrons, which may exist in many excited states in which one electron remains in the 1s subshell and the other is in the 2s or 2p subshell. The states are further separated into singlet states (electron spins opposite) and triplets (electron spins parallel). Allowed transitions are shown as solid arrows and forbidden transitions by dashed arrows.

No two electrons in an atom can occupy the same quantum state.

24.2 ATOMIC STRUCTURE

This means that no two electrons can have exactly the same combination of quantum numbers n, l, m_l, and m_s. This is a very general principle, and has been applied successfully on larger scales than single atoms. It also determines the behavior of entire classes of particles, known as fermions and bosons (discussed in Chapter 26).

We can determine the electron energy level structure of an atom by considering each electron independently. We recall that each electron tends to be in its lowest possible state, and then apply the exclusion principle as we assign quantum number combinations. The lowest possible value of the principal quantum number is $n = 1$. As we have seen, for $n = 1$ the only allowed value of the orbital quantum number is $l = 0$, which means that m_l must also be zero. We can still have $m_s = +\frac{1}{2}$ and $m_s = -\frac{1}{2}$, so there are two possible combinations of quantum numbers that are unique. Thus, an atom can have only two electrons in the $n = 1$ state at one time.

Now if we consider the next value of the principal quantum number, $n = 2$, we find that 8 unique combinations of quantum numbers are possible (Example 24.2). The value of l can be 0 or 1; for $l = 1$, we can have $m_l = -1, 0,$ or $+1$; and for each of the four allowed combinations of l and m_l we can have $m_s = +\frac{1}{2}$ or $m_s = -\frac{1}{2}$. Similarly, for $n = 3$, there are 18 possible electron states. In general, for every value of n, there are $2n^2$ possible states.

Thus, the exclusion principle makes it rather simple to deduce how many electrons are in each of the principal energy levels. To calculate the energy of each level is more difficult, requiring the solution of the Schrödinger equation. As we noted earlier, determining the energy levels for a complex atom is often done experimentally.

Example 24.3

List the quantum numbers of the ground states of helium ($Z = 2$) and carbon ($Z = 6$).

In helium there are two electrons. Both may go into the state ($n = 1$, $l = 0$), since one can have $m_s = +\frac{1}{2}$ and the other $m_s = -\frac{1}{2}$, satisfying the Pauli exclusion principle. The quantum numbers are indicated in the table.

Helium ($Z = 2$)

n	l	m_l	m_s
1	0	0	$+\frac{1}{2}$
1	0	0	$-\frac{1}{2}$

Carbon has six electrons. The first two can have ($n = 1$, $l = 0$) as with helium, but the additional electrons must go into higher states to satisfy the Pauli principle. Thus, the third and fourth electrons go into the ($n = 2$, $l = 0$) state, with $m_s = +\frac{1}{2}$ and $m_s = -\frac{1}{2}$, respectively. The remaining two electrons go into the ($n = 2$, $l = 1$) state. The choice of m_l and m_s is determined in practice by rules for how spins line up as successive electrons are added. The horizontal lines separate electrons having common values of (n,l).

	Carbon ($Z = 6$)		
n	l	m_l	m_s
1	0	0	$+\frac{1}{2}$
1	0	0	$-\frac{1}{2}$
2	0	0	$+\frac{1}{2}$
2	0	0	$-\frac{1}{2}$
2	1	1	$+\frac{1}{2}$
2	1	1	$-\frac{1}{2}$

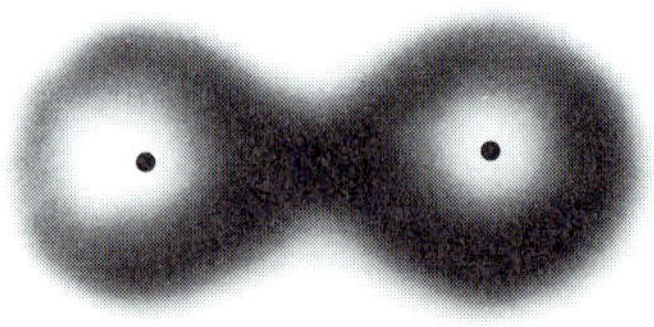

Figure 24.12

A diatomic molecule, H_2, in which two hydrogen nuclei are bonded together by the mutual orbit of their two electrons.

The Periodic Table

The identity of an atom is determined by the number of protons in its nucleus (that is, the **atomic number** Z), which is equalled by the number of electrons (unless ionization has occurred). As we have now seen, the electrons are distributed in energy levels according to the quantum mechanics principles we have just discussed. Thus, for each chemical element, the electron energy level structure in the ground state is fixed.

The number of electrons in the outermost energy state, and the strengths of their bonds to the atom, determine how an atom will react chemically. For example, a diatomic molecule (one consisting of two atoms bound together) forms when one or more electrons from one atom enter states in which they orbit both nuclei (Fig. 24.12). The likelihood that this process will occur when the two atoms encounter each other depends on how easily one atom can give up an electron, and how likely the other is to receive it. Thus, the chemical reactivity of an element is a function of the arrangement of electrons in its outermost energy state.

Chemists speak in terms of the **valence** of an atom, which is a designation of the number of electrons or electron spaces available in the outermost level (the electrons in the outermost level are called valence electrons). For example, a valence of $+1$ means that the atom can easily give up one electron to be shared with another atom in forming a molecule. Similarly, a valence of -2 means that an atom has two available spaces for electrons in its outermost level, and can therefore react easily with another atom that can give up one or two electrons (that is, an element with a valence of $+1$ or $+2$).

Without knowing much of atomic structure, the Russian chemist D. Mendeleev in the latter part of the nineteenth century arranged the known elements into a table according to their chemical behavior. This is the **periodic table**, in which the elements are arranged in rows and columns (Fig. 24.13). The elements are arranged across each row according to the type of reaction they undergo, and the rows are arranged vertically according to atomic mass. Since Mendeleev's time the number of known elements has increased, so the table

Periodic Table of the Elements*

Group I	Group II						Transition elements							Group III	Group IV	Group V	Group VI	Group VII	Group 0
H 1 1.0079 $1s^1$																			He 2 4.0026 $1s^2$
Li 3 6.941 $2s^1$	Be 4 9.01218 $2s^2$													B 5 10.81 $2p^1$	C 6 12.011 $2p^2$	N 7 14.0067 $2p^3$	O 8 15.9994 $2p^4$	F 9 18.9984 $2p^5$	Ne 10 20.179 $2p^6$
Na 11 22.9898 $3s^1$	Mg 12 24.305 $3s^2$													Al 13 26.9815 $3p^1$	Si 14 28.0855 $3p^2$	P 15 30.9738 $3p^3$	S 16 32.06 $3p^4$	Cl 17 35.453 $3p^5$	Ar 18 39.948 $3p^6$
K 19 39.0983 $4s^1$	Ca 20 40.08 $4s^2$	Sc 21 44.9559 $3d^14s^2$	Ti 22 47.88 $3d^24s^2$	V 23 50.9415 $3d^34s^2$	Cr 24 51.996 $3d^54s^1$	Mn 25 54.9380 $3d^54s^2$	Fe 26 55.847 $3d^64s^2$	Co 27 58.9332 $3d^74s^2$	Ni 28 58.69 $3d^84s^2$	Cu 29 63.546 $3d^{10}4s^1$	Zn 30 65.39 $3d^{10}4s^2$			Ga 31 69.72 $4p^1$	Ge 32 72.59 $4p^2$	As 33 74.9216 $4p^3$	Se 34 78.96 $4p^4$	Br 35 79.904 $4p^5$	Kr 36 83.80 $4p^6$
Rb 37 85.468 $5s^1$	Sr 38 87.62 $5s^2$	Y 39 88.9059 $4d^15s^2$	Zr 40 91.224 $4d^25s^2$	Nb 41 92.9064 $4d^45s^1$	Mo 42 95.94 $4d^55s^1$	Tc 43 (98) $4d^65s^1$	Ru 44 101.07 $4d^75s^1$	Rh 45 102.906 $4d^85s^1$	Pd 46 106.42 $4d^{10}5s^0$	Ag 47 107.868 $4d^{10}5s^1$	Cd 48 112.41 $4d^{10}5s^2$			In 49 114.82 $5p^1$	Sn 50 118.71 $5p^2$	Sb 51 121.75 $5p^3$	Te 52 127.60 $5p^4$	I 53 126.905 $5p^5$	Xe 54 131.29 $5p^6$
Cs 55 132.905 $6s^1$	Ba 56 137.33 $6s^2$	57–71‡	Hf 72 178.49 $5d^26s^2$	Ta 73 180.948 $5d^36s^2$	W 74 183.85 $5d^46s^2$	Re 75 186.207 $5d^56s^2$	Os 76 190.2 $5d^66s^2$	Ir 77 192.22 $5d^76s^2$	Pt 78 195.08 $5d^96s^1$	Au 79 196.967 $5d^{10}6s^1$	Hg 80 200.59 $5d^{10}6s^2$			Tl 81 204.383 $6p^1$	Pb 82 207.2 $6p^2$	Bi 83 208.980 $6p^3$	Po 84 (209) $6p^4$	At 85 (210) $6p^5$	Rn 86 (222) $6p^6$
Fr 87 (223) $7s^1$	Ra 88 226.025 $7s^2$	89–103‖	Rf 104 (261) $6d^27s^2$	Ha 105 (262) $6d^37s^2$	106 (263)	107 (262)		109 (266)											

Symbol — Cl 17 — Atomic number
Atomic mass — 35.453
$3p^5$ — Electron configuration

Lanthanide series‡

| La 57
138.906
$5d^16s^2$ | Ce 58
140.12
$5d^14f^26s^2$ | Pr 59
140.908
$4f^36s^2$ | Nd 60
144.24
$4f^46s^2$ | Pm 61
(145)
$4f^56s^2$ | Sm 62
150.36
$4f^66s^2$ | Eu 63
151.96
$4f^76s^2$ | Gd 64
157.25
$5d^14f^76s^2$ | Tb 65
158.925
$5d^14f^86s^2$ | Dy 66
162.50
$4f^{10}6s^2$ | Ho 67
164.930
$4f^{11}6s^2$ | Er 68
167.26
$4f^{12}6s^2$ | Tm 69
168.934
$4f^{13}6s^2$ | Yb 70
173.04
$4f^{14}6s^2$ | Lu 71
174.967
$5d^14f^{14}6s^2$ |

Actinide series‖

| Ac 89
227.028
$6d^17s^2$ | Th 90
232.038
$6d^27s^2$ | Pa 91
231.036
$5f^26d^17s^2$ | U 92
238.029
$5f^36d^17s^2$ | Np 93
237.048
$5f^46d^17s^2$ | Pu 94
(244)
$5f^66d^07s^2$ | Am 95
(243)
$5f^76s^2$ | Cm 96
(247)
$5f^76d^17s^2$ | Bk 97
(247)
$5f^96d^07s^2$ | Cf 98
(251)
$5f^{10}6d^07s^2$ | Es 99
(252)
$5f^{11}6d^07s^2$ | Fm 100
(257)
$5f^{12}6d^07s^2$ | Md 101
(258)
$5f^{13}6d^07s^2$ | No 102
(259)
$6d^{0}5f^{14}7s^2$ | Lr 103
(260)
$6d^15f^{14}7s^2$ |

*Atomic mass values averaged over isotopes in percentages as they occur on earth's surface.
**For many unstable elements, mass number of the most stable known isotope is given in parentheses.

Figure 24.13

is now more complex, but his basic arrangement has not changed. Normally the entry for each element includes the atomic weight as well as the atomic number Z. Also sometimes included is a notation showing the arrangement of electrons in the various levels.

We have seen how the exclusion principle governs the number of electrons in each principal quantum state, and how the quantum numbers of each electron is each state are determined. The energy of each state is primarily a function of the quantum numbers n and l, because there is no distinction on the basis of m_l unless a magnetic field is applied, and because the energy difference between states with different values of m_s is negligible for most practical purposes. Thus, the electrons of a complex atom can be grouped into energy levels so that within each group the electrons have the same value of n and l, and therefore the same energy (except for the very small difference corresponding to different values of m_s). The electrons having the same value of n are said to belong to the same **shell,** whereas those within a shell having equal values of l belong to the same **subshell.** The shells and subshells are designated by letters chosen during the early days of spectroscopy.

The lowest-lying ($n = 1$) shell is called the **K shell,** and has only 2 electrons in it, both having $n = 1$ and $l = 0$. Both belong to the same subshell, called the $1s$ subshell. The next shell ($n = 2$), is called the **L shell,** and it has two subshells, since there are two possible values of l. The $l = 0$ subshell is again called the $2s$ subshell, while the $l = 1$ subshell is the $2p$ subshell. There can be 2 electrons in the $2s$ subshell, and up to 6 in the $2p$ subshell, as we already know. The third shell ($n = 3$) is the **M shell,** with 2 electrons in the $3s$ subshell ($l = 0$, as before); 6 in the $3p$ subshell ($l = 1$); and up to 10 in the $3d$ subshell ($l = 2$). (The letters s, p, and d stand for sharp, principal, and diffuse appearances of the resulting spectra under some conditions. After d, the subshells are named in alphabetical order.) Table 24.2 summarizes the arrangement of shells and subshells, and the number of possible electrons. Elements with completely filled shells (and sometimes subshells) tend to be chemically stable; the total numbers of electrons for such elements are often called **magic numbers.**

The **configuration** of electrons in an atom is a designation of the subshells that are present and the number of electrons in them. In the standard notation, the leading numeral indicates the value of n; this is followed by a letter indicating the subshell, which has a superscript showing the number of electrons in that subshell. Thus, the configuration of the single electron of a hydrogen atom in its ground state is $1s^1$, because the electron is in the $n = 1$ state, with $l = 0$ (so it is the s subshell), and there is only 1 electron in this subshell, so the superscript is 1. For helium, the configuration is almost identical, except the superscript becomes 2: the ground-state configuration is $1s^2$. The next element is lithium, with 3 electrons. The configuration is $1s^2 2s^1$, indicating that there are 2 electrons in the s subshell of the $n = 1$ state, and 1 in the s subshell of the $n = 2$ state. For sodium, which has 11 electrons, the configuration is $1s^2 2s^2 2p^6 3s^1$, indicating that 2 electrons are in the s subshell of the $n = 1$ state; 2 are in the s subshell of the $n = 2$ state; 6 are in the p subshell of the $n = 2$ state; and 1 is in the s subshell of the $n = 3$ state.

The configurations of the inner subshells are often known because the subshells are full, so for many atoms the configuration of the outermost subshell is all that is given. Thus, for sodium we might say that the configuration is

24.2 ATOMIC STRUCTURE

Table 24.2
Electrons in shells and subshells*

Shell	(n)	Subshell (l)	Symbol	Maximum electrons in subshell	Total number of electrons in all shells
K	1	0	1s	2	2
L	2	0	2s	2	4
	2	1	2p	6	10
M	3	0	3s	2	12
	3	1	3p	6	18
	3	2	3d	10	28
N	4	0	4s	2	30
	4	1	4p	6	36
	4	2	4d	10	46
	4	3	4f	14	60

*Beyond Z = 18, the subshells fill out of order (4s before 3d, 5s before 4d, and so on). Consult the Periodic Table (Appendix 5 or Fig. 24.13) for details.

simply $3s^1$. This could also be written as (Ne) + $3s^1$, since neon has a closed shell, $1s^2 2s^2 2p^6$. If the inner shells are not filled (as discussed below), it may be necessary to give the configuration of the outermost two or three subshells.

An element whose outermost subshell is an *s* subshell, and which has just one electron in that shell, gives it up easily in chemical reactions (the lone outer electron is somewhat shielded from the positive charge of the nucleus by the electrons in the inner shells, so it is not tightly bound). Such elements are called **alkali** elements, and they form the far left column of the periodic table. Thus, in this column all the elements have configurations of ns^1, where *n* is the principal quantum number. As we have seen, sodium has this configuration, so it is an alkali, listed in the far left column. In the far right column, the elements have completely filled subshells, either $1s^2$ or np^6. This means that the electrons are tightly bound, and that the atom can not easily accommodate any additional electrons. These elements are therefore **inert,** because they do not generally react with other elements. In the second column from the right, however, the outermost subshell is one electron short of being filled (np^5; the valence is -1). These elements, called the **halogens,** react easily, accepting one electron in the process. Halogens react particularly easily with alkalis, forming stable products called **salts** (ordinary table salt, NaCl or sodium chloride, is in this category).

Generally, the sequence of increasing electron energy levels is consistent with the sequence of electron configurations. That is, the configurations of lowest electron energy are the 1s configurations, the next are the 1s2s configurations, next are the 1s2s2p configurations, and so on. This consistency does not hold up throughout, however. For example, we can continue to fill subshells in order through the 3p configurations, but then the next subshell to be filled is not the 3d configuration, as we might expect, because the 4s configuration has slightly lower energy than 3d. (The degree to which an inner shell "screens" outer electrons from the attractive force of the nucleus depends in part on the symmetry of the electron clouds, which depends in turn on the exact quantum state of the electrons and the number of electrons in a shell.) Therefore the subshells are filled out of sequence. In potassium (K), which has atomic number $Z = 19$, the 1s, 2s, 2p, 3s, and 3p shells are filled, accounting for 18 of the electrons, but the nineteenth electron is in the 4s subshell, not the 3d. Thus, potassium has a single electron in its outermost subshell, which is an s subshell, making potassium an alkali. The next element, calcium (Ca), has two 4s electrons. Following this, the 3d subshell begins to fill, so that element 21, scandium (Sc), has a single 3d electron in its outermost subshell.

In some atoms, the energy states associated with different subshells can be so close that the subshells overlap. For example, in element 24, chromium (Cr), it is energetically preferred for the configuration to be $1s^2 2s^2 2p^6 3s^2 3p^6 3d^5 4s^1$, so that the 4s subshell is one electron short of being filled, even though in elements 20 through 23 the 4s subshell is filled. In these atoms, the configurations of the outermost subshell and any unfilled subshells must be given to fully specify the arrangement of electrons; thus, for chromium the abbreviated notation is $3d^5 4s^1$. Some entire groupings of elements in the periodic table are characterized by this kind of overlapping of subshells. The **rare earth** or **lanthanide elements** have filled outermost $6s^2$ subshells and therefore react similarly, even though the configurations of the inner 4f and 5d shells vary quite a bit. This series of elements, as well as the **actinide elements** (each having a filled outermost $7s^2$ subshell) is usually listed separately along the bottom of the periodic table. In the center of the table is a large group called the **transition elements,** again having gaps in their inner shell configurations but similar outermost subshells (these occupy columns 3 through 12 of the table, having outermost s subshells, with either one or two electrons, and unfilled spots in an inner subshell).

The periodic table is a very useful tool, both for understanding atomic structure and for predicting the chemical behavior of an element. As we have seen, the table can be understood in terms of quantum mechanics, and can be partially constructed using the simple rules provided by the exclusion principle. To calculate energy levels in detail and to predict exactly the order in which subshells are filled requires difficult quantum mechanical calculations, particularly for the more complex elements.

Example 24.4   Write the electron configurations for the elements carbon ($Z = 6$) and silicon ($Z = 14$).

Both have Z smaller than 19, so we need not worry about subshells filling out of sequence. For carbon, there are 2 electrons in the s subshell of the K shell ($n = 1$), filling that shell; there are 2 electrons in the s subshell of the L shell ($n = 2$), and the remaining 2 are in the p subshell of the L shell. Therefore the configuration is $1s^2 2s^2 2p^2$ or simply $2p^2$ in abbreviated form.

Silicon has its K and L shells filled, accounting for 10 of its electrons, and the last 4 are in the ($n = 3$) M shell—2 in the s subshell and 2 in the p subshell. Thus, the configuration of silicon is $1s^2 2s^2 2p^6 3s^2 3p^2$, or $3p^2$ for short.

Example 24.5   The transition element cadmium (Cd) has $Z = 48$, and its configuration is $4d^{10}5s^2$. Summarize the number of electrons in each shell and subshell.

We start with the K shell and work outward, assuming each shell in order is filled until we reach those listed in the configuration. Therefore, the K shell has 2 electrons, both in the $1s$ subshell; the L shell has 8 electrons, two in the $2s$ subshell and 6 in the $2p$ subshell; and the M shell has 18 electrons, 2 in the $3s$ subshell, 6 in the $3p$ subshell, and 10 in the $3d$ subshell. In the N shell ($n = 4$), there are 2 electrons in the $4s$ subshell, 6 in the $4p$ subshell, and 10 in the $4d$ subshell, filling that subshell. Next, however, instead of being in the $4f$ subshell, the remaining 2 electrons are in the $5s$ subshell.

24.3 Changes in Quantum Numbers: Atomic Spectra

We have already learned a good deal about the formation of spectral lines by atoms when an outer electron makes a transition between energy states. Now we see that when a transition occurs, one of the quantum numbers of the electron changes and alters the energy state of that electron. So far we have only discussed a rather limited case, where the principal quantum number n changed for a single, outer electron. In this section we will examine other possible kinds of transitions, including some that can occur in molecules.

Energy Level Diagrams and Selection Rules

We already have some familiarity with the concept of an energy level diagram, in which the energies associated with electron states are depicted as horizontal lines. Now we are armed with a new notation in which we can specify the state (n,l) of an electron. In a typical energy level diagram, the configurations of the states are given (Fig. 24.14). For a complex atom with many levels, some of which are close together, it is often necessary to display the levels side by side.

So far when we have discussed the formation of spectral lines, we have said little about the probability that a transition between a given pair of energy levels will occur. This did come up indirectly when we mentioned metastable excited states, but there is more to be said.

Let us consider emission lines first, since their formation does not require a source of photons. When an electron is in an excited state, its natural tendency is to drop to a lower state, giving up a photon in the process. The quickness with which it does so, however, depends on the probability of the transition, which depends in turn on the relationship between the wavefunctions associated with the upper level the electron is in and the lower one to which it might jump. A detailed calculation is required to find the exact probability for a given transition, but it turns out that a few relatively simple rules can be

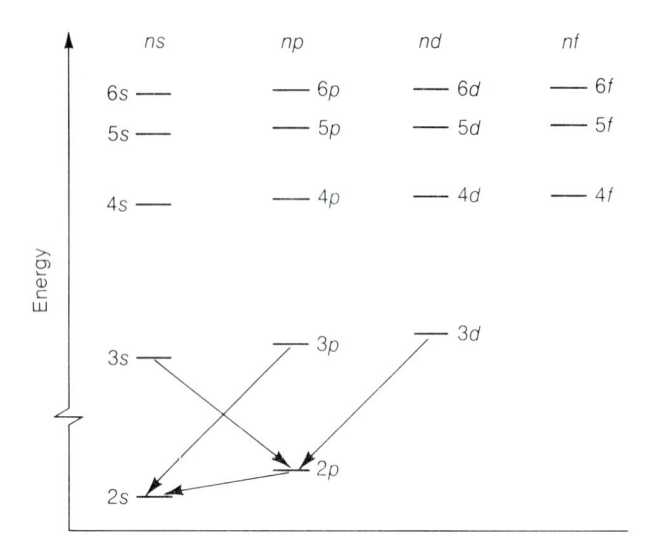

804 CHAPTER 24 ESSENTIALS OF QUANTUM MECHANICS

Figure 24.14

Energy levels for a lithium atom. Lithium has three electrons, with ground state $1s^2 2s^1$ (labelled 2s here). The excited states have the 2s electron raised to a higher level. Allowed transitions between several low states are shown by arrows.

applied; these are called **dipole selection rules,** because the photon emitted in the transition between upper and lower quantum states may be viewed as radiation from an oscillating dipole charge distribution in the electron cloud.

These selection rules are stated in terms of the changes in quantum numbers of the electron that jumps:

$$\Delta l = \pm 1 \tag{24-8a}$$

$$\Delta m_l = 0, \pm 1 \tag{24-8b}$$

$$\Delta s = 0. \tag{24-8c}$$

In other words, the orbital quantum numbers must change by ± 1 ($3p \to 2s$ or $2p \to 1s$, for example, are allowed, whereas $3d \to 2s$ or $2s \to 1s$ are not), and the total spin of the electron cloud must remain unchanged. With these and a few other rules, it is possible to decide which transitions between states in an energy level diagram are likely, and therefore which lines will appear in the spectrum. The same rules apply for upward transitions as for downward ones, so lines that are forbidden by dipole selection rules are unlikely to appear as either absorption or emission lines.

Example 24.6 Draw an energy-level diagram for the $n = 1$, 2, and 3 states (n,l) of hydrogen. Neglect the fine-structure splitting due to spin effects. Show with arrows all dipole allowed transitions.

24.3 CHANGES IN QUANTUM NUMBERS: ATOMIC SPECTRA

Since there is only one electron with spin $s = \frac{1}{2}$, the selection rule $\Delta s = 0$ is automatically satisfied, and we need only look for transitions with $\Delta l = \pm 1$; that is, between neighboring columns. Figure 24.9 shows the levels (n,l). The allowed transitions are: $3s \to 2p$, $3p \to 1s$, $3p \to 2s$, $3d \to 2p$, and $2p \to 1s$. Note that the $2s$ state cannot decay to the ground ($1s$) state via dipole radiation because $\Delta l = 0$ (it decays via a much slower process).

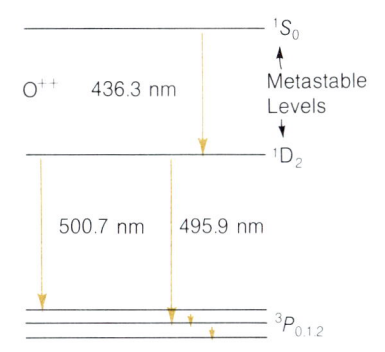

Figure 24.15

Formation of a forbidden emission line by twice-ionized oxygen. Collisions in interstellar gas clouds excite electrons to metastable levels which eventually decay by emitting photons in the visible and infrared wavelengths. (The latter arise from the transitions between fine-structure levels in the ^3P state).

O^{++} 436.3 nm $^1S_0$
Metastable Levels
$^1D_2$
500.7 nm   495.9 nm
$^3P_{0,1,2}$

The term **forbidden line** is used in spectroscopy, but in fact any transition has some probability of occurring. The so-called forbidden lines are never seen in the spectra of gases under ordinary conditions in the laboratory, because some other process will normally change the state of the electron before there is time for the excited state to radiate and the spectral line to form. An electron in a metastable excited state will often absorb a photon and move to a higher state, or be forced into some other state by a collision, before it has time to make a spontaneous downward transition. Similarly, an electron is unlikely to make a forbidden upward transition because it will normally absorb a photon corresponding to some other, more probable transition before it makes the improbable one.

There are situations in which forbidden lines are observed, however. The first was during the 1920s when unidentified emission lines were observed from certain hot gas clouds in space. These lines were initially thought to result from a new element called "nebulium." It was eventually established, however, that these gas clouds contained twice-ionized oxygen ions (oxygen with six electrons instead of the usual eight) which were being bombarded by energetic electrons. In the resulting collisions, the oxygen ions are excited into metastable levels 2 eV to 5 eV above the ground state, from which they slowly decay (the lifetimes of these states are about 30 s) by emitting forbidden photons (Fig. 24.15). The density in the gas clouds is so low that the forbidden transitions actually occur more frequently than collisions between ions, so forbidden lines appear in the emission spectra of these clouds.

In another very important case for astronomy, the strongly forbidden transition between the $m_s = +\frac{1}{2}$ and the $m_s = -\frac{1}{2}$ levels of the ground state of hydrogen is observed. This transition is so improbable that the lifetime of the electron in the upper of the two states is several million years, but there are so many hydrogen atoms in space, and the average density is so low, that the emission line resulting from the transition is readily observed. Its wavelength is quite long, 21.1 cm, in the radio portion of the electromagnetic spectrum, because the energy difference between the two levels is very small (9.4×10^{-25} J $= 5.9 \times 10^{-6}$ eV). In this case, the upper state is populated by occasional collisions between atoms.

Fluorescence and Phosphorescence

When an electron is in an excited state and spontaneously moves back to the ground state, it is most likely to do so in a series of steps, rather than dropping all the way down in a single transition (Fig. 24.16). If the electron was excited by absorbing a photon, then the net effect is to convert a single photon into

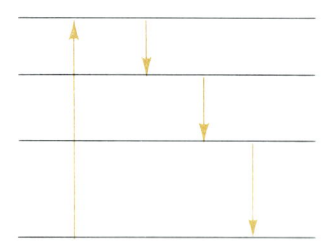

Figure 24.16

Fluorescence. A high-energy photon excites an electron to the uppermost level. Then the electron drops back to the ground state in a series of steps and emits a photon each time.

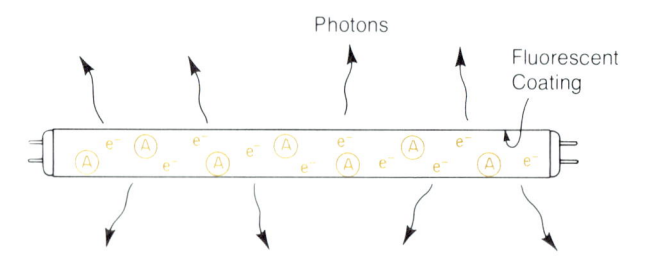

Figure 24.17

The fluorescent tube. Electrons accelerated inside the tube by the applied voltage collide with gas atoms and excite them to emit ultraviolet photons. These photons strike the fluorescent coating and create secondary photons in the visible part of the spectrum.

two or more of lower energy. The wavelengths of the new photons must therefore be longer than that of the initial photon that caused the excitation. This conversion of short-wavelength photons into longer-wavelength ones is called **fluorescence**.

Probably the most familiar form of fluorescence to most people is the remarkable effect of using "black lights", which are simply ultraviolet lamps. Many common substances, such as articles of clothing, appear to glow with a luminous deep violet color. Certain minerals (including one called fluorite) fluoresce strongly when exposed to ultraviolet light. The nature of the fluorescence from a mineral or crystal can help in identifying it or the impurities in it. The fluorescence from minerals can have any color, but always the color is remarkable for its purity and depth. Certain paints use materials that fluoresce under ordinary lighting conditions to produce a glowing effect.

As might be expected, fluorescence is used in a fluorescent light tube (Fig. 24.17). An applied voltage accelerates electrons inside the tube so that they collide with gas atoms, causing excitations that lead to the production of ultraviolet photons. These photons in turn excite electrons in a fluorescent coating on the inside of the tube, causing it to glow in visible wavelengths. This mechanism is quite distinct from that of an ordinary incandescent bulb, in which a resistive filament is heated and glows.

The process known as **phosphorescence** is essentially the same as fluorescence, except that the emission of light continues for a while after the source of exciting photons has been turned off. This happens if an excited electron must pass through a metastable state on its way back to the ground state. It lingers in the metastable state for as long as a second or more before resuming its journey with the emission of a photon. Thus, secondary photons will be emitted for a while after the excitation stops.

Phosphorescence is commonly used in labelling watch dials, because a weak glow will persist for some time after the watch has been exposed to light. Even if the lifetime of the metastable state is only a few seconds, some excited electrons will persist for minutes or even an hour after exposure to light, so the watch dial will continue to glow for some time.

Inner-Shell Transitions: X-Ray Emission

The transitions that produce spectral lines in visible and ultraviolet wavelengths generally involve changes in the principal quantum number of an electron in the outermost subshell of an atom's structure. This is because the outer electrons are effectively screened from the positive electric charge on the nucleus, so they are not tightly bound. The energy differences between levels tend to be of the order 10^{-19} to 10^{-18} J, or a few eV, corresponding to photons whose wavelengths are in the ultraviolet or visible part of the spectrum.

The electrons in the inner shells are much more tightly bound to the nucleus, however, so that as electrons make transitions between these shells, the photons emitted or absorbed have very high energies and short wavelengths. Normally the inner shells are filled, so there are no vacancies into which an electron can move, and transitions of inner-shell electrons do not occur. In a very hot gas exposed to x-ray photons, however, an inner-shell electron can be knocked out, leaving a vacancy that is soon filled by an outer electron that drops down. As it does so, it can emit a photon whose energy is equal to the difference in energy between the inner and outer shells, typically several hundred to several thousand eV. The wavelengths of these photons are in the x-ray portion of the spectrum.

Very hot gases contain atoms that have lost many electrons in collisions with energetic electrons. This state of ionized matter is called a **plasma**. Collisions between ions and energetic electrons can also produce highly excited energy levels of these ions, which decay to the ground state by emitting x-ray photons. Thus, very hot plasmas (temperatures 10^6 K and higher), or gas exposed to x-rays which ionize inner-shell electrons, will both produce an x-ray emission spectrum.

It is possible, using a simple approximation, to estimate the energy, hence the wavelength, of a transition from $n = 1$ to $n = 2$ for an inner-shell electron. The electrons in the innermost subshell (the 1s subshell of the K shell) are essentially unshielded from the positive charge on the nucleus. Therefore, it is not too poor an approximation to apply the Bohr model to these electrons. It turns out to work fairly well to assume that the binding force is proportional to $(Z - 1)^2$ rather than Z^2, because there are two electrons in the innermost subshell, and the negative charge on one of them in effect cancels one unit of positive charge on the nucleus as "seen" by the other electron. The Bohr relation (Eq. 23–13) for photon wavelength can now be modified by substituting $(Z - 1)$ for Z:

$$\frac{1}{\lambda} = R(Z - 1)^2(1 - 1/2^2) = 0.75R(Z - 1)^2 \qquad (24\text{–}9)$$

where R is the Rydberg constant, and where we have explicitly set $n = 1$ for the lower state and $n' = 2$ for the upper, since the approximation is valid only for transitions between these two states. A commonly-observed x-ray emission line appears in the spectrum of molybdenum (Mo) that is bombarded by high-energy electrons; this line is due to the transition of an L-shell electron into the K shell, following the removal of a K-shell electron by a collision. Our simplified theory (substituting $Z = 42$ into Eq. 24–9) predicts $1/\lambda = 1.38 \times$

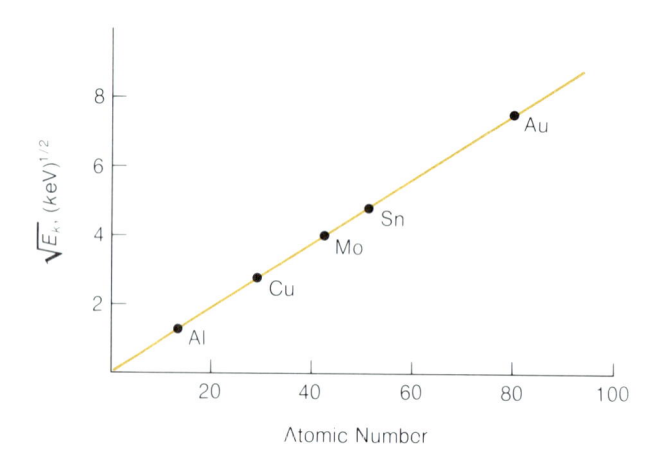

Figure 24.18

The relation between Kα transition energy and atomic number. A simplified adaptation of the Bohr theory to transitions between the L and K shells predicts that the transition energy E_K is proportional to Z^2, which is verified by this experimental plot.

10^{10} m^{-1}, or $\lambda = 7.23 \times 10^{-11}$ m $= 0.0723$ nm. The observed wavelength of this line is 0.071 nm.

The spectral line produced by the transition between the *L* shell and the *K* shell is usually called the *K*α line for a given element. The energy associated with this transition, as we have seen, is proportional to $(Z - 1)^2$. This was discovered in 1914, when a plot was made of the square root of the transition energy $\sqrt{E_K}$ versus *Z*, and a straight line was the result (Fig. 24.18). Such a plot, with observed values of E_K, was used by H. Mosely to determine the values of *Z* for several elements.

It is worth noting that when a gas is hot enough to produce x-ray spectral lines, the gas is normally a very highly ionized plasma. If collisions are energetic enough to free the tightly bound inner-shell electrons, they are surely energetic enough to free many of the less tightly bound outer electrons also. For elements of high *Z*, the wavelengths of the x-ray spectral lines do not depend very strongly on the state of ionization, because the energies of the inner levels are not strongly affected by the loss of some of the electrons in the outer shells. For small *Z*, however, most or all of the outer shell electrons may be lost at 10^6 K temperatures, and then the x-ray line spectrum depends more strongly on the state of ionization.

Example 24.7 Estimate the wavelength of the *K*α line of oxygen ($Z = 8$).

This is a straightforward application of Eq. 24–9:

$$\frac{1}{\lambda} = 0.75R(Z - 1)^2 = (0.75)(1.097 \times 10^7 \text{ m}^{-1})(7^2) = 4.03 \times 10^8 \text{ m}^{-1}$$

24.3 CHANGES IN QUANTUM NUMBERS: ATOMIC SPECTRA

Thus, the wavelength is

$$\lambda = 2.5 \times 10^{-9} \text{ m} = 2.5 \text{ nm.}$$

One of the x-ray emission lines from the sun's corona is produced by oxygen ions that have lost six electrons, leaving only two; the wavelength of this line is 2.2 nm, not far from our calculated value. Another observed coronal line is due to oxygen that has lost seven electrons, leaving only one. The wavelength of this line is 1.9 nm, farther from the value we have found here, but precisely what we would calculate using the Bohr model (Eq. 23–13) with $Z = 8$, since there is no shielding effect for this hydrogen-like ion.

Infrared and Radio Spectral Lines

Physical processes can create spectral lines in any portion of the electromagnetic spectrum. We have already learned how atoms can create visible and ultraviolet lines as well as x-ray lines. Furthermore, we have found that infrared spectral lines arise when transitions occur between relatively high excited levels, as in the Paschen and Brackett series of hydrogen, discussed in the previous chapter. Similarly, most other elements have infrared spectral lines.

Infrared lines are also produced when electrons make transitions between fine-structure levels (Fig. 24.15); that is, states with the same values of n, l, and s but different values of the total angular momentum $(l + s)$. These transitions are forbidden because selection rules call for l to change in an allowed transition, but in low-density gases, such transitions do take place via other radiative processes. Typically a collision causes an electron to jump to an excited fine-structure level, and then, given enough time before an additional collision takes place, a spontaneous downward transition will occur. The energy differences between fine-structure levels are typically a fraction of an eV, corresponding to wavelengths in the infrared portion of the spectrum.

Let us now see how radio spectral lines are formed. As in infrared lines, radio lines can represent changes in the principal quantum number or of fine structure levels. First, to see how changes in n can involve photons in the radio portion of the spectrum, consider a transition between two very high energy levels in an atom. Suppose, for example, that a transition occurs between the $n = 101$ state and the $n = 100$ state in hydrogen. According to the Bohr formula, the inverse wavelength is given by

$$\frac{1}{\lambda} = R(1/100^2 - 1/101^2) = 21.61 \text{ m}^{-1}$$

and the wavelength is 0.046 m = 4.6 cm. This is in the radio portion of the electromagnetic spectrum.

It is unlikely to find absorption lines due to such transitions, because to populate such highly excited states requires imparting just enough energy to the electron to excite it, but not so much that ionization occurs, which only takes a little more energy (just 0.0013 eV in the example just cited). Furthermore, the lifetime of an electron in such an excited state is very short, so that in a collection of atoms very few at any moment would be in a state where a 4.6 cm photon could be absorbed.

It is quite easy to produce emission lines between very high-lying states, however. In any ionized gas, there are always electrons and ions recombining with each other, and the electron will probably be captured in a very highly excited state when this occurs. It then drops down to the ground state rapidly, but, because of the relative transition probabilities, it will often do so in a series of small steps, with $\Delta n = 1$. Therefore these radio emission lines are often called **recombination lines.**

A second mechanism for producing radio emission lines has already been mentioned; this is the 21 cm spin-flip transition in hydrogen atoms, created when the electron spontaneously changes its spin quantum number from $m_s = +½$ to $m_s = -½$. Similar transitions can occur in other elements, but the conditions are rarely right for observing them. Because the transition is strongly forbidden, a large quantity of atoms must be present for the line to be strong, because at any moment only a very small fraction of the excited atoms will emit. Hydrogen is by far the most abundant element in the universe, so the 21-cm line is observed, but the only other element in which a spin-flip transition has been observed is carbon, the fourth most abundant element (after hydrogen, helium, and oxygen).

Radio spectral lines can also be formed by molecules, which have energy levels that lie very close together. This is discussed in the next section.

24.4 Molecular Bonds and Molecular Spectra

Since most matter on the Earth is composed of molecules rather than individual atoms, we now complete our chapter on quantum mechanics with a discussion of molecular bonds and spectra. Quantum theory applies to both these areas, and both play major roles in chemistry, biology, medicine, and other areas of science.

Molecular Bonding

When two atoms approach one another, the electron cloud of each is distorted by the electrostatic forces of the protons and electrons in the other. At close distances, the electrons can orbit both nuclei, resulting in a sharing of electrons (an electron cloud that is shared by both nuclei is called a **molecular orbital**). Alternatively, an electron can be transferred from one atom to the other. These two possibilities correspond to the two basic types of strong chemical bonds: covalent and ionic. As we shall see, most bonds are intermediate between these two extremes.

A **covalent bond** is formed by sharing electrons between two nuclei. We can show how these bonds form by discussing the simplest molecule, hydrogen (H_2), consisting of two hydrogen atoms bonded together. This molecule has two nuclei, each consisting of a single proton, and two electrons forming a covalent bond. The electron spin orientations are crucial in forming the bond. As illustrated in Fig. 24.19, the electron probability distributions for parallel spins and antiparallel spins are dramatically different. When the two electrons'

24.4 MOLECULAR BONDS AND MOLECULAR SPECTRA   811

spins are parallel (both $m_s = \frac{1}{2}$), the Pauli exclusion principle ensures that they do not occupy the same quantum state by keeping them as far apart as possible; no bond is formed. On the other hand, when the spins are antiparallel ($m_s = \frac{1}{2}$ and $m_s = -\frac{1}{2}$), the electrons can occupy different quantum states in the 1s sublevel, and a covalent bond is formed as the electron cloud concentrates between the two nuclei. The bond is simply the electrostatic attraction of the two positively charged nuclei toward the negatively charged electron cloud in between. Note that quantum effects, including the exclusion principle, are key in understanding the covalent bond. Common examples, in addition to H_2, are the diatomic molecules O_2 and N_2, and to a partial degree the more complex species CO, H_2O, and CO_2.

The pure covalent bond is a nearly equal sharing of electrons between two nuclei. In contrast, the **ionic bond** results from an unequal distribution of electrons. For example, in a molecule of table salt (sodium chloride, NaCl), the valence electron of the sodium atom is "stolen" by the chlorine atom to form an ionic bond. A glance at the periodic table helps illustrate how this works. Sodium is an alkali, with atomic structure $(1s^2 2s^2 2p^6)3s^1$, with an extra 3s electron outside a closed shell; chlorine is a halogen $(1s^2 2s^2 2p^6 3s^2)3p^5$, lacking one 3p electron to completely fill its outer shell. Thus, when the sodium atom gives up its 3s electron to the chlorine atom, both atoms can fill their outer shells. To really see why the sodium and chlorine atoms interact in this way requires a quantum mechanical calculation of the energies. It turns out that the molecule has a lower energy than the sum of the energies of the individual atoms, and therefore the reaction is energetically favored when the two atoms get together. As a result of the electron transfer, the chlorine atom has a net negative charge, and the sodium atom has a net positive charge; hence the name ionic bond. Again, it is the electrostatic force between the two ions (Na^+ and Cl^-) that holds the molecule together. The role of quantum mechanics lies in understanding why closed shells are energetically favorable.

As mentioned earlier, most molecules involve aspects of both covalent and ionic bonds, particularly when the atoms are different (CO and H_2O, for example). The shared electrons in these molecules spend more time near one atomic nucleus than the other, resulting in a **polar molecule,** with a partially ionic character. One part of a polar molecule has an excess positive charge and another part has an excess negative charge, which gives the molecule a charge separation (that is, a **dipole moment**).

The covalent and ionic bonds are the extremes of strong chemical bonds. Weaker bonds form from residual electrostatic attractions in polar molecules, or from polarization of electron clouds when two atoms are brought close together. The strength of a bond is best characterized by the potential energy stored in the electrostatic fields. This energy is called the **bond energy,** and represents the energy that must be put into the molecule to break the bond. For strong covalent and ionic bonds, the bond energies are between 1 and 5 eV; for weak bonds the energies range from 0.03 to 0.3 eV. From these energies we can estimate the latent heats of vaporization or fusion.

Weak bonds are important in biology and medicine. A bond energy $E_b = 0.03$ eV corresponds to a temperature $T = E_b/k = 350$ K. Thus, collisions between molecules moving at moderate speeds can break these weak bonds and open up new reaction pathways such as biochemical reactions of the very large molecules involved in enzyme reactions, proteins, and DNA replication.

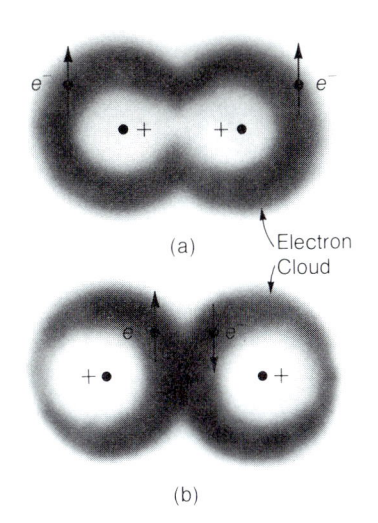

Figure 24.19

Formation of a covalent bond. This shows the electron probability distribution for two atoms (a) when the electron spins are parallel and there is no bond; and (b) when the electron spins are antiparallel, and a covalent bound is formed by the electron cloud concentrated between the two nuclei.

FOCUS ON PHYSICS

Bonds in Biology

Explaining the nature of the chemical bond was one of the first practical uses of quantum mechanics. By donating or sharing electrons, atoms can form stable molecules held together by electrostatic forces. These strong chemical bonds are either **covalent** or **ionic**, depending on the degree to which the electrons are shared equally or unequally.

The primary characteristics of bonds are their strength and their length. The bond strength is usually expressed as the binding energy needed to break the bond, which is the same as the depth of the potential energy well (Fig. 24.21). The bond length, typically between 0.5 and 1.5 Å, is given by the separation between the two atoms when they are in equilibrium at the bottom of the potential well. The energies of strong covalent or ionic bonds are between 1 and 5 eV (Table 24.1.1), and are often expressed in kcal/mole. One kcal is 4,180 J or 2.61×10^{22} eV and one mole is 6.022×10^{23} atoms, so 1 kcal/mole corresponds to a bond energy of 0.0433 eV/bond.

Two other characteristics of molecular bonds are the **valence**, equal to the maximum number of bonds an atom can make, and the **bond angle** between two bonds from a given atom. A single pair of shared electrons is known as a single bond. Atoms with valences greater than one can form single, double, or triple bonds, indicated in drawings by one, two, or three lines between atoms. For instance, oxygen has a valence of 2 (it needs two electrons to fill its outer 2p shell) and can form at most two covalent bonds, as in the water molecule H_2O. Carbon, with a valence of 4, needs four electrons to close its 2p shell. When it forms four single covalent bonds, as in the methane molecule CH_4, the bonds form a tetrahedron with bond angles of

Table 24.1.1

Bond*	Energy (eV)
H – H	4.48
C – C	3.53
N – N	1.66
O – O	1.51
C – N	3.00
O – H	4.77
C – O	3.53
C = C	6.32
C = N	5.85
N = N	4.23
C = O	7.49
O = O	4.16
C ≡ C	8.31
C ≡ O	11.08
N ≡ N	9.74

*Single line denotes single bond; two and three lines denote double and triple bonds.

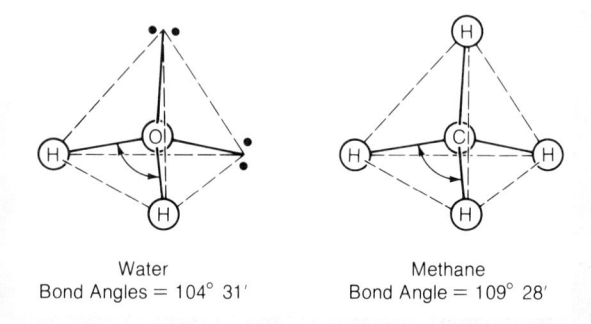

Figure 24.1.1

The molecules of water (H_2O) and methane (CH_4), showing characteristic bond angles.

Water
Bond Angles = 104° 31'

Methane
Bond Angle = 109° 28'

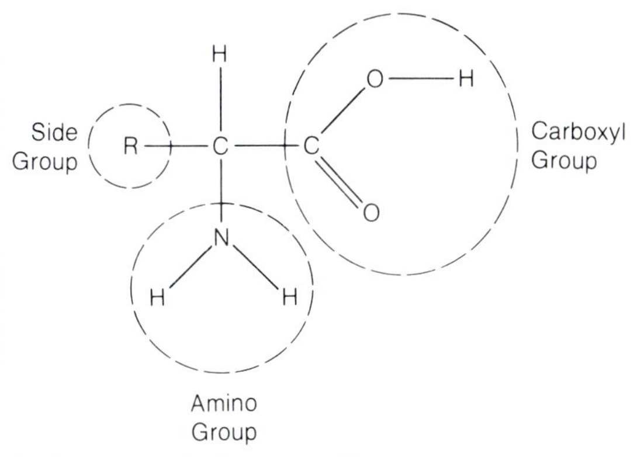

24.4 MOLECULAR BONDS AND MOLECULAR SPECTRA

Figure 24.1.2

Chemical structure of an amino-acid, showing the basic amino group (NH₂), carboxyl group (COOH), and the side group (R), which may be one of 20 variations.

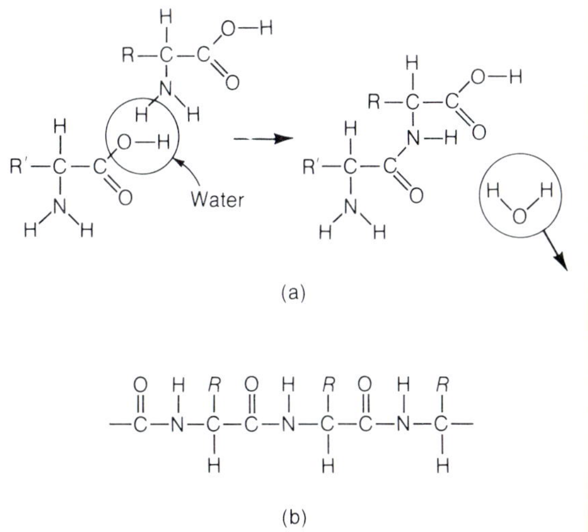

Figure 24.1.3

Peptide bond is formed when a water molecule (H₂O) splits off (a) and a link is formed between the NH₂ group of one amino-acid and the COOH group of the second. A polypeptide chain consists of hundreds of such links between amino-acids (b).

109° (Fig. 24.1.1). Bond angles play an important role in how groups of molecules combine. Double and triple bonds are quite rigid and permit little rotation of groups of atoms about the bond axis.

In addition to covalent or ionic bonds, atoms also form weak bonds due to the attraction of the dipole moments of molecules. The dipoles can be permanent (Table 14.2), or they can result from the charge separation that occurs when a polar molecule's electron cloud approaches an electrically balanced (non-polar) molecule. These weak bonds are known as **Van der Waals bonds,** and if one of the atoms involved is hydrogen, they are called **hydrogen bonds.** Biochemists have discovered that weak bonds are responsible for holding together the long chains of molecules in hemoglobin (the oxygen-carrying agent in blood). Hydrogen bonds are responsible for the cross links in the double helix of DNA (the molecule responsible for reproduction and the genetic code). These **macromolecules** or chains of molecules have molecular weights ranging from 100,000 into the millions.

The structure of covalent and hydrogen bonds is of utmost importance in biology. Nearly all the molecules on which life is based are formed from just six atoms: carbon (C), nitrogen (N), oxygen (O), hydrogen (H), phosphorus (P), and sulfur (S). Proteins, which are the basis of most enzymes, cells, and DNA, are macromolecules formed from smaller sub-units of nitrogen-containing molecules called **amino acids** (Fig. 24.1.2). The symbol R is called the **side group** and may be a single hydrogen atom (H), a methyl group (CH₃) or a more complicated group. There are 20 common amino-acid molecules, whose names are given three-letter abbreviations (*ala* for alanine, *gly* for glycine, etc.) The amino acids can condense into long chains, linked by covalent bonds between the —NH₂ group of one amino-acid and the —COOH group of another. This link is called a **peptide bond** (Fig. 24.1.3), and the proteins formed by hundreds of such links are called **polypeptide chains.**

Each protein differs in the number and order of amino-acid links. The primary structure is determined by the linear sequence of C, N, or O

(continued)

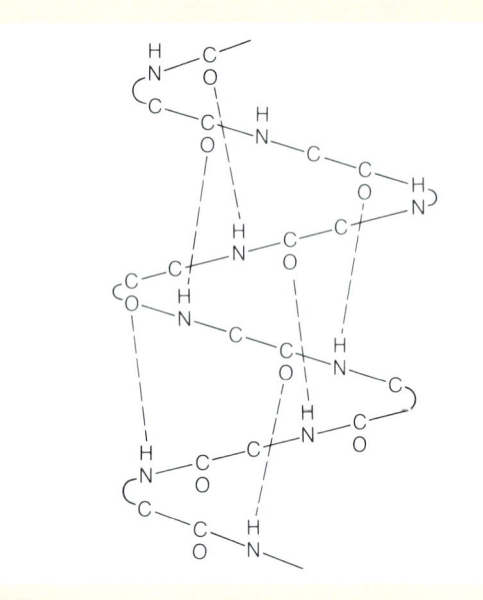

Figure 24.1.4

The alpha helix of a protein, showing the hydrogen bonds (dashed lines). The side groups are not shown for clarity.

atoms which form the backbone of the polypeptide chain. Finding the **amino-acid sequence** of protein molecules in tissue cells can give clues to evolutionary relationships between species. The secondary structure of the protein is determined by the shape of the polypeptide chain, which is set by the bond angles between the C, N, and O atoms in the backbone of the chain. One of the most common shapes is the alpha-helix (Fig. 24.1.4), first proposed by the American chemist Linus Pauling on the basis of bond angles and x-ray crystallography (Chapter 20). The amino-acids are arranged along a spiralling helix, and the structure is stabilized by cross-links, weak hydrogen bonds between H and N or O atoms in the peptide groups along the spiral. The pattern repeats itself every five turns over a distance of 27 Å, and there are 3.6 R-groups per turn (18 R-groups per pattern).

We conclude our discussion with DNA, the car-

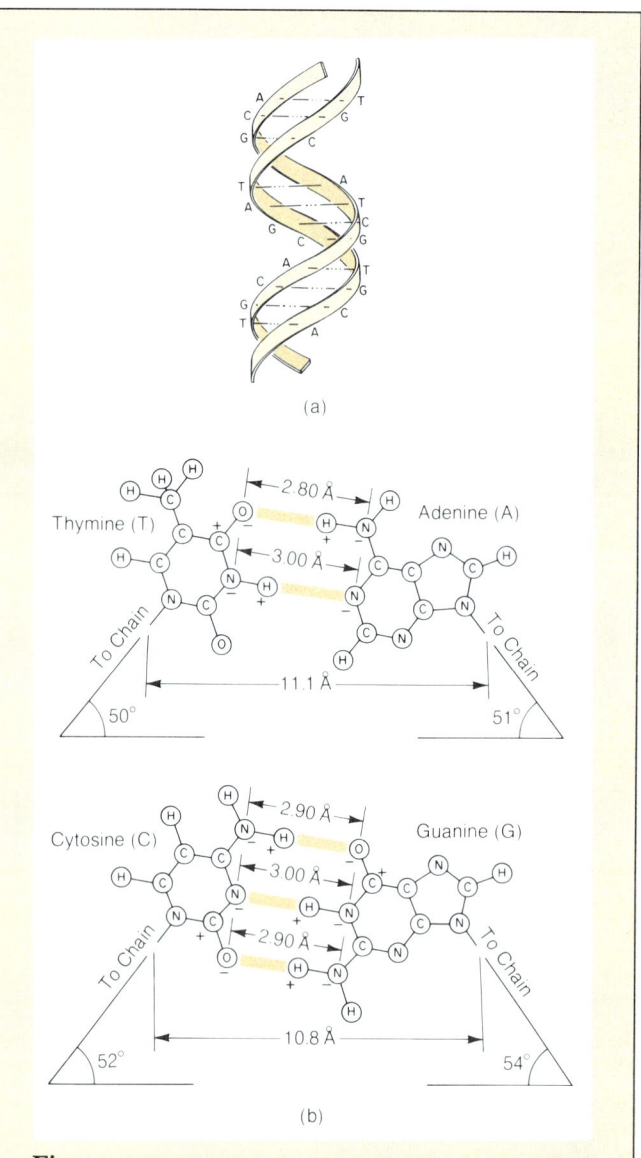

Figure 24.1.5

(a) A portion of the DNA double helix, showing matching base pairs (A–T and G–C). (b) A view of the hydrogen bonds between the base pairs, showing bond lengths in Ångstroms. (After J. D. Watson, Molecular Biology of the Gene, 3rd ed., W. A. Benjamin, Menlo Park, 1976.)

24.4 MOLECULAR BONDS AND MOLECULAR SPECTRA

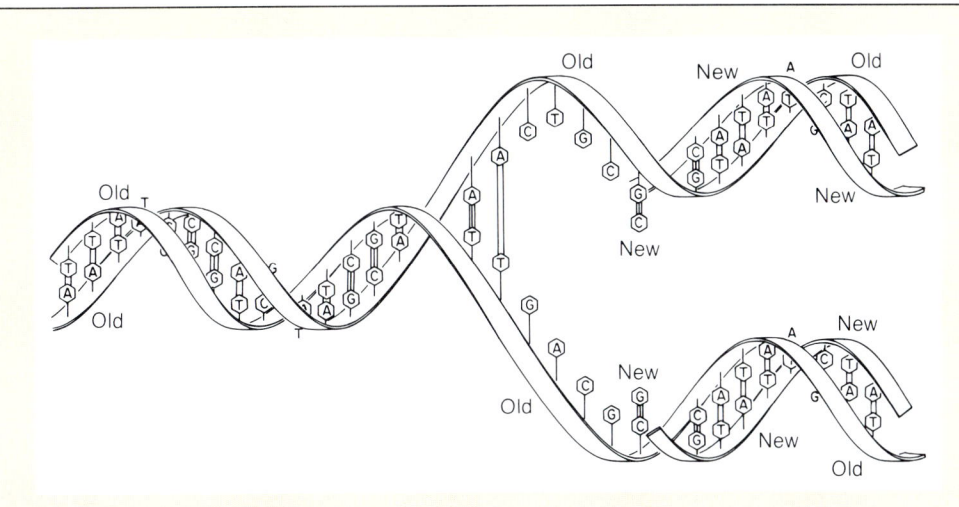

Figure 24.1.6

rier of the genetic code. Life and its continuation require that cells be able to carry out specific tasks and reproduce this capability from generation to generation. To ensure the reliability of this process, nature has encoded the reproductive information in chromosomes, which are composed of genes. Each gene contains the information required to produce a specific protein molecule. As shown by F. Crick and J. D. Watson in 1953, the principal molecular material in the chromosome is DNA, or deoxyribonucleic acid, which consists of a double helix of sub-unit molecules, cross-linked by hydrogen bonds. The sub-units are known as **nucleotide base pairs**, of which there are four—adenine (A), guanine (G), cytocine (C), and thymine (T). Because of the shapes of these base pairs, it turns out that adenine will only form a hydrogen bond to thymine, and guanine will only hydrogen bond to cytocine (Fig. 24.1.5). Thus, A is always opposite T along the helix, and G is opposite C.

When the cell reproduces itself, the double helix unravels at one end (Fig. 24.1.6), leaving the base pairs free to find new mates. Because the only matchings that will occur are A–T and G–C, a new double helix will regenerate from stray base pairs floating about in the cellular fluid. In this manner, genetic information is duplicated and passed on to the next generation. The error rate in this process is miniscule, but it can occur as a spontaneous mutation or as a result of chemical damage from radiation.

The science that grew out of genetics, biochemistry, and molecular physics is called **molecular biology**, and a more complete discussion of the genetic code, protein synthesis, and the role of chemical bonding would take us far afield. Several excellent books exist for further reading.

Further reading:

Watson, J. D., *The Double Helix*, Atheneum, New York, 1969.

Watson, James, D. *Molecular Biology of the Gene*, 3rd ed., W. A. Benjamin, Menlo Park, 1976.

Lehninger, A. L. *Bioenergetics: The Molecular Basis of Biological Energy Transformations*, 2nd ed., W. A. Benjamin, Menlo Park, 1971.

Example 24.8 The latent heat of vaporization of water is 539 kcal/kg. Estimate the bond energy in H₂O liquid (assume 6 bonds per molecule to nearby molecules).

We know that 1 mole of H₂O (containing roughly 6×10^{23} molecules) has a mass of 18 grams (the total number of protons and neutrons in two hydrogen atoms and one oxygen atom is 18). Thus, in 1 kg of water the number of bonds N is

$$N = \left(\frac{1 \text{ kg}}{0.018 \text{ kg}}\right)(6 \times 10^{23})\left(\frac{6 \text{ bonds}}{\text{molecule}}\right) = 2 \times 10^{26} \text{ bonds}.$$

The bond energy, equal to the latent heat divided by the number of bonds in 1 kg, is

$$E_b = \frac{(539 \text{ kcal})(4180 \text{ J/kcal})}{(1.6 \times 10^{-19} \text{ J/eV})(2 \times 10^{26} \text{ bonds})} = 0.07 \text{ eV/bond}.$$

Molecular Spectra and Harmonic Oscillators

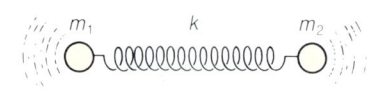

Figure 24.20

Molecular vibrational energy. The bonds holding nuclei together in a molecule are elastic, so the nuclei can vibrate about their equilibrium positions with spring constant k. Their energies are quantized.

A molecule, two or more atoms bonded together, can contain energy not available to individual atoms or ions. In addition to the energies of the electrons in their allowed states, a molecule has vibrational energy and rotational energy. The vibrational energy is due to the fact that the individual atoms can oscillate about their equilibrium position, much like point masses attached by springs (Fig. 24.20). The vibrational energy alternates between kinetic and potential forms as the "spring" is stretched and compressed. The rotational energy is the internal kinetic energy due to the motion of a molecule as it rotates about one or more axes. Both kinds of energy are imparted to molecules by collisions.

Both vibrational and rotational energy are quantized, just as the electron energy is. That is, a molecule can be in one of a series of discrete vibrational energy states, but not in any state in between. Similarly, it can have rotational energy only in specific, discrete amounts. To fully specify the energy state of a molecule requires stating not only the electron quantum numbers we have already discussed, but also a vibrational quantum number v and a rotational quantum number J.

The quantized vibrations of a diatomic molecule like O₂ or CO may be modelled as a **simple harmonic oscillator** (see Chapter 9). If the two atoms have masses m_1 and m_2, and are separated by a "spring" with force constant k (Fig. 24.20), then the natural angular frequency of oscillation is

$$\omega_0 = \sqrt{\frac{k}{m}} \qquad (24\text{--}10)$$

where

24.4 MOLECULAR BONDS AND MOLECULAR SPECTRA

$$m = \frac{m_1 m_2}{m_1 + m_2}$$

is the effective mass of the two atoms loading the spring. The cyclic natural frequency v_0 is related to the angular frequency by

$$\omega_0 = 2\pi v_0$$

since there are 2π radians per cycle. The period of oscillation is

$$T_0 = \frac{2\pi}{\omega_0} = \frac{1}{v_0}.$$

The potential energy stored in a spring is $\frac{1}{2}kx^2$, when the spring is extended a distance x from its equilibrium position (which for a molecule is just the bond length or normal separation between atoms). The overall energy of a bound molecule is negative, meaning that energy must be supplied to break the molecule apart. Vibrations of the atoms represent small positive amounts of energy, thus effectively reducing the magnitude of the overall negative energy (of course, sufficiently energetic vibrations could completely overcome the bonding energy and break the molecule apart). For a bound molecule, we say that the atoms are in a **potential well** (Fig. 24.21), which has a parabolic shape $\frac{1}{2}kx^2$. The energy levels that are created in this well by different vibration frequencies are quantized:

$$E(v) = \hbar\omega_0\left(v + \frac{1}{2}\right), \quad v = 0, 1, 2, \ldots \quad (24\text{--}11)$$

where v is the **vibrational quantum number**. The ground state corresponds to $v = 0$, and the higher energy states ($v = 1, 2, 3, \ldots$) are the harmonics or overtones. Notice that the energy levels are half-integral multiples of the fundamental energy $\hbar\omega_0$. These energies and their associated wavefunctions are shown in Fig. 24.21. (Note that the wavefunctions extend beyond the boundaries of the potential well. This effect is due to the uncertainty in the particle's position and is called **tunnelling**. We will return to this effect in Chapter 25 when we discuss tunnelling in nuclear reactions and α decay.)

Two points are worth emphasizing about the harmonic oscillator model of the molecule. First, the fundamental quantum of energy is

$$\hbar\omega_0 = \hbar\sqrt{\frac{k}{m}}.$$

Thus, if one can determine the "spring constant" k and effective mass m of a molecule, the vibrational energy levels are easily computed. Second, even when the molecule is in its lowest vibrational state ($v = 0$), the quantum theory still predicts (Eq. 24–11) that it has a vibrational energy $\frac{1}{2}\hbar\omega_0$. This residual energy is known as the **zero-point energy**, and is a direct consequence of the uncer-

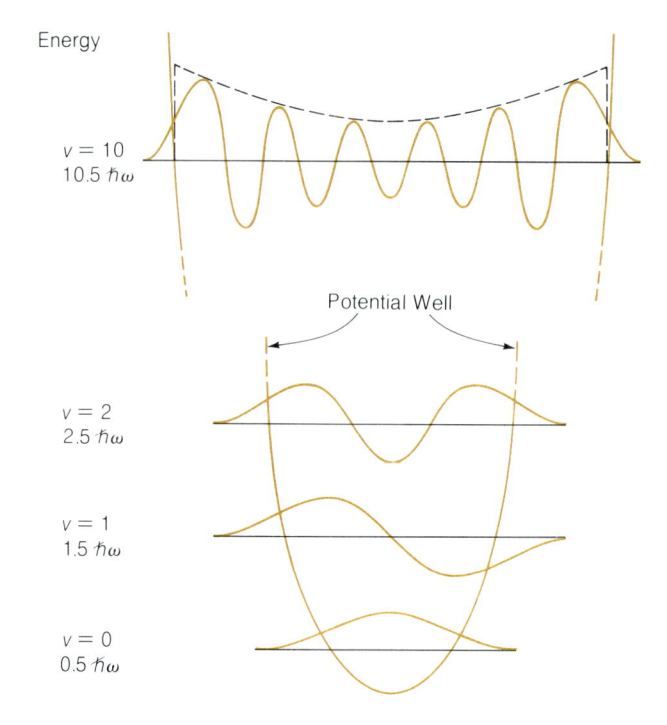

Figure 24.21

Energy levels and corresponding wavefunctions for a harmonic oscillator confined to a parabolic potential well, ½kx². The lowest three states (v = 0,1,2) and v = 10 are shown. Notice that the wavefunctions extend beyond the boundaries of the potential well—an effect called tunnelling.

tainty principle. Basically, it is impossible to completely stop the motion of the oscillator, since if the motion were zero, the uncertainty in position Δx would be zero, resulting in an infinitely large uncertainty in momentum (since $\Delta p \simeq \hbar/\Delta x$). The zero-point energy represents a sharing of the uncertainty in position and the uncertainty in momentum. The energy associated with the uncertainty in momentum gives the zero-point energy.

Example 24.9 The diatomic molecule carbon monoxide (CO) has a harmonic oscillator potential well corresponding to a spring constant $k = 1.9 \times 10^3$ N/m. Evaluate the molecule's natural vibration frequency ω_0, compute the vibrational energy-level splitting $\hbar\omega_0$, and find the wavelength of the $v = 4$ to $v = 0$ vibrational transition.

The effective mass of the molecule is

$$m = \frac{m_1 m_2}{m_1 + m_2} = \frac{(12)(16)m_H}{12 + 16} = 6.86 m_H$$

24.4 MOLECULAR BONDS AND MOLECULAR SPECTRA

since C and O have masses 12 and 16 times the hydrogen mass m_H, respectively (here we are using mass numbers instead of the more precise atomic masses, which are explained in Chapter 25). The natural frequency is therefore

$$\omega_0 = \sqrt{\frac{k}{m}} = \sqrt{\frac{(1.9 \times 10^3 \text{ N/m})}{(6.86)(1.67 \times 10^{-27} \text{ kg})}} = 4.07 \times 10^{14} \text{ rad/s}.$$

The vibrational energy is

$$E_0 = \hbar\omega_0 = \frac{(6.626 \times 10^{-34} \text{ J} \cdot \text{s})(4.07 \times 10^{14} \text{ rad/s})}{2\pi}$$
$$= 4.29 \times 10^{-20} \text{ J} = 0.268 \text{ eV}.$$

The wavelength of the transition from $v = 4$ to $v = 0$ corresponds to an energy difference of $4\hbar\omega_0$. Thus, we may use the formula relating photon energy to wavelength,

$$E = h\nu = \frac{hc}{\lambda}$$

to find the wavelength,

$$\lambda = \frac{hc}{E} = \frac{(6.626 \times 10^{-34} \text{ J} \cdot \text{s})(3 \times 10^8 \text{ m/s})}{(4)(4.29 \times 10^{-20} \text{ J})}$$
$$= 1.16 \times 10^{-6} \text{ m} = 1.16 \text{ μm}$$

which is in the infrared part of the spectrum.

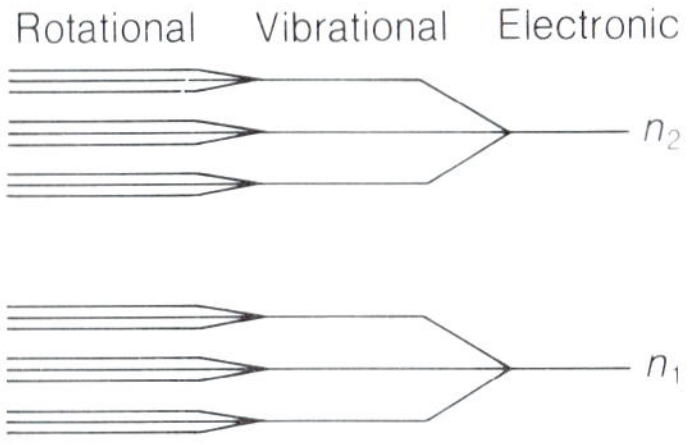

Figure 24.22

Energy levels in a molecule. The electronic energy states in a molecule consist of multiple vibrational states, which consist in turn of multiple rotational states.

The rotational energy of diatomic molecules is also quantized, with energies

$$E(J) = \left(\frac{\hbar^2}{2I}\right)J(J + 1), \quad J = 0, 1, 2, \ldots \quad (24\text{--}12)$$

where J is the **rotational quantum number** and I is the moment of inertia of the molecule about an axis perpendicular to the line between atoms. This formula results from quantizing the angular momentum of rotation. (The details are discussed in Special Problems 2, 3, and 4 at the end of this chapter.)

The energy difference between electronic states is usually much greater than between vibrational states, which in turn are more widely spaced than rotational states. An energy-level diagram becomes very complex, involving not only the electronic energy levels, typically separated by a few eV, but also vibrational levels within each electronic level separated typically by a few tenths of an eV, and rotational levels within each vibrational level with separations of perhaps 10^{-3} to 10^{-2} eV (Fig. 24.22). As with most atoms, the transitions of electrons between electronic energy states result in the emission or absorption

Figure 24.23

A photograph of a molecular spectrum. Transitions between the rotational levels within each vibrational level create the regular banded appearance of the spectrum.

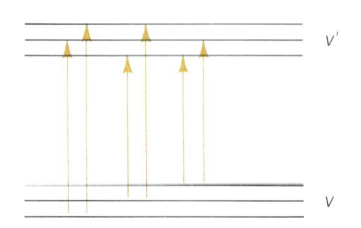

Figure 24.24

A vibration band. Each vibrational state v consists here of three rotational states (there would be many more in a real molecule). Arrows indicate the allowed transitions. The series of closely-spaced lines forms a band in the spectrum.

of photons in the visible or ultraviolet portion of the spectrum. Transitions between vibrational states usually fall in the infrared portion of the spectrum, and rotational transitions are commonly found in the radio (microwave) portion of the spectrum. Because each electronic level has so much fine structure, and because the fine levels are usually very regularly spaced, in a typical visible-wavelength molecular spectrum each band is composed of many lines very close together (Fig. 24.23).

Selection rules require that the vibrational and rotational quantum numbers change when a molecule makes a transition between principal quantum states. Thus, for a given change in n, there are bands for various changes in the vibrational quantum number v. These bands are regularly spaced in the spectrum, because the vibrational states within an electronic level are regularly spaced. Each band consists of many lines, each due to a different change in rotational quantum number J. Fig 24.24 indicates the transitions that are possible between energy levels.

The complex spectra of molecules are very useful in deducing their structure, since knowledge of energy levels can be translated into knowledge about the potential wells. Clearly, it would be very difficult to calculate all the possible energy states from quantum mechanical considerations alone; this has been done for only a very few species, such as molecular hydrogen (H_2) or carbon monoxide (CO).

Perspective

The addition of quantum mechanics to our repertoire, even though we have been able to discuss it only qualitatively for the most part, has added much to our comprehension of atomic structure and spectroscopy. We have learned something of the powerful science that in a sense underlies all others because it allows the analysis of the very structure of matter itself. At the same time, we have seen how spectral lines are formed in nearly all parts of the electromagnetic spectrum, and we have an idea of how elements can be identified by analyzing spectra.

We have left one part of the atom unexplored, and a portion of the electromagnetic spectrum as well. The next chapter presents a discussion of the nucleus, and the emission of gamma rays.

SUMMARY TABLE OF FORMULAS

de Broglie wavelength:
$$\lambda = \frac{h}{p} = \frac{h}{mv} \tag{24-1}$$

Quantization of electron angular momentum:
$$m_e v_n r_n = \frac{nh}{2\pi} = n\hbar \tag{24-2}$$

Electron energy levels in the Bohr model:
$$E_n = -\frac{2\pi^2 e^4 mk^2}{h^2 n^2} = \frac{-13.6 \text{ eV}}{n^2},$$
$$n = 1, 2, 3, \ldots \tag{24-3}$$

Quantization of orbital angular momentum:
$$L = \sqrt{l(l+1)}\hbar, \quad l = 0, 1, \ldots, (n-1) \tag{24-4}$$
$$L_z = m_l \hbar, \quad m_l = 0, \pm 1, \pm 2, \ldots, \pm l \tag{24-5}$$

Quantization of spin angular momentum for electron:
$$S = \sqrt{s(s+1)}\hbar, \quad s = \frac{1}{2} \tag{24-6}$$
$$S_z = m_s \hbar, \quad m_s = \pm \frac{1}{2} \tag{24-7}$$

Dipole selection rules:
$$\Delta l = \pm 1 \tag{24-8a}$$
$$\Delta m_l = 0, \pm 1 \tag{24-8b}$$
$$\Delta s = 0 \tag{24-8c}$$

Wavelength for K-shell transition:
$$1/\lambda = R(Z-1)^2(1 - 1/2^2) = 0.75R(Z-1)^2 \tag{24-9}$$

Angular frequency of a simple harmonic oscillator:
$$\omega_0 = \sqrt{\frac{k}{m}} \tag{24-10}$$

Quantization of molecular vibrational energy:
$$E(v) = \hbar\omega_0\left(v + \frac{1}{2}\right), \quad v = 0, 1, 2, \ldots \tag{24-11}$$

Quantization of molecular rotational energy:
$$E(J) = \left(\frac{\hbar^2}{2I}\right)J(J+1), \quad J = 0, 1, 2, \ldots \tag{24-12}$$

THOUGHT QUESTIONS

1. In the wave mechanics picture of the atom, what is oscillating? What determines the wavelength?

2. How do the wavelengths of typical (0.1–1 kg) masses moving at 10 – 100 m/s compare to the size of atoms? Why are the quantum mechanical effects not noticeable?

3. The ionization energy of helium ($Z = 2$) is 24.58 eV, while that of lithium ($Z = 3$) is 5.39 eV. Explain the difference.

4. Why do fluorine and chlorine have similar chemical properties?

5. Why are helium, neon, argon, krypton, etc. called inert gases? Explain these properties according to the shell model of their configurations.

6. In a quantum energy level with high angular momentum quantum number l, what is rotating? Is the electron farther or nearer the nucleus, on the average, than a level with $l = 0$?

7. Do the probability distributions of an s and p electron overlap? Can an s electron ever find itself farther out from the nucleus than a p electron?

8. Explain the importance of the Pauli exclusion principle for atomic structure and the chemical properties of elements.

9. Define the following terms: configuration, subshell, energy state, spin.

10. What is electron screening? How does it affect the order of shells and the energies of outer electrons?

11. Why are "forbidden lines" seen only in low-density gases?

12. What happens to the proton and electron in hydrogen when it emits the 21-cm radio line?

13. In a fluorescent light, what provides the excitation energy of the upper level? How does this radiation differ from that produced in an incandescent bulb?

CHAPTER 24 ESSENTIALS OF QUANTUM MECHANICS

14 How does an atom or ion emit an x-ray line? Infrared or radio lines?

15 Explain why a diatomic molecule has many more energy levels, and thus a richer emission-line spectrum, than an isolated atom.

PROBLEMS

Section 24.1–Wave Mechanics

- **1** How many de Broglie wavelengths of an electron fit into the circumference of the nth Bohr level of hydrogen?

- **2** A particle of mass m is constrained to move in a one-dimensional box of length L. In wave mechanics, the possible modes correspond to standing waves with nodes at the box ends.
  (a) Show that the nth mode satisfies $(n\lambda/2) = L$, where λ is the particle's wavelength (b) Show that the KE of the nth mode is $KE = n^2h^2/8mL^2$.

- **3** Estimate the ground state energy ($n = 1$) of the hydrogen atom by approximating it by an electron confined to a box of width 10^{-10} m.

- **4** What is the ground state energy of a 100 g ball confined to a racquetball court 5 m in width?

- **5** Estimate the energy of a nuclear ground state. Assume that a proton $m_p = 1.67 \times 10^{-27}$ kg is confined to a box of size 2×10^{-15} m.

Section 24.2–Atomic Structure

- **6** List the (n,l)-subshells possible for $n = 4$.

- **7** What values of m_l are possible for a $3d$ state?

- **8** List the $n = 3$ subshells of hydrogen. How many quantum states are possible?

- **9** How many different states (n,l,m_l,m_s) are possible for an electron with $n = 4$?

- **10** List the quantum numbers for the ground state of aluminum.

- **11** What is the magnitude, L, of the orbital angular momentum in the $n = 4$, $l = 3$ state of hydrogen? How many Zeeman states does this level split into when a magnetic field is applied?

- **12** What is the maximum number of electrons that can occupy a $5f$ state?

- **13** List the "magic numbers" of electrons that complete subshells up to $n = 3$. Show that the total number of electrons that fill a shell (all nl subshells for a given n) is equal to $2n^2$.

- **14** Draw energy level diagrams for neon ($Z = 10$) and sulfur ($Z = 16$), showing the occupancy of subshells.

- **15** Give the complete electron configuration of gold ($Z = 79$).

- **16** Give the generic configuration type for inert gases, alkali metals, and halogens. How are the three types related?

- **17** In which columns of the periodic table would one find elements with the following configurations: (a) $1s^22s^22p^63s$, (b) $1s^22s^22p^63s^23p^5$, (c) $1s^22s^22p^4$.

- **18** What are the valences of silicon ($3p^2$), copper ($3d^{10}4s$), and zinc ($3d^{10}4s^2$)?

- **19** Write out the electron configurations of the man-made element Californium ($Z = 98$).

- **20** Rubidium ($Z = 37$), niobium ($Z = 41$), and molybdenum ($Z = 42$) each have one $5s$ electron. Write their configurations and explain why they are different chemically.

- **21** Many periodic tables list Lawrencium ($Z = 103$) as the last element. Write out its complete electron configuration and the probable configurations of the next two man-made elements ($Z = 104$, $Z = 105$).

Section 24.3–Atomic Spectra

- **22** Which of the following radiative transitions are allowed according to dipole selection rules? If not allowed, explain why.
  (a) $2p \rightarrow 1s$, (b) $2s3p \rightarrow 2s^2$, (c) $3s3d \rightarrow 3s^2$, (d) $2p^34d \rightarrow 2p^4$

- **23** Is the radiative transition $1s^22s2p \rightarrow 1s^22s^2$ allowed according to dipole selection rules? Why or why not?

- **24** Four configurations of the oxygen atom ($Z = 8$) are given below:
  (1) $1s^22s^22p^33s$, (2) $1s^22s^22p^5$, (3) $1s^22s^22p^4$, (4) $1s^22s^22p^33p$
  (a) Which is the ground state?
  (b) Which may be excited by an allowed radiative transition from the ground state?
  (c) Which may be excited by an allowed radiative transition from an inner shell?

(d) Which may not decay to the ground state by an allowed transition?

•• 25 Draw an energy level diagram for the hydrogen atom, showing all (n, l) substates up to and including $n = 4$. Using the dipole selection rules, label with arrows the 3 possible allowed transitions responsible for the first Balmer line ($n = 3 \to 2$) and the 5 transitions responsible for the first Paschen line ($n = 4 \to 3$).

••• 26 Figure 24.25 shows five energy states of sodium vapor ($Z = 11$), with ground state configuration $1s^2 2s^2 2p^6 3s$.
 (a) Label with arrows all *downward* allowed radiative transitions.
 (b) If the 4p level is selectively excited by an ultraviolet laser, what fluorescent lines would one expect to detect? Give their wavelengths in nm.

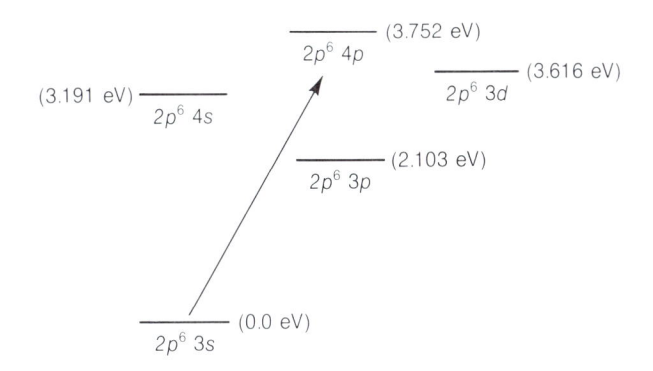

Figure 24.25

•• 27 Figure 24.26 shows three energy levels for a hypothetical substance. The radiative transition $3 \to 2$ is allowed; that from $2 \to 1$ has a long lifetime. To make a laser of this material, which level should be pumped? Which transition will produce a laser?

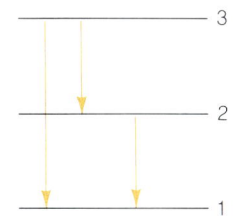

Figure 24.26

•• 28 Estimate the energy of the inner shell K-line of uranium ($Z = 92$).

•• 29 Estimate how much energy is required to ionize the inner (K-shell) electron from iron, lead, and xenon.

•• 30 A prospector brings a piece of metal ore into a bar, claiming it is gold. Suspecting it may be copper or iron, the bartender exposes it to a burst of energetic electrons and measures an x-ray emission line of energy 8000 eV. Should the bartender let the prospector buy drinks? What type of ore is it likely to be?

••• 31 What are the wavelengths (cm) of:
 (a) The $n = 60 \to 59$ line of atomic hydrogen?
 (b) The $n = 108 \to 107$ line of hydrogen-like He^+?
 (c) The $n = 200 \to 199$ line of hydrogen-like carbon, C^{+5}?

••• 32 A radio astronomer has a telescope and electronic receivers capable of measuring radio recombination lines of hydrogen with $\Delta n = 1$ and frequencies near 4.3×10^9 Hz. Which upper level n is involved in these lines?

•• 33 The oxygen atom has its ground state configuration, $2p^4$, split into a number of fine structure states. Figure 24.27 shows the lowest three levels. The radiative transition probabilities between levels are $A(2 \to 1) = 9 \times 10^{-5}$ s^{-1} and $A(3 \to 2) = 1.7 \times 10^{-5}$ s^{-1}. These are the probabilities per second that the upper level emits a photon.
 (a) What are the wavelengths of the emission lines, $2 \to 1$ and $3 \to 2$?
 (b) In what spectral ranges do these lines fall?
 (c) In a low-density gas, how long would an electron be, on average, in each of the upper levels 2 and 3 before radiating to the ground state?

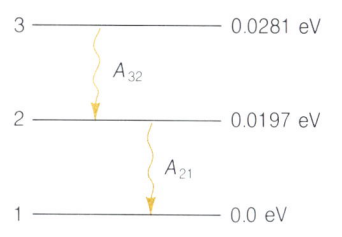

Figure 24.27

•• 34 For an oxygen atom in Problem 33, how hot would a gas need to be before a free particle had a substantial chance of exciting level 2 collisionally? (Use the rule of thumb that the *average* thermal energy of a free particle, kT, must be about 30% of the excitation energy for high-energy particles in the tail of the velocity distribution to excite the higher level by collisions.)

Section 24.4 Molecular Bonds and Spectra

35 To melt water ice (H_2O) requires a latent heat of fusion 80 kcal/kg. Assuming 6 bonds per molecule to nearest neighbor molecules, estimate the bond energy in eV.

36 The bond energy between C and H in hydrocarbon fuels is 4.28 eV/bond; between C and C it is 3.60 eV/bond. How much heat (kcal) is released in burning 1 kg of ethane (Fig. 24.28)?

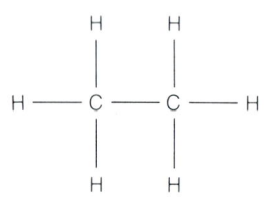

```
       H   H
       |   |
   H — C — C — H
       |   |
       H   H
```

Figure 24.28

37 Using the C-H bond energy of 4.28 eV, estimate how much heat is released in burning 1 kg of methane (CH_4).

38 Carbon monoxide (CO) releases 84 kcal/mole when burned. What is the CO bond energy in eV?

39 Two solid spheres of radius R and mass M are composed of molecules with mass m, bound by energy E per molecule. The spheres approach one another head-on, each with velocity v. Derive a formula for the critical velocity V_c such that the KE of the collision completely disrupts the solid (breaks all the bonds). What is V_c for iron spheres ($m = 56m_p$, bond energy $E_b = 4.3$ eV)?

40 The first vibrationally excited level ($v = 1$) of the molecule H_2 lies about 0.5 eV above the ground state ($v = 0$).
  (a) At what wavelength will radiative transitions between these levels occur?
  (b) How hot would molecular gas need to be for collisions to excite this level? See Problem 34.

41 The H_2 molecule may be idealized as two protons of mass m_p held together by a spring with force constant k. If the first vibrationally excited level ($v = 1$) has an energy of 0.5 eV, and if the "effective mass" on the spring is 0.5 m_p, calculate the natural vibrational frequency ω_0 (rad/s) and the spring constant k (N/m). (Recall that vibrational energy is quantized: $E(v) = \hbar\omega_0(v + 1/2)$.)

42 Estimate the energy of the $v = 1$ state of O_2. Assume that the spring constant k is the same as for H_2 (Problem 41), and that the effective mass on the spring is larger by a factor of 16, the ratio of oxygen mass to hydrogen mass.

SPECIAL PROBLEMS

1. Zeeman Effect

When an atom is placed in a magnetic field B, the interaction between the electron's current I and the field creates a small splitting of energy levels known as the Zeeman effect (Section 24.2). Imagine that a single electron of mass m and charge e moves around the nucleus with velocity v at radius r (Fig. 24.29). The plane of the orbit makes an angle ϕ with **B**. Suppose that the electron's mass and charge are spread out uniformly around the circumference, $2\pi r$, of its orbit, forming a current loop.
  (a) Show that the electron's current (coulombs/s) is $I = (ev/2\pi r)$.
  (b) As shown in Chapter 16 (Eq. 16–8), the torque τ on the loop due to magnetic forces is $\tau = (IAB)\cos\phi = (IAB)\sin\theta$. Using techniques of calculus, one can show that the potential energy associated with the work done against this torque is PE = $(IAB)\cos\theta$. Show that we may write this energy as

$$PE = (eLB/2m)\cos\theta = (e/2m)L_zB$$

where L is the total orbital angular momentum of the electron and $L_z = L\cos\theta$ is the component of L along the z-axis aligned with **B**.
  (c) Using the quantization rules for L_z (Eq. 24–5) derive a formula for the energy splitting between different states of magnetic quantum number m_l. Estimate this energy splitting for a field of 1 Gauss.

2. Diatomic Molecules

A diatomic molecule may rotate about an axis perpendicular to the line between its two atoms and passing through their center of mass (Fig. 24.30). The masses, distances, and velocities of the two atoms are labelled.
  (a) Show that $m_1R_1 = m_2R_2$ and that $(v_1/R_1) = (v_2/R_2) = \Omega$, where Ω is the rotational angular velocity of the molecule.

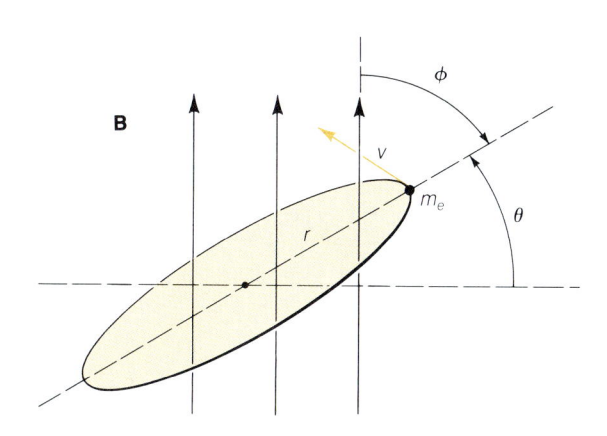

Figure 24.29

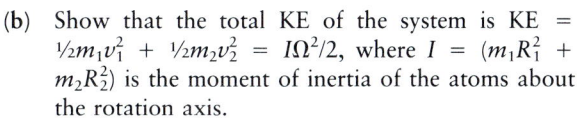

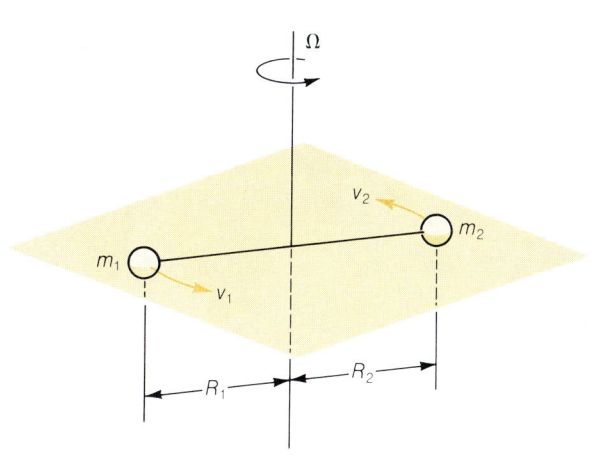

Figure 24.30

(b) Show that the total KE of the system is KE = ½$m_1v_1^2$ + ½$m_2v_2^2$ = $I\Omega^2/2$, where $I = (m_1R_1^2 + m_2R_2^2)$ is the moment of inertia of the atoms about the rotation axis.

(c) Show that the angular momentum of the molecule is $L = m_1v_1R_1 + m_2v_2R_2 = I\Omega$, and that the KE may therefore be written as KE = $L^2/2I$.

(d) Using the usual quantization rules for the magnitude of the angular momentum, $L = \sqrt{J(J+1)}\,\hbar$, where $J = 0, 1, 2, \ldots$ is the rotational quantum number, show that the quantized energies of the molecule are given by the formula (Eq. 24–12), $E(J) = (\hbar^2/2I)J(J+1)$.

3. Molecular Hydrogen Rotation

Using the results of Special Problem 2, estimate the energy of the first excited rotational state ($J = 1$) of H_2. (First compute the moment of inertia I of two protons about a rotational axis as in Fig. 24.30. Assume that the protons are separated by a bond length 7.7×10^{-11} m.)

4. Carbon Monoxide Rotation

Using the rotational energy formula $E(J) = (\hbar^2/2I)J(J+1)$, estimate the wavelength of the $J = 5 \rightarrow 4$ rotational transition of carbon monoxide (CO). Assume a bond length of 10^{-10} m between C and O atoms.

SUGGESTED READINGS

"The spectra of streetlights illuminate basic principles of quantum mechanics." J. Walker, *Scientific American*, Jan 84, p 138, **250**(1)

"The isolated electron." P. Ekstrom, D. Wineland, *Scientific American*, Aug 80, p 104, **243**(2)

"High-energy collisions between atomic nuclei." W. C. McHarris, J. O. Rassmussen, *Scientific American*, Jan 84, p 58, **250**(1)

"Fifty years of matter waves." H. A. Medicus, *Physics Today*, Feb 74, p 38, **27**(2)

"It might as well be spin." S. A. Goudsmit, *Physics Today*, Jun 76, p 40, **29**(6)

"Heisenberg and the early days of quantum mechanics." F. Bloch, *Physics Today*, Dec 76, p 23, **29**(12)

"Is the moon there when nobody looks? Reality and the quantum theory." D. Mermin, *Physics Today*, Apr 85, p 38, **38**(4)

"Let the elements teach periodic law." J. F. Hyde, *The Physics Teacher*, Dec 75, p 538, **13**(12)

Thirty years that shook physics. George Gamow, Doubleday, NY, 1966

The strange story of the quantum. Banesh Hoffman, Dover, NY, 1959

Mr. Tompkins in Wonderland. George Gamow, Cambridge University Press, New York, 1940

Great scientific experiments. Rom Harre, Phaidon Press, Oxford, 1981 (the Stern-Gerlach experiment)

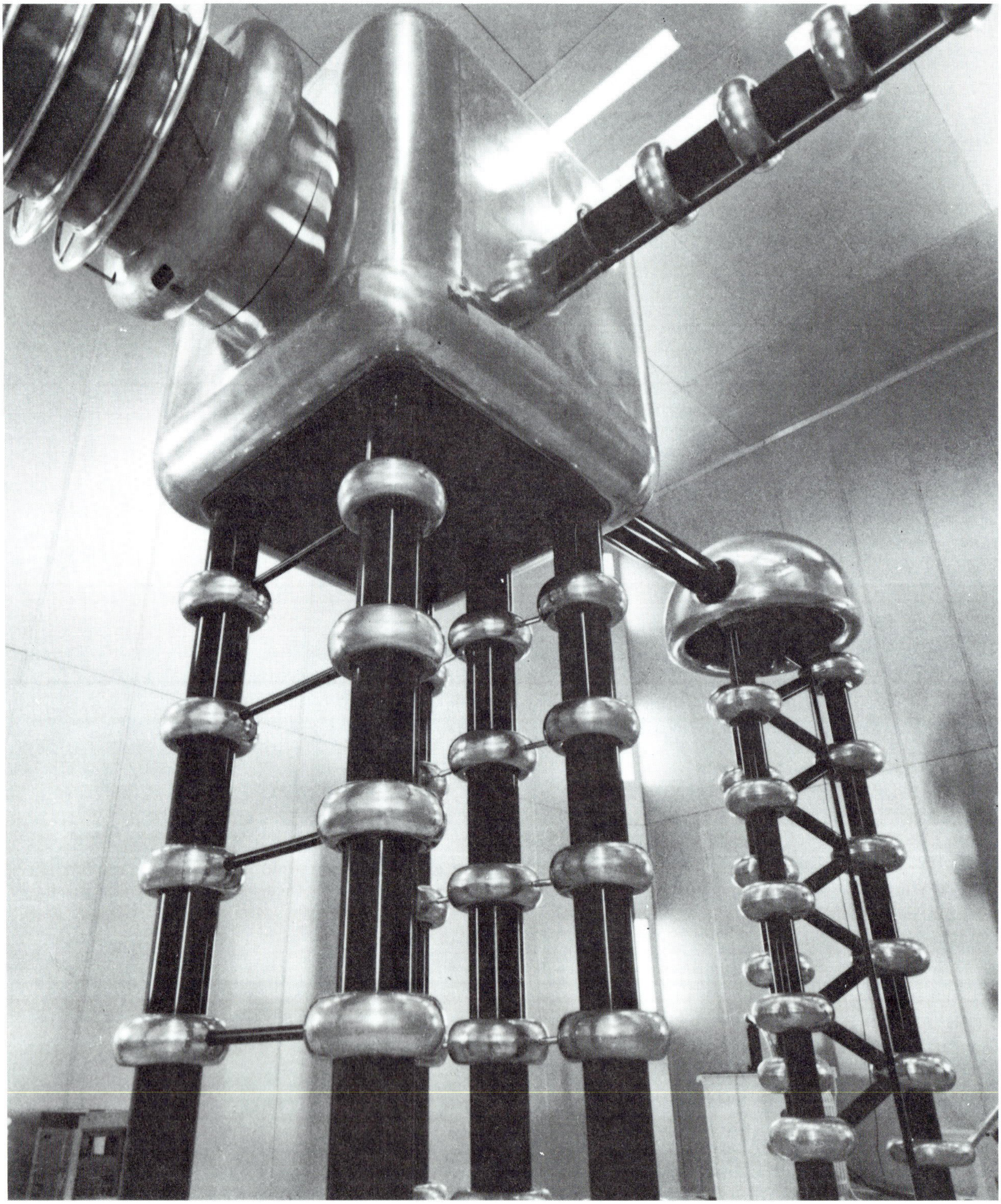

Chapter 25

The Nucleus and Nuclear Energy

Between 1890 and the 1920s, many developments occurred in understanding the electron energy level structure of atoms. At the same time, similarly remarkable progress was made in discovering the properties of nuclei. The earliest clues came from the study of radioactivity, the spontaneous emission of subatomic particles in some elements. The pioneering work in this area was done by the French physicist Henri Becquerel and by Pierre and Marie Curie, who found that radioactive substances emit three distinct types of "rays" having different abilities to penetrate solid materials. (Two of the three, alpha-rays and beta-rays, are beams of high-energy particles; the third, gamma-rays, are very high-energy photons, as we now know.)

In 1911 the experiments of Rutherford led him to deduce the existence of the nucleus. Soon Rutherford and then Moseley carried out experiments which led to understanding the atomic number Z and its role in establishing the identity of the elements. In 1932 Chadwick discovered the neutron. Within the next decade, the first experiments with colliding nuclei and nuclear reactions took place, and by the end of the 1930s a fairly complete picture of the nucleus had been developed. Since then, a great deal has been learned, mostly through the use of massive machines to accelerate particles to high energies.

In this chapter we will examine the general properties of the nucleus, the nature of radioactivity, the interactions of nuclei in reactions, and the machines used both to explore the nucleus and to exploit its potential as an energy source.

25.1 Properties of the Nucleus

Great progress in understanding the nature of atomic nuclei took place during the first decades of this century. The empirical advances in nuclear physics, unlike those in atomic physics, were not followed by a complete theory with the rise of quantum mechanics. There still is not a wholly satisfactory theoretical understanding of nuclear structure, although rapid progress has again been seen in the past decade. Some of this progress will be described in the next chapter, in our discussions of elementary particles. For now, however, we will describe the experimental evidence for the properties of the nucleus.

Nuclear Structure

It was suspected from early experiments that a nucleus consists of at least two types of smaller particles: one, the **proton**, which carries a positive charge equal to the negative charge of the electron; and another, the **neutron**, with no electrical charge. Collectively these particles are called **nucleons**.

The number of protons in a nucleus, as we have seen, is the **atomic number Z**, a unique characteristic of an element. The total number of nucleons is the **mass number A**. The atomic number and the mass number together are sufficient to completely specify the composition of a nucleus, although sometimes we speak of the **neutron number N**, which is simply the difference $(A - Z)$ between the mass number and the atomic number. Any nucleus, with its characteristic values of Z and A, is called a **nuclide**. A single element may have more than one possible nuclide, and the various forms are called **isotopes** of that element. The isotopes of an element all have the same number of protons (the same Z), but different numbers of neutrons (N), and thus different mass numbers (A). In the most abundant form of hydrogen, for example, $Z = 1$ and $A = 1$ (that is, the nucleus is just a single proton), but there are less common forms with $A = 2$ (**deuterium,** with one proton and one neutron in the nucleus), and with $A = 3$ (**tritium,** with one proton and two neutrons). So long as there is only one proton in the nucleus, it is still hydrogen (Fig. 25.1).

Most elements exist in two or more different isotopic forms, although usually one isotope is much more abundant than the others. The relative abundance of each isotope in a sample is a function of the history of the material under consideration and, as we shall see, the measurement of isotope ratios can provide valuable information on the age or the formation of the sample.

A special notation is used by physicists to specify a nuclide. The usual

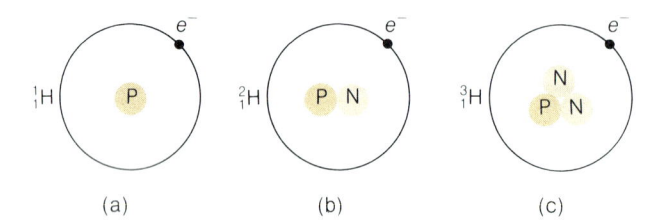

Figure 25.1

The isotopes of hydrogen: (a) ordinary hydrogen, (b) deuterium, (c) tritium.

25.1 PROPERTIES OF THE NUCLEUS

chemical symbol indicates the element, such as H for hydrogen, He for helium, and so on. A leading superscript gives A, and a subscript indicates Z. In this notation, the three forms of hydrogen are ^1_1H, ^2_1H, and ^3_1H (hydrogen, deuterium, and tritium, respectively). Another common element having two or more well-known isotopes is carbon, which is most commonly $^{12}_6C$ (usually called "carbon-12", even though the superscript 12 precedes the C), but also is found as $^{13}_6C$ and $^{14}_6C$ (carbon-13 and carbon-14).

The composition of a nucleus can be changed, either by the emission of particles in radioactive decay, or by undergoing a nuclear reaction. Both of these processes are described later in this chapter. The heaviest naturally-occurring element is uranium ($Z = 92$, A as high as 238 in natural isotopes), but an additional fifteen elements have been created in reactors or accelerators. The largest nuclei tend to be unstable. They spontaneously emit particles (that is, they are radioactive), which changes them into other elements, so they do not last for a very long time. Therefore if any natural elements beyond $Z = 92$ once existed on the earth, they have long ago disappeared.

The sizes of nuclei are estimated from scattering experiments, as originally done by Rutherford. Whereas he used helium nuclei, modern experimenters use electrons as the scattering particles. These experiments show that the radius of a nucleus is approximately proportional to the cube root of the mass number ($R = (1.2 \times 10^{-15}\,m)A^{1/3}$), just what would be expected if the nuclei are spherical and the nucleons themselves are not compressed when they combine in a nucleus. There is essentially no empty space between them. Thus, all nuclei have approximately the same density, a remarkably high value of about 10^{17} kg/m^3.

The masses of nuclei are derived from **mass spectrometer** experiments. A mass spectrometer injects ionized gas at a known velocity into a region of a uniform magnetic field (Fig. 25.2). The nuclei follow a curved path due to the force exerted on them by the magnetic field, and the radius of curvature is a function of the ratio of charge to mass of the particles. The particles are allowed to strike a detector (such as a piece of photographic film), so that their point of impact can be measured and the radius of curvature determined. This is the same sort of experiment that was used by J. J. Thomson in 1897 to demonstrate the existence of the electron; the discovery of isotopes was later made in this way as well.

Although it is possible to express the masses of nuclei in standard units such as kilograms, it is often more convenient to use other units instead. The most common is the **unified atomic mass unit** (**u**), which is related to the average mass per nucleon. More specifically, the unified atomic mass unit is defined as precisely one twelfth of the mass of the most common form of carbon, $^{12}_6C$. Therefore, the mass of a carbon-12 nucleus is exactly 12.0000 u. In this system, the mass of a proton is 1.00728 u, that of a neutron is 1.00867 u, and that of an electron is 0.000549 u. In the SI system, 1 u = 1.660×10^{-27} kg. Often in periodic tables, or other listings of basic physical data about the elements, the atomic mass in u is given. Examination of such a table shows that the atomic masses are not simply integral multiples of 1 u, as we might naively expect. The reasons for this will become clear in later sections of this chapter. According to convention, the masses of heavier nuclei are given for the neutral atom and therefore include the mass of the Z electrons.

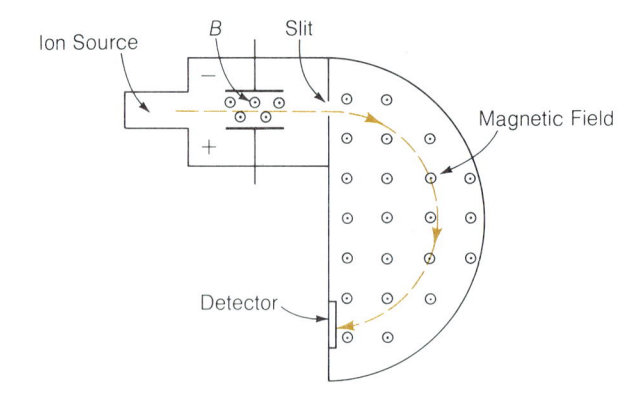

Figure 25.2

The mass spectrometer. Atoms are ionized and injected into a chamber with perpendicular electric and magnetic fields, so that only those ions with a certain velocity pass through to the slit without being deflected. For constant speed, the radius of curvature of the ions due to the magnetic field in the semicircular chamber depends on particle mass. Hence the ions are dispersed according to mass as they strike the photographic film. The equations governing the motions of these charged particles are discussed in Chapter 16.

Nuclear masses are also sometimes expressed in terms of their equivalent energies. Recall from our discussion of relativity that mass and energy are equivalent (according to the equation $E = mc^2$) and therefore in principle they can be interchanged. When we discuss nuclear reactions we will see how this exchange may occur. For now we simply point out that masses are sometimes expressed in terms of energy, usually in MeV (10^6 eV). In these units, the mass of a proton is 938.3 MeV, that of the neutron is 939.6 MeV, that of the electron is 0.511 MeV; and 1 u = 931.50 MeV.

The relative abundances of the elements depend on their history of formation, as do the isotope ratios, as we have already mentioned. The history will not be fully discussed until the next chapter. Appendix 6 at the end of the book contains information on the abundances of the elements, and Appendix 7 lists masses, abundances, and other properties for selected isotopes.

Nuclear Binding Energy

In Chapter 3 of this text, we mentioned that there are four known forces in nature. These are the gravitational force, (the earliest known), the electromagnetic force, (which we have also discussed at considerable length), and the strong and weak nuclear forces. We are finally at the point where these latter two forces come into play.

We approach the question of the force holding a nucleus together by

25.1 PROPERTIES OF THE NUCLEUS

considering the energy required to separate the nucleons to an infinite distance. This is consistent with the convention we have used before: when discussing orbits, we defined gravitational potential energy so that its value is zero when two bodies are infinitely far apart, and similarly we have established the electron energy levels so that the energy is zero when the electron is completely free of the atom. We say that two bodies in orbit, or an electron and a nucleus, have **binding energy** equal to the energy that must be added to separate them. The binding energy is positive by definition, because energy must be added to separate them. When the nucleons are bound, their total energy is negative.

How do we measure the binding energy of an atomic nucleus? We note that the mass of any nucleus is *less* than the sum of the masses of the nucleons of which it is made. For example, a $^{12}_{6}$C atom has 6 protons, 6 neutrons, and 6 electrons, whose masses total 12.09894 u, yet we have seen that the mass of this atom is actually 12.00000 u. We conclude that the missing mass has been converted into binding energy. If we substitute m = 0.09894 u into Einstein's formula, we find that the energy equivalent is $E = mc^2 = 1.48 \times 10^{-11}$ J = 92.2 MeV. Similarly, the mass difference between a $^{4}_{2}$He atom and the sum of the mass of its constituent particles is 0.03038 u (use Appendix 7), implying that the binding energy is 4.53×10^{-12} J = 28.3 MeV. Notice that we have used six figures following the decimal point in the masses to maintain accuracy.

The binding energy calculated by comparing masses as we have just done is the *total* binding energy, because it represents the entire amount of energy that holds a nucleus together. A related quantity is the **average binding energy per nucleon,** which is equal to the total binding energy divided by A, the mass number. For $^{12}_{6}$C, we find that the average binding energy per nucleon is 7.7 MeV, and for $^{4}_{2}$He it is 7.1 MeV. The average binding energy per nucleon increases with mass number for light nuclei, then is roughly constant for nuclei with mass numbers of 20 or more (Fig. 25.3). The lowest value, 1.1 MeV, is found for deuterium, and the peak, just under 9 MeV, occurs for nuclides with A between 50 and 60 (the elements near iron in the periodic table). The relative peak for $^{4}_{2}$He indicates that this nucleus is unusually tightly bound compared to others with low values of A. This is why helium nuclei, also known as alpha-particles, are sometimes released intact in radioactive decay (recall that Rutherford used alpha-particles in his scattering experiments).

Besides calculating the total binding energy and the average binding energy per nucleon, it is also interesting to consider the energy required to remove a single nucleon from a nucleus. We choose nitrogen as our example. From Appendix 7, we see that the mass of a $^{14}_{7}$N atom is 14.003074 u, while that of a $^{13}_{7}$N atom is 13.005738 u. The sum of the masses of the $^{13}_{7}$N atom and a free neutron is 14.014403 u, 0.01133 u greater than the mass of a $^{14}_{7}$N atom. Therefore the energy required to remove a single neutron from a $^{14}_{7}$N atom is the energy equivalent of 0.01133 u, which is 10.6 MeV. This is substantially greater than the average binding energy per nucleon for $^{14}_{7}$N, which is 7.5 MeV. We will discuss the implications of this for the stability of a nucleus in the next section.

Before we do so, however, we consider the nature of the force that holds nucleons together in a nucleus. We know there is a force at work, because we know there is substantial binding energy; that is, work is required to separate

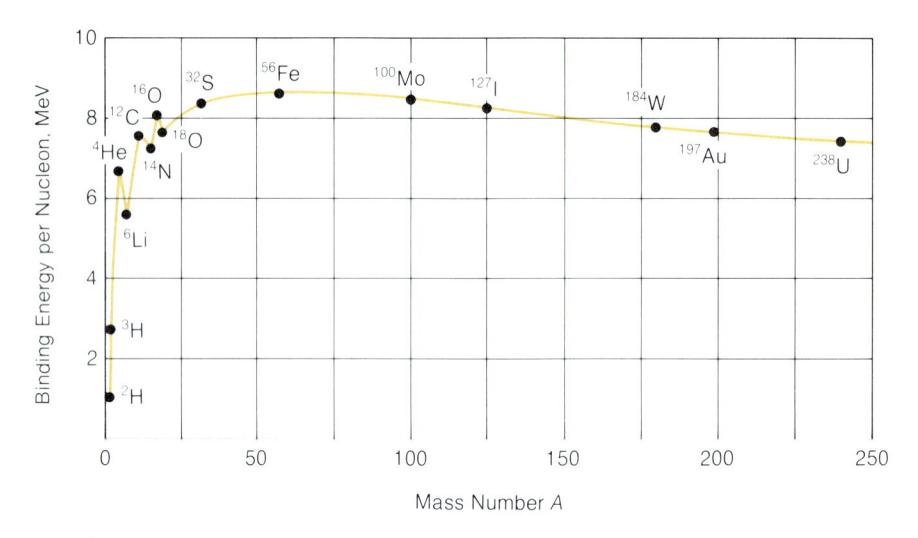

Figure 25.3

The average binding energy per nucleon. This quantity indicates the relative stability of nuclei. Note the sharp peak (far left) at A = 4, for the helium nucleus $^4_2$He, also known as the α-particle. The high binding energy per nucleon of this nucleus explains why it is often ejected intact during radioactive decay.

the nucleons. Furthermore, the force holding a nucleus together must overcome the repulsive electrostatic force that the protons exert upon each other. Hence the binding force must be much stronger than the electrostatic force, and it is called the **strong nuclear force.** The exact nature of this force is not fully understood. It is known that it acts only over very short distances (within about 10^{-15} m), so it does not vary simply as the inverse square of the separation between particles, as the gravitational and electrostatic forces do.

The strong nuclear force acts between all nucleons, whether neutrons or protons. The force is diminished by the repulsive electrostatic force between protons, however, so that in a nucleus with many protons, the binding force is weakened and the nucleus may be unstable. As atomic number Z increases, the number of neutrons required to maintain stability becomes larger than Z. Even so, there is a limit on how many protons a nucleus can have and still be stable; this limit is 83 (the element bismuth). Therefore, all nuclides with Z greater than 83 are unstable. We will discuss the implications of this in the next section.

Example 25.1

Calculate the total binding energy, the average binding energy per nucleon, and the energy required to remove one neutron from $^{16}_8$O.

The total binding energy is the energy equivalent of the mass difference between a single $^{16}_8$O atom, which is 15.994915 u, and the sum of the masses of 8 protons, 8 neutrons, and 8 electrons, which is 16.131920 u. (In Appendix 7, the mass of the electrons is included with the neutral atom, so we take 8

25.2 RADIOACTIVITY

times the mass of the $_1^1$H atom plus 8 times the mass of a neutron.) Thus, the mass difference is 0.137005 u, corresponding to a total binding energy of

$$E = mc^2 = (0.137005 \text{ u})(1.66 \times 10^{-27} \text{ kg/u})(3 \times 10^8 \text{ m/s})^2$$
$$= 2.045 \times 10^{-11} \text{ J} = 127.6 \text{ MeV}.$$

The average binding energy per nucleon is therefore

$$\frac{E}{16} = 8.0 \text{ MeV}.$$

The sum of the masses of a $_8^{15}$O atom plus a free neutron is 16.011730 u, which is 0.016815 u greater than the mass of a single $_8^{16}$O atom. Therefore the binding energy for a single neutron in a $_8^{16}$O atom is

$$E(\text{neutron}) = mc^2 = (0.016815 \text{ u})(1.66 \times 10^{-27} \text{ kg/u})(3 \times 10^8 \text{ m/s})^2$$
$$= 2.51 \times 10^{-12} \text{ J} = 15.7 \text{ MeV}.$$

Problems like these can be simplified by noting that there is a straightforward conversion factor from the atomic mass unit u to energy: $E = mc^2 = (1.6606 \times 10^{-27} \text{ kg})(2.99792 \times 10^8 \text{ m/s})^2/(1.6022 \times 10^{-19} \text{ J/eV}) = 931.50$ MeV/u.

25.2 Radioactivity

As mentioned earlier in this chapter, radioactivity was discovered (in 1896) by H. Becquerel, and explored in detail after the turn of the century by the Curies. This pioneering work came before the structure of the atom was known; indeed, the existence of the nucleus was not established until 1911. Becquerel found that a mineral called pitchblende could stimulate phosphorescence in the complete absence of any light source. The Curies were able to isolate two elements, named radium and polonium, which were even stronger sources of phosphorescence. The term "radioactivity" was applied, because it was clear that some kind of radiation was emitted by these materials. The experiments of the Curies established that radioactivity was an intrinsic property of certain substances, not dependent on external conditions.

The Three Modes of Decay

Following the work of the Curies, other experimenters, including Rutherford, found three distinct types of radiation. This was established by finding the sign of the electrical charge on the emitted particles (by observing the direction in which their paths were bent by magnetic fields; Fig. 25.4), and by measuring the distances the ray penetrated through various substances such as lead. The

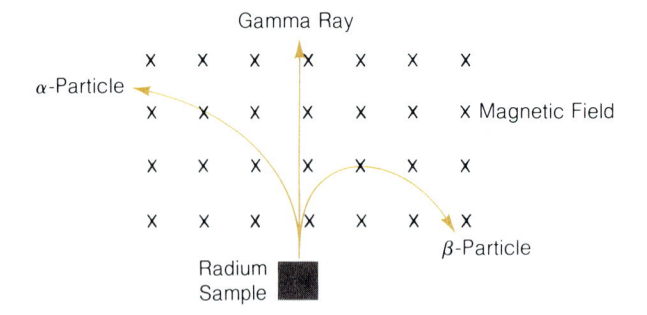

Figure 25.4

Types of radiation. A magnetic field (directed into the page) forces electrons and α-particles to follow curved paths in opposite directions, while γ-ray photons travel in a straight line.

three types of radiation were named alpha-rays, which consist of positively charged particles, beta-rays, which are negatively charged, and gamma-rays, which have no charge. We now know that the first two are actually helium nuclei (α) and electrons (β), while the γ-rays, which penetrate the farthest through solids, consist of very high-energy photons.

Measurements of nuclear mass and charge, before and after the emission of an α-particle, showed that such a particle has mass number $A = 4$ and charge $Z = 2$. This is a nucleus of $^{4}_{2}\text{He}$. A typical α-particle emission, observed by early experimenters, is the decay of radium-226. The radium nucleus has $Z = 88$ and $A = 226$, and the products are an α-particle (helium nucleus) and the remainder of the nucleus, a new nucleus with $Z = 86$ and $A = 222$. This is the element radon. In standard notation, this decay is represented as

$$^{226}_{88}\text{Ra} \rightarrow {}^{222}_{86}\text{Rn} + {}^{4}_{2}\text{He}.$$

This example shows that when an α-decay occurs, a new element is formed. Such a change is therefore called a **transmutation.**

We may wonder why the particular combination of nucleons that constitute an α-particle should be emitted, rather than some other combination. The answer lies in the average binding energy per nucleon, discussed in the preceding section. There we saw that the average binding energy per nucleon for a $^{4}_{2}\text{He}$ nucleus is substantially greater than it is for other relatively small nuclei. Thus, an α-particle is unusually tightly bound, and can stay together under some circumstances where a large nucleus is unstable and splits apart.

Spontaneous decay occurs because the binding energy is positive; that is, the mass of the parent nucleus ($^{226}_{88}\text{Ra}$) is *greater* than the sum of the masses of the products. Hence it is energetically favored for the large nucleus to split. In the decay of radium-226, the mass of the parent particle is 226.025402 u, whereas the sum of the mass of radon and helium is 222.017570 + 4.002603 u = 226.020173 u. The mass difference is 0.005229 u, which corresponds to an energy of 4.87 MeV; this represents excess energy that is

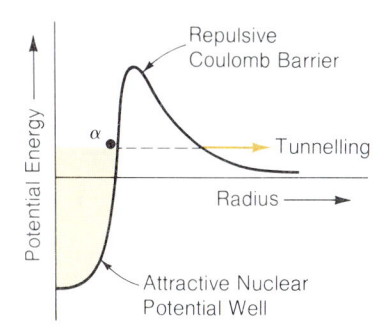

Figure 25.5

Schematic potential curve of an α-particle near a nucleus. Classical physics predicts that the α-particle cannot surmount the potential barrier and escape the nuclear forces. The uncertainty in the position of the particle allows a small chance for the α to "tunnel" through the barrier.

released when the decay occurs. In radium decay, the energy goes into kinetic energy of the products, mostly the α-particle. If all the kinetic energy were gained by the α-particle, its velocity would be $v = \sqrt{2E/m} = 1.5 \times 10^7$ m/s, or about 5 percent of the speed of light.

The process of α-decay is a prime example of **quantum mechanical tunnelling** (first mentioned in Section 24.4). As shown here, the α-particle has enough energy to escape the nucleus, but to do so it must first overcome the short-range nuclear forces. The problem is best illustrated by the potential energy curve of the α-particle near the nucleus (Fig. 25.5). Imagine that the α-particle is separated from the nucleus and approaching it. Outside the nucleus, the potential energy curve rises as $1/r^2$ because of the electrostatic repulsion (this is often called the **Coulomb barrier**). As the α-particle approaches closer to the nucleus, the curve turns and even becomes negative because of the attractive nuclear force. The depth of this potential well is related to the nuclear binding energy. To escape the nucleus, the α-particle must get through the potential barrier. According to classical laws of physics, this is impossible. Because of the uncertainty principle, however, the α-particle cannot be localized completely within the nucleus; its wavefunction spills over outside the nucleus, and there is a small but finite chance that it will suddenly find itself outside. Physicists say that the α-particle has tunnelled through the barrier. It sounds like magic, but once again quantum mechanics appears to be a better description of nature than classical physics.

In considering α-decay, we have implicitly assumed that energy is conserved in a radioactive decay (more specifically, we have assumed that mass and energy together are conserved, since the two are equivalent and interchangeable). Our assumption that the energy that was lost by the parent particle must appear in some other form (kinetic energy of the α-particle in this case) is fully verified by experiment. Furthermore, no spontaneous reactions occur when the mass (that is, the total energy) of the parent is less than that of the products. This also supports our assumption of energy conservation, for if this decay did occur, energy would have to be created somehow. Later we will discuss reactions for which the products have more mass than the initial particles, but we will find that energy must be supplied.

There are other conservation laws that apply to radioactive decays, as we are about to see. For example, the total number of nucleons remains constant, even though some may change their identities. Further, the total electrical charge is constant. Angular momentum is also conserved. It is significant that the seemingly elementary postulates that certain quantities are conserved are so general that we now find them applicable on scales even smaller than that of the atom.

In β-decay, the emitted particle is an electron (hence electrons are sometimes called β-particles). In this case, the total number of nucleons remains constant, but one of the neutrons is replaced by a proton. This satisfies the conservation of charge, because the negative charge on the released electron is balanced by the positive charge on the newly created proton. It is as if a neutron spontaneously turned into two particles, a proton and an electron (in fact, a free neutron does this in a typical time of about 10.8 minutes), and the first experiments to isolate neutrons were very difficult because of this spontaneous decay.

An example of a β-decay is offered by the radioactive isotope of carbon, $^{14}_{6}C$. When an electron is released, one of the 8 neutrons in the nucleus is converted into a proton, so that the new nucleus has 7 protons and 7 neutrons. This is $^{14}_{7}N$, the most common form of nitrogen, and the decay may be written

$$^{14}_{6}C \rightarrow {}^{14}_{7}N + {}^{0}_{-1}e$$

where the symbol $^{0}_{-1}e$ stands for an electron, with negligible mass number (this is sometimes replaced by simply writing e^-, but the more formal notation makes keeping track of the total charge easier).

In energy units, the total mass of the parent is 14.003241 u, while that of the products is less, 14.003074 u. (We have been using total atomic masses throughout, which are the mass values listed in Appendix 7; thus, the mass of the emitted electron is included in the listed mass for $^{14}_{7}N$, even though when the decay of $^{14}_{6}C$ occurs, this electron is a free particle, rather than being bound to the $^{14}_{7}N$ nucleus.) The mass difference is 0.000167 u, corresponding to an energy of 0.156 MeV.

If the decay of carbon-14 is observed, it is found that the kinetic energies of the emitted electrons are not constant and are often less than 0.156 MeV, in apparent violation of the conservation of energy. Furthermore, the linear and angular momenta of the product particles are not conserved. This dilemma was finally resolved in 1930, when W. Pauli proposed that a new kind of particle, very difficult to detect, was emitted during β-decay. To satisfy the conservation laws, this particle needed to have kinetic energy and momentum (including angular momentum in the form of spin), but no charge and, strangely enough, no mass (that is, no rest mass). The needed particle was named the **neutrino** (for "little neutral one") by E. Fermi, who in 1934 worked out a detailed theory for β-decay. We will have more to say about neutrinos in the next chapter, when we discuss elementary particles. For now, we need only be aware that both theory and experiment has confirmed Pauli's original suggestion, and that we must include neutrinos when we discuss β-decays. In complete form, then, the decay of $^{14}_{6}C$ is

$$^{14}_{6}C \rightarrow {}^{14}_{7}N + {}^{0}_{-1}e + \bar{\nu}$$

where the final symbol represents the neutrino (technically an antineutrino, as we shall see later).

There are many examples of β-decays, always involving parent nuclides that have an imbalance of neutrons over protons. Another closely related decay occurs when the number of neutrons is much less than the number of protons. In these reactions, a proton is converted into a neutron, and a positively charged particle called a **positron** is emitted. A positron has the same mass as an electron, but an opposite charge. It is called an **antiparticle** to the electron for this reason (more on antiparticles in the next chapter). This kind of decay is also called β-decay.

In β-decay the weak nuclear force comes into play. This force is the only interaction between a neutrino and other particles, because a neutrino is believed to have no mass, no charge, and no strong interactions. The existence of the weak nuclear force was first postulated by Fermi when he developed the theory of neutrinos.

25.2 RADIOACTIVITY

A third type of reaction involving an electron is the reverse of the β-decay. Sometimes a nucleus can capture one of its orbiting electrons, allowing a proton to be converted into a neutron. This is called **electron capture**, and, since it usually involves one of the K-shell electrons, those closest and most tightly bound to the nucleus, it is sometimes called **K-capture**.

A γ-decay is analogous to the spontaneous emission of a photon by an electron as it drops from an excited state into a lower one. It is not an electron, however, but a nucleon, that makes a downward transition. As we might have expected from quantum mechanics, the nucleons in a nucleus can exist only in discrete energy states. A nucleus tends to stay in its lowest, or ground state, but can be excited to higher levels by collision or decay (or other nuclear reactions). The nucleus will then return to the ground state, emitting a photon whose energy is equal to the difference in energy levels of the nucleus. These differences are very large, a fraction of an MeV or more, so the emitted photon has very high energy. Thus, a γ-ray is emitted, carrying away energy and angular momentum but no charge.

The emission of γ-rays often accompanies γ-decays or electron captures in which a product nucleus is left in an excited state from which it quickly makes a transition to the ground state. If the excited level is metastable, then the emission of the γ-ray photon may be delayed. While in the excited state, the nucleus is called an **isomer**.

The three types of radioactive decays can be written in generic terms, through their effects on atomic number Z and mass number A of the "parent" and "daughter" nuclei:

$$\begin{aligned}
\alpha\text{-decay:} \quad & (Z,A) \rightarrow (Z-2, A-4) + {}^{4}_{2}\text{He} \\
\beta^{-}\text{-decay:} \quad & (Z,A) \rightarrow (Z+1, A) + e^{-} + \bar{\nu} \\
\beta^{+}\text{-decay:} \quad & (Z,A) \rightarrow (Z-1, A) + e^{+} + \nu \\
\gamma\text{-decay:} \quad & (Z,A)^{*} \rightarrow (Z,A) + \gamma
\end{aligned}$$

where the asterisk indicates the nucleus is initially in an excited state. These formulas are useful in finding the decay products.

Example 25.2 Uranium-232 ($^{232}_{92}$U) undergoes α-decay. What is the resulting nuclide, and what are the kinetic energy and velocity of the emitted α-particle, if it receives all the released energy?

If an α-particle is emitted, then Z must decrease by 2 and A by 4 to conserve charge and nucleon number. Thus, the product nucleus must have Z = 90 and A = 228. Element 90 is thorium (Th), so the product is thorium-228. The decay is written symbolically as

$$^{232}_{92}\text{U} \rightarrow {}^{228}_{90}\text{Th} + {}^{4}_{2}\text{He}.$$

The mass of the parent nuclide is 232.037130 u, and the sum of the masses of the products is 228.028715 u + 4.002603 = 232.031318. Thus, the quantity of mass converted into kinetic energy of the α-particle is 0.005812 u, corresponding to an energy of

$$E = mc^2 = (0.005812 \text{ u})(931.5 \text{ MeV/u})$$
$$= 5.4 \text{ MeV}.$$

If this energy is entirely in the form of α-particle kinetic energy, then the velocity of the particle is

$$v = \sqrt{\frac{2E}{m}}$$

$$= \sqrt{\frac{2(8.67 \times 10^{-13} \text{ J})}{(4.002603 \text{ u})(1.66 \times 10^{-27} \text{ kg/u})}} = 1.6 \times 10^7 \text{ m/s}.$$

Example 25.3 The isotope of hydrogen called tritium ($^3_1$H) is unstable and undergoes β-decay. Write the equation for this decay. What is the kinetic energy of the emitted electron if it receives all the energy released?

We know that an electron is emitted, and that in the process a neutron is converted into a proton. Hence the remaining nucleus must have 2 protons and 1 neutron. Therefore the reaction is

$$^3_1\text{H} \rightarrow {}^3_2\text{He} + {}^{\ 0}_{-1}\text{e} + \bar{\nu}$$

(an antineutrino is always emitted during β⁻-decays).

The mass of the tritium atom is 3.016049 u, whereas the sum of the masses of the products is 3.016029 u (note that the e^- is included in the mass of the $^3_2$He atom), so the mass converted into energy is 0.000020 u, which corresponds to an energy of

$$E = mc^2 = (0.000020 \text{ u})(931.5 \text{ MeV/u})$$
$$= 0.0186 \text{ MeV}.$$

Thus, the maximum kinetic energy that the emitted electron could have is 0.0186 MeV or 18.6 keV.

Example 25.4 When radioactive neon-23 ($^{23}_{10}$Ne) undergoes beta-decay, forming sodium-23 ($^{23}_{11}$Na), it sometimes leaves the resulting sodium-23 nucleus in an excited state 0.44 MeV above the ground state. When the excited nucleus drops to the ground state (which it does very quickly), a γ-ray is emitted. What is the wavelength of the γ-ray?

We know that for a photon $E = hc/\lambda$, so solving for λ yields

$$\lambda = \frac{hc}{E}$$

$$= \frac{(6.63 \times 10^{-34} \text{ J} \cdot \text{s})(3 \times 10^8 \text{ m/s})}{(7.04 \times 10^{-14} \text{ J})} = 0.0028 \text{ nm}.$$

25.2 RADIOACTIVITY

Decay Probabilities and Half-Lives

An unstable nucleus can be likened to an electron in an excited state; it has the potential to make a spontaneous transition to another state, but when it will do so cannot be predicted for any individual nucleus. As for electron transitions, we speak of the probability that a nucleus will decay, and we can make accurate predictions only for large numbers of nuclei.

In α- or β-decay, the identity of a nucleus changes. Therefore, if we start with N unstable nuclei, N decreases steadily with time. It is found that the rate of decrease at each instant t is proportional to the number of nuclei present at that time and to the average number of decays per nucleus per second (which is the reciprocal of the average lifetime of a nucleus in the unstable state). This factor, which depends only on the nuclide, is called the **decay constant** λ. Then the number of decays per second is

$$\frac{\Delta N}{\Delta t} = -\lambda N \qquad (25\text{--}1)$$

where the minus sign indicates that N is decreasing with time. This is an example of a general class of equations that can be readily solved (using calculus), leading to

$$N = N_0 e^{-\lambda t} \qquad (25\text{--}2)$$

where N is the number of nuclei remaining at time t, and N_0 is the number originally present (at $t = 0$). The symbol e in this case is the base of the natural logarithm (Appendix 2), approximately $e = 2.71828\ldots$. The decrease in N is an **exponential decrease**, because the rate at any time depends on the number of nuclei left at that time, and the solution has the exponential form given in Eq. 25–2.

If we solve Eq. 25–2 for the fraction that remain at t, we find

$$\frac{N}{N_0} = e^{-\lambda t}.$$

Notice that for any fractional change N/N_0, the value of t is the same, regardless of the initial number of nuclei present. If, for example, we set $N/N_0 = 1/2$, we find that

$$\frac{1}{2} = e^{-\lambda T_{\frac{1}{2}}}$$

where $T_{\frac{1}{2}}$ is called the **half-life** of the nuclide. No matter what quantity N_0 we start with, half of that number will decay in a time $T_{\frac{1}{2}}$ (Fig. 25.6). Thus the half-life is a convenient measure of the radioactivity of an unstable nuclide. To see how $T_{\frac{1}{2}}$ is related to the decay constant λ, we solve this expression for $T_{\frac{1}{2}}$ (using a calculator or a table of exponential functions):

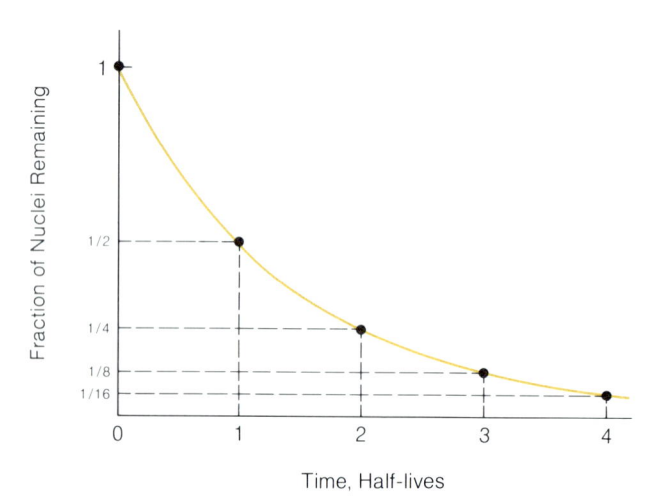

[Graph: Fraction of Nuclei Remaining vs Time in Half-lives, showing exponential decay from 1 at t=0 to 1/2, 1/4, 1/8, 1/16 at t=1,2,3,4 half-lives]

Figure 25.6

Exponential decay curve. The number of nuclei remaining is halved during each half-life. Thus, the greatest change in the number occurs during the first half-life, and the amount of change decreases with each succeeding half-life.

$$T_{\frac{1}{2}} = \frac{\ln 2}{\lambda} = \frac{0.693}{\lambda} \tag{25-3}$$

As an example, consider the decay of $^{14}_{6}C$, which has a half-life of $T_{\frac{1}{2}} = 5730$ yr. From Eq. 25–3, we see that the decay constant is $\lambda = 0.693/T_{\frac{1}{2}} = 3.83 \times 10^{-12}$ s^{-1}, so the average lifetime of a $^{14}_{6}C$ nucleus is $1/\lambda = 2.61 \times 10^{11}$ s $= 8270$ yr. (This is the same as the time required for the number to diminish by a factor of e.)

In time t, the fraction of radioactive nuclei present can be given in units of the half-life. For example, in 11,460 years (two half-lives), $(1/2)(1/2) = 1/4$ of a sample of $^{14}_{6}C$ nuclei will remain, and in 5730 years more, only $(1/2)(1/2)(1/2) = 1/8$ will remain. In general, the fraction remaining after time t is

$$\frac{N}{N_0} = \left(\frac{1}{2}\right)^{t/T_{\frac{1}{2}}}. \tag{25-4}$$

For example, after 10,000 years, the fraction of $^{14}_{6}C$ nuclei left in a given sample is $(1/2)^{10,000/5730} = (1/2)^{1.745} = 0.298$.

Implicit in Eq. 25–4 is a technique for measuring the age of a material that contains a radioactive isotope. Suppose we know the initial quantity of unstable nuclei in a sample, and at some later time we measure how many are present (such measurements are discussed in the next section). To write an expression for the age t, we take the natural logarithm of Eq. 25–4 and solve for t:

25.2 RADIOACTIVITY

$$t = -1.443 T_{\frac{1}{2}} \ln\left(\frac{N}{N_0}\right) \qquad (25\text{--}5)$$

Thus, if we encounter a sample in which we determine that only 1/16 of the original $^{14}_{6}C$ remains, then we find that the age of the sample is $t = -1.443(5730 \text{ yr})\ln(1/16) = 22{,}900 \text{ yr}$, which is equal to $4T_{\frac{1}{2}}$, as we would expect. Using this technique to find the age of something is called **radioactive dating**.

It is not always easy to know what the original concentration of radioactive nuclei was in a sample. We have cited carbon-14 in this discussion, because it is believed that the ratio of $^{14}_{6}C$ to the more common $^{12}_{6}C$ in the Earth's atmosphere has remained quite constant over the history of the atmosphere. All forms of plant life absorb carbon dioxide (CO_2) from the atmosphere, as do plant-eating animals indirectly. So as long as an organism is alive, it maintains a fixed ratio of $^{14}_{6}C$ to $^{12}_{6}C$ in its composition. When it dies, however, it stops taking in CO_2, so from that time on, its $^{14}_{6}C$ content decreases while its $^{12}_{6}C$ content remains constant. Therefore the original quantity of $^{14}_{6}C$ in a fossilized plant can be determined from the amount of $^{12}_{6}C$ it contains, and its current quantity of $^{14}_{6}C$ can be measured directly from its present level of radioactivity. Then Eq. 25–5 can be solved for t, the age of the fossilized plant. There are uncertainties in this technique, primarily because the atmospheric $^{14}_{6}C$ to $^{12}_{6}C$ ratio may have varied over time, but nevertheless carbon-14 dating is one of the most powerful techniques available for finding the ages of dead organisms.

It is sometimes possible to determine the original concentration of an unstable nuclide by measuring the present concentration of the decay product. This is not easily possible for $^{14}_{6}C$, because the product is $^{14}_{7}N$, a common form of nitrogen already abundant. There are elements, however, most of them heavy elements found in rocks, for which the decay product is not thought to be formed by any other process, or at least the quantity resulting from decay can be determined. Some isotopes of uranium, for example, decay into forms of lead that do not otherwise occur, so the present-day ratio of uranium to lead (in the appropriate isotopes) in a rock shows the original quantity of uranium. The original and present amounts of uranium may then be substituted into Eq. 25–5 to find the age of the rock. Uranium (in the form $^{238}_{92}U$) is especially useful for this, because its half-life is 4.5×10^9 yr, comparable to the age of the earth. Other useful nuclides for radioactive dating of rocks include $^{40}_{19}K$ (half-life 1.25×10^9 yr), $^{87}_{37}Rb$ (half-life 49×10^9 yr), and $^{235}_{92}U$ (half-life 0.7×10^9 yr).

Often the product of a decay is itself unstable, and further decay will occur. There may be several steps before a stable product is finally created and the sequence stops. The decay of uranium into lead is an example; this does not occur in one step, as implied by our discussion of radioactive dating, but instead involves intermediate steps. A sequence of nuclides that results from the decay of a single unstable one is called a **decay series**. A series can have branch points (Fig. 25.7), because some unstable nuclei may decay in more than one way. The relative number of decays along the different branches is determined by

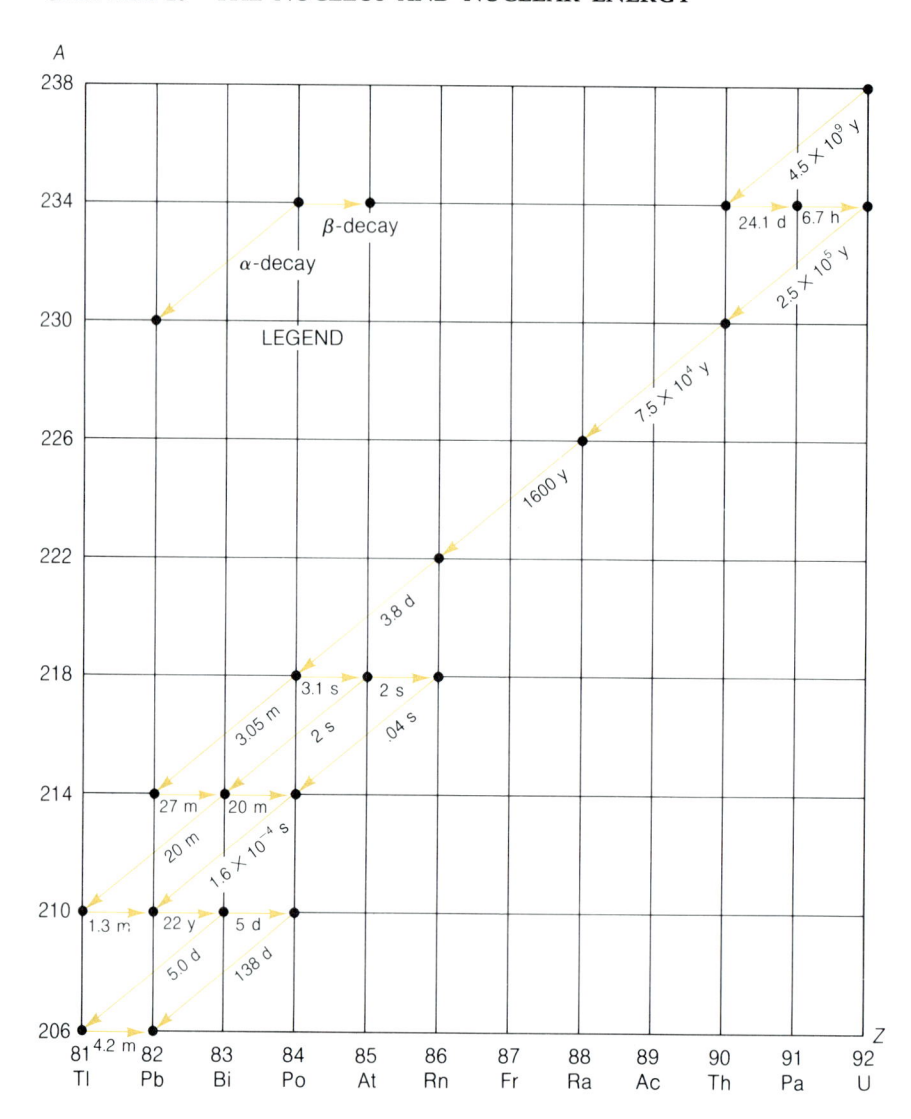

Figure 25.7

Radioactive decay series. Here $^{238}_{92}U$ decays through a series of intermediate stages to $^{206}_{82}Pb$. The diagonal arrows represent α-decays, in which A changes by 4, and the horizontal arrows represent β-decays, in which A is constant but the atomic number Z increases by 1. Note branch points and parallel paths, with half lives indicated.

the relative decay rates. The most favored decay series for the transmutation of $^{238}_{92}U$ into $^{206}_{82}Pb$ is

$$^{238}_{92}U \rightarrow ^{234}_{90}Th \rightarrow ^{234}_{91}Pa \rightarrow ^{234}_{92}U \rightarrow ^{230}_{90}Th \rightarrow ^{226}_{88}Ra \rightarrow ^{222}_{86}Rn \rightarrow ^{218}_{84}Po \rightarrow$$
$$^{218}_{85}At \rightarrow ^{218}_{86}Rn \rightarrow ^{214}_{84}Po \rightarrow ^{210}_{82}Pb \rightarrow ^{210}_{83}Bi \rightarrow ^{206}_{81}Tl \rightarrow ^{206}_{82}Pb.$$

25.2 RADIOACTIVITY 843

We see here that a series may include a large number of steps. Notice that one of the steps is the formation of radium-226 ($^{226}_{88}$Ra), one of the radioactive nuclides first isolated by the Curies. The half-life of this nuclide is about 1600 yr, so sufficient time has passed since the earth formed that no original radium-226 should be present. We conclude that all the present-day radium is a product of the decay of uranium-238 or possibly other isotopes.

We note in passing that radioactivity is an important source of heat in the earth's interior. It has been known for a long time that portions of the interior are molten, but it was not immediately understood why. Eventually it was realized that energy released by radioactivity was responsible (see Special Problem 2 at the end of this chapter).

Example 25.5 Suppose we have a sample of $^{13}_{7}$N, which is radioactive (β^+ emitter) with a half-life of 600 s. If the sample has a mass of 2.0×10^{-4} kg, and is pure initially, what is the decay rate and heat release rate 15 minutes later? What is the fraction that is still in the form of $^{13}_{7}$N?

At time $t = 15$ min $= 900$ s, from Eq. 25–4 the fraction of $^{13}_{7}$N nuclei remaining is

$$\frac{N}{N_0} = \left(\frac{1}{2}\right)^{t/T_{\frac{1}{2}}} = (0.5)^{900 \text{ s}/600 \text{ s}} = (0.5)^{1.5} = 0.354.$$

(Note that we get the same answer if we use Eq. 25–2 and Eq. 25–3).

We need to know N_0 to answer the other questions. If m is the total mass of the sample, and m_N the mass of a $^{13}_{7}$N atom, we find

$$N_0 = \frac{m}{m_N} = \frac{2.0 \times 10^{-4} \text{ kg}}{(13.005738 \text{ u})(1.66 \times 10^{-27} \text{ kg/u})} = 9.26 \times 10^{21}.$$

Therefore the number of $^{13}_{7}$N nuclei remaining at $t = 900$ s is

$$N = 0.354 N_0 = (0.354)(9.26 \times 10^{21}) = 3.28 \times 10^{21}.$$

The rate of decay at time t is found from Eq. 25–1, using Eq. 25–3 to substitute for λ:

$$\frac{\Delta N}{\Delta t} = -\lambda N = \frac{-0.693 N}{T_{\frac{1}{2}}} = \frac{-0.693(3.28 \times 10^{21})}{600 \text{ s}} = -3.8 \times 10^{18} \text{ s}^{-1}.$$

Therefore after 15 minutes, radioactive decays are occuring at a rate of nearly 4×10^{18} s^{-1}; the minus sign shows that the number of $^{13}_{7}$N nuclei is decreasing at this rate.

The heat release rate is the product of the reaction rate and the energy released per reaction,

$$^{13}_{7}\text{N} \rightarrow {}^{13}_{6}\text{C} + {}^{0}_{1}e + \nu.$$

The energy is the difference between the masses of $^{13}_{7}N$ (13.005738 u) and $^{13}_{6}C$ (13.003355 u), which equals 0.002383 u = 2.22 MeV. Technically, only 1.20 MeV is released promptly; the remaining 1.02 MeV comes when the positron annihilates itself with a nearby electron to yield two γ-rays of energy $m_e c^2 = 0.511$ MeV. The net heat release rate is therefore

$$(4 \times 10^{18}\ s^{-1})(2.22 \times 10^6\ eV)(1.6 \times 10^{-19}\ J/eV) = 1.4 \times 10^6\ J/s$$
$$= 1.4\ \text{Megawatts!}$$

Example 24.6

Suppose that $^{206}_{82}Pb$ is present in rocks *only* as the decay product of $^{238}_{92}U$. The half-life for this decay is 4.5×10^9 yr (this includes all of the steps in the decay series). How old is a rock in which the ratio of $^{238}_{92}U$ to $^{206}_{82}Pb$ is 4.3? Ignore any of the original $^{238}_{92}U$ that may be in the form of intermediate nuclides in the decay series (this is a very small fraction, because the half-lives for all the intermediate decays are much shorter than 4.5×10^9 yr).

For simplicity of notation we set the present number of $^{238}_{92}U$ nuclei equal to N_U, and the present number of $^{206}_{82}Pb$ nuclei equal to N_{Pb}. N_0 stands for the initial number of $^{238}_{92}U$ nuclei. We know that N_0 is equal to the sum of N_U and N_{Pb}, and we know that $N_U/N_{Pb} = 4.3$. Therefore the fractional quantity of the original $^{238}_{92}U$ present now is

$$\frac{N_U}{N_0} = \frac{N_U}{N_U + N_{Pb}} = \frac{N_U}{N_U + \dfrac{N_U}{4.3}} = \frac{1}{1 + \dfrac{1}{4.3}} = 0.811.$$

Now we use Eq. 24–5 to find the age of the rock:

$$t = -1.443 T_{\frac{1}{2}} \ln\left(\frac{N}{N_0}\right) = -1.443(4.5 \times 10^9\ \text{yr})\ln(0.811) = 1.4 \times 10^9\ \text{yr}$$

Biological Effects of Radioactivity

When a charged particle passes through matter, it is likely to collide with atoms and molecules and ionize some of them. The process is direct for α-particles and β-particles because they can exert electric forces on electrons and give them sufficient energy to separate them from their parent atoms or molecules. Neutrons, x-rays, and γ-rays, which have no electrical charge, also cause ionizations by colliding with nuclei and disrupting them, creating α- or β-particles, or ionizing atoms by the photoelectric effect. Since α- and β-particles typically have energies of 0.1 to 10 MeV, whereas ionizations of atoms and molecules in cells require only 10 to 50 eV, a single α-particle or β-particle can produce hundreds or thousands of ionizations.

Radiation damage to metals, electronic circuits, and other materials can be a serious problem. The strength of support structures can be diminished, chemical composition can be altered, and computers can malfunction. For example, radiation damage to the walls of nuclear reactors and to space satellites is significant. Even the presence of large numbers of high-energy charged particles in the earth's magnetosphere can cause interference with electronic circuits in space vehicles.

Ionization in living tissue has several harmful effects. Highly reactive molecular ions may be created, for example, which can disrupt normal chemical activity such as the exchange of nutrients and waste products by cells. Radiation can break chemical bonds and deposit large amounts of heat, resulting in serious radiation burns. The β-rays are much more penetrating than α-rays. They can pass through thin sheets of metal or moderate lengths in air. The α-rays are usually stopped by short pathlengths of air or thin paper, but their ionizing power is greater. The γ-rays are 10 to 100 times more penetrating than β-rays, but do not produce as much ionization. As a rough guide, the ionizations produced by α- and β-particles and by γ-rays are in the ratio 10,000 : 100 : 1.

Probably the most serious effect of radiation is the damage to the large protein molecules, DNA and RNA, that are responsible for replication of gene patterns in the reproduction of organisms. Radiation can also destroy the cells in bone marrow that produce red blood cells. Ionization in living tissue would have to be very widespread to seriously harm a plant or animal, but a single damaged DNA molecule can disrupt the reproduction process of a cell and trigger abnormal reproduction leading to cancer, or it may cause mutations in later generations.

Radioactivity has some beneficial roles in biology as well. Minute quantities of radioactive substances such as tritium ($^{3}_{1}$H), $^{14}_{6}$C, or positron emitters can be introduced into organisms and then traced, allowing doctors or research scientists to learn how fluids are transported internally, how certain amino acids are synthesized, or how some chemicals are passed through the body, brain, or food chain. Radiation therapy is used to control or stop the growth of cancer by killing the cells that are reproducing abnormally. The radioactive source or tracer may be a β-emitter, such as $^{60}_{27}$Co, $^{32}_{15}$P, or $^{131}_{53}$I or a beam of photons (x-rays or γ-rays) or π-mesons (subatomic particles which interact strongly with nuclei). Radio-isotopes have the same chemistry as the stable isotope of an element, so the assimilation or concentration of the radio-isotope in certain regions is unaffected by the radioactivity. For example, radioactive $^{131}_{53}$I is useful in studies of the thyroid gland because its assimilation is rapid and its effects are short (half-life of 8.04 days).

It is thought that radiation from space, in the form of **cosmic rays**, is at least partially responsible for the mutations that have aided the evolution of life on earth. Cosmic rays are charged particles, the vast majority of them electrons, protons, or $^{4}_{2}$He nuclei from the sun, so they can interact with matter just as α-rays or β-rays do. The earth's magnetic field prevents most cosmic rays from reaching the ground by trapping them in radiation belts, zones of ionized gas high above the atmosphere. Some do get through, however, particularly at high altitudes or when the radiation belts are disrupted by solar flares or periodically over the age of the earth when its magnetic field reverses.

Measurement of Radioactivity

The level of radioactivity from a source is usually measured in terms of the number of events (decays) per second. A conventional unit is the **curie** (**Ci**), 3.70×10^{10} events/s. In biological applications especially, the **dosage**, the amount of energy deposited by radiation, is measured instead. Historically, the unit used for measuring dosage was the **roentgen** (**R**). This unit is defined in terms of the number of ions produced per unit volume: 1 R of radiation produces 2.08×10^{15} ion pairs/m^3 in air at standard temperature and pressure (an ion pair consists of a positive ion and a negative electron, the products of one ionization). A more general and practical unit has largely superseded the roentgen today; this is the **rad**, the quantity of radiation that deposits 10^{-2} J of energy per kilogram of any absorbing material. The rad is a useful measure of dosage, because it applies to any radiation absorbed in any material. (Note that the dimensions of the rad are joules per kilogram.) For example, the radiation source might deposit more energy in dense bone than in human tissue.

Even the rad does not fully characterize radiation effects, however. As we have seen, one rad of α-radiation produces far more ionization than one rad of β-rays. Radiation experts apply a correction factor, the **relative biological effectiveness** (**RBE**), defined as the equivalent number of rads of x-rays or γ-rays that would produce the same biological damage (see Table 25.1). By definition, the RBE of x-rays or γ-rays is 1.0; for α-particles, it may be as high as 10 to 20. Thus, the product of dosage in rads and RBE yields the most useful unit, called the **rad equivalent man**, or simply the **rem**,

$$\text{rem} = (\text{rad})(\text{RBE}). \qquad (25\text{–}6)$$

With the RBE correction factor, the rem places all types of radiation on a comparable scale, according to its effects on an average human.

The average dose of natural radiation at sea level is about 0.1 rem/year, mostly from cosmic rays and natural radioactivity in rocks, soil, and cement (this exposure rises to 0.2 rem/year at 6000 feet elevation). Some regions, particularly locales near mine tailings or waste sites, are subject to higher radioactivity levels. Employees who work with radio-isotopes are generally

Table 25.1

Relative biological effectiveness of radiation (RBE)

Radiation type	RBE
X-rays or γ-rays	~1
β-rays (electrons)	~1
Slow neutrons	3–5
Fast protons and neutrons	~10
α-particles ($^4_2$He)	10–20
Heavy ions	10–20

25.2 RADIOACTIVITY

required by laws to limit their exposure (film badges provide a convenient way to record one's total exposure). A U.S. Government-recommended limit is 0.5 rem/year. A lethal dose of radiation varies from 300 to 1000 rem in a single exposure, although radiation sickness (severe nausea, hair loss, and drop in white blood cells) can occur at much lower levels. Long-term exposure to low-level radioactivity does not have the same ravaging effects as a single large dose, but our knowledge of the long-term effects is not good, and a prudent person will keep his or her exposure as low as possible.

Example 25.7 A 0.01 Curie source of radioactive uranium ($^{236}_{92}U$) is ingested by a 60 kg person. If all the α-particles are absorbed, calculate the disintegrations per second and the dosage in rad/hr and rem/hr.

From the definition of the Curie, the uranium undergoes $(10^{-2})(3.7 \times 10^{10}$ decays/s) = 3.7×10^8 decays/s. Each decay,

$$^{236}_{92}U \rightarrow {}^{232}_{90}Th + {}^{4}_{2}He$$

releases $(236.045562 - 232.038054 - 4.002603)u = 0.004905$ u $= 4.57$ MeV $= 7.31 \times 10^{-13}$ J of energy. In one hour (3600 s), the body will absorb an energy of

$$\frac{(3.7 \times 10^8 \text{ decays/s})(3600 \text{ s/h})(7.31 \times 10^{-13} \text{ J})}{(60 \text{ kg})} = 0.016 \text{ J/kg} \cdot \text{h}.$$

Since 1 rad equals 0.01 J/kg, the dosage is 1.6 rad/h. The RBE of α-rays is 10–20, so the rem dosage, equal to the dosage in rads times the RBE, is between 16 and 32 rem/h. The person would absorb a lethal dose of 400 rem in less than 25 hours.

FOCUS ON PHYSICS

The Effects of Low-level Ionizing Radiation

As industrial and military uses of nuclear energy have developed over the past forty years, public awareness and concern about the possible harmful effects of radiation have grown apace. Although the damage to living organisms from high levels of radiation is well documented, there is considerable

(continued)

uncertainty about the effects of the very low levels that are so much under discussion today. It is very complicated to determine the risk due to low-level radiation.

For many people living in the United States, natural sources of radiation are at least as important as man-made sources. Ionizing particles from space and from natural radioactivity in rocks and soil result in an average annual body dose of 90 millirem per year throughout the United States. (There is some debate over whether some locales, such as high-altitude regions, have higher natural levels, but the most recent studies show that for the predominantly indoor lifestyles of most of us, the level is about the same throughout the country.) The level of natural radiation is largely beyond anyone's control, and must be accepted.

Most of the current interest and concern is about man-made radiation, but the levels reported are in most cases no greater than natural levels. The average dose in the United States due to artificial sources is 100 millirem per year, nearly the same as the average dose due to natural sources. For most people, medical diagnosis contributes the greatest part of the man-made dosage; a chest x-ray, for example, gives 40 millirem.

One of the most widespread causes of above-average doses is a natural one: the radioactive isotope radon-222 (and other isotopes into which it decays). This isotope, present in soils, can cause annual lung doses as high as 3,000 millirem, depending on the materials and design of the home (underground houses are especially vulnerable). There is great variation about this level, so even higher doses occur. Since natural sources of radiation are common and unavoidable, it is ironic that there has been little public concern about elevated natural exposure to radiation.

The principal sources of public concern, whether the source of radiation is man-made or natural, are long-term or delayed effects, primarily cancer. It is clear that large doses of ionizing radiation (in excess of hundreds of times the natural background) cause increased rates of incidence of cancer (although it is ironic that one of the most effective and widely used treatments for cancer is ionizing radiation, which can kill cancerous cells).

What is not yet clear is whether or how low doses cause cancer.

Most of what is believed about the low levels of radiation is based on what is known about high levels. It is very difficult to carry out direct observations or experiments that show a cause-and-effect relationship between low levels and harmful effects. The main reason is that it may be years or decades from the time of exposure until the harmful effect appears. In the meantime, many other factors that may contribute to cancer can come into play, so the results are ambiguous at best. At least 1,500 chemical and physical agents are known to produce cancerous effects similar to the effects of radiation. It is not known how all these potential causes of the disease interact with each other, but many causes of cancer are recognized. Low levels of radiation may contribute to causing cancer, but that other factors may be necessary as well. Unfortunately, the long timescales and possible multiplicity of causes make statistical studies very complex, so that different people studying the same data can come to remarkably different conclusions.

It is interesting that no more than 3% of cancer cases in the United States can be plausibly attributed to ionizing radiation from all sources, yet concern over radiation as a carcinogen has become dominant in the minds of many people. Other sources (such as smoking, for example, which is responsible for 30% of all cancers) receive much less attention relative to their importance.

Perhaps what is most bothersome about radiation as a source of harmful effects is its invisibility and the general ignorance and even hysteria about its nature and effects. The ignorance (but in most cases not the hysteria) extends into the scientific community, because of the difficulties of making unambiguous tests of low levels of radiation as causes of cancer. Hanging in the balance is the resolution of the so-called energy crisis, a problem that could be solved by nuclear energy if satisfactory ways are found to overcome its dangers, both real and perceived.

Because there is so much confusion and contradiction concerning the sources and effects of low-level radiation, we give a brief list of references used in preparing this article.

Gertz, S. M. and M. D. Loudon (eds), Statistics in the Environmental Sciences, *American Society for Testing and Materials Special Technical Publications,* No. 845, 1984.

Harley, J. H., Environmental Radioactivity—Natural, *Reports of the National Council on Radiation Protection and Measurements,* No. 5, 1983.

Natural Background Radiation in the United States, *Reports of the National Council on Radiation Protection and Measurements,* No. 45, 1975.

Exposure From the Uranium Series with Emphasis on Radon and its Daughters, *Reports of the National Council on Radiation Protection and Measurements,* No. 77, 1984.

Evaluation of Occupational and Environmental Exposures to Radon and Radon Daughters in the United States, *Reports of the National Council on Radiation Protection and Measurements,* No. 78, 1984.

Principles for Limiting Exposure of the Public to Natural Sources of Radiation, *Annals of the International Council on Radiation Protection* (Pergamon Press: Oxford), 1984.

Report of the National Institutes of Health Ad Hoc Working Group to Develop Radioepidemiological Tables, *National Institutes of Health,* Publication No. 85-2748, United States Department of Health and Human Services, 1985.

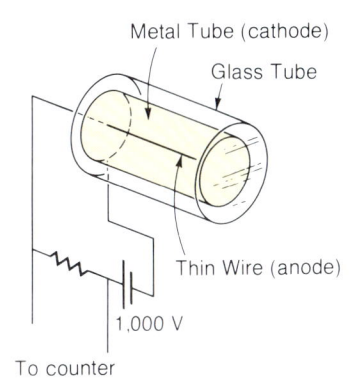

Figure 25.8

The Geiger counter. A high voltage is maintained between the central wire and the metal tube. When a charged particle enters the chamber and ionizes a few gas atoms, the freed electrons quickly cause more ionizations. An avalanche of electrons reaches the central wire, creating a pulse that can be amplified, displayed and recorded.

Several devices are used to detect and measure radioactivity. Some are as simple as a strip of film that will darken upon exposure to radiation, and others are quite sophisticated. Probably the best known is the **Geiger counter,** in which a charged particle entering an enclosed tube causes ionization in a gas kept under an applied potential difference (Fig. 25.8). The applied voltage plays two roles. First, it keeps the atoms in the gas excited nearly to the point of ionization. Therefore little additional energy is required for the incoming particle to trigger an ionization. Second, after ionization, the potential accelerates the electron so that it collides with other atoms, causes more ionizations, and creates more free electrons. Thus, when a single charged particle enters the tube it creates a burst of electrons. This burst is collected on the anode, where it creates an electrical signal that can be recorded or used to produce an audible click.

In a **scintillation counter,** atoms in a light-tight enclosure (usually an opaque crystal such as sodium iodide, NaI, or certain plastics) are excited by incoming particles so that they emit photons (Fig. 25.9; this emission process is sometimes called scintillation). These photons are then detected by a photomultiplier tube, a device that detects light by the photoelectric effect. Photons cause the ejection of electrons from a surface called a **photocathode,** and then a voltage accelerates them to an anode where they are detected electronically. Scintillation counters are usually more sensitive to incoming charged particles than Geiger counters, particularly for the detection of γ-rays, which are uncharged and therefore not readily detected by Geiger counters. The material that scintillates can be a liquid or solid, either of which is more dense than the gas used in a Geiger counter, so the probability that an incoming particle will interact with an atom and create a detectable signal is greater.

The paths of individual α-rays or β-rays can be traced in two types of instruments. In one, called a **cloud chamber** (Fig. 25.10), gas is cooled nearly to the point of condensation. When a charged particle passes through the

850 CHAPTER 25 THE NUCLEUS AND NUCLEAR ENERGY

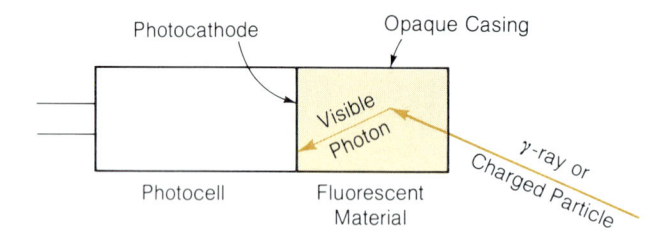

Figure 25.9

The scintillation counter. A γ-ray or charged particle creates visible-light photons in a material (often a NaI crystal) that fluoresces in visible wavelengths. The secondary photons strike the cathode of the photocell and are detected as an electrical impulse.

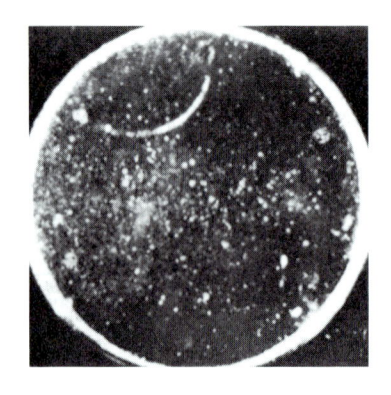

Figure 25.10

Particle tracks in a cloud chamber.

chamber, gas molecules are ionized along its path, and the vapor condenses on the ionized molecules. This produces a track of tiny bubbles that can be photographed. In the second type of device, called a **bubble chamber,** a liquid is heated near its boiling point, so that vaporization along the path of a charge particle creates bubbles that can be photographed. The bubble chamber is generally more sensitive than the cloud chamber because the liquid it uses is denser and more likely to interact with an incoming particle. In either device, a magnetic field may be applied so that the incoming particle follows a curved path whose radius provides information on the charge-to-mass ratio of the particle.

In another type of particle-tracing instrument, a **spark chamber,** a charged particle creates secondary electrons when it passes through a series of closely-spaced conducting plates with large potential differences between adjacent pairs (Fig. 25.11). The electrons cause horizontal arcing across the gaps between plates, making the path of the particle visible and allowing it to be photographed.

A number of other, more specialized devices are used and new developments are occurring. For example, semiconductors have recently been effectively used to detect ionization caused by incoming charged particles and high-energy x-rays.

25.3 Nuclear Reactions

The radioactive decay processes we have just discussed are examples of spontaneous nuclear reactions. Nuclear reactions are said to occur any time nuclei undergo changes as a result of the loss or gain of nucleons, or the change of nucleons from one form into another. Reactions in which nuclei split into smaller ones are **fission** reactions; those in which nuclei merge to form larger ones are **fusion** reactions. If energy is transferred from binding energy to other forms, such as kinetic energy and γ-rays, a reaction is said to be **exothermic.** All spontaneous decays must be exothermic, as discussed in the preceding

25.3 NUCLEAR REACTIONS

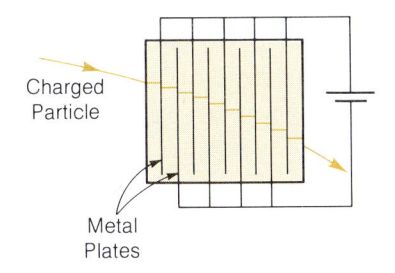

Figure 25.11

The spark chamber. Adjacent metal plates have large potential differences, so that horizontal spark discharges occur between them when a charged particle passes through, leaving a trail of ions.

section. There can also be exothermic reactions that are very unlikely to occur spontaneously, but which will take place readily if some energy is supplied to overcome an initial barrier. In this section we will discuss such reactions. In **endothermic** reactions there is a net loss of energy because more is required to make the reaction occur than is produced. These reactions are observed only in particle accelerators or under extremely energetic astrophysical conditions, which we will only mention in the next chapter.

Fission Reactions

During the years before World War II, scientists in several countries were exploring the properties of the nucleus by experimental techniques. The study of spontaneous decay processes continued, but attempts were being made to trigger nonspontaneous reactions. The usual method was to bombard a substance with particles by exposing it to a radioactive material (recall that Rutherford's scattering experiments used α-rays). The Italian physicist Enrico Fermi realized that neutrons might be more effective than positively-charged particles such as α-rays, since neutrons are not repelled by the positive charge on the nucleus.

Fermi was actually attempting to create new elements by fusion when he set the stage for the first fission reactions. He bombarded the heaviest natural element, uranium, with neutrons in hopes of creating even heavier elements (which he apparently succeeded in doing). When a group of German scientists led by Otto Hahn followed up on Fermi's experiments, they realized that the products of the reactions contained nuclei smaller than uranium. Later experiments showed that when a large nucleus gains neutrons, it also gains energy and becomes highly excited. An excitation of a single nucleon would most likely be followed by a drop to the ground state and the emission of a γ-ray photon. For uranium, however, the excitation is much greater, and many of the nucleons are set into motion. The entire nucleus oscillates and changes its shape much like a liquid droplet (Fig. 25.12); the result can be a fission reaction in which the nucleus splits into fragments. Many different products may be formed in the process; all that is required is that the sum of the masses of the products be less than the mass of the initial nucleus, so that the fission reaction is exothermic. This is generally possible for nuclei with mass numbers A greater than 80, which are on the downward-sloping portion of the binding energy curve (Fig. 25.3), since by splitting into smaller, more tightly bound fragments the nucleus can release its extra binding energy. In all of the possible reactions, free neutrons are released along with the larger nuclei. These new neutrons can collide with other uranium nuclei, causing further reactions which produce more neutrons, and so on, in a chain reaction.

In a typical uranium fission, the reaction starts with $^{235}_{92}$U and a neutron, whose addition creates the very short-lived isotope $^{236}_{92}$U, which then splits and releases neutrons. The steps in this reaction may be written

$$^{1}_{0}n + ^{235}_{92}U \rightarrow ^{236}_{92}U \rightarrow ^{141}_{56}Ba + ^{92}_{36}Kr + 3\,^{1}_{0}n$$

where the symbol $^{1}_{0}n$ represents a neutron. These particular fragment nuclei,

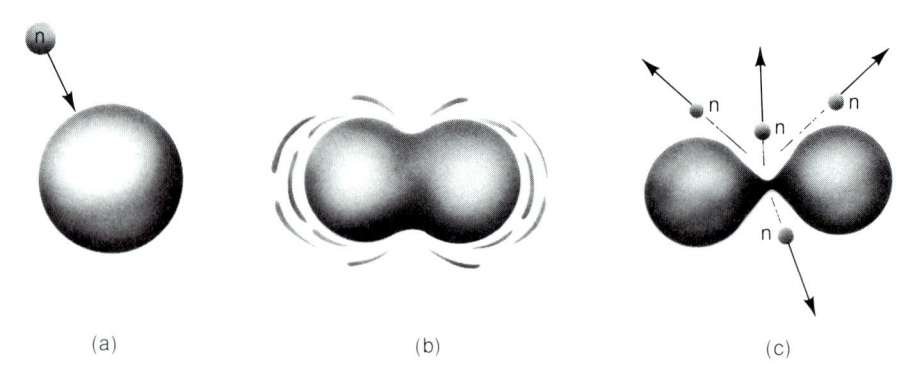

Figure 25.12

A fission reaction. A large nucleus (a) becomes excited when struck by a slow neutron. The nucleus becomes distorted and undergoes oscillations (b). It splits into fragments (c) and releases free neutrons.

barium-141 and krypton-92, are typical products, but many others may result instead.

A large quantity of energy is released in a fission reaction. As an example, we can calculate the yield in the reaction

$$^{1}_{0}n + ^{235}_{92}U \rightarrow ^{90}_{38}Sr + ^{136}_{54}Xe + 10\, ^{1}_{0}n$$

which is another possible result of the interaction of a neutron with uranium-235 (we have not shown the intermediate step $^{236}_{92}U$). The total mass of the reactants is 1.008665 u + 235.043924 u = 236.052589 u, and the total mass of the products is 89.907738 u + 135.907214 u + 10(1.008665 u) = 235.901602 u. The difference is 0.150987 u, corresponding to a released binding energy of 141 MeV, an enormous quantity. The energy released depends on the reaction products. The average energy per $^{235}_{92}U$ disintegration is about 200 MeV. One gram of $^{235}_{92}U$ would release about 8×10^{10} J of energy (for comparison, a medium sized city uses 3×10^{13} J of energy each year). Even if this energy were released gradually, over a period of a year for example, the power output it represents is substantial (a steady release of this amount of energy in one year corresponds to a power of 2.5 kW). Table 25.2 gives the relative energy contents of various phenomena.

Once the possibility of fission reactions was known, and the potential for energy production was appreciated, it became important to try to sustain a chain reaction. This was a very complex business, because a release that was too rapid could be destructive. Fermi was the first to construct a reactor in which controlled fission reactions took place (Fig. 25.13).

The minimum condition for a sustained reaction is that at least one product neutron from each fission go on to cause another fission (Fig. 25.14). An average of 2.5 neutrons per reaction is produced by the fission of uranium-235. Not all of these will interact with other uranium nuclei, however, because some may escape through the surface of the uranium sample. If there is a sufficient quantity of uranium-235, however, then on the average each free

25.3 NUCLEAR REACTIONS

Table 25.2
Energy equivalents

	Energy*	Energy (Joules)*
Energy of hydrogen 21-cm line	6×10^{-6} eV	9.5×10^{-25}
Hydrogen ionization	13.6 eV	2.18×10^{-18}
Electron rest mass ($m_e c^2$)	511 keV	8×10^{-14}
One U-235 fission	188 MeV	3×10^{-11}
Proton rest mass ($m_p c^2$)	938 MeV	1.5×10^{-10}
Athlete sprinting (KE)		3000
Car at 60 mi/h (KE)		10^6
One kilowatt-hour		3.6×10^6
One ton TNT explosive		5×10^9
One barrel fuel oil		6×10^9
One megaton H-bomb		5×10^{15}
Earthquake (8.0 magnitude)[a]		2.5×10^{16}
Hurricane (KE of winds)		10^{17}
Krakatoa volcano (1883)		10^{18}
Tambora volcano (1815)		8×10^{19}
US annual energy consumption		10^{20}
World fossil fuel reserves (estimated)		3×10^{22}
Supernova explosion of star		10^{44}
Radio galaxy (KE of hot gas)		10^{54}
Rest mass of universe (estimated)		10^{69}

*One electron volt (eV) equals 1.6×10^{-19} J. The symbols keV = 10^3 eV; MeV = 10^6 eV.
[a]Richter earthquake scale: Magnitude = $0.667 \log_{10}(E/2.5 \times 10^4$ Joules)

Figure 25.13

Fermi's first reactor.

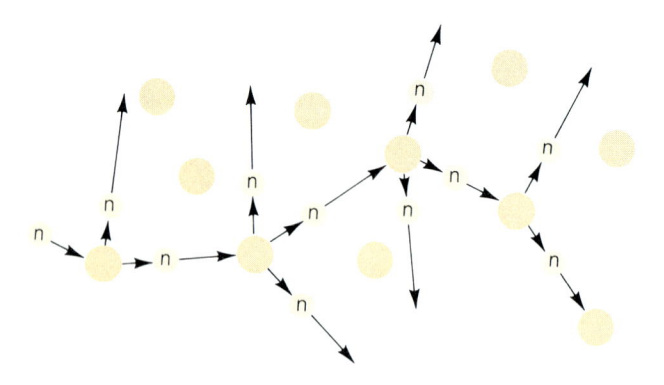

Figure 25.14

Self-sustaining reaction. Each time a nucleus is hit by a neutron, a fission reaction occurs, releasing additional neutrons. If, on average, each fission produces at least one neutron that causes another nucleus to fission, then the reaction is sustained.

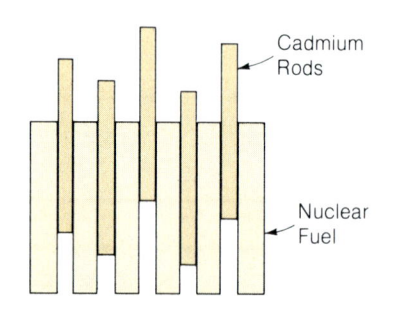

Figure 25.15

Control rods. Rods made of cadmium can be moved in and out between fuel elements, controlling the number of free neutrons, and moderating the reaction rate.

neutron will encounter a uranium nucleus so the reaction will be sustained. The minimum amount for this to happen is called the **critical mass**. Once this mass is reached (its value depends on the purity of the $^{235}_{92}U$ and other factors), the number of free neutrons grows exponentially. If not controlled, the enormous energy of fission reactions can be devastating. This is how an atomic fission bomb, such as those used at the end of World War II, works.

The first fission chain reactions that took place were controlled, however. Fermi's reactor, first tested in 1942, used graphite and cadmium to absorb neutrons. The cadmium was in the form of movable control rods, which absorbed enough neutrons to inhibit the reaction entirely (Fig. 25.15). When these rods were pulled out, neutrons began to interact with uranium-235 nuclei, producing more neutrons, and a sustained reaction was under way. We will discuss nuclear reactors more fully in the latter part of this chapter.

Fusion Reactions

The possibility that fusion reactions might occur was explored theoretically in the 1920s and 1930s, partly in response to the great mystery surrounding the Sun's unknown source of energy. It was known that the Sun was over four billion years old, and that no chemical or mechanical energy source could produce the quantity of energy the Sun must have released in its lifetime. Spurred by this and the mass-energy equivalence embodied in Einstein's theory of relativity, many scientists pondered the possibility that nuclear fusion reactions might be the Sun's internal energy source. By the late 1930s, a fairly complete theory of solar nuclear reactions had been developed. In 1952, the

25.3 NUCLEAR REACTIONS

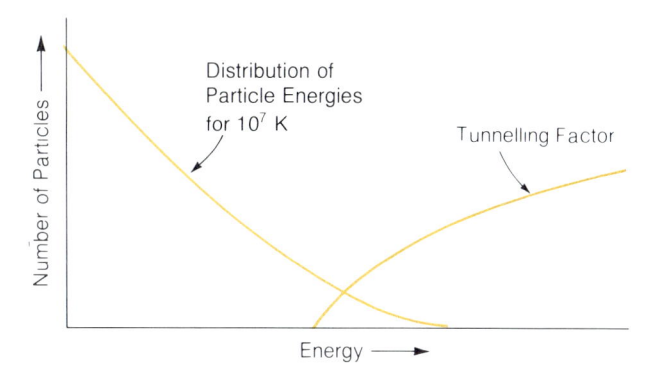

Figure 25.16

Fusion probabilities. Even at very high temperatures (10^7 K in the Sun), very few nuclei have sufficient kinetic energy to overcome the electrostatic repulsion due to their like charges. Reactions occur at a low rate, due to quantum mechanical tunnelling of particles through the potential energy barrier.

first fusion bomb (hydrogen bomb) was exploded. Since the 1960s, scientists at many laboratories have pursued the development of a controlled fusion reaction. Since 1980, laboratory fusion reactions have been performed, but do not yet produce more energy than is required to create them.

In a fusion reaction, two relatively light nuclei come close enough for the strong nuclear force to bind them into a single, larger nucleus. For this to happen, the nuclei must collide at very high speed to overcome the Coulomb barrier (Fig. 25.5), because otherwise their positive electrical charges would keep them too far apart for the nuclear binding force to act. Consequently, fusion reactions require very high temperatures.

Even at very high temperatures such as those inside the Sun (about 10^7K), the potential energy barrier created by the electrostatic force is greater than the typical particle kinetic energy (Fig. 25.16). Thus, except for quantum mechanical tunnelling (see Section 25.2) these reactions would not be possible. The location of a nucleus at any time is not perfectly defined, but instead follows a probability distribution represented by its wave function. This means that there is a finite probability that a nucleus can penetrate the potential energy barrier, even if the kinetic energy of the nucleus is less than required to go over the barrier. Because tunnelling is a low-probability event, and because the average particle kinetic energy is well below the energy needed to overcome the potential barrier, only a very small fraction of collisions between nuclei result in reactions. Therefore not only high temperature is needed, but also a high density of particles, so collisions can be sufficiently frequent.

One of the simplest fusion reactions is two nuclei of hydrogen forming a deuterium nucleus:

$$^1_1\text{H} + ^1_1\text{H} \rightarrow ^2_1\text{H} + ^{\ \ 0}_{+1}e + \nu$$

where $_{+1}^{0}e$ represents a positron and ν a neutrino (these last two particles are required to conserve charge and momentum). The mass of the reactants is 2.015650 u, and that of the $_{1}^{2}H$ is 2.014102 u. The difference is 0.001548 u, corresponding to an energy release of 1.44 MeV. Some energy is carried away by the neutrino, and the rest becomes kinetic energy of the positron and the deuteron (the deuterium nucleus) and the energy released when the positron annihilates with a nearby electron, releasing $2m_ec^2$, or 1.02 MeV.

The energy in the Sun is produced by a sequence of reactions, called the **proton-proton chain**, of which the fusion of two hydrogen nuclei is the first step. Following this, the deuteron combines with another hydrogen nucleus:

$$_{1}^{2}H + _{1}^{1}H \rightarrow _{2}^{3}He + \gamma.$$

The energy released is 5.49 MeV. Some energy is retained by the γ-ray photon, and the rest goes into kinetic energy of the $_{2}^{3}He$ nucleus. Next, the $_{2}^{3}He$ nucleus combines with another:

$$_{2}^{3}He + _{2}^{3}He \rightarrow _{2}^{4}He + 2_{1}^{1}H$$

In this step, the energy released is 12.86 MeV, which goes into kinetic energy of the product particles. Here the reaction sequence stops, so the net effect is that four hydrogen nuclei are merged to form a helium nucleus, releasing a total of 26.72 MeV of energy in the process (recall that the first two steps have to occur twice each to produce two $_{2}^{3}He$ nuclei for the final step). If the total energy that could be produced in the Sun by this sequence of reactions is calculated, it is found that there is sufficient energy supply available to power the Sun for about 10^{10} years, about twice its current age. Thus, fusion reactions provide a plausible power source for the Sun and stars (see Special Problem 3).

Fusion reactions are thought to be responsible for the origin of most of the chemical elements in the universe. The proper conditions of particle density and temperature for fusing protons and neutrons to make $_{2}^{3}He$, $_{2}^{4}He$, and $_{3}^{7}Li$ are thought to have prevailed during the first 3 or 4 minutes of the lifetime of the universe (see Special Problem 4). The universe appears to have begun in a very hot, highly compressed state, from which it has been expanding ever since; this is discussed in some detail in Chapter 26. Heavier elements, such as carbon, nitrogen, and oxygen are formed by fusion in the hot interiors of stars. Even heavier elements, such as iron and beyond (gold, silver, lead, etc.) are believed to form in rapid neutron-capture reactions during the fantastically hot conditions that prevail when stars explode. We will discuss the origin of the elements at greater length in the next chapter.

Controlled nuclear fusion reactions would be a very attractive source of energy for society, but there are severe technical difficulties. The advantages over fission reactions are that neither the reactants nor the products need be radioactive (although they often are), that there is no danger of a runaway reaction (the reaction would just stop in the event of any malfunction), and that the reactants are extremely plentiful and cheap. Active research toward controlled nuclear fusion is under way in many places; this will be discussed a bit more in the next section.

25.4 Particle Accelerators and Nuclear Reactors

Progress in understanding nuclei and nuclear processes has been accompanied by the development of a variety of massive, expensive machines. Some analyze properties of nuclear and subatomic particles; others extract useful energy from nuclear reactions. In this section we briefly describe the principles of nuclear technology.

Accelerators

The first successful device for accelerating subatomic particles to very high speeds was the **cyclotron,** developed in 1930 by E. O. Lawrence and M. S. Livingston. In a cyclotron, charged particles are accelerated by potential differences and constrained to move in a circular path by magnetic fields. The charged particles (usually protons) move in two semicircular cavities called "dees," with the region of accelerating potential between them (Fig. 25.17). This potential must alternate directions in phase with the particles. The frequency of revolution of the particles is found by setting the magnetic force qvB equal to the centripetal force mv^2/r, and solving for v:

$$v = \frac{qBr}{m} \qquad (25\text{-}7)$$

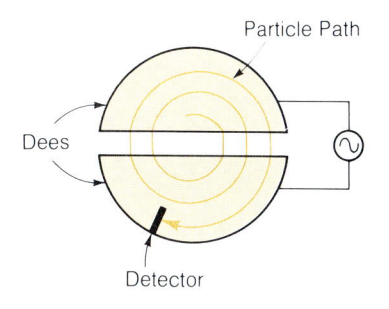

Figure 25.17

The cyclotron. A charged particle is constrained by a magnetic field inside the dees to follow a circular path. Its velocity is increased by a potential difference each time it crosses the gap between the dees; therefore the radius of its path increases.

where q is the charge on the particles, B is the magnetic field strength (assumed to be uniform), r is the radius of curvature (essentially equal to the outer radius of the dees), and m is the particle mass. The period for each trip around is $T = 2\pi r/v$, and the frequency is the inverse of the period.

$$\nu = \frac{v}{2\pi r} = \frac{qBr}{2\pi rm} = \frac{qB}{2\pi m}. \qquad (25\text{-}8)$$

Notice that this value depends only on the charge and mass of the particle and the strength of the magnetic field, but not on the radius of the orbit. Therefore the frequency of the alternating potential need not be adjusted as the particles accelerate. This is not true if the particles reach relativistic velocities, however, because then their masses increase and the frequency decreases. Then sophisticated control systems are needed to adjust the frequency of the applied voltage.

In rarefied, ionized gas in space or the earth's upper atmosphere, charged particles undergo centripetal acceleration in the presence of a magnetic field, just as in an accelerator. The frequency of these natural motions is called the **cyclotron frequency.**

As the velocity v of a particle in a cyclotron is increased by many trips through the gap between the dees, the radius increases until the particle strikes the outer wall or a target placed at a certain radius. The speed, hence the

Figure 25.18

(a) The ring accelerator at Fermi National Accelerator Laboratory in Batavia, Il. (b) A portion of the underground Fermilab accelerator ring, showing superconducting magnets.

kinetic energy, of the particle depends on the radius, so the energy of the particles when they strike the target can be both known and controlled.

25.4 PARTICLE ACCELERATORS AND NUCLEAR REACTORS

Rather than adjusting the frequency of the applied voltage as the particles become relativistic, it is also possible to adjust the strength of the magnetic field. The value of B can be increased to compensate for the increase in mass as the particles accelerate. Accelerators which adjust the magnetic field are called **synchrotrons**. The highest particle energies are achieved by synchrotrons, and these machines can be enormous. The largest to date is the CERN accelerator in Switzerland, which has a radius of 1.1 km and produces particle energies as high as 400 GeV (a GeV is 10^9 eV); the accelerator at the Fermi National Accelerator Laboratory in Batavia, Illinois (Fig. 25.18), has a radius of 1 km and will soon exceed 1000 GeV. A much larger (20,000 GeV) facility is being planned for a U.S. site not yet selected; funding and construction may begin in a few years.

Rather than a single gigantic magnet, large synchrotrons employ a series of smaller magnets located along the circumference of the circular path, with the accelerating voltages applied in the spaces between magnets. Particles must be introduced into such an accelerator with an initially high velocity, since they must begin with a large radius of curvature, so the particles are accelerated first by a smaller accelerator.

Particle energies as high as a few tens of MeV can be achieved by a cyclotron; such energies are comparable to typical nucleon binding energies and are used to study nuclear composition and energy levels. To explore the properties of more fundamental particles (such as those discussed in the next chapter), however, requires far greater energies. Furthermore, to show fine

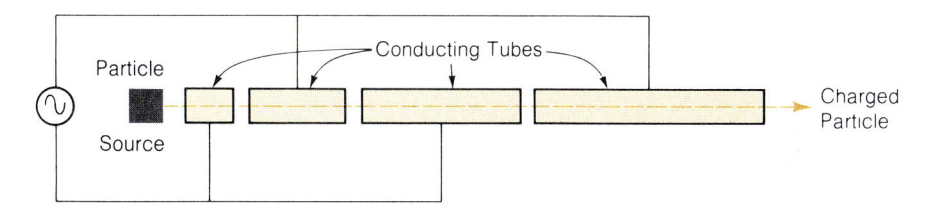

Figure 25.19

A linear accelerator. A charged particle is accelerated by the potential differences in the gaps between conducting tubes. The voltage must be alternated in synchronization with the passage of the particles through the gaps, so the potential difference encountered by the particle at each gap has the same sign. As the particle accelerates, the lengths of the tubes increases to keep the travel time between gaps constant.

details of nuclear or particle structure with scattering experiments requires a very short wavelength of the scattering particles. According to de Broglie's equation, $\lambda = h/p$, a large momentum is needed. Large synchrotrons currently accelerate particles to energies as high as 10^6 MeV (or 1000 GeV). Even higher energies are expected in the next few years.

Electrons are sometimes used as the impacting particles in scattering experiments, and they are often accelerated in **linear accelerators**. These devices (sometimes used for positive ions as well) employ a series of conducting tubes of alternating potential, so that a charged particle is accelerated each time it crosses a gap between conductors (Fig. 25.19). The length of the accelerator

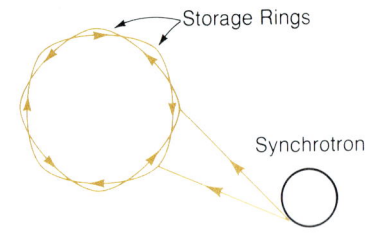

Figure 25.20

Storage ring beam collider. Particles are injected at high energy (from the synchrotron) into two rings where they may circulate for a long time under the control of magnets. Where the rings intersect, the particles collide. (Based on the Intersecting Storage Rings at CERN, located near Geneva, Switzerland.)

determines how high the particle speeds can get, but for the low-mass electrons, relativistic velocities can be reached quickly.

When very high-energy impacts of particles are desired, a class of devices using **colliding beam** accelerators may be employed (Fig. 25.20). Both the target and impacting particles are accelerated so that the total energy of the collisions is the sum of the energies of the two types of particles. Usually a single accelerator produces both types of beams, by accelerating first one particle, then the other, and injecting them into **storage rings,** which are circular chambers where magnetic fields are used to maintain particle velocities for long periods (up to several hours). Once the two types of particles have been injected into the storage rings, they collide where the rings intersect.

In all accelerators, but especially those which reach highly relativistic velocities, energy is lost in the form of electromagnetic radiation. Because the charged particles are being accelerated, they emit photons by a process that has come to be called **synchrotron radiation**. From the point of view of accelerating the particles to some desired energy level, this is a disadvantage, but on the other hand, such radiation can be useful. The spectrum intensity is nearly constant over a wide range of photon energies, and there is substantial x-ray and ultraviolet radiation, so synchrotron radiation is a useful source of intense, high-energy beams of photons.

Like cyclotron motion, synchrotron radiation also occurs in nature, but was not discovered until after the development of man-made devices. In high-energy objects in the universe, electrons accelerated to relativistic velocities by strong magnetic fields are constrained to spiral around the field lines (Fig. 25.21) and emit photons. This intense continuous radiation is highly polarized. Observation and analysis of such radiation has become a major tool for astrophysicists who study such objects as exploded stars (Fig. 25.22), radio galaxies and quasars.

Example 25.8

Suppose a certain nuclear reaction requires incident protons with an energy of 8.5 MeV. Can this be achieved by a cyclotron whose maximum radius is 0.3 m, if the magnetic field is 1.5 T? What is the cyclotron frequency?

The first question requires finding the particle kinetic energy when the accelerated particles reach the maximum radius. We will assume that the protons are not moving at relativistic speeds. The particle velocity is given by Eq. 25–7, so the energy is

$$\begin{aligned} KE &= \frac{1}{2}mv^2 = \frac{(qBr)^2}{2m} \\ &= \frac{[(1.6 \times 10^{-19} \text{ C})(1.5 \text{ T})(0.3 \text{ m})]^2}{2(1.67 \times 10^{-27} \text{ kg})} \\ &= 1.55 \times 10^{-12} \text{ J} = 9.7 \text{ MeV}. \end{aligned}$$

Therefore this small cyclotron can produce sufficiently energetic protons.

The cyclotron frequency, from Eq. 25–8, is

$$\nu = \frac{qB}{2\pi m} = \frac{(1.6 \times 10^{-19} \text{ C})(1.5 \text{ T})}{2\pi(1.67 \times 10^{-27} \text{ kg})} = 2.3 \times 10^7 \text{ Hz}.$$

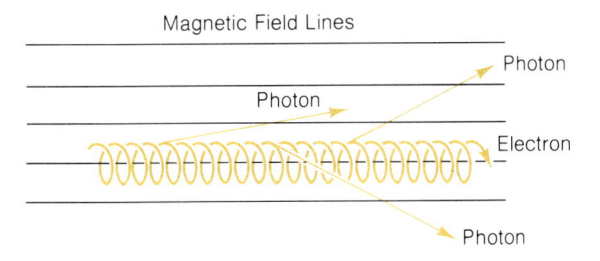

25.4 PARTICLE ACCELERATORS AND NUCLEAR REACTORS   861

Figure 25.21

Synchrotron radiation. As a charged particle (usually an electron) moves at relativistic velocity in a magnetic field, it is forced to spiral around the field lines. The acceleration due to the magnetic field causes the emission of photons.

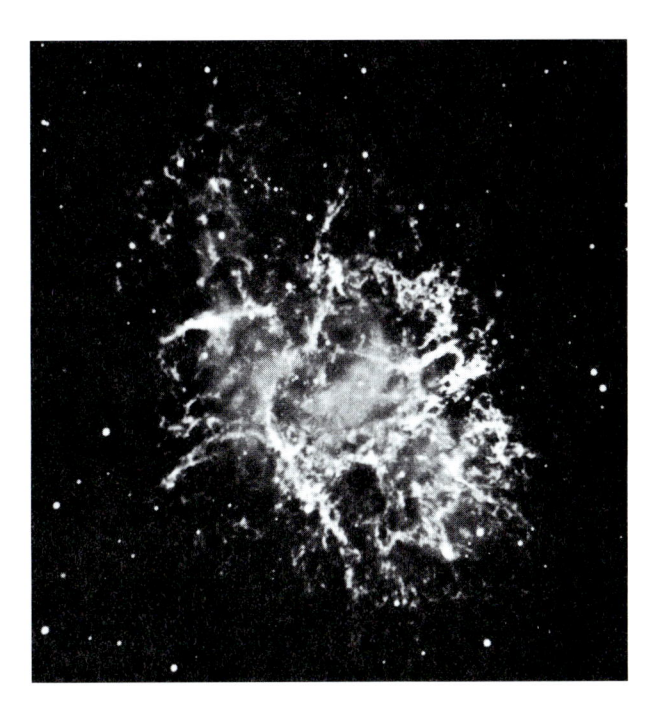

Figure 25.22

The Crab Nebula. This remnant of a star that exploded in 1054 A.D. radiates vast quantities of energy by the synchrotron process.

Nuclear Reactors

The basic principle of a fission reactor has already been discussed: a sufficient quantity of an appropriate nuclear fuel is assembled so that critical mass will be reached when control rods are pulled out. As discussed earlier in this chapter, $^{235}_{92}U$ is a suitable fuel, and has been widely used in reactors. Another fuel is

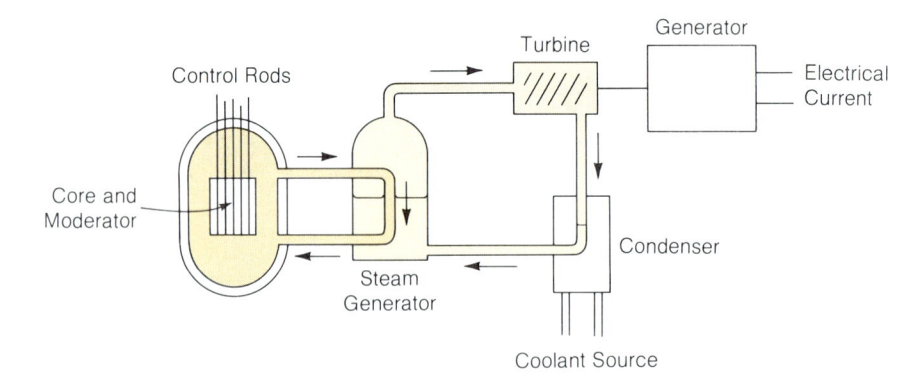

Figure 25.23

Nuclear power plant. The fission reactions in the core heat a fluid (water or liquid sodium), which is circulated through a chamber of water and creates pressurized steam. The steam passes through a turbine, which operates the generator, and is then recondensed into water and recirculated to the steam generator.

$^{239}_{94}$Pu, an isotope of plutonium that does not occur in nature but which is a by-product of $^{235}_{92}$U fission. Thus, plutonium fuel can be produced in a **breeder reactor** (see Special Problem 1 at the end of this chapter), one in which uranium fission occurs (the $^{239}_{94}$Pu actually is formed by the addition of neutrons to $^{238}_{92}$U, which is always present with the $^{235}_{92}$U). Breeder reactors have had difficulties in achieving effective production of energy, however.

The goal in operating a fission reactor is to balance the rate of neutron production with the rate at which neutrons are absorbed by control substances or allowed to escape, so that on the average only one neutron produced in each fission reaction causes another fission reaction.

The core of a reactor typically consists of nuclear fuel surrounded by a control substance which absorbs neutrons (Fig. 25.23). This substance may be graphite or heavy water (water containing deuterium in place of hydrogen). The control substance, sometimes called the moderator, slows the neutrons so they have a greater chance of reacting with nuclei before escaping. (An important discovery was that slow neutrons are more easily captured by fissionable nuclei.) Embedded within the core are a number of control rods, made of a substance such as cadmium that effectively absorbs neutrons. When the control rods are in place, they completely inhibit chain reactions, but when pulled out, they can be used to adjust the quantity of free neutrons in the core to achieve the desired steady reaction rate.

Heat produced in the core boils water, and the steam drives turbines attached to electrical generators. The advantages of producing electrical power from fission reactions are that uranium is an efficient and plentiful source, and nuclear power may be "cleaner" than traditional fossil-fuel energy sources, since it does not add pollutants to the atmosphere. The disadvantages are: (1) storage of radioactive waste products having long half-lives, (2) thermal pol-

25.4 PARTICLE ACCELERATORS AND NUCLEAR REACTORS

lution of rivers which carry away excess heat (see *Focus on Physics* article in Chapter 13), (3) the slight possibility of core melt-down, (4) a danger that human error or equipment malfunction could result in a leakage or theft of radioactive materials; and (5) the enormous cost of licensing and constructing a nuclear power plant. For these reasons, nuclear power generation is highly controversial today, and society is faced with a difficult trade-off between the need for power and the dangers associated with fission plants.

The possibility of creating sustained, controlled nuclear fusion reactions is tantalizing because if it proves feasible, many of the problems associated with fission plants may be eliminated. The difficulty, as mentioned earlier in this chapter, is that fusion requires extreme physical conditions. It turns out that temperatures high enough to overcome the Coulomb barrier are not prohibitively difficult to create. The difficulty lies in containing a hot, ionized gas (plasma) for sufficient time, so that the fusion energy released exceeds the energy used to heat the plasma. The break-even point is reached when the plasma density n (particles/m^3) and the confinement time τ (s) satisfy the relation

$$n\tau \geq 3 \times 10^{20} \text{ m}^{-3} \text{ s}. \tag{25-9}$$

In a fusion bomb, for example, extremely high temperature and density are created by exploding a small fission bomb to trigger the fusion reaction. This presents obvious difficulties in a reactor, so instead the strategy must be to work with lower densities and seek a means of containing the superheated material for a sufficiently long time.

One proposal is to use intense laser beams to vaporize small pellets of fusionable material, momentarily creating extremely high temperatures and densities where the beams strike and compress the pellet (Fig. 25.24). Another possibility is to use magnetic fields to confine a plasma as it is heated and pressurized until reactions can occur (Fig. 25.25). The magnetic fields keep the plasma away from the solid walls, whose impurities would pollute the plasma and cool it by radiating heat. The major effort in magnetic confinement has taken advantage of an ingenious device called a **tokamak** (Fig. 25.26). The tokamak is a doughnut-shaped tube in which powerful magnetic fields from electromagnets are twisted as they wrap around the tube. Charged particles in the plasma spiral around the magnetic field lines and are kept bottled up inside the tube. The goal is to heat, compress, and confine the plasma for a sufficient time to satisfy the fusion requirement, Eq. 25–9.

Several fusion reactions may be used to produce energy in a sustained reaction. Probably the most likely candidate, called D-T, is one in which a deuterium nucleus (D) merges with a tritium nucleus (T).

$$^3_1\text{H} + ^2_1\text{H} \rightarrow ^4_2\text{He} + ^1_0 n$$

The D-T reaction releases 17.6 MeV. It requires a somewhat lower temperature than some of the other possibilities, and it has a higher energy output. Eventually other reactions (D-D, for example, in which two deuterium nuclei fuse)

864 CHAPTER 25 THE NUCLEUS AND NUCLEAR ENERGY

(a)

(b)

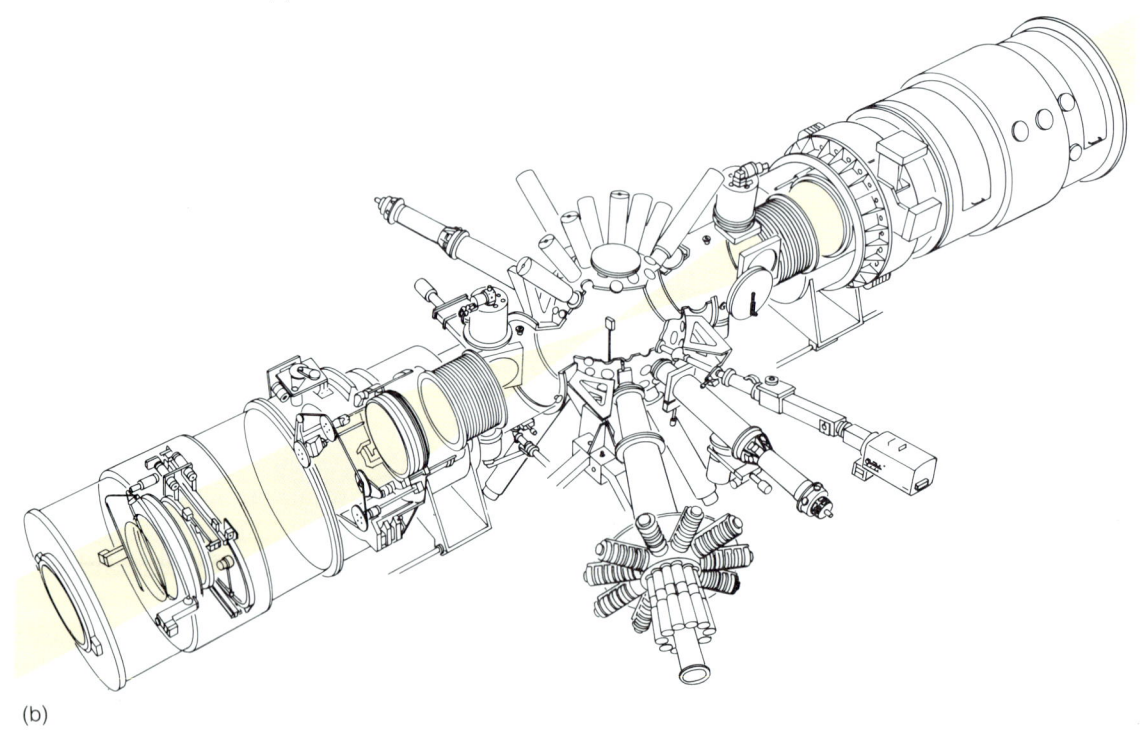

Figure 25.24

NOVA laser fusion device at Lawrence Livermore Labs, U.Cal. Several powerful laser beams focus at a single point, providing sufficient energy to cause nuclear fusion in pellets of nuclear fuel.

25.4 PARTICLE ACCELERATORS AND NUCLEAR REACTORS

Figure 25.25

Magnets for magnetically confined plasma fusion. This is a large pair of superconducting electromagnets constructed at the Lawrence Livermore Laboratory for magnetic confinement of hot plasmas at the densities and temperatures required for sustained nuclear fusion.

would need to be explored, however, because tritium is scarce. The D-D reactions are

$$_1^2H + {_1^2H} \rightarrow {_1^3H} + {_1^1H} \text{ (4.0 MeV)}$$

$$_1^2H + {_1^2H} \rightarrow {_2^3He} + {_0^1n} \text{ (3.3 MeV)}.$$

Other possibilities include

$$_1^3H + {_1^3H} \rightarrow {_2^4He} + 2\, {_0^1n} \text{ (11.3 MeV)}$$

$$_1^2H + {_2^3He} \rightarrow {_1^1H} + {_2^4He} \text{ (18.3 MeV)}.$$

The United States and other nations are engaged in major programs to develop controlled nuclear fusion, using the tokamak and other devices. Significant progress has been achieved so far, with several research groups actually succeeding in producing very brief fusion reactions. The breakthrough—when the energy produced exceeds that required to create the energetic particles and immense magnetic fields—may be reached by 1990. Even so, it will be substantially longer before fusion power becomes generally available.

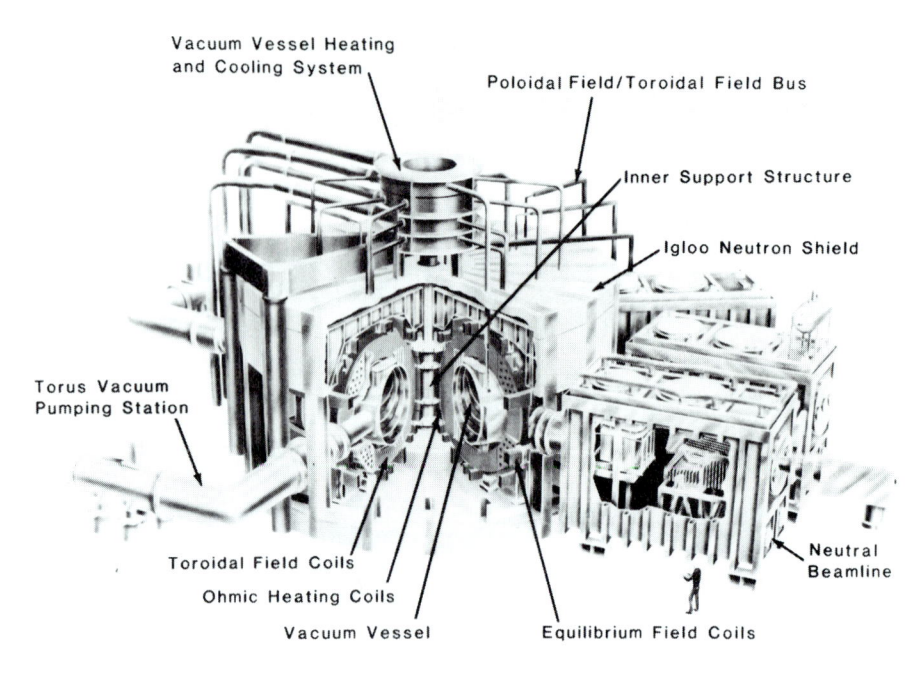

Figure 25.26

A tokamak. The path followed by particles inside the vacuum vessel keeps them away from the walls of the chamber.

Perspective

We are almost finished discussing modern physics, having only briefly treated some of the most fascinating aspects of current research into nuclear structure. Even so, we have seen that a great deal is known and that much has been learned about how to take practical advantage of nuclear properties, even though a complete theoretical understanding of the strong nuclear force has not yet been attained.

Modern work in nuclear physics has capitalized on recent advances in the understanding of subnuclear particles and the strong nuclear force. These are part of elementary particle physics. Therefore, before we complete this text, we will discuss one more area, a topic that may at first seem to be an odd combination of subjects. Lumping together our discussions of elementary particles and of the overall organization and destiny of the universe makes sense when we realize how much of the origin of the universe is embodied in what is being learned today about the most fundamental of elementary particles.

SUMMARY TABLE OF FORMULAS

Radioactive decay rate:
$$\frac{\Delta N}{\Delta t} = -\lambda N \tag{25-1}$$

$$N = N_0 e^{-\lambda t} \tag{25-2}$$

Radioactive half-life:
$$T_{\frac{1}{2}} = \frac{\ln 2}{\lambda} = \frac{0.693}{\lambda} \tag{25-3}$$

Fraction of radioisotope remaining after time t:
$$\frac{N}{N_0} = \left(\frac{1}{2}\right)^{t/T_{\frac{1}{2}}} \tag{25-4}$$

Age of a radioisotope sample:
$$t = -1.443 T_{\frac{1}{2}} \ln\left(\frac{N}{N_0}\right) \tag{25-5}$$

Radioactive dosage:
$$\text{rem} = (\text{rad})(\text{RBE}) \tag{25-6}$$

Particle velocity in a cyclotron:
$$v = \frac{qBr}{m} \tag{25-7}$$

Cyclotron frequency:
$$\nu = \frac{v}{2\pi r} = \frac{qB}{2\pi m} \tag{25-8}$$

Fusion condition:
$$n\tau \geq 3 \times 10^{20} \text{ m}^{-3}\text{ s} \tag{25-9}$$

THOUGHT QUESTIONS

1. What is the evidence that the nucleus of an atom is confined to such a small volume?

2. What is the evidence for the strong nuclear force?

3. What are the definitions of isotope, element, nuclide and isomer?

4. What are the differences between the three basic types of radioactivity: alpha, beta, and gamma rays?

5. Explain the differences between fission and fusion. What is the importance of the binding energy curve in determining which process will release energy?

6. The chemical atomic weights (atomic masses) for most elements in the Periodic Table are not whole numbers (integers). Why?

7. Can hydrogen, deuterium, or helium emit an alpha particle?

8. Why must fission reactions release neutrons to be useful in producing sustainable large amounts of energy?

9. What two basic physical conditions must be present, either in a star or in the laboratory, to produce fusion?

10. Explain the role of hydrogen and deuterium in fusion in stars.

11. Does the neutrino have a mass? Does it carry energy and momentum?

12. Describe three devices or methods for detecting charged particles, such as beta or alpha particles.

13. Explain why it is easier to split a nucleus with neutrons than with protons.

14. Explain the concept of half-life. Why, after many half-lives, is there still a small amount of a radioisotope left?

15. Discuss the advantages and disadvantages of radioactivity. Include such topics as medicine, energy generation, and archaeological dating.

PROBLEMS

Section 25.1 Nuclear Properties

- **1** For the isotope of lead, $^{207}_{82}$Pb, how many protons, neutrons, and electrons are in a neutral atom?
- **2** The isotopes of uranium range from U-227 to U-240. What is the range in neutron numbers?
- **3** The two stable isotopes of carbon are C-12 and C-13, with natural abundances 98.89% and 1.11% respectively. What is the chemical (abundance-weighted average) atomic mass of a typical sample of carbon?
- **4** If a nucleus has atomic number Z, atomic weight (mass number) A, and mass $M(Z,A)$, including electrons, show that the binding energy per nucleon is

$$\frac{E_B}{A} = \left(\frac{c^2}{A}\right)[ZM_H + (A-Z)M_n - M(Z,A)]$$

where M_H is the mass of a neutral hydrogen atom and M_n the mass of a neutron.

- **5** Calculate the binding energy per nucleon of hydrogen, deuterium, and tritium.
- **6** What are the binding energies per nucleon of gold ($^{197}_{79}$Au) and uranium ($^{235}_{92}$U)?
- **7** What is the binding energy of the last strongly bound neutron in $^{56}_{26}$Fe? Consider the difference in mass between $^{56}_{26}$Fe and the sum of $^{55}_{26}$Fe and $^{1}_{0}n$.
- **8** Calculate the total binding energy, the average binding energy per nucleon, and the energy required to remove one neutron from $^{13}_{6}$C.
- **9** Iron exists in four stable isotopes: $^{54}_{26}$Fe (5.80%), $^{56}_{26}$Fe (91.72%), $^{57}_{26}$Fe (2.20%), and $^{58}_{26}$Fe (0.28%) (natural abundances are in parentheses).
  - (a) Calculate the chemical (abundance-weighted average) atomic mass of iron.
  - (b) Calculate the binding energy per nucleon (in MeV) of each isotope.
  - (c) Why is $^{56}_{26}$Fe the most abundant isotope?
- **10** How many nuclei are contained in a 5-gram pure-gold coin?
- **11** What is the average density of nuclear matter (kg/m^3)? Compare to the average density of a neutron star of 1 $M_\odot$ = 1.99 × 10^{30} kg and 10 km radius.
- **12** What is the mass of an electron in atomic mass units (u)?
- **13** In a fission reaction, a $^{235}_{92}$U nucleus sometimes splits into $^{141}_{56}$Ba and $^{92}_{36}$Kr nuclei, and three free neutrons. What are the radii of the particles? Is the total volume of nuclear matter approximately the same before and after?

Section 25.2 Radioactivity

- **14** A nucleus with atomic number Z and mass number (atomic weight) A has a mass $M(Z,A)$, including electrons. Show that this nucleus is unstable and emits the following particles, if certain criteria are satisfied:
  - (a) Proton emission, if $M(Z,A) > M(Z-1,A-1) + M_H$
  - (b) Neutron emission, if $M(Z,A) > M(Z,A-1) + M_n$
  - (c) Alpha particle emission, if $M(Z,A) > M(Z-2,A-4) + M_\alpha$.

  where M_H, M_n, and M_α are the masses of the neutral hydrogen atom, a free neutron, and an α-particle ($^{4}_{2}$He atom).

- **15** Show that a nucleus of mass $M(Z,A)$ is unstable for β^- emission if $M(Z,A) > M(Z+1,A)$, and unstable to β^+ emission (positron emission) if $M(Z,A) > M(Z-1,A) + 2m_e$, where m_e is the electron (positron) mass.

- **16** Derive a criterion similar to those in Problem 14 for a nucleus of mass $M(Z,A)$ to be unstable to spontaneous fission into two fragments of mass $M(Z_1,A_1)$ and $M(Z_2,A_2)$, plus a number N of free neutrons.

- **17** Using Problems 14 and 15, determine whether tritium ($^{3}_{1}$H) is unstable for decay by emission of neutrons, β^-, and β^+ particles.

- **18** One of the first radioisotopes used in medicine was radium, $^{226}_{88}$Ra, which is an α-emitter. What is the energy released? What daughter (decay product) isotope is formed?

- **19** In the walls of a well-known California university, the cement contains radioactive potassium-40 ($^{40}_{19}$K). What is the energy of the β^- particle emitted?

- **20** The curve of binding energy per nucleon shows peaks at $^{4}_{2}$He, $^{12}_{6}$C, and $^{16}_{8}$O, reflecting the stability of groups of 1, 3, and 4 α-particles. Using nuclear mass data, show that $^{8}_{4}$Be is unstable to decay into two α-particles. Why are groups of 3 and 4 stable?

- **21** Nuclear data show that there exist no stable isotopes with mass number A of 5 or 8. Using helium (Z =

2), lithium (Z = 3), beryllium (Z = 4), and boron (Z = 5) show which decays are likely to destroy their isotopes with A = 5 or A = 8.

••• 22 One of the radioactive nuclei discovered by Marie Curie was polonium-210, $^{210}_{84}$Po, which is unstable for α-emission.
(a) What is the decay product?
(b) How much energy (MeV) is released?
(c) Is the decay product stable?
(d) What is the velocity of the α-particle?

•• 23 The radioactive isotope strontium-90 was common during the early 1960s, during atmospheric nuclear tests. What is the decay product of the $^{90}_{38}$Sr β$^-$-decay? How much energy is released?

•• 24 Write the reaction equations (including Z and A) and identify the daughter nuclides (decay products) for the following:
(a) A $^{14}_{7}$N nucleus captures an α-particle.
(b) A $^{13}_{7}$N nucleus emits a positron and neutrino.
(c) A cobalt-56 nucleus emits a positron (β$^+$) and neutrino.

Section 25.2 Decay Probabilities and Half Lives

• 25 After 6 half-lives, what fraction of a substance is left?

• 26 What is the decay constant, λ, of $^{239}_{94}$Pu, whose half-life is 24,000 yrs?

• 27 In a dead part of a bristlecone pine, the carbon-14: carbon-12 ratio is a factor 0.6 of its original value. How old is the tree?

•• 28 Laboratory tests can detect a carbon-14: carbon-12 ratio as low as 8×10^{-15}. The normal ratio in living matter is 1 part in 10^{12}. What is the oldest sample that can be measured by the carbon-14 technique?

•• 29 In a sample containing two isotopes of the same element, having decay constants λ_1 and λ_2, the ratio of abundances changes with time. Show that if the ratio initially (t = 0) is f_0, the ratio after time t is

$$f(t) = \frac{N_2(t)}{N_1(t)} = f_0 e^{-(\lambda_2 - \lambda_1)t}.$$

•• 30 A sample of uranium ore contains an isotope ratio uranium-235: uranium-238 = 1:138.5. The half-lives are $t_{235} = 7.1 \times 10^8$ yr and $t_{238} = 4.5 \times 10^9$ yrs. If the sample contained equal amounts when it was formed, how old is it? (Refer to Problem 29.)

•• 31 A 10 g sample of the radioactive tracer iodine-123 is stored on a hospital shelf for 5 days. If its half life is 13 h, how much is left?

••• 32 A radioactive waste dump contains 10 kg of radio-isotopes of average nuclear mass 100u and 100 year average half-life. How many decays per second are occurring after 250 years?

• 33 The first large atmospheric nuclear tests were made around 1950. If large amounts of radioactive strontium-90 ($T_{\frac{1}{2}}$ = 28.8 yr) and cesium-137 ($T_{\frac{1}{2}}$ = 30.2 yr) were injected, what fraction of each remains today?

• 34 The decay constant of radium-226 is 1.37×10^{-11} s^{-1}. What is its half-life?

•• 35 What is the activity (decays per second) of a 1 g sample of cobalt-60 whose half-life is 5.27 yr? How many Curies?

•• 36 A sample of uranium-238 registers 10^5 counts per second in a Geiger counter. Assuming the counter is 50% efficient at counting decays, how many nuclei are present in the sample? (Its half-life is 4.5×10^9 yr.)

••• 37 The $^3_1$H isotope of hydrogen, called tritium, is unstable to β$^-$-decay with a half-life of 12.36 years.
(a) Identify the daughter nuclide and find the energy of the β$^-$ emitted.
(b) What is the power emitted (watts) by a 1 g pure sample?
(c) What is the power emitted after 100 years?

•• 38 Tritium ($^3_1$H) is not naturally occurring, but is made by cosmic ray collisions with atoms high in the atmosphere, from where it is washed down to earth by rain. If an art historian discovers that the ratio of tritium to ordinary hydrogen contained in a watercolor painting is 0.1 times that in a modern painting, how old is the painting? The half-life of tritium is 12.36 years.

••• 39 Consider the decay series of $^{238}_{92}$U down to $^{206}_{82}$Pb discussed in the text.
(a) How many alpha particles ($^4_2$He) were emitted per nucleus of U-238?
(b) If the helium nuclei are trapped inside a sample of granite, containing 100 g of U-238 originally (with no helium), what mass of $^4_2$He should be present today, 2.5 billion years after formation? The half-life of U-238 is 4.5×10^9 yrs.

Section 25.3 Measurement of Radiation

• 40 A radioactive waste pile contains 100 kg of $^{60}_{27}$Co, with 5.27 yr half-life. How many Curies are present?

• 41 A 10^{-8} Curie sample of radioactive phosphorus-32 is injected into the bloodstream as a tracer. If a de-

42 What is the activity (in Curies) of 1 gram of $^{226}_{88}$Ra, an alpha emitter of 1600 year half-life?

43 A 0.02 µCi sample of radioactive $^{32}_{15}$P (half-life 14.28 days) is injected into tissue as a tracer. How many days must one wait before the counting rate in a Geiger counter is 100 counts per second? Assume the counter intercepts 50% of the emitted beta particles and counts them with 90% efficiency.

44 The Earth's crust is 17 km thick, on average, has an average density of 2800 kg/m³, and contains 1 part per million (by mass) of radioactive uranium-238.
(a) Calculate the volume and mass of the crust, assuming it to be a thin shell. The Earth's radius is 6378 km.
(b) Find the total power generated (Joules/s) in the crust by $^{238}_{92}$U α-decays.
(c) Compare the heat flow from radioactivity to the heat falling on the Earth's surface in sunlight. The flux of sunlight at the distance of the Earth is about 1360 Watts/m².

45 The average dose of radiation at sea level is 100 mrad (millirads) per year, but as much as 180 mrad at the elevation of Boulder, Colorado (5400 feet). How much energy is absorbed by a 60 kg person in each location per year?

46 When a cosmic ray passes through the atmosphere, it can ionize a molecule and produce an ion and free electron (together, known as an ion pair). If 35 electron volts (about 5.6×10^{-18} Joules) is required to produce one ion pair, calculate the dose (rads) when air absorbs 7 Roentgens (7 R) of cosmic rays. (Recall that 1 R produces 2.08×10^{15} pairs/m³ in air at standard temperature and pressure, with density 1.29 kg/m³.

47 Refer to Problem 46. Show that a dosage of 1 R of cosmic rays is equivalent to the absorption of about 0.012 Joules/m³ in air. Thus, verify that 1 R is approximately equal to 1 rad for cosmic rays.

48 A lethal dose of radiation for most humans is 400 rem. How many rads is this if the radiation is γ-rays (RBE = 1)? How many rads if it is alpha particles (RBE = 15)?

49 Compute the energy absorbed (Joules) by a 60 kg person when exposed to 1000 rem of γ-rays (RBE = 1). Compare this energy with that released when a 60 kg person falls 1 m.

50 A source of $^{60}_{27}$Co emits gamma rays of energy 1.33 MeV and 1.17 MeV in approximately equal numbers. A 65 kg patient is exposed to a 500 Curie source, located 5 m away. The cross-sectional area exposed to the source is 1 m², and the patient absorbs 50% of the emitted γ-rays.
(a) How many gamma rays are absorbed each second?
(b) Calculate the total dose (rads) in a one minute exposure.

51 What is the mass of a 1 µCi source of $^{40}_{19}$K? Of $^{14}_{6}$C?

52 How many rads of slow neutrons (RBE = 4) would do as much radiation damage to humans as 100 rads of x-rays (RBE = 1)?

53 A 150 gram sample of powdered thorium ore is measured with a Geiger counter, which registers 200 counts/s. If the Geiger counter measures 10% of the total counts, how many grams of pure thorium-232 are present? How many Curies?

54 In a certain city at high altitude, the average yearly radiation background consists of 150 mrad cosmic rays, 30 mrad in x-rays or γ-rays, and 4 mrad from alpha particles (RBE = 15). How many rems will a person receive each year, on average?

55 A wasteful city burns 2 billion kg of garbage annually, most of which is carbon matter. If the natural abundance of carbon-14 is 1 part per 10^{12} by mass, and if the garbage contains mostly "fresh" organic material in which the carbon-14 has not beta-decayed, compute the total amount (Curies) of carbon-14 put into the atmosphere each year.

Section 25.4 Nuclear Reactions

56 Fill in the blanks for the missing particle in the following reactions:
(a) $^{13}_{6}$C + $^{1}_{1}$H → _____ + γ
(b) _____ → $^{17}_{8}$O + β$^+$ + ν
(c) $^{235}_{92}$U + $^{1}_{0}$n → _____ + $^{141}_{55}$Cs

57 Calculate the maximum energy of the e^- emitted in the decay of a free neutron ($n \to p^+ + e^- + \bar{\nu}$).

58 How much energy is required to initiate the "inverse beta reaction," $p^+ + e^- \to n + \nu$?

59 Calculate the efficiency of nuclear hydrogen burning: find the energy released in fusing 4 $^{1}_{1}$H nuclei into one $^{4}_{2}$He nucleus, relative to the rest mass energy, $4M_H c^2$.

60 A radioactive isotope with initial atomic number Z_0 and mass number A_0 decays through a series of steps, emitting N_α alpha particles and N_β beta (e^-) particles. Show that the final values are:

$$A = A_0 - 4N_\alpha \text{ and } Z = Z_0 - 2N_\alpha + N_\beta$$

SPECIAL PROBLEMS

Calculate the number of alphas and betas emitted in the transmutations of
(a) $^{238}_{92}U$ down to $^{206}_{82}Pb$,
(b) $^{232}_{90}Th$ down to $^{208}_{82}Pb$,
(c) $^{235}_{92}U$ down to $^{207}_{82}Pb$.

- 61 Calculate the energy released in the fusion reaction, $n + ^{238}_{92}U \rightarrow ^{239}_{92}U$.

•• 62 Calculate the energy released in the fission reaction, $n + ^{235}_{92}U \rightarrow ^{141}_{56}Ba + ^{92}_{36}Kr + 3 ^{1}_{0}n$.

••• 63 Calculate the mass of the $^{235}_{92}U$ in the 1945 atomic explosion of 12 kilotons equivalent of TNT. Assume an average energy yield of 200 MeV per fission, and use the conversion: 1 kiloton TNT = 5×10^{12} J.

- 64 Calculate the energy released in the fusion reaction of deuterium and tritium, $^{2}_{1}H + ^{3}_{1}H \rightarrow ^{4}_{2}He + ^{1}_{0}n$.

- 65 Calculate the energy released in the fusion of two deuterium nuclei, $^{2}_{1}H + ^{2}_{1}H \rightarrow ^{3}_{2}He + ^{1}_{0}n$.

•• 66 In the hot cores of exploding stars, silicon nuclei can fuse to form nickel, $^{28}_{14}Si + ^{28}_{14}Si \rightarrow ^{56}_{28}Ni$. The nickel is radioactive and beta decays to cobalt, which in turn beta decays to iron via the following reactions:

$^{56}_{28}Ni \rightarrow ^{56}_{27}Co + \beta^+ + \nu$, and
$^{56}_{27}Co \rightarrow ^{56}_{26}Fe + \beta^+ + \nu$.

Most of the iron in the universe is made this way. Compute the energies released in the Si-Si fusion, and in the two positron (β^+) decays.

SPECIAL PROBLEMS

1. Breeder Reactors

In one version of a breeder reactor, neutrons released from the fissionable uranium-235 fuel are used to convert the more common uranium-238 into plutonium-239, thus providing new fissionable fuel. The following reactions occur:

1. $^{238}_{92}U + ^{1}_{0}n \rightarrow ^{239}_{92}U$
2. $^{239}_{92}U \rightarrow ^{239}_{93}Np + \beta^- + \bar{\nu}$ ($\tau = 23.5$ min)
3. $^{239}_{93}Np \rightarrow ^{239}_{94}Pu + \beta^- + \bar{\nu}$ ($\tau = 2.33$ days)

(a) Calculate the energy generated in each step.
(b) Using the mean lifetimes for beta decays (given in parentheses), find the maximum energy generation rate (watts/kg) of uranium-238.
(c) What is the efficiency of energy released by this process, relative to the rest mass energy of uranium-238?
(d) A core sample (1 gram) of uranium contains equal amounts of uranium-235 and uranium-238. If 180 MeV per uranium-235 is released, and plutonium-239 is created, calculate the total energy generated.

2. Radioactive Heating of Earth's Crust

The earth's core and crust are heated by the decay of radioactive nuclei. Granitic rock in the crust contains the following amounts of radioisotopes by mass: uranium-238 (4×10^{-6}), uranium-235 (2.8×10^{-8}), thorium-232 (1.8×10^{-5}), and potassium-40 ($^{40}_{19}K$) (4.2×10^{-6}).

(a) Compute the number of atoms of each isotope contained in a kg of granite.
(b) Calculate the energy released in the radioactive decay series of each isotope, ending in a stable form:

$^{238}_{92}U \rightarrow ^{206}_{82}Pb + 8 ^{4}_{2}He + 6\beta^-$
$^{235}_{92}U \rightarrow ^{207}_{82}Pb + 7 ^{4}_{2}He + 4\beta^-$
$^{232}_{90}Th \rightarrow ^{208}_{82}Pb + 6 ^{4}_{2}He + 4\beta^-$
$^{40}_{19}K \rightarrow ^{40}_{18}Ar + \nu$ (K-capture)

(c) Find the average heat production rate (watts/kg) in the rock.
(d) How long would it take the earth's crust (total mass 2.4×10^{22} kg) to rise 100 C° in temperature if no heat escapes? Assume that granite has a specific heat of 740 Joules/kg/C°.

3. The CNO Nuclear Burning Cycle in Stars

The basic CNO cycle of nuclear hydrogen burning in the cores of massive stars consists of the following reactions:

1. $^{12}_{6}C + ^{1}_{1}H \rightarrow ^{13}_{7}N + \gamma$
2. $^{13}_{7}N \rightarrow ^{13}_{6}C + \beta^+ + \nu$ ($\tau = 870$ s)
3. $^{13}_{6}C + ^{1}_{1}H \rightarrow ^{14}_{7}N + \gamma$
4. $^{14}_{7}N + ^{1}_{1}H \rightarrow ^{15}_{8}O + \gamma$
5. $^{15}_{8}O \rightarrow ^{15}_{7}N + \beta^+ + \nu$ ($\tau = 178$ s)
6. $^{15}_{7}N + ^{1}_{1}H \rightarrow ^{12}_{6}C + ^{4}_{2}He$

(a) By summing the particles before and after the cycle, show that the net result is $4 ^{1}_{1}H \rightarrow ^{4}_{2}He + 2 \beta^+ + 2 \nu + 3 \gamma$.
(b) Show that $^{12}_{6}C$ is a catalyst—it is neither created nor destroyed.
(c) Using nuclear mass data, calculate the energy released in each of the six steps above.
(d) In the CNO cycle, energy cannot be generated any faster than the time it takes for the two beta decays

(lifetimes in parentheses). What is the shortest cycle time?
(e) What is the maximum energy generation rate per unit mass (watts/kg) in the CNO cycle? (Consider the mass and energy involved in converting four hydrogen nuclei to one helium, and assume that carbon-12 has an abundance of 10^{-3} by mass.)

4. Nucleosynthesis in the Early Universe

About 3 or 4 minutes after the big bang, matter and radiation in the cosmological fireball have cooled to slightly under 10^9 K, so that nucleons can bind together. Because the universe is still younger than the half-life of a free neutron (10.8 min), both protons and neutrons participate in the fusion reactions. Further discussion is given in Section 26.4.

(a) The key first reaction is to form deuterium by p-n fusion, $^1_1H + ^1_0n \rightarrow ^2_1H + \gamma$. How much energy is released by this reaction? How is it different from the first reaction in the proton-proton chain in stars?

(b) Primordial helium is formed by either of the following reaction pairs:

$$^2_1H + ^1_1H \rightarrow ^3_2He; \text{ then } ^3_2He + ^1_0n \rightarrow ^4_2He$$
$$^2_1H + ^1_0n \rightarrow ^3_1H; \text{ then } ^3_1H + ^1_1H \rightarrow ^4_2He$$

What are the energies released in these pairs?

(c) Further fusion reactions to form heavier nuclei are difficult, because there are no stable nuclei with mass number $A = 5$ or 8. For example the following attempts to form heavier nuclei fail: $^4_2He + ^1_1H \rightarrow ^5_3Li$; $^4_2He + ^4_2He \rightarrow ^8_4Be$, because the lithium-5 and beryllium-8 nuclei are not stable. Show, however, that the following reaction is energetically favored:

$$^4_2He + ^3_2He \rightarrow ^7_4Be$$

The beryllium-7 decays to lithium-7 via K-capture, so that small amounts of lithium-7 can form this way. Further reactions are essentially non-existent.

SUGGESTED READINGS

"Scintillation counters." G. B. Collins, *Scientific American*, Nov 53, p 36, **189**(5)

"The bubble chamber." D. A. Glaser, *Scientific American*, Feb 55, p 46, **192**(2)

"The discovery of fission." O. Hahn, *Scientific American*, Feb 58, p 76, **198**(2)

"The spark chamber." G. K. O'Neill, *Scientific American*, Aug 62, p 36, **207**(2)

"Tokamak approach in fusion research." B. Coppi, J. Rem, *Scientific American*, Jul 72, p 65, **227**(1)

"Fusion power by laser implosion." J. L. Emmett, J. Nuckolls, L. Wood, *Scientific American*, Jun 74, p 24, **230**(6)

"A natural fission reactor." G. A. Cowan, *Scientific American*, Jul 76, p 36, **235**(1)

"Gamma-ray line astronomy." M. Leventhal, C. J. MacCallum, *Scientific American*, Jul 80, p 62, **243**(1)

"The engineering of magnetic fusion reactors." R. W. Conn, *Scientific American*, Oct 83, p 60, **249**(4)

"Recollections of Rutherford and the Cavendish." S. Devons, *Physics Today*, Dec 71, p 39, **24**(12)

[Magnetically confined fusion issue]. *Physics Today*, May 79, **32**(5)

"Neutrons in science and technology." D. Allan Bromley, *Physics Today*, Dec 83, p 30, **36**(12)

"Atomic physics with synchrotron radiation." B. Crasemann, F. Wuilleumier, *Physics Today*, Jun 84, p 34, **37**(6)

"Reaching ignition in the tokamak." H. P. Furth, *Physics Today*, Mar 85, p 52, **38**(3)

"The SSC: A machine for the nineties." S. L. Glashow, L. M. Lederman, *Physics Today*, Mar 85, p 28, **38**(3)

"To cleave an atom." A. P. Lightman, *Science 84*, Nov 84, p 103, **5**(9)

The discovery of subatomic particles. Steven Weinberg, W. H. Freeman, New York, 1984

A source book in physics. William F. Magie, Harvard University Press, Cambridge, 1985

Mr. Tompkins explores the atom. George Gamow, Cambridge University Press, New York, 1942

Great scientific experiments. Rom Harre, Phaidon Press, Oxford, 1981 (Rutherford and the artificial transmutation of the elements)

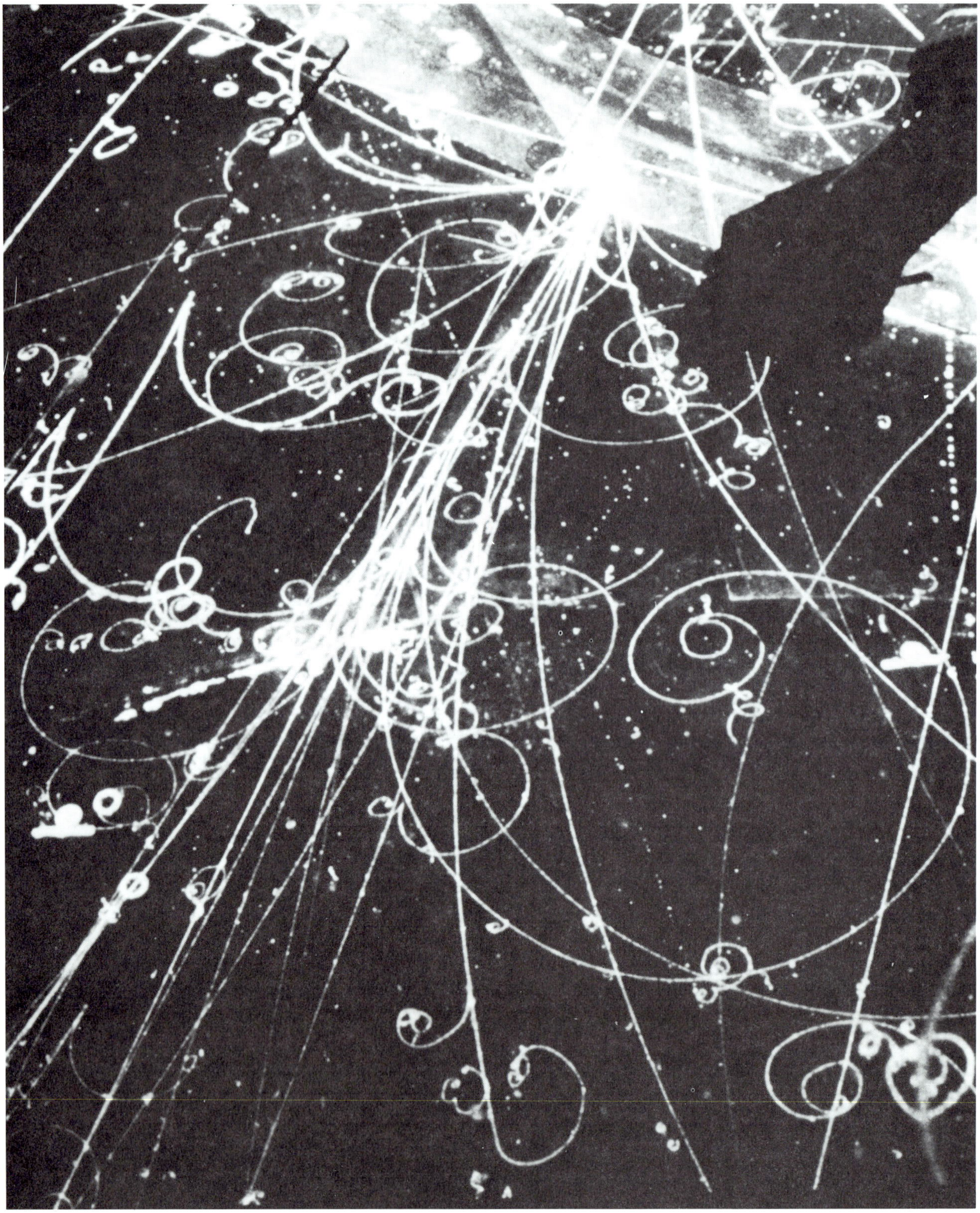

Chapter 26

Elementary Particles and Cosmology

By the 1930s, a fairly simple picture of atomic structure had been developed, in which there were just three basic subatomic particles. Besides the proton, the neutron, and the electron, however, it soon became evident from studies of cosmic rays and nuclear decays that there were others, such as the positron and the neutrino, whose existence was implied by the conservation laws in β-decay. Before long, the family of **elementary particles** had grown to hundreds of members, and even today it is not entirely clear how many there may be. Many of these particles are unstable and live fleeting lives before decaying. Faced with such a menagerie of shifting entities, theoretical physicists have begun to study ways of grouping these particles into families and constructing some of them from a small number of truly fundamental constituents called quarks and leptons.

The detection of elementary particles usually involves a process for converting energy into mass in the form of particles. For many particles, the energies required do not occur naturally on the earth, so it is necessary to create them in particle accelerators, where high-energy collisions yield product particles whose charge and mass can be deduced by one of the methods described in Chapter 25. Relatively massive new particles have been discovered in recent years, following the development of new particle accelerators with enhanced energy capabilities. Accelerators capable of energies twenty times higher than those currently available are now being designed to search for fundamental particles predicted by new theories for unifying the forces of nature.

It may not be obvious that a link exists between elementary particles and cosmology, the study of the nature and evolution of the universe. There is, however, an intimate connection: the present state of matter in the universe is the direct result of elementary particle formation in its earliest times, when sufficient energy densities existed. We will discuss this toward the end of this chapter.

26.1 Particle Exchange and Field Forces

In the early chapters of this text we alluded to the four fundamental forces of nature: gravitation, the electromagnetic force, and the strong and weak nuclear forces. We have described them as field forces, capable of acting at a distance, and possibly transmitted by the continuous exchange of particles. In this section we explore that idea more fully.

Photons and the Electromagnetic Force

In classical electrodynamics the force between two charges arises because of an invisible **electromagnetic field** transmitted across space. The field concept was invented to explain "action at a distance": two objects exerting forces on each other without physically touching. (We know now that "touching" actually makes little sense, because atoms are mostly empty space). In classical field theory, a force between particle 1 and particle 2 arises because of the field that particle 1 creates at the position of particle 2, and vice versa. This concept has been applied to both electromagnetism and gravity, the two long-range forces in nature.

But what about the strong and weak nuclear forces, which are short-range, and govern the interactions between nuclear and other subatomic particles? Are these forces due to fields also? In trying to answer this question, theoretical physicists have developed a new, more fundamental way of understanding action at a distance that involves **exchange particles**. To see how this works, we resort to a crude analogy of two hockey players moving along parallel

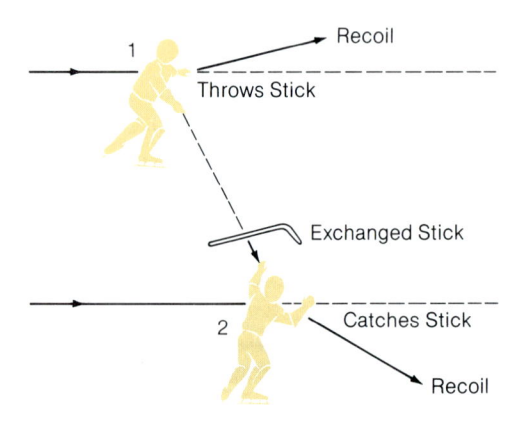

Figure 26.1

Exchange particle for a repulsive force. Two ice hockey players skate along parallel paths. One throws his hockey stick to the other, transferring momentum and causing both players to recoil. The hockey stick is like the virtual photon in electrostatic repulsion.

26.1 PARTICLE EXCHANGE AND FIELD FORCES

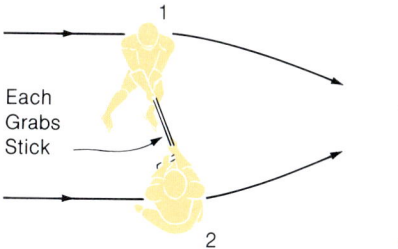

Figure 26.2

Exchange particle for an attractive force. Here each hockey player momentarily pulls on the stick, so that the two players are deflected toward each other. The stick is like the exchange particle for an attractive force.

Figure 26.3

Electrostatic repulsion. Two electrons are deflected away from each other by the exchange of a virtual photon.

paths (Fig. 26.1). One throws his hockey stick at the other, recoiling slightly because of momentum conservation. When the second skater catches the stick, he too recoils; the entire sequence of events looks like a repulsive force, mediated by the exchange of a "particle" which carried momentum from one skater to the other. To see how an attractive force might arise from exchange, we must imagine the two skaters exchanging the stick, each grabbing onto an end for a short time. Their paths would move toward each other (Fig. 26.2).

In quantum field theory, all forces are the result of the exchange of particles. For the electromagnetic force, the first to be explained as a particle exchange force, the photon (γ) mediates. For example, two electrons exert a repulsive Coulomb force on each other (Fig. 26.3) through the exchange of a photon, which transfers momentum and energy. Because the photon is absorbed by the second electron shortly after it was emitted, it is called a **virtual photon**, in contrast to the real photons that we see and detect at great distance. A more fundamental difference is that the virtual photon violates energy conservation. It is created spontaneously with energy ΔE out of nothing (physicists say "from the vacuum"). According to the classical laws of physics, this cannot happen. But in quantum mechanics the uncertainty principle allows energy conservation to be violated for a short time interval $\Delta t \approx \hbar/(\Delta E)$. As long as the energy is conserved after this time, we can regard the virtual particle exchange as a small fluctuation of energy that is entirely consistent with quantum mechanics.

We can now understand the relation between the range of a force and the mass of the exchange particle. The range is simply the distance d travelled by the virtual particle during the brief time interval $\Delta t \approx \hbar/(\Delta E)$ allowed by the uncertainty principle. If we assume that the particle has mass m, then $\Delta E = mc^2$ (it costs energy to create a particle). We also assume that the exchange particle travels at nearly the speed of light c. Therefore,

$$d \approx (c\Delta t) \approx c\hbar/\Delta E \approx \hbar/mc. \qquad (26\text{-}1)$$

This length $\hbar/mc$ is called the **Compton length** of the particle, and it represents the range of the force carried by an exchange particle of mass m. Notice that the photon, with zero rest mass, corresponds to an infinite range for the force (the electromagnetic force).

Mesons and the Strong Nuclear Force

In 1935 the Japanese physicist H. Yukawa, using the electric force as an analogy, predicted the existence of a new exchange particle that was responsible for the strong nuclear force. This new particle, called the **meson**, was expected to have a rest mass because the strong nuclear force has a short range.

Example 26.1 Using the Compton wavelength, estimate the mass of the meson needed as an exchange particle for the strong nuclear force.

Yukawa knew that the strong force has a range comparable to the size of the nucleus; that is, that $d \approx 1.5 \times 10^{-15}$ m. If the exchange particle obeys Eq. 26–1, its estimated mass is

$$m \approx \frac{\hbar}{dc} \approx \frac{(6.63 \times 10^{-34} \text{ J} \cdot \text{s})}{2\pi(1.5 \times 10^{-15} \text{ m})(3 \times 10^8 \text{ m/s})} \approx 2.3 \times 10^{-28} \text{ kg}$$

which corresponds to a rest energy of 130 MeV.

In the 1930s particle accelerators capable of producing collisions with sufficient energy to create mesons did not yet exist, so the meson was first sought in reactions caused by high-energy cosmic rays reaching the earth's surface from space (these were mentioned in Chapter 25). A new particle with nearly the correct mass (106 MeV rest energy) was indeed found, but it did not interact with matter through the strong nuclear force, and this disqualified it as the particle predicted by Yukawa. The new particle was named the **mu meson** or **muon**, usually designated μ^-, and it turned out to be related to the electron.

Further experiments with cosmic rays finally detected the Yukawa particle in 1947. The new particle was called the **pi meson** or **pion**, and it was soon created in laboratory experiments as well. It was found to have three possible charge states: $+1$, -1, or 0 (in units of the electron charge). The charged states, denoted π^+ and π^-, have rest masses of 140 MeV, while the neutral particle, the π^0, has a rest mass of 135 MeV. Example of reactions found to produce these particles are

$$p + p \rightarrow p + p + \pi^0 \tag{26-2a}$$

$$p + p \rightarrow p + n + \pi^+ \tag{26-2b}$$

26.1 PARTICLE EXCHANGE AND FIELD FORCES

$$n + n \rightarrow n + n + \pi^\circ \tag{26-2c}$$

$$p + n \rightarrow p + p + \pi^-. \tag{26-2d}$$

It may appear that conservation of mass is violated in these reactions, but recall that it is the total energy that is conserved, not just the portion of it that is in the form of mass. The excess energy required to provide the mass of the pions comes from the kinetic energy of the colliding protons. We discuss particle production in Section 26.2.

The verification of Yukawa's prediction confirmed that the strong nuclear force, like the electric force, is due to continuous particle exchange. Furthermore, the short range of the strong force compared to that of the electric force is a consequence of the uncertainty principle and the fact that the pion has mass, whereas the photon does not.

In the past twenty years, high-energy physicists have discovered many other mesons associated with nuclear matter and the strong force. It now appears that these mesons are composed of even more fundamental particles called **quarks**. The strong force between quarks is believed to be governed by the exchange of exotic particles called **gluons** (they provide the nuclear "glue"). We will further discuss the theory of quarks and gluons in a later section.

It is appropriate to summarize the current view of the four forces according to the particle exchange theory (Table 26.1). The strong nuclear force is by far the strongest, although its range is limited to about the diameter of the nucleus because its exchange particle has a finite Compton length (finite rest mass). Electromagnetism is the second-strongest force, with an infinite range because the photon has zero rest mass (the Coulomb force decreases in strength as $1/r^2$, but we still consider this long-range). The weak force has a short range and is considerably weaker than the strong or electromagnetic forces. It is mediated by exchange of one of three **intermediate vector bosons,** symbolized by W^+, W^-, and Z°. These particles were hypothetical until 1983, when a group of European physicists at CERN found them in the debris left from collisions of protons and antiprotons. These particles have rest masses about a hundred times that of the proton ($m_W c^2 = 81 \pm 3$ GeV; $m_Z c^2 = 93 \pm 3$ GeV). The 1984 Nobel Prize in physics was awarded to the leaders of the

Table 26.1
The four forces in nature

Force	Relative strength	Exchange particle	Range[a], m	Typical lifetime[b], s
Strong	10	Mesons (gluons)	10^{-15}	10^{-23}
Electromagnetic	10^{-2}	Photon (γ)	∞	10^{-16}
Weak	10^{-7}	W^+, W^-, Z°	10^{-18}	10^{-10}
Gravitational	10^{-45}	Graviton	∞	∞

[a]Range of force determined by Compton length, $\hbar/mc$, of exchange particle
[b]Lifetime of unstable particles decaying through action of these forces

discovery group, for confirming the particle exchange model for the weak force.

The gravitational exchange particle is called the **graviton,** and by analogy with the other long-range force, it must have zero rest mass. The graviton is to gravity as the photon is to electromagnetism. Thus, if we ever detect gravitational radiation, we will have detected the graviton. Gravity is so weak a force that its quantum exchange nature will be exceedingly difficult to detect.

Example 26.2 Physicists have discovered a short-lived Δ-particle which decays with a lifetime of 10^{-23} s. What type of interaction is involved? What is the maximum distance over which the Δ can interact?

From Table 26.1, we see that the lifetime of the Δ-particle suggests a strong interaction. At the speed of light, the Δ travels

$$d = (3 \times 10^8 \text{ m/s})(10^{-23} \text{ s}) = 3 \times 10^{-15} \text{ m}$$

which is about the diameter of the nucleus. The Δ provides a link between different parts of the nucleus, and thus contributes to the many meson exchange interactions that hold the nucleus together. All such strongly-interacting particles are called **hadrons,** and are thought to be composed of quarks.

Unification of Forces

The success of the particle exchange theory for electromagnetism and the weak interaction has led researchers to expect that all four forces in nature might someday be "unified". Unification means that all forces might be manifestations of the same phenomenon—that there really is just one force in nature. Physicists have always valued simplicity, yet their theories have become increasingly mathematical. Einstein spent the last twenty years of his life working on unified field theory, but was unsuccessful.

In the 1970s a breakthrough occurred when two physicists, Steven Weinberg and Abdus Salam, created a theory, based on previous work by Sheldon Glashow, which unified two of the forces, the electromagnetic and the weak forces, into a single **electroweak force.** At low energies, the electromagnetic and weak forces are certainly quite different (see Table 26.1). The exchange particles are quite distinct: the photon (γ) for the electromagnetic force; and the three vector bosons (W^+, W^-, and Z°) for the weak force. What Weinberg and Salam showed was that at sufficiently high energies (or temperatures) these four exchange particles were part of a single, indistinguishable family (called a **group of gauge bosons**). Weinberg and Salam estimated that this unification occurred at energies greater than about 100 GeV, which roughly corresponds to the limit of present-day particle accelerators. Thus, when CERN experimenters found the W^+, W^-, and Z° particles with masses in the range 80–

26.1 PARTICLE EXCHANGE AND FIELD FORCES

Table 26.2
Stable particle table*

	Particle	Symbol	Anti particle	Rest mass MeV	B	Q	L_e	L_μ	L_τ	S	Mean lifetime (s)	Typical decay modes*
Gauge Bosons	Photon	γ	Self	0	0	0	0	0	0	0	Stable	
	Weak boson	W^+	W^-	80800	0	+1	0	0	0	0		$e^+\nu_e$
	Weak boson	W^-	W^+	80800	0	−1	0	0	0	0		$e^-\bar{\nu}_e$
	Weak boson	Z^0	Self	92900	0	0	0	0	0	0		$e^+e^-, \mu^+\mu^-$
Leptons	Electron	e^-	e^+	0.511	0	−1	+1	0	0	0	Stable	
	Muon	μ^-	μ^+	105.66	0	−1	0	+1	0	0	2.2×10^{-6}	$e^-\bar{\nu}_e\nu_\mu$
	Tau	τ^-	τ^+	1784	0	−1	0	0	+1	0	3.4×10^{-13}	$e^-\bar{\nu}_e\nu_\tau$
	Neutrino(e)	ν_e	$\bar{\nu}_e$	0(<4.6 × 10⁻³)	0	0	+1	0	0	0	Stable	
	Neutrino(μ)	ν_μ	$\bar{\nu}_\mu$	0(<0.50)	0	0	0	+1	0	0	Stable	
	Neutrino(τ)	ν_τ	$\bar{\nu}_\tau$	0(<164)	0	0	0	0	+1	0	Stable	
Mesons	Pion	π^0	Self	134.96	0	0	0	0	0	0	0.83×10^{-16}	$\gamma\gamma$
	Pion	π^+	π^-	139.57	0	+1	0	0	0	0	2.60×10^{-8}	$\mu^+\nu_\mu$
	Pion	π^-	π^+	139.57	0	−1	0	0	0	0	2.60×10^{-8}	$\mu^-\bar{\nu}_\mu$
	Kaon	K^+	K^-	493.67	0	+1	0	0	0	+1	1.24×10^{-8}	$\mu^+\nu_\mu, \pi^+\pi^0$
	Kaon	K^-	K^+	493.67	0	−1	0	0	0	−1	1.24×10^{-8}	$\mu^-\bar{\nu}_\mu, \pi^-\pi^0$
	Kaon	K^0_S	$\bar{K}^0_S$	497.67	0	0	0	0	0	+1	0.89×10^{-10}	$\pi^+\pi^-, \pi^0\pi^0$
	Kaon	K^0_L	$\bar{K}^0_L$	497.67	0	0	0	0	0	+1	5.2×10^{-8}	$3\pi^0, \pi^+e^-\bar{\nu}_e$
	Eta	η^0	Self	548.8	0	0	0	0	0	0	7×10^{-19}	$\gamma\gamma, 3\pi^0$
Baryons	Proton	p	$\bar{p}$	938.28	+1	+1	0	0	0	0	Stable	
	Neutron	n	$\bar{n}$	939.57	+1	0	0	0	0	0	898 ± 16	$pe^-\bar{\nu}_e$
	Lambda	Λ^0	$\bar{\Lambda}^0$	1115.6	+1	0	0	0	0	−1	2.6×10^{-10}	$p\pi^-, n\pi^0$
	Sigma	Σ^+	$\bar{\Sigma}^-$	1189.4	+1	+1	0	0	0	−1	0.8×10^{-10}	$p\pi^0, n\pi^+$
	Sigma	Σ^0	$\bar{\Sigma}^0$	1192.5	+1	0	0	0	0	−1	6×10^{-20}	$\Lambda^0\gamma$
	Sigma	Σ^-	$\bar{\Sigma}^+$	1197.3	+1	−1	0	0	0	−1	1.48×10^{-10}	$n\pi^-$
	Xi	Ξ^0	$\bar{\Xi}^0$	1315	+1	0	0	0	0	−2	2.9×10^{-10}	$\Lambda^0\pi^0$
	Xi	Ξ^-	$\bar{\Xi}^+$	1321.3	+1	−1	0	0	0	−2	1.64×10^{-10}	$\Lambda^0\pi^-$
	Omega	Ω^-	Ω^+	1672	+1	−1	0	0	0	−3	0.82×10^{-10}	$\Lambda^0 K^-, \Xi^0\pi^-$

*Stable (or long-lived) particles, with baryon number(B), charge(Q), lepton numbers(L), strangeness(S), and typical decay modes ($e^-\bar{\nu}_e\nu_\mu$ means $\mu^- \to e^- + \bar{\nu}_e + \nu_\mu$). Source: "Review of Particle Properties" (*Rev. of Mod. Phys.*, 56, Apr 84)

100 GeV (see Table 26.2), this was taken as a strong indication that the unification picture had merit.

A major effort in theoretical research today is to develop a theory that will unify the electroweak force with the strong force. Some attempts have been made which predict a strong-electroweak unification at energies of about 10^{15} GeV. Just as in the electroweak unification theory, physicists believe they may have identified a family (or group) of 24 **gauge particles** which appear identical above a certain energy. If these ideas are correct, then at energies greater than 10^{15} GeV, the photon, the three intermediate vector bosons, the gluons, and 12 new particles called **X-bosons** become identical. In that ultra-high-energy world, the strong nuclear, weak nuclear, and electromagnetic forces become one, governed by a single theory which has acquired the name **Grand Unified Theory** (GUT). This is something of a misnomer, since one force (gravity) has not been included. A tiny minority of physicists hold onto the hope that one day even gravity will be included.

Energies approaching 10^{15} GeV are far beyond the capabilities of any particle accelerator even dreamed of today. Thus the GUT theories are probably untestable in the laboratory for the foreseeable future. Nevertheless, there is one laboratory which might have the required energies. That is the early universe, a fraction of a second after the beginning of its expansion. By studying the remnants of this expansion (usually called the **big bang**), we might be able to discover some relic of that high-energy, unified era. For this reason, many particle physicists and cosmologists have begun to look into ways of testing these theories through astronomical observations. We will return to this subject in Section 26.4.

26.2 Particle Interactions

We have alluded to reactions in which particles are created from energy, and we have discussed the interaction of particles with matter as a requirement for their detection. In this section we specify what we mean by these terms, and we find that elementary particles can be classified according to their interactions.

Pair Production

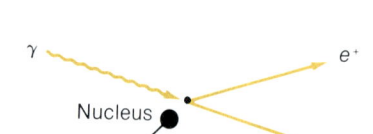

Figure 26.4

Pair production. A γ-ray with energy greater than $2m_ec^2$ (1.022 MeV) creates an (e^+, e^-) pair from its excess energy. This occurs near the nucleus, which absorbs some of the photon momentum.

We have said that a particle can be created when an amount of energy equivalent to its rest mass is available. For example, the mass of an electron corresponds to an energy of 0.511 MeV, so we could have electron production if this quantity of energy were available. Suppose we have a photon with this energy (this would be a γ-ray photon, with wavelength 0.024 Å). If this photon were transformed into a single electron, however, charge would have to materialize from somewhere. Therefore, a second particle, having an equal but opposite charge to that of the electron, must be formed as well. Thus, pairs of particles and **antiparticles** are formed. Whereas a 0.511 MeV photon cannot produce a free electron, a 1.022 MeV photon can produce a pair of particles, one an electron (e^-) and the other a positron (e^+), which has the same mass as the electron but opposite charge. This is called **pair production** (Fig. 26.4). Pair production can be detected on a photographic emulsion exposed to γ-ray photons (Fig. 26.5), where a pair of particle tracks, corresponding to equal masses but opposite charges, emanates from a single point. Pair production is triggered by the nearby presence of another particle, such as a nucleus, that absorbs some of the momentum of the photon.

In theory, every particle has an antiparticle. The first antiparticle to be discovered was the positively-charged electron, or **positron**, whose existence was implied by the early studies of nuclear reactions, as discussed in Chapter 25. The positron was found in cosmic rays by Carl Anderson in 1932. The next to be found were the μ^+ (antiparticle of the muon μ^-) and the π^+ and π^- pions (the neutral pion π^0 is its own antiparticle). The proton's antiparticle, the **antiproton**, designated $\bar{p}$, was produced in a new high-energy accelerator in 1955.

An antiparticle has the opposite charge of its corresponding particle, but the same (positive) mass (for example, the e^+ has positive charge and the e^- has negative charge; the proton has positive charge while the antiproton has

26.2 PARTICLE INTERACTIONS

Figure 26.5

Pair-production particle tracks, showing a particle pair bending in opposite directions in a magnetic field.

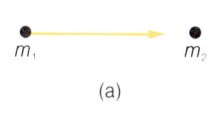

Figure 26.6

Particle production. In the lab frame, a high-energy particle (m_1) collides with another (m_2) at rest. After the collision, particles m_1 and m_2 and the newly created particles move slowly. The total energy, including rest mass, of the new particles is E.

negative charge). If a particle and its antiparticle are ever brought close together, they will annihilate into pure energy (mostly photons). Antiparticles are denoted by placing a bar over the particle's standard symbol (for example $\bar{p}$, $\bar{n}$, and $\bar{\nu}$), or the distinction is understood by the opposite charge (for example e^+ versus e^-, or π^+ versus π^-). In a few cases, where the charge is zero, the particle and its antiparticle are identical (for example π°, η°, and γ).

Particle-antiparticle pairs can also be created in high-energy collisions if the colliding particle has sufficient excess kinetic energy to create the rest mass of the particle pair, which is $2mc^2$, where m is the mass of one of them. We say that the colliding particle's kinetic energy must exceed the **production threshold energy** (KE_{th}). Consider Fig. 26.6. A high-energy particle of mass m_1 collides with a second particle (at rest) with mass m_2. After the collision, new particles of total rest energy E are created. If we write equations representing the conservation of energy and momentum (using special relativistic expressions), we find that the threshold energy is

$$KE_{th} = \left(\frac{m_1 + m_2}{m_2}\right)E + \frac{E^2}{2m_2c^2} \qquad (26-3)$$

where E is the difference between the rest energy of the product particles and the initial particles. The threshold energy is the *minimum* energy that particle 1 must have to produce the new particles.

Example 26.3 Evaluate the threshold kinetic energy that a proton must have when it collides with a second proton at rest and produces an additional $(p,\bar{p})$ pair.

A proton and an antiproton each have rest mass energy $m_p c^2 = 938$ MeV. The desired reaction is

$$p + p \to p + p + (p + \bar{p})$$

where the newly-created pair (in parentheses) has a total rest mass energy of $2m_p c^2 = 1876$ MeV. We therefore set $E = 2m_p c^2 = 1876$ MeV and $m_1 = m_2 = m_p$ in Eq. 26–3 to find

$$\text{KE}_{\text{th}} = \left(\frac{2m_p}{m_p}\right) 2m_p c^2 + \frac{(2m_p c^2)^2}{2m_p c^2} = 6m_p c^2 = 5630 \text{ MeV}.$$

Notice that the threshold kinetic energy is $6m_p c^2$ and not just the $2m_p c^2$ needed to create the rest mass of the $p,\bar{p}$ pair. The reason for this difference is that energy is not an absolute quantity in special relativity; instead, like space and time, it depends on the frame in which it is measured. In the frame moving with the $p,\bar{p}$ pair, $2m_p c^2$ is the required energy. In the laboratory frame (in which particle 2 is stationary), however, the energy required is $6m_p c^2$. The Bevatron at Berkeley was expressly designed in 1955 to accelerate protons to 6.3 GeV, slightly above threshold for $p,\bar{p}$ pair production.

The opposite process to pair production, matter-antimatter annihilation, occurs spontaneously whenever a particle and its antiparticle come together. Two photons (or other particles) are formed whose total energy is equal to the sum of the rest masses of the two particles. Gamma-ray photons from space have been observed with energies (0.511 MeV) which correspond to the electron-positron annihilation. One of the intriguing questions facing astrophysicists is whether there now is or ever was a large quantity of antimatter in the universe. If so, it would be difficult to distinguish it from matter, because the interaction of light with antimatter would be the same as that of normal matter; an anti-hydrogen atom, for example, consisting of a $\bar{p}$ and a e^+, would have the same energy level structure and therefore the same spectrum as a normal hydrogen atom. Only if large quantities of matter and antimatter came together, creating annihilation and the release of vast quantities of energy, would the presence of antimatter become obvious.

Conservation Laws

We have seen that energy and charge are conserved in particle interactions. Analysis of reactions has shown that there are other conservation laws as well. It has been found, for example, that a nuclear reaction can occur only if the total number of nucleons remains constant. It is possible to think of reactions in which energy and charge are conserved, but the number of nucleons would

26.2 PARTICLE INTERACTIONS

change, yet no such reactions take place. Nucleons (that is, protons and neutrons) are members of a class of particles called **baryons**. Studies of reactions have shown that the baryon number is always conserved. The baryon number of ordinary nucleons (or baryons) is $B = +1$, whereas for antinucleons (antibaryons) it is $B = -1$. Thus we see that the reaction

$$p + n \rightarrow p + n + p + \bar{p}$$
$$B = 1 + 1 = 1 + 1 + 1 - 1,$$

where the baryon number is conserved, is possible, whereas the reaction

$$p + n \rightarrow p + p + \bar{p}$$
$$B = 1 + 1 \neq 1 + 1 - 1,$$

for which the baryon equation is not conserved, is not possible. In the first of these reactions, the total baryon number is $+2$ on both sides of the equation, whereas in the second, it is $+2$ on the left but only $+1$ on the right. Baryon number conservation is another way of understanding why new particles are produced in pairs. The particle-antiparticle pair has zero net baryon number, so it does not change the initial baryon number. Table 26.2 lists some of the most common baryons, together with some of their quantum numbers and decay modes. Note that the antiparticle of each has baryon number -1 (for example $\bar{p}, \bar{n}, \bar{\Lambda}, \Xi^+$).

Example 26.4 Does the reaction

$$\Omega^- \rightarrow \Xi^- + \pi^\circ$$

conserve charge and baryon number?

Charge is obviously conserved; both sides have charge -1. Both Ω^- and Ξ^- are baryons ($B = +1$), while the π° meson has $B = 0$. Thus baryon number is also conserved. It turns out that this reaction is very slow (that is, strongly inhibited) because it violates another conservation law, called **strangeness**, which we will discuss in the next section on particle stability.

Besides baryon conservation, there is another class of particles (actually three subclasses) whose number is conserved in particle decays and interactions. These particles, called **leptons**, are light particles such as the electron (e^-), muon (μ^-), and the recently-discovered tau particle (τ^-), plus their associated neutrinos (ν_e, ν_μ, and ν_τ). Each of these six leptons has an antiparticle. Table 26.3 lists the six known leptons and six antileptons, grouped into three subclasses (or families) known as electron-type, muon-type, and tau-type. Notice that we now distinguish three different types of neutrinos. Leptons have zero baryon number; they do not interact by the strong force, but rather by the weak or electromagnetic forces. We will see in Section 26.3 that the strong force does not affect leptons because they are not composed of quarks.

Table 26.3
Leptons and their families

	Leptons	Anti-leptons
e-type	e^-, ν_e	$e^+, \bar{\nu}_e$
μ-type	μ^-, ν_μ	$\mu^+, \bar{\nu}_\mu$
τ-type	τ^-, ν_τ	$\tau^+, \bar{\nu}_\tau$

To see how lepton conservation works, think about baryon conservation. Every lepton or antilepton is assigned a **lepton number**: $+1$ for particles; -1 for antiparticles. Non-leptons are assigned 0. Lepton number must be conserved separately for each subclass. Thus there is an electron lepton number (L_e), a muon lepton number (L_μ), and a tau lepton number (L_τ). For example, $L_e = +1$ for an electron (e^-) and its associated neutrino (ν_e); $L_e = -1$ for a positron (e^+) and its antineutrino ($\bar{\nu}_e$); and $L_e = 0$ for all other particles, including muons and taus. For muon-type leptons, $L_\mu = +1$ for μ^- and ν_μ; $L_\mu = -1$ for μ^+ and $\bar{\nu}_\mu$; $L_\mu = 0$ for all others. The tau lepton numbers are assigned similarly.

Example 26.5 Analyze the following possible decays for the muon μ^-. Which are possible according to conservation of L_e and L_μ? Recall that μ^- has $L_e = 0$ and $L_\mu = +1$.

$$\mu^- \to e^- + \nu_\mu$$

The final state has $L_e = +1$ for the e^- and $L_\mu = +1$ for the ν_μ. Thus, even though muon lepton number is conserved, the electron lepton number is not.

$$\mu^- \to e^- + \bar{\nu}_\mu$$

The final state has $L_e = +1$ for the e^- and $L_\mu = -1$ for the $\bar{\nu}_\mu$. In this case, neither L_e nor L_μ is conserved.

$$\mu^- \to e^- + \nu_e$$

The final state has $L_e = +2$ and $L_\mu = 0$. Again, neither L_e nor L_μ is conserved.

$$\mu^- \to e^- + \bar{\nu}_e + \nu_\mu$$

This final state has $L_e = 0$, since the L_e of the e^- and the $\bar{\nu}_e$ cancel to zero. The L_μ of the ν_μ is $+1$. Thus, both L_e and L are conserved, and this is the way in which the muon decays. This decay is the primary evidence that both L_e and L_μ must be conserved in lepton decays.

26.2 PARTICLE INTERACTIONS

Particle Stability

Many particles can decay spontaneously, releasing energy or other particles. The length of time a particle typically exists before decaying depends on the type of interaction it can undergo. For example, particles subject to the strong nuclear interaction normally decay very rapidly, having typical lifetimes of only about 10^{-23} s. Note that this time is similar to the range of the strong interaction (10^{-15} m) divided by the speed of light (3×10^8 m/s). These decays all take place over very small distances.

Even though a particle may interact via the strong nuclear force, it may be prevented from decaying quickly. We may view such particles as being in a kind of metastable state, so that decay by the strong interaction is somehow forbidden, much as certain electron transitions in an atom are forbidden. The stability of many particles is due to the conservation laws we have already discussed. A proton does not decay, for example, because there is no particle less massive than the proton that has a baryon number $B = +1$, so no possible decay could conserve the baryon number. (Some new unified theories of strong and electroweak forces do predict that protons might violate baryon conservation and decay to $e^+ + \pi^°$, with a 10^{32} year lifetime, but experiments have thus far failed to confirm this prediction.)

Another class of particles is inhibited from certain decays because they violate **strangeness (S)**, a new quantum number and conservation law developed in 1955 by Murray Gell-Mann and K. Nishijima to explain the strange behavior of particles such as the K-mesons and the Λ, Σ, Ξ, and Ω baryons. These "strange" particles, as they were called, had lifetimes far longer than the typical strong interaction (10^{-10} to 10^{-8} s instead of the usual 10^{-23} s). In addition, the strange particles appeared to be produced in pairs (this was called **associated production**). For example, the reaction

$$\pi^- + p \rightarrow K^° + \Lambda^°$$

occurs readily when pions collide with protons ($\Lambda^°$ and $K^°$ are the strange pair), whereas the reaction

$$\pi^- + p \rightarrow K^° + n$$

has never occurred because the strange $K^°$ is not paired with another strange particle.

The strangeness conservation law works just like the previous laws we have discussed: charge, baryon number, or lepton number. Unless the interaction or decay conserves strangeness (S), the reaction is forbidden or strongly inhibited. In this latter respect, strangeness conservation differs from conservation of charge, baryon number, or lepton number, because strangeness is only a rigid conservation law for strong decays (lifetimes of order 10^{-23} s), and not for weak decays (lifetimes 10^{-10} to 10^{-8} s). This explains the abnormally long lifetimes for the strange baryons (see Table 26.2). In Section 26.3, we shall see that the strangeness quantum number is actually carried by the strange quark. That is, the strange mesons and baryons all contain one or more strange-type quarks.

888 CHAPTER 26 ELEMENTARY PARTICLES AND COSMOLOGY

As seen in Table 26.2, the $\Lambda°$, Σ^+, $\Sigma°$, and Σ^- particles are assigned strangeness $S = -1$ (they all contain *one* strange quark, which for historical reasons was assigned strangeness $S = -1$). The $\Xi°$ and Ξ^- have $S = -2$ (they contain two strange quarks), and the Ω^- particle has $S = -3$ (it contains three strange quarks). The K^+ and $K°$ mesons have $S = +1$, while the K^- and $\overline{K}°$ mesons have $S = -1$. Note that the antiparticles have the opposite strangeness of their corresponding particles (thus the $K°$ and K^+ have $S = +1$, because $\overline{K}°$ and K^- have $S = -1$).

Example 26.6   Which of the following decays violate strangeness? How much energy is released to the decay products?

$$K^- \rightarrow \mu^- + \overline{\nu}_\mu$$

This decay violates strangeness because the K^- has $S = -1$, yet neither μ^- nor $\overline{\nu}_\mu$ is a strange particle ($S = 0$). The reaction does occur, however, by a strangeness-violating weak interaction (lifetime 1.24×10^{-8} s). Note that muon lepton number is conserved. The energy released is the difference in rest energy between the K^- and μ^-: $(493.67 - 105.66)$ MeV $= 388$ MeV. The neutrino is massless.

$$\Omega^- \rightarrow \Lambda° + K^-$$

Strangeness is violated (Ω^- has $S = -3$, while $\Lambda°$ has $S = -1$ and K^- has $S = -1$). The decay occurs via the weak interaction, with lifetime 0.82×10^{-10} s and releases energy $(1672 - 1115.6 - 493.7)$ MeV $= 62.7$ MeV to the $\Lambda°$ and the K^-.

$$\Sigma^- \rightarrow n + \pi^-$$

This also violates strangeness (Σ^- has $S = -1$; n has $S = 0$). This weak decay has a lifetime of 1.48×10^{-10} s and releases energy $(1197.3 - 939.6 - 139.6)$ MeV $= 118$ MeV to the n and the π^-.

$$\Sigma^- \rightarrow \Xi^- + \pi°$$

This decay violates strangeness conservation, but it cannot occur at all since the Ξ^- is more massive than the Σ^-. A particle can decay only to products that are less massive.

If a particle is prevented from decaying via the strong interaction, it may exist long enough to decay by a less likely process. The electromagnetic interaction will cause decay in about 10^{-16} s, so this will be the minimum lifetime of a particle that does not decay by the strong interaction but is subject to the electromagnetic interaction. If the electromagnetic decay is forbidden, then the particle can live for at least 10^{-10} s, the typical time needed for decay via the

26.2 PARTICLE INTERACTIONS

weak interaction. It is not known whether particles that are stable against decay by the strong, electromagnetic, and weak interactions can decay through the gravitational interaction; if so, the lifetimes are enormous, and no such decay has yet been detected (some work on development of a GUT predicts that protons might decay with a lifetime of about 10^{32} years, but this is attributed to a new, superweak interaction, rather than to the gravitational interaction).

Even if a particle is subject to spontaneous decay via an allowed process, the probability of decay, hence the lifetime of the particle, depends on the details of the possible reaction. An important factor is the amount of energy that will be released in the decay, which corresponds to the mass difference between the initial particle and the products. If the difference is small, the probability of the decay is low, and the particle may last considerably longer than the nominal time associated with this interaction. Thus, the π^+ meson has a lifetime of 2.6×10^{-8} s, slightly longer than typical decays via the weak interaction. Other weakly-interacting particles may last for much longer times; this explains why radioactive β-decay half-lives may be years or even billions of years.

The very short-lived particles, with lifetimes of 10^{-20} s or less, are impossible to observe directly. Their fleeting existence is inferred from high-energy particle collision experiments in which the mass and energy of the initial and product particles is measured. If a very short-lived particle is formed and then decays during a collision, the secondary particles produced by the decay will have a total quantity of mass and energy equal to the mass and energy of the transitory particle. Therefore in an experiment where many collisions take place, and the distribution of product energies is measured, there will be a peak in this distribution at the energy corresponding to the rest energy of the temporary particle (Fig. 26.7). Such a peak is called a **resonance**, in analogy with many other processes where a single energy or frequency is favored. The transitory particle itself is usually referred to as a resonance in the experiment because it scarcely exists as a particle.

By the uncertainty principle, we may estimate the lifetime of a resonance particle. Since $(\Delta E)(\Delta t) \approx \hbar$, we set Δt equal to the lifetime $\tau = \hbar/\Delta E$.

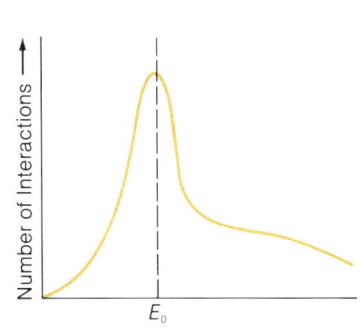

Figure 26.7

Particle scattering resonance. A peak in the distribution of product particle energies indicates that an unstable particle is formed whose rest energy is E_0. The unstable particle decays very rapidly, yielding product particles whose total energies are equal to the energy of the parent particle.

Example 26.7 The width of the energy peak in the distribution of $\eta°$ meson formation is 0.88 keV. What is its lifetime?

Using the uncertainty principle,

$$\Delta t = \tau = \frac{\hbar}{\Delta E} = \frac{(6.626 \times 10^{-34} \text{ J} \cdot \text{s})}{2\pi (0.88 \times 10^3 \text{ eV})(1.6 \times 10^{-19} \text{ J/eV})}$$
$$= 7.5 \times 10^{-19} \text{ s}.$$

This lifetime is typical for the electromagnetic decay

$$\eta° \rightarrow 2\gamma$$

although the $\eta°$ also decays via the strong interaction.

Classification of Particles

One of the goals of modern particle physics is to make some sense out of the hundreds of elementary particles that have been discovered since the 1940s. One way to make progress is to classify the particles according to their interactions (summarized in Table 26.2). The hope is that by seeing the family structure, we may understand the physical laws.

Recall that there are four fundamental forces in nature: the strong, electromagnetic, weak, and gravitational forces. Some progress has been made in unifying these forces, but the unification only applies at ultrahigh energies. Any particle having mass interacts gravitationally (actually, according to general relativity, even massless particles such as photons respond to gravity, as long as they have energy). All charged particles interact electromagnetically, and if a reaction involves photons, that is an indicator that the electromagnetic force was involved. Likewise, the appearance of neutrinos signals the weak interaction, while reactions involving nuclear matter generally implicate the strong force. We are now ready to classify.

We define four general classes of particles: (1) gauge bosons, (2) leptons, (3) mesons, and (4) baryons. The gauge bosons are the photon (γ) and the three intermediate vector bosons (W^+, W^-, and Z^0) responsible for the weak force. These four particles form a family or group, which unifies the electromagnetic and weak forces at high energies into a single electroweak interaction. The photon is massless, while the weak bosons have mass energies between 80 and 100 GeV.

The leptons were discussed in an earlier section. There are six known leptons (e^-, μ^-, τ^-, ν_e, ν_μ, ν_τ) and their corresponding antiparticles. They interact via the weak or electromagnetic forces, but not via the strong force. Leptons are believed to come in three types: electron-type (e^-, ν_e); muon-type (μ^-, ν_μ), and tau-type (τ^-, ν_τ), plus antiparticles. The muon and tau particle are essentially heavy versions of the electron, and they both can decay into the less massive electron, which is itself stable against further decay. The neutrinos associated with each type are thought to be massless and stable (there has been some recent suggestion, however, that neutrinos may have a very tiny mass). Since neutrinos have no charge and do not interact via the strong interaction, they pass easily through nuclear matter. To detect them at all requires many tons of material.

The last two classes, mesons and baryons, are the strongly interacting particles, generically called **hadrons**. The distinction between mesons and baryons comes from their intrinsic spin angular momenta (see Chapter 24); mesons have integral spin (0, 1, 2, . . ., in units of $\hbar$), while baryons have half-integral spin (1/2, 3/2, . . ., in units of $\hbar$). Another way to describe the two is that baryons are like nucleons (p, n, and heavier particles), while mesons are produced in collisions of nuclear matter (pions, K-mesons, etc.). As we have mentioned before, both baryons and mesons are believed to be composed of quarks and antiquarks.

We mention, finally, that in addition to the four classes of particles described above, there is another distinction that has far-reaching implications for particles in large numbers. Because of symmetry properties rooted in quantum mechanics, particles behave differently according to their spin. Particles with integral spin are called **bosons** (for example photons, pions, and *K*-mesons) and tend to clump together into the same quantum state. The laser is an example of bosons (photons) clumping together into a single, coherent state. In contrast, particles with half-integral spins (for example protons, neutrons, and electrons) are called **fermions** and obey the Pauli exclusion principle. All fermions obey this principle, and the protons and neutrons in a nucleus are prohibited from occupying the same quantum state in the same way that electrons in atoms are.

26.3 Quarks and Gluons

We have mentioned the need for ever higher energies in particle accelerators, as new particles with large rest masses are sought. Another motivation for developing very high-energy particles beams is that finer and finer structural details can be resolved in scattering experiments. The de Broglie wavelength is inversely proportional to particle momentum; for particles travelling at nearly the speed of light, as in accelerators, the particle momentum is $p \approx E/c$, so the de Broglie wavelength is

$$\lambda = \frac{h}{p} \approx \frac{hc}{E} \qquad (26\text{-}4)$$

where E is the particle energy. As particle energies increase and the de Broglie wavelength becomes smaller, the resolution of scattering experiments becomes finer. In Rutherford's early experiments, he had just sufficient resolution to detect the presence of nuclei in gold foil, but modern high-energy accelerators are capable of resolving detailed structure within a nucleus as fine as a few hundredths of a fermi (that is, structure on scales in the 10^{-17} m range). Even higher resolutions will become available with the development of new accelerators.

When modern scattering experiments probe the structure of elementary particles, a major contrast emerges between leptons and hadrons. No internal structure, nor indeed any measurable physical extent, is found for the six leptons (and their antiparticles). Hence this structure is smaller than the best resolving powers yet achieved. On the other hand, hadrons are found to have measurable physical size, and some have internal structure as well, consisting of isolated dense centers.

These findings suggest that hadrons are not truly elementary particles, but are instead composed of subordinate particles (Fig. 26.8). This suggestion has been verified, at least tentatively, by the results of beam-collision experiments, and is the basis of the **quark** theory. Murray Gell-Mann and George Zweig proposed in 1963 that certain symmetries among hadrons could be explained if baryons were composed of quark triplets (QQQ) and mesons were composed of quark-antiquark pairs ($Q\overline{Q}$). Their theory initially involved three quarks (the name "quark" itself is said to come from a line in James Joyce's book,

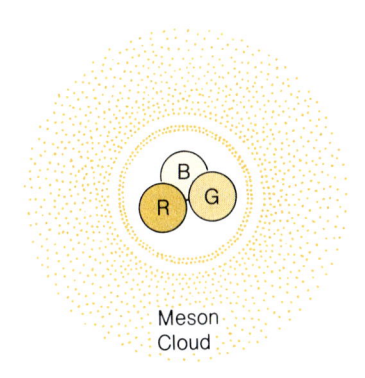

Figure 26.8

A nucleon (p or n), showing 3 quarks surrounded by a meson cloud. The quarks are bound by the strong force, mediated by an exchange of gluons between the "color charges" carried by the quarks (here, red, green, and blue).

Finnegan's Wake, but quark also means "nonsense" in German), but today we have evidence for six distinct types of quarks (providing a nice symmetry with the six types of leptons). Thus, there are thought to be only 16 truly elementary particles: the six leptons (and their antiparticles), the six quarks (and their antiparticles), and the four gauge bosons.

If we are to have six distinct types of quarks, then we must have at least six unique combinations of quantum numbers. All the quarks have spin 1/2 (in the usual units of $\hbar$) and baryon number $+1/3$, so they cannot be distinguished by spin or baryon number. They do have two possible electric charges, either $+2/3$ or $-1/3$ of the electron charge, but this does not provide enough combinations to account for even the three quarks first envisioned. Hence new quantum properties were inferred.

Even before the quark theory was proposed, there was already evidence for one new quantum property. As we have seen, experiments in the 1950s had shown that a new quantum property, strangeness, must be invoked to explain the lifetimes of certain particles. The addition of strangeness and a conservation law to go with it satisfactorily explained the 1950s experimental results. An additional quantum property, called **charm**, and a fourth quark were added in 1974 to explain a newly discovered meson which did not fit the old three-quark scheme. By convention, a strange quark has strangeness -1, and a charmed quark has charm $+1$; the antiquarks of each have opposite signs. Both strangeness and charm are conserved in strong and electromagnetic interactions, but not in weak interactions.

When the number of known leptons grew from four to six in the late 1970s, the expectation of symmetry between leptons and quarks led to the suggestion that there are six quarks as well, as we have already mentioned. This required two new quantum properties, called **top** and **bottom** by some, and **truth** and **beauty** by others. The six types of quarks, often described as the six **flavors**, are summarized in Table 26.4. Each has a name (for example up, down) and is designated by a letter (u or d, for example). There is now indirect experimental evidence, in the form of new heavy mesons, favoring the existence of a fifth and sixth quark, the bottom and top quarks. The two newest quantum properties, bottomness or beauty and topness or truth, apparently are also conserved in strong and electromagnetic interactions, but not in weak interactions.

According to the quark theory, mesons consist of quark-antiquark pairs, and baryons consist of three quarks in combination. A pion (π^+) is composed of an up quark and a down antiquark, the combination $u\bar{d}$. The total charge is $2/3 + 1/3 = 1$, as it should be. An antiparticle of the π^+ (the π^-) is composed of the antiparticles of the quarks that make up the π^+; that is, the π^- consists of $\bar{u}d$, and has a charge of -1. A proton, a stable hadron in the baryon subclass, consists of three quarks in the combination uud, and a neutron has three in the form udd. It is easy to use Table 26.4 to verify that these combinations produce the correct charge (and other quantum properties) for the proton and the neutron (see the Special Problems at the end of this chapter for further discussion). The quark theory very nicely predicts the observed properties of all of the known hadrons. Furthermore, all of the allowed combinations of up, down, and strange quarks, as well as a few mesons and baryons with charm and bottomness, have been observed in experiments. It remains

26.3 QUARKS AND GLUONS

Table 26.4
Properties of quarks*

Name	Symbol	Charge	Baryon number	Strangeness	Charm	Bottomness (beauty)	Topness (truth)
Up	u	$+2/3$	$1/3$	0	0	0	0
Down	d	$-1/3$	$1/3$	0	0	0	0
Strange	s	$-1/3$	$1/3$	-1	0	0	0
Charmed	c	$+2/3$	$1/3$	0	$+1$	0	0
Bottom	b	$-1/3$	$1/3$	0	0	$+1$	0
Top	t	$+2/3$	$1/3$	0	0	0	$+1$

*All quarks have spin ½, and come in three colors (say, red, green and blue). Antiquarks have opposite charge, baryon number, strangeness, charm, bottomness, and topness.

to be seen whether all of the additional combinations provided by the inclusion of charmed, top, and bottom quarks will be found also.

We have said that hadrons are made up of quarks, and that hadrons as a class undergo the strong interaction. In the modern theory, called **quantum chromodynamics,** the strong force between quarks is mediated by the exchange of additional gauge particles called **gluons,** of which there are eight types. By analogy with the electromagnetic force, in which the photon is the exchange particle between positive and negative electric charges, the eight gluons are the exchange particles between the three types of strong charge. **Strong charge,** usually called **color** in this theory, comes in three types, called red, green, and blue (based on an analogy with the primary colors of light), but of course they have nothing to do with the colors seen by the eye. Whereas only one photon is needed in the electromagnetic force (there is only one type of electric charge, which can be positive or negative), eight gluons are required to account for the possibilities of exchange between the three color charges in the strong force.

All real particles are colorless. By analogy with color combinations in light, they are composed either of equal mixtures of red, green, and blue (that is, quark triplets QQQ, called baryons), or they are equal mixtures of a colored quark and its antiquark (for example a red quark and an anti-red quark, $Q\overline{Q}$, called a meson). That real particles are colorless has tremendous predictive power; it helps explain why the only strongly interacting particles are baryons (QQQ) and mesons ($Q\overline{Q}$), and why the other possible combinations (QQ, $\overline{QQ}$, or even single quarks Q) are not observed. These particles would have an unbalanced color. Gluons, the exchange particles that change the colors of quarks, are bicolored. They come in eight varieties to account for all the color combinations.

The Grand Unified Theory, the ultimate goal of many physicists, must include a detailed description of the behavior of quarks and gluons, for they appear to hold the key to the fundamental nature of forces. The next few years will be exciting times in this area of research, because important theoretical developments have been emerging, and because powerful new accelerators are scheduled to go into operation soon.

26.4 Matter in the Universe

Cosmologists are those who study the properties of the universe as a whole, and attempt to understand its origin and evolution. In their studies, they are unavoidably faced with the question of explaining the present observed state of matter. Some questions, such as the ultimate origin, may never be answered; but others, such as the reasons for the observed distribution of elementary particles or of chemical elements, may be answerable. In this section we briefly outline progress that has been made in these directions.

The Origin of Matter

As we learned in Chapter 21, the universe apparently began in an initially hot, dense state, and has expanded from that state ever since (Fig. 26.9). The scenario of universal expansion from a compressed, hot state is called the **big bang**. Now we have learned that mass and energy are interchangeable, and that elementary particles can be created from sufficient quantities of energy. It follows that matter in the universe today originated from the creation of particles from energy during the earliest stages of the universal expansion. With this assumption, elementary particle theory has been remarkably successful in explaining the present-day composition of the universe. It has proven possible to reconstruct the sequence of events as the universe progressed from an early state where it was filled only with radiation and elementary particles to one where matter, in the form of nuclei and atoms, dominated.

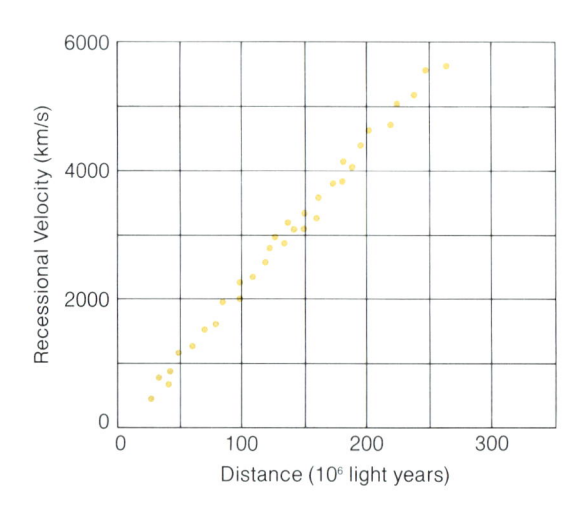

Figure 26.9

Expansion of the universe. Distant galaxies are receding from us with a velocity proportional to their distance (v = H_0d, where H_0 is the Hubble constant). This relationship, discovered by E. Hubble in 1929, shows that the universe is expanding.

FOCUS ON PHYSICS

The Cosmological Redshift and Universal Expansion

One of the most startling discoveries of the twentieth century was the expansion of the universe. This led to the big bang theory, and new ideas about the origin of galaxies, elements, and perhaps even elementary particles. The discovery began in the 1920s when the American astronomer V. M. Slipher realized that the spectral lines of distant galaxies are Doppler shifted to longer (redder) wavelengths. This effect is known today as the cosmological redshift.

The redshift is measured by recording the spectrum of a galaxy and measuring the wavelength of an emission or absorption line. If the galaxy is moving away at a velocity v that is much smaller than the speed of light c, then (as described in Chapter 10) all wavelengths of light are shifted by an amount $\Delta\lambda$ given by

$$\frac{\Delta\lambda}{\lambda_0} = \frac{(\lambda_{\text{obs}} - \lambda_0)}{\lambda_0} \simeq \frac{v}{c}$$

where λ_{obs} is the observed (redshifted) wavelength and λ_0 is the emitted wavelength in the rest frame of the galaxy (that is, the laboratory wavelength). For example, a galaxy receding at 1% of the speed of light ($v/c = 0.01$) will have all wavelengths lengthened by 1%. An emission line which normally appears at 500 nm will be observed at 505 nm. A convenient way to write the Doppler shift formula is

$$\lambda_{\text{obs}} = (1 + z)\lambda_0$$

where $z = \Delta\lambda/\lambda_0$ is a dimensionless number known as the redshift factor. In this example, $z = 0.01$.

Until the 1960s, most observed galaxies had redshift factors of only a few tenths or less. Improved observation techniques have allowed fainter galaxies, some with larger redshifts, to be measured. Beginning in the 1960s a new class of objects called quasars has been observed to have enormous redshift factors, as large as 2 or 3 (the current record-holder is the quasar PKS 2000-330, with $z = 3.78$). A value of z larger than 1 would seem to imply a value of v larger than c, according to the formula given above, but for large velocities a relativistic redshift formula must be used (see Chapter 23, Special Problems 3 and 4). The correct equation is

$$z = \frac{\Delta\lambda}{\lambda_0} = \sqrt{\frac{1 + v/c}{1 - v/c}} - 1.$$

Thus, the quasar with redshift factor $z = 3.78$ has a velocity of recession of $v = 0.916c$ (over 90% of the speed of light!).

In the 1920s the American astronomer Edwin Hubble undertook a systematic study of the redshifts of galaxies, and found that the more distant galaxies are moving away with higher velocities, and that the velocity of recession is proportional to the distance. Mathematically, this relationship is expressed as

$$v = H_0 d$$

where d is the distance and H_0 is the Hubble constant. The equation is known as the Hubble law. The best estimate today of the Hubble constant is about 75 km/s/Mpc. This means that the velocity of recession increases by 75 km/s for every increase in distance of 1 Mpc. (Mpc stands for megaparsec, or 10^6 parsecs. A parsec is an astronomical unit equal to 3.26 light-years or 3.09×10^{16} m.) In Figure 26.1.1, the slope of the line is the Hubble

(continued)

constant H_0 (the reason that the value of H_0 is uncertain today is that the observations are difficult and there is a lot of scatter in data points). As discussed in Special Problem 5 at the end of this chapter, the age of the universe can be estimated by taking the reciprocal of the Hubble constant. (This is equivalent to estimating the travel time by knowing the rate of travel and the distance covered.) The time $1/H_0$, known as the Hubble time, is between 10 and 20 billion years, depending on the value of H_0.

One of the profound implications of Hubble's discovery is that all of the galaxies in the universe, or at least the material from which they formed, must once have been in the same place. Thermodynamic considerations suggest that the universe was a glowing hot fireball at that time. The expansion from that initial state has become known as the big bang. The thermal radiation that filled the early universe is still present, but redshifted into the microwave portion of the spectrum where it has been detected by radio astronomers.

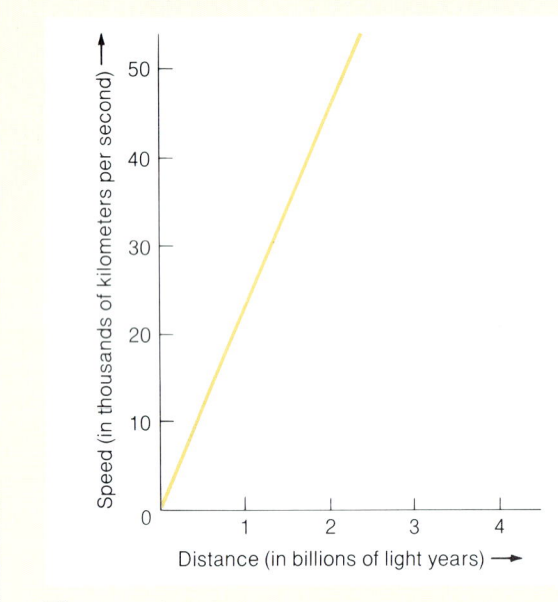

Figure 26.1.1

There remain uncertainties, mostly concerning the very earliest instant and the process by which a (supposedly) more or less uniform early universe produced the present fragmented distribution of matter. One of the most important uncertainties regarding the initial moments arises from the difficulty of starting the process of creating matter. If we postulate that the universe began as pure energy, then we need to understand how and when the fundamental particles (that is, quarks and leptons) were created. Because this is not known, many theorists begin their calculations at a very short time (about 10^{-4} s) after the start of the expansion, and assume that a certain number of fundamental particles were already present at that time.

Another puzzle is the apparently large imbalance between the quantity of matter and antimatter in the universe. Simple conservation laws and symmetry arguments might lead us to expect that particles and their antiparticles should be produced in equal numbers, yet in today's universe normal matter dominates by an enormous factor. The explanation for this is not yet complete, but current Grand Unified Theories include the notion that certain interactions can be asymmetric, resulting in a net conversion of antiparticles to particles.

Let us ignore for now the unknown processes that led to an early universe consisting of elementary particles, and consider what happened next. First, we introduce without explanation one result of general relativistic calculations describing the early universe, namely, that the temperature of the early universe is inversely proportional to the square root of its age,

26.4 MATTER IN THE UNIVERSE

Table 26.5
Events and eras in the universe

Age of universe	Radiation Temperature	Energy	Event	Era
15 billion yrs	3 K	10^{-3} eV	Today	Era of Galaxies
1 billion yrs	100 K	10^{-2} eV	Galaxies form	Era of photons and atoms
500,000 yrs	3000 K	1 eV	Plasma recombines into atoms	
3–4 min	10^9 K	1 MeV	Formation of stable Nuclei	Era of nuclei
10^{-6} s	10^{13} K	1 GeV	Formation of hadrons from quarks	Particle Era
10^{-12} s	10^{15} K	100 GeV	Weak and EM force unify	Quark-Lepton Era
10^{-35} s	10^{28} K	10^{15} GeV	Strong and electroweak forces unify	Electroweak Era
10^{-43} s	10^{32} K	10^{19} GeV	All four forces unify	GUT Era
0 s	∞	∞	Big bang	Quantum Gravity Era

$$T \approx \left(\frac{1.5 \times 10^{10} \text{ K}}{t^{1/2}}\right)\left(\frac{2}{n}\right)^{1/4} \quad (26\text{-}5)$$

where t is the age of the universe in seconds, and n is the number of fundamental particle types present. By "age", we mean the length of time since the beginning of the universal expansion. By "temperature" we mean the Kelvin temperature that characterizes the radiation, that is, the temperature of a blackbody that would produce the spectrum of the radiation. (See the discussion of blackbody radiation in Chapter 22.) In the earliest stages of the expansion, the radiation and the particles were in equilibrium, so this temperature also described the particles. At age $t = 1$ s, the temperature was about 10^{10} K; at $t = 100$ s, the temperature was 10^9 K, and so on.

Given the temperature history and the known energies of elementary particles, it is possible to calculate the times at which various particles began to appear out of the vacuum in the early universe (Table 26.5). It is proposed by theorists working toward the Grand Unified Theory that all four forces were unified at the earliest times (before $t = 10^{-43}$ s), when the temperature was 10^{32} K. This means that the four forces were equal and indistinguishable from each other (for example, the density was so high, or spacetime so tightly curved that the gravitational force between two particles was as strong as the strong nuclear force). Because of asymmetries that developed as the temperature decreased, the gravitational force at $t = 10^{-43}$ s became weaker than the other three. When the temperature dropped further, to about 10^{28} K (the equivalent of 10^{15} GeV energy) at $t = 10^{-35}$ s, the strong nuclear force became distinct from the unified electroweak force. Even later, at $t = 10^{-12}$ s (temperature 10^{15} K or 100 GeV energy), the weak and electromagnetic forces became distinct. At each step, new classes of elementary particles emerged from the primordial medium (see Table 26.5).

Between steps, the particles and antiparticles were in equilibrium with the radiation field, that is, their energy alternated between the forms of rest mass and radiation. Each type of particle stayed in equilibrium with the radiation only as long as the typical photon had more than enough energy to provide the energy for the formation of the particle-antiparticle pair. Therefore, as the universe cooled and photon energies decreased, a threshold temperature was reached for the production of each type of particle pair. As the universe cooled to lower temperatures, those particles no longer formed. Thus, the universe would have a net population of each kind of particle only if there was an imbalance between it and its antiparticle. This was brought about by the asymmetries in one or more of the four types of interactions that we have already mentioned.

At $t = 0.02$ s (temperature 10^{11} s), during the "particle era", the most common particles were photons, electrons, and neutrinos, and their antiparticles, but some protons and neutrons were also present. The proportions of these particles changed as the expansion and cooling continued, and when the production threshold for e^+, e^- pairs was passed at $t = 14$ s (temperature 3×10^9 K), positrons became much less common. Some nuclear fusion reactions began, but the temperature was still too high for these fusions to produce even simple nuclei such as deuterium ($^2_1$H) in any great numbers. Deuterium in particular was significant, because it is an intermediate stage in the formation of heavier nuclei such as helium (see Special Problem 4 in Chapter 25), yet deuterium is easily destroyed at a temperature of 3×10^9 K. When the temperature cooled to less than 10^9 K, deuterium began to survive long enough to allow subsequent reactions to produce some heavier nuclei, including not only helium ($^3_2$He, $^4_2$He), but also small amounts of other light elements such as lithium. By the age of about one hour, when the universe had cooled enough so that nuclear reactions had substantially stopped, and the primordial composition of the universe was established, the "era of nuclei" began. Now the universe consisted primarily of hydrogen (about 74% by mass) and helium (about 26%), with only traces of other chemical elements.

In the next section, we briefly outline the formation of heavier elements, but first we comment on the later development of the state of matter in the early universe. After the end of nuclear reactions and for nearly a million years after the expansion began, not much happened. The universe continued to expand and cool, and the equilibrium between radiation and matter continued. Now the interaction was Compton scattering of photons by electrons; thus, the temperature of the radiation and matter remained equal.

About 500,000 years after the expansion began, however, another very significant threshold was passed, the onset of the "era of atoms". It then became cool enough (temperature around 3000 K) for electrons and hydrogen nuclei (protons) to combine to form atoms and survive for a while, rather than being immediately ionized by the absorption of high-energy photons. When the uni-

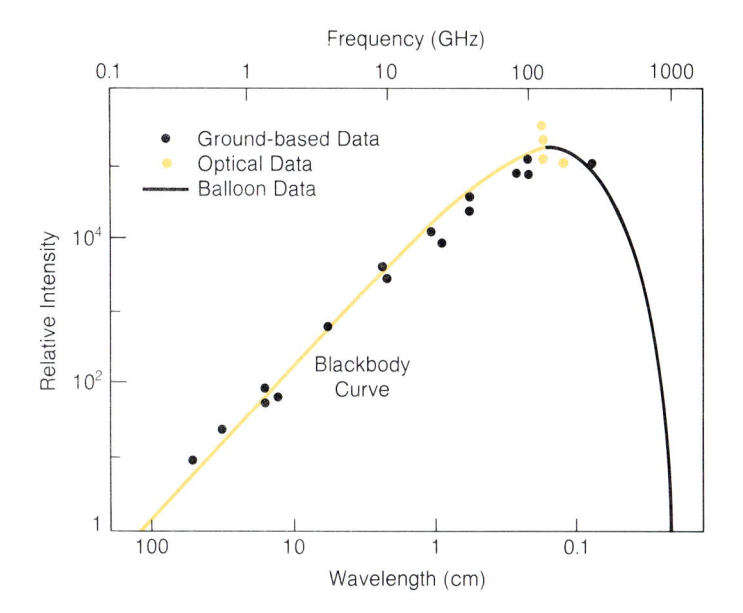

Figure 26.10

Spectrum of the microwave background radiation.

verse was in an ionized state (plasma), it was completely opaque to the transmission of light. A photon could not travel far before Compton-scattering off a free electron. The disappearance of free electrons into atoms suddenly allowed light to travel great distances across space. Because atoms have discrete energy levels and therefore cannot absorb light at all wavelengths, the appearance of atoms suddenly meant that there were photon wavelengths for which the matter in the universe was transparent. The radiation and matter in the universe "**decoupled**" from one another, and were no longer in equilibrium. Since that time, the number of photons in the universal radiation field has been essentially constant, while its temperature has continued to cool. Today we find that the universal radiation field has a characteristic temperature of about 3 K (so that its wavelength of peak intensity is in the microwave part of the spectrum, near 1 mm; Fig. 26.10). This radiation, variously called the **microwave background radiation,** or the **three-degree background radiation,** is a remnant of the big bang. Together with the expansion of the universe, this radiation is the most compelling evidence favoring the big bang theory.

After the time of decoupling, the matter in the universe has followed its own course, separately from the radiation. Stars, galaxies, and clusters of galaxies eventually formed, as did smaller aggregates of material such as planets. The story of how all of this came about is complex, is partially understood, and is the subject of an entire discipline known as astrophysics.

The Present State of Matter

Since we have described the formation of a few of the simplest elements according to elementary particle theory, for completeness we now sketch the later development of heavier elements. Here we will draw upon knowledge gained in Chapter 25, where we discussed nuclear physics.

As we learned in our earlier discussions, very high temperatures are needed in fusion reactions for collision energies to overcome the repulsive Coulomb force created by the like electrical charges on nuclei. A high density is also required to produce reasonably high reaction rates, because even at high temperature, only a small fraction of the collisions between nuclei result in reactions. As we have just seen, conditions of high temperature and density during the early expansion of the universe set the stage for nuclear reactions to produce a few simple elements. These reactions stopped too quickly to lead to the formation of heavier elements, however, so we must look elsewhere to find the origin of these species.

The cores of stars are very hot and dense because of gravitational compression (Fig. 26.11). In the center of the Sun, the temperature is about 1.5×10^7 K, and the density is about 1.6×10^5 kg/m^3. These conditions permit nuclear reactions. In stars more massive than the Sun, even higher temperatures and densities exist, and a greater variety of reactions is possible.

We mentioned (in Chapter 25) a reaction sequence called the **proton-proton chain**, in which four hydrogen nuclei combine, in several steps, to form one nucleus of helium ($^4_2$He). This is the predominant process in the Sun and other stars of similar or lower mass. In stars more massive than the Sun, a different reaction sequence, called the **CNO cycle**, occurs. The CNO cycle has the same result as the proton-proton chain, converting four protons into a single helium nucleus, but it does so at a higher rate (reaction rates are highly temperature-dependent) and it uses $^{12}_6$C as a catalyst. The steps in the CNO cycle are listed in Table 26.6. (See also Special Problem 3 in Chapter 25).

The conversion of hydrogen into helium is the first nuclear stage in the lifetime of a star. The process continues as long as hydrogen remains in the core of a star, but stops when the core hydrogen is depleted. The lowest-mass stars may never undergo any further reaction stages, but stars more massive than the Sun generally start new reactions after the hydrogen-burning stage. When the hydrogen reactions stop, the core of a star collapses to greater density and temperature because its source of internal pressure is gone, while the star's outer layers expand roughly 100 times in size to become a **red giant**. As the collapse proceeds, eventually sufficiently high temperatures are reached so that a new reaction involving helium may begin. The most favored is the **triple-alpha reaction**, in which three helium nuclei (alpha-particles) merge to form carbon ($^{12}_6$C).

As the core of a star is converted to carbon, other reactions may occur, depending on how high the temperature becomes as the core compresses further. Very massive stars may undergo a number of additional reaction stages, changing core composition each time, whereas a star of only moderate mass may stop its nuclear evolution with the completion of the triple-alpha stage. The types of reactions that follow the triple-alpha reaction are **alpha-capture reactions**, in which nuclei are converted to heavier species ($^{16}_8$O, $^{20}_{10}$Ne, $^{24}_{12}$Mg, etc.) by the addition of α-particles.

The heaviest elements that can be formed by energetically favored reactions are iron and nickel (recall our discussion of exothermic and endothermic reactions in Chapter 25). Thus, alpha-capture and other reactions build up the

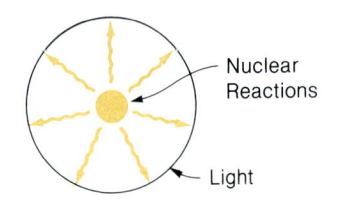

Figure 26.11

Stellar nuclear reactions. In the core of a star, the density and temperature are sufficient to sustain nuclear fusion reactions. These reactions occur in several steps, building up a core of heavy elements from an initial composition dominated by hydrogen.

26.4 MATTER IN THE UNIVERSE

Table 26.6
Basic steps in the CNO cycle

1. $^{12}_{6}C + ^{1}_{1}H \rightarrow ^{13}_{7}N + \gamma$
2. $^{13}_{7}N \rightarrow ^{13}_{6}C + e^{+} + \nu_e$
3. $^{13}_{6}C + ^{1}_{1}H \rightarrow ^{14}_{7}N + \gamma$
4. $^{14}_{7}N + ^{1}_{1}H \rightarrow ^{15}_{8}O + \gamma$
5. $^{15}_{8}O \rightarrow ^{15}_{7}N + e^{+} + \nu_e$
6. $^{15}_{7}N + ^{1}_{1}H \rightarrow ^{12}_{6}C + ^{4}_{2}He$

result: $4\,^{1}_{1}H \rightarrow ^{4}_{2}He + 2e^{+} + 2\nu_e + 3\gamma$

core composition of a massive star until iron has formed. Thereafter, any further fusion reactions are endothermic, and will therefore not occur unless large quantities of energy are available. With no more nuclear fuel to burn, the core of the massive star collapses to form a neutron star or black hole, while the outer layers explode as a **supernova**, releasing some 10^{44} J of energy as high-velocity ejecta of heavy elements. The formation of iron is the end of nuclear processing in stable stars, and we might expect that there is a relatively high abundance of iron in the universe. This is indeed so (Fig. 26.12)

Additional reactions must still occur, because we know that isotopes as heavy as $^{238}_{92}U$ are found in rocks. There are two classes of **neutron-capture** reactions that are responsible for the formation of elements heavier than iron. Free neutrons can be captured by nuclei, thereby creating nuclei heavier than those near the iron peak of the nuclear binding energy curve. Inside a normal star, there are few free neutrons available (they come only as by-products of certain reactions in red giant stars, such as $^{4}_{2}He + ^{13}_{6}C \rightarrow ^{16}_{8}O + ^{1}_{0}n$), so the neutron-capture process is very slow. Typically in these **s-process reactions** (s for slow), the neutron-captures are slow enough that the newly-formed nucleus undergoes β-decay, converting a neutron into a proton. Thus, the s process creates a sequence of elements whose neutron numbers grow only rather slowly. In the **r-process reactions** (r for rapid), neutrons are available in large quantities, and usually the newly-formed nuclei do not have time to undergo beta-decay before additional neutrons are added. The new nuclei that are built up have higher proportions of neutrons (relative to protons) than do the s-process elements.

The large numbers of neutrons needed for r-process reactions are available for short periods during stellar explosions, particularly the **supernovae** when massive stars complete all possible stable reactions and collapse catastrophically (Fig. 26.13 and Fig. 25.22). Thus, many of the elements beyond the iron peak that are found on the Earth were formed in supernova explosions.

Supernovae are also responsible for distributing throughout space the elements formed in stellar cores by stable fusion reactions. As succeeding generations of stars form, evolve, and die, the overall composition of the universe gradually is enriched with heavy elements. Today the cosmic composition of the universe is still dominated by hydrogen and helium, but there are relatively high abundances (about 2% by mass) of the elements formed most readily in stellar cores (such as carbon, nitrogen, and oxygen) and the elements near iron

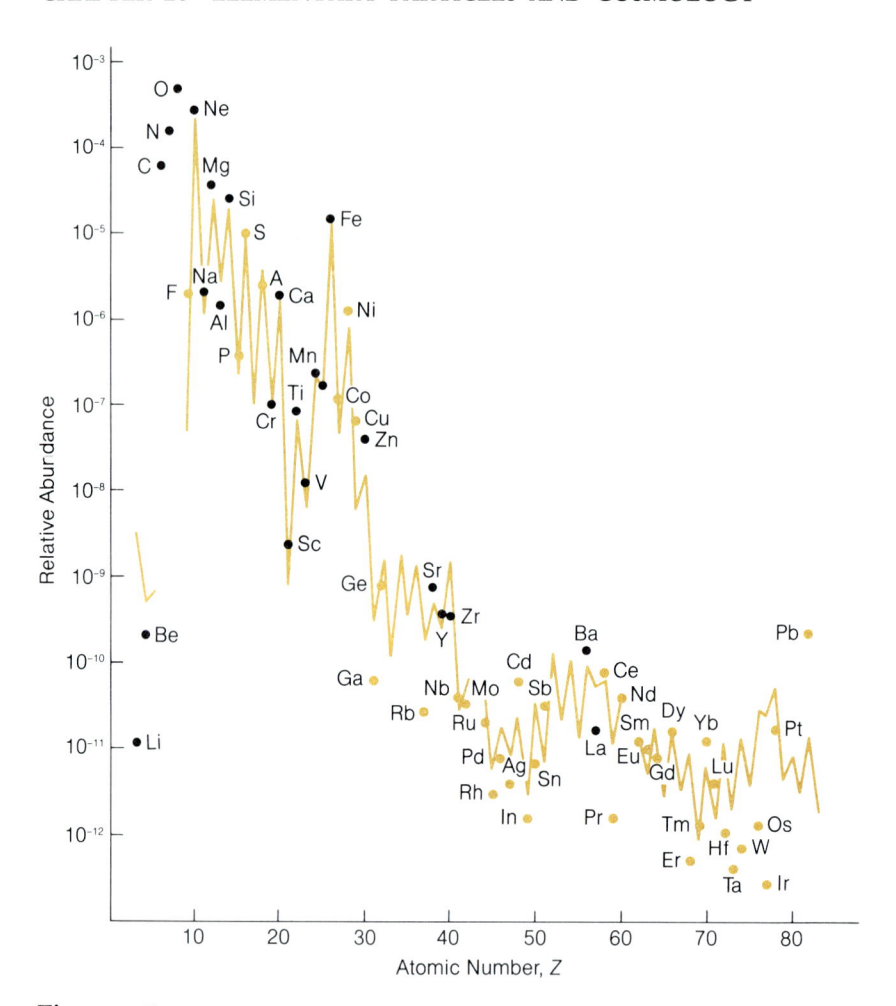

Figure 26.12

Cosmic abundances of the elements. Hydrogen is 1.0 and helium is 0.1 on this scale. All the other elements have been created in stellar nuclear reactions. The distribution has a peak at iron (Fe) because it is the heaviest element produced in exothermic reactions in stars. The dots represent abundances on the Sun's surface, and the solid line is from the Earth and meteorites.

in the periodic table (Fig. 24.13). Presumably as the universe continues to age, more and more of its mass will be converted from simple elements to heavy ones by stellar processing.

The Future of Matter

It is possible to speculate about the future of the universe, with our knowledge of the processes that are gradually altering its chemical composition, and our

Figure 26.13

Supernova in a distant galaxy.

understanding of elementary particle theory. Of course, the final outcome depends on whether the universe is open or closed (see Chapter 21), for this will determine whether it will simply keep expanding and cooling, or will eventually contract to a new era of high temperature and density. If the universe is closed, so that it does finally contract to a new state of compression, we can expect it to reach a state where matter and radiation are again in equilibrium at very high temperatures. In this state, often called the "big squeeze", the conditions of the early expansion will be reproduced. Elementary particle physics can tell us what will happen for a time after this, until all four forces are again unified and the universe consists of pure radiation. It is presently impossible to say what will happen next, but many have speculated that a new expansion would follow, repeating the entire cycle. The period for a cyclic universe, the approximate time for expansion and contraction, is about 10^{11} years.

If, on the other hand, the universe is open (or flat), then there will be no recompression, just a continued expansion and dissipation. Much of the hydrogen will be converted to other elements in about 10^{14} years and new star formation will effectively cease. Gradually matter will become locked up in dead stars and widely dispersed gas and dust. Galaxies themselves will dissipate due to random escapes of stars by about 10^{19} years. By age 10^{32} years, all of the protons in the universe may have had time to decay (by a very low-probability process called for in some modern Grand Unified Theories), and the universe will consist of only free electrons, positrons, black holes, photons, and neutrinos. Black holes themselves can decay by quantum mechanical processes, and all of the black holes will be gone by age 10^{100} years. Thus, the final state of the universe may be a cold near-vacuum, containing electrons, positrons, and very low-energy photons and neutrinos.

CHAPTER 26 ELEMENTARY PARTICLES AND COSMOLOGY

Perspective

It is ironic and fitting that we have completed our discussions of physics by being led from a study of the most minute particles to an understanding of the grandest phenomenon of all: the evolution of the universe itself. In a way, the breadth and scope of this final chapter represent the role of physics in the world of science. Physicists deal with natural phenomena on all scales. We have found during our studies that basic physics underlies every natural science, that physics encompasses all science the way elementary particles comprise all material objects.

The broad background provided by this text has provided insufficient depth in any single area to allow a student to go forth and solve all problems in the real world. What it has done, we hope, is provide the student with the tools to learn how to deal in depth with specific areas. We hope further that the background has provided the intellectual stimulation that accompanies any new understanding, and that the student has enjoyed knowing more about how things work. That is, after all, part of the benefit of being a scientist.

SUMMARY TABLE OF FORMULAS

Compton length:
$$d \simeq \frac{\hbar}{mc} \tag{26-1}$$

Reactions producing pi-mesons:
$$p + p \rightarrow p + p + \pi^\circ \tag{26-2a}$$
$$p + p \rightarrow p + n + \pi^+ \tag{26-2b}$$
$$n + n \rightarrow n + n + \pi^\circ \tag{26-2c}$$
$$p + n \rightarrow p + p + \pi^- \tag{26-2d}$$

Threshold energy for particle production:
$$KE_{th} = \left(\frac{m_1 + m_2}{m_2}\right)E + \frac{E^2}{2m_2c^2} \tag{26-3}$$

de Broglie wavelength for relativistic particles:
$$\lambda = \frac{h}{p} \simeq \frac{hc}{E} \tag{26-4}$$

Temperature of the universe at age t:
$$T \simeq \left(\frac{1.5 \times 10^{10} \text{ K}}{t^{\frac{1}{2}}}\right)\left(\frac{2}{n}\right)^{1/4} \tag{26-5}$$

THOUGHT QUESTIONS

1. Why does neutron decay ($n \rightarrow p + e^- + \bar{\nu}_e$) involve the weak interaction?

2. Show that the ratio of a particle's total energy E to its rest mass energy m_0c^2 equals the Lorentz factor $\gamma = 1/\sqrt{1 - (v^2/c^2)}$. (See Chapter 21.)

3. What is the relation between a particle's total energy E, its kinetic energy T, and its rest mass energy m_0c^2 in special relativity?

4. Show that in neutron decay (Question 1), a down quark is transformed into an up quark.

5. Show that when a pion at rest decays into two photons ($\pi^\circ \rightarrow \gamma + \gamma$), the two photons must have the same energy to conserve both momentum and energy.

6. When a particle and its antiparticle annihilate at rest into two photons (for example, $e^+ + e^- \rightarrow \gamma + \gamma$ or $p + \bar{p} \rightarrow \gamma + \gamma$), show that the energy of one of the

photons equals the rest mass energy of one of the original particles.

7 Why, in the quark theory, do only baryons and mesons interact via the strong interaction?

8 Describe the differences between a particle and its antiparticle. Discuss mass, charge, baryon number, lepton number, etc.

9 Why is the lifetime of an unstable particle related to the uncertainty in its energy?

10 Define: lepton, baryon, hadron, meson, photon, neutrino, quark.

11 What are the similarities and differences between photons and gluons?

12 Why may the early universe be the best laboratory for testing new theories of particle physics at ultrahigh energies?

13 Discuss the type, forces, and interactions of the constituents of matter, beginning with solids, liquids, and gases (molecules and atoms) and proceding to smaller and smaller units.

14 In the early universe, at temperatures of 10^{12} K, what are the fundamental or predominant particles?

PROBLEMS

- 1 What is the total energy of an electron whose kinetic energy is 1.022 MeV?
- 2 What is the Lorentz factor γ of the electron in Problem 1?
- 3 How much energy is released in the decay of a π^+ meson, $\pi^+ \to \mu^+ + \nu_\mu$?
- 4 What are the energies of the photons released in the decay of a $\pi°$ meson at rest?

$$\pi° \to \gamma + \gamma$$

•• 5 What are the baryon numbers of reactants and products in the following reactions?
(a) $p + p \to p + p + p + \bar{p}$
(b) $p + n \to p + p + \pi^-$
(c) $n \to p + e^- + \bar{\nu}_e$
(d) $p + \pi^- \to K° + \Lambda°$
(e) $\Sigma^+ \to p + \pi°$
(f) $\Xi° \to \Lambda° + \pi°$
(g) $\Omega^- \to \Xi° + \pi^-$
(h) $\Lambda° \to p + \pi^-$

•• 6 What are the lepton numbers of the reactants and products in the following reactions?
(a) $\pi^+ \to \mu^+ + \nu_\mu$
(b) $n + \nu_e \to p + e^-$
(c) $K^+ \to \mu^+ + \nu_\mu$
(d) $\mu^+ \to e^+ + \nu_e + \bar{\nu}_\mu$

•• 7 What are the values of strangeness of the reactants and products in the reactions?
(a) $\Lambda° \to K^- + n + \pi^+$
(b) $\Lambda° \to p + \pi° + \pi^-$
(c) $\Xi° \to \Lambda° + \pi°$
(d) $\Omega^- \to \Lambda° + K^-$
(e) $\Lambda° \to K^- + p$

- 8 How much energy is released when an electron and its antiparticle, the positron, annihilate into gamma rays ($e^+ + e^- \to \gamma + \gamma$)?
- 9 How much energy is released when a proton and its antiparticle annihilate ($p + \bar{p} \to \gamma + \gamma$)?
- 10 The energy uncertainty of an unstable particle is 10 MeV. What is its approximate lifetime from the Heisenberg uncertainty principle?
- 11 The energy uncertainty of the ψ/J particle resonance is 0.070 MeV. What is its lifetime?
- 12 The energy uncertainty of the Δ resonance when pions scatter off protons is about 100 MeV. Estimate the lifetime of the Δ resonance particle.

•• 13 Two protons, each travelling at the same speed, approach each other, collide, and produce a $\pi°$. (The reaction is $p + p \to p + p + \pi°$.) What minimum threshold kinetic energy must the protons have?

•• 14 When high-energy cosmic ray protons collide with stationary interstellar protons (contained in hydrogen gas in space), they produce pions via the following reactions:

$$p + p \to p + p + \pi°$$
$$p + p \to p + n + \pi^+$$

(a) Show that both baryon number and charge are conserved.
(b) Compute the energy threshold (see Eq. 26–3) needed by the cosmic ray p to produce the pion in each reaction.

•• 15 Which of the following reactions are allowed according to the conservation laws of charge, baryon

number, lepton number, and strangeness? For the forbidden reactions, state which law or laws are violated.
(a) $n + p \rightarrow \pi^+ + \pi^- + \pi^0$
(b) $n + p \rightarrow p + p + \bar{p}$
(c) $n + \bar{\nu}_e \rightarrow p + e^-$
(d) $\pi^+ \rightarrow \mu^+ + \nu_\mu$
(e) $\Lambda^0 \rightarrow p + \pi^0 + \pi^-$
(f) $\Lambda^0 \rightarrow K^- + n + \pi^+$
Recall that the antiproton, $\bar{p}$, has the opposite charge from the proton p.

16 If the cosmic rays in Problem 14 collide with neutrons (in interstellar helium nuclei, for example), which of the following reactions are allowed? For those forbidden, state which conservation law is violated.

$$p + n \rightarrow p + n + \pi^0$$
$$\rightarrow p + p + \pi^-$$
$$\rightarrow p + n + \pi^0 + \pi^0$$
$$\rightarrow p + n + \pi^+ + \pi^-$$
$$\rightarrow p + p + \pi^- + \pi^0$$
$$\rightarrow n + n + \pi^+ + \pi^0$$

17 The charged pions, π^+ and π^-, produced when cosmic rays strike protons or neutrons in the upper atmosphere (see Problems 14 and 16), decay into muons via the reactions,

$$\pi^+ \rightarrow \mu^+ + \nu_\mu \text{ and } \pi^- \rightarrow \mu^- + \bar{\nu}_\mu.$$

The neutral pions decay into two photons ($\pi^0 \rightarrow \gamma + \gamma$).
(a) What are the energies of the two photons resulting from π^0 decay? (Assume the π^0 is at rest.)
(b) Show that lepton number is conserved in the π^+ and π^- decays.
(c) Explain why, on the basis of conservation of both muon lepton number L_μ and electron lepton number L_e, two neutrinos (ν_μ and ν_e) are released when a μ^+ or μ^- decays according to the reactions

$$\mu^+ \rightarrow e^+ + \bar{\nu}_\mu + \nu_e \text{ or } \mu^- \rightarrow e^- + \nu_\mu + \bar{\nu}_e.$$

18 To create an antiproton in a particle accelerator, one must create enough energy in the center-of-mass frame to produce both a new proton and antiproton ($p + p \rightarrow p + p + p + \bar{p}$).
(a) Using the concept of baryon number, explain why the antiproton must be accompanied by a proton. This is why antiparticles are created as particle-antiparticle pairs.
(b) Using the energy threshold formula given in the text (Eq. 26–3), find the minimum kinetic energy needed to create the p-$\bar{p}$ pair by colliding a fast p with a stationary target p.
(c) Compare your answer for KE in part b with the rest energy of the p-$\bar{p}$ pair. Why was the extra energy required?

19 Consider the following particle reactions and state which ones violate conservation of charge, baryon number, and lepton number (e and μ type).
(a) $n \rightarrow p + e^- + \nu_e$
(b) $p + p \rightarrow p + p + \bar{p} + e^+$
(c) $\pi^+ \rightarrow \mu^+ + \nu_\mu$
(d) $\mu^- \rightarrow e^+ + \nu_e + \nu_\mu$
(e) $e^- + e^+ \rightarrow \gamma + \gamma$
In these reactions, p is a proton (p^+) and $\bar{p}$ is an antiproton with a negative charge.

20 Complete the following reactions with a appropriate particle or nucleus, using conservation of charge (Z) atomic weight (A), and lepton number L. Electrons and positrons have negligible atomic weight compared to the baryons and nuclei. Photons and neutrinos are massless.
(a) $^1_1H + ^2_1H \rightarrow \underline{} + \gamma$
(b) $^4_2He + ^3_2He \rightarrow \underline{} + \gamma$
(c) $^{12}_6C + \underline{} \rightarrow ^{13}_7N + \gamma$
(d) $\underline{} \rightarrow ^1_1H + e^- + \bar{\nu}_e$
(e) $^1_0n + e^+ \rightarrow ^1_1H + \underline{}$

SPECIAL PROBLEMS

1. Quark Theory

In the original quark theory, baryons were composed of three quarks (QQQ), which came in three types: up (u), down (d), and strange (s). One group of such quark triplets is called the first octet and contains the following:

$$p = (uud) \qquad \Xi^{\circ} = (ssu)$$
$$n = (ddu) \qquad \Sigma^{\circ} = (uds)$$
$$\Sigma^{-} = (dds) \qquad \Lambda^{\circ} = (uds)$$
$$\Xi^{-} = (ssd) \qquad \Sigma^{+} = (uus)$$

Note that both Λ° and Σ° have the same quark constituents (uds), although their mixture is different. From Table 26.4 (in the text) showing the baryon number, charge, and strangeness of the quarks, find the total charge, baryon number, and strangeness of the eight particles in the first octet.

2. Mesons

In quark theory, strongly interacting mesons are composed of quark-antiquark pairs ($Q\overline{Q}$).
(a) Show that because a quark is paired with an antiparticle, a meson must have zero baryon number.
(b) Using Table 26.4, confirm the charge of the following mesons:

$$\pi^{+} = u\overline{d} \qquad K^{\circ} = d\overline{s}$$
$$\pi^{-} = \overline{u}d \qquad \underline{K}^{\circ} = \overline{d}s$$
$$K^{+} = u\overline{s} \qquad K^{-} = \overline{u}s$$
$$\pi^{\circ} = u\overline{u} + d\overline{d}$$

(c) Show that K^{+} and K^{-} are antiparticles of each other; show likewise for K° and $\underline{K}^{\circ}$ and for π^{+} and π^{-}.
(d) What is the strangeness of each meson in part b?

3. Exotic Quarks

In the current version of quark theory, there are six quarks (Table 26.4).
(a) Give an example of the quark constituents of a charmed baryon.
(b) Give an example of a meson containing truth or beauty.
(c) Give an example of the quark constituents of a strange antibaryon.

4. Quark Transformations

In the following reactions involving a transformation of one baryon into another, describe which quark or quarks change type. (Refer to Special Problems 1 and 2 for quarks).
(a) $n \rightarrow p + e^{-} + \overline{\nu}_{e}$
(b) $\Lambda^{\circ} \rightarrow p + \pi^{-}$
(c) $\Sigma^{+} \rightarrow p + \pi^{\circ}$
(d) $\Omega^{-} \rightarrow \Xi^{\circ} + \pi^{-}$

5. Expansion of the Universe

Before doing this problem, please read the *Focus of Physics* insert in Chapter 26 on the Hubble constant H_0 and the Doppler shift $z = \Delta\lambda/\lambda_0$.
(a) Astronomers find a distant galaxy with its spectral lines shifted to the red by 10%; its redshift factor z is 0.1. What is the galaxy's velocity according to the nonrelativistic Doppler shift formula?
(b) If the galaxy's distance is determined to be 1.3×10^9 light years (400 Mpc), what is the Hubble constant H_0? Give your answer in "astronomers units," km/s/Mpc.
(c) If 1 Mpc = 3.09×10^{22} m, show that your value for H_0 in part b may be written in more basic units of s^{-1}. (You will need to convert the units of km and Mpc to a common unit.)
(d) Remembering that $d = vt$, use Hubble's law, $v = H_0 d$, to show that the time a galaxy has been travelling is equal to the reciprocal of the Hubble constant, $1/H_0$. What is this Hubble time in years, for your value of H_0 in part c?
(e) If the expansion of the universe has been slowing down (galaxies were receding more rapidly in the past), how would your answer in part d be affected? Is the actual age of the universe greater or less than $1/H_0$?

SUGGESTED READINGS

"Quarks with color and flavor." S. L. Glashow, *Scientific American*, Oct 75, p 38, **233**(4)

"The search for new families of elementary particles." D. B. Cline, A. K. Mann, C. Rubbia, *Scientific American*, Jan 76, p 44, **234**(1)

"Will the universe expand forever?" J. R. Gott III, J. E. Gunn, D. N. Schramm, B. Tinsley, *Scientific American*, Mar 76, p 62, **234**(3)

"The confinement of quarks." Y. Nambu, *Scientific American*, Nov 76, p 48, **235**(5)

"Supergravity and the unification of the laws of physics." D. Z. Freedman, P. van Nieuwenhuizen, *Scientific American*, Feb 78, p 126, **238**(2)

"Heavy leptons." M. L. Perl, W. T. Kirk, *Scientific American*, Mar 78, p 50, **238**(3)

"The cosmic asymmetry between matter and anti-matter." F. Wilczek, *Scientific American*, Dec 80, p 82, **243**(6)

"A deep-sea neutrino telescope." J. G. Learned, D. Eichler, *Scientific American*, Feb 81, p 138, **244**(2)

"The decay of the proton." S. Weinberg, *Scientific American*, Jun 81, p 64, **244**(6)

"Superheavy magnetic monopoles." R. A. Carrigan, Jr., W. P. Trower, *Scientific American*, Apr 82, p 106, **246**(4)

"The future of the universe." D. A. Dicus, J. R. Letaw, D. C. Teplitz, V. L. Teplitz, *Scientific American*, Mar 83, p 90, **248**(3)

"The structure of quarks and leptons." H. Harari, *Scientific American*, Apr 83, p 56, **248**(4)

"Dark matter in spiral galaxies." V. Rubin, *Scientific American*, Jun 83, p 96, **248**(6)

"Particles with naked beauty." N. B. Mistry, R. A. Poling, E. H. Thorndike, *Scientific American*, Jul 83, p 106, **249**(1)

"The large-scale structure of the universe." J. Silk, A. Szalay, Y. B. Zel'dovich, *Scientific American*, Oct 83, p 72, **249**(4)

"The inflationary universe." A. H. Guth, P. J. Steinhardt, *Scientific American* May 84, p 116, **250**(5)

"The hidden dimensions of spacetime." D. Z. Freedman, P. van Nieuwenhuizen, *Scientific American*, Mar 85, p 74, **252**(3)

"Elementary particles and forces." C. Quigg, *Scientific American*, Apr 85, p 84, **252**(4)

"The search for proton decay." J. M. LoSecco, F. Reines, D. Sinclair, *Scientific American*, Jun 85, p 54, **252**(6)

"Cosmology and elementary particle physics." M. S. Turner, D. N. Schramm, *Physics Today*, Sep 79, p 42, **32**(9)

"The birth of elementary particle physics," L. M. Brown, L. Hoddeson, *Physics Today*, Apr 82, p 36, **35**(4)

"The early universe and high-energy physics." D. N. Schramm, *Physics Today*, Apr 83, p 27, **36**(4)

"The evolution of SLAC and its program." W. K. H. Panofsky, *Physics Today*, Oct 83, p 34, **36**(10)

"The SSC: A machine for the nineties." S. L. Glashow, L. M. Lederman, *Physics Today*, Mar 85, p 28, **38**(3)

"The shadow universe." D. Overbye, *Discover*, May 85, p 12, **6**(5)

"Inventing the beginning." A. Sandage, *Science 84*, Nov 84, p 111, **5**(9)

"Sensing the ripples in space-time." M. Bartusiak, *Science 85*, Apr 85, p 58, **6**(3)

"Detecting next to nothing." G. Taubes, *Science 85*, May 85, p 58, **6**(4)

The first three minutes. Steven Weinberg, Basic Books, New York, 1977

The moment of creation. James S. Trefil, Scribners, New York, 1983

Appendix 1
Fundamental Constants

	Symbol	Approximate Value	Best Value
Speed of light	c	3.00×10^8 m/s	2.99792458×10^8 m/s
Gravitational constant	G	6.67×10^{-11} N·m²/kg²	6.6720×10^{-11} N·m²/kg²
Gravitational acceleration	g	9.81 m/s²	9.80665 m/s²
Electron rest mass	m_e	9.11×10^{-31} kg = 0.511 MeV/c²	9.109534×10^{-31} kg 0.5110034 MeV/c²
Proton rest mass	m_p	1.67×10^{-27} kg = 938.3 MeV/c²	$1.6726485 \times 10^{-27}$ kg 938.2796 MeV/c²
Neutron rest mass	m_n	1.67×10^{-27} kg = 939.6 MeV/c²	$1.6749543 \times 10^{-27}$ kg 939.5731 MeV/c²
Atomic mass unit	amu	1.66×10^{-27} kg = 931.5 MeV/c²	$1.6605655 \times 10^{-27}$ kg 931.5016 MeV/c²
Avogadro's number	N_A	6.02×10^{23}/mol	6.022045×10^{23}/mol
Gas constant	R	8.31 J/mol·K	8.31441 J/mol·K
Boltzmann constant	k	1.38×10^{-23} J/K	1.380662×10^{-23} J/K
Absolute zero	0 K	−273 °C	−273.15 °C
Atmospheric pressure	1 atm	1.01×10^5 N/m²	1.01325×10^5 N/m²
Triple point of water		+273 K	+273.16 K
Electron charge	e	1.60×10^{-19} C	$1.6021892 \times 10^{-19}$ C
Permittivity of free space	ϵ_0	8.85×10^{-12} C²/N·m²	$8.85418782 \times 10^{-12}$ C²/N·m²
Permeability of free space	μ_0	$4\pi \times 10^{-7}$ T·m/A	$12.5663706144 \times 10^{-7}$ T·m/A
Planck's constant	h	6.63×10^{-34} J·s	6.626176×10^{-34} J·s
	$\hbar = h/2\pi$	1.05×10^{-34} J·s	$1.0545887 \times 10^{-34}$ J·s
Stefan-Boltzmann constant	σ	5.67×10^{-8} W/m²·K⁴	5.67032×10^{-8} W/m²·K⁴
Rydberg (infinite mass nucleus)	R_∞	1.10×10^7 m⁻¹	1.097373177×10^7 m⁻¹
Rydberg (Hydrogen)	R_H	1.10×10^7 m⁻¹	1.096775854×10^7 m⁻¹
Rydberg energy (Hydrogen)	I_H	13.6 eV	13.598457 eV
Bohr radius	r_0	0.529×10^{-10} m	$0.52917706 \times 10^{-10}$ m
Proton/electron mass	m_p/m_e	1836	1836.15152
Astronomical unit	AU	1.50×10^{11} m	1.49599×10^{11} m

Appendix 2
Review of Mathematical Techniques

While it is assumed that the student has had sufficient mathematics to understand and solve problems in this text, a review of general techniques may be helpful for those whose previous mathematics courses were taken long ago, or for those unfamiliar with certain methods that come into play. This review is by no means comprehensive, but instead concentrates on a few key areas.

Numerical Accuracy and Scientific Notation

In physics and other sciences, we are concerned with expressing and manipulating numbers that represent physical quantities. This means that there are certain conventions and rules for expressing numbers so that their representation is realistic. It also means that we have to deal with numbers that can be very large or small, so that the normal way of writing them can be very cumbersome.

The most important factor in determining how we should write a number representing a physical quantity is the accuracy with which the physical quantity is known. This is usually determined by the degree of uncertainty in an experimental measurement. If, for example, we measure the mass of an object on a balance, we may typically find that we can determine the result accurately to the nearest one-hundredth of a kilogram, but no better. Then the mass of the object would be written with only two digits after the decimal point. It is often an important part of an experiment to determine how accurate the measurements are. This depends on the nature of the experiment and the measurement apparatus, as well as on the statistics of random errors.

We can characterize the accuracy of a number as the ratio of the uncertainty to the measured value itself. If, for example, we find that the mass of an object is 13.41 ± 0.03 kg, we say that the mass is known to an accuracy of 0.03 kg/13.41 kg = 0.0022 or 0.22 percent. Expressing an accuracy in this fashion can provide a better intuitive feeling for how precise an experiment is than to just say what the uncertainty is in absolute units (such as kilograms).

In solving problems involving experimental values, it is important to keep in mind how many **significant figures** there are. The number of significant figures is the number of digits in the value that are well known from the measurement. In the mass determination example above, where the value is 13.41 ± 0.03 kg, the number of significant figures is 3. In calculations involving this value, we would write the number as 13.4 kg. Furthermore, we would only give three significant figures for the answer to the problem, because to express it more accurately would be misleading. The result of a calculation can only be as accurate as the *least* accurately known value in the calculation. Thus, for example, if we were to find the density (the mass divided by the volume) of the object whose mass is 13.4 kg, we would express the density to only three significant figures, even though the volume might be known to much better accuracy than that.

In dealing with very large or very small numbers, we find it convenient to use **powers of ten notation,** sometimes called **scientific notation.** In our system of numbers, every number can be expressed as a multiple of some power of ten. The number 1,420, for example, is the same as 1.42 times 1000, which is equal to 10^3. Thus, this number can be expressed as 1.42×10^3. Similarly, the number 0.0032 is the same as 3.2/1,000, or $3.2 \times 1/1000$. The number 1/1000 can be written as $1/10^3$ or 10^{-3}. Thus, 0.0032 can be written as 3.2×10^{-3}.

There is no obvious advantage in writing numbers like 1,420 or 0.0032 in this notation, but consider much larger or smaller numbers. The distance from the Earth to the Sun, for example, is about 149,000,000 km, and the value of the gravitational constant G is 0.0000000000667 N·m^2/kg^2. It becomes cumbersome to write out all the digits for these numbers. Furthermore, the zeros are not significant digits, so there is no loss of information if we do not write them all. In scientific notation, numbers like these become much easier to handle, being written as 1.49×10^8 km and 6.67×10^{-11} N·m^2/kg^2.

To convert a large number to scientific notation, place the decimal point to the right of the first significant digit, then count how many places you had to move the decimal to the left to put it there. The number of places is the power to which ten is raised. Thus, in the Sun-Earth distance, we moved the decimal eight places to the left in order to place it between the 1 and the 4, so the exponent of ten in the result is 8. For small numbers, the same rule applies, except that now the decimal is moved to the right and the exponent is negative.

Numbers written in scientific notation can be added or subtracted only if they have the same exponent. Thus, we could add 6.7×10^8 and 1.8×10^8 to get 8.5×10^8, but to add 6.7×10^8 and 4.1×10^6 we would first have to convert one number to the same power of ten as the other (in this case we could write 6.7×10^8 as 670×10^6 and do the addition). Multiplying and dividing numbers in scientific notation is always possible, regardless of the exponent. To multiply, we just multiply the preceding numbers together, then add the exponents. For example, the product of 6.8×10^{22} and 1.3×10^{-5} is $(6.8)(1.3) \times 10^{22 + (-5)} = 8.8 \times 10^{17}$. Division is done in an analogous fashion; we divide the preceding numbers and subtract the exponents. Thus, if we want to divide 4.48×10^{14} by 7.21×10^6, we get $(4.48)/(7.21) \times 10^{14 - 6} = 0.621 \times 10^8$, or 6.21×10^7.

In these examples we have been careful to express the result with the same number of significant digits as the input numbers. In writing a number in scientific notation in the first place, we include only the significant digits, and we observe the same rules described above in deciding how many significant digits to preserve in the answer.

If we want to raise a number in scientific notation to a power, or take a root of it, this is relatively simple. We just raise the preceding number to the power, and multiply the exponent by the power. Thus, for example, the cube of 4.1×10^{11} is $(4.1)^3 \times 10^{3(11)} = 69 \times 10^{33} = 6.9 \times 10^{34}$. To find a root of a number, we recognize that this is the same as raising the number to a power that is a fraction. That is, for example, the square root of a number is that number raised to the ½ power; the cube root is the number raised to the ⅓ power, and so on. Then the same rule applies as for raising the number to any other power. The cube root of 9.23×10^{12} is therefore $(9.23)^{1/3} \times 10^{12/3} = 2.1 \times 10^4$.

Algebra

Algebra consists of carrying out numerical operations on equations, so that unknown quantities can be found. Thus algebra is the principal basis for solving problems in physics, because what we do is to write equations based on physical laws relating physical quantities, and then solve for the unknown quantities of interest. The numerical operations involved in this text are usually just the basic ones: addition, subtraction, multiplication, and division, although others, such as taking square roots or raising terms to some power, may be used also. Even in higher-level physics courses, where calculus is used, the basic task is to solve equations algebraically. The difference is that the operations that may be applied to equations are more complex and varied than we use here.

The basic premise in algebra is that an equation remains an equation if the same operation is applied to both sides of the equality. As a very simple example, if we have the equation

$$x = 14.8,$$

then it is also true that

$$x + 4 = 14.8 + 4.$$

Similarly, if we have the equation

$$7y = 21,$$

we can divide each side by 7 to find

$$y = 3.$$

The same basic principle applies to more complex

equations. For example, suppose we have

$$\frac{x}{4} + 23.5 = 12x,$$

and we want to find the value of x. We can start by multiplying both sides by 4 to get

$$x + 94 = 48x,$$

then subtracting x from both sides to find

$$94 = 47x$$

This is easily solved by dividing both sides by 47, yielding

$$x = 2.$$

In this example, we have deliberately carried each step separately, for the purpose of illustration. As the student becomes more familiar with the technique, it becomes easy to perform several steps at once.

It is a general rule that there must be one equation for each unknown quantity in a problem, or else the unknowns cannot be found. In the examples above, we had only one unknown, and needed only one equation to find it in each case. There are examples, however, where two or more unknowns are to be found, so that an equal number of equations is needed. It is a physics problem to find the necessary equations, but then an algebra problem to solve them. Here we will restrict ourselves to only two equations and two unknowns. Suppose, for example, that we have the equations

$$x + 3y = 32,$$
$$2x - 4y = 24.$$

A good first step is to solve one of the equations for one unknown in terms of the other. From the first equation, we can write

$$x = 32 - 3y.$$

Now we can substitute this in place of x in the second equation:

$$2(32 - 3y) - 4y = 24.$$

Notice that now we have a single equation in a single unknown, y. We multiply out the terms on the left:

$$64 - 6y - 4y = 24,$$

and then combine:

$$64 - 10y = 24.$$

Subtracting 64 from both sides yields

$$-10y = -40,$$

or, dividing both sides by -10,

$$y = 4.$$

Now we can substitute 4 for y in either of the original equations to find x. If we use the first equation, we get

$$x + 12 = 32$$

which, upon subtraction of 12 from each side, yields

$$x = 20.$$

The answer would have been the same if we had substituted $y = 4$ into the second equation, and then solved for x.

It sometimes happens that an unknown is raised to some power in an equation. In many cases the solution is basically the same as in the above examples, because all that is needed is to perform a simple operation on both sides. For example, suppose

$$x^2 + 5 = 54.$$

To solve this, first subtract 5 from both sides to get

$$x^2 = 49,$$

and then take the square root of both sides:

$$x = \pm 7.$$

Notice that there are two possible answers to the equation.

Things are not quite so simple, however, if more

than one power of the unknown is involved. For example, suppose we have the equation

$$x^2 + 2x = 8.$$

There are two methods for solving such equations. One is to rearrange it as an equality with zero (by subtracting 8 from both sides in this case), and then finding the multiplicative factors:

$$x^2 + 2x - 8 = 0.$$

We recognize that the left side is the product of

$$(x - 2) \times (x + 4).$$

In order for the product of these two terms to be zero, one or the other of the terms must be zero. Thus, there are two possible solutions to this equation:

$$x - 2 = 0,$$

which leads to

$$x = 2;$$

and

$$x + 4 = 0,$$

which yields

$$x = -4.$$

Few **quadratic equations**, as those involving the square of the unknown are called, are as simple as the two types we have just illustrated. There is, however, a general formula for solving all equations of the form

$$ax^2 + bx + c = 0.$$

The formula is

$$x = \left(\frac{-b \pm \sqrt{b^2 - 4ac}}{2a} \right).$$

Note that, because of the $\pm$ sign, there are always two possible solutions of quadratic equations. Physically, only one of them may be realistic, and it sometimes requires judgement to decide which one is correct and which one is impossible.

In the last equation cited, if we apply the quadratic formula, where $a = 1$, $b = 2$, and $c = -8$, we find that the solutions are

$$\begin{aligned}
x &= \left[\frac{-2 \pm \sqrt{(2)^2 - 4(1)(-8)}}{2(1)} \right] \\
&= \left[\frac{-2 \pm \sqrt{4 - (-32)}}{2(1)} \right] \\
&= \left(\frac{-2 \pm 6}{2} \right) \\
&= 2 \text{ or } -4.
\end{aligned}$$

These are the same solutions we found by the factoring method. Equations involving the square of the unknown are particularly common in problems having to do with uniformly accelerating bodies, such as falling bodies, since the displacement in those cases is proportional to the square of the time (Chapter 2).

Trigonometry

Trigonometry is an algebraic method of solving for unknowns in geometric problems. The basic principle is that two triangles having equal angles are **similar,** meaning that the relative lengths of their sides are the same. Thus, in the two triangles below, which have equal angles, the ratios of the lengths of the sides are the same; that is, $a/b = a'/b'$, $b/c = b'/c'$, and $c/a = c'/a'$.

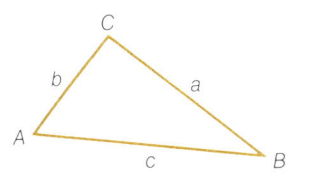

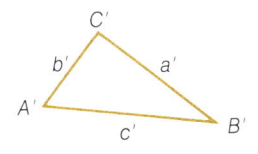

The sum of the angles in a triangle is always 180°. Thus, if two of the angles are known, the third one is also known, since it must equal 180° minus the sum of the other two. In order to establish that two triangles are similar therefore requires that only two of the angles are known to be equal; if this is true, then the third angles must also be equal. In trigonometry we often restrict ourselves to special triangles called **right triangles** that have one angle equal to 90°. For two right triangles to be similar requires only that one of the acute angles of one triangle equal one of the acute angles of the other. For example, the two right triangles below are similar, because angle A equals angle A'.

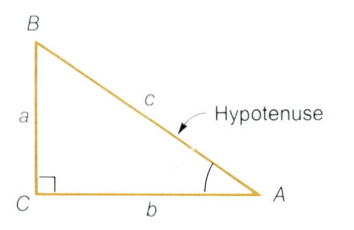

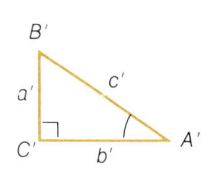

Because similar triangles always have the same relative lengths of their sides, it is possible to tabulate these relative lengths, so that if the angles are known, the relative lengths of the sides could be found in tables. This is impractical in general, because there is an infinite variety of triangles, but if we restrict ourselves to right triangles, then the number of different types of triangles becomes manageable. This is the essence of trigonometry.

Consider the two right triangles below:

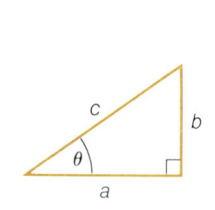

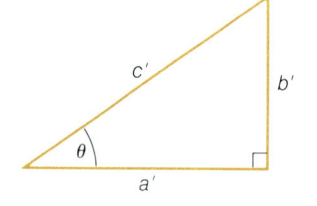

These are similar triangles, because the angles indicated as θ are the same in each. Therefore the ratio of sides b/c in the first triangle is equal to the ratio of sides b'/c' in the second. For *any* right triangle with an angle equal to θ, the ratio of the side opposite that angle to the long side (the hypotenuse) will always be the same. This ratio is called the **sine** of angle and it can be found in tables of sines (modern scientific calculators provide trigonometric functions such as sines as well, so tables are rarely used today).

Similarly, the ratio of the side adjacent to a given angle to the hypotenuse (the ratio of a/c or a'/c' in the triangles above) is called the **cosine**, and is also tabulated or calculated as a function of angle. A third standard function is the **tangent**, which is the ratio of the opposite side to the adjacent side (b/a or b'/a', above).

As a simple example of a problem requiring the use of trigonometry, consider the situation sketched below.

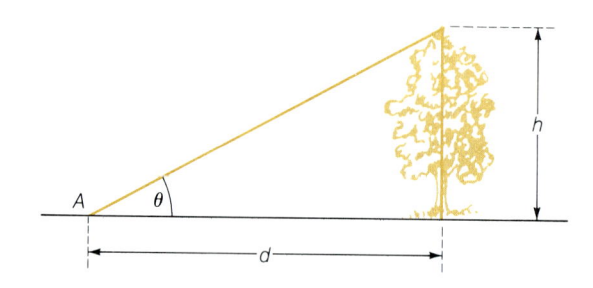

It is desired to measure the height of the tree, but this is difficult to do directly. Suppose, therefore, that the distance d from the tree's trunk to point A is measured, and the angle θ toward the top of the tree from A is also measured. Then the height h of the tree is found from the relation

$$\tan\theta = \frac{h}{d}.$$

In this expression the values of θ and d are known, and it is a simple matter to solve for h.

So far we have discussed only angles less than 90°, but trigonometric functions are defined for other angles as well. The reference circle below helps illustrate the sign conventions.

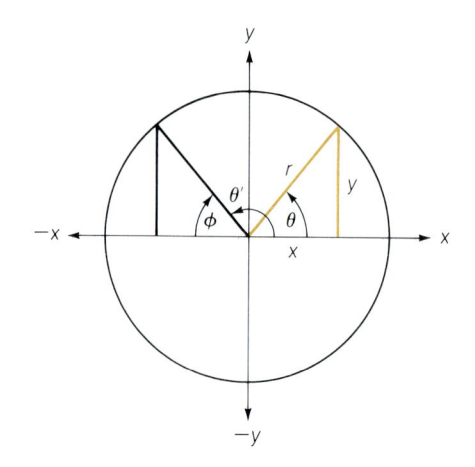

For any point on the circle, we can draw a right triangle with legs x and y and hypotenuse r (the radius of the circle). Then the trignometric functions of the angle θ (measured from the x-axis) are $\sin\theta = y/r$; $\cos\theta = x/r$, and $\tan\theta = y/x$. In the upper right quadrant of the circle, x, y, and r are positive, so for angles between 0° and 90°, the trigonometric functions are all positive.

In the upper left quadrant, where the angle is shown as θ' above, x is negative but y and h are positive, so the cosine and tangent are negative while the sine is positive. Note that the angle ϕ is equal to 180° − θ'. Therefore we say that, for an angle that is greater than 90° but less than 180°,

$$\sin\theta = \sin(180° - \theta),$$
$$\cos\theta = -\cos(180° - \theta),$$

and

$$\tan\theta = -\tan(180° - \theta).$$

Now let us consider the lower right quadrant. Here x and r are positive but y is negative, so the cosine is positive but the sine and tangent are negative. From this we see that

$$\sin(-\theta) = -\sin\theta,$$
$$\cos(-\theta) = \cos\theta,$$

and

$$\tan(-\theta) = -\tan\theta.$$

In the lower left quadrant, we can consider the angle to be either between 180° and 270° or between −90° and −180°. In this quadrant both x and y are negative (but r is positive), so we know that the sine and cosine are negative while the tangent is positive. It is left as an exercise for the student to determine the expressions for the functions in this quadrant.

Given these relations, we can now find the trigonometric functions for any angle, positive or negative, from 0° to 180°. Functions of any larger angles can be found as well, keeping in mind the reference circle. Calculators usually make all the necessary conversions, so the angle can simply be entered, but in using trigonometric tables, it is often necessary to convert to the appropriate angle between 0° and 90° first, keeping in mind the sign conventions.

There are several useful relationships among trigonometric functions. They can all easily be derived from simple plane geometry, but we will not do so here. We simply list a few of them, with figures to illustrate the meaning:

$$\sin^2\theta + \cos^2\theta = 1 \text{ (Pythagorean theorem)}$$

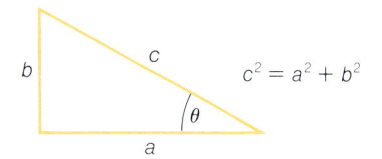

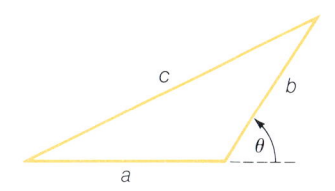

$$c^2 = a^2 + b^2$$

$$c^2 = a^2 + b^2 + 2ab\cos\theta \text{ (law of cosines)}$$

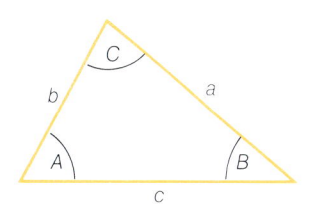

$$\frac{a}{\sin A} = \frac{b}{\sin B} = \frac{c}{\sin C} \text{ (law of sines)}$$

There are other useful relations between trigonometric functions, but they are not encountered in this text.

Logarithms

In some physical situations, we find that the relationship between two quantities takes on the following form

$$x = A^y,$$

where A is a standard base number. We can define the logarithm with respect to that base as follows:

$$\log_A x = y.$$

APPENDIX 2 REVIEW OF MATHEMATICAL TECHNIQUES

Trigonometric Table

Angle in degrees	Angle in radians	Sine	Cosine	Tangent	Angle in degrees	Angle in radians	Sine	Cosine	Tangent
0°	0.000	0.000	1.000	0.000					
1°	0.017	0.017	1.000	0.017	46°	0.803	0.719	0.695	1.036
2°	0.035	0.035	0.999	0.035	47°	0.820	0.731	0.682	1.072
3°	0.052	0.052	0.999	0.052	48°	0.838	0.743	0.669	1.111
4°	0.070	0.070	0.998	0.070	49°	0.855	0.755	0.656	1.150
5°	0.087	0.087	0.996	0.087	50°	0.873	0.766	0.643	1.192
6°	0.105	0.105	0.995	0.105	51°	0.890	0.777	0.629	1.235
7°	0.122	0.122	0.993	0.123	52°	0.908	0.788	0.616	1.280
8°	0.140	0.139	0.990	0.141	53°	0.925	0.799	0.602	1.327
9°	0.157	0.156	0.988	0.158	54°	0.942	0.809	0.588	1.376
10°	0.175	0.174	0.985	0.176	55°	0.960	0.819	0.574	1.428
11°	0.192	0.191	0.982	0.194	56°	0.977	0.829	0.559	1.483
12°	0.209	0.208	0.978	0.213	57°	0.995	0.839	0.545	1.540
13°	0.227	0.225	0.974	0.231	58°	1.012	0.848	0.530	1.600
14°	0.244	0.242	0.970	0.249	59°	1.030	0.857	0.515	1.664
15°	0.262	0.259	0.966	0.268	60°	1.047	0.866	0.500	1.732
16°	0.279	0.276	0.961	0.287	61°	1.065	0.875	0.485	1.804
17°	0.297	0.292	0.956	0.306	62°	1.082	0.883	0.469	1.881
18°	0.314	0.309	0.951	0.325	63°	1.100	0.891	0.454	1.963
19°	0.332	0.326	0.946	0.344	64°	1.117	0.899	0.438	2.050
20°	0.349	0.342	0.940	0.364	65°	1.134	0.906	0.423	2.145
21°	0.367	0.358	0.934	0.384	66°	1.152	0.914	0.407	2.246
22°	0.384	0.375	0.927	0.404	67°	1.169	0.921	0.391	2.356
23°	0.401	0.391	0.921	0.424	68°	1.187	0.927	0.375	2.475
24°	0.419	0.407	0.914	0.445	69°	1.204	0.934	0.358	2.605
25°	0.436	0.423	0.906	0.466	70°	1.222	0.940	0.342	2.748
26°	0.454	0.438	0.899	0.488	71°	1.239	0.946	0.326	2.904
27°	0.471	0.454	0.891	0.510	72°	1.257	0.951	0.309	3.078
28°	0.489	0.469	0.883	0.532	73°	1.274	0.956	0.292	3.271
29°	0.506	0.485	0.875	0.554	74°	1.292	0.961	0.276	3.487
30°	0.524	0.500	0.866	0.577	75°	1.309	0.966	0.259	3.732
31°	0.541	0.515	0.857	0.601	76°	1.326	0.970	0.242	4.011
32°	0.559	0.530	0.848	0.625	77°	1.344	0.974	0.225	4.332
33°	0.576	0.545	0.839	0.649	78°	1.361	0.978	0.208	4.705
34°	0.593	0.559	0.829	0.675	79°	1.379	0.982	0.191	5.145
35°	0.611	0.574	0.819	0.700	80°	1.396	0.985	0.174	5.671
36°	0.628	0.588	0.809	0.727	81°	1.414	0.988	0.156	6.314
37°	0.646	0.602	0.799	0.754	82°	1.431	0.990	0.139	7.115
38°	0.663	0.616	0.788	0.781	83°	1.449	0.993	0.122	8.144
39°	0.681	0.629	0.777	0.810	84°	1.466	0.995	0.105	9.514
40°	0.698	0.643	0.766	0.839	85°	1.484	0.996	0.087	11.43
41°	0.716	0.656	0.755	0.869	86°	1.501	0.998	0.070	14.30
42°	0.733	0.669	0.743	0.900	87°	1.518	0.999	0.052	19.08
43°	0.750	0.682	0.731	0.933	88°	1.536	0.999	0.035	28.64
44°	0.768	0.695	0.719	0.966	89°	1.553	1.000	0.017	57.29
45°	0.785	0.707	0.707	1.000	90°	1.571	1.000	0.000	∞

Two bases A are often used. One is the number 10, so that the logarithm of x is the exponent to which 10 is raised to yield x. Logarithms based on 10 are called **common logarithms.** As a simple example, the common logarithm of 100 is 2, because $10^2 = 100$. Fractional values of logarithms are also possible; for example, the logarithm of 6 is 0.778, because $10^{0.778} = 6$ (actually, the logarithm has more decimal places, but typically three provide enough accuracy for most problems).

The second base for logarithms is the "natural" number $e = 2.718\ldots$. This number, given by the sum of the series $1 + 1/1 + 1/(1)(2) + 1/(1)(2)(3) + \ldots$, arises in certain exponential functions (see the next section of this appendix). A **natural logarithm,** one based on e, is designated by ln instead of log; there we write

$$ln\ x = y.$$

The natural logarithm of 6, for example, is $ln\ 6 = 1.792$. Natural logarithms are rarely encountered in this text.

Either type of logarithm obeys certain rules. For example,

$$\log\ (ab) = \log\ a + \log\ b,$$

or

$$ln\ (ab) = ln\ a + ln\ b.$$

It should be easy to see why this is so, because we know that

$$(A^a)(A^b) = A^{a+b},$$

where A is either 10 or e in this case.

Similarly,

$$\log\left(\frac{a}{b}\right) = \log\ a - \log\ b,$$

and

$$ln\left(\frac{a}{b}\right) = ln\ a - ln\ b.$$

From the rule for logarithms of products we can find the logarithm of a quantity that is raised to some power. First, consider the logarithm of x^2. This is the same as the logarithm of $(x)(x)$, which we know is

$$\log(x)(x) = \log\ x + \log\ x = 2\ \log\ x$$

In general, this leads to

$$\log\ (x^n) = n\ \log\ x,$$

where n can be any exponent, integer or fraction, positive or negative.

The values of logarithms can be found in tables, although today it is usually easier to use a scientific calculator. If tables are used, then it is often necessary to apply the rule for the logarithm of a product or a quotient, because normally values are tabulated only for numbers between 1 and 10. If the log of 20 is needed, for example, we would note that $20 = (10)(2)$, so that

$$\log(20) = \log\ (10) + \log(2),$$
$$= 1 + 0.301$$
$$= 1.301$$

Similarly, the logarithms of numbers smaller than 1 can be found using the rule for logarithms of quotients. For example, suppose we need to find $\log(0.0046)$. We recognize that $0.0046 = 4.6/1000$, so that

$$\log(0.0046) = \log(4.6) - \log(1000)$$
$$= 0.663 - 3$$
$$= -2.337.$$

Sometimes we know the logarithm of a quantity, and want to find the quantity. Then we must find the **antilogarithm,** which is done by raising 10 (or e, in the case of natural logarithms) to a power equal to the logarithm. For example, if we know that $\log\ x = 2.147$, then

$$x = 10^{2.147} = 140.3$$

(if tables are used, it will probably be necessary to rewrite this as

$$x = (10^2)(10^{0.147}) = (100)(1.403) = 140.3,$$

because tables are usually compiled only for numbers between 1 and 10.)

Physical situations in which common logarithms are involved include those having to do with human senses such as hearing or sight, because it is found that the human response is logarithmic. That is, what the ear perceives as uniform gradations in sound intensity, and what the eye sees as uniform gradations in brightness, are actually proportional to changes in intensity by powers of 10. An example consequence of this is the definition of the unit of sound intensity called the **decibel**:

$$\beta = 10 \log(I/I_0),$$

where β is the intensity in decibels, I is the actual intensity of the sound, and I_0 is a reference intensity (see Chapter 10). The range from sound that is barely perceptible to sound that is loud enough to damage the ears is the range from 0 to 120 dB, or a factor of $10^{120/10} = 10^{12}$ or one trillion. A similar relationship is used by astronomers to measure brightness of stars as seen by the eye.

Logarithms are also useful in expressing or depicting a quantity that varies over a very large range. An example of this is the **Richter scale**, a system for measuring the energy released in earthquakes. Each division on the scale represents an increase in energy by a factor of ten. The scale goes from 1 to 10, so the full range of energies, from smallest to greatest, is 10^9 or one billion. A magnitude 6.5 earthquake, for example, is $10^{2.1} = 126$ times more powerful than a magnitude 4.4 earthquake. Thus, by expressing a simple number between 1 and 10, it is possible to represent numbers ranging over a factor of a billion. It is often useful to plot graphs of the logarithms of quantities that vary widely, so that the broad range can be easily seen graphically.

Exponential Functions

There are many physical situations where a quantity varies as a power of itself. A good example is population growth in cases where the growth rate is a fixed percentage of the existing population. Another example is the half-life of a radioactive substance, which is the time required for half of the substance to decay.

An exponential growth or decay rate can be expressed as

$$x = x_0 A^n,$$

where A is the base rate of change per unit n. Often A is a percentage or fractional change that occurs in some interval of time, and n is the number of such intervals that elapse. For example, if the population is growing at a rate of 2% per year, then the population after n years is

$$P = P_0(1.02)^n.$$

Here P_0 is the population in some reference year (the present, for example), and the 1.02 term represents the fact that the population is multiplied by this factor every year.

It is interesting to actually calculate how much the population increases in a few years of exponential growth. Suppose we assume a 2% growth rate per year for 50 years. Then we have

$$P = P_0(1.02)^{50}.$$

This is most easily solved by using logarithms. First, divide both sides by P_0:

$$\frac{P}{P_0} = (1.02)^{50},$$

then take the common (or natural) logarithm of both sides:

$$\log\left(\frac{P}{P_0}\right) = 50 \log(1.02) = 50(0.0086) = 0.43.$$

Now find the antilogarithm of both sides:

$$\frac{P}{P_0} = 10^{0.43} = 2.69.$$

In only fifty years the population has increased by more than a factor of 2½! This illustrates that exponential growth can produce enormous changes in modest times (for fun, calculate the value after a few years of a bank deposit that earns at a fixed interest rate, say 5% per year. Then compare this growth in value with the loss in value due to an annual inflation rate of, say, 7% per year)

When a quantity is decreasingly exponentially, we often call it an exponential decay. The half-life of a radioisotope is a good example, where we have relation

$$N = N_0(0.5)^n,$$

where N_0 is the initial abundance of the isotope, the (0.5) term represents the fact that the quantity is reduced to one-half of its initial value during each half-life, and n is the number of half-lives that have elapsed. For example, radium 226 (^{226}Ra; see Chapter 25) has a half-life of 1600 years. Suppose we want to know what fraction of some initial quantity of radium is left after a period of 10,000 years (= 6.25 half-lives). We have

$$\frac{N}{N_0} = (0.5)^{6.25} = 0.0131.$$

(again, the use of logarithms simplifies this kind of calculation). The amount of radium remaining after 10,000 years is about one percent of the original quantity.

The graph of an exponential growth or decay always has a characteristic appearance, such as shown in our two examples of population growth and radioactive decay.

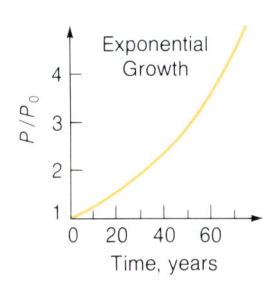

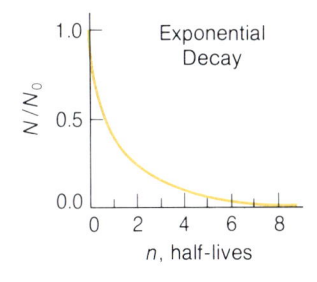

Binomial Expansion

It sometimes happens that we need to raise a complex term to some power, as for example, in the form $(1 + x)^n$. This can be a tedious operation to carry out, but the solution has a general form that we can simply adopt:

$$(1 + x)^n = 1 + \frac{nx}{1} + \frac{n(n-1)x^2}{1(2)} + \frac{n(n-1)(n-2)x^3}{1(2)(3)} + \ldots + x^n$$

This is still a complex expression, but in certain circumstances it can be greatly simplified. If x is much less than 1, for example, then we can ignore the terms that contain x^2 or higher powers of x, and write

$$(1 + x)^n \simeq 1 + nx.$$

To see that this is correct, suppose we want to calculate

$$(1 + 0.001)^4.$$

Using the formula for binomial expansion yields an expression with five terms,

$$(1 + 0.001)^4 = 1 + 0.004 + 0.000006 + 0.000000004 + 0.000000000001$$
$$= 1.004006004001.$$

We see that this is, to a high degree of accuracy, the same as

$$(1 + 0.001)^4 \simeq 1 + nx = 1 + 4(0.001) = 1.004.$$

It is often advantageous to recognize when an expansion can be approximated in this manner; sometimes algebraic manipulation is required in order to write the expression in the appropriate form.

Appendix 3
Units and Conversion Factors

Units	Dimensions*	SI (MKS)	CGS	FPS	FSS
Length	L	m	cm	ft	ft
Mass	M	kg	g	lb	slug
Time	T	s	s	s	s
Velocity	L/T	m/s	cm/s	ft/s	ft/s
Acceleration	L/T^2	m/s^2	cm/s^2	ft/s^2	ft/s^2
Force	ML/T^2	kg m/s^2 = Newton	g cm/s^2 = dyne	lb ft/s^2 = poundal	slug ft/s^2 = lbwt
Momentum, Impulse	ML/T	kg m/s = N s	g cm/s = dyne s	lb ft/s = pdl s	slug ft/s = lbwt s
Energy, Work	ML^2/T^2	kg m^2/s^2 = N m = Joule	g cm^2/s^2 = dyne cm = erg	lb ft^2/s^2 = ft pdl	slug ft^2/s^2 = ft lbwt
Power	ML^2/T^3	kg m^2/s^3 = J/s = Watt	g cm^2/s^3 = erg/s	lb ft^2/s^3 = ft pdl/s	slug ft^2/s^3 = ft lbwt/s
Pressure	M/LT^2	kg/m s^2 = N/m^2 = Pascal	g/cm s^2 = dyne/cm^2	pdl/ft^2	lbwt/ft^2

*Dimensions M(mass), L(Length), T(time), Q(charge)

Electricity and Magnetism (SI units only)

Electric Charge	Q	Coulomb
Electric Current	Q/T	Coulomb/sec = Ampere
Electric Potential	ML^2/QT^2	Joule/Coulomb = Volt
Resistance	ML^2/Q^2T	Volt/Ampere = Ohm
Magnetic flux	ML^2/QT	Joule/Ampere = Weber = Tesla · m^2
Magnetic field	M/QT	Weber/m^2 = Tesla = 10^4 Gauss
Inductance	ML^2/Q^2	Weber/Ampere = Henry
Capacitance	Q^2T^2/ML^2	Coulomb/Volt = Farad

APPENDIX 3 UNITS AND CONVERSION FACTORS

Conversion Factors

Length

1 km = 1000 m
1 m = 100 cm = 1000 mm
1 cm = 10 mm = 0.01 m
1 mm = 0.001 m
1 μm = 10^{-6} m
1 nm = 10^{-9} m
1 Å = 10^{-10} m

1 inch = 2.54 cm
1 ft = 30.48 cm
1 mi = 1.609 km = 5280 ft
1 nautical mile = 1.852 km = 6076 ft

1 cm = 0.3937 in
1 m = 39.37 in = 3.28 ft
1 km = 0.6214 mi

Mass

1 kg = 1000 g = 2.2046 lb = 0.06852 slug
1 lb = 453.59 g = 0.45359 kg = 0.03108 slug
1 oz(troy) = 31.103 g
1 oz(avoirdupois) = 28.349 g
1 slug = 32.174 lb = 14.594 kg
1 amu = 1.6606×10^{-27} kg

Volume

1 liter = 1000 cm^3 = 1.057 quart = 61.02 in^3 = 0.03532 ft^3
1 m^3 = 1000 liter = 35.315 ft^3
1 ft^3 = 7.4805 U.S. gallon = 0.02832 m^3 = 28.32 liter
1 U.S. gallon = 231 in^3 = 3.7854 liter
1 British gallon = 277.4 in^3 = 1.20095 U.S. gallon

Time

1 solar day = 24 hr = 1440 min = 86,400 s
1 yr (calendar) = 365.25 day = 8766 hr = 525,960 min = 3.15576×10^7 s

Speed

1 km/h = 0.2778 m/s = 0.6214 mi/h = 0.9113 ft/s
1 mi/h = 1.467 ft/s = 1.609 km/h = 0.4470 m/s
1 ft/s = 0.305 m/s = 0.682 mi/h
1 m/s = 3.28 ft/s = 3.60 km/h = 2.237 mi/h
1 knot = 1.151 mi/h = 0.5144 m/s

Force

1 lbwt = 4.448 Newton = 32.17 poundal
1 Newton = 10^5 dyne = 0.2248 lbwt
1 U.S. short ton = 2000 lbwt
1 long ton = 2240 lbwt
1 metric ton = 2205 lbwt

Energy

1 Joule = 10^7 erg = 0.7376 ft lbwt = 0.23885 cal = 9.4788×10^{-4} BTU
1 calorie = 4.1868 Joule = 3.968×10^{-3} BTU
1 Calorie = 1 kcal = 1000 calories
1 BTU = 1055 Joule = 0.293 Watt-hr
1 kWatt-hr = 1000 Watt-hr = 3.60×10^6 Joule = 859.8 calorie = 3412 BTU
1 ft-lbwt = 1.356 Joule = 0.3239 calorie = 1.285×10^{-3} BTU
1 eV = 1.602×10^{-19} Joule

Power

1 Watt = 1 J/s = 10^7 erg/s = 0.2389 cal/s
1 horsepower = 550 ft-lbwt/s = 0.178 kcal/s = 745.7 Watt
1 kWatt = 1000 Watt = 1.341 horsepower = 737.6 ft-lbwt/s = 0.9478 BTU/s

Pressure

1 Pascal = 1 N/m^2 = 10 dynes/cm^2 = 9.869×10^{-6} atm = 2.089×10^{-2} lbwt/ft^2
1 lbwt/in^2 = 6895 Pascal = 51.71 mm Hg = 27.68 in. water
1 Torr = 1 mm Hg = 1.316×10^{-3} atm = 133.3 N/m^2
1 atm = 760 Torr = 1.013×10^5 Pascal = 14.7 lbwt/in^2 = 406.8 in. water = 29.9 in. Hg

Angles

1 radian = 57.30° = 57° 18'
1° = 0.0175 rad
1 rev/min = 0.1047 rad/s = 2π rad/min

Appendix 4
Table of Planetary Data

Planet	Sidereal period (yr)	Semi-Major[1] axis of orbit a(AU)	Orbital eccentricity	Mass[1] (M_e)	Radius[2] (km)	Oblateness (ϵ)	Bond[3] albedo	Rotational period
Mercury	0.2408	0.3871	0.2056	0.0558	2439	< 0.029	0.06 ± 0.01	58.65 d
Venus	0.6152	0.7233	0.0068	0.8150	6052	< 0.001	0.75 ± 0.02	243 d
Earth[4]	1.0000	1.0000	0.0167	1.0000	6378	0.0034	0.31 ± 0.01	$23^h 56^m 04^s$
Mars	1.8809	1.5237	0.0934	0.1074	3384	0.0059	0.15 ± 0.01	$24^h 37^m 22^s$
Jupiter	11.8623	5.2026	0.0485	318.05	71,398 ± 100	0.0637	0.343 ± 0.032	$9^h 55^m 30^s$
Saturn	29.458	9.5547	0.0556	95.147	60,330 ± 60	0.09	0.342 ± 0.030	$10^h 39.4^m$
Uranus	84.01	19.2181	0.0472	14.54	26,145 ± 30	0.024 ± 0.003	0.34 ± 0.02	16 ± 1 h
Neptune	164.79	30.1096	0.0086	17.23	25,225 ± 30	0.021 ± 0.004	0.29 ± 0.02	18.2 ± 0.4 h
Pluto	248.5	39.44	0.250	0.0023:	1445:	?	0.4 ± 0.1	6.387d

[1]These masses based on earth mass, 5.976×10^{24} kg. One astronomical unit (AU) equals 1.495979×10^{11} m.
[2]The radius quoted is R_e = equatorial radius. The polar radius $R_p = R_e/(1 + \epsilon)$ and the mean radius $\bar{R} = R_e/(1 + \epsilon)^{1/2}$ where ϵ is the oblateness.
[3]Bond albedo refers to fraction of energy in solar spectrum reflected from full disk of planet.
[4]Moon has $M = 7.35 \times 10^{22}$ kg, $R = 1738$ km, $a = 384,000$ km.

Appendix 5
Periodic Table of the Elements*

Group I	Group II						Transition elements							Group III	Group IV	Group V	Group VI	Group VII	Group 0
H 1 1.0079 $1s^1$																			He 2 4.0026 $1s^2$
Li 3 6.941 $2s^1$	Be 4 9.01218 $2s^2$													B 5 10.81 $2p^1$	C 6 12.011 $2p^2$	N 7 14.0067 $2p^3$	O 8 15.9994 $2p^4$	F 9 18.9984 $2p^5$	Ne 10 20.179 $2p^6$
Na 11 22.9898 $3s^1$	Mg 12 24.305 $3s^2$													Al 13 26.9815 $3p^1$	Si 14 28.0855 $3p^2$	P 15 30.9738 $3p^3$	S 16 32.06 $3p^4$	Cl 17 35.453 $3p^5$	Ar 18 39.948 $3p^6$
K 19 39.0983 $4s^1$	Ca 20 40.08 $4s^2$	Sc 21 44.9559 $3d^14s^2$	Ti 22 47.88 $3d^24s^2$	V 23 50.9415 $3d^34s^2$	Cr 24 51.996 $3d^54s^1$	Mn 25 54.9380 $3d^54s^2$	Fe 26 55.847 $3d^64s^2$	Co 27 58.9332 $3d^74s^2$	Ni 28 58.69 $3d^84s^2$	Cu 29 63.546 $3d^{10}4s^1$	Zn 30 65.39 $3d^{10}4s^2$			Ga 31 69.72 $4p^1$	Ge 32 72.59 $4p^2$	As 33 74.9216 $4p^3$	Se 34 78.96 $4p^4$	Br 35 79.904 $4p^5$	Kr 36 83.80 $4p^6$
Rb 37 85.468 $5s^1$	Sr 38 87.62 $5s^2$	Y 39 88.9059 $4d^15s^2$	Zr 40 91.224 $4d^25s^2$	Nb 41 92.9064 $4d^45s^1$	Mo 42 95.94 $4d^55s^1$	Tc 43 (98) $4d^55s^2$	Ru 44 101.07 $4d^75s^1$	Rh 45 102.906 $4d^85s^1$	Pd 46 106.42 $4d^{10}5s^0$	Ag 47 107.868 $4d^{10}5s^1$	Cd 48 112.41 $4d^{10}5s^2$			In 49 114.82 $5p^1$	Sn 50 118.71 $5p^2$	Sb 51 121.75 $5p^3$	Te 52 127.60 $5p^4$	I 53 126.905 $5p^5$	Xe 54 131.29 $5p^6$
Cs 55 132.905 $6s^1$	Ba 56 137.33 $6s^2$	57–71‡	Hf 72 178.49 $5d^26s^2$	Ta 73 180.948 $5d^36s^2$	W 74 183.85 $5d^46s^2$	Re 75 186.207 $5d^56s^2$	Os 76 190.2 $5d^66s^2$	Ir 77 192.22 $5d^76s^2$	Pt 78 195.08 $5d^96s^1$	Au 79 196.967 $5d^{10}6s^1$	Hg 80 200.59 $5d^{10}6s^2$			Tl 81 204.383 $6p^1$	Pb 82 207.2 $6p^2$	Bi 83 208.980 $6p^3$	Po 84 (209) $6p^4$	At 85 (210) $6p^5$	Rn 86 (222) $6p^6$
Fr 87 (223) $7s^1$	Ra 88 226.025 $7s^2$	89–103‖	Rf 104 (261) $6d^27s^2$	Ha 105 (262) $6d^37s^2$	106 (263)	107 (262)		109 (266)											

Symbol — Cl 17 — Atomic number
Atomic mass** — 35.453
$3p^5$ — Electron configuration

Lanthanide series‡

| La 57
138.906
$5d^16s^2$ | Ce 58
140.12
$5d^14f^16s^2$ | Pr 59
140.908
$4f^36s^2$ | Nd 60
144.24
$4f^46s^2$ | Pm 61
(145)
$4f^56s^2$ | Sm 62
150.36
$4f^66s^2$ | Eu 63
151.96
$4f^76s^2$ | Gd 64
157.25
$5d^14f^76s^2$ | Tb 65
158.925
$5d^14f^96s^2$ | Dy 66
162.50
$4f^{10}6s^2$ | Ho 67
164.930
$4f^{11}6s^2$ | Er 68
167.26
$4f^{12}6s^2$ | Tm 69
168.934
$4f^{13}6s^2$ | Yb 70
173.04
$4f^{14}6s^2$ | Lu 71
174.967
$5d^14f^{14}6s^2$ |

Actinide series‖

| Ac 89
227.028
$6d^17s^2$ | Th 90
232.038
$6d^27s^2$ | Pa 91
231.036
$5f^26d^17s^2$ | U 92
238.029
$5f^36d^17s^2$ | Np 93
237.048
$5f^46d^17s^2$ | Pu 94
(244)
$5f^66d^07s^2$ | Am 95
(243)
$5f^76d^07s^2$ | Cm 96
(247)
$5f^76d^17s^2$ | Bk 97
(247)
$5f^96d^07s^2$ | Cf 98
(251)
$5f^{10}6d^07s^2$ | Es 99
(252)
$5f^{11}6d^07s^2$ | Fm 100
(257)
$5f^{12}6d^07s^2$ | Md 101
(258)
$5f^{13}6d^07s^2$ | No 102
(259)
$6d^05f^{14}7s^2$ | Lr 103
(260)
$6d^15f^{14}7s^2$ |

*Atomic mass values averaged over isotopes in percentages as they occur on earth's surface.
**For many unstable elements, mass number of the most stable known isotope is given in parentheses.

Appendix 6
Elemental Abundances

Z	Element	Symbol	Abundance[a,b]
1	Hydrogen	H	1.00
2	Helium	He	0.10
3	Lithium	Li	1.0×10^{-11}
4	Beryllium	Be	1.41×10^{-11}
5	Boron	B	8.51×10^{-10}*
6	Carbon	C	4.90×10^{-4}
7	Nitrogen	N	9.77×10^{-5}
8	Oxygen	O	8.13×10^{-4}
9	Fluorine	F	3.02×10^{-8}*
10	Neon	Ne	1.0×10^{-4}
11	Sodium	Na	2.14×10^{-6}
12	Magnesium	Mg	3.80×10^{-5}
13	Aluminum	Al	2.95×10^{-6}
14	Silicon	Si	3.55×10^{-5}
15	Phosphorus	P	2.82×10^{-7}
16	Sulfur	S	1.62×10^{-5}
17	Chlorine	Cl	1.86×10^{-7}*
18	Argon	Ar	3.80×10^{-6}
19	Potassium	K	1.32×10^{-7}
20	Calcium	Ca	2.29×10^{-6}
21	Scandium	Sc	1.3×10^{-9}
22	Titanium	Ti	1.05×10^{-7}
23	Vanadium	V	1.0×10^{-8}
24	Chromium	Cr	4.68×10^{-7}
25	Manganese	Mn	2.82×10^{-7}
26	Iron	Fe	4.68×10^{-5}
27	Cobalt	Co	8.32×10^{-8}
28	Nickel	Ni	1.78×10^{-6}
29	Copper	Cu	1.62×10^{-8}
30	Zinc	Zn	3.98×10^{-8}
31	Gallium	Ga	7.59×10^{-10}
32	Germanium	Ge	4.27×10^{-9}
33	Arsenic	As	2.45×10^{-10}*
34	Selenium	Se	2.24×10^{-9}*
35	Bromine	Br	4.27×10^{-10}*
36	Krypton	Kr	1.62×10^{-9}*
37	Rubidium	Rb	3.98×10^{-10}
38	Strontium	Sr	7.9×10^{-10}
39	Yttrium	Y	1.74×10^{-10}
40	Zirconium	Zr	3.63×10^{-10}
41	Niobium	Nb	2.57×10^{-11}*
42	Molybdenum	Mo	8.32×10^{-11}
43	Technetium	Tc	—
44	Ruthenium	Ru	6.92×10^{-11}
45	Rhodium	Rh	1.32×10^{-11}
46	Palladium	Pd	4.90×10^{-11}
47	Silver	Ag	1.91×10^{-11}*

[a]Abundances from N. Grevesse, *Physica Scripta*, Vol T8, p 49 (1984)
[b]Listed abundances are solar, unless marked with * (those values are meteorite abundances). Elements with missing values are radioactive with short half-lives.

APPENDIX 6 ELEMENTAL ABUNDANCES

Z	Element	Symbol	Abundance[a,b]
48	Cadmium	Cd	7.24×10^{-11}
49	Indium	In	4.57×10^{-11}
50	Tin	Sn	1.0×10^{-10}
51	Antimony	Sb	1.0×10^{-11}
52	Tellurium	Te	1.78×10^{-10}*
53	Iodine	I	3.24×10^{-11}*
54	Xenon	Xe	1.55×10^{-10}*
55	Cesium	Cs	1.32×10^{-11}*
56	Barium	Ba	1.35×10^{-10}
57	Lanthanum	La	1.66×10^{-11}
58	Cerium	Ce	3.55×10^{-11}
59	Praseodymium	Pr	5.13×10^{-12}
60	Neodymium	Nd	2.19×10^{-11}
61	Promethium	Pm	—
62	Samarium	Sm	6.3×10^{-12}
63	Europium	Eu	3.2×10^{-12}
64	Gadolinium	Gd	1.32×10^{-11}
65	Terbium	Tb	1.6×10^{-12}
66	Dysprosium	Dy	1.3×10^{-11}
67	Holmium	Ho	3.2×10^{-12}*
68	Erbium	Er	8.5×10^{-12}
69	Thulium	Tm	1.0×10^{-12}
70	Ytterbium	Yb	1.20×10^{-11}
71	Lutetium	Lu	1.3×10^{-12}*
72	Hafnium	Hf	7.6×10^{-12}
73	Tantalum	Ta	8.1×10^{-13}*
74	Tungsten	W	1.29×10^{-11}
75	Rhenium	Re	1.8×10^{-12}*
76	Osmium	Os	2.82×10^{-11}
77	Iridium	Ir	2.24×10^{-11}
78	Platinum	Pt	6.3×10^{-11}
79	Gold	Au	6.6×10^{-12}*
80	Mercury	Hg	1.9×10^{-11}*
81	Thallium	Tl	6.6×10^{-12}*
82	Lead	Pb	7.94×10^{-11}
83	Bismuth	Bi	5.1×10^{-12}*
84	Polonium	Po	—
85	Astatine	At	—
86	Radon	Rn	—
87	Francium	Fr	—
88	Radium	Ra	—
89	Actinium	Ac	—
90	Thorium	Th	1.0×10^{-12}
91	Protactinium	Pa	—
92	Uranium	U	3.2×10^{-13}*
93	Neptunium	Np	—
94	Plutonium	Pu	—
95	Americium	Am	—
96	Curium	Cm	—
97	Berkelium	Bk	—
98	Californium	Cf	—
99	Einsteinium	Es	—
100	Fermium	Fm	—
101	Mendelevium	Md	—
102	Nobelium	No	—
103	Lawrencium	Lr	—

Appendix 7
Selected Isotopes†

(1) Atomic number Z	(2) Element	(3) Symbol	(4) Mass number, A	(5) Atomic mass‡	(6) Percent abundance, or decay mode if radioactive	(7) Half-life (if radioactive)
0	(Neutron)	n	1	1.008665	β^-	10.8 min
1	Hydrogen	H	1	1.007825	99.985	
	Deuterium	D	2	2.014102	0.015	
	Tritium	T	3	3.016049	β^-	12.36 yr
2	Helium	He	3	3.016029	0.00014	
			4	4.002603	≈100	
3	Lithium	Li	6	6.015121	7.5	
			7	7.016003	92.5	
4	Beryllium	Be	7	7.016928	EC, γ	53.3 days
			8	8.005305	2α	0.067 fs
			9	9.012182	100	
5	Boron	B	10	10.012938	19.8	
			11	11.009305	80.2	
6	Carbon	C	11	11.011433	β^+, EC	20.3 min
			12	12.000000	98.9	
			13	13.003355	1.1	
			14	14.003241	β^-	5730 yr
7	Nitrogen	N	13	13.005738	β^+	9.97 min
			14	14.003074	99.63	
			15	15.000108	0.37	
8	Oxygen	O	15	15.003065	β^+, EC	122 s
			16	15.994915	99.76	
			18	17.999160	0.200	
9	Fluorine	F	18	18.000937	β^+, EC	109.8 min
			19	18.998403	100	
10	Neon	Ne	20	19.992435	90.51	
			22	21.991383	9.22	
11	Sodium	Na	22	21.994434	β^+, EC, γ	2.605 yr
			23	22.989767	100	
			24	23.990961	β^-, γ	14.97 h
12	Magnesium	Mg	24	23.985042	78.99	
13	Aluminum	Al	27	26.981541	100	
14	Silicon	Si	28	27.976927	92.23	
			31	30.975362	β^-, γ	2.62 h
15	Phosphorus	P	31	30.973762	100	
			32	31.973907	β^-	14.28 days
16	Sulfur	S	32	31.972070	95.02	
			35	34.969031	β^-	87.2 days
17	Chlorine	Cl	35	34.968852	75.77	
			37	36.965903	24.23	

‡The masses given in column (5) are those for the neutral atom, including the Z electrons.

APPENDIX 7 SELECTED ISOTOPES

(1) Atomic number Z	(2) Element	(3) Symbol	(4) Mass number, A	(5) Atomic mass‡	(6) Percent abundance, or decay mode if radioactive	(7) Half-life (if radioactive)
18	Argon	Ar	40	39.962384	99.60	
19	Potassium	K	39	38.963707	93.26	
			40	39.963999	β^-, EC, γ, β^+	1.25×10^9 yr
20	Calcium	Ca	40	39.962591	96.94	
21	Scandium	Sc	45	44.955910	100	
22	Titanium	Ti	48	47.947947	73.8	
23	Vanadium	V	51	50.943962	99.75	
24	Chromium	Cr	52	51.940509	83.79	
25	Manganese	Mn	55	54.938047	100	
26	Iron	Fe	54	53.939612	5.8	
			55	54.938296	EC	2.7 yr
			56	55.934939	91.72	
			57	56.935396	2.2	
			58	57.933277	0.28	
27	Cobalt	Co	56	55.939841	β^+, EC	77.7 days
			59	58.933198	100	
			60	59.933819	β^-, γ	5.272 yr
28	Nickel	Ni	56	55.943124	EC	6.10 days
			58	57.935346	68.3	
			60	59.930788	26.1	
29	Copper	Cu	63	62.939598	69.2	
			65	64.927793	30.8	
30	Zinc	Zn	64	63.929145	48.6	
			66	65.926034	27.9	
31	Gallium	Ga	69	68.925580	60.1	
32	Germanium	Ge	72	71.922079	27.4	
			74	73.921177	36.5	
33	Arsenic	As	75	74.921594	100	
34	Selenium	Se	80	79.916520	49.6	
35	Bromine	Br	79	78.918336	50.69	
36	Krypton	Kr	84	83.911507	57.0	
			92	91.926270	β^-, n	1.84 s
37	Rubidium	Rb	85	84.911794	72.17	
			87	86.909187	27.83 (β^-)	4.9×10^{10} yrs
38	Strontium	Sr	86	85.909267	9.86	
			88	87.905619	82.58	
			90	89.907738	β^-	28.8 yr
39	Yttrium	Y	89	88.905849	100	
			90	89.907152	β^-	64.0 h
40	Zirconium	Zr	90	89.904703	51.5	
41	Niobium	Nb	93	92.906377	100	
42	Molybdenum	Mo	98	97.905406	24.1	
43	Technetium	Tc	98	97.907215	β^-, γ	4.2×10^6 yr
44	Ruthenium	Ru	102	101.904348	31.6	
45	Rhodium	Rh	103	102.905500	100	
46	Palladium	Pd	106	105.903478	27.3	
47	Silver	Ag	107	106.905092	51.84	
			109	108.904757	48.1	

APPENDIX 7 SELECTED ISOTOPES

(1) Atomic number Z	(2) Element	(3) Symbol	(4) Mass number, A	(5) Atomic mass‡	(6) Percent abundance, or decay mode if radioactive	(7) Half-life (if radioactive)
48	Cadmium	Cd	114	113.903357	28.7	
49	Indium	In	115	114.903880	95.7; β^-	4.4×10^{14} yr
50	Tin	Sn	120	119.902200	32.4	
51	Antimony	Sb	121	120.903821	57.3	
52	Tellurium	Te	130	129.906229	33.8; β^-	2.4×10^{21} yr
53	Iodine	I	123	122.905594	EC	13.1 h
			127	126.904473	100	
			131	130.906114	β^-, γ	8.04 days
54	Xenon	Xe	132	131.904144	26.9	
			136	135.907214	8.9	
55	Cesium	Cs	133	132.905429	100	
			137	136.907073	β^-	30.17 y
56	Barium	Ba	137	136.905812	11.2	
			138	137.905232	71.7	
			141	140.914363	β^-	18.3 min
57	Lanthanum	La	139	138.906346	99.911	
58	Cerium	Ce	140	139.905433	88.5	
59	Praseodymium	Pr	141	140.907647	100	
60	Neodymium	Nd	142	141.907719	27.13	
61	Promethium	Pm	145	144.912743	EC, α, γ	17.7 yr
62	Samarium	Sm	152	151.919729	26.7	
63	Europium	Eu	153	152.921225	52.2	
64	Gadolinium	Gd	158	157.924099	24.8	
65	Terbium	Tb	159	158.925342	100	
66	Dysprosium	Dy	164	163.929171	28.2	
67	Holmium	Ho	165	164.930319	100	
68	Erbium	Er	166	165.930290	33.6	
69	Thulium	Tm	169	168.934212	100	
70	Ytterbium	Yb	174	173.938859	31.8	
71	Lutecium	Lu	175	174.940770	97.39	
72	Hafnium	Hf	180	179.946545	35.2	
73	Tantalum	Ta	181	180.947992	99.998	
74	Tungsten (wolfram)	W	184	183.950928	30.7	
75	Rhenium	Re	187	186.955744	62.60, β^-	4.5×10^{10} yr
76	Osmium	Os	191	190.960920	β^-, γ	15.4 days
			192	191.961467	41.0	
77	Iridium	Ir	191	190.960584	37.3	
			193	192.962917	62.7	
78	Platinum	Pt	195	194.964766	33.8	
79	Gold	Au	197	196.966545	100	
80	Mercury	Hg	202	201.970617	29.65	
81	Thallium	Tl	205	204.97440	70.5	
82	Lead	Pb	206	205.974440	24.1	
			207	206.975872	22.1	
			208	207.976627	52.4	
			210	209.984163	α, β^-, γ	22.3 yr
			211	210.988735	β^-, γ	36.1 min
			212	211.991871	β^-, γ	10.64 h
			214	213.999798	β^-, γ	26.8 min

APPENDIX 7 SELECTED ISOTOPES

(1) Atomic number Z	(2) Element	(3) Symbol	(4) Mass number, A	(5) Atomic mass‡	(6) Percent abundance, or decay mode if radioactive	(7) Half-life (if radioactive)
83	Bismuth	Bi	209	208.980374	100	
			211	210.987255	α, β^-, γ	2.14 min
84	Polonium	Po	210	209.982848	α, γ	138.38 days
			214	213.995176	α, γ	163 μs
85	Astatine	At	218	218.008684	α, β^-	
86	Radon	Rn	222	222.017570	α, γ	3.8235 days
87	Francium	Fr	223	223.019733	α, β^-, γ	21.8 min
88	Radium	Ra	226	226.025402	α, γ	1.60×10^3 yr
			228	228.031064	β^-	5.75 y
89	Actinium	Ac	227	227.027750	α, β^-, γ	21.77 yr
90	Thorium	Th	228	228.028715	α, γ	1.9131 yr
			230	230.033127	α	7.54×10^4 yr
			232	232.038054	100, α, γ	1.41×10^{10} yr
			234	234.043593	β^-	24.10 d
91	Protactinium	Pa	231	231.035880	α, γ	3.27×10^4 yr
92	Uranium	U	232	232.037130	α, γ	68.9 yr
			233	233.039628	α, γ	1.592×10^5 yr
			234	234.040946	0.0055 (α)	2.45×10^5 yr
			235	235.043924	0.72; α, γ	7.038×10^8 yr
			236	236.045562	α, γ	2.342×10^7 yr
			238	238.050784	99.275; α, γ	4.468×10^9 yr
			239	239.054289	β^-, γ	23.5 min
93	Neptunium	Np	236	236.046550	EC, β^-	1.2×10^5 yr
			237	237.048167	α	2.14×10^6 yr
			239	239.052933	β^-, γ	2.35 days
94	Plutonium	Pu	239	239.052157	α, γ	2.41×10^4 yr
			242	242.058737	α	3.76×10^5 yr
95	Americium	Am	243	243.061375	α, γ	7.37×10^3 yr
96	Curium	Cm	245	245.065483	α, γ	8.5×10^3 yr
			247	247.070347	α	1.56×10^7 yr
97	Berkelium	Bk	247	247.070300	α, γ	1.4×10^3 yr
98	Californium	Cf	249	249.074844	α, γ	351 yr
			251	251.079580	α	8.9×10^2 yr
99	Einsteinium	Es	252	252.082944	α, EC	1.29 yr
			254	254.088019	α, γ, β^-	275 days
100	Fermium	Fm	253	253.085173	EC, α, γ	3.0 days
			257	257.075099	α	100.5 days
101	Mendelevium	Md	255	255.091081	EC, α	27 min
			258	258.098570	α	56 days
102	Nobelium	No	255	255.093260	EC, α	3.1 min
			259	259.100931	EC, α	≈58 min
103	Lawrencium	Lr	257	257.099480	α	0.65 sec
			260	260.105320	α	3 min
104	Rutherfordium (?)	Rf	261	261.108690	α	65 sec
105	Hahnium (?)	Ha	262	262.113760	α	34 sec
106			263	263	α	0.8 sec
109			266	266		5 ms

†Data from *Chart of the Nuclides*, 12th ed., 1977, and from *Handbook of Chemistry and Physics*, 66th ed., 1985.

Appendix 8
Answers to Odd-Numbered Problems

CHAPTER 1

1. 9.54 m, 17.5° from vertical.
3. 6.71 cm
5. 127 m, 0 m
7. −1.21 km (west); 55.86 km (north)
9. spiral
11. $x = 0.866$ m; $y = 0.500$ m
13. 127 m
15. $x = -28.3$ km; $y = 28.3$ km
17. 5.83 mi; 350 mi/h
19. $d = (r_2 \cos \theta_2 - r_1 \cos \theta_1)^2 + (r_2 \sin \theta_2 - r_1 \sin \theta_1)^2$
21. 1 yard = 0.914 m; 9.4%
25. 12,576,000 m; 7816 mi
27. 0.0636 m^3; 350 kg
29. 4 m^3, 0.98 m, 8800 lb

CHAPTER 2

1. 90.9 km/h
3. 10 m/s, 9.1 m/s, 7.7 m/s
5. 0.38 min
7. 19.3 m/s
9. 3.82 km/h
11. 1672 km/h
13. 0.1 h, 0.5 km
15. 20.7° W of S; 129 km/h
17. 5 km; 3.76 min; #2
19. 0 km/h; 0.36 s
21. 5 m/s^2; 55 m/s
23. 4.17 m/s^2; 208 m
25. #1 (30 m/s, 15 s); #2 (36 m/s, 14.1 s)
27. 0.001 s; 4080g
29. 1 m/s^2
35. 3×10^4 m/s; 2.4×10^{-3} m/s^2, 3.8×10^4 m/s
37. 123 m; 49 m/s
39. 123 m
41. 4.99 s
43. 9.04 s
47. 37.3 m/s
49. $h = (V_0^2 \sin^2\theta_0)/2g$; $x = V_0^2 \sin(2\theta_0)$; $t = (2V_0 \sin\theta_0)/g$
51. 17.7 m/s; 2.52 s
53. 4.04 s; 12.8 m

CHAPTER 3

1. 2000 N
3. 10.2 kg
5. 28 g; 16,440 N, 3696 lbs
7. 46 N
9. -6.57 N; -0.22 m/s^2
11. 10 N
13. 4.47 m/s^2; 894 N
15. 0.2 m/s^2
17. 980 N
19. 0.2 m/s^2
21. 14
23. 28.8 m/s
25. 250 N
27. 0.98 m; 13,720 N
29. $m(g - a)$
31. $a = m_2 g/(m_1 + m_2)$
33. 2.5×10^{14} m/s^2; 2.28×10^{-16} N
37. 9.23 m/s^2
39. 20 N; 2 m/s; 13.3 N
41. $k_1 = 0.836$ kg/m; 250 N
45. 0.025 m/s (toward #2)
47. 400 N; 4000 m/s^2; 0.01 s
49. 1470 N; 0.13 s
51. 16.7°
53. 0.2
55. 1.02 m
59. $(m_2 - \mu m_1)/(m_1 + m_2)$
61. 25.5 m
63. No slip
65. Object with $\mu (\times 2)$
67. 70.7 N; 0.126; 132 N
69. 0.56 g
71. 5.05 m/s
73. 1.25 g
75. 11.8°

CHAPTER 4

1. 7.07 N toward SE
3. 412 N
5. 612 N
7. 214 N, $F_x = 175$ N, $F_y = 123$ N, $F = T$
9. 76.2 N
11. $T_1 = 1598$ N, $T_2 = 1568$ N
13. 60°, 396 N, 396 N
15. $T_1 = T_4 = 71.7$ N, $T_2 = T_3 = 87.9$ N
17. 1470 N
19. $F_1 = F_2 = 1470$ N
21. 186 kg
23. 6000 N
25. 113 N, 103 N
27. 65.6 kg
29. 0.289
31. 7.17 m
33. 8.03 kg
35. 3.92×10^5 N·m
37. $\theta = 0°$, 40,000 kg
39. 1.29 m from front
41. 70 N·m, 20 N·m
43. > 1.2 m
45. 0.30
47. 296 N
49. 68.6 N
51. $a \leq \mu g$, no slip
53. 71.9 N
55. 7000 N
57. d_2/d_1, $(d_1 + d_2)/d_1$, $d_1/(d_1 + d_2)$
59. 0.882 N
61. 2768 N
63. 18.4°
65. 21.8°
67. 274 N, 412 N
69. 1.08 m/s^2
71. No

CHAPTER 5

1. More massive
3. 139 N
5. 600 kg m/s, 600 N
7. 12 tons
9. 6
11. 58 kg m/s, 2600 N/m^2, 2.1×10^5 N
13. 16.7 m/s
15. No
17. 605 drops/m^2/s, 0.037 N
19. 10^4 kg
21. 19.23 m/s at 14.04°, 21.1 m
23. 60,000 N·s, 60 m/s
25. 312 lbs
27. 39.2 m/s, 0.082 m
29. 14,700 J
31. 4.18×10^7 J, 1.61×10^8 J, 1.21×10^8 J
33. 40.6 J
35. 4041 J
37. 1455 J
39. 6.04×10^5 J
41. 3.7×10^6 J
43. 2×10^{11} J
45. 5×10^{-6} J, 5 J, 1.8×10^8 J, 2.7×10^{33} J, 6.2×10^{40} J, 1×10^{53} J
47. 500 m/s
49. 10 m/s
51. 9490 J
53. 7.19×10^5 J
55. 6.93×10^6 J = 1660 Cal
57. 140 m/s, 2000 m
59. 8.86 m
61. 5.85×10^5 J
63. $2h$
65. $\sqrt{2gh}$, $2.5r$
67. $(m^2v^2/2g)/(M + m)^2$
69. 30.1 m
71. 1.24×10^8 J, 0.24 gal
73. all way, 12.5 m
77. 839 N
79. 22.9°, yes, if $\mu_s > 0.42$
81. 363 W = 0.48 hp
83. 3.6×10^6 J = 861 Cal
85. 741 N, 14.1 s
87. 387 W
89. 5×10^9 W
91. 2750 hp
93. 5.6 W
95. 2
97. $\left[\dfrac{1}{2\sin(\theta/2)}\right]$
99. 5, 2.25, 45%
103. 0.135
105. 11.3°
107. 44.1

CHAPTER 6

1. $\pi/6$, $\pi/4$, $\pi/2$, π, 3.752 rad
3. 0.796 rad
5. 2092 ft
7. 18.85 m/s
9. 0.105 rad/s, 0.00175 rad/s
11. 2×10^{-7} rad/s, 29.7 km/s
13. 1430 ft
15. 0.628 rad/s
17. 4×10^{23} m/s
19. 0.84 m/s, 16.8 rad/s
21. 8.38 rad/s^2, 83.8 rad/s
23. in m/s: (0.266, 0.531); (0.359, 0.718); (0.622, 1.245)
25. 23.9 rev.
27. 6.97 s, 1.93 rev.
29. 460 rpm, 3.77 rad/s^2

31 4.77 revs
33 4.19 m/s, 70.2 N
35 1.27 mi, 0.244 m/s^2
37 5.9×10^{-3} m/s^2, 3.5×10^{22} N
39 5.04g
41 102 m
43 8.46 rpm
45 5.8–21.6 mi/h
47 15.65 m/s
49 7340 m
51 No, gravity
53 1372 N
55 833 kg m^2
57 0.012 N
59 1.6 and 1.8 kg m^2, Wheel A
61 41.2 s
63 0.377 kg m^2/s, 2.37 J
65 0.12
67 3.17 rad/s
73 3.33 kg m^2, 0.32 s
75 290 m
79 16 rev/s
81 0.51 m/s, 2.78 rad/s
83 790 m/s
85 4.33 rev/s
87 226 kg
89 1.58 m/s, 0.13 J
91 4.93×10^6 J
93 $v' = 2v$, $1.5Mv^2$ from tension work
95 1737 s

CHAPTER 7

1 $GM/R^2 = 9.8$ m/s^2
3 7.86×10^{-10} N
5 8.24×10^{-8} N, no
7 1.63×10^{-9}
9 1936 s
11 0.38g, 2.50g, 1.07g
13 same
15 $3R_e$
17 353.5 yr
19 2.08 lt yr
21 0.707 yr
23 1000 AU from pair, 63,200 yr
25 $v = (GM/R)^{1/2}$, $P = 2\pi(R^3/GM)^{1/2}$
27 17x
29 6.34×10^{-4} yr
33 $1.3 \times 10^{11} M_\odot$
35 0.128″, 0.203″, 0.421″
37 0.999
39 Pluto 49.41, 29.65 AU; Neptune 30.32, 29.80 AU
41 0.967, 35.29 AU
43 1.63×10^8 m/s
45 297 km/s
47 0.893 km/s
51 0.905, 8.6 yr, 4.2 AU
53 29.8 km/s, 36.5 km/s, 42.1 km/s, 51.6 km/s
59 $(\pi/2)(R^3/2GM)^{1/2}$
61 10.3 km/s
65 5.46 yr, 5.46 revs
67 6.52×10^{-6} m/s^2
69 5.9×10^{-6}
71 3.22×10^{-6} m/s^2 (a,b), 2.43×10^{-6} m/s^2 (c)

CHAPTER 8

1 2.3×10^{17} kg/m^3
3 $Z = 6$, $A = 13$; 8,16; 92,238
5 118
7 10^{46}
9 29 grams
11 2700 kg/m^3

13 1000 times smaller ratio
15 0.13 m
17 5000 N/m
19 10^3 N/m, 3.4 ± 0.4 cm
21 980 N/m
23 90 lbs
25 3950 N/m
27 180 lbs
29 1333 N/m^2
31 9.8×10^4 N/m^2, 4.9×10^{-6}
33 2×10^6 kg
35 0.014
37 5×10^4 N
39 10^7 N/m^2, 10^{-3}
41 5×10^5 N (112,000 lbs)
43 9×10^{10} N/m^2
45 10^{10} N/m^2
47 1 cm
49 0.2
51 1.73
53 3×10^{-21} N/m^2/atom
55 2700 N
57 10^{-7} rad
59 1.5×10^7 N/m
61 2.1 ratio
63 65 atm = 6.5×10^6 N/m^2
65 0.1
67 10^{10} N/m^2
69 2×10^7 N/m^2
71 7 km

CHAPTER 9

1 100 N/m
3 6.53×10^{-4} m
5 2.6×10^8 yr, 1.2×10^{-16} Hz
7 0.05 Hz
9 1.6×10^{-3} s, 4×10^7 m/s
11 4.63 h
13 2.67×10^{-9} Hz, 1.28×10^{-10} Hz
15 3 cm
17 55.1 J
19 0.0703 m, 0.704 m/s, -6.89 m/s^2
21 1.5 kg
23 1.29 Hz
25 0.317 s
27 9.87 J
29 1.55 m/s
31 21.2 cm
33 0.707
35 6, 1.26 s, 0.796 Hz
37 $(1/2\pi)(g/A)^{1/2}$
39 0, $2\pi/3$, $4\pi/3$, . . .
41 $(0.206 + n\pi)$ s, $(1.365 + n\pi)$ s, $n = 0, 1, 2, . . .$
43 x: 1, 9, 17 s, . . . ; v: 7, 15, 23 s, . . .
45 add 198 kg
47 both 4.41 J
49 0.993 m
51 0.352 Hz
53 uniform rod
55 1/6 as long
57 0.16 g
59 12,756 km above
61 4.3×10^4 N/m
63 2.61×10^3 N/m, 1.29 Hz
65 5×10^{-3}
67 0.21 s
69 2.01 s
71 3.56 Hz
73 208 N/m, 0.257 Hz
75 $gA\rho(\Delta x)^2$, $2gA\rho$, $(1/2\pi)(2g/L)^{1/2}$

CHAPTER 10

1. 16.7 m/s
3. 3300 Hz
5. 3×10^8 m/s
7. 4 kg/m
9. 3.33 s
11. 459 m
13. 1.70 mm, 7.66 mm
15. 3 cm
17. $d = (349T)$ m
19. 1400 W/m^2
21. 6.28×10^7 W
23. 2.3×10^{-2} W/m^2, 6×10^4
25. 250 m, 2.94 m, 1.77×10^{-5} W/m^2
27. 346/350
29. 1.4 times
31. 110 m/s
33. $\lambda' = \lambda(1 - v_s/v)$
35. $\Delta \nu = +148$ Hz
37. 17.8 m/s
39. 313 m/s
41. $(\Delta\lambda/\lambda) = \pm(v_s/c)$
43. 516.7 Hz, 514.3 Hz, yes
45. 440 Hz
47. steel/iron = $\sqrt{2}$, 20°, 14°
49. $d < 1.98$ m
51. $\theta_i \le$ arc sin (v_i/v_r), total reflection
53. 8
55. 4 Hz
57. 2.21 Hz
59. 100 Hz, 3
61. 15.8 Hz, 0.0632 s
63. 1200, 1800, 2400, 3000 Hz
65. 44
67. 567 Hz
69. 30 min
71. 50 dB
73. 10^{-6} W/m^2
75. 10
77. 10^6
79. 5×10^{-5} J/s
81. 28.2 N/m^2, 3×10^{-4} atm
83. 0.688 m
85. 275 Hz, 413 Hz
87. 0.062 Hz, $n = 160$–320
89. 3300 Hz, $n = 1,3,5$
91. Mach 1.0 and 0.83
93. 602 m/s, Mach 1.74
95. 1.5×10^{38} kg

CHAPTER 11

1. 1.031×10^5 Pa, 1.018 atm, 1.031 bar, 1031 mbar, 775.2 torr
3. 1.3×10^{-3} g/cm^3, 1.27×10^{-2} N
5. 6580 atm
7. 5070 m, yes
9. 7.7×10^4 N
11. 10 g/cm^3
13. 0.831 m
15. 133 Pa/torr
17. 2.5×10^{-7} atm
19. 1 cm
21. 87.2 N
23. 10^3 cm^3, 4 g/cm^3
25. 2.1 m
27. double work
29. r(water) 2.47 times larger
31. 0.49 m, no
33. 12 N/m^2

35 $(3Ad/2\pi r^3)$, $[(3d/r) - 1]A$, $r/d = 6$, $-(\gamma A/2)$
37 decrease r by factor $\sqrt{10}$
39 2.2×10^{-3} m/s
41 50:1
43 0.459 m down from top
45 2 m
47 1.19 times smaller
49 9.9 m/s
51 1.88×10^5 N/m^2
53 235 m^2
55 $(8\pi\eta LQ^2/A^2)$
57 0.3 N s/m^2
59 3.6×10^{-3} N s/m^2
61 1.2 km
63 $(2r^2g/9\eta)(\rho_s - \rho_f)$
65 1.5×10^7, turbulent
67 $(2\rho_w r^2 g/9\eta) = 30$ m/s

CHAPTER 12

1 $T_F = (9/5)T_C + 32$
3 427°C, 800°F
5 $-40°C = -40°F$
7 37°C
9 $T_F = (9/5)T_K - 459.7$
11 $-442°F$
13 83°C
15 0.27 mm
17 $-48°C$
19 103°C
21 61°C
23 5.53×10^{-4}, 1.66×10^{-3} (°C)$^{-1}$
25 0.96 ml
27 1.8×10^5 m^3
29 238 g
31 3.35×10^{25}, 55.6 moles
33 0.366 l
35 74.4 m^3
37 2.52 cm^3
39 8.4×10^4 g
41 1.38×10^{-23} J/K
43 28.8 g, 2.2×10^{22} N$_2$, 5.1×10^{21} O$_2$
45 86.8 atm CO$_2$, 3.2 atm N$_2$
47 0.044 atm, 9.37 kg
49 3.7×10^{16} kg
51 7.8 m^3
53 570 g each
55 40%
57 Solid to liquid to gas
59 273.16°C, 4.58 mm Hg
61 1 BTU/lb°F = 1 cal/g°C
63 20 J, 50%
65 133 J/kg°C
67 5.5×10^4 J
69 1.67×10^6 J
61 2.26×10^6 J, 3.33×10^6 J
73 38.8 cal/g
75 16.7°C
77 3.88×10^4 J/l, 1.44×10^{-18} J
79 509 J
81 1600 W, same
83 10^4 W, yes
85 decrease R by ½
87 0.24 W, 295 K
89 $R_2/R_1 = (W_2 e_1/W_1 e_2)^{1/2} (T_1/T_2)^2$
91 4 m
93 T increases by 1.32 or R increases by 1.73

CHAPTER 13

1. 2.69×10^{25}
3. 405 m/s
5. 5 mm
7. 13 u, yes
9. 237 m/s
11. 60 u
13. 73.6 u
15. 0.991
17. 1.29 km/s, 0.389 km/s, $v_p \propto m^{-1/2}$
19. 12 km/s, 35 km/s, 1580 km/s
21. 278 m/s, 482 m/s, 0.018 s
23. 7.8×10^9 s
25. diffusion 2×10^5 faster
27. 2.5×10^{-3} m/s
31. 1.1×10^{-7} m/s, 1.1×10^{-11} m²/s
33. 4 mm
35. 0.5 m/s
37. Increase U by 3
39. 1.6×10^{21}
41. 3.35 moles
43. 10^4 J
45. C_P = 9.96 cal/mole K, C_V = 7.97 cal/mole K
47. $i = 3$
49. $i = 7, 9/7$
51. 2.42×10^{-21} J/degree
53. 7.5×10^6 J
55. 3000 K
57. 212 K
59. 5.22, 1.85
61. 500 K
63. 16°C, 2×10^5 J

CHAPTER 14

1. -3.6×10^9 N, attractive
3. 1.14×10^{-21} C
5. 4.5×10^{10} N/C
7. 1.72×10^{11} N/C
9. 3.8×10^{25} C/m³
11. 6.24×10^{11} N/C away from midpoint
13. weaker by 1/9
15. 2.45×10^{-12} C
17. 2.57×10^{-7} N
19. 9.65×10^4 C
21. 6.1×10^{-4} N
23. 0.674 m from 10 C, between charges
25. center of square
27. 7.5×10^4 N, 127 Mg
29. 2.11×10^{11} N
31. 25
33. 1.64×10^5 m
35. 0
37. 0, smaller by 1/15
39. 2.3×10^{39}
41. 1 V/m = 1 J/C/m = 1 N/m
43. 6.4×10^4 V
45. 6 V
47. 9×10^{10} V
49. 0.24 m from 10 C
51. 2.56×10^3 V
53. 9:1
55. 3×10^{17} C
57. 9.37×10^7 m/s = 0.3 c
59. 1.39×10^7 m/s
61. 1.2×10^{-5} C
63. 5.65×10^{-5} m²
65. 1.13
67. 5×10^{-6} J
69. 2.66×10^{-7} C
71. 5
73. 8.85×10^{-11} F
75. 7.2×10^{-5} J

CHAPTER 15

1. 3600 C
3. 1.08×10^5 C
5. 900 s
7. 7.8×10^{17}
9. 9.65×10^4 A
11. 2.33×10^{-5} m/s
13. 204 Ω
17. 0.015 A, 0.225 W
19. 1.15×10^{-3} A, 0.132 W
21. 661 W
23. 13 A, 8.82 Ω
25. 44,000 W
27. $R_2/R_1 = 2/3$
29. 1.27×10^{11} Ω
31. 20 A
33. 1.25×10^5 Ω
35. 1.67 A
37. 500°C
39. $T = 1/\alpha$ (°C)
41. 2.33×10^{-2} (C°)$^{-1}$
43. 8.35×10^{-3} $\Omega \cdot$m, 2.51×10^{-2} $\Omega \cdot$m, 0.1 (C°)$^{-1}$, semiconductor
45. 100 W bulb
47. $59.00
49. 11.9 W
51. 674 W
53. 1.0 Ω
55. 2 V
57. 15 Ω
59. 0.789 Ω
61. 1.20 A (5Ω), 0.667 A (2,3,4Ω)
63. 4 V across each
65. 0.923 Ω
67. 1.636 Ω
69. 8 μF
71. 1.33 μF
73. 0.857 μF
75. 1.56 μF, 2.33×10^{-5} C and 11.65 V (2 μF), 9.99×10^{-6} C and 3.33 V (3 μF), 1.33×10^{-5} C and 3.33 V (4 μF)
77. 5×10^{-5} Ω
79. 0.1 Ω

CHAPTER 16

1. 1.5×10^{-5} N
3. parallel or anti-parallel to I
5. infinite
7. 2.5×10^{-3} N/m
9. 3.2×10^{-21} N, along equator
11. 7.7×10^{-6} N, 0
13. 1.92×10^{-14} J
15. qBr, no, no
17. 8.39 Hz, 9.5×10^3 m, 0.119 s
19. 8.53 cm
21. 10^{-7} N/m
23. 0.577 A, 1.73 A
25. 1.0 m outside wire
27. 2×10^{-6} T, 4×10^{-6} N/m
29. 0.796 A
31. 4×10^5
33. 4.7×10^{-6} T
35. 0.2 T
37. 56.5 T
39. 0.45 Wb, 0.423 Wb
43. 3.95×10^{-5} Wb, 3.95×10^{-4} V
45. 3750 A/s
47. 0.4 A
49. B^2L^2v/R, $P = (BLv)^2/R$
51. 10^5 V, 10^5 A
53. $\Phi = (\pi r^2 B) \cos[\theta_0 \cos(\sqrt{g/L}\,t)]$.

APPENDIX 8 ANSWERS TO ODD-NUMBERED PROBLEMS

CHAPTER 17

1. 160 A
3. 12 A, 10.8 A
5. 18 V
7. 76.4 Hz
9. $N_s/N_p = 100$
11. 5 V, 3
13. 22.4 kV
15. 8.33 MW, 41.7 kV, $N_p/N_s = 348$
17. 25
19. $N_s/N_p = I_p/I_s = 417$
21. $I_p/I_s = 1/2$
23. 0.25 m
25. 79.6 A/s
27. $PE = \tfrac{1}{2}LI^2$
29. 79.6 Hz
31. 503 Hz
33. 2.21×10^{-5} F
35. 0.009 V
37. $L_{eq} = L_1 + L_2$
39. 12.9 Ω
41. 5.03 Ω
43. 76.5°, 90°, 66.6°, 30.5°
45. 2.51 Ω, 37.2°
47. 0.452 Ω
49. ∞, ∞, 65 Hz
51. $I/I_0 = e^{-t/RC}$
55. 0.318 A, 377 Ω, 89.85°
57. 0.159 Hz
59. 37.8 Ω, 3.04 A, 87°, 9.25 W
61. 1.2 mF
63. 704 pF
65. 0.1 A

CHAPTER 18

1. 1.26×10^9 A/s
3. -0.126 m/s
5. $(\epsilon_0 E^2 \pi R^2 d/2)$, $(\epsilon_0 \pi R^2 V/d)(\Delta V/\Delta t)$
7. $(\mu_0 \epsilon_0 R^2/2rd)(\Delta V/\Delta t)$
9. 13.6 V/m
11. 36.9 pF
13. 1.496×10^{21} Å, 1.496×10^{20} nm
15. 1 Å, 0.1 nm
17. (a) $4 \times 10^{14} - 7.5 \times 10^{14}$ Hz
    (b) $\leq 3 \times 10^8$ Hz
    (c) $3 \times 10^{16} - 3 \times 10^{18}$ Hz
19. 1.5×10^{-18} m, $10^{-3}\, R_{nucl}$
21. 3×10^8 m/s
23. 8.3 min
25. 1.19 h
27. 10^6 m/s
29. 4.6×10^{14} Hz
31. 10^{-3} J/m³
35. 1.59×10^{-3} T
37. c
39. 38.7 V/m, 1.29×10^{-7} T
41. 1.5×10^{18} m = 160 lt yr
43. 5.6×10^{28} J
45. 0.173 V/m, 5.8×10^{-10} T
47. 27.4 V/m, 9.2×10^{-8} T
49. 6.3×10^7 W/m²
51. 4.53×10^{-6} N/m²
53. 2.53×10^{-5} H, 3.91×10^{-10} F
55. 3.98×10^{-7} W/m², no line-of-sight

CHAPTER 19

1. 1.25
3. 2.0, 1.5×10^8 m/s
5. 1.42
7. 0.497 mm, $\Delta y = (2 \text{ mm})(\tan \theta_i - \tan \theta_r)$
9. 47.3°
11. 2.0
13. 81.1°
15. 51.5°
17. 7 m down wall
19. $2 \arcsin(1.54 \sin\theta) - 2\theta$
21. -30 cm, -12 cm; 60 cm, 12 cm
23. $h_i = -5.33$ cm, $d_i = -13.3$ cm
27. 2 m
29. 0.5
33. 13.3 cm
35. 1.5, 0.15
37. 1.38
39. $h_i = -4$ cm, $d_i = -10$ cm
41. 1, -2
43. 1.67 m
45. 1.556
47. $f = (2.5 \text{ cm})/(n - 1)$
49. 17.6 cm, 5.68 D, 0.417
51. 30 cm
53. 0.383 mm^2
55. $f/3.13$
57. -2.0 D, -0.5 m
59. 6 D
61. -2.08 D
63. 12.5 cm
65. 5 cm, 6x
67. 0.0139 cm
69. 1000x, $f_e = 1.25$ cm, $f_o = 0.39$ cm
71. -20x
73. -200x, 1 m

CHAPTER 20

1. 8×10^{-6} m
3. 33 cm
5. 2×10^{-5} m
7. 3 mm
9. 1.5 m
11. 5×10^{-7} m, yes
13. 2.67 m
15. 2.44 nm/mm, 1.16 nm/mm, 0.725 nm/mm
17. 5 m
19. 19.6 arc sec, yes
21. same (0.126 arc sec)
23. 1.22×10^{-4} m
25. 4.5×10^{-3} cm
27. 1.22 m
29. 6.6×10^{-3}
31. 6 mm diameter
33. maxima 16.9°, 35.6°, 60.8°; minima 8.4°, 25.9°, 46.7°
35. 0.625 mm
37. $2dn$
39. 1.08
41. 250 nm
43. 400 nm, 600 nm
45. 385 nm, 96 nm
47. 500 nm
49. 605.780211 nm, 1.817×10^{-4} m
51. 1.99
53. 61.0°, 29.0°, 67.5°
55. $1 \le n_2 < 1.73$
57. 7.5×10^{-6} W/m^2
59. (a) 0.866 E_0, 0.750 I_0; (b) 0.750 E_0, 0.563 I_0; (c) 0.250 I_0
61. (a) 90° $- 2\alpha$; (b) $-30°$, driver away from sun; (c) 22.3°, driver toward sun, 0.31°

APPENDIX 8 ANSWERS TO ODD-NUMBERED PROBLEMS

CHAPTER 21

1. Same place, 13.9 m forward
3. 120 mi/h
5. 5.908 and 5.895 s
7. 3.19×10^{-2} s
9. 15.15, 0.9978c
11. Coyote 5.03, 50.3 yr; roadrunner 7.09, 70.9 yr
13. 2.03 yr
15. 2.6×10^{-3} s
17. 71.4 m
19. 13.2 m
21. 0.5c
23. $\theta' > \theta$
25. 2.78×10^{-27} kg
27. 0.14c
29. $2.66 \times 10^{-5}:1$
31. per. 94.1 m/s, ap. 1.58 m/s, 4.92×10^{-14}
33. v_2/v_1, 1.424:1
35. 1.14×10^{-14} g
37. 3.02×10^{-29} kg
39. 3.67×10^{-10} kg
41. 4.64×10^{-14} g
43. 6.1×10^{17} kg/s, 1.9×10^6 yr
45. 3.45×10^{-10} J, 1.94×10^{-10} J
51. 5.09×10^{-19} kg m/s
55. 0.882c
57. $v_2 = 0.976c$, $v = 0.8c$

CHAPTER 22

1. 2.90×10^{-6} m
3. 9.94×10^{-19} J, 6.2 eV
5. 2.44 eV
7. 12.4 keV
11. 7×10^{31} ph/s
13. 10^{45} ph/s
15. 505 nm
17. 0.25 eV
19. 2.76 eV
21. 40.5 keV, 0.01485 nm
23. 0.5 MeV, $P_x = 8.05 \times 10^{-22}$ kg·m/s
25. 79.9°
27. 1.54×10^{-34} m
29. 3.88×10^{-10} m
31. 2.6×10^{-11} m
33. c/v
35. X-rays 4.96 keV, neutrons 0.013 eV
37. 5.34×10^{-3} nm
39. 2.1×10^{-14} m
41. 6.3×10^7 m/s, 21 MeV
43. $(KE) = (nh/2L)^2/2m$
45. 200 MeV
47. 10^8 Hz
49. 38.7 rad/s, 4.1×10^{-33} J, 2.3×10^{-18} m, 9×10^{-17} m/s

CHAPTER 23

1. v/RB
3. 4.6×10^{-14} m
5. 47
7. 121.567, 102.572, 97.254 nm
9. 1876 nm, infrared
11. 364.7, 820.6, 1459 nm
13. $r_n = (n^2\hbar^2/m_e k e^2)$
15. 4348

17 5.29×10^4, 7×10^{-15}
19 2186 km/s = 4.9×10^6 mi/h
21 -1.51 eV
23 13.6 eV, 3.4 eV, 0.54 eV
25 $n \geq 6$
27 $r_1(\mu) = r_1(e)/207$, $E_1(\mu) = 186 E_1(e)$
29 $r_n = (n^2\hbar^2/mkZe^2)$, $E_n = -(mk^2Z^2e^4/2n^2\hbar^2)$, $v_n = (kZe^2/n\hbar)$
31 2.03×10^{-12} m, -9194 eV
33 6895 eV
35 12, magnesium
37 A-436.5 nm, B-500.9 nm, C-496.1 nm
39 4.56×10^{-6}
41 121.5 nm, 102.5 nm, 97.2 nm

CHAPTER 24

1 n
3 38 eV
5 50 MeV
7 $0, \pm 1, \pm 2$
9 32
11 $\sqrt{12}\,\hbar$, 7 states
13 2, 4, 10, 12, 18
15 Xe + $4f^{14}6s^15d^{10}$
17 $3s$ (I), $3p^5$ (VII), $2p^4$ (VI)
19 Rn + $5f^{10}6d^07s^2$
21 Lr = Rn + $5f^{14}6d^17s^2$, then $6d^27s^2$, $6d^37s^2$
23 yes (l = 1)
27 pump 3, 2 $\to$ 1 laser
29 8.5 keV, 8.9 keV, 38 keV
31 0.960 cm, 1.41 cm, 1.00 cm
33 63 μm and 148 μm, far-IR, 1.11×10^4 s, 5.88×10^4 s
35 0.01 eV/bond
37 2.5×10^4 kcal
39 $v_c = (2E/m)^{1/2}$, 3836 m/s
41 7.6×10^{14} rad/s, 486 N/m

CHAPTER 25

1 82 p and e, 125 n
3 12.0111 u
5 0, 1.11, 2.83 MeV/nucleon
7 -11.2 MeV
9 55.8468 u; 8.736, 8.790, 8.770, 8.792 MeV; from ^{56}Ni beta-decays in stars (^{28}Si + ^{28}Si)
11 2×10^{17} kg/m^3, 5×10^{17} kg/m^3
13 7.41×10^{-15} m (U), 6.25×10^{-15} m (Ba), 3.96×10^{-15} m (Kr), 1.2×10^{-15} m (n), yes
17 n stable, β^- unstable, β^+ not possible
19 1.31 MeV
21 $^5_3$Li $\to$ $^4_2$He + p, $^8_4$Be $\to$ 2 $^4_2$He, etc.
23 $^{90}_{39}$Y, 0.553 MeV
25 1/64
27 4220 yr
31 0.0167 gram
33 0.42, 0.44
35 4×10^{13} decays/s, 1.1×10^3 Ci
37 $^3_2$He, 18.6 keV, 1.06 W, 3.9×10^{-3} W
39 8, 4.3×10^{-3} kg

41	111 cts/s	55	8.94 Ci
43	24.8 d	57	0.782 MeV
45	0.06 J, 0.11 J	59	0.0071
49	600 J, 588 J	61	4.81 MeV
51	0.143 g, 2.24×10^{-7} g	63	0.73 kg
53	0.49 gram, 5.41×10^{-8} Ci	65	3.27 MeV

CHAPTER 26

1. 1.533 MeV
3. 33.91 MeV
5. 2, 2, 1, 1, 1, 1, 1, 1
7. $(R,P) = (-1,-1), (-1,0), (-2,-1), (-3,-2), (-1,-1)$
9. $2m_p c^2$
11. 9.4×10^{-21} s
13. 67.48 MeV
15. Only d and f allowed
17. 67.48 MeV, L_e and L_μ are conserved
19. Only c, e, allowed

Index

aberration, *see lense, aberrations of*
absolute temperature scale, 365–367
absolute zero, 367
absorption coefficient, 772
absorption lines, 758–759
acceleration, 8, 34–44
   angular, 158–159
   average, 35–36
   centripetal, 160–161
   of gravity, 8, 39–44
   instantaneous, 35–36
   related to force, 60–61
   tangential, 159
   uniform, 36–37
accelerators, *see particle accelerators*
accelerometer, 175–176
airplane wing, 346–347
air resistance, 73–74, 128, 352
alpha decay, 833–835
alpha particles, 758, 827, 833–835, 837, 851
alternating current (AC), *see electric current*
alternator, 553–554
amino acids, 812–815
ammeter, 497–498
ampere (unit), 449, 478, 530–531
Ampere, A. M., 528
Ampere's law, 527–531
amplitude, 257
amplitude modulation (AM), *see AM radio*
AM radio, 606–607
Angstrom (unit), 597
angular acceleration, 158–159
angular magnification, 645–647
angular momentum, 170–173
   classical, 170–173, 174–176
   conservation of, 170–173, 195
angular quantities, 153–173
angular size, 616
angular velocity, 155–158
anode, 464
antielectron *(see positron)*
antimatter, 836, *see also antiparticles*
antiparticles, 836, 882–884
Archimedes, 337
Archimedes' principle, 337–340

architecture, 95–97
area moment of inertia, 236, 238
Aristotle, 6, 57, 59
astigmatism, 639, 641
astronomical unit (AU), 190
atmosphere (unit pf pressure), 332
atmospheric pressure, 332
atomic bomb, 854, 863
atomic mass, 228–229
atomic mass number, 229
atomic mass unit, 229, 829
atomic number, 228, 828
atomic spectra, 758–773, 803–810
atomic structure, 228–229, 757–769, 791–803
   Bohr model, 761–769
   of complex atoms, 796–803
   of hydrogen, 761–767
   planetary model of, 758
   quantum mechanics of, 791–803
atoms, 227–230, 757–769, 791–803
average binding energy per nucleon, 831–833
Avagadro's number, 373

back emf, *see electromotive force, back*
back torque, *see torque, back*
balance, *see equilibrium*
Balmer, J. J., 759, 769
Balmer formula, 759
Balmer series, 759, 764–765
bank gap, 736
Banks, W., 68–69
bar (unit), 332
barycenter, 193
barometer, 332–333
baryons, 885–886
battery, 464–465
beats, 307, 315
beauty, 892–893
Becquerel, H., 827, 833
Bernoulli's equation, 345–352
beta decay, 833–836
beta particle, 835
bicycle ergometer, 139–140
big bang, 715–716, 894–904

binary stars, 196–197, 214–215
binding energy, 830–833
   in atoms, 830–833
   of molecules, *see bonds; also molecules*
   of nuclei, 830–833
   related to radioactivity, 830–833
binomial expansion, 703
biomechanics, 64–65, 68–69, 138–140
birefringence, 686–687
blackbody radiation, 612, 730–732
black hole, 215, 223, 714–715
Bohr, N., 760, 761–764, 787
Bohr model of atomic structure, 761–769
Bohr radius, 763
Boltzmann, L., 409
Boltzmann constant, 376
bonds, 810–816
   in biology, 812–815
   covalent, 810–811
   ionic, 811
   weak, 811–815
bone, 235–236
bosons, 879
bottomness, *see beauty*
Boyle, R., 372
Boyle's law, 372
Brackett series, 760, 765
Bradley, J., 599–600
Bragg, W. H., 738
Bragg, W. L., 738
Brahe, T., 7
breaking point, 231
breeder reactor, 862
Brewster's angle, 685
brilliant cut (for gemstones), 622
British Thermal Unit (BTU), 384
BTU, *see British Thermal Unit*
bubble chamber, 850
bulk modulus, 243, 309
Bunsen, R., 730, 759, 769
buoyancy, 337–340
buoyant force, 337–340

calorie, 138
Calorie, 138, 150
calorimeter, 138

calorimetry, 138
camera, 642–645
candela (unit), 12
Canterna, R., 64, 68, 139
capacitors, 461–464
    energy stored in, 461–464, 588
    impedance of, 563–564
    reactance of, 563–564
    in combination, 495–496
capillarity, 341–343
carbon dating, see radioactive dating
Carnot, S., 428–429
Carnot engines, 429–430
cathode, 464, 571
cathode ray tube (CRT), 473, 571–573
Cavendish, H., 223
Celsius temperature scale, 365–367
center of gravity, 88, 91–93
center of mass, 88, 193–194, 223
centigrade temperature scale, 365–367
centrifugal force, 165–168
centrifuge, 166–167
centripetal acceleration, 160–161
CERN accelerator, 859
cgs system of units, 62
Chadwick, J., 827
characteristic curves, transistors, 576–577
charge, see electric charge
Charles, J., 372
Charles' law, 372
charm, 892–893
chromatic aberration, 649
circuits; see electric circuits,
circuit breaker, 503
circular motion, 153–176
clocks, 270–272
cloud chamber, 849, 850
CNO cycle, 871–872, 900
coil, see inductor,
colliding beam accelerators, 860
collisions, 115, 118–120, 131–133, 151, 227
    elastic, 131–133
    inelastic, 151
color (in quark theory), 893
commutator, 533, 552–553
compass, 516
compressibility, 244
Compton, A. H., 738
Compton formula, 739
Compton wavelength, 739–740
Compton scattering, 738–740
computers, 578–580
conduction band, 736
conductivity, thermal, see thermal conductivity
conic sections, 197–199
conservation laws, 118–121, 129–133
    of angular momentum, 171–172

in elementary particle interactions, 884–886
    of energy, 129–133, 141
    of momentum, 118–121
conservative force, see force, conservative
contact lenses, see lenses, eyeglass
continuity, equation of, 344–345
convection, 391–393, 396–398
Conversion of units, 12, 19–20
coordinate systems, 16, 75–76
    polar, 198
Copernicus, N., 6–7
coriolis force, 397
cosmic rays, 845
cosmological redshift, 895–896
cosmology, 715–716, 872, 894–904
coulomb (unit), 449
Coulomb, C., 448
Coulomb barriar, 835
Coulomb's law, 448–449
covalent bond, see bonds
critical angle, 621–622
critical height, 239
critical mass, 854
critical point, 380
cross product, 89, 540–541
CRT, see cathode ray tube,
crystals, 230
Curie, M. and P., 827, 833
current, see electric current,
cyclotron, 857–859
cyclotron frequency, 523, 857
cyclotron radius, 523

damping, 273–274
dating, radioactive, 840–841
Davisson, C. J., 742
de Broglie, L., 741, 787–788
de Broglie wavelength, 741–742, 787–788
decay, radioactive, see radioactivity,
decay constant, 839
decibel, 311
defects of the eye, see eye, defects of
deformations, see strain
degrees of freedom, 425
density, 333–334
deoxyribonucleic acid (DNA), 813–815, 845
de Sitter, W., 715
deuterium, 229, 828, 898
dew point, 379
diamond, 622
dichroism, 687
dielectric constant, 451
differential gravitational force, see tidal force
diffraction, 302–303, 662–674
    of light, 662–674

by single slit, 670–674
    x-ray, 692, 738–739
diffraction grating, 668–670
diffraction limit, 673–674
diffusion, 418–422
diffusion constant, 418–419
dimensional analysis,
diode, 553, 575
    zener, 575
diopter, 636
dipole, electric, 459–460
dipole moment, 459, 811
dipole selection rules, see selection rules
direct current (DC), see electric current,
dispersion, 619, 658–659, 667
displacement, 27–28, 32–33, 37–38
displacement current, see electrical current, displacement
distance, 27–28
DNA, see deoxyribonucleic acid,
domains, magnetic, see magnetic domains
Doppler effect, 297–302, 326, 784, 785
dosimetry, 846–847
double slit experiment, see Young's double-slit experiment
drag, 352, 355–357
drag coefficient, 352, 355–357
dynamics, 27, 57–83
dynamo, see electric generators
dyne, 62

earth
    convection in, 396–398
    internal structure, 295, 326, 396–398
    magnetic field of, see magnetic field, earth
    seismic waves in, see waves, earthquake
earthquakes, 245–247
eccentricity (orbital), 198
eddy currents, 555–556
efficiency
    of heat engines, 429–430
    of simple machines, 136–138
Einstein, A., 697, 699–700, 712–713, 715, 717, 734, 761, 880
elasticity, 227, 229–231
elastic limit, 231
elastic modulus, 233–234
    table of, 234
electric charge, 445–447
    in atom, 446
    conservation of, 447
    on electron, 449, 473
    induced, 446
electric circuits, 478, 487–503
    AC, 477, 559–570
    analysis of, 488–502, 504–505
    DC, 477, 487–503
    household, 502–503
    in the human body, 512

integrated, 578, 578–580
and Kirchhoff's rules, 488–492
LRC, 564–567, 581, 586
RC, 559–560
resonant, 569–570
time constants of, 560
electric current, 12, 477, 478–487
alternating, 477
direct, 477
displacement, 593
measurement of, 478–479, 483, 497–498
electric dipole, 459–460
electric energy, 455–464
stored in electric field, 461–464
electric field, 452–455
energy stored in, 457–458
electric field lines, 454–455
electric flux, 594
electric force, 447–452
electric generators, 551–554
electric motors, 554–555
electric potential, 455–461
of dipole, 459–460
of point charge, 458
electric power, 556–558
in AC circuits, 567–569
production of, 551–554
transformation of, 558–559
transmission of, 556–558
electrolysis, 465
electromagnet, 518–520
electromagnetic force, 706–707, 876–878
electromagnetic induction, *see magnetic induction*
electromagnetic spectrum, 596–597
electromagnetic waves, *see waves, electromagnetic*
electromotive force (emf), 477–478
back, 554–555
induced, 532–536
electron, 228
charge on, 449
discovery of, 757
energy states of, 761–769
electron cloud, 790–792, 795
electron microscope, *see microscopes*
electron spin, 794–796
electron volt (eV), 457
electroplating, *see electrolysis*
electroscope, 446, 473
electrostatic force, *see electric force,*
elementary particles, 875–904
accelerators of, *see particle accelerators*
classification of, 890–891
and conservation laws, 884–886
interactions of, 882–884
resonances, 889
stability of, 887–889
table of, 881

elements, 228–229, 828–830 (*see also periodic table*)
cosmic abundance of, 902
formation of, 894–902
ellipse, 190, 197–198
emf, *see electromotive force*
emission coefficient, 772
emission lines, 730
emissivity, 731
energy, 124–133
conservation of, 129–133, 141, 383
in EM waves, 601–605
electric, 455–464
equipartition of, 425
equivalence to mass, *see special relativity*
and heat, 382–389
ionization, 771–772
kinetic, 124–126
orbital, 201–203
of a photon, 732
potential, 126–127, 200–201
rotational, 173–176, 778, 819–820
of simple harmonic motion, 258–261
thermal, *see energy and heat*
vibrational, 816–820
of waves, 293–297
and work, 127–128
energy equivalents, table of, 853
energy levels, *see energy states*
energy level diagrams, 803–804
energy states
in atoms, 761–769
in molecules, 816–820
engines, *see heat engines*
entropy, 427, 441
equation of state, 372–376, 398–399
equilibrium, 85–111
thermodynamic, 414
equipartition of energy, 425
equipotential surface, 460–461
erg, 121–122
escape speed, 205–207
exchange particles, 876–878
excitation, 771–772
of atom, 771–772
of nucleus, 837
exclusion principle, 796–797
exercise and body metabolism, 138–140, 150, 406
expansion of the universe, *see universal expansion*
eye, 637–642
defects of, 640–642
structure of, 637–638
eyeglasses, 639, 641

Fahrenheit temperature scale, 365–366
farad, 461
Faraday, M., 461, 532

Faraday's law, 535–536
farsighted eye, *see hyperopia*
Fermi, E., 836, 851, 853
Fermilab, *see Fermi National Accelerator Laboratory*
Fermi National Accelerator Laboratory, 858–859
ferromagnetism, 515
fiber optics, *see optical fibers*
Fick's law, 418
field (*see also electric field, gravitational field, magnetic field, field forces*), 452
field equations, 715–716
field forces, *see force, field*
fine structure, 792
fission, *see nuclear reactions*
Fizeau, H. L., 600
flow of fluids, 343–357
laminar, 343–352
streamline, 355–357
turbulent, 353–355
fluids, 331–363
fluorescence, 805–806
flywheel, 174–175
FM radio, 606–607
focal length, 626, 631–632
focal point, 626, 631–632
focal ratio, 643
forbidden lines, 805
force, 10, 57–58, 880–882
and acceleration, 60–61
adhesive, 341–342
buoyant, 337–340
centrifugal, *see centrifugal force*
centripetal, *see centripetal force*
conservative, 127–128
drag, *see drag*
electric, 447–452
electromagnetic, 706–707, 876–878
in equilibrium, 85–99
field, 58, 191, 880–882
friction, 69–73
gravitational, *see gravitational force*
mechanical, 58
net, 58
normal, 70
restoring, 230, 255–257
transmission of, 66–69
unification of, 880–882, 897–898
units of, 62–63
force constant, 231, 257
force diagram, 58
Foucault, J., 600
four-dimensional space-time, *see spacetime*
Fourier analysis, 274
Fourier, J., 274
frames of reference, *see reference frames*
Franklin, B., 445
Fraunhofer, J., 730

free-body diagram, 58
frequency, 156–157, 257
    carrier, 606
    of rotation, 156–157
    of vibration, 257
frequency filters, 587
frequency modulation, *see FM radio*
friction, 69–73, 75–76, 127–128
    coefficients of, 70
fusion, heat of, 386
fusion reaction, *see nuclear reactions*

Galilean relativity, *see relativity, Galilean*
Galileo, G., 8–9, 57, 59, 74, 189, 267, 270, 599, 699
galvanometer, 497, 526–527
gamma decay, 837–838
gamma rays, 837–838
gas constant, *see universal gas constant*
gases, 371–379, 409–436
    change of phase, 379–382, 386–389
    heat capacities of, 384–386, 422–426
    ideal, 372–376
gas laws, 372–376
gauss (unit), 521
gaussmeter, 525
Gay-Lussac's law, 372
Geiger counter, 849
Gell-Mann, M., 891
geodesy, 680–682
geosynchronous orbit, 209
general relativity, 712–719
    and cosmology, 715–716
    principle of equivalence in, 713
    tests of, 717–718
generator, *see electric generator*
Germer, L. H., 742
geysers, 362
Glashow, S., 880
glasses, eye, *see eyeglasses*
gluons, 893
gram, 62
Grand Unified Theory (GUT), *see unified field theory*
graphical analysis, 32–34, 47–49, 117, 123–124
grating, diffraction, *see diffraction grating*
gravitation, law of, 189–192
gravitational force, 189–192
gravitational potential energy, *see energy, potential*
gravitational redshift, 718
graviton, 880
gravity, 39–44, 62, 880
    acceleration of, 39–44
    center of, *see center of gravity*
Greek science, 5–6
guage bosons, *see bosons*

GUT, *see unified field theory*
gyroscope, 170, 174–176

hadrons, 890
Hahn, O., 851
half-life, 839–844
    and radioactive dating, 840–841
Hall effect, 524–525, 546
Hall, E. H., 525
Halley's comet, 203–204
harmonic motion, *see simple harmonic motion*
harmony, musical, *see music*
heart, 362
heat, 382–398
    and energy, 382–389
    transport of, 389–398, 399–400
heat capacity, 422–426
heat engine, 427–433
    efficiency of, 429–431
    pollution, *see thermal pollution*
heat of combustion, 388
heat of fusion, 386
heat of vaporization, 386
heat pump, 431–433
heat transfer, *see heat, transport of*
Heisenberg, W., 747
Heisenberg uncertainty principle, *see uncertainty principle*
heliocentric theory, 6–9
henry (unit), 561
Henry, J., 532
Hipparchus, 6–7
holography, 777
Hooke, R., 231
Hooke's law, 231, 256–257
horsepower, 133
Hubble, E. 716, 894–896
Hubble constant, 895–896
Huygens, C., 270–271, 615, 616, 661, 662, 746
hydroelectric generators, 552–553
hydrogen, 229
hyperbola, 199
hyperfine splitting, 794
hyperopia, 638–639, 641

ideal gas law, 372–376
images, 625–653
    formed by plane mirrors, 625–626
    formed by spherical mirrors, 626–630
    formed by lenses, 631–634
    real, 628
    virtual, 625–626
impedance, 562–564
impulse, 115–117
inductance, 560–562
    mutual, 561
    self, 561

inductor, 562–564, 588
inertia, 8, 59–60
    law of, 59
    moment of, 169
inertial guidance, 175–176
infrared radiation, 809–810
infrasound waves, 310
integrated circuits, *see electric circuits, integrated*
interference, 304–308, 312–314, 663–682
    of light waves, 663–682
    of particles, 789–790
    of sound waves, 304–308, 312–314
    by thin films, 675–678
interference filter, 677–678
interferometers, 678–682
interferometry, very long baseline, 680–682
internal reflection, *see total internal reflection*
internal resistance, *see resistance, internal*
ionic bond, *see bonds*
ionization, 771–772, 784
isomer, 837
isothermal process,
isotopes, 228, 828

joule (unit), 20, 121–122
Joule, J., 480
Joule's law, 480
Jupiter, 204–205, 223

K-capture, 837
kelvin (unit), 367
Kelvin temperature scale, 365–367
Kepler, J., 7–8, 189, 194
Kepler's laws, 7–8, 10, 189–190, 194–197, 762
kilocalorie, *see Calorie*
kilogram, 11, 12, 61
kinematics, 27–55
kinesiology, 64
Kirchhoff, G. 488, 730, 759, 769
Kirchhoff's rules
    for electric circuits, 488–489
    for spectroscopy, 769–771

laminar flow, 343
lasers, 773–779
    helium-neon, 775–776
    and holography, 777
    and nuclear fusion, 863–865
    ruby, 775
latent heat, 386
Leibniz, G. W., 10
LeMaitre, G., 715
length contraction, relativistic, *see special relativity*
lens, 631–653
    compound, 651–652

contact, 641
converging, 636
diverging, 636
of eye, 638, 640–642
eyeglass, 639, 641
focal length of, 632
magnification of, 633–634
negative, 636
positive, 636
power of, 636–637
lens equation, 632–634
lens-maker's equation, 635–637
Lenz's law, 533–534, 541
leptons, 885–886
lever arm, 89
light, 596–597, 615–653, 661–688
as electromagnetic wave, 596–597
infrared, 597
interference in, 663–668, 674–682
photon theory of, 729–741
polarization of, 682–686
speed of, 597–601, 616–620, 706–707
ultraviolet, 597
visible, 596–597
wave properties of, 661–668
wave theory of, 734
light pipe, see optical fibers
lightning, 466–468, 512
linear accelerator, 859
lines of force, 454–455
Livingston, M. S., 857
Lorentz, H. A., 705
Lorentz factor, 702
Lorentz transformation, 705
luminosity, 394, 754
Lyman series, 760, 765

machines, 134–138
Mach number, 318
magnetic confinement (nuclear fusion), 863
magnetic domains, 516–517
magnetic dynamo, 537–539
magnetic field, 515–540
of a current loop, 518–519
of a dipole, 548–549
of the earth, 516, 537–539
path of charged particle in, 423–524
of solenoid, 518, 529–530
of straight wire, 527–529
of a toroid, 548
magnetic flux, 534–536
magnetic force, 520–524
magnetic induction, 531–536
magnetic mirrors, 547–548
magnetic monopole, 518
magnetic permeability, 528, 530
magnetic poles, 516, 518
magnetic quantum number, 792–794
magnetism, 515–517

electro-, 518–520
permanent, 516–517
magneto, 552
magnification, 629–630, 633–634, 645–647, 651–653
angular, 645–647
of lens, 633–634
of magnifying glass, 645–647
of microscope, 651–653
of mirror, 629–630
of telescope, 647
magnifying glass, 645–647
magnifying power, 646
Marconi, G., 605
maser, 778–780
mass, 59–62
mass-energy equivalence, see special relativity
mass increase, relativistic, see special relativity
mass number, 828–829
mass spectrometer, 547, 829–830
matter, origin of, 894–904
Maxwell, J. C., 409, 592, 599, 699
Maxwell distribution of molecular speeds, 413–418
Maxwell's equations, 592
mechanics, 8
mechanical advantage, 135–136
Mercury, 717
meson, 702–703, 878–880
metabolism, 139–140
metal detectors, 587–588
meter, 11, 12
Michelson, A. A., 600–601, 678, 699
Michelson interferometer, 678–682, 692, 699
Michelson-Morley experiment, 699, 705, 717
microscope, 651–653
compound, 651–653
magnification of, 651–652
scanning electron, 744–745
transmission electron, 743–744
Millikan, R. A., 473, 734
mirror equation, 628–630
mirrors, 625–630
focal length of, 626
parabolic, 649
plane, 625
spherical, 626–630
used in telescope, 649–650
MKS system, see System International
molar heat capacity, 423
mole, 373
molecular biology, 812–815
molecular bonds, 810–816
molecular spectra, 778–780, 816–820, 824–825

molecules, 229, 810–820, 824–825
in space, 778–780
moment of inertia, 169, 171
momentum, 113–121
angular, 170–176
conservation of, 118–121, 170–176
moon, 193, 211–212
Morley, E. W., 699
Mosely, H. G. J., 808, 827
motion
circular, 153–176
graphical analysis of, see graphical analysis
Newton's laws of, see Newton, laws of
projectile, 39–44
rotational, 88, 177–178
translational, 88
uniform circular, 153–158
uniform linear, 29–31
uniformly accelerated, 36–44
mu meson, 702–703, 878–880
muon, see mu meson
music, 314–317
musical instruments, 315
musical scales, 315, 316–317, 325
mutual inductance, 561
myopia, 638, 641

natural frequency, see resonant frequency
nearsighted eye, see myopia
Neptune, 217, 223
neutrinos, 836
neutron, 228
neutron number, 828
neutron star, 215
newton (unit), 20, 62
Newton, I., 9–10, 58, 189–191, 194, 661–662, 699, 729–730, 762
Newton's laws of motion, 58–66
Newton's rings, 675–677
nuclear binding energy, 830–833
nuclear decay, 833–844
nuclear energy, 830–833, 861–866
levels of, 837–838
released in reactions, 850–856, 861–866
nuclear force, 830–833
and binding energy, 830–833
strong, 58, 832
weak, 58
nuclear power, 861–866
nuclear radiation, 827, 833–850
activity of, see radioactivity
alpha, 827, 833–835
beta, 833–836
gamma, 837–838
damage to biological organisms, 844–849
detection of, 846–850
dosimetry for, 846–847

nuclear reactions, 850–856
  in bombs, 854, 863
  chain, 851–854
  energy released in, 850–856
  fission, 851–854
  fusion, 854–856
nuclear reactors, 861–866
  breeder, 862, 871
  fission, 861–863
  fusion, 863–866
nuclei, 827–866
  half-lives of, 839–840
  masses of, 829–830
  radioactive decay of, 833–844
  size of, 829
  stability of, 839–840
  structure of, 827–830
nucleon, 828
nucleosynthesis
  in early universe, 872, 894–899
  in stars, 900–901
nuclide, 828

Oersted, H. C., 517
ohm (unit), 480
Ohm, G. S., 481
ohmmeter, 499
Ohm's law, 481
optical activity, 687–688, 693
optical density, 616
optical fibers, 622–624
optical instruments, 637–653, 673–674
optic axis, 686–687
optics, 615–653
orbital quantum number, 793
orbits, 193–210, 216–217
Orion nebula, 779
oscillations, see vibrations
osmosis, 421–422
osmotic pressure, see pressure, osmotic
ozone, 754–755

pair production, 882–884
parabola, 199
partial pressure, 376–379, 399
particle accelerators, 857–861
  colliding beam, 860
  cyclotron, 857–859
  linear, 859
  synchrotron, 858–859
particle exchange, see exchange particles
pascal (unit), 331
Pascal's principle, 335–336
Paschen series, 760, 765
Pauli, W., 796, 836
Pauli exclusion principle, see exclusion principle
pendulum, 8, 266–272
  physical, 269–272
  simple, 266–268, 270–272
peptide bonds, 813
period, 156–157, 257
  of vibration, 257
periodic table, 798–803
permeability
  magnetic, see magnetic permeability
  of membranes, 421
permittivity, 448–449
phase
  in AC circuits, 564–567
  changes of, 379–382
  of simple harmonic motion, 263–266
phase angle, 565
phase diagram, 379–380, 382
phasor diagram, 564–565
phosphorescence, 806
photocathode, 849
photocell, see photomultiplier tube
photoelectric effect, 734–736
photomultiplier tube, 734
photon, 700, 732,
  as exchange particle, 876–878
  momentum of, 754
photon counting, 754
piezoelectricity, 247
pi meson, 878–879
pion, see pi meson
Planck, M., 732, 761
Planck's constant, 732
plasma, 807
Plato, 6
Pluto, 217, 604–605
polar molecules, 811
polar moment of inertia, 242
polarization, 682–686
  circular, 685–686
  plane, 682–685
polarizing angle, 685
pollution, see environmental pollution
positron, 836
potential difference, 455–461
potential energy
  elastic, 127
  electric, 455–461
  gravitational, 127, 200–201
potentiometer, 501–502
power, 133–134
Poynting vector, 603
precession, 170
presbyopia, 640
pressure, 331–336
  atmospheric, 332
  blood, 333
  of gas, 332–336, 410–412
  gauge, 332–333
  measurement of, 332–333
  osmotic, 421–422, 441
  partial, see partial pressure
  units and conversions, 331–332
  vapor, see vapor pressure
principal quantum number, 791–792
principle of equivalence (general relativity), 713
Priestly, J., 448
probability distribution, 789–791
problem solving, 19–20, 45–46, 47–48, 75–76, 177–178, 216–217, 276–277, 398–400, 504–505, 540–541, 581, 719–720
proper time, 701
proteins, 812–815
proton, 228
proton-proton chain, 856
Ptolemy, 6–7
pulsar, 214
pyramid, 150
Pythagoras, 6, 11

Q factor of a resonant circuit, 570
quantum, 732
quantum mechanics, 787–820
quantum number, 763, 791–796
  magnetic, 792–794
  orbital, 792
  principal, 763, 791–792
  spin, 794–795
quantum theory, 734, 787–820
  of atoms, 761–769
  of blackbody radiation,
  of light, 732, 734
quarks, 879, 891–893, 907
quasar, 785

r-process reactions, 901
rad, 846
rad equivalent man (rem), 846
radian, 154
radiation
  alpha, 827, 833–838
  beta, 835–836
  blackbody, see blackbody radiation
  energy transport by, 393–395, 405–406
  gamma, 837–838
  natural, 848
radiation damage, 844–849
radiation therapy, 848
radio, 605–607, 612, 809–810
radioactive dating, 840–841
radioactive decay, 150
radioactive heating (earth's crust), 871
radioactive tracers, 845
radioactivity, 228
rainbow, 658–659
random walk, 419, 440
Rayleigh criterion, 673
ray model of light, 616
ray tracing, 616

RBE, *see relative biological effectiveness*
reactance, 562–564
reactions, nuclear, *see nuclear reactions*
reactors, *see nuclear reactors*
recombination lines, 810
rectifier, *see diode*
redshift, cosmological, *see cosmological redshift*
reference circle, 262
reference frames
   inertial, 697–698
   noninertial, 698
reflecting telescope, 647–650
reflection, 302–303, 619–621, 625–630, 661–662
   law of, 303, 620, 661–662
   total internal, 621–624
   of sound waves, 313
reflection grating, 669
refraction, 302–304, 616–621, 658, 661–662
   double, *see birefringence*
   in air, 658
   index of, 616–619
   law of, 303, 619–621, 661–662
   and Snell's law, 303, 619–621
   by thin lenses, 631–634
relative biological effectiveness (RBE), 846–847
relative humidity, 378–379
relativity, 697–719
   Galilean, 697–699
   general, *see general relativity*
   special, *see special relativity*
   tests of, 717–718
relativity principle (Galilean), 698
relaxation oscillations, 274
rem, *see rad equivalent man*
resistance, 479–487
   internal, 494–495
resistivity, 482–487, 513
resistors, 479, 482–487
   in combination, 492–494
   shunt, 497
resolution, 673
   of electron microscope, 744
   of human eye, 674
   of microscope, 673–674
   of telescope, 673–674
resolving power, 673–674, 744
resonance, 274–276 (*see also waves, standing*)
   in AC circuits, 569–571
   in mechanical systems, 274–276, 308
   in particle collisions, 889
Reynold's number, 353–355
rheostat, 487
ribonucleic acid (RNA), 812–815, 845
right-hand rule, 90, 518

RNA, *see ribonucleic acid*
Roche limit, 212–213, 224
rock, 245–247
rockets, 44–45, 117–118, 150, 205–206
Roemer, O., 599
roentgen (unit), 846
rotational quantum number, 819–820
Rutherford, E., 758, 761, 827, 829, 831, 833
Rydberg constant, 759
s-process reactions, 901
sailboats, 347
Salam, A., 880
satellites, 208–210
Saturn, 213, 223
scalar quantities, 13–14
scales, musical, 315, 316–317
Schrodinger, E., 789
Schwarzschild radius, 714
science
   history of, 5–10
   philosophy of, 3–5
scientific method, 4–5
scintillation counter, 849, 850
sedimentation, 352
seismic waves, *see waves, earthquake*
selection rules (spectroscopy), 803–806
semiconductor, 485, 573–578, 736–737
   energy bands in, 574, 736–737
   n and *p* types, 574
shear modulus, 240
shear stress, *see stress, shear*
shells, atomic, 800–803
SHM, *see simple harmonic motion*
shock waves, 317–319
significant figures, 12–13
simple harmonic motion, 255–272
simple harmonic oscillator, 256, 816–820
single-lens reflex (SLR) camera, *see camera*
sinusoidal motion, 264
SI units (Systeme International), 11–12
Slipher, V. M., 895
slug, 62
Snell's law, 303–304, 619–621
Socrates, 6
solar cells, 737–738
solar constant, 604
solar energy, 604, 737–738
solar sails, 754
solenoid, 518, 529–530
solids, 229–247
solid state electronics,
sonic boom, 317–319
sound waves, 309–319
space diagram, 58
spacetime, 712
spark chamber, 850
special relativity, 699–711, 719–720
   Doppler effect in, 785

   length contraction in, 704–706, 725–726
   mass-energy equivalence in, 708–710, 831
   mass increase in, 706–707
   postulates of, 700
   simultaneity in, 703–704
   tests of, 717
   time dilation in, 700–704, 725, 726
specific gravity, 334–335
specific heat, 384–386
spectral lines, 730, 758, 764–767, 769–771, 803–810
spectrograph, 669
spectrometer
   light, 669
   mass, *see mass spectrometer*
spectrum
   absorption, 758
   band (molecular), 816–820
   blackbody, 730–731
   electromagnetic, 596–597
   emission, 730, 758
   molecular, 816–820, 824–825
   visible light, 596–597
speed, 29
sphygmomanometer, 333
spin quantum number, 794–796
spring, vibration of, *see simple harmonic motion*
standard temperature and pressure (STP), 374
Stefan-Boltzmann constant, 394, 731
stimulated emission, 773, 774
Stoner, P. B., 95
strain, 232–247 (*see also stress*)
strangeness, 885, 887
stress, 233–248
   linear, 232–235
   shear, 239–243
   volume, 243–244
stress-strain curve, 234
strong nuclear force, *see forces, nuclear*
subshell, atomic, 800–803
superconductor, 485
superposition theorem, 304–305
supersonic speed, 317–319
surface energy, 340
surface tension, 340–343
synchrotron, 859
synchrotron radiation, 860–861

tachyons, 707
telescopes, 647–651
   astronomical, 647–651
   Galilean, 648–649
   magnification of, 647
   reflecting, 649–651
   refracting, 647–649
television, 571–573, 607, 612

temperature, 365–368
temperature scales, 365–367
tensile strength, 67–69
tension, 66–69
terminal velocity, 74
Thales, 6
thermal conductivity, 389–391
thermal expansion, 368–372
    coefficient of, table, 369
thermal pollution, 434–435
thermal stress, 370
thermionic emission, 571
thermistor, 485
thermodynamic equilibrium, 414
thermodynamics, 409–436
    first law of, 422–426, 383–384
    second law of, 426–427, 428
thermometers, 367–368
thin-film interference, 675–678
thin lenses, *see lenses*
Thomson, G. P., 742
Thomson, J. J., 757, 829
threshold frequency, 734
tidal force, 210
tides, earth, 211–212
time dilation, *see special relativity*
topness, *see truth*
torque, 88–99
    back, 555
torr, 332
Torricelli, E., 349
Torricelli's equation, 349
total internal reflection, *see reflection, total internal*
transformer, 558–559
transition probability, 772–773
transition temperature, 485
transistors, 575–578
transmutation, 834
triple-alpha reaction, 900
triple point, 380
tritium, 229, 828
truth, 892–893
tunneling (quantum mechanical), 817, 835
turbulence, 353–357

u, *see unified atomic mass unit*
ultimate strength, 234–235
ultrasound, 310
ultraviolet catastrophe, 732–733
uncertainty principle, 746–750, 790

unification of forces, *see unified field theory*
unified atomic mass unit (u), 829
unified field theory, 880–882, 897–898
universal expansion, 894–896, 907
universal gas constant, 374
universal law of gravitation, *see gravitation*
universe, *see cosmology*
Uranus, 223

vacuum tubes, 570–573
valence, 798
valence band, 736
Van der Waals bonds, 813
vapor, *see gases*
vaporization, latent heat of, 386
vapor pressure, 377–379
vector bosons, *see bosons*
vectors, 13–18
    addition of, 14–18
    components of, 16–18
    resultant, 14
    subtraction of, 14–18
velocity, 29–34
    angular, 155–158
    average, 31–32
    of electromagnetic waves, *see light, speed of*
    instantaneous, 31, 35
    relative, 30–31
    relativistic, addition of, 710–712
    uniform, 29–31
    of waves, see *waves, velocity of*
Venturi tube, 347
vibrational energy, 816–820
vibrations, 255–277
    of atoms and molecules, 271–272, 283, 816–820
    damped, 273–274
    forced, 274–276
virtual images, *see images, virtual*
virtual photon, 877
viscosity, 350–352
    coefficient of, table, 351
voltage, 456
voltmeter, 498–499

watt (unit), 133
waves, 285–319
    continuous, 287
    earthquake, 288, 294–295, 326
    electromagnetic, 591–607

    energy transport by, 601–605
    formation of, 594–596
    information transport by, 605–607
    speed of, 597–601
    interference of, 304–308, 312–314
    longitudinal, 288
    and particles, 740–745
    periodic, 287
    plane, 287–288
    radio, 605–607
    sound, 314–319
    standing, 306–308, 312–314
    transverse, 287–288
    velocity of, 289–292
wave function, 788–791
wave intensity, 293–297
wavelength, 287
    de Broglie, 741–742, 787–788
    of particles, 741–742
    of visible light, 596–597, 666–668
wave nature of light, *see light, wave theory of*
wave nature of matter, 741–743
wave-particle duality, 740–743
wave velocity, *see waves, velocity of*
weber, 534
weight, 62, 191–192
    and exercise, 138–140
Weinberg, S., 880
Wheatstone bridge, 499–500
Wien's law, 730–731
Wollaston, W. H., 730
work, 121–123
    relation to energy, 127–128
    of rotation, 173–176
work-energy principle, 127–128
work function, 734–735

x-rays, 807–809

Young, T., 663
Young's double-slit experiment, 663–668
Young's modulus, *see elastic modulus*
Yukawa, H., 878
Yukawa particle, 878

Zeeman, P. 793
Zeeman effect, 793–794, 824
zener diode, *see diode, zener*
zero-point energy, 817
Zweig, G., 891

Chapter 9

Fig. 9.1 © Pam Hasegarva/Taurus Photos. **Fig. 9.14** © Bettman Newsphotos, UPI-Reuters Photo Libraries. **Fig. 9.15** Sandia National Laboratories.

Chapter 10

Fig. 10.1 © Fundamental Photographs. **Fig. 10.27a** Sandia National Laboratories. **Fig. 10.27b** Dr. A. C. Charters, Jr., *Album of Fluid Motion*, The Parabolic Press.

Chapter 11

Fig. 11.15 © 1974 ONERA photograph, Werlé, *Album of Fluid Motion*, The Parabolic Press. **Fig. 11.17b** T. J. Mueller, photo by F. N. M. Brown, Notre Dame, *Album of Fluid Motion*, The Parabolic Press. **Fig. 11.20** © Russel A. Thompson/Taurus Photos. **Fig. 11.21a** ONERA photograph, Werlé, 1980; © Thomas Corke and Hassan Nabib: all from *Album of Fluid Motion*, The Parabolic Press.

Chapter 12

Fig. 12.4 Golden Gate Bridge Highway and Transportation District. **Fig. 12.9** National Center for Atmospheric Research/National Science Foundation. **Fig. 12.15** © Peter Angelo Simon/Photo Take.

Chapter 13

Fig. 13.8 Department of Surgical Pathology, Stanford University. **Fig. 13.1.1** Jeff Becker.

Chapter 14

Fig. 14.1 © Peeter Vilms. **Fig. 14.1.1** Greg Gilbert/Seattle Times.

Chapter 15

Fig. 15.4 Richard Megna/Fundamental Photographs. **Fig. 15.5** Theodore P. Snow. **Fig. 15.6** Richard Megna/Fundamental Photographs.

Chapter 16

Fig. 16.3 Photo by R. K. Mishra, General Motors Research Labs. **Fig. 16.9** Daniel Brody/Stock, Boston.

Chapter 17

Fig. 17.35 General Electric Company. **Fig. 17.36** Sandia National Laboratory.

Chapter 18

Fig. 18.1 Ronald Rosberg, The Phone Co., Burlingame, CA.

Chapter 19

Fig. 19.1 American Institute of Physics, Niels Bohr Library. **Fig. 19.5** © Roy King. **Fig. 19.15** Courtesy of AT&T Bell Laboratories. **Fig. 19.1.1** Courtesy of AT&T Bell Laboratories. **Fig. 19.19** Perkin-Elmer Corporation. **Fig. 19.44** Courtesy of Palomar Observatory, California Institute of Technology.

Chapter 20

Fig. 20.10 Courtesy of Professor Philip M. Rinard/American Journal of Physics, Jan. 1976, Vol. 44, No. 1. **Fig. 20.12** © Fundamental Photographs. **Fig. 20.17** © Lester V. Bergman.

Chapter 22

Fig. 22.9a Natsu Uyeda, Institute for Chemical Research, Kyoto University. **Fig. 22.9b** Porter, K. R. and Bonneville, M. A.: *Fine Structure of Cells and Tissues*. Lea and Febiger, Philadelphia, 1973. **Fig. 22.12a, b** J. Pickett-Heaps, Department of Molecular, Cellular, and Development Biology, University of Colorado.

Chapter 23

Fig. 23.1.1 Lick Observatory photograph. **Fig. 23.1.2** Lick Observatory photograph.

Chapter 25

Fig. 25.10 California Institute of Technology. **Fig. 25.13** Stagg Field Stadium, University of Chicago, Argonne National Laboratory. **Fig. 25.18** Fermilab, Batavia, IL. **Fig. 25.19b** Fermilab, Batavia, IL. **Fig. 25.22** Lick Observatory photograph. **Fig. 25.24** University of California, Lawrence Livermore Laboratory. **Fig. 25.25** University of California, Lawrence Livermore Laboratory. **Fig. 25.26** Department of Energy, Washington, D.C.

Chapter 26

Fig. 26.5 California Institute of Technology, Pasadena, CA. **Fig. 26.13** Palomar Observatory, California Institute of Technology.

Periodic Table of the Elements*

Group I	Group II				Transition elements									Group III	Group IV	Group V	Group VI	Group VII	Group 0
H 1 1.0079 $1s^1$																			He 2 4.0026 $1s^2$
Li 3 6.941 $2s^1$	Be 4 9.01218 $2s^2$													B 5 10.81 $2p^1$	C 6 12.011 $2p^2$	N 7 14.0067 $2p^3$	O 8 15.9994 $2p^4$	F 9 18.9984 $2p^5$	Ne 10 20.179 $2p^6$
Na 11 22.9898 $3s^1$	Mg 12 24.305 $3s^2$													Al 13 26.9815 $3p^1$	Si 14 28.0855 $3p^2$	P 15 30.9738 $3p^3$	S 16 32.06 $3p^4$	Cl 17 35.453 $3p^5$	Ar 18 39.948 $3p^6$
K 19 39.0983 $4s^1$	Ca 20 40.08 $4s^2$	Sc 21 44.9559 $3d^14s^2$	Ti 22 47.88 $3d^24s^2$	V 23 50.9415 $3d^34s^2$	Cr 24 51.996 $3d^54s^1$	Mn 25 54.9380 $3d^54s^2$	Fe 26 55.847 $3d^64s^2$	Co 27 58.9332 $3d^74s^2$	Ni 28 58.69 $3d^84s^2$	Cu 29 63.546 $3d^{10}4s^1$	Zn 30 65.39 $3d^{10}4s^2$			Ga 31 69.72 $4p^1$	Ge 32 72.59 $4p^2$	As 33 74.9216 $4p^3$	Se 34 78.96 $4p^4$	Br 35 79.904 $4p^5$	Kr 36 83.80 $4p^6$
Rb 37 85.468 $5s^1$	Sr 38 87.62 $5s^2$	Y 39 88.9059 $4d^15s^2$	Zr 40 91.224 $4d^25s^2$	Nb 41 92.9064 $4d^45s^1$	Mo 42 95.94 $4d^55s^1$	Tc 43 (98) $4d^55s^2$	Ru 44 101.07 $4d^75s^1$	Rh 45 102.906 $4d^85s^1$	Pd 46 106.42 $4d^{10}5s^0$	Ag 47 107.868 $4d^{10}5s^1$	Cd 48 112.41 $4d^{10}5s^2$			In 49 114.82 $5p^1$	Sn 50 118.71 $5p^2$	Sb 51 121.75 $5p^3$	Te 52 127.60 $5p^4$	I 53 126.905 $5p^5$	Xe 54 131.29 $5p^6$
Cs 55 132.905 $6s^1$	Ba 56 137.33 $6s^2$	57–71‡	Hf 72 178.49 $5d^26s^2$	Ta 73 180.948 $5d^36s^2$	W 74 183.85 $5d^46s^2$	Re 75 186.207 $5d^56s^2$	Os 76 190.2 $5d^66s^2$	Ir 77 192.22 $5d^76s^2$	Pt 78 195.08 $5d^96s^1$	Au 79 196.967 $5d^{10}6s^1$	Hg 80 200.59 $5d^{10}6s^2$			Tl 81 204.383 $6p^1$	Pb 82 207.2 $6p^2$	Bi 83 208.980 $6p^3$	Po 84 (209) $6p^4$	At 85 (210) $6p^5$	Rn 86 (222) $6p^6$
Fr 87 (223) $7s^1$	Ra 88 226.025 $7s^2$	89–103‖	Rf 104 (261) $6d^27s^2$	Ha 105 (262) $6d^37s^2$	106 (263)	107 (262)		109 (266)											

Symbol — Cl 17 — Atomic number
Atomic mass** — 35.453
$3p^5$ — Electron configuration

Lanthanide series‡

La 57 138.906 $5d^16s^2$	Ce 58 140.12 $5d^04f^26s^2$	Pr 59 140.908 $4f^36s^2$	Nd 60 144.24 $4f^46s^2$	Pm 61 (145) $4f^56s^2$	Sm 62 150.36 $4f^66s^2$	Eu 63 151.96 $4f^76s^2$	Gd 64 157.25 $5d^14f^76s^2$	Tb 65 158.925 $5d^04f^96s^2$	Dy 66 162.50 $4f^{10}6s^2$	Ho 67 164.930 $4f^{11}6s^2$	Er 68 167.26 $4f^{12}6s^2$	Tm 69 168.934 $4f^{13}6s^2$	Yb 70 173.04 $4f^{14}6s^2$	Lu 71 174.967 $5d^14f^{14}6s^2$

Actinide series‖

Ac 89 227.028 $6d^17s^2$	Th 90 232.038 $6d^27s^2$	Pa 91 231.036 $5f^26d^17s^2$	U 92 238.029 $5f^36d^17s^2$	Np 93 237.048 $5f^46d^17s^2$	Pu 94 (244) $5f^66d^07s^2$	Am 95 (243) $5f^76d^07s^2$	Cm 96 (247) $5f^76d^17s^2$	Bk 97 (247) $5f^96d^07s^2$	Cf 98 (251) $5f^{10}6d^07s^2$	Es 99 (252) $5f^{11}6d^07s^2$	Fm 100 (257) $5f^{12}6d^07s^2$	Md 101 (258) $5f^{13}6d^07s^2$	No 102 (259) $6d^05f^{14}7s^2$	Lr 103 (260) $6d^15f^{14}7s^2$

*Atomic mass values averaged over isotopes in percentages as they occur on earth's surface.
**For many unstable elements, mass number of the most stable known isotope is given in parentheses.